The Life and Works of John Napier

John Napier (1550–1617), inventor of logarithms and numerous other mathematical concepts and calculating devices, as detailed in this work. The portrait (by an unknown artist) was donated to the University of Edinburgh by Margaret, Baroness Napier (d. *1706) in the late seventeenth century and subsequently passed to the Scottish National Portrait Gallery at an unknown date before returning to the University in 1983. Oil on canvas, 43½" × 38". (Courtesy of the University of Edinburgh.)*

Brian Rice • Enrique González-Velasco
Alexander Corrigan

The Life and Works of John Napier

Springer

Brian Rice
Horsham
West Sussex, United Kingdom

Alexander Corrigan
School of Divinity
The University of Edinburgh
New College, Mound Place
Edinburgh, United Kingdom

Enrique González-Velasco
Department of Mathematical Sciences
University of Massachusetts at Lowell
Lowell, Massachusetts, USA

ISBN 978-3-319-53281-3 ISBN 978-3-319-53282-0 (eBook)
DOI 10.1007/978-3-319-53282-0

Library of Congress Control Number: 2017931974

Cover illustration courtesy of Lord Napier and Ettrick

Printed on acid-free paper

This Springer imprint is published by Springer Nature
The registered company is Springer International Publishing AG
The registered company address is: Gewerbestrasse 11, 6330 Cham, Switzerland

CONTENTS

ILLUSTRATIONS

I REMEMBER as a child seeing the portrait of my remarkable ancestor (or my great—to the power of 12—grandfather), the Scottish theologian and mathematician John Napier, hanging in my parents' home. It always fascinated me, and my late father frequently told me that he was a very clever man and that I must study my maths really well!

To the vast majority of people it seems incredible that someone could find a way to simplify multiplication and division, mathematical functions that seem so fundamental that they could not possibly be simplified any further. Yet this is exactly what he did with his invention of Logarithms, to the benefit of all those involved in the complex computations required for astronomy, navigation, and engineering etc.

He developed an ingenious but simple method of multiplying and dividing using small rods, commonly called "Napier's bones", which remained popular throughout the 17th century and beyond. He also invented numerous other devices for simplifying arithmetical calculations, in particular his Promptuary, though none of them ever became popular, probably due to the very success of his own logarithms, which rendered them unnecessary. Nowadays, modern technology means we take such calculations for granted.

Napier was also known for his strong religious views. Many stories have been told about him and some locals even thought he was in league with the Devil. However, Logarithms are undoubtedly Napier's greatest legacy.

THE LIFE AND WORKS OF JOHN NAPIER

Napier was a truly remarkable person and his work was highly valued by many illustrious astronomers and mathematicians of his generation. I am privileged to know that I am related to this great man, who will forever have a place of honour alongside the greatest and most famous names in the history of science.

I am delighted that the three authors of this splendid book have brought all of Napier's publications together for the first time, some 400 years after his death. The greater accessibility this provides to Napier's works can only help to enhance his fame.

NAPIER and ETTRICK

December 2016

PREFACE

THIS work was conceived by a descendant of John Napier to celebrate the quadcentenaries of the publication of Napier's invention of logarithms in 1614 and his death in 1617. The tercentenary of the publication of the *Descriptio* in 1614 was marked by an International Congress held in Edinburgh, but the quadcentenary was celebrated much more modestly. This was largely because by 2014 the everyday use of logarithms had been displaced by the electronic calculator, introduced in the 1970s. Whole generations have subsequently grown up without ever seeing log tables at school, so that interest in the subject is naturally confined to a much narrower audience. Yet logarithms played a pivotal role in the development of western science and they are still used to measure many natural phenomena. Napier's key part in their discovery should not be forgotten but, as one recent biographer wrote, Napier is "probably the greatest unknown Scot".[1]

Clearly if Napier had not invented logarithms when he did, someone else would have done so shortly thereafter, because the development of mathematics had reached that particular stage throughout Europe. However, that does not diminish Napier's achievement, particularly when one considers that he was working alone, in a remote and periodically war-stricken country, and that he was devoid of the academic support he could have expected had he been based at Oxford, Cambridge, or any of the other major centres of learning in Europe.

However, logarithms are not Napier's only claim to fame. He published three works in his own lifetime; another was published two years after his death by one of his sons while the last was not published for another two centuries. In order of publication, Napier's five works are:

1. *A plaine discovery of the whole revelation of Saint Iohn*, Robert Waldegrave, 1593.
2. *Mirifici logarithmorum canonis descriptio*, Andreæ [Andro] Hart, 1614.
3. *Rabdologiæ seu numerationis per vigulas libri duo*, Andreas [Andro] Hart, 1617.[2]

---

[1] Francis Shennan, *Flesh and bones, the life, passions and legacies of John Napier*, Napier Polytechnic, 1989, (p. xv).

[2] Napier died on 4th April 1617 while the *Rabdologiæ* was entered for copyright protection at the Stationers' Company in London on 22nd May of the same year. However, there

4. *Mirifici logarithmorum canonis constructio*, Andreas [Andro] Hart, 1619.

5. *De arte logistica Joannis Naperi, Merchistonii Baronis libri qui supersunt*, The Maitland Club, 1839.

As indicated by their titles, the latter four works were published in Latin, the common language for scholars within Europe at that time. However, for this work it was determined to publish them in English, so that all of Napier's titles would be available in one work and in English for the first time. The aim has also been to replicate the layout of all five of Napier's books as closely as practically possible.

For the sake of brevity these five works are referred to throughout this work as the *Plaine Discovery*, *Descriptio*, *Rabdologiæ*, *Constructio* and *De Arte Logistica*.

The *Plaine Discovery* has been reprinted using Napier's original orthography and with the contemporary use of the long *s* (ſ, or *ſ* in italics). For easy reference the headlines to each page have been abbreviated to show the Proposition or Chapter numbers or other subsection name. The revised edition of 1611 has been reproduced since it includes an additional chapter in which Napier responded to some of the criticisms made of the first edition. The main text of the revised edition was almost identical to the original 1593 edition, except for the addition of extensive marginal notes and the omission of the original's Dedicatory Epistle to James VI (later James I of England), which is reproduced in Appendix 1. There were three reasons for Napier to increase the marginalia:

1. To further support his arguments by providing scriptural references to which he was either alluding or which he felt corroborated the point

2. To give credit to and recommend further reading to demonstrate that his arguments were in line with those of noted scholars

3. To provide the reader with a short summary of his argument or point for ease of reference

Thus this work makes available the full content of all editions of the *Plaine Discovery* printed in Napier's lifetime for the first time in a single volume.

For the mathematical works the authors have used the earliest available translations wherever possible. The *Descriptio* was translated by Edward Wright (1561–1615) with Napier's full approval and with a new introduction written by Napier specifically for the translation. This edition was published within Napier's lifetime and it has been used for the text in this work. However,

was no equivalent to the Stationers' Company in Scotland, so it is uncertain exactly when *Rabdologiæ* was published in Edinburgh and whether Napier saw a copy before he died.

Wright abbreviated the logarithmic values in the tables from a maximum of eight places to seven places, so that for this edition the original tables from the first Latin edition have been used. The values in these tables were entered into spreadsheets by two persons working independently and then compared to identify any differences, so it is believed that there are few (if any) errors. However, very occasionally a number is not legible in the original table and a judgement has had to be made regarding the value to be entered. Since these logarithms have no modern day use, such judgements and any possible errors are of no great significance.

The *Rabdologiæ* was translated by the New Zealand classical scholar William Frank Richardson (1938–2004) and published by MIT Press in 1990. They have kindly agreed to his translation being reproduced here. However, a small number of corrections have been made by Enrique González-Velasco to improve the translation and the formatting of the text has been altered to more closely reflect that of Napier's original work.

The *Constructio* was translated and published in 1889 by William Rae Macdonald (1843–1923) and his translation has been used, with some changes to the formatting.

Finally, the *De Arte Logistica* was translated by the English mathematician William Francis Hawkins (1906–1990) as part of his PhD thesis submitted to the University of Auckland in 1982 during his retirement in New Zealand. Despite extensive searches it has not been possible to trace any relatives to obtain permission for its use. Should any of them identify themselves, the originator of this work would be delighted to hear from them. Again Enrique González-Velasco has amended the translation in parts and reformatted Hawkins' original typescript to more closely follow the original prepared for the press by Mark Napier and published in 1839.

This work also includes three introductory chapters covering Napier's life, his single religious work and his four mathematical works. Each of these chapters is complete in itself but this has inevitably resulted in the occasional repetition. However, it was felt that this was a necessary compromise to allow the full story on each of these major cross-disciplinary themes to be told without the use of distracting cross references.

However, a major difficulty when researching Napier's mathematical output is that almost all his manuscripts were lost in a fire in 1801.[3] This limits any detailed analysis of Napier's method of working to a study of his published works.

The internet provides much general information on Napier, some of which is of an academic nature, occasionally with appropriate references. However,

[3] See p. 50.

much is of a more popular nature, often narrating how Napier was a "warlock", "sorcerer" or "necromancer" but these latter claims are provided with few, if any, references.

The documentary evidence shows that this characterisation of Napier is based on events in his life that have been distorted or exaggerated over time by repetition. Napier was an alchemist, a normal interest for an educated person of his time, but this should not be confused with more outlandish claims, however diverting or amusing they may be.

ACKNOWLEDGEMENTS

THE originator of this work was fortunate in being able to identify two colleagues who were experts in their respective fields to assist with its preparation.

Enrique González-Velasco is a Professor at the Department of Mathematical Sciences, University of Massachusetts Lowell, who has specialised in the history of mathematics, though his previous major work, *a Journey Through Mathematics*, covers a much broader time period than this work, which concentrates on a single individual. He has prepared an introduction to the mathematical aspects of Napier's works, whilst also reviewing and correcting (where necessary) the four translations used as source material for this work.

By a very fortunate chance, Alexander Corrigan was undertaking a PhD on Napier's sole religious work, the *Plaine Discovery*, when this work was first proposed so he was particularly well qualified to undertake the commentary on it. It was rather appropriate that he was doing this at the University of Edinburgh, Napier's home city.

All three of the authors have worked on this project part-time, two of them while also holding down full-time jobs which have taken up much of their time, while the third was engaged in preparing his PhD. Therefore the gestation period has been far more protracted than originally envisaged and they each thank their respective spouses and partners for their forbearance.

The authors thank Dr. Garry Tee for his generous assistance and encouragement. He reviewed various parts of the mathematical works, whilst also providing numerous useful leads regarding other aspects of Napier and his works. Tee supervised William Francis Hawkins' PhD on Napier's mathematical works and they jointly arranged for a copy of Napier's Promptuary to be built at the University of Auckland in 1979.

Particular thanks are due to Dr. Athol Murray, former Keeper of the Records of Scotland, for clarifying the records relating to the acquisition of Merchiston by the Napiers in the fifteenth century and for explaining the Scottish meaning of a barony.

The translation of the *Rabdologiæ* is taken from William Frank Richardson's *Rabdology* published by MIT Press which is reproduced here with their permission.

The originating author is particularly indebted to Lynsey Halliday at the

National Library of Scotland for the considerable trouble she took in answering numerous queries regarding references and in tracking down original copies for some of the illustrations of Merchiston Tower. Chris Fleet of the Library's Map Department kindly advised on the maps in this work, as well as answering questions relating to contemporary views of Edinburgh.

The photograph of the Madrid Promptuary was taken by Raúl Fernández Ruiz of the Museo Arqueológico Nacional, Spain (N.I. 52698).

Thanks are due to Wilma Mackie for typing the *Plaine Discovery* and to Jonathan Webley of Words & Numbers for typesetting it, a challenging task to both of them.

Richard Samways undertook the translation of various Latin texts, as well as correcting Mark Napier's transcription of Napier's *Secrett Inventionis*, while Carole Noble translated the French texts by James Hume and Pierre-Simon Laplace from her mother tongue into English. Hugh Watson kindly volunteered to redraw many of Napier's original line drawings for this work.

The authors gratefully acknowledge the financial support provided by The Strathmartine Trust in the preparation of this work.

THE
STRATHMARTINE
TRUST

A NOTE ON ORTHOGRAPHY

TWO of Napier's works reproduced here were first published in English within his lifetime, using contemporary spelling conventions. However, in an age before dictionaries or standardised spelling, authors adopted their own spelling rules, largely based on the sound of the words. The result was that spelling was much more variable than is now considered acceptable and some words may seem strange to the modern eye. These differences have been faithfully replicated in this work. The most common spelling differences include:

The use of the long *s* instead of an *s* at the start or in the middle of a word. However a double *s* in the middle of a word generally (but not always) used a single long *s* followed by a normal *s*, while a long *s* almost never replaced the last *s* of a word. In roman font the long *s* looks very similar to an *f* with the crossbar only to the left of the vertical or with no crossbar at all, but the difference is quite noticeable in italic font:

expreſſe = express	*ſelfſame = selfsame*
himſelfe = himself	*poſſeſsion = possession*
ſatisfaction = satisfaction	

The use of *u* for *v* and of *v* for *u* since these two letters were not considered as being distinct. Generally *v* was used at the start of a word and *u* within the word, regardless of sound:

euen = even	deceiued = deceived
vnloued = unloved	vtterance = utterance

Similarly, *i* and *j* were considered as scribal variants of the same letter:

Iohn = John	conioyne = conjoin
iubilee = jubilee	iudgement = judgement

However, the printer used an uppercase *J* instead of an *I* in some unusual and seemingly random instances in the *Plaine Discovery*, including the first person pronoun—*Jtalia* for *Italy*, *Jt is knowne* for *It is known*, *Neuertheles J haue* for *Nevertheless I have*, etc.

The letter *y* was often used as a substitute for an *i*:

fyrste = first	voyces = voices
fynde = find	poyson = poison

Sometimes a word included an *e* at the end, which is no longer used, and the final consonant might also be doubled:

ſinnes = sins	ſumme = sum
terme = term	logarithme = logarithm

Words were often spelled inconsistently, occasionally on the same page, though this was sometimes due to the printer altering words to create a justified margin:

ſhee or ſhe	maner or manner
tearme or terme	ſelfſame or selfe-ſame

Diacritical marks were frequently used above vowels to mark a missing *m* or *n*, a device often used by printers to achieve an even column of text. The marks used by Napier's printers varied considerably; sometimes what could be a straight horizontal bar (a macron) but more commonly a wavy one (a tilde). Tildes have been used throughout this work for the sake of consistency and the convenience of the reader:

euẽ = even	preſẽt = present
foũd = found	cõprehẽded = comprehended

Typographical errors by printers were far more common in Napier's time than would be permitted now. Letters and numbers might be printed upside down (an entry for $21^\circ\ 39'$ in Napier's original logarithm tables of 1614 suffered this fate) or omitted altogether (*vide* $7^\circ\ 20'$ and $22^\circ\ 58'$). Numbers were sometimes transposed, so that 93 might be printed as 39, and it was very easy for 6 or 9 to be printed upside down. Other examples include:

iuferior = inferior	prodnced = produced
cmoming = coming	ſnbverſion = subversion

Finally, Napier used the + and − signs throughout his mathematical works but it is clear that his printers were not familiar with them and often used symbols of inconsistent size and, for the minus sign, of inconsistent length. This, combined with the printer's use of rules to guide the eye across the page, often makes it difficult to determine whether a minus sign or a rule was intended in the original text. The transcription in this work resolves this issue.

JOHN NAPIER

Brian Rice

The portrait of John Napier owned by the Napier family. The artist and date are unknown, though it is clearly almost identical to those owned by the University of Edinburgh and the Scottish National Portrait Gallery. In the top right corner is ÆTATIS 66 (Aged 66), indicating that the portrait was painted the year before Napier's death. The motto in the escrol reads In Prudentia et Simplicitata (*literally* In Prudence and Simplicity). *Underneath the escrol are a serpent and a dove, from Matthew, Chap. 10, v. 16*: Beholde, I sende you forthe as shepe amonge wolves. Be ye therfore wyse as serpentes and innocent as doves. *Oil on canvas*, 44½" × 38". (*Courtesy of Lord Napier and Ettrick*.)

JOHN NAPIER

JOHN Napier's fame rests on his invention of Logarithms, which were first revealed in 1614 with the publication of *Mirifici Logarithmorum Canonis Descriptio* (*Description of the Marvellous Canon of Logarithms*).[4] Their invention was driven by the need to reduce the laborious work involved in astronomical calculations, though logarithms soon found many other applications with navigators, scientists, engineers, surveyors and actuaries.

The influence of logarithms was immediate and wide-ranging throughout Europe, even reaching China by 1653 where they were introduced by a Jesuit priest,[5] and they were crucial to the development of both astronomy and applied mathematics during the following centuries. As early as 1635 the mathematical writer Edmund Wingate (1596–1656), who introduced logarithms into Paris in 1624, could state that:[6]

> For as much as amongst many inventions, that concerne the Mathematicks, none can be found comparable to this of the Logarithmes, the worthy labours of thofe Learned men which have endevoured to advance it, are to be prized accordingly.

while the distinguished mathematician and astronomer Pierre-Simon Laplace (1749–1827) pointed out that Kepler had the benefit of logarithms "which, by reducing to a few days the work of several months, doubles, so to speak, astronomers' lives and spares them from the errors and the distaste inherent to long calculations".[7] The consequence for science was far reaching since "Newton was essentially dependent on the results of Kepler's calculations, and these calculations might not have been completed in Kepler's lifetime but for the aid that the logarithms afforded".[8]

Nor did it take long for the word *logarithm* to appear in English literature. In Act I, Scene VI of Ben Jonson's comedy, *The Magnetic Lady*, first performed

---

[4] Napier used the term *Canon* in what is now the obsolete mathematical sense of a table, typically of sines, tangents etc.

[5] Denis Roegel, *A reconstruction of Smogulecki and Xue's table of logarithms of numbers (ca. 1653)*, http://locomat.loria.fr.

[6] Edmund Wingate, *Logarithmotechnia; or the construction and use of the Logarithmeticall Tables*, Miles Flesher, 1635, p. A3.

[7] Pierre-Simon Laplace, *Exposition du système du monde*, Cercle-Social, 1796, Book 5, Chap. 4, p. 266.

[8] *Napier and the invention of logarithms*, a paper read by Prof. George A. Gibson before the Royal Philosophical Society of Glasgow on 11th February 1914, Proceedings of the Royal Philosophical Society of Glasgow, Vol. 45, p. 35.

in 1632 but not published until 1641, the following lines occur:

> Sir *Intereſt*, as able a Philoſopher,
> In buying, and ſelling! has reduc'd his thrifte,
> To certaine principles, and i' that method!
> As hee will tell you inſtantly, by *Logorythmes*,
> The utmoſt profit of a ſtock imployed:

However, to many of his contemporaries Napier's best known work was his religious book, *A Plaine Discovery of the whole Revelation of Saint Iohn*, a polemic against the Roman Catholic Church which was published in 1593. To most modern readers the issues behind the religious schism of this period are so remote from their daily concerns as to be barely intelligible, and Napier's work now strikes many readers as largely incomprehensible. The fact that Napier's forecast that the end of the world would occur between 1688 and 1700 proved to be incorrect did not help the credibility of his work to subsequent generations. However, when it was first published it proved to be so popular that it was soon translated into three languages in Europe.

Such a diverse range of subjects may seem strange to a modern audience, but this was a time when philosophy, religion, astronomy and science were not the distinct disciplines they are now considered to be.

Napier lived during one of the most turbulent yet interesting times in Scottish history. Martin Luther wrote his *Ninety-Five Theses on the Power and Efficacy of Indulgences* in 1517, the catalyst for the Protestant Reformations which subsequently swept through Europe. Religious divisions became the principal issue of the day, dividing families and country alike, and the Scottish Parliament formally broke from Rome in 1560 when Napier was ten years old. In 1561 the Catholic Mary Stuart returned from France after the early death of her husband, Francis II of France, to reign over the country of her birth. She married her first cousin Lord Darnley in 1565 but he was murdered two years later. Her third marriage in 1567 to the Earl of Bothwell (which was conducted by Napier's uncle) lasted only a few months before Mary was forced to abdicate and Bothwell fled Scotland. He was subsequently imprisoned for ten years in a Danish castle where he died insane in 1578.

By the time of Mary's accession, Scotland was largely, but by no means completely, Protestant in outlook. After Mary's abdication her son was taken from her and brought up as a strict Protestant. Scotland erupted into civil war and two Regents were murdered and another died in office before Mary's son, James VI, came of age. The fourth and last Regent was executed shortly thereafter.

Despite these decidedly unpropitious circumstances, James succeeded his mother to the throne. He not only survived a number of attempts on his own life but, on the death of Elizabeth I in 1603, he became James I of England and subsequently proclaimed himself King of Great Britain.

Napier himself was a man of many parts. Besides his religious and mathematical talents, he dabbled with the invention of weapons of warfare, agricultural improvements and a search for treasure, while he also gained a reputation for sorcery. His strong religious beliefs led him to become involved in the governance and promotion of the Reformed Church through its General Assembly, though he never became involved with the Royal Court or with politics. His personal life involved two marriages and twelve children who survived to adulthood.

NAPIER'S ANCESTRY

Napier was born into a distinguished family whose principal home was Merchiston Tower, located 1¼ miles south-west of Edinburgh Castle. The family were closely involved with much of the turbulent politics of their times and of Napier's antecedents, one was killed at the Battle of Sauchieburn, another at Flodden and a third at Pinkie.

The first Laird of Merchiston was Alexander, a successful merchant and Provost of Edinburgh in 1437.[9] Merchiston was originally part of the Crown demesne (domain) and was pledged to Alexander sometime prior to 1438

[9] A *Laird* is a courtesy title for a member of the Scottish landed gentry, roughly equivalent to an English squire, whose position depended on the ownership of a substantial estate. This title was confused by some English contemporaries, who were probably unfamiliar with the term *Laird*, and by many subsequent historians with *Lord*. For a very early example see Henry Briggs's letter to the Rev. James Ussher on p. 33 written in 1615 and Bretnor's *A new almanacke ... for 1618* (fn. 80 on p. 43).

However, Laird is not a title of nobility and it clearly ranks below the peerage. Napier often signed himself as *Fear of Merchiston*, meaning that he held the property in Fee, owning the fee-simple (absolute possession) of Merchiston as opposed to being a life-renter. A typographical error in the 1645 edition of the *Plaine Discovery* printed this as *Peer of Merchiston* which can only have further confused Napier's actual position as a member of the landed gentry with one in the peerage.

Napier was also described as *Barone Merchistonii* on the title page of the *Descriptio* and the *Rabdologiæ* and he was often and correctly referred to as a baron by his contemporaries. However, this term was used in the Scottish sense of a person who held land or property from the crown but not as a baron in the peerage. Three of Napier's direct ancestors had been knighted (a life title) and his son was to be raised to the peerage as a Baron (a hereditary title) by James VI after he became James I of England. Ironically neither title was offered to Napier who proved to be the most distinguished of them all.

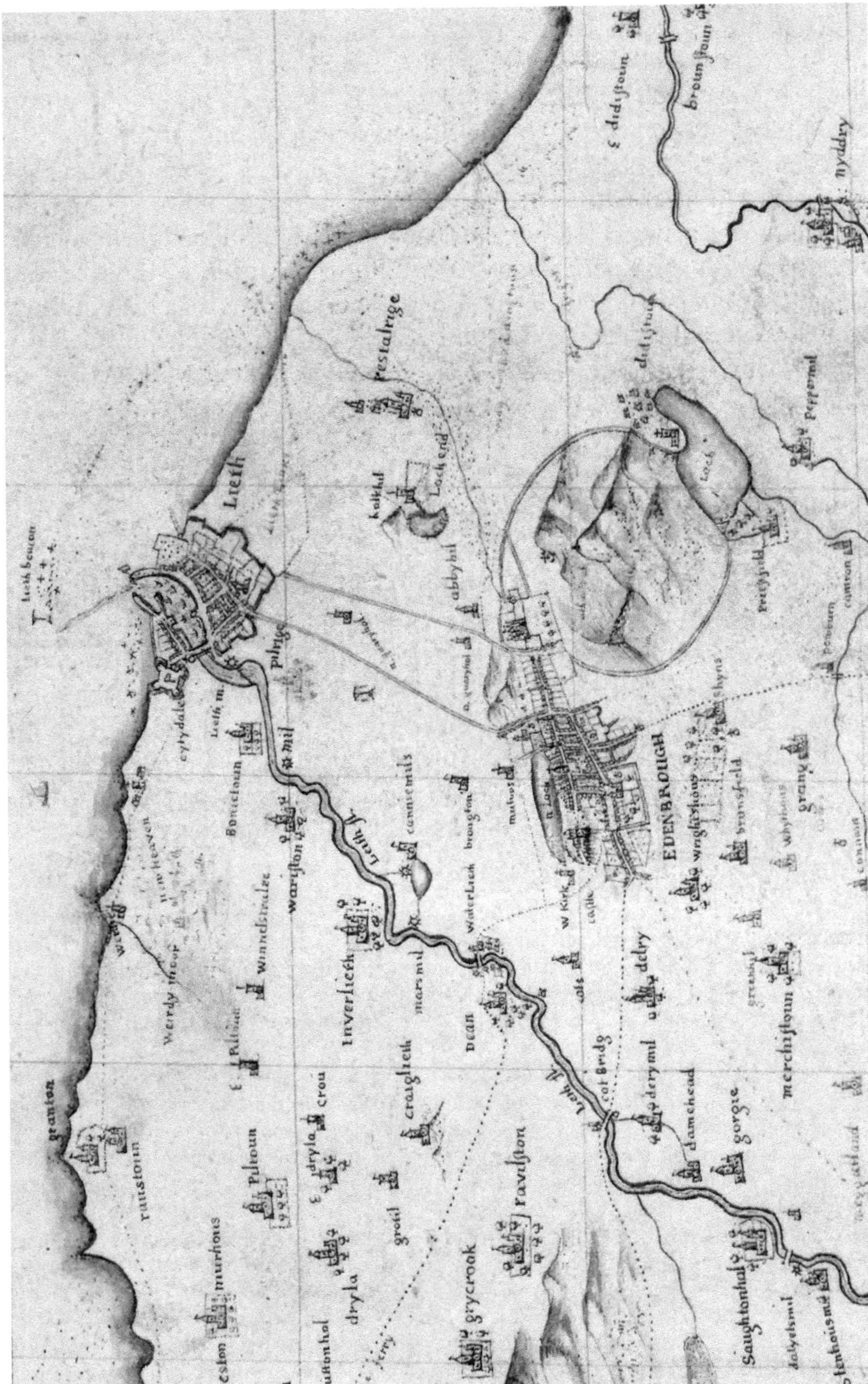
Lieth
EDENBROUGH
granton
Inverlieth
merchistoun
craiglieth
Saughtonhal
Pilloun
dryla
grycrook
ravelston
Dean

by James I of Scotland in wadset (mortgage) for a loan from Alexander. However, James I was assassinated in 1437 and the loan was never repaid so Alexander acquired outright possession in 1438.[10] The Napier family built Merchiston Tower later in the fifteenth century.

Alexander died in 1454 and was succeeded by Sir Alexander Napier (*d.c.* 1473), the second of Merchiston. He pursued a successful state career under two monarchs, becoming Comptroller to James II within his father's lifetime. When James II came of age, he executed the two leading traitors who had placed his mother under house arrest at Stirling Castle during his youth, and the forfeited estates of one and the position of another were given to Alexander Napier who had been wounded in defending the Queen. He was Provost of Edinburgh in 1455. By a commission under the privy seal, dated at Edinburgh on 24th February 1464, "Sir Alexander Napar of Merchamston" and others were appointed searchers of the port and haven of Leith, in order to prevent the exportation of gold and silver and in 1473 his name occurs as a Parliamentary Commissioner for "searching of the money". After the premature death of James II, Sir Alexander was one of those sent to negotiate the marriage between James III and the King of Denmark's daughter and he was subsequently appointed to the office of Master of the Household.

Sir Alexander not only succeeded his father to the estate at Merchiston but he also held lands from the Crown called the Pulterlands, to which was attached the hereditary office of *Pultriæ Regis* or *King's Poulterer*. This required Sir Alexander to provide an annual gift of poultry to the king whenever required. These lands lay near the village of Dean, barely a mile west of Edinburgh. The office remained in the family until John Napier, the inventor of Logarithms, sold it and the associated lands in 1610 for 1,700 marks to "Williame Nisbet

---

Facing page. *A small portion of a manuscript map of Midlothian surveyed and drawn by John Adair, a leading surveyor of his time. It is dated 1682. The port of Leith lies to the north-east of Edinburgh while "Merchiftown" is shown to the south-west. Also identified on this map is "W[est] Kirk" (just west of Edinburgh), also called St Cuthbert's, the parish church of Merchiston where Napier is buried. Also shown are "Restalrige" (east of Edinburgh), the home of Sir Robert Logan with whom Napier signed a contract to search for treasure at Fast Castle, and "Dean" (west of Edinburgh), the location of the lands held from the crown called the Pulterlands, to which was attached Napier's hereditary office of* Pultriæ Regis *or* King's Poulterer. *(Courtesy of the National Library of Scotland.)*

[10] See Appendix 9 (p. 964) for further details regarding the acquisition of Merchiston by Alexander Napier.

of Deane, his airis maill and assignais quhatsomevir".[11]

John Napier (*d.* 1488), the third of Merchiston belonged to the royal household in his youth and at various times was Provost of Edinburgh. Three successive Lairds of Merchiston held this office on various occasions during a period of half a century. John was killed at the Battle of Sauchieburn in 1488, fought between the followers of James III (who was killed) and a large group of rebellious Scottish nobles, nominally under the leadership of James III's own son, who then became James IV.

Archibald (*d.* 1521/2), fourth of Merchiston, succeeded to the estate. Little is known about him except that he married three times, on each occasion to well-established Scottish families.[12] He was present at the Battle of Flodden which took place in Northumberland in 1513. Although he managed to escape the carnage which saw the flower of the Scottish nobility brutally cut down, his eldest son fell, as did James IV, the last monarch in the British Isles to die on the field of battle.

Archibald was succeeded by Sir Alexander (*d.* 1513), fifth of Merchiston, who had received the lands of Merchiston from his father prior to the latter's death. However, he was slain at Flodden within the lifetime of his father but not before Merchiston was erected as a barony for Sir Alexander by James IV in 1512.[13]

Sir Alexander was therefore succeeded by his son Alexander (*c.* 1509–1547), sixth of Merchiston, while an infant. He had to obtain a dispensation from the Pope to marry his cousin. Alexander spent much time abroad, partly to undertake a pilgrimage to Amiens where the reputed head of John the Baptist was held, but also due to his delicate health. However, he returned to Scotland only to fall at the Battle of Pinkie in 1547, on the banks of the River Esk near Musselburgh, only three years before Napier was born.

---

[11] National Records of Scotland, PA2/18, f. 18r. K. M. Brown et al eds, *The Records of the Parliaments of Scotland to 1707*, St Andrews, 2007–2014, 1612/10/61.

[12] Records in the National Records of Scotland (GD430/101) state that "in 1509 there was erected for Archibald Napier of Merchiston the barony of Edinbellie Napier, comprising lands in the sheriffdoms of Stirling, Dumbarton and Perth".

[13] Scottish baronies were baronies by tenure, meaning that a Scottish baron was not a member of the peerage but rather a tenant-in-chief holding lands from the crown in free barony (*in liberam baroniam*), with special baronial powers exercised in the barony court. Up to Robert I's reign (*d.* 1329) the terminology was applied quite loosely but from his reign onwards baronies required specific creation (erection) by the king or his recognition that they already existed.

The barony court had both judicial and local government functions, as well as being part of the administration of a landed estate. From the 15th century barons ceased to sit in the Scottish Parliament as of right (though they could be elected as members for their shire) unless they were created lords of parliament, the lowest rank of the Scottish peerage. However, barons were still regarded as minor nobility.

Sir Archibald Napier (*c.* 1533/4–1608), seventh of Merchiston, was the eldest son of Alexander. At the time of his father's death, Archibald was only fourteen years old but in or before 1549 he married Janet Bothwell. Archibald was only 16 when his eldest son, John Napier the inventor of logarithms, was born in 1550 at Merchiston.[14]

Given their position in Scottish society the Napier's inevitably had numerous distinguished and influential connections, many acquired by marriage but others by opportunity. Archibald, 5th Duke of Argyll, made Sir Archibald Napier a Depute-Justice when he was appointed Lord Justice-General. The Lord Justice-Clerk, Sir John Bellenden, was a cousin of Sir Archibald's wife and when, in March 1566, the Edinburgh magistrates, including Sir Archibald, were put on trial for releasing a prisoner who had committed "slauchter", the prosecutor objected to Bellenden on the grounds that "he and the Lard of Merchameftonis wyfe wes fifter and brethir bairnis, and that thair wes bairnis betnix the said Lard and his fpous":[15] When Mary, Queen of Scots, returned to Scotland in 1561 Bellenden was made a Privy Councillor, becoming closely involved in the events that eventually led to her downfall and flight to England in 1568.

Sir Archibald was knighted in 1565 and appointed to the Mastership of the Mint in about 1582, taking sole charge of the all the mines and minerals in Scotland until his death. It is recorded that in 1604 he "went to London, to treat with the English commissioners anent [regarding] the cunzie, who, to the great amazement of the English, carried his business with a great deal of dexterity and skill".[16]

[14] The first mention of Napier's birthplace appears to be in David Stewart, Earl of Buchan, & Walter Minto's *An account of the life, writings and inventions of John Napier of Merchiston* (1787, R. Morison & Son, p. 10), in which Merchiston is stated as being his birthplace with the footnote "As appears by an inscription on his portrait, engraved by old Cooper, from an original painting". This is a reference to Richard Cooper Senior (1701–1764) who sold a mezzotint of Napier as a loose item (rather than being bound in a book) sometime in the 1730s. It was clearly based on the portrait now owned by the University of Edinburgh. However, the Latin inscription on this, Cooper's only known engraving of Napier, makes no reference to his birthplace, merely stating in Latin "The most noble, by birth and then by his writings, John Napier, a Scot, Baron of Merchiston, the most celebrated inventor of logarithms etc, died on 3rd April, 1617. Aged 67." (Nobilifsimus, cum genere tum Scriptus, Ioannes Napierus, Scotus, Merchistony Baro: Celeberrimus logarithmorum Inventor, &c. Obijt April 3, 1617. Ætat: 67). Despite this lack of proof, Merchiston is quite likely to have been John Napier's birthplace, given that his father was in full possession of the property at the time of Napier's birth. Mark Napier's *Memoirs of John Napier* (William Blackwood, 1834, p. 56) merely states that he was born at Merchiston without providing any source for this information.

[15] Robert Pitcairn, *Ancient criminal trials in Scotland from AD 1488 to AD 1624*, William Tait, 1833, Vol. 1, Part 2 (1584–1596), p. 443.

[16] Sir Robert Douglas & John Philip Wood, *The peerage of Scotland*, Archibald Constable,

---

The Napier name has been variously spelt, particularly in the distant past. His first marriage contract of 1572 and his contract with Sir Robert Logan dated 1594 to search for treasure at Fast Castle are both signed in his own hand Johne Neper while his *Secrett Inventionis* of 1596 is signed in his own hand as Jo Nepar. The title pages to both the 1593 and the 1611 editions of the *Plaine Discovery* use Iohn Napeir, as does the Dedicatory Epistle to James VI prefixed to the 1593 edition. However, the 1645 edition uses the modern form of John Napier. The 1616 English edition of the *Descriptio* uses Iohn Nepair for both the title page and the dedication to Prince Charles. However, in the same work Henry Briggs' name is spelt Brigs, Brigges and Briggs, indicating that the spelling of personal names was regarded as being of no great importance at the time. A letter to Napier's son dated 1608 is signed Jhone Napeir, as are all the deeds signed after that date. This list obviously excludes the Latin variant of Napier's name, Ioanne Nepero, consistently used in the *Descriptio* of 1614, *Rabdologiæ* of 1617 and *Constructio* of 1619. His children used almost every possible variation except Napier itself, while previous generations generally signed themselves Neper. Therefore the current form of Napier is relatively recent.

Sir Archibald Napier, later the 1st Lord Napier, recorded the origin of the Napier name in a document sent to relatives in 1625.[17] It states that Donald Lennox, second son of an Earl of Lennox, particularly distinguished himself during a battle. Afterwards the king declared that Donald had "Na-pier" [No peer] and commanded him to change his name accordingly. This ancient connection with the Lennox family is borne out by the Napiers having always carried the Lennox coat of arms but with the cross engrailed to make the necessary heraldic difference for a cadet branch. No other Napier family carries anything resembling these arms. An alternative derivation of the family's name is from *Naperer*, the keeper of the table linen in the Royal household, but this seems less likely given the similarity of the Napier arms to those of the Lennox family and the strict rules regarding the creation of coats of arms.

EDUCATION AND MARRIAGE

Perhaps the most influential of Napier's relations was Janet Napier's brother, Adam Bothwell (*c.* 1527–1593), who pursued a successful career in both

---

1813, p. 288.

[17] John Hutchins, *The history and antiquities of the county of Dorset*, W. Boyer & J. Nichols, 1774, Vol. 2, p. 46. This document is reproduced in Appendix 2 (p. 948).

church and state, eventually becoming Bishop of Orkney, though he only visited the place twice. He officiated at the highly controversial marriage of Mary, Queen of Scots, to the 4th Earl of Bothwell (no relation) only three months after the murder of Darnley, Mary's second husband, in which the Earl of Bothwell was strongly suspected of being involved. He also anointed the infant James VI.

Scotland's three universities were in a very poor condition at this time, both educationally and structurally, with St Andrews having an "entrenched medieval curriculum".[18] In 1560 Adam Bothwell wrote to Sir Archibald Napier regarding the education of the ten year old John Napier:[19]

> I pray you, Schir, to send your sone Jhone to the schuyllis; oyer to France or Flandaris; for he can leyr na guid at hame, nor get na proffeit in this maist perullus wordle, that he may be savet in it, that he may do frendis efter honnour and proffeit as I dout not bot he will.

In 1563 Napier became a student at St Salvator's College at St Andrews University, aged thirteen or fourteen.[20] However, Napier's mother, Janet, died shortly afterwards on 20th December 1563. It is clear from her will that Napier boarded within the college and under the care of the Principal because the statement of debts due at her death reads "item, to Johnne Rutherfurde for hir sonnis burde, auchtene [eighteen] pundis."

In 1563, the year Napier matriculated, a petition was presented to Queen Mary:[21]

> "in the name of all that within this realm ar desyrous that leirning and letters floreis", stating that the patrimony of some of the foundations in the colleges, particularly those of St Andrews, was wasted, and that several sciences, and especially those that were most necessary, the tongues and humanity, were very imperfectly taught in them, to the great detriment of the whole lieges, their children and posterity; and praying that measures should be taken to remedy these evils.

However, the chaotic state of Scotland prevented any effectual reform taking place until 1579. Therefore, it is hardly surprising that Napier, whose academic talents must have already been noticed, cannot be found in the

[18] Steven J. Reid, *Humanism and Calvinism, Andrew Melville and the Universities of Scotland 1560–1625*, Ashgate, 2011, pp. 22 & 24.

[19] Mark Napier, *Memoirs of John Napier*, p. 67.

[20] St Andrews, founded in 1413, was the first university to be established in Scotland while the College of St Salvator was founded in 1450.

[21] Thomas M'Crie, *The life of Andrew Melville*, William Blackwood, 1819, Vol. 1, p. 238.

college records after his matriculation. If Napier had remained at St Andrews his name would appear in the list of Determinants (Bachelor of Arts) for 1566 and of Masters of Arts for 1568, but it is not amongst them and it seems that only the basis of his education was laid at St Andrews.

In fact it was a common practice for anyone who had any pretensions to becoming learned to follow a few years of study at home with some years in a foreign university, as is borne out by some of Napier's contemporaries who subsequently distinguished themselves. It should also be noted that the Bishop of Orkney had previously advised exactly this course of action for his nephew. Napier's excellence in theology, his working knowledge of Greek and Latin, as well as his mathematical ability, all indicate that he received an excellent education.

The earliest specific mention of Napier's subsequent education appears to be by George Mackenzie (1669–1725) who wrote over a century later that:[22]

> Our Author had no sooner finiſhed his Studies in Philosophy at St Andrews, but he was ſent to his Travels by his Parents; and having ſtayed for ſome Years in the low Countries, France and Italy, he returned to his Native Country, and applyed himself closely to the Study of the Mathematics, in which he excelled all the Mathematicians of his Age.

At this time the Huguenots were well represented at the University of Paris, then one of the pre-eminent universities in Europe, and there were also long-standing political connections between Scotland and France, so it is quite possible that Napier attended there for at least part of the time he was abroad. Numerous other authors have subsequently speculated on which university or (more likely) which universities he attended. Besides Paris they have included Flanders, Italy (which seems unlikely due its strongly Catholic nature) and Geneva but the records are far too sparse to make any definitive statement regarding the location or nature of Napier's subsequent education.

He was certainly at home in 1571 (having just come of age) when the preliminaries for his marriage to Elizabeth Stirling were arranged at Merchiston. In December 1571 the respective fathers of the couple entered in to an undertaking in which Napier was invested with the Baronies of Edenbellie and Merchiston while Sir James Stirling agreed to pay a tocher (the marriage portion which a wife brings to her husband) of 3,000 marks. A contract was signed on 23rd February 1572[23] and on 2nd April 1572 a deed

[22] George Mackenzie, *The lives and characters of the most eminent writers of the Scots nation*, William Adams, 1722, Vol. 3, p. 519.

[23] Mark Napier, *Memoirs of John Napier*, p. 130.

A portrait of Napier's first wife, Elizabeth Stirling, daughter of Sir James Stirling of Keir. They married in 1571/2 when she was aged 16 but she died in 1579 having given birth to two sons. The eldest, Archibald Napier, became the 1st Lord Napier in 1627. The shield in the top right corner displays the arms of Napier of Merchiston on the left and those of Stirling of Keir on the right. 29½" x 24½". (Courtesy of Lord Napier and Ettrick.)

was signed at Merchiston by John Napier and Elizabeth Stirling preliminary to their marriage.[24]

Finally, a Royal Charter was issued on 8th October 1572 granting to Napier and his wife the lands of Edenbellie and Gartness in conjunct fee (jointly), while Napier also received "the lands of Merchiston with its tower and the Pultrielands", amongst others.[25] However, "The life-rent of all the lands save those in conjunct fee was reserved to Sir Archibald and his wife". In summary, Napier became the owner of most of the lands mentioned but the income they generated was retained by his parents. In fact his father, Sir Archibald, married for the second time at almost the same time as Napier married. Sir Archibald had six children by this marriage so Napier's half siblings would all have been of about the same age as Napier's own children, though this is less surprising when one recalls that Napier was only some sixteen years younger than his father.

After the death of his first wife, Elizabeth, in 1579, Napier married Agnes, daughter of Sir James Chisholm of Cromlix, by whom he had ten further children. The date of their marriage is not known but all the generally available sources state that their second son, Robert (who subsequently became his father's literary executor) was born in 1580. This would appear to be incorrect, given the well documented date of Elizabeth's death only a year earlier.

Napier succeeded his father in 1608 and there then followed a long drawn-out dispute with his half-brother Alexander due to "some settlements made to his prejudice by his father, in breach of a promise made, in 1586, in presence of some friends of the family, not to sell, wadset [loan], or dispose from his son John, and his heirs, the lands of Over Merchistoun".[26] These disputes seem to have been resolved by June 1613 when Napier was "retoured [returned] heir of his father, in the lands of Over Merchistoun".

CIVIL WAR

In 1568 civil war broke out in Scotland between the supporters of Mary, Queen of Scots, and those of the infant James VI, who was born at Edinburgh Castle in 1566. Mary had been forced to abdicate in favour of her son in 1567 and she fled to England in 1568 to seek the protection of Elizabeth I, while

[24] Sidney Lee (ed), *Dictionary of National Biography*, Macmillan & Co, 1894, Vol. 40, p. 60.

[25] This Royal Charter still exists and is held by the National Archives of Scotland (GD430/135/1/1).

[26] Sir Robert Douglas & John Philip Wood, *The peerage of Scotland*, Archibald Constable, 1813, p. 291.

Scotland was ruled by a series of Regents on behalf of the infant King. In 1571 Edinburgh came under siege for almost two years, the city being held by the Queen's faction against the besieging King's forces.

Sir Archibald's name does not appear in any of the retaliatory decrees of forfeiture which both parties issued against each other from their respective Parliaments. However, he seems to have incurred the displeasure of the Queen's party at this time, probably because he took little active part in the civil war, and on 18th July 1571 "Naper of Merchingstoun knycht wes tane, and brocht to Edinburgh castell be the laird of Mynto and his cumpany".[27] (Pollock, p. 233).

As his brother-in-law, Adam Bothwell, the Bishop of Orkney, had written to him a few years before, he was now "sett amiddis betwix twa grete inconvenientis" for the King's faction had kept him prisoner in his own house under heavy securities in 1568, when a dreadful plague broke out in Edinburgh and those affected were driven onto the Borough Muir near Merchiston, while in 1571 the Queen's faction summoned him to Edinburgh for not remaining in the vicinity of the town.[28]

The civil war seems to have entered a particularly violent phase in 1572, with Merchiston periodically being closely involved because of its strategic location:

> The regent and the kingis auouraris haid ftuffit [reinforced] the houffis of Craigmillar, Merchingftoun, Sclatfurd, Reidhall, Corftorphin, and the college thairof, and the abbay, with all places about the toun of Edinburgh. This wes done for the ftoping of the victuallis and vtheris neceffaris for furneifing [provisioning] of the toun of Edinburgh, [...] the saids regentis fettand furth be his proclomatioun, that quhatfumeuer [who so ever] come with any furneiffing to Edinburgh, the man to be hangit, and the woman drownit, (Pollock, p. 291) [...] And als [also] all inhabitouris within twa

[27] This and subsequent details regarding Merchiston and Sir Archibald Napier during the civil war are taken from the Pollock MS (Anon, *A diurnal of remarkable occurents that have passed within the country of Scotland since the death of King James the Fourth till the year MDLXXV, from a manuscript of the sixteenth century, in the possession of Sir John Maxwell of Pollock, Baronet*, The Bannatyne Club, 1833). There are similar references to these events in David Calderwood, *The history of the Kirk of Scotland*, The Wodrow Society, 1842–49, Vol. 3, pp. 212 & 213.

[28] Adam Bothwell's letter is quoted in Mark Napier's *Memoirs of John Napier*, p. 110. It appears that the lands surrounding Merchiston were often used for plague victims. On 11 March 1602 Napier brought a complaint against the provost and baillies of Edinburgh who had caused "build scheillis [rude huts] and ludgeis [temporary shelters] to their seik personis infectit with the pest [bubonic plague] upoun the said complenaris yairdis [enclosed grounds] of his proper lands of the schenis [Sciennes, an area south of Edinburgh, barely a mile north-east of Merchiston]". (Privy Council Register, vi., p. 359.)

> [two] myles to Edinburgh wer conſtranit to leave thair houſſis and lands, to that effect Edinburgh ſould haue na furneiſſing, and damnit poore men and women to the deid for inbringing of victuallis to Edinburgh. (Pollock, p. 293)

On 5th May 1572 the Queen's troops left Edinburgh to besiege Merchiston. They captured the outworks and finally the entire castle, with the exception of its "dungeoun, in the quhilk [which] wes certaine ſuddartis [soldiers] in Leith; the hail houſſis wes ſpoulgeit and brunt, to have ſmokit the men of the dungeoun out" (Pollock, p. 295) to which the Regent's garrison had retreated as being impregnable. However, the King's party in Leith were well aware of the strategic importance of Merchiston and marched in force to raise the siege, but this was not the last attempt by the Queen's forces to take Merchiston. On 10th June soldiers from Edinburgh besieged Merchiston again, bombarding it with a cannon and making breaches in its walls, until the Leith men came to their rescue. They captured many of the Queen's men, though the latter managed to get their cannon back to Edinburgh.

Sir Archibald seems to have been released from his confinement in Edinburgh and escaped forfeiture because on 3rd July 1572 he was summoned to appear before the queen on "pain of rebellioun and putting of him to the horne [proclaimed an outlaw]". (Pollock, p. 303).

Therefore, having been compelled to seek refuge elsewhere, he was now summoned by the Queen's faction to appear before them at the very time that the King's forces had garrisoned Merchiston with their soldiers. However, it is unlikely that Archibald or his son were at Merchiston for any length of time during the civil war since it was occupied by the King's troops. Instead he probably retired to his estates in the Lennox or Menteith.[29]

RELIGION

Napier was only nine years old when John Knox (1514–1572) returned to Scotland in 1559, the country of his birth, to lead the Protestant Reformation. The latter half of the 16th century was marked by bitter religious dissension and Napier was an uncompromising Protestant. From an early age he had no love for the Church of Rome, as he explained in the prefix to his *A Plaine Discovery of the whole Revelation of Saint Iohn*:

[29] The Lennox was an historic district, which included Loch Lomond, located about eight miles north of Glasgow, while Menteith was another historic district which lay between the River Teith and the Forth in south Perthshire.

> And although I haue but of late attempted to write this ſo high a worke, for preuenting the apparant danger of Papiſtrie ariſing within this Iland, yet in trueth, it is no few yeares ſince firſt I beganne to precogitat the ſame: For in my tender yeares and barneage in Sainct-Androes at the Schooles, hauing on the one parte contracted a louing familiaritie with a certaine Gentleman, &c. a Papiſt: And on the other pairt, being attentive to the Sermons of that worthie man of God, *Maister Christopher Goodman*,[30] teaching vpon the Apocalyps, I was ſo mooued in admiration, againſt the blindnes of Papiſts, that could not moſt euidently ſee their ſeuen hilled citie Rome, painted out there ſo liuely by Saint Iohn, as the mother of all ſpirituall whoredome, that not onely bursted I out in continual reaſoning againſt my ſaid familiar, but alſo from thenceforth, I determined with my ſelfe (by the aſsistance of Gods ſpirit) to employ my ſtudie and diligence to ſearch out the remanent myſteries of that holy booke.

The immediate cause for Napier to start on work on the *Plaine Discovery* seems to have been the arrival of the Spanish Armada in 1588 when he "was conſtrained of compaſſion, leauing the Latine, to haſte out in Engliſh this preſent worke, almoſt vnripe, that thereby, the ſimple of this Iland may be inſtructed, the godly confirmed and the proud, and fooliſh expectations of the wicked beaten downe".

The hierarchical authority of the Roman Catholic Church was formally abolished in favour of Reformation by the legislation of the Scottish Reformation Parliament in 1560 but that certainly did not mean that Roman Catholicism promptly disappeared from Scotland. Nor was the form of the newly emerging reformed church finally settled. The reformers wished to abolish Bishops, Dioceses and Parishes and enforce a Presbyterian system led by lay Elders but James VI viewed the Bishops as natural allies to the monarchy and prevaricated in his support for the Reformation.[31] In addition, James' contacts with Catholic Spain alarmed many Scots. Having commenced his work in 1588, Napier proposed to address his commentaries to James but the latter was absent at the time, on his expedition to marry Anne of Denmark in Oslo

[30] Christopher Goodman (1520–1603) was a reforming clergyman and close friend of John Knox, who fled from England to Geneva in 1554 during the reign of the Roman Catholic Mary I of England. In 1558 he published a tract bitterly attacking Mary and the rule of women in general, which did little to endear him to Elizabeth I who succeeded Mary later the same year. He re-joined Knox in Scotland in 1559 and was appointed to St Andrews the following year, preaching throughout Scotland. He returned to England in 1565 but was never much in favour due to his acrimonious temperament and uncompromising views. He obtained a living at Alford, near Chester, but was subsequently deprived of it for non-conformity.

[31] Presbyterians believe in the laity (the people) electing their pastors and elders while the Anglican Church appoints bishops and priests to serve the laity.

in November 1589, and they were not published until 1593, in the *Plaine Discovery*.

Napier had little to do with the court and its intrigues but as a fervent Protestant he was certainly involved in the religious dissension of the time. The first notice of this dates to 1588 when he was chosen by the Presbytery of Edinburgh as one of their Commissioners to the 54th General Assembly of the church. Here it was agreed that the "the presbytereis therof to give in the names of Papist excommunicats, and mainteaners of them to their commissioner".[32]

In late 1592 incriminating papers were found on George Kerr as he was about to sail for Spain. They included blank documents signed by prominent Catholic nobles which were, supposedly, to be filled in by others to encourage a Spanish invasion of Scotland. This affair became known as the Spanish Blanks Affair and one of those implicated was Sir James Chisolm, Napier's own father-in-law and an ardent Catholic.

In September 1593 a convention was established in Fife to consider the matter and Napier, who was a delegate, was a signatory to a proclamation which declared Sir James, along with five others:[33]

> in name and authoritie of the Lord Jesus Christ, did cutt the saids persons from their communioun, and delivered them to Satan, to the destructioun of their flesh, that the spirit might be safe, if so it pleased God to returne and reclame them by true repentance; otherwise, to their just condemnatioun everlasting;

The convention also appointed commissioners to seek an audience with James VI to personally state their views regarding the issue. Napier was one of the six chosen for this meeting, the others including James Melville, the Moderator of the Church, and James Maxwell of Calderwood. The "affemblie of Fyff was bittterlie inveyt againft" by the King who was incensed by the way in which they had acted and stated that he would not acknowledge the convention, which was convened without his authority, nor recognise the commissioners sent by the convention.[34] Melville replied that he "ftude honeftlie be it, faying it was in trew and vpright hartes, with all dewtie and reuerence to his Maieftie for preventing of imminent euill and danger to his Stat, Relligion, and Countrey."[35] Eventually James VI agreed to treat with the Commission-

---

[32] David Calderwood, *The history of the Kirk of Scotland*, Wodrow Society, Vol. 4, 1843,p. 683.

[33] Idem, 1844, Vol. 5,1844, p. 268.

[34] Idem, p. 271.

[35] James Melville, *The Diary of Mr James Melvill 1556–1601*, The Bannatyne Club, 1824, p. 208.

ers on a personal basis. They submitted their petition in writing and the king agreed to hold a convention to consider the matter further.[36]

Not even family connection could divert Napier from what he considered to be his duty to the reformed church.[37]

Napier wrote a Dedicatory Epistle to James VI for the first edition of the *Plaine Discovery* published in 1593. However, it was not included in the 1611 edition, by which time James VI of Scotland had also become James I of England. It clearly demonstrates Napier's strong views on what he believed the King should do to protect the Reformed church. Could any Prince, he argued, destroy the Roman Catholic church if he did not purge his own house, family and court? Could he purge his own house if he did not purge himself by private meditation with his God? Napier did not restrain himself when addressing the King on this issue, requiring him:

> in the meane time, vntill the reformation of that Idolatrous seate, to be preparing and purging your M. own ſeat and kingdome from all the enemies of that cauſe: [. . .] Therefore, Sir, let it be your M. continuall ſtudy (as called and charged thereunto by God) to reforme the vniuerſall enormities of your country, and firſt (taking example of the princely prophet David) to begin at your M. owne houſe, familie and court, and purge the ſame of all ſuſpicion of Papiſts and Atheiſts or Newtrals.

The main text the *Plaine Discovery* is divided into two parts, as described by Napier:

> THE FIRST AND INTRODUCTORY TREATISE, CONTAINING a ſearching of the true meaning of the Revelation, beginning the diſcouerie thereof at the places moſt eaſie, and moſt euidentlie knowne; and ſo proceeding from the knowen, to the proouing of the vnknowne, vntil finallie, the whole grounds thereof be brought to light, after the manner of propoſitions.

---

[36] Mark Napier records this incident in his *Memoirs of John Napier* but typically overstates Napier's involvement by stating that James VI's "timorous heart must have quaked at the sight of the unflinching moderator of the church and the majestic Merchiston" and the King "turning fiercely to Merchiston and Calderwood . . . " (p. 165 & p. 166). Melville's diary does not record any such dramatic detail.

[37] Despite this complete divergence of religious allegiance between Napier and his father-in-law, it is recorded that "On 28th of July 1593 [some two months before the Convention] he [Napier] became surety in £ 5000, that Sir James Chisholm, his father-in-law, should do nothing to hurt the King, the realm, or the true [Protestant] religion, during his [the King's] absence abroad". (Sir James Balfour Paul, *The Scots Peerage founded on Wood's edition of Sir Robert Douglas' Peerage of Scotland*, David Douglas, 1909, Vol. VI, p. 418 & Privy Council Register, v, p. 610). Similarly, on 30 April 1601 Napier became cautioner for his father's brother, Andrew Napier, "touching the mass which was said in his house" (Privy Council Register, vi, p. 632).

This first part consists of thirty-six Propositions, to each of which is added a few pages of "proof" by Napier, dealing with Rome and Catholicism. There then follows:

> THE SECOND and principall treatiſe wherein (by the former grounds) the whole Apocalyps or Reuelation of S. JOHN, is paraphraſticallie expounded, hiſtoricallie applyed, and temporallie dated with notes on euerie difficultie, and arguments on each Chapter.

This second part covers each of the twenty-two chapters of the Book of Revelation verse by verse in the first column. Napier paraphrased each verse in the second column and deduced a historical application in the third. Each chapter is followed by Napier's own 'Notes, Reasons and Amplifications'.

Napier stated that he wrote the *Plaine Discovery* in English instead of Latin so that 'hereby, the ſimple of this *Iland* may be inſtructed, the Godly confirmed and the proud, and fooliſh expectations of the wicked, beaten downe'. This enabled a much wider audience to read it in Britain, besides allowing easier access to the original text for the modern reader. It is interesting that Napier deliberately set out to use English that would be understood by both his English and Scottish audience, as he explained:

> also ſeeing herein purpoſely I preaſſ not to follow the particular ornat tearmes, of neither Scottes nor Engliſhmen: but rather contrarilie for both their inſtructions, I vſe ſo much as I can, theſe words and tearmes, that be more common and ſenſible to them both, then proper or ornate to any one of them.

Normally the use of English would not have encouraged the wider dissemination of the *Plaine Discovery* throughout Europe, where Latin was the common medium. However, the work was so well thought of that it was soon translated, passing through many editions in Dutch (1600 & 1607), French (1602, 1603, 1605 & 1607) and German (1611, 1612, 1615 & 1627) as well as three further English editions (1594, 1611 & 1645).[38]

In the first edition of the *Plaine Discovery* Napier promised to publish a more erudite Latin version sometime in the future but after eighteen years he had still not started on it. By then the various translations had been widely distributed throughout Europe and many, both Protestant and Catholic, had commented on the work. Therefore, Napier published a new edition in 1611, again in English. He opened the new text with an explanation for its necessity:

[38] William Rae Macdonald (trans), *The construction of the wonderful canon of logarithms by John Napier translated from Latin into English with notes and a catalogue of the various editions of Napier's works*, William Blackwood & Sons, 1889, pp. 109–128

And ſeeing there are certain wel-affected brethren, who not in the ſpirit of arrogancie and contention, but in all ſobrietie and meekenes, haue craved of me the reſolution of ſome doubts, ariſing vpon my Treatise of the Revelation: therefore, for diſcharging my duetie, I haue thought good to write a Reſolution of their doubts, and to inſert the ſame in this Treatiſe vpon the Revelation, for the better ſatisfaction of their reaſonable deſire, and inſtruction of others, meek and zealous perſons, whom the like doubts might hinder. As to the contentious and arrogant reaſoners, I leaue them to the mercie of the Lord.

SUPERSTITION, ALCHEMY AND INVENTION

Since his death Napier has been periodically cast in some works as a sorcerer who was in league with the Devil. However, this was an age when sorcery and witchcraft were brutally proscribed, not least by James VI who took a keen interest in the subject and whose reign saw the extensive persecution of witches. Indeed, James VI wrote an 80 page book on witchcraft (*Dæmonologie*, 1597) and became personally involved in some of the trials of alleged witches. Napier's rank in society would have been no protection against any such charge.

One story states that Napier used a black cock as his "familiar" (assistant) and, in particular, of using it to catch a thief.[39] Having noticed some of his property was missing and suspecting his servants, Napier told each of them that the cock would crow when they stroked it, if they were the thief. He then instructed each of them to enter a darkened room to stroke the cock which had previously been coated in soot. The cock remained silent throughout the ceremony but the hands of one of the servants were found to be free of any soot, indicating that he was too frightened by his theft to stroke the cock. However, one cannot help suspecting that this legend is in some way connected with the hereditary office of *King's Poulterer* (*Pultrie Regis*) which descended to Napier, having been held in the family for many generations.

Another story relates how Napier was annoyed by pigeons eating the seed in his fields so he saturated the seed with alcohol and was able to collect the drunken pigeons next morning. In Mark Napier's time, a field in front of Merchiston was pointed out as the scene of this exploit and he recorded that it was then still called "the Doo [Dove] Park".[40]

Mary, Queen of Scots, employed a Frenchman, Claude Nau, as her confidential secretary and at the time of the Queen's imprisonment at Loch Leven

[39] Mark Napier, *Memoirs of John Napier*, p. 214.
[40] Idem, p. 215.

Castle in 1567–68, he recorded in his diary of Napier's father, Sir Archibald, that:[41]

> The Laird of Markyston, who had the reputation of being a great wizard, made bets with several persons to the amount of five hundred crowns, that by the 5th of May her Majesty would be out of Lochleven.

In fact she escaped on 2nd May but there is no record of whether Sir Archibald was able to collect his winnings. It is a combination of these and other similar stories that has led some authors to conclude that Napier was involved in sorcery though none of them provide the slightest supporting evidence.

However, like many of his well-educated contemporaries (and Sir Isaac Newton some eighty years later) Napier did practise alchemy, something that was considered perfectly legitimate at the time and openly acknowledged. This interest in alchemy was shared by Napier's second son by his second marriage, Robert (who went on to become Napier's literary executor), who wrote a small manuscript on the subject.

Napier also signed an agreement with Sir Robert Logan (*c.*1555–1606) of Restalrig, a knight with a very dubious reputation, to search for treasure at his seat at Fast Castle, located on a promontory in Berwickshire jutting into the North Sea, about 40 miles east of Edinburgh.[42] Logan was described by one of his contemporaries as "ane godless, drunkin, and deboshit [debauched] man".[43] In June 1594, only a month before the contract with Napier was signed, Logan was "ordained to be denounced rebel" by the High Court of Justiciary for failing to appear before them and hand over some of his household servants for whom he was responsible. Robert Gray, a Burgess of Edinburgh, had been severely assaulted while on his way to Berwick by two of Logan's servants and three accomplices who:[44]

---

[41] Rev. Joseph Stevenson, *The history of Mary Stewart, from the murder of Riccio until her flight into England, by Claude Nau her secretary*, William Paterson, 1883, p. 80.

[42] Robert Logan was a supporter of Mary, Queen of Scots during the siege of Edinburgh and was implicated in the Gowrie Conspiracy of 1600 in which the 3rd Earl of Gowrie allegedly attempted to abduct James VI and hold him at Fast Castle. Logan's involvement only came to light in 1608, after his death, and his body was exhumed in 1609 and exhibited in court when the sentence of forfeiture was pronounced against him, the family losing all their estates. Restalrig is located about a mile north-east of Holyrood Palace, the family residing at Lochend Castle.

[43] Quoted without attribution in William Anderson, *The Scottish Nation; or the surnames, families, literature, honours, and biographical history of the people of Scotland*, A. Fullarton, 1862, Vol. 2, p. 688.

[44] Robert Pitcairn, *Ancient criminal trials in Scotland from AD 1488 to AD 1624*, The Bannatyne Club, 1833, Vol. 1, Part 2 (1584–1596), p. 335.

Fast Castle by Henry Bright (1810–1873). The castle belonged to Sir Robert Logan with whom Napier signed a contract to search for treasure within its grounds. The castle was largely derelict when this view was painted in the 19th century but now almost nothing is left standing. Oil on canvas, 24" x 36". (*Courtesy of the National Trust for Scotland.*)

> not onlie reft [robbed] and fpuilzeit [plundered] fra him nyne hundreth and fiftie pundis money quhilk [which] he had vpoun him, bot alfwa, maift cruellie and barbarouflie invadit and perfewit him of his lyffe, hurte and woundithim in the heid, and ftraik him with diuers vtheris bauch ftraikis vpoun his body, to the grite danger and perrill of his lyffe.

The contract between Napier and Sir Robert states that there were reports that treasure was "heid and hurdit up secritlie, quilk [which] as yit is on fund be ony man". Napier would "do his utter and exact diligens to serche and sik out, and be al craft and ingyne that he dow, to tempt, trye, and find out the sam" and that Napier would receive "the just third pairt of quhatsoewir poiss [hoard] or heid treasour the said Jhone sall find".[45]

[45] This transcript is taken from the facsimile of the original document reproduced in Mark Napier's, p. 220. The present location of the original is unknown but it is not held by the National Library of Scotland nor the National Records of Scotland. A facsimile of the original document and a full transcript are reproduced in Appendix 3 (pp. 949–951).

EDYNBURGUM.

With the exception of Logan's signature, the entire contract is written by Napier himself. The contract also specifically states that he should be guaranteed a safe conduct back to Edinburgh, escorted by Logan. Clearly Napier was under no illusions regarding Robert Logan's character. Whether or not Napier actually visited Fast Castle in pursuit of the treasure is unknown but the fact that the contract was not destroyed, as it states it should be at the conclusion of the search, suggests that the search did not take place. However, it is clear that the agreement did not proceed amicably because in 1596 Napier inserted a clause in his tenancy agreements forbidding his lessees from subletting to or engaging anyone by the name of Logan.[46]

Two years later Napier wrote a manuscript entitled *Secrett Inventionis proffitabill and necessary in theis dayes for defence of this Iland, and withstanding of strangers, enemies of Godes truth and Religion.* The original of this document is still extant and it sets out a number of ideas that bear an uncanny resemblance to modern day machine guns, tanks and submarines:[47] Napier's summary of his *Secrett Inventionis* provides the outcome of his inventions but does not reveal how his devices could be made to work. This approach was typical of the period and even the greatest of his inventions, logarithms, was first published in the *Descriptio* as a tool to be followed by rote. It was not until the *Constructio* was published five years later that the method of calculating logarithms was revealed.

On 30th January 1597 Napier was granted a Precept of Sasine (an instrument conferring a privilege) by James VI for his invention of a "machine for

---

Facing page. *A view of the city of Edinburgh drawn by Wenceslaus Hollar (1607–1677) in about 1641, though it is not certain if Hollar ever actually visited Edinburgh. It was published in Matthaus Merian's* Theatrum Europaeum *in 1649. The view shows Edinburgh looking towards the north. Edinburgh castle is shown to the left (west) with Holyrood Palace a mile to the right (east) and Calton Hill in the centre, beyond the town. The port of Leith is shown in the distance with the island of Inchkeith on the right and the Firth of Forth beyond. Merchiston Tower is located outside this view to the south-west of the Castle, almost exactly the same distance from the Castle as is Holyrood. The book was begun in 1635 by Matthaus Merian the Elder and upon his death in 1650 it was continued by his son, Merian the Younger, resulting in a total of 21 volumes to this huge work. 7.5" × 13.5".*

[46] Mark Napier, *Memoirs of John Napier*, p. 223.

[47] Lambeth Palace Library, MS 658, f 58. A facsimile of the original document and a full transcript are reproduced in Appendix 4 (pp. 952–954).

drawing water which in a single revolution brings up water more productively than buckets, pipes or any other machine previously used in Scotland, whence waterlogged coal pits may be quite recovered with small expense [. . .] Napier and his heirs male and their spouses were granted the sole privilege of erecting and using such machines for all time in the future, with power to erect them wherever in realm necessary; preventing others from using them without Napier's licence, under penalty of 2,000 merks Scots, half to be paid to the king and half to Napier".[48]

Although this device sounds very much like the well-known Archimedes screw there are insufficient details within the Precept to be able confirm this or to determine how the machine worked. However, it must have had some advantage over the previous technology to be granted what was effectively a patent.

FAMILY ISSUES

Archibald, John Napier's eldest son by his first marriage, attended Glasgow University from 1593. After completing his studies he did not travel abroad, as many of his well-connected contemporaries would have done, but almost immediately joined the household of James VI. Here he enjoyed a successful career in the service of his sovereign, including time spent in London when James VI became James I of England. He became Gentleman to the Bedchamber and a Lord of Session, eventually being raised to the peerage as the 1st Lord Napier in 1627. However, in 1598, at the very start of his career, he was awarded a patent by James VI for manuring the land with salt to increase the yield. Certainly, Scotland was no stranger to poor harvests resulting from poor weather. In 1595 it was recorded that "During all this yeir thair was a great scant of cornes, and exceiding great derth [. . .] and manie poor people deit for hungar, and sum of better estait had na better conditioun",[49] while 1596 proved to be little better.

Of course it was widely known that too much salt kills most plants so its application required particular care. Archibald published a small pamphlet in 1598 explaining in great detail how the soil should be prepared before it was manured with salt, how much salt should be used for different types of soil and crops, as well as when this should be done.[50]

---

[48] National Records of Scotland, GD430/150.

[49] John Colville and Thomas Thomson, *The historie and life of King James the Sext being an account of the affairs of Scotland from the year 1566 to the year 1596, with a short continuation to the year 1617*, The Bannatyne Club, 1825, p. 355.

[50] Most of this small work was published by Mark Napier in his *Memoirs of John Napier*

Clearly much thought and experimentation had gone into the patent to prove its efficacy so it is doubtful that Archibald, who was only about 23 at the time, was responsible for this. As Mark Napier suggests, it is much more likely that it was the work of Napier himself, despite the patent lauding Archibald as being an expert in the subject. It was probably given to Archibald as a present so that he could benefit from any royalties it generated:[51]

Napier's father, Sir Archibald, married for a second time at almost the same time as Napier was married for the first time. Sir Archibald's new spouse was Elizabeth Mowbray and they had a further six children, including three sons, Alexander, Archibald and William who were of about the same age as Napier's own children. However, the sons proved to be rather wild. Archibald and William wished to join an equally restless uncle, Francis Mowbray, in the Low Countries who had been outlawed for running a rapier through a certain William Schaw. The fact that Mowbray was also a Catholic consorting with known Papists in exile did little to endear him to the Protestant Sir Archibald and his son John.

In August 1600 the young Archibald was riding in the border country between Scotland and England, an area well known for its lawless Border Reivers (raiders) who indiscriminately stole cattle and personal goods from both sides of the border, regardless of their victim's nationality. Archibald's horse was either stolen or strayed and while searching for it he met a certain Scott of Bowhill who took Archibald's questions about his horse's whereabouts as a personal affront. Archibald managed to extricate himself from the situation but later that evening Scott waylaid him in a lonely spot and Archibald killed his adversary during the ensuing fight. It appears that no attempt was made to bring Archibald to justice since he had been provoked into the attack but on 8th November Scott's brothers waylaid Archibald as he was riding alone to his home in Edinburgh and viciously slew him near the Palace of Holyrood House. The diarist Robert Birrel (fl. 1567–1605) recorded that:[52]

> The aucht [8th] of November, Archibald Napier slaine by fyve of the name of Scot and Thomas Crichton, ryding hame to hes awen hous to the Wowmit.[53]

(p. 284). The original pamphlet was printed by Robert Waldegrave, a prolific printer in London and then to James VI in Edinburgh, but no copies appear to have survived. It was reprinted in *Archaeologia Scotica or Transactions of the Society of the Antiquaries of Scotland*, William Blackwood, 1818/22, Vol. II, Part 1, p. 154. Extracts from this document are reproduced in Appendix 5 (p. 955).

[51] Mark Napier, *Memoirs of John Napier*, p. 283 quoting Privy Seal 70, 22nd June 1598.

[52] Thomas Dalyell, *Fragments of Scotish history*, Archibald Constable, 1798, p. 52.

[53] Woolmot, near Edinburgh.

Sir Archibald sought redress from James VI who was equally shocked by the event. Both Alexander and William proposed revenge for their brother's death but were mortified to find that their proposal only produced a severe rebuke from their father, Sir Archibald, and from Napier himself. The sons regarded any inaction on their part as reflecting dishonour on themselves and therefore felt that they should quit Scotland. Mark Napier reproduced two letters that Sir Archibald and John Napier wrote, the first appearing to be an unsent draft, possibly in Napier's hand, while the second, certainly in Napier's hand, was signed by both of them and sent.[54] These letters are of particular interest since they are rare survivors relating to Napier's personal life, almost everything else having been lost. The first starts:

> SONEIS, – We greit you with sick commendatiouns as your folifull deid deserves, be ministering unto us new occaisoun of gritter displessour. Ye lament onlie your brothir's deid, but we lament bayth his deid and your foleis, and the mair, that as he was in his tyme uncounsalabill, sua ye appeir to follow the samyn cours. God mak ye end better. We haif ressavit your letter, or rather you fuillische gudenycht, by the quhilk [which] ye wald bring upoun us the occasioun of this your vane interpryse; allegeing us to be slak in the revengeing of the lait injurie doen be our enemeis, sayinge ye proponit [propose] dyvers wayis to haif reparit the wrange done, bot we would nawyis prosecute your devices, bot objectit in the contrair. Judgement, wit, and experience craveis ressouning in materis of grit importance befoir conclussioun.

The second letter is somewhat less forthright but still makes it quite clear that Sir Archibald and John Napier strongly opposed any revenge. It was sent from "Merchistoun this last of November 1600" and, after suggesting that any course of action would have to be reviewed by the whole family, their relations and friends, it continues:

Facing Page. *A lithograph of "Merchiston" viewed from the north-west and drawn sometime between 1770 and 1778. It is therefore the earliest view that was published, though first publication did not take place until 1825, 13 years after the self-taught artist, John Clerk died* (*John Clerk,* Etchings chiefly of views in Scotland 1773–1779, *Bannatyne Club, 1825 and 1855*). *Merchiston Tower was where John Napier was probably born in 1550, published his invention of logarithms in 1614 and died in 1617. Merchiston Tower still exists, largely unchanged from this view but now surrounded by the concrete of Napier University. 5.75" × 8.5".* (*Courtesy of the National Library of Scotland.*)

[54] Mark Napier, *Memoirs of John Napier*, p. 303.

> Bot put the caiss it wer sa, is that the occasione of your oncumlie depairting? nay surlie, altho ye mak it the only pretext thereof, for it is knawin to God, to the world, and to your awin consciencis, that this lang season, before the said slaughter and lett pretext of dishonor, ye heff socht all occasionis, pretextis, and causes ye culd, to heff departit away. We wold wisch you, in the feir of God, to behaif yourselfis mair plainlie and obedientlie to us your parentis than to be making sik pretextis to colour your inobediences, quherby [whereby] ye do that in you lyis, to bring us, that are your parentis, to the graif with sorrow, as also to heipe on yourselfis the wratht of God therfor. Ye tak in mynd deiplie the reveng of the said murther! and na marvell, for sa do we all; but quhy [why] neglek the originall pairt of the tragedie? forgetting how your umquhil [late] brother became ane inobedient and stubborn contemptnair of father and mother. . . .
>
> Be yours loving parentis,
>
> S^r^ AR^d^ NEPER.
>
> JHONE NEPER.

Records of the High Court of Justiciary show that criminal proceedings were started by the Napiers against the Scott brothers in January 1601 "for the Slauchter of the ſaid vmq^le^ [former] Archebald; be the quhilkis [which] lettres the ſaidis perſonis ar denunceit our ſouerane lordis rebellis, and put to the horne [proclaimed an outlaw].[55]

However, no further entry appears to exist, and it is quite likely that no redress was obtained by the family. The monarchy was always short of funds and it would not have been the first time that James VI issued a royal pardon in exchange for a contribution to his depleted coffers.

LOGARITHMS

By a fortunate chance it is known that in 1594 the Danish astronomer Tycho Brahe (1546–1601) was indirectly advised of Napier's work on logarithms prior to their publication. As it turned out, Napier's other commitments, as well as the sheer labour of the computations themselves, delayed their publication for some twenty years until *Descriptio* was issued in 1614.

Sir Archibald Napier's colleague in the office of Justice-Depute was Sir Thomas Craig (*c.* 1538–1608) of Riccarton. Napier had become a close friend of the latter's younger brother, John Craig (*d.* 1620).[56] Craig and Brahe communicated with each other in a very friendly manner, with Craig addressing

[55] Robert Pitcairn, *Ancient criminal trials in Scotland from AD 1488 to AD 1624*, The Bannatyne Club, 1833, Vol. 2, Part 2 (1600–1609), p. 339.

[56] Mark Napier in his *Memoirs of John Napier* (p. 361) describes John Craig as Thomas

Brahe as his "Honoured friend" and signing himself "Your most affectionate, John Craig, Doctor of Philosophy and Medicine". The first recorded letter from Craig to Brahe commences "About the beginning of last winter that magnificent man Sir William Stuart delivered to me your letter and the book you sent."[57] The date is not given but a mathematical work of Brahe's in the library of Edinburgh University has a manuscript dedication in Latin on the first blank leaf:[58]

> To that most celebrated man, and one endowed with varied and excellent learning, Doctor John Craig, practising medicine in Edinburgh in Scotland, and most skilled in mathematics, I send this as a gift, Tycho Brahe, and this he wrote in his own hand at Uraniburg the 2nd of November in the year 1588.[59]

This book is clearly that referred to in Craig's letter to Brahe which must have been written, therefore, in the early part of 1589. Craig had long wanted to visit Brahe, as he subsequently makes clear in his response to Brahe:[60]

> In truth from the time when the most learned Wittich[61] returned from Denmark, he acclaimed you to such an extent, that from that time I have held you in love and honour, and five years ago [*i.e.* in 1584] when I was to sail to your homeland through your Porthmus, which he had recommended, I had personally longed and determined to do so, but contrary winds having blown up and Norwegian rocks lying in the way before your Urania [Uraniborg], my desires were with regret disappointed.

Craig's third son but this has since been shown to be incorrect – he was Thomas' younger brother. John Craig was a proficient mathematician, having matriculated at the University of Frankfurt an der Oder in 1573, and was Professor of Mathematics and Logic there when he graduated MD from the University of Basle in 1580. He returned to Edinburgh in about 1582 where it seems he concentrated on his medical practice, eventually being appointed Principal Mediciner to James VI in 1603. He moved from Edinburgh to London shortly after James acceded to the English throne. (*Oxford Dictionary of National Biography*, Oxford University Press, 2004).

57 Rud. Aug. Noltensii [Rudolf August Nolte], *Commercium litterarium clarorum virorum*, Brunsuigæ [Braunschweig/Brunswick], 1737–38, p. 1. None of the letters are dated.

58 This copy of Tycho Brahe's *De mundi aetherei recentioribus phaenomenis liber secundus* (Uraniburgi, 1588) is still held in the University of Edinburgh Library under Shelf-mark Dd.5.50.

59 Uraniborg is located on the small island of Hven, in the narrow Øresund Strait between present day Denmark and Sweden. Hven was under Danish rule until 1658.

60 Rud. Aug. Noltensii [Rudolf August Nolte], *Commercium litterarium clarorum virorum*, Brunsuigæ [Braunschweig/Brunswick], 1737–38, p. 2.

61 Paul Wittich (*c.* 1546–1586) was a German mathematician and astronomer who studied at the universities of Leipzig, Wittenberg and Frankfurt an der Oder. He briefly worked with Tycho Brahe at Uraniborg in about 1580 before being employed by Landgraf Wilhelm IV of Hessen-Kassel.

Only a few months later, in August 1589, James VI was betrothed to the fourteen year old Anne of Denmark but she became storm-bound in Norway when trying to sail to Scotland for her marriage. On hearing this James immediately sailed to Oslo where they were married in November 1589. He then visited Brahe at Uraniborg for one day in March 1590 before returning to Scotland. Mark Napier in his *Memoirs of John Napier* (p. 365) states that Craig accompanied James VI to Norway and probably encouraged the king to visit Brahe. However, Craig was not appointed as James' Principal Mediciner until 1603 and there appears to be no independent proof that he accompanied the king on this voyage or that he ever met Brahe. On the other hand, it is almost certain that Craig knew James VI long before his appointment as Principal Mediciner and it is therefore quite possible that Craig did indeed accompany the king to Uraniborg. Either way, it is clear that Brahe had been told of the forthcoming benefit of logarithms by his correspondent Craig, because in a letter written in 1624 by Kepler to his friend Krüger [62] he comments on those who had done most to advance mathematics. In the process Kepler reveals that Napier was working on logarithms at least twenty years prior to their eventual publication: [63]

> Moreover, I consider nothing to be superior to Napier's method: even though a certain Scotsman [John Craig], in a letter written to Tycho in the year 1594, already expressed hope of that wonderful Canon.

The *Descriptio* was first published in 1614 and an early enthusiast of logarithms was Henry Briggs who was eleven years younger than Napier. [64] His letter of 10th March 1615 to James Ussher clearly shows that he was very ex-

[62] Peter Krüger (1580–1639), sometimes Peter Crüger which in turn was Latinised as Petrus Crugerus, graduated from Wittenberg but spent the rest of his life in Danzig (now Gdańsk) where he became Professor of Mathematics and of Poetry at the Academic Gymnasium. He published treatises on many scientific subjects and contributed to the advancement of trigonometry, geography and astronomy.

[63] Michael Gottlieb Hanschius [Hansch] (ed), *Epistolæ ad Joannem Kepplerum mathmaticum Cæsareum scriptae insertis ad easdem responsionibus Kepplerianis, quotquot hactenus reperiri potuerunt*, Frankfurt, 1718, p. 460.

[64] Henry Briggs (1561–1630) was a "Geometry-reader" (mathematician) who attended St John's College, Cambridge, graduating in 1581. He became interested in navigation and astronomy, collaborating with Edward Wright. In 1596 he became the first Professor of Mathematics at the newly founded Gresham College in London and in 1619 he was appointed Savilian Professor of Geometry at Oxford University, residing at Merton College until his death.

cited by Napier's work.[65] After speaking of various other matters he wrote:[66]

> *Napper*, Lord of *Markinſton*, hath ſet my Head and Hands a Work, with his new and admirable Logarithms. I hope to ſee him this Summer if it pleaſe God, for I never ſaw [a] Book which pleaſed me better, or made me more wonder.

Shortly before Briggs arrived in Scotland in the summer of 1615, Napier was prepared for the visit by the presence of John Marr, a mathematician attached to the household of James I. Marr described the meeting in some detail to William Lilly who included it in his posthumously published autobiography, though the statement that on first meeting "almost one quarter of an hour was spent, each beholding other with admiration, before one word was spoke" must be considered an exaggeration.[67]

Having travelled to Edinburgh and stayed a month at Merchiston in 1615, the pair clearly got on well and had much to discuss so Briggs repeated his visit in 1616. In 1617 he again prepared to travel north but Napier died before the arrival of the summer travelling season. However, Briggs published *Logarithmorum Chilias Prima* in that year and *Lucubrationes et Annotationes in Opera Posthuma J Neperi* in 1619.

No sooner was the *Descriptio* published in 1614 than it found an able and enthusiastic translator in Edward Wright. In the Foreword to Wright's translation his son, Samuel, states that his father "gaue much commendation to this worke (and often in my hearing) as of very great uſe for Mariners". This must have been in the first year of its publication, for in that or the following year Wright sent Napier his translation for review and Samuel states that "ſhortly after he had it returned out of *Scotland*, it pleaſed God to call him away afore he could publiſh it." Therefore the task devolved upon Samuel Wright, with the assistance of Henry Briggs who was calculating his improved logarithmic tables. The translation was printed in London by Nicholas Okes in 1616. Samuel Wright noted his father's clear understanding that logarithms would be of particular value to navigators when he wrote his Foreword, which

---

[65] James Ussher (1581–1656) was a highly influential theologian and a prolific scholar who became Archbishop of Armagh and Primate of all Ireland in 1625. Most famously, he used the Bible to "calculate" that the beginning of the world occurred on the night preceding Sunday 23rd October 4004 BC.

[66] Richard Parr (ed), *The life Of the Most Reverend Father in God, James Usher, Late Lord Arch-Bishop of Armagh, Primate and Metropolitan of all Ireland. With a Collection of Three Hundred letters, between the said Lord Primate and most of the Eminentest Persons for Piety and Learning in his time, both in England and beyond the Seas*, Nathanael Ranew, 1686, Letter XVI, p. 36 of the letters.

[67] See Appendix 6 (pp. 956–957) for Lilly's description of the first meeting between Napier and Briggs.

was dedicated to the East India Company whose "imployment of ſo many Mariners in ſo many goodly and coſtly ſhips, in long and dangerous voyages, for whoſe vſe (though many other wayes profitable) this little booke is chiefly behoouefull".

Napier's concluding words to the 1616 English translation of the *Descriptio* accurately summarises the enormous value of logarithms:

> Now therfore it hath been ſufficiently ſhewed that there are Logarithmes, what they are, and of what vſe they are: for with helpe of them, we haue both demonſtratiuely ſhewed and taught by examples of both kindes of Trigonometrie, that the Arithmeticall ſolution of any Geometricall queſtion may moſt readily bee performed without trouble of Multiplication, Diuiſion, or extraction of roots.

On the last page of the tables in some copies of the original Latin edition of the *Descriptio*, but not in the translation nor in any other edition, is the following interesting "Admonition" from Napier himself:[68]

> Seeing that the calculation of this table, which ought to have been perfected by the labour and pains of many calculators, has been finished by the operation and industry of one alone, it is not surprising if many errors have crept into them. I beseech you, benevolent readers, pardon these, whether caused by the weariness of computation or an oversight of the press; for, as for me, declining health, and weightier matters have prevented my adding the last finish. But if I shall understand that the use of this invention proves acceptable to the learned, I will, perhaps, shortly give (God willing) the philo sophy, and method either of amending this Canon, or of constructing a new one upon a better plan; so that through the diligence of many calculators, a Canon more highly finished and accurate than the work of a single individual could effect, may at length see the light.
>
> *Nothing is perfect at its birth.*

Facing page. *An etching of "Marchiston Tower" viewed from the north-east and drawn in 1789. The two storey building with a sloping roof in the re-entrant between the two legs of the L and the large windows would not have existed in Napier's time. In addition the top floor did not exist at this time since the two pitched roofs would have been much shallower, covering a stone barrel vaulted ceiling to the floor below. 4.75" × 6.75".* (*Francis Grose*, The Antiquities of Scotland, *S. Hooper, 1791*).

[68] Mark Napier states in his *Memoirs of John Napier* (p. 412) that he had seen copies of the original 1614 edition of the *Descriptio* which did not include this Admonition. This is confirmed by William Rae Macdonald in *The construction of the wonderful cannon of logarithms* (William Blackwood & Sons, 1889, p. 86).

It must have been early in 1614 that the *Descriptio* was published because Edward Wright died in November 1615, but by then he had completed the translation, sent it to Napier in Scotland and received the revision back before his death. The translation was then published by Samuel Wright under the guidance of Briggs. The work opens with a Dedication to Prince Charles, James I's eldest son:

> Seeing there is neither ftudy nor any kinde of learning that doth more acuate and ftirre vp generous and heroicall wits to excellent and eminent affaires: and contrariwife that doth more deiect and keepe downe fottifh and dull mindes then the Mathematickes. ... and efpecially feeing this new course of *Logarithmes* doth cleane take away all the difficultie that heretofore hath beene in mathematicall calculations, (which otherwife might haue been diftaftfull to your worthy towardneffe) and is fo fitted to helpe the weakneffe of memory, that by meanes thereof it is eafie to refolue moe Mathematical queftions in one houres fpace, than otherwife by that wonted and commonly receiued manner of Sines, Tangents, and Secants, can bee done euen in a whole day. ... For what can bee more delightfull and more excellent in any kinde of learning then to difpatch honourable and profound matters, exactly, readily, and without loffe of either time or labour.

Napier's own Preface follows:

> Seeing there is nothing (right well beloued Students in the Mathematickes) that is fo troublefome to mathematicall practife, nor that doth more moleft and hinder Calculators, then the Multiplications, Diuifions, fquare and cubical Extractions of great numbers, which befides the tedious expence of time, are for the most part fubiect to many flippery errors, I began therefore to confider in my minde, by what certaine and ready Art I might remoue thofe hindrances.

In the original Latin edition Napier adds a verse of his own, which is not given in the translation:

IN LOGARITHMOS
Quae tibi cunque sinus, tangentes atque secantes
Prolixo preaestant, atque labore grav:
Absque labore gravi, et subito tibi, candide Lector,
Haec Logarithmorum parva tabella dabit.

[REGARDING LOGARITHMS
Whatever sines, tangents and secants
present you with after lengthy and heavy labour:
Fair Reader, this little table of logarithms will give you,
without serious toil, and immediately.]

Although logarithms in the form published by Napier were almost immediately superseded by the more practical variant to Base 10, it is remarkable how accurate they were. Macdonald has shown that the method employed by Napier was capable of producing logarithms such that "the true logarithm will always be very nearly equal to the logarithm found by Napier's method of computation less a six hundred billionth part".[69]

However, Napier made an error in his calculations which adversely affected the accuracy of his final figures. Biot identified the source and published it in his abstract of Mark Napier's *Memoirs of John Napier*, stating that he had:[70]

> detected that there is in fact a slight error of this kind [of calculation], a very slight error, in the last term of the second progression which he [Napier] forms prepartatory to the calculation of his table. Now all the subsequent steps are deduced from that, which infuses those slight errors that have been remarked.

After Napier's death in 1617 and having been appointed the first Savilian Professor of Geometry at Oxford in 1619, Briggs devoted himself to the computation of a completely new table of logarithms in the form agreed with Napier before the latter's death, but to a much higher degree of accuracy. He also developed a number of new methods for calculating the values, each designed to simplify this extremely laborious task. The result was published as *Arithmetica Logarithmica* in 1624, which listed the logarithms of 30,000 numbers (1–20,000 and 90,001 to 100,000) to 14 decimal places. In his Preface to this work, published seven years after Napier's death, Briggs makes it clear that it was Napier who proposed the new format to Base 10:[71]

> That these Logarithms differ from those which that illustrious man, the *Baron of Merchiston*, published in his Canon Mirificus, must not surprise you. For I myself, when expounding publicly in *London* their doctrine to my auditors in *Gresham* College, remarked that it would be much more convenient that 0 should stand for the Logarithm of the whole sine, as in the Canon Mirificus, but that the Logarithm of the tenth part of the same whole sine, that is

[69] William Rae Macdonald (trans), *The construction of the wonderful cannon of logarithms by John Napier translated from Latin into English with notes and a catalogue of the various editions of Napier's works*, William Blackwood & Sons, 1889, p. 90.

[70] Jean-Baptiste Biot's abstract was translated from French and published as an article (Abstract of the Memoirs of John Napier of Merchiston by M. Biot) in *The Edinburgh new philosophical journal*, Adam & Charles Black, 1836. Vol. XX, October 1835 – April 1836. The article starts on p. 255 but the accuracy of Napier's tables is discussed on p. 285.

[71] Henricvs Briggivs [Henry Briggs], *Arithmetica Logarithmica*, Gvlielmvs Iones [William Jones], 1624, p. A3. The translation is from Mark Napier's *Memoirs of John Napier*, p. 409.

to say, 5 degrees, 44 minutes, and 21 seconds should be 10,000,000,000. Concerning that matter, I wrote immediately to the author himself; and, as soon as the season of the year and the vacation time of my public duties of instruction permitted, I took journey to *Edinburgh*, where, being most hospitably received by him, I lingered for a whole month. But as we held discourse concerning this change in the system of Logarithms, he said, that for a long time he had been sensible of the same thing, and had been anxious to accomplish it, but that he had published those he had already prepared, until he could construct tables more convenient, if other weighty matters and his frail health would suffer him so to do. But he conceived that the change ought to be effected in this manner, that 0 should become the Logarithm of unity, and 10,000,000,000 that of the whole sine; which I could not but admit was by far the most convenient of all. So, rejecting those which I had already prepared, I commenced, under his encouraging counsel, to ponder seriously about the calculation of these tables; and in the following summer I again took journey to *Edinburgh*, where I submitted to him the principal part of those tables which are here published, and I was about to do the same even the third summer, had it pleased GOD to spare him to us so long.

After falling out with the new King of Denmark, Brahe left Uraniborg in 1597 but it was not until 1599 that he settled near Prague and was appointed as Imperial Court Astronomer to Rudolf II (1552–1612), King of Hungary, Croatia and Bohemia, Archduke of Austria and Holy Roman Emperor. Here he and his assistant Longomontanus returned to their observations. Kepler (1571–1630) joined them in 1600 and in the following year he was presented by Brahe to his patron Rudolph II. Following Brahe's sudden death in 1601 Kepler was appointed as Imperial Mathematician. This was the beginning of Kepler's long connection with his famous *Rudolphine Tables* which were not published until 1627. These tables involved an enormous number of calculations but Kepler had completed most of those related to the astronomical discoveries now referred to as Kepler's Laws before he learnt of the invention of logarithms.[72] Kepler was in correspondence with many of his scientific contemporaries throughout Europe and in a letter dated 11th March 1618, four years after the publication of the *Descriptio*, he wrote (in Latin) to his friend Schickard.[73]

[72] The sheer quantity of calculations undertaken by Kepler is detailed in Francis Jeffrey (ed), *The works of John Playfair Esq*, Archibald Constable, 1822, Vol. 2, p. 181.

[73] Wilhelm Schickard (1592–1635) graduated from the University of Tübingen and became a Lutheran minister before being appointed Professor of Hebrew at the same university. He was later appointed there as Professor of Astronomy but his broad interests included mathematics and surveying. He invented a mechanical calculator that incorporated Napier's

A view of Merchiston Tower painted and glazed on a china plate. It was given to Nigel, 14th Lord Napier, in 1958 as a wedding present by Col. Charles Napier, the son of Col. John (Jack) Scott Napier, the 2nd son of Francis 10th Lord Napier. Jack Napier (1848–1938) was the last Napier to own Merchiston, having inherited it from his mother on her death in 1911. The painting used to hang in Merchiston Tower and the inscription on the frame in which the plate is held states that it dates from c. 1850. 5¼" dia. (Courtesy of the Dowager Lady Napier and Ettrick).

Bones which he described in a letter to Kepler. Therefore this machine, which was destroyed in a fire before it could be delivered to Kepler and was not replaced, predated the generally accepted first invention of a mechanical calculator by Blaise Pascal (1623–1662). However, there is some doubt as to whether Schickard's machine would have been fully functional. He and his entire family died of the bubonic plague.

After detailing the various difficulties and resources of trigonometry, he stated:[74]

> A Scottish baron has appeared, whose name escapes me, but he has proposed some wonderful method by which all necessity of multiplications and divisions are commuted to mere additions and subtractions; nor does he make any use of a table of sines. However, he still requires a table of tangents and in some cases the variety, frequency and difficulty of additions and subtractions exceed the labour of multiplication and division.

This statement by Kepler is rather obscure since it seems to indicate that although he had seen a demonstration of logarithms before writing this letter to Schickard, Kepler had not fully understood how they were used (and therefore probably had not used them himself) since "a table of tangents" (*At tamen opus est ipsi Tangentium Canone*) is not required. Nor can Kepler's concluding sentence be considered an accurate assessment of logarithms.

This was Kepler's first impression of Napier's work but later in the same year he happened to read *Cursus mathematici practici volumen primum*, published in 1618 and written by his former assistant Benjamin Ursinus, in which Napier's tables were reproduced for the first time on the Continent.[75] Kepler soon recognised the potential of logarithms and he wrote a long letter to Napier in 1619, publishing it as a generous Dedication to Napier in his *Ephemerides* for 1620, shortly before he published *Harmonices Mundi.* Despite Kepler's wide correspondence throughout Europe, it is clear that he was not aware that Napier had died two years previously.[76]

Kepler became an enthusiastic promoter of Napier but without having any description of how Napier generated his logarithms, he set out to create his own using a geometrical procedure, the results of which were published in the *Chilias logarithmorum* in 1624. Kepler also dedicated his *Ephemerides* to 'The Illustrious Baron', writing:[77]

---

[74] Michael Gottlieb Hanschius [Hansch] (ed), *Epistolæ ad Joannem Kepplerum mathmaticum Cæsareum scriptae insertis ad easdem responsionibus Kepplerianis, quotquot hactenus reperiri potuerunt*, Frankfurt, 1718, p. 672.

[75] Charles Coulston Gillispie, *Dictionary of scientific biography*, Charles Scribner's Sons, 1974, Vol. IX, p. 613.

[76] Ioanne Kepplero, *Ephemerides novæ motuum coelestium ab anno vulgaris æræ MDCXVII / Ephemeris motuum coelestium ab annum incarnationis verbi MDCXX*, Iohannes Plancvs, 1617–1620, sig. N recto. A translation of the greater part of the Kepler's letter to Napier is given in Appendix 7 (pp. 958–960).

[77] Joannis Kepleri, *Supplementum chiliadis logarithmorum*, Marpurgi, Casperis Chemlin, 1625, p. 113.

> When I arrived in Upper Germany in the year 1621, and everywhere discussed Napier's logarithms with experienced mathematicians, I discovered that those whom age had blessed with sound judgement but whose ready competence it had diminished, were hesitant to accept this class of numbers in place of their accustomed table of sines. They said that it was shameful for a professor of mathematics to get childishly excited over any summary scheme of numbers, and for the time being to agree to use a form of calculation without any proper demonstration, which at some future time might drag you into error when you least expected it. They complained that Napier's demonstration rested upon the illusion of a certain motion of geometry, whose slippery and changeable nature was no good as a firm basis for reasoned demonstrations. This was why I then immediately devised the initial stages of a proper demonstration, which later, when I returned to Linz, I worked out more carefully, mostly from the beginning of 1622.

A further and conspicuous confirmation of Kepler's admiration for Napier was included in the allegorical frontispiece to his *Rudolphine Tables*. On the roof of the temple there is a symbolic figure adorned by a halo emblazoned with the digits 69314.72.[78] This is the logarithm of sine 30°, as published in Kepler's own *Chilias logarithmorum* of 1624, his re-calculation of Napier's logarithms. Napier's published value in 1614 was 6931469.

CALCULATING TOOLS

As part of the computation of his logarithmic tables, Napier invented a number of calculating tools which he described in the *Rabdologiæ*. Initially Napier had not considered any of them worth publishing but having shared his inventions with his friends, they were becoming known both in Scotland and abroad and were in danger of being appropriated by others. Alexander Seton, Earl of Dunfermline, was then Lord High Chancellor of Scotland, and a friend of Napier. At his instigation Napier collected the most important of his inventions into a Latin digest which was published in 1617 as the *Rab dologiæ*. In his introduction Napier states that he was encouraged to publish his "mechanism and use of the rods [. . .] "lest they should be published in the name of another. He also states:[79]

---

[78] For a full explanation of the symbolism see Mikael Rågstedt, "Kepler and the Rudolphine tables", *Bulletin (New Series) of the American Mathematical Society*, Vol. 50, No. 4, October 2013, pp. 629–639.

[79] This is a fresh translation prepared for this work and is not the one by Richardson used in the copy of *Radbologiæ* published elsewhere in this work.

TABULÆ RUDOLPHI ASTRONOMICÆ
COPERNICUS.
TYCHO BRAHE.
HIPPARCHUS.
PTOLEMAEUS.
MYSTER.COSMO. ASTR.P.OPTICA COM.MARTIS EPIT.AST.COP.
INSULA HUENNA DANIÆ.
Georg. Celer sculpsit Norimbergæ

> I published of late years the Canon of Logarithms, wrought out by myself a long time ago, which, casting aside the natural numbers, and the more difficult operations performed by them, substitutes in their place others affording the same results, by means of easy additions, subtractions, bisections, and trisections. Of which Logarithms, indeed, I have now found out another species much superior to the former, and intend, if God shall grant me longer life, and the possession of health, to make known the method of constructing, as well as the manner of using them. But the actual computation of this new Canon, I have left, on account of the infirmity of my bodily health, to those versant in such studies; and especially to that truly most learned man, Henry Briggs, public professor of geometry in London, my most beloved friend.

The simplest, best-known and certainly the most popular of Napier's devices for everyday use is commonly referred to as Napier's Bones.[80] These are essentially multiplication tables set out on small square sectioned rods about 3 inches long which can be easily used for multiplication and, less easily, for division. With certain additional rods they can also be used for extracting square and cube roots though, as David Bryden states:[81]

Facing page. *The elaborate frontispiece to Kepler's* Tabulae Rudolphinæ *of 1627. It was designed by Kepler and almost every feature has a symbolic meaning, Under the Hapsburg eagle, which is dropping coins to those below, is* Urania, *the muse of astronomy, hastening forward on a chariot. Surrounding the dome are six scientific muses and the third from the left is* Logarithmica. *The twelve sides to the temple reflect the twelve signs of the zodiac whilst the columns supporting the roof range from simple logs at the rear (in the past) to elegant classical columns at the front, reflecting astronomical progress. They are inscribed with the names of distinguished astronomers from the past. Brahe is shown pointing to the heavens, asking the seated Copernicus* Quid si sic? (What about this?). *Also included are Galileo's telescope, Copernicus' heliocentric system and the elliptical orbit of a planet. The front panel in the base of the temple depicts Hven, the island where Brahe carried out most of his star measurements, while in the panel to the left Kepler is seated with a few coins from the Hapsburg emperor on the table (Courtesy of the Syndics of Cambridge University Library).*

---

[80] The earliest reference to the term "Napier's Bones" occurs as early as 1617 in Thomas Bretnor, *Bretnor, 1618, [a new almanacke] and prognostication for 1618* (no publisher and unpaginated). Recommending recently published mathematical works, Bretnor states "And also the *Logarithmes* in English, and an excellent Treatise of *Rabdologia*, or Arithmeticall Rods, invented and published by the learned Lord of *Merchiston*, which Instruments or Rods (commonly called *Napier's Bones*) are made by *Nathaniel Gosse* in Hosier-lane".

[81] D. J. Bryden, *Napier's bones, a history and instruction manual*, Harriet Wynter, 1992, p. 21.

> Contemporary instructions for the use of Napier's rods for extracting square and cube roots do not provide an explanation for the procedures to be followed. Furthermore, in the case of extracting cube roots, key stages in the procedure in the method by rote are more or less glossed over. The beginner would be bemused rather than enlightened. There is no apparent rhyme or reason as to why, for example one number is tripled and multiplied by the square of another and then added to another partial product. The need to undertake various side computations introduced further complexities. There is an arcane sense of mystery about the use of the instrument for use other than in basic multiplication. Certainly the procedure for extracting square roots can be blindly followed, but that for cube roots is much more demanding. Here, the tyro, who is kept blissfully ignorant of the theoretical basis on which the procedure is grounded, has ample opportunity to introduce a whole series of errors.
>
> Nevertheless, in the hands of competent arithmetician, John Napier's calculating rods could indeed be used for more than just simple multiplication. Indeed, no lesser mathematical luminary than Isaac Newton recommended their use for squaring a term in a series of repetitive computations.

However, probably the most powerful of Napier's devices is what he termed the Promptuary of Multiplication. The name Promptuary comes from the Latin *promptuarium*, meaning storehouse, and the machine involves a more complex version of Napier's Bones, allowing the answer to a multiplication to be read through apertures once the number to be multiplied has been set out in vertical strips and the multiplier set out in horizontal strips.

For centuries the Promptuary was little known or appreciated, only Napier's instructions in Latin for its construction being readily available. However, the only known Promptuary that is anywhere near contemporary with Napier was re-identified in 1988 in the Museo Arqueológico Nacional in Madrid by Erwin Tomash.[82]

This re-discovery was at least partially due to the first English translation of the *Rabdologiæ* by William Francis Hawkins and the subsequent interpretation by Garry Tee of the Appendix, which described the design and operation of the Promptuary, to show how it worked.

In 1979 Tee and Hawkins supervised the building of a replica by technicians at the University of Auckland and they found that, after some practice, they

[82] William Francis Hawkins, *The first calculating machine (John Napier, 1617)*, 1988, Annals of the History of Computing, Vol. 10, No. 1, pp. 37–51. Erwin Tomash, *The Madrid Promptuary*, 1988, Annals of the History of Computing, Vol. 10, No. 1, pp. 52–67. However, Felipe Picatoste did include details of the Madrid Promptuary in a rare book published in 1878 (*Abaco neperiano ó rabdológico del Siglo XVII, que se conserva en el Museo Arqueológico Nacional*, Museo Español de Antigüedades, Madrid, Vol. IX, 1878, pp. 51–80.)

could each obtain a 20-digit product in 45 seconds, including selecting the strips from their container and packing them away again. Napier himself wrote that a pair of 10-digit numbers could be multiplied on the Promptuary in one minute. Elsewhere Briggs clearly states that Napier worked alone on the calculation of logarithms but, even though he was clearly very proud of his Promptuary, there is no indication that he used it during his calculations.[83]

PUBLICATION

Printing and publishing were tightly controlled trades during the early modern period, the two activities often being synonymous. In England the Stationers' Company, which was founded in 1403 and received its Royal Charter in 1557, had a monopoly of the industry, setting and enforcing regulations until 1710 when an Act of Parliament established statutory copyright regulated by the courts, rather than through the Stationers' Company. The usual arrangement was for an author to come to a private agreement with a printer/publisher who then registered and held the copyright with the Stationers' Company.[84]

However, the Scottish publishing industry, which was far smaller than the English one, was regulated much more diffusely, through a system of copyright, licence and censorship. Similar to the arrangement in England, most Scottish authors sold their works outright to a printer/publisher. Copyright was normally granted by the crown and confirmed by Privy Seal or an act of Privy Council but only for a defined numbers of years, Scottish copyright being regarded as a limited privilege rather than an absolute property. Completely separately, all books also required a licence for publication, to confirm that the material had been checked and approved, but enforcement was largely carried out by the Burghs through their bailies and magistrates.

The records for Scottish copyright are very incomplete, so it is not known if Napier or his printer/publisher applied for copyright on any of his titles. It is quite possible that the *Plaine Discovery* was covered by copyright, since its subject matter would have been considered very marketable in such a religiously contentious age. However, Napier's mathematical works would have had a much more limited audience and, since there was a charge for registering copyright, Napier and his publisher may have considered such a payment

[83] In his Preface to the Reader in the *Descriptio* of 1616 Henry Briggs wrote about "this little Table of Logarithmes being first begun, and finished by the charge and paines of the honourable Authour alone". See also the "Admonition" from the 1614 edition of the *Descriptio* quoted on p. 35 above.

[84] Many of the details regarding the Scottish printing and publishing trade are taken from Alastair J. Mann, *The Scottish book trade 1500–1720*, Tuckwell Press, 2000.

unnecessary.

The Scottish press was noted for a strong line of presbyterian dissent at this time and many printers fell foul of the government's changing censorship policies at one time or another. Of course those who published anti-government material, or religious material which did not conform to the prevailing orthodoxy, did not seek a license for these works. Therefore they often found themselves imprisoned and forced to pay a surety for their future good behaviour.

Scottish printers were also notable for using Roman type for their religious works, rather than the heavy Gothic or Blackletter typefaces favoured on the Continent and in England at this time.

Napier published three works in his lifetime, with a further work appearing shortly after his death, using three different publishers. His first, the *Plaine Discovery*, was published in 1593 by Robert Waldegrave (*c.* 1554–1604). Between 1578 and 1588 Waldegrave was located in London, printing mainly religious works by reformers including those by Martin Luther, John Calvin and John Knox. His publication in 1584–5 of the puritan *Booke of the form of Common Prayer* was banned by the English Star Chamber. Waldegrave was twice jailed for such publications and eventually his press and type were seized for his anonymous publication of a religious work that was deemed unacceptable by the authorities. He continued to publish dissenting religious works from a covert press established at a series of friend's houses in England before he moved to Edinburgh in 1590. There he continued to be a prolific printer, publishing over 100 titles, again mainly religious, including three by James VI himself, as well as Napier's *Plaine Discovery*. Waldegrave was appointed King's Printer in 1590 and returned to London shortly after James' accession to the English crown in 1603 but died the following year.

In 1594 John Norton published an almost identical version of the *Plaine Discovery* in London, though it was probably printed by Robert Waldegrave in Edinburgh. Although Norton hailed from Shropshire, he established himself as a publisher and bookseller in Edinburgh in about 1586 but moved to London when James VI acceded to the English throne in 1603. There he held the position of King's Printer in Latin, Greek and Hebrew. He developed extensive contacts overseas and by 1600 he was trading at the twice yearly Frankfurt book fair where he purchased books for the libraries of both James I and Thomas Bodley. The Norton family were powerful landowners in Shropshire and John Norton was a prominent member of the Stationers' Company, being Master in 1607, 1611, and 1612.

In 1611 a further edition of the *Plaine Discovery* was published by Andro Hart in Edinburgh and also by John Norton (1556/7–1612) in London. The two editions are almost identical and it is very likely that they were both printed by Andro Hart in Edinburgh.

Napier's next two works, *Descriptio* (1614) and *Rabdologiæ* (1617) were published by Andro Hart (*d.* 1621), Scotland's wealthiest and most successful printer of this period. As with Waldegrave, Hart held strong religious beliefs, being a fervent presbyterian. He started printing on his own account in 1608 and his most celebrated work was his edition of the Geneva Bible, published in 1610, even though he never held the position of King's Printer. This version was particularly noted for its accuracy and by the middle of the 17th century it had become the working text for many Dutch printed, English language editions destined for sale in Scotland.

The Netherlands was the home of a flourishing and largely unrestricted Calvinist printing industry and they frequently undertook commissions for publishers in Scotland. Hart had strong connections with the Low Countries and became one of the leading importers of English language works printed in the Netherlands. He even purchased an exclusive right to do so from James I in 1614 but this privilege was soon struck down by the Privy Council.

The first English language edition of *Descriptio* (1616) was published by Nicholas Okes (*d.* 1645) in London, the year before Napier's death. Okes became a Freeman of the Stationers' Company in 1603 and set up on his own account in 1607. He printed works on a wide variety of subjects but is now largely noted for his publication of Renaissance drama. However, Okes was frequently in trouble with the authorities for printing controversial works without official approval. There can be little doubt that the choice of a London publisher, rather than one based in Edinburgh, was due to Samuel Wright overseeing his father's translation through the press. Both father and son lived in London.

The first edition of *Constructio* was printed in 1619, two years after Napier's death. This was arranged by Robert, Napier's second son, who lived near Edinburgh and Andro Hart was again chosen as the printer and publisher.

NAPIER'S DEATH

Napier suffered from poor health for a number of years prior to his death but the details are unknown, except for the fact that he suffered from severe gout. In a complaint to the Privy Council by "Johnne Naper of Marcheistoun" dated 28th April 1613, Napier, his Baillie and three tenants complained about their treatment by the Grahams of Boquhopple:[85]

[85] David Masson (ed), *The Register of the Privy Council of Scotland*, H. M. General Register House, 1891, Vol. 10 (AD 1613–1616), p. 41.

> Thomas Grahame of Bochopill, "having consavit [conceived] ane deidlei haitrent [hatred] and malice aganis the said laird of Marcheistoun," has sought by all indirect means to execute his private grudge against the said laird. More especially, the said Grahame, "knowing that the said laird is havelie diseasit with the pane of the gute [gout], and is unable to repaire to the saidis landis for halding of courtis thairon, and doing of justyice to his poore tennentis who labouris the same," intends by open violence to lay the said lands waste.

His Dedication to Prince Charles in the *Descriptio* of 1614 states "[I] that am now almost spent with sicknesse", while in the *Rabdologiæ*, published in 1617, Napier wrote that:

> I have now discovered a greatly superior species of *Logarithms* and have decided (if God will grant me life and health for a little longer) to publish the method of creating them and also their use, but because of the poor state of my bodily health I have left the actual computation of the new canon to men versed in this type of pursuit.

Napier made out his final will on 1st April 1617, three days before his death and in it he states that:[86]

> I, Johne Naipper of Merchinstoun, being sick in bodie at the plesour of God, bot haill in mynd and spereit, and knawing nathing mair certane nor death, and the tyme and manner thaiof maist uncertane, and unwilling to dispose upon my wurdlie effairis, and to be dischairgit of the burding and cair thereof, sua that at the plesour of Almichtie God I may be reddie to abyd his guid will and plesour quhen it sall pleis him to call me out of this transitory lyfe.

Napier's ill health was further mentioned in his will, which states that:[87]

> *Sic subscribitur* Jon Naipper above-writtine, with my hand at the pen led be the nottaris under-writtine, at my command, in respect I dow not writ myself for my present infirmitie and seiknes.

Napier died aged sixty-seven, on 4th April 1617.[88] At one time it was generally thought that Napier was buried at St Giles church (now a cathedral)

[86] Mark Napier, *Memoirs of John Napier*, p. 430.

[87] Idem, p. 431.

[88] Mark Napier states that Napier died on 4th April 1617 and this is the date on the memorial to John Napier erected in St Cuthbert's church in 1842 (in which Mark Napier must surely have taken a leading part). This date is clearly based on the Testament to Napier's will which he made out on 1st April 1617. Mark Napier transcribed this in his *Memoirs of John Napier* (p. 427) as "Jon Naipper of Merchinstoun, within the parochine

in Edinburgh, due to an old but non-specific memorial to the Napier family located there that reads S. E. P. / FAMILLÆ NEPERORVM INTERIVS / HIC SITUM / EST (*The tomb of the Napier family is situated within*). This memorial is now located in a niche on the outside wall of the cathedral, to the east of the north door. However, Mark Napier, in his *Memoirs of John Napier* (p. 426), reproduced a letter written by Prof. William Wallace in 1831 which pointed to a treatise on trigonometry by a Scotsman, James Hume of Godscroft, Berwickshire.[89] The latter clearly states (in French) that Napier "was buried in the Church of St Cuthbert, outside Edinburgh's West Port".[90]

This was the parish church for Merchiston but it was rebuilt in 1775, though the steeple was not added until 1789, and it was subsequently rebuilt again in 1894, while retaining the previous tower and steeple. Thus, there is no

of Sanctcuthbert and schirefdome of Edinburgh, the tyme of his deceis; quha [who] deceist upon the fourt day of Appryle the yeir of God i^{m} vi^{c} and sevinteine yeiris".

However, George Crawfurd's *The Peerage of Scotland* (Printed for the Author, 1716, p. 363), published more than a century prior to Mark Napier's work, states that "He died in a good advanc'd Age, on the 3d *April* aged 67" with a footnote in Latin which reads "Charta penes [Papers in the possession of] Francis D. Napier [6th Lord Napier]". George Mackenzie's *The lives and characters of the most eminent writers of the Scots nation* (William Adams, 1722, Vol. 3, p. 523) states that Napier died on "the third of April the same Year, in the 67th Year of his Age". Sir Robert Douglas' *The Peerage of Scotland* (R. Fleming, 1764, p. 509) merely states that 'He died anno 1617'. David Stewart, Earl of Buchan, & Walter Minto's *An account of the life, writings and inventions of John Napier of Merchiston* (1787, R. Morison & Son, p. 18) gives "3^{d} of April, O.S." Burke's *Peerage, Baronetage & Knightage* (Burke's Peerage & Gentry, 2004, 107th edn) gives 3rd April, as do other similar modern works, all presumably following Crawfurd.

It is known that Francis, 8th Lord Napier (1758–1823), compiled a summary of the family's charters and private papers to create an early genealogical account of the family (Mark Napier, *Memoirs of John Napier*, p. viii). This formed the core for the Napier entry in Wood's edition of Douglas' Peerage (*The Peerage of Scotland*, Archibald Constable, 1813) in which Wood also states that Napier "died on 3d April 1617, in the 67th year of his age". It is unclear why this minor discrepancy arose and why it still persists, when the evidence for 4th April seems so incontrovertible.

However, as David Stewart & Walter Minto indicate, this date is subject to further correction since it pre-dates the change from the Julian to the Gregorian calendar which occurred in Great Britain when the eleven days from 3rd to 13th September were omitted in 1752 (Catholic Europe had changed their calendar in 1582, with Protestant countries following at various later dates). Up to 1751 Britain also counted the start of the year from 25th March, but then changed to 1st January.

[89] *Archeaologia Scotica or Transactions of the Society of the Antiquaries of Scotland*, Wiilliam Blackwood, 1831/57, Vol. IV, Part 1, p. 213; Paper read on 9th May 1831 by Prof. William Wallace *On the burial place of John Napier of Merchiston, the inventor of Logarithms*.

[90] James Hume, *Traité de la trigonométrie, pour résoudre tous les triangles rectilignes et sphériques*, N. et J. de la Coste, 1836, p. 136.

sign of Napier's grave, though a much later memorial on an internal wall commemorates him.[91]

Such an early confirmation of Napier's burial place, from a fellow Scot and mathematician, has been accepted as correct ever since, though the exact location of his grave is still unknown. However, given Napier's social status, it is almost certain that Hume was correct in stating that he was buried within the church itself, rather than in the churchyard. Hume also states (in French) that Napier suffered from gout, further confirmation that Hume was well informed about Napier, writing that:

> The inventor was a Sire of great name and, whose posterity is today in possession of great respect in the kingdom, who, being of age and greatly affected by gout, could not do much else but devote himself to science, mainly mathematics and logistics, which pleased him enormously and, with great effort, created his tables of Logarymes, printed in Edinburgh in the year 1614.

Napier's will shows that he left a number of large landed estates to his decendants.[92] He also left a mass of papers, including his mathematical treatises and notes, all of which came into the possession of Napier's third son (the second by his second marriage), Robert (1580–1655), who became his father's literary executor. Encouraged by the positive reception that the *Descriptio* had received, Robert undertook the task of seeing the *Constructio* through the press and it was finally published in 1619.

Subsequently these manuscripts passed to Col. Robert Milliken-Napier (1765–1808) of Culcreuch, one of Robert Napier's direct descendants who adopted the additional surname of Milliken under the terms of his maternal grandfather's will. His wife stored them in a garret in their house near Kilbarchan, Renfrewshire, before going abroad to join her husband but during their absence the house caught fire in 1801 and was largely destroyed. Napier's manuscripts and other relics were entirely lost, including a portrait of Napier and a Bible signed by him, all of them a great loss to future historians.

---

[91] This memorial seems to have been erected at the instigation of Prof. William Wallace, Emeritus Professor of Mathematics at the University of Edinburgh. A copy of a printed letter from him dated 26th November 1838, calling for a meeting to be held at the Royal Society of Edinburgh to discuss the erection of a memorial to Napier, is pasted in a copy of Mark Napier's *Memoirs of John Napier* held by the National Library of Scotland (Shelf-mark L.C.1224). No doubt interest in Napier was re-awakened by Mark Napier's work published in 1834.

[92] A transcription of the will is given in Mark *Napier's Memoirs of John Napier*, pp. 427–431. The original will is held by the National Records of Scotland, CC8/8/49.

Fortunately, two of Napier's treatises had previously been given to Francis, 6th Lord Napier (*c.* 1702–1773), by William Napier of Culcreuch, probably at the time Francis was considering writing a life of his ancestor, and they subsequently lay in the Napier charter chest, unknown to the world. They passed by descent to Francis, 8th Lord Napier (1758–1823) who, as he noted on a blank leaf, "finding them in a neglected state amongst my family papers, I have bound them together, in order to preserve them entire".

The two surviving documents were entitled "*The Baron of Merchiston, his booke of Arithmeticke, and Algebra*; *for Mr Henrie Briggs, Professor of Geometrie at Oxforde*."[93] They appear to have been written for Henry Briggs after the latter had been appointed to the Savilian chair in 1619.

They were subsequently prepared for the press by Mark Napier and published in 1839 as the *De Arte Logistica*. The originals were beautifully written in the hand of Robert Napier who concludes by stating that "Ther is no more of his algebra orderlie sett doun".

EDINBURGH & GARTNESS

By far the best description of Edinburgh during Napier's lifetime was written by Fynes Moryson, an experienced traveller who visited the city in 1598, providing an interesting insight into Napier's home town:[94]

> From the Kings Pallace at the Eaſt, the City ſtill riſeth higher and higher towards the Weſt and conſiſts eſpecially of one broad and very faire ſtreet, (which is the greateſt part and ſole ornament thereof), the reſt of the ſide ſtreetes and allies being of poore building and inhabited with very poore people.

Napier's association with Merchiston is well known but he also lived for part of his married life at Gartness, in the parish of Killearn, Stirlingshire, about 60 miles by road west of Edinburgh and only five miles from Loch Lomond. However, none of the few letters from Napier quoted in Mark Napier's *Memoirs of John Napier* is dated from Gartness and the hamlet hardly features in the existing records relating to Napier. An early record of

93 The present whereabouts of this manuscript is unknown.

94 Fynes Moryson, *An itinerary written by Fynes Moryson Gent. first in the Latin tongue, and then translated by him into English*: *containing his ten yeeres travell throvgh the twelve domjnions of Germany, Bohmerland, Sweitzerland, Netherland, Denmarke, Poland, Jtaly, Turky, France, England, Scotland, and Ireland*, John Beale, 1617, Part I, Booke 3, Chap V, p. 273. Further extracts relating to Edinburgh are included in Appendix 8 (pp. 961–962).

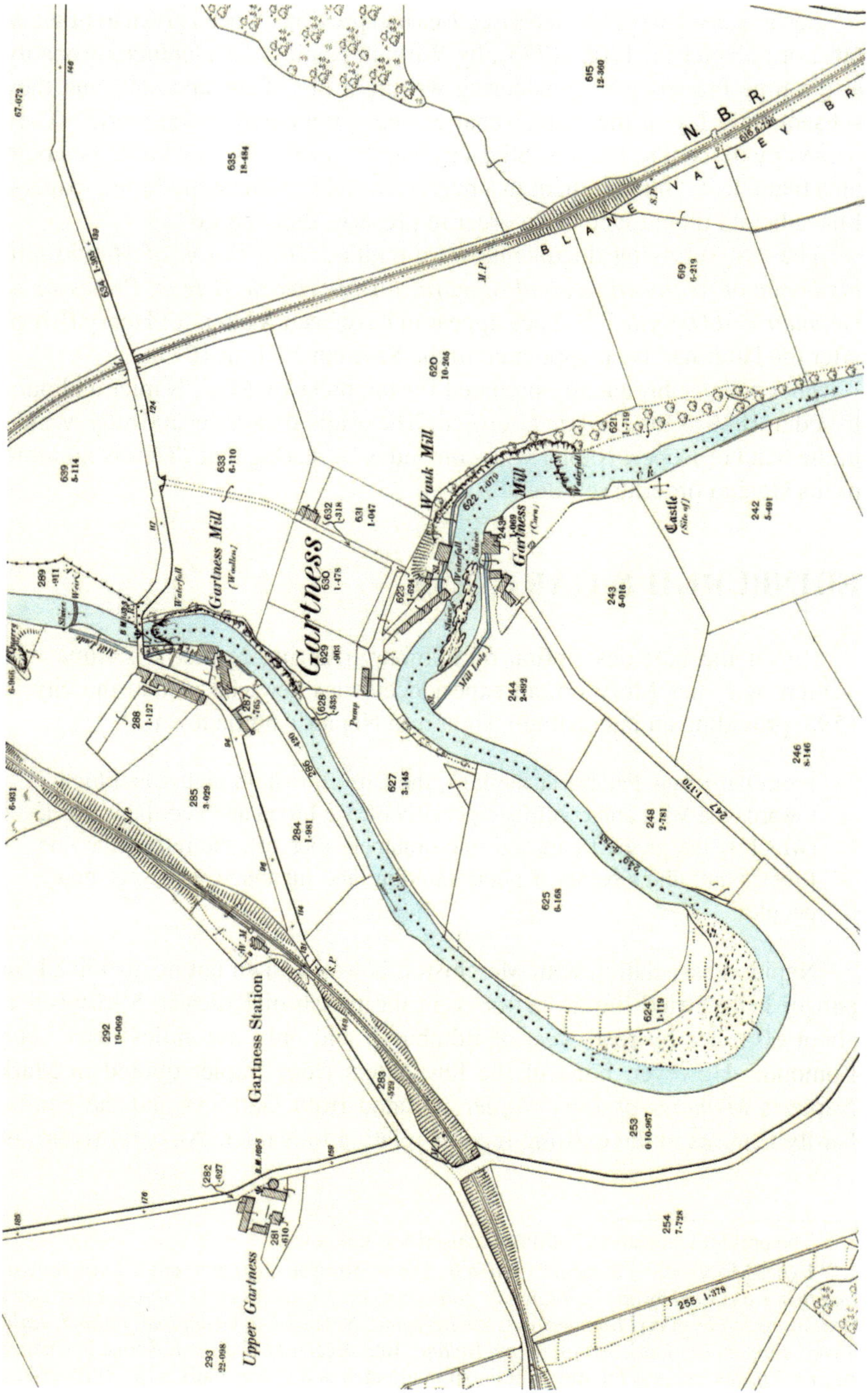
Gartness
Gartness Mill
Wauk Mill
Gartness Station
Upper Gartness
N. B. R.
BLANE VALLEY
Castle
Mill Lade
Waterfall
Pump

the area, albeit written some 200 years later, states:[95]

> *Gartneſs.* – The *Pot of Gartneſs*, in the water of Enrick, has, for its beauty and ſingularity, always attracted the attention of ſtrangers. It is a deep linn, ſhaped like a caldron or pot, whence the origin of the name. It is occaſioned by the fall of water over a rock that lies acroſs the river; the fall is not perpendicular, but is interrupted by three or four breaks.

This was part of the Napier property in the Menteith and, assuming that Napier lived there for a large part of the time until his father's death in 1608, much of Napier's work on logarithms must have been undertaken at Gartness. This is confirmed by the Rev. David Ure's record of local folklore gathered when putting together the history of the parish of Killearn:[96]

> Adjoining the mill [of Gartness] are the remains of an old houſe in which JOHN NAPIER of Merchiſton, inventor of the logarithms, reſided a great part of his time (for ſome years) when he was making his calculations. It is reported, that the noiſe of the caſcade being conſtant, never gave him uneaſineſs, but that the clack of the mill, which was only occaſional, greatly diſturbed his thoughts. He was, therefore, when in deep ſtudy, ſometimes under the neceſity of deſiring the miller to ſtop the mill, that the train of his ideas might be interrupted. He uſed frequently, in the evening, to walk out in his night gown and cap. This, with ſome things which to the vulgar appeared rather odd, fixed on him the character of a *warlock*. It was firmly believed, and currently reported that he was in compact with the devil; and that the time he ſpent in ſtudy was ſpent in learning the *black art*, and holding converſations with *Old Nick* [the Devil]. But John Napier was not the only great man who,

---

Facing page. *A portion of the Ordnance Survey 25" to the mile map for Stirlingshire, (Sheet 020.07), revised ca. 1896 and published in 1898. The Parish of Killearn is on the east side of the Endrick Water while the Parish of Drymen is on the west side. The site of Gartness Castle, Napier's home before he moved to Merchiston Tower, is shown on the west bank, just south of Gartness itself. Three mills are shown, the most northerly being marked as 'Gartness Mill (woollen)', the middle one as 'Wauk [fulling] Mill' and the most southerly as 'Gartness Mill (corn)'. It is known that the clack of one of these disturbed Napier's studies and that he sometimes requested the miller to stop work, though it is not known which one.* (*Courtesy of the National Library of Scotland.*)

[95] Sir John Sinclair, *The statistical account of Scotland drawn up from the communications of the ministers of the different parishes*, William Creech, 1795, Vol 16, p. 107.
[96] Ibid., p. 108.

in days of ignorance, was fuppofed to be a wizard; nor were the vulgar in this neighbourhood the only people who believed in witchcraft.

Napier himself remarks in his notes upon the first chapter of the Apocalypse in his *Plaine Discovery* that, "long garments or gownes were of olde, and to this day worne of Doctors and Senatours, to represent grauitie and wifedome" and he clearly felt quite entitled to wear them himself.

Some authors have stated that Napier built his home at Gartness, perhaps based on the fact that a stone with the date 1574 carved into it was found in the south gable of a store-house adjoining the mill.[97] However, they provide no further evidence to support this conclusion and the property might equally well predate Napier's arrival and been improved when his family moved there.

MERCHISTON TOWER

Over the years, Merchiston has been referred to as a castle, a tower and as a house.[98] None of these designations is necessarily wrong but architecturally the building is of the classic form known as a tower house. They were built throughout Europe during the Middle Ages as combined defensive structures and residences for local chiefs, often in wild or barren areas which did not warrant the expense of building and maintaining a proper castle. They were particularly popular in Scotland and Ireland where, like Merchiston Tower, they were often built to a familiar L plan.

Alexander, the first Laird of Merchiston acquired Merchiston by wadset (mortgage) from James I sometime before 1438 but the king was assassinated in 1437 and the loan was never repaid.[99] However, it is not known exactly when Merchiston Tower was built. The consensus suggests that it was built later in the fifteenth century, either by Alexander himself, who died in 1454,

[97] Although the site is recorded on Ordnance Survey maps as a castle, it was not marked as such on Joan Blaeu's *Atlas Novus* of 1654. Blaeu does mark Fast Castle as a castle, for instance, but not Merchiston. Therefore Gartness was probably closer to being a fortified house than an extended castle complex. This view is supported by the Rev. David Ure's previously quoted description of the site as containing "the remains of an old house".

[98] Norman Dixon in *The place names of Midlothian* (A PhD thesis awarded by the University of Edinburgh, May 1947) states that the first recorded use of the name Merchiston appears in 1264 (p. 68). He contends that name comes from a combination of the Old Welsh personal name of Merchiaun with the familiar Old-English suffix *tūn*, meaning an enclosure or farm (pp. 20, 68, 234 & 247). He also indicates that the correct pronunciation is mə;rkistən (Mer kis t'n, not Mer chis t'n), with a hard *k* (p. 68), a pronunciation still used by people living in the area.

[99] See Appendix 9 (pp. 963–973) for further details of on the acquisition of Merchiston by Alexander Napier, its history and restoration.

or by his son, Sir Alexander, the second laird who died in 1473. The first two surviving references to the tower relate to an instrument "Done at Merchamston, 18 April 1491" and another document "4 May 1495 ... Done in the tower at Merchamston, near the burgh of Edinburgh".[100]

When the Napiers first built Merchiston, and for many centuries thereafter, it would have stood almost alone in the surrounding countryside. Edinburgh itself was situated between the Castle to the west and Holyrood House to the east, with all the houses and shops located close to the long street running the length of the sloping ridge connecting the two. South of Edinburgh there was the low lying land of the South Loch, now the Meadows, while even further south lay the relatively high ground of the Borough Muir, bounded at its western edge by Merchiston Tower. The ground slopes away from Merchiston in all directions except the east, providing clear views to the north and the south, so the tower occupied a strategic position, commanding the south-western approaches to Edinburgh.

Merchiston Tower was sold by the Napier family in 1647 and became a domestic house and then, for almost a century, a school. The tower was virtually abandoned from 1930 and ownership passed to Edinburgh Corporation in 1935. It was extensively restored by the Corporation between 1958 and 1964, a process which revealed much about its previous history, as detailed in Appendix 9.

NAPIER'S DESCENDANTS

Napier's eldest son, Archibald Napier (*c.* 1575–1645) and ninth laird of Merchiston, was knighted in 1616, created a baronet in March 1627 and was raised to the peerage as Lord Napier of Merchiston in May of the same year. He married Margaret Graham, daughter of John Graham, 4th Earl of Montrose and sister of James Graham, 1st Marquess of Montrose. Archibald was Gentleman of the Privy Chamber to James VI and accompanied him to London for his coronation in 1603, subsequently becoming Treasurer-Depute for Scotland from 1622–31, during which time James I was succeeded by Charles I. However, Scotland was soon to be plunged into war again. Archibald was a strong supporter of Charles I, whose religious reforms were strongly opposed in Scotland by the Covenanters, a presbyterian movement. This led to a civil war in Scotland from 1644–5 as the Royalists, led by the Marquess of Montrose, fought the Covenanting government. Montrose won a series of

[100] Gordon Donaldson (ed), *Protocol book of James Young 1485–1515*, Scottish Record Society, 1952, p. 99 & p. 174.

THE TOWER OF MERCHISTON IN THE FIFTEENTH CENTURY

The form of the barmkin and details of the work above parapet level are conjectural

PROBABLE APPEARANCE
OF SOUTH FRONT,
EIGHTEENTH CENTURY

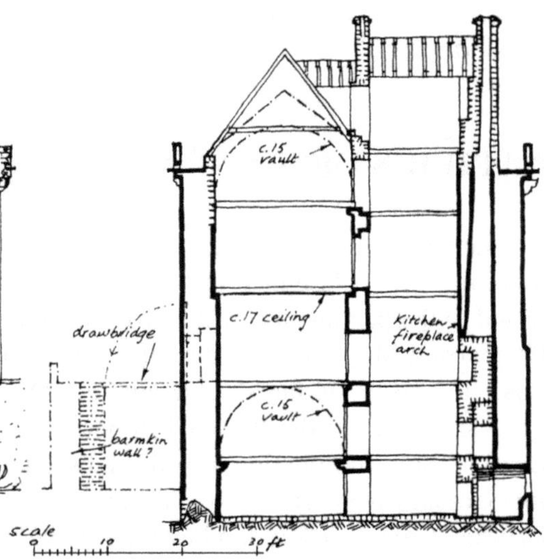

SECTION (S TO N) OF
TOWER AND JAMB

Conjectural views of Merchiston Tower in the 15th and 18th centuries. The section shows the original barrel vaulted stone ceilings to the 1st and 4th floors. (*The illustrations on pp. 56 & 57 are from Stuart Harris, The Tower of Merchiston,* The Book of the Old Edinburgh Club, *Published for the Members of the Club, 1962 & 1969.*)

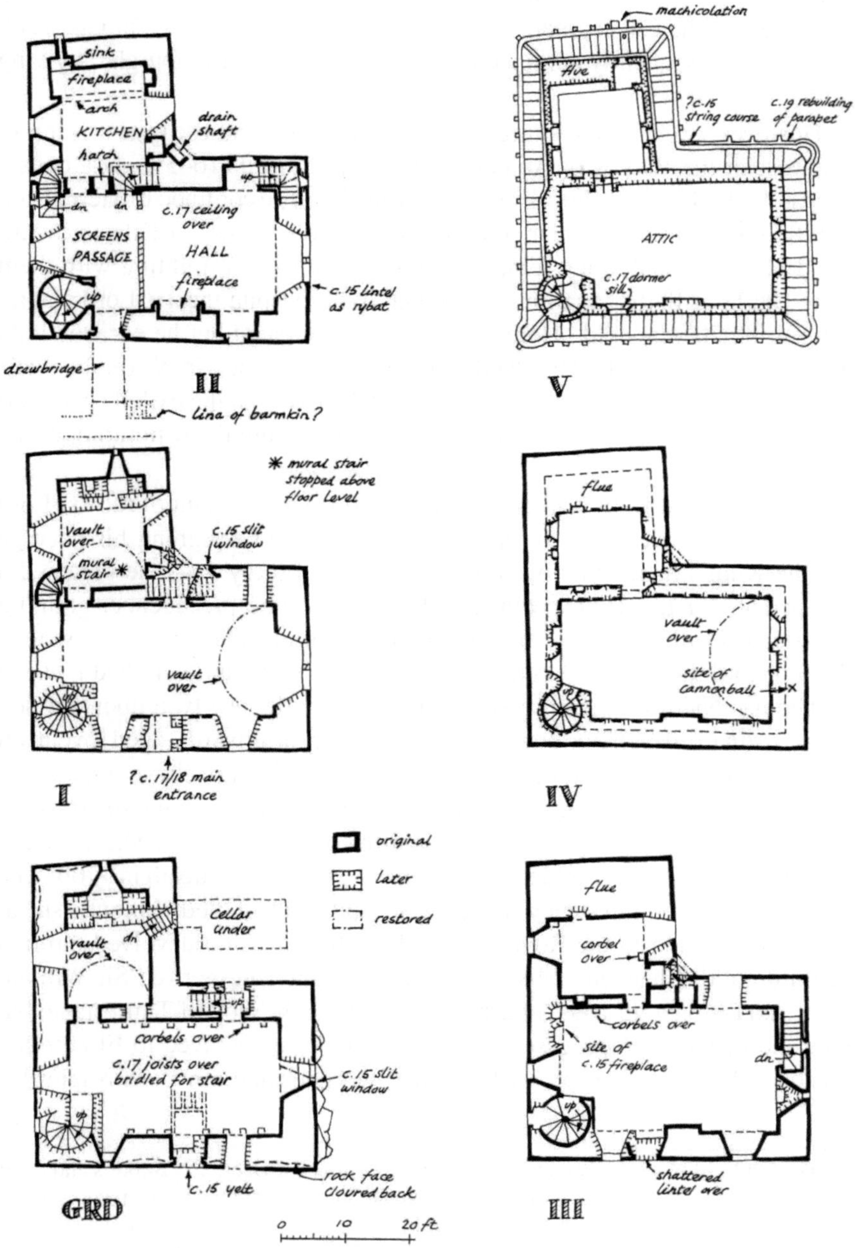

Floor plans of Merchiston Tower. See Appendix 9 for further details of the tower's restoration. (The illustrations on pp. 56 & 57 are courtesy of the Old Edinburgh Club.)

rapid victories but Archibald and his son had already been imprisoned in Edinburgh by the Covenanters in 1641 because of their support for Charles I. They were subsequently confined to their apartments in Edinburgh on giving a surety.

However, Archibald's son ignored the injunction and escaped to support Montrose at the Battles of Auldern (1644) and Laford (1645). As a consequence the 1st Lord was imprisoned again and his wife confined to Merchiston until they were freed by their son after Montrose's victory at Kilsyth in August 1645. Archibald then joined Montrose and his son, fighting with them at the Battle of Philiphaugh in September 1645, despite the 1st Lord Napier being some 70 years old. Here Montrose was defeated but he escaped with the two Napiers into the Highlands where Archibald died in November. His son, Archibald the 2nd Lord Napier, left Scotland to join the exiled Charles II in Holland where he died in 1660, a few months before the restoration of the king.

Archibald, 3rd Lord Napier, was the last male descendant of the 1st Lord and being unmarried, he had taken the precaution of resigning his peerage into the King's hands in November 1676. In February 1677 he obtained a new patent, for the same peerage and with the same precedence, granting the title of Lord Napier to him, his heirs or failing that, his three sisters. The 3rd Lord died in 1683 but, since his eldest sister Jean, who had married Sir Thomas Nicolson, had predeceased him, the title devolved upon her son Thomas, 4th Lord Napier. However he died shortly afterwards, aged seventeen and unmarried in 1686, so the title devolved to the 3rd Lord's younger sister (the 4th Lord's aunt), Margaret, who had married John Brisbane (*d.* 1684) in 1676. Thus Baroness Napier became the fifth holder of the title. She had three children but they had all predeceased her by the time of her death in 1706. The eldest son had died in infancy while the second son had died without issue at sea, whilst serving in the Royal Navy. Therefore the title devolved through the third child, Elizabeth, Mistress of Napier. She had married Sir William Scott, 2nd Baronet, in 1699, eldest son of Sir Francis Scott of Thirlestane, but since she had also predeceased her mother, the title devolved to Elizabeth's eldest son Sir Francis Scott, 3rd Baronet, who became 6th Lord Napier.[101]

[101] The 6th Lord Napier, as numbered in modern editions of Debrett's, Burke's and other historical works, is numbered as the 5th Lord Napier in older editions of such works, since the latter ignore the fact that his grandmother, Margaret, 5th Baroness Napier, held the title in her own right, albeit as a female, on the death of the 4th Lord Napier in 1686. This mis-numbering was carried on through subsequent generations until quite recently but it can be a fertile ground for confusion when cross-referencing old histories with more recent ones, particularly as many of the Napiers shared common Christian names over numerous generations.

Since then the title has devolved through successive elder sons to the present holder Francis, 15th Lord Napier, 6th Baron Ettrick.

In the nineteenth century the family produced numerous distinguished military figures including Admiral Sir Charles Napier (1786–1860), who was known as "Black Charlie" due to his swarthy appearance, and General Sir Charles Napier (1782–1853). In 1843 the latter conquered the province of Scinde (Sindh) in India (now south-east Pakistan) and it is said that he sent a one-word despatch to Lord Ellenborough, the Governor General of India, reading *Peccavi*, Latin for 'I have sinned'. In fact, this apocryphal story arose from a joke published in 1844 in the magazine *Punch*.[102]

Another distinguished descendant, was the diplomat and administrator Francis, 10th Lord Napier (1819–1898). He was offered an elevation in the peerage for his work as Governor of Madras but he preferred to keep his precedence in the Scottish peerage as 10th Lord Napier and took the additional title of 1st Baron Ettrick in the English peerage. This also enabled him and his descendants to sit in the House of Lords by right, unlike Scottish peers who only sat there after the Union of 1707 as a matter of courtesy.

As stated above, Archibald Napier was created a Baronet in 1627 but this title became dormant in 1683 on the death of Archibald, the 3rd Lord and 3rd Baronet, who had no male heirs. However, it was successfully claimed in 1817 by the 3rd Lord Napier's heir male general, Sir William Milliken-Napier the 8th Baronet, and the present holder is his descendant Sir John Napier, 14th Baronet.

[102] The *bon mot*, much quoted when the Napier family are being considered, is often attributed to Gen. Napier. In fact it was a joke that started with a remark by a 17 year old schoolgirl, Catherine Winkworth, which she was encouraged to send to the magazine *Punch* who then published it as a seven line report under the heading "Foreign Affairs" (18th May 1844, Vol. 6, p. 209). The origin of the Peccavi story was reported by Catherine's cousin many years later in a short notice sent to *The East and the West, a quarterly review for the study of missions* (October 1907, Vol. 5, pp. 467–8).

THE NAPIER FAMILY TREE

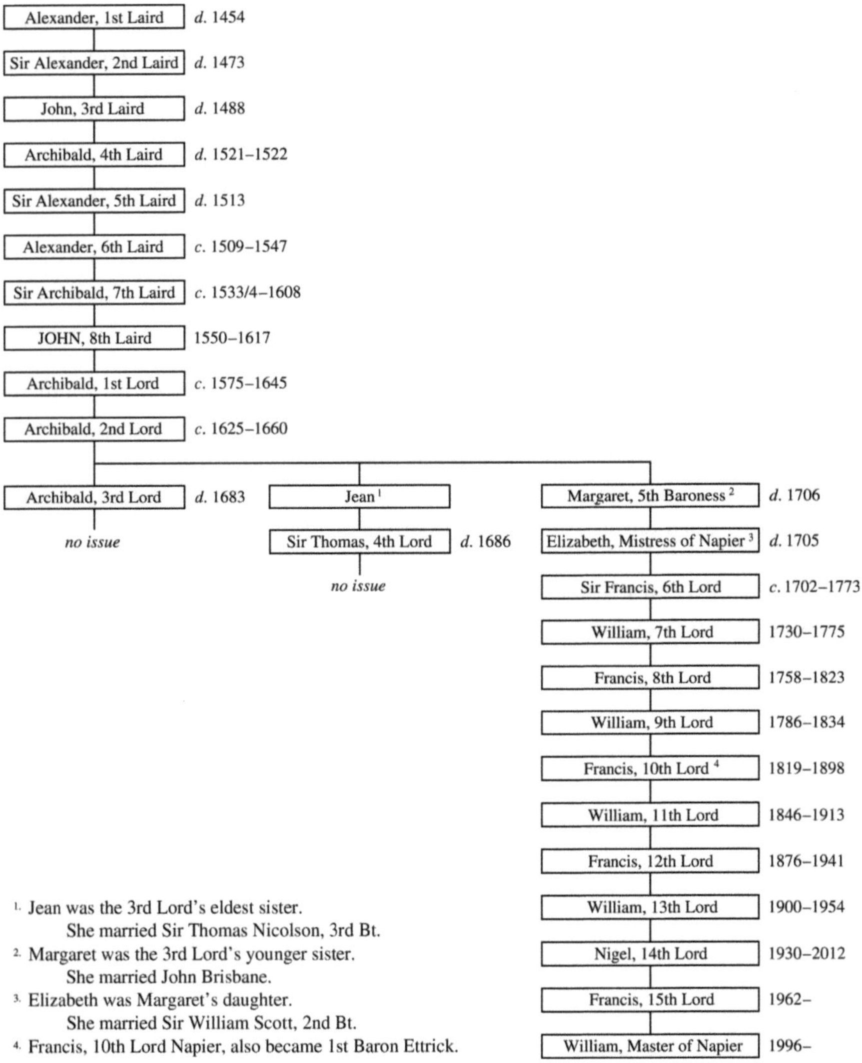

1. Jean was the 3rd Lord's eldest sister. She married Sir Thomas Nicolson, 3rd Bt.
2. Margaret was the 3rd Lord's younger sister. She married John Brisbane.
3. Elizabeth was Margaret's daughter. She married Sir William Scott, 2nd Bt.
4. Francis, 10th Lord Napier, also became 1st Baron Ettrick.

This tree is based on Sir Robert Douglas & John Philip Wood, *The Peerage of Scotland* (Archibald Constable, 1813); Anon (Archibald Scott Napier), *A history of the Napiers of Merchiston ... compiled from old records*, (J & E Bumpus, 1921) and *Burke's Peerage, Baronetage & Knightage* (Burke's Peerage & Gentry, 2004, 107th edn).

REVEALING THE *PLAINE DISCOVERY*

Alexander Corrigan

A PLAINE DIS-couery of the whole Reue-lation of Saint IOHN: set downe in two treatises: The one searching and prouing the true interpretation thereof: The other applying the same paraphrastically and Historically to the text.

SET FOORTH BY IOHN NAPEIR L. of Marchistoun younger.

WHEREVNTO ARE annexed certaine Oracles of SIBYLLA, agreeing with the Reuelation and other places of Scripture.

EDINBVRGH
PRINTED BY RO-bert Walde-graue, prin-ter to the Kings Ma-iestie. 1593.
Cum Priuilegio Regali.

REVEALING THE *PLAINE DISCOVERY*

DESCRIPTION AND KEY FEATURES OF THE *PLAINE DISCOVERY*

JOHN Napier's *A Plaine Discovery of the Whole Revelation of St John* was first published in 1593/4, when Napier was 42 or 43 years of age and the edition reproduced in this volume appeared in 1611.[1] Napier's life, personality, worldview and mathematical achievements cannot be understood fully unless one understands something of this, his first published work. In the early modern period, Christianity provided people in Europe and the British Isles with a framework for understanding oneself, the rest of humanity, the universe, and everything in it. Christian faith was the most dominant force in the majority of people's lives and governed every intellectual pursuit, including those that might now be classified as 'the sciences', driving people to make discoveries about the natural world to attain a closer relationship with its creator. The *Plaine Discovery* is an important text for anyone with an interest in Napier and his works because it explains the faith that governed his intellectual, public and private lives. His hopes and fears, and those of his audiences, were reflected in the book, and the first shoots of his interest in mathematics that blossomed two decades later in the logarithms may be seen.

Napier must have been aware of his prodigious talent for mathematics many years before he published the logarithms. The *Plaine Discovery* was

---

Facing page. *Title page of* A Plaine Discovery, *1593, courtesy of the National Library of Scotland.*

[1] The title page of the first edition provided 1593 as the year of publication but Napier's *Episle Dedicatorie* (see below pp. 945–947) to James VI carried the more specific date, 'the 29. Daye of Ianuar. 1593.' When Napier wrote that first edition Scotland began each new year on 25th March, only adopting 1st January as New Years Day in 1600. As far as he was concerned, 1593 began on 25th March 1593 in the new style (N.S.) and what he called 'January 1593' occurred nine months after that. Therefore, Napier's 29th January 1593 (O.S.) corresponds to January 1594 (N.S.). Later in 1594 a version was published by John Norton in London, which contained minor corrections. In 1611 two further versions were published, again in Edinburgh and London respectively. These editions were more extensively revised, especially in terms of type-setting, the addition of marginal scriptural references in the first treatise and an additional treatise addressing criticisms of the earlier editions. See Macdonald, W. R., pp. 109–28 & 166–7 for a list of the various editions.

distinguished from the many texts that influenced it by the prominence of mathematics in Napier's understanding of the meaning of prophecies from the Book of Revelation, the most challenging book of the Christian canon. He believed his calculations gave his chronological system more gravitas and precision than those that preceded it. The invention of logarithms may be Napier's most famous achievement but the *Plaine Discovery* must not be overlooked. It provides the best source of information about Napier the man and his attitudes to the world in which he lived, as well as the beliefs that provided a foundation for his later achievements.

The greatest part of the *Plaine Discovery* was a commentary on the Book of Revelation, the final text of the Christian Bible. Napier saw the events foretold in the Revelation as occurring in chronological order and explaining key events throughout the history of Christianity from the baptism of Christ to the end of the world. Other sixteenth century writers had constructed 'chronologies of salvation history' and John Bale had even based a chronology on the Revelation in his seminal *The Image of Both Churches* of 1545.[2] However, Napier's chronology was unlike anything that had gone before because of its explicit and detailed nature, and the prominent role it played in his polemic. For Napier, a sense of chronological development was a defining characteristic of the Revelation and was vital to his work because of his emphasis on predictions about the future. His belief that the world would end around a century into the future was an important reason for this, as events at the end of the chronology represented the fulfilment of humanity's relationship with God. He also predicted that the final battle between good and evil foretold in the Revelation would soon occur and would include a great war between Protestants and Roman Catholics. He portrayed everything through which he and other Christians had lived and were experiencing as leading up to these events and his chronology was intended to show the progression of humanity towards their collective destiny.

The *Plaine Discovery* was comprised primarily of two treatises. The first contained 36 'Propositions', or introductory claims that were intended to support and provide evidence for assertions made in the second treatise, Napier's

[2] Bale, J., *The image of bothe churches after reulacion of saynt Iohan the euangelyst*, S. Mierdman (Antwerp, 1545). Bale's work was ground-breaking and was the first of its kind in English but it was influenced by earlier works such as Robert Barnes' *Uitae Romanorum Pontificum* (Basileae, 1535). Since the Reformation, several works had discussed Christian history in such a way as to bolster the image of the Protestant Churches and present Roman Catholicism as corrupt, evil and oppressive, including the seminal *Magdeburg Centuries*. Napier's work would have beeen influenced by a varietyof such sources, often indirectly through scholars like Bale. However, Napier does not appear to have been influenced by John Foxe's famous *Actes and Monuments* and the *Plaine Discovery* was distinguished from earlier works by the mathematical precision of its chronology.

detailed commentary on the Revelation. The introductory propositions had several important aims. They explained the mathematical bases of Napier's system of dating the age of the world and his predictions about the date of the apocalypse. Some of these issues were complex and Napier could not have given them adequate attention had he attempted to incorporate them into his second treatise, the tabular format of which demanded brevity. Napier was not simply explaining his methods; he was also attempting to justify them. He was the first Protestant writer in English to attempt to date the Second Coming and end of the world and he knew that doing so was controversial. Consequently, he had to demonstrate that his conclusions were in line with proper Christian faith and practice. As was *de rigueur* amongst sixteenth century polemicists, Napier used frequent references to scripture to achieve this end. This was even more pronounced in the 1611 editions, to which Napier added numerous references to biblical verses in the margins of his first treatise.[3] The 'propositions' were also intended to provide a framework for the polemical claims in the second treatise. For example, the idea that the pope was Antichrist was a central feature of Napier's worldview and the chronology that formed the basis of his commentary. Accordingly, his 26th proposition, which sought to demonstrate the papacy's role as the representative of Satan on the earth, was his longest. This structure, where evidence for Napier's polemical claims was presented at the beginning of the book meant that those assertions could be treated as proven facts in the commentary on the Book of Revelation.

That second treatise was primarily organised into between two and four columns on each page. The first column was Napier's own version of the Book of Revelation, derived from the 1560 Geneva translation and the Vulgate. The second was a 'paraphrase' of the scriptural text and the third was Napier's historical application of prophecies in the Revelation to events in world history. A fourth, narrow column was occasionally included to identify the period of time, associated with the prophecies from the Revelation, in which the events being described belonged. In the sixteenth century, humanist scholars began to prioritise ease of use over aesthetics in their published works. The placement of analytical comments beside the biblical text meant that most of the information the author wished the reader to consider was before their eyes, on the same page, so that leafing back and forth between biblical text and commentary was no longer necessary. Napier wanted the *Plaine Discovery* to appeal to the widest audience possible, claiming to have published the work in English, as opposed to Latin, so that, "the fimple of this *Iland* may be inftructed."[4]

[3] See below, pp. 106–82. All quotations from that work are taken from the edition of the *Plaine Discovery* reproduced in full below, henceforth *PD* in footnotes.

[4] *PD*, p. 101.

He determined "not to follow the particular ornat tearmes, of neither Scottes nor Englifhmen", so as not to alienate English readers, or those in his home country, and translated most of the Latin quotations into English for the benefit of those who could only read their native language.[5]

However, the system of tables that Napier employed was not only intended to assist the reader; it also suited Napier's polemical and rhetorical aims. The third column, Napier's 'historical application' of the biblical prophecies, will probably be of greatest interest to modern scholars. This was where the most dramatic polemical claims were made and where Napier's mathematical skills were brought to bear. However, that column could not have been so effective if the second column had not existed; Napier's paraphrase of the scriptural text allowed him to subtly manipulate its meaning before making dramatic claims about historical or contemporary figures in his third column. In short, the columns allowed Napier gradually to impose layers of meaning onto the biblical text, which made his interpretation appear less controversial or surprising. For example, Napier's assertion that 'Babylon' in the Revelation referred to Rome might seem something of a logical leap.[6] However, in his 'paraphrase', Napier explained that the text did not literally mean Babylon but its successor as, "the mother of al fpiritual whoredome, and abominable idolatrie of the world."[7]

The *Plaine Discovery* was first published by Robert Waldegrave, printer to James VI, who published many works that James patronised and those he authored personally. The first edition was dedicated to King James, but the dedicatory epistle was absent in the 1611 editions. The book was relatively small; both the 1593 and 1611 editions were in octavo format and consequently it was lightweight and portable, and could easily have been carried around or read from one's hands. It utilised several technical innovations from humanist scholarship that had become popular in the sixteenth century. The tabular form of the second treatise reflected the most up-to-date humanist scholarly techniques that existed at the time. Petrus Ramus had popularised the technique during the mid-sixteenth century and it was employed in the 1560 Geneva Bible, on which Napier based his translation of the Revelation and which influenced the *Plaine Discovery* in numerous ways. Also like the

[5] *PD*, p. 101. In fact, the finished work was written in language far closer to English than Scots, reflecting Napier's anglophile tendencies. Also on p. 95, in the epistle to the reader of the 1611 version Napier discussed 'our English or Scottifh tongue', implying he regarded them as two dialects of the same language.

[6] The idea that 'Babylon' meant 'Rome' had been expressed by many Protestants before Napier and would have been familiar to some of Napier's readers. However, he explained and defended the interpretation in unusual detail.

[7] *PD*, p. 298, Rev. 17:5.

version of the bible, the *Plaine Discovery* was printed in Roman font, distinguishing it from many contemporary works which were published in the more ornate Blackletter. Napier summarised the Book of Revelation and his interpretation of it in tables and a poem. This made the work highly accessible and they may have been included to win favour with James VI, who frequently patronised poets and wrote a good deal of verse himself.[8]

HISTORICAL CONTEXT

EXILE, REFORMATION AND CIVIL WAR

The *Plaine Discovery* was a response to events in the British Isles and on the continent through which Napier and his contemporaries and forebears had lived. In a matter of decades, the discovery of the Americas, the rise of the Ottoman Empire, the Protestant Reformation and the accompanying erosion of the Holy Roman Empire and Spain's dominance in Western Europe changed European's philosophical, geographical and political understanding of the world forever. So dramatic were these events that the world seemed to be coming to an end. In turn, that increasingly prevalent apocalyptic worldview shaped how individuals and groups viewed events relating to and effecting them more directly. In July 1553 the young Protestant king of England, Edward VI died and his Catholic half-sister Mary Tudor succeeded to the throne. In the months that followed she began to persecute leading Protestants and reverse the reforming legislation enacted during the reigns of her brother and father, Henry VIII, restoring Roman Catholicism as the state religion of England. In 1554 many English Protestants fled to the continent where they viewed the apostasy of England and the burning of their co-religionists by Mary within an eschatological framework.[9]

In November 1554 John Knox arrived in Frankfurt after being invited to minister to the city's English congregation of Protestant 'Marian exiles'. That congregation soon divided into two factions, comprising those who were willing to use the 1552 English Book of Common Prayer and those, like

[8] See Graigie, J. (ed.), *The Poems of James VI of Scotland*, Vol. II, Scottish Text Society (Edinburgh, 1958).

[9] Eschatology is a concern with 'end times' or 'end things' and is a facet of all major religions. However, it has played an especially prominent role in Christianity, especially during times of trauma.

Knox, who were not. During this time, modern Germany lay within the Holy Roman Empire, and Frankfurt, despite being a free city that had embraced the Protestant Reformation in 1533, had to maintain good relations with the Emperor. Knox's opponents in the city worried that his fiery denunciations of Catholic monarchs would encourage them to move against Frankfurt and that Mary's vengeance would include increased persecutions of Protestants in England. Knox was asked to leave the city after controversially comparing Charles V to the often-demonized Roman Emperor Nero. He, along with some of the English exiles who desired a more 'Reformed' form of worship than the Book of Common Prayer, travelled to Geneva, which was governed according to the uncompromising religious policies of John Calvin.[10]

At Geneva Knox and his good friend Christopher Goodman ministered to a congregation that was hard-line in its reforming agenda. They had fled England rather than recant their faith and even rejected the standard liturgical text of their home country under the last Protestant monarch, Edward VI. They would have absorbed and contributed to the atmosphere of a city that housed some of the most ambitious Protestant Reformers of the time and was a hotbed of religious scholarship. They worked on a new form of worship, a metrical psalter and a bible, hoping to one day use these texts as the bedrock of a new, Reformed Church in England.

In Scotland, political and religious tensions were also on the rise and 1558 was a pivotal year for Anglo-Scottish relations and Protestants in and from the British Isles. On 24th April Mary, Queen of Scots was married to Francis, Dauphin of France, sparking fear among many Scots that their country would come under total French rule. On 17th November Mary Tudor died and the Protestant Elizabeth I was immediately named as her successor. The Marian exiles celebrated by dedicating what became known as the Geneva Bible to the new queen whilst Protestants in Scotland were emboldened and hoped for a political alliance with their neighbour to the south.

John Knox returned from the continent the following year and began preaching inflammatory sermons, sparking iconoclastic riots, the first and most famous of which was at Perth. Mary was in France and her mother, Mary of Guise, acting as regent, unleashed a swift and brutal military response, which helped turn public opinion against her and in favour of the Protestant Lords of the Congregation. These Protestant nobles, with Knox as their chief mouthpiece and political agitator, fought to overthrow the Guise regime and make Protestantism the new state religion. In January 1560, an English fleet landed in the Firth of Forth, north of Edinburgh, and reinforced

[10] See Dawson, J. E. A., *John Knox*, Yale University Press (New Haven, 2015), Chapter 7, for the best account of Knox's 'troubles' at Frankfurt.

the Protestant Party, thereby securing their military victory. In February, the Lords of the Congregation and the Duke of Norfolk, acting as Elizabeth I's representative, signed the Treaty of Berwick, which promised English support to protect the new regime from French interference. From 10th July, what has become known as the 'Reformation Parliament' enacted legislation that made Scotland a Protestant country and declared the influence of the papacy over the country's church to be over.[11]

Although England had been restored to a Protestant country under Elizabeth I, to her royal court the word 'Geneva' itself suggested seditious extremism. Knox was never accepted by Elizabeth because of his poorly-timed tract *The First Blast of the Trumpet Against the Monstrous Regiment of Women*, which employed scripture to argue that female monarchy was against the natural, divinely mandated order.[12] The text was aimed primarily at Mary Tudor, but Knox's mistake was to attack the legitimacy of all female rulers. Nor was the Geneva Bible officially recognised in England. Despite these problems, Knox's prominent role in Scotland's Reformation gave him influence over the new church and he introduced a form of worship in line with the Genevan model. The Presbyterian structure of the new Kirk was also based on Calvin's. Napier's devotional life from an early age was formed within this context.

However, the impact of the exile and its legacy on the *Plaine Discovery* was even more direct. The single greatest influence on that work was the 1560 Geneva Bible, the first Bible to be printed in Scotland. In 1579, an act of parliament decreed that all families of adequate wealth must own a copy and it was therefore the standard religious text of Napier's life.[13] Although he claimed to have used the Vulgate to produce his commentary, there is no doubt that Napier primarily employed the Genevan translation.[14] It is also certain that Napier's apocalyptic thinking was heavily influenced by the marginal notes of that Bible, which were polemically charged and revealed their authors' preoccupation with eschatological, themes.[15] Furthermore, it was Knox's close friend, the English exile Christopher Goodman, whom Napier personally credited with firing his fascination with the Book of Revelation:

[11] For a more detailed explanation of these events see, Dawson, J. E. A., *Scotland Re-Formed: 1488–1587*, Edinburgh University Press (Edinburgh, 2007), Chs. 8 & 9.

[12] Laing, D. (ed.), *The Works of John Knox*, Vol. 4, Woodrow Society (Edinburgh, 1846), pp. 349–422.

[13] Brown, K. M. et al. (eds.), *The Records of the Parliaments of Scotland to 1707* (St Andrews, 2007–2014), 1579/10/25 (accessed 12/04/2013).

[14] *PD*, p. 103.

[15] For a detailed analysis of the Geneva Bible's influence on Napier, see Corrigan, A., *John Napier of Merchiston's* Plaine Discovery: *A Challenge to the Sixteenth Century Apocalyptic Tradition*, University of Edinburgh Doctoral Thesis (Edinburgh, 2014), Ch. 9.

> in my tender yeares and barneage in Sainct-Androes at the Schooles, hauing on the one parte contracted a louing familiaritie with a certaine Gentleman, &c. a Papiſt: And on the other pairt, being attentiue to the Sermons of that worthie man of God, *Maiſter Chriſtopher Goodman*, teaching vpon the Apocalyps, I was so mooued in admiration, againſt the blindnes of Papiſts, that could not moſt euidently ſee their ſeuen hilled citie Rome, painted out there so liuely by Saint Iohn, as the mother of all ſpirituall whoredome, that not onely burſted I out in continual reaſoning againſt my ſaid familiar, but alſo from thenceforth I determined with my ſelfe (by the aſsiſtance of Gods ſpirit) to employ my ſtudie and diligence to ſearch out the remanent myſteries of that holy booke: as to this houre (praiſed be the Lorde) I haue bin doing at all ſuch times, as conveniently I might haue occaſion[16]

Goodman's views had been shaped, to a great extent, by his experiences as a Marian Exile and Napier's life and work might have been very different had it not been for that sermon. Men like Goodman and Knox saw themselves as prophets and Napier's claim that his unlocking of scriptural secrets with 'the aſsiſtance of Godſpirit' is another example of how he followed their model in many aspects of his life. Despite never feeling the call to ministry that dominated their lives he believed he was a prophet and that the *Plaine Discovery* was, at least in part, divinely inspired.

Scotland's Reformation had been facilitated to a great extent by armed conflict including the ousting of a ruling Catholic regent by force, which coloured Napier's view of religion and especially his view of the 'end times'. Moreover, the break with France and new political alliance with Protestant England, led to a conception that the British Isles were united in politics and religion for the first time, creating a sense of shared destiny. This, combined with the countries' geographical separation from the Roman Catholic countries of mainland Europe, fostered an atmosphere of both confidence in the island nations' favoured status with God and fear of foreign Catholic influence, which would be enhanced in subsequent decades.

Mary of Guise died in June 1560 and Francis II of France died in December. The widowed Mary, Queen of Scots returned to rule her native country. Despite her devotion to Roman Catholicism, she tolerated the new Protestant parliament and Kirk, and was granted religious concessions including a private Roman Catholic Mass at the Palace of Holyrood. However, the peculiar conditions of a Catholic queen ruling over a Protestant nation led to political and religious tension. On 29th July 1565, Mary married her cousin Henry Stewart, Lord Darnley, who also had a claim to the English throne. This obvious attempt to secure the succession displeased Elizabeth and relations with

---

[16] *PD*, p. 100.

England deteriorated. Mary's attempts to re-catholicise the nobility and her new husband's brash personality did not help in Scotland. Darnley openly boasted of personally reinstating the mass and of his award of the order of St Michael, conferred upon him by a visiting French delegation. Fear that France would soon conquer Scotland with the cooperation of the monarchy spread. In the Protestant backlash that followed, one of Mary's servants was murdered and Darnley met a similar fate eighteen months later on 9th February 1567 (N.S.).

Mary failed to conform to the expectations of a grieving widow, attending a wedding of one of her servants. Soon after she married the Earl of Bothwell, who had been accused of orchestrating Darnley's murder, and suspicion fell on her. Once again, Scotland was split into two camps: those loyal to the queen, and those who demanded her removal. The latter side commanded more military might and forced Mary to abdicate on 24th July 1567. Her son was one year old at his coronation as James VI of Scotland on 29th July, in a ceremony performed by Napier's maternal uncle, Adam Bothwell, Bishop of Orkney. During the early years of James' minority, the country was ruled by a Protestant regent, Mary's illegitimate half-brother, James Stewart, Earl of Moray. Sporadic fighting occurred in the intervening years following Mary's escape from Lochleven Castle in 1568, especially after Moray's assassination in 1570, and what happened next directly affected the Napier family.

In 1571 the country was in the grip of a civil war that was increasingly concentrated around the capital, Edinburgh, which during the early modern period had emerged as the legal and administrative centre of Scotland. The queen's supporters had garrisoned the city and the regent's troops attempted to starve them out and force a surrender. Merchiston was important because it lay on the main route into the city from the south, and could be used by the queen's party to supply the town, or by English troops to approach from the south in support of the regent. The regent's troops acted first, occupying the Napiers' home in that year. On 18th July John Napier's father, Archibald, was summoned to Edinburgh Castle by his cousin, William Kirkcaldy of Grange, who bombarded Merchiston with cannon fire in an attempt to capture the tower. Archibald tried to avoid committing to either side and may have taken his children to stay at the family's holdings in the Menteith or with John at his marital home for their safety.[17] Further skirmishes took place at Merchiston over the summer of 1572, although the Regent's supporters continued to hold it until their ultimate victory in this civil war and Archibald ultimately aligned himself with the victorious side.

Modern research has shown that religious devotion did not necessarily

[17] Napier, M., pp. 110 & 135.

decide affiliation with either the queen's or regent's side but accounts in subsequent years tended to portray the King's supporters victory as a triumph of Protestantism over Catholicism.[18] Thus, the religion to which Napier subscribed and the Church to which he belonged were coloured by warfare and political intrigue.[19] These events in Napier's youth shaped his experience of religion and current affairs and it is no surprise that his theological work reflected this. The rise and decline of earthly empires and kingdoms were central to his chronological system and the final battle between good and evil foretold in the Revelation was not metaphorical or to occur only in heaven but would include a real war between Protestants and Roman Catholics.

CONTEMPORARY CRISES

During the early modern period, fears about warfare and foreign powers were often expressed in the kind of apocalyptic language that typified the *Plaine Discovery*. However, in the years closely preceding the publication of the work, two key events occurred that would raise the intensity of these preoccupations to new heights. Napier's early religious influences like Adam Bothwell may have been religious moderates but by the late 1580s he was closely associated with more ambitious, uncompromising thinkers whose political and religious beliefs were unified under the banner of Presbyterianism. In 1588, the Spanish Armada failed to land an invasion force against England. The fact that many Spanish ships were wrecked by storms on the coasts of Scotland and Ireland enabled English observers to portray their victory as the result of divine providence. Protestants in the British Isles became even more convinced that God was on their side. However, the size of the Spanish fleet and rumours of further invasion attempts also created fear, which was exacerbated by the isolation of Britain from the European mainland. In December 1592, in this atmosphere of fear and tension, tempered with confidence in the perfect truth of Protestantism, the Spanish Blanks Affair came to light. A group of Roman Catholic nobles had attempted to smuggle blank parchments, which they had undersigned, out of Scotland, to be filled in by Jesuits, which

[18] Donaldson, G., *All the Queen's Men: Power and Politics in Mary Stewart's Scotland*, Batsford (London, 1983) showed that many factors determined allegiance to either side, and that many people changed sides during the civil war.

[19] For more on the role and place of Merchiston in the civil war, see Napier, M., *Memoirs of John Napier of Merchiston: His lineage, life, and times, with a History of the Invention of Logarithms*, pp. 131–40, & Pollock M. S., reproduced in Thomson, T., *A diurnal of remarkable occurrents that have passed within the country of Scotland since the death of King James the fourth till the year M.D.LXXV*, The Bannatyne Club (Edinburgh, 1833).

amounted to a *carte blanche* for a Spanish invasion of Scotland.

Napier discussed the Armada several times in the *Plaine Discovery*, integrating it into his chronology as evidence for the waning power of Antichrist and imminence of Protestantism's final victory over Roman Catholicism. He perceived the Armada as a watershed in the history of the British Isles, and portrayed the Spanish Blanks and proselytizing activities of Jesuits as a continuation of the precedent set in 1588:

> this new infolencie of Papifts arifing about the 1588 year of God, and dayly increfing within this *Iland* doth fo pitie our heartes, feeing them put more truft in Iefuites and feminarie Priests, then in the true fcriptures of God, and in the Pope and king of *Spaine*, then in the King of Kings [20]

Napier repeatedly called Britain an 'iland'. Like many of his contemporaries, he was presenting the Atlantic Archipelago's isolation from Europe as evidence that God wanted Scotland and England to unite as allies under a single church. The united countries would provide a beacon of hope to the oppressed Protestants of the continent and help defeat the Antichrist in the impending final battle.[21] Nowhere were Napier's views on the consequences of these events and his views on the political future of the British Isles made clearer than in the *Episle Dedicatorie* to James VI, in the first edition of the *Plaine Discovery*.

That text was a direct correspondence with his king that revealed Napier's views on monarchy and the events that had recently transpired. In the dedication, Napier tasked James with safeguarding the spiritual and temporal future of Scotland, believing that kings and the political establishment were best-suited to lead reform, and that their subjects would follow their example. He implied that there was special meaning in the Book of Revelation and his commentary for all kings and urged James, "to begin at your M. owne houfe, familie and court, and purge the fame of all fufpicion of papifts, and Atheifts or Newtrals".[22] He combined rebuke with flattery, appealing to an elevated image of kingship by comparing James to King David. He also integrated James into his chronology, casting him as one of the ten horns of the beast in Revela-

[20] *PD*, p. 101. See MacDonald, A. R., 'James VI and the general Assembly, 1586–1618' in Goodare, J. & Lynch, M. (eds.), *The Reign of James VI*, John Donald (Edinburgh, 2000), pp. 170–85, for a brief discussion of the fear created by the Counter-Reformation in Scotland.

[21] Dawson, J. E. A., 'Anglo-Scottish Protestant culture and integration in sixteenth-century Britain' in Ellis, S. & Barber, S. (eds.), *Conquest and Union: Forging a Multi-National British State*, Longman (London, 1995), pp. 87–114, p. 104.

[22] 1593 edition, see below page 946.

tion 17, which represented ten kings, who originally sided with the Antichrist but ultimately followed Christ and helped destroy the whore of Babylon.[23]

This combination of flattery and remonstrance is unsurprising, given the events that preceded the composition of the Epistle, which was dated 29th January, 1593 but was almost certainly written in 1594 according to the modern system of dating.[24] The epistle reflected the Presbyterian party's desire for James to act decisively against the Catholic nobility in Scotland, meting out harsh punishments to the Spanish Blanks conspirators. In the months following the discovery of the 'blanks' ministers and general assembly members urged the king to act and, on 25th September 1593, the General Assembly impatiently excommunicated the plotters, who included Napier's father-in-law from his second marriage, Sir James Chisholm of Cromlix. The fact that Napier sat on the delegation that excommunicated the plotters, in his capacity as the Presbytery of Edinburgh's Commissioner to the General Assembly, reveals a commitment to the Reformed Protestant cause that transcended familial loyalties. The king was furious that the conspirators had been excommunicated without his consent and complained that "the ministers were cruell, and as they sought blood they should have it."[25] The broader issue of royal authority was at stake; James wanted to exercise greater control over excommunication in particular and the Kirk in general. Napier's standing amongst his colleagues in the Kirk is highlighted by the fact that he was commissioned, along with James Melville and Patrick Galloway, with meeting with James and petitioning for his support for harsh punishments for the plotters. However, their pleas were unsuccessful. James was personal friends with one of the Plotters, the Earl of Huntly, and tended to pursue policies of appeasement. In late 1593 he effectively pardoned the conspirators, on the proviso that they never engage in seditious activities again.[26]

Napier's subsequent epistle to James VI was something of a desperate final attempt to encourage James to submit to the Kirk's demands and act decisively against the perceived Roman Catholic threat in Scotland, especially among the nobility. It reveals Napier's attitudes and those of his fellow Presbyterians and shows that the Spanish Blanks Affair had hit close to home for both Napier and his king. The scandal seemed to prove that the Roman Catholic threat to Scotland was real and imminent and this was was reflected in the

[23] 1593 edition, see below p. 946. Napier appealed to the royal couple as a unit by casting Denmark, the home-country of James' bride, as another horn.

[24] See above, fn. 1.

[25] Calderwood, *History*, Vol. 5, p. 269 & Napier, M. p. 162.

[26] James' attitudes and motivations are a highly complex issue. For more information see Macdonald, A. R., *The Jacobean Kirk, 1567–1625: Sovereignty, Polity and Liturgy*, Ashgate (Aldershot, 1998), p. 39.

urgent language used in Napier's epistle. That short text was therefore well-suited to open the *Plaine Discovery* because the book was distinguished by its sense of apocalyptic urgency that was unprecedented in the English language. However, Napier's epistle failed to persuade James to punish the Catholic nobility and it was absent from subsequent editions of the work.

CONTENT OF THE *PLAINE DISCOVERY*

NAPIER ON ROMAN CATHOLICISM

Denunciation and vilification of the papacy and its church were the most dominant themes of the *Plaine Discovery*. Every part of the work was loaded with anti-Catholic language and imagery, from the dedicatory and introductory epistles, to the tables that summarised the Revelation and the poems included by Napier. He employed Roman Catholicism as a foil for Protestantism, attempting to remind Protestant readers of the evils of the alternative to remaining constant in their faith and urging Catholics to repent immediately. He was seeking to provide his audience with a scapegoat, on which to focus all of their fears and hatred. However, demonization of Roman Catholicism was not merely a key *raison d'être* of the *Plaine Discovery*, Napier portrayed it as the reason for the Book of Revelation itself, which Napier understood as a divine warning to humanity that they might identify and understand their enemy in the last age of human history. The destruction of Rome was presented as an essential part of the apocalyptic narrative he claimed was unfolding and he believed Protestants in the British Isles and their king might soon have to literally take up arms against the forces of Antichrist, led by the pope. This conflict would follow the precedent of the Spanish Armada, in which foreign tyrants were simply puppets of the papacy, which was cast as the supreme leader of all Roman Catholics:

> Thefe mightie marriners with their great Galliaffes, Venetians, Spanifhes, & other of that Antichriftian flote, who hitherto, and lateft of all in this 1588. yeare of God, haue bene fo readie at euery nod of their Apoftatik ftep-mother, *Rome*, to haue overwhelmed vs poore and true 'Christiã profeffors.[27]

Napier employed arguments that would have been familiar to his domestic audience. He attacked Roman Catholic traditions and doctrine as evil black

[27] *PD*, p. 314. 'Apoftatik' referred to the apostasy, rejection of religious faith, with which Napier charged Rome.

magic and connected every evil figure in the Revelation in some way to Rome. A detailed analysis of all his arguments could not be accommodated in a work of this size. Instead, this chapter considers as an example Napier's presentation of the papacy as Antichrist.[28]

Portraying the papacy as Antichrist was not new. Martin Luther had made the claim as early as 1520 and pre-Reformation scholars like John Wycliffe (1320–84) had also made the connection. There were a number of Protestant works that did so in English, including the 1560 Geneva Bible and John Bale's *Image of Both Churches*, from which Napier derived his idea of a false Church of Antichrist, pitted against the real Church of God, which was often hidden from view. However, Napier brought his own unique style to his arguments, which were unusually detailed and set out to prove beyond all doubt that the papacy was the Man of Sin and Antichrist, foretold in 2 Thessalonians and Revelation respectively.

Napier's explanation of the papacy as Antichrist was unusually thorough compared to the works that influenced it, and this was consistent with the detailed and systematic approach taken in the *Plaine Discovery* in general. Not only did it provide an enemy on which to focus, it answered an important question. If God was on the Protestants' side and they were his true church, why was so much of the world opposed to them? One answer was that God had ordained that his chosen people must endure persecution and minority status in the Old Testament, when the Israelites were enslaved in Egypt and kept in captivity in Babylon. These examples showed that this suffering was temporary because God had ultimately delivered the Israelites from oppression. Napier's reading of the Revelation was intended to reinforce this image and refine it for a new era. The scriptural text showed that Antichrist's dominance over the world was a predestined but temporary part of God's plan, culminating in eternal rewards for his chosen and eternal punishment for their enemies.

In Napier's commentary, the idea of the papacy as Antichrist was treated as an accepted fact that provided a foundation and evidence for some of his more controversial and novel conclusions. However, he first went about 'proving' the papacy's nature in the longest of his introductory propositions: "The Pope is that only Antichrift, prophecied of, in particular."[29] He argued that there are many things that oppose Christ's teachings and example and are therefore 'antichrist' but that there is one that leads and unifies them all. Napier referred to 2 Thessalonians, which foretold the coming of a Man of Sin who would seek to pervert the teachings of Christianity. He argued the text meant that this

[28] See Corrigan, 2014, Ch. 4 for a detailed analysis of these themes.
[29] *PD*, pp. 149–59.

Man of Sin was also the Antichrist and must emerge from within the Church and could not be an external aggressor: "he needs be a latent, and not a patent enemie, that the Spirit of God ſo carefully points out: and ſo no *Turke*, *Jew* nor *Pagan*, yea, no plaine Wolfe muſt hee be: but a Wolfe in Lambs skin, euen an Antichriſt vnder the name of a Chriſtian."[30] He identified false miracles used by the Roman Catholic Church and claimed these were tools of Antichrist, designed to trick good Christians into worshipping him. These false miracles were widely denounced by Protestants and much of Napier's audience would have been predisposed to this type of thinking.

Another of his arguments appears highly dubious to the modern observer, but probably had more currency in the late sixteenth and early seventeenth centuries. This was a list of 17 examples of the ways in which the papacy supposedly did the opposite of Christ, making it 'antichrist'. For example, "Christ ſhed his blood once, that we may reign ſpirituallie. The Pope ſheddeth our blood daylie, that he may reigne temporallie".[31] This section seems unplanned and rambling, which sets it apart from the rest of the book, which suggests fervour on Napier's part. He evidently believed what he was writing and his words seem emotionally charged.

Napier's analysis of 1 John 2:22 appeared more thoughtful. Here he displayed his talent for anticipating possible criticism. The scriptural text had: "Who is a lyer, but he that denyeth that Iesus is Christ? the same is the Antichrist that denyeth the Father and the Sonne."[32] Clearly though, the pope does not deny Christ but Napier argued,

> there be two denials, the one in mouth and profeſsion plainely, and that way the Pope ſeemeth a Christian . . . The other deniall is in heart and deede tacitly . . . the Pope denieth Chriſt, in that he eſtabliſheth other mediators . . . Wherefore, in effect the Pope denieth the Sonne, and conſequently both the Father & the Sonne.[33]

Again, the idea that Christ is the only mediator between humanity and God, and that saints served no such function was central to Protestant doctrine and Napier would have expected his audience to be receptive to such an argument.

---

[30] *PD*, p. 160.

[31] *PD*, p. 156.

[32] Unless otherwise stated, all biblical quotations outwith the Book of Revelation are taken from the 1560 Geneva Bible: Whittingham, W. et al., *The Bible and Holy Scriptures conteyned in the Olde and Newe Testament. Translated according to the Ebrue and Greke, and conferred with the best translations in diuers languges. With moste profitable annotations vpon all the hard places, and other things of great importance as may appeare in the epistle to the reader*, Rouland Hall (Geneva, 1560), STC/1019:06. That was the translation upon which Napier primarily relied and would have used for study and devotion.

[33] *PD*, pp. 157–8.

Napier's assertions about the papacy were typified by his uncompromising polemical style that differentiated the *Plaine Discovery* from works that preceded it. For example, John Bale used the word 'antichrists' to denounce a wide range of people and religious practices.[34] However, by using the word as a proper noun exclusively to describe the papacy, Napier increased its impact immeasurably. In the *Plaine Discovery*, he portrayed the papacy as an almost supernatural, monolithic institution that embodied absolute evil and whose rise and ultimate destruction were predicted in scripture.[35]

THE CHRONOLOGY

Napier's chronology of salvation history was more detailed and was more integral to its author's polemical arguments, especially those directed at Roman Catholicism, than any that had been devised previously. It reflected Napier's interest in mathematics, and his skills were employed in its unusually rigid and precise structure. It identified events, in chronological order, as fulfilments of the prophecies of the Book of Revelation. The prophecies most emphasised in the chronology were typically regarded as allegories for events that would occur in the physical world. To Napier, the structure of the Revelation itself suggested a fixed pattern within which those prophecies would be fulfilled incrementally. Revelation Chapters 6 to 8 described the opening of seven seals that bound up the scroll of God's Revelation to John being opened one by one. In Chapters 8 to 11 seven angels blew trumpets to announce key events in the unfolding apocalyptic narrative. Chapter 16 was concerned with the pouring out of seven vials, or in some translations bowls, representing the wrath of God against the followers of antichrist. Finally, Revelation 14 related the appearance of angels, which made important announcements. These prophecies and Napier's interpretation of historical and contemporary events that fulfilled them formed the basis of his chronological scheme.

The chronology began with the opening of the first seal, which Napier claimed occurred in 29 A.D., which he identified with the baptism of Christ, and continued to his predicted date of the Second Coming, Day of Judgement and destruction of the world between 1688 and 1700. The concept of time passing was vitally important throughout the work, but Napier presented these final occurrences as different aspects of the same event and abandoned his

[34] See Corrigan, 2014, Chapter 7.

[35] This elevated conception of the papacy's power was common among sixteenth century Protestants. See Weiner, C. Z., 'The beleaguered isle: a study of Elizabethan and early Jacobean anti-Catholicism', *Past & Present* 51, 1971, pp. 27–62.

chronology after the Second Coming. Near the end of Chapter 20 he wrote,

> And we are ſure that then the ſecond and laſt reſurrection should bee, wherein generallie all men ſmall and great, that euer died, shall riſe againe, and appeare before that iustice ſeat of God, and the regiſter bookes of all mens conſciences bee opened vp[36]

Then, at the very beginning of Chapter 21 he wrote,

> Then I foreſaw that the heauens and the earth ſhould be renewed to their vttermoſt perfection: for this imperfect eſtate of the heauens & the earth that now is, ſhall diſſolue and melt away, and there ſhall be no ſea, or raging inundations of worldly troubles.[37]

Thus, when Napier employed one term like 'second cõming' or 'day of judgement' he intended his audience to think generally about the 'end times' when Christ would return, judge humanity, punish the wicked, save the elect and the world would be destroyed and renewed, all within a very short space of time.

The events that fulfilled the prophecies of the seven seals, trumpets and vials, and four 'thundering angels jubilees' occurred at uniform intervals of seven, 245 and 49 years. These figures were built upon the method of interpreting 'days' in prophetic texts as 'years' and the concept of *Shmita*. This was a sabbatical interpretation of years, in which every seventh year was seen as a Sabbath, just as every seventh day was a Sabbath.[38] Napier began explaining the system in his first Proposition, which illustrates its importance to his work. He used Numbers 14:34 to justify substituting days for years: "After the number of fourtie daies, in which ye ſearched out the land, the yeare ſhall be counted for a day".[39] However, he went even further than using scripture to defend his methods. He asserted that the 70 weeks described in Daniel 9:24 represented 490 years and that this prophecy accurately predicted the birth of Christ.[40] Damnation awaited any who dared deny his system because to do so was to deny Christ:

> vpon neceſsitie of ſaluation, all chriſtians muſt confeſſe, in the ſeuentie weekes of DANIEL, a day to be taken for a yeare, extending in the whole to 490. yeares; otherwiſe, that prophecie of the Meſsias comming would not fall vpon the juſt time of Chriſts comming, as neceſſarily it ought to doe.[41]

---

[36] *PD*, p. 325.
[37] *PD*, p. 331–2.
[38] See Genesis 2:2–3, Exodus 23:12 & Deuteronomy 5:13–14.
[39] *PD*, p. 107. Napier's translation of the Vulgate.
[40] Ibid., 70 weeks × 7 days = 490 days, interpreted as years.
[41] Ibid.

NAPIER'S CHRONOLOGY OF SALVATION HISTORY

Event in the Revelation	Scriptural Reference	Napier's Date	Page in *Plaine Discovery*	Napier's Historical Interpretation
1st Seal	Rev 6:1	29 A.D.	p. 216	Baptism of Christ. Gospel of Matthew written.
2nd Seal	Rev 6:3	36 A.D.	p. 217	Christians persecuted. Gospel of Mark written.
3rd Seal	Rev 6:5	43 A.D.	p. 217	Global famine. Gospel of Luke written.
4th Seal	Rev 6:7	50 A.D.	p. 218	Gospel of John written.
5th Seal	Rev 6:9	57 A.D.	pp. 218–9	Nero's temporal power increased.
6th Seal	Rev 6:12	64 A.D.	p. 219	Nero persecuted Christians, committed incest & murder.
7th Seal	Rev 8:1	71 A.D.	pp. 227–8	Persecution of Christians suspended under Flavian Dynasty.
1st Trumpet/ Vial	Rev 8:7/ 16:2	71 A.D.	pp. 229/ 287–8	Effeminate & tyrannical Roman Emperors.
2nd Trumpet/ Vial	Rev 8:8–9/ 16:3	316 A.D.	pp. 229–30/ 288	Constantine shifted imperial seat from Rome to Constantinople.
3rd Trumpet/ Vial	Rev 8:10–11/ 16:4-7	561 A.D.	pp 230/ 288–9	Rise of Islam & apostasy of Christians in the Near East.
4th Trumpet/ Vial	Rev 8:12/ 16:8–9	806 A.D.	pp.230–1/ 289–90	Church corrupted by Islam in the East & Papacy in the West. Charlemagne divided Holy Roman Empire between his sons.
5th Trumpet/ Vial	Rev 9:1/ 16:10–11	1051	pp. 233/ 290	Rising power of Islam.
6th Trumpet/ Vial	Rev 9:13/ 16:12–13	1296	pp. 236/ 290–1	Unification of formerly disparate Islamic peoples.
7th Trumpet/ Vial	Rev 11:15/ 16:17–21	1541	pp. 251–2/ 291–3	Protestant Reformers active & successful.
1st Jubilee	Rev 14:6	1541	p. 276	Protestant reformers bringing truth of Gospels to light.
2nd Jubilee	Rev 14:8	1590	pp. 276–7	Fall of Rome as new Babylon. Military successes of Protestant against Catholic states.
3rd Jubilee	Rev 14:9–10	1639	p. 277	Final defeat of Rome.
4th Jubilee	Rev 14:14	1688–1700	p. 278	Second Coming of Christ, God's Judgement, Destruction of World, Creation of New Heaven and New Earth.

Exodus 23:10–11 decreed that, just as every seven days was to be a Sabbath, every seventh year was to be a year of rest, when fields were to lie fallow and the poor could gather any unharvested crops. Leviticus 25:8 commanded that every seventh Sabbath, or every 49th year, was to be a celebratory Jubilee year, which is why Napier's 'thundering angels' jubilees' were every 49 years. These periods formed the basis of Napier's scheme and had already been used for centuries by scholars to calculate the age of the world, by counting backwards through the Old Testament. This supposed mathematical precision allowed Napier to portray his scheme as perfectly accurate and probably impressed his contemporaries in an age when historical data were not easily verified. However, to the modern observer, this chronological precision is the *Plaine Discovery*'s greatest failing because its fixed pattern of dating forced Napier to falsify when many events occurred, and also to include certain dates that correlated to nothing in particular.

SEVEN SEALS

> And they ſung a new ſong, ſaying, Thou art worthie to take the Booke, and to open the ſeals thereof, becaus thou was killed, & haſt redeemed vs to God, by thy blood out of every kinred, and tongue, & people and nation.
>
> **Revelation 5:9**

> After, I beheld when the Lamb had opened one of the ſeales, and I heard one of the four beaſtes ſay, as it were the noiſe of thunder, Come and ſee.
>
> **Revelation 6:1**

In Napier's commentary, the seven seals in Revelation 6 and 8 were interpreted as being opened every seven years, from the Baptism of Christ to the end of the persecutions of Christians under the Roman Emperor Vespasian in 71 A.D.[42] Napier's basis for the Baptism of Christ occurring in 29 A.D. derived from Luke 3:1–2:

> Now in the fiftenth yere of the reigne of Tiberius Cesar, Pontius Pilate being governour of Iudea, and Herode being tetrarch of Galile, and his brother Philippe tetrarch of Iturea and the countrey of Trachonitis, and Lysanias the tetrarch of Abilina, (When Annas and Caiphas were the high Priests) the worde of God came vnto John, the sonne of Zacharias, in the wilderness.

[42] *PD*, Christ's baptism at p. 228 seventh seal at pp. 233–4.

Augustus Caesar died in 12 A.D., which may be regarded as the first year of his successor's reign. The first four seals also coincided with the writing of the four Gospels, though modern scholarship has disproved this and tends to date Mark as the earliest, whereas Napier identified Matthew as the earliest.[43] The precision of Napier's system is emphasised by a comparison to Bale's *Image*. For example, in Bale's chronology, one may infer that the fourth seal was opened at some point in or around the fourth century A.D.[44] In Napier's scheme it was stated explicitly as opening in 50 A.D.

Napier celebrated the church at this time as uncorrupted by the evils that he believed typified the Roman Catholic church. He presented the opening of the seals, and associated writing of the gospels, as acts of divine mercy, intended to comfort the persecuted Christians of the Roman Empire. This was consistent with Protestant humanist philosophy, which regarded the early church as an ideal, uncorrupted by invented traditions, to which Christianity should return. Even at this early stage in his chronology, he emphasised the supposed evils of Roman Catholicism, in relation to this position:

> the deuill was ſo ready, that how ſoon the true word of God was ſowen by the primitiue Church, incontinently he raiſed vp the papiſticall antichriſt to deuour & confound the ſame with falſe gloſſes, & ſiniſter interpretations.[45]

The fifth and sixth seals were identified with Nero. Although 57 and 64 A.D. correlated to no specific event by which Napier could prove that emperor's evil nature, Napier was bound to these years by his system and connected them to the start of his reign, which was actually 54 A.D., and his persecution of Christians. However, Napier was skilled at wording his arguments to make them fit with the biblical text and he interpreted Revelation 6:12, "And I beheld when he had opened the ſixt ſeale, & loe, there was a great earthquake, and the ſunne was as black as ſack-cloth of haire, and the Moone was like bloode", as meaning that Nero had, "vtterly darkened and extinguiſhed the light of the Goſpel".[46] Similarly, the seventh seal was accompanied by, "ſilence in heauen about halfe an houre", and Napier interpreted this as a temporary suspension of persecution of Christians during the Flavian Dynasty.[47]

---

[43] *PD*, p. 216 & Barclay, W., *The Gospel of Mark*, Saint Andrew Press (Edinburgh, 2001), p. 1: "it is agreed by nearly everyone that it [the Gospel of Mark] is the earliest of all the gospels".

[44] Ibid., ff. 78*v*-*r*.

[45] *PD*, p. 258.

[46] *PD*, p. 219.

[47] *PD*, p. 228.

SEVEN TRUMPETS & VIALS

> And I ſaw the ſeuen Angels which ſtood before God, and to them were giuen ſeuen trumpets.
>
> **Revelation 8:2**

> So the firſt Angel blew the trumpet, and there was haile and fire mingled with blood, & they wer caſt into the earth, and the 3 parte of the trees was burnt, and all green graſſe was burnt.
>
> **Revelation 8:7**

> And one of the four beaſts gaue vnto the ſeauen Angels, ſeauen golden vials, full of the wrath of God, which liueth for euermore.
>
> **Revelation 15:7**

> And the firſt went, and powred out his vial vpon the earth, and there fell a noyſome and a grieuous ſore vpon the men, which had the marke of the beaſt, & vpon thẽ which worſhipped his image.
>
> **Revelation 16:1**

Napier believed that the seven vials in Revelation 16 and 18 simply reinforced the trumpets in Chapters 8, 9 and 11. This idea of 'recapitulation' had been used to varying degrees in many earlier works on the Revelation since its initial usage by Victorinus of Pettau in the third century.[48] It was employed by John Bale and, in the case of the *Plaine Discovery*, meant that the chronology in Chapters 16 and 18 repeated that in Chapters 8, 9 and 11. From this point on, the word 'trumpets' will be used to mean both trumpets and vials.

Like the seals, the trumpets occurred at regular intervals, in this case every 245 years. This was half of the 70 weeks of Daniel, or 490 years, which Napier presented as the period for which God allows earthly institutions and kingdoms to remain dominant. Napier derived his interpretation of these 'Great mutations of Empires' from a work called *Carion's Chronicle*, though that book did not explain the system in nearly as thorough or systematic a way as the *Plaine Discovery*.[49] The period denoted by the trumpets was the longest in Christian history and was concerned primarily with corruption of the Church and growing dominance of the papacy as Antichrist, ending with the new hope that came in the form of the Protestant Reformations.

---

48 Backus I., *Reformation Readings of the Apocalypse: Geneva, Zurich and Wittenburg*, OUP (Oxford, 2000), pp. 135 & xiii.

49 Funke, J. (ed.) & Lynne, W. [A Wouter van Lin] (trans.), *The thre bokes of Chronicles, whyche John Carion (a man syngularly well sene in the Mathematycall sciences) Gathered with great diligence of the beste Authours that haue weitten in Hebrue, Greke or Latine*, Mierdman (London, 1550), STC 4626, f. ciiii*r* on Constantine and the papacy. The book of Daniel contained prophecies about kingdoms, and refers to a time during the Babylonian Captivity, hence the connection between the text and empires.

Again, many of the dates were approximate or incorrect. For example, Constantine's transfer of his imperial seat from Rome to Constantinople was a gradual process that took place from around 306 to 330 A.D. when the city was consecrated. However, Napier claimed that Constantine transferred his imperial seat in 316 A.D. because this was 245 years after his previous date of 71 A.D.[50] 316 was critically important to Napier's system because it represented a major step forward in the temporal power of the papacy, which, over the centuries, increased its authority and cruelly oppressed Christendom.[51] Similarly, the fourth trumpet in 806 highlighted the rise of the Holy Roman Empire, which Napier portrayed as the beast whose wound was healed in Revelation 13:3.[52] Again, this was vital to his demonization of the Roman Catholic Church. In his reading of history, the Holy Roman Empire of Charlemagne and his descendants was not a new empire, but the same idolatrous, Roman Empire that had killed Christ, represented in the Book of Revelation by a monstrous beast. When the beast was wounded, the Western Roman Empire fell but the Holy Roman Empire was simply the same beast, healed and restored to temporal dominance. However, Napier was also careful to portray these emperors, along with all Roman Catholic secular leaders, as puppets of the papacy. They were not just tyrants, but agents of Satan on the earth, led by the Antichrist whose throne was in Rome.

The fifth and sixth trumpets reflected the other great source of fear among Protestants in Napier's lifetime, the Islamic World. Napier asserted that the power of Islam was increasing in 1051 and that four races of Islamic peoples, '*Turks*, *Tartarians*, *Saracens*, and *Arabians*' united after 1296.[53] Although these dates were rather vague in terms of how they related to world events, the Ottoman Empire did emerge around the turn of the fourteenth century. The view that Revelation 9:14, "loose the 4 Angels, which are bound in the great riuer Euphrates", referred to four Islamic groups was fairly common and Napier had derived his interpretation from the 1560 Geneva Bible.[54] He cast Muslims as Gog, to the papacy's Magog, two opposing but evil forces described in Revelation 20:7–10 whose warfare and oppression threatened the true church but whose destruction was assured as part of God's predestined plan for humanity.[55]

[50] *PD*, p. 239–40.

[51] Napier's stance on the so-called 'Donation of Constantine' is a highly complex issue, see Corrigan, 2014, pp. 130–2, 155–6 & 193–6.

[52] *PD*, pp. 146 & 266–7.

[53] *PD*, pp. 110 & 236.

[54] Indeed, the Geneva Bible named the same groups as Napier. See Geneva Bible, Rev. 9:14, note *a*.

[55] *PD*, pp. 323–4.

Finally, the seventh trumpet heralded the Protestant Reformation which, for Napier, was one of the most important turning points in human history. God's true Church was no longer hidden and oppressed by the visible Roman Church, but active and forceful and the balance of power began to shift away from the forces of Antichrist in favour of the elect. Unfortunately for Napier, he could not identify an important event that occurred in 1541, the year dictated by his system, and he simply claimed that, by this point, the Reformation and some of its key players were active and confident.[56]

THUNDERING ANGEL'S JUBILEES

> Then I ſaw another Angel flie in the middes of heauen, hauing an euerlaſting Gospel, to preach vnto them that dwel on the earth & to euerie nation and kinred & tongue, and people. Saying with a loud voice, fear God, and giue glorie to him: for the houre of his iudgemẽt is come, and worſhippe him that made heauen and earth, and the sſea, and the fountaines of water.
>
> **Revelation 14:6-7**

The period of the four Jubilees, occurring every 49 years, was special for Napier's scheme because it represented a shift from allegorical interpretations about historical and contemporary events, to predictions about the future. It was the most positive period for Napier because it included the victory of God's true church over the church of Antichrist. He reiterated the importance of reformers around 1541, including Luther, Calvin and Melanchthon, linking their efforts to ostensibly bring the truth of scripture to light, to Revelation 14:6: "Then I ſaw another Angel flie in the middes of heauen, hauing an euerlasting Gospel, to preach vnto them, that dwel on the earth, and to euerie nation, and kinred and tongue & people."[57] The second Jubilee heralded the fall of Babylon, interpreted by Napier as meaning Rome and he claimed it had occurred in 1590, a mere three years before the *Plaine Discovery* was published. This speaks volumes about Napier's view of the age in which he lived. He believed that the defeat of the Spanish Armada in 1588 and the succession of France's first Protestant king in the following year, which ended the dominance of the house of Guise, were the result of divine providence demonstrating God's favour towards English and Scottish Protestants.[58] The

[56] *PD*, p. 276.

[57] *PD*, p. 276.

[58] Henry of Navarre actually converted to Roman Catholicism in 1593. Napier had clearly heard about this by 1611 and changed his description of the king from 'a Proteſtant' to 'one named a proteſtant' in the editions of that year. See *PD*, pp. 276–7.

Guise dynasty had controlled not only France, but Scotland under the Regent Mary of Guise. Thus, domestic affairs, including the Scottish Reformation that had occurred with the ousting of Mary were presented as important on a cosmic scale, reassuring the reader that everything they had experienced mattered and was part of God's plan. The events seemed to confirm that the final defeat of the Roman church and the institutions that supported it, by the Protestant nations of Europe was at hand. If that were true, it was inevitable that Protestants would soon receive their eternal reward for embracing true religion and suffering under the oppression of tyrannical Catholic forces.

The fall of Babylon was one of the last occurrences before the Second Coming and Napier moved to predicting dates in the future. He saw the Armada as evidence that the balance of temporal and spiritual power was shifting in favour of Protestants and predicted the ultimate destruction of Rome in 1639. Napier used violent and horrific imagery to encourage Protestants to remain constant and Roman Catholics to convert. Those who continued to worship the beast would, "be tormented eternallie in the bitter ſulphurious fire of hell, in the preſence of Christ and all his holie Angels."[59] The final 'angel' which appeared 49 years later was Christ appearing to reap humanity and gather the elect, before the reprobate were cast into hell.[60] The Bible does not cast this figure as an angel, but Christ, or one, 'like vnto the Son of man'.[61] Napier probably did so to bring the total of angels listed in the Revelation to the mystical number seven which, through his preoccupation with Sabbaths and Jubilees, formed the basis of his chronology.

THE MILLENNIUM OF SATAN'S BINDING AND ANTICHRIST'S REIGN

Two other important numbers in the *Plaine Discovery* were 1260 and 1000. These referred to the number of years for which the Antichrist was permitted to reign and the millennium for which Satan was bound. The idea that these things would occur at the same time may seem illogical but Napier argued that God had displayed his mercy by binding Satan during the reign of his representative on earth, "leaſt on both sides, Gods Church were vtterly extinguiſhed".[62] Napier believed that, when Revelation 20:3 stated that Satan was bound "that he ſhould deceiue the people no more," he was being prevented from harming

[59] *PD*, p. 277.
[60] Ibid.
[61] *PD*, p. 278.
[62] *PD*, p. 177.

only the outward, visible church of Antichrist and not the true, hidden Church of God, "which euer he troubleth".[63]

The number 1260 occurred many times in the Books of Daniel and Revelation, the two most important prophetic texts of the Christian canon, which usually referred to a period of days. Napier interpreted these as years and cited Revelation 12:6 as evidence of the true church being forced into hiding by the church of Antichrist: "And the womã fled into the wildernesse, where shee hath a place prepared of God, that there they should feed her there a thousand, two hundreth and threescore daies."[64] In Napier's Chronology, this period lasted from the so-called 'Donation of Constantine', "as is alleged betwixt the yeare of Christ 300. & 316. to the year of God 1560. which time the notable decay of his kingdome began."[65] These dates were a clever selection. The first was connected to the rise to dominance of the papacy in the Western Roman Empire and Napier saw it as a hugely significant historical turning point. The period of Antichrist's reign saw gradual corruption spread throughout Christendom via the use of idolatry in the Roman Catholic Church until the Protestant Reformation. Napier's usage of 1560 as the end of Antichrist's reign elevated the Scottish Reformation to universal significance, again appealing to his domestic audience.

Similarly, the binding of Satan occurred around the year 300, when Constantine ended the persecution of Christians by the Roman Empire. Constantine and his successors, "(except a fewe of short raign) maintained Christianisme to the abolishing of Sathans publike kingdome: and therefore, saye wee, this yeare Sathan is bound."[66] Thus, Napier's view of this emperor was ambivalent; his actions were connected both with the binding of Satan and the beginning of Antichrist's reign. Satan was released after the sixth trumpet was blown in around 1296 and this coincided with the release of Gog and Magog. However, something monumentally evil had to have coincided with the release of Satan and Napier selected something that supported his conclusions perfectly. He claimed that Boniface VIII's introduction of Jubilees to encourage pilgrimage to Rome coincided with the release of Satan. This underscored Napier's hatred of Roman Catholic doctrine because the purpose of these pilgrimages was to achieve remittance of one's sins. The idea that no human actions could affect one's salvation and that one was saved by faith alone is one of the central aspects of Protestant belief that differentiates it from

---

63 *PD*, pp. 322 & 176.

64 *PD*, pp. 258–9. For other scriptural references to 1260, see Rev. 11:2–3; Dan. 7:25 & Dan. 12.7.

65 *PD*, p. 152.

66 *PD*, p. 174.

Roman Catholicism. In fact, Martin Luther's objection to the selling of indulgences, which supposedly allowed people to spend less time in Purgatory, was generally considered the first act of the Reformation. However, Napier was not only presenting himself in the tradition of Luther and emphasising his doctrinal differences to the papacy, he was also attempting to avoid any allegations that the Jubilees that formed the basis of his entire system were connected to the hated doctrine of Purgatory.

However, the *Plaine Discovery* did not emphasise the millennium described in Revelation 20:6, when "the Prieſtes of God and of Chriſt ... ſhall reigne with him a thouſãd year."[67] Napier interpreted this number figuratively, as meaning, "for euer, and euer", after the apocalypse.[68] This belief in a figurative millennium was called 'Amillennialism' and was opposed to 'Premillennialism', or Chiliasm, which posited that the millennium would occur on earth in the future and 'Postmillennialism', which identified it in the past. This stance was necessary for a number of reasons. Napier predicted that the world would end in around a century and a millennium of temporal bliss could not therefore exist in the future. Nor could the millennium have occurred in the past because Napier's view of history was so negative. Since Christ's ascension, the Church had been typified by a steady decline into tyranny and corruption. Logically then, the millennium would have to occur in the next world. However, the most important reason was probably that Chiliasm had been tainted by the Anabaptist Revolt at Münster in 1535–5. This Protestant sect had seized Münster and declared the city the New Jerusalem, believing that Christ would return and rule the world from this city for 1000 years, with the community by his side. The city was besieged by a force led by the expelled Prince-Bishop Franz von Waldeck and the ensuing panic, violence and bloodshed shocked Europe. For decades, Protestants had to distance themselves from these beliefs due, if nothing else, to political expediency.

NAPIER'S CALCULATIONS ON THE END OF THE WORLD

Napier's fascination with mathematics found an outlet in his calculations of the approximate date of the end of the world, based on numbers and various books of the Bible and the Talmudic Prophecy of Elias. This prophecy stated that the world would endure for a total of 6000 years: 2000 before the law or under the law of nature, 2000 under Mosaic Law and 2000 under the law of

[67] *PD*, p. 323.
[68] Ibid.

Christ.[69] This was combined with Matthew 24:22: "And except those dayes shulde be shortened, there shulde no flesh be saued: but for the electes sake those dayes shalbe shortened."[70] This told Napier that God's predestined plan for humanity was to spare the suffering of the elect by bringing about the end of the world before the year 2000. He then produced two possible years during which the second coming would occur, conceding that it might have happened at any time between them. He calculated the earliest date in this range by reaffirming that the first 'Thundering Angel's Jubilee' had occurred in 1541 and adding three more jubilees at intervals of 49 years, which made the final year 1688.[71]

His calculations that provided the later date of 1700 were more complex. Daniel 12:11–12 read:

> And from the time that the dayly sacrifice shal cease, & the abhomination put to desolation shal be 1290 daies. Blessed is he that waiteth and commeth to the thousand, thre hundreth and fiue and thirtie dayes.

Again, Napier interpreted these 1335 days as years and argued that the abomination and sacrifice prophesied meant the attempt to rebuild the temple of Jerusalem under the Emperor Julian in 365 A.D.[72] They were brought to desolation when an earthquake destroyed the partially restored building and 1335 years after this date was 1700. Again though, Napier's dates were incorrect. The building works were abandoned in the year of Julian's death, 363 A.D. Finally, he interpreted Revelation 14:20 as also referring to a number of years: "blood came out of the Wine-preſſe vnto the horſe bridles, by the ſpace of a thouſand and ſixe hundreth ſtades or courſes."[73] He claimed that the Revelation was written between 97 and 99 A.D and 1600 years after this also provided an approximate date around the end of the seventeenth century.

Napier's discussion of this issue is intriguing for what it reveals about his interest in mathematics but it also reveals something of his perception of the religious context in which he lived. The *Plaine Discovery* was written in forceful language and Napier's conclusions were presented as incontrovertible facts. However, he knew how controversial dating the apocalypse would be and he went to great lengths to justify his actions regarding this issue. His main justification was that he was only attempting to predict the approximate

[69] Napier's usage of Elias was inspired by *Carion's Chronicle*.

[70] Geneva Bible & *PD*, p. 117. Mark 13:20 has similar content.

[71] *PD*, p. 120. 1541 + 49 + 49 + 49 = 1688.

[72] *PD*, p. 126.

[73] *PD*, p. 127. A 'stade' or 'stadium' is equal to a 'furlong', which Napier had in the 1593 edition.

age of the end, and not the exact date, which he conceded was forbidden by scripture. Mark 13:32 had: "But that day and houre knoweth no man, no, not the Angels which are in heauen, nether the Sonne him self, saue the Father." However, Napier believed that this simply meant that Christ did not know the exact date as opposed to the approximate time. After all, Revelation 12:12 stated that the Devil knows his time is short, and it would be blasphemous to argue, "that Chrift fhall be ignorant of that, which the Deuill doeth know."[74] He claimed that the Bible contained so many clues about the end that God must have wanted his Church to know when it was approaching:

> he giues vs diuers indices & foretokẽs, which he could not, nor would not have forewarned, if hee had bene vtterly ignorant of the time thereof, or yet minded that we fhould neuer haue fore-knowne the age, and appearant yeares thereof, although the precife day and houre be only knowne to God.[75]

Napier believed that because he was living in the last age, the rules that had previously governed prophecy no longer applied. He read Acts 1:7 literally. When Jesus said, "It is not for you to know the times, or the seasons, which the Father hathe put in his owne power", he was referring only to the disciples and did not necessarily mean all of humanity. So desperate was Napier to justify his prefiguring of the end that he went so far as taking liberties with the Biblical text. Although Napier primarily employed the Geneva translation, when justifying his attempts to predict the end he employed the Vulgate's version of Daniel 12:4, which suited his polemical agenda: "Seale the booke till the appointed time, manie fhall goe to & fro, and knowledge fhall be encreafed."[76] The Geneva Bible had, "seale the boke til the end of the time". Napier was determined to express his views and was not afraid to challenge the *status quo*. The officially recognised Bible in Scotland, which was the most authoritative book in his lifetime had forbidden prophesying the end of the world and had expressly forbidden the use of the Prophecy of Elias and making predictions about the future.[77]

---

[74] *PD*, p. 123.

[75] *PD*, p. 121.

[76] *PD*, p. 123.

[77] A marginal note to 1 Thessalonians 5:1 had, "beware of all dreames & fantasies of men which wearie themselues & others in searching out curiously the time that the Lord shal appeare, alledging for them selues a vaine prophecie, and moste falsely ascribed to Elias that 2000 yere before the lawe, 2000 vnder the Lawe and 2000 after the law the worlde shal endure." An annotation to 1 Thessalonians 5:20 'Despise not prephecying' defined this prophesying as preaching and not predictions about the future, which by implication were to be despised. Annotations in 2 Thessalonians condemned predictions concerning the date of the Second Coming as the work of false prophets.

FINDING NAPIER IN THE *PLAINE DISCOVERY*

John Napier is a mysterious figure because so few of his personal papers have survived and his reputation has been dominated by rumour and innuendo. However, the *Plaine Discovery* offers an insight into its author, not least because it expressed his religious views, which were such a dominant force in people's lives in the sixteenth century. Napier's condemnation of 'Newtrals' in the *Episle Dedicatorie* was in keeping with an attitude that prevailed in the sixteenth century and that was especially pronounced in the works of Knox and the Marian exiles.[78] They believed that absolute good was pitted against absolute evil in every aspect of human activity and expression so that a compromise or neutral position on spiritual matters was impossible. The language of absolutes was present throughout the *Plaine Discovery* and illustrates Napier's view of his fellow Protestants and anybody he felt was a threat to the spiritual welfare and political stability of the British Isles. Moreover, it reveals his self-perception as a messenger of God's revealed truth.

In the preface, *To the Godly and Christian Reader*, Napier explained something of his religious experiences as a youth in St Andrews, revealing that Christopher Goodman's sermons had made a significant impact on him. The work also reveals his Reformed doctrinal views, which emphasised justification by faith alone, election and predestination in God's providential relationship with humanity. He also repeatedly argued against the Roman Catholic doctrine of transubstantiation by arguing that Christ's body could not be in the host because it was with the Father in heaven.[79] This argument is definitely Calvinist as opposed to Lutheran, which supposed that Christ was omnipresent and therefore in the communal bread as much as anywhere else.

The *Plaine Discovery* betrays a concern with the world in which its author lived and he used fear sparked by international events to achieve his polemical ends. Demonization of Roman Catholicism was the central aim of the work and Napier was willing to go to great lengths to this end. His use of evidence was not judicial and he included every piece of material he could find to support his arguments, even when this meant he contradicted himself.[80] He was convinced of his prophetic status and that his work was not an interpretation but simply reported the truth. To dispute him was to dispute God and be damned. These facets of his personality emerged most markedly in the short

[78] See above, fn. 24

[79] *PD*, pp. 190–1

[80] For example, Napier deliberately used the Greek letters 'stigma', ϛ and the similar looking 'final sigma', ς interchangeably when calculating the meaning of the Number of the beast, but elsewhere tacitly admitted that they were not the same. See *PD* pp. 168–9 & Corrigan, 2014, pp. 170–3.

treatise he added to the 1611 edition, *A Resolvtion of Certaine Doubts*.[81]

By 1611 Napier was feeling more defensive than he had been in 1593. Unlike the first edition, the 1611 version was not dedicated to James VI, who had been crowned James I of England in 1603, nor was it published by the royal printer. Not only had James steadfastly refused to deal with the Catholic lords in the years following the Spanish Blanks Affair in a way that satisfied Napier, his religious policies had marginalised the Presbyterian radicals with whom Napier was aligned. James had also written works that denounced Scotland's Protestant Reformation as the result of sedition and advocated a greater degree of royal authority over the ecclesiastical arena than Napier would have supported.[82] In his additional treatise Napier responded to criticisms that had been made against the first edition but did so in a way that merely reiterated his positions and denounced those that would question him.

At first glance the *Resolution* appears to address Napier's critics in a cordial tone. He wrote that he had written the piece because, "there are certaine wel-affected brethren, who not in the ſpirit of arrogancie and contention, but in all ſobrietie and meekenes, haue craved of me the reſolution of ſome doubts".[83] However, he also implied that there were critics with whom he was refusing to engage: "As to the contentious and arrogant reaſoners, I leaue them to the mercie of the Lord."[84] This leaves no evidence about what criticisms might have been voiced, or by whom. Instead, it simply dismissed his critics as evil and implied that God's judgement against them would be harsh. Once more, this reflected his dualistic analysis of people, their thoughts and deeds as either absolutely good or absolutely evil.

His responses, or *Resolutions*, were generally effective, serving to explain points about which people had been confused. However, the sixth stands out as revealing Napier's intransigence, even when his critics had made a good case. In Napier's Chronology, the Donation of Constantine, a document in which the emperor allegedly transferred control of the Western Roman Empire to the papacy, was a major turning point. It represented the beginning of the papacy's dominance over the temporal, as well as the spiritual arenas and secured its status as Antichrist. However, since Lorenzo Valla's *On the Donation of Constantine* was published in 1517, the document had been widely considered a forgery. Although some disputed Valla's claims, they were so popular that

[81] *PD*, pp. 351–82.

[82] James VI, *The True Law of Free Monarchies*, Robert Waldegrave (Edinburgh, 1598) & *Basilikon Doron*, Robert Waldegrave (Edinburgh, 1599).

[83] *PD*, p. 356.

[84] Ibid.

it is impossible that Napier would not have known that the provenance of the *Donation* had been called into question and his failure to acknowledge this in his first edition was a glaring omission. Protestants had widely celebrated Valla's work because it called into question the centuries-old claim of the papacy to command the secular and ecclesiastical arenas. However, it played such an important role in his chronology that Napier could not bear to concede any of Valla's critiques. For example, he argued that Valla's claim that the Latin of the Donation was anachronistic meant nothing, as Valla might have been analysing a later copy.[85] However, the most telling of his arguments was that his chronology proved that the Revelation had foretold the usurping of temporal power by the church, which led to its corruption. Referring to Revelation 17:17 he wrote:

> *For God hath put in their* [hearts] *to fulfil his wil, and to doe with one confent, for to giue their kingdome vnto the beafte, vntil the words of God be fulfilled.* Ye fee then that Chriftian kings muft giue their kingdomes to the beaft, Shall we then thinke that the Emperour fhal be free from this decree of God?[86]

Napier believed that he was a prophet and that his commentary on the Revelation was, at least in part, divinely inspired. Describing his attempts to uncover the mysteries of the Revelation he wrote:

> And fo, when after a long time fpent, with little knowledge, I (iustlie difpairing of mine owne habilitie) became truly forowfull and humble in heart then it pleafed God, (to whome be the only glory) to giue me that grace to efpy in fhorte time, that wonderfull ouerture which in long time before I could not confider.[87]

God had revealed the meaning of the text to Napier and to dispute his interpretations was to dispute God.

LEGACY

The *Plaine Discovery* did not spark a wave of interest in predicting the date of the Second Coming in Napier's own country. However, the work was popular and appears to have been widely discussed and disseminated in the years immediately following its publication. The fact that it was printed in England in 1594, so soon after the first edition, suggests it was instantly popular.

85 *PD*, p. 374.
86 *PD*, p. 375.
87 *PD*, p. 100.

Even Bishop William Cowper, whose views on ecclesiastical polity differed immensely from Napier's, described him as, "worthily renowmed, as peere-lesse indeed, and specially for his great pains taken vpon this Booke, out of rare learning, and singular Ingene", in his commentary on the Revelation.[88] By the time the 1611 edition appeared in print, the work had been translated into French and Dutch, while the first German edition was published in that year. In subsequent decades, several editions were produced in these languages. Napier's original plan to produce a Latin version never came to fruition because the popularity of the English versions and translation into several European languages rendered it unnecessary.[89]

The greatest impact of the *Plaine Discovery* in the seventeenth century was probably felt in England. Joseph Mede's *Clauis Apocalyptica* of 1627 contained complex diagrams and calculations that displayed a significant debt to Napier's mathematical method of interpreting prophecy.[90] The idea that the world would end between 1688 and 1700 found new currency in the 1640s during the English Civil Wars and another English edition of the *Plaine Discovery* was published in 1645. These conflicts were deeply traumatic and, once again, people reacted by believing that the world was coming to an end and that they were engaged in a prelude to the final battle between good and evil predicted in the Revelation. Napier's prophetic statements were also condensed and combined with astrological predictions about the coming year in annual pamphlets known as the *Bloody Almanack*.[91] Of course, when Napier's predictions failed to come true, the *Plaine Discovery* fell from favour, and enlightenment scholars like Baden Powell denounced the work for bringing Christianity into

[88] Williamson, A. H., 'Number and national consciousness: the Edinburgh mathematicians and Scottish Political culture at the union of the crowns' in Mason, R. A., (ed.), *Scots and Britons: Scottish political thought and the union of 1603*, Cambridge University Press (Cambridge, 1994), pp. 187–212, p. 198 & Cowper, W., *The workes of Mr William Cowper late Bishop of Galloway: Now newly collected into one volume. Whereunto is added a comentary on the Reuelation neuer before published. Also an alphabeticall table for the finding out the principall heads contained in euery booke*, Iohn Budge (London, 1623), p. 822.

[89] *PD*, p. 101.

[90] For Napier's probable influence on Thomas Brightman's *Apocalypsis apocalypseos*, (Frankfurt, 1609), see Williamson, A., 'Empire and Anti-Empire: Andrew Melville and British Political Ideology, 1589–1605' in Mason, R. A. & Reid, S. J. (eds.), *Andrew Melville (1545–1622): Writings, Reception, and Reputation*, Ashgate (Farnham, 2014), pp. 75–99, p. 80 & Cogley, R. W., '"The Most Vile and Barbarous Nation of all the World": Giles Fletcher the Elder's The Tartars Or, Ten Tribes (ca. 1610)', *Renaissance Quarterly*, Vol. 58, No. 3 (2005), pp. 781–814, p. 789.

[91] For example, see *A bloody almanack: foretelling many certaine predictions which shall come to passe this present yeare 1647. With a calculation concerning the time of the day of judgement, drawn out and published by that famous astrologer. The Lord Napier of Marcheston*, I. Coe (London, 1647), Thomason/59:E.371[3].

disrepute.[92] The twentieth century saw many scholars downplay the importance of the theological and metaphysical works of early modern scholars like Isaac Newton and Napier in an attempt to cast them as wholly rational, scientific thinkers. However, since the 1980s scholars have increasingly valued such texts for what they reveal about their authors and the world in which they lived.

CONCLUSION

On first inspection, the *Plaine Discovery* may appear to many modern observers to be a standard sixteenth century polemical text. The majority of the arguments and imagery Napier employed would have been familiar to his audience. Indeed, many of his ideas, such as the pope being Antichrist, would have been shared by his audience before they ever picked up the book. However, certain aspects of the work were ground-breaking and would have had a dramatic impact on the reader. For example, the imposing of dates, even approximate, onto predictions about the Second Coming was virtually unheard of and remained highly controversial after the *Plaine Discovery* was published.[93]

Furthermore, the very nature of the chronology, the fact that it strove for mathematical precision and formed the basis of the entire work set it apart from works like Bale's *Image*, and every other chronological work that had gone before, in which dates were approximate and could only be inferred. The emphasis on mathematics in a theological work demonstrates that the kind of hard and fast divisions between intellectual pursuits that are now taken for granted simply did not exist in the minds of early modern intellectuals. Napier felt at ease using his mathematical talents to illuminate the hidden secrets of scripture, just as men like Isaac Newton employed theology along with

[92] Powell, B., *An Historical View of the Progress of the physical and mathematical Sciences, from the earliest ages to the present time*, Longman, Rees, Orme, Brown, Green & Longman (London, 1834), pp. 190–5 cit. Napier, M. p. 181, n. Baden Powell's assessment of Napier is discussed in Almond, p. 56.

[93] Even Napier's close friend Robert Pont demurred from predicting the end of the world in his tract which borrowed heavily from Napier and recommended the reader to the *Plaine Discovery*. See *A newe treatise of the right reckoning of yeares*, R. Waldegrave (Edinburgh, 1599), STC/349:07. Robert Rollock actively condemned the attempt in his *Lectures vpon the first and second Epistles of Paul to the Thessalonians: preached by that faithfull seruant of God M. Robert Rollock, sometyme minister of the Euangell of Iesus Christ, and rector of the Colledge in Edinburgh*, Robert Charteris (Edinburgh, 1606), STC/21281. See Corrigan, 2014, Part IV.

alchemy and natural philosophy to reveal the secrets of creation. Napier used every and all means at his disposal to achieve the polemical ends for which he was striving. The several editions of the *Plaine Discovery* and Napier's return to and defence of his contentions in the 1611 editions illustrates the lifelong prominence of theological and, more specifically, eschatological concerns in his mind. Those concerns never faded as his mathematical works increasingly consumed his spare time. The fact that the *Plaine Discovery* was notable for its emphasis on mathematical calculations shows how remarkable an intellect Napier must have had. His interest in mathematics as a tool for making the world a better place and understanding God's creation more perfectly could be seen throughout all of his published works.

A PLAINE DISCOVERY OF THE WHOLE REVELATION OF S. IOHN

A

PLAINE DISCOVERY, OF THE WHOLE REVELATION of S. IOHN: set *downe in two treatises: the one searching and* proving the true interpretation thereof: The other applying the same paraphrasticallie *and Historicallie* to the text.

SET FOORTH BY IOHN NAPEIR *L. of Marchiston. And now revised, corrected* and inlarged by him.

WITH A RESOLVTION OF *certaine doubts, mooved by some well-* affected brethren.

WHEREVNTO ARE ANNEXED, CER- taine Oracles of SIBYLLA, agreeing *with the Revelation and other* places of Scripture.

LONDON,

Printed for IOHN NORTON. 1611.

To the Godly and Chriſtian Reader.

ALthough the nature of the trueth be of ſuch force and efficacie, that after it is hearde by the ſpirituall man, it is immediatlie beleeued, credited and embraced: yet the naturall man is ſo infirme, and weake, that his beleefe, muſt bee ſupplied by naturall reaſons and euident arguments: Wherefore, many learned and godly men, of the Primitiue Church, haue gathered out diuers pithie and forcible, naturall and Philoſophicall arguments, to prooue and confirme the Chriſtian faith thereby: As in the 1. Cor. 15.36 Paul the learned and godly teacher of the Gentiles, perſwading them to confeſſe the reſurrection of the dead, induceth a marveilous pithie and familiar argument, by a natural compariſon of ſeede ſowne in the ground, that firſt muſt die and be corrupt in the earth, and then doth it quicken vp and riſe againe after an other forme, than it was ſown into: And likewiſe other learned Doctors of the primitiue Church writing to the Ethnicks, who ſturred at the Virgins conception and at Chriſts divinitie, reaſoneth with them on this maner: ſaying, Your gods (as ye beleue) hath converſed with many women among you and haue begotten many children, who haue wrought no miracles: & how can ye that ſo beleeue, deny vs, that our great God hath begotten one Sonne, in whom divinitie, and humanitie are conjoyned, ſeeing your eies and forefathers, haue ſeen ſo many and divine miracles wrought by him, & in his name? And ſo moſt wiſely vſed they theſe Gentiles own opinions and arguments against themſelues, which mooued the malicious Apoſtate *Iulian* the Emperour, to diſcharge from Chriſtians, the Schooles and learning of Philoſophie, yeelding the reaſon, becauſe ſaith he, *Propriis pennis ferimur*, and as by that meanes many of the moſt learned Gentiles in theſe daies, were either mooued neceſſarilie to confeſſe the Chriſtian Religion to be true, pure and holie: or then at the leaſt were made ſo aſtoniſhed, that they were not able to write or teache againſt it. So likewiſe, now in this ſecond ariſing of the Evangelical trueth, from the horrible Antichriſtian darkneſſe: it is not enough that this divine Reuelation which diſcouereth the Antichriſt be onely by ſimple aſſertion interpreted, or hiſtoricallie applyed to the confirming of their zeale, that already knowes and deteſts that man of ſinne and ſonne of perdition, but alſo to the effect the ſauourers of his errors may either be conuerted or then their mouthes bridled from calumniating the trueth, it is needfull that ſuch neceſſarie reaſons and ſure arguments or at the leaſt, ſuch notable tokens be induced, as may make that interpretation vndenyable.

And therefore being of purpoſe (by the aſsiſtance of the Almightie) to expound and open vp the myſteries of this Revelation, by a two-folde diſcourſe, the one Paraphraſtical, the other hiſtoricall, both confronted together: I haue thought good before that worke, to premit by way of introduction, a reſoning for the inveſtigatiõ of the true ſenſe and meaning of euerie notable myſterie thereof, and to ſet the ſame in forme of Propoſitions, as neere the analytick or demonſtratiue manner, as the phraſe & nature of holy Scriptures will permit: and that for the better ſatisfaction of theſe naturaliſts, of whom happilie, God may either make true ſpirituall profeſſors, or at the leaſt, hereby bridle their opprobrious mouthes from calumniating the vndoubted trueth, when they ſee the naturall coherence of Gods myſteries ſo marueilouſly reuealing one another, that thereby the trueth of God, the hiſtory of his Church and perſon of the Antichriſt are detected, and the very Antichriſtians induced neceſſarilie, either to confeſſe themſelues to bee hereby diſcouered or then muſt they deny the Scripture it ſelfe. And although I haue but of late attempted to write this ſo high a worke, for preuenting the apparent danger of Papiſtrie ariſing within this Iland, yet in trueth, it is no few yeares ſince firſt I beganne to precogitat the ſame: For in my tender yeares and barneage in Sainct-Androes at the Schooles, hauing on the one parte contracted a louing familiaritie with a certaine Gentleman & c. a Papiſt: And on the other pairt being attentiue to the Sermons of that worthie man of God, *Maiſter Chriſtopher Goodman*, teaching vpon the Apocalyps, I was ſo mooued in admiration, againſt the blindnes of Papiſts, that could not moſt euidently ſee their ſeuen hilled citie Rome, painted out there so liuely by Saint Iohn, as the mother of all ſpirituall whoredome, that not onely burſted I out in continual reaſoning againſt my ſaid familiar, but alſo from thenceforth I determined with my ſelfe (by the aſsiſtance of Gods ſpirit) to employ my ſtudie and diligence to ſearch out the remanent myſteries of that holy booke: as to this houre (praiſed be the Lord) I haue bin doing at all ſuch times, as conveniently I might haue occaſion: But (to confeſſe the trueth of the infirmitie of man to the glorie of God) I found euer during all that longtime, more fruit in one houres ſobrietie, prayer and humble meditation, than in a thouſand daies of curious or preſumptuous inquiſition: yea, the more ſubtillie I ſearched, the more darknes I found, in ſo much, as curious inquiſition rather diſcouraged me by finding out of doubts, then profited me by finding any reſolution thereof: And ſo, when after long time ſpent with little knowledge, I (iuſtlie deſpairing of mine owne habilitie) became truely ſorowfull and humble in heart then it pleaſed God, (to whome be the only glory) to giue me that grace to eſpy in ſhorte time, that wonderfull ouerture which in long time before I could not conſider: So that then I beganne to take vp matters by their right beginning, and by moſt eaſie grounds, And to proſecute out the reſt by their coherence in order, as is here ſet downe to the full refolution of all the former doubts. After the which althogh (greatly reioycing in the Lord) I began to write thereof in Latnie: yet I purpofed not to haue fet out the fame fuddenlie, and far leffe to haue written the fame alfo in Englifh, til that of late, this new infolencie of Papifts arifing about the 1588. year of God, and dayly increfing within this Iland doth fo pitie our heartes, feeing them put more truft in Iefuites and feminarie Priefts, then in the true fcriptures of God, and in the Pope, and king of *Spaine*, then in the King of Kings: that, to preuent the fame, I was conftrained of compafsion, leauing the Latine, to hafte out in Englifh this prefent worke, almoft vnripe, that hereby, the fimple of this *Iland* may be inftructed, the Godly confirmed and the proud, and foolifh expectations of the wicked, beaten downe. And where as after the firft edition of this booke in our Englifh or Scottifh tongue, I thought to haue publifhed fhortlie the fame in Latine (as yet God willing I minde to doe) to the publike vtilitie of the whole Church. But vnderftanding on the one part, that this work is now imprinted, & fet out diuerfe times in the French & Dutch tongs, (befide thefe our Englifh editions) & therby made publik to manie. As on the other part being aduertifed that our papiftical, adverfaries wer to write larglie againft the faid editions that are alreadie fet out. Herefore I haue as yet deferred the Latine edition, till hauing firft feene the aduerfaries obiections, I may infert in the Latin edition an apologie of that which is rightly done, and an amends of whatfoeuer is amiffe. Whatfoeuer therefore through hafte, is here rudelie and in bafe language fet downe, I doubt not to be pardoned thereof by all good men, who confidering the necefsitie of this time wil efteeme it more meete to make hafte to preuent the rifing againe of Antichriftian darknes within this Iland, then to prolong the time in painting of language: Efpeciallie alfo, feeing herein purpofely I preaffe not to follow the particular ornat tearmes, of neither Scottes nor Englifhmen: but rather contrarilie for both their inftructions, I vfe, fo much as I can, thefe words and tearmes, that be more common and fenfible to them both, then proper or ornate to any one of them, Here then (beloued Reader) thou haft this work divided into two treatifes, the firft is the faid introduction and reafoning, for inveftigation of the true fenfe of euery chiefe Theologicall tearme and date contained in the Revelation, whereby not onely is it opened, explained, and interpreted, but alfo that fame explanation and interpretation is proved, confirmed, and demonfstrated, by euident proofe and coherence of fcriptures, agreeable with the euent of hiftories: The feconde is the principal treatife, in the which the whole Apocalyps, Chapter by chapter, Verfe by verfe, and fentence by fentence, is both Paraphrafticallly expounded and Hiftorically applyed. This fetond and principall treatife, keepeth this order, that firft before euery chapter is premitted the argument, not of the literall fenfe of the chapter but of the true meaning and interpretation of the fame. Thereafter (feeing all the chapters of the Apocalyps doe either deſcribe the viſion, or then expreſſe the prophecie) if the preſent chapter be deſcriptiue, euery page thereof is divided in two columnes: in the firſt wherof is ſet the text by verſes, in the ſecond columne: is ſet the Paraphraſticall expoſition, anſwering verſe by verſe to the text of the firſt columne: but if the preſent chapter be propheticall, then is euery page thereof divided in three columnes, the firſt containeth the Text by verſes, the ſecond the Paraphraſticall expoſition, thereof verſe by verſe, the third alſo agreeable therewith by verſes containeth the hiſtorie or hiſtoricall application, in the which, and by the which, the ſame prophecie is, or ſhall be performed. And if any chapter is partly deſcriptiue, partly propheticall, then where it is propheticall, there are three columnes, the firſt is the Text the ſecond, is the Paraphraſticall expoſition of the text. The third, is the hiſtoricall application and accompliſhment thereof. But where in that ſame chapter any parte is deſcriptiue, then are there forenent that part onely two columnes, whereof the firſt is the Text, the ſecond is common to both the Paraphraſtical columne, & to the hiſtoricall columne, and therefore it paſſeth through, occupying the roome of both the ſecond and third columnes, ſo that when yee read the ſecond columne, ye ſhall reade it with the ſecond columne, and when againe ye read the third columne, yea ſhall alſo reade it with the thirde columne: for it is ſuch a common thing as agrees to bee reade with either of them. Furthermore, there is two drawne lines betwixt the Paraphraſticall columne, & the Hiſtoricall, within the which lines there is ſet downe the dates of times, by ſeals trumpets, viales, thunders and yeares of God, anſwering, to the preciſe time, in the which euery Prophecie thereof was, or ſhall be performed. Moreouer, incaſe any part of the text may be thoght to be wrong tranſlated, any part of the Paraphraſe wrong expounded: or finallie, any part of the hiſtorie wrong applied, ouer briefly handled, not ſufficientlie prooued, or hauing neede of annotations: I haue therefore forenent euery doubtfull and needful place of the Text, Paraphraſe, and Hiſtory ſet markes of the ordinarie letters of *a, b, c,* & c. In the Paraphraſe, and after the end of the whole chapter haue ſubioyned Notes, Reaſons, and Amplifications marked coreſpondentlie with the like letters of *a, b, c,* & c. By the which Notes and amplifications, not onely the brevitie of the Paraphraſe and hiſtorie is ſupplied, but alſo by reaſons, proofes, and arguments of ſcripture therein inſert, the whole ſmaller doubts thereof are reſolued. And as toward the greater doubts, they are referred to the firſt treatiſe wherein they are at length and demonſtratiuely prooued. So the note here in the ſecond treatiſe, doeth onely quot the propoſition of the firſt treatiſe, that proues that preſent aſſertion. And becauſe this whole work of Reuelation concerneth moſt the diſcouerie of the Antichriſtian and Papiſticall kingdome, I haue therefore (for remoouing of all ſuſpition) in all hiſtories and prophane matters, taken my authorities and cited my places either out of Ethnik auctores, or then Papiſticall writers, whoſe teſtimonies by no reaſon can be refuſed againſt themſelues but in matters of diuinitie, doctrine and interpretation of myſteries (leauing all opinions of men) I take me onely to the interpretation and diſcouerie thereof, by coherence of ſcripture, and godly reaſons following therevpon: which alſo not only no Papiſt, but euen no Chriſtian may iuſtly refuſe. And foraſmuch as our ſciptures herein are of two ſortes, the one our ordinary text to the other extraordinary citations, In our ordinarie text, I follow not altogether the vulgar Engliſh tranſlation but the beſt learned in the greek tong, ſo that (for ſatisfying the Papiſts) I differ nothing from their vulgar text of S. *Ierome*, as they call it, except in ſuch places, where I proue by good reaſons, that he differeth frõ the Original Greek. In the extraordinary text of other ſcripturs cited by me I follow euer *Ieromes* latine tranſlation, where any controuerſie ſtands betwixt vs & the Papiſts, & that moueth me in diuers places to inſert his very latine text, for their cauſe, with the iuſt Engliſh thereof, for ſupply of the vnlearned. And hereof it commeth that our quotations are according to the ſections & verſes of *Ieromes* tranſlation, printed in Antuerp by *Plantine*, ageeing alſo with the great concordances of *Robert Staphane*. In other texts not controuerted the Engliſh text it alone is ſufficient in a Scottiſh or Engliſh treatiſe, & therefore omit we the Latine. Further as concerning my citations of ancient & vnſuſpect writters, I haue choſen the ſame out of the moſt old & faithful copies, chiefly thoſe that be found in old Popiſh libraries and imprinted by Popiſh Printers, but as for the new copies and latter editions, there is amalicious ordinance obſerued in theſe latter daies, by the which, ye Friers of *Louane*, & other Papiſtical correctors eiketh paireth, & peruerteth al good & rare books, cauſing firſt one ſort of them to be imprinted vuder their teſtimoniall, ſo that as that, ſame teſtimonial therein imprinted beareth, nothing is left therein that may be contrarious to the Romane Church although the Authour by his true edition had neuer ſo much inweighed againſt Rome. Thereafter all the remnant Popiſh Printers do follow there new peruerted Coppies, in their latter editions without any teſtimoniall, as being warranted by the ſaide firſt testimonial & this is moſt euident to be ſeen by their booke intituled *Index Expurgatorius* Praying therefore all good men to beware of theſe, We deſire earneſtly with our hearts ſo to proceede hereinto, that trueth may come to light, and that ſuch groundes may be laide, as thereby the ignorant and ſimple Reader may be beſt inſtructed, the godly and learned Chriſtian moſt ſurely confirmed, and the arrogant and obſtinate calumniator rathereſt confounded and put to ſilence. But forſomuch as this our good intention and godly purpoſe, doeth alwayes proceed of a very tender and fraile veſſell, and that as all liquors, (how precious ſoeuer) doth take ſome taſte of their veſſels: So this holy work may in ſome things (though not eſpied by my ſelfe) tafte of my imperfections. Therefore humbly I fubmit thefe imperfections whatfoeuer, to the gentle correction of euery wife and difcreete perfon, who in the motion of Gods fpirite iudgeth vprightly, without envie or partialitie. Praying all good men to haue me apardoned of whatfoeuer is amiffe: for although I haue not done herein perfectly as I would, yet zealoufly as I could, knowing that the poore widowes mite was acceptable vnto the Lord: for euery man hath not gold, filuer, filks, & purple to offer to the work of the Sanctuarie: to me (as faith *Ierome*) it fhall be much, If I may purchafe wooll, or flockes, to offer to that holy work. And furely, this that I haue, how fmall foeuer it be, till God enlarge me with more, I offer it gladly vnto the glory of God and edification of his true Church. To God therfore the difpofer of this, and all other Godly workes and meditations, who liueth and reigneth eternally in Trinitie and Vnitie, be glory praife laude and thanks, for euer and euer, *Amen*.

The booke this bill fends to the beaft,
Crauing amendement now in heaft.

GOD *firft to* IOHN *in* Pathmos *me prefents,*	Apoc.1.1.9
Who fent me fyne the feuen Kirkes vntill:	Apo-1.11.
As foorth I foore with the two Teftaments,	Apo.11-3.7.
Gods truth to teach, in witnefsing his will:	Apoc 11.7.
Thou bloudie Beaft, vs cruellie did kill,	Apoc.11.3.
In fack of fchifmes fyling vp our fenfe,	Apoc. 11.8.
Our corps vnkend then ftonifhed lay ftill,	Apo 11.3.9.
Til feuentie years eighteen tims paffed hence:	Apo. 11.11 & 14.6.7.
But now fince comd is vntil audience	Apo. 11.12.
Gods word from heauen the voyce of veritie,	Apo.11.11.
Quickning thefe corps with true intelligence,	Apoc.11.2 and 15.8.
So long fuppreft by thy fubtilitie:	Apo. 10.11.
I plaine proclaime and prooue by prophecie,	Apo.17.9.
That thou, O Rome, *raifd vp on hilles feuen,*	Apoc.17.18 & 11.8 note (m) thereof.
Citie fupreame and feat of Sodomie:	Apo. 17.6 & 18.24.
Vnder whofe reigne our Lord to death was driuen,	Apoc. 17.5.18 & pro. 23.
And many Martyrs rudelie rent and riven:	Ap.17. & 18.
Art Heire and Eroy to great Babylon,	Ap.17.1.3.6.
Whereby her name here God hath to thee giuen,	Ap.17. & 18.
Thou whore that fitteft the bloudie beaft vpon:	Apoc.18.2.9.8.
Thy dayes are done, thy glory now is gone.	Apo.14 & 17.
Burnt fhall thou be, and made a Den of Deuils.	Apoc. 14.14.15.
Flee from her then my flocke, leaue her alone,	Propofi. 23.
Left that ye be partaker of her euils:	
For doth at hand approch the latter day,	Apocal.11.14.
When Chrift his Church fhall reigne with him for ay.	Apoc.11.15.

A Table of the Conclusions introductiue to the Revelation, and prooved in the firſt Treatiſe.

The firſt Treatise, is an introduction to the knowledge of the Revelation, expounding by proofe and demonſtration the meaning of

- Dates and chieſe reckenings hid vnder tearmes,
  - Vulgare and vſed ſuch as
    - generally, a day for a year, a week for ſeven yeares, a moneth for 30. years, and a yeare of 7'eares, or three hundreth & three ſcore yeares, proved in the 1. propoſition.
    - particularly 42. moneths 1260. propheticall daies : three great dayes and an half and a time, times, and half a time, prooved to be all one date, propoſition 15. and every one of them to meane a thouſand, two hundreth aud three ſcore Iulian yeares, propoſition 16.
  - New and ſtrange, where with is dated
    - the hiſtorie of S. Iohn his daies, vnder the tearms of 7, ſeales, whereof, the firſt begins, An Chri. 29. prop. 8. the laſt An. Chriſti. 71. propoſition 7. and every one of them indureth at the leaſt ſeven yeares, as is proved by the ninth propoſition.
    - The prophecy to the worldes end, whereof.
      - The progres is dated by 7. trumpets, or 7. vials which both ar proved to be one propofition 2. and everie one of them doeth endure 245. yeares, gathered by the third and fourth propoſition, and proved by the fift propoſition: and the firſt beginneth anno Chriſti. 71. propoſition, 6. and the laſt beginneth anno. Chriſt.1541. and ends An. 1786. at the fartheſt prop. 10.
      - The ende and laſt daies, are dated by 7. thunders, & 7. An'gels of Gods harveſt, both prooved to bee one, prop, 11. & to beginne Anno Chriſt 1541. prop. 12. & every one of the firſt three that goeth before the day of iudgement, to indure 49. yeares, prop. 13. making the latter day to fall betwixt the years of Chriſt 1688. & 1700. by prop. 14.
- Affaires and chiefe matters concerning
  - Gods ſervants, whoſe true, and holie
    - Church with her contents is tearmed Gods Temple in heavẽ, Pro. 20 and a woman clad with the Sunne, prop. 22.
    - Religion is tearmed God his throne in heauẽ, pro. 17 whereof the written word is teſtified by the two witneſſes prooved prop. 21. to be the two Teſtaments, whereof
      - The twenty four books of the olde Teſtament, & 24. Writers therof, are called ẏͤ 24. Elders propoſi. 18.
      - The four Evangels of the new Teſtament are called the foure beaſtes, propoſition. 19.
  - Gods greateſt enemies, as
    - Sathan who is bounde 1000. years, beginning An. 300. prop. 34 yet bond but only from raiſing of vniverſal warres, prop. 35
    - Wicked kingdoms which are ſet out vnder the tearmes of
      - Two nations, Gog and Magog prooved (by the thirtie and two propoſition) to bee the Papiſtes & Mahometans, whoſe armies are all one with thoſe of the fixte trumpette or viall, as is proved in the 33. propoſition.
      - Two Beaſts
        - A two horned beaſt, ſignifying the Antichriſt only by the twentie fiue propoſitiõ, who is proved to be the Pope by the twenty ſixe propoſuiõ, whoſe raign of a thouſand, two hundreth ſixtie yeares beginneth Anno Chriſti 300. or 316. at the fartheſt, propoſition, 36.
        - A ten horned beaſt, ſignifying by the twentie foure propoſ. on, the whole Latine or Romane Empire, with the Antichriſt deſcẽding of the fame, of which Roman or Latine Empire, the ſeat and Metropolitan Citie Rome, is meaned and repreſented by Babylon in the twentie three propoſition. To this tenhorned beaſt and Roman or Latine Empire (as is proved in the twentie ſeuen propoſition) appertaineth
          - An Image, prooued by the 28. propoſition to be theſe degenerate princes that only in name and shewe, are Romane Emperours.
          - A mark, prooved by prop 30. to be the league of fervitude profeſſed to the Romanes by their ſubiect; noted afterwarde by the Pope with viſible marks, which by the 32. prop. are $\chi\widetilde{\varrho}\varsigma$ & croſſes of al kinds.
          - A name, prooved by the twentie nine propoſition, to be the name of the Latime or Romane Empire, $\lambda\alpha\tau\epsilon\iota\nu o\varsigma$
          - A number of 666. aſſigned by the text.

Theſe Propoſitions are ſet in this table after their natural order, but in the Treatiſe following, they are placed demonſtratiuely, to the effect everie propoſition may be prooved by the former propoſitions.

THE
FIRST AND
INTRODVCTORY
TREATISE, CONTAINING

a ſearching of the true meaning of the Revelation, beginning the diſcouerie thereof at the places moſt eaſie, and moſt euidentlie knowne; and ſo proceeding from the knowen, to the proouing of the vnknowne, vntil finallie, the whole grounds thereof be brought to light, after the maner of propoſitions.

THE FIRST PROPOSITION.

In propheticall dates of daies, weekes, moneths and yeares, euery common propheticall day is taken for a yeare.

THis Propoſition is prooued, by appearance, by a law, by practiſe, and by neceſsitie. By appearance, becauſe the common weeke is called a weeke of daies, as in Geneſis (according to IEROMS tranlation) *Imple hebdomadam dierum Fulfil a weeke of daies.* And the common yeere is called a yeare of daies, as *adbue duo anni dierum. Within two yeares of daies:* and *Poſt duos amios dierum: after two yeares of daies*: Inferring thereby, that there is another weeke of yeares, containing ſeauen years, as the common weeke containes ſeauen daies. And another yeare of yeares, containing 360. yeares, as the common yeare (not intercalar) among the Hebrewes and Grecians, contained †360.daies. †Secondly by the Leuitical law, (which is the figure of all prophetical veritie) it was inſtituted, that as in the common weeke, ſixe daies were for labour, and the ſeauneth for reſt: ſo ſhould there alſo be a weeke of yeares, in the which the land ſhould be ſixe yeares laboured, and reſt the ſeuenth: Making ſo the common weeke of daies, to reſemble the figuratiue weeke of yeares, and ſo conſequently, euery day to reſemble a yeare. Thirdly, by the prophetical cuſtome and practiſe,

1. Reaſon by appearance.
Gen. 29.27

Ordinarie Weeke,
Iere. 28.3.11.
1. Mach. 1.30.
Ordinarie yeare.
Extraordinarie weeke.
Extraordinarie yeare.
†Ioſephus Scaliger deemen.temp.
†2. *By a Law.*
Exod. 23
Leuit. 25

3. By practiſe.

euery day reprefented a yeare: as in Numb. *Iuxta numerum 40. dierum, quibus confideraftis terram, annus pro die imputabitur, & 40 annis recipietis, &c. After the number of fourtie daies, in which ye fearched out the land, the yeare fhall be counted for a day, and in fourtie years ye fhall receaue, &c.* And in Ezech. *Ego autem dedi tibi annos iniquitatis corum numero dierum, &c. And I haue giuen thee the yeares of their iniquitie by the number of daies.* And afterwards, *Et affumes iniquitatem domus Iuda 40. diebus, diem pro anno, diem inquam, pro anno dedi tibi. And thou fhalt beare the iniquitie of the houfe of Judah fourtie dayes a day for a yeare, euen a day for a yeare, I haue given thee.* Fourthlie, vpon necefsitie of faluation, all Chriftians muft confeffe, in the feuentie weekes of DANIEL, a day to be taken for a yeare, extending in the whole to 490. yeares, otherwife that prophecie of the *Mefsias* comming, would not fall vpon the juft time of Chrifts comming, as neceffarily it ought to doe. So then, a propheticall day is a yeare, the weeke feuen yeares, the moneth thirtie yeares (becaufe the Hebrew and Grecian moneth hath thirtie daies) and confequently, the propheticall yeare is 360. yeares. Befides this common prophetical day, there is an other great and extraordinary day, where of hereafter fhall be fpoken.

Num. 14.34.

Ezec 4.5.6

4. Upon necefsitie.

Dan. 9.24

IOfEPHUS *Scaliger deemend. Temp. Of this great daye, read propofition 15.*

2. PROPOSITION.

The feuen Trumpets of the 8. 9. and 11. chapters, and the feuen Vials of the 16. chapter are all one.

THis is prooued, in that both the feuen Vials, and alfo the feuen Trumpets are the feuen laft plagues: The feuen Vials, in that by the text they are called, *The feuen laft plagues*: thefe fame being hereafter called, *The feuen golden Vialles, and the feuen vials of the wrath of God.* As to the feuen Trumpets, the laft conteineth the day of judgement, as teftifies the Angels

1 Reafon, Accord of time.

Apoc. 15.1 & 7.

Apoc. 16.1

oath, ſwearing, *Quod tempus non crit amplius, ſed in diebus vocis ſeptimi Angeli, cum coeperit tuba canere, conſummabitur myſterium magnum Dei, &c. That there ſhal be no more time, but in the daies of the ſeuenth Angel, when he ſhall blow the trumpet, the great myſterie of God ſhall be finiſhed.* Which myſterie, PAVLE to the Corinthians, maketh the latter day and reſurrection, ſaying, *Ecce, myſterium dico vobis, & c. Behold, I declare vnto you a myſterie.* And againe, *In momento, in ictu oculi, in noviſsima tuba (canet enim tuba) & mortui reſurgent: In a moment, in the twinkling of an eye, at the Laſt Trumpet, (for the Trumpet ſhall blowe) and the dead ſhal riſe.* And ſeeing then the ſeuen trumpets follow one an other in order, in the eight, ninth, tenth and eleuenth chapters, and the laſt containeth the day of judgement and generall reſurrection: Therefore muſt the ſeuen trumpets be alſo the ſeuen laſt plagues, and conſequently, they & the ſeuen vials muſt be all one. Moreouer, for confirmation hereof, they agree in their principall termes: the ſecond trumpet with the ſecond Viall: the third trumpet with the third Vial: the fourth trumpet with the fourth Vial: the ſixt trumpet with the ſixt Vial, the ſeuenth trumpet with the ſeuenth Vial: So, that thereby wee may be ſure, and conclude both thoſe trumpets, with thoſe Vials & alſo the reſt of the trumpets, with the reſt of the Vials, reſpectiue in purpoſe, meaning, time, & in all other circumſtances, to be one and the ſelfe ſame thing.

Apoc. 10.6

This great myſterie of God is the Reſurrection. 1. Cor. 15.51.

2. Agreeance in tearmes & circumſtãces

3. PROPOSITION

The ſtarre and locuſts of the fift trumpet, are not the great Antichriſt and his Cleargie, but the Dominator of the Turkes and his armie, who began their dominion, in anno chriſti 1051.

BY the former propoſition, the fift Vial is al one with the fift trumpet: But in that Vial (ſaith the text) there aroſe ſuch a plague againſt the ſeate of that Antichriſtiã beaſt, that his kingdome was darkned, and they did gnaw their tongues for ſorrowe. So that this may no waies be himſelfe that aroſe againſt himſelfe, but rather ſome other godleſſe tyrant like him, whom we ſhall prooue to be the Apoſtate *Mahomet*, and his locuſtes the Turkes: Firſt, by the name of their Chieftanes: ſecondly, by the ſpace of their oppreſsion done to pretended Chriſtians in their firſt dominion: thirdly, and laſt of all, by all the tokens and circumſtances contained in the text. As to the firſt, they ſhall haue (ſaith the text) their king, whoſe name ſhall be in Hebrewe *Abaddon*, and in Greeke *Apollyon*, & in Latine (as S. IEROM tranſlateth it) *Exterminans*, and in Engliſh a deſtroyer, or a waſter. But ſo it is, that trying from language to language the names of princes, yee ſhall only finde both their temporall and ſpirituall kings names to ſignifie the ſame, that *Abaddon* in Hebrew, and *Apollyon* in Greeke doeth: for their temporal king is called *Turca,* which is aſmuch to ſay, as a Waſter or deſtroyer, as teſtifieth *P. Melancton* in *Carion* his *Chronicles. And their ſpiritual kings name *Mahomet**, ſignifieth *delens*, a deſtroyer or waſter: and beſide that, it ſignifieth alſo a Meſſenger or Angel, moſt agreeable with this text, where he is alſo called, the Meſſenger or Angel of the depths. Therefore, theſe locuſtes by the name of their Chieftaine agreeth certainly to be the Turkes. †Secondly, as to the ſpace of their oppreſsion done to pretended Chriſtians. Theſe locuſts (ſaith the text) had power giuen them not to kill but to vexe thoſe who had not the ſeale of God in their fore-heads fiue moneths, which being prophetically taken, (becauſe this is a prophecie) maketh juſt an hundreth and fiftie yeares, as is prooued by the firſt propoſition. And ſo long continued the Turkes in oppreſsing & ſpoyling the pretended Chriſtians, before the Empire of the *Tartarians* aroſe, to wit, from the daies of *Zadok* their firſt dominator. *Anno Chriſt.* 1051. to the time that *Changius Chan* of *Tartarie*, did begin in *anno Chriſti*, 1201. that Empire

Hier, de int nominum.
1. Reaſon, agreeance of names.

Both their temporall kings name Turca, *is* Abaddõ *a deſtroyer.*

Chro. lib. 1 ſol. 8. & lib. 5. ſol. 107.
Hier. de int. nominiũ.
And their ſpirituall kings name Mahumetes is Abaddon, *a deſtroyer.*
†2. *Reaſon Accorde of time.*

3. Accord of circumſtances. Read IOH. BALEYS *in the end of his*

which ſubdued them. So then by the time of their oppreſsiõ done to pretended Chriſtians, theſe locuſts agree to be the Turkes. Thirdly, and laſt of all, the whole circumſtances and tokens of the text, agrees moſt conueniently with the *Turke*, as at length our paraphraſtical & hiſtorical diſcourſe ſhal prooue. Wherfore, the *Star*, mentioned in the fift trumpet that before had fallen downe from heauen, and his *Locuſts* that aroſe, muſt needs be the *Mahomet*, who fel frõ his former Chriſtiã profeſsion, & became an Apoſtate, and out of the ſmoke of his hereſie, ſtirred vp the Turkes to be his armie.

preface vpon the Image of both the Churches, citing Cuſpinianus *and* Biblianders *teſtimonie, that Mahomet granted Chriſt, and ſpeaketh truely of him, and yet made defection from him.*

4. PROPOSITION.

The Kings of the Eaſt, or foure Angels, ſpecified in the ſixt trumpet or ſixt vial. cap. 9. & 16. are the foure nations, Mahometanes beyond and about Euphrates, who began their empire by Ottoman, in the year of Chriſt, 1296 or thereabout.

FOr proofe hereof, there needeth no other argument, than the perfect concord of the whole tokens of the texts, cap. 9. & ca. 16. agreeing ſo in euery point with that hiſtorie, that no other hiſtory can be applied thereto. For firſt, cap. 9. they are called Angels, that is to ſay, Meſſengers euill or good, executers of Gods will: So were theſe *Mahometans* meſſẽgers ſent of God, to ſcourge the Chriſtians falling away. Secondly, cap. 16. they are called Kings of the Eaſt: & ſo are theſe Mahometans indeed Kings of the Eaſt. Thirdly, they were in number foure (ſaith the text) ſo were there of theſe *Mahometanes* foure chiefe nations, to wit, *Turks, Tartarians, Saracens*, and *Arabians*: and foure imperial or royal families, to wit *Aſsimbeis, Candelors, Caramans*, and *Ottamans* Fourthly, they were hitherto bound (ſaith the text) about the riuer *Euphrates*, & ſo haue they their reſidence about *Euphrates*, where they were ſo bound and tyed 100. years, by ciuill & inteſtine warres, betwixt the *Turks* & *Tartarians*, that they

1. Reaſon, Mahumetans angels or meſſengers.

2. Mahumetanes Kings of the Eaſt.

3. They are foure nations: & 4. families

4. Bounde at Euphrates.

came not by weſt Euphrates to make cõqueſt, til after this yeare of Chriſt 1296. afterwhich time, the whole Mahometik empire ouer the moſt part of the ſaid foure nations, came frõ the other three families, in the only hand of *Ottomannus* their firſt Emperour. Fiftly, theſe were appointed (ſaith the text) at this time to ſlay the third part of men. And ſo did they, what by corporal, and what by ſpiritual death, ſlay, cõqueſt & poyſon with hereſie, all *Aſia* & much of *Europe*, euẽ the large third part of the world. Which propertie can be attributed to none other hitherto, but to the ſaid Mahometike armie, which exceeded far, any armie heard of in Chriſtiã hiſtories. And therfore, ſixtly, are they numbred to ſo great a multitude in the text, to wit, two hundreth thouſand thouſands horſemen. Seuenthly, as out of the mouths of the deuil, the Antichriſtian beaſt, & this fals *Mahomet*, proceeded out three vncleane ſpirits, ſtirring vp the world to battell, as the Text ſpecifies: So, by the hiſtory it is found, that out of their mouths proceeded ſuch inſpirations of the deuil, and deuiliſh exhortations, that ſtirred vp (in fiery wrath, in ſmoking hereſie, and in ſulphurious and bitter auarice) the whole world to battell: that is to ſaye, theſe Mahometanes, Emperours of the whole *Orient*, againſt the Antichriſtian beaſt ſupreame head of the *Occident,* to fight in *Armageddon*: that is to ſay, to fight in the mountaine of the choſen fruite, and in the mountaine of the Euangel euen in *Aſia minor*, and *Europe*, and in other lands of the Chriſtians, and among Gods elect fruit, & choſen profeſſors of his Euangel, troubling them on both ſides, which at length and orderly ſhall be found in our paraphraſticall & hiſtorical Diſcourſe, together with all the reſt of the ſmaller circumſtances contained in the text. So, then by theſe perfect properties we conclude, the ſixt trumpet or Vial, to be ſpoken of the great Empire of theſe foure nations Mahometanes, which began about the yeare of Chriſt 1296. vnder their firſt Emperour *Ottomannus*.

5. Appointed to ſlay.

6. Their great multitude.

7. Their three diueliſh exhortations.

Hier, de. int. nomin.

Concluſion

5. Proposition.

The ſpace of the fift trumpet or vial, containeth 245. yeares, and ſo much alſo, euery one of the reſt of the trumpets or vials doe containe.

SEing at the powring out of the fift vial, or ſounding of the fift trumpet, the Turkes began their dominion vnder *Zadok* in Anno Chriſti 1051. by the third propoſition: and by the fourth propoſition, at the ſounding of the ſixt trumpet, and powring out of the ſixt Viall, the whole foure nations Mahometanes began their great Empire about the yeare of Chriſt 1296. Therefore, the fift Trumpet or Viall indured from the 1051. yeare, to the 1296. yeare, which is the ſpace of 245. yeares. Now, that euery one of the reſt of the Trumpets or Vials doeth containe the ſame ſpace, it is probable by theſe reaſons. Firſt, becauſe in euery diſtribution, æqualitie is moſt apparant and probable, and ſo theſe ſeuen ages, termed by trumpets and vialles, appeare to haue no reaſon, why one ſhould be longer and another ſhorter, but rather all alike. Secondly, as in thoſe 245. yeares, the effects of the fift Trumpet and Viall were performed (as now is prooued) ſo by our paraphraſticall and hiſtoricall Diſcourſe, you ſhall finde, that in the former 245. yeares, the effect of the fourth Trumpet and Vial, and in the 245. yeares before theſe, the effect of the third Trumpet or Viall was performed: and conſequently, the effects of euery one of the reſt of the trumpets or Vials were performed in their owne ſpace of 245. yeares, ingendering ſo in theſe intervals of 245. yeares a perfect harmonie and analogie betuixt the prophecie and hiſtorie. Thirdly, becauſe euery one of theſe ages are tearmed and diſtinguiſhed by the ſounds of trumpets here, as the *Iubilies* were in* *Leviticus*. For there, at the end of euery 49. yeare, and beginning of the 50. yeare the trumpets ſounded, making betuixt euery ſounding thereof 49. yeares preciſely for a Iubilie intervall without any oddes, as the name *Iubileus* betokens,

1. Reaſon Equalitie.

2. Performance of effects.

3. Their diſtinguiſhing by ſoũd of trumpets.
Leuit. 25.9

Of this, reade our Solution of doubts, concerning the

which fignifieth a Trumpet or Cornet: Therefore, muft thofe ages alfo containe certaine whole Iubilie intervalls, precifely without any oddes: otherwife, apparantly, they would not haue bene limited here by the foundes of Trumpets, as the Iubilies are. Then, if euery one of thefe ages doeth containe foure Iubilies, which is 196. yeares, or fixe Iubilies, which is 294. yeares, the one fhall be fo fewe, and the other fhall be fo manie, that the hiftorie could not agree thereto: Therefore, fiue Iubilies, which is 245. yeares (as the middeft betuixt extremities) agreeth exquifitlie. Fourthlie, it is founde in hiftories, that great mutations of Empires followed neere by 245. years one after another: as, *Ierufalem* was deftroyed Anno 71. Anno 316, or thereby, *Conftantine* tranfported the Imperiall feate from *Rome* to *Byzantium*, which is *Conftantinople*. And in *Rome*, Pope *Sylvefter*, is faid then to haue begunne the Papifticall kingdome. Anno 561, *Narfes* recouering *Rome*: and after many burnings, repairing it, contracted peace with the *Oftrogottes*. About the yeare of Chrift 806, *Charlemaigne* receiued the newe Empire of Germanes and Romanes. Anno 1051, *Zadok* began the firft dominion of the Turkes. Anno 1296, or foone after, *Ottoman* began the firft Empire of the foure nations Mahometanes. Laft of all, about the yeare 1541. arofe our true profeffors againft the kingdome of Antichrift. And euery one of thefe great mutations followed neere by 245. yeares one after another. And this mooueth *Carion*, and other Hiftoriographers, to affirme by diuers obferuations, that fiue * hundreth yeares (as groffelie they thinke, or rather more exactlie, 490. yeares) is *fatalis periodus imperiorum: the fatall period of Empires*, for that it is about twife 245. yeares. Fiftlie, 245. yeares, is the juft halfe of the great Iubilie, or feuentie weekes of *Daniel*, containing 490. yeares: And becaufe the world would not ftand whollie feuen of thefe great Iubilies: therefore S. *Iohn* by the Spirit of God diuideth feuen halfes of the great Iubilie,

Jubilie.
Heir. de int. nominũ

4. *Great changes euery 245. yeares.*

Carion. lib. 4. de Henr. 4. Imp.
490. years the fatall terme of Empires.

5. 245, yeares are the iuft half of Daniels 70. weekes.

among the ſeuen Trumpets or vials. Sixtly, there are bookes of the Iewes, containing (as they alledge) doctrines proceeding from the mouthes of the Patriarches, affirming euery great Angel of ſeuen, to rule the world 490. yeare, which wee (wanting the warrant of Scripture) can neither affirme nor condemne. Alwaies (if it be ſo) the halfe thereof, to wit, 245. yeares is taken in ſtead of the whole, becauſe (as is ſaid) the time to come to the worlds end, would not containe ſeuen of theſe whole diuiſions and to containe fewer diuiſions than ſeuen, that had bene repugnant to the propheticall maner, that diuides all by ſeuen: as ſeuen ſeales, ſeuen trumpets, ſeuen vials, ſeuen thunders, &c. So then, for concluſion, euery Trumpet or Viall endureth 245. yeares.

6. The ancient opinion of Jewes.

6. PROPOSITION.

The firſt Trumpet or Viall began at the Iubilie, in Anno Chriſti, 71.

SEing by the third Propoſition, the fift trumpet or viall began in Anno 1051. And that by the fift propoſition euery trumpet or viall containeth 245. years, it muſt needs follow by juſt count, that the firſt trumpet or viall began in Anno 71. The ſecond in An. 316, The third in An, 561. The fourth in anno 806. The fift (as is already proued) in An, 1051. The ſixt alſo in An. 1296. *T*he ſeuenth in An. 1541. And that this 71. yeare of Chriſt, and conſequentlie, the fift yeare of euery trumpet or Viall was Iubilie, you may conſider by the diſccurſe of *Joſephus Scaliger de emendarione temporum.*

Prouen by induction.
71. yeare, a Jubilie yeare.
More heereof, read in our Solution of doubts.

7. PROPOSITION.

The Laſt of the ſeuen ſeales, and firſt of the ſeuen Trumpets or vials, begin both at once, in Anno 71.

THis appeareth euidently, for euen by the text (cap. 8.) the ſeuenth ſeale being opened, that verie houre, the ſeuen Angels receiued their trumpets to blow, for hauing ſaid in the firſt verſe, *and when he had opened the ſeuenth ſeale, there was ſilence in heauen about halfe an houre:* hee ſubjoyneth immediatly (verſ. 2.) *And I ſaw the ſeuen Angels, &c. receaue their ſeuen Trumpets.* And further, after he hath ſhewed (verſ. 5) how Chriſt powred out the fiery coales of his wrath, in reuenge of the blood of his Saints, (which doubtleſſe, was vpon *Ieruſalem*, which ſhead it) incontinent he ſubjoyneth (verſe 6. and 7.) that then the ſeuen Angels prepared themſelues to blow. *And the firſt blew his trumpet, &c.* So therefore, juſtlie at this deſtruction of *Ieruſalem*, in Anno Chriſti 71. make wee both the laſt ſeale, and firſt Trumpet or Viall to beginne. And this is alſo confirmed by the former, calculations, which come backward from the fift Trumpet or Viall preciſely, to the yeare of this deſtruction.

The reaſons hereof, read in our Solution of doubts.

8. PROPOSITION.

The firſt ſeale beginneth to be opened, in Anno Chriſti 29. compleat.

THis is prooued by this reaſon. The opening of the firſt ſeale (cap. 6.) and out-going of one riding vpon a white horſe, &c. is interpreted to be the word of God, in the 19. chapter, verſ. 13. which paſſed out victoriouſlie, conquering and piercing our hearts with the arrowes of feruencie and godly zeale. This behooued onely to beginne at that time that Chriſt was baptized, and began to preach and open vp the ſealed doctrine of our ſaluation, which was in the end of the 29. yeare of the age of Chriſt: and about the beginning of his thirtieth yeare (as *Luke* teſtifieth) So, then conſequently, the firſt ſeale beginneth to be opened about the end of the ſaid 29. yeare of the age of Chriſt.

Gods word victorious.

Chriſts preching, opened the ſeales of ſaluation.

Luk. 35.23

9. PROPOSITION.

Euery Seale muſt containe the ſpace of ſeuen yeares.

THe firſt Seale beginning to be opened in Anno Chriſti 29. as by the former propoſitiõ is prooued, we ſay, that the ſecond ſeale beginneth in Anno Chriſti 36. The third beginneth in Anno. 43. The fourth in An. 50. The fift, in An. 57. The ſixt in An. 64. And finallie, the ſeuenth beginneth in Anno 71 proceeding euer æqually by ſeuen yeares internall, for theſe reaſons. Firſt, becauſe in all diſtributions, æqualitie is moſt apparant and probable: and truth it is, that the firſt ſeale could not begin at the terme and yeare of Chriſt 29. prooued by the former propoſition, and the ſeauenth ſeale at the tearme of the yeare of Chriſt 71. prooued by the ſeuenth Propoſition, and proceed æquallie: vnleſſe that euerie ſeale containe ſeuen yeares. Secondly, becauſe in proceeding, and giuing ſeuen yeares to euery ſeale, ye ſhall finde the effect of euery ſeale to be performed within the ſeuen yeares of that ſeale: And ſo the harmonie to be perfect betuixt theſe ſeales and the juſt hiſtorie, as in our paraphraſtical and hiſtoricall Diſcourſe ſhall appeare more at large. Thirdly, becauſe all compts of yeares within a Iubilie, were reckoned among the Leuites and Prophets† by weekes of yeares, euery weeke containing ſeuen yeares, as the Iewes doe obſerue yet vnto this day: and this ſpace of ſeales opening, proceeding from the yeare of Chriſt 29, to the Iubilie in Anno. 71. is leſſe than a Iubilie intervall, for it is but 42. odde yeares: and therefore the diuiſion ought of neceſsitie to fall by weekes of yeares, or by ſeuen yeares. Fourthly, the Angel in DANIEL† reckoneth 70. weekes of yeares betuixt the commandement to build *Ieruſalem*, and the latter deſtruction of *Ieruſalem*, including the *Meſsias* comming. And this Angell calleth his words cloſed & ſealed, *Sermones clauſi & ſignati: words cloſed and ſealed*: whereby juſtly theſe 70. weekes may be

Induction

1. Reaſon, equalitie probable.

2. Performance of effects in euery ſeuen yeares.

3. Reckoning Within a Jubily is made by weekes of yeares.
† Exod. 23.
Leuit. 25.

4. Theſe ſeales are a part of Daniels ſealed weekes.
Dan. 9.24.

† Dan. 12.9.

called ſealed weeks. Now, becauſe theſe 42. odde yeares of the ſeales are certainlie a part of the ſeuentie ſealed weeks of DANIEL: Therefore, we may juſtlie, affirme theſe ſeales to be ſealed weekes, and ſo conſequentlie euery one of them to containe ſeuen yeares.

10. PROPOSITION.

The laſt trumpet and Vial beginneth anno Chriſti 1541. and ſhould end in anno Chriſti, 1786.

SEing by the Propoſition, the fift Trumpet or Viall began in anno 1051. And by the fift propoſition, euery Trumpet or Viall containeth 245. yeares, it muſt needs follow that the ſeuenth Trumpet or Vial began in Anno Chriſti 1541. and conſequently if ſhould end 245. yeares after: which is in Anno 1786. Not that I meane, that that age, or yet the world ſhall continue ſo long, becauſe it is ſaid, † that for the Elects ſake, the time ſhal be ſhortned: but I meane, that if the world were to endure, that ſeuenth age ſhould continue vntill the yeare of Chriſt 1786.

By induction.

Mat. 24.22

11. PROPOSITION.

The ſeuen Thunders, whoſe voices are commanded to be ſealed, and not written [cap. 10.4] are the ſeuen Angels, ſpecified cap. 14. verſe 6.8.9.14.15.17.18.

THis tenth chapter doeth agree with the twelth of DANIEL almoſt in euery poynt: there doth he likewiſe ſhew, that thoſe verie ſelfe-ſame myſteries are ſealed, yet but for a time, ſaith the Angel, in theſe words, *Tu autem* DANIEL, *daudeſermones, & ſigna librum vſque ad tempus ſtatutum, &c, But thou, O* DANIEL, *cloſe vp the Words, and ſeale the booke, vntil the appoynted time.* And againe, he ſaith, *Vade* DANIEL, *quia clauſi ſunt ſignatique ſermones, vſque ad præfinitum tempus. Goe thy*

Dan. 12.4. and 9.13.

way, DANIEL, *for the words are cloſed and ſealed vntil the appoynted time*. This time that theſe myſteries were cloſed, appeareth to be the time of darkneſſe alreadie paſt, vnder the Antichriſtian errors, now hope we in God, that thoſe myſteries are able to be found out, ſeeing that time is expired. Then to our purpoſe, let vs conferre the tokens of the ſeuen thunders, with the tokens of the ſeauen Angels, ſpecified in the fourtenth chapter, and we ſhall finde them to agree. For firſt, at the comming of the ſeuen thunders, Chriſt doeth offer the open booke of his trueth. (So ca. 14.) the firſt of the ſeuen Angels bringeth the euerlaſting Goſpel, and openly preacheth out the ſame. Secondly, cap. 10, Chriſt by the mightie voice of his word, as a roaring Lion in the mouthes of his preachers, proclaimed the truth: then came the ſeuen thunders: ſo alſo (cap. 14.) the voice of Gods heauenly elect is heard as the mightie ſound of waters, & as thunder (ver. 2.) Then came orderlie the ſeuẽ Angels (ver. 6.8.9.14.15.17 & 18) Thirdly (cap. 10) the ſeuen thunders are ſealed vp as a myſterie. So likewiſe (ca. 14) thoſe ſeuen Angels are preparers and reapers of Gods great harueſt, and what myſterie is more ſealed than that, as teſtifieth † *Marke*, ſaying, *De die autem illo vel hora, nemo ſcit, neque Angeli in coelo, neque Filius, niſi Pater: But of that daye and houre knoweth no man, no not the Angels which are in heauen, neither the Sonne but the Father:* And *Paule* to the *1. Cor* ſaying, *Ecce, myſterium dico vobis, &c. Beholde, I ſhew you a myſterie, &c*. Fourthly, (cap. 10.) the thunders are commanded to be ſealed, and not to be writtẽ plainly. So (ca. 14) they are ſo ſealed, that they are neither named duelie with their former name of thunder: neither are they placed in their due place, which ſhould haue bene in the tenth chapter, but here are called Angels, and placed in the fourtenth chapter. Fiftlie (chap. 10) they be called thunders, and yet are ſealed: So (chap. 14) though they be ſealed with the name of Angels,

Myſteries no longer cloſed.
Reaſons of Analogie.

1. The Booke and euerlaſting Goſpel opening.
Apoc. 10.2 8.9.10.11
Apoc. 14.6
2. The mightie voice of both.

3. They both are called myſteries.

Mar. 13.32

1. Cor. 15.51

4. They ar both ſealed vp.

5. The one called thũders, and a thunder goeth before the other.

yet for a token that they be the ſeuen thunders, he deſcribeth the ſound of great thunder going before them. Then for concluſion, in reſpect of the harmonie betuixt the ſeuen thunders and ſeuen Angels of Gods great harueſt, we conclude them both to be one.

Apoc. 14.2

Concluſion

12. PROPOSITION.

The firſt of the ſeuen thunders, and the ſeuenth and laſt Trumpet or vial, begin both at once in Anno 1541.

IN the eleuenth chapter it is ſaid, that at the blaſt of the ſeuenth Trumpet, the kingdome of the world becommeth Chriſt his kingdome: that is, the kingdome of the Antichriſt, & all other prophane kingdomes ſhall fall, and Chriſt his kingdome ſhall be ſpread and enlarged ouer all: And this can no other waies come to paſſe, but by the preaching of the Euangel, which was of newe opened vp, and preached at the comming of the firſt Angel, whom the Text ſaith, to haue †*Euangelium æternum, vt euangelizaret ſedentibus ſuper terram: An euerlaſting Euangel, to preach vnto them which dwel vpon the earth.* Which Angell, by the former propoſition, is prooued to be the firſt of the ſeuen thunders: Therefore, it followeth conſequentlie, that the new reſtoring of the Euangel, the ſeuenth Trumpet or Vial, & the firſt thunder, or thundring Angel, come all at once. And therefore, by the tenth Propoſition, they began in Anno Chriſti 1541. For confirmation hereof, after the end of the ſixt Trumpet, completed in the ende of the ninth chapter: in the beginning of the tenth chapter, where the ſeuenth Trumpet appeareth to beginne (the ſixt being newlie ended) ye ſhall finde there declared, how Chriſt proffereth the open booke, & manifeſt doctrine of the Euangel, & then immediatlie that verie time, while as Chriſt by his Miniſters, preached and proclaimed out that manifeſt word, as with the mightie voice of a Lion:

Apoc. 14.6

The reſtoring of the Euangel, Anno 1541. beginneth the ſeuenth trũpet or vial, & the firſt thunder or Angell.

incontinent (I ſay) that verie time (verſe 3. the ſeuen thunders began to vtter their voice: And ſo the beginning of the voice of the ſeuen thunders, or voice of the firſt thundering Angel commeth at once with the completing of the ſixt Trumpet, and beginning of the ſeuenth: although the very latter blaſt of the ſeuenth Trumpet be not then come.

13. Proposition.

Euery one of the firſt three thundering Angels containeth a Iubilie, and then the laſt foure all at once compleateth the day of iudgement.

THe firſt part of this propoſition, is at the command of the Angel in DANIEL, and in the *Reuelation* ſo ſealed & cloſed, that be no part of thoſe texts may it be gathered, how long euery thunder or Angel of the great harueſt doeth follow after other: yet after the cuſtome, both of the Leuites and Prophets, † who diuide all great dates by Iubilies, and ſmall dates by weekes of yeares, we judge moſt apparantlie, this laſt age to be diuided by Iubilies: * and ſo euery thunder or Angel of theſe three that are ſaid in the text (ca. 14. ver. 16.8 & 9.) to preceed Gods great harueſt, to containe 49. yeares, ‡ beginning the firſt (by the former propoſition) at the yeare of Chriſt, 1541. who (verſe 6. and 7) preached his euerlaſting Euangell to the yeare of God, 1590. † Which yeare, beginneth the ſecond thundring Angel (verſe 8.) and continueth proclaiming the finall decaie and fal of *Babylon*, to the yeare of Chriſt, 1639 Which yeare, the thirde thundering Angel beginneth, who (verſe 9.) is ſtiled the third Angel, and he continueth exhorting and threatening theſe of the laſt dregs of the Antichriſt to repentance, to the yeare of Chriſt, 1688. Where the fourth thundring Angel, euen Chriſt himſelfe (verſ. 14. & 16.) enters actually vnto his great harueſt, to gather vp his elect, at the warning of the fift Angel, euen Gods holie Spirit (verſ. 15) Like as contrarilie, the ſixt

Dan. 12.9.
Apoc. 10.4

†Great dates divided by Jubilies.
** Each Angell of the firſt three thunders to contein 49 yeares.*
‡ 1541. 1. thunder in this Jubily the Euangel preched
†1590. 2. thũder proclameth the fal of Babilõ to the year 1639.
3. thunder threatning to repentance to the yeare, 1688.

The great harueſt gathering in of his elect.

thundering Angel, executor of Gods fiery and fierce wrath (mentioned verſ. 17. and 19) is admoniſhed and warned (verſe 18) by the ſeuenth Angel commander thereof, to cut downe and tread out the vines of the earth, in the wine-preſſe of Gods wrath: And ſo by theſe laſt foure Angels or thunders (to wit, by Chriſt himſelfe, and by his whole miniſtering Angels, vnder the figure of theſe laſt Angels) all the whole great day of Gods harueſt, and latter judgement is accompliſhed.

Deſtruction of the reprobat.

14. PROPOSITION.

The day of Gods iudgement appeares to fall betuixt the yeares of Chriſt, 1688. and 1700.

ALthough it be ſaid in *Marke*, that the day of judgement & houre thereof, none doth know, yea, not the Sonne, but the Father only: yet let none be ſo baſe of iudgemẽt as to conclude thereby, that the yeare or age thereof is alſo vnknowne to Chriſt, or vnable to be knowne any waies to his ſeruãts, by reaſon that firſt in that ſame part of *Mark*, Chriſt letteth vs not only ſee, that he knew the age & yeares, neere the which that day ſhould fall, but alſo to the effect, that wee may likewiſe fore-know the ſame, he giues vs diuers indices & foretokẽs, which he could not, nor would not haue fore-warned, if hee had bene vtterly ignorant of the time thereof, or yet had minded that we ſhould neuer haue fore-knowne the age and appearant yeares thereof, although the preciſe day and houre be only knowne to God. Secondlie, although the Spirit of God hath hitherto concealed theſe myſteries from them whom the knowledge thereof might haue endamnaged: yet that prooueth not, that the ſame ſhall be hid from vs, to whom the knowledge thereof might bring repentance and amendement: for as the foreknowledge of death, to him that were to liue long, might make the forke-knower negligent of his dutie to God, & careful to prouide inordinatly

Mar. 13.32
Reaſons that the latter day may neerly, (though not preciſely) be knowen.
1. Chriſt teacheth the foretokens.

2. Reaſon, becauſe that foreknowledge might now mooue repentance & amendment.

for his long life here: Where through God hath made the houre of death vncertaine till it aporoach. Euen ſo, if the foreknowledge of the latter day had bene granted to men any waies long before it come, that long aſſured continuance of the world, foreknowne by them ſo long before, had made them to become more carefull *per fas & nefas*, for their families and poſterities, that were long to ſtande, than for that heauenly kingdome, that was long to be delayed. And therefore, was that myſterie juſtlie by the prouidence of God, cloſed from our predeceſſors: but cõtrarilie, ſo ſoone as that day beginneth to approach, God by his Scriptures, ſhall make the age and yeares thereof to be manifeſted, as a ſpurre in his mercie, to mooue the elected ſinners to repentance, and a teſtimonie in Gods juſtice, againſt the hard-hearted miſ-beleeuers, continuing in ſinne. Thirdly, and for confirmation of the former, Chriſt teſtifieth, that his comming ſhal be like vnto the floud of *Noah*, before the which, the world being admoniſhed, both by *Noahs* preaching, and by the viſible building of the Arke, would not the more repent, nor amend their liues, till vnawares they were all deſtroyed: & is God now otherwiſe then he hath bene? *Is he* (ſaith the ſcripture) *a man, that he ſhal repent that now, which he did then*, and forbeare the like? Nay, but contrarily, he ſhall make his owne to be fore-ſeene of this time, aſwell as of that time, ſeeing it is now as neceſſarie, that Gods Miniſters exhort vs to repentance, of that coldneſſe of charitie, and al other vices that abound in this laſt age, and to terrifie vs with the certaine and aſſured approaching of the latter day, and deſtruction of the world by fire, aſwel as the longſome building of the Arke, the deeds, geſture, and apparant preaching of *Noah*, was a fore-warning 120. yeares to the olde world, that for their vnlawfull luſts, & other their horrible vices, they ſhould be deſtroied by water. Fourthly, that the yeares or age of the latter day is not vnknowne to Chriſt, nor euer vnſearchable to his ſeruants,

Why death & the latter day are concealed till they approach.

Mat. 24.37

3. It ſhal be fore-warned as Noahs flood was.

1. Reg. 15.29.

Gen. 6.3.
1. Pet. 3.19
4. Daniel prophecieth that in thir latter daies, knowledge of myſteries ſhall abound.

is certaine by *Daniel*, to whom, although it is ſaid, †*Vade* DANIEL, *quia clauſi ſunt ſignatique ſermones hi: Goe thy way* DANIEL, *for theſe words are cloſed and ſealed*: † As to the Apoſtles, *Non eſt veſtrum ſoire tempora: It is not for you to knowe the times*, for that their time was far from the latter day, yet, ſaith DANIEL of the time of reuelation, *Signa librum ad tempus ſtatutum, multi pertranſibunt, & multiplexerit ſcientia, Seale the booke till the appointed time, manie ſhall goe to and fro, and knowledge ſhall be encreaſed*: meaning, when as the due time beginneth to approach, theſe dates ſhall be knowne, for knowledge ſhall then abound. And againe he ſaith, *Impie agent impii, neque intelligent omnes impii, porro docti intelligent. The Wicked ſhall doe wickedly, and none of the wicked ſhall haue vnderſtanding, but the wiſe ſhall vnderſtand.* Fiftlie, this age or apparant years of the latter day, are neither vnſearchable to Chriſts ſervants, nor vnknowne to himſelfe, ſeeing it is euident, that the Deuils haue a certaine fore-knowledge thereof, in that they ſaid to Chriſt, †*Veniſti hue, ante tempus torquere nos? Art thou come hither, to torment vs before the time?* meaning, that although they are continually reſerued in feare and dread, yet (as *Peter* and *Iude* teſtifie) their chiefe tormẽt is not vntil the latter day, which they knew was not to be at that time. Again, it is ſaid in the *Reuelation, Letamini cœli, &c. Væ aut ens terræ & mari, quia deſcendit Diabolus ad vos, habens iram magnam, ſciens quod modicum tempus habet. Therefore, reioyce yee heauens, &c. and woe to the inhabitants of the earth, and of the ſea, for the Deuil is come downe vnto you, which hath great wrath, knowing that he hath but a ſhort time.* So then, ſeeing the Deuill hath great wrath in the latter daies, and doeth know that his time is ſhort, ſhall we ſay, that Chriſt ſhall be ignorant of that, which the Deuill doeth know. Sixtly, to what effect were the Prophecies of DANIEL, and of the *Reuelation* giuen to the Church of God, and ſo many dates of yeares,

† Dan. 12.9

† Act. 1.7

Dan. 12.4.

Dan. 12.10

5. Seing the deuils doe know there of, Chriſt and his Elect cannot be ignorãt.

†Mat. 8.29

2. Pet. 2.4.
Iude 6.

Apo. 13.12

6. In vaine were it prophecied, if it were not to be vnderſtood.

and circumſtances of time, fore-ſhewing the latter day, contained thereintill, if God had appointed the ſame to be neuer knowne or vnderſtood before that daye come. Therefore, aſſuring our ſelues, that all theſe Prophecies of the latter day ſhall be knowne and manifeſted to Gods Church, before Chriſt come to judgement: let vs conferre all theſe prophecies and propheticall figures thereof together, and we ſhall finde thẽ come ſo neere to one time, that verily the matter is wonderfull. And firſt (beginning at the Symbole of the Sabboth) it is thought by the moſt learned, that the ſixe daies of labour, weekly obſerued, doth meane & bear the ſymbole of 6000. years that mankind ſhall indure the trauels and cares of this world: and that is confirmed by *Peter*, who ſpeking of the day of iudgement, noteth, that a *thouſand years ſhal be as one day in Gods ſight:and a day as a thouſand years*: And ſo conſequently, the ſixe dayes of worke, to repreſent ſixe thouſand yeares: after the which ſixe thouſand yeares of worldly trauels and cares, then ſhall come our æternall Sabboth & reſt, in the glorie of heauen, ſignified by the ſeuenth daies reſt: For that truely, there is no inſtitution appoynted by God to *Moſes*, which (beſides the ciuill commoditie) had not alſo a ſpirituall figure: And ſure it is, that no figure appeareth more conſonant with theſe ſixe daies of worke, and ſeuenth day of reſt, than the preſent miſeries of this world, and æternall Sabboth hereafter. Secondly, and agreeable with the former, there is a ſentence of the houſe of *Elias* referued in all ages, bearing theſe wordes: *The world ſhal ſtand ſixe thouſand yeares, and then it ſhall be conſumed by fire*: two thouſand yeares voide or without lawe, two thouſand yeares vnder the law: & two thouſand yeares ſhall be the daies of the MESSIAS: And for our offences, which ſhall be manie and great, ſhall theſe yeares lacke which ſhall lacke. Thus farre ſaieth *Elias*: Now, the tearme of theſe 6000. yeares doeth expire, about the 2000. yeare of Chriſt, which terme, both by the ſaid ſaying of

Concluſion of the poſsibilitie to fore-know.
Inquiſitions of that foreknowledge. The firſt is the Symbole of the Sabboth.

2. Pet. 3.8.

2. The ſentence of the houſe of Elias.

Elias, and by Chrifts owne faying in *Matt.* appeareth to be fhortned. † And therfore, thirdly, by *Daniel* wee fhall approch nearer that tearme: for he in his 12. chap intreating both of the firft refurrection (which is the refurrection from the Antichriftian errors) and of the fecond refurrection, which is the generall refurrection from the dead: fuch an Angel, as afterwards appeareth to IOHN (*Reu. 10*) teaching him concerning the latter day, doth here appeare to DANIEL, fwearing folemnly a tearme of prophetical daies to euerie one of thofe queftions, doubted of by the Prophet: of which, the greateft tearme is expreffed to be 1335. propheticall daies, which (by the I. propofition) meaneth 1335. yeares, whofe beginning is determined in thefe words, *And frõ the time that the daily facrifice fhal ceafe, and the abomination put to defolation* (or made defolate) *fhal be 1290. daies: bleffed is he that awaiteth and abideth to the 1335. day.* As if he would fay, from that time that both the Iewifh daily Sacrifice fhal vtterlie ceafe, and the abominable rites of the Gentiles be abolifhed, to the compleating of the great refurrection from the Antichriftian errours, and vtter decay of his kingdome, fhal be 1290. yeares, and then bleffed fhall he be that abideth and remaineth conftant in the fmall time, that fhall be betwixt that and the latter day, which fhall fall in the 1335. yeare, from the taking away of the faid Iewifh ceremonies, and Gentiles fuperftitions. But fo it is, that thefe were neither actually taken away by Chrifts pafsion, neither in the deftruction of I*erufalem*, neither other-waies elfe, till miraculouflie in the daies of *Iulian* the Apoftate, Anno Chrifti 365. This Apoftate, hauing in defpight of Chrift, ordained the Iewes to builde vp and repaire the Temple of *Ierufalem* on the one part, and on the other part, hauing fent his Legates to offer Sacrifice in the chiefe Ethnicke Temple of the *Gentiles* in *Delphos*, and to confult with that *Oracle* of *Apollo* there, God fent his thunder from aboue, and earthquake from beneath, and thereby over-

†3. Daniel dateth both the refurrections
Mat. 24.22
Dan. 12.

Apoc. 10.5 6.7.

Dan. 12.11

Carion. Chro.lib. 3

whelmed both thoſe chiefe Temples* about one time, ſo vtterly, that to this day al the Iewiſh daily Sacrifice of the one, & the abominable Ethnick ſuperſtitions of the other, haue ceaſed, and bene put to vtter confuſion and deſolation. But if (after the opinion of ſome learned men) this text doeth meane in the original Hebrew, not the *Abomination* to be put in *Deſolation*, but contrarilie, † a deſolating and waſting *Abomination* to be *ſet vp*, together alſo, with the taking away of the former and foreſaid daily Sacrifice: In this caſe appearingly, the ſame date is not the leſſe eſtabliſhed: † for who was a more horrible and waſting *Abomination*, than was the foreſaid *Julian*, that blood-thirſtie Apoſtate, together with his deteſtable, idolatrous and magicall decrees, which publikelie he erected and ſet vp, to bring *Chriſtianiſme* to vtter *deſolation*. So that howſoeuer this *Abomination* be expounded, either paſsiuely, to *be made deſolate*, as the Gentiles chiefe abominable and idolatrous *Oracle* of *Apollo* at that time was made deſolate, or actiuely, that the *Abomination* ſhould make a waſting or *deſolation*, and that that waſting *Abomination* ſhould be then erected, as the abominable Tyrant and Apoſtate *Iulian* was promoted Emperour, and his cruell decrees of perſecution at that time ſet out: Euen alwaies in that verie ſame 365. yeare, all the foreſaid accidents occurred: for in that one yeare both firſt the foundations of the *Iewiſh* Temple of daylie Sacrifice: as ſecondly (in the paſsiue ſenſe) the ſaid Temple of the Ethnicks *abominations*, in *Delphos*, were by thunder and earthquake both deſtroyed as thirdly (in the actiue ſenſe) that a-abominable *Julian*, and his decrees for deſolating & devouring Gods ſeruants, were then ſet vp. Therfore from this yeare of Chriſt 365. we are commanded to reckon 1335. years which falleth in Anno 1700. And then (ſaith DANIEL, verſ-6) ſhall be the end of all wonders, euen the wonderous and great day of the Lord: But yet, as the foreſaid prophecie of *Elias*, defined by whole thouſands, is by DANIEL more narowly

**The temples of Jeruſalem & Delphos deſtroyed at one time anno 365. & the dailie ſacrifice ceaſed, and abominable Ethnik ſuperſtitions, then were put to deſolation.*

† A deſolating abomination ſet vp.

**Juliã Apoſtat, and his deteſtable decrees were a waſting abomination.*

1. Occurrance.

2. Occurrance.

3. Occurrance.

1700. the ende of all wonders.

defined by 17. hundreth. So it appeareth that the laſt prophecie of the Revelation ſhould moſt narrowlie of all point out the ſame by odde yeares: for the which, now wee proceed to that laſt Prophecie of the *Revelation.* Fourthly therefore, it is reaſoned in the former propoſitiõ, that euery thundering Angell of three, contained 49. years, and then commeth the great Day of the Lord: and by the 12 Propoſition, the firſt thundring Angel, began in An. 1541. to which adde thriſe 49. yeares, which is 147. yeares: and ſo by that account, the latter day wil fal in An. Chriſti, 1688. Wherfore, appearingly betwixt this 1688. yeare, according to the *Revelation*, and the 1700. year, according to DANIEL, the ſaid latter day ſhould fal. And for further confirmatiõ hereof, there is a number put in the end of the 14 ch. of the *Revelation*, which appeares to be a date of the latter day: for that chapter altogether ſpeaketh of the Lordes great harueſt, & latter judgement. There (ſaith *John*) the blood came out of the Wine-preſſe vnto the horſe bridles, by the ſpace of a thouſand and ſixe hundreth ſtades or courſes: as if appearingly, hee would meane metaphoricallie, as wine may bee thought to flowe from the preſſe, or the blood of ſlaine men in a fielde, to aſcend to the horſe bridles: ſo, æternallie ſhall the torment of the wicked aſcende, after that a thouſand and ſixe hundreth yeares be accompliſhed: For, theſe *Stades* agree well to meane yeares, ſeeing a *ſtade* is that race or courſe, that one may be thought to runne with one breath, before hee beginne to renewe his breath againe: as one yeare is that race or courſe, that the Sunne maketh in a circuit, before he beginne to renewe his circuit againe. Now, counting therefore, a thouſand and ſixe hundreth yeares, from the time that this was written, which was about the 97. yeare of Chriſt, as *Euſebius* in his Chronicle ſaith, or in the ende of the raigne of *Domitian*, as *Irenæus* ſaith, which was in Anno Chriſti 99. The end of the count ſhall fall about the yeare of Chriſt 1697. or the yeare 1699. which is

4. Reaſon, by the ages of the thũders.

The latter day betwix 1688 and 1700.

A confirmation hereof.

betuixt the ſaid terme 1688. and 1700. And ſo the difference of theſe dates is but final, and if the time of hiſtories were ſurely written and obſerued, it may be thought, that the difference would be either leſſe, or vtterlie nothing at all.

The terms of this account, are 1697. or 1699.

15. PROPOSITION.

Dan. 7.25.
Dan. 12.7.
Apo. 11. & 12.
Apoc. 13

The 42. moneths, a thouſand two hundreth and threeſcore prophetical daies, three great daies and a halfe, and a time, times, and a halfe a time mentioned in Daniel, & in the Reuelation, are all one date.

Everie moneth among the Grecians, contained thirtie daies preciſely, as witneſſeth *Ioſephus Scaliger*, and ſo conſequentlie, fourtie two moneths are æqual to a thouſand two hundreth, & threeſcore daies. Alſo, three great daies and an halfe, are æquall to them for theſe reaſons. † Firſt, by reckoning, becauſe three daies and an halfe, after the propheticall manner (prooued in the firſt propoſition) is three yeares and an halfe: and then counting (after the Græcian manner) twelue moneths in the yeare, and thirtie daies in euerie moneth, theſe three yeares and a halfe (called in the text, three daies and a halfe) will be fourtie two moneths, or a thouſand, two hundreth and threeſcore daies juſt. Secondly, they muſt be all one, for that in purpoſe they agree: for it is ſaid, that the Antichriſtian beaſt at his riſing, ſlew the two *Witneſſes* of God (which hereafter are prooued to be the two Teſtaments) and their dead carcaſes (or naked letter) lay three daies and a halfe, and then were they reuiued, and at that time of their reviuing, came a great decaie on the Antichriſtian citie: So that both by this text, and other good reaſons, it appeareth that theſe *Witneſſes* of Gods trueth, lay dead and ſilent, and their teſtimonie neither buried in our breaſts, nor ingraued in our hearts, during all the daies of the Antichriſt. And ſo the time of their

lib.de emen dat. tempor.

Thre great dayes and a halfe, are 1260. propheticall dayes.
†1 Reaſon, by reckonning.

2. One in purpoſe.
Apoc. 11.7
Read propoſition 21.

lying dead, and the time of the Antichrifts raigne, to be all one date. Now, the time that they lay dead, is called three daies and an halfe, and the time that the Antichrift raigned, and oppreffed Gods Church, is called 42. moneths: wherfore, three great daies & an halfe, are al one with 42. moneths, or 1260 daies: It refteth then, to prooue a time, times, and halfe a time, to be likewife equal to them, which muft needs befor thefe reafons. Firft, becaufe it is faid (cap. 12.6., that the *Woman* (which hereafter is prooued to fignifie the Church of Chrift) fled into the wilderneffe, where fhe was nourifhed 1260. daies: And againe (verfe 14) it is faid, that fhee fled away in the *wilderneffe, where fhe was nourifhed for a time, times, and halfe a time*. So then, *a time, times, and halfe a time*, and *1260*. daies muft be both one. Secondly, this *time, times, and halfe a time*, or three times & an halfe, correfpondent with the forefaid three great daies and an halfe, are æquall to 1260. daies for this caufe. A time taken fimplie and without figure, meaneth a yeare, as in *Daniel, Nebucadnetzars* feuen times that he remained a beaft, are taken for feuen yeares: Thẽ three times and a halfe, are three yeares and an half, which being counted, will make 42. moneths, or 1260. daies: And fo finally, for conclufion, feeing the raigne of the Antichrift, his blafpheming of God, and opprefsion of Gods Saints, the treading of fpirituall *Ierufalem* vnder feete, the prophecying of Gods witneffe vnto the true and invifible Church in humble maner, and their lying dead and filent, vnto the outward vifible pretended Church. And the flying of the Spoufe & Church of Chrift to the wildernes, & becomming invifible, be matters concurrent and adjunct, fuch as apparantly muft at one time beginne together, continue together, and end together. Therefore, moft neceffarily the dates that the text attributeth to their continuance, to wit, 42. moneths. 1260. propheticall daies, three great daies and an halfe, and a time, times, & halfe a time, muft be all one date.

Apoc. 11.2
Apoc. 13.5.
A time times, and half a time are 1260. prophetical dayes.
1. Reafon: the church hid 1260. dayes, and againe, a time times and halfe a time.

A time fimplie taken is a yeare.
Dan. 4.13.20.29.

Conclufion

Matters alwayes one, muft haue their dates one.

16. PROPOSITION.

The 42. moneths, 1260. propheticall daies, three great daies and a halfe: And a time, times, and halfe a time, ſignifieth euery one of them, 1260. Iuliane yeares.

SEing theſe dates are prooued by the former propoſition to be æquall, therefore neceſſarily what one is, all is, then wee come to theſe 1260. daies, which wee ſay, muſt needes be either vnderſtood ſo many natural dayes, or elſe ſo many propheticall daies, which are prooued to be yeares in the firſt proproſition: But naturall or common daies they cannot be, for theſe cauſes. Firſt, when the Prophets in number, or great dates meane of naturall dayes, they vſe to adjoyne the definition thereof, by morning and euening, defined in *Geniſ.* cap. 1. as *Daniel* doeth chap. 8. verſe 14. and 26. But that it is not done here, this being a prophecie: and therefore, theſe are not naturall and common daies, but conſequently, are propheticall daies. Secondly, it is not apparant, that ſo manie Prophets would haue written ſo much: for only a raigne of 1260. of common daies, which is but three years and a halfe, ſeeing many cruell tyrants and hereticall Emperours, that did raigne longer ouer Gods Church, were not prophecied of in particular, as *Nero, Domitian*, and others. Thirdly, if theſe were naturall and common daies, then ſhould the Antichriſt raigne, but 1260. common daies, which is three yeares and an halfe: but contrarilie, it is certaine, by the *Reuelation*, cap. 20.4.5. that that the Antichriſtian beaſt had authoritie, and did raigne at the leaſt a thouſand yeares, martyring the ſaints of God, and perſecuting them that reuerenced not him and his *marke, &c.* but ſpirituallie raigned, and liued as true Chriſtians theſe thouſand yeares, and that the reſt of the people ſhould lie dead and drunken in his errours, vntill the ende of theſe thouſand yeares: and ſhould not while then, beginne to riſe in the firſt reſurrection, which is, to riſe from their errors.

Reaſons why thir dayes are yeares.

1. Not defined as natural daies.

2. not aparant that the raigne is ſhorte where of ſo much is written.

3. the Antichriſt muſt (at the leaſt) raigne 1000 years

So then confequently, thefe daies (which exceede 1000. yeares) cannot be 1260. of common daies, but 1260. propheticall daies, which are prooued by the firft propofition to be 1260. years: and fo apparantly, the Antichrift raigning mightily 1000. yeares: the remnant, 260. yeares are attributed to the falling and decreafing of his kingdome, making therby his whole raigne 1260. yeares, to the which, feeing by the former propofition, three great daies & an halfe or three times and an halfe is æquall, it muft needes be, that euery great day, or euery time, muft not fimplie fignifie a common yeare, as *Nebuchadnezars* times did, but muft fignifie a prophetical time, or yeare, prooued by the firft propofition to be a yeare of yeares, or 360. yeares precifely. For confirmation, wherof, it is to be vnderftood, that the firft inftituters of times, to wit, the *Caldeans, Grecians* and Aftrologes in their directions do agree with this defcription of time: for they divide the *Equinoctial* into 360. degrees, and attribute a yeare for euery degree of their directions, whereby the whole time of the great revolution or direction of the whole *Equinoctiall*: will be 360. yeares, and confequently, three of thefe great times and a halfe, or three times and a halfe, revoluing of the whole *Equinoctiall*, will containe 1260. yeares. And fo *Daniel*, writing in the Chaldee tongue to the Church of God, then refident chiefly among the *Chaldeans*, and S. Iohn alfo writing in the Greek tongue to the Church of God in his time refident chiefly among the *Grecians*, obferued the Chaldean, and Grecian manners of counting times, as being a vulgar compt then amongft them. But now, although it is prooued, thefe dates to bee 1260. yeares: yet, forafmuch as 1260. of Grecian yeares, are but 1242. *Julian* yeares, and 8. moneths, or there-about: and 1260. *Julian* yeares, are 1277, and an halfe of Grecian yeares, making therby, nere 18. yeares of difference. It refts therefore, to prooue what kind of yeares thefe be. Thefe (we fay) are common *Julian* yeares, for two caufes: Firft, although the

Conclufion 1260 days are 1260 yeares.

A great day or great time is a yeare of yeares.
Dan 4.13 20.29

The diverfitie of yeares united.

Grecian common yeare contained but 12. moneths and 30 dayes in euery moneth, yet do they adjoyne certaine intercalar dayes, which doeth make euery yeare ouerhead to conteine 12. moneths, fiue dayes and a quarterwhich is 365. dayes and a quarter: and ſo conſequentlie, are ouerhead equall with our common *Julian* yeare. Secondly, among the Hebrue Prophets, where a day is taken for a yeare, although the common yeare contain but 12. moneths, yet almoſt euery third yeare, they adjoyned an intercalar moneth, by doubling the moneth *Adar*, which made their Hebrew yeares ouerhead æquall alſo with our *Julian* yeares, as teſtifieth *J of. Scal. de emenda. temporum.*

Hitherto are the difficulties of dates reſolued: now followeth the reſolution of the principall termes and matter.

17. PROPOSITION.

The deſcription of the throne of God in the fourth chapter, is not the deſcription of the maieſtie of God in heauen, but of his true religion, where in he is authoriſed and ſits in the throne among his holy elect on earth.

BEcauſe it is ſaid in the text, that this throne is ſet in heauen: therefore, ſome thinks this to be a viſion of Gods glorie in heauen, but that can no waies be, for theſe reaſons. Firſt, becauſe heauen, for the moſt part prophetically, is taken for Gods heauenly Elect or true Church vpon earth. Secondly, becauſe it were ſuperfluous curioſitie for vs to know any farther of Gods heauenly eſtate, and glorie of his Maieſtie, than the ſimple points of our ſaluation. Thirdly, becauſe the Scripture teſtifies, that no pen can deſcribe, nor wit comprehend, the glory of Gods majeſtie in heauen. Fourthly, becauſe the foure *Beaſts* and 24. *Elders*, who here are coherent members of this throne, confeſſeth themſelues

Reaſons why,

1. Becauſe heauen is takẽ for our true church on earth.
2. Superfluous curioſitie.

3. Impoſſible to bee deſcribed or comprehended.
Ap. 5.9.10
Apoc. 6.6.

to *raigne vpon earth, and that Chrift hath redeemed them with his blood. Fiftly, becaufe it is faid, that among thẽ, euẽ among thefe four *beafts*, arofe a †famin and dearth of Barley and Wheat. Therefore this throne cannot be Gods throne in heauen, but muft needs be his throne vpon earth among his heauenly elect here: & confequently, are either his Church or true religion: but his Church is not properlie his throne and feate, but rather are thefe ouer whom he fits. Therefore, this throne muft needs be his truth & true religion, wherein he fitteth, abideth, & making his refidence, therein is authorized and inthronized here vpon earth, among his heauenlie elect feruants.

4. The affeffors of this throne are earthly mẽ

5. They are fubiect to hunger and dearth.

Conclufion, This throne is Gods trueth, and true Religion.

18. PROPOSITION.

The 24. Elders, are the 24. bookes of the old Teftament, with the 24. writers and [*metonymice*] *all the true profeffours thereof.*

THefe 24. Elders, being prooued by the former propofition to be vpon earth, becaufe that the glory of the whole throne (wherof they are one coherent parte) is vpõ earth, we fay now further, that they doe reprefent the 24. bookes of the old Teftament, writers and profeffours therof, for thefe reafons. Firft, *ab officio*, becaufe thefe ancients are faid in the † text, to glorifie God day and night, and what thing in earth is God more glorified by, than by his Scriptures and holy wrytings, &c? whereof thefe be the firft. Secondly, becaufe in name they doe agree, for thefe 24. are called the antients, fo are thefe 24. books called the old Teftament. Thirdlie in number they do agree, for thefe antiẽts are 24. fo there be 24 authentick books of the old Teftament, nominat by *Jerome, in prologogaleato*, & 24. writers therof, nominat by *Zanchius* & by vs, *Ap. 4.* note *d.* thereof. Fourthly, & finally, whatfoeuer is fpokẽ in the *Reuelatiõ*, in name & be halfe of any of the 24, Elders: the fame

Apo. 4.10.11.
1. reafon ab officio.

2. accord in names.

3. accord in their number.

4. accord in fpeach and teftimonie.

ſhal ye find ſpecially writtẽ in one of theſe 24. books of the old Teſtament, as particularly ſhall be noted in their dew place of our principall diſcourſe. As to the ſecond part of this propoſition, that vnder the name of theſe 24. ancients, not only the 24. bookes, and 24. writers therof, but alſo all the true profeſſours thereof be (*metonymice*) included, it is certain otherwaies the only books could not ſay, that Chriſt hath redeemed them, and that they raigne vpon earth. Ap. 5.9.10.

19. PROPOSITION.

The foure beaſts are the foure Evangels with al the true writers and profeſſours thereof.

THat theſe foure beaſts are on earth, is alſo proued by the 17. propoſition, in reſpect the glory of the whole throne, (whereof they are one coherẽt part) is vpon earth. We ſay now further, that they do repreſent the four Euangels for theſe reaſons. Firſt, *ab officio*, for that theſe foure beaſts doth here decore the throne, & neuer † ceaſed frõ praiſing God day & night. And what thing on earth doth more adorne Gods true throne & Chriſtiã religion, thẽ doth theſe four Evãgels & their true profeſſors, who neuer ceaſe frõ praiſing God continuallie? Secondly, they agree in number, for there be foure beaſts, ſo are there four Evangels. Thirdly, in their particular and diſtinct titles or faces, they agree, hauing conſideration of the cuſtome of the ancients, that vſed for to intitle their books according to the beginning thereof, as in Hebrue, *Geneſis* is called *Bereſchith*, becauſe it beginneth *Bereſchith bara Elohim*, &c. & ſuch like other books: as we alſo entitle our booke of lawes, *Regiam Maieſtatem*, becauſe it beginneth ſo. Then in compariſon, the face of theſe *Beaſts*, are compared to the titles or beginnings of theſe bookes: becauſe, as men or beaſts are readilieſt knowne & diſtinguiſhed by their faces: ſo are bookes by their titles, and beginnings. And

Reaſones

1. ab officio
Apoc 4.8

2. accord in number.
3 accord in their titles.

to come in particular, like as in the *Revelation*, the *Beaſtes*, and in EZECHIEL, their faces were, one like a *man*, another like a *Lyon*: the third, like a *Bullocke*, the fourth, like an *Eagle*: So, of theſe foure Evangeliſts, *Matthew* beginneth his firſt *face* or leafe, at the Genealogie of Chriſt, as he is a man: and *Marke* beginneth his firſt *face* or leafe at the voyce, crying (like a roaring Lion) in the wilderneſſe, *Prepare the Way of the Lord*, & c. And *Luke* beginneth his firſt *face* or leafe, at *Zacharias* offering incenſe (as it were a *Bullocke*) at the Altar. And *John* beginneth his firſt *face* or leafe, at the high and divin eſſence or Chriſts God-head, flying ſo high in his ſtile, that he is compared to an *Eagle*. Fourthlie, in their order of prioritie, according as they firſt wrot, they agree with *Ezechiels* firſt order, where the firſt was a mans face, that is *Matthew*, who firſt of all wrote, and that in Hebrew. The ſecond was a Lions face, and that is *Marke*, who was the ſecond that wrote, and that in greeke. The third in *Ezechiel*, is a Bullockes face, & that was *Luke*, who was the thirde that wrote, and that in Greeke. The fourth was the *Eagles face*, who is *John*, that wrote the fourth Euangel, and that in Greeke. And ſo the order of their firſt editions, agrees preciſelie with *Ezechiel* his order. Fiftlie, their order of tranſation or edition in Greeke, doeth agree with the order that here *S. Iohn* (who wrote both in Greeke, & to the Greeke Churches) ſetteth thẽ into: to wit, The Lion, *Marke* wrote firſt of al in Greeke. The *Bullocke, Luke* wrote ſecond in greeke: Then was the *mans* face, that is, *Matthewes* Evangel, tranſlated in greeke, who now is the third, and before was the firſt. Afterwardes, laſt of all, the *Eagle, Iohn* (as before) wrote the fourth Euangel. And ſo correſpondẽtly, as *Ezechiel*, the Hebrew Prophet, writing to the Hebrews, preferreth *Matthewes* Hebrew Euangel, for that it was firſt written. So Saint IOHN, a prophet, writing to the Grecians, in Greeke, ſpake of *Matthewes* greeke Euangel, and ſetteth it in the third order, for

Matthew a mans face.
Mark a Lyons face

Luke a bullockes face.
John an Eagls face

4. Reaſon accord in order.

5. Their accorde in their order of edition in Greeke.

Apoc. 6.1.35.7

that it came after both *Markes* and *Lukes*. Sixtlie, the precife time of their firft writing, and occafion why they did write thefe foure Euangelles, agreeth with the hiftory and time of the foure firft Seales, where their firft comming is mentioned. For firft, when the word of God paffed out on the white horfe in the firft feale, betuixt the yeare of Chrift 29. and 36. then (faith the text) one of the foure beafts faid, *come and fee*, &c. that is, *Matthew* the Apoftle, who came that time, and wrote his firft Euangel in Hebrew, exhibiting the fame to be feene of the *Jewes* exprefly, for converting them, that Gods word of his new couenant might victorioufly go out among them, to conquer and ouercome (as faith the text.) Then in the fecond feale, when *S. Steuen* and S. *James* were martyred, and great perfecution rofe againft the Church of God, *viz* betwixt the 36. and 43. yeare of Chrift, then (faith the text) the fecond Beaft faith, *Come and fee*: that is, the fecond Euangell was at that time fet forth by *Marke*, to be feene and read, in comfort of the afflicted Church. Afterward in the third feale, when hunger and dearth arofe, betwixt the 43. and 50. yeare of Chrift, the third Beaft faith, *Come and fee*: and this is *Luke*, whofe Euangel came to light at that time, and was fet foorth to be feene, and alfo writeth of the fame dearth, in his booke of the Acts of the Apoftles. Laftly, in the fourth feale, when deadly herefies arofe againft the Divinitie of Chrift, betwixt the 50. and 57. yeare of Chrift. Then (faith the text) came the fourth Beaft: and fo indeed at that time did *John* write the fourth Euangell, beginning at the defcription of Chrifts Diuinitie, † expreffelie againft the faid Hereticks. Seuenthly, thefe foure Beafts wings, eyes, and all their other tokens and circumftances, contained both in *Ezechiels* prophecie, and here in the *Reuelation*, doeth fo properlie agree with the foure Euangelifts, that they cannot be fo conueniently attributed to any other, as fhall be fhewed at length in the principall Treatife. Eightly, it

6. Accord with the hyftorie and prophecie in time and purpofe.

Act. 11.28.

† *Cyrillus in his preface vpon S. John, & Irenaus cõtra herefes lib. 3. ca. 11*

7 The accord of all their properties and tokens.

8. Reafon for the cõmoditie refulting hereof.

appeareth by *Hierome* and others learned, that thefe indices and tokens, fo aptlie agreeing to thefe foure Euangelifts, and to their *faces*, haue confirmed the ancient Fathers of the Church, both to receive thefe foure Euangels true and certaine, as alfo to repell many falfe Euangels, written in the name of *Thomas, Matthias, Bartholomæus*, and other Apoftles. Alfo the Euangels of *Bafilides, Apelles*, and *Nicodemus*. For *Luke* teftifieth, that many pretended to write Euangels. Then certainely, feeing there could not, nor cannot redounde to the Church of God fo great a commoditie, by interpreting thefe foure *Beafts* any other waies, we haue great occafion, by all thefe forefaid reafons, to repofe vs alfo vpon this interpretation. As to the fecond part of this propofition, that vnder the name of thefe foure *Beafts* or foure Euangels, both the writers and the profeffors thereof are *metonymice* included: it is certaine, otherwife (as in the difcourfe of the former propofition, is declared) they could not fay, that Chrift hath redeemed them, or yet that there arofe* famine among them, for neither can that redemption, nor fuch famine properlie come to any but to men.

Luke 1.1.

Proofe of the fecond part.

Apoc. 5.9.
Apoc. 6.6.

20. PROPOSITION.

Gods Temple, although in heauen, is alfo taken for his holy Church among his heauenly Elect vpon earth, and [*metonymicè*] *for the whole contents thereof.*

IN figuratiue fpeakings, heauen is taken for Gods holie Elect vpon earth, whom he hath preordained to inherite heauen, as is alreadie fhewed by the 17. propofition, wherfore, by the throne of God in heauen, is meaned his trueth and true religion in earth among his heauẽly Elect here: So in this place alfo wee fay, that Gods *Temple* in heauen, is more properly his Church among his heauẽly Elect vpon earth, than among the glorified bodies in heauẽ: for that glorified number, called the † *new Jerusalem*,

Proofe of the firft part.
Apoc. 21.22.

are ſaid properly, to haue no diſtinct *Temple* among thẽ, for *John* ſaith therof. *Et templũ non vidi in ea*, &c. *I ſaw no temple therin* &c. So, no *Temple* being there, this *Tẽple* properly muſt be amõg Gods heauenly Elect vpon earth, and conſequently his holie and true Church. As to the ſecond part of this propoſition, it agreeth verie well with Chriſts ſpeaking, ſaying of the materiall Temple, *Qui iurat in altari, iurat in eo & in omnibus quæſuper illud ſunt, & quicunque iuraverit in tẽplo, iurat in illo & in eo qui habitat in ipſo: Whoſoeuer ſweareth by the altar, ſweareth by it and by all things vpon it, and whoſoeuer ſweareth by the Temple, ſweareth by it, and by him that dwelleth in it.* Euen ſo likewiſe, vnder the name of this ſpiritual *Tẽple*, is not only meant Gods true Church among his heauenly Elect vpon earth, but alſo is meaned *metonymice* the whol cõtents ther of, to wit, Gods truth & true religion: yea (althogh leſſe properly) euẽ the very Majeſty of God himſelf becauſe he dwelleth in his Church, and they in him: for proofe whereof, though *Iohn* ſaith firſt, *I ſaw no temple therein*, becauſe there is no diſtinct Church properlie, where all is the Church, yet doeth he ſubjoyne immediatlie, *Dominus enim Deus omnipotens templum corum eſt & Agnus: For the Lord God almightie and the Lambe are the temple of it*: calling the Deitie to be their Church that are glorified, or rather to them in ſtead of a Church: although more properly the Deitie is the thing contemplate, than the *Temple* it ſelffe. From this Temple in this ſenſe, that is, from the majeſtie of God, are ſaid to goe out the laſt Angels of Gods great harueſt and latter judgement, to gather in his Elect, and to deſtroy the wicked. And ſthe *Temple* of God in heauen, is properlie taken for his heauenly elect Church vpon earth, and for their true doctrine, profeſsion, and religion: yea, and (although leſſe properlie) euen for the Dietie itſelff. Herevpon followeth this corollar, agreeable with al the former ſenſes: that when this *Temple* is ſaid to be open, then it meaneth Gods true Church

Proofe of the ſecond part.
Math. 23.20.21.

The temple metonymice for the contẽts thereof.

Apo. 14.15 17.
A corollar
The open temple is the Church to bee viſible, & contra.

to be viſible, his trueth, and true religion preached and opened vp, and finallie, the Majeſtie of God to be knowne and reuealed by that preaching of his truth. And when againe, that *Temple* is cloſed or filled with fume, it ſignifieth Gods true Church to lurke, and become inviſible, his trueth, and true religion, and knowledge of the Deitie and divine Maieſtie to be obſcured, darkened, and cloſed vp. As in confirmation hereof, it is ſaid, † That the T*emple* of the Tabernacle of the Teſtimonie was open in heauen, and the ſeuen Angels of the laſt plagues come out of the Temple, and afterwards no man was able to enter into the Temple againe till theſe ſeuen plagues were fulfilled: which agreeth preciſely with the euent after all the former ſenſes, to wit, that the true Chriſtian Churches, profeſſors of Chriſt his true teſtimonie, were viſible and patent, their true doctrine and Chriſtian religion opened vp, and the majeſtie of God thereby made knowne, and manifeſted in the daies of the Apoſtles: and afterwards, vpon neglecting their teſtimonie, and vpon the contempt of the manifeſtation and opening vp thereof, there proceeded all the whole plagues of the ſeuen laſt ages, during the which, the fume of Gods wrath was ſo great, that by the Antichriſt he made his true Church lurke and become inviſible, and his trueth and true religion, and knowledge of the divinitie to be obſcured and darkened: ſo that none might viſiblie enter or haue acceſſe to the knowledge thereof, till by the ſeuenth & laſt plague, that the Antichriſts kingdome began to fall, and then after 1260. years darkenes, was that true Temple of God opened & made manifeſt, as is plainely ſaid in the † ſeuenth Trumpet or age. So then, this Temple, the opening, cloſing, and opening againe thereof, agrees ſo well with the Eccleſiaſtical hiſtorie in all points, after the former ſenſes, that wee muſt conclude, by this Temple of God in heauen, to be meaned his holie Church among his heauenly elect vpon earth, with the whole contents and pertinents

A confirmation & inſtance.
Apoc 15.5.8.

The Antichriſtian Church viſible, and true church inviſible, and latent 1260. years

The true temple opened.
Ap. 11.19.

thereof, to wit, his trueth, true religion, and (after a maner) the verie Deitie it felffe.

21. PROPOSITION.

The two witneſſes mentioned [*Reu. II*] *are the two Teſtaments, and* [*metonymicè*] *the whole true profeſſours thereof.*

THis is proued by their number, name and office and by the whole circumftances of the text. Firft, by their number, they are two witneffes: fo are they two Teftaments. Secondlie, by their name, they are called *Witneſſes*: fo in that language of Latine, wherein they haue bene vfed moft, thefe 1200. years and more, they are called *Teſtamenta* from the word *Teſtis*, which is to fay, a *Witneſſe*, as being witneffes of Gods will: all other doubtful teftimonies of men being refufed, as teftifieth Chrift, faying, **I receiue not the teſtimonie of men*, &c. But afterward he faith, *Search the Scriptures for they are they which teſtifie of me.* Thirdly concerning their office, (as Oliues) they annoint vs Kings and Priefts to God: And (as Lanternes) they illuminat vs with knowledge of the true way of faluation. So that who (as the text faith) would peruert them, God fhall confume him with the æternall fire of hell: therefore it is faid, Ier. 23.29. *Is not my word euen like a fire* (faith the Lord) *and like a hammer that breaketh the ſtone?* And Ier. 5.14. *I will put my words into thy mouth like a fire, and this people ſhall be as wood, & it ſhall deuoure them.* Laftly, the whole circumftances of the text (which here for breuitie are omitted, and are amplified in the principall Treatife) doe fo harmonically agree with thefe two Teftaments, that neceffarily, they be the two *Witneſſes* here made mention of. As concerning the witnefsing of men, becaufe it is faid, *Omnis homo mendax: Euerie man is a liar:* And againe. *Non ab homine teſtimonium accipio: I receiue not the record of man*: Therefore, no men are fimply

Reaſons

1. By their number.
2. By their name.

Ioh. 5.34.39.

3. By their office.

Ier. 23.29.

Ier. 5.14.

Laſt reaſõ accorde of circumſtãces.

Pfal 116.11.

Rom. 3.4.
Ioh. 5.1.34.

to be accepted in place of thefe *Witneffes*. Yet notwitftanding, fo far as men do profeffe thefe Teftaments, and do purely preach the fimple doctrine thereof, in that cafe thefe men, are *metonymice* included vnder thefe Teftaments, for that their worde is not their owne word, but the word of thefe two Teftaments.

22. PROPOSITION.

The Woman clad with the Sunne [*chap. 12*] *is the true Church of God.*

THis is proued, firft, by the Song of *Salamon*, wher *Salamon*, bearing the figure of Chrift, his beloued Woman and Spoufe, throughout all that Canticle, beareth the figure of Chrifts Church. Secondly fpirituall *Hierufalem*, which is Chrifts Church by diuerfe Scriptures, is alfo called Chriftes Spoufe in the *Reuelation*. Thirdly, bodily marriage, is by Saint *Paul* called a Symbol, and a Sacrament of the vnion of Chrift, and his Church: wherby the husband reprefenteth Chrift, and the Woman efpoufed reprefenteth the Church. Fourthly, in the whole Scriptures, idolatrie being called fpirituall whoredome, neceffarily the true worfhipping of God is reprefented be perfect Spoufage, and the true Church that worfhippeth him, is his Spoufe, and fo the Church of God is figured by a Woman. Laft of all, the whole tokens of this Woman, contained in the text fo viuely and perfectly agree with Chriftes Church (as is declared in the principall treatife) that neceffarily wee muft conclude, this Woman to meane the true Church of God.

1. Reafon, by Salamõs Canticle.

2. Chrifts Church called Chrifts fpoufe.
Apoc. 21
Eph. 5.23.
3. the fymbol of mariage.

4. Idolatry called Whordome

Laft the accord of tokens.

23. PROPOSITION.

The whore who in the Reuelation is ftiled fpiritual Babylon, is not reallie Babylon, but the very prefent citie of Rome.

Apoc. 14.16.17.18
Cap. 19.

IN the former propoſition was deſcribed the holie Spouſe of Chriſt, here is to be deſcribed the filthie Whoore of Sathan, there that Ladie, who is adorned with the Sunne, Starres, and heauenly vertues: here that Adultreſſe, who glorieth in gold, ſiluer, precious ſtones, and worldly pleaſures: there ſhe who is perſecuted by the Dragon, here ſhee vnto whom the Dragon doeth giue authoritie: There ſhe who is chaſed into the wilderneſſe, and hath no lodge to hide her in: here ſhe who impireth aboue all people, and is the Metropolitane citie of the world. And finallie, ſeeing in all things this Whoore or whooriſh *Babylon*, is contrarious to Chriſts holie Spouſe, let vs, and all Chriſtians trie her out, as our deteſtable and deadlie enemie, and ſee what *Babylon* ſhe is. Wee ſay then, that this *Babylon*, is not that reall *Babylon* of *Chaldee*, but *Rome*, for theſe reaſons. Firſt, for that this *Babylon* is called † *myſterium Babylon*, that is to ſay, myſticall or figuratiue *Babylon*: Therefore, it is not *Babylon* it ſelfe in *Chaldee* Secondlie, that olde *Babylon* in *Chaldee* was deſtroyed long before Saint *Iohn* wrote this, as was prophecied by *Eſay*, by *Ieremie*, and by *Dauid*, and neuer roſe to authoritie: But this *Babylon*, euen when Saint *Iohn* wrote, raigned ouer the † Kings of the earth: Therefore, it is not olde *Babylon*, but meaneth *Rome*, which at that time, and a thouſand and foure hundreth yeares after that, had Empire ouer the whole earth. Thirdly, this myſticall *Babylon*, is ſaid by the text, to be ſet vpon ſeuen hilles, and therefore it is *Rome*, which is called *Septicollis*, as VIRGIL teſtifieth, *ſaying*,

Apoc. 17.5

1. Reaſon. Not Babylon, but myſtical Babylon.

2. Old Babylon of old deſtroyed.

Eſa. 13.17.
Iere. 51.11
Pſal. 136.
Ap. 17.18.

This Babylon yet raignand.

3. This Babylon ſeauen-hilled, Rome ſeuẽ hilled.

Apoc. 17.9

Septemque una ſibi muro circundidit arces:

Within the walles of that citie, contained are ſeuen hils hie

Virgil, lib. 2 Geor, & lib 6 æneid.

And *Propertius* ſaith,

Septem urbs alta iugis, toti quapræſidet orbi,

On hils ſeuen that city ſtands, that hath empire aboue all lands.

Propert. lib 3. eleg. 10.

Againe, Saint † *Ierome libro Epiſtolarum*, in diuers places approoueth the ſame: and *Eutropius* in

parte, 2, tra. 5, Epiſt, 59 & part. 3,

his Hiſtorie, & *Publius Victor de deſcript one urbis Rome*, and diuerſe other doe name theſe ſeuen hilles this way, as *Palatinus, Cœlius, Tarpeius* (otherwiſe called *Capitolinus*) *Aventinus, Quirinalis, Viminalis*, and *Eſquilinus*. Fourthly, becauſe the manner of Families, Kings, and Kingdomes, is to entitle themſelues by the name of the firſt Conquerours, or moſt notable perſons, as the whole *Romane* Emperours were called *Cæſars*, according to the name of the firſt Emperour, at the leaſt, firſt founder of the Romane Empire *Julius Cæſar*. So *Rome* hauing obtained the Monarchie ouer the whole kingdomes of the earth, whereof the *Babylonians* were the firſt Conquerours, therefore, juſtlie is called *Babylon*, as ſucceeding in her place. Fiftlie, *per Jcona, ſeu a ſimili*, for that in idolatrie, pride, couetouſneſſe, and exceeding crueltie againſt the Saints of God, *Rome* was nothing inferiour to *Babylon*. Sixtlie, becauſe by vſe and cuſtome of ancient and learned writers, *Rome* is very often called *Babylon*, as by *Jerome lib. Epiſtolar*. Alſo by *Tertullian* in his booke againſt the *Jewes*, and *lib. 3*. againſt *Marcion*. And for final confirmation hereof, al they that would proue *S. Peter* to haue remained at *Rome*, ſay, that by *Babylõ* in the end of the firſt epiſtle of *Peter*, is meant *Rome*, from whence he then wrote, as *Jerome* ſaith, in his Prologue, vpon *Marke*, and in *Catalogo Scriptorum Eccleſiaſticorum*, in theſe words, *Et Petrus in Epiſtola primaſub nomine Babylonis, figuraliter Romam ſignificans: And Peter in his firſt Epiſtle, vnder the name of Babylon ſiguratiuely ſignifying Rome.*

And ſo for aſſured certainty, this whoriſh and myſticall *Babylon* is verily *Rome*. But ſome ſeeking ſophiſtical ſubterfuges will ſay, that this *Babylon* meaneth olde Rome, and not this preſent Citie of Rome, that now is, as though both were not one, euen builded both within one wall, and vpon theſe ſelfe ſame ſeuen mountaines: and finallie, one ſelſfe-ſame towne, both in name, ſeate, and gouernement, and in all things elſe, onely that where

tract. 8. Epiſt. 30.

Names of the 7 hilles

4. Rome called Babylon as heire and ſucceſſour to Babylon.

5. Be their reſemblance in all vices.

6. Rome vſually ſtiled Babylon.
part, 2. tract 6 Epiſt, 80 & part 2. tract. 5 Epi. 9. & part. 3 tract. 8. Epi. 30.

Concluſion

one houſe was ruined, deſtroyed, or burned by the *Gothes*, and other enemies, another within the ſame wals is builed vp againe. Of theſe then it is to be demanded, if olde Rome hath bene ſo deſtroyed, that it is not to be found, or that neuer inhabitant hath dwelt in it ſince that time, neither euer ſo much as a candle hath ſhined therein hitherto, as is prophecied of *Babylon* in the Revelation, chap. 18. But the contrarie being true: Therefore, it is not onely this olde Rome, but all whollie Rome that is called *Babylon*, and as yet, reſteth to be that way for euer deſtroyed.

Not olde Rome but this Rome reſts to be deſtroied.

Not a repared deſtruction, but an vnreparable yet to come prophecied.

24. PROPOSITION.

The great ten-horned beaſt, is the whole bodie of the Latine Empire, whereof the Antichriſt is a part.

For proofe hereof, the moſt notable tokens aſſigned by the text to this Beaſt, doeth onely agree to the *Latine* or *Romane* Empire. For firſt (ſaith the Text) the Woman that ſate vpon this Beaſt, is the great Citie, that ſitteth ouer the Kings of the earth, So the chiefe ſeate and citie, of the *Latin* or *Roman* Empire, is that great citie Rome, that had Empire ouer all the kindomes of the earth. Secondly, ſaith DANIEL, *There ſhall foure chiefe kingdomes ariſe vpon earth, vnder the figure of foure beaſts.* Whereof (by plaine interpretation hee ſayeth) the fourth beaſt that had theſe ten hornes, is the fourth kingdome of the earth: And ſo it is, that the firſt great kingdome or Monarchie being of the *Babylonians*: The ſecond of the *Medes* and *Perſians*: The third, of the *Græcians*. † The fourth and laſt, is certainely the Monarchie of the *Latins*, or *Romans*: and therefore, that fourth beaſt which both there in DANIEL, and here in IOHNS *Reuelation*: hath ten hornes, muſt neceſſarilie be the *Romane* or *Latine* Empire. Thirdly, this *Beaſt*, whereon the ſaid whooriſh *Woman*, or *Babylonical citie* ſitteth,

Apo, 17 18
1. Reaſon. Rome is the great citie. &c.
Dan. 7.17.

2. The fourth beaſt ten-horned is the fourth kingdome.
Dan. 7.13.

† The reaſones of this Romane Monarchie are to be read in our ſolution of doubts.

3. The ſeuen heades and ſeuen hilles.

hath feuen heads, which the text † interpreteth to be feuen *mountaines*: Euen fo *Rome*, the chiefe citie or *Metropolitane* feate of the *Latin* Empire, is fet vpõ feuẽ hils, nominated and prooued in the third reafon of the former propofition. Fourthly, there be here of feuen Kings (faith the text) that is, feuen forts of royal gouernements: for a *King* is oft taken for a whole race of Gouernours of one fort, as in *Daniel* cap. 7. So had *Rome* feuen royal gouernments, to wit, *Kings*, *Confuls, Dictators, Decem-virs, Tribuns, Emperours*, and *Popes*: Of which (faith the text) fiue are fallẽ, one is, & another is not yet come. Euen fo was it indeed that before S. *Johns* daies, thofe fiue gouernmẽts, *viz.* the *Kings: Confuls, Dictators, Decem-virs*, and *Tribunes* were paft, and in his dayes raigned the *Emperours*, & the *Popes* were not come to the gouernement while after his dayes. Fiftlie, (faith the text) this *Beaft* that Saint *John* did fee, was, and is not, and fhall rife of low eftate, and fhall goe into decay, *Was* (faith hee) becaufe the *Romane* Empire, & gouernment of *Emperours*, was before that time, that he faw this viſiõ. *Js not* (faith he) becaufe the time wherin he faw this vifion, the Emperor *Domitian* was dead, & the Empire vaked, a fhort time, for the next Emperor *Nerva Cocceius*, was not yet creat, as appeareth by *Irenæus*, who faith that S. *John* faw thefe vifions, in the end of the raigne of *Domitian*. Then (faith he) He *fhall rife of lowe eftate*: Euen fo arofe there a new Emperour, not of the noble ancient blood of Romanes, but of bafe eftate, to wit, the faid *Nerua Cocceius*, who was the firft ftranger, or Barbarian *Emperour*: This (faith he) *fhall goe into decay*: that is, fhall goe fhortlie into decay: for otherwife fimplie, all flefh muft goe at fometime into decay: And fo was it indeed, that this *Nerua* goeth fhortly into decaie: For he liued Emperour but one yeare, three moneths, and nine daies, and then he died: Againe, the *Beaft* (faith that fame text) *Was, and is not, and yet is*, that is to fay, there were Emperours before S. *Iohn* faw thefe vifions, none when

† Apo. 17.9

4. The feuen kings. Apoc 17.

The 5 that were fallen emperours then extant.

Popes then vncomde

5. Reafon & 5. token

Tefte Aurelio victor & Achille P. Gaffaro & Dione Nicæo

he ſaw them, and yet when hee ſaw them, the Empire ſtood, and the gouernment by Emperours was vnaboliſhed: for, incontinent ſtarted vp *Nerua*, and manie moe Emperours after him: ſo that the Empire might be be ſaid at that time, in a maner not to be and in a maner to be. Sixtlie (ſaith the text) the Beaſt that was, and is not, is euen the eight, and is one of the ſeuen: and that meaneth, that that royall gouernement of the Empire by Emperours, which was immediatly before S. *Iohn* ſaw theſe viſions, and vaked when hee ſawe them, by the deceaſe of DOMITIAN that ſame gouernment ſhal be the eight gouernement, and was alſo one of the ſaid ſeuen gouernments: and euen ſo was it, that after the Romane Emperours (who was the ſixt gouernement) were aboliſhed, by the *Hunnes, Gothes*, and *Vandalles*, and thereafter, *Rome* ſeuenthly gouerned by the Popes: then eightly, ſtart vp by *Charlemaign* a newe race againe of Emperours, intituled of *Germanes* and *Romanes*. So that evidently the Emperours were the eight eſtate that gouerned *Rome*, and were alſo one of the former ſeuen, euen the ſixt gouernment thereof, as is alreadie ſaid. Seuenthly (ſaith the text) one of theſe ſeuen heades receiued a deadlie wound, and that deadlie wound was heald: So the chiefe of the ſeuen gonernements of the Romane Empire, to wit, the Eſtate of Emperours was cut off and aboliſhed by the *Hunnes, Gothes*, and *Vandalles*, and vaked from the daies of *Auguſtulus*, Anno Chriſti 475. vntill that *Charlemaign* came, in Anno 806. At which time, the eſtate of Emperours was of new repaired, and whollie made vp againe. Eightly, this beaſt hath ten crowned hornes, that is, ten Kings (ſaith the text) which when Saint *Iohn* wrote, had not receiued their kingdomes, but ſhould receiue their kingdome at an houre after the beaſt, & then (ſaith he) ſhall they giue their power againe to the Beaſt, and all at once ſhall make warre with the *Lambe*: but at length (ſaith hee), theſe ſhall hate and deſtroy that harlot beaſt: Euen ſo came it to paſſe of the Roman Empire, that long after S. *Johns*

Sixt reaſon & his tokens.

7. The deadlie wound healed.

Emperours reſtored.

8. Reaſon His tenne hornes and ten Kings declared, Apoc. 17 note l. therof.

dayes when that old Empire began to fall, and the eftate of Emperours to be abolifhed, (as is faide) immediatly thereafter, fprang thereof ten Chriftian Kings, who againe gaue their power, afsiftance and fortification, to the next Governours of the newe Latine Empire, which was to the Pope, authorizing him, as their fupreame head, with whome they concurred, fighting againft Chrift, as hereafter fhall be fhewed: But at length fhall thefe tenne Chriftian Kings, deftroy that fpirituall harlot and idolatrous Empire, as (praifed be God) *England* and *Scotland*, with fome others haue already begun. Ninthly, that beaft is the Latine Empire, and the chiefe feat thereof is Rome: becaufe in the text, both the plaine name of the Latine Empire is figuratiuely expreffed, & the figuratiue name of *Rome* is plainely expreffed. As to the firft, how the plaine name of the Latine Empire, which is $\lambda\alpha\tau\epsilon\iota\nu o\varsigma$, is figuratiuely expreffed, read the 29. propofition following. As to the fecond, how the name of *Babylon* (which is plainly faid to be the figuratiue name that was written in the Womans fore-head that fate on the beaft) is the perfect, certaine and accuftomed figuratiue name of that citie Rome, that fitteth ouer the Latine Empire, is moft euident by the former propofition.

9 Reafon. The plane name of the Empire is figuratiuelie and the figuratiue, name is plainly expreffed. Apoc. 17.5.

Finally, all the reft of the fmaller circumftances of this tenne-horned beaft, contained in the text, doe fo properly and fitlie agree with the Latine Empire as in the principall treatife fhall be found at length) that we muft neceffarile conclude, that felfe-fame Beaft to meane and fignifie the whole Roman Empire.

Proofe of the fecond parte.

Refteth then the fecond part of this Propofition, to prooue the Antichrifts Kingdome to be a part of the great *Romane* or *Latine* Empire. Seeing then it is prooued, that this whole ten-horned beaft fignifieth the whole Romane Empire. Neceffarilie, it that one part of this beaft fignifieth, muft alfo be a part of the *Romane* Empire: But one part of this beaft fignifieth the Antichrift: Therefore the

Antichrift muft needs be a part of member of the whole *Romane* Empire. That one part of this beaft, to wit, † the litle blafphemous horne thereof in *Daniel*, and the blafphemous mouth thereof in the *Revelation*, doeth fignifie the Antichrift: it is certaine by the tokens fet downe in the text, wherein it is faid, that that *blafphemous* †*mouth and* † *horne fhould fpeake proude blafphemie againft God, and againft his Saints, and againft them that be in heauen: And fhould change times and lawes, and fhould make warre againft Gods Saints*, and ouercome them, and fubdue them vnder his tyrannie, for the fpace of 1260. propheticall daies, prooued to be yeares in the 1. and 16. propofition. So this muft neceffarily be the Antichrift, that hath fo wicked and fo long a raigne: And confequentlie, the Antichrift muft rife of the Romane Empire, and be a member thereof. Secondly, his raigne muft be of that Romane Empire, for that he muft fit in fpiritual *Babylon*, which by the former propofition, is prooued to be Rome. And fo Rome fhould become the denne of all diuelifhneffe and Antichriftian errours, as in the principall Treatife vpon the 17. and 19. chap. fhall be fhewed.

1 reafone
† Read the proofe here of in our folution of doubts.
Ap. 13.5.6
Dan. 7, 25
2 reafon.

25. Proposition.

The two horned Beaft, is the Antichrift and his kingdome, it alone.

P*Aul* defcribing the Antichrift and his kingdome, among certaine fpeciall notes to know him by, giueth many of thefe fame tokens of this two horned beaft: to wit: heere in *John*, this beaft hath two hornes like the *Lambe*: that is, double power, fpirituall and temporall: fo there in *Paul*, † he fheweth himfelfe to be like God, and hath temporall power, whereby he is aduerfarie to Gods feruants: and fpirituall power to worke wonders, albeit lying wonders. Secondly (faith *John*,) he fhal fpeak like the *Dragon*: fo faith *Paul*, he fhall come by the working of *Sathan*, &c. Thirdly, (faith *John*,) he

2. Theo 2.
Apo. 13.11
1. Reafon is the beafts conformities with Chrift in fhew, and with the dragon in effect is the fecond reafon.
† 2. Theo. 2.4.9.
Apo. 13.13 14.

wrought great wonders and ſignes, and deceiueth men thereby: So in *Paul*, he came with power, and ſignes, and lying wonders & in al deceaueableneſſe of vnrighteouſneſſe. Whereby (doubtleſſe) that man of ſinne, and ſon of perdition, euen the very Antichriſt, whome *Paul* their deſcribed, is the very ſame two-horned beaſt, mentioned here by *John.* † And for confirmation hereof, the properties of this two-horned beaſt (cap. 13) are the ſelfe-ſame properties of the falſe Antichriſtian Prophet (cap. 19) For this two-horned Beaſt is ſaid to worke great miracles, before the firſt tenne-horned beaſt, and to deceiue them that dwelt vpon the earth, whom hee cauſed to worſhip the image of that firſt beaſt and to receiue his mark. Euen ſo, the falſe Prophet, which muſt needes be the Antichriſt, is ſaid alſo to worke miracles before that firſt beaſt, wherby he deceiued thẽ that receiued that beaſts marke, and worſhipped his Image. Wherefore neceſſarilie, as of this two-horned beaſt, and of that falſe Antichriſtian Prophet, all the tokens be one: ſo moſt alſo conſequentlie they them-ſelues be both one.

3. His lying wonders.

† Their ſẽblable properties. Apoc. 13.14.15. Ap. 19.20.

26. PROPOSITION.

The Pope is that only Antichriſt, prophecied of, in particular.

AMongſt the pluralitie of Antichriſts, ſpoken of in the Scripture to bee generallie in all ages: there is one particular Apoſtatik kingdome, who is the chiefe and principall of all Antichriſts, and is that great Antichriſt, whom *Paul* calleth the *man of ſinne, and ſonne of perdition, adverſarie to God, and an extoller of himſelſe aboue all that is called God*, with diverſe other epithets contained, *2. Theſ 2.* Leauing therefore all other ſmaller Antichriſts, this great Antichriſt, and chiefe head of all Antichriſts, is hee whom here we haue to trie out, whõ (for remouing of al doubts) we ſay cannot be the *Mahomet*, neither any *Turke*, *Jew* or

One chiefe antichriſt.
This Antichriſt is not the Mahomet, and why,

Ethnick. Firſt, becauſe it is not apparant, that the Spirit of God, would trauel ſo carefully, to point them out vnto vs, by dark myſteries, and ſecret ſigns to be the Antichriſt, who when they come, doe with moſt cleare confeſsion, grant and advouch thẽſelues to be Antichriſtians, as all *Jewes*, *Turkes* and *Ethnicks* do plainely confeſſe: and therefore their owne confeſsion is ſufficient to knowe them by, and all further ſecrete tokens, were ſuperfluous. Then muſt he needs be a latent, and not a patent enemie, that the Spirit of God ſo carefully points out: and ſo no *Turke, Jew* nor *Pagan*, yea, no plaine Wolfe muſt hee be: but a Wolfe in a Lambs skin, euen an Antichriſt vnder the name of a Chriſtian. Secondly, he muſt ſit, ſaith *Paul.* In the Church of God: Therefore an alledged Chriſtian muſt he be: for no *Turke, Jewe*, nor *Pagan* hath other ſitting, than in their owne Temples: theſe be not Gods temples: onely the ſocietie of Chriſtians is Gods temple: of their number therefore muſt the Antichriſt call himſelfe: and among them muſt he ſit, as a pretended Chriſtian: and conſequently muſt he be no *Iew*, *Turke*, nor *Ethnick*. Thirdly, a part muſt his raigne be of the *Latin* Empyre, as is prooued (*Propoſition* 24.) and one pretending friendſhip thereunto, yea, he muſt ſit in that ſpiritual *Babylon*, prooued to be *Rome*: Then muſt he not be the *Mahomet*, nor any *Turke*, or other *Ethnicke*, but a pretended Chriſtian Prince. Finally, and in a word, this Antichriſt is no *Turke, Jew*, nor avouched *Ethnick*, becauſe on the one part, all the notes and tokens of the Antichriſt, giuen vs by the ſcriptures, are moſt evident properties of the Pope, and doth altogether agree to him. And on the other part the ſpecial points of doctrine & converſatiõ of Chriſt and his Apoſtles, are altogether turned to the contrarie into his perſon: Him therefore ſhall we proue both by the ſyntheſis or ſympathie, that his properties hath with the Antichriſtiã tokẽs, as by the antitheſis or antipathie, that he hath ẅ Chriſt, to be the very ſelf great antichriſt prophecied of in particular.

1. Reaſon. No advouched adverſarie is this Antichriſt.

The antichriſt muſt be a latent enemie. Mat. 7.15. & ac. 20.29

2. He muſt ſit in the Church of God. 2. Theſſ 2.

3 His raign muſt be a part of the latine Empyre. The popes properties agreeing to the Antichriſts, & repugnant to Chriſts. Simpathies betuix the Pope and antichriſt.

And firſt, to account his notable Sympathies with the Antichriſt. Rome (vnder the name of Babylon) is the Antichriſts ſeate: ſo in Rome doth the Pope ſit. In the † Church of God, muſt the Antichriſt ſit: ouer Chriſtian Churches doeth the Pope ſit, profeſsing himſelfe their ſupreame head. The Antichriſtian kingdome muſt be a part of the *Latin* Empire, and therof muſt it proceid (propoſition 24) ſo is the Popes kingdome a part of the latine or *Romane* empyre, for in *Rome* doeth he ſit. DANIEL neere the end of the 11 chap. ſpeaking inwardly and occultly of the Antichriſt, & outwardly of his ſymbol *Antiochus* ſheweth that he ſhall fixe his tent and dwelling betuixt the ſeas, and ſhall worſhip the god *Mauzim*, whom his fathers knew not. So doth the Pope make his reſidence at Rome, & in Italie, betuixt the *Hadriatick* & *Mediterran* ſeas, & hath erected a new and vnknown *Artolatrik* Idol, which he calleth (*Miſſam.*) Chriſt ſhewes that there ſhal ariſe falſe Antichriſtian Prophets, who ſhall ſay, there Chriſt is in the deſert, or he is in the ſecret places, &c. So the Pope and his Clergie ſaith, that there Chriſt bodily is that hoſte, which ſometime they carrie in proceſsion in deſert and barren fields, to bleſſe the fruits therof, at other times they cloſe it vp againe in their ſecreet box beſide their altars. The Antichriſt ſpeaking lies through hypocriſie, ſhal † forbid all his to marie, & ſhall command abſtinence from certaine meates: ſo doth the Pope vnder hypocriſie, and lying pretext of chaſtitie and abſtinence, forbid all his Cleargie to marrie, and to eate fleſh in Lent, or on Frydaies. The falſe Antichriſtian Prophets ſhall priuilie bring in damnable errours, thereby denying the Lord that redeemed them, and through couetouſneſſe, they ſhall with fained wordes, make marchandiſe of Chriſtians (as ſaith *Peter*) and ſhall ſell their ſoules, as ſaith the *Revelation*: ſo hath the Popes Clergie brought in pardons and indulgences, whereby they promiſe remiſsion of ſinnes, and the kingdome of Heauen, priuilie and in

1. Rome the antichriſtes ſeat.
2. Ouer chriſtiã churhes muſt the antichriſt ſit.
† 2 Theſ. 2.
Lib. 5. ſexti can. fœl cis. in gloſſ& lib. 3 ſexti tit. 16. can. periculoſo
3 the Antichriſts kingdome.
4. The antichriſts dwelling and artolatrie.
Mat 24.26
5 Reaſon. Many Chriſts heere and there are the many hoſts heere and there.
6 Mariage forbidden and choiſe made of meates.
1 Tim. 4.3.
7 Inducing of damnable errours.
2. Pet. 2.
Apo. 18 13

effecte thereby, denying CHRIST to bee their onelie Redeemer, and ſo haue they through covetouſneſſe vnder fained pretext of religion, made merchandiſe of poore Chriſtians, ſelling by ſuch deuiliſh wares, euen their ſoules to the Deuill. The Antichriſt, the ſonne of perdition ſhall extoll himſelfe againſt all that is called God, ſo doeth the Pope ſtyle himſelfe God, and cauſeth Emperours and Kings to kiſſe his feete. And Pope *Alexander* the third trod with his feet vpon the neck of the Emperour *Frederick Barbaroſſa*, his clergy ſinging, *Super aſpidẽ & Baſiliſcum ambulabis & Leonẽ & Draconem pedibus conculcabis*: In *Daniel*, that blaſphemous Antichriſtian horne, that ſprang of the fourth monarchie, and in the *Reuelation*, that blaſphemous mouth that was giuen to the *Romane* beaſt, had the ſaints of God vnder his hand, impiring ouer them and perſecuting them 1260. prophetical daies, proued by the firſt and 16. *Propoſitions*, to be 1260. yeares: ſo the Popes kingdome which is of the *Romane* empire, hath had power and authoritie ouer all Chriſtians, & of theſe that were moſt true Chriſtians, ſuch as *John Hus*, *Hieronymus Pragenſis*, *Hieronymus Savonarola*, & of many thouſand moe, hath he bene a perſecuter theſe 1260. yeares, euen from the time that † *Syflueſter* the firſt, and firſt Pope vnmartyred receiued his three crownes and large patrimonie of the Emperour *Conſtantin*, as is alledged betwixt the yeare of Chriſt 300. and 316. to the yeare of God 1560. which time the notable decay of his kingdome began. † The Antichriſts comming (ſaith *Paul*) is by the working of *Satan*, with all power, and ſignes, and lying wonders in all deceiuableneſſe. So of Popes, 22. knowne (ſaith *Platina* and others) † came to the Popedome, being abhominable Necromancers, and excerciſed their mightie imperiall power ouer whole Chriſtianitie, ſubdewing, and ouer-whelming the moſt mightie Princes and Emperours thereof: And whereas *Paul* ſaith, that he ſhall worke with ſignes and lying wonders in

8. Extolling himſelſe aboue al Gods.
2. Theſſ. 2.
Diſt. 96 cauſatis.
euiden ter.

9. His tyrannie thir 1260. years
Dan 7.25.
Ap. 13.5
† Diſt. 96
10 His ſorcerie and lying wonders,

2. Theſ. 2.
† *Read this and the vnſuſpect writers therof in ſpeculo pontificum roman. editum, Anno 1584. per Stephanum Szegedinum pannonium*

all deceaueablenes: So hath the Pope aduãced his credit, deceiuing the ſimple ſort, by his fained fables, & alleadged miracles of the Legendary, &, *de vitis Patrũ*, & by ſuch juglings, as the holy blood of *Wailes*, & the maide of *Kent* in *England*, and the fained miracles of *Lawret* in *Scotland*, and lateſt of all, by ſuch as the *Spaniſh* goddeſſe, with her fiue bloody wounds, Prioreſſe of the Abbay of *L'Annunciata*, in this laſt 1588. year of Chriſt now priſoner there, as a deceitful witch. And ſhortly all the whole indices and tokens of the Antichriſt, contained in the *Revelation* (beſide theſe before ſpecified) doe moſt perfitlie agree to the Pope, his ſeate, kingdome & cleargie, as ſhall be particularlie diſcuſſed in our principall Diſcourſe, being heere for breuitie omitted. It reſteth then to ſhew a certaine of the notable contrarieties betwixt Chriſt and his alleadged Vicar, the Pope, whereby he ſhall bee found not his Vicar, but his adverſarie: yea, not a Chriſtian, but the Antichriſt.

The holy blood of wales.
The maid of kent.
Miracles of Lawret.
Maria l'Annuntiata in Spane.

Antipathies betuix the Pope and Chriſt.

And firſt (ſaith *Matthew*) Chriſt came ſo ſoberlie, & ſo meekelie, that he would not haue brokẽ a bruiſed reed, The Pope came proudly, ouerwhelming the moſt mightie Empires: He changed the kingdom of *France* from *Chilperike*, the lawfull king to *Pipin* the Popes confederate: he extinguiſhed the king of the *Longobards*, and brokes of their kingdome to this day: he bereft the *Exarchat* of *Ravenna* from the Emperour of the *Orient*, he trode with his feet vpon the neck of *Frederik Barbaroſſa*, Emperour of the *Occident* Chriſt would not worſhip the deuill, for al the kingdomes of the world, as teſtifieth *Matthew*: † Twenty two Popes bound themſelues ſlaues to the Deuil for euer, to be made Popes, as teſtifieth *Platina* and other vnſuſpected writers. Chriſt *caſteth out the merchands out of the Temple, the Pope eſtaliſheth his merchants in the Temple to ſell his pardons, maſſes, and indulgences. ‡ Chriſt waſhed his poore Diſciples feet: The moſt rich and mighty Emperors kiſſe the Popes feet, & he doth tread on

Mat 12.20
1 his proud comming.
Depriuing the king of france.
Extinguiſhing the kings of Longobards
Bereauing the Exarchat of Ravenna.
Treading on the emperors neck
2. His deuiliſh bond
*Mat 21.12
3. His Church merchãdiſe
† *The writers hereof are cited in ſpeculo Romanorum pontificum as is aforeſaid.*

their necks. Chrift honoured marriage in † *Cana* of *Galilie*, had fome married Apoftles, and commanded all to marrie, that could not liue chaft: The Pope forbiddeth his cleargie to marrie, though they were neuer fo filthie whore-mongers. Chrift forbiddeth whoredome of all kinds, Of Popes, 13. were adulterers, three were common brothellers, foure were inceftuous harlots, eleuen were impoyfoned with vile Sodomie, feuen were whore-mongers, and erecters of brothel-houfes. Finally, one was a whore, and died of childe in open procefsion, and all thefe are named in their owne hyftories. God forbiddeth that the pryce of an harlot be ‡ offered in his fanctuarie. The Pope receiued of euery whoore in his brothels weekly, a Iulian pennie, which will amount fome yeares, to 40000. ducats by yeare. Chrift payed tribute to the Emperour: The Pope bereft the Emperour of both landes and tribute within *Jtalie*. Chrift had not of propertie, fo much money of his owne, as to pay his *tribute pennie*. The popes propertie is faide at this houre, to be 18000. ducates euery day, † alfo Pope *John* of *Avinion*, left in ftore 25. milliõs of gold. Chrift grants libertie of all meats, faying that *nothing entereth in the mouth that defileth the foule*, & commanding vs to eat of all thing, that is † fold in the market. The Pope forbiddeth flefh in Lent, Friday, and other his fafting times. God miflyketh the *obferuing of dayes, moneths, and yeares*. The Pope commanded to obferue his feftiuall and holy dayes, feptuagefes, Lẽtes, and yeares of *Jubilies*, & indictions. † Chrift is the *only Mediatour betwixt God and man*. The Pope maketh, and canonizeth his owne faints, whom he calleth Mediatours betwixt God and vs. Chrift is both *God and man*. The Pope is ftyled neither God nor man. *The fowles ef the aire, haue their nefts, and the beafts haue their dens* (faith Chrift) *but the Son of man hath not where to repofe his head*, the pope hath ftore of towres, caftels, and princely palaces. Chrift threatneth *Peter* for drawing his fword: the Pope vfurpeth both the

‡ Iohn 13.5
4. The kiffing of his feet.
† Iohn 2.
Mat. 8.14.
1. Cor. 7.2.
5. His forbidding of marriage.
6. Adultery & inceft of Popes
The Pope Joanna a womã died of child.
‡ Deu. 23.18.
7. *Whorifh tributes.*
8 The popes vfurping againft the Emperour.
† Mat. 17,27
9. Chrifts povertie, the Pops riches.
Palmerius.
10. His forbidding of meats.
Mat. 15.11
1. Cor. 10 verfe 25.
11 His obferuing of fuperftitious dayes.
Galat. 4.10
12. He maketh many mediators.
1. tim, 2.5
13. The Pope is ftiled neither God nor man.
In prologo Clementina rnm in gloffatore.
† Mat. 8.20
14. The Pope hath ftore of palaces.

temporall and ſpirituall ſword: for proofe whereof † Pope *Boniface* the 8. ſhewed himſelf, the one day in his pontificall apparel of a Biſhop, the next day clad in armour as an Emperour, bearing before him two ſwords, proclaiming theſe words, *Ecce duo gladii hic*, *Behold heere the two ſwordes*. God forbiddeth in his ‡ ſecond commandement, not only the worſhipping of images, but alſo to make any ſo vile figures, to repreſent his ſo incomprehenſible Maieſtie. The Pope and his Cleargie, not only make they vile images to repreſent God & his Saints, but alſo they command to reverẽce thẽ, to come in pilgrimage, barefooted & bare-headed kneeling before them praying, ſuch prayers to them, as are intituled in their bookes, A prayer to the image of Chriſt, a prayer to the Croſſe of Chriſt, and ſuch others. God hath commanded to engraue theſe ten commandements on the poſts of our houſes. and on all patent parts: yea, and in our hearts. The Pope and his Cleargie leaueth out the ſecond commandement, & to make them ſeeme ten, they divide the laſt commandement into two, and teacheth them that way to the people. Chriſt cõmandeth † vs not to make much babling, nor oft repetition of prayers as the Heathen. The Pope commandeth to repeat ouer your prayers, as to ſay fiue *Pater noſters*, fiue *Avees*, and to ſay a *Trental of Maſſes*, which is 30 Maſſes, and ſo to obſerue a number, as the Witches doe, & as Ovid ſaith of the Witch *Medea* in theſe words,

15. Vſurpeth both the ſwords.
† Ma. 26.52
† Pope Boniface his oſtentation.
16. He erecteth Images and idolatrie.
‡ Exod. 20.4.
Deut 11.18 19.20
17 The pope blotteth out the ſecond cõmandemẽt which forbideth idolatrie.
† Mat 6.7
18 babling and magicall repititions of prayers.

Verbaque ter dixit placidos facientia ſomnos,
Que mare turbatum, quæ flumina concita ſiſt ant.
Thriſe from her lips did words redound, which moueth ſleepe and ſlumbering ſound.
And ſtayeth ſtormes of roaring ſeas, and furious flouds doe ſet at eaſe.

Lib. Metamorph 7.

And for better proofe hereof, they muſt haue beads to keep their count with. And to ſpeake briefly of

19. His religion of bodelie and externall rites.

their doctrin in one word, Chriſt teacheth a religion wholly in ſpirit, and void of al external ceremonies, except his two ſacramentall ſeales. The Pope hath conſtitute a religion, conſiſting wholly in bodily rits and full of all externall ſuperſtitions, partly *Ethnick*, partly *Jewiſh*, which Chriſt hath vtterly aboliſhed. Finally, both Chriſt was mocked with a Crowne of thornes, and the Pope contrarilie honoured with a threefold imperiall crowne. As alſo, Chriſt ſhed his blood once, that we might raigne ſpirituallie. The Pope ſheddeth our blood daylie, that he may raigne temporallie: And ſo although in al points of doctrin & cõverſation, Chriſt & he, be altogether contrarie, yet for the breuity of this treatiſe, we muſt pretermit the reſt at this preſent, truſting theſe foreſaid to be ſufficient to inſtruct the ſound judgement of him, whoſe heart God hath mollified as to the reprobat, the words of Chriſts owne mouth: yea, the Law nor the Prophets, neither finally the riſing of the dead to teach them, could anie wayes convert them. And yet for the more ſuperabundant certaintie of this matter, we will alſo remoue certain doubts of Scripture, which might moue the ſimple ſort (by adhearing to the literall ſenſe) to beleeue that all the Antichriſts that were to come, were already come, & paſt in the dayes of the Apoſtles, or that might any waies ſeeme to purge the Pope from being the Antichriſt.

20 The Pope honoured with a threefold crowne of golde.
21. His daylie tyrannie in martyrdome.

Luk 1 6.

Explications of texts alleadged for the Pope.

And firſt, where it is ſaide in S. *John.* Theſe are the laſt dayes, and ye haue heard that the Antichriſt ſhal come: Euen now are there many Antichriſts, and againe ſaith he, *but this is the ſpirit of the Antichriſt, of whõ ye haue heard that he ſhould come, & now he is in the world.* It is to be anſwered hereunto, that the meaning of the firſt text is not, that that preſent time, but the whole age after Chriſt, is the laſt age & dayes of the world, in the which, the great Antichriſt ſhould come. And euen then, ſaith he, there were many Antichriſts, not that he ſaith, that the great Antichriſt was any of them, but rather, that he meaneth, that there were

Firſt text.
1. Ioh. 2.18
Second text.
1. Iohn 4.3

Expoſition of the firſt.

ſo many ſmall Heretiks and Antichriſtes riſen vp, that thereby the comming of the great Antichriſt was portended: for *Paul* makes this place plaine, where he ſpeakes in expreſſe words that the myſtery of the Antichriſts iniquitie was euen then begun to worke, but himſelfe ſhould not come, while he who helde ſhould yet holde ſtill a while, and then be diuided or remooued out of the way: meaning, that the myſterie of his iniquitie, and ſpirit of his errours, was then begun to worke in theſe ſmall Antichriſts, fore-runners of the great Antichriſt, vnto whoſe errours (ſaith † he) becauſe the world hath attended, and hath left the ſpirit of veritie, whereby they ſhould haue bene ſaued therefore, God ſhal ſend vpon thẽ the efficacie of errours, that they may beleeue lies, that is to ſay, God ſhall ſend the great Antichriſt to work his miracles of lies, &c. As to the ſecond text, it is not ſpoken of the perſon of the great Antichriſt, but plainely by the text is meant, of the ſpirit of his errours, or (as *Paul* calleth it) *the myſterie of his iniquitie*, which euen then was in the world and wrought in his fore-runners, whom S. *John* calleth alſo antichriſts: So be theſe texts, rightly conceiued, it is gathered, that many ſmal Antichriſts (in whom wrought the ſpirit and myſterie of the great Antichriſt) were come in the daies of the Apoſtles, but yet not the great Antichriſt himſelf. Moreover, incaſe it be doubted how the Pope can be this Antichriſt, ſeeing ſaint *Iohn* ſaith, that the Antichriſt denieth the Father and the Sonne, whereas apparantly the Pope denieth neither of them.

2. Theſ. 2.

2. Th. 2.11

2. Th. 2.11

Expoſition of the ſecond.

1. Ioh. 2.22

Two denials, In mouth & in heart.

To this it is anſwered, there be two denials, the one in mouth and profeſsion plainely, and that way the Pope ſeemeth a Chriſtian, and not to denie Chriſt otherwiſe hee would not be credited of Chriſtians, neither could he be a meete Mediatour betwixt the Dragon and them, as the Antichriſt muſt bee. The other deniall is in heart and deede tacitly, and that way the Pope hath denied altogether, both the Father and the Sonne: for who ſo denieth the Sonne, the ſame (ſaith the ſame text immediatly after) denieth alſo the Father, and conſequently them both: but ſo it is that the Pope denieth Chriſt, in that he eſtabliſheth other mediatours of his canonized Saints, betwixt God and vs, and vindicats to himſelfe power to remit ſinnes, and ſo craftilie maketh Chriſt but a cipher, without either office of mediation, or of remiſsion of ſinnes. Wherefore, in effect the Pope denieth the Sonne, and conſequently both the Father and the Sonne. And beſides this former reaſon, is not this certaine, that the 22. *Necromantik* Popes already ſpoken of, could not be *Necromancers*, vnles they had by priuat and tacit paction, renounced both the Father and the Sonne, and became ſlaues to the deuill for euer? Alſo hath not diuers of them bene eſpied & noted with Atheiſme, ſuch as Pope *Leo* the tenth, who when Cardinal *Bembus* was in commoning with him, of certaine Evangelical hoſtories, made this godles anſwere. *Quantum nobis & noſtro cœtui profuerit ea de Chriſt ofabula ſatis eſt ſæculis omnibus notum. Jt is knowne evidently to all ages, how much this fable of Chriſt, hath bene profitable to vs and our Clergie.*

1. Ioh. 2.23
His denial of the ſonne and conſequently of the Father.

Chriſt made a cipher.

22. Necromantik Popes, deniers of God.

An atheiſt ſentence.

Of this godleſſe number, there were eight Popes nominated in true regiſters, beſide the foreſaide Necromantick Popes, and others vnſpied Atheiſts. Finallie, the very name of Antichriſt αντιχριϛος hath two ſignifications, to wit, *Vicarius Chriſti*. Chriſts lieuetenant, and *adverſarius Chriſti*, Chriſts enemie: both the which doe bewraie the Pope, & moſt viuely deſigne him: for as τυραννος ſignified of olde times a King, and now degeneratly, is taken for a tyrant or oppreſſour, becauſe that Kings oftimes become oppreſſours: euen ſo is αντι in the Greeke taken both for a lieutenant, or Vice-gerent, in Latine, *pro vicem gerente aut vicario*, as Matth. 2.22. where it is ſaid that *Archelaus* did raigne in ſteede of his father *Herod*, he vſeth the words ἀντι τοὺ Ηρωδου. Likewiſe ϛρατηγος is taken for a captaine or

Eight Atheiſts.

αντιχριϛος both for vicarius chriſti & adverſarius chriſti.

αντι taken pro vicario

Mat. 2.22

gouernour, αντιϛρατηγος for his lieutenant or depute, as on the contrarie part αντι alſo is taken for an enemie of vſurper of a mans place, euen both for *pro* and *contra*, becauſe that a deputie, lieutenant, or place-bearer ofttimes vſurpes and retaineth his place by rebellion, and therefore in the firſt ſenſe, αντιχριϛος is as much to ſay as Chriſts lieutenant: In Latin, *Vicarius Chriſti*: Which name the Pope chalengeth to himſelfe, profeſsing him thereby to be the Antichriſt in that ſenſe. In the other ſenſe, the very effects of his doctrine, life, & conuerſation, prooueth him to be alſo the Antichriſt, an enemie to Chriſt, and vſurper of his place, as learnedly that worthie man *Wolfgangus Muſeulus* doeth obſerue & note. Here then haue we after foure diuers maners diſcouered the Antichriſt: firſt, by proouing him to be neither, *Jew, Turke*, nor other *Ethnick*, but a pretended Chriſtian Prince. Secondly, by applying of the ſpeciall tokens of the Antichriſt preciſely to the Pope. Thirdly, by declaring the notable contrarieties in doctrine and converſation, betwixt Chriſt and the Pope. Fourthly, and laſt of all haue wee expounded and remooued certaine doubts, which would ſeeme to excuſe the Pope, and purge him frõ being the Antichriſt: ſo that nothing needs more to be ſaid, if it pleaſeth God to beſtow his ſpirit vpon the auditour: otherwiſe in vaine doth the Sun clearly ſhine to the blind mould-warp. The Lord opẽ the eyes of all men, & inſpire the harts of Princes (as in the *Revelation* is promiſed) to deſtroy that idolatrous ſeat, & bring the Antichriſtian kingdome thereof to an end.

αντι taken in the contrair ſenſe euen both for pro and contra.

αντιχριϛος in greike is in latine the Popes proper ſtyle vicarius chriſti.

αντιχριϛος ane enemie in effect. An foure fold diſcouerie.

1

2

3

4

Apo. 17.16

27. PROPOSITION.

The Image, marke, name, and number of the beaſt, are of the firſt great Romane beaſt, and whole Latine empire vniverſallie, and not of the ſecond beaſt, or Antichriſt alone in particular.

Some hath mixed confufedly, the tokens of thefe two Beafts through other, and fhaue thought to applie this *Image, marke, name and number* to the Antichrift in particular: but here fay wee, the fame doeth appertaine to the great ten horned Beaft, and whole Latine Empyre in generall, and not onely to the Antichrift in particular, and that wee prooue by thefe reafons. Firft it is faid, † that in the time of the firft *vial*, which was betwixt the year of Chrift 71. & 316. *there fel a great plague on them that had the marke of the beaft, and on them that worfhipped his image*. But at this time, the fecond Beaft or great Antichrift was fcarcely come: as by the 9. fynthefis of the former propofition is fhowne, and fhall be more at length in the 36. propofition. Therefore the Image and marke, muft be of the firft great *Romane* beaft or *Latin* empire, and not of the fecond beaft. And where it is faid † that the fecond Beaft and onely Antichrift, caufed to make the image of the firft great *Romane* Beaft, and caufed all men to receaue his marke, and that none might buy or fell, that had not the marke or the name of the Beaft, or the number of his name, it followeth not thereby that there was no *Jmage, marke, name, nor number of the beaft*, before the Antichrift came, but rather, that the fecond beaft or Antichrift, caufed to repaire them after they were almoft abolifhed, and that he did confirme and approoue them of new, as in the next propofition is declared, Secondly, this *marke, image, name and number of the name*, appertaines to the firft great *Roman beaft*, becaufe it is plainely † faid, that *the image is of the great Roman beaft which receiued the deadly wound*: & therafter it is faid (or at the leaft may be gathered by the text,) that not onely the *image* and the *marke*, but alfo the *name* and *number* doeth appertaine, to one felfe-fame beaft. And therefore neceffarilie *the image, marke, name, and number of the name*, muft al appertaine to the great *Roman beaft and Latin Empyre*, and nowaies to the fecond Antichriftian beaft, nor to the Antichrift onely.

1 Reafon
Apoc 16.2

Apoc. 13.14.
16.17.18.

2. Reafon.
Apo 13 14

Apo. 19.20
Apo. 13 14.
16.17.18.

28. PROPOSITION.

The Image of the Beaſt, is theſe degenerate Princes, that in name onely were called Romane Emperours, and were neither Romans of blood, nor Emperours of Magnanimitie.

SEing that great ten-horned beaſt, is the *Romane* Empyre (by the 24. propoſition) and (by the former propoſition) this Image, is the image of that Empyre, and (by the text) this image muſt haue a ſpirit, and muſt ſpeake as the Antichriſt inſpires him. What elſe can this liuely and ſpeaking Image then be, but a man? yea, euen ſuch a man, as doeth repreſent and beare the figure of that Romane Empyre: and therefore (as ſaith the text) doe all men worſhip and reuerence him. This image then is of two ſorts, the one were theſe vnworthie Emperours, (not of the ancient blood of *Romanes*) in whom the empyre did decay. Of theſe there were in the daies of this firſt *Vial*, as is ſaid in the diſcourſe of the former propoſition: the laſt of theſe were extinguiſhed by the *Hunnes, Gothes, and Vandalles*. The other forte were the new Emperours, called the Emperours of *Germanes* and *Romanes*, of whom *Charlemaigne* was the firſt. Theſe alſo were but a figure of the firſt olde Empire, and were but only *Roman Emperours* in name & neither of their anciẽt blood, neither hauing ſuch authority as they. Theſe did the Pope crown & inaugurat as Emperors, or rather as images of the firſt empire. Theſe inſpired he with the ſpirit of his errours, and made them to ſpeake as he willed them, and then confirmed he their empyre, and proclaimed them *Emperours*, & cauſed al men to reuerence them, as is ſaide in the † text. So then for concluſion, all the properties of the beaſts image, agree ſo fitly with theſe degenerate Emperours, both of the firſt ſort & laſt ſort (which neither were *Romanes* of blood, neither *Emperours* in effect) that we can call none other, than them to be the Images of the ten horned *beaſts* or *Roman* and *Latin* empyre.

A ſpeaking Jmage.

Jmages of two ſorts.
Firſt ſort.

Second ſort Emperours of Germans and Romans.

Theſe Images did the Pope authorize & made them to ſpeak as he liſt.
Apo. 13.14 15.

29. PROPOSITION.

The name of the beaſt expreſſed by the number of 666. [*cap. 13*] *is the name λατεινος onely.*

MAny hauing miſtakẽ this text, haue judged this number to be the number of the Antichriſts name in particular, as if it were the number of the ſecond *Beaſts* name, which we haue prooued by the 27. propoſition to be of the firſt *beaſts* name. Here then ſay we, that name is λατεινος, for theſe reaſons. Firſt, becauſe the name of the beaſt is prooued (by the ſaid 27. propoſition) to be the name of the ten-horned *Roman beaſt*, or *Latine Empyre* in general, and not of the Antichriſt onely, and ſo it muſt either be *Romanus* or *Latinus*, but of theſe two, *Latinus* is the eldeſt ſtyle: for King *Latinus* (from whom that people were called *Latini*, and their countrey called *Latium*) was long before King *Romulus*, of whom the Citie was called *Rome*, and the people thereof *Romanes*. Secondly, it muſt be the number of a mans name (ſaith the text) ſo is *Latinus* the name of a man, euẽ the name of one of their firſt kings. Thirdly, foraſmuch as the *Grecians* had a cuſtome in their myſteries and Oracles to obſerue the number of names, as ye ſhall finde in diuers parts of *Sibylla*: And as in that countrey, the name of the flood Νειλος is celebrated as holie, becauſe it containeth the number of the daies of the yeare 365. as *Gaſparus Peucerus* vpon *Carion* teſtifieth, *Chro. lib. 4.* where he deſcribeth the *Cattes* and *Heſſes*. Therefore, Saint *John* (obſeruing the cuſtome of them to whome hee writeth) ſaieth † that the number of the Beaſt, or rather (as he termed it in the former 17 verſe) the number of the Beaſts name, is 666. And euen ſo it is, that λατεινος contained the number of 666. for λ is 30. α is 1. τ is 300. ϵ is 5. ι is 10. ν is 50. o is 70. and ς is 200. which altogether make ſixe hundreth three ſcore and ſixe. Therefore, λατεινος, is the verie name of the Beaſt, meaned by the ſaid number. And for better confirmation of this forme of

It is the firſt beaſts name and number & is λατεινος
1. Reaſon. Latinus their eldeſt king & eldeſt ſtile.

Ap. 13.17 18.
2. Jt is the number of a mans name: So is latinus a man.
3. The greciã cuſtom to account of names be their number.

Apo. 13.18.
The letters of λατεινος maketh the number 666.

interpretation, *Irenæus Martyr* (who was Difciple to *Polycarpus*, the Difciple of S. *John*) about the end of the fift booke *de hærefibus*, approoueth this maner of interpretation, faying, *Et teftimonium perhibentibus his, qui facie adfaciem viderunt Johannem, & ratione docentibus, quoniam numerus nominis Beftiæ, fecundum Græcorum computationem, per literas, quæ in eofunt, fixcentos habebit & fexagintafex, &c. And thefe bearing teftimonie, that face to face haue feene Saint John, and by reafon teaching, how that the number of the beafts name, doeth containe fixe hundreth three fcore and fixe after the Grecian reckoning, by the letters that are therein.* And thus far faith he generally, concerning the maner how this text fhall be interpreted, as appearingly he had his warrant, receiued mouth by mouth from Saint Iohn. But as to the particular names, although hee reckoneth out λατεινος, with other of his owne conception, yet becaufe he had not fuch particular warrant mouth by mouth thereof, he leaues off to fpeake determinatly thereof: But by the euent now fallen out, wee may more conftantly affirme this fame. For conclufion, feeing by the 24. propofition, the firft great ten-horned beaft is the *Romane* Empire, whereof (by the 27. propofition) the number of his name, is the number of 666. and that (by the text) the fame muft be the number of a mans name: Therfore, this number of 666. cannot be yeares, as fome doe interpret it. nor αρνουμε, for αρνουμαι, which is to denie, nor ἐκκλησια ἰταλικα, neither *Dic Lux*: becaufe although thefe containe the number of 666. yet thefe be neither the names of a man, neither the names of the *Latine* or *Romane* Empire, as is prooued, that the name of the faide number ought to bee.

Jrænaus cõfirmeth this interpretation.

Epilogue of the former reafons and conclufion thereupon.

† *This nõber is not of yeares, nor of* αρνουμε, *nor* εκκλησια ἰταλικα, nor DIC LVX, *but of a mans name.*

Alfo, this name is neither τειταν nor μαομετες, becaufe, although they both anfwere the faid number, as likewife are the names of a man or mafculine, yet are they not the names of the firft beaft, or *Romane* Empire. And therefore that name

† *Neither* of τειταν, *nor of* μαομετες, but *of the great imperiall beaft* x *man.*

neceſſarily muſt be $\lambda\alpha\tau\epsilon\iota\nu o\varsigma$, as being not onely the name of a man, but alſo the proper name of the firſt beaſt: as thirdly containing in it the ſaid number 666. conforme to the three properties and tokens thereof, warranted by the text.

30. PROPOSITION.

The marke of the Romane beaſt, is that inviſible profeſsion of ſeruitude and obedience, that his ſubiects hath profeſſed to his Empire, ſince the firſt beginning thereof, noted afterward by the Pope, with diuers viſible markes.

IN the ninth cap. of *Ezech.* it is ſaid, that the Angel marked Gods Elect with *Jeruſalem*, with a marke on their fore-heads: the like is ſaid in the *Reuelation*: yet in no hiſtorie of theſe daies, it is hard that any ſuch viſible marke was ſet vpon the faithfull: wherethrough that marke muſt of neceſsitie meane nothing elſe, but that profeſsion of ſeruice and obedience towards God, which with bolde faces wee aduouch as conſtantly and openly, as if we were viſibly marked therewith. And ſemblablie, on the contrary part, this marke of the *Roman* beaſt or *Latin* Empire, is nothing elſe, but that profeſſed ſeriutude, obedience, confederacy, or concurrence, which the ſubjects thereof haue avouched to obſerue thereunto, And for confirmation hereof, it is ſaid in the Revel. that in the daies of the firſt *vial*, there fel a great plague on them that receiued the *beaſts* marke, and worſhipped his Image: but at that time, (which was betuixt the yeare of God 71. and 316) there was no viſible mark receiued by thẽ, nor that can be attributed vnto thẽ of that empyre, nor any difference betwixt them & the enemies of the empire, but onely their foreſaid profeſsion, which therefore doubtles muſt be called in that text, the marke of the empyre. But as to the ſecond part of this propoſition, howe ſoone the Antichriſt, that two-

Ezech. 9.4.
Apoc. 7.34

† *Profeſsion and obedience to God, is his publike marke.*

Obedience to the beaſt is his firſt marke.
Apoc. 16.2

Firſt markes inviſible.
Second marks viſible.

horned Beaſt came, he cauſed all men to receiue a marke, which the text deſcribes ſo ſpecially, with ſuch notes, as it muſt needes be a more notable and viſible marke, than the ſimple profeſsion only: for the former marke of ſimple profeſsion, is but ſimply called the marke of the *beaſt* (*Reuel.* 16) but the markes that fell in the time of the Antichriſt, are not onely ſaid to be marked in their fore-heads or right hands: but alſo the forme & ſhape of the marke is expreſſed by the text, where the ſame marke is ſaid to be, the *marke of the Beaſts name or number thereof*: as if he would ſay, there are markes taken out of the beaſtes name, or rather number therof, which all his ſubjects ſhall beare or weare ſometimes on their foreheads, ſometimes in their hands: which, what they be, followeth in the next propoſition.

Apo. 16.2

Ap. 14.11.

Apo. 15.2.

31. PROPOSITION.

The viſible markes of the Beaſt, are the abuſed characters, of χρ̃ϛ and croſſes of al kindes, taken out of the number of the firſt beaſts name.

THeſe viſible markes, whatſoeuer they be, haue fiue notes or tokens, aſsigned by the text to know them by. Firſt, they are called the *markes of the Beaſt*, Reuel. 13.17. and 14.9. that is to ſay, markes of the firſt and great Romane beaſt, as is prooued by the 27. Propoſition. Secondly, & more eſpecially, they are called *markes of that beaſts name*. Reu. 14.11. Thirdly, and moſt eſpecially of all, they are called the markes of the number of that beaſts name, in theſe words, Reuel. 15.2. *And theſe who obtained victorie of the beaſt, and of his Image, and of his marke of the number of his name*, word by word out of the original greeke: bearing, *καὶ ἐκ τοῦ χαράγματος αὐτου, ἐκ τοῦ ἀριθμου ῗου ὀνὸματος αὐτου*, And ſo this victorie is not ſpoken of his marke, (*and*) of the number of his name, as ſome tranſlates, but of his

Fiue tokẽs.

1. Marks of the beaſt

2. Marks of the beaſts name.

3. Marks of the number of the beaſts name.

marke of the number of his name, meaning ſo expreſſelie, that his marke, is a marke of the number of his name, that is to ſay, is deriued or compoſed of the number of his name, but the greeke number of his name is, χξϛ, As expreſly ſaith the text, *Re*. 13.18. Therfore, in theſe numeral letters χξϛ are theſe marks cõtained, & this we haue very ſpecially for the third tokẽ. Fourthly, theſe marks muſt be deuiſed, enioyned, and appoynted: by the ſecond Beaſt or Antichriſt, as proues the text. Revel. 13.16. Fiftly, theſe markes muſt be vniuerſall vpon all men, rich and poore, free and bound, and that no man might vſe his traffick or marchandiſe, or lawfull affaires, who had not receiued theſe marks as teſtifieth the text, *Revel.* 13.16.17. Then, vnto whatſoeuer markes all theſe fiue tokens doe agree, the ſame are the Antichriſtian markes: But vnto the abuſed characters of χ͂'ρϛ, and croſſes of all kinds, doth all theſe fiue tokens agree: for firſt χρ͂'ϛ, is juſtly called the mark of the firſt Beaſt & *Roman* Empire for two cauſes: the one for that all theſe of that Empire, are entiſed vnder colour of the Name of Chriſt, to reverence that character, being nevertheleſſe, as it is abuſed, nearer to the name of the Antichriſtian and *Latin* kingdome, nor to the name of Chriſt, as herafter ſhal be ſaid. The other cauſe is, for that theſe marks of χ͂'ρϛ, ar gathered of theſe letters, χξϛ, which are the number of the name of that *Latin* kingdom, and *Roman* Beaſt (as hereafter more ſpecially shall be proued) Therfore, juſtly ar the marks of χ͂'ρϛ, called the marks of the firſt *Roman* Beaſt, agreeing with the firſt token. Secondly, and agreeable with the former becauſe (by the 29. propoſition) that Beaſts name is λατεινοϛ, whoſe numerall letters, χξϛ repreſents & yeelds the marks of χ͂'ρϛ, (as more ſpecially immediatly ſhal be ſaid,) Therefore χ͂'ρϛ, agreeth with the ſecond tokẽ, to be called the mark of that beaſts name. Thirdly, and according to the third token, this mark χρ͂ϛ, is contained

In χξϛ are theſe markes to be ſought.

4. They are marks appointed by the ſecond antichriſtian beaſt.

5. Markes vniverſall to all men.

Theſe markes be χρ͂ϛ and croſſes of al kinde.

1. Reaſon χρ͂ϛ the marke of the firſt beaſt for two cauſes 1 cauſe.

2. cauſe.

2. Reaſon is their agreance with the ſecond token.

3. Agreable to the 3 token.

vnder the numerall letters of that beaſts name $\chi\xi\varsigma$, and is hereby perfitely repreſented, the two extream letters reſpectiuely agreeing in one, and the ſmall diſagreeance of the middlemoſt: to wit, betwixt ξ and $\rho^{\sim\prime}$ to be vpon very neceſsitie: for where as S. *John* heere ſpeaketh expreſly of numerall letters, he could haue no one numerall letter to repreſent both ρ and his crowne, this way ρ, more like it nor ξ, whereof the vpper part repreſents the crowne, the nether part the figure of ρ. Moreouer, there is yet greater affinitie betwixt ξ & ρ, for ρ or rather $\grave{\rho}$ after this forme in the Greek, and ξ in this Beaſts language and common written letter in Latin is σ which is all one in figure, and ſo the markes of $\tilde{\chi}\rho\varsigma$ agree with the third token.

4. Reaſon. Accord with the fourth token.

Then fourthlie the marke $\tilde{\chi}\rho\varsigma$ is deuiſed, and taken vp by the ſecond Beaſt, the Pope & his Cleargie, to repreſent as they ſay, the Name of Chriſt thereby, but rather the contrary, as followeth hereafter.

5. Reaſon agreing with the fift token.

Fiftly, and according to the fift token, the Pope and his Cleargie ordaineth vniuerſallie all theirs (whom only they call Chriſtians) to reverence the very naked figure of $\tilde{\chi}\rho\varsigma$ and bare character thereof, as hauing a divine vertue in it, transferring and diſtracting our zealous reverence from Ieſus Chriſt our Sauior, who is God with vs, to ſuch bare characters, and to ſuch dead & naked wordes in an vnknowen language, as IH˜S † xp˜s † SOTER † IMMANVEL †: And therefore appoint they all men to beare and weare this foreſaid character $\tilde{\chi}\rho\varsigma$ with ſuch like other bare names, in their rings, amulets, pendaries, breaſt-plates, tablets, *Principioes*, and *Agnus Deies*: Abuſing it in way of ſorcerie, as in a thouſand Papiſticall charmes and cõjurations, not only in the Cleargies priuie books but alſo in their publike ſeruice called *Exorciſmus*, is to be ſeene.

Concluſion vpon $\tilde{\chi}'\rho\varsigma$ Reſts to proue the like of croſſes.

So then, ſeeing theſe fiue tokens doe agree with $\tilde{\chi}'\rho\varsigma$ it followeth neceſſarily, that that abuſed character is one of the Antichriſtian markes. It remaineth then to prooue the like concerning their croſſes of all kindes.

1.Reaſone. hath three cauſes.

And firſt the croſſe is juſtly called the marke of the firſt great *Romane* beaſt, &

Latin Emperours, for three caufes, the one becaufe that the Emperour and all his fubjects generally, are fome-times marked therewith, & beareth the fame, as fhall be faid. The other caufe, becaufe the Emperour *Conftantine* illuded by a croffe fhadow in the clouds, taketh vp at the devife & perfwafion of *Sylvefter* the firft, and firft Antichriftian Pope, the mark of the croffe, and that the rather becaufe (as thefe fabulous hiftories report) his mother *Helena* was faid to haue found Chrifts croffe, and fo juftly may the croffe be called, the marke of the firft beaft and Roman Emperour.

1. caufe.

2. caufe.

The laft caufe, why the croffe is called the marke of the *Roman* Beaft, or *Latin* Emperour is, for that it is taken out of his name, euen out of the three numerall letters thereof, as prefently you fhall heare. So then the firft token of the Antichriftian marke, agreeth with the croffe. Then agreeable to the fecond token: forfomuch as by the 27. propofition λατεινος is that beafts name, at the leaft vndoubtedly, the numerall letters of his name are χξϛ, out of the which are gathered croffes of all kindes (as immediatly fhall be faid.)

3. Caufe.

The fecond reafon. Agreance with the fecond token.

Therefore thefe croffes, are juftly called the markes of this Beafts name. Then thirdly according to the third token, in the faid numerall lerters χξϛ, are contained croffes of diuers kindes, both in figure and in name: In fuch maner, as both the *Greek* Churches & *Latin* Churches, are certified thereof: for among the *Grecians* the forme of χ is a croffe, and among the *Latines* ξ which is X, is alfo a croffe, and thefe bee croffes afidewayes called S. *Andrewes* croffe: Therefore the third and laft letter is a double letter of abbreuiation, containing thefe two letters ς τ. wherof the laft is called a headles croffe, likeft to the croffe which they cal our *Lords* croffe: And fo alwaies & of euery letter hereof arifeth the figure of the croffe. As to the names of the croffe, in Latine *crux*, and in greeke ϛαυρος, it is not pofsible for them to be expreffed vnder fo fewe letters more viuely, nor they are here, for here haue

Third reafon agreing with the 3 token.

The greik X the Latin X are S. Androwes croffe.

τ the headles croffe.

The greeke and the latine names of the croffes found in χξϛ.

you their chiefe letters, euẽ both their capital & their final: for by χ and ξ, haue you in latine C. & X. making *crux*, or C. H. & X. making barbarouſlie C H R V X, as I haue ſeen & heard it ſo in print & pronounced. And again by the laſt letter ϛ haue ye the greeke name of the croſſe ϛαυρος likewiſe expreſſed, by his capital and finall letters. So then according vnto the third token in χξϛ, which are the greeke number or numerall letters of the beaſts name, are contained croſſes of diuers kindes, both by figure and by name evidently expreſſed, both to the orientall or greeke Churches, and to the occidentall or Latine Churches, and what is more evident? Fourthly, & agreeing to the fourth tokẽ, theſe croſſes (I mean the abuſes therof) were deviſed by the ſecond *Beaſt*, the Pope and his Clergie, as magicall markes of ſorcerie, gree by gree, to with-drawe that godly meditation that we ought to haue of Chriſts paſsion, and that reuerence that we ought to haue towards Chriſt, for ſuffering death for our ſakes, and to beſtowe the whole merite, honour and vertue thereof, firſt, to the materiall Croſſe, whereon hee ſuffered, and then to bare figures and markes thereof, and laſt of all, to other curious ſophiſticall figures & marks, ſuch as  which they call the croſſe of *Jeruſalem*, and ſuch as this croſſe  which is to be ſeene vpon diuers old Church walles. Which figures are ſo degenerat gree by gree from croſſes to compaſſes, crooks and rammald, that altogether they haue no reſemblance of Chriſts paſsion. Finally, and according to the fift and laſt token: theſe croſſes were ſo vniuerſally vſed, (that beſide their daylie croſsings with their right hand on their fore-heads which they call *ſauing*) all maner of perſones behooued to beare and weare, euen viſible markes thereof, at apointed times on their fore-heads, or in their hands, ſuch as to be croſſed with aſhes on their fore heads on *Aſhwedneſday*, and to beare Palme croſſes in their hand on *Palmeſunday*, otherwiſe they were thought to be *Heretickes*, and curſed from the cõmunion of

Fourth reaſon agreing to the 4 token. Chriſts due honor with drawen from him and giuen to impertinent figours

Fift reaſon agreeable to the 5 token of vniuerſality

Fore heads croſſed on Aſhwedneſday. Palme croſſes in their hands on Palme ſunday.

Chriſtians: And theſe excommunicates (as ſaith the Bull of Pope *Martin* the fift, *Fol.* 134) ſhall not be permitted to keepe houſes or lodging, or to make any blocke or bargaine, or to occupy any traffike or trade of marchandize, or to haue any ſocietie with Chriſtians. Then ſeeing theſe croſſes agree now with al theſe fiue tokẽs of the Antichriſtiã marke, as a litle before χ̃ρς was proued to do. Therfore, neceſſarilie theſe marks of χ̃ρς and croſſes of all kinds, muſt be the beaſts viſible markes. Let no man grudge hereat nor grieuouſly take it, obiecting that we haue made theſe to be the marks of the Antichriſt, which mẽ haue hitherto deuiſed (as they ſay) to repreſent the name of Chriſt, and his paſsion. For certainly theſe ſame thinges, which not men, but God hath deuiſed and appointed, howe ſoone they are miſuſed, they are called by the Spirit of God, abhominatiõs. Appointed not God the braſẽ Serpẽt to be erected, as health ful, but after that it was worſhipped it was deſtroied by † *Ezechias* as damnable? Appointed not God the Sacrifices & incenſe offerings, & yet ſaith he in *Oſee, Miſerieordiam volui & non ſacrificium: I deſired mercy and not Sacrifice*: And in *Eſay, Ne afferatis vltra ſacrificium fruſtra: incenſum abominatio eſt mihi: Bring no more Sacrifice in vaine: Jncenſe is an abomination vnto me*? Seeing the Lord then doth reiect thoſe ordinances which himſelfe hath inſtituted, and calleth them, *Abomination*, when they are abuſed, how much more ought we to reiect theſe ſophiſtical marks of χ̃ρς and croſſes of all kindes, ſuch as theſe ✝ which they call our Lords croſſe, and ✠ Saint. *Georges* croſſe, and X S. *Andrewes* croſſe, and ☩ the croſſe of *Jeruſalem* and ⊛ the croſſe of Dedication, beſides the Popes croſſe and manie Croſſes moe, and eſteeme them as abhominable Antichriſtian badges, whereas we ſee them not onely deuiſed by men, but alſo abuſed by them in way of ſorcerie and magicke: imputiug vnto theſe naked figures, a vertue and ſanctitude in their charmes and exorciſms, as though, the whole

Concluſion againſt croſſes.

Præoccupation.

4 Reg. 18.4

Oſee. 6.7.

Eſa. 1.13.

vertue of Chrift and his pafsion, were transferred ouer into them. And although they fay, that they haue devifed χ˜′ρϛ, to reprefent the name of Chrift χριϛος thereby, and croffes of fundrie kindes, to reprefent his pafsion therewith: yet the Spirit of God lets vs fee euidently by this text, that he doth reject and caft ouer thefe their magicall characters and forceries to themfelues againe, as liker to the numeral letters of their name, than to his name, euen liker to χξϛ than to χριϛος and fo let none doubt, but thefe fore-faid figures, as they are vfed, or rather abufed, are the very markes and tokens of the beaft, agreeing with the fore-faid tokens of the text, in all points and properties. So that if any would demande what is the marke that thefe numerall letters do defigne? thefe very letters capitally, do make the anfwere. *χ. χαραγμα. ξ, του ξυλου, ϛ. του ϛαυρου. Character ligni crucis*: the marke of the tree of the croffe.

Note to which of thefe, thefe marks and croffes be likeft.

32. PROPOSITION.

Gog is the Pope, and Magog is the Turks and Mahometanes.

THis is proued three waies, to wit, firft by the fignification of the names, fecondly, by comparifon, and thirdly, by their linage: As to the firft, *Gog* doth fignifie *Tectum*, that is to fay, *Covered*: and *Magog* (as *Carion* doeth teftifie) fignifieth *De Gog*, or *Detectum*, that is to fay, *Difcouered*, as many learned doe affirme, though others efteeme it rather to meane *de tecto, of a couering*, than *detectum, difcouered.* So then, there being in thefe latter dayes, only two great notable and mighty enemies to God: the one, the Pope, a couered enemie and a diffembling Chriftian, the other the *Turkes* and *Mahometanes* difcovered and open profeffed enemies: It muft needs be, that, that fecret enemy *Gog*, is the *Pope*, and the open enemie *Magog*, is the *Mahometanes*. As to the fecond, like

Three reafons.

1 Their names.
Chro. lib. 1 fol. 21.

Two enemies a couered and a difcouered, the Pope and the Turke.

2. Reafone. Comparifon,
1 Chro. 5 4

as by comparifon, *Gog* was an *Iſraelite* borne, the *Jſraelites* then beeing the Church of God. And as *Magog* was a borne *Ethnicke*: ſo is the Pope borne among Chriſtians, which now are Gods Church: And the *Turkes* and *Mahometanes* are borne *Ethniks*. And again, as *Gog* was of the tribe of *Reuben*, who was accurſed for inceſtuous whooredome. So is the Pope chief of that *Romiſh* ſeat, which is called that ſpirituall whoore, and accurſed Idolatreſſe, with whom the Princes of the earth haue committed Idolatrie, and ſpirituall whooredome. As to the third, concerning their linage: *Ezechiel* defineth *Gog* to be the chief Prince of *Tubal* and *Meſhech*: then read *Jerome de interpretatione nominum*, &c. vpon *Jtalia* and *Tubal*, and ye ſhall find of *Tubal* to be deſcended, *Jberi* or *Iberes*, which he calleth the *Spaniards* or *Italians* in *Europe*, & likewiſe of *Meſhech* to be deſcẽded the *Cappadocians* in litle *Aſia*: which all hauing embraced the Chriſtiã faith, did afterward, nil they, wil they, accept the Pope as their ſupreme head, or chief Prince: & ſo *Gog* the chief prince of *Tubal* & *Meſhech*, muſt be the Pope, ſeeing he is ſupream head of theſe *Tubaliſts* in *Europe* & *Meſhechiſts* in *Aſia*. And as cõcerning *Magog*, he was the ſonne of *Iapheth*: of this *Magog*, deſcended the whole *Scythians*, which now are the *Turkes* and *Mahometans*, according to the opinion of all writers. Then by theſe three reaſons (by and beſides the whole notes and tokens of the 20. chap. contained in the principall treatiſe) it appeareth euidently that *Gog* is the Pope, and *Magog* the *Turke*, or *Mahemetick Emperour*,

Gen 10.2

Gen. 49 4-
Apoc. 17 2

3 Reaſone Their linage.
Ezech. 38. and 39.

Gen 10.2.
Concluſion

33. PROPOSITION.

The armies of Gog and Magog [*chap. 20.*] *are all one with the two adverſarie armies of the ſixt Trumpet, or ſixt Viall.*

THis is ſufficiently prooued, by the harmonie of the whole tokens aſsigned by the Text of

Harmonies and tokens.

the *Trumpets* and *Vialles*, and by the text of *Gog* and *Magog*. For thefe of the fixt *Trumpet* and *Vial*, dwelt beyond and about *Euphrates*, and were kings of the Eaft: fo here doeth *Magog*, for that he is proued lately, to meane the *Mahometanes*. They are there called in the fixt *Trumpet* foure, as indeed they are foure families, and foure nations *Mahometanes*: Here likewife, are they tearmed the people of the foure quarters or corners of the earth. There were thefe foure nations bound at *Euphrates* til their due time, & then loofed to make warfare: Here is the Deuill bound from ftirring them vp or entifing them, till their due time, and then is loofed to ftirre them vp to battell. There are thefe *Mahometanes* alone reckoned, to two hundreth thoufand thoufandes of horfemen: Here are both the Armies, euen both *Gog* and *Magog*, compared to the fea-fand in number. There, doeth three diuelifh infpirations (called fpirits of Deuils) gather & ftirre vp all thofe people to battel: here, the Diuel himfelf is loofed to ftir thẽ vp to battell. There, doe they conveene to fight in *Armageddon*, which is the mountaine of the Euangell, and landes of the Elect and chofen people: Here, do they compaffe the whole lands, companie, and armie of Chriftians, and their elect cities. And finally, thefe foure of that fixt Trumpet or Viall, are prooued by the fourth propofition, to be the foure nations *Mahometanes*. The fame is this *Magog*, prooued to be by the former propofition. And confequently, the armies of *Gog* and *Magog*, whom *Sathan* here in this 20. cap. doeth ftirre vp to warre-fare, are the felfe-fame armies of Papifts & Mahometanes, that made great wars in the daies of the fixt Trumpet or Viall.

Apoc. 20.
Their dwelling,

Their number of nations.

Bondage and loofing
Their multitude.
Divelifh inftigation.

Hier de int. nominum.
The place of their battell.

Proued be the 4 and 32 propofitions.

Conclufion

34. PROPOSITION.

The thoufand yeares that Sathan was bound [Revel. 20] began in An. Chrifti 300 or therabout.

For proofe hereof, it is euident by hiſtories, that after the continuall and ſuccefsiue tyrannie of Ethnick Emperours, and laſt of *Diocletian*, (who in one moneth made ſeuenteene thouſand Martyrs) there aroſe about this 300. year of Chriſt *Conſtantine* the great, a Chriſtian and baptized Emperour, who, and whoſe ſucceſſors (except a fewe of ſhort raigne) maintained Chriſtianiſme, to the aboliſhing of Sathans publike kingdome: and therefore, ſaye wee, this yeare Sathan is bound. Secondly, ſhortlie after this time, was the firſt publike and generall godlie councel, holden by the Chriſtians at *Nice*, in the which the Apoſtolical beleefe was publiſhed, the authentick Scriptures authorized: and finally, the true Chriſtian religion ſo receiued, that all Sathans outward oppoſition was baniſhed, and his publike tyrannie and kingdome ouercome: although yet, by his Lieutenant the Antichriſt, euen then began his difsimulate and hypocriticall kingdome. Thirdly, ſeeing by the former propoſition, *Gog* and *Magog* are the armies of the ſixt trumpet and Viall: and theſe (by the fourth propoſition) were looſed about the yeare of God 1296. to make warres. Therefore, about the yeare of God 1296. or rather (as hiſtories preciſely report) about the yeare of God 1300. were the armies of *Gog* & *Magog* looſed, & ſo Sathã was then looſed, to ſtir them vp to battell: from the which 1300. yeares deduce the thouſand yeares, that Sathan lay bound, and it will conſequently follow, that Sathan was firſt bound in Anno Chriſti 300. Fourthly, and for confirmation of the former, the text ſaith, that how ſoone the Deuill is looſed, he paſſeth foorth to ſtirre vp and ſeduce, theſe papiſticall & Mahometick armies of *Gog* and *Magog*, to ſtrife and warre-fare. But in that 1300. yeare, began (by Sathans inſtigation) that proude ſtrife betuixt them for ſupremacie, both of them chalenging to themſelues the Empire of the whole earth: for Pope *Boniface* the eight, inſtituted the firſt Iubilie that yeare, and clothed himſelf

1. Reaſon Chriſtianiſme did riſe and Sathanes tyrannie reſtrained.

2. Godly councels and good ordinances. Sathanes hypocriticall kingdome ariſing.

3. be a count of yeares. Sathan bound Anno 300. and looſed An. 1300.

Apoc. 20.7.
4 Reaſone the proud & ambitius oftẽtatiõ of the Pope one the one part.

the one day in the Pontificals of a Biſhop, & the other day in the robe-royal of an Emperor, & hauing borne before him two ſwords, proclaimed theſe words, *Ecce duo gladii hic*: meaning therby, that he was ful Monarch & more thã Monarch: For, therby he vſurped to him ſelf vniverſally both ſpiritual & temporal power. Like wiſe, on the cõtrarie part, that ſame yeare, the chiefe nations *Mahometanes* crowned *Ottoman* the great, their firſt Emperour: And ſo betwixt theſe contrarie Empires, euer from thencefoorth followed out ſuch vniverſal and terrible warres, that the Deuil (doubtles) from that 1300. yeare of God foorth, ran looſe among them, ſtirring vp theſe warres, and conſequently, he lying bound 1000. yeares afore, behooued to begin that bondage, in Anno 300.

The pride and contentiõ betwix the Pope on the one ſide and Mahometans on the other ſide begun about the yeere 1300

35. PROPOSITION.

The Deuils bondage a thouſand yeares [*cap. 20.*] *is no waies elſe, but from ſtirring vp of vniuerſal warres among the nations.*

SOme by this bondage of Sathan, haue ſought out to find the viſible Church of God a thouſand yeares pure and vncorrupt, or beleeued that the true Church ſhould enjoy 1000. yeares peace: But the plaine contrary appeareth here by the fourth and fift verſes, where it is meant expreſly, that in the time of theſe 1000. yeares, ſhould the Antichriſtian beaſt raigne, and ſhould ſlay and behead thoſe, that worſhipped not him, nor reuerenced his marke and image, but liued and raigned with Chriſt Ieſus, as true Chriſtians, all theſe 1000. years: and that all the reſt of the people lay dead in his errours: and roſe not there from, till theſe thouſand yeares were expired. And ſo, ſeeing both by this text, theſe Ethnick-like Antichriſtians raigned in this time, as alſo (by the chap. 11, 2.) they poſſeſſed the outward viſible Church: Therefore, could neither the true inviſible Church enjoy peace vnder the Antichriſt,

Sathanes publike tyrannie reſtraned, the Antichriſts privat tyrannie ariſeth.

neither yet could that viſible Church be pure and incorrupt during theſe thouſand yeares, but rather after the ſaid 1000. yeares, becauſe it is meant (verſe 5.) that the reſt, who were dead in Antichriſtian errours ſhould not reuiue, or riſe there-from, till after theſe thouſand yeares. And beſide this, it is not ſaid, that the Deuil is bound from troubling the Church of God (which euer he troubleth) but from ſeducing or deceiuing the Gentiles or nations, which for the moſt part, are taken in the Scriptures in an euil part, for the idolatrous, wicked, and lewde people of the worlde, to whom God here granteth theſe thouſand yeares peace, becauſe they (Revel. 11.2) doe poſſeſſe the outward and viſible Church, and haue the true inviſible Church lurking vnder their feete, which by their continuall trouble, might alſo be troubled. Then reſteth the affirmatiue part of this propoſition, to wit, this bondage of *Sathan*, to be from ſtirring vp of theſe nations, to any vniverſal warre-fare, while the end of theſe thouſand yeares. For proofe whereof, when he was looſed, then (ſaith the text) he paſſed forth to ſtir vp the whole nations of the foure quarters of the earth, *Gog* and *Magog*, to warre-fare and vniverſall blood-ſhed: whereas contrarily, by hiſtories it is found, that al the 1000. yeares præceeding, there were no ſuch hudge imperial armies, nor terrible blood-ſhed as now, whẽ the whole Papiſts & pretended Chriſtians on the one ſide, againſt the whole *Mahometan*s on the other ſide, euen the whole *Occident* againſt the whole *Orient*, vnder the names of *Gog* and *Magog*, were convened in number as the ſand of the ſea, as the text ſaith. So that in compariſon hereof, al ſmall ciuil warres, skirmiſhings, and incurſions, that occurred during theſe thouſand years, were thought nothing but peace, and conſequently thoſe terrible and vniverſal warres, that fell out from this time foorth, were the onely troubles and ſeditions, from entiſing and raiſing whereof, the text meaneth, that *Sathan* was bound and reſtrained all the former 1000. yeares, and not from ſtirring vp of errors,

Note to what end the D*evill is bound.*

His actions (after his looſing) ſhewe wherefore he was bound.

Papiſts & Occident againſt Mahometans and Orient.

Concluſion

herefies, martyrdome, & provinciall warfare.

36. PROPOSITION.

The 1260. yeares of the Antichrifts univerfall raigne ouer Chriftians, begins about the yeare of Chrift 300. or 316. at the fartheft.

THis is prooued, by reafons of Gods apparant mercy, and of the Deuils fubtiltie, and by diuers neceffarie reafons of Scriptures confirmed by hiftories. And firft, feing on the one part, it was meet for the iniquities of the world, that fometime a generall Antichrift fhould come: but on the other part, whileft as Sathan ranne at libertie, and his kingdome, publikely flourifhed by the tyrannie of *Ethnick* Emperours, by whom Gods Church was extreamely perfecuted: It was vnmeete at that time, that the Antichrift fhould alfo rife: leaft on both fides, Gods Church were vtterly extinguifhed: and therefore it is not apparant, that God in his mercie would fuffer the Antichriftian kingdome to rife, vntil the Empyre of *Ethnike* Emperours were firft remooued, and Chriftian Emperours eftablifhed in their place: and fo Sathans publike tyrannie bound and reftrained. But this came to paffe, in Anno 300. that the firft race of Chriftian baptized Emperours, is begunne in *Conftantine* the great: & Sathans publike and imperiall tyrannie fuppreffed. Therefore, at this time it is moft apparant, that God in the temperature of his juftice & mercie, would rathereft fet vp the Antichrifts kingdome, and herewith agreeth the faying of Paul, that *he that holdes, muft holde ftill a while, till he be taken out of the way or remooued, and then muft that man of finne or Antichrift come*: meaning, the wicked Emperours that held the Empyre, muft hold ftill the fame till they be remooued, and then fhould the Antichrift ftart vp: & fo the one enemie of God to giue place to the other, and not both to ftande. † Secondly, by the Diuels craft and fubtiltie (herein

1. Reafone. the midft betwixt Gods Juftice and mercie.

2. Theff. 2.

† 2 Reafon. The devils craft to tyrannife diffimulatlie when he was reftraned publiklie.

permitted by God) it is very apparant, that while as his publike tyranny, that he vſed by his *Ethnick* Emperours, was newly extinguiſhed, and himſelfe in a maner, to be bound and reſtrained for 1000. yeares, he would in this his bondage from publike tyrannie, rather eſtabliſhe the Antichriſt, as his Lieutenant, to vſe diſsimulate tyrannie, rather than at any other time: but this his bondage and Chriſtianiſme of Emperours, beganne about this 300. yeare, by the 34. propoſition. Therefore, about this 300. yeare, it appears rathereſt that the deuil by his ſubtilty, hath ſubſtituted the Antichriſt, and made him to beginne his kingdome. Thirdly, theſe 1260 yeares of the Antichriſts raigne, or three greate daies and a halfe, that the witneſſes or teſtaments of God lyeth dead (which by the fifteenth propoſition are all one date) being ended then (ſaieth the text) *ſhall the tenth parte of the Antichriſtian citie fall, euen the tenth part of all the Abbacies, Monaſteries, Nunries, and Papiſtical policies be deſtroyed.* But ſo it came to paſſe, that neere about the yeare of GOD 1560. theſe Papiſticall policies were deſtroyed in England, Scotland, and in ſome partes of Germanie, France and other countries, euen their large tenth part, from the which 1560. yeare, deduce 1260. for the yeares of the Antichriſts vniverſall reigne, ſo reſts 300. yeares to be the yeare of Chriſt, about the which the Antichriſt beganne his kingdome. Fourthly, *Paul* teſtifieth, that *thoſe that hold, muſt hold ſtil a while til their ſeate be tranſported, diuided, or remooued out of the way, and then ſhal the Antichriſt reueale and open vp his empire, for he ſhall ſit* in their ſeat, *ouer the Church of God, extolling himſelfe as God, &c.* And ſo it came to paſſe, that the *Romane* Emperours ſate that time, & held the empyre of *Rome*, til *Conſtantine*, Emperour thereof, tranſported the ſeate of the empyre to *Bizantium*, which he called *Conſtantinople*, leauing the ſeate of *Rome* to Pope SYLVESTER the firſt and his ſucceſſours, who was the firſt Pope or Biſhoppe that raigned in *Rome* vnmartyred, and the firſt that

Third reaſon. By deduction of yeares.

Ap 11.13.

2 Theſſ. 2.4.7.
4. Reaſon The Emperiall ſeat tranſported.

euer poſſeſſed patrimonie, rent or dignitie, from which time, their kingdome euer hath encreaſed vntill our daies, and therefore, juſtlie about his time, which beganne in the yeare of God 313. or thereby, eſtabliſh wee the beginning of the Antichriſtian and Papiſtical raigne. Fiftly, becauſe it is ſaid by *Daniel* & by the *Reuelation*, that he muſt all the ſaid ſpace, haue power and authoritie ouer the Saints of God, and by *Paul* in the ſaid text, that he muſt be extolled againſt all that is called God. Therefore muſt wee begin our count, at the time of the Popes firſt power & authoritie, euẽ his very firſt coronation, and this was about the yeare of God 316. or thereby: for then *Conſtantine* Emperour, gaue to the ſaid Pope *Sylveſter* the towne of *Rome*, and great landes and dominions there-about, and gaue vnto him the triple Crowne, to be crowned therewith, in token that he made him ſupreame head ouer all the Churches in *Aſia, Africa*, and *Europe*, as his gift, contained in the decrees, Diſtinct. 96. at length proportes, the date whereof is, *Data Romæ. 3. Calend. April. a Conſtantino Auguſto quarto conſule, & Gallicano quarto conſule*, which date repugneth in it ſelffe, for wee finde by common writers that *Conſtantine* his fourth conſulſhip, and *Licinius* his fourth conſulſhip, were together *in anno Chriſti* 315. or according to *Haloander in anno 316.* But we finde neuer, where *Conſtantine* and *Gallicanus* were conſulles together, neither yet that *Gallicanus* was foure times Conſull: Whereby, not onely the date of this gift is vncertaine, what yeare of God it befell, but alſo, by this repugnance, and diuerſe other reaſons, ſome thinke this whole gift to be falſe and fained, yet alwayes in this purpoſe be it true or falſe, it is al one matter: for ſeing they acclaime it as true, and thereat make their beginning: it pertaineth not to vs to reaſon here, whether their beginning was truelie giuen, or falſlie vſurped, but to make the beginning of our compt, where they begin their kingdome: to wit, at the ſaid gift, falling betwixt the yeares of God 300. and

Dan. 7.25, Apoc. 13.5

5. His authoritie & power then begunne.

Conſtantins gift to Pope Sylveſter.

Alwayes bee it true or ſalſe, thereat they begin their account of primacy by indictions.

316. whereas the precise yeare thereof, is (as is said) vncertaine. Sixtly, *Hermannus Gigas* writteth, that at the time of that donation there was a voyce heard in the aire, saying, *Hodie venenum infusum est Ecclesiæ*. Also *Maleolus de nobilitate*, and *Nauclerus lib. 2. generatione 11*. saieth that the same words at that same time were written by an hand vpon the wal of the Lateranne palace, in al mens presence, which miraculous voyce and hand, apparantly establisheth and maketh plain to vs, that at this time began the horrible and detestable kingdome of the Antichrist, as a poysonable venome vnto the Church of God. Seventhly, at this time began the first Antichristian marke, proued by the 31. proposition, to be the abused marke of the crosse, which now was induced among Christians, by the fabulous alleageance of two fained miracles: the one that Queene *Helen* the mother of *Constantine*, admonished by an heauenly vision, passed, & did find that very real crosse whereon our Lord suffered: the other that *Constantine* her sonne, fighting against *Maxentius*, saw appeare in the aire, the figure of a crosse with these wordes, *Jn hoc signo vinces, by this marke thou shalt ouercome*, with which marke and inscription, the *Portugall* ducat and some other coines of late are imprinted. And therefore here, where the first publike and visible Antichristian marke beginnes, esteeme we iustlie the Antichristian kingdome to beginne. Eightlie, about this time, beginnes the account of two sorts of *Indictions*, in place of the olde *Olympiads*: the one, *Indictio Constantiniana*, which beginneth *anno Christi*. 312. the 24. of *September*, at which time the said Emperour *Constantine* ouercame *Maxentius*: The other is *Indictio pontificia* beginning the first day of Ianuarie *Anno* 313. which was also the first yeare of Pope *Sylvester* the first, & first Pope vnmartyred, euẽ that very day & year beginneth the commõ account of the papisticall kingdome, by *Indictions*, and so proceedeth for

6. A marveilous voyce and vision.

7. The first antichristian marke then began

The first fained miracle.

The second false miracle.

8. Reasone the Indictions.

First sorte of Indictions,

Second sort of Indictiõs

Ottoman. Iose. Scal. de emendat. temporum.

15. yeares, and then beginneth againe the firſt yeare of a newe *Indiction*, keeping thereby, the exact memorie of the current yeares of their kingdome, within every fiftene, and dating therewith all their papiſtical evidences, even like the Iewes, who keeped compt of the od and current yeares of their *Iubilies*, from their conqueſt of the land of *Canaan*: The number of whole *Indictions*, and of whole *Iubilies*, being more eaſie to keepe account of, nor their current yeares. Therefore, iuſtlie at this firſt yeare of this *Indiction*, and firſt account of their kingdome, beginning at new yeare day in *Anno* 313. which yeare alſo the ſaide *Sylveſter* began, ought wee to begin the Antichriſtian kingdome. So by theſe eight former reaſons, we conclude, that betwixt the yeere of Chriſt, 300. and 316. hath the Antichriſtian and Papiſticall reigne begunne, reigning vniverſally and without any debatable contradiction. 1260. yeares, and ſo (as is ſaid) about the yeare of God. 1560. beganne the firſt publike decay of their policie, and the open repining againſt their kingdome to their confuſion, euer more and more, till firſt their rents: then their citie of *Rome*: laſt of al, by Chriſts owne comming their latter dregges be vtterly extinguiſhed and exterminate, as in the principall treatiſe, vpon the 17 and 18 Chapters ſhall be declared.

The progreſſe of the decay of Rome,

So ends this demonſtratiue reſolution of all difficulties of the Reuelation, firſt of al dates and times, and laſt of the principall termes and matters, as to the meaner termes and ſmaller matters they are interpreted in the notes of the principall treatiſe.

CONCLVSION.

THen for concluſion, by theſe interpretatiue propoſitions, followeth foure things marueilous and notable. Firſt, that the interpretation of euery part of the Reuelation, is acceſſorie or conſectarie to other: that is to ſay, it is ſo chained and linked together that euery myſterie opens other to the diſcouerie of the whole. Secondly, that the firſt halfe of the booke is orderly, that is to ſay, it containeth in order of time the moſt notable accidents that concerneth Gods Church, frõ the time of Chriſts Baptiſme ſucceſsiuelie to the latter day. Thirdly, that euerie hiſtorie prophecied is limited or dated with his own number of yeares. Fourthly and laſt of al, that whatſoeuer hiſtorie is more orderlie and ſummarlie, than plainly ſet downe in the firſt orderlie parte of the booke: the ſame is repeated, interpreted, or amplified in the laſt part of the booke: which therefore we call the amplificatiue parte of the booke, diuiding the whole Reuelatiõ according to the table following before we proceid to the principal mater.

Foure conſectaries in the propoſitions.
1. Each part conſectarlie openeth other.
2. That the firſt part of Apocal. is orderlie.
3. Euerie prophecie is dated.
4. The breuitie of the firſt part ſupplied & amplified in the ſecond part.

A TABLE DEFINITIVE AND

Diviſiue of the whole Reuelation.

THe *Reuelation*, is a diſcouerie of thoſe notable matters that concerne the Chriſtian Church, from her beginning to the latter day, ſet out by viſions, which Chriſt hath ſhewed to his Apoſtle *Iohn* : And it containeth

- A particular admonition, which as an Epiſtle, is directed vnto the ſeuen Churches of Litle Aſia, and it containeth
  - The preface, which ſheweth the director, the receiuer, & to whome the meſſage is directed, ſet out ſpiritually, cap. I.
  - The meſſage directed particularly to the Churches of Epheſus, Smyrna Perganus, Thyatira, cap 2. And to Sardis, Philadelphia & Laodicea, cap. 3. And it containeth
    - The reproofe of certaine points of their defection, threatening them to amendement
    - The commendation of certain of their godly doings, exhorting them to perſeuerance.
- A generall diſcourſe of all alterations, which generally doe concerne or intereſt the Chriſtian Church, from her beginning to the latter day, and it containeth
  - An orderly part, which deduceth ſucceſsiuely, briefly, & in due order of time all alterations concerning the Chriſtian Church : and it containeth
    - A hiſtorie of thoſe thinges which were accompliſhed before they were written, expreſſed cap. 4. 5. 6. 7. & beginning of the 8. vnder the termes of ſeven ſeales.
    - A prophecie, forewarning all notable accidentes alterations and changes, that concerne the chriſtian Church, betuixt S. Iohn his dayes and the latter day : and it is
      - Secular, which forewarneth ſuch mutations of earthlie kingdomes, and commotions of Empires as any way concer neth the Chriſtian Church, & this is done in the reſt of the 8. chap. & in the whole 9. chapter, vnder the termes of ſix of the ſevẽ Trumpets. Hereafter followeth in the tenth chapter, an introduction to the ſeuen thunders, or ſeuen thundering Angels, which are myſteries of the ſeuenth trumpet and latter day.
      - Eccleſiaſticall, which ſimply and altogether forwarneth the onely progreſſe of God's true church, and this is done Chap. 11.
  - Ane amplificatiue part, wherein are more largely ſet downe and expounded thoſe greateſt and moſt notable maters, which wer ouer ſummarlie, briefly, or obſcurely expreſſed in the orderly part, & this forewarneth the eſtate of the Elect,
    - on earth by commotions
      - Eccleſiaſticall, wherin are deſcribed, the Antichriſtian beaſtes ſtrife with Gods Church cap. 12 and that Beaſt his tokens and qualities cap 13 and the progres of his falling, and the riſing of the Evangell to the latter day, cap. 14. and the interpretation of the Beaſtes tokens cap. 17. & the decay of his citie and ſeat, Cap. 18. and the triumph alſo of Gods Church, cap. 19. And finallie is expounded briefly, cap. 20. how Gods publik & priuat enemies, ſtriue against other, and both againſt his church : but at length to their owne wracke and confuſion eternally, Amen.
      - Seculare, which are introduced by preface, ca. 15 & dilated in order, cap. 16. vnder the termes of ſeuen Vials.
    - in heauen.deſcribed cap. 21 and 22. vnder the terme of new Ieruſalem

THE SECOND

And principall treatiſe wherein [by the former grounds] the whole Apocalyps or Reuelation of S. IOHN, is paraphraſticallie expounded, hiſtoricallie applyed, and temporallie dated with notes on euerie difficultie, and arguments on each Chapter.

CHAP. I. THE ARGVMENT.

Theſe three firſt Chapters, being in way of Epiſtle, particularlie directed to the ſeuen Churches of ASIA *minor: he declareth in this firſt Chapter (in maner of preface) who is the director of this meſſage, who are the bearers thereof, and thirdlie, to whome the meſſage is directed: confirming the veritie of all by an heauenlie viſion, whereof in the end of the Chapter he yeeldeth the interpretation.*

The paraphraſticall Expoſition.

THis is a booke (*a*) reuealed by IESVS CHRIST, whereby God appointed him to ſhew foorth and open vnto his ſervants, things that muſt ſhortly bee performed in the laſt daies, the which Chriſt hath ſent his (*b*) Angels to ſhew & reveale to his beloued Apoſtle (*c*) IOHN,

2 Who alſo is an Evangeliſt, and therby beareth the record of the word of God, and of the hiſtory of our ſaluation in Ieſus Chriſt, and of all things that he hath ſeene, both then, and at this time.

3 Bleſſed is he, that reading and hearing the words of this Prophecie, doeth obſerue and keepe, truely the meaning and contents thereof: for ſhortly, & in this laſt age of the world muſt the ſame be performed.

4 I *Iohn*, vnto the ſeven Churches which are in litle *Aſia*, wiſh grace and peace from God the Father, who is, and euer was, and who ſhall come, and abide for euer: and from that holy, and [*d*] one Spirit of God, whoſe ſeuen-fold vertues and perfections appeare, euer miniſtring before the

The text.

1 The reuelation of Jeſus Chriſt which GOD gaue vnto him, to ſhew vnto his ſeruants things which muſt ſhortlie bee done: which he ſent, and ſhewed by his Angel vnto his ſeruant John.

2 Who bare record of the word of God, & of the teſtimonie of Ieſus Chriſt, and of all things that he ſawe.

3 Bleſſed is he *that readeth, and they that heare the words of this prophecie, and keepe thoſe things which are written therein for the time is at hand.*

4 John to the ſeuen Churches which ar in Aſia, Grace be *with you, and peace from him, Which is, and Which was, & Which is to come, and from the ſeuen Spirites, which are*

throne of God.
5 And from Iefus Chrift, God and Man, who in his owne perfon, hath borne a faithfull teftimonie of the refurrection of the dead, and of all that euer died, hee is the [*e*] firft borne, and begotten before all eternitie, and is the firft fruits of the refurrection, euen the firft among the dead that arofe vnto heauen: & he is King & Lord ouer all the princes of the earth: Vnto him therefore, that fo hath loued vs, that he hath wafhed vs from our finnes, by fhedding his blood to the death for vs
6 And thereby hath made vs as Kings, to raigne with God his Father and as Priefts, to offer vp our daylie prayers and laudings to him, euen vnto that Lord Iefus (I fay) be glorie and dominion for euermore, So be it.
7 Behold, he fhall come to judgement in the [*f*] clouds, and all men fhall fee him with their eyes, yea, euen they which wounded and crucified him, and the wicked of all the kinreds of the earth, fhall quake and mourne before him, euen verily fo fhall it be.
8 I am the [*g*] firft, I am the laft, euen the firft, without any beginning, and the laft without any ending, faith the Lord God, who prefẽtly is, hitherto euer hath bene, and fhall come, & abide in all time hereafter, euen for euer and euer: for I am the Almightie.
9 And I *John*, your brother in Apoftlefhip, and a partakar of tribulation with you, for the kingdome of heauen, through patience in Iefus Chrift was fugitiue in one of the yles of the *Cyclades*, called [*h*] *Pathmos*, for profefsing the worde of God, and for bearing the true teftimonie of Chrift Iefus.
10 And there I was rauifhed in fpirit, vpon that day of the week that our Lord rofe frõ death, which is the [*i*] Sunday: then heard I behind me a mighty voyce, terrible, and loud as a trumpet,
11 Saying, I am the beginning, and the ending, yea, the firft before al beginning, and the

before his Throne:
5 And from Jefus Chrift, which is a faith full witneffe, and *the firft begotten of the dead, and Prince of the kings of the earth, vnto him that loued vs, and wafhed vs from our finnes in his blood,*
6 And made vs kings and Priefts vnto God, euen his father, to him be *glorie and dominion for euermore, Amen*
7 Behold, he cõmeth with clouds, & euerie eie fhal fee him: yea, *euen they which pierced him through: and all kinreds of the earth, fhall waile before him, enen fo, Amen.*
8 J am Alpha & Omega, the beginning & the ending, faieth the Lorde, which is and which was, & which is to come, euen *the Almightie.*
9 I Iohn, euen your brother, & companion in tribulation, and in the kingdome & patience of Iefus Chrift, was in the yle called Pathmos, for the word of God, & for the witneffing of Iefus Chrift,
10 And I was rauifhed *in fpirit on the lords day, and heard behind me a great voice, as it had bin of a trumpet.*
11 Saying, I am Alpha & Omega the firft and the laft: and that which thou feeft, write in a booke and fend it to the feuen Churches which are in Afia vnto Ephefus & vnto Smyrna,

laſt after all ending: All therefore, that I let thee ſee, write in a booke, and ſend the ſame to the ſeuen Chriſtian Churches, which are in litle *Aſia*, to wit, to the Churches of *Epheſus*, of *Smyrna*, of *Pergamus*, of *Thyatira*, of *Sardis*, of *Philadelphia*, and of *Laodicea*.

& vnto Pergamus & vnto Thyatira, & vnto Sardi and vnto Philadelphia, and vnto Laodicea,

12 Then I turned back to ſee him, whoſe voyce I had heard ſpeaking to me: and when I had turned, I ſawe ſeuen ſhining Candleſtickes, bearing the type and figure of thoſe ſeauen [*k*] Churches.

12 Then I turned back to ſee the voice, that ſpake with me: & whẽ I was turned, I ſaw ſeuen goldẽ candleſticks,

13 And in the middeſt of theſe ſeuen, Chriſt himſelfe in his Deitie, vnder the [*l*] likeneſſe of his Humanitie, arrayed as in [*m*] long garments, betokening, grauitie and wiſedome: and in ſigne of [*n*] diligence and readineſſe, hee was girded with the golden, and precious girdle of veritie, juſtice, and power.

13 And in the middes of the ſeuen candleſticks, one like vnto the Sonne of man, clothed with a garment downe to the ſeete, and gird about the pappes with a golden girdle.

14 His head and white haires, whiter than wool or ſnow, importeth his [*o*] honourable and reuerent antiquitie and age from all beginning: his fierie eyes doe pearce the ſecrets of all hearts, and terrifie his enemies.

14 His head & haires were *white as white wool, & as ſnow, & his eies* were *as a flame of fire,*

15 His feete are firme and ſtable to ſtand, as braſen pillers in defence of his Elect, and are ſwift and terrible as a conſuming fire, againſt the wicked, and his voice by his preachers is loude and open, as of a multitude of [*p*] people and waters.

15 And his ſeete like vnto fine braſſe, burning as in a fornace, and his voice as the ſound of many waters.

16 And he preſerued and keeped in his mightie hand of Protection, the holy [*q*] teachers and inſtructers of theſe ſeuen Churches, who are ſtars of heauen, yeelded the light of Gods trueth among them, and out of his mouth, by his teachers, proceded the ſword of [*r*] true doctrine two edged, conquering in his elect on the one ſide, and ſubdewing and deſtroying his enemies on the other ſide, and in the beautie of his face ſhined (brighter than the Sunne) the bright light of æternall life.

16 And he had in his right hande ſeuen ſtarres: and out of his mouth went a ſharpe two edged ſworde: and his face ſhone, *as the Sunne ſhinneth in his ſtrength.*

17 And when I ſaw his heauenlie majeſtie, I

17 And when I ſaw him,

(lacking ftrength) fell at his [*f*] feete as dead: then extended he his mercifull hande, and gentle protection ouer me, comfortably faying vnto me, be not afraide of me, I am thy Lord, without beginning, and without ending.

18 I rofe to life, and doe liue, though I [*t*] was flaine to redeeme you, yea, now beholde, I liue for euermore, Amen: and I haue power ouer death and hell, to open and clofe the fame at my will.

19 Write vp therfore, thefe things which thou haft feene by vifion, both concerning thefe things that [*u*] are, and thefe things which fhall come to paffe hereafter.

20 And be fure, the feuen ftarres, which by vifion thou didft fee in my right hand, and the feuen golden candlefticks, are myftically to be vnderftood (as before is faid) euen the feuen ftarres doe meane the [*x*] feuen meffengers or minifters, which God hath fent to inftruct thefe feuen Churches and the feuen candlefticks which appeared to thee, doth meane thofe fame feuen Churches.

I fell at his feete as dead: then he laide his right hand vpon mee, faying vnto me, Feare not: I am the firft and the laft,

18 And am aliue, but J was dead: & beholde, J am aliue for euermore, Amen: and I haue the keies of hell and of death.

19 Write the things which thou haft feene, and the thinges which are, and the thinges which fhall come hereafter.

20 The myfterie of the feuen ftarres, which thou faweft in my right hand, and the feuen golden candle-ftickes, is this, *The feuen ftarres are the Angels of the feuen Churches: and the feuen candleftickes, which thou faweft, are the feuen Churches.*

Notes, Reafons, and amplifications.

(*a*) As God hitherto never left his Church deftitute of doctrin to inftruct them, and of prophecies to fore-warne them: so here hath his Maieftie prouided, al the former bookes hitherto of the new Teftament, to containe the Chriftian doctrine: & this book of *Reuelation*, to containe the whole chriftian Prophecie to the worlds end. *Apocalyps* then is a Reuelation, and therefore, is this booke fo intituled: for that, by the will and ordinance of God, Chrift hath revealed the fame, word by word, as it is written in the text, to his Difciple *Iohn*, and now doth daily by his holy Spirit, reueile more and more the meaning thereof, to vs in this our age, for the which it hath bene chieflie inftituted.

(*b*) Here is this booke defcribed to proceede from God, & come to the feuen Churches by foure degrees: Firft, God giueth the fame to Chrift: Secondly, Chrift fendeth his Angel, or Angelles bearing commifsion: Thirdly, the Angel inftructeth *Iohn*: Fourthlie, *Iohn* writeth the fame to the feuen

Churches. Not meaning hereby, that the Angelles onely are appearers to *Iohn*: but ſometime the Father, as chap. 4. ſometime the Sonne, as in this chapter: ſometime the holie Spirit, as chap. 14.15. and other places: at which times, for confirmation of the ſmaller viſions: the Trinitie in theſe great viſions, ſheweth his vnſpeakable glorie, but in other ſmaller viſions, concerning things to come, the Angels (as ſayeth this text) are made reporters, ſhewers, and expoſiters, as chap. 17.1.7. and diuers other parts, as plainely appeareth cap. 22.8.16. For hee ſaieth not here, the Angels are ſent to ſhewe all the viſions, or viſions of Gods glorie, for it was the Trinitie it ſelfe that ſhewed it ſelfe, and appeared in his owne viſions, and not the Angels, as hereafter (note *l*) ſhall be ſhewed: But he ſpeakes here, of propheticall viſions, which Chriſt ſaith here, he ſent his Angels to ſhewe.

(*c*) Some haue vnadviſedly impugned the authoritie of this booke, by diuers friuolous conjectures: & among the reſt, becauſe *Iohn* hath ſo often herein named his name, thinking it not agreeable to the Apoſtolike grauitie, to caſt vp ſo oft his owne name: and therefore, eſteeme they rather this booke to be ſet out in his name by others than by himſelfe, for that hee doeth not the like in his Euangell. But contrarie to their opinion, in all the Scriptures, the writers of doctrine, doe not ſo often expreſſe their names, as the writers of prophecie doe: for the doctrine receiues the authoritie of the owne preſent veritie, and of the Spirit of God, but the Prophecy receiues the authority by the author, who is knowne to be a Prophet, whereas otherwiſe, the booke could not be knowne to be true before the performing thereof, conſider hereupon, *Daniels* prophecie, where ye ſhal find his name more than threeſcore times repeated: yea, in the firſt perſon, ye ſhal find (*I Daniel*) repeated about nine times, in the 7.8 9. & 10, chapters, wheras *Iohn* repeateth his name but fiue times in all this booke. Hereby then no man can gather, but that this book was written by that ſelfe-ſame *Iohn*, who was both Chriſts beloued Diſciple, and an Euangeliſt. And therefore, ſtileth hee himſelfe here the ſeruant of Chriſt, who hath borne witneſſe of Chriſt Ieſus, of theſe things which he ſaw, as being one of the foure Euangeliſts that wrote the hiſtorie of Chriſt, which with his eyes he ſawe.

(*d*) There is but one holie Spirit, whoſe chiefe functions and graces are deuided into ſeuen: whereof reade the note (*e*) of the fourth chapter. For the better comfort therefore of the ſeuen Churches, to ſupplie their beleefe and weake faith, hee letteth them vnderſtand, that there is ſuch a compleat perfection in Gods Spirit, that it hath as many functions and graces to conſerue them with, as are agreeable to their number of ſeuen.

(*e*) For two cauſes is Chriſt called the firſt borne of all that died: Firſt, becauſe in the Leuiticall Law, the firſt borne, & firſt fruites, or primices, were conſecrated and offered vp vnto the Lord, and ſeeing of all men that euer tooke life and died, onely Chrift is that firft fruite, that is offered vp an acceptable Sacrifice to God the Father, as PAVLE teftifieth, 1. *Cor.* 15.20. Therefore, he is called the firft borne, euen in his humanitie figuratiuely, although really many others were borne neere foure thoufand yeares before him. Secondly, in confideration of his Deitie, euen of that worde, that is God, and long after tooke flefh: euen in confideration (I fay) of the eternitie thereof, becaufe it was before all beginning: Therefore, is Chrift (who is that word) called the firft borne among the dead, *Col.* 1.15.17.18. Although not the worde or Deitie that tooke flefh, but the flefh it felfe died: and fo in this fenfe, the being firft begotten is imputed to Chrifts Deitie, and eternall word, and to die, is imputed to his flefhlie humanitie, and fo to Chrift, as he is fully both God and man, both thefe properties of being the firft borne, and of dying, are juftlie appropriate, according to the diuers refpects aforefaid.

(*f*) Euen in the clouds he afcended from hence into heauen, Act. 1.9.10.11. and that way fhall he returne to iudgement: read hereupon cap. 14. note (*m*) thereof.

(*g*) Of the greeke A, B, C. (in which language this booke was written) *Alpha* is the firft letter, and *Omega* is the laft: and fo God ftiling himfelf the firft letter and the laft of that A, B, C, he both thereby meaneth, as alfo incontinently fubioyneth the meaning thereof, faying, *I am the beginning and the ending.*

(*h*) *Pathmos* is one of the yles, that are called *Cyclades*, lying in *Mare Aegeo*, be-Weft the coaft of *Afia minor*, as *Strabo* witneffeth: Herein hid S. *Iohn* himfelf from the great tyrannie and perfecution of *Domitian*, where he abode vntill *Domitian* died, and that yeare *Anno Chrifti*, 97. or 99 he wrote this booke in that yle: as appeareth by *Irenæus* and *Eufebius*,

(*i*) The day of reft and godly exercifes, while Chrifts death was celebrated vpon the Saterday, which of olde was the Iewifh Sabboth day, but in memorie of Chrifts refurrection, and victorie ouer death, the Apoftolike Church tranfported the exercifes of that day to the Sunday, which was the day of the weeke, that Chrift arofe vpon, and therefore is called the Lords day: the which day, the people (appearingly conveening to ferue God) are defired by PAVLE, I. Cor. 16.2. to giue contributions to the poore, & there he calleth it the firft day of the week: And fo indeed, where the Saterday is the feuenth and laft day of the weeke, the Sunday falles to be the firft, and is vfed for our Chriftian Sabboth and day of the Lord.

(*k*) That thefe feven Candleftickes doe meane thefe feuen Churches of litle *Afia*, is plaine by the end of this fame chapter. Furthermore, how the olde Leuiticall Candleftick was alfo a figure hereof, is to be faid in the fourth chapter, note (*e*) thereof.

(*l*) Becaufe the text calleth this midmoft not the Son of man, but like vnto the Son of man: Therefore, ſome may thinke this not to be Chriſt, but an Angel bearing the type and figure of Chriſt, whom Chriſt had deputed verſ. I. to ſignifie and repreſent theſe things to *Iohn*, and ſo this Angel, not to be the Sonne of man, but like vnto the Sonne of man: But (after mine opinion) God, who in his ſecond commandement, will haue vs to repreſent him by no forme, will alſo repreſent himſelfe to vs by none other, but by himſelfe, leaſt we otherwaies being abuſed, ſould giue the honour of God to Angels. And certaine it is, that neither wil God be the authour, nor good Angels the inſtruments, ſo to illude mankinde: And for confirmation then, that this mid moſt is Chriſt, & no Angel, nor other bearing the type of Chriſt: it is euident, that hereaſter he calleth himſelfe by the ſtiles of Chriſt the *firſt* and the *laſt, who liueth & was dead*: and finally (cap. 2.18.) he calleth himſelf the Sonne of God: And theſe titles (I am ſure) none in heauen will accept nor vſurpe, except Chriſt onely, to whom they appertaine. But then may you demande, why doeth not Chriſt call himſelfe really the Sonne of man, but like vnto the Sonne of man? The anſwere hereof, making much againſt Tranſubſtantiation, is therefore to be deduced at length. I ſay then, that ſo long as Chriſt was in this worlde, appearing corporally to man, he ſtiled himſelfe the Sonne of man, as hauing his fleſh and humane body converſant with vs: but contrarily, both before his incarnation, as alſo ſince his aſcenſion, when euer in all the Scriptures he did viſibly appeare to his Saints, his viſible ſhape & apperance, is not called the Sonne of man, but the likeneſſe of the Son of man, *quaſi Filius hominis*, or *ſimilis filio hominis*, as in *Dan.* 7.13. & *Dan 10. Reu*, 14.14. and in this place, or whereſoeuer elſe: & yet by the whole properties of theſe ſelfe-ſame texts, that ſame that appeared in viſion, is knowne to be Chriſt, as this ſame, who here is called the likenes of the Sonne of man, or like vnto the Son of man, is hereaſter in expreſſe tearmes, called the Sone of God: neither maketh this any repugnance or contradiction, to the eſſential nature of Chriſt, who is really both the Son of God, & the son of man, being ſpoken of Chriſt, either before his incarnation, or ſince his aſcẽſion: for firſt, becaus his humanity could not be really ſeen of *Daniel*, before it was conceiued of the Virgin: therefore, *Daniel* ſaw Chriſts Godhead, euen that Word that is God, ſaw he tranſfigured (for his capacity) into the likenes of a man, or likenes of the son of man, but not to become man, nor the son of man, till hee tooke fleſh of the Virgin. The like tranſfiguration is oft read to be alſo of Angels in the likenes of men, but not in men. Alſo, ſecondly, ſince Chriſt his aſcenſion into heauen, his humanitie hauing reſidence onely there, his Deitie being both there, and alſo euery where: that Deitie (I ſay) euẽ after Chriſts aſcenſion, as before his incarnation, is ſaid here to appeare, not in his humanitie, as the Son of man, but in the likeneſſe of the Son of man, for that his humanitie, whereby he is the Sonne of man, is not here, but in heauen, and his Godhead which is here, taketh not the flefh of the Sonne of man, but the vifible likeneffe of the flefh of man, by fuch a notable difference, either in the vifion, or in the spirit of the Prophet, that he knew it to be Chrift in his Deitie, and not in his humanitie, but bearing the figure and likeneffe of his humanitie. Some vnadvifedly will thinke this, to meane a feparation of Chrifts Deitie from his humanitie. But contrarily, we fay, that Chrift his Dietie, conjunct in heauen with his humanitie, doeth both make his chiefe refidence with the fame, and doeth alfo flowe there-from ouer all the whole world infinitly, as in comparifon, the light beames and fpiritual vertues of the Sunne, are conjunct in heauen with the corporall and vifible bodie of the Sunne, wherein they haue their chiefe refidence, & do alfo flow from thence vnto euerie place: Alfo, the fpirituall vertue and fight is in the body of the eye, as his chiefe feate, and doeth likewife flow from thence towarde all things vifible: Can any man fay, that the beame fubfifteth it felfe alone, and is feparate from the Sunne, or that the fight fubfifts it felfe alone, and is feparate from the eye, feeing the bodie of the Sunne is the verie locall feate of light, and the eye of the fight? No more make we Chrifts Deitie alone to be feparate from his humanitie: the more that the Deitie is vniverfall and euery where, and the humanitie locally in heauen: for the Deitie (being euery where) is alfo in heauen with the humanitie, though it likewife from thence extendeth it felfe, and floweth ouer euery place. So that, for conclufion, the Divinitie of Chrift, that heere appeared to S. *Iohn*, brought not down his humanitie from heauen, for then he had called that humanity that appeared, the Son of man, as (cap. 2.18) he calleth his Divinitie the Son of God: but that Divinity tooke only, in prefence of the Prophet, and for his inftruction, a fhape like a man, which therefore, he calleth not the Sonne of man, but the likenes of the Son of man. Hereof then followeth, that feeing Chrift commeth no otherwife into the worlde fince his afcenfion, than he came before his incarnation, that is to fay, by his Deitie, and that his Deitie fheweth not his humanitie, here on earth to his Saints, but onely the likeneffe thereof (as faith the text) and fo brings not downe his real and naturall bodie from heauen while the latter day: therefore, his real bodie can never be in the Hofte, nor tranfubftantiate therein. Further hereof, reade note (*f*) following, and note (*d*) of the fecond chapter.

(*m*) Long garments or gownes were of olde, and to this day worne of Doctors and Senatours, to reprefent grauitie and wifedome, as may be feene, that euen in thofe dayes the Scribes, Pharifes, and Doctors of the Law did vfe, or rather abufe them: & therefore, are reproached of Chrift, *Mar.*12.38. But here by the right vfe and wearing thereof, Chrift doth typicallie reprefent, his wifedom and perfect grauitie.

(*n*) Girding is a figne of readineffe: read *Luke*, 12.35. and *Luke* 17.8. fpecially, readineffe in execution of iuftice, *Efa.*11.5. *Iuftitia cingu lum lum-*

borum eius: Iuſtice ſhall be the girdle of his loynes, conioyned with power and ſtrength, as *Pſal.* 17.33.40. The girdle therfore, that repreſenteth theſe excellent properties, is ſaid to be of gold: that is to ſay, precious, and honorable, as likewiſe is to be ſaide, *Revel.* 15.6 note (*e*).

(*o*) White haires import alſo gravitie, wiſedome, and age, worthie and due of all reverence, as ſaith the Poet, *Ovid. Faſt. 5.*

Magnafuit quondam capitis reverentia cani.

In old times men much honoured, The old and graue white-haired head.

And ſeeing theſe properties of grauitie, wiſedome, and honour, as alſo of antiquitie: belong to Chriſt, he therefore, for our capacitie, is deſcribed with white haires.

(*p*) Waters here may either be literallie taken, or prophetically, as *Revelat*, 17.15. for people: alwayes meaning here the voyce of God to be lowd, open and publique.

(*q*) Theſe ſeuen ſtarres are in the end of the chapter, interpreted to be ſeuen Angels or meſſengers, which are the miniſters of the ſeuen Churches.

(*r*) As the tongue and euill ſpeaking of the wicked, is called the ſword of their mouth, *Iob.*5.15, *Pſal.*56.5. and *Pſal.*58.8. and *Prov.* 5.4. whereby the godly are trapped, deceiued, and oftentimes deſtroied: So alſo *à contrariis*, this two edged ſword of Chriſt, in his teachers mouths, is the doctrine of trueth and true diſcipline, whereby the wicked are confounded, and the godly are conquered, & broght in into Chriſts kingdome, *Apoc.* 19.15. And for further confirmation that this two edged ſword is the true doctrine of God, read *Paul* to the *Hebrewes*, 4.12. ſaying, *The worde of God is liuely, and mightie in operation, and ſharper then any two edged ſword*, and *Pſal.* 148.6 *And a two edged ſword in the hands of the Saints*

(*ſ*) This viſion is agreeable in this, and almoſt in all points with the viſion of the man that was ſeene by DANIEL, chap. 10.5. and therefore, as this is Chriſt, ſo is alſo he whom there *Daniel* ſawe: but where he calles him a mã according to the ſhape he did beare, he againe incontinently interpreteth him, neither to be a man, neither the Sonne of man, but like to the ſimilitude of the Sonne of man. verſ. 16. and like to the viſion or figure of a man, ver. 18. & this example of *Daniel*, confirmes our former note *l*, where wee prooued Chriſt, neither before his incarnation, nor after his aſcenſion, to appeare in his humanitie to his Saints, but euer in his Deitie vnder a figure, like to his humanitie: his humane bodie ſince his aſcenſion, abiding in heauen vntill the latter day: reade hereupon, *l*, aboue written, and the note, *d*, chapter 2. following.

(*t*) Here may ſome contrarie to the foreſaid notes, *l*, and *ſ*, induce a ſophiſme, ſaying, he who was dead, and reuiued eternally appeared to *Iohn*: But Chriſt in his humanitie died, and reuiued againe eternally: Therefore, Chriſt in his humanitie appeared vnto *Iohn*. For opening the deceit of this caption: the *ſubiest* of the aſſumption is Chriſt alone, his *attributum* is to die in his humanity, and to reuiue againe eternally, and therefore neither this his humanitie, nor any part of this *attributum* ought to be repeated in the concluſion, but onely the *ſubiectum* Chriſt, with the *attributum propoſitionis* after this forme: Hee who was dead and liueth eternallie appeared vnto *Iohn*: But Chriſt died in his humanitie, and reuiued againe eternallie. Therefore, Chriſt appeared vnto *Iohn*. And to the effect that the vulgar capacities may vnderſtand theſe fraudes: This is, as one would ſay in a familiar example, He who caried this booke to you: wrote the ſame: But on horſebacke. I caried this booke to you. Therefore on horſebacke I wrote this booke: Whereas the right argument ſhould be this waies diſpoſed: he who caried this book to you wrote the ſame: but I caried this booke to you on horſeback, or rather onely ſimply: but I caried this booke to you: therefore, I wrote this booke. Praying therefore, the ſimple to beware of theſe and the like ſophiſmes, I thought good in this due place, to yeelde this one by way of example.

(*u*) Hereby, and by the writing of the booke within and without mentioned cap. 5.1. appeares plainely, that not onely there are prophecies of things to come, written in this *Reuelation*, but alſo hiſtories of ſuch things paſt, as may be a briefe introduction to the prophecie, as ye ſhall find (chap. 6.) of the ſeuen ſeales, which were all performed, before they were written.

(*x*) Seeing in the Greeke, all meſſengers of God are termed by Angels, therefore muſt theſe Angels of the ſeven Churches without all queſtion, meane theſe miniſters or meſſengers, that God hath ſent to teach them, as is proved immediatly by the next chap. where S. *Iohn* is commanded to write to the Angell of the Church of *Epheſus*, of *Smyrna*, &c. This neceſſarilie, is to the miniſters of the Churches, and not to the Angels that S. *Iohn* ſhould write.

CHAP. II. THE ARGVMENT.

The general preface to the ſeuen Churches, being premitted in the firſt chap. now in this chap. followes out the particular meſſages, direct to foure of the ſaid churches commending their well-doing, and exhorting perſeuerance, and reprouing their defection, and crauing their amendement.

The paraphraſticall Expoſition.	The Text.
VNto the Paſtour or Miniſter of the Church of *Epheſus* writ thou & ſhew that theſe things ſaith the Lord Ieſus, that holdeth the	*VNto the Angell of the church of Epheſus: write, Theſe things ſaith he, that holdeth the ſeuen ſtars*

paſtors of theſe ſeuen Churches vnder his merciful hand and protection and is preſent himſelf in the middeſt of theſe his ſeauen holie Chriſtian Churches.

2 I know thy [*a*] workes, O Church of *Epheſus*, and the trauel, and patience thou haſt for my ſake, and how thou deteſteſt, and reproueſt all that doe euill, and how thou haſt tried out the falſe Apoſtles, and haſt found them diſſembling liers.

3 And thou haſt endured trouble patientlie, and for my names ſake, haſt ſtill continued in all godly exerciſes, without fainting or falling backe.

4 Neuertheles, I haue to reproue thee, that thou haſt made [*b*] defection from that charitable loue, that thou ſometimes had.

5 Remember therefore of that defection, repent and amend the ſame, by taking vp againe thy former charitable doing, or els I wil come ſhortlie in my iuſtice, and remoue my Church from among you, vnleſſe you ſpeedilie amend.

6 But in this I praiſe thee, that thou abhorreſt the common harlotry and confuſion of the heriticks [*c*] *Nicolaitans*, which alſo are abhominable in my ſight,

7 Let him that hath iudgement vnderſtand, what Chriſt, [*d*] ſpiritually appearing, ſpeaketh vnto his Churches: To him, ſaith he, that ouercommeth his tentations, ſhall I giue the foode of euerlaſting life, which he ſhall enioy in the ſpirituall [*e*] paradiſe of heauen, with God the Father æternallie.

8 And vnto the paſtour or miniſter of the church of the *Smyrniãs*, writ: theſe things ſaith the Lorde Ieſus who is without beginning, and without ending, who ſuffred [*f*] death for mans ſake, and now liueth æternally.

9 I know thy works, thou Church of the *Smyrnians*, and the trouble that thou endureſt, and both thy povertie in ſpirite, and contemning of

in his right hand & walketh in the middeſt of the ſeuen golden Candleſticks.

2 I know thy works & labor, & thy patience, & how thou canſt not forbeare thẽ which are euil, & haſt examined thẽ which ſay they are Apoſtles & are not & haſt found them liers.

3 And thou haſt ſuffered, & haſt patience and for my names ſake haſt laboured, and haſt not fainted.

4 Neuertheles J haue ſomwhat *againſt thee becauſe thou haſt left thy firſt loue.*

5 Remember therefore frõ whẽce thou art fallen, & repent, & do the firſt works: or els J wil come againſt thee ſhortly, & wil remoue thy candleſticke out of his place, except thou amẽd

6 But this thou haſt, that thou hateſt the works of the Nicolaitans, which I alſo hate

7 Let him that hath an eare hear what the ſpirit ſaith vnto the churches: To him that ouerc ommeth, will I giue to eate of the tree of life which is in the mids of the paradiſe of God.

8 And vnto the Angel of the Church of the Smyrnians write, theſe things ſaith hee that is firſt, & laſt, which was dead, and is aliue.

9 J know thy works, and tribulation, and pouertie (but thou art rich) and I know *the blaſphemie of thẽ which ſaye they are Jewes,* worldly riches, but thou haſt [*g*] treaſure abiding thee in heauen: I knowe alſo the blaſphemous intiſements that certain diſſembled Iewes make vnto you, but they are the very devils diſciples, and Synagogue of Sathan.

10 Feare not therefore theſe perſecutions, which you ſhall ſuffer: for beholde, I fore-ſhewe you, it ſhall come to paſſe, that the devil ſhall ſtir vp his tyrants to perſecute and impriſon you, that yee thereby may bee tempted, and abide triall, and that perſecution ſhall laſt ten [*h*] yeares: but be thou faithful euẽ to the death of the bodie, & thou ſhalt be crowned with the glorie of æternall life.

11 Let him that hath judgement, vnderſtand what Chriſt, [*i*] ſpirituallie appearing, ſpeaketh vnto his Churches: he that abideth conſtant, ouercomming his tentations, ſhall be deliuered from the [*k*] eternall death and damnation of the ſoule.

12 And to the Paſtour of the Church of *Pergamus*, write: Thus ſaith the Lord Ieſus, out of whoſe mouth proceedeth the two edged ſword of true doctrine and ſharpe threatnings.

13 I know thy works O Church of *Pergamus*, and what wicked companie you dwell among, euen among thoſe where Sathã hath ſet his throne and yet you worſhip my Name, and haue not denyed the true Chriſtian faith, in the dayes of the greateſt perſecution, when your [*l*] Paſtour *Antipas*, my faithfull martyr was ſlaine by thoſe tyrants among you, in whome Sathan dwelleth.

14 But I haue certaine offences to lay to thy charge, becauſe thou entertaineſt them, who teach to eat things ſacrificed to idols, & commit whoredome, as *Balaam* did [*m*] teach *Balac*, to put out their beautiful whoores of *Moab*, with their gliſtering ſacrifices to idols, and allured the children of *Iſrael* both to eat of their idolothyts, and alſo to commit fleſhlie whooredome.

15 Euen ſo you entertaine them among you,

and are not but are *the Synagogue of Satan.*

10 Feare none of thoſe things, which thou ſhalt ſuffer: beholde, it ſhall come to paſſe, that the Deuill ſhall caſt ſome of you into priſon, that yee may be tried, & ye ſhal haue tribulation ten daies: be thou faithful vnto the death, and J will giue thee the crowne of life.

11 Let him that hath an ear, heare what the ſpirit ſaith to the churches. He that ouercõmeth, ſhall not be hurt of the ſecond death.

12 And to the Angel of the Church which is at Pergamus, write, This ſaith hee, which hath the ſharpe ſworde with two edges.

13 J know thy works and where thou dwelleſt, euen *where Satãs throne is, & thou keepeſt my name, and haſt not denied my faith, euen in thoſe days when Antipas my faithfull martyr was ſlaine among you, where Satan dwelleth.*

14 But J haue a few things againſte thee, becauſe thou haſt there them that maintain the doctrine of Balaam, which taught Balac to put a ſtumbling blocke before the children of Jſrael, that they ſhould eat of thinges ſacrificed vnto idoles, and commit fornication.

15 Euen ſo haſt thou them, that maintaine the doctrine of thẽ Nicolaitans, which thing J hate.

that maintain the like doctrine of the *Nicolaitans*, who confufedly make wiues, and al other things to be common, whofe doctrine is abhominable in my fight,

16 Repent and amend thy felfe hereof fhortly, or els I will fuddainlie come againft thee, & will ouerthrow them with the devouring fworde of my threatnings.

17 Let him that hath judgement vnderftand, what Chrift, fpirituallie appearing, fpeaketh vnto his Churches: To him (faith he) that overcommeth his tentations, will I giue my felfe, as food to his foule, who am the true, fpirituall and invifible [*n*] bread of life, that came from heauen: And as the ancients gaue a [*o*] white ftone in token of victory and freedome: fo fhall I, who am the true corner ftone, giue my felfe to dwell in him, & he in me, euẽ that ftone fhal he haue, which is intituled by my newe and fauing name of Iefus, which none [*p*] doeth rightly know, but hee that receiueth faluation thereby.

18 And vnto the Paftour of the Church of *Thyatira*, write: Thefe things fayeth the [*q*] Sonne of God, the fierie flame of whofe eyes feeth the fecreets, and terrifieth the thoughts of all hearts: and whofe feete are firme and ftable to ftande, as brafen pillars, in the defence of his Elect.

19 I know thy workes, thy charitable loue, thy true feruice, thy liuely faith, & thy great patience, and that thy good & holy works do increafe, and grow moe in the end, than they were in the beginning:

20 Notwithftanding, I haue certain offences to lay to thy charge, in that thou fuffereft the divelifh woman *Jezabel* (who calleth her felfe a Prophetefse) to teach and feduce my feruants entifing them both to the filthy whooredome of the body, and to [*r*] fpirituall whoordome with idols, in eating of their Ethnick facrifices, and idolothytes.

21 And I haue forborne her for a feafon, giuing

16 Repent thy ſelfe, or elſe J will come vnto thee ſhortlie, and will fight againſte them, with the ſworde of my mouth.

17 Let him that hath an eare heare what the ſpirit ſaith vnto the Churches. To him that ouercommeth, will I giue to eat of the Manna that is hid & will giue him a white ſtone, and in the ſtone a newe name written, which no man knoweth, ſauing he that receiueth it.

18 And vnto the Angell of the Church which is at Thyatira write, theſe things ſaith the ſonne of God, which hath his eies like vnto a flame of fire and his fietelike fine braſſe.

19 J know thy works and thy loue and ſeruice, and faith, & thy patience, & thy works, & that they are *moe at the laſt than at the firſt:*

20 Notwithſtanding I haue a few things againſt thee, that thou ſuffereſt the woman Jezabel, who calleth her ſelfe a Propheteſſe to teach & to deceiue my ſeruants, to make thẽ commit fornication, and to eate meates ſacrificed vnto idols.

21 And I gaue her ſpace

her leaſure and ſpace to repent, and amend her whooriſh idolatrie, but ſhe repented not.

22 Beholde, I ſhall cauſe her [*ſ*] buske and ſet out all her idolatrous entiſements, and thoſe that goe awhooring after her idolatries ſhall be caſt with her, into moſt extreame affliction: vnleſſe they repent them of their wicked workes.

23 And her [*t*] idolatrous children ſhall I deſtroy to the death: and all the Churches ſhall know that I the Lord, am he, which ſearcheth out the deſires of the mind, and ſecretes of the heart, and I wil reward euerie one of you according to your deſeruings.

24 And vnto you I ſay, euen vnto the reſt of them that are of *Thyatira*, ſo many of you as haue not embraced this diveliſh doctrine, nor learned that profound knowledge (as they terme it) of Sathan, I will lay vpon you no [*u*] burthen of other traditions.

25 But that which I haue already teached you, that keepe and hold faſt and ſure, till I come againe, to judge the world.

26 For to them that do ouercome their tentation, & perſeuere in good workes, keeping my precepts to the end, will I giue power to raigne and triumph with me ouer all the wicked [*x*] nations.

27 And I in them, and they in me, ſhall ſubdue theſe wicked vnder the ſcepter of our ſeruitude, and as potſhardes, ſo ſhall theſe be broken and deſtroyed, euen of me ſhall they haue power, as I haue receiued that power of God my Father.

28 And I ſhall giue vnto them the light of the Evangel, and ſhal adorne them with the morning beames, of the Sonne of juſtice.

29 Let him that hath judgement vnderſtande what Chriſt, ſpirituallie appearing, ſpeaketh vnto his Churches.

to repent of her fornication, and ſhee repented not.

22 Behold J wil caſt her into a bedde, and them that commit fornication with her into great affliction, except they repente them of their works.

23 And J wil kil her children with death: and all the Churches ſhal know, that J am he which ſearch the reines and heartes: and J will giue vnto euery one of you according vnto your works.

24 And vnto you I ſay, the reſt of them of Thyatira, as many as haue not his learning, neither haue knowne the deepnes of Sathan, (as they ſpeake) I will put vpon you none other burthen.

25 But that which you haue already, hold faſt till I come.

26 For he that ouercommeth, and keepeth my works vnto the end to him will I giue power ouer nations,

27 And he ſhall rule them with a rod of iron and *as the veſſels of a potter ſhal they bee broken: euen as I haue receiued* that power of my father.

28 And J wil giue vnto him the morning ſtarre.

29 Let him that hath an eare heare what the Spirit ſaith to the Churches.

Notes, Reaſons, and Amplifications.

(*a*) Such is the mercie of God, that theſe good workes that we acknowledge, to be done by his holy Spirit, working in vs, he imputes them to be our workes. But contrarily, let the preſumptuous man aſſure himſelf, that if he eſteem theſe good works to be of himſelfe, GOD in his iudgement ſhall let him finde, to his confuſion, that onely God is good (Marke. 10.18) and that no goodneſſe is in man, *ibid.* And that all fleſh is ſubdued to ſinne (Rom. 3 9.10) Woe therefore, to him who otherwiſe preſumptuouſlie iudgeth.

(*b*) Hereby learne, that a great number of good works are not ſufficient, but God requireth at our hand, that we preaſſe by all meanes poſsible, to attaine to a perfection in everie good worke, as Chriſt teſtifyeth, (Matth. 5.48) ſaying, *Be ye alſo perfect, as your Father in heauen is perfect.*

(*c*) The *Nicolaitans* were heretickes, affirming wiues, and all things to be common. Hereof read more amplie the note (*m*) following.

(*d*) This Spirit that ſpeaketh to *Iohn*, and to the Churches, is neither an Angel, neither the perſon of the holie Spirit, but Chriſt himſelfe: for this ſame Spirit that ſpeaketh, calleth himſelfe immediatlie the Firſt and the Laſt, and he who died, & liued againe, (verſ.8.) and the Sonne of God (verſ. 18) Where note, that he therfore, being Chriſt, calleth here himſelfe a Spirit, becauſe he appeared not to *Iohn* in his humanitie, but ſpirituallie in his Deitie, vnder a ſhape, like to the Sonne of man (as ſaith the text, cap. 1.13. and not in his perfect ſhape of the Sonne of man: Of this read the notes (*l*) and (*s*) of the firſt chapter.

(*e*) Paradiſe literallie in the Scripture, is that part in the earth, wherein *Adam* was firſt placed: But there is a ſpirituall taking of Paradiſe, called the Paradiſe of God, taken for that ioye in heauen, which the elect ſoules, and glorified bodies doe poſſeſſe: read thereupon Luk 23.43. and 2. Cor. 12.4.

(*f*) Chriſt, in the text of the former note *d*, calling himſelfe a Spirit, becauſe he appeared not to *Iohn*, by his humanitie (as there is ſaid) yet doeth Chriſt here ſpeake of his humanitie, how it died, and roſe againe, although his Deitie vnder an humane ſhape, and not his humanitie it ſelfe, did then appeare to *Iohn.*

(*g*) How gladlie worldlie pouertie is to be imbraced for riches in heauen, Chriſt teacheth, Mat. 19.21. and Mat. 6.19.20.21.

(*h*) How among the Prophets, daies doe meane yeares, is prooued by our firſt propoſition, and ſo theſe ten daies are ten yeares. euen the laſt ten yeares of *Traianus* raigne, wherein was great perſecution, beginning that yeare, that *Ignatius*, teacher of *Antiochia*, and this S. *Iohn* his diſciple was martyred, *Anno Chriſti*, 109 and continuing vnto the year of Chriſt 119, that *Hadrian* the Emperour began his raigne.

(*i*) How this Spirit that ſpeakes, is ſimply no spirit, but Chriſt, who is both fleſh and Spirit, conſider the notes *f* and *d* aboue written, and *l* and *ſ* chap. I.

(*k*) Here, and Apoc. 20.6.14. and Apoc. 21.8. mention is made of the ſecond death, meaning two deaths, to be euen the death of the bodie, which oftentimes in the Scriptures is called a ſleep: & the death of the ſoule, which is eternall damnation. This ſame diſtinction vpon the cõtrary part ye ſhall finde in the ſcripture of the ſoul, or life, taking it in the firſt ſignification for the ſoul & life that liueth here: and in the ſecond ſignification for the ſoul & life that liueth eternally, as Mat. 10.38. & 16.25. Alſo two reſurrectiõs you ſhall finde mentioned in the Scriptures, the one in this world from ſins and errours, the other at the latter day, when the fleſh ſhal riſe again. Apoc. 20. Cõſider therefore, the diſtinctiõs of theſe three, death, life, & reſurrectiõ, when in the scripture they occur.

(*l*) *Antipas* (ſaith *Aretas*) was the paſtour & miniſter of this ſame Church of *Pergamus*, by whoſe example of martyrdome, Chriſt here exhorteth the reſt of that Church to conſtancie.

(*m*) By, and beſides the common Eccleſiaſticall hiſtories, which do deſcribe the hereſie of the *Nicolaitanes*, to be, in that they eſteeme wiues & al things common: Here S. *Iohn* doth alſo ſet out their confuſed opinion and double whoordome, by example and compariſon to the wicked Prophet *Balaam*, who deuiſed a way to entiſe the children of Iſrael to periſh by offending God, both in corporal and ſpirituall whoordom. Num. 25.1. and Num. 31.16. So then doubtles, the *Nicolaitans* hath bene a ſect: that vſed their wiues and women as common, and ſpared not to eate the offrings made to idoles, expreſſely againſt *Paul*, 1, Cor. 7. and 8. and 10.

(*n*) That this hid Manna is Chriſt Ieſus, who is our ſpirituall food, all Chriſtianes muſt confeſſe. So then, the bread and the wine in the Sacrament, beeing publike and viſible ſignes, patent to our eyes, cannot be that inviſible and hid Manna, which is Chriſt: and ſo conſequentlie, by transſubſtantiation, the bread and wine do not become Chriſts body and blood: but contrarily, that viſible Manna, and corporal food of bread and wine, that is taken with our hands, and incorporat in our bodies, and nouriſheth our bodies in this tranſitorie life, betokeneth only, and repreſenteth in the Sacrament, that the glorified and inviſible Manna Chriſt Ieſus who is aſcended, and hid in heauen, apprehended of vs by faith, and adioyned to vs ſpiritually, doth nouriſh our ſoules to eternal life. And ſo, for concluſion, that patent Manna, is not this hid Manna by transſubſtantiation, as the Papiſts do iudge: but contrarilie, the viſible Manna, doth repreſent the inviſible, and aſſureth vs of the full fruite and fruition thereof,

(*o*) *Aretas* reporteth, that the ancients were accuſtomed to giue a certain white ſtone to him, that did get the victory in their playes, and games: and therefore, iuſtly here may be taken for a ſigne of victory, that Chriſt doeth giue vs ouer our tentations. Moreouer, among the ancients, they that clenſed or abſolued an accuſed perſon, did caſt in a white ſtone: and they that filed or convicted him, did caſt in a black ſtone, as *Ovid* teſtifieth *lib. 15. Metamorph.* in theſe words,

Moserat antiquis, niveis atriſque lapillis,
His damnare reos, illis abſolvere culpa.
Sentence was giuen in ancient times, by ſtones blacke and white,
The blacke convicted men of crimes, the other did them quite.

So, in that ſenſe, this white ſtone may well be taken for the abſolution of our ſinnes, which Chriſt giueth to al that victoriouſly perſevere in his faith. In the third ſenſe, himſelfe is called the corner ſtone, Mat. 21.42. And therefore, ſaith the text, There is written thereon a new name, or rather (as chap. 3. 12.) my new name: euen that corner ſtone repreſenteth Chriſt Ieſus. For as *Iehova* is his old name common to the whole God-hood, in the old teſtament, ſo is Chriſt Ieſus his new name proper to his perſon alone in the new Teſtament. Where note in Prophecies, that it is no abſurditie to take a Propheticall figure in diuerſe agreeable ſenſes as Apoc. 17.9, the ſeuen heads are both taken for ſeuen hils, and for ſeuen kings and gouernements: read that text itſelfe.

(*p*) Whether by the former note, this white ſtone meaneth Chriſt himſelfe, or that victorie and abſolution, that proceedeth from him: ſure it is, that by Chriſt onely obtaine we Chriſt, and all his benefites: and the name of Chriſt and Chriſtians receiue wee thereby, dwelling ſo inſeparablie in him, and hee in vs, that none doe perfectly taſte how ioyfull a thing it is to receiue him, and be participant of his ſaluation, except thoſe that receiue the ſame. This name (Apoc. 19.12) none is ſaid to know but himſelfe: meaning none to know Chriſt, but the whole bodie of Chriſt, who dwell in him, and he in them: for Chriſt and his members are called Chriſt. 1. Cor. 12.12.

(*q*) How this ſonne of God, who alſo ſuffred fleſhlie death, verſ. 8. is in the verſes 7.11.17. and 29. called a Spirit, without any repugnance: reade thereupon the notes, *d*, and, *f*, aboue-written.

(*r*) By this text and the conference hereof with the former 14. and 15. verſes, it is evident, that this *Iezabel* was a falſe *Nicolaitan* Propheteſſe, teaching the verie doctrine of *Balaam* and *Nicolaus.*

(*ſ*) As the bed is the execution place of whoordom, ſo metaphoricallie, to lay this falſe Propheteſſe in her bed, is as much to ſay, as to ſet her gloriouſly at her altar and ſacrifices, among her idols, where is meaned, that if any there, following her idolatries, doe commit ſpirituall whooredome, and eat of her idolothytes, they ſhall periſh with her for euer: Where note, that in the Scripture whoordome is oft taken in a ſpiritual meaning for idolatrie:

as Deut. 31.16, Iudga. 17. and infinit other places. This bed may alſo meane care-bed, or the bed of ſicknes and affliction.

(*t*) As the children of the true Church are thoſe, who haue bene foſtered vpon the word of life, and ſhall euer retaine the ſame, ſo they that without recouerie, are poyſoned with heretical doctrin, are here called the children of theſe heretikes that haue teached them.

(*u*) Here doth our Sauiour expreſſely note how ſweet a yoak his yoake is, as Matth. 11.30. & how to adde vnto the pure & naked word of God, is ſo burthenous, that he himſelfe will adde no farther traditions thereunto, till his latter comming. Seeing Chriſt thẽ wil adde no more than his ſimple word, by what authoritie doeth men, vnder pretext of Catholikes & Chriſtians, adde to the Chriſtian religion infinite traditions and ceremonies of mens inventions, againſt the which Chriſt ſpeaketh both here, and in Marke 7. and Luke 11. and in other places.

(*x*) Though *to gouerne al Nations with a Scepter of yron &c.* bee a propertie onely aſsigned to Chriſt in all the Scriptures, yet in his mercie, he ſo is in vs, & we in him, that therby we are made heires & partakers with him in his kingdome & glory, & ſo conſequẽtly alſo in this preheminence ouer the wicked nations. Of this vnion & communion promiſed to vs, read Ioh. 14.20. & Ioh. 17.21. and Mat. 19.28 and 25.34. Rom. 8. 17. & 1. Cor. 6.3. Apoc. 3.21. And therefore, in this text, to the ſame ſentence immediatly are ſubioyned theſe words, *Euen as I receiued of my father*: meaning, that he giues vs power to reigne ouer the wicked, as he hath receiued that power of his father, &c. And ſo I allow not this lection: *Even as I receiued of my Father, ſo wil I giue him the morning ſtarre.*

CHAP. III. THE ARGVMENT.

The former foure Churches of Epheſus, Smyrna, Pergamus, and Thyatira, being in the former chapter admoniſhed to perſevere in their wel-doings, and amend their euill-doings, our Saviour in this chapter, proceedeth with the like admonition to the last three Churches of Sardis, Philadelphia and Laodicea.

The paraphraſticall expoſition.

AND vnto the Paſtour of the Church of *Sardis*, write: Theſe things ſaith the Lord Ieſus, who hath the ſeven-fold graces of his holy Spirit, vnited in God-hood with him, and thereby holdeth

The text.

ANd write vnto the Angell of the Churche, which is at Sardi, theſe things ſaith hee, that hath the ſeven Spirits, of God, the Paſtours of your ſeven Churches vnder his protection. I know thy works, O *Sardis*, thou beareſt the [*a*] name, that thou liueſt in me by liuelie faith: but contrarilie, thou and thy faith, are but dead.

2 Awake from thy ſlouthfulneſſe, and by godly exerciſes ſtrengthen and quicken vp againe, the ſmall ſparke of faith, which as yet is in thee, but is in danger vtterlie to be extinguiſhed for I haue no wayes found thy works to bee perfect before God.

3 Remember therefore, the doctrine that you receiued, and heard, and cleaue faſt thereunto, and repent thy ſweruing therefra: for if thou wilt not repent, & bee wakened from ſinne, I ſhall come as a [*b*] theeſe vnawares vpon thee, and ſhall ſpoyl thee of all my benefites, and thou ſhall not knowe the houre nor time when I come.

4 Nevertheleſſe, I know yee haue a certain in *Sardis* with you, knowne to me by name, which yet haue not defiled their bodies, and theſe ſhal walk with me in the purity and regeneration of the body, for I haue accepted them as worthie thereof.

5 He that overcommeth his tentations, ſhall bee cloathed with a pure and glorified bodie, and I will not blot out his name from among the liuing, but I will advouch him by name, as mine owne before [*c*] God, and his holy Angels.

6 Let him that hath iudgmẽt, vnderſtãd what Chriſt ſpiritually appearing ſpeaketh vnto his Churches.

7 And vnto the Paſtour of the Church of *Philadephia*, write: Theſe things ſayeth the Lord Ieſus,who is holy and true in all his ſayings, and hath the key of the ſpiritual [*d*] houſe of DAVID, euen his holy Temple & church, to open vp the true doctrin thereof which none thereaſter ſhall be able to obſcure: & to cloſe vp the myſteries thereof, which none, but by him, ſhall be able againe to open vp.

8 I know, O Church of *Philadelphia*, thy good workes: Behold I haue made to thee an open acceſſe to the knowledge of mine Evangell, and

and the ſeven ſtars. J know thy works for thou haſt a name that thou liueſt, but thou art dead.

2 Be awake & ſtrẽgthen the things which remaine, that are readie to die: for I haue not found thy workes, perfect before God.

3 Remember therefore, how thou haſt receiued and heard, and hold faſt, and repent: If therfore thou wilt not watch, I will come on thee as a thiefe, and thou ſhalt not knowe what houre I wil come vpon thee.

4 Notwithſtanding *thou haſt a few names yet in Sardi, which haue not defiled their garments: and they ſhall walke with me in white, for they are worthie.*

5 He that ouercommeth, ſhal be clothed in white aray, & I will not put his name out of the booke of life, but J will confeſſe his name before my Father, and before his Angels.

6 Let him that hath an eare, heare what the Spirit ſaith vnto the Churches.

7 And write vnto the Angel of the Church which is of Philadelphia, theſe things ſaith he that is holie & true which hath the key of Dauid, which openeth and no man ſhutteth, and ſhutteth and no man openeth, none ſhall be able to cloſe vp or hide the ſame from thee: for thou art strong [*e*] in a part, and haſt firmelie kept mine holy word, and haſt neuer denyed my Name.

9 Behold, I wil make theſe pretended Iewes, who are indeed the [*f*] Sinagogue of Sathan, and are no true Iſraelits, but doe teach lies: Beholde, (I ſay) I ſhall make theſe, that now would ſeduce and trouble thee, to come, and fall downe penitent [*g*] at thy feet, yeelding reuerence vnto thee and worſhip vnto mee, when as they ſhall know that I haue loued thee.

10 And becauſe thou haſt kept my precepts & holy word, with patience and conſtancie, therefore, I will deliuer thee in the time [*h*] of triall and perſecution, which ſhall come ſhortly vpon the Chriſtians throughout all the world, whereby their conſtancie ſhall be tried, in all the quarters of the earth.

11 Beholde, it ſhall not be long to my laſt comming, hold therefore, and ſticke faſt to that [*i*] doctrin that thou haſt receaued, leaſt thy glorious rewarde and triumphant crowne be giuen to another.

12 Him that here ouercometh his tentations, will I promote hereafter to beare ſpeciall rule in that heauenly Temple and kingdome of God and he ſhall no more ſwerue from me nor wander aſtray, for I will intitle him by name [*k*] of a godlie man, and by the name and ſtile of the citizens of Gods heauenlie kingdome and new *Jeruſalem*, euen of that glorious eſtate, that God from the higheſt heauen, ſhall vouchſafe to powre downe on his elect: and finallie, I will ſtile and name him by mine owne newe name, a true Chriſtian.

13 Let him that hath iudgement vnderſtand, what Chriſte, ſpirituallie appearing ſpeaketh to his Churches.

14 And vnto the Paſtour of the Church of the *Laodiceans* write thou, Theſe things ſaith [*l*] veritie it ſelfe, euen the Lord Ieſus the moſt faithfull and true witneſſe, the [*m*] original by whom all the

8 I know thy workes: beholde, I haue ſet before thee an open doore and no man can ſhute it: for thou haſt a little ſtrength, and haſt kept my word, and haſt not denied my name.

9 Beholde I will make them of the Synagogue of Sathan, which call themſelues Iewes and are not, but doe lie: behold, I ſay, *I will mak them, that they ſhall come and worſhip before thy feete, and ſhall know that I haue loued thee.*

10 Becauſe thou haſt kept the worde of my patience, therefore, I will deliuer thee from the houre of tentation which will come vpon all the worlde, to trie them that dwell vpon the earth.

11 Beholde I come ſhortlie: holde that which thou haſt, that no man take thy crown

12 Him that ouercommeth will I make a pillare in the Temple of my God, & he ſhall goe no more out: and I will writ vpon him the name of my God, and the name of the citie of my God, which is *the newe, Jeruſalem, which commeth down out of heauen from my God, and* I will write vpon him *my newe name.*

13 Let him that hath an eare, heare what the Spirit ſaith vnto the Churches.

creatures of God were created.

15 I know thy workes, O Church of *Laodicea*, that thou art neither ſimplie ignorant, nor zealous & feruent, according to thy knowledge, I would thou wereſt either cold by [*n*] ignorance, or feruent by knowledge:

16 But becauſe thou art betwixt them luke-warme, and neither a ſimple ignorant nor a feruent Chriſtian: it muſt therefore come to paſſe, that I caſt thee out from theſe whom I profeſſe with my mouth to bee mine.

17 For thou vanteſt that thou art rich, and increaſed with worldlie goods, and ſuppoſeſt that thou haſt neede of no other thing, but thou knoweſt not, how in trueth thou haſt but a wretched and miſerable ſoule, and art poore and indigent of all heauenly treaſure, and art blinded with worldly affections, and ſeeſt not the filthie nakedneſſe of thy ſinnes.

18 I counſell thee therefore, by thy prayers and dailie ſupplications, to buye and purchaſe at my hande, the fine and [*o*] golden treaſures of ſpirituall graces, that you may bee made rich in heauen, and ſute by prayers to be clad with the [*p*] merit, and innocencie of my paſsion, that being clothed therewith, the [*q*] filth of thy ſinnes be neuer diſcovered: adioyne alſo to thy dull ſenſes my [*r*] cleare doctrine, that you may ſee the way of life.

19 As many as I loue, I doe rebuke and chaſten them, bee yee therefore feruent and zealous, and amend your former life.

20 Beholde, I ſtande euer readie knocking at the doore of your conſcience: who ſo yeeldeth vnto mine admonitions, and opens his heart and minde, by penitence and remorſe, I will certainly dwell in him, and will communicate all thinges with him, and he with mee.

21 To him that overcommeth his tentations, will I grant power to [*ſ*] ſit with me in my kingdome, partaker of my glorie, euen as I ouercame ſinne

14 And vnto the Angell of the Church of the Laodiceans write, Theſe thinges ſaith Amen, the faithfull and true witneſſe, the beginning of the creatures of God.

15 I know thy works that thou art neither cold nor hote: I would thou wereſt colde or hote.

16 Therefore, becaus thou art luke-warme, and neither colde nor hote, it will come to paſſe that I ſhall ſpew thee out of my mouth.

17 For thou ſayeſt, I am rich and increaſſed with goods, and haue neede of nothing, and knoweſt not how thou art wretched and miſerable, and poore, and blinde and naked.

18 J counſell thee to buy of me golde tryed, by the fire that thou mayeſt be made rich, & white raiment, that thou maieſt be clothed and that thy filthie nakednes do not appeare: & annoint thine eies with eie-ſalue, that thou maiſt ſee.

19 As manie as J loue, I rebuke and chaſten: be zealous therefore and amend.

20 Beholde I ſtande at the doore and knock Jf any man heare my voice, and open the dore, I wil come in vnto him, and will ſuppe with him, and he with mee.

21 To him that ouercommeth. Will I grant to

and death, and doe ſit with my father in his kingdome and glorie.

22 Let him that hath iudgement vnderſtand what Chriſt, ſpiritually appearing, ſpeaketh vnto his Churches.

ſit with me in my throne, euen as I ouercame, and ſit with my Father in his throne.

22 Let him that hath an eare hear, what the Spirit ſaieth to the Churches.

Notes, Reaſons, and Amplifications.

(*a*) Hereby is to be noted, in what eſtate & danger of death ſtand all hypocrites, and theſe that in name onely are called Chriſtians, whereas God contrarily craveth of vs, that ſo far as we may, wee ever aſpire to perfection in effect, and not in name.

(*b*) Of this metaphore and compariſon of Chriſt his ſuddaine comming, and the comming of a thiefe, read his owne words, Mat. 24.43. Luke 12.39. & 1. Theſ. 5.2. & 2. Pet. 3.10. and Apoc. 16.15. thereby exhorting vs to continuall watching and attentiueneſſe.

(*c*) This is a comfortable ſpur to mooue vs publikely, boldly & with conſtancie to the death, to profeſſe and confeſſe the name of ſo worthie a Lord: ſeeing he before God and his Angelles doeth not ſpare by name to confeſſe vs, although we be moſt vnworthy creatures. Where, note the vnſeparable communion betuixt Chriſt and vs, he doeth here confeſſe our name: and verſe 12. hee writes on vs his name: reade, *k*, following.

(*d*) Hereby plainly doeth it appeare, that the Prophecie of *Eſay* 22.20, of *Eltakim*, hath both a literal & ſpiritual meaning: A literall, whereby it is prophecied, that *Eliakim* the ſonne of *Hilkia*, ſhould beare the charge ouer the houſe of *David*, as was performed 4. Reg. 18.18. A ſpirituall meaning, prophecied and meaned of Chriſt the true *Eliakim*, that is to ſay, the God of the reſurrection, or God that roſe againe (*ex Hieron. de interpretatione nominum*) who deſcended of *Eliakim*, Luk. 3 31. Whoſe father *Hilkia* is here in *Luke*, called *Melcha*. Elſe were there two ſeveral *Eliakims* deſcended of *David*. In this ſenſe, Chriſt the true *Eliakim* is ſpirituallie prophecied of in *Eſay*, & here expounded in this text, to bear the keies of the houſe of *David*, &c. meaning, that he beares the ſpiritual gouernement and cenſure of his Church, both concerning iudiciall power, as alſo to open and cloſe, &c. as is before ſaid in the paraphraſe.

(*e*) This litle ſtrength or ſtrongneſſe in a part, being appearingly no great ſtrength, that God had giuen them for their talent: is neuertheleſſe greatly commended of God, becauſe they haue gained there-with, and neither haue loſt it, nor the profit and gaine therof, but haue put it to profitable vſurie, according to the parable, Matth. 25. And therefore, moſt of all the ſeuen Churches, this Church is alwaies commended of God, and in nothing reproued.

(*f*) Theſe appearingly were ſome counterfeit Chriſtians, & alledged Iewes, mentioned before, cap. 2.2.9. who purpoſed with their hereſies to haue ſeduced theſe Churches from the trueth.

(*g*) This is a reuerence to be done by them to the Church of *Philadelphia*, and an adoration and worſhipping made at their feet of *Philadelphia*, not to them by way of idolatrie, but to God: & there fore, is it ſubioyned in the text, *and they ſhal know that I haue loued thee*: meaning that they vnderſtanding that I loue thee ſhal there fore reuerence thee, and worſhip mee at your feet, that I may alſo loue them: Otherwiſe it were idolatrie, and not only would not be cauſed by God, but alſo would be forbidden by him, and by his Angels, as is done Revel. 19.10. and Reuel. 22.8.9.

(*h*) This was (appearingly) that great and vniverſal perſecutiõ, during the laſt ten yeares of *Traianus* raigne, mentioned before, Apo. 2.10, and note, *h*, thereof.

(*i*) No rites, ceremonies, traditions of men, nor other burthen, but my ſimple word and ſweete yoke, as before (cap. 2.24.25. and note, *u*, thereof) is ſaid at length.

(*k*) This is an vn-remooueable marke, a perpetual title, & conſtant profeſsion, whereby euen to the death wee are marked, named, and ſtiled by the name of God, godly men, & by the name of his true Church, true profeſſors: yea, & finally, by that name that our Sauiour (who was before al eternitie) had but newly receiued when he was called Ieſus or Saviour, & the only Chriſt: wee are therefore (after his name) called Chriſtians, and made participant of that ſalvation ſignified by that new name, IESVS, in the Newe Teſtament of his blood.

(*l*) *Amen* moſt commonly is taken for confirmation of a true and certaine thing, but likwiſe it ſignifieth truth & veritie it ſelf, & in that ſenſe it is here taken, & is put as an epithet & ſtile of Chriſt who is the fountaine of all trueth, and trueth it ſelfe, as he teſtifieth, Ioh. 14.6. *I am the way, the trueth, and the life*: and this ſtile of *Amen*, or trueth, is ſet before, being purpoſed in the following text to ſhew immediatly of his faithful and true teſtimonie & witneſſing.

(*m*) Hereof read Coloſſ. 1.15.16.17. where ye ſhal find, that God by Chriſt created all things viſible & inviſible, & that Chriſt was before all things, euen before all eternitie, and therefore, is here called the beginning and original of all the creatures of God.

(*n*) Of men colde in true religion or yet, exceeding hot & feruent in a wrong religion before their calling, there are infinite examples, that God hath made thẽ after their calling notable inſtruments in his Church, & zealous profeſſors: But how dangerous the eſtate of theſe luke-warme people is, who, after their calling, are neither hote nor colde, but knowing the truth, do notwithſtãding liue in a careles mediocrity & idle ſecuritie, Chriſt teſtifieth, Luke. 12.37. Matth. 11.20 Alſo of the terrible danger of theſe, read Heb. 10.26.

(*o*) Gods word, wiſdome & doctrin, is compared and preferred to golde or ſiluer, fined in the fornace: read Prov. 8.10.11 & Pſal. 12.7, So, by this golde ſo fined, is meant the ſpirituall riches and treaſure of theſe and all other godly benefites.

(*p*) Theſe white garments, are not onely our mortified fleſh, and regenerate bodies, but euen vnder the termes of garments *Synecdochicè* expreſſed, is meaned that vniuerſall puritie, both of bodie and ſoule, that wee obtaine, being waſhed by the innocent blood of Chriſt Ieſus, as is proued by theſe, who (Apoc. 7. had long white robes, who (ibid. verſ. 13. and 14.) are ſaid to haue waſhed their robes, and made the ſame white in the blood of the Lambe Chriſt Ieſus: and wee know by the Chriſtian trueth, that not particularly our robes, yea, not our fleſh onely, but vniverſally, both in bodie and ſoule, we are waſhed in that blood, and obtaine ful remiſsion of all our ſinnes.

(*q*) Nakednes of it ſelfe is no villanie, as before the fal, *Adam* had no ſhame, but a ſinful nakednes is indeed vile & ſhamful, which as God corporally hid to *Adam* with garments of skins after his fall (Gen. 3.21) So Chriſt ſpiritually hid the ſame, by his white garments of innocencie in his death & paſsion, wherby our offences are remitted, hid, and waſhed cleane away.

(*r*) This ſight and eie-ſalue is ſpiritually taken (as in Chriſts parable. Mat. 7.3.4.5. and many other places) for the iudgement and foreſight of the mind, and helps thereof, ſuch as ſtudie, diligence. and exerciſe, which leads vs the right way of ſaluation: and can therefore, no waies be literallie imputed to the corporal ſight.

(*ſ*) Here is a compariſon, wherof we vile creatures, are not worthie, to him be glorie, that glorifieth the vnworthie: Then as I (ſaith Chriſt) haue obtained victory ouer all tentations, & therefore, raigne with my Father: ſo ſhal ye alſo (if ye obtain victory ouer your tentations) raigne with mee eternally. The cauſe here of, is becauſe we are vnited in him, and he in vs, as the Father is in him, and he in the Father: read chap. 2.26. and note *x*, thereof.

And here endeth the firſt part of this booke, concerning the particular admonitions to the ſeuen Churches. Now followeth the general diſcourſe to the worlds end.

CHAP. IIII. THE ARGVMENT.

This viſion the ſpirit of God doth premit, and ſet before the whole prophecie and hiſtorie following, as a moſt neceſſarie thing, to know the true throne of God and his word here in this world, and theſe authenticke bookes and ſcriptures, which teach the ſame, here ſet out in this chapter, before he proceed to declare the tyrannie that the Antichriſt, and his prophane princes are to vſe a gainſt that holy word and conſtant profeſſours thereof, to the end of the world.

The paraphraſtical expoſition.

AFter this I beheld, and loe, the doore of heauenly knowledge was opened vnto me, and [*a*] firſt, I was called, as by the mightie & fearfull trumpet of Gods Spirit, to ariſe from earthly cares, and affections, to high and heauenly contemplations, that I might ſee thoſe things that heereafter ſould come to paſſe.

2 Then was I immediatlie rauiſhed in ſpirite, and did beholde and contemplate the heauenly [*b*] throne and temple of God, ſet here among his faithfull, and in that throne, God by his word erected in honour, and ſet in authoritie.

3 And this maieſtie of God (though in him ſelfe he be pure, bright and ſhining) yet in our dull apprehenſion it ſeemeth darke and mixed with myſteries, as the Iaſper ſtone is mixed with vaines and ſpottes, which our eies can not pearce: neuertheleſſe, that weake apprehenſion is euer increaſing flouriſhing, and continuing greene, as the Iaſper, holding freſh, and recent the memorie euer of Chriſts blood and paſsion, repreſented by the bloodie and fleſhie hew of the Sardine, and this inferior throne of Gods maieſtie, is circled ſo about with the couenant of his [*c*] protection, as a rainebowe, that whoſoeuer ſhall approach to his Maieſtie, are vnder the wings of his protection, for this couenant of protection, proceedeth from the throne & maieſtie of God, as the *Smaragdus* hath his beginning and growing from the *Iaſper teſte Cardano de variet. lib. 5. cap 18. de Jaſpide*.

The text.

AFter this I looked, and behold, a doore was open in heauen, and the firſt voice which I heard, was as it were of a trumpet, talking with mee, ſaying, Come vp hither, and J will ſhewe thee things which muſt bee done hereafter.

2 And immediatlie, I was rauiſhed *in the ſpirite, and beholde, a throne was ſet in heauen, and one ſate vpon the throne.*

3 And he that ſate, was to looke vpon, like vnto a Jaſperſtone, and a Sardine, and there was *a rainebow round about the throne in ſight like to an Smaragde.*

4 And round about the throne were foure and twentie ſeates, and vpon the ſeates, J ſawe foure and twenty Elders ſitting, clothed

4 And round about, and on euerie ſide is this Majeſtie of Gods trueth and true religion, compaſſed, adorned, and ſet out by the honourable thrones & authorities of the [d] twẽty foure authentick books of the old Teſtament, as wiſe Senators, ſitting and conſulting in Gods cauſe, clothed all in puritie, for that in thẽ is found no lie, and crowned with victorie, for conquering, and enlargeing Gods kingdome.

5 Foorth of this throne and Majeſtie of Gods trueth, and againſt the contemners thereof, proceedeth forth ſhewings of tempeſtuous troubles, threatnings of warres, and finallie the thundring out of all maner of plagues, in Gods iuſtice: before this throne and Maieſtie of God, brightly ſhineth the [e] ſeuen ſpecial graces of Gods only Spirit, communicated to all the faithfull.

6 Before the eyes of his Majeſtie, the raging ſeas and ſtormes of this world, are made fixed and ſtable as glas, that his elect (though perhaps they ſlide) yet they ſhall not be drowned therein: yea theſe drumlie ſeas of worldlie tyrants are made cleare as criſtall, and the ſecrets of their hearts diſcloſed. This holy throne of Gods trueth is adorned within and without, with the perfect teſtimony of his [f] foure Euangeliſts, repleate with the eies of diſcretion, eſpying before them the prophecies of the *Meſſias* to be fulfilled in Chriſt, and eſpying after them that this our *Mesſias* ſould reigne for euer.

7 The firſt Euangeliſt *Mark* begins his firſt face or leafe, at the voice, crying like a roaring Lyon in the wilderneſſe, *Prepare the way of the Lord. &c.* The ſeconde Evangeliſt *Luke*, beginnes his firſt face or leafe, at *Zachariah* his offering incenſe [as it were a bullocke] at the altar, &c. The third Euangeliſt *Matthew* hath his firſt face or leafe, of the genealogie of Chriſt as hee is man. The fourth Euangeliſt *Iohn*, beginnes his firſt face or leafe, at the high and diuine eſſence of Chriſts Godhoode, flying ſo high in his ſtile, that he is compared to an Eagle.

8 Theſe foure as with ſixe winges, betokening threefold haſte and expedition proceede ſwiftlie in

in white raiment, & had on their heads crowns of gold.

5 And out of the the throne proceeded lightenings, and thunderings, and voyces, and there wer ſeven lamps of fire, burning before the throne, which are the ſeven ſpirits of God

6 And before the throne, there was a ſea of glas, like vnto criſtal and in the mids of the throne, & round about the throne were *foure beaſtes full of eies before and behinde,*

7 And the firſt beaſt was like a lion and the ſecond beaſt like a calfe and the third beaſt had a face as a man, and the fourth beaſt was like a flying Eagle.

8 And the foure beaſts had each of them ſixe wings about him, and they were full of eies within, and they ceaſed not day nor night, ſaying, Holie, holie, holie Lord God almightie Which was, and which is, & which is to come.

9 And when thoſe beaſts gaue glorie, and honour, and thankes to him that ſate on the throne, which liueth for euer their functiõ, opening the hiſtory of Chriſt and making the ſame patent to the eies of the whole worlde: ſo that day nor night, theſe neuer ſtay from vttering Gods praiſe, ſaying, *Holie, holie, holie Lord God almighty, which was, which is, and which is to come.*

9 And as theſe Euangels and their profeſſonrs doe vtter this glorie: honour, and thankes to God that liueth for euer, for that hee redeemed vs in Chriſt.

10 The foure and twentie bookes of the olde teſtament by the true profeſſors thereof, kneele down in Gods preſence, and worſhip him that liueth for euer and euer, acknowledging him to be the authour of their triumphant victorie, and glorious crownes ſaying,

11 Thou onely, O Lord, art worthy to receiue glorie, honour & power of all thy creatures, for that thou haſt created all things, and at thy pleaſure & will they ſubſiſte, and to expreſſe thy glorie, they haue beene created.

and euer.

10 The foure & twentie Elders fel down before him that ſate on the throne, and worſhipped him, that liueth for euermore, and caſt their crownes before the throne, ſaying,

11 Thou art worthie, O Lord to receiue glorie, and honour, and power: for thou haſt created al things & for thy wils ſake they are, & haue bene created.

Notes, Reaſons, and Amplifications.

(*a*) Before any acceſſe to heauenly knowledge, we muſt firſt leaue off all worldly affections.

(*b*) How & why this throne, that here is ſaid to appeare in heauen, is Gods heauenly trueth and true religion, wherby he ſitteth enthronized and authorized among his Saints vpon earth is proued at length in our 17. propoſition.

(*c*) The Rainbow was ſhewed to *Noah*, as a couenant of Gods protection, that the world ſould no more be deſtroyed by water again. Geneſ. 9.15. and ſo generallie in myſteries, it is taken for the couenant of protection.

(*d*) Of theſe foure and twenty Elders, and how they mean the foure and twentie authentike bookes of the olde Teſtament, read the 18. propoſition. Theſe doeth S. *Hierome in prologo Galeato* nominate this way, *Geneſis, 1. Exodus. 2. Leviticus. 3. Numeri. 4. Deuterom. 5. Ioſhua 6. Iudges. 7. Ruth. 8. Samuel. 9. Kings. 10. Chron. 11. Ezra. 12. Heſter. 13. Iob. 14. Pſalmes. 15. Prouerbs. 16. Eccleſiaſtes 17. Song of Salomon. 18. Eſay. 19. Ieremies prophecie. 20. Ieremies Lamentations. 21. Ezechiel. 22. Daniel. 23.* the twelue ſmal Prophets. 24 Theſe ſame 24. ancients, meaning alſo the 24. writers of theſe ſame books are nominate by *Zanchius* this way. *Moyſes. 1. Iob. 2.*

Iofbua, or his conveyer *Eleazar* 3. *Samuel* writer of the Iudges, *Ruth*, and 24. chapters of the firft booke of *Samuel.* 4. *David. 5. Salomon 6.* thefe two wrot the *Pfalmes*, the *Prouerbs, Ecclefiaftes, Cantica cantic.* And they with other Prophets wrote the reft of *Samuel* and the *Kings. Ezras* wrote the bookes of *Efdras, Nehemia* and *Chronicles 7. Mardocheus 8.* wrote Efter: *Ifaie 9. Ier. 10. Ezech. 11. Daniel 12.* The 12. final Prophets makes out the number of 24. compleet.

(*e*) Thefe feuen Lampes, or feuen Spirits, illuminating Gods throne, are the feuen fpecial and principal gifts, graces, and functions of Gods onely Spirite, befides the innumerable fmaller benefits therof: of which reade Rom. 12. and 1. Corinth. 12. And this feptenarie partition hereof, to be receiued of olde, is evident by the Hymne, *Veni Creator Spiritus*, where after it is faid, *Tu feptiformis munere, &c.* In thefe feuen Lampes, or feuen Candlefticks, cap. 1. or feuen grained Candlefticke, fpecified in Exodus, chapter 25. and chapter 37. is alfo a further fymbole or figure contained, to wit, both of the feuen Churches of *Afia* to whom Saint *Iohn* wrote (as alfo of the feuen Churches that PAVLE wrote vnto. And of the one and twentie Epiftles and remnant bookes of the new Teftament, figured by the one and twentie cuppes of the graines, of that Leviticall Candlefticke, with the remanent bowles and flowres thereof: Which Churches, bookes, and Epiftles are planted and framed, by the feuen-folde grace of Gods onely Spirit, and being the firft lights and lanterns of the primitiue church are therefore here called feuen lampes, whereby the throne of Gods word on earth, is illuftrated and adorned.

(*f*) That thefe foure beafts be the foure Euangelles, is prooued by our 19. propofition, including alfo *metonimycè*, the true profeffours thereof: but how EZECHIEL agrees thereto: & alfo with this Text is to be declared. EZECHIEL therefore, cap. 1. feeth thefe beafts after a more ample maner, different from this Text in circumftance, but not in effect. For firft, in Ezech. they haue foure wings, meaning fpeede in their future comming: here haue they every one fixe wings, meaning much greater fpeed & diligence, in their prefent function. Secondly, there euery one of them, haue all their foure faces, and the wings of euery one ioyned with others: meaning that everie Evangelift agreeablie teacheth of Chrift, borne as a man, facrificed and flaine as a bullocke, victorious ouer finne and death as a Lyon: and finallie, rifing vp and afcending to heauen as a flying Eagle. And both thefe four patent faces, and chiefe heads of Chriftianifme, as alfo euery fmaller head & point of doctrine, of any one Evangelift to be contained in the others, and fo one to be all, and all foure to be as one, as true witneffes fould be, and therefore, are their wings ioyned, that where one flewe, all flewe, tending all to one purpofe and fcope: But here in the Revelation, though their faces are feparate, there is no repugnance: for they are foure diftinct witneffes indeed, but here alfo, are they agreeable with Ezechiel, in that by this text, they tend all one way, ſaying al in one voice, *Holie, holie, holie, &c.* Thirdly, there they haue wheeles with them, &c. that is, the armie & congregations of Gods elect, going the waies they go, eſchewing the waies they eſchew, repoſing on the grounds whereon they repoſe, euen on Chriſt the ground-ſtone, for that the Spirit is all one, that is in them, and in vs Gods congregation, euen that Spirit of God that indited thoſe foure beaſtes and Evangelles, inſtructeth vs his Armie and Chariots, to fight out valiantly our ſpirituall battels. So on the other part, the Revelation agreeth herewith, ſpeaking (chapter 5.11.12.13) of infinite number of saints and creatures, that with theſe Evangeliſts praiſed God. For concluſion, theſe and al other appearing varieties, & ſuppoſed repugnance of ſcriptures in circumſtance, are meere harmonies, and perfect agreements in effect.

CHAP. V. THE ARGVMENT.

As God neuer left his Church deſtitute of Prophets and prophecies, to comfort and forewarne them against all troubles: So here is deſcribed how Chriſt his Church and throne of his trueth, and true Religion, hath this booke of prophecie adioyned: Wherein is ſet out, what great griefe and dolour it were to the Elect, if they lacked this fore-ſight of the good ſucces, & proſperous end of their miſeries: for the which cauſe, Chriſt here hath revealed this booke of prophecie to his Elect Church: like as on the other part, his whole Saints and elect congregation, doth render him praiſe and thanks therfore for euer.

The Paraphraſtical Expoſition.

AND I conſidered, that onle in the mercifull hand of the Almightie that ſitteth in the throne of his trueth, ſtandeth the [*a*] revelation of all things done hitherto, or that ſhall occurre and come to paſſe hereafter, and that theſe myſteries were hid and cloſed from the world, as if they [*b*] were ſeuen-foldly ſealed.

2 And I perceiued, that although the moſt mightie Angels of heauen were inquiſitiue, and ſhould openlie proclaime, who were worthie to open theſe ſealed myſteries, and to make the ſame patent:

3 They ſould finde no creature in heauen, earth,

The Text.

AND J ſaw in the right hande of him that ſate vpon the throne, a booke written within, and on the backeſide, ſealed with ſeuen ſeales,

2 And I ſawe a ſtrong Angel, which preached with a lowde voice, who is worthy to open the booke and to louſe the ſeales thereof.

3 And no man in heauen, nor in earth, neither vnder the earth, or vnder the earth, worthie of able to open the ſame, or to fore-ſee them:

4 For the which cauſe, I (for the whole elect) greatly mourned and lamented, that none was found worthie to open theſe Evangelical myſteries of our redemption, and Chriſtian hyſtorie following thereupon.

5 Till by one of the twentie foure bookes of the olde Teſtament, euen [*c*] by *Geneſis* (chap. 49.8.9.10.) wee receiued firſt this comfort, that out of the tribe of *Juda* (which was the root and ſtocke, whereof *David* and *Chriſt* ſprang) ſhould ariſe a victorious Lion ouer al his enemies, who ſhould be the true *Shiloh* and our *Meſſias*, in whom all nations ſhould bee bleſſed, and ſhould be the ladder, Gen. 28.12 by which wee ſhould attaine to heauen, and all heauenly myſteries: ſo that no ſecrets, although they were ſeuen-foldlie ſealed, could be latent or hid from him.

6 And beholde, among theſe of Gods true Temple and holy religion, euen his foure Evangeliſtes, and true profeſſours of the foure and twentie bookes of the olde Teſtament, this came to paſſe, and was performed in Ieſus Chriſt our Lambe, who was ſacrificed for vs, and ſeeming vtterly dead, did riſe againe, aud liueth eternally, in [*d*] ſeuen-fold power and providẽce of the only Spirit of God, which is diſperſed ſeverallie amongſt the Elect, throughout the whole world.

7 For he is now come fleſhlie into this world, and hath receiued at the handes of his Father, perfect knowledge & revelation of the whole myſteries of our redemption, and hiſtory following thereupon.

8 Which as that immaculate Lambe receiued, and began to open and preach: his whole Elect of the old and new Teſtaments, proſtrate themſelues before his Maieſtie, and reioycing as with timbrels, doe offer vp their vials ful of the heauenly ſmel of their holy prayers.

9 Singing gladly the joyful [*e*] ſõg of our new redemption in his blood, confeſsing this their Lord,

was able to open the booke neither to look thereon.

4 Then I wept much, becauſe no man was found worthie to open and to reade the Booke, neither to looke thereon

5 And one of the Elders ſaid vnto me, weep not: beholde the Lyon which is of the tribe of Iuda, the roote of Dauid, hath obtained to open the booke, and to looſe the ſeauen ſeales thereof.

6 Then I beheld, and loe, in the middes of the throne, and of the four beaſts, and in the mids of the Elders, ſtoode a-Lamb as thogh he had bene killed, which had ſeuen hornes, and ſeuen eies, which are the ſeuen Spirits of God, ſent into all the world.

7 And he came, and tooke the Booke out of the right hand of him, that ſate vpon the throne.

8 And whẽ he had taken the book, the foure beaſts, and the foure and twentie Elders, fel down before the Lamb hauing euerie one harps and golden Vials ful of odours, which are the prayers of the Saints.

9 And they ſung a new ſong, ſaying, Thou art worthie to take the Booke, and to open the ſeals thereof, becaus thou was killed, & haſt to be only worthie to receiue the doctrin of our redemption, and to open vp the myſteries, and whole hiſtorie following thereupon, for that euen he it is, that hath died for vs, and hath [*f*] redeemed vs in his blood, and hath reconciled vs to God, and of al the tribes, tongues, kinreds, & nations, he hath gathered vs together.

10 And made vs Kings, and Prieſts to God, and euen with him ſhall wee raigne, while wee are yet in this [*ff*] earth.

11 And I conſidered the ioyfull voyce alſo of infinit of Gods holie Angels, that ſtand awaiting on the throne of Gods trueth, and in defence of his holy religion, and doctrine of the new and olde Teſtaments, euen thouſand thouſands beheld I,

12 Profeſsing openly, that immaculat Lamb, that was crucified to be worthie to receiue power, and riches, wiſedome, and ſtrength, with honour, glorie and praiſe.

13 And all the creatures which are in heauen, and on the earth, and vnder the earth, and in the ſea, euen all that are in them, vnderſtood I to yeeld euery one in his owne kind, the like praiſe, honour, glory, and power vnto that Almightie, that ſitteth in the throne of his trueth, and vnto this his immaculate Lambe for euermore.

14 And as the foure Evangeliſts yeelded thereto their teſtimonie, the twentie foure Bookes of the olde Teſtament, euen all the true profeſſours thereof (vnderſtanding their Prophecies and expectation of the *Meſsias*, now compleate) doe proſtrate themſelues, and doe worſhip him therefore, that liueth for euer and euer.

redeemed vs to God, by thy blood out of every kinred, and tongue, & people and nation.

10 And haſt made vs vnto our God, kings, & Prieſts, and we ſhall raigne on the earth.

11 Then I beheld, & I hard the voice of many Angels round about the thron, & about *the beaſts, and the Elders, and there* were *thouſand thouſands,*

12 Saying with a loud voyce, Worthie is the Lamb, that was killed to receiue power, & riches, and wiſdome, and ſtrength, and honor, & glory, and praiſe.

13 And al the creatures which are in heauen, & on the earth, & vnder the earth, and in the ſea, & al that are in them, heard I, ſaying, Praiſe, & honor, & glorie, and power be *vnto him, that ſitteth vpon the throne, & vnto the Lambe for euermore.*

14 And the foure beaſts ſaid: Amen, & the foure and twentie Elders ſell downe, and worſhipped him that liueth for euermore.

Notes, Reaſons, and Amplifications.

(*a*) That by a book is meaned Prophecies, Revelations, and admonitions, conſider by the tenth chapter hereof: and Ezech, 3.1. But in that it was written without & within, it ſignifieth that it contained patent hiſtories of things in

S. Iohns dayes, & ſecret prophecies of things to come after his dayes, among the Chriſtians, as evidently appeareth, Revel. 1.19. and note *u*, thereof: the like ye ſhall find in Ezech. 2.10.

(*b*) That each of theſe ſeales doth contain ſeuen yeares, is prooued by our 9. propoſition: but why, and in what reſpects they are called ſeales, appeareth to be theſe cauſes: Firſt, becauſe (as propoſition 9. is ſaid) they are a part of *Daniels* weeks, which indeed, are myſticall and ſealed weekes, as ſaith the Angell, Dan. 12.9. *Theſe words are cloſed and ſealed.* Therefore, iuſtly may both theſe weeks and theſe ſeales be called, ſealed weeks, or weeklie ſeales. secondly, becauſe in the time of theſe ſeals opening, containing the ſpace of 42. yeares, the whole ſeales of our ſaluation were ſufficiently opened hy the doctrine of Chriſt and his Apoſtles, preached in that time (as *Paul* teſtifieth, Rom. 10.18) throughout all the whol world, and in that ſpace the Iewes, and other enemies were brought to confuſion. Thirdly, this ſpace of theſe 42. yeares, is iuſtly termed by ſeals, becauſe that ſhort ſpace, in compariſon of the reſt of the yeares vntill the latter day, which are contained in this booke, is as the time of the opening of the ſeales of letters, in reſpect of the longer time in reading the ſame. Fourthly, as ſeales muſt firſt be opened, ere the letter be read, ſo theſe accidents of theſe 42. firſt years, being firſt in time, muſt firſt be performed before the reſt follow.

(*c*) Although this be *Geneſis*, in the which this comfortable promiſe of the *Meſsias* is firſt made to vs, yet all the reſt of the 24. Ancients do alſo prophecie and forewarne the ſame, which moueth *Iohn* here not to ſay, only one Elder, but, one of the Elders, as if he would ſay, the firſt of them: & afterward, he repeats the ſame, for all the reſt of the Elders, in the 8. and 9. verſes: where they ſay al in one voice, *Thou art worthy to take the booke, & open the ſeales thereof, &c.* In which ſaying of this Elder ye ſhall not ſturre, although ye find not theſe words mẽtioned word by word in *Geneſis*, for Chriſt & his Apoſtls cite not their teſtimonies of the old Teſtamẽt alwaies word by word, as ye ſhall find more amplie in the note *g*. Ap. 7

[*d*] Among the prophets, commonlie hornes are taken for kings, Kingdomes, and powers, as in Dan. 8. and Pſal. 74.4.5. And eyes are taken for wiſedome, prouidence, and foreſight, as in Eſai 35.5. and Mat. 7. meaning hereby the princely powers, and wiſe foreſight, that is in Gods holy spirit: but of their number of ſeuen, read *e*, in the former chapter.

[*e*] That this ſong is the ſong of our new redemption, is proued by the words following, to wit: *And thou haſt redeemed vs &c.*

[*f*] [*ff*] That this, for this text and other cauſes, is a viſion of Gods throne, not in heauen, but on earth, is proved by our 17. propoſition.

CHAP. VI. THE ARGVMENT.

In theſe two former chapters, being deſcribed how God adornes the throne of his trueth and true religion, both with the preſent doctrine of ſalvation, and with the propheticall admonitions of all things needfull to bee foreknowne. The firſt being taught be the 24 Elders, and four beasts, and other books and lamps of the new Testament: This booke of prophecies was adioined in the ſecond place, whereof the firſt part historicall, now in this chapter, beginnes at Chriſts baptiſme, An Chriſti 29. and continueth from thence 42. od yeares, euen to the deſtruction of Ieruſalem, in the yeare of Iubilie in anno Chriſti 71. vnder the termes of ſeven ſeales, whereof the hiſtory of ſix ſeales is orderlie ſet downe in this chapter: The reſt, to wit, how from the ſeventh ſeale, in the year of Chriſt 71. continueth the ſeven trumpets or vials, to the yeare of Chriſt, 1541. and how from thence to the latter day, cotinueth the ſeuen thunders, in the chapters hereafter following ſhall be declared. Now firſt of the ſeuen ſeales.

Hiſtorical application.

FIrſt Chriſt in his compleat age of 29. yeares is baptized, beginnes to open and preach the Goſpel. In theſe ſeuen years *Matthew* writs his Hebrue Euangell, and publiſheth it.

2 Chriſt & his Diſciples paſſe out preaching to all nations. Act. 2.5 the teſtimonie of the goſpell, and true word of God, which by Chriſtes doctrine, miracles, death and glorious victorie ouer death, ſo touched and mooued the heartes of all the faithfull, that ſodainlie it ſpread,

Years of Chriſt.

29 Sabboth year

The weeke of the first Seale.

Paraphraſe.

THereaſter I ſaw when Chriſte opened the ſealed doctrine of our redemptiõ, in this firſt ſealed week. One [*a*] of the four Euangeliſtes in the mightie ſtile of Gods thundring Spirite ſet out the ſame, wiling vs to approch & cõſider it

2 And as wee beheld and conſidered that Evangel, lo, the [*b*] pure & holy teachers and Apoſtles, are directed forth, *Mat. 28.29. Mar 16.15* bearing with them Gods word, ſhoting the arrowes of zeale to pierce all godlie hearts:

The Text.

A*Fter, I beheld when the Lamb had opened one of the ſeales, and I heard one of the four beaſtes ſay, as* it were *the noiſe of thunder, Come and ſee.*

2 *Therefore I beheld and loe, there* was *a white horſe, & he that ſate on him, had a bow and a crowne was giuen vnto him, and he went forth conquering*

encreaſed, and triumphantlie it got victory ouer all the world.

3 Nowe after this firſt ſeede of the Evangell once ſowen, beginneth perſecution, about the ſix and thirtieth yeare of Chriſte Within theſe ſeconde ſeuen yeares *Marke* writeth the ſecond Evangel: in comforte of the afflicted Church.

4 And Saint *Steuen* ſuffered martyrdome for Chriſtes ſake, and *Iames* the brother of *Iohn*, likewiſe was beheaded (Act. chap. 7. and chap. 12) for even then, and from thence foorth, the *Romain* magiſtrates, and Synagogue of the Iewes neuer ſtinted from perſecuting and putting to death by fire and ſworde, all true Chriſtians.

5 Notwithſtanding this perſecution, the Evangel ever more and more, is opened & encreaſed, ſo that within theſe third ſeuen years beginning in the yeare of Chriſt, 43. *Luke* alſo wrote the third Evãgel & publiſhed it. At this this time there fell a great famine vniverſally.

6 Whereof *Agabus.* prophecied, and foretold to the Euangeliſts and Diſciples, and whole Church, Act. 11. and they contributed among them, and gaue ſuccour to their poore brethrẽ: in this

36 A ſabbaoth or year of reſt.

The weeke of the ſecond ſeale. 43

A Sabbaoth

The weeke of the third ſeale

crowned with victorie, for it paſſed ouer all the world, ouercõming and conquering.

3 And as Chriſt by his Apoſtles, opened the doctrine of our redemption, in the ſecond ſealed week, the ſecond Evangeliſt ſpak out his goſpell, that all men might come & ſee it

4 And that time proceded [*c*] tyrannous & bloody Magiſtrats, ſitting in the ſeat of tyrannie, to whom power was giuen to perſecute Gods Saints on earth and to depriue them of peace by cauſing al men to kil thẽ, abuſing ſo the ſworde of iuſtice, that God had giuen them.

5 And while as the third ſealed weeke, (after Chriſt began to open his doctrin) came, the [*d*] thirde Evangeliſt wrote foorth his Evangel, that all men might come and ſee the ſame: then beholde there came a deadly plague of famine, and therewith ſuch ſcarſitie, that victuals were weighed by weight.

6 So that it was

that he might ouercome.

3 And when hee had opened the ſecond ſeale I heard the ſecond beaſte ſay, Come and ſee.

4 And there went out another horſe, that was *red, & power was giuẽ to him that ſate thereon, to take peace from the earth, and that they ſhould kill one another, & there was giuen vnto him a great ſword.*

5 And when he had opened the third ſeale, I heard the third beaſt ſay, Come and ſee. Then I behelde, & loe, a blacke horſe, & hee that ſate on him, had ballances in his hand

6 And I heard a voice in the middes of the foure beaſts ſay a meaſure of wheat for dearth, the buſhell of wheate was ſold for fiue accuſtomed prices, and the buſhell of bareley for three: of this dearth ſheweth *Suetonius* in *Claudio Cæſare*, and *Dion*, and other: but of the skarſity of other victualles we read not.

7 Then after the 50 yeare of Chriſt, in this fourth ſeuen yeares. S. *Iohn* wrote the fourth Evangel and publiſhed it, beginning at the deſcription of Chriſts Diuinitie,

8 Expreſſelie againſt certaine Heretikes, for there aroſe manie in thoſe dayes, as *Cerinthus, Nicolaus*, and *Ebion*, and (in the Acts chapt. 13. and chapt. 15.) *Elymas* the ſorcerer, and certaine Phariſaicall heretickes, and others procuring God his plagues againſt *Iudea*, and other partes of *Aſia*, where they reſorted.

9 Then in the fift ſeuen yeares, euen the yeare of Chriſt 57. aroſe the tyrant *Nero*, on whom and on whoſe bloodie ſeate, and ſucceſſours, the blood of Saint *Steuin* & Saint *Iames*, martyred by his predeceſſours, and

foreſpoken of, and came to paſſe amõg the midds of the Evangeliſts and Diſciples of the primitiue Church, that the meſure of wheat was ſold for no leſſe than fiue ordinarie prices, and the meaſure of barley for three ordinarie prices but there was no skarſitie of wine nor oyle.

50 A ſabbaoth

The weeke of the fourth ſeale

7 And when the fourth ſealed weeke (after Chriſt firſt opened his doctrine) came: the fourth Evangeliſt ſhewed forth his Euangel, that all men might come and ſee it.

8 And beholde, at that time came in [*e*] heriſies among the people bringing therewith the death of the ſoule, & procuring hel & eternal damnation: and through that quarter of the earth, where theſe hereſies raigned, God gaue them over to the ſworde, to famine, to death, and to the [*f*] tyrants of the world.

57 A ſabbaoth.

9 And whẽ the fift ſealed weeke (after Chriſt firſte opened his word) came, then (ſome newe tyrant being riſen) there appeared proſtrate before the [*ff*] iuſtice ſeate of God, the

a pennie, and three meaſurs of barlie for a pennie, and oyle, and wine hurt thou not.

7 And when hee had opened the fourth ſeale I hard the voic of the fourth beaſt ſay, Come and ſee.

8 And I looked and beholde, a pale horſe, and his name that ſate on him was Death, & Hell followed after him, and power was giuen vnto them ouer the fourth parte of the earth, to kill with the ſword and with hunger and with death, & with the beaſtes of the earth.

9 And whẽ he had opened the fift ſeal, I ſawe vnder the altar the ſoules of them, that wer killed for the worde of God & for the teſtimonie which they maintained.

10 And they cried with a

other Saintes martyred by himſelfe,

10 Cryed vp to the heavens to God, to reuẽge their blood, vpon theſe tyrant emperors of this bloody Empire, but as God is euer holie and iust,

11 So within a ſhorte ſpace after this, euen whẽ *Peter* & *Paul*, and others of gods ſeruãts had likewiſe ſuffered martyrdom vnder the ſame tyrant *Nero*, then powred out GOD his great vengeance on him, his ſeate, familie, and ſucceſſours, as followeth,

12 To wit, after the yeare of Chriſt, 64. euen in the ſixt ſeuen yeares, a great change and defection came in all eſtates of the *Roman* Empire: for now *Nero* raiſed great perſecutions againſt the true ſprituall profeſſors, to haue vtterly darkened and extinguiſhed the light of the Goſpel, as alſo hee became a vile inceſtuous tyrant and paricide, a polluter of his whole Empire, and eſtates thereof. with tyrannie, and with the blood of GOD his Saints, *Peter, Paul*, and others.

13 And in that tempeſtuous perfection *Demas* (2. Tim. 4.10) and others weake and feeble profeſſors fel from

The weeke of the fift Seale

64 A Sabbaoth.

The weeke of the ſixt Seale.

ſoules of them that had bene martyred for the word of God, and for the teſtimonie of Chriſt Ieſus, which they did bear.

10 Crying vehemently to God, holie & true, to judge and revenge their bloode, on theſe that had ſhead the ſame vpõ the earth.

11 Theſe hath God clothed with his [*g*] innocencie, & willeth them to abide a litle while, vntil the number of their brethren and fellow-ſeruãts that muſt likewiſe ſhortly be martyred for the Evangel were alſo fulfilled.

12 And whẽ the ſixt ſealed weeke, (after Chriſt firſt opened his worde) came, then is the great empire of the earth ſhaken, & the Spiritualitie and bright light of the Goſpel darkened with perſecution, [*h*] & ſecular princes & Magiſtrates now are become bloody murtherers of Gods Saints.

13 And the weak and feeble profeſſors ſhaken, by that tempeſtuous perſecution, fell (as ſhaken fruit) from the true profeſsion, to

loud voice, ſaying How long, Lord, holie and true doeſt not thou iudge and auengeour blud or them that dwell on the earth?

11 And long whit robes wer giuen vnto euery one, & it was ſaid vnto them, that they ſhould reſt for a little ſeſon vntil their fellow ſeruants and their brethrẽ that ſhuld be killed, euẽ as they were, were fulfilled.

12 And I beheld when he had opened the ſixt ſeale, and loe, there was a great earthquake, and the ſunne was as black as ſack-cloth of haire, and the Moone was like bloode

13 And the ſtarres of heauẽ fell vnto the earth as a figtree caſteth her greene figges, when it is ſhaken of a the true Church, to earthly ambition.

14 So that the true Church of God is forced to lurk in ſecret, for revenge whereof, and of the blood of other martyrs ſhead by other Emperours before: God moueth the kingdomes of *France*, *Spaine*, & Ile of *Britane* to revolt from this tyrant.

15 And theſe kingdomes with the Princes, Tribunes, Conſuls, Senators and people of *Rome*, fled from his tyrannie, and left him and all his race of the *Cæſars* blood.

16 And choſed of a new blood the *Proconſull* of *Spaine, Galba*, to be their Emperour againſt *Nero*, wherupon *Nero* (after he had burned his towne of *Rome*, murthered his Senators, mother, & chiefe kinſmẽ) horribly ſlayeth himſelfe.

17 That ſame yeare the Emperours *Galba, Otho* and *Vitellius* cruellie murther ech other with manie thouſand Romanes, in ſigne and token of Gods angrie face and vnreſiſtable vengeance.

earthlie ambition.

14 And Gods Church lurketh, as a cloſed letter: wher fore, euen now God ſtirred vp great kingdomes of [*k*] maine landes, and Ilands to revolt frõ the obedience of that Empire.

15 And the kings that wer vnder that great empire of the earth, with the Princes, Tribunes, and members thereof, ſmal and great, free and bond, hid and abſented thẽſelues, among obſcure people from their Emperours preſence, and fled vnder the protection of the new headesmen, of mightie barbarous kingdomes.

16 Crauing the princes of theſe Barbar kingdomes, to raigne ouer them, and to couer them from the horrible tyrannie of theſe Emperours, who repreſented the face of Gods wrath and of Chriſt Ieſus.

17 For the greate day of their wrath, and time of revenge is come, and none is able to with-ſtand the ſame.

mightie wind.

14 And heauẽ departed away, as a ſcrolle when it is rolled and euerie mountaine and Jle were mooued out of their places.

15 And the kinges of the earth and the great mẽ, & the rich men and the chiefe captaines, and the mightie men, & euery bondman, and euery free-man, hid them ſelues in dennes, and among the rocks of the mountaines.

16 And ſaid to the mountaines & rocks Fall on vs, and hid vs from the preſence of him that ſitteth on the throne and from the wrath of the Lambe.

17 For the gret day of his wrath is come, and who can ſtand?

Notes, Reaſons and Amplifications,

(*a*) That this beaſt is *Matthew*, the order of time prooueth for that in that time, he wrote the firſt Euangel in Hebrue, and therefore is *Matthew* (vnder the terme of a mans face) firſt in order by *Ezechiel* (as is ſaid in our 19. propoſition) and though he be third in *Iohns* former viſion (Ap. 4) for that his Greeke Evangel was third in order, yet here muſt the account be made by his firſt Hebrew Evangel, becauſe it was that work that firſt paſſed out vnto the Iews and Hebrew Church, & began that Chriſtian victory, whereof the text here ſpeaketh. The reſt read in the ſaid 19 propoſition.

(*b*) That by this rider on the white horſe, is meaned the word and teſtimonie of God which his ſeruants do beare and profeſſe, is evident by the 19 chapter hereof.

(*c*) That this red hewe meaneth bloodſhed, read *Eſay* 63.1 & how perſecution muſt follow Chriſts doctrine read *Mat.* 10.34.

(*d* This third beaſt agrees to be *Luke* two waies both for that he wrot his Evangel that time as alſo for that he writes that hiſtorie in his book of the Acts. chap. 11. ver. 28.29.30.

(*e*) This muſt needs be falſe doctrine and hereſies, which (beſide temporall puniſhments) procured both death and hell.

(*f*) That theſe beaſts be earthly kings and tyrants, read Ezech. 34. and Dan. 7. and Dan. 8.

(*ff*) It were vnproper here, and chap. 8.5. and chap. 9.13. & chap. 14.18. to interpret this altar to be Chriſt, ſeeing Chriſt was the ſacrifice, that was offered vpon the altar of Gods iuſtice, for full ſatiſfaction of his wrath. Therefore as the *Ark* & couerture thereof, doeth repreſent Gods mercie ſeat, Exod. 25. So muſt the altar repreſent his iuſtice ſeat, for that the altar was the place of fire & bloodſhed.

(*g*) This is a metaphore taken frõ them, that by the raſh furie & ouerthrow of their enemies, haue bene ouertrode, and their garments pudled and defiled and afterward knowne guiltles, are reſtored to cleane garments, betokening their innocency: reade thereupon, Apoc. 7.14.

(*h*) The Sun for the moſt part is taken for the ſpiritual eſtate, be it pure or corrupt, as in Eſay 24.23. & Eſa. 30.26. and here & in diuers other places: And ſometimes alſo, it is taken for thoſe things good or euill, that are adioyned to the ſpirituall eſtate, as, Apo. 12.1. for the light of the Evangel, and as in Marke 4.6.17. for perſecution: whereof read the note *i*, in the ſeuenth chapter following. As touching the Moone, ſhee is ſometime taken for the changeable world, as Apoc. 12.1 but ofteſt for the ſecular eſtate thereof, as here, & in Eſa in the verſes afore ſaid: & that both becauſe of the great mutabilitie of that eſtate, as alſo becauſe al ſecular lawes and ciuil ordinances, are borrowed from the ſpiritual Law of God, and miniſtred to the people as the Moone borroweth her light from the Sunne, and reflexeth the ſame to the earth.

(*i*) That ſtarres betoken Gods Miniſters, read Dan. 12.3. and Apoc. 1. and notes *q*. and *x*, thereof. And that heauen be tokeneth Gods Church here on earth, conſider by Apoc. 19. conferred with the premiſſes.

(*k*) Mountaines to meane kingdomes and kings, is euident by Eſai chap. 2. where he expreſſeth the kingdome of Chriſt, to become erected aboue all kingdomes by terme of *mountaines*. Further the kingdomes of the *Chaldeans* and *Idumeans* are called mountaines, Ier. 51.25. and Ezech 35.3.

CHAP. VII. THE ARGVMENT.

God hauing in the fift ſeale aboue written, promiſed the revenge of his Martyrs blood, and in the ſixt ſeale begun the ſame vpon the Romane Emperours, conſenters thereunto: and being of purpoſe to doe the like againſt the Iewes, that were executers thereof: Here in this chapter he ſetteth out how before that deſtruction of the Iewes, in this next plague, and alſo before the whole plagues, that are to come both on Iew and Gentile, in the ſeuen ages following, he will firſt of his tender care, reſerue and marke his owne of euery tribe of Iſrael, ſo that theſe with infinite of the Chriſtian Gentilles ſhall reioyce in that deliuery, and ſhall giue praiſe therefore to God.

The Paraphraſticall expoſition.

AFterward I conſidered, that on all the four quarters of the earth God had his holie [*a*] Angels ſtaying the Spirits of his wrath, frõ blowing foorth as yet their vengeance on maine landes, Ilandes, or people of the world.

2 And I perceaued Eaſterlie [*b*] towarde *Ieruſalem*, that Chriſte the great Angell, and *day-ſpring from on high*, who hath the [*c*] ſeales and badges of ſaluation, did by the mightie voice of his worde, commande theſe Angels, who had power ouer the ſaid deſtroying Spirits of the earth: and Ilandes thereof.

3 Saying, Let ye them not hurt the landes, Ilandes, nor the people of the world, till we haue ſealed and put the marke of Gods

The text.

AND after that J, ſawe foure Angels ſtand on the foure corners of the earth, holding the foure windes of the earth, that the winds ſhuld not blow on the earth, nether on the ſea, nether on any tree.

2 And I ſaw another Angell come vp from the Eaſt, which had the ſeale of the living God, & he cryed with a loud voice to the four Angels, to whom power was giuẽ to hurt the earth and the ſea,

3 Saying, hurt ye not the earth, neither the ſea, neither the trees, til we haue ſealed the ſeruants of our God, in

protection [*d*] patently, vpon all his true ſeruants.

4 And I heard the number of thoſe that out of [*e*] all the tribes of *Iſraell*, Chriſt had ſealed with the Profeſsion of his Evangel, and marke of his protection, to be an [*f*] hundreth fourtie foure thouſand.

5 To wit, of the tribe of *Juda* were marked [*f*] twelue thouſand Chriſtians. Of the tribe of *Reuben* were marked twelue thouſand Chriſtians. Of the tribe of *Gad* were marked twelue thouſand Chriſtians.

6 Of the tribe of *Aſher* were marked twelue thouſand Chriſtians, Of the tribe of *Nephtalim*, were marked twelue thouſand Chriſtians. Of the tribe of *Manaſſes* were marked twelu thouſand Chriſtians.

7 Of the tribe of *Simeon* were marked twelue thouſand Chriſtians, of the tribe of *Leui* were marked twelue thouſand Chriſtians. Of the tribe of *Iſſacar* were marked twelue thouſand Chriſtians. Of the tribe of *Zebulon* were marked twelue thouſande Chriſtians.

8 Of the tribe of *Joſeph* twelue thouſand Chriſtians: finallie of the tribe. Of *Beniamin* were marked twelue thouſand Chriſtians.

9 Then of the Gentiles I conſidered the Congregation to be out of number, euen of all kinreds, people uations and tongues, who ſtanding conſtant before this throne of Gods treuth, and in preſence of Chriſt Ieſus clad with the white garmẽts of innocencie of heart, and hauing the ioyful palmes of righteouſneſſe and vpright doing in their handes,

10 Publikelie profeſſed with their mouthes, that all ſaluation commeth from God, that ſits on the thron of his trueth, & from his Son Chriſt Ieſus.

11 Like as all the whole Angels and heauenlie armie of God, that awaites about that throne of Gods trueth, in defence thereof, and of

their fore-heads:

4 And I heard the number of them, which were ſealed, and there were *ſealed an hundreth and foure and fourtie thouſand of all the tribes of the children of Iſrael.*

5 Of the tribe of Iuda were ſealed twelue thouſand. Of the tribe of Reuben were ſealed twelue thouſand. Of the tribe of Gad, were ſealed twelue thouſand.

6 Of the tribe of Aſher wer ſealed twelue thou ſand. Of the tribe of Nepthali wer ſealed 12 thouſãd, Of the tribe of Manaſſes were ſealed twelue thouſand.

7 Of the tribe of Simeõ wer ſealed 12 thouſand, Of the tribe of Leui wer ſealed 12 thouſand, Of the trib of Jſſacher wer ſeled 12 thouſãd, Of the tribe of Zabulon were ſealed 12 thouſand.

8 Of the trib of Ioſeph were ſealed twelue thouſand. Of the tribe of Beniamin were ſealed twelue thouſand.

9 After theſe thinges I beheld, and loe, a great multitud, which no mã could number of al natiõs & kinreds, & people & tongues ſtood before the throne, and before the Lamb, clothed with long whit robs, & palmes in their hands.

10 And they cryed with a loud voice, ſaying Saluation commeth *of our God, that ſitteth vpon the throne and of the Lamb.*

11 And all the Angeles ſtood rounde about the throne and about *the Elders, and the four beaſts, & they fel before* the holie doctrine of the twentie foure bookes of the olde Teſtament, and foure Evangeliſtes proſtrated themſelues before Godes throne, and worſhipping him cõfirmed the ſame.

12 Yelding praiſe and glorie, wiſdome and thanks, with honour, power, and might vnto our God for euermore, ſo be it.

13 And as [g] *Eſay* one of the 24. Elders writing (cap. 63.1.) *who is this that commeth from Edom in red garments, from Bozra, he is glorious in his apparell, &c.* I iudged alſo by him, to know what they were that were arayed in theſe white robes.

14 So I found in him (verſ9.) that theſe be Gods holie Iſrael, whom he hath brought out of trouble, and redeemed them, and (as chap. 1.18.) hath made them white as ſnowe, and cleane as woll, though they had bene red as ſcarlet.

15 And that becauſe (verſ. 16.) *they ceaſed to doe euill, and learned to doe good, &c.* and ſo ſerued God continually, & haue God dwelling among them.

16 Theſe ſaith he (chap. 49.10) ſhal no more haue [*h*] hunger, nor thirſt in their ſouls, nor the burning Sunne, or the [*i*] heat of perſecutiõ to wither thẽ

17 For Chriſt, which is in the mids of his throne & true profeſſours ſhall guid them and lead them to drinke of the liuelie doctrine and fountaines of his ſpirituall waters, and (chap 25 8.) ſhall wipe away the teares of dolour from their eies, and giue them ſpirituall comfort.

the throne on their faces & worſhiped god.

12 Saying, amẽ, praiſe & glory, & wiſdom, & thanks & honor, & power & might, be *vnto our God for euermore Amen.*

13 And one of the Elders spake, ſaying vnto me, What are theſe which are arayed in long white robes, and whence came they?

14 And I ſaide vnto him Lord thou knoweſt: and he ſaid to me, Theſe are they which came out of greate tribulation, and haue waſhed their long robes and haue made their long robes white in the blood of the Lambe.

15 Therefore are they in the preſence of the throne of God, and ſerue him day & night in his Temple, and he that ſits on the throne will dwel among them

16 They ſhall hunger no more, neither thirſt any more, neither ſhall the ſunne light on them nor anie heat.

17 For the Lamb which is in the mids of the throne, ſhall gouerne them and ſhal lead thẽ vnto the liuelie fountaines of waters, and God ſhall wipe away all teares from their eies.

Notes, Reaſons, and Amplifications.

(*a*) Correſpondent to the foure quarters of the earth, the prophet here by a propheticall phraſe and conformitie of language, ſetteth downe foure Angels, and foure executers of Gods wrath, whereas they are innumerable. The Iewes alſo haue bookes of antiquitie, dividing the government of the earth among foure great Angels, and vnder them to be many inferiour Angels: But theſe aſſertions being more curious then certaine, or pertinent to vs to know, we leaue this, and returning to the expoſition of the text, what theſe foure Angels and winds be. We ſay, the foure Angels appeare by the text, to bee good Angels, bearing command ouer theſe winds, whom we eſteem to be euil angels, executers of Gods wrath, at command of the other: for although theſe angels (verſ. 2) are forbidden to hurt the earth, yet that inferreth not them to be euill Angels, more than where the like is ſaid to God, *Lead vs not into tentation*: but contrarily, in ſo far here as of themſelues, they ſtay theſe winds and blaſts of Sathans tyranny, to hurt the earth, they muſt needs be good Angels, and the winds euil Angels. And whereas theſe firſt foure are by the text, ready to hurt, or not to hurt the earth, according as they be commanded of Chriſt that great Angell. Yet for as much as that hurt ſeemeth to be executed not by them, but (verſe 1.) by theſe winds, whom they haue power to looſe or bind: Therefore, apparantlie, theſe foure Angels be the good Angels of God, and the foure winds, the ſpirits of Satan and executers of vengeance.

(*b*) This day-ſpring or riſing in the Eaſt, may two wayes be vnderſtood, to wit, either as an epithet of Chriſts, as Zach. 3.8. and Luke 1.78. called by *Hieroms* tranſlation, *Oriens ex alto*, &c. Or then it may be vnderſtood, that Chriſt aroſe vp, & abode Eaſterlie aboue *Ieruſalem*, to ſeale his Elect Iſraelites there, with his protection: For *Ieruſalem* lieth Eaſterlie, both from *Pathmos*, where Saint *Iohn* wrote, as alſo from the Churches of *Aſia* to whom he wrote.

(*c*) This marke is a compariſon from ſheep-heards, who as they marke their ſheep, ſo Chriſt our ſheepheard, marketh vs with the marke of his Profeſsion and protection: alluding here to the Symbole of the Paſſe-ouer. Exod. 12.22. The like heereof read in Ezech. 9.

(*d*) Chriſt marketh vs with no ſecret marke, but patentlie, in our foreheads, meaning thereby, that vowablie and with bolde faces, wee muſt profeſſe the marke of Chriſtians. Saying with S. Paul, *I am not aſhamed of the Goſpel, &c.* Rom. 1.16. For who denieth him here, he will denie them in heauen. Matth. 10, 33 and 2. Tim. 2.12.

(*e*) Among theſe tribes, *Dan* is left out, and in his ſtead is put in the tribe of *Ioſeph*, meaning *Ephraim, Joſephs* yongeſt ſonne: For *Manaſſes, Ioſephs* eldeſt ſonne, is put in for himſelfe. Why *Dan* is ſo left out, the reaſon appeareth to be, that, that tribe hath bene more accurſed than the reſt: for by the Spirit of God, it is called Gen. 49.17. a ſerpent, or an Adder, and is called a Lyons whelpe, Deut. 33.22. and for their golden calues and great idolatrie, appeareth by the Prophet *Amos*, 8.14. that they fell, and neuer roſe againe, and ſo could not be participant of this Chriſtian marke.

(*f*) The Spirit of God here expreſſeth the number of an hundreth fourtie and foure thoufand, to wit, twelue thoufand of euerie tribe, not that euery tribe was equall, or that the number was fo in whole thoufands without oddes: as if there were neither more, nor fewer, as though we would addict the Spirit of God to refolue and fatisfie our curiofities: but the meaning of the Spirit of God hereinto, is to fatisfie our comfort, and not our curious fcrupulofitie, giuing to vs Gentiles, and to the Chriftian Iewes, that comfort, that whereas we rafhlie feare that we be left alone to ferue Chrift, and that none moe than we of *Ifrael* are chofen, as *Elias* iudged 3. Reg. 19.18. and Rom. 11.3.4.5. God will let vs know, that he hath fealed about twelue thoufand of euerie tribe: where note for certaintie, that in fpeaking to men, and in accounting of men, the Spirit of God fpeaketh after the phrafe of men, in reiecting the oddes, and in fpecifying the perfect or whole number, that is neereft the trueth, inftead of the trueth. And as to this number, no maruell in deede it is, though in the long time of thefe feven plagues following, there be about 144000. of the Ifraelites elected Chriftians of men women, and children, feeing euen befides women and children, and impotent men, there were at once of Ifraelites in the onely dayes of *David*, thirteene hundreth thoufand fighting men. For here is neither meaned that thefe were all at this one time, neither that they were altogether marked at this time, but hereby he meaneth, this to be a time of the beginning of troubles, from the which, alwayes he fignifieth hereby, that he will exeme his owne, not only by preferuing them fucceffiuely, as they are borne in the world, but alfo by bringing their parents for their fakes, through thefe troubles, that of them they may be procreated. Of thefe read further, Apoc. 14.1. and note *a*, thereof.

[*g*] Let none doubt that the 24. Elders do fignifie the 24. booke of the old Teftament: although thefe wordes, here fpoken by one of the 24 Elders, be not found word by word, and altogether in any of the bookes of the old Teftament: for the teftimonies cited by Chrift and his Apoftles, are not al to be found, word by word, in the old Teftament: as the words cited in Matth. 27.9. out of *Ieremie*, are not to be found together in any one part of Ieremie, and in *Zach*. 11.12. that fentence is found nereft, but not word by word. Alfo that Song, which in the Revel. 15.3. is called the Song of *Mofes*, is not to be found word by word in *Mofes* Song, neither in any of the fiue bookes of his *Pentateuchon*. The like is to be fene in divers other places of Scripture: For God [as is faid in the former note] travels to fatisfie our comfort, but not our curiofity. For, although not fcrupuloufly in words, yet fubftantially in fentences, all fpoken here by this Elder, is contained in *Efay*. For firft, in defcribing *who is he that was clad in red garments &c.* (*Efay 63.*) It followeth of necefsity tacitly, that he fheweth what it is alfo to be clothed in white, feing in this cafe they are contrarie: For whereas he interpreteth, that, to be clothed in red, meaneth blood-fhed, and flaughter: to be clothed

in white garments, muft meane puritie, innocencie, cleannes, and holineffe: againe yet more amplie, (Efay 1.18.) is expreffed what is meant by cleane wooll, white as fnow, and by red crimofin and skarlet, whereas hee faith, *though thy finnes were as crimofin and red as skarlot, they fhal be made white as fnow, and cleane as wool.* So then none can denie, but here *Efay* interpreteth fufficientlie, that to be white as fnow, or to be clothed in white garments or in cleane wooll, is to be purged, and made cleane, and holy from finne. And in like maner, as toward the reft of the fentences fpoken here by this Elder, they are all expreffely contained in *Efai*, as they are quoted in the Paraphrafe, which feing none of the other Elders, or 24. bookes of the old Teftament do containe, Therefore without all doubt, *Efay* is this Elder here meant of.

[*h*] This hunger and thirft cited by *Iohn* here out of *Efay*, meaneth the hunger and thirft of our foules for the word of God, as is plainly interpreted in *Amos*, 8.11.12,

[*i*] Though the Sunne for the moft part, is taken for the light of the Gofpell, and for the fpirituall eftate it felfe: yet, when the burning heat thereof, is fet oppofit againft the feed of the Gofpel, and liuely fprings of true doctrine (as here) then doth the Sunne meane the heat of perfecution, this is plainly prooued by the fourth chapter, and 6. verfe of Mark, and by the 17. ver. thereof that interpreteth the fame. Read more hereof, Apoc. 6. note *h*, thereof,

CHAP. VIII. THE ARGVMENT.

God in the former chapter, hauing firft marked his Elect Ifraelits, now in this proceedeth to the deftruction of Ierufalem, and reprobat Ifraelites, in revenge of their tyrannie vfed againft Chrift and his feruants, at which deftruction of Ierufalem, endeth the historie bygone, of the fealed weeks, and beginneth the prophecie then to come, of the feuen trumpets, euery trumpet containing the fpace of 245. yeares, wherein, are orderly expreffed, the notable accidents and alterations, that wer to come among the Chriftian kingdomes, in a maner to the worlds end. In this chapter, then is intreated of the laft of the feuen feales, and firft four of the feuen trumpets, as followeth.

Hiftoricall application.	*an. Chrift.*	*Paraph. expofition.*	The text.
IN the feuenth feuen yeares, beginning in the yeare of Chrift 71. came	71. a Jubile	ANd as the feaventh fealed week (after Chrift opened the	*ANd when he had opened the*

the Emperors ſauourers of Chriſtians *Veſpaſian* and *Titus* under whome the Church of God had peace a litle while.

2 And here begins the ſeuen ages to the worldes end, this 71. yeare being Iubilie.

3 And was that year in the which Chriſt Ieſus procured the revenge of his blood, & the blood of his holie martyrs vpon the synagogue of the Iews (vpon whom the ſaid innocent blood had hitherto craved vengeance) So that now (by the interceſsion of Chriſt) God in his iuſtice, ſtirreth vp the ſaid Roman Emperors, *Veſpaſian* and *Titus* againſt them.

4 And the continuall prayers of poore martyrs, craving the revenge of their blood now taketh effect.

5 For the ſaid *Titus* accõpanied with many thouſand Romans, incloſeth the whol I*ewes,* celebrating the Paſſeouer within *Ieruſalem*, where they are beſieged and ouercome their citie deſtroyed, innumerable ſlaine, and the reſt are taken, and ſolde as bond-ſlaues, abiding without land, libertie, or gouernment to this day. At this time God thundered out againſt them, all manner of plagues of warres both externall and inteſtine, of famine and of peſtilence, & the great Roman Empire, Monarch of the earth, is

The opening of the ſeuenth ſeale

Evangel) came, God his Church had peace a litle while.

2 And beholde, here the [*a*] Angels of the ſeuẽ ages, begin to receiue the ſeuen trumpets of Gods wrath.

3 And the [*b*] great Angell Chriſt Ieſus, came and ſtood as our high Prieſt before the iuſtice ſeate of God, hauing the office of interceſsiõ: and many ſweet acceptable praiers wer made to him of all Gods martyrs and Elect, that he by his [*c*] mediation might offer vp the ſame before that iuſtice ſeat of Gods throne.

4 And the ſweete requeſt of theſe praiers were accepted, & granted by God the Father, through the procurement & interceſsion of Chriſt Ieſus.

5 So that Chriſt tooke the cuppe of Gods vengeance, and filled the ſame with the fierie wrath of his iuſtice, and powred it out vpon the tyrantes of the earth, murtherers of the Saintes of God: & there came great thunderinges of Gods vengeance vpon them

ſeaventh ſeal, there was ſilence in heauen about halfe an houre.

2 And I ſaw the ſeuen Angels which ſtood before God, and to them were giuen ſeuen trumpets.

3 Then another Angel came and ſtood before the Altar, hauing a golden cenſer, & much odours were giuen vnto him, that hee ſhould offer with the prayers of all Saints vpon the golden Altar, which is before the throne.

4 And the ſmooke of the odours with the prayers of the Saints, went vp before God, out of the Angels hand.

5 And the Angel tooke the cenſer, and filled it with fire of the altar and caſt it into the earth, and there were voices, & thunderings, & lighteninges, now ſtired and occupied in this deſtruction of the Iewes:

6 At which deſtruction begins the account of the ſeuen ages to come to the worlds end.

7 In the firſt of the which ages, to wit, frõ this 71 yeare of Chriſt, to the 316. yeare, the effeminate & ſoft Emperors, *Macrinus, Heliogabalus, Philippus, Gallus Voluſianus, Lieinius & Gallienus*, by their great ſloth and coldnes: and the fierie & fierce Emperours, *Domitiã, Cõmodus, Didius, & Maximinus*, by their furie & raſhnes, mixt with the tyrannie of the thirty tyrants that then raigned, as vnder them began the firſt decay of the Roman Empire, During this time the miniſters of Gods word, chiefly in *Europe*, are perſecuted: for euẽ of *Rome* 33 Biſhopes, ſuccesſiuely after other were martyred, & the weake in faith were conſtrained to make defection.

8 Hitherto the Empire of *Rome* decaying beginneth the ſeconde age in this 316. yeare of Chriſt, ſoone after the which, the Emperour *Conſtantine* tranſporting the ſeat therof to *Conſtantinople*, the whole Peninſul of *Cethim* which is interpret the gathering or toſsing of the

Seale and progreſſe of the firſt age or trumpet

316 A Iubilie

The ſecond age

with forth-ſhewings and threatninges of all maner of tempeſtuous plagues, and the great Empire of the earth was then aloft, and greately commooued.

6 At this inſtant, the Angels of the ſeuen ages, hauing the ſeuẽ laſt plagues, mak thẽſelues, readie to blowe out the ſame.

7 And the Angels of the firſt age blew forth their plagues, and there were certaine effeminate Emperours, colde as haile & other furious & fierce as fire and bloodie tyrantes among them, who all were caſt in the Empire of the earth. At this time the conſtant profeſſours of Chriſt, thorowe the thirde parte of the worlde, were perſecuted, and the weake in faith vtterlie, decaied withered, and fell away therefrom.

8 And the Angels of the ſeconde age threatned out Gods iudgement, and there came a multitude of people, fierce as fire, who entered vpon ſome [*d*] Peninſull of the ſea, or landes named by the ſea, and they conſumed

and earth-quake.

6 Then the ſeuen Angelles, which had the ſeuen trumpets, prepared themſelues to blowe the trumpets.

7 So the firſt Angel blew the trumpet, and there was haile and fire mingled with blood, & they wer caſt into the earth, and the 3 parte of the trees was burnt, and all green graſſe was burnt.

8 And the ſecond Angel blew the trumpet, & as it wer *a greate mountaine, burning with fire, was caſt into the ſea, and the thirde parte of the ſea* ſeas, & now called *Italie* is left a pray to the *Hunnes*, *Gothes, Vandals*, and other fierce nations.

with the edge of the ſword the third part of that Peninſull and ſea coaſt.

became blood.

9 Who deſtroyed *Rome*, and ſubdued all Italie, ſlaying & ſpoiling of great and ſmal euen euery third man.

9 And the thirde parte of the people of that peninſull who dwelt there, were ſlaine, and deſtroyed, euen from the poore [*e*] Marriner and inhabitant to the great Pylots, & gouernours

9 And the 3 part of the creatures which wer in the ſea and had life dyed, & the third part of ſhippes were deſtroyed.

10 Then in the yeare of Chriſt 561. begines the third age, in the which *anno* 591. (as the *Arabians* counts *per an nos Hegira*) there aroſe the horrible *Mahomet*, who firſt Profeſsing Chriſt to be that moſt high prophet, thereafter made defection, & throgh pride & glorie to raigne, hee with the aid of *Sergius* a Monke, deuiſed the damnable and bitter doctrine of the *Alkaron*, and by the aſsiſtance of the *Saracens*, he began his conqueſt *anno* Chriſti 623.

561 A Iubilie

The third age

10 And the Angels of the third age, blew out Gods thretnings, and there fell from the true church a notable & [*f*] great Apoſtate, burning in pride, and wrath, and hee entred and ſowed his hereſie in the thirde parte of the worlde, amongſt the (*g*) liuelie fountaines of the pure doctrine of the Euangell.

10 Then the third angel blew the trumpet, and there fell a great ſtar from heauen burning like a torch, and it fell into the third parte of the riuers, and into the fountaines of waters.

11 And ſpred that bitter hereſie through all *Aſia*, impoiſoning therby the ſouls of all theſe of *Aſia*, that came vnder his dominion, and receiued the ſame hereſie.

11 And hee may iuſtlie be called (*h*) gal or wormwood to the ſoules, for hee made the doctrine of the thirde parte of the world bitter as wormwood, ſo that many ſoules thereby died euen all that ſpiritually dranke of that doctrine, becauſe it was horrible and bitter hereſie.

11 And the name of the ſtar is called wormwood: therefore the third part of the waters became wormwood, and many men died of the waters, becauſe they were made bitter.

12 Thereafter in the fourth age, beginning in the year of Chriſt. 806. this

806

12 And the Angelles of the fourth age blew forth Gods wrath and

12 And the fourth Angel blew

erronious doctrine of the *Mahomet*, ſo on the one part increaſed, & on the other part ſuch hereſies, and new traditions are crept in, & authoriſed by the Pope, that both Eccleſiaſticall doctors, ſecular magiſtrats, and ſimple profeſſours, are polluted and darkened with the *Mahometick* & papiſticall errours, ſo that the third part of the light of Gods truth that ſhined ſomtime in the primitiue Church, is now extinguiſhed, the oriental or Greciã Churches being ſeduced by the *Mahomets*, & the occidental or Latin Churches, by the Pope of *Rome*,

A Iu-bilie.

The fourth age or trumpet.

the third part of the ſpirituall Preachers, ciuill Magiſtrates, and Chriſtian profeſſors, were ſo plagued with theſe hereſies, and were ſo blinded with mens traditiõs, that they loſte that light of Gods trueth that ſometime ſhined in them, and neither the Oriental [*i*] or eaſterly Church where the day ſprings) nor the occidẽtal or weſterly Church, (where the night glooms) haue (by the 3. part) that bright light of Chriſts pure Evangell that they were wont to haue.

the trumpet, & the third part of the Sun was ſmitten, & the thirde part of the moone, & the thirde part of the ſtarres, ſo that the thirde part of thẽ was darkned, & the day was ſmitten, *that the third part of it, could not ſhine, and likewiſe the night.*

13 This muſt needs ſpeedily procure Gods true Church to lurk in patience, & abide inviſible, till God in his iuſtice & iuſt iudgement, haue brought moſt terrible plagues and vengeance vpõ the world, for theſe errors & vniverſal defectiõs, whereof the revẽging plagues ſhal immediatly follow & fal out in theſe three next ages yet reſting.

13 Whereupon I beheld, and foreſaw Gods ſpeedie warning as the [*k*] flight of an Eagle, going through the whole congregation of his Elect, proclaiming in his terrible threatnings, 3. fold [*l*] wo & wrack, yet to com therfore, vpon the wicked worldlings, by the executers of the plagues of the three ages following

13 And I beheld, and heard one angel flying through the mids of heauen ſaying with a loud voice, Wo wo, wo, to the inhabitants of the earth, becauſe of the ſounds to come of the trumpet of the three angels, which wer yet to blow the trumpets.

Notes, Reaſons, and Amplifications.

(*a*) That at this yeare of Chriſt. 71. beginnes both the laſt of the ſeuen ſeales, and the firſt of the ſeuen trumpets, appears plainely by this text, and is prooued by our ſeuenth propoſition, and how thereafter, euery trumpet containeth the ſpace of 245. yeares, or fiue *Iubilies* is alſo proued by our fift propoſition.

(*b*) This other Angel muſt be Chriſt, for none ſtandeth before the Altar of Gods iuſtice, as high Prieſt and mediator, making intercefsion for vs, except only Chriſt, as in 1. Tim. 2. verſ. 5. and Hebrews 7.8.9. chapters, We muſt not doubt hereof, though here he is ſimplie called but an Angell, for the like is found in diuerſe ſcripturs, as in Malach. 3. verſe 1. Chriſt is called the Angell of the Couenant, though he was more than an Angell, as contrarily S. *Iohn* the Baptiſt, is there called the Angell that goeth before him, in S. *Ieromes*, tranſlation, althogh he be leſſe than an Angel, as teſtifieth Matthew 11.11.

[*c*] Theſe orations and prayers, though generally they meane all our prayers, which indeed Chriſt offered vp before God his Father, yet in this place is ſpecially meant, the prayers made for revenge of the late martyrdomes ſpecified cap. 6, in the fift ſeale, which prayers, there are delayed, and here taketh effect, as by the text, paraphraſe, and hiſtorie thereof doth evidently appeare.

(*d*) That the ſea is taken for the ſea coaſts or peninſuls, read *Eſas* 21.1, *by the burthen of the deſert ſea*, meaning the burthen of the deſert land of *Arabia* beſide the ſea: alſo *Eſay* 23. ver. 4. by the *ſea and ſtrength of the ſea*, is meant the peninſull of *Tyrus*: and againe in *Eſay* 66. ver. 19. *Italie* (called there *Tubal*) is named among the ſea countreyes. Finally, anent the ſignification of *Gethim*, read *Ierome de interpretatione nominum.*

(*e*) That ſhippes be taken for their marriners, it is certaine: Alſo wheras the ſea is taken for the ſea coaſt (as by the former note) it is probable that the ſhippes betoken the inhabitants of that coaſt and governors thereof, as in *Ezechiel*, 27. verſ. 25. by *Ieromes* tranſlation, *Naves maris principes tui, &c. The ſhippes of the ſea are thy princes*.

(*f*) That ſtarres be taken for miniſters or profeſſours of God his trueth, euen after their defection or apoſtaſie, you may ſee both here by this *Mahomet*, and the monk *Sergius* his colleague, as by the Turks hereafter, *Rev*. 9. ver. 1. Of ſtarres generally taken for true teachers, read *Dan*. 12. ver. 3. and Rev. 1. notes *q*, & *x*, therof, *Reve*. 6.13. and note *i*, therof, and other places.

(*g*) What theſe floods and fountaines be, conſider by the hunger and thirſt ſpoken of before, *Revel*. 7. ver. 16 and 17. And note *h*, therof, and confer there with *Ieremie*. 2.18. ſpeaking of the drinking of the bitter floods of the *Egyptians* and *Aſſyrians*, alſo *Iohn* 7.37.38 and *Revel*. 21.6.

(*h*) That this bitter worme-wood meaneth the doctrine of falſe Prophets, how ſweet ſoeuer it ſeemes, read *Ieremie* 23.15. & how bitter and vile the *Mahomets* doctrine is, they may conſider by his *Alkaron*.

(*i*) In theſe Churches of the Oriental and Occidentall empires, where God ſometimes was truelie honoured (as *David* pſal. 113.3. prophetically foretelleth) at this time aboundeth idolatrie, hereſies and many vaine traditions: of which, read the Eccleſiaſtical hiſtories.

(*k*) There is ſo ſmall difference in the greeke, betwixt the orthographie of an Angel & of an Eagle, [$\gamma\gamma$ being excepted, which by blot or wearing may be defaced] that ſome old copies hath the one, and ſome hath the other, alwayes to them both, agreeth the interpretation of ſpeedy warning, or haſtie meſſage.

[*l*] The whole viſible face of the Orientall and Occidentall Churches, hauing now made defection: He notes by wo, wo, wo as by an emphaſis, how great and terrible ſhall bee the remanent three plagues that are to come vpon them therefore.

CHAP. IX. THE ARGVMENT.

Hauing deſcribed the eſtate of the Empire, Theſe foure firſt ages: and lately how both the Orientall and Occidentall empyres, haue loſt a great part of that ſpiritual light that firſt they had: Now he procedeth in this chapter to the fift and ſixt ages, wherein is deſcribed how horriblie they are puniſhed for their defection, & how neuertheles they abide obſtinate and impenitent, till in the remanent chapters, the reformatiõ of a great part, and ruine of the remanent enſue.

Hiſtorical application.	*An. Chriſt.*	*Paraph. explication.*	The text.
NOw, in the fift age, beginning in the yeare of Chriſt 1051. proceeded yet one further horrible plague, for euẽ the foreſaid *Mahomet*, who	1051 A Iubilie The	ANd the angels of the fift age, blew forth Gods threatnings & plagues, & I perceiued that former [*a*] Apoſtat, who before had fallen from God his true Church, to earthly pride and ambition receiued yet more power by his profeſſours after him, to open vp againe the infernall pit of his deuiliſh errours.	*ANd the fift Angel blew the trũpet, and I ſawe that ſtar which had fallen from heuẽ vnto earth & to him was giuẽ the key of the bottomleſſe pit.* 2 *And he*

2 And he by his profeſſours and doctrin opened the very way to the endles deepe of hell, and out of that his deuiliſh fornace, there aroſe ſuch darknes of errours, that darkened greatly both the light of the Evangel and temporall eſtates.

3 So that nowe are ſprung vp the Turkes, and become new *Mahometiſts*, and oppreſſe all Chriſtian landes, riſing at laſt ſo high in conqueſts, and ſo deep in hereſie, that their ſect is ſpred ouer in all places.

4 Yet compell they as then, no man to become *Mahometiſts*, but ſets more their intent to reaue and ſpoile thoſe that were vnder the *Romane* and *Papiſticall* empire.

5 And not being of power to deſtroy them and conquere *Europe*, they cõtinued ſo ſpoiling and reving, from the daies of *Zadok* their firſt *Dominatour, in anno Chriſti*. 1051. vnto the time that *Changius, Chan* of *Tartarie*, began his Empyre in *anno* 1201, which was the ſpace of an hundreth and fiftie years, beginning firſt to enter in *Europe* as friends in ſupply of *Baſilius*, Emperour againſt the *Bulgarians*. his rebeles, afterward, they proceded waſting and ſpoyling ouer all parts.

fift age or Trumpet

3 So that now, newly againe of this deuiliſh darknes and hereſie, is ſprung vp new [*b*] *ſwarmes* of that ſect, waſters of al lands, hauing power in the end with their [*c*] *ſtings* of hereſie, to infect a great part of the earth.

4 Yet did God ſo bridle them, that as yet they were not able to ouercome the conſcience of the very tendereſt & moſt weake Chriſtians on earth, neither any that were as greene plants, or as fruitfull trees in Chriſts vineyarde, but onely to trouble and annoy theſe, that making defectiõ from Chriſt profeſſed ſubiection to the antichriſtian empire.

5 Yea, euen towards theſe alſo were they ſo brideled, that they ſhuld not [*d*] vtterly deſtroy them, but onely oppreſſe them for the ſpace of [*e*] 150. yeares: but this

opened the bottomles pit, & there aroſe the ſmoke of the pit, as the ſmoke of a great fornace, and the ſun and the air wer darkned by the ſmoke of the pit.

3 And there came out of the ſmoke Locuſtes vpon the earth, and vnto them was giuen power, as the ſcorpions of the earth haue power.

4 And it was commanded thẽ that they ſhould not hurte the graſſe of the earth, neither any greene thing, neither any tree but only theſe men which haue not the ſeale of God in their fore heades

5 And vnto them was commanded, that they ſhould not kill them, but that they ſhould be vexed five monethes, and that their paine ſhuld be as the pain that commeth of a ſcorpion whẽ he hath ſtung a man.

6 So that now their poore, oppreſſed and depauperate captiues would wiſh to die, yet doe they referue them aliue for rowling in their gallies and other feruitudes.

their ſmall oppreſsion beginning firſt ſoftlie, as the ſting of a *ſcorpion* proceedeth at length to incurable wrack and exterminion.

6 And in theſe daies men whom they haue thus vexed and depauperat would wiſh to die but they ſhall be reſerued aliue in torment.

6 Therefore in thoſe daies, ſhal men ſeeke death, & ſhall not finde it, & ſhall deſire to die, and death ſhall flee from them,

7 And the form of the locuſts was *like vnto horſes prepared vnto battel, and on their heades* were, *as* it were *crowns, like vnto gold, & their faces* were *like the faces of men.*

7 And thoſe oppreſſour were as barded horſes, proceeding to battell with ſtrength boldenes, and agilitie, and ſo triumphed in oppreſsion, that they raigned as crowned Kinges, enriched with their prayes: In the firſte face they ſeemed friendly as men ſhould be.

8 And they had hair as the hair of women & their teeth were as the teeth of lyons,

8 And ſmiled alluringlie, as with womens attyre, yet they deuoured and waſted all landes, as with the teeth of Lyons.

9 And they had habergiõs like to habergiõs of yron, & the ſoũd of their wings was *like the ſound of of charets when many horſes run vnto battel.*

9 They were armed on the moſt ſtrongeſt maner, and there ſpeedie haſt made a noiſe as chariots of warre.

10 And they had tailes like vnto ſcorpiõs, & their wer ſtings in their tailes & their power was to hurt men fiue moneths.

10 But their tail, end, and finall intention was to oppreſſe and ſubdue all men, and after that ſubiection, to impoyſon them with their ſtings of hereſie, and they had power to torment & oppreſſe men the ſaid 150. years.

11 And they had ſpirituallie raigning ouer the mthe ſaid *Mahomet*, whoſe name ſignifieth a

11 And they had ouer them their king, euen a [*f*] meſſenger of the Devill, whoſe name

11 And they

meſſenger, or deſtroyer, & ouer them temporallie raigned the great *Turk*, which likewiſe ſignifieth the ſame: for *Turca* is interpreted a waſter or deſtroyer.

12 Theſe at this time wrought greate woe to the *Roman* Empire, but afterward, theſe ſame with others wrought yet more woe.

13 For in the ſixt age (begun in the yeare of Chriſt, 1296,) it pleaſed God in his wrath to ſtirre vp.

14 The foure nations *Mahometiſts*, that dwelled beyond and about *Euphrates*, to wit, the *Saracens*, *Turks, Tartarians*, and *Arabians*. who al being confederate together in one lawe of the *Mahomet*, and vnder one great Emperor *Ottoman* began euen then their firſt Empire.

15 Theſe with huge Armies came againſt both the Empires of the Orient and occident, ouerthrowing, ſlaying & ſubduing all *Aſia*, and much of *Europe*, euen about the third parte of the world,

16 For their number was infinite,

17 And themſelues and their horſes armed moſt ſtronglie, like irefull proude oppreſſours haſting to their pray profeſsing with their mouthes theſe 3. ſpeciall pointes of their

1296 A Iubilie.

The ſixt age or trumpet.

(trying from languag to languag) ſhal be found to ſignifie a (*ff*) deſtroyer or waſter.

12 The firſt of the great woful plagues is paſt, and beholde the two laſte are to come ſhortlie.

13 And the Angels of the ſixt age blew forth Gods threatnings, & a warning came at all quarters from the throne of Gods iuſtice,

14 To theſe Angels of this 6 age, to looſe the 4 [*g*] great natiõs which as Angels, & meſſengers of Gods wrath, abode at the floud *Euphrates*.

15 And theſe foure nations (which here dwelt in readineſſe againſt this preciſe time & houre, to ſlay the third part of men) now raiſed their armies:

16 And the number of their horſmẽ were [*h*] 200 000.000. & I heard that nũber to be certain

17 So I ſawe, in ſpirite, theſe horſes, and their horſmẽ armed, as men burning with fierie wrath, puft vp with gliſtring pride, & tending for fearfull

haue a King ouer thẽ, which is the Angel of the bottomleſſe pit, whoſe name in Hebrew is Abaddon, and in greeke he is named Apyllyon.

12 One woe is paſt, and behold, yet two woes come after this.

13 Then the ſixt Angel blewe the trũpet, & J heard a voice from the 4. hornes of the goldẽ altar which is before God,

14 Saying to the 6. Angel, which had the trũpet, looſe the 4. Angels, which are boũd in the great riuer Euphrates.

15 And the four Angels were looſed, which were prepared at an houre, at a day, at a moneth, & at a yeare, to ſlay the 3 parte of men.

16 And the nũber of horſmen of warre were 20 thouſand times tenne thouſand for I heard the number of them

17 And thus J, ſaw the horſes in a viſion, & them that ſat law, to wit, their vnquenchable ire, againſt al chriſtian profeſſours: the eſtabliſhing of the *Mahometicke* hereſie and doctrin vniuerſallie, and finally their bitter auarice, to reaue and conquere all the worlde to thẽſelues as appertaining to them by right, for that they are the ſonnes of *Abraham*,

18 And they being on theſe three reſpects puft vp ouerran ſlew & oppreſſed all *Aſia*, & much of *Europe*.

19 Conquering mightilie before them with their armie, and leaning behinde thẽ their prieſtes and falſe preachers to ſeduce.

20 Neuertheleſſe the reſt of the Roman Empire reformed not themſelues, nor amended their workes, neither left they their vngodlie & deuiliſh worſhipping of alleadged Saintes, neither their filthie idolatrie, and worſhiping the ſenſeleſſe images made of euerie mettall and materiall:

deſtruction riding fiercely forward to deuour their pray as Lyõs, Profeſsing plainlie with their mouthes, their fierie wrath, their dark and ſmokie hereſie, & their horrible & bitter avarice againſt all Chriſtians:

18 And in their [*i*] ire hereſie, & avaritious oppreſsiõ, which they profeſſed, they invaded, ſlew & deſtroied the 3. part of all the Empire.

19 For theſe horſemen had power before them to conquere and deuoure vp all that come in their teeth & power, alſo behind them becauſe the Deuill, that old ſerpent followed in their tailes by the mouthes of their falſe teachers, to ſting and ſeduce all whom they had ouercome.

20 Neuertheleſſe the reſt of the people of the Empire, that were not ouercome by theſe plagues, repented not of their [*k*] vngodly workes, nor abſtained from [*l*] diueliſh worſhiping of [*m*] images of gold, of ſiluer, of braſſe, of ſtone and of timber,

on them hauing fierie habergiõs & of Iacinth, of brimſton, & the heads of the horſes were as the heads of lyons, & out of their mouthes went out fire & ſmoke and brimſton.

18 Of theſe three was the third parte of men killed: that is, *of the fire and of the ſmoke, and of the brimſtone, which came out of their mouthes.*

19 For their power is in their mouthes, and in their tailes: for their tailes were like vnto ſerpẽts and had heades where with they hurt.

20 And the remnant of men which were not killed by theſe plagues, repented not of the workes of their hands, that they ſhould not worſhip deuils, and idoles of golde, and of ſilver, & of braſſe, and of ſtone, and of woode, which neither can ſee,

	which neither can ſee, nor heare nor moue.	*neither heare nor goe:*
21 Neither left they off to murther all them that ſpake againſt their errours: but continued in their charmes, and exorciſmes, & in their vild Sodomie, and horrible whoredome, and epented not to dereiue ſimple men, in cealing from thẽ both their goods and their ſouls, for ſoul-maſſes, diriges, pardons, indulgences, & ſuch other trifles.	21 Neither repented they of their [*n*] tyrannie & martyrdomes, neither of their [*o*] conjuring charms, or witchcraft neither of their filthy and [*p*] vnſpeakable whoredome, neither yet of their [*q*] theeuiſh taking of goods from ſimple men.	*21 Alſo they repented not of their murther, and of their ſorcerie, neither of their fornication, nor of their theft.*

Notes, Reaſons, and Amplifications.

(*a*) Now doth the Spirit of God repeat here the fall of the *Mahomet*, not that he now fell (for as is ſaid, he fell in the time of the thrid trumpet) but meaning that his old fall, his *Akaron* and profeſſours that he left behind him, haue made the way patent to the Turks, now at this time, to fall alſo from the trueth, and follow him. Conferre here with our third propoſition, and note following, by the which, both theſe ſtarres appeare to be one, and to meane the *Mahomet*, and his locuſts to be the Turkes, and for confirmation that this ſtarre, is that ſame ſtarre that before had fallen, the Greeke terme of falling, expreſſed in the text, is πεπτωκότα, in the preterit time, which had fallen, and not πιπτοντα in the preſent time, which falleth. Meaning thereby expreſly: that S. *Iohn* ſaw that in this fift age, the *Mahomet* (who before had fallen from God in the third age) now by his ſucceſſours, and doctrine, doth open vp the gates of hell more manifeſtly, in ſtirring vp theſe Turks by his hereſie.

(*b*) That theſe Locuſts be the Turks, it is plaine: for that as Locuſts doe deſtroy and waſte euery where, and are plagues appointed by God to that effect. Exod. 10. So are the Turks both in name and effect, waſters and ſpoylers, as ſhall be ſhewed in the note *f*, following.

(*c*) That theſe, who with pleaſant flatteries profeſſe good, and yet in effect are rebels, and ſeducers from God, are compared to ſcorpions, read Ezech. 2.6. and how euil women for their faire reſemblance and wickedneſſe in effect, are called ſcorpions, reade *Eccl.* 26.10 So theſe Turks are deſcribed here with mens faces, & womẽs haire, & ſcorpions tails, becauſe they firſt entred in *Europe* humainly, & proceded ſince with all deſpite & oppreſsion. Alſo for that the ſtings of ſcorpiõs ſeem firſt nothing, & after become incurable, as the ſmal cõpany of Turks becam almoſt a monarchy

[*d*] In hiſtories is more mention made of the *Turks* oppreſsiõs, and ſpoiles done againſt Chriſtians in that age, than of any conqueſt or vtter deſtruction of Chriſtians, as *Iohn Baptiſta Egnatius*, ſpeaking *de origine Turcarum*, ſaith, *Vagi diſpalatique quò ve cuique ſors affust, latrocinantes magis quàm belligerantes provincias vaſtarunt Theſe going in vaging and wandering whereſoeuer their chance was, they waſted countries, as theeues, rather then as warriors*. And as to their dominion, how frõ the firſt year of their firſt Dominator *Zadok*, in the year of Chriſt, 1051. to the time that the *Tartarians* ſubdued thẽ, in An. 1201. ar iuſt 150. years, diuers hiſtories do declare

[*e*] That theſe fiue prophetical moneths, are an hundreth and fiftie yeers, read the firſt propoſition, with the ſecond reaſon of our third propoſition.

[*f ff*] This flitting from *Abaddon* in Hebrew to *Apollyon* in greek ſignifieth, that his name is neither *Abaddon* nor *Apollyon*, but that we ought to deſcend from language to language, vntill we find the name of a Governor, that ſignifieth the ſame that *Abaddon* & *Apollyon* do in Hebrew and Greek. And ſo it is, that Abaddon, & Apollyon do ſignifie a Deſtroyer or waſter (*Ierome do interpretatione nominum*) And again, on the other part, *Mahometes* ſignifieth *ex eodem) Delens*, a *Deſtroyer*, or, *Nuncius*, a *Meſſenger*, agreeable with both the ſtiles that this text giueth him, to wit, both a deſtroier, & alſo, the Angel or meſſenger of the bottomles pit. Moreouer the like is to be ſaid of their temporal kings name *Turca*, which (*ex Carionis chronico, fol. 8. lib. 1. & lib. 5. fol. 107.*) ſignifieth *idem quod vaſtator*, a Waſter or deſtroyer.

[*g*] That all meſſengers and executers of Gods affairs, either of favour or wrath be called Angels, it is common in al the Scriptures: for *Angelus* ſignifieth a meſſenger: & for proofe that theſe four of the ſixt trumpet be no ſpirituall Angels, the ſame in the ſixt viall (Apoc. 16.12.) are called the Kings of the Eaſt, for ſo are theſe *Mohametans* indeed.

[*h*] Albeit this number of two hundreth thouſand thouſand is not ſo expreſſed to ſatisfie our curioſitie, as if there were no oddes (as I wrot, Apoc. 7. and note *f*, thereof) Yet certainlie it hath bene nere this number, for that he ſubioyneth for the more certaintie again, *And I heard the number of them, and saw the horſes in a viſion, &c* Agreeable hereto, it is reported that the great Turke contemptuouſlie ſent to the Emperour of Romanes, a Camel or a Dromadarie loadẽ with wheat, with this vow by meſſage, that he ſould bring againſt him, as many fighting men, as there were grains of wheat therein. This armie hereafter (Apo. 20,8) is compared to the ſea ſand in number,

[*i*] Of theſe three points of their Profeſsion, read their *Alkaron*, & confer

therewith the hiſtories, for of their ire againſt the name of Chriſtians, euen from *Mahomets* firſt beginning, read *Iohn Baptiſt Egnatius* vpon *Heraclius* Emperour, ſaying, *Reſponderunt impii,* [meaning, *Mahometiſts*] *non prius. Romanis pacem ſe daturos, quam Chriſtum abnegaſſet Imperator. Theſe wicked made anſwer, that they would giue no peace to the Romans, til firſt their emperour renounced Chriſt.* The ſecond concerning their diligence to ſpread their hereſie is certain. The third, concerning their avaritious claiming the title of the whole world, that is conteined expreſly in their law, that they being the ſons of *Abraham*, and of *Sara* (as they alledge) the land is promiſed to them by God, Gen. 13.16. & therevpon *Mahomet* promiſeth the kingdome of heauen to all them, that fights valiantly in that quarrell. Yet theſe alledged *Saracens*, are the children of *Abraham* and *Hagar*, that is to ſay, *Hagarens* or *Iſmaelits*, & not of *Abraham* and *Sara.* But as to the Turks wee find no certaintie that any wayes they are the ſonnes of *Abraham*, yet claim they that promiſe, as if they were the true ſonnes of *Abraham* and *Sara.* And ſo in theſe three points of their hereſie & vſurping of kingdomes, do they invey againſt Chriſtians.

[*k*] Theſe of the Romane Empire regarded not what euil and vitious works they did, ſeing they were perſwaded, that for litle coſt, they would get indulgence, pardons, holy water, and light pennance, which would relieue and purge their ſinnes.

(*l*) Of earthly hypocrites: who diſſembled their euill life, as ſome hiſtories report, the Pope made his Saints: And thogh ſome of theſe be deuils, condemned for their hypocriſie, yet hath the Pope and his Cleargie ordained all theſe, whom they canonized to be worſhipped. Moreouer, they are ſaid here to worſhip the deuil, in that they worſhip not God, as he hath cõmanded them but after their own imaginations.

(*m*) What crucifixes and other Idoles of golde ſiluer and other materiales were both priuatly in their oratories, and publikly in Churches adored, the whol world knoweth. If they will ſay, they worſhipped them not, but vſed them as a remembrance: O *Ananiaſſes* and *Zaphiraſſes*, lyars againſt their owne ſoules, came they not bare-footed, bare-legged, bare-headed, growflings, and kneeling with their hands holden vp to theſe Images, and with their faces and eyes direct to them, making ſuch prayers and ſupplications as in their very books are intituled, a praier to the image of Chriſt, and a prayer to the croſſe of Chriſt, and ſuch others. How are they then able to deny, but that they prayed euen to theſe very images, yea, and for the greater deceit to ſimple ſouls, leſt it ſhould be knowne that idolatrie any wayes was ſin, their Cleargie took away the ſecond commandement out of their treatiſes and diuided the laſt commandement into two, to make out ten. And likewiſe, they held the whol Bible in Latin vntranſlated, leſt it ſhould be knowen to the people, that God in it forbad the worſhipping of images: Repining thereby againſt their knowledge & conſciẽce, without any repẽtance as yet: as ſaith the text.

[*n*] Amidſt the great invaſions that theſe *Mahometans* made at this time againſt the papiſts. The Pope neuertheles ſpared not, euen then to murther the poore ſaints of God, ſuch as *Iohn Huſſe*, I*erome* of *Praga, Ierome Savonarola*, and many moe that opponed againſt his indulgences, and others his wicked traditions.

[*o*] What charms euery commõ Curat had for the fellone, the farſie, the worme, for blood-ſtemming, for lecherie, and for other horrible purpoſes, all their familiars, or that hath ſene their works and books knowes. Alſo in *Platina*, and in other credible hiſtoriographers, ye ſhall find the legend of 22. popes or thereby, that wer moſt horrible *Necromancers*, like as there is one part of their ſeruice that is called *Exorciſmus*, conteining ſuperſtitious & magical inchantments and coniurations of oyl, of ſalt, of holy water, of bels, beads, *Principioes. Agnus Deis*, and other ſuch like. And further, euerie high Altar is erected with ioints, hair, clouts cloths, threeds and worſeds of diuers hews, and diuers other reliques of dead men, laid and builded behind the ſaid Altars, as they do teſtifie whoſe eies hath ſene the ſame. And leſt they ſhuld doubt thereof who hath not ſene it, the collects of the Maſſe do beare record of the ſame, granting theſe reliques to be there as in the maſſe booke *ſecundum vſum Sarum, in ſervitio beatæ Mariæ, orationibus: Propitiare quæſumus Domine*, and, *Præſta quæſumus omnipotens*: And again, *infeſto reliquiarum, oratione præſta quæſumus omnipotens, &c.* and *lectione prima, Hodie dilectiſsimi, &c.* ye ſhall find theſe clauſes contained. *Quorum reliquiæ in hac continentur Eccleſia*, that is to ſay, *whoſe reliques are here contained in this church.*

(*p*) What whoredome and vile Sodomie is vſed both in Rome, and likewiſe euery where, among the Popes, Cardinals & whole cleargie, not onely their innumerable baſtards teſtifies, but alſo their lawes priuilie permitting Prieſts to haue concubines, but not wiues, (Diſtinct. 34) though ſome late glozers (excuſing the mater) define theſe concubines to be wiues, contrary vnto the text. Further hereof read in *ſpeculo Romanorum pontificum*, ſet out by *Stephanus Szegedinus pannonius*, An. 1584. from the 91. page, to the 111. citing and quoting the vnſuſpect writers of the Legend of licentious popes, whereof thirteene were adulterers, three were common brothellers, 4. were inceſtuous harlots, 11. were impoyſoned with vile Sodomie, 7. were whooremongers and erecters of brothel houſes, whereof euery whoore payeth weekly a Iulian penny to the Pope, which diverſe years will extend to 40000, ducats by year. And all theſe beſide the ſhe pope called *Ioane*, who was firſt called *Gilberta* an Engliſh woman, borne in the town of *Lyn*, who being two years and an halfe pope, at laſt dieth in deliuerie of child openly vpõ the high ſtreit of *Rome*, going in publik procefsiõ by the way to the Church of S. Iohn *de Lateran*, & fo made her whordom & filthines of her feat, patent, In memory whereof, al popes euer fince going in procefsion, efcheweth & goeth about that *Lateran* way, where their fhe predeceffour died.

(*q*) It is more nor notorious what great Abbacies & bifhopricks from fimple princes, what lands and yearly rents from landed men, what money and goods from men of all eftates, hath bin deceitfullie ftolne for indulgences, pardons, remffsions of fins, trigintals, foule-maffes, dirigies, deprofundis, & other fuperftitions, ftealing thereby not onely mens goods, but even their fouls, as hereafter Rev. 18.13 and note *i*, thereof, and 2. Pet. 2.3. Thefe bee like the Scribes and pharifies, *who devour vp poore widowes houfes vnder pretence of long prayers*. as is faid in *M*atthew 23.14.

CHAP. X. ARGVMENT.

Now orderly fhould follow the feventh age, which the fpirit of God beginneth not here to defcribe, after the ftyle of the other fix: but firft fetteth out a vifion confonant & relatiue, with the vifion wherewith he beganne to defcribe the eftate of the primitiue church, and that becaufe Chrift Iefus, that came then in the flefh, and fpread the true light of his Evangell to the primitiue Church, is now (after long darknes and horrible defection) returned by his Spirit to reftore and open vp again the fame light and trueth to vs, in this laft age. And therefore he ftaieth here the imperial or fecular hiftorie of this feventh age, till the Ecclefiafticall hiftorie be alfo farre proceeded. In this vifion firft is defcribed the fpirituall returning of Iefus Chrift to vs, with the open word of his truth. Secondly, that now the feven thunders or Angels of Gods great haruef t, are comming, where Chrift maketh his folemne vow, that in this feventh age the world fhall end. Last of al, how (vnder the name of the Prophet) the whole Church at this time, fhould rightly conceiue in their hart the true underftanding and meaning of this Revelation, and fhould preach and publifh with their mouths the fame, more plainly of new again to the whole world.

The Paraphraftical expofition.

AND now I faw another, euen the great Angel of the covenant, [*a*] Chrift Iefus by his holy Spirit defcending frõ heaven, whõ as now the world thoght cloudy, & knew not: ouer his head, ftood the fign & memorie of his

The text.

AND J faw another mighty Angel come down from heauen, clothed with a clud, & the rainbow vpon his

protection, & in the beautie of his face, ſhined the bright light of eternal life, ſtanding conſtant as a pillar in defence of the elect, & going as fierce as fire to confound the wicked.

2 And he offered now at hand, the open trueth and revelation of theſe [*b*] Evangelical and propheticall myſteries, and ſubdued at laſt vnder his feet al empires, both of ſea and land.

3 And he cried, & publikely preached repentance by the mouths of his Preachers, as the fearfull voyce of lions: & as that doctrin was preached the ſeuen terrible thunders, [*c*] preparers & reapers of Gods great harueſt) begã to forewarn their meſſage.

4 And when they had ſhewed forth and fore-told the ſame, I thought to haue written it plainly, but (that being ſo great a myſterie) I was cõmanded from heauẽ to [*d*] ſeal & cloſe vp the ſame in dark lãguage (as cap. 14.1 haue don) & no to writ it plainly here

5 And Chriſt Ieſus that great Angel, Dominatour now ouer ſea and land, ſolemnly voweth, as one holding vp his hands to [*e*] the heavens,

6 And ſweareth by him that liueth for euer and euer, who made the heauens, and all that is therein, and the earth, and al that is therein, & the ſea, and al that is therein, that [*f*] time ſhal be no longer:

7 But in the dayes of this ſeuenth age, euen [*g*] after the Angels of this age blow foorth the plague of Gods laſt wrath, the Antichriſt ſhall be deſtroyed, and Chriſts kingdome onely erected, and the great [*h*] myſterie of Gods latter iudgement, ending the prophecies of all prophets, ſhall be performed.

8 Then was I warned againe, by that former voice of God, to go now and receiue this cleare inſtruction of theſe myſteries, evangelicall and propheticall, at the hands of Chriſt Ieſus Dominatour ouer ſea and land.

9 And I paſt & craued, earneſtly of him to reueile now to mee, (in the name of the whole

head, and his face was as the Sunne, and his feete as pillars of fire.

2 And he had in his hand a litle booke open, and he put his right foot vpon the ſea, and his left on the earth,

3 And cried with a loud voice, as when a lion roareth: and when he had cryed, ſeuen thunders vttred their voices

4 And when the ſeuen thũders had vtred their voyces, J was about to writ: but I hard a voice from heauen, ſaying vnto mee, Seale vp theſe things which the ſeuen thunders have ſpoken, and write them not.

5 And the Angell which I ſaw ſtand vpon the ſea, and vpon the earth, lift vp his hand to heauen.

6 And ſware by him that liueth for euermore, which created heauen, and the things that therein are, and the earth, and the thinges that therein are, and the ſea & the thinges that therein are, that time ſhould be no more:

7 But in the daies of the voice of the ſeuenth Angel, after he ſhal blow the trumpet, euen the myſterie of God ſhall be finiſhed, as he hath declared to his ſeruants the Prophets.

8 And the voice which

Church) this revelation and knowledge, & Chriſt granting this our petition, commanded vs all now in this age, to receiue this patent reuelation and knowledge, and deeplie to ſwallow it vp, and ſettle it in our heartes, and it ſhall, ingender within vs the Spirite of his [*i*] zeale, & bitter indignation, deteſting horriblie all impietie, but to preach it out with our mouths, & disburthen our conſciences thereof, it ſhall be moſt ſweete & comfortable for our ſoules.

10 Then (the true light being now come) we receiue this perfecte and cleare inſtruction at the handes of Chriſt Ieſus, and ſettle the ſame deeplie in our breaſts, and the preaching and conſtant Profeſsion thereof with our mouthes, becomes moſt pleaſant and delectable to our ſouls: but when we doe deepelie ponder the terrible plagues of God, prophecied therein, to come for our impietie, it moueth our bodies to quake, and our ſouls to be full of bitter anguiſh.

11 And Chriſt Ieſus by his ſpirit reueiled vnto me, that it behoued me, that is to ſay, the whole miniſtrie in me, to (*k*) preach forth this preſent reuelation againe in this laſt age, and plainlie open the prophecie thereof to all nations, people, tongues, kingdomes, and kinges thereof.

I heard from heauen, ſpake vnto me againe, and ſaid, Goe, and take the little booke, which is open in the hand of the Angel which ſtandeth vpon the ſea, and vpon the earth.

9 So J went vnto the Angel, and ſaid to him Giue me the little book And he ſaid vnto mee, Take it, and eate it vp, & it ſhal make thy bellie bitter, but it ſhall be in thy mouth as ſweete as honie.

10 Then I took the little booke out of the Angels hand, and ate it vp, and it was in my mouth as ſweete as honie: but when I had eaten it, my bellie was bitter.

11 And hee ſaid vnto me, thou muſt prophecie againe among the people and nations, and tongues, and to manie kings.

Notes, Reaſons, and Amplifications

(*a*) This Angell, who here appeareth to *Iohn*, is alone with him whome *Daniel* ſaw in the ſimilitude of a man, Dan, 10.5. continuing to *Daniels* 12. chapter, verſe 6.7.) as is euident by their common ſignes and tokens, and for the ſame cauſe, he who appeareth in the middes of the ſeven Candleſtickes, Apoc. 1.13. is alſo all one with them: For by Dan. 10.16.18. he is ſtiled, *One like the ſimilitude ofthe ſonne of man*, or like the appearance of a man. So Apoc. 1.13. he is ſtiled, *One like vnto the ſonne of man*. In Daniel he is girded with a golden girdle: So is he alſo in Apoc. 1. In both places his face did ſhine, as here alſo it ſhineth: In all theſe places his feet were as pillers of fire, and of fiery and burniſhed braſſe, and his voyce terrible, as of a multitude of people or of manie waters, or of a roaring lyon. Againe, both there in *Daniel* and Apoc. 1. his eies burne as a flame of fire, or as burning lampes: both *Daniel* and *Iohn* fall down at his feet through feare & he againe doth cõfort them both: Further both in Dan. 12.6.7. & here in this chapter, he ſtanding vpon the waters, holds vp his hands and ſweareth the end of times. So that without all doubt, that ſimilitude of a man that there appeared to *Daniel.* and Apoc. 1.13. appeared to *Iohn*, and this Angell here mentioned, muſt all three be one. But ſeing he, who (Apoc 1.13) appeared, to *Iohn*, is proued there, note *l*, thereof to be Chriſt, ſpirituallie appearing: Therefore, both he, who there in *Daniel* appeared, and this Angell, who heere appeares, is alſo Chriſt. And for confirmation hereof, none hath this book in heauẽ or earth or is worthie to open the ſame, except Chriſt our Lambe, Apoc 5.2.3.6. Therefore, he who here hath the ſame open in his hand muſt be Chriſt.

(*b*) Of this booke of Reuelation (whereof Chriſt here beareth the type in his hand) onlie the hiſtoricall part, to wit, the ſeuen ſeales thereof, is opened by Chriſt and his Apoſtles to the primitiue Church, in that, that part was fulfilled in their dayes. But the text doth neuer ſhewe, that the booke within is opened vp and offered patent to vs, till now in this our ſeuenth age, wherein Chriſt here (verſ. 2.) not onely opens all the propheticall myſteries thereof, but alſo (ver. 9.) offereth the ſame to the Church vnder the perſon of the prophet: as thirdlie ordaineth them to preach the ſame ouer againe, openlie and patently to the whole world, where-through, without all queſtion, this is the time of knowledge euen preſentlie, of which ſaieth *Daniel* chap. 12.4. *Claude ſermones, & ſigna librum vſque in tempus ſtatutum. Cloſe vp theſe ſpeaches, and ſeale the booke vntill the time appointed.*

(*c*) How ſoone in the beginning of this ſevẽth age, (about the year of Chriſt, 1541.) the light of the Evangel, and of theſe prophecies againſt the Antichriſt beginnes to be reuiued & preached out by Ieſus Chriſt, in the mouthes of *Luther, Caluin*, and other his miniſters, euen that very year begin the ſeuen thunders of the latter day, and reapers of Gods great harveſt, to expreſſe their prophecies, euerie one of the firſt foure following other, by the ſpace of fourtie and nine yeares, as at length is ſhewed in our 12. and 13. propoſitions.

(*d*) At the command and voice of God, the prophecie of theſe thunders is ſo ſealed and cloſed, that none could before the due time eſpie, either what they meaned, or where they were writtẽ. But the compleate time of Reuelation, and knowledge, ſpoken of by *Daniel.* (chapter, 12.4, 9.13) ſo ſeemeth by the dates contained both there in *Daniel*, and here in the *Reuelation*, to be come. And in this open booke, that Chriſt in this text proffereth to be ſo manifeſted, that we may certainlie conclude now, that this, nor no other myſterie hereof is vnſearchable: what therefore, theſe thunders be, is declared in our 11. propoſition, to wit, that they be the ſeuen reapers of God his great haruest, mentioned Apoc 14. No maruell then, though ſo greate a myſterie be commanded to be cloſed at the leaſt, during the dayes of the Antichriſtian tyrannie that theſe, that raigne in iniquitie might continue ſtill therein till vnawares, they be ruined (Matth. 13.13 14.15.) like as contrariwiſe at length and chieflie to the elect, the ſame is commanded to be opened, and not ſealed, Apoc. 22.10. And ſo, though euen vnto the wicked, I with many moe, now write this patentlie', yet God hath prepared a vaile for their eies, that ſeeing they ſhall not ſee till they goe into deſtruction.

[*e*] This is a compariſon and ſimilitude, taken from the geſture of men when they make a ſolemne oath: and not meaning that God ſweareth: for, the ſmalleſt worde of his mouth is as ſure as an oath, yea, ſo ſure, that not one jot thereof ſhall periſh, (Matth. 5.18.) The ſame oath you ſhall finde in Dan. 12.7. Alſo the like Gen. 22.16. The mention of this oath is but put here for the greater Emphaſis, to ſupplie our weake credit and beleefe.

(*f*) Then let the Pope and all earthlie Princes, looke for no longer delay, but that in this ſeuenth age alreadie begunne in the year of Chriſt, a thouſand fiue hũdreth fourty one, the *corner ſtone* Ieſus Chriſt, ſhal become a mountaine, & ſhal deſtroy al temporall kingdomes and raigne for euer, Eſay. 2.2.3. Dan. 2.44. Pſa 2.9. and before the 1786. year of Chriſt yea euẽ ſooner, to wit when we ſee the tokẽs & plagues of this age ſpent and paſt, then ſhall immediatly come, the day of Gods dreadfull iudgment, as more in ſpecial is declared in our 10.13. & 14 propoſitions, and in chap. 14 following. With this vow conferre the like, mẽtioned in Daniel. 12.7.

[*g*] The vulgar text ſaith here, *Quum cœperit tuba canere conſummabitur myſterium magnum*, that is, *When he begins to blow the trumpet*, &c. But the originall Greek may rather import, *After he ſhall blow the trumpet*: for the word ὅταν may more iuſtlie be taken for *After* rather than for *immediatly* or, *incontinently when*, &c. as is to be ſene in Mar. 4.32. where ὅταν is taken for along time after, and not *inſtantly*: for there it is not meaned, that the ſeed which is ſowen, doeth *inſtantly* riſe vp, &c. and Iohn 8.28. by the worde ὅταν he meaned not, that inſtantly after the crucifying of Chriſt, they ſhould know him truely, but rather after a certaine progreſſe of time from his paſsion. We therefore here iuſtly diſaſſent from the vulgar tranſlation, which maketh the beginning of the ſeuenth blaſt to be the end of the world, where contrarily, Apoc. 11.19. and Apoc. 16.18.19.20.21. there is mention made of ſome worldly commotions, betwixt the beginning of the ſeuenth blaſt or outpowring of the ſeuenth viall, and the latter day.

[*h*] What this myſterie is, and how it is the myſterie of the day of iudgement, and generall reſurrection, read 1. Cor. 15. whole chapter, ſpecially ver. 51. *Behold, I ſhew vnto you a myſtery*, &c,

[*i*] It is certain, that the knowledge of Gods myſteries, mortifieth in vs

all worldly ioy, and maketh vs to conceiue horrible bitternes againſt worldly impietie, ſuch is the vehement zeale that the Spirit of God worketh in our bowels, as ye ſhall find in Ezech. 3.14. This bitternes alſo may be taken for the danger that it is to preachers, to conceall and hide vp Gods myſteries, and due admonitions within their bowels, & not to disburthen their conſcience thereof, by preaching out the ſame, as may be ſeene of *Ionah*, chap. 1 and *Paull* teſtifieth the ſame, 1. Cor. 9.16. ſaying, *We were vnto me if I preached not the Goſpell.* And Iere. 20.9. *His worde was in my heart, as a burning fire ſhut vp in my bones, and I was weary with forbearing, and I could not ſtay.* But contrariwiſe, vpon the other part, what ſweetnes and comfort commeth by preaching foorth the ſame, read Pſal. 118.103. The like ye ſhall find in Ezech. 3.3.

[*k*] This prophecying ouer again by the open booke, meaneth, that beſides *Iohn*s firſt prophecying and writing of this Revelation, in cloſed and darke language, it muſt be prophecied, preached and opened ouer againe by the whole Church in the Name of *Iohn*, in theſe our daies of the ſeuenth age, according to the perfect ſenſe, meaning, and interpretation thereof. The like is ſaide of *Daniel* chap. 12,13. That he muſt riſe and ſtand vp againe after manie yeares.

CHAP. XI. THE ARGVMENT.

The Spirit of God hauing hitherto orderlie prophecied the imperiall or ſecular hiſtorie vnto this ſeuenth age or trumpet, and in the former chapter hauing ſhewed that in this ſeuenth age the world will end: Therefore now in this chapter before that end come, hee beginnes againe, and enters to the ecleſiaſticall hiſtory of the progreſſe of Gods word and true profeſſors therof, proceding briefly therein, vntill that hiſtorie be as farre proceeded as the other: And then doth he introduce the trumpet of the latter day, after the very manner and ſtyle of the other ſix trumpets before expreſſed.

PARAPHRASE.

AND there was power and knowledge giuen me ſtraightlie, as with a metwand, to meaſure the eſtate of thinges to come: ſo Chriſt the great Angell of the couenant aſsiſting me, commanded me to ariſe from all earthlie affections, and to prophecie now the preciſe meaſure of times, that God hath carefully appointed ouer his true inward and inviſible Church his holie religion, and

The Text.

THen was giuen me a reed like vnto a rodde, and the Angel ſtood by ſaying, riſe and mette the Temple of

all the true profeſſours thereof.

2 But as for the [a] outward ſuperſtitious and viſible face of the pretended Church, it muſt be reiected from God, and no care, meaſure, nor account had by him thereof, becauſe it muſt be giuen ouer to Antichriſtian and Idolatrous people.

who ſhal ſubdue his holy [b] Church and ſpirituall *Ieruſalem*, and treade it vnder foote [c] 1260, years.

3 Yet in the meantime ſhall GOD giue that grace, that his two Witneſſes, the [d] olde & newe Teſtamentes ſhall be authorized, and ſhal preach and prophecie theſe (e) thouſand two hundreth and threſcore yeares, though not in there due apparel but diſguiſed with the obſcuritie (f) of mens traditions.

for ſince the year of God 303 or thereby, the pope and his clergie haue vendicate to themſelſe the ſupremacie ouer all Chriſtiã churches as granted be Conſtantine the Emperour to Pope *Sylveſter*, & to thir our daies even 1260. yeares haue poſſeſſed the ſuperſtitious, outward and viſible Church of Chriſtians, treading down and ſuppreſsing all that truelie opponed againſt their abuſes.

3 Seeming to authorize the old and newe Teſtamẽtes, & to preach therof theſe 1260. yeares yet not rightlie interpreting them, but obſcuring thẽ with mens traditions & coloured gloſſes.

An Chr 300

The Eccleſiaſticall.

4 Theſe teſtamentes be as two (g) Oliues, to annoint vs Kings and Prieſts to our God, and as two [g] Lanternes to illuminate vs with the true light of his holie Spirit Theſe abide and ſtande before God with his elect vpon earth.

5 If any hereticke would peruerte theſe the [h] fire of godly zeale, proceeding from the mouthes of their profeſſors, ſhall confound theſe heretickes. And if anie Atheiſt would vtterlie ſubvert theſe Teſtaments, they ſhall be deſtroyed and die eternally.

6 Theſe haue ſuch authoritie, that the heauen ſhall be cloſed, and no raine ſhall deſcend from thence, and tyrannie and bloodſhed ſhal be among the people of the earth, and the worlde ſcourged with all kinde of plagues, dureing all the dayes that theſe Teſtaments and profeſſors thereof doe ſo prophecie and threaten vs for our iniquities: yea

God, and the altar, and them that worſhip therein.

2 But the court which is without the Temple caſt out, and mette it not, for it is giuen vnto the Gentiles, & the holy Citie ſhall they tread vnder foote two and fourtie moneths.

3 But I will giue power *vnto my two witneſſes, and they ſhall prophecie a thouſand, two hundreth, and threeſcore daies clothed in ſackcloth.*

4 Theſe are two oliue trees, and two candleſticks ſtanding before the God of the earth.

5 And if any man will hurte them, fire proceedeth out of their mouthes, and deuoureth their enemies: for if any man would hurt thẽ, thus muſt he be killed.

6 Theſe haue

euen ſo oft, as they liſt to pronounce Gods wrath againſt vs for that cauſe,

7 Now after the primitiue Church had receiued and admitted theſe teſtaments and confirmed the ſame in the coũcel of *Nice*, anno *Chriſti*, 322. euen from thencefoorth the Romane and Papiſticall empire hath troubled thoſe ſcriptures of God, with humane traditions and ſuperſtitions, and ſo hath preuailed that the true ſenſe & meaning thereof, was as dead out of the world.

8 And this dead letter did lie obſcured in latine throughout all the Empire of *Rome*, that great citie, that is more vile in all ſtinking Sodomie tha *Sodom* it ſelfe, and a more tyrannical Empire ouer Gods Church then *Egypt* was, for in it Chriſt is daylie crucified in his mẽbers, yea, euẽ in *Rome* was Chriſt himſelfe crucified, in that he ſuffered death vnder that Empire and vnder *Pontius Pilat*, one of the magiſtrates thereof.

9 And all the *Roman* & *Papiſtical* empire, had of thoſe Latine & il gloſed Bibles patent in their Churches and ouſes during the ſaide thouſand two hundreth and ſixtie years of their reign all the which time their *Legendarie*, and other fables were more ingraued in their hearts, than theſe teſtamẽts of Gods trueth.

Hiſtorie and eſtate of Gods word and Scriptures during the Anti-chriſt his raigne,

7 And when theſe two Teſtaments are compleitly written out, and their teſtimonie authorized the (*i*) Antichriſtiã Monarch, riſing from the deuil, ſhal ſtriue againſt them, and ſhall ouercome them and ſhal depriue them of their ſpirituall meãing & liuely ſenſe

8 And their dead letter ſhal ly as dead carkeiſes patentlie, throughout all the [*k*] juriſdiction of that great Antichriſtian citie which ſpirituallie is [*l*] as *So*dom, in all ſtinking and vnnatural lecherie, and as [*l*) *Egypt* in tyrannie and oppreſsion ouer the Saints of God, foreuen there, Chriſt our Lord, in manie reſpects is continually (*m*) crucified and put to death.

9 And they of all tribes people, tongs and nations, ſhal ſe this dead letter, lying as dead carkeiſes the ſaid ſpace of (*n*) ane thouſand two hundreth and ſixtie yeares, and ſhall not ſuffer the ſame to be duely buried, ſetled, & ingraued within their

power to ſhutte heauen, that it raine not in the daies of their prophecying, & haue power ouer waters, to turne thẽ into blood & to ſmit the earth with all maner of plagues, as often as they will.

7 And when they haue finiſhed their teſtimonie, the beaſt that commeth out of the bottõles pit ſhal make war againſt thẽ and kille them

8 And their corps ſhall lie in the ſtreets of the great city, which ſpirituallie is called Sodom & Egypt where our Lord alſo was crucified.

9 And they of the people and kinreds, and tongues & gentils ſhal ſee their corps thre daies and a halfe and ſhall not ſuffer their carkeiſes to be put in graves

10 Yea, they reioyced when the fearfull threatninges of Gods worde were hid, and the Popes plauſible pardons opened vp and offered to them and made mery in thier Chriſtmas and Eaſter feaſts fẽding for gladnes their propines and gifts at new-yeares-day & hanſel Monday: wher as if they had knowne the fearfull ſentence of the Scriptures, they had not ſo reioyced in mens traditions and earthly pleaſures.

11 But now are theſe 1260 yeares expired in *anno chriſti*, 1560 or ther about, & the Scriptures of God are liuely and quicklie interpreted & reſtored to their true eſtimation and honour and thouſands converted dayly from their former papiſtry, to the true feare of God.

12 Hearing it openlie preached by the moſt true profeſſors of God, that theſe his holy Teſtaments haue authoritie, and authentikenes aboue al decrees, writs, and traditions of men, and therfore (how dark & deſpiſed ſoeuer they ſeemed before) are only to be brought in as witneſſes of Gods holie truth: and now this authoritie of theſe

breaſts,

10 For all that dwel in earthly pleſures ſhall be merie to ſee Gods worde ſo ſmothered, and ſhall reioice together, and ſend gifts and propines euerie one to another: For theſe two teſtaments would haue preached againſt ſinnes and prophecied their deſtruction and haue troubled the glad eſtate of theſe that dwelt in earthlie pleaſures.

11 But after the ſaid a thouſand two hundreth and ſixtie yeares, the liuelie ſpirit of interpretation from God ſhall quicken them vp, and they ſhall bee authorized and ſet vp in eſtimation, and the feare of God ſhall come vpon the heartes of all men that ſhall then rightlie ſee & conſider them.

12 For by the voyce and preaching of Gods heauenly and true Church, they were choſen & called vpon, to aſcend vp to them, and be promoted aboue all humane traditions, and they aroſe vp, and were promoted to their due honor and authority, among them of

10 And they that dwell vpon the earth ſhall reioice ouer thẽ and be glad, and ſhall ſende giftes one to another, for theſe two prophets vexed thẽ that dwelt on the earth.

11 But after 3 dayes & an half the ſpirit of life comming *frõ God, ſhall enter into them, and they ſhall ſtand vp vpon their feete, and great feare ſhall come vpon them which ſawe them.*

12 And they ſhall heare a great voyce frõ heauen ſaying vnto thẽ, Come vp hither, And they ſhal aſcend vp to heauen in a cloud, and their enemies ſhall ſee them.

holie Teſtaments is become ſo cleare & euident, that the very papiſts do ſee and perceiue the ſame, and muſt needs admit their Teſtimonie.

13 At this time this Romane and papiſticall Empire, is greatlie ſhaken, for the tenth part thereof, euen all their Temples, Abbaies Frieries, Nunries, and other their cities, and ciuill policies through all *England, Scotland*, & diuers parts of *Germanie* are pulled down, and deſtroyed: And of their Biſhops, Abbots, priors, Chãnons, Monks, Friers and prieſts, a great number depriued of their rents, and liuings, and the teſt repenting their idolatrous papiſtrie, become true profeſſors of God.

14 This papiſticall kingdom ſuffered great troble by *Turkiſh* dominators, and far greater by the *Mahometan* Emperours, and theſe troubles are paſſed, but now thirdlie, and greateſt of all, theſe Miniſters of Gods trueth, ariſing in this age, ſince the yeare of God 1541, ſhall bring it to vtter wracke.

15 For that yeare began the ſeuenth age wherein the laſt days approching all the empire and kingdomes of the worlde muſt be troden down. & Gods & his ſon Chriſtes Eternall kingdome

that heauenly and true Church, while as yet they were but cloudy, dark, & vnknown to the worlde: then ſhall their enemies the Antichriſtiãs, now begin to ſee and perceiue their authoritie.

13 And the great empire of the erth ſhall quake and be commoued: and the [*o*] tenth part of the Antichriſtian Synagogues, & buildings ſhall be ruined, and in that commotion [*p*] ſeuen thouſand of thoſe men ſhall be depriued of their rents wheron they liued, and the reſt of theſe ſhall repent, and feare the Lord, and ſhal giue glorie to the God of heauen.

14 The ſecond of the great wofull plagues, that that empire ſhould ſuffer is paſt, and this is the third and laſt, that approacheth at hand:

15 And the Angels of the ſeuenth and laſt age blewe out the plagues of Gods wrath, and then were there great voices of gladnes, ſolemnlie hearde throughout all the Church of God, rejoicing, & ſaying, Now all the kingdoms

13 And the ſame houre ſhal there be a great earthquake, & the tenth parte of the citie ſhall fall, and in the earthquake ſhal be ſlain in number ſeuen thouſand: and the remnant ſhal be afraid, and giue glory to the God of heauen.

14 The ſecond woe is paſt, and behold the third woe will come anone.

15 And the ſeuenth Angell blew the trumpet, and there were greate voyces in heauen, ſaying, The kingdomes of this worlde are our Lordes, and his Chriſtes and hee ſhall reigne for euermore.

16 Then

only erected.

16 And Gods holie ſcriptures which lay as dead the time of the Antichriſtian raigne, now being reſtored to their former authoritie and honour, do ſtirre vp the harts & mouthes, of all true profeſſors, to humble themſelues, and to yeelde their due praiſe and honour to God almightie, ſaying,

1541

Here beginneth the ſeuenth trumpet, in which the world endeth

of the earth [*q*] are throwen down, & that holy kingdom of God, and his ſon Chriſt Ieſus (who reigneth for euer and euer) only ſet vp.

16 And the 24. bookes of the olde Teſtamẽt (dead & vnmẽtioned ſince the dayes of the primitiue church) now appeares before God in their due and accuſtomed place and eſtate, and proſtrating themſelues, yeelde by the mouths of their profeſſors, praiſe & worſhippe to God, ſaying,

the foure and twenty Elders which ſate before God on their ſeates fell vpon their faces and worſhipped God.

17 Saying, wee giue thee thankes, Lord GOD almighty, which art, & which waſt, and which art to come: for thou haſt receiued thy great might, & haſt obtained thy kingdome.

18 And the gentiles wer angry, and thy wrath is come, and the time of the dead that they ſhould be iudged and that thou ſhouldeſt giue reward vnto thy ſeruãts the Prophetes, & to the ſaints, & to them that feare thy name, to ſmall and great, and ſhouldeſt deſtroy thẽ which deſtroy the earth.

19 Then the tẽple of God was opened

17 O our Lord God almightie who was, who is, and who now ſhall returne to judgement, wee yeelde thee moſt harty thankes, for that now thou haſt taken vpon thee thy great and mightie power, and ſtrength, and doeſt now onely reign ouer all.

18 And the wicked nations are angrie thereat, but thy wrath is come vpon them, and the time of the generall judgement is come, that the deade muſt riſe, and thy holy ſeruants the propletes, and Saintes, and all men both great and ſmall that haue feared thy name, muſt receiue the reward of eternall life: but contrarilie, thoſe muſt be rooted out, and caſt into eternall damnation, who haue ſeduced the world, and polluted the earth with their abominations.

19 And vntill that day (now ſhortlie approaching) the true temple & Church of God ſhall be made [*r*] patent, vouchable, viſible and open to all his heauenlie and true elect, and among theſe of that true Church & heauenlie

elect, ſhal Chriſt the true arke of our couenant, and atonement with God the Father bee ſeene, knowne, and vnderſtoode, preached out, and conſtantlie advouched. In that time ſhall there be great ſhewinges foorthe of tempeſtuous troubles, great noiſe, rumors and threatninges of warres, and great trouble and commotions vpon the Empire of the earth. And the colde and tempeſtuous [*s*] haile of partiality, ſelfloue, and lacke of charity, ſhal extinguiſh and quench out that ardent zeale, that we owe towards God and the ferueut loue and charitie, that we owe to our neighbour.

in heauẽ and there was ſeene in his temple the Arke of his couenant: and there were lightenings & voyces, & thundrings, & earthquakes, & much haile.

Notes Reaſones and Amplifications.

(*a*) The temple of *Ieruſalem*, contained three parts, to wit, the vtter parte, where all men reſorted, the inner place, called the holy place, & the innermoſt part, called the holieſt of all. Theſe two holy and holieſt parts that are within, repreſents here the true, latent and inviſible Church, and the vttermoſt & moſt patent part, repreſents the outward and viſible pretended Church, with the externall rits and gliſtering ſuperſtitions thereof: this vtter court is without the temple, if yee call onelie the holie and holieſt places to be the Temple: but if yee call the whole worke to be the temple, then is this vtter courte within the temple, as a parte of the whole: & therefore ſome read, *atrium quod eſt intra templum*, and ſome reade *atrium quod eſt extra templum, &c*. Alwaies in this viſible and outward face of the Church, repreſented here by this outward court *Paul* doth mean, that the Antichriſt muſt ſit 2 *Theſ* 2.4. and therefore, it is commanded here to be caſt out, and no care, meaſure, nor account thereof to be taken of God.

(*b*) The holy citie is not one only citie, but is ſpiritual *Ieruſalem* euen the true Church & Spouſe of Chriſt Ieſus generally throgh the whole world, as *Reuel*, 21 appeareth.

(*c*) That theſe 42 moneths are 1260. yeares, is prooued by our 1.15. & 16. propoſitions, and that this account of 1260 years beginnes about the yeares of Chriſt. 300. or 316. is prooued by our 36. propoſition. and that this time, the Antichriſt muſt ſitte ouer the Church of God, reade 2 Theſ. 2. and our ſolution of doubts therevpon: and that ſo doeth the pope profeſsing himſelfe the head of the Church, read lib. 5. *Sexti tit 9. de pœnis in fine, Canone fœlicis in gloſſa, & lib. 3, Sexti Tit. 16. Can periculoſo, in gloſſa litera, K prima. Item in prœmio ſexti*, hereof reade alſo the 2. ſyntheſis of our 26. propoſition.

(*d*) That the old and new Teſtaments be theſe two witneſſes that here teſtifies of God is prooued by our 21 propoſition.

(*e*) That 1260. dayes meanes here 1260. yeares is prooued by our 1.15. and 16. Propoſitions: and that they begin *anno Chriſti.* 300. or 316. or thereby, is declared in our 36 propoſition.

(*f*) Call to mind how in the dayes of Papiſtrie, none durſt vnder the paine of their liues haue the Bible in vulgar language, but all the ſcripturs wer hidden vp from the vulgar people in Latin language, and wer hidden vp from ſimple ſchollers by wrong gloſſes and falſe interpretations, as who woulde hide vp and diſguiſe a prince in ſackcloth. Of theſe wicked gloſſes and falſe tranſlations, wreſted by the original Hebrue and Greeke, there is diuers to be ſeene in that age, ſuch as euen now, and lateſt of all the *Rhemiſh* Teſtament.

(*g*) There is mention made of theſe in Zach. 4. where the Prophet ſpeking of *Zerubabel*, & of *Iehoſhua* the ſon of *Iehozadak*, their repairing vp of the materiall Temple introduceth in that viſion a figure of theſe two Candleſticks, and of theſe two Oliues or Teſtaments that yeelde ſpirituall oyle, both to the Lampes of the ſeuen primitiue Churches, as alſo now to vs to kindle our Lampes againe anew, againſt our bridgroomes comming to iudgement. Matth. 25.1.

[*h*] How the enemies of God and his worde are confounded with the fire of godlie zeale, is teſtified pſal. 20.9.10. and Ier. 5.14. and 23.29 and 1. King. 19.10.14. and 2. King, 1.10.12. Alſo, how manie blaſphemous heretickes haue beene confounded by the mouthes of Gods miniſters, and afterward died moſt horrible deaths, in hiſtories may be ſeene.

[*i*] That this beaſt is the Roman and papiſticall Empire, by the whole indices and tokens thereof, is moſt clearlie declared Apoc 12. and 13 and 17. and euidentlie proued by our 24. propoſition.

[*k*] This great citie and ſtreetes thereof (though chiefly it meaneth the Antichriſtian citie, which God willing, both hereafter ſhall be, and is heretofore propoſition 23. prooued bo be *Rome*) meaneth no one citie, but all the landes cities, and places of the world, ſubiect to the Antichriſtian Empire: Euen like as contrarilie, ſpirituall *Ieruſalem*, meaned all thoſe through the worlde that were of the true Chriſtian Empire, as is before ſaid in the note [*b*] hereof.

[*l*] This citie beeing called here, not ſimplie *Sodom* nor *Egypt*, but ſpiritually, like as hereafter, Apoc. 17,5. the ſame is called not ſimplie *Babylon*, but myſticallie *Babylon*: meaneth thereby, that, that citie is not named literallie *Sodom Egypti*, or *Babylon*, but that becauſe it is polluted with their vices, and like to them it is: therefore, ſpirituallie compared to them.

[*m*] Chriſt is ſaid to be crucified in this Antichriſtian citie of *Rome*, for four cauſes: Firſt, for that as the Iewes did put Chriſt to death, who is Gods true incarnat Word, & depriued him of this earthly life: So the Antichriſt here hath

ſlain Gods holy Word, and Scriptures, by depriuing them of their true ſenſe, which is their ſpirituall life, and the execution of this ſlaughter procecdeth from *Rome*. Secondlie, wherſoeuer Chriſt is truelie preached & afterward deſpiſed (as in *Rome*) he is ſaid there to be crucified, Heb 6. verſ 6. Thirdly, whereſoeuer Chriſts mẽbers are crucified and afflicted, there Chriſt eſteemeth himſelfe to be crucified and afflicted, Mat. 25.40.45, But in *Rome* many martyrs haue ſuffered death. Therefore, &c. Fourthlie, becauſe (by the note *k*, preceeding) the whole Antichriſtian and Romane Empire, is comprehended vnder the name of this chiefe metropolitan city thereof, which is *Rome*, as teſtifieth, *Ovid. faſt. 2.*

Gentibus eſt aliis tellus data limite certo,
Romanæ ſpatium est Vrbis & orbis idem.

But vnder the Roman Empire, and by *Pontius Pilate*, one of the magiſtrats thereof was Chriſt crucified: Therfore, in this city (which ſpecially meaneth *Rome*) is Chriſt figuratiuely ſaid to be crucified.

[*n*] That theſe three daies & an halfe, are 1260. years, is proued by our 15. & 16. propoſitions, & that theſe years begã betwix the years of Chriſt 300. and 316. is declared in our 36. propoſition.

[*o*] In conſideration of the ten dominions that were vnder the Roman and Papiſticall Empire, it may well be eſteemed, that the Abbeyes, Monaſteries, and Papiſticall places of *England*, and *Scotland*, with the addition of many in *Germanie*, and ſome in *France*, that were ruined and pulled down, about the time of the yeare of Chriſt 1560. were no leſſe than the tenth part of all the Antichriſtian citie or papiſticall policie.

[*p*] At this time, the tenth part of the papiſticall Empire being reformed, and the idolatrous policies deſtroyed, it is ſpecified here that ſeuen thouſand were ſlain, that is, they were degraded of their dignities & honours, and ſpoyled of their rents, whereon they liued, for as hereaſter, Apo. 13.15. to giue life to the image of the beaſt, is nothing els, but to confirme, authorize, & inaugarat it, whereby it may enioy office, honour and rent, So here, to ſlay, is as much as to degrade & depriue them of their offices, honours, and rents, whereon they did liue. As to the preciſe number of ſeuen thouſand, the like and like number ye ſhall find, 3. Reg. 19.18. and Roman. 11.4. which we muſt not thinke to be ſo preciſely ſpoken, as if there were no oddes: for ſuch reaſons, as you ſhall reade Apoc, 7. note *f*, thereof, and Apoc. 9. note *h*, thereof.

[*q*] That in this age (which began in the year of Chriſt, a thouſand fiue hundreth fourtie and one, and endeth at the furtheſt, in the yeare of Chriſt, a thouſand, ſeuen hundreth, foureſcore and ſix,) the Roman and papiſticall Empire, and all tyrannicall kingdomes ſhall be ſuppreſſed, & Chriſts kingdome only erected: & the laſt iudgement, ending all prophecies ſhall come, is man-

ifefted by the oath & vow, made Ap. 10,6. of which, reade the notes *e*, and *f*, thereof.

[*r*] How this true temple of God, was patent in the dayes of the true primitiue Church, & then clofed vp againe, to thefe dayes is faid in the difcourfe of the 20. propofition, and fhal be confirmed hereafter, Revel. 15.8. and note *h*, thereof. Whereby they notablie erre, that think the Church of God euer to be patent, publik, & vifible, feing by the former verfes 2.3.7. and 8. it muft be trod downe by the Antichrift, 1260. yeares, til at length here, ver. 19. a litle before the latter day, it be made publike, patent, and open to all men.

[*ſ*] This cold haile of partialitie, iniquity, felfe-loue, and lacke of charitie, is not only here the laft token, that preceeds the day of iudgement, but alfo it is the very caufe why the day of iudgement fhal be haftened before the faid 1786. yeare come, as Matth. 24.12.22 teftifieth: Of this felfe-loue and partialitie of the later daies teftifieth *Paul*, 2, Tim. 3.2. The remanent & perfect hiftorie of the latter day is here omitted, and remitted to the 14 chapter, as fhall be difcuffed there at length, and was alreadie noted, Reuelat. 10. note (*d*) thereof, and in our progreffe vpon the 11. propofition is mentioned. And fo endeth the orderlie part of this *Revelation*, beginning at Chrifts baptisme, and continuing vntill the latter day.

Here followeth the amplificatiue part of the Revelation of the Evangelist S. IOHN, *containing a repetition, amplification and expofition of ſuch notable things, as before are but ſummarlie intreated of in the former part.*

CHAP. XII. THE ARGVMENT.

The Spirit of God, for our full inſtruction, minding to repeat, amplifie & expound in this laſt part of the booke, ſuch matters as hath bene ſummarlie and obſcurelie treated of in the former part, of the which this laſt Eccleſiaſticall hiſtorie being the chiefe: He therefore beginneth in this twelfth chapter againe, to expound and repeat the ſame after the maner of a battell betwixt Chriſts Church, holy Spirit, Word, and true profeſſours on the one part: and the deuill, the Antichriſtian and Romane Empire on the other part. In which is deſcribed the ſtrife and malice that ſathan by his Antichriſt, executed againſt the viſible, Apoſtolik & primitiue Church, which firſt conceiued, imbraced, and preached forth Gods word, and againſt that holy word, which they preached foorth. So that the true Church became latent and inviſible, and God kept and reſerued his holy word in puritie, in the hearts of that inviſible, and true latent Church, by meanes of his holy Spirit working in them, whereupon followed out a ſpirituall battell, the ſpirit of Sathan ſtriving in the mindes of the Elect by falſe doctrine, and other entiſements againſt the Spirit of God. But Sathan ſeing that he prevailed not in that ſpirituall battell, againſt the ſoules of the elect number, he raiſeth by his Antichriſt perſecution corporally, againſt the bodies of euery particular profeſſour.

Hiſtoricall application.

IN the daies of Saint *Iohn*, the true Primitiue Church of Chriſt flouriſhed, the Evangell increaſed, ſhining as the Sun, ſo that the profeſſours thereof, deſpiſed the glorie and dignities of this world, caſting down all their rents at the feet of the Apoſtles, to be diſtributed in cõmõ to the whol Church Act. 4.35. hauing no other crown of glorie, than the croſſe of chriſt, and doctrin of his Apoſtles, Galat. 6.14.

2 Altogether bent in their hearts to cõceiue and contemplat Gods true word, and in their mouthes and workes to teach the ſame, and expreſſe the fruites thereof

3 But ſodainly after the primitiue Church, the deuill ſtirred vp his inſtruments, euen the mightie Empire of *Rome* that great citie, that is ſituate vpon the ſeuen notable hilles *Palatinus, Cœlius, Tarpeius, Auentinus, Quirinalis, Viminalis, & Eſquilinus*, and hath theſe ten Chriſtian kingdomes, *Spaine, France, Lombardie, England* Scotland, *Denmarke,*

Paraphraſt. expoſition.

ANd there was ſhewed to me the great & marveilous viſion of Chriſt his heauenlie [*a*] ſpouſe & true Church, ouerſhadowed eueriewhere, with the bright light of his Euangel, ſhining as the Sun, & ſhe contemned and trode vnder her feet, all the glories of this world, which areas mutable as the Moone, and ſhe was crowned with the triumphant doctrin of the twelue Apoſtles.

2 And within her bowels ſhe cõceued and meditated [*b*] Gods true worde, & with her mouth preached and proclaimed out the ſame, & was preaſſed by the Spirit of God to yeeld forth and teach the ſame as a woman is preſſed to yeeld foorth her birth.

3 Then on the contrarie parte, I beheld amongſt Gods heauenlie and holy elect and loe,

The text.

AND there appeared a great wonder in heauen: A woman cloathed with the Sunne, and the Moone was *vnder her feete, and vpon her head a crown of twelue ſtarres.*

2 And ſhe was with child and cried, trauelling in birth and was pained readie to be deliuered.

3 And there appeared another wonder

S*weden*, The kingdome of the *Gothes*, in *Italie*, and of the *Hunnes*, in *Pannonia*, or *Hungary*, and the *Exarchat of Ravenna*, confederate with it, & haue their Empir gouerned by feuen eftates all ruling fuccefsiuely after other as kings. to wit. *1 Kings 2 Confuls 3. Dictators 4. Decemvirs 5. Tribuns, 6. Emprors, 7. Popes.*

4 By the meanes of the which Romane and Papifticall Empire, the deuill hath drawne after him, much of Europe & of *Afia*, euen the large thirde part of the world caufing their paftours, and minifters to leaue their heauenlie doctrin, and to reach Antichriftian Apoftafie, for loue of dignities, rents, and earthly benefices: for the deuill was fo ready, that how foon the true word of God was fowen by the primitiue Church, incontinently he raifed vp the papifticall antichrift to deuour & confound the fame with falfe gloffes, & finifter interpretations.

5 But contrary to their expectation, that pure doctrine of the Evangel, preached by the primitiue Church, was referued by God among the invifible Church, till at length, it fhould rule ouer all the kingdomes of the earth, and fo God preferued his worde in puritie.

6 But his vifible Church

the deuil [*c*] that horrible and bloodie dragon, (in the perfon of that Empire whofe feate and citie ftandeth vpon feuen hilles, and hath vnder it tenne confederate kings, and hath had feven princely eftats gouerning it)

4 With the poyfõ of his taile, drew after him the third part of Chriftians, from their heauenlie contempla-tion, to earthly affectiõs and this dragon, and his members awaited vpõ Gods Church and holie Spoufe, that preached and expreffed out the fruit of Gods word, that how foone fhe had preached out that word, he might chocke the fame, and devoure the fruite thereof.

5 And fhe deliuered and expreffed foorth her fruit, and increafe of the true [*d*] incarnat word, Chrift Iefus, who fhall rule and governe all nations vnder the fcepter of his fervitude: And God accepted this fruite of his holy Word vnder his protection, and authorized it among his holie elect.

6 Whereof the vifible

in heauen: for behold, a great red dragõ, hauing feuen heads and tenne hornes and feuẽ crownes vpon his heads:

4 And his taile drew the third part of the ftars of heauen, and caft them to the earth. And the dragõ ftood before the woman, which was ready to be deliuered, to devour her child when fhee had broght it forth.

5 So fhee brought foorth a man childe, which fhould rule all nations with a rod of yron: and her fon was taken vp to God, and to his throne.

6 And the

was baniſhed and made inviſible among a few godly perſons, who ſuccesſiuely after other intertained the trueth all the dayes of the Antichriſt, from the daies of Pope *Sylveſter* the firſt, to thoſe our dayes, euen 1260 yeares.

7 For at the *Pentecoſt* after Chriſts aſcenſion, was the holie Spirit and third cõforter ſent down to dwell in our mindes, and therein to fight and withſtand al the priuate tentations and motions of the deuill, vſed againſt the Chriſtian church. For the deuil hauing firſt ſeduced the outward, viſible & Antichriſtian Church, now ſtriueth againſt the Spirit of God, in the inward mindes of the inviſible Church,

8 But he by his tentations and intiſements is no waies able to ouer come the mightie force of Gods Spirit, working in our Spirites, far les to poſſeſſe the ſouls of the faithfull and heauenlie elect.

9 So that now in Chriſt his paſsion that craftie Seducer and accuſer, & his members, hauing no power againſt theſe elect ſouls doeth moſt cruellie extend his tyrãnie againſt their fleſhlie bodies onely.

number, euen Gods true Church and ſpouſe was chaſed away, and [*e*] remained inviſible and ſolitarie among certaine priuate perſons, predeſtinated and elected of God where ſhe was nouriſhed, and foſtered with their godlie praiers and ſecret meditatiõs during the ſpace [*f*] of 1260 yeares.

7 Hereupon proceeded a great and heauenlie ſtrife, for God, the [*g*] holie Spirite and third comforter by his miniſtring Angels and godly inſpirations, fought againſt the Deuill, that olde dragon, who with his wicked ſpirits and tentations fought and withſtood the ſpirit of God within vs.

8 But Sathan preuailed not, neither obtained anie place amongeſt Gods heauenly & faithfull elect to ouercome them,

9 And ſo that wicked dragon & olde ſerpent who iuſtlie is called an accuſer and odious, who deceiues all the whole world (now reſtrained from ouercomming the faithfull & heauenly elect ſouls) is caſt out,

womã fled into the wilderneſſe, where ſhee hath a place prepared of God, that they ſhould feed her there a thouſand two hundreth and threeſcore daies.

7 And there was a battell in heauẽ, Michaell and his Angelles fought againſt the dragon, and the dragon fought and his Angels.

8 But they preuailed not, neither was their place found any more in heauen.

9 And the greate dragon, that olde ſerpent called the deuill and Sathan was caſt out, which deceiueth al the world he was euen *caſt into the earth, and his Angels were* he, his mẽbers and inſtrumentes hauing only permiſſion to tempt, trie, and perſecute their earthlie bodies.

caft out with him.

10 Therefore, with vs and for our cauſe all the Angels of heauen doe yeeld praiſe and thanks to God eternall, ſaying,

10 For which delivery of our ſoules there wer voices of thankesgiuing and great ioy among al the Angels of heauen ſaying,

Now is their ſaluation in heauen prepared for Gods Elect, and the ſtrength and kingdome of our God, and power of his Sonne Chriſt Ieſus, is enlarged ouer all: for the accſuer of Elect ſouls (who are euen as brethrẽ vnto vs his angels) is caſt down & diſplaced: for continually ſince the fall of the firſt *Adam*, he accuſed them therefore in Gods juſtice.

10 Then I heard a loude voice faying now is faluation in heauen, & ftrength, and the kingdome of our God, and the power of his Chrift, for the accufer of our brethren is cast down, which accufed them before our God day and night.

11 But now they haue ouercome this their accuſer, in the blood of the ſecond *Adam*, Chriſt Ieſus, and in the conſtant teſtimonie they did beare of him, not regarding their liues, but beeing contented to ſuffer martyrdome for his ſake.

11 But they ouercame him by the blood of the Lamb, & by the word of their teftimony, & they loued not their liues vnto the death.

12 Therefore, rejoyce all ye godly and celeſtiall creatures: but woe be vnto you, that do fixe your felicity on earthly glorie & affections, and do plunge your ſelues in the muddie ſeas, and tumbling cares of this world, for the deuil is ſent downe amongſt you, raging in great furie, knowing his time to be ſhort, for that the world draweth neere to an end,

12 Therefore, rejoice ye heavens, & ye that dwell in them: Wo to the inhabitantes of the earth, & of the fea, for the Devil is com down vnto you, which hath gret wrath knowing, that hee hath but a fhort time.

13 So then, Sathan ſeing himſelf ſecluded frõ anie further power againſt the elect, but only a litle power ouer the earthly bodies of the true Church, ſtirreth vp the Biſhopes of *Rome* to vſurpe temporall dignitie and Antichriſtian ſupremacie, aboue all the

13 Now Sathã perceiuing himſelfe ſo ouer-come, that hee had no more power but ouer the earthlings [as being prince of this worlde] he in the perſon of the Antichriſt perſecuted the true and viſible

13 And when the dragon fawe that he was caft vnto

vifible Churches, whereby thefe feuen Churches of *Afia* that S. *Iohn* wrote to, and other Apoftolicall Churches, were ftill perfecuted by that Roman Church, till their whole vifible number were faine to imbrace the *Roman* Apoftafie, in fted of that true word, that fomtime was preached among them.

14 So God from thefe daies forth, viz. from the yeare of Chrift 316. hath withdrawen his vifible Church from open affemblies to the hearts of particular godly men, where fuccefsiuelie it abode in puritie, invifibly all the daies of the Antichriftian raigne, euen to thefe our daies the fpace of 1260. yeares.

15 Then this deuilifh antichrift, & Roman feat fearing left the fmall fparke of Gods invifible Church fhould fomtime reviue, and extinguifh their fynagogue, preached & fet out fuch plaufible herefie, as might trap and allure the invifible Church, fuch as *merits, & workes of fupererogation*, deferuing (as they fay) more then heauen, with *Indulgences*, & *pardons* for fins, althogh they fhould deferue worfe than hell.

16 Which, with many more plaufible herefies, their vifible Church hath wholly embraced: yet ever fome of the true invifible Church haue

primitiue Church & fpoufe of Chrift, who had procreated & broght forth the noble fruit and increafe of Gods Word.

14 But God armed her with the fwift [*h*] winges of his protectiõ, whereby fhee became his invifible Church, and fledde from the open affembly to the priuate heartes of particular godly men, as to her deferte place, where fhe (as is faid before) was nourifhed with godlie meditations, for the fpace of [*i*] a thoufand two hundreth & threefcore yeares free from the poyfon of that dragons herefie.

15 Then fpewed Sathan out of his mouth, a flood of (*k*) falfe doctrin & herefie, that hee might feduce God his true Church therewith.

16 And the earthly & vifible Church of the Antichrift, did fo drink vp and fwallow euery drop of that doctrin, that the true invifible Church, did abide pure, and did not get one drop of that poyfon, that Sathan had fpewed out.

17 And the deuill raged in ire againft that invifible Church, &

the earth, hee perfecuted the woman which had brought foorth the man child.

14 But to the woman were giuen two winges of a great eagle, that fhee might flie into the wilderneffe, into her place, where fhe is nourifhed for a time, and times, and halfe a time, from the prefence of the ferpent.

15 And the ferpent caft out of his mouth, water after the woman like a flood, that he might caufe her to be carried away of the flood.

16 But the earth holpe the woman, and the earth opened her mouth, & fwallowed vppe the floode, which the dragon had caft out of his mouth.

17 Then the Dragon was wroth with the woman, &

remained in their puritie, as teſtifieth *Catalogus teſtium veritatis.*

17 Whereupon the devil, in that *Romiſh* Antichriſt, is ſo inraged, that whereſoeuer particularly theſe diſperſed childrẽ of Gods true inviſible Church, their workes & bookes are apprehended, they are al burnt together, and perſecuted with fire and ſword.

18 And theſe things did S. *Iohn* foreſee, and further, as followeth, ſtanding on the ſeaſhore of *Pathmos.*

proceeded in perſecuting the reſidue of her children, which from thence-forth doe obſerue the commandements of God, and doe beare truly and conſtantlie, the teſtimonie of Ieſus Chriſt.

18 And I ſtood on the ſea (*l*) ſhoare of *Pathmos*, beholding the viſion which followeth.

went and made warre with the remnant of her ſeed, which keepe the commandments of God, & haue the teſtimonie of Ieſus Chriſt.

18 And I ſtood on the ſea ſand.

Notes, Reaſons, and Amplifications.

(*a*) That this woman clad here with the Sunne, &c. is the true Church of God, is proued by the 22. propoſition.

(*b*) How the true Church euer travelled to procreat moe children to God, in whom Chriſt is continually conceaued, formed, & ſpirituallie borne, ye ſhal read the *Galathians*, 4.19. where *Paul* calleth the *Galathians* his litle children, of whom he travelled again in birth, that Chriſt may be formed & ſhaped in thẽ: whereby it is evident, that the Church of God is this *Woman*, & Chriſts true word is the fruit and *birth*, that the formeth and produceth into the hearts and bowels of the Elect.

(*c*) That this Dragon is the deuill, it appeareth plainly by the 9. verſe hereof following, and that this ſhape and figure of his, of ſeuen heads and ten hornes, and that the 1260. yeares of his raigne, are the proprieties and notes of the Latine or Roman empire, is proued by the 24. propoſition. Hereby then is meant, that the deuill, in this ſtrife againſt Gods Church, vſed or clothed himſelfe with the perſon of the Roman or Latin empire.

(*d*) This man-childe that ruled all nations with a rod of iron, is called (pſal. 2.) Chriſt the Sonne of God, and (Revel. 19.13.) he is called *the word of God*, which are both one, ſeeing that word took fleſh, & was incarnat, Iohn 1.14. Gods Church then, & true ſpouſe, is ſaid to produce this birth of this man-child, when they preach foorth the true worde of God, and his ſonne

Chriſt Ieſus, reade hereof Gal. 4.19. how Chriſt muſt be formed and ſhapen in vs, &c.

(*e*) Seeing by the opinion of *Methodius* & *Primaſius*, and diuers old writers, theſe notes of this woman agree not with the Virgin *Marie* in particular, but with the whole Church of Chriſt in generall: Therefore, and by theſe words of this text, and by our 16 propoſition, it is certaine, that the Church of Chriſt muſt lurke & abide inviſible, 1260. yeares, vnder the Antichriſt, where through they are deceiued, that think the true Church to be euer viſible, patent and publike, or yet the publike and viſible Church to bee euer Gods true Church

(*f*) That 1260. dayes in prophecies, ſignifie 1260. years is prooued by our 1. and 16. propoſitions, & that theſe years beginne at *Sylveſter* the firſt, about the yeare of Chriſt, 316. is prooued in the 36. propoſition. And that all this time, euer ſome of the true and inviſible Church were preſerued, who deteſted the Pope & entertained Gods trueth, is to be ſeene in the booke intituled, *Catalogus teſtium veritatis.*

(*g*) That both in *Daniel* 10.13,21. and *Daniel* 12.1. and alſo in this place, *Michael* is taken for one of the perſones of the Trinitie, euen either for the Sonne or for the holy Spirite, it is certaine, by the propertie conteined in theſe texts for there is *Michael* called one of the firſt Princes, and our Prince, and ſo is both Chriſt and the holy Spirite and firſt Prince before all beginning, and the laſt after all ending, euen α. and ω. and is alſo prince ouer vs, yea, he is a prince for vs that are his elect, who (as ſaith *Daniel* there) ſtandes for *Daniels* people, euen for the true and ſpirituall Iſraelits, and ſo ſtandeth both Chriſt in heauen as our mediatour with God, and the holy Spirite on earth with vs, as our defender and gouernour, Againe with the Name of *Michael* which is to ſay *who is lyke God*, or otherwiſe, *Deus percutiens a beating or ſtriking God* [*ex Hieronymo de interpretatione nominum.*] doth both the perſone of Chriſt and the holy Spirite agree, for both with the father are one God, and none is like to that God, whereof all the three perſons beate and ſtrike downe Sathans tyranny, and fighte for vs. The queſtion therefore is, which perſone of the Deitie doth *Michael* ſignifie in theſe forenamed textes: though ſome think he meane Chriſt, for the properties aboue written agreeing to Chriſt: yet in reſpect (as is ſaid) the ſame alſo doeth agree with the holy ſpirite and in ſome points agreeth not to Chriſt. Therefore (in my iudgement, *Michael* meaneth the holie Spirit, and not the perſon of Chriſt: for ſeeing the man or one like vnto the Sone of man that appeared to *Daniel*, (10.) is prooued Apoc. 1. note, [*ſ*] thereof, and Apoc. 10, note (*a*) thereof, to be vndoubtedly Chriſt: and that it is he only whoſe voice there (Dan, 10.6.9.) is deſcribed and heard, and who there after (Dan. 12.6.) alanerlie is deſired to ſpeake, and conſequentlie muſt be he, and no third ſuppoſit Angel, who (Dan. 10.13.21.) ſpeaketh and ſayeth that *Michael* helped him. How can it be that Chriſt there ſaith that *Michael* helped him, except that *Michael* be taken for the holie Spirite? For the Deitie diſtinguiſhed in three perſons [although eſſentiallie inſeparable] euerie perſon is ſaid in the Scriptures, to helpe one another, *as the Father by the Sonne created the world* (Col. 1.16. and Iohn 1.3.) The Sonne ſent by the father, redeemed the world. (Luke 4.43.) The holie Spirite ſent both from the Father and the ſonne comforteth, defendeth, and regenerateth his elect of the worlde: And the Angels are not ſaid to helpe, but to ſerue and miniſter to the Deitie, But contrarilie to diuide one perſon of the Deitie in two names and to make Chriſt, named with the one name to, helpe Chriſt, named with the other name, ſemeth altogether improper: Wherby, *Michael* that helped Chriſt, appeareth euidentlie not to be Chriſt, but the holie Spirit. and this I ſpeake of theſe forenamed Texts onelie and not of *Michael*, in the Epille of *Iude*, where hee repreſenteth an Archangel, who durſt not himſelfe, but referred it to God to rebuke Sathan, neither of other parts of Scripture, where *Michael* is the proper name of a man. More of this reade in our ſolution of doubts.

(*h*) Theſe Eagles winges doe meane Gods mightie protection becauſe (Den. 32.11) *As the Eagle flattereth ouer her yong, and ſtretcheth out her wings and carieth them on her ſhoulders: ſo the Lord* hath alſo kept and preſerued his owne. Furthermore, theſe two Eagles wings may be taken for the two Teſtaments, the olde and the new, with the which Gods true and inviſible Church clothes her ſelfe and therewith flieth from the Antichriſtian ſynagogue, and aſſemblie, to her priuate meditations.

(*i*) That this *time, times, & halfe a time*, meaneth a thouſand two hundreth and three ſcore years, is prooued by our 16. propoſition, beginning that date about the yeare of Chriſt 316. by the 36 propoſition.

(*k*) That theſe Waters and floods mean doctrin, conſider by the note *g*, of the 8. chap. hereof

(*l*) Some read this Text to be ſpoken of the Dragon, in theſe words, *'And it ſtood vpon the ſea ſhore*, and ſo it may be interpreted to mean, that Sathan ſtandes ready betwixt ſea & land (according to the 12. verſe hereof) to oppreſſe them both: that is, to ſeduce al thoſe, that fixe and repoſe their felicitie on earthly pleaſures and eaſe on the one part, and theſe on the other part, who as the waltering waues of muddie ſeas play the tyrants, reauing to themſelues the kingdoms of the world: But I rather aſſent to thoſe other, who read. *And I ſtood on the ſea ſhore*, as if S. *Iohn* would ſignifie to vs, that he ſtood on the ſhore or ſea coaſt of the Ile of *Pathmos*, while as he ſaw this viſion following of the beaſt ariſing out of the ſea. The like is ſaid in Dan. 10.4. that he ſtood on the bank of the riuer *Tigris*, where he ſaw his viſion.

CHAP. XIII. THE ARGVMENT.

Hauing in the former chapter deduced the continual ſtrife betwixt Chriſts Church & holy Spirit, on the one part, and the Deuill and his inſtruments on the other part: In this chapter he reneweth a more ſpecial deſcription of theſe chiefe inſtruments, whoſe perſon the deuill is to vſe in all times hereaſter, to perſecute Gods Church withall, to wit, of the whole Latin Empire in generall, and of the Antichriſtian kingdome in particular, ſet out reſpectiuelie vnder the figure of two beaſts, the firſt ten-horned, the laſt, two-horned, with al their notes & tokens, agreeing moſt exactly with the Roman, and Papiſticall kingdomes as followeth,

Hiſtorical application.

And among the ambitious tyrants and tumultuous people of the world, aroſe the great *Latin* or *Roman* Empire, whoſe chiefe ſeat and metropolitan citie *Rome*, is ſet vpon theſe ſeuen hils, *Palatinus, Cœlius, Tarpeius*, otherwiſe called, *Capitolinus Aventinus, Quirinalis, Viminalis*, and *Eſquilinus*, and hath bene gouerned, by theſe ſeuen princelie governements, *Kings, Conſuls, Dictators, Decem-virs, Tribuns, Emperours & Pops*, Afterward it was divided into ten chiefe kingdoms, to wit, *Spain, France, Lombardie, England, Scotland, Denmark, Sweden*, the kingdome of *Goths* in *Italie*, & of the *Hunnes* in *Pannonia* or *Hungary*, and the *Exarchat* of *Ravenna*, and overthwart theſe ſeuen hils, & notable parts of that Empire, wer ſet vp idolatrous temples, images, pillars, Piramydes,

Paraphr. expoſition.

AND I foreſawe, that out of the waltering [*a*] waues and muddie ſeas of Tyrantes, there ſhould ariſe a great and mightie (*b*) Empire, whoſe principall ſeate and chiefe Citie, is placed vpon ſeuen (*c*) hilles, and ſhould bee governed by ſeuẽ princely governements: and (*d*) tenne ſeuerall parts of the ſame Empire are to bee divided, whereof there ſhal be tenne crowned Kings, and euerywhere vpon theſe ſeuen hilles, ſhall bee monumentes of oſtentation, with proude and blaſphemous ſuperſcriptions, and titles, ſet vp by theſe princelie and royall Gouernours.

The text.

And J ſaw a beaſt riſe out of the ſea, hauing ſeuen heads, and ten hornes, and vpon his hornes, were *tenne crownes, and vpõ his heads the name of blaſphemie.*

& monuments of great oftentatiõ & blafphemous fuperfcriptions and titles, fuch as *Diis Manibus, Minervæ, Fortunæ, Romæ-Fortunatæ, Romæ Victrici, Romæ æternæ, Plutoni, Veneri Priapa*, & infinit others of old erected befide alfo infinit Papifticall idols of new.

2 This Empire fucceeding to the former three Empires and monarchies, was

2 And this Empire that I forefaw, was

like the Græcian Monarch, in that it fwiftly as with the fpeed of a [*e*] Leopard, fhortlie ouer ran the whole world, and like the Monarchie of the *Meedes* and *Perfians*, who mightily marched forward, as on Bears feet, treding down al other kingdoms, and like the Monarchie of the *Babylonians*, who (as with the mouths of Lyons) devoured vp innocent blood, without meafure:

So that at length by their deuilifh works of Idolatrie, auguries, forceries, oracles of falfe gods, and by bloodfhed it pleafed our God, to fuffer thefe Emperours to conquer & poffeffe the whole world.

And the devill gaue him his power, and his throne, that is, he made him [*f*] Prince of the world with full authoritie.

2 And the beaft which J faw was like a Leopard, and his feet like a Bears, & his mouth as the mouth of a lion: & the dragon gaue him his power, & his throne, & great authoritie.

3 But afterward, the *Hunnes Gothes*, and *Vandalles*, fubverted that gouernement of Emperours, and abolifhed them, where-through the Empire vaked, and lay dead more than three hundreth yeares. Yet, afterward rofe new Emperours againe, intituled of *Germanes* and *Romanes*, of the which *Charle-maigne* was the firft, and in them was the Roman Empire in a manner

3 And I forefaw one of his chiefeft [*g*] gouernements, as it were vtterlie deftroyed, and abolifhed: but that loffe was mended again, & that gouernment wholly reftored, and all men marueiled at the power & profperitie of that Empire, & gaue ouer themfelues as flaues & feruants vnto it.

3 And J faw one of his heades, as it were *wounded to death: but his deadly wound was healed, and al the world wõdred and followed the beaft.*

reuiued and reſtored againe. The great fœlicity of this Empire aſtoniſhed all men, who therefore called it *Imperium ſacrum perpetuum, & ſemper auguſtum*, and not only became ſubiect to it,

4 But alſo for loue of it, they worſhipped the deuill, in the idols of *Mars, Minerva*, Ladie *Fortune*, and others, and did beare ſuch reverence to that Empire, that by protection of falſe gods, they thought the ſame without all compare, and for euer invincible.

5 In this Empire ariſeth the pope, who proudly alleadgeth himſelfe to be Gods Vicar, and that he can remit ſinnes, and exeme from hell, and ſuch other blaſphemies againſt God and he began his mighty power and kingdome, by pope *Sylveſter* the firſt, about the year of Chriſt three hundreth and ſixteene, and hath raigned a thouſand two hundreth and threeſcore yeares,

6 Not onely blaſpheming God and his name but alſo boaſting that he hath heauen to ſell for money, and can make Saintes in heauen of his owne canonizing, to whom we muſt pray as our mediatours.

7 And whoſoeuer would not embrace theſe & other his errors, but meaned to ſerue God truely, of theſe did he perſecute and put to death manie a thouſand

4 And they worſhipped the [*h*] deuill that gaue ſuch preheminence to that Empire, and they honoured that Empire, eſteeming none like thereunto, or able to make warre againſt it.

5 Then was there adioyned to that Empire, a [*i*] fals mouth, euen an Antichriſtian Prophet, that ſpake foorth proudly things of great oſtentation, & blaſphemies againſt God, and he receiued a mightie power of doing, for the ſpace of [*k*] 1260. yeares.

6 During the which time, the Empire', opened this his Antichriſtian mouth whereby he taught blaſphemous doctrine againſt God, and againſt his holie name and heauenlie Tabernacle, and againſt al the Saints that dwell in heauen.

7 And God ſuffered him to make warre againſt his elect ſervants and to ſubdue them

4 And they worſhiped the dragõ which gaue power vnto the beaſt & they worſhipped the beaſt, ſaying, who is like vnto the beaſt? who is able to warre with him?

5 And there was giuen vnto him a mouth, that ſpake great things, and blaſphemies, and power was giuen vnto him to do two and fourtie moneths.

6 And he opened his mouth vnto blaſphemie againſt God, to blaſpheme his Name and his Tabernacle, and them that dwell in heauen.

7 And it was giuen unto him to make warre with the ſaintes and to ouercome them, and power was giuen him ouer every

extending his tyrannical power ouer al Chriſtian kinreds and people.

8 Whereby wordlie men, addicted to earthlie pleaſures, reuerenced him as God (diſtinct. 96 Can, *Satis euidenter*) kiſſing his feet, & adoring him ſo that in him the name of *Rome* and whole Roman Empire is eſteemed famous and holden in great eſtimation and reuerence of all.

his power extended ouer euerie kinred, tongue and nation.

8 Therefore, all wordlie men dwelling in earthly pleaſures, ſhall giue ſuch homage and reuerence as appertaineth to God, vnto that Empire, euen all vnder his tyrannie, ſo that theſe, whoſe names are not regiſtred among the elect number of Chriſt Ieſus our Sauiour, who was predeſtinate to [*l*] ſuffer death for vs, before the foundations, of the worlde were laide.

kinred tongue & nation

8 Therefore all that dwel vpõ the earth ſhal worſhip him, whoſe names are not writtẽ in the booke of life of the lamb, which was ſlaine from the beginning of the world.

9 But who hath underſtanding, let him conſider Gods iuſt iudgement.

9 If any man haue an eare, let him heare.

10 Whereby theſe tyrants ſhall go at length into captiuitie, becauſe they haue led others captiue: and ſhall be killed with the ſworde, becauſe they haue killed others with the ſword and this yeeldeth occaſion of comforte, patience, and perſeverance in faith to Gods ſervants,

10 If anie leade into captiuitie, he ſhal goe into captiuitie, if any kill with a ſword, he muſt be killed by a ſworde, here is the patience and the faith of the Saints.

11 And to returne againe to the particular deſcription of the Popes kingdome, it aroſe of the ſaid greate Empire, and contained a two folde power, ſpiritual and temporall: ſignified by their two horned Miter, and by the two ſwordes that Pope *Boniface* the eight cauſed to bee caried before him, ſaying *Ecce duo Gladii hic*, while as he was clade the one day in the Pontificall of a Biſhop, the next day in armour like an Emperour, vſurping ſo double power, but teaching ſuch deuiliſh doctrine as before is ſaid.

11 And I conſidered [*m*] by it ſelfe and aparte, the deſcription of this Antichriſt, euen how his kingdome was, in a manner another kingdome comming vp out of the former great and ſtabliſhed Empire of the earth, and that hee had double power, ſpirituall and temporall, euen [*n*] Prieſthood & kingdom as Chriſt hath, yet in his doctrine teacheth he diveliſh errours.

11 And I beheld another beaſt comming vp out of the earth which had two hornes like the Lambe, but he ſpake like the dragon.

12 And he wrought what he lifte [the Emperours yet ftanding] with no leffe authoritie and power than the Emperour himfelfe, yea he eftablifhed and inaugurated again of new, that decayed ftate of Emperours, that now ftarted vp againe by *Charlemaigne*.

13 And he wrought many wonders partely falfe as thefe of the legendary, partely by forcerie throwing fire out of his fleues, as fent from heauen, partely by his ordinarie ceremonies of curfing, vfed with the fetting vp on hie of light torches, and burning candles, and then throwing them downe headlong to the ground, extinguifhing them before al the people,

14 Terrifying fimple men to obey him, who thought their foules wer vndone for euer, by thefe lying terrours and deceits, which he vfed againft both the people, and Princes of the Roman Empire: Alfo, it was the Pope that caufed *Charlemaigne* & his fucceffours new Emperours to be confirmed and eftablifhed, after that, that eftate of Emperors had vaked more then three hundreth yeares, by the warres of the *Hunnes, Gothes*, and *Vandalles*: and

12 And he did all that he lift in prefence of the greate Empire, with no leffe power then that firft great Empire was wont to do and he compelled all men to doe homage, reuerence & be fubiect to that great empire, whofe gouernment newlie [*o*] was ftart vp againe, and his former decaies and loffes amended.

13 And this Anti-Chrift, wrought many devilifh and [*p*] deceitfull wonders, fo that he made [*q*] fires of diuers kinds to be throwne downe from on high to the ground in prefence of the whole people, as if the fire of God his wrath were at his commãd, to be broght down from heavẽ.

14 And he deceiued, and terrified them that dwelled on the earth, by thefe lying fignes and terrors which God fuffered him to vfe, in prefence of all thofe of the great Empire: and hee [*r*] caufed all thofe that dwel on the earth, to renew & make vp again a new forme of Emperours, or rather images of the former great Empire, whofe imperiall

12 And he did all that the firft beaft could doe before him, and he caufed the earth and them which dwell therein to worfhippe the firfte beafte, whofe deadly wound, was healed.

13 And he did greate wonders, fo that he made fire to com down from heauen on the earth in the fight of men.

14 And deceiued them which dwell on the earth, by the fignes, which were permitted to him to doe in the fight of the beaft, faying to them that dwell on the earth, that they fhould make the image of the beaft, which had the wound

ordained Princes, Electours, who ſhould chooſe the King of Romans.

15 Who yet was not Emperour but as a dead figure, Image, or ſhew of the Empire, till he folẽnlie receiued the Popes injunctions, and did ſweare to maintaine his profeſsion, and then did the *P*ope crowne, inaugurate and confirme him Emperour of *Germanes*, and *Romanes*: And when the Emperours is ſo ſworne, and confederate with the pope, then who euer doe not reuerence them & obey their common decrees wil be purſued of them both and of all their confederates, to death, by fire and ſword, as rebels, apoſtates, and heritikes.

16 Moreouer, the pope cauſed all his ſubiects of Chriſtendome, to receiue and beare in their hands and foreheads markes & tokens, repreſenting the name of the *Latin Empyr* ſuch as the marks of χρ~'ς and croſſes of ſundry ſorts.

17 Elſe were they not thought to be Chriſtians but excommunicate, and by the Bul of Pope *Martin* the fift theſe wer not permitted to vſe any trafique or trade of merchãdize that would not ſubiect themſelues to the ſeate of *Rome*, by marking thẽſelues thereto in bearing croſſes on their foreheades

governement had bene deſtroied and aboliſhed before by the ſword, but now is repared and reſtored againe.

15 And God ſuffered this Antichriſt to [*ſ*] inſpire thoſe Emperours (or rather images and formes of the great Empire) with the ſpirit of his errours, ſo that they ſhould ſpeake and profeſſe his doctrine, and then ſhould he confirme, crown and inaugurat theſe counterfeite Emperours with ſuch power, that now all men muſt honour and reuerence them, and yeelde them worſhippe and obedience, vnder paine of their liues.

16 And he cauſed all men of al degrees and qualities, to receiue and beare viſible [*t*] markes or liueries of the Empire, ſome times in their hands, ſomtimes on their foreheads.

17 And he diſcharged all men of doing their lawfull affaires, or vſing their trade of merchandize, except thoſe, that doe beare and profeſſe the marke or name of the Empire, or the figure of the number of his name.

of a ſword, and did liue.

15 And it was permitted vnto him to giue a ſpirit vnto the image of the beaſte, ſo that the image of the beaſt ſhuld ſpeak and ſhould cauſe that as manie as would not worſhip the image of the beaſt ſhould be killed.

16 And he made all both ſmal and greate, rich and poore, free and bound to receiue a marke in their right hands, or in their fore-heads.

17 And that no man might buy or ſell ſaue he that had the marke of the name of the beaſt or the number of his name.

on Aſh-wednesday, & palme croſſes in their hands on Palm-ſunday or by profeſsing their name, that is to ſay, to beleeue as the *Latine* Church beleeues, or to reuerence magically the croſſes or naked figure of χρ̃′ς.

18 Which who wil wiſelie conſider, ſhall finde more like to χξς which is the nũber of the name λατεινος,) than to the name of Chriſt, for this λατεινος is the name of that king, from whome that Empire is called the *Latine Empire*, and the number of that name λατεινος is 666. expreſſed in Greeke by theſe numerall letters χξς

18 This is a matter, which requireth wiſdome: let him therefore that hath skil and knowledge reckon the literall number of the name of this Empire, for the ſame number is alſo the number of one of the chiefe [*u*] heads-men, and firſt Princes thereof, and the letters of that name mak the number of ſixe hundreth three ſcore and ſixe.

18 Here is wiſdome, Let him that hath wit, count the number of the beaſt for it is the nũber of a man, and his number is ſix hundredth, threeſcore and ſixe.

Notes, Reaſons, and Amplifications.

[*a*] Seeing (Apoc. 17.15.) waters generally are interpreted to meane all people and nations: the ſea whereout this kingdome and the foure kingdomes in (Dan. 7.) ariſeth, muſt needs be the tumultuous people and tyrants of the world, amongſt the which, and of whom, the monarchies ſtart vp, conforme to *Eſay*, 57.20. *the wicked are like the raging ſea that cannot reſt, &c.* and Ezech. 26.3 *I will bring vp many nations againſt thee, as the ſea, &c.*

[*b*] That beaſts be interpreted kings, kingdomes, empires, and Monarchies, read Dan. 7.17. and Dan. 8.20.21. and that this beaſt is the Roman kingdome and empire in particular, is prooued by our 24. propoſition.

[*c*] That theſe ſeuen heads, are interpreted ſeuen hilles of the great citie, and ſeuen ſorts of Kings or gouernours thereof, read Apoc. 17.9.18. and how Rome is that ſeuen-hilled city, gouerned by ſeuen royall and princely gouernements, is proued by the 23. propoſition, and in our foreſaid 24. propoſition.

[*d*] How this empire of Rome, became diuided into ten kings or Kingdomes, compared to hornes, read Apoc. 17.12. and the ſaid 24. propoſition.

[*e*] *Daniel* cap 7. defcribing the foure monarchies, compareth the firft (which was of the *Babylonians*) to a Lyon. The fecond (which was of the *Medes & Perfians*) to a Beare. The third which was of the *Grecians*) to a Leopard: but becaufe to al thefe, fourthly fucceded the great Latin empire: Therefore Saint *Iohn* compofeth here the fourth beaft, which reprefenteh the fame of all the former three, fignifying thereby, that this Empire lacked no inftrument of crueltie, that any of the former had.

[*f*] What throne & feat can this be els, that the deuill giueth to the Romans, but that which he promifeth to giue to Chrift, Mat. 4.9. to wit, euen the Empire of the world, as being the Prince of this world, Ioh. 12.31. and Eph. 6.12. and Col. 2.15. and fo indeed it pleafed God to permit the Romans by their deuilifh auguries, oracles, forceries and bloodfhed, to obtain the whole monarchie of the earth, to the wrack of other godles people, and at length of themfelues.

[*g*] The heads, Apoc. 7.19. being two wayes interpreted, to wit for feauen mountaines, and for feuen kings or royall gouernments. The wounding of one of thefe feuen heades muft neceffarilie, rather mean the decay of one of the feuen princely gouernments thã, any thing concerning any of the hilles, fpecially feeing experience of the event hath declared, that the moft fpeciall of thefe feuen gouernments, to wit the Emperors, wer abolifhed by the war-fare of the *Hunnes, Gothes, & Vandals*, & vaked from the dayes of *Augu*f*tulus*, in the yeare of Chrift. 475. vntil the comming of *Charle-maign* in the 806. yeare, in whofe perfon and fucceffors, that deadly wound of the Empire was healed, and the eftate of Emperors renued. Wherevpon reade *Iohannes Bapti*f*ta Egnatius Lib. 3. Romanorum Principum, in Carolo Magno.*

(*h*) Who can deny but thefe Romans, adorers of thefe Idols, in whom the deuill actually fpake by oracle, in that doing, worfhipped the Deuil, feeing the moft fimple fort of Idolatrie, is nothing els, but the worfhiping of deuils and are adiunct. Apo. 9.20.

(*i*) This mouth of the Empire, euen the Pope, that Antichriftian prophet, here receiues his beginning of the *Latin* empire, as a member thereof, but afterward (ver. 11.) he becomes a feuerall King, and is particularly defcribed: hereof read the 24. 25. and 26. Propofitions.

(*k*) That 42. moneths be 1260. yeeres, and begin about the 300. or 316. yeare of God, reade our 1. 16. and 36. propofitions.

(*l*) This flaying of the Lambe from the beginning, meaneth the certaintie of Godes predeftination, in that the decree of God preordinate from the beginning, fhall as furelie take effect, as if it had bene alreadie performed from the beginning. *Arethas*, reads this wife, *Whofe names were not written from the beginning of the world in the booke of life of the Lamb, who was* f*laine.* Agreeable with the like, contained Apoc. 17.8. Alwaies, feeing both the faued number as alfo the forme of redemption are from the beginning appointed and preordinate by God, therefore both thefe readings may ftand.

(*m*) After he hath defcribed the Pope as a member, euen the mouth of the great *Roman* Empire: now defcribeth he him more particularlie & apart, as hauing obtained a particular kingdom of his owne: hereof read not *i*, with the 24. 25. and 26. propofitions

(*n*) *Hornes*, to fignifie powers and kingdomes, reade the note *d*, Apoc. 5.

[*o*] This deadly wound note *g*, is declared to be the decay of the Imperiall eftat, which now being ftart vp again in *Chairle-maigne* the Pope proclaimes him Emperour, and willeth all men to reuerence him.

[*p*] Of thefe miracles and wonders, wrought partlie by forcerie, partlie by iugling, and partly but falfly alledged, ye fhall finde diuers hiftories: hereof read the tenth fynthefis of our 26, propofition.

[*q*] Of thefe fierie fignes, we reade two fortes the one is by inchantment, fuch as Pope *Gregory* the feuẽth did practife, who firft was a monke and was called *Hilde-brand* which is to fay, a *helth-brand*, or rather a *hel-brand*, and being afhamed of that name, did change the fame but not the meaning thereof, for as Cardinal *Benno* writes *invita Hildebrandi*, in his booke intituled, *Vita & gefta Hildebrandi qui Rom. Pontifex factus, Gregorius 7 dictus eft, defcripta per Bennonem Cardinalem & Archiprefbyterum Romanæ ecclefiæ. Cum vellet manicas fuæs difcutiebat, & in modũ fcintillarũ ignis difsiliebat & eis miraculis oculos fimpliciũ veluti figno fanctitatis ludificabat: & quia diabolus per paganos Chriftum publicè perfequi non poter at per falfum monachum fub habitu monaftico, fub fpecie religionis, nomen Chrifti fraudulenter fub vertere difponebat.* that is *when he lift he did fhake downe his fleeues, and fire did flie out, in maner of fparkes, and by thefe miracles as by a figne of holineffe he deluded the eyes of the fimple people: And becaufe the deuil might not perfecute Chrift openly by Pagans, therefore by this falfe Monk vnder a Monks habite and fhew of religion, he fraudfullie preaffed to fubuert Chrifts name.* The other forte is by ceremonies of terrour, as when the Pope or his cleargie do caft down fierie torches, and Candles from on high, at the time of their curfing, as though they would meane, that the fire of Gods wrath were likewife readie at their command, to bee fhaken out of heauen, as out of Hildebrands fleeue againft them whome they curfe, And by thefe horrible fignes and terrours they haue fo terrified Princes that they made them to loofe their kingdomes and due obedience of their fubiects, till they behoued to ftoupe vnder their feete, and craue pardon and abfolution. What kingdoms thereby the Pope hath cowped and what rents and dignities he hath conquered, is to be feene in hiftories.

[*r*] The pope politiklie for his eafe, caufeth thefe new counterfeit Emperors: to be made by princes Electors, that thefe Emperors of his owne making may not onlie augment his Ecclefiafticall rents, but alfo may be readie, as

his ſlaues, to fight his battels, & reuenge all his quarrels, as is to be ſeene in manie & diuerſe hiſtories, elſe ſhal the Pope depriue them and creat other Emperors in their place: And ſo hath he puiſſance indeede to doe, for that her hath hitherto ſo pinched the empire, that hee hath the ſubſtance thereof, and they little more than the bare name, image or ſhadow therof: and therefore are they called images of the beaſt as being in a manner but counterfait Emperours: Reade hereupon our 28. propoſition.

[*s*] The king of Romanes and apparent Emperour choſen by the Princes Electors is neuer the more Emperour, vnleſſe, being ſworne to maintaine the Papiſticall ſeate, he be after ward inaugurate and crowned by the Pope: for manifeſtation of which ſupremacie, Pope *Cœleſtine,* crowning the Emperour *Henrie* with his hand, hee afterward with his foote threw downe the crowne from his head, ſaying he had authoritie both to make Emperours and to depoſe them at his pleaſure.

[*t*] That theſe viſible marks are the abuſed characters of χρ͂ς, and croſſes of all kindes ſuch as † ✠ × ⊞ ⊛ and ſuch other, and how theſe be rather deriued from the beaſts name λατεινος, than from the name of Chriſt, read the 31. propoſition.

(*u*) That the name of the beaſt is λατεινος, or *Latin Empire*, and how that name reſulteth of the number 666. or χξς, reade the 29. propoſition, & how the beaſts viſible markes are deriued therefrom, read the 31. propoſition.

CHAP. XIIII. THE ARGVMENT.

Vnder the figure of 144000. Iſraelites marked (Apoc. 7) with the marke of Gods protection, before the troubles of the ſeuen ages, and now appearing in that ſame full number after theſe troubles, reioycing with Chriſt on mount Sion: the ſpirite of God expreſſeth to vs the deliuerie of his church from all the tyrannie of the antichriſt, & reuiuing thereof again in ful nũber which being the firſt reſurrectiõ, he procedeth frõ thence orderly to the ſecond reſurrection, and Gods great harueſt, dividing the whole time to that great day, as alſo that dayes work, among ſeuen Angels, whom extraordinarlie here the prophet hath inſert, and not in the tenth chapter, where they ſhould haue bene, for the better concealing the myſtery thereof to theſe dayes, which before theſe our dayes was not meet to be knowne, as may be gathered by Dan. 12.4. 8. 9. 13. and Apoc. 10. 4 and note (d) *thereof, conferred with our 11. propoſition, and with the ſecond and fourth generall reaſons in our 14. propoſition.*

Hiſtoricall application. *Paraphraſt. expoſition.*

THen I beheld, and conſidered the Lambe of God, Chriſte Ieſus, to ſtand on the mount of his inheritance, euen among his elect Church, and with him, thoſe [*a*] hundreth, fourtie & foure thouſands Iſraelits, who before (Apo, 7) were marked patently, with the marke and name of God their protectour.

2 And I heard further, proceeding from Gods Church, the voice and ſound of many (*b*) people and nations, and the thundring & threatning voyce of Gods true Miniſters, and I heard the joyful voyce of their reioycing in God.

3 And theſe Iſraelites, ſung not the old ſong that their Fathers did, of the Meſsias to come: but contrarily in the preſẽce of God, & before his profeſſors of the old and new Teſtaments, they gladly ſung the ſongs & hymnes of their newe redemption in his blood, already ſhead: and no [*c*] moeof Iſrael could learne to reioyce and triumph in that newe redemption, but onely thoſe hundreth, fourtie foure thouſands, which thereby were redemed and ſelected from among the earthly Iſraelits.

4 Theſe are they, which are neither [*d*] whoremongers corporallie with women, nor ſpirituallie with idoles, for they are chaſt and cleane from both: Theſe follow Chriſt, whereſoeuer hee directeth them: theſe are redemed and choſen out from worldly men, being that portion of the [*e*] fruit of the world, that is dedicated to God, and to his Sonne Chriſt Ieſus.

5 And in their mouths is found no guile nor deceit, for Chriſt hath purged their ſinnes, ſo that they appeare without ſpot before the Maieſtie of God.

The text

Then I looked, & loe, a Lambe ſtood on mont Sion, & with him, a hũdred fourty & four thouſande, hauing his Fathers Name written in their fore-heads.

2 And I heard a voice frõ heauen, as the ſound of manie waters, and as the ſound of a greate thũder: & J heard the voyce of harpers, harping with their harpes.

3 And they ſung as it were *a new ſong before the thron and before the foure beaſts, & the elders, and no man coulde learne that ſong, but the hundreth fourty and foure thouſand, which were bought from the earth.*

4 Theſe are they which ar not defiled with womẽ: for they ar virgins, theſe follow the Lamb whitherſoeuer he goeth, theſe are boght from men, beeing the firſt fruits vnto God, and to the Lambe.

5 And in their mouths was found no guile: for they are without ſpot before

6 Thefe Ifraelites with many Gentiles being hitherto hidlings, preferued by God in thefe fix ages vnder the Popes tyrannie. Now in the feuenth age, about the 1541. yeare of Chrift, God ftirred vp openly, his firft publike minifters & feruants *Luther, Calvin, Melãcthon* and others, to preach out during all this Iubilie the perfite veritie of the Gofpel publikly, which before was hid and obfcured vnder the Antichriftian raigne, fo that now the true Church lurketh no more, but Gods trueth to all the world is made patent by thefe minifters.

The feuenth age begins. 1541 *The firft thundering Angels Iubilie.*

6 Then after Chrifts fpirituall prefence, for the defence of thefe hitherto: Now came another mighty Meffenger, euen the [*f*] firft of the feuen thundering Angels of the latter day, & he, in the mouthes of God his Minifters, did publifh the euerlafting Evangell of Chrift Iefus, and preached foorth the fame openlie, to all them which dwelt vpon the earth, of euerie Nation, Kinred, tongue and people.

the throne of God.

6 Then J faw another Angel flie in the middes of heauen, hauing an euerlafting Gofpel, to preach vnto them that dwel on the earth, & to euerie nation and kinred, & tongue, and people.

7 Teaching them all vniuerfallie, by their publike and open preaching to feare God onlie, and leaving all other fuperftitious worfhipping, to giue glorie onelie vnto him, for the [*g*] houre draweth now neere in the which all fhall be iudged, and therefore, to leaue their Antichriftian Idolatrie, and only to worfhippe him, that made the heauen, earth, waters, and all that in them is.

7 Saying with a loud voice, fear God, and giue glorie to him: for the houre of his iudgemẽt is come, and worfhippe him that made heauen and earth, and the fea, and the fountaines of water.

8 About the end of which Iubilie, and beginning of this next Iubilie, begun in Anno. 1590. God hath fhewed marueilous indices, that the Empire of *Rome* & Papiftical kingdome fhall fhortelie fall: the Antichriftian & *Spanifh* flote is deftroyed, the late king of *France,* Duke of *Guize*, and his brother Papifts, and committers of the *Parifiã* maffacre all murthered by other, one

Iubily 1590 *The fecond thundring Angels*

8 Then came the fecond of the feuen thundering Angels, declaring euen prefentlie, the horrible fall and decay to become of that figuratiue h) *Babylon*, & great Antichriftian citie, becaufe fhee had entifed all nations hitherto, to committe

8 And there followed another Angel, faying, it is fallen, it is fallen Babylon the great citie: for fhee made al nations to drinke of the wine of the wrath of her fornication.

named a proteſtant nowe made king of *France*: So that before the ende of this Iubilie (God willing) *Rome* and the whole papiſticall kingdome thereof ſhall be ruined.

Iubilie

idolatrous whoredome with her and ſo made them to drinke of the cuppe of God his eternall wrath and indignation,

9 So that in the thirde Iubilie beginning in An. 1639. the miniſters and meſſengers of God, ſhall God willing haue only the latter dregges of that *Romiſh* ſeat to admoniſh & threaten by open intimation, that if any bee, in whoſe heartes the reuerence of that ruined Empire and counterfeit Emperors defẽders therof is ſuperſtitiouſlie in veterate, or if anie now be marked with Chriſme or croſſe on hand or head

1639 *The third thundering*

9 Afterward comes [*i*] the third order of thundering or threatning Meſſengers, openly and loudelie threatening, that if any bee now, that doe worſhip or reuerence that Antichriſtian Empire or counterfeiters thereof, or that receiues the olde ſuperſtitious markes thereof on their fore heade or in their handes.

9 And the third Angel followed them, ſaying with a loud voice, Jf anie man worſhippe the beaſt and his image, and receiue his *marke in his foreheade or on his hand.*

10 The ſame ſhall drinke of the horrible vengeance of Godes wrath yea, of his vengeance and wrath, pure and vnmixed with anie pointe of his mercie: and he ſhall be tormented eternallie in the bitter ſulphurious fire of hell, in the preſence of Chriſt and all his holie Angels.

11 And the griefe of their torment ſhall encreaſe for euer and euer: and

10 The ſame ſhall drinke of the wine of the wrath of God, yea of the pure wine which is powred into the cup of his wrath and he ſhall be tormẽted in fire and brimſtone, before the holie angels, & before the Lambe.

their torment ſhal neuer ende that reuerence that *Romane*, and Papiſticall ſeate or obey the Princes maintainers thereof, be it *Spaine* or *Germanie*, or who euer els, that maintaining that ſeate doeth counterfeit the Emperour, and whoſoeuer receiues in honour the ſuperſtitious markes of χρ῀ϛ and croſſes of all kinds which are deriued

Angels Iubilie

they ſhall haue no ſlackning day nor night, who reuerence that Antichriſtian empire or counterfeit Emperors thereof, or whoſoeuer receiueth the ſuperſtitious markes that are [*k*] deriued and compoſed of the

11 And the ſmoke of their tormẽt ſhall aſcend euermore, and they ſhall haue no reſt day nor night, which worſhippe the beaſt and his image, and who ſoeuer receiueth the printe of his name,

12 Here is the patience of the Saints, here are they that keepe the

from $\chi\xi\varsigma$ which is the number of $\lambda\alpha\tau\epsilon\iota\nu o\varsigma$, the name of the Latin Empire

name of that Empire,

12 And this eternall puniſhment of the wicked, doeth miniſter occaſion of patience to the Saintes of God, and to theſe that obſerue his commandements, and are the faithfull ſeruants of Chriſt Ieſus

13 And now doeth the prophete *Iohn* at the commandement of Gods Spirite write & publiſh thoſe now to be happie that haue died in the Lord, for that [*l*] incontinent (ſaith Godes Spirite) after this Iubilie, ſhall they reſt from their troubles, & the rewarde of their workes (deſerued in Chriſt) ſhall now followe them.

14 And beholde now and loe (the fourth Iubelie being begunne in Anno a thouſand ſixe hundreth eightie eight.) bright clouds ſhall appeare, and (*m*) thervpon Chriſt the ſonne of man comming vnto iudgement, as the fourth of the thundering Angeles having on his head the glorious crowne of eternall victore, and in his hand the (*n*) ſickle of ſharpe diligence, to gather in his elect with.

15 And for the fift of the thundering Angels came another euen the (*o*) thirde perſone of the Trinitie, who proceeding out from the (*p*) Maieſtie of the father proclaimed with a loude voice to Chriſt that ſate vpon the cloudes, ſaying, Enter to now with diligence, and gather in thine elect: for the houre of the greate harueſt is now come, and the elect people of the world are readie and ripe, exſpecting their reward.

16 And Chriſt, comming in the clouds entered to as a reaper, with the ſharp ſickle of diligence, to gather in whollie his elect, from the earth, and they were then all gathered vp from the earth.

17 And then came another (euen the ſixt thundering Angel) executer of God his wrath from the maieſtie of God, appointed alſo diligentlie (as with a ſickle) to gather in the wicked.

18 And the ſeuenth and laſte thundering Angell came out from the iuſtice ſeate of God, hauing power, concerning Gods fierie and fierce wrath and he

commandements of God and the faith of Ieſus.

13 Then J heard a voice from heauen, ſaying vnto me, write, bleſſed ar the dead which die in the Lord, for that incontinent hereaſter (ſaith the ſpirit) ſhal they reſt from their labours and their works follow them.

14 And J looked and behold, a white cloude, and vpon the cloud one ſitting like vnto the Son of man, hauing on his heade a golden crowne, and in his hand a ſharpe ſickle

15 And another Angel came out of the Temple, crying with a loude voice to him that ſatte on the cloude. Thruſt in thy ſickle and reap: for the time is come to reape: for the harueſt of the earth is ripe.

16 And hee that ſat on the cloude, thruſt in his ſickle on the earth, & the earth was reaped.

17 Then another Angel came out of the Temple which is in heauen, hauing alſo a ſharpe ſickle.

18 And another Angel came out frõ the Altar, which had

proclaimed with a loude voice to the former Angell, that was readie as one with a ſickle, to gather in the reprobate, ſaying: Enter to worke diligentlie, as a reaper with his ſickle and gather the heapes of the wicked of this world, for their [*q*] iniquitie is fullie ripe.

power ouer fire & cryd with a loud cry to him that had the ſharp ſickle, & ſaid, Thruſt in thy ſharp ſickle, & gather the cluſters of the vineyard of the earth: for her graps are ripe.

19 And that ſixte Angel entered vpon thoſe people of the earth, as a reaper with his ſickle, and cut downe the wicked growing in all earthlie wickedneſſe, and caſt them into hell, the greate tormente place of God his fierce wrath.

19 And the Angel thruſt in his ſharpe ſickle on the earth, and cut downe the vines of the vineyard of the earth, & caſt them into the great wine-pres of the Wrath of God

20 And their they were tormented, farre from the companie of Gods true citizens, and as the blood of an infinite ſlaughter, may be thought to flowe vp to the bridles of horſe in a fielde [*r*] a thouſand ſix hundreth ſtades off: euen ſo ſhall their terrours and eternall torment in hell, beginne to ouerflow al meaſure, after the outrunning of a [*r*] thouſand ſix hundreth yeares.

20 And the winepreſſe was troden without the citie, & blood came out of the winepres, vnto the horſe-bridles, by the ſpace of a thouſand and ſix hundreth furlongs.

Notes, Reaſons, and Amplifications.

(*a*) Of theſe hundreth fourtie foure thouſand Iſraelites, which the Spirit of God ſignified to S. Iohn, in the beginning of the troubles of the ſeuen ages, that he would preſerue and marke to himſelfe, there lackes not one, but they are all exhibited now, after theſe troubles, merrie and reioycing with Chriſt their redeemer, as Chriſt ſaith, Iohn 6.39. and 17.12. *Hec est enim voluntas eiusqui miſit me Patris, vt omne quod dedit mihi, non perdamex eo, ſed reſuſcitem illud in noviſſimo die. This is the Fathers will which hath ſent mee, that of all which he hath giuen me, I ſhould looſe nothing, but ſhould raiſe it vp again at the laſt day:* And, *Quos dediſti mihi, cuſto diui, & nemoex ijs perijt, &c. Theſe that thou gaveſt me, haue I kept, and none of them is loſt. &c,*

(*b*) As in the ſeuenth chapter hee hath expreſſed and fore-warned the marking, preſeruing and electing of theſe 144000. Iſraelits, & of the Gentiles of all nations, an innumerable people. So here for the Iſraelits, he repeats that ſelfe-ſame number of 144000. and for the Gentiles, he repreſents them vnder the ſound of many waters, & great thunders, for Rev. 17.15. waters are interpret people, nations, and tongues, and thunders are thoſe preachers, of whatſoeuer nations, that boldlie threaten out Gods iudgements againſt the wicked, for the which, Chriſt called Iohn and Iames, the ſonnes of thunders. Marke 3.17. and the ſeuen Angels, that here threaten the deſtruction of this Babylon, and day of iudgement, are therefore (Revel. 10.3.4.) called the ſeuen thunders.

(*c*) This doubtleſſe, is literally ſpoken of Iſrael, of whom he wold mean, that, boaſt as they will, that they are the ſeed of Abraham and choſen people of God, yet ſhall no moe of them than theſe 144000, be ſaued, that becomes Chriſtians: whereas of the Gentils, the number that ſhal be ſaued, Rev. 7.9. ſhall be infinite, and their ſound ſhalbe (here ſaith he ver. 2) as the ſound of many waters. But concerning the preciſnes of this number, read Apo. 7. note *f*, thereof.

[*d*] This is not ſo preciſely meaned, but that both the whoremonger and idolater, whom God afterward hath indewed with repentance and amendement, hauing faith in Chriſt Ieſus, ſhal be reſtored to puritie in his blood: for, how oft ſoeuer a ſinner doth repent him of his iniquities, the Lord will put his iniquitie out of his mind, *Yea, though thy ſinnes* (ſaith God Eſay 1.18) *were as skarlet yet ſhall I make thee white as ſnow, and cleane as wool.*

(*e*) As the firſt fruits in the Leviticall law were offered to God, and the reſt diſponed vpon by the people: So are the elect onelie dedicate and appointed for God, and therefore, they onely are called his firſt fruites: and the great multitude reprobat, and reject.

[*f*] That this Angell (although he calleth him another Angell, that is to ſay, another than the Lambe) is no ſecond, but is the firſt of theſe, which Apoc. 10.3.4. are called the ſeuen thunders, appeareth hereaſter by verſe 9. and note *i*, following, and by the 11. and 12, propoſitions: and therefore this Angels time falleth betwixt the 1541. and the 1590. yeare of Chriſt.

[*g*] That at this time the houre of Gods latter judgement drawth nere, euen within an hundreth and ten yeares from the end of this firſt Angels time, is to be ſeene by our 13. and 14. propoſitions.

[*h*] That vnder the name of figuratiue *Babylon*, is meaned *Rome* read the 23. propoſition. And as *Rome* it ſelfe is here comparable to *Babylon*, in that ſhe ſucceded to her Empire and impietie, ſo alſo here, & in the 18. chap hereaſter, doth the prophet S. *Iohn* in deſcribing the fal of this *Babylonicall Rome*, borrow the very words of *Eſay* & *Ieremie*, deſcribing the fal of

Babylon it ſelf. Hereof read the note *b*, of the 18. chap. Further, how before the end of this Iubilie, ending in Anno 1639. *Rome*, appearingly is to be deſtroyed, and the kingdome thereof to fall, the marveilous beginning of this ſame Iubilie ſeemeth to portend: for nere to the beginning hereof, in the 88. 89. and 90. yeares of God, God hath by the tempeſt of his winds, miraculouſly deſtroied the hudge and monſtruous Antichriſtian flote, that came from *Spaine*, againſt the prefeſſours of God in this poore Iland: Againe, God hath ſtirred vp one of the chiefe murtherers of the Saints of God in *Paris,* euen the late King of *France*, to murther the Duke of *Guize,* and the Cardinall his brother, ſpeciall deviſers of that cruell Maſſacre. Then further that mighty God hath ſtirred vp a deſperat papiſticall Frier, to change liues with that bloody King. So that by the ſword, and mutuall blood-ſhed of papiſts among themſelues, the right of the crown of *France* is now fallen into the hands of the king of *Navar* who pretending himſelfe to haue bene a proteſtant, the Church of true proteſtants vnder him, hath thereby had reſt hitherto. And with theſe miraculous accidents hath this Iubilie begun, hoping in God before the end thereof, to heare that whole papiſticall city and kingdome of *Rome* vtterly ruined: For theſe premiſſes were as vnlikely before thoſe three yeares.

(*i*) Becauſe this text calleth this Angell the third Angel: therefore, neceſſarily the Angel preceeding him in the 8. verſe, muſt be the ſecond: And again, the Angel preceding that ſecond in the 6. verſe, muſt be the firſt Angel, although the text adioines to that firſt Angel, the name of *other*, as being another than the Lamb, mentioned before, verſe 1. For the Lamb is euer ſpiritually preſent, and aſsiſtant with his Church: But that other was the firſt meſſenger, that commeth in the laſt age: and this, contained in this 9. verſe, is the third, & Chriſt himſelfe, by his bodily comming to iudgement, ſhall be the fourth. Read more hereof in our 13. propoſition.

(*k*) Whereas in other parts of this book, this marke is but ſimplie called the *beaſts marke*, as doubtfull what it is like, or from whence it is deriued: here ſaith he plainlie, it is the *marke of the beaſts name*: as if thereby he would plainly let vs underſtand, that, that marke is ſome wayes deriued from the name of the beaſt λατεινος, to wit, euen from the three numerall letters thereof χξϛ aſsigned by the text, as more euidentlie doth appeare by the next chapter verſe 2, and note *c*, thereof, where that marke is expreſly called, the marke of the number of the beaſts name, and therfore theſe marks, are the marks of χρ̃ϛ and croſſes of all kindes, as is proued by our 31. propoſition.

(*l*) Some read here, *Bleſſed are they that incontinent hereaſter die in the Lord, for that (ſaith the Spirit) they ſhal reſt from their labours*: but ſeeing certainly, they are alwaies bleſſed that die in the Lord, be it heretofore, or hereaſter, I therefore aſſent rather to them that reade *Bleſſed are they that die in the Lord, for that incontinent hereaſter (ſaith the ſpirit) they ſhall reſt from their labours*, and this varietie commeth by the placing of the point before the adverbe of time, or after the ſame in the Greeke.

(*m*) This cloud is meant literallie, for in a cloud Chriſt paſſed out of his diſciples ſight, and that ſame way (ſaith the Angels, Act. 1.9.10.11) ſhall he returne to judgement: hereof read Math. 24.30 and Mat. 26.64. the like is ſpoken, Revel. 1.7. and note *f*, thereof.

(*n*) This hooke or ſickle mentioned here, and in Ioel 3 13. and mar. 4.29. can mean no other thing, than that earneſt & ſharp diligence, that Chriſt ſhall vſe in gathering in his elect: by compariſon, as ſhearers or reapers with their hookes, gather in their corne, & not that Chriſt ſhall vſe any material inſtrument in this work of the latter day: yea, no other than that, that God did vſe at the firſt creation of all things, which was, *Dixit & factum eſt*, euen the vertue and puiſſance of his word.

(*o*] The preciſe day & houre of Gods great iudgement, being both vnknowne to Chriſt, ſo far as he is man, & alſo to the Angels of heauen, as Chriſt teſtifieth in Marke 13.32. Therefore, this Angel, that is the advertiſer of Chriſt, muſt needs be the holy Spirit the great meſſenger of God, & God himſelf, by whom Chriſt is made fore-ſeene, and priuie to all myſteries, as being conjunct & vnit perſons in one God-hood.

(*p*) This temple that is ſaid here to be in heauen, is by the Rev. 21.22. & by the 20. prop. nothing elſe than the maieſtie of God: from whome the holy Spirit going out figuratiuelie, as it were, after the manner of men, beareth this meſſage of the latter day to Chriſt, & ſo are expreſſed three ſeveral functions, correſpondent to the three perſons of the God-hood. The Father directeth the meſſage of his judgement, the holie Spirit beareth it vnto the Sonne, the Sonne then executeth it. And this orderlie progreſſe is ſet down, but for ſupplie of our weake underſtanding.

(*q*) This Vine is the wicked people of the world, which the Lord hath planted to bring foorth good fruit, but they haue brought foorth euill and corrupt works, of whom ſaith Eſay 5.2. *I waited that they ſhould produce good vines, & they brought foorth ſowre grape*s. That is to ſay, in the ſeuenth verſe thereof: I waited that they ſhould giue indgement, and behold, they wrought iniquitie, and I looked for iuſtice, and behold the complaint.

(*r*) Seeing by the Spirit of God, there is nothing idlely or in vain expreſſed. Theſe whole premiſſes being ſpoken of the latter day, it appears this date of 1600. to be the date of that great day, elſe aparantly, this great preciſe number of 1600. would not haue bene put here more than any other nũber, & this doth chiefly appeare becauſe this number agreeth ſo neere with other propheticall dates of the latter day, falling about the yeare of Chriſt 1699. as at more length is ſhewed in our 14. Propoſition.

CHAP. XV. THE ARGVMENT.

The Spirit of God, intending to repeat the prophecie of the ſeauen ages, now vnder the termes of Vials, which before were expreſſed by ſeuen trumpets, to the effect, the one may be a commentary to diſcipher the other, firſt ſetteth down this chapter as a preface thereunto, containing in it chiefly two godly inſtructions. The one of Gods mercy, that theſe laſt plagues being imminent & ready, do not the more fall out, till God haue firſt eſtabliſhed ſuch in ſecurity, amidſt the raging ſeas, and conſuming fire of worldly perſecution, as refuſing the Antichriſtian errours, reioyce triumphantly in God. The ſecond inſtruction, is of Gods iuſtice, that while as his temple is moſt patent, and his trueth publikly preached in the dayes of the primitiue Church, foorth of the open prædication of the Evangell, proceeds both Gods horrible plagues, vniverſally upon the contẽners thereof. As alſo, it pleaſeth Gods maieſtie, to giue power & permiſsion to the Antichriſt, to obſcure the true contemned light of the Evangell, till the end of theſe plagues.

The Paraphraſticall expoſition.

AND I behelde againe, further of theſe great and heauenly maruels, and conſidered Gods meſſengers of the ſeuen latter plagues readie, whereby in them the wrath of GOD might be fulfilled.

2 Then I perceiued and ſaw, how firſt God would make for his owne, the flowing [*a*] waues, and raging tyranny of the world calme, and firme as glaſſe, though mixed with a litle fire of perſecution, and thoſe that had victoriouſlie attained to the true (*b*) knowledge, which is the antichriſtian empire, who are the counterfait emperors thereof, which is his marke, that comes [*c*] of the number of his name, would he make firmelie and conſtantly to ſtand in all theſe rageing ſtormes, as ſtable and ſure with perfite gladnes,

3 Singing and reioycing with Moſes and the Prophets of God, in the old Teſtament, and with Chriſt and his diſciples of the new Teſtament.

The text.

AND I ſaw another ſigne in heauen, greate and marueilous, ſeuen Angels hauing the ſeuẽ laſt plagues: for by thẽ, is fulfilled the wrath of God.

2 And I ſaw as it were, *a glaſſie ſea mingled with fire and them that had gotten victorie of the beaſt, and of his image & of his marke of the number of his name, ſtand at the glaſſie ſea, hauing the harps of God.*

3 And they ſung the ſong of Moſes the ſeruant of God, and the ſong of the Lambe, ſaying, great and marueilous are thy workes, Lord God Almightie: iuſt and true are *thy waies, King of Saints.*

4 *Who ſhall not feare thee, O Lord, and glorifie thy Name? for thou onely* art *holie, and*

all nations ſhall come and worſhip before thee: for thy iudgements are made manifeſt.

5 Thereaſter I conſidered, and behold the true Temple of God, euen his holie Church was viſible and patent, and the Goſpel publikely preached, among his heauenlie elect of the primitiue Church.

6 And vpon the patencie of that true Church & [*d*] opening vp of the goſpell and true chriſtian religion, there flowed out vpon the diſobedient contemners thereof, the ſeuen great plagues, powred out in the ſeuen laſt ages, by the Angels of God, who are clad in [*e*] puritie and innocencie, and glorioufly girded with iuſtice, power and readines.

7 And *Luke* one of the (*f*) foure evangeliſts (ending here his acts of the Apoſtles) giues place from this foorth, to theſe ſeuen Angels, to proſecute out their prophecie of theſe plagues, which they are to powre foorth of the precious cups of the juſtice of God euerlaſting.

8 And during that time, ſhall the true Church of God, and holy Chriſtian religion be (*g*) obſcured and darkened with Antichriſtian errours, which the majeſtie and power of God hath pleaſed to permit & ſuffer, for the cōtempt of his word, ſo that thereby, there ſhal be no true viſible Church, nor ſincere doctrin publikely opened, that any may haue acceſſe vnto, till theſe (*h*) ſeuen plagues bee powred out, by theſe Meſſengers & Angels of God, againſt the wicked of the world.

5 And after that I looked, and behold, the temple of the Tabernacle of teſtimonie was open in heauen.

6 And the ſeauen Angels cam out of the temple, which had the ſeuen plagues, cloathed in pure & bright linen, and hauing their breaſts girded with golden girdles.

7 And one of the four beaſts gaue vnto the ſeauen Angels, ſeauen golden vials, full of the wrath of God, which liueth for euermore.

8 And the Temple was full of ſmoke from the glory of God, and from his power, and no man was able to enter into the Temple, till the ſeauen plagues of the ſeuen Angels were fulfilled.

Notes, Reaſons, and Amplifications.

(*a*) This glaſſie ſea, is the world, wherein, although the wicked bee drowned ſometime, with the ſtormie waues and troubles, ſometime againe, with the pleaſures & voluptuouſnes thereof: yet doeth God a litle quiet the rage thereof, and make it ſo firme and ſtable to his owne, that though they ſometime ſlide thereon, and be grieued with the fire of perſecution, yet ſhall neither the raging ſtormes thereof, on the one part, nor the alluring pleaſures thereof, vpon the other part, ouer-whelme or drowne them, hereof read the Paraphraſe of the Revel. 4.6.

(*b*) The text here ſpeaketh of victory, and ouercomming the *beaſt, his image, and marke of the number of his name.* Though the ouercomming the beſt, might import a battell or ſtrife: yet it were vnproper to ſay, they ſtroue and ouercame the marke of the number of the beaſts name. This victory then, appeares to meane a ſpirituall ſtrife by knowledge, to diſcerne the trueth and true meaning of theſe, from the wrong meaning, and ſo to ouercome the difficultie of theſe myſteries, that knowing their true meaning, we may victoriouſly follow the trueth.

[*c*] Many hauing miſtaken this text, haue not followed the originall Greeke, but thinking to make the matter more ſenſible, haue turned it to the wrong ſenſe: for ſome haue here interponed the conjunction, *and*, ſaying, of his marke, [*and*] of the number of his name: theſe, by interponing that conjunction, *and*, would meane the beaſts marke to be a ſeueral thing, no waies pertinent to the number of the beaſts name. Others, not finding there that conjunction, *and*, which is *καὶ* in the Greeke, doeth apparantly iudge theſe words [*Of his marke of the number of his name*] to bee put *Appoſitiuè*, as to ſaye, *Iohn an Apoſtle*, and ſo doeth iudge this mark & this number to mean both one thing, & therefore leaues out the one, to wit, the marke, & this is done in the vulgar latine tranſlation in theſe words: *Et eos qui vicerunt beſtiam & imaginem eius & numerum nominis eius &c.* that is, *and they who haue gotten victory of the beaſt, and of his image, and of the number of his name, &c.* but betwixt theſe extremities, whereof the one addeth, and the other paireth the text, we holde the text it ſelfe as the true middeſt, which word by word according to the greeke originall & as both T. Beza in the Latine, and we here in the engliſh haue expounded the ſame, doeth import not the victory of the beaſts marke, and of the number of his name, but of the beaſts marke of the number of his name, *καὶ ἐκ τοῦ χαράγματος αὺτου, ἐκ τοῦ ἀριθμου τοῦ ὀνὸματος αὐτου*: inferring expreſſely thereby that the beaſts mark comes of the number of his name & therfore, haue we cited this text in our 31. propoſitiõ, as a ſure ground & authoritie to trie this mark by. Reade further hereof in the ſaid prop.

(*d*) That this Temple meanes Gods holy Church and whole contents thereof, euen his very trueth & true religion amongſt his holy and heauenly elect vpon earth, is ſhewed in our 20. propoſition. And here is to be noted, that euer the more patently that this temple is opened, and the more openly that Gods truth is publiſhed, the greater euer are the plagues that floweth out from thence, if it be not duly imbraced, and reuerently accepted' as experience both here, after the daies of the primitiue Church, and alſo in all ages proueth, and ſhal, without all queſtion, be alſo prooued vpon vs in this preſent age, except more ſpeedy repentance come nor yet appeareth.

(*e*) That fine linnen, is puritie, innocencie and iuſtification, read the Reuel. 19.8. and how this girding is a ſigne of readines, read Luke 12.35. and Luke 17.8. and ſpecially meaneth readines to iuſtice, as in Eſay. 11.5. *Iuſtitia cingulum lumborum eius, Iuſtice ſhall be the girdle of his loynes*, &c. and that with great power, as Pſal. 17.33 40. So that the girdle of theſe properties, for their excellencie, is ſaid here to be of golde: that is to ſay, precious and glorious, as is ſaid, Apoc. 1.13. note (*n*) thereof.

(*f*) Where it is ſaid here, that this beaſt or Evangeliſt, gaue the ſeuen Vials to theſe ſeuen Angels; it is not literallie ſo to be takẽ, but after a propheticall and figuratiue maner of ſpeach: as if hee would ſay, that, that Evangeliſt *Luke* remitted, & gaue ouer the writing of the reſt of his Eccleſiaſtical hiſtorie, and entred to, theſe ſeuen Angels to prophecie out, where he left off: The like phraſe is commonly vſed, in ſaying, Thou haſt deliuered him thy penne: that is, Thou haſt giuen place to him to write the reſt. Here then note, that if either the foure beaſts were wrong interpreted, to be the foure Evangeliſts, or yet the ſeuen trumpets, or ſeuen vials were not both one, or at the leaſt, beganne not about the yeare of God, 71. and ſo our 2. 6. and 19. propoſitions were impugned, then would there ſome repugnance appeare here: but in that there is no repugnance here (for the Actes of the Apoſtles end at the remaining of *Paule* captiue two yeares in *Rome*, in free-ward, where ſhortly after he ſuffred death, in the end of *Nero* his raign, which approacheth within a yeare or two to the ſaid 71. yeare of Chriſt). Therefore, the ſaid propoſitions are hereby confirmed & vndoubtedlie approoued.

(*g*) Some doe tranſlate this to be the ſmoke and reeke of Gods maieſtie, & of his power, that hindereth vs to enter into his Temple, which is not probable, but rather, as *Ierome* & other learned tranſlate it, a ſmoke, not of Gods maieſtie, nor of his power, but from Gods maieſtie, and from his power, as being the ſmoke of Antichriſtian errours, which neuertheleſſe floweth from the permiſsion and power of Gods maieſtie, for the puniſhment of the contemners of his trueth. The like is ſaid in the Lords prayer, *Lord, lead vs not into temptation*. and Amos 3.6. *Is there any euil in the citie, which the Lord hath not done*: not that the Lord tempteth vs, or doth any euil in the citie directly, but only, that, that temptation and euil-doing doeth flow from the permiſsion and ſufferance of his maieſtie, and his almightie power, as permitting, in his iuſt wrath, his ſlaues, the devil, the world, and the fleſh, to ſtir vp all euill in vs: And ſo, this ſmoke that hindereth vs from entering into Gods Temple, is not his ſmoke, but the ſmoke of Antichriſtian errours, which neuertheles floweth from the permiſsion of Gods maieſtie and power.

(*h*) This abiding of Gods Temple ſo long latent and obſcure, is moſt certaine: For, during the moſt part of the firſt age, by diuers heretikes, forerunners of the great Antichriſt: & during the ſecond, third, fourth, fift, ſixt, and beginning of the ſeuenth ages, by the Romane Antichriſt himſelfe, the true Temple of God, and light of the Goſpell was obſcured, till by the ſeuenth plague, in this ſeuenth & laſt age, his kingdome began to fall, ſo that henceforth to the worlds end, ſhall Gods true Church and holie religion be patent, open, and publikelie knowne. Hereof read the paraphraſe of chap. 11.19. and note (*r*) thereof, with the diſcourſe of our 20. propoſition.

CHAP. 16. THE ARGVMENT.

Hauing præmitted the former chapter, as a preface to the prophecie of the ſeauen ages, nowe to bee repeated vnder the tearmes of vials. In this chapter hee proſecutes out, in due order of time, the plagues of the whole ſeuen vials, correſpondent and agreable in purpoſe, ſenſe, and almoſt in tearmes with the ſeuen trumpets before expreſſed, whereby moſt ſurely, the one may be vſed as a commentary to the other.

Historical application.

AT the appointment of God, there are ſeuen ages to the worlds end, in euery one of the which, there are ſeueral plagues poured foorth, for the iniquities of the world.

2 In the firſt age, from the yeare of God, 71. to the 316. yeare, the Empire of *Rome* and *Roman* Monarchie beganne to decay: For, the ſubiects of that Empire, and all who obeyed the degenerate Princes thereof, were ſometime plagued and oppreſſed by their tyrannie, and by the tyrannie of the thirty tyrants, that then raigned vnder thẽ: Others, ſomtime were oner-run

Anno Chriſti. 71 *The firſt Viall or age*

Paraphraſtical expoſition.

AND I heard the mightie decree of God, directed from the Temple of his trueth, to his miniſtring Spirits of the ſeuen ages, cõmanding them to paſſe, and powre out the cuppes of his wrath of theſe ſeuen ages, vpon the [*a*] people of the earth.

2 And his meſſengers of the firſt age paſſed, and powred out the cups of God his wrath vpon the great Empire of the earth, & there came a cruell and horrible plague vpon al people

The text.

ANd I heard a great voice out of the temple, ſaying to the ſeuen *Angels, Goe your waies, and poure out the* ſeuen *vialles of the wrath of God, vpon the earth.*

2 And the firſt went, and powred out his vial vpon the earth, and there fell a noyſome and a grieuous ſore vpon the men, which had the marke of

by ftrangers, through the effeminate lenitie of thefe Emperours.

3 Wherethrough, and becaufe in the feconde and next age, beginning about the yeare of God, 316. the Emperour *Conftantine* tranfported the imperiall feate to *Conftantinople*, it came to paffe, that the towne of *Rome*, and whole peninfull of *Cittim*, which is (*ex Hieronymo de interpretat. nominum*) interpreted the gathering or tofsing of the feas, and nowe is called *Italie*, is left a praie to the *Huns, Gothes*, *Vandales*, and other barbarous nations, who flewe and deftroyed of the inhabitants of that countrey at their plefure.

4 Then begins in the yeare of God, 561. the third age, in the which, the wicked *Mahomet* arofe & taught the damnable doctrine of his *Alkaron*, among the Chriftian Churches of the *Orient*, where-through thefe Eafterly countries became altogether corrupt with his bloodie herefie.

5 Herein doeth God his iuftice and moft iuft iudgement appeare: for that men had then contemned the true doctrine.

6 And fhed the blood of Gods feruants, therefore

316 *The fecond Viall*

561 *The third Viall*

that profeffed obedience to it, and vpon them that reverenced that Empire, & the degenerat Princes therof.

3 And the meffengers of the fecond age, powred foorth their cups of God his wrath vpon a certaine (*b*) fea coaft, peninful or land, named by the fea: & that land became full of blood-fhed and murther: and all the people that fometimes liued wealthilie in that Peninful, were (for the moft part) flain and deftroyed.

4 And the meffengers of the third age powred out their cups of Gods wrath (*c*) vpon the wholfom flouds aud liuelie fountains of pure doctrin, and the doctrine became corrupt, tending al to (*d*) bloodfhed and murther.

5 And I perceiued the Meffengers and Minifters of the true doctrin extolling Gods iuftice, faying, Thou Lord, who is, and euer was holy, art iuft, and haft iudged righteoufly in this caufe.

6 For thefe people haue fhead the blood

the beaft, & vpon thẽ which worfhipped his image.

3 And the fecond Angell poured out his viall vpon the fea, and it became as the blood of a dead man: and euerie liuing thing died in the fea.

4 And the thirde Angell powred out his viall vpon the riuers & fountaines of waters and they became blood.

5 And J hearde the Angell of the waters fay, Lord, thou art iuft which art, & which waft, & holy, becaufe thou haft iudged thefe thinges.

6 For they fhed

did God fende among them this *Mahometicke* doctrine, which cruellie exhorteth men to make war and bloodfhed, & falfly promifeth the kingdome of heauen to all thofe, that valiantly fight in their quarrels.

7 And this, all that contemplate the myfteries of Gods iuftice, vnderftande to bee iuftlie deferued at the hands of the almightie, who is true and righteous in all his iudgements.

8 Then beginneth in Anno 806. the fourth age, about the which time, the Spiritualitie & Papiftical Cleargie, induced many contentious arguments, and fophifticall reafonings, as the captious diftinctions of *δουλεια* & *λατρεια* excufing their idolatrie, and new erection of Imagerie, eftablifhed lately before, in the feuenth councell of *Nice* againft certaine Godlie Emperours of *Greece*, who had demolifhed Imagerie: Alfo, the curious fophifmes and contentious arguments of tranfubftantiatiõ to excufe their *αρτολατρειὰ* and adoration of the bread.

9 And with thefe, and fuch like fires of contentions, they kindled vp men in ftrife, fmothering the trueth, and blafpheming the name of God, by calling it God, that was not God, and perfeuered in thefe errours, without repenting, or glorifying of the name of

806

The fourth Viall

of thy holie Saints and Minifters, & therefore haft thou fuffered them fpirituallie to drinke bloody doctrine, exhorting blood-fhed, for they were no better worthie.

7 And I cõfidered frõ the Sanctuary, euen from among them, that entered into the fecret meditatiõ of Gods myfteries, the fame alfo confeffed and approued, faying, O Lord God Almighty, true and righteous are thy iudgements.

8 And Gods Meffengers of the fourth age, powred out the cup of Gods wrath, againft the [*e*] fpirituall Preachers: & God fuffred them to ftir vp arrogant opinions, and to vex & grieue men with their fophifmes, and furious fire of their hot contentions.

9 And men were marueiloufly chafed, and puft vp with [*f*] contentious arguments, defending blafphemies againft the Name of God, who only had power to reveale the trueth, and to remoue thefe plagues of bitter contention, neither

the blood of the Saints and prophets, & therefore haft thou giuẽ them blood to drinke: for they are worthie.

7 And J hearde another out of, the Sanctuarie fay, euen fo Lord God Almightie true and righteous are thy iudgements.

8 And the fourth Angell poured out his viall on the Sun, & it was giuen vnto him to torment men with heate of fire.

9 And men boyled in great heat, and blafphemed the name of God, which hath power ouer thefe plagues, and they repented not to giue him glorie.

the true & almightie God.

10 Then the fift age being begun in An. 1051. That yeare *Zadok* was made firſt Dominatour of the *Turkes*, whoſe dominion aroſe dailie ſo gret, & ſo diminiſhed the *Roman* Empire, that they of the Empire are greatly aſtoniſhed and grieued.

11 And yet continue they ſtill in their former impieties, idolatries and ſchiſmes, without repenting the ſame, or crauing pardon of God for their wicked workes.

12 Afterwarde, begins the ſixt age, in *An.* 1296. About this time, euen in *An.* 1300. *Ottoman* being crowned the firſt *Mahometan* Emperour, it pleaſed God in his wrath, to ſtirre vp from that time foorth the foure nations Mahometans, *Turks, Tartarians, Saracens*, and *Arabians*, that dwel beyonde and about *Euphrates*, with the ſaid *Ottoman* and his vnder-kings of the Eaſt countries, to come ouer *Euphrates*, and raiſe warre againſt them of the whol Roman Empire.

13 At this time, the Devil ſtirres vppe falſe teachers on both the ſides, euen the Legates of the Pope and Roman ſeat on the one ſide, and the doctours of the Mahometick hereſies,

1051 repented they therof, that they might glorifie God.

10 And the meſſengers of the fift age, powred out their cups of God his wrath, againſt the Antichriſtian & imperial thron, and other more [*g*] puiſſant

The fift Viall.

1296 Conquerors, obſcured and ſtayned their monarchy, ſo that they orped & gnaſhed their tongues for ſorrow.

11 And yet inſiſted, blaſpheming the God of heauẽ euen the more becauſe of theſe plagues and griefes, and repented not thẽſelues of their wicked works.

12 And the meſſengers of the ſixt age, powred out their cups of Gods wrath, toward the people that remained about *Euphrates*, that theſe Kings and Princes of the *Eaſt*, may make ready way ouer *Euphrates*, & take journey againſt the people of the *Weſt*.

13 So I conſidered out of the deuiliſh [*h*] mouths of the great Empire, and of the falſe Prophet, to proceed three ſpeciall exhortations & wicked entiſements, as

10 And the fift Angel powred out his viall vpon the throne of the beaſt, and his kingdom waxed darke, and they gnew their tongues for ſorrow.

11 And blaſphemed the God of heauen for their paines and for their ſores, and repented not of their works.

12 And the ſixt Angel powred out his viall vpõ the great riuer Euphrates and the water therof dryed vp, that the way of the kings of the Eaſt ſhould bee prepared.

13 And I ſawe three vncleane ſpirites, like frogs, come out of the mouth of the Dragon, on the other ſide, to ſeduce and ſtirre vp their ſubiectes of the whole worlde, three waies: to wit, in moſt furious wrath and cruelty, the one againſt the other. Secondly, to ſtand obſtinately in their hereticall doctrine. Thirdly, to be illuded with vaine hope, to obtaine avaritiouſlie, both the whole kingdomes of the worlde, and alſo the kingdom of heauen hereafter, if they fight valiantly in theſe their holie warres, as they call them.

14 And by theſe deuiliſh intiſementes, confirmed ſome-times with lying wonders, they ſtirred vp all the princes of the world by Eaſt, and about *Euphrates*, againſt thoſe that are by Weſt, in continual warfare all this ſixt age.

15 So hath the Almightie brought this great & ſudden vengeance vpon the Papiſticall Chriſtians for their defection. Bleſſed are they that abide conſtant, for God ſhal deliuer them from ſhame & confuſion.

16 And this wicked multitude of the whol world, euen the Roman
1541 & Papiſtical Princes frõ the weſt, and *Mahometick* Princes frõ the eaſt, conveened, & fought many battels in Chriſtendom, both in *Aſia* where S. *Iohn* had ſent this revelation vnto the ſeuen Chriſtian Churches thereof, and in

The ſixt age or Viall,

vile vermin, or frogs, creeping from their pool,

14 Theſe be [*i*] deuiliſh inſpirations, entiſing men by lying wonders and entered into the hearts of all the Princes of the world, to entiſe & ſtirre them vp to make warre together, in this time of the horrible vengeance of Almighty God.

15 Behold (ſaith the lord) I come at vnawares to puniſh: bleſſed is hee that abideth, praying, and watching from ſin, and keepeth his clothings of righteouſneſſe, leaſte hee beeing ſtripped thereof, al his horrible iniquities appeare diſcouered, to his eternal ſhame and confuſion.

16 And theſe people ſhal gather themſelues together, to fight in the [*k*] mountaine of gods Evangel, euẽ in the lande of his choſẽ fruit, & Chriſtian elect people.

17 And the Angels of the ſeventh age, powred forth their vials of gods wrath, againſt the volving ayre, euen the changeable eſtate of things, & there paſſed out a mighty decree &

and out of the mouthe of the beaſt, and out of the mouth of the falſe prophets.

14 For they are the Spirits of Deuils, working miracles, to goe vnto the Kinges of the earth, and of the whol world, to gather them to the battell of that great daye of God almightie.

15 (Behold, I come as a thiefe. Bleſſed is hee that watcheth and keepeth his garments, leaſt hee, walke naked, & men ſee his filthineſſe.)

16 And they gathered themſelues together into a place, called in Hebrew Armagedon.

17 And the ſeuenth Angell powred out his viall into the aire, and there ſuch parts alſo of Europe, where other Apoſtles had preached.

17 Laſt of all, beginnes the ſeuenth age, in *Anno Chriſti*, 1541. in the which age the world ſhal ende.

18 In the meane time there ſhall be great and tempeſtuous troubles of all ſortes, and ſpeciallie, the great Romane Empire, and Papiſticall kingdome ſhall (GOD willing) quite be ouerthrowne, and the Monarchie thereof ſo vtterlie deſtroyed, as neuer Monarchie was, ſince the beginning of the world.

19 At this time the whol people that remain in Chriſtendome, ſhal either profeſſe Proteſtantrie, Papiſtrie, or a pretended Chriſtian neutrality: for as touching the publike Profeſsion of any Ethnick religion, there ſhall be none within Chriſtendome: then ſhal that Papiſticall ſeat and citie of *Rome*, in Gods wrath, and iuſt iudgement bee deſtroyed.

20 *England*, *Scotland*, and other Ilands ſhal revolt from her, and leaue her Papiſtrie: her mighty & confederat kingdomes of

The ſeuenth Viall, and laſt age.

ordinance from Gods throne, and Temple of his trueth among his faithfull, ſaying (*l*) al is done, euen all mortal things, in this age ſhal end

18 And vntill that end, there ſhal be great ſhewes, and foretokens of tempeſtuous troubles, rumours of warres, and Gods iuſt iudgements thundered out, & ſuch a commotion ſhall come in the great Empire of the earth, (which ſeemeth more ſtable than the earth) that there came [*m*] neuer ſo great a commotion thereof, ſince firſt men wer vpon earth.

19 And the whole people that [*n*] dwelt in Chriſtendom, became of three ſorts of opinions, and all the Ethnick religions were aboliſhed, & then the great (*o*) Antichriſtian city, ſucceſſour of *Babel*, came in remembrance before the Almightie, to giue her now to drinke in his iuſtice, the cup of his wrath, and bitter indignation.

20 And euerie Iland revolted frõ her, & her mighty kingdomes

came a loude voice, out of the Temple of heauen from the throne, ſaying, Jt is done.

18 And there were voices, and thundrings, & lightnings, & there was a greate earth-quake, ſuch as was not ſince men were vpon the earth, euen ſo mightie an earth-quake.

19 And the great city was divided into three parts, and the cities of the nations fell: & great Babylon came in remembrance before God, to giue vnto her the cup of the wine of the fiercenes of his wrath.

20 And euery Ile fled away, and the

Germanie, *France*, *Spain*, and others: ſhall either be ruined and diſſolued themſelues, or then ſhall diſſolue their league with her.

21 In this time, zeale to God and charitie to our nighbors ſhal grow cold, and ſelfe loue ſhall abound among all people, whereby the moſt part ſhall neglect & deſpiſe God, and follow their auarice, ſelfe loue, and partial affections for vndoubtedly, to the worlds end, theſe vices ſhall euer encreaſe.

of maine land were diſſolued.

21 That time the (*p*) cold & tempeſtuous hail of ſelfeloue and lacke of zeale and charity, fell downe as from on high, and in abundance among the people of the world, ſo that they deſpiſed & blaſphemed God throgh their coldnes of charity & zeal, for that plague of cold zeale doeth euery-where increaſe.

mountains wer not found.

21 And ther fell a great haile like talents out of heauen vpon the men, & men blaſphemed God becauſe of the plague of the haile: for the plague thereof was exceeding great.

Notes, Reaſons, and Amplifications.

(*a*) The earth, in al the Revelation, is not only taken for the people of the earth in generall, but for the Antichriſtian & vniverſal Empire of the earth, euen the *Romane* Monarchie in ſpecial, as appeareth here plainly by theſe ſeuen plagues, which here are ſaide to be powred foorth vpon the earth, and fell indeed againſt the *Romane* Empire, & chieflie, where he ſaith, the firſt Vial was powred forth vpon the earth, he ſubjoyneth incontinent, *and there fel a noyſome and grieuous plague on them that did beare the mark* of that Antichriſtian beaſt or Empire: as meaning by the falling of that plague on the earth, the falling therof on the empire of the earth and conſequentlie, the earth meaneth the Empire of the earth, or *Romane* Monarchie: and therfore, earthquake doth euer mean commotions of the Empire, both actiuelie and paſsiuelie, as you ſhall find through all our Paraphraſe, to agree with the hiſtorie.

(*b*] Why iuſtly we interpret the ſea to meane *Italie*, reade the note *d*, Apoc. 8.

[*c*] That fountains & floods mean doctrin, read the note g, Ap. 8

(*d*) This doctrin is iuſtly called bloudy, ſeeing the *Mahomet* preched and proclaimed warre againſt all men, and promiſed the Kingdome of heauen to them, that fight valiantly in his cauſes.

(*e*] That the Sunne for the moſt part, meaneth the ſpirituall eſtate, read Apoc. 6. note *h*, thereof.

(*f* This heat that the Cleargie, at this time, both chafed themſelues, and others with, agrees verie wel, by hiſtories of that time to be both the contentious arguments, & opinions, conteined in our hiſtoricall application, wherby they troubled & vexed Chriſtians, as alſo the pride and ire of the Pope & his Clergie, which at this time roſe ſo hie, that they by the hand of *Charlemaigne*, bereft the *Exarchat*, frõ the Emperor of the *Orient*, & the kingdom of *Lombardie*, frõ the *Lombards*, and do brook a great part of both to this day, & then, & euer ſince, haue ſtirred vp the world in ſtrife

(*g*) This plague, that here makes againſt the *Romane* ſeat, in this fift Vial, proueth the plague of the fift trumpet, not to meane the Pope [as ſome beleue] but rather the Turke, as in our 3. propoſition is proued more at length.

[*h*] Theſe three vncleane Spirits, that here are ſaid to proceed foorth of the mouthes of the Deuill, of the Antichriſtian beaſt, and of the falſe Prophet, are called (Apoc. 9.17.18.) three plagues of fire, ſmoke, and brimſtone, that proceeded out of the mouthes of the Mahometanes: where-through of neceſsitie, they muſt meane ſuch three common heades of doctrine, as both the Pope and the Mahomet doe ſeduce their ſubiects with. Such as, firſt, their ire and tyrannie: for, the Pope proclaimed havocke of all Mahometick blood. So do the Mahometans with Chriſtians. Secondly, hereſie: For the Pope preaſeth by all meanes poſsible to ſpread his papiſticall hereſie vniverſally: ſo doe the *Mahometans* their wicked hereſie of their *Alkaron*. Thirdly, bitter & ſulphurious avarice: For either of thẽ aſpired to be Monarch of the whol world. The Pope charging the Empire, and his vnder-kings to fight his holy wars, (as he tearmeth them) for the augmenting of S. *Peters* patrimony. The *Mahomet* charging his *Saracens* to take vp the whole kingdoms of the world, as promiſed to thẽ by God (Gen. 13.16,) as being the ſonnes of *Abrahã*, euẽ *Agarens* indeed, and *Saracens* (as they think) by adoption, & therefore, promiſing the kingdome of heauen to thẽ that fight valiantly in that quarrel: and ſo for concluſion, theſe deuiliſh plagues, and inſpirations of ire, hereſie, and avarice, do ſeeme to be the three plagues and deuilith intiſements, that here both mooueth Papiſticall & Mahometicke kings to this great bloodſhed and war-fare.

(*i*) As God is in the trueth, and the trueth in him, and he is the truth (Ioh. 14.6.) So (*â contrariis*) the Deuil is in deceit, and deceit in him, and he is deceit, and that ſo inſeparablie, that where the one is, there alſo is the other. Whereupon, theſe wicked and deceitful intiſements and three inſpirations of the Deuil, are iuſtlie called by the text, *ſpirits of Deuils*, becauſe the Deuill goeth with them, and they with him coniunctly, in token of the which his perſonal preſence with them, hee confirmeth them oftentimes with lying wonders, and deceitful miracles, as ſaith the Text. Of theſe falſe Papiſtical miracles, and *Mahometical enthouſiaſmes*, diuers are to be read in their hiſtories.

(*k*) That *Armageddon* ſignifieth the mountaine of the Euangel, or of the elect fruit, read *Hieromede interpretatione uominum*. There is no reaſon here to read *Geddon arma*, as ſome do, vnder pretẽce of the Hebrew maner of reading backward: for if the Hebrew maner were here to be obſerued, not only would the laſt word be firſt read, but euen likewiſe the laſt letters firſt, and euery letter backward in his owne order, making *Noddegamra*, and not *geddon arma*.

(*l*) Of this conſummation of all things, read the note (*q*) Apoc. 11.

(*m*) There came never ſo great a change among the wicked Monarchies of the earth: for euer as one was deſtroyed, another as wicked aroſe, but now ſhal al wicked kingdomes be diſſolued, and ſhal giue place vnto the eternal kingdome of Chriſt.

(*n*) This great citie, hauing neither the epithet of Holie, nor contrarilie, any Antichriſtian epithet ioyned to it, and yet apparantly by the text, including no profeſſed Gentiliſme, muſt therefore, (as appeares) meane Chriſtendome, which neither is altogether Gods holie citie and ſpiritual Hieruſalem, neither yet altogether Antichriſtian, but partly of both: but hauing no publike Gentiliſme profeſſed in it: for there is no religion publikelie profeſſed in it, but Chriſtian: yet by the craft of the deuil, the ſame is rent in peces, euẽ in three factions: one being Proteſtãts, another Papiſts, the third, ignorãt Neutrals, alike to both: who al call themſelues Chriſtians: But of theſe three, the Papiſtical ſeate being the author both of their owne wilfull errours, as alſo of the ignorant peoples Neutralitie and doubtfulnes of faith, ſhall therefore (in the next verſe) goe to ruine.

(*o*) That this Babylon, that here is mentioned, is Rome, is prooued by our 23. propoſition: by destruction of the which Romane ſeate, here prophecied, occaſion of perſeverance is miniſtred to the proteſtants, penitence to the Papiſts, and of perfect certaintie to the doubtfull and wauering Neutrall: which becauſe theſe do not the more hereof acknowledge, & thankfullie accept, but for the moſt part, become all cold as haill (as in the next verſe is ſaid) therefore doth the Lord ſhorten the latter dayes, for finall end of all iniquitie.

(p) Of this cold and tempeſtuous haile of ſelfe-loue, & lack of zeale & charitie, read the note *ſ*, Apoc. 11. Here note, that profeſſed Chriſtians, beeing of three ſorts, Proteſtants, Papiſts, & Neutrals: the Papiſts being the Authors, both of their owne wilfull errours, and of the Neutrals ignorant errours, are therfore in the former verſe, firſt puniſhed: but becauſe, neuertheles, theſe ignorant Neutrals do encreaſe & augment, daylie in ſelf-loue, & their zeale and charitie euer freezeth colder & colder to the worlds end, therefore, ſhal the Lord ſhorten the latter day, as appeareth by the teſtimonies cited in the ſaid note *ſ*, Apoc. 11.

CHAP. XVII. THE ARGVMENT.

The moſt ſpecial intention of the Spirit of God, in this holy worke, being couertly to diſcouer the Antichriſt (couertlie I ſay, leaſt before the time, and to Antichriſtian perſons it being known, had miniſtred occaſion to their tyrannie, through care of their owne ſtanding, to haue deſtroied this whole Revelation) now, vnder this couerture, and locking vp of his ſecrets, God hath bene ſo beneficiall to his own, that he prouideth the means, where by this moſt needfull diſcouerie of the Antichriſt, ſhal be in due time made patent to them: among the which means, this chapter is the key and chiefe: For here, in expreſſe tearms, the Spirit of God hath interpreted the great Antichriſtian and Idolatrous whoore, to be that citie, that is ſet as Miſtres ouer the kingdomes of the earth: and the ſeuen heads of the beaſt, whereon ſhe ſitteth to be her ſeuen mountaines, whereby (beſide many other tokens herein contained) that imperiall and ſeuen-hilled citie of Rome is moſt liuelie expreſſed, and only deſigned, and all true Chriſtians thereby certified, that in it muſt the Antichriſt ſet his ſeat and dwelling.

Hiſtoricall application.

NOw, the laſt of the ſeuen ages being come, the Spirit of God letteth vs evidentlie ſee the deſtruction approaching of the great AntiChriſtian citie, and Idolatrous Empire of *Rome*, which hitherto hath empired ouer all nations.

2 And with whome all the princes of the earth haue bene entiſed to goe ſpiritually a whoring, both of olde, after the Pagan Idols of *Mars, Minerua*, lady *Fortune*, and others, as alſo of late, after the images of our Ladie, the Crucifixe, *S. Paul*, S. *Helen*, S. *Margaret*, S. *Sylveſter*

Paraph. expoſition.

AND there cam, & communed with mee, the [*a*] laſt of the Angels, which did beare the plagues of the ſeuen latter ages, ſaying, Approch, & I ſhal ſhew vnto thee, the deſtruction of the great Antichriſtiã city [*b*] and idolatrous whore, who impires over many [*c*] nations.

2 And with whom the kings of the earth haue bin entiſed to cõmit idolatrie, and all that dwell vnder her earthly empire, are drunken with

The text.

THẽ there came one of the ſeuẽ angels, which had the ſeuen vials, & talked with me, Come, I wil ſhew thee the damnation of the great whore that ſitteth vpon many waters.

2 With whom haue committed fornication the kings of the earth, and the inhabitants of the earth are

Pope, and infinit moe papiſticall idols, and haue cauſed all people to dote with theſe fond ſuperſtitious errors.

3 And thoſe, whoſe ſpirit God indued with true iudgement, may perceaue, howe that citie of Rome hath broked a bloody Empire, and hath vſurped proude, glorious, and blaſphemous Antichriſtian titles, calling her ſelfe of olde, *Romæ æterna, fœlix, invicta*, &c. and her Empire *ſacroſanctum & perpetuum;* and of new, *S. Peters chair, the Apoſtolike ſeate, & head citie of the Church.* Shee is builded vpon ſeven hils, and hath bene governed ſuccefsiuely, by ſeven princely gouernments, & aſsifted by the ten Chriſtiã kings, her confederats, all hereaſter nominated.

4 And this imperial citie inioyeth all the pompe & riches of the worlde: her pillers are of marble, her tapeſtries of ſilkes, her ſilerings overgilt with gold, her Cardinals and rulers clothed in fine skarlet, red & purple: their temples & idols decored with gold, pearles, & precious ſtones: for all the worlde payes tribute vnto her, but ſhe communicated nothing to thẽ again, but her poiſonable doctrin, and vile allurements of idolatrie.

5 And though her common name be *Rome*,

the (*d*) doctrin of her abominable errours.

3 Then rauiſhed he my ſpirit in (*e*) ſolitarie contemplations: and I perceiued, vnder the figure of a [*f*] womã a citie, ſet ouer a cruell & (*g*) bloodthirſtie Monarchie, full of glorious and blaſphemous (*h*) ſtyles, which is ſituate vpon ſeuen (*i*) mountaines, and ruled by ſeuen princelie (*k*) gouernements, vnder the figure of ſeuen heads, and afterwarde divided into (*l*) tẽ vnder-kings vnder the figure of ten hornes.

4 And this whoriſh citie, was clade with all princelie (*m*) riches of purple, skarlet, gold, precious ſtones, and pearle, exhibiting and propining (as out of her hande) to the whole world the venoumed drinks of her poyſonable doctrine & idolatrous hereſie,

5 And ſhe bare the title & ſtile of (*n*) myſtical

drunken with the wine of her fornication.

3 So he caried me away in to the wildernes in the Spirit, and I ſaw a woman ſit vpon a skarlet coloured beaſt, full of names of blaſphemy, which had ſeuẽ heads & ten horns.

4 And the woman was arayed in purple & skarlet, & gilded with gold & precious ſtones, and pearles, & had a cup of golde in her hand full of abominations, and filthineſſe of her fornication.

5 And in her fore-head, was *a name written A myſterie,*

yet is ſhe often ſtiled glorioufly, by her Clarkes, after her grand-mother, *Babylon*, whoſe Empire ſhee brooketh, & ſo is as mother of all vice, & fountaine of all filthines.

6 And we may daylie ſee, how that cruel citie hath ſupped the blood of many thouſands of Gods martyrs, ſo that it is wonderful to behold, how ſhee doth play the Antichriſt in al her tokens & behauiour.

7 Which the Spirit of God, in theſe latter dayes letteth vs evidẽtly vnderſtãd, to agree in all points with the Antichriſtian citie, Empire, ſeuen mountaines, ſeuen princelie governments, and finally with the ten vnder kings thereof.

8 For ſtood not that Empire or Monarchie of Rome, being gouerned by Emperours, before S. *Iohn* ſawe theſe viſions? Did it not vake by the deceaſe of *Domitian*, whẽ hee ſawe theſe viſions? Roſe not the next Emperour *Nerva Cocceius*, not of the ancient blood, or honourable ſtock of the Romans, but of baſe linage? As teſtifieth *Aurelius Victor, in deſcriptione Nervæ*, euen hee was the firſt ſtranger or barbarian Emperour: and then died hee not ſhortly? to wit,

Babylõ meaning literally her ſucceſſour the mother of al ſpiritual whoredome, and abominable idolatrie of the world.

6 And I ſaw this city, as drunken with the blood of Gods ſeruants, & martyrs of Ieſus Chriſt, and I (*o*) wondered marveiloufly with great admiration, when I conſidered her.

7 Then ſaid the Angell vnto mee, why wondereſt thou? for I ſhall make plaine vnto thee, & to the whol church in thee, the perfect interpretation, what is meaned by that woman, and by that imperial beaſt or Monarchie, which vpholdeth her, who hath the ſeuen heads and ten hornes.

8 That imperiall [*p*] beaſt that thou didſt ſee, was that Monarchie which hitherto raigned, & now vaketh through the deceaſe of the Emperor thereof, and immediatly ſhall the next Emperor therof ariſe of low degree, & baſe linage to that Monarchie, & ſhortlie thereaſter he ſhall die, & the people of the earth,

great Babylon the mother of whordomes, & abominations of the earth.

6 And I ſaw the woman drunken with the blood of Saints, & with the blood of the martyrs of Jeſus: & when I ſaw her, J wondered with a great maruel.

7 Then the Angel ſaid vnto me, Wherefore marueileſt thou? I will ſhewe thee the myſterie of the woman, and of the beaſte that beareth her, which hath ſeuen heades, and ten hornes.

8 The beaſt that thou haſt ſeene, was, and is not, and ſhall aſcend out of the bottomles pit, & ſhal goe into perdition, and they that dwel on the earth ſhall wonder (whoſe nams are not written in the booke of life, from the foundatiõ of the worlde) when after he had raigned onely one yeare, three moneths, and nine daies: and finally, doth not the wicked worlde haue in great reuerence and admiration thefe Emperours, whofe eftate (as is faid) ftood, before Saint *Iohn* fawe thefe vifions, and though by deceafe of *Domitian*, it vaked whẽ thefe vifions appeared, yet in effect ftood it ftill vnabolifhed, *Nervæ* and others ftill fucceeding.

9 And befide thefe infallible tokenes, doeth not the reft of the greater circũftances alfo more plainly agree: for is not *Rome* fituate vpon the feuẽ hils of *Palatinus*, *Cœlius*, *Tarpeus* (otherwife called *Capitolinus*) *Auentinus*, *Quirinalis*, *Vinimalis* & *Efquilinus* & hath not that citie & empire therof, bene gouerned by thefe feuen princely gouernementes *Kings, Confulles, Dictators Decem-virs, Tribunes, Emperours*, and *Popes*.

10 Whereof the gouernement of thefe fiue, to wit, *Kinges*, *Confuls*, *Dictators*, *Decem-virs, and Tribunes*, were paft before S. *Iohns* daies. The fixt which is the ftate of Emperors gouerned in his daies, & the feuenth ftate, which is of Popes came not while after his daies: and fince they came they haue gouerned 1260. yeares, a time but fhorte in Gods fight, yea fhort in refpect of the

euen they whofe names are not frõ the beginning of the world, regiftred into the booke of life, fhall haue in great (*q*) admiration, reverence & eftimation: this great monarchie, that hitherto ftood, now vakes, and (*r*) yet doeth ftand vnabolifhed, for that moe Emperors thereof do inftantly fucceed.

9 And now followeth the mind, and meaning of the Spirit of God, wherein confiftes the knowledge & interpretatiõ hereof. Thefe feuen heads which appeared to thee, are certainly to be interpreted feuen mountains, where upon that whorifh citie is fet, & they do alfo meane the feuẽ princely (*ſ*) governmẽts thereof.

10 Whereof fiue are paft, & the fixt therof now ftands & the feuenth governement is not yet come, & when it fhall come, it muft rule a [*t*] certain fpace.

they beholde the beaſt that was, & is not, & yet is

9 Here is the minde, that hath wiſdome. The ſeuẽ heads ar ſeuen mountaines, whereon the woman ſitteth: they are alſo ſeuen kings.

10 Fiue are fallen, and one is, and another is not yet come: and when hee commeth, hee muſt continue a ſhort ſpace.

contrary eternall kingdome of Gods Elect.

11 And ſo that imperiall ſtate, that reigned before S. *Iohn* wrote, to witte, the gouernement by Emperours [after it had vaked from the daies of *Auguſtulus*, *in anno*. 475. more then three hundred years) was renued again in *Charlemaign in Anno*. eight hundreth and ſix and became the eight *Roman* gouernement, like as it had beene before the ſixt, but now ar the *Cæſars* and imperial eſtate, worne to leſſe than a priuate Kingdome.

12 And there ar ariſen of their decayed empyre, ten chief Chriſtiã kings: to wit, of *Spane*, *France*, *Lombardie England*, *Scotlãd*, *Denmark*, *Swaden*, of the *Hunnes* or *Hungarians* in *Hungarie*, of the *Gotthes* in *Italie* & the *Exarchat* of *Ravenna*, who were not in S. *Iohns* daies, nor begunne to reigne while the empyre began to decline.

13 Theſe whollie agreeth in one Papiſticall faith, & Idolatrous Profeſsion: and therefore, with all their might, force & ſtrength, maintained the Pope, who had the ſeuenth gouernment of *Rome*.

14 And concurring with him, they perſecuted & martyred all true Chriſtiã profeſſors, who opponed againſte their papiſticall abuſes, but doubtleſly, Chriſt Ieſus king of kings,

11 And ſo that ſtate of monarchie, & imperiall gouernement, which latelie was, and now vaketh, that ſame imperiall gouernement (being heereaſter long ſuppreſſed) ſhall bee againe the eight gouernement, like as before it hath beene one of the ſeuen, and at length ſhall it goe into final deſtruction.

12 And as for the tenne hornes which appeared to thee, theſe are ten vnder-kings, or confederate kings, who as yet haue not begun to raign, but ſhal ſtart vp as kings immediatlie [*u*] after that, that monarchie, or imperiall ſtate beginnes to fall.

13 Theſe ſhall agree vpon one faith and opinion, and ſhall giue ouer againe their force, power, and aſsiſtance, to the next governors of that monarchie.

14 Theſe wil aſſiſt that Antichriſtian monarchie, fighting [*x*] againſt Chriſt Ieſus and his true ſeruants, but in the end, Chriſt ſhal ouercome them, for they are but earthlie Kings, and he is Lord of lords, and King

11 And the beaſt that was and is not, is euen the eight, and is one of the ſeuen, and ſhall goe into deſtruction.

12 And the ten hornes that thou ſaweſt, are ten Kings, which yet haue not receiued a kingdome, but ſhal receiue power as kings at one houre after the beaſt.

13 Theſe haue one minde, and ſhall giue their power & authoritie, vnto the beaſt.

14 Theſe ſhall fight with the Lambe, and the Lambe ſhall ouercome them, for he is Lord of Lordes and king of

ſhall now ſhortly ouercome them, either cõverting or ſuppreſsing thẽ (as hee hath already begun) and ſhall erect aboue all his kingdome of his holy people.

15 And ſo, al-though *Rome* hath borne hitherto the empire & ſupremacie over all kinreds, people and nations.

16 Yet now ſhall theſe ten Chriſtian kings, her confederats beginne to hate her, and renounce her Papiſticall ſuperſtitions, depriue her of her dignities and honors, appropriate & annexe her rich benefices and rents, to their owne patrimonies: and at length, ſome of them ſhall ſpoyle that verie citie it ſelfe, of all her coſtly ornaments, & burne vp with fire her edifices, and deſtroy the ſame for euer.

17 For although God made theſe princes, firſt to aſsiſte and authorize her with one conſent: yet nowe is the time come, in the which God hath decreede them to revolt from her and deſtroy her.

18 Who hitherto hath bene the Metropolitane Citie, that hath empired ouer all the whole

of kings, & his ſeruants are by him called and choſen, and found faithfull in al their doings.

15 And expounding yet further vnto me: theſe waters (ſaith he) that appeared to thee, ouer the which, is that whoriſh citie ſet, are to be interpreted people, kinreds, nations, and tongues, ouer the which ſhe empires

16 And theſe ten vnder-kings (which by hornes appeared to thee) ſhal afterward begin (y) to hate that Idolatrous citie, and ſhall (y) make her bare & naked of her honours & dignities, and ſhal (y) eat vp her patrimonie & rents, and at length (y) deſtroy her-ſelfe with fire.

17 For God had put into their hearts to work at all times his will: euen that firſt they ſhould giue ouer their whol power might and ſupremacie of their kingdomes to her till at laſt the time were come, in the which God had predeſtinat thẽ to revolt from her.

18 And be ſure, that richly decked whore which appeared to thee, meaneth nothing els,

kings, & they that are on his ſide are called, & choſen, and faithfull.

15 And hee ſaid vnto me, the waters which thou ſaweſt, where the whore ſitteth, are people & multitudes & nations, and tongues.

16 And the ten hornes which thou ſaweſt vpon the beaſt, are they that ſhal hate the whore, and ſhall make her deſolate, and naked, and ſhall eate her fleſh and burne her with fire.

17 For God hath put into their harts to fulfill his will, & to doe with one conſent, for to giue their kingdome vnto the beaſt, vntill the wordes of God be fulfilled.

18 And the woman which thou ſaweſt, is the great citie,

kingdomes and countries of the world.	but the Metropolitan & idolatrous citie, which vſurpeth the empire over the whol kingdomes of the earth.	*which reigneth over the kings of the earth.*

Notes, Reaſons, and Amplifications.

[*a*] This one of theſe ſeuen Angels, both mentioned here, & in the Apoc. 21.9. agreeth rather to be the laſt, euẽ the Angel of the laſt age, as being he in whoſe time, both in this place that Antichriſtian whore was deſtroyed, & in that place Gods holy *Hieruſalem*, & true Spouſe and Church decored and decked: & therefore, of all the ſeuen, hee ſemeth meeteſt to deſcribe theſe things to *Iohn*, which God had appointed him in effect to execute.

[*bf*] That an idolatrous citie is called in the Scriptures, a whore it is common, For ſaith Eſai. 1.21. *How is the holy citie become a whoore*, &c. And how more particularlie this decked whoore, and filthie woman, is interpreted to be that citie, that doeth empire ouer the kings of the earth, read the laſt verſe of this chapter, and ſo it muſt needs meane *Rome*, the onlie chiefe Metropolitan citie of the Empire.

[*c*] Waters are plainly interpreted by the Angell, to be people, and nations, ver. 15.

[*d*] Of potable liquors: wine & water, the cleare waters & wel ſprings, are euer taken in the Scriptures in the beſt parte, for the wholeſome, pure, true, and moſt ſimple doctrine of the Evangell, which our ſoules continuallie thirſt after: But contrarilie, wine, for that it is more alluring, delectable, and provoketh drunkenneſſe: therefore, in all the Scriptures, it is commonlie taken in an evill part for, idolatrous, licentious, and voluptuous doctrins, provoking Gods ire & fierce wrath, Read Apoc. 14.8.10. The like is in diuers other places.

(*e*) Except we be ſolitarily, & as in a deſert, diſtracted & withdrawne from worldly ſocietie and partiall affaires, it is hard to haue our Spitit and perfect iudgement ripe and ready to diſcerne the trueth of ſpiritual matters.

(*f*) Hereof read the note *b*.

(*g*) As al red colours, betoken bloodſhead in the Scriptures (as is ſaid Apoc. 6.4. note *c*.) and skarlet is a more glorious red than ſimple red: So betokeneth it, this Roman kingdom not only ſimplie to be blood-thirſtie, but euen to glorie in their blood-ſhead and murthers. And further, beſide the ſpiritual meaning of theſe colours, the Princes of that Empire both firſt and laſt, euen both Emperours and Popes, were corporallie & reallie clothed in purple and skarlet, in their Pontificals, as is to be read in the *Decret. diſtinct.* 96. where *Conſtantin* the Emperor giues his imperial garments of purple & skarlet, called therein, *Chlamydem purpuream, & tunicam coccineam,* to Pope *Sylveſter* the firſt, and his ſucceſſors to be worne by them, as they doe to this day.

(*h*) Theſe names of blaſphemie, that here are ſaid, that the beaſt is full of, are ſaid (Apoc. 13.1) to be vpon the ſeuen heads of the beaſt. The trueth is therefore, that theſe proud, glorious and preſumptuous ſtiles and ſuperſcriptions, are not only ſet vp in ſundry monuments vpon divers places of theſe ſeuen hils, which are her ſeuen heads, but alſo through all the whole bodie of the beaſt, euen in every part of that Empire, are there infinite of theſe Temples, Idols, and other monuments erected, bearing ſuch proude and blaſphemous ſuperſcriptions and titles dedicatories, as *Diis manibus, Fortunæ, Plutoni, Veneri, Priapo.* And even at *Muſsilbergh*, among our ſelues in *Scotland*, a foundation of a Roman monument lately found (now vtterlie demoliſhed) bearing this inſeriptiõ dedicatorie, *Apollini Granno Quintus Lucius Sabinianus Proconſul Aug.* Alſo there is to be ſeene in their coines the like vniverſally, euen titles of the pride & vain-glorie of Rome, as in the daies of *Conſtantius* and *Maximianus*, a coine printed with this inſcription, *Romæ æternæ prat.* In the daies of *Licinius Iunior*, a coine with this inſcriptiõ, *Romæ æternæ*, with the figure of Rome, ſitting as a glorious ladie. In the daies of *Priſcus Attalus*, a coine with this inſcription, *Invecta Roma æterna conob.* Of which writeth *Adolphus Occo, libro Numiſmatum Romanorum.* Of this their blaſphemous ſtile of eternity, read S. *Ierome, lib. Epiſt. parte 2. tract. 2. Epiſt. 18. queſtione 11. Litera Z.* in theſe words. *Vnde ſecundum Apocalypſim Ioannis, in fronte purpuratæ meretricis, ſcriptum eſt nomen blaſphemiæ, id eſt Romæ æternæ*, that is to ſay, *Whereby, according to the Reuelation of S. Iohn in the forehead of the skarlet whoore', there is written a name of blaſphemie, which is Eternal Rome.*

(*i*) Theſe ſeuen heades are plainely interpreted by the Angell hereafter, verſ. 9. to be ſeuen mountaines, and ſeuen gouernments or princely eſtates.

(*k*) Read theſe heads, taken for kings, and kings for any princely gouernours in verſe 9. following, and note *ſ.* thereof.

(*l*) Now haue we to trie, what theſe ten hornes do meane: firſt by *Daniel* 7.24. and Rev. 17.12. theſe be ten kings. Secondly, their kingdomes muſt be part of the Roman beaſt, & fourth Monarchy as both appeares, Dan. 7.7.23.24. Reu. 12.3. and 13.1. & 17.3.7. as alſo becauſe that whole ten horned beaſt, meaning the whole *Roman* Empire by our 24. prop. what els can the parts of that beaſt ſignifie, than parts of that Empire, & conſequentlie the ten crowned hornes of the beaſt ten kings and kingdomes out of the Empire? Thirdly by ver. 12. hereof and note (*u*) following, appeares that thefe ten kings arofe when the *Roman* gouernment by Emperours decayed, and when the Pope beganne to clame temporall dignities, which all occurred betwixt the 300. and 450. year of Chrift by our 36 prop. and approued hiftories. Fourthly by the 13 and 14 verfes, appeares, that thefe fhall be of one religion, and fhal therefore authorize and giue their power and afsiftance to the beaft, that next governed that Roman feate, even to the Pope (the Emperours then being decayed). Fiftly, though (as is faid) thefe ten hornes came at once with that Papifticall beaft, & litle Antichriftian horne, yet they rofe firft to their kingdomes before this Papifticall horne rofe to his Monarchie, and then he fuppreffed three of them, which teftifieth, Dan. 7.24. faying, *Another* (even that Antichrift) *fhal rife vp after them* (to wit after thefe ten kings) *and he fhal fubdue three kings*, and fo the Antichrift that commeth at one time with them, rifeth not to an eftablifhed Monarchy till after their rifing, yea, till three of them were fallen: for the tearm in the original Hebrwe or Chalde doeth not meane there of his firft comming or fimple beginning to rife, but of his eftablifhing or confirming of his rifing: for the fame tearm that here is put for this rifing, is exprefly put in Iob. 22.28, & 2. Chro. 7.18 & diuers other places, for eftablifhing & confirming. So by thefe notes & tokens, thefe ten appear to be the kings of *Spain, France, Lombardy. England, Scotland, Denmarke, Swaden*, of the *Hunnes* or *Hungarians* in *Pannonia*, now called *Hungarie*, of the *Gothes* in *Italy*, and the *Exarchat* of *Ravenna*, all ftart vp with the Pope, betuixt the 300. and 450. yeare of God, As to the kingdomes of *Arragon, Granad, Portugal, Caftile, Legeon, Tolet, Galice,* and *Nauarre*, they are onely inferiour kingdomes, partly fubiect to the Empire, partly pendicles of *Spain*, & are not kingdoms of old. The other kingdome of *Nauarre* is alfo of late, and now is all one with *France. Sicil, Boheme*, and *Naples*, for the moft part are of the Empire, and are neither olde nor free kingdomes: *Burgundie* is of *France*. and was onely a fmall kingdome, or rather dukedome. *Norway* is but of late a diuided kingdome from *Denmarke*, and is now conioyned againe with *Denmarke*. *Ireland* is conioyned with *England*. *Polland* is but of late erected in a kingdome. *Cyprus* is a kingdome both late, fmall, and feldome in Chriftians hands. As to the *Vandals* and *Vifegothes*, that people, as appeares, came from S*waden*, and the Northren countries, and began their kingdome in Spain, *Anno Chrifti 382.* by *Modogiflaus* their king: & therefore iuftly, of thefe *Vandals* and *Goths* whatfoeuer, wee comprehend their goers out vnder *Spaine, Italie*, and *Hungarie*, where whey came, and their remainers at home, wee comprehend vnder *Swaden*, whofe king now ftileth himfelfe, *King of* Swaden, *Gothes and Vandals*, as in the circumfcription of their coines is to be feene. And fo apparantly, remaine onely the faid ten kingdomes to be chiefly the beaftes ten hornes, with which the Popes litle horne and fmal kingdome comming at one time, was afterward confirmed and eſtabliſhed by *Pipin, Charlemaign*, and *Ludovicus Pius*, and then aroſe it greateſt of all: and by his craft, and meanes of his confederates, ſuppreſſed the kingdomes of the *Goths*, *Longobards* and *Exarchat*, and brooketh their lands within *Italy* to this day: But the other ſeuen kings yet extant, with the people alſo of theſe three kingdomes, ſhall (God willing) one day repay that Antichriſtian ſeat with fire ſword, and vtter deſtruction, as appeareth by the 16 verſe hereof and other places of this booke.

[*m*] Of the great pomp & riches of this towne, read *Hierom. part 2. tract. 5. Epiſtola 59 litera, G.* where he concludeth, *Aurata ſua tecta magis velint aſpicere quàm cœlum*, that is, *they had rather beholde their gilded ſilerings than the heavens*: & what marvel is it that it be rich, ſeeing almoſt the whole world haue bene tributaries to it, about 2000. yeares, including the time of pardons, as being the moſt welthy tributes. And as in all things they are glorious, ſo alſo in their tributes, they appoynted that the ſame ſhould bee brought in earthen pots, and the pots broken in a certaine place of Rome, where, by the great quantity of broken pots, there is waxed a hill, called *Monte teſtaceo*. And this haue they done for their glorious name & oſtentation, which (confirming this text) beareth recorde of the great riches that hath bene brought to them from all the world.

(*n*) How and why *Rome* if ſtiled *Babylon* myſtically or figuratiuely, is ſhewed at length in our 23. propoſition.

[*o*] Here ſaith Saint *Iohn*, hee wondered at the beaſt, and hereaſter (verſe 8) he ſaith, that they, *whoſe names were not written in the booke of life wondred at the beaſt*, not making any repugnance, neither making himſelfe to be of the wicked number: for there is a diſtinction in wondering, according to the qualitie of the wonderer: the godly ſeeing the pompe and glory of the world, wonder at the vanitie and impietie thereof, with great griefe of minde, deteſting the ſame: and ſo S. *Iohn* here wondered at the whoriſh citie: But the wicked contrarily, ſeeing hereaſter (ver. 8) the great pompe, wealth, and worldly felicitie of the Romane Emperours, they wondered thereat, not deteſting the ſame, but rather are rauiſhed in admiration, with reverence, obedience & feare of them, as if they were gods, and therefore it is ſaid (Apoc. 13.3.4) that the *whole world wondred at the beaſt, and worſhipped the beaſt, and the dragon that gaue her power.*

[*p*] It is ſo notable among the Prophets, that a beaſt, generally meaneth a King, kingdome, Empire, or Monarchie (as may bee prooued by Dan. 7. where hee calleth his foure beaſtes, the foure Kingdomes or Monarchies of the earth) That here S. *Iohn* ſuperfluouſly interpreteth not, this beaſt to meane a Monarchie in general, but entereth to the particular deſcription therof, ſo plainly, that he lets vs evidently ſee it, to be not only in general a Monarchie,

but even in particular, that ſelfe-ſame ten-horned beaſt which *Daniel* calleth the fourth Kingdome or monarchie: and ſo conſequently, to be the Roman Empire, as is prooved in our 24. propoſition at more length.

(*q*) What manner of wondring this is, read before in the note *o*.

(*r*) Some read here, *The beaſt that was, and is not*, onely: but I rather approoue thoſe, that read, *the beaſt that was, and is not, and yet is* for although this ſeeme a repugnance, yet it is none: for the Empire or imperial eſtate, at that time that S. *Iohn* wrote, might iuſtly be ſaid, not to be; ſeeing the Emperour *Domitian* was dead, and the next Emperour *Nerva Cocceius* not yet received: & yet in a manner, the Empire or imperial eſtate that time *was*, becauſe the governement of Emperours ſtood as yet that time vnaboliſhed. And ſo in divers maners, and in divers reſpects, the imperial eſtate at that time, both was and was not, without repugnance: whereof read the 5. reaſon of our 24. propoſition.

[*ſ*] A King, in the ſcripture, is a general name for any ſupreame governour, be he King, Monarch, Emperour, or Pope, as in *Daniel* chap. 7. all the Kings of one Monarchie, are called but one King: whereby the foure Monarchies are called foure kings: And againe, *Moſes* being but a governour, is called a king. Deut. 33.5. and ſeeing hereby, that not onely a king is taken for whatſoeuer ſupream governour, but alſo ſeing how many ſoever ſupream governours there be of one race or ſort, they are all called but one king. Therefore, we iuſtly eſteeme theſe ſevẽ kinges here mentioned, to be the ſeven princely, royal, and ſupreme governments of Rome, ſeeing in all the reſt of their circumſtances, they agree ſo well with the text: Hereof is ſpoken in the fourth reaſon of our 24. propoſition.

(*t*) It were vnproper, to meaſure the brevity & length of this time that the ſeventh Papiſtical & Antichriſtiã government doth ſtand, according to our fantaſie, as to cal it long, that men thinke long, for ſo indeed, ſhould that Antichriſtian raigne be counted long, but more properly the spirit of God (in whoſe ſight a thouſand years are but as one day) here, in compariſon of the æternity of the Chriſtiã kingdom of Chriſt Ieſus, accounteth the contrary kingdome, euen the Antichriſtian kingdome, to be but ſhort, for that truely 1260. yeares are nothing in reſpect of æternitie. Of a longer time called but ſhort, read Apoc. 22.6.7.10.12.20,

[*u*] Here doe wee followe *Ieromes* vulgare tranſlation, *Post beſtiam*, that is, *After the beaſt*, agreeing ſufficiently with the Greek, $\mu\epsilon\tau\alpha\ \tau o\upsilon\ \theta\eta\rho\iota o\upsilon$, for verily ſo came it to paſſe, that after the Empire of Rome decayed, and the imperiall gouernment was aboliſhed by the *Hunnes, Gothes, Vandales*, and ſuch others: Then immediatly, and at once (betuixt the yeares of God, 300. & 450) ſtart vp theſe ten kings. The other tranſlation of. T. Beza and others, who tranſlate, *vna hora cum beſtia, At one houre with the beaſt*, may alſo very

well ſtand with the word $\mu\epsilon\tau\alpha$ which ſometimes ſignifieth *after*, ſometimes *with.* And ſo in that ſenſe, although after the beaſt, euen after the fall of the Emperours (who in their time were that Roman beaſt and Latine Empire) the ten kings came and began to raigne, yet alſo this comming of theirs, was at one houre (that is to ſay immediatlie) with the beaſt: to wit, they came and aroſe immediatly with the pope, who ſince the fall of Emperors hath euer bin that *Roman* beaſt and poſſeſſed the Empire thereof, and therefore ſaith the text, in the next verſe: *Theſe* (to wit tenne kings) *haue one counſell and ſhall giue ouer their ſtrength and power to the beaſt*, meaning to the Pope, who ſince this decay of Emperors haue broked for the moſt part the ſtile of *pater patrum or patriæ* vnder the abbreiuat name of Pa. Pa, together with the rent, honors and ſeat of this *Latine* Empire, and *Romane* beaſt.

[*x*] Theſe be the wars (called the holy warres) which the Pope, his vnder-kings and confederats, did raiſe againſt thoſe kingdoms princes, and particular profeſſors, that following the truth, would any waies mean to repine againſt his Antichriſtiã authoritie: But the Spirit of God here, diſcouereth their hypocritical tearmes of holy warres, affirming ſtedfaſtly, them to fight againſt Chriſt, and that at length to their æternal confuſion.

(*y*) Here doeth the Angel declare to vs, by whom, and after what maner the Antichriſtian and Idolatrous Citie of Rome ſhal be deſtroyed. The doers thereof ſhall be (ſaith he) the ſame tenne Chriſtian Kings, who before had maintained her. The manner of this deſtruction of Rome, ſeemeth to be threefold. Firſt (ſaith the text) they ſhall hate her, and leaue her deſolate: that is to ſay, [as appeareth by *Paule* 2. Theſſ. 2.8. and Apoc. 14.6.] The trueth of Gods word and holie Evangel ſhall be firſt publikely preached, and by the Maieſtie of the comming therof, and two edged ſword of that holie Word, many of theſe ten Chriſtian Princes, their kinreds, and people ſhall be converted from their Antichriſtian and Papiſtical ſuperſtitions, and ſhall hate the Romane ſeat, turne backe from it, and leaue it deſtitute of their aſsiſtance, and this is her firſt wracke. Secondly [ſaith the text] *They ſhal eate her fleſh:* that is, theſe ten Chriſtian kings, who before had enriched her, and made her ſat with great benefices, rents and dailie caſualities, now hating her, ſhall eate vp theſe rents and rich benefices, and liue thereon themſelues: and both theſe two wrackes are already come, though not as yet altogether perfected. Reſteth yet the laſt, whereof, thirdly (ſaith the text) *they ſhal burne her with fire*: which appeareth not to be figuratiuely taken, but literally, by the next chapter, verſes 8.9.16.17.18.21. by which it appeareth, that at length, Princes ſhal make warre againſt the verie citie of *Rome*, take it captiue, ſpoile it, and finallie, ſo burne it with fire, that it ſhall be waſte for ever, as at more length in that chapter we ſhall now deduce.

CHAP. XVIII. The Argvment.

In this chapter coherently with the former, is pronounced the foreſentence and doome of deſtruction, againſt the ſeate and citie of Rome, for her wicked demerites, with an exhortation to all true Chriſtians, to leaue and renounce her: and expreſſe commandement to all men to put hand to the execution of Gods vengeance againſt her, for her pride. And laſt of all, is ſet forth the great horrour, and anguiſh of her familiars, for the ſuddaine and terrible ſpoliation, demoliſhing, burning, and deſtruction of her for euer, with the triumph of Gods whole true ſeruants therefore, after ſo plaine a ſtile, that the Paraphraſe and hiſtorie (or rather Prophecie) thereof, may bee conſolidate in one Propheticall paraphraſe.

The Prophetical Paraphraſe.

AFterward came the (*a*) ſecond of the ſeven thundering Angels from heauen, with great power and Maieſtie, lightening all the earth with the beames of his glorie:

2 And he pronounced with a loud voice the doome and ſentence following, ſaying, now ſhal fal, yea, even nowe ſhal fall that mightie imperial ſeate of [*b*] *Rome*, ſucceſſour of great *Babylon*, and ſhe ſhal become a deſert dwelling of [*c*] Devils and Fayries, and a ſolitarie wilderneſſe, for the repaire of wilde and ravenous ſowles.

3 Becauſe ſhe hath poyſoned al nations, with the venemous drinkes of her idolatrous doctrine: and all the kings and governours of the earth, haue followed her a whooreing after idolles: and the Marchants of the earth [*d*] both ſpiritual and temporal, haue bene inriched through her voluptuous and delicate traffique.

4 Then heard I againe, a ſpirituall and heavenly voice, charging all God his true, holie and elect Chriſtians, to depart from that Roman ſeat, & leaue all her ſuperſtitions, leaſt they being participant of her offences were alſo deſtroyed with her.

5 For her ſinnes haue craued vengeance from the heauens, and the Lord hath called to minde her

The Text.

A*ND after theſe things, J ſaw* another *Angel come down from heauẽ, hauing gret power, ſo that the earth was lightened with his glory.*

2 And hee cried out mightely, with a loude voice, ſaying, It is fallen, it is fallen, Babylon that great city, *& is become the habitatiõ of devils, & the hold of al foul ſpirits, & a cage of every vncleane & hatefull bird.*

3 For all nations haue drunk of the wine of the wrath of her fornicatiõ, & the kings of the earth haue committed fornication with her, & the marchants of the earth are waxed rich of the aboundãce of hir pleſures.

4 And I heard another voice from heauen, ſaye, Go out of her my people, that yee be not partakers in her ſins, & that ye receiue not of her plagues.

iniquities.

6 Commanding therefore all men to (*e*) render vnto her ſuch wracke, as ſhe hath deſerved at their hand: yea, and to double the revenge vpon her, that her workes haue deſerved, and as ſhe hath hitherto entiſed thẽ to drink in the cuppe of Gods wrath, ſo they nowe to execute Gods wrath twofolde againſt her.

7 And ſo much as ſhee hath vaunted and gloried in her ſelfe, and hath lived wealthilie in all ſenſualities and pleaſures, ſo much the more to adde vnto her torment and dolour. For in the preſumption of her heart, ſhe hath ſaid, I ſit (*f*) Queene and Miſtreſſe over the whole worlde, and I ſhall be deſtitute of no voluptuous pleaſure, neither ever ſhall I ſee any ſorrowe.

8 Therefore, at once ſhall God powre out on her the plagues of death & mourning, for the (*g*) ſlaughter of her inhabitants, and ſhee ſhall be incloſed and effamiſhed: finally, taken captiue, and burnt with fire: for almightie is the Lord, that hath given that ſore ſentence againſt her.

9 The Kings, Cardinals, and rulers of the earth, who haue followed her in whooring ſpiritually after Idols, and haue led a careles life in all her worldly pleaſures, ſhall now mourne and bewaile her, and ſhall become dolorous and ſorrowfull for her ſake, when they ſhall vnderſtãd that ſhe is conſumed & burnt vp with fire, & deſtroied for ever.

10 They dare not approach to aſſiſt her, for feare to be wracked with her, but abſenting themſelues, ſhall mourne in their hearts, ſaying: Alas, alas, our chiefe citie *Rome*, that great & mightie ſucceſſour and daughter of Babel, for vnawares is come thy deſtruction.

11 The Merchants of the earth, crafts-men and traffiquers with her wares, ſhal lament and mourne for her deſtructiõ, for their delicate wares ſhal not hencefoorth be ſo ſaleable.

12 Even their rich and coſtly Idols, chalices,

5 For her ſinnes are come vp vnto heauen, & God hath remembred her iniquities.

6 Rewarde her even as ſhee hath rewarded you, and giue her double according vnto her workes: and *in the cup that ſhee hath filled to you, fill her the double.*

7 In as much as ſhee glorified her ſelfe, and lived in pleaſure, ſo much giue yee to her torment and ſorrow: for ſhee ſaith in her heart, J ſit, being a Queene, and am no widow, and ſhall ſee no mourning.

8 Therefore ſhal her plagues come at one day, death, & ſorrow, & famin, & ſhe ſhal be burnt with fire: for ſtrong is the Lord God, which will condemne her.

9 And the kings of the earth ſhal bewaile her, and lament for her which haue committed fornication, and liued in pleaſure with her, when they ſhall ſee the ſmoke of her burning.

10 And ſhal ſtand farre off for fear of her torment, ſaying, Alas, alas, the great city Babylon, the mighty city, for in one houre is thy iudgement come.

11 And the merchants of the earth ſhall weepe and waile ouer her, for no man buyeth their ware any more.

Paxes, and crucifixes, threefolde crownes, myters, Bifhops-ftaues, and rings of golde, and filver fet with pearles, and all manner of precious ftones: furplices of fine linnen, copes, maffe-clothes, & other veftments of purple, of filk, & of skarlet, filerings, tabernacles and other ornaments in Churches, and Church-places of all manner of [*h*] wel-carved, and long-lafting wood, Hofte-boxes, and relique-buftes of Ivorie or fine Woode, offering ftockes and diverfe veffelles for Chrifme, oyle, falt, and holy water, of fine timber, fnuffers, and incenfors of braffe, copper, or yron, founts, and holy water ftones of Alabafter, Marble, or Iafper.

13 Yea, their confections of cannel, & fuggered fpices, for their Chriftmas and Eafter feafts, their delicate perfumes, odours, & coaftly oyntments: for their Pontificals, their incenfe, rofet and frankincenfe, to burne in Churches, wine for their drink, and for daily maffes, oyle for their extreame vnction, fine flowre for their Hoftes and maffe bread, wheat and plentie of all maner of cornes, with ftore of kyne & fheepe, for their daylie fuftentation: horfes, mules, & mulets, with coftly decored chariots for their journeying, Moores, Pages, Lackies, and feruants to ferue them, and finally, the pretended (*i*) merchandife of the foules of men, from hell and purgatorie, by foule-maffes, dirigies, pardons, indulgences, remifsions of finnes, and fuch foule-wares for lucre.

14 And the delicate and wealthie pleafures that thy foule lufted after, are departed from thee, and all thy profitable rents, and glorious dignities are taken from thee, and fhall neuer hereafter bee reftored to thee.

15 Thefe Marchants and fellers of thefe wares, which hitherto were made wealthie thereby, dare not afsift her, but abfenting themfelues left they be wracked with her, fhall bewaile and lament her,

16 Saying, Alas, alas, that mightie citie, which was decored with fine linnen, & with abullemẽts, pales, hangers & tapeftries of filke, purple &

12 The ware of gold and filver, & of precious ftone, & of pearles and of fine linnen, and of purple, and of filk & of skarlet, & of al maner of Thyne wood, & of all veffels of Jvorie, & of al veffels of moft precious wood, and of braffe, and of yron, and of marble.

13 And of cynamon and odours, and oyntments, & frankincenfe, and wine, and oyl, and fine flowre, and wheat, and beafts, and fheepe, and horfes, and chariots and feruants, and foules of men.

14 (And the apples that thy foule lufted after, are departed from thee, and al things that were fat and excellent, are departed from thee, & thou fhalt find them no more)

15 And the marchãts of thefe things which were waxed rich, fhall stand a far of from her, for fear of her torment, weeping and wailing,

16 And faying, Alas alas, the gret citie, that was clothed in fine linnen, and purple, and skarlet, & gilded with gold, and precious ftone and pearles.

17 For in one houre fo great riches are com to defolation. And euerie fhipmafter, and all the people that occupie fhips, and fhipmen, and whofoeuer trauell on the

skarlet and had her (k) filerings ouer-gilt with gold, and was richly dect and replenifhed with all manner of precious ftones and pearle:

17 For at an inftant is fhee depriued, and fpoyled of all thefe riches, and all her mariners, pylats, & gouernours of her great (*l*) fhippes, and huge Galliaffes, and euerie fhipmaifter, that tranfporteth her wares, ftanding a farre off for feare to be wracked,

18 Vnderftanding her to be burnt vp with fire, and vtterlie deftroyed, fhall lament, faying, what citie on earth, was fo mightie as this great citie?

19 And they fhall be grieued and disfigured with forrow and anguifh, and fhall crie out, weeping and wayling, faying, Alas, alas, that mightie and great citie, by whom, all wee her fhip-mafters and marriners were inriched, and made wealthie, through the coftlines of her delicat wares, for in an inftant is fhee brought to defolation.

20 But contrarily, O Elect Church of God, Apoftles & holy Prophets reioyce and triumph yee, for the Almightie hath iudged and revenged your caufe againft her.

21 Then a mightie Angel fhewed me the comparifon of a great milfton, caft into the fea, declaring that with fuch fuddentie and violence fhould thou, O *Rome*, the mightie citie, fucceffour and daughter of great *Babylon*, be brought to vtter confufion, fo that thou fhall neuer (*m*) arife nor recouer thereafter.

22 And the triumphant melodie of thy Organes, and belles, hymnes, and canticles, bummes, timbrels, and rigols, fhall neuer bee hereafter heard, founding in thy quiers, nor in thy houfes or habitations, fhall any maner of crafts-men be found dwelling: yea not fo much as the found of any mill or quernes fhall be heard grinding corne towards thee.

23 For there fhall not dwell in thee, fo much as one to light a torche or candle to fhine in thee, and farre leffe fhal it be heard, anie to marrie, or to be married in thee, or to procreat, or replenifhe

fea, fhal ftand a far off,

18 And cry when they fee the fmok of her burning, faying, What city was like vnto this gret citie.

19 And they fhal caft duft on their heads, & crie weeping and wailing, and fay, alas, alas, the great citie, wherein were made rich al that had fhips on the fea, by her coftlines: for in one houre fhe is made defolate.

20 O heauen reioyce of her, and ye holy Apoftles and Prophets: for God hath giuen your iudgement on her.

21 Then a mighty angel took vp a ftone, like a great milfton, & caft it into the fea, faying, with fuch violence fhal the great citie Babylon be caft, and fhall bee found no more.

22 And the voyce of harpers, & muficians, and of pipers, & trumpeters fhal be heard no more in thee, and noe crafts-man, of whatfoeuer craft he be, *fhal be found any more in thee and the found of a milftone fhall be heard no more in thee.*

23 And the light of a candle fhall fhine no more in thee: and the voice of the bridegrom and of the bride, fhall be hearde no more in hee: for thy merchants were the great men of the earth: & with

with people thy ruined walles, for thy merchants (chiefe ſelfers of indulgences, and ſoule-wares) became Cardinals and Biſhops, and great men vpon earth, and with thy alluring ſuperſtitions, all people and nations were blinded and deceiued.

24 And that miſerable citie was founde and conuict of God, to be the author of the blood-ſhed and martyrdome of all the Chriſtians, & true profeſſors that were ſlaine throughout the whole earth.

thine inchantments were deceiued all nations.

24 And in her was found the blood of the Prophets, and of the ſaints, and of all that were ſlaine vpon the earth.

Notes, Reaſons, and Amplifications.

(*a*) We interpret the Angel, that here pronounceth this decay of this ſpirituall Babylon, to be the ſecond of the ſeuen thundering Angels, becauſe that ſame Angel proclaimed that ſame decay to fall in his time. Apoc. 14.8 and ſo conſequently, wee looke (God willing) for the performance hereof, in this preſent Iubilie, begunne alreadie *in anno 1590.* and ending *in anno 1639.* or thereby.

(*b*) This *Babylon* is not literally *Babylon* itſelfe, for that *Babylon* was deſtroyed by the *Medes* ad *Perſians*, according to the prophecies of *Eſay* 13.17. and *Ieremie* 51.11. and *Pſal.* 137. and that, long before S. *Iohn* wrote this, and never roſe in authority again, but this *Babylon* is figuratiue or myſtical *Babylon*, ſo called (Apoc. 17.5.) for that this *Babylon* ſucceedeth to the monarchie of the former, as the eroy to her grandmother, & is certainly the citie of *Rome*, as is prooved by our 23 propoſition, and by the whole tokens of the 17. chapter hereof, reade alſo hereupon, Apocal. 14. note *h*.

(*c*) The ſpirit of God, deſcribing the horrible decay of real *Babylon*, Eſa. 13.21. for the greater aggravating of Gods iuſtice, & to mooue vs to conceaue the greater horrour, not only ſaith ſimply it ſhall be a deſert, or ſuch a deſert or wildernes, as no tame beaſt or birde dare reſort into, for ravenous beaſts and ſowles of reaſe that abide there: but even that it ſhall be a place ſo vtterly elraich (as wee tearme it) that alſo no man dare make reſidence there, for feare of devils, fairies and ſpirits of illuſions: and, as the Rabbins interpret that text, *the ſaltabunt Fauni, Satyri, ſeu damones piloſi, &c*. The like is ſaid here of this myſtical *Babylon*, or *Babylonical Rome*, that it ſhal become a dwelling of Devils, and a holde or reſidence of vncleane ſpirits, not that the Devils are as yet annexed to any certaine reſidence, or that they like rather to goe dwell in deſerts, than to abide among men to ſeduce them, but that the ſpirit of God would haue vs his faithfull to deteſt that citie as abhominable, wherein Devils ſhall dwell and poſſeſſe the ſame as their due patrimonie.

(*d*) Theſe marchants appeare hereafter plainly by their wares, to meane both ſeculare marchants, ſellers of all delicate wares to the *Romans*, as alſo their Eccleſiaſtical merchants, prieſts, pardoners, and Legats, that ſell their indulgences, pardons, and ſoulewares vnto the whol world: conſider this by the 12. & 13. verſes hereof.

[*e*] Now Princes, potentats, and mightie men on earth, go to, & execute Gods wrath againſt this godles citie, for here you haue a ſpecial warrant and earneſt commande to revenge that cauſe, that is, not onely your owne cauſe, but alſo it is the cauſe of the Almightie, by the which ye are certified here, that that work ſhal go wel with you, & ſhal proſper in your hands. Where note, that this doubling of revenge is ſpoken by a Prophetical *Emphaſis*, as meaning thereby, that we ought without pitie, rueth, & mercie to proceed with all poſsible extremitie againſt that devilliſh ſeat, to the vtter extirpation thereof, and not meaning that any puniſhment in this world that poſsiblie may be executed by men, can be ſo much as equivalent with the ſingle, and far leſſe can be double to her wicked demerits: where-through, in the ende of the next chapter, ſhe and hers are caſt into hell fire eternallie, as a full revenge.

(*f*) Here doeth the ſpirit of God deſcribe & paint Rome no otherwiſe, then ſhe in her olde monuments & coynings doth paint and ſet out her ſelfe: to wit, aftet the forme of a luſty Lady, clad in no mourning weede, nor widowes apparrel, but gloriouſlie and richlie deckt, and decored with all delectable and coſtly ornamẽts and therewithal, ſitting in-throned in her throne, as Lady & miſtreſſe over the whole world, with her glorious titles, & blaſphemous inſeriptions, *Roma æterna, felix, invicta*, and ſuch others, whereof reade the note *h*, Apoc. 17.

(*g*) Here haue we inſerted certaine clauſes gathered out of other parts of this booke, opening the maner of the deſtructiõ of *Rome*, that the prophecie may be the more plaine. And firſt, where wee interpret her mourning to be for the ſlaughter of her inhabitants (beſide alſo for the ruine of her kingdome) we do not coniecture it without a warrant, that there ſhall be great ſlaughter vpon her inhabitants, for which they ſhall mourne, becauſe the former plague (expreſſed in this text) that ſhould fall on them, is death. Againe, by Apoc. 13.10. it is doubtleſſe meant, that they ſhall be killed by the ſword, becauſe they haue killed others with the ſword. Secondly, becauſe the text ſpecifies, that they ſhall be effamiſhed and burnt with fire, it betokeneth a ſiege and incloſing of them, and that ſhe ſhall be taken captiue, and therefore haue wee alſo inſerted theſe clauſes agreeing with the ſaid warrant. Apoc. 13.10. wherein it is likewiſe meant, that they ſhall be led captiue, becauſe they haue led others captiue.

(h) This $\theta \upsilon \iota \nu o \nu$, is the wood *Thyia* which *Theophraſtus* reporteth to be a long-laſting, & vncorruptible timber: hereof mentioneth *Plinius* lib. 13. cap. 16. And with this timber, Temples in old times were decored and repleniſhed.

(i) After the ſpirit of God had reckoned out certaine bodilie wares which the *Romane* ſeat accuſtomed moſt to buy, now concludes he with theſe ſoule-wares, which they and their marchãts vſed chiefly to ſell, whereby, they were ſo enriched, that thereby they were made able to buy all the former coſtly wares. Theſe ſoule-wares are pardons, indulgence, remiſsions of ſinnes, trentalles, ſoule-maſſes, and diriges, with ſuch godles ieſtes, whereby they promiſe for money, to deliver vs from hell and purgatorie, and to canonize vs ſaints in heauen, and make vs interceſſors for our brethren on earth, and conſequently, Chriſts interceſsion to be needles. O blaſphemie intollerable, that maketh rich men [whom Chriſt ſaith ſhall hardly come in the kingdome of heauen] to become ſaints for money, and ſo tacitly to depriue our ſaviour of his office of interceſsion and onely mediation. Doth not *Peter* ſpeake truely of theſe his falſe pretented ſucceſſors (2. Epiſt. cap. 2) ſaying, they ſhal bring in damnable errors, denying the Lord that redeemed them, and through fained words ſhall they make marchandiſe of men, to wit, (as here is ſaid) by ſelling both their poore peoples ſoules, and their own ſoules to the Devill for worldly lucre. Here-with confer the note *d*, preceeding.

[*k*] Of theſe golden & gilded ſilerings & great riches of *Rome*, read the note *m*, in Apoc. 17. with divers olde writers, deſcribers of the pompe of *Rome*.

(*l*) Theſe mightie marriners with their great Galliaſſes, Venetians, Spaniſhes, & other of that Antichriſtian flote, who hitherto, and lateſt of all in this 1588. yeare of God, haue bene ſo readie at every nod of their Apoſtatik ſtepmother *Rome*, to haue overwhelmed vs poore and true Chriſtiã profeſſors. Seeing now and perceaving by daily experience, that God fights for vs, & dailie advanceth our cauſes more and more to their deſtruction, ſhall now ſhortly faint and fall backe, from aſsiſting that cauſe any more, onely ſtanding a far and bewailing that irreparable deſtruction that God is ſending vpon that godleſſe citie, and in their hearts lamenting the decay of their owne proſperitie, without making any farther debate or encounter in defence of that Romiſh ſeate.

(*m*) Let none beleeue, that this deſtruction, deſolation, & burning of *Rome*, is any of theſe deſtructions which are already performed by the *Hunnes, Gothes, Vandales*, or others in time paſt, for theſe deſtructions were ſoone repared, but this deſtruction (ſaith the ſpirit of God before, verſe 14. & here verſes 21.22.23.) ſhall be ſuch, that never ſhall be repared againe, that citie never found nor founded againe, her dignities & rents never reſtored to her, and finally, no inhabitant to dwel there, or ſo much as to light a candle in that execrable and accurſed citie againe.

CHAP. 19. THE ARGVMENT.

Following out the exhortation and Commandement of Gods ſpirit in the former chapter, verſe 20. here in this chapter, the whole Church of God, underſtanding that Antichriſtian enemie of theirs, the Romane ſeate to be ſhortly brought to ruine, and themſelues now to be ſteadfaſtlie cowpled with their Lord and Spouſe, Chriſt Ieſus, doth therefore reioyce & magnifie God. Thereaſter followeth a briefe deſcription of the word of God, and how the ſame doeth now obtaine victorie ouer all the enemies, that are conveened againſt it: and how all falſe Prophets, and chiefe members of that Antichriſtian empire, are condemned to the vnquencheable fire of hell for euer, as the latter and iuſt rewarde of their demerits, beſide that bodilie vengeance that commeth vpon the carkaſses of all their aſsiſters.

Paraphraſticall expoſition.

HEreafter I conſidered, and behold, the voyce of the whol multitude, and Congregation of Gods true Church, ſaying, (*a*) let vs prayſe the Lord, ſaluation, and glorie, honor and power, belongeth to the Lord our God for euer.

2 For true and righteous are all his proceedings, and hee hath iuſtly condemned that idolatrous citie and ſpirituall whore, which poyſoned and infected all people on earth, with her Idolatrous doctrine, and God hath righteouſlie avenged the blood-ſhed, and cruell martyrdome of his Saints and ſeruants, committed by her tyrannie.

3 And againe, and againe, inceſſantlie continued they, ſaying, Let vs praiſe the Lord, and the grief of her torment ſhall reſound to the heauens, and neuer ſhall haue an ende.

4 And al the true profeſſours of the olde and newe Teſtaments, proſtrated themſelues and worſhipped God, who raigned in his throne among them, ſaying, even ſo, O Lord, let vs praiſe thee for ever.

The text.

ANd after theſe things, J heard a greate voice of a greate multitude in heauen ſaying, Hallelu-iah ſaluation, and glory, & honour, and power be *to the Lord our God.*

2 For true and righteous are *his iudgemẽts for he hath condemned the great whore, which did corrupt the earth with her fornication, and hath auenged the blood of his ſeruants* ſhed *by her hand.*

3 And againe they ſaid, Halleluiah: and her ſmoke roſe vp for euermore.

4 And the foure and twentie elders, and the foure beaſts fel

5 Then did the voice of Gods Spirit, proceeding from the throne of his trueth, exhorte all his elect ſervants, both ſmall and great, that feare him, to render him praiſe for ever.

6 And with that was hearde the great dinne and mightie voyces of all the huge multitude of God his (*b*) people, as the reſounding noyſe of manie waters, and of his true Teachers and Prophets, thundering out Gods praiſes, ſaying, Let vs praiſe God, for our Lord God Almightie hath prevailed, and ſhal raigne hencefoorth for evermore.

7 Let vs be glad and reioyce, and giue to him the praiſe, that hath nowe vouchſafed to vnite his Sonne, that immaculate Lambe, Chriſt Ieſus, in ſpiritual [*c*] ſpouſage with vs, his holie Church, and now is our Church, purged from Antichriſtianiſme, & readie decked for him, as his true Spouſe.

8 And hee hath propined and preſented her with gifts and ſhining garments of puritie and innocencie, betokening that perfect purity and righteouſneſſe, that God hath granted to euery one of her particular ſaints and profeſſours.

9 Then ſaid the Angell vnto mee, publiſh thou, and ſhew foorth vnto all men, how bleſſed they are, that are called of God, to bee participant of that holy (*d*) Supper and marriage banquet, whereby wee are vnited and eſpouſed to Chriſt Ieſus, and he ſaid moreouer, all that I haue ſaide vnto thee is of [*e*] God, and ſhall ſurely come to paſſe.

10 And I fell downe at his feete to haue [*f*] worſhipped him: but hee rebuked me, ſaying, Beware thou do it not, for I am but a ſeruant of God, as thou art, and as one of thy brethren, bearing the teſtimonie of Chriſt Ieſus: worſhip nothing except God onelie: for wee that doe teſtifie of Chriſt Ieſus, are not God, but Spirits of his prophecie.

11 At this time the Church of GOD ſhall be patent and viſible, and therein ſhall bee perfect and ſincere profeſſours, and that Profeſsion that they

downe, and worſhipped God, that ſat on the throne, ſaying, Amen, Halleluiah.

5 Then a voyce came out of the throne, ſaying, Praiſe our God, all ye his ſeruants, and ye that feare him, both ſmall and great.

6 And I heard like a voyce of a great multitude, and as the voyce of manie Wateres, and as the voyce of ſtronge thunderinges, ſaying, Halleluiah: for our Lord God Almightie hath reigned.

7 Let vs be glad, and reioyce, and giue glorie to him: for the mariage of the Lambe is come, & his wife hath made her ſelfe readie.

8 And to her was granted, that ſhe ſhould bee arrayed with pure fine linnen, & ſhining: for the fine linnẽ is the righteouſnes of ſaints.

9 Then he ſaid vnto me, Write, Bleſſed are *they which are called vnto the Lambs ſupper And he ſaid vnto me Theſe wordes of God are true.*

10 And J fell before his feete, to worſhip him, but he ſaide vnto me, See thou do it not: J am thy fellow ſeruãt and one of thy bretheren, which haue the teſtimonie of Jeſus. Worſhip God. For the doe beare on them ſhall bee of [*g*] Chriſt Ieſus, who is called the faithfull and true incarnat word of GOD, who doth judge with juſtice, and fighteth righteouſlie.

12 And with the fierce and fierie eyes of foreſight doeth terrifie his enemies, and his head is crowned with manie victories ouer them, and hee hath a name, which none knoweth but himſelfe, euen none doth know Chriſt Ieſus, the very true incarnat Word of God, but they who are his myſtical bodie, euen they who are in him, and hee in them.

13 His apparell was beſprinkled with the blood that he did ſhead for our redemption, [*h*] as alſo with the blood of his enemies, whom he hath troden downe: and he is named that Word of God, that was before all beginning, and now hath receiued fleſh.

14 And all the whole Armie and Elect Congregation of his true Church followed his will and precepts, truelie journeying, and trauailing through this world in their mortified and regenerate fleſh, cloathed with the garments of his puritie and righteouſneſſe.

15 And out of his mouth proceeded true doctrine, piercing and ſubdewing on euerie ſide, the heartes of all Nations, and vnder the Scepter of his ſervitude doeth he governe and ſubdew them: for he alſo it is, who doeth (*i*) preſſe and tread down theſe wicked and reprobate people, that drinke vp iniquitie, and provoke the vengeance of almightie God.

16 And hee beareth in his badge, the ſtile & title of KING OF KINGS, AND LORD OF LORDS.

17 Then I ſaw a mightie (*k*) Angel who did awaite vpon the Eccleſiaſticall eſtate, and aſſayres of the Goſpel, proclayming openly and patently, to all Warriours, Souldiours, and other ravenous people, that dwelled among them of the true Church of God, forewarning them, that they ſhould conveene & gather themſelues together, againſt that time, which almightie God had appoynted for that greate

teſtimonie of Jeſus, is the Spirit of prophecie.

11 And J ſaw heauen open, and beholde, a white horſe, & he that ſate on him was called faithfull and true, and hee iudgeth, and fighteth righteouſlie.

12 And his eyes were *as a flamme of fire, and on his head* wer *many crownes, and he had a name written, that no man doeth knowe, but himſelfe.*

13 And he was cloathed with a garment, dipt in blood, and his name is called, THE WORD OF GOD.

14 And the warriors which were in heauen followed him vppon white horſes, cloathed with fine linnen, white and pure.

15 And out of his mouth went a ſharpe ſword, that with it he ſhould ſmite the heathen: for hee ſhall rule them with a rod of iron for hee it is that treadeth the winepreſſe of the fiercenes & wrath of Almightie God.

16 And he hath vpon his garment, and vpon his thigh a name writtẽ, THE KING OF KINGS, AND LORDE OF LORDS.

17 And J ſaw an Angel ſtand in the ſun who cried with a loude voice ſaying to all the ſpoyle, ſlaughter, and blood-ſhed, which they muſt eate vp and devoure.

18 And that then they ſhould ſhead the blood, and eate vp the rents, and ſpoyle the riches of Papiſticall kings and Cardinals, and of their riding-knights and Captaines, ſubverting them, horſe and man: & finally ſhould devoure vp and deſtroy all, both free and bond-man, both great and ſmall, that were of that ſect.

19 And I fore-ſawe that the greate Romane Monarch, and his confederat kings, and their great armies had euer convened together to fight and rebell againſt that true word of God, that his profeſſours did beare, and againſt all the true Miniſters thereof.

20 And therefore alſo were taken theſe Monarches, and their (*l*) *Pſeudopropheticall* Popes, who had counterfaited miracles, and made lying wonders before them of the Empire, to deceiue all thoſe who did profeſſe obedience vnto that Empire, bearing the badges and markes thereof, and them that reverenced the counterfait Emperours thereof: and theſe two, euen ſtraight after this (*m*) mortall life, were caſt down to hell, there to be tormented for euer, with the burning fire and ſulphurious flammes thereof.

21 And the reſt of their Armie, and partakers were altogether ouercome, vanquiſhed, and ſuppreſſed by that true and ſincere doctrine, which did proceed from out of the mouth of Chriſt Ieſus, that incarnate and holy Worde, which all true profeſſours doe moſt conſtantly beare teſtimonie of, and profeſſe, and all gredie and ravenous people, reavers, ſpoylers, and auaritious men, were ſatisfied aboundantlie with the rich ſpoyles, Church-rents, and benefices of theſe antichriſtian people.

fowles that did flie by the middeſt of heauen, Come and gather your ſelues together vnto the ſupper of the great God.

18 That ye may eate the fleſh of Kinges, and the fleſh of high Captaines, & the fleſh of mightie men, & the fleſh of horſes, and of them that ſit on them, and the fleſh of al free-men and bond-men, & of ſmall and great.

19 And I ſaw the beaſt and the kings of the earth, and their warriours gathered together, to make battel againſt him that ſate on the horſe, and againſt his ſouldiers.

20 But the beaſt was taken, and with him that falſe Prophet, that wrought miracles before him, whereby he deceiued them that receiued the beaſtes marke, and them that worſhipped his image. Theſe both were aliue caſt into a lake of fire, burning with brimſtone.

21 And the remnant were ſlaine with the ſworde of him that ſitteth vpon the horſe, which commeth out of his mouth, & al the fowles were filled full with their fleſh.

Notes, Reaſons, and Amplifications.

(*a*) That *Halleluiah* is interpreted, *Praiſe the Lord*, read *Ierome de interpretatione nominum.* This word, as alſo *Amen*, and *Selah*, and diuers other Hebrue words, as being *Vocabula artis*, euen notable Theologicall tearmes, that in fewe ſyllables doe containe great matter, are therefore in all tranſlations left vninterpreted, for that S. *Iohn* himſelfe writing in Greek, doth neuertheles retaine them in Hebrew, yet in the Paraphraſe we thinke meet to leaue nothing vnexponed.

(*b*) Waters to meane people, and Thunders to mean Preachers, read in the note *b*, Apoc. 14.

(*c*) Of this Spouſe and ſpouſage, read the note *a*, Apoc. 12. and our 22. propoſition.

(*d*) The parable of this marriage banquet of Chriſt Ieſus, is at length expreſſed, in Mat. 22. Hereof the Supper of the Lord is the perfite type and Symbole.

(*e*) Here is the certainty of theſe prophecies, threefoldlie confirmed: firſt, in that they are revealed by an Angel. Secondlie, in that hee teſtifieth that they are of God. And thirdly, for that in expreſſe tearmes hee affirmeth them to be true and certaine.

(*f*) O miſerable Papiſts, why go ye bare-footed, & bare-legged on pilgrimage, and do fall kneeling & groueling before your idols adoring either them, or yet at the beſt, thoſe Saints whom they repreſent, ſeeing a more glorified Saint then any of your Saints (whom for the moſt part, your ſelues make and canonize) euen an holy Angel, who neuer was a ſinfull man, neuer fel, nor made defection at any time from God, as men Saints, a thouſand thouſand times haue done, but euer abode perfect & holy, here doeth abhorre and deteſt all kneeling and worſhipping before him, greatlie rebuking Saint *Iohn*, for meaning & preaſſing thereunto, and in expreſſe tearmes, forbidding the ſame, and yeelding the godly reaſon, that none ſhould bee worſhipped, but God only. Away, away with your captious diſtinctions of δουλεὶα and λατρεὶα, and all ſuch viſardes, as excuſe idolatrie, for here, the pure, naked, and ſimple worde of GOD, ſpoken by the mouth, and vttered by the example of this Holie Angell of GOD, and of God his holie Prophet Saint *Iohn*, doeth detect all your hypocriticall ſophiſmes, The Lord open your eyes, for in this text, among manie moe, doeth the light clearelie ſhine. The ſame example for the more aggreaging hereof, is repeated ouer againe in the Revel. 22.8.9.

(*g*) That this, that rideth on the white horſe, is the true and incarnate worde of GOD, euen CHRIST IESVS, appeareth plainelie by the thirteene and ſixteene verſes following, and by the whole harmonicall tokens of this text, and of Apocal. 6.2. conferred alſo with the note *d*, of Apoc. 12.

(*h*) Seeing Chrift, the incarnat Word of God, treadeth the wine-preffe of Gods wrath, in the 15. verfe; therefore doeth this blood wherewith his garments are befprinkled, beare the type of the blood of his enemies, which he hath fhead in the wine-preffe of Gods juftice, as appeareth by Efay 63.3. befide alfo the figure of that blood, which he fhead out of his bodie for our redemption.

(*i*) To treade the wine-preffe of God his wrath, doeth here not onelie meane, to execute God his latter judgement of eternall damnation, againft the wicked, as it doeth particularly meane, in the end of the fourteenth chapter: but it doeth meane generally the execution, of all God his judgements, againft offenders. And in that general fenfe, how Chrift our righteous judge and revenger, doeth treade this preffe, reade Efay, chapter 63. and Lamentations, chapter 1. verfe 15.

(*k*) This Angell certainlie beareth the type of all God his Minifters in this age, who ftanding in the Sunne, that is, ftanding, and awayting on the Gofpell and Ecclefiafticall affaires, openlie and patently proclaimed the deftruction of *Babylon*, whereby thofe Ravens, and rauenous fowles, that flie by the middeft of Heauen, that is, the ravenous and greedy people, that dwel among them of the true Church, fhall bee couragioufly inanimated, to put hande to worke, and fpoyle that wicked Citie, and eat vp all their riches. So that their greedy auarice in this cafe, fhall aduance the work of the Lord in a part, although the applying thereof to the vfe of the true Church, were more the duetie of thefe temporall auaritious perfons.

(*l*) Here is to be feen, what it is to worldly Princes to maintaine and afsift falfe teachers, for here is not only the Pope, that falfe illuder of thofe of the Empire, throwne downe headlong to hell, but alfo thofe Monarches and Emperours that afsifted him, and fuch other falfe teachers in this world, in the world to come fhal participat with them in eternall torment. And that this falfe prophet here mentioned, is one with the two-horned Antichriftian beaft, and confequently is the Pope, is prooued in our 25. and 26. propofitions.

(*m*) This lake burning with fire and brimftone, where thefe imperiall beafts, and *Pfeudoprophetical* Popes ar placed, are torments induring for euermore (as faith S. *Iohn*, Apoc. 20.10) & are the fecõd death, euẽ the death of the foul, (Apoc. 20.14. & Apoc. 21.8.) and therefore is it neither a temporall punifhment, neither the fire of Purgatorie, but the eternall flamming fire of *Gehenna*, out of the which there is no redemption. And heere note againfte Purgatorie, that the Pope himfelfe and his imperiall princes, deuifers and maintainers thereof, haue, by this text, no leafure giuen them to refort thither, but ftraight-way, while they are skarfe dead, to hell muft they goe the neereft way, and therefore faith the text here, *And thefe two were aliue caft into that lake &c.* as if he would fay, skarfe is their foule and life gone foorth of their bodies, but it is as ſoone in hell, without any intermiſsion or mediat place. So that they and their ſect hereafter in experience ſhall find no other Purgatory to them, but the eternall flãming fire of hell, nor to the elect, but the blood of Chriſt, although now they debate Purgatorie by fire and ſword.

CHAP. XX. THE ARGVMENT.

There beeing onely two ſpeciall kingdomes enemies to God, the one Papiſticke, a coloured enemie, the other Mahometicke, a profeſſed enemie, and hauing in the third, fourth, fift, and ſixt trumpets or vials, exponed and prophecied orderly, the eſtate and increaſe of the Mahometicke kingdome. And in many of theſe chapters lately paſt, hauing deſcribed the Romane and Papiſticall kingdome, nowe in this chapter S. Iohn ſheweth, that after a thouſand yeares of peace granted to the Romane and Papiſtical Empire, in the ende the Devil is looſed to ſtirre vp moſt cruel warres betwixt theſe two mightie kingdomes, Papiſticall and Mahometicke, in the revenge of the blood of holie martyrs, ſhead in the meane time: and how afterward, men beginne to reviue and ariſe from the blinde errours and deadly ſleepe of Antichriſtianiſme. And laſt, how by Gods trueth, theſe godleſſe kingdomes, ſhall be extinguiſhed, and Chriſt ſhal come to iudge all men according to their demerits.

Hiſtorical application.

NOwe it pleaſed the mercie and providence of God, by his holy Spirit, and heavenlie Angels, to reſtraine the tyrannicall power and vniverſall rage of Sathan.

2 So that after the continuall and ſuccеſsiue tyrannie of heathen *Roman* Emperours, & laſt of *Diocletian*, who in one moneth ſlewe 17000. martyrs, aroſe at length the firſte publike Chriſtian Emperour, *Conſtantine* the great, about the yeare of

Paraphraſtical Expoſition

ANd I ſaw an Angel deſcẽding from heauen, who had power of God, to open and cloſe the myſteries of hell, & to reſtrain by his hand, the power of the Devil, in the chaines of Gods feare.

2 And hee foreſhewed mee, howe God had appointed to take that olde (*a*) dragon and venemous ſerpent, euen the deuill, that odious

The text

ANd I ſawe an Angell come downe from heauẽ, hauing the keye of the bottomles pit and a great chain in his hand.

2 And hee tooke the dragon, that olde ſerpent, which is the deuil & Satan, and he bound

Chriſt, 300. from which time to the daies of Pope *Boniface* the 8. and of *Ottoman* the great, and firſt Emperour of Mahometans, An. 1300. even the ſpace of a thouſand yeares, the Devil and his raging tyrants of this worlde, were reſtrayned from that high degree of vniverſall tyrannie, that they both before and after vſed.

3 For God ſo dantoned them, and abaſed the power of Sathan, that hee might well ſtir vp particular and provincial warres, but vniverſal wars among the people and whole nations of the world, that is to ſay, betweene Monarch and Monarch there was none raiſed during al theſe 1000. yeares, & after theſe thouſand yeares, ending in An. 1300. it pleaſed God to permit Sathan, by his inſtruments, to raiſe vp moſt terrible & vniverſall wars for a ſeaſon.

4 Now in the beginning of theſe 1000. years, a new kinde of ſupream government, & iudiciall authoririe ſtart vp, to wit, the papiſtical governmẽt, by Pope *Sylveſter* the firſt, in Anno 315. to whom, and whoſe ſucceſſours, *Conſtantine* gaue both kingdome and iudiciall power over all Chriſtian Churches: this at length came to that tyrannie, that they martyred all thoſe that truely profeſſed Chriſt,

Sathan, & ſhould reſtrain him in bondage (*b*) a thouſand yeares.

3 And ſhoulde depoſe him from his (*c*) higheſt tyrannicall power, to his loweſt eſtate, and in ſuch wiſe ſhould he be reſtrained and incloſed, and the cloſure ſo confirmed of God, that he ſhould (*d*) no further ſtirre vp deceitfully the Heathen nations, to vniverſall (*e*) warfare, till theſe thouſande yeares were expired, and then ſhould the deuill be looſed, to ſtirre vp great wars among them, for a ſeaſon.

4 At this time hee let mee ſee, that there ſhould be (*f*) new ſeats, and ſupreame authorities erected, and theſe that were erected in authoritie, ſhould haue judiciall power giuen them, ſo I perceiued many, whom they had judged, martyred, and condemned to death, for bearing the teſtimonie of Ieſus Chriſt, and

him a thouſãd years

3 And caſt him into the bottomles pit, and he ſhutte him vp, and ſealed the doore, *vpon him, that he ſhould deceiue the people no more, till the thouſãd years were fulfilled for after that he muſt be looſed for a little ſeaſon.*

4 And J ſaw ſeates: & they ſat vpõ them, and indgemẽt was giuen vnto them, and I ſaw *the ſoules of them that wer beheaded for the witnes of Ieſus, and for the worde of God, and which did not worſhippe the beaſte, neither* and preached the worde of God, and all that reverenced not that *Romane* ſeate, nor theſe counterfeit Emperours or images of the Empire, whom they inaugurated, neither were marked with Chriſme, or Croſſes on their fore-heads or hands, for even then was there a latent & inviſible Church, that lived privilie as true Chriſtians, triumphing & raigning over theſe Papiſticall abuſes, during all theſe thouſand yeares.

5 But the reſt, I meane the whole outward viſible Church laye whollie as dead, and corrupted with Papiſticall errours, and beganne not to be raiſed vp nor quickened by the worde of life, till after the yeare of God, a thouſand three hundreth, when that all theſe thouſande yeares were outrun, and then began moe and moe to riſe daily from their former Antichriſtian errours: this

for Profeſsing the true worde of God, and which did not reverence the Antichriſtian Empire, nor the counterfait Emperours thereof, neither had profeſſed his ſuperſtitious markes, or borne the ſame on their fore-heads, or in their handes, and theſe liued and raigned as true Chriſtians [*g*] during al theſe thouſand years.

5 But the reſt of the people, that lay [*h*] dead in Antichriſtian errors, aroſe not therfrõ, to embrace the worde of life, till theſe thouſand yeares were runne out: this

his image, neither had takẽ his marke vpon their foreheades or on their handes, & they liued and reigned with Chriſt a thouſande yeares.

5 But the reſte of the dead mẽ ſhal not liue againe, vntill the thouſand yeares be finiſhed: this is the firſt reſurrection.

6 Bleſſed and holie is he, that hath parte in the firſte reſurrection: for on ſuch the ſecond death hath no power: but they ſhall be the Prieſtes of God and of Chriſt, and ſhall reigne with him a thouſãd year.

7 And when the thouſand yeares ar expired, Sathã ſhal be looſed out of his

riſing from Antichriſtian errors, is the [*i*] firſt reſurrection, that goeth before the reſurrection of the dead.

6 Happie is he, and bleſſed, that riſeth truelie from theſe errours, and ſo is participant of that firſt reſurrection, for vpon them the ſecond death, which is the death of the ſoule, ſhall haue no power, but they ſhall become Prieſts and Sacrificers of their pure and holy prayers vnto God and his Sonne Chriſt Ieſus, and ſhall raigne with him ſpiritually, the [*k*] thouſand yeares of the great Sabboth, which is to ſay, for euer, and euer.

7 But theſe former 1000. yeares being expired in An. 1300. even that verie yeare Pope

7 And when the (*l*) former thouſand yeares ſhall be (*m*) outrun, the deuill ſhal be looſed of

Boniface the 8. beginning the firſt Iubilie, was clothed the one day in his Popelike-pontificall, & the next day in an Emperours Robe-royal, & bearing before him two ſwords, ſaying, *Ecco duo gladii hic*, thereby he vendicated to himſelfe both the ſpirituall and temporall empire over all the quarters of the earth: on the contrary part, was *Ottoman* the great, who aſpiring to bee Monarch, was crowned that ſame yeare Emperour over all the foure nations *Mahometans*.

8 And ſo from this forth the Devill ſtirred vp betwixt theſe two pretended Monarches and their ſucceſſors, moſt terrible warres, by ſuch vniverſall multitudes and innumerable Armies, as hath not bene heard of before.

9 And their chiefe skirmiſhes and battels, were fought beſides the landes of Chriſtians, in *Europe* & in *Aſia minor*, and thereabout, where God had firſt planted his holy Citizens, even theſe ſevẽ Churches that S. *Iohn* writeth vnto: But (God willing) at lẽgth ſhall theſe Papiſticall and *Mahometane* kingdomes, both enemies to God, be deſtroyed by the power & force of Gods word.

10 And Sathan, that great deceiver of the world ſhall be caſt into hell fire eternallie, where alſo all

his foreſaid bondage,

8 And ſhal paſſe out to deceiue the people that dwell in al the arthes, or foure quarters of the earth, euen (*n*) both *Gog*, and *Magog*, ſtirring vp the one aganſt the other to battel, & ſhall gather them in number, as the (*o*) ſea ſand.

9 And they ſhal gather together in the plain of the earth, toward and about thoſe countries, (*p*) cities and dwellings, where Chriſtians, and Chriſts holy citizens had made their reſidence: but at length by the fiery tongues of the Apoſtolike doctrin and flamming fire of God his worde, from heauen, ſhall both *Gog*, and *Magog*, be ouercome.

10 And the deuill that ſhal deceiue thẽ and ſtirre them vp to warrefare, ſhall bee caſt into the eternall flamme of *Gehenna*, where alſo the great Emperours of the earth, and the falſe Antichriſtian Prophets, ſhall bee tormented day & night for euer-more.

11 For I ſaw that at this time ſhould the true & ſincere juſtice ſeat of God appear, and

priſon.

8 And ſhal go out to deceiue the people which ar in the foure quarters of the earth, even *Gogge and Magog, to gather them together to battel whoſe number* is *as the ſande of the ſea.*

9 And they went vp into the plaine of the earth, which compaſſed the tentes of the Saints about, and the beloued citie: but fire came downe from God out of heavẽ, & devoured thẽ.

10 And the devil that deceived them, was caſt into a lake of fire & brimſtone, where the beaſt & the falſe Prophet ſhall be tormented even day & night for evermore.

11 And J ſawe a greate white throne, and one that ſat on it, from whoſe face the godles *Romane* & *Mahometick* Emperours, & other wicked Princes, together with al falſe Prophets *Papiſticals*, *Mahometicks*, & others, ſhall terriblie bee tormented day and night for evermore.

11 When Chriſt in his maieſty ſhal come to iudgment, for then certainlie ſhall heaven and earth be renewed.

Chriſt our iuſt iudge ſitting theron, at whoſe preſence this heauen & earth ſhall then vtterlie (*q*) melt away, and al things bee renewed.

fledde awaye both the earth and heauen, & their place was no more found.

12 And I ſaw the dead, both great & ſmal, ſtand before God, and the books wer opened, & another booke was opened, which is the booke *of life, and the dead were iudged of thoſe things which were written in the books, according to their works.*

12 And we are ſure that then the ſecond and laſt reſurrection ſhould bee, wherein generallie all men ſmall and great, that euer died, ſhall riſe againe, and appeare before that iuſtice ſeat of God, and the (*r*) regiſter bookes of all mens conſciences bee opened vp, and laide abroad, and the great (*r*) regiſter of God his predeſtination, & booke of life ſhall be opened, and made patent, and the dead ſhal bee judged according to their (*ſ*) workes, written and regiſtred in their conſciences

13 And the ſea gaue vp her dead, which were in her, and death & hell delivered vp the dead, which were in them: & they were iudged, every man according vnto their workes.

13 And the waltering world, by his raging ſeas, and ſtormie tempeſts inveighing againſt our fleſh, ſhal render againe all that haue died by it, & the deuill by his deadly and deſperat griefes, and infernall tentations of the mind, ſhall render againe all that hath died by him, and euery man ſhall bee judged according to his workes.

14 And death and hell were caſte into the lake of fire: this is the ſecond death.

14 And the (*t*) infernall ſpirits and damned deuils, who are as death and hell it ſelfe, ſhall bee caſt in the burning flame, and vnquencheable fire thereof: this is the ſecond and euerlaſting death.

15 And whoſoeuer was not found written in the booke of

15 And whoſoeuer is not contained in the regiſter of Gods mercifull predeſtination, and booke of life, ſhal alſo be caſt in with them, in that eternall and vnquencheable fire.

life, was caſt into the lake of fire.

Notes, Reaſons, and Amplifications.

(*a*) The former part of the hiſtory of this Dragon Sathan, is to be found in the twelfth chapter hereof: to the which, this chapter is to be ſubioyned.

[*b*] To wit, a thouſand yeares of the 1260. yeares, that the Church abode latent and inviſible, as at more length is opened in our 34. Propoſition.

[*c*] This deepe wherein Sathan is now cloſed, is not as yet the deepe pit, and endles paine of Gehenna, for as yet, he is but retained in the chaines of darknes, till the latter day, when he ſhal be damned æternally to the deepe of hell [2. Pet. 2.] But this hell or deepe, called in Greeke ἀβύσσος, and in Latine *abyſſus* is ſomtime takẽ for the depth of the earth, or of waters locallie: as Pſal. 105.9. Sometime for lowe eſtate and baſe degree, as both here, and Apoc. 17.8. where the imperial beaſt riſeth out of this hel or deep and goeth thereafter to decay, but out of hell there is none that riſeth, ſpeciallie, no emperour or man kinde: wherefore in this text (*abyſſus* meaning but a lowe degree and baſe eſtate) the Devil is but thirled and reſtrained from his vttermoſt tyrannie, and not at this time bound in Gehenna: for as al the ſcriptures teſtifie he hath, ever had & ſhal haue to the worlds end that freedom & libertie to goe tempting mankinde, ſeeking as a rageing Lyon whom hee may devoure: reade further hereupon in our 35. proſition.

[*d*] It is ſaid in the Apoc. 12. [where the former part of this hiſtorie is ſet out] that the Church of God ſhall be perſecuted and chaſed into the deſert by this Dragon, and made inviſible 1260. yeares: and therefore now alluding to that, he ſaith here, that Sathan is bound, that he ſhall not ſeduce any farther: to wit, any farther than is ſaid alreadie in that 12. chapter: and ſo in this chapter he is not reſtrained ſimplie from perſecution, and from al maner of ſeduction, ſuch, as particular warres, and martyrdomes, and ſowing of hereſies, but from ſtirring vp of vniverſall warres, as is prooved in the ſaid 35. propoſition.

(*e*) Since Chriſts daies to this day, the Church of God wanted never either ſchiſmes or perſecution: therefore, this bondage of Sathan for 1000. yeares, is onely (as is ſaid) from ſtirring vp of vniverſall warres, as is prooved in the ſaid 35. propoſition.

(*f*) Becauſe the iudgement ſeate, & iudiciall throne of Chriſt followeth ſhortly, verſe 11. and that with him his elect ſhall ſit (Apoc. 2.26.27. and note *x*, thereof, and Apoc. 3. verſe 21. & note *ſ*, thereof.) Therefore this throne

& authoritie that here is erected, appeares not to be of the elect ſoules, but rather vpon the contrarie part to be the throne of the Antichriſtian beaſt, who here doth martyr theſe Saints of God, that raigned againſt antichriſtianiſme theſe thouſande yeares profeſsing Chriſt in this worlde, and ſhall therefore eternallie raigne with Chriſt in his throne hereafter verſ. 11.

(*g*) Although during theſe 1000. yeares and more, the true Church lurked in the wildernes (as Apoc. 12.6.14. is ſaid) yet was not that Church ſo inviſible, but the Dragon in the Antichriſts perſon in all ages, eſpied ſome of the children of the true latent Church, and martyred them: as both here, and Apoc. 12.17. is mentioned. So this confirmes our former aſſertion, where we ſay, the Devil here is not ſimplie bound, but in a maner, as we prooue in our ſaid 35. propoſition. Moreover, hereof is gathered, that ſeeing (by this text) they that worſhipped not the Antichriſtian beaſt, his image, or marke wer perſecuted 1000. yeares, therefore neceſſarily the Antichriſt raigned 1000. yeares at the leaſt: and conſequently, theſe who ſuppoſe him to raigne but three yeares and an halfe, are deceived, as is prooved in our 16. Propoſition.

(*h*) The former martyrs and profeſſors being the inward & inviſible Church (Apoc. 11.1.) Theſe whole remnant here ſpoken of, are the vtter court & viſible pretended Church called in that chapter Gentiles, that is, Idolatrous Ethnicks, & *Iohn* there ver. 2 is forbidden to account or regard them, for here ſaith he, they lye all dead and drunken in Antichriſtian errors, for the ſpace of 1000. yeares: and ſo indeed from the daies of Pope *Sylveſter* the firſt, to the daies of Pope *Boniface* the eight: that 1000. yeares the ſeat of *Rome* having no match nor encounter, ſuffred never any to be ſeen vouchable or viſible of the true church: but there about, and from thence foorth ſuch hote warres fell betwixt the empire of *Rome* and the *Mahometicke* empire, that vnawares, divers true profeſſors openly and vowably did ariſe and ſtart vp, ſuch as *Iohn de rupe ſciſſa, an. 1240. Gulielmus de Sancto amore, an. 1260. Arnoldus de villa nova & Petrus Caſsiodorus Italian, an. 1302. Dantes Aligherius & Gulielmus Occham ann. 1321. Franciſcus Petrarcha 1374. Ioannes Wicklevus, an. 1390. Ioãnes Huſſe & Hieroymus de Praga Bohemians* and martyrs, anno 1415. And then did ariſe by their doctrine a viſible reformed Church in *Bohemia*, and thereafter by *Luther* in *Germanie*, and ſo ever moe and moe viſible Churches and publike teachers are riſen from Papiſtical and Antichriſtian errors, even to this day: and all theſe were ſometime dead in theſe former Antichriſtian errors, but [praiſed be God] are now revived by the ſpirit of God in this firſt reſurrection.

[*i*] That there be two reſurrections, and two deathes, & what they be is plaine in the ſcriptures. The firſt reſurrection is the riſing from errours, of which ſaith *Salomon, Prou.* 24.16. Seuen times in a day falleth the iuſt man, and yet riſeth againe: & this firſt reſurrection is chiefly meant, from

antichriſtian errors, as here, and in *Daniel*, 12.2. wherein faith he, many ſhal riſe, &c. whereas the ſecond reſurrection is the general reſurrectiõ of the dead, wherein not onely many, but euen all, ſhall ariſe againe: as to the two deathes they are alſo plaine. The firſt, is the death of our fleſhlie bodies here, which the ſcripturs call oft a ſleepe: the ſecond death is the endleſſe dying of body and ſoule in hel fire eternallie: reade hereof, Apoc. 2. ver. 11 and note (*k*) thereof.

[*k*] After our exemption (here ſpoken of) from the ſecond death and after our æternall kingdome and prieſthood with Chriſt, this reigne of 1000. yeares in this place, cannot be temporall or definite: for by the grounds of Chriſtian religion that reigne is æternall, how then here a thouſand yeares is taken for æternitie, yee may conſider by the compariſon of the ſix dayes of labour, & the Sabboths reſt, with the ſix thouſand yeares that we muſt labour in this world, and thereafter haue eternal reſt whereof we haue ſpoken in our 14. propoſition, and ye ſhall finde, that although euery day of weeklie worke daies meaneth (according to 2. Pet. 3.8.) 1000. yeares definitly: yet the Sabboths 1000. yeares repreſents æternitie. And ſuch like doubtles, though the other 1000. yeares. of this chapter are definitly to be taken for 1000. common yeares: yet theſe 1000. yeares of our reigning with Chriſt, after wee are freed from the ſecond death, muſt infinitely be taken for æternity and for the eternal Sabboth. Where note, that by this text literally and definitlie taken, refulted the great errour of *Cerinthus* and his ſecte of *Chiliaſts* or *Millenaries* who thought our raigne with Chriſt to be on earth, and temporall for a thouſand yeares, and we then againe to die, and ly dead another thouſand yeares, and ſo about by viciſsitudes, as did of olde the *Platonickes*, and of newe in a manner the *Originiſts*. Further, ſome alſo by the miſtaking of this text, ſuſpected the authoritie of this whole Reuelation: but to the true Chriſtian conceiuer hereof, both is the authoritie of this booke confirmed, and the heriſie of the *Millenaries* refelled.

(*l*) Though in the former verſes by the laſt note 1000. yeares is there prooued to meane æternitie: yet theſe thouſand yeares heere ſpoken of, are the ſame thouſande yeares literallie and definitlie ſpoken of in the verſes, ſecond third fourth and fyfte for all theſe thouſande yeares are ſpoken of thinges temporall, while we are yet in this worlde, and agreeth hiſtoricallie with theſe ſame thouſande yeares that beginnes about the yeare of Chriſt three hundreth, and ends in the yeare of Chriſt a thouſand three hundreth, as is proued in our thirtie foure propoſition, but yeares, times, and numbers ſpecified of the worlde to come (as verſe 6.) muſt meane æternitie, ſeing after the latter day, there ſhal be no day, yea no year, time nor numeral deſtinction thereof, but æternitie without meaſure, Apoc. 10.6. And therefore be ſure this preſent text here, is the progreſſe and large deduction of that looſing of Sathan, which was but ſummarly touched, and forewarned in the end of the third verſe, and ſo this looſing is all one with that looſing.

(*m*) This text doth confirme our former aſſertion, whereby wee aſſirme Sathans bondage theſe thouſande yeares, to be onely from raiſing of vniverſall tyrannie and warres, for proofe now whereof when here he is looſed, he raiſed now vniverſall warres, by inanimating of Gods publike enemie *Magog*, with his Princes of the *Orient*, againſt Gods latent enemy, and Antichriſt *Gog*, with his whole Princes of the *Occident.* and ſo came it to paſſe that very laſt yeare of theſe thouſand, euen the yeare of God 1300. that Pope Boniface the eight chalenged, to himſelfe in his triumphant Iubilie, the double Monarchie, both ſpirituall and temporall ouer the whole worlde. And contrarilie, that ſame yeare *Ottoman* the greate, was crowned the firſt Emperour of the *Mahometans*: and ſo from thence foorth betwixt theſe, and their ſucceſſours, there followed out continuall and vniverſall warres of the whole *Orient*, againſt the whole *Occident*, by armies of innumerable people.

(*n*) That *Gog* is the *Romane* and Papiſtick Empire, and *Magog* the *Mahometicke* Empire, read and conſider the 32. propoſition.

(*o*) The *Mahometicke* horſemen only of this armie, being (Apoc. 9.16.) reckened to two hundreth thouſand thouſand horſemẽ, no maruel thogh here both the Armies, *Mahometicke & Papiſtick*, yea both horſemen and footmen thereof be more comparable to the ſea ſand than to any definite number: further concerning this matter, read Apoc. 9.16. and note (*h*) thereof.

[*p*] Conſider the progreſſe of *Ottomans* warres, and his ſucceſſors, Emperours of *Mahometanes*, & ye ſhall find the greateſt part of all their battels, both inteſtine, & alſo againſt the Popes forces, to haue bene fought among Chriſtians, and Gods holy Citizens. For *Ottoman* fought, and did win *Bithynia*, & much of the coaſte ſide at *Pontus Euxinus*, in the countrey of *Aſia minor*, where theſe ſeven Churches lay, that S. Iohn writeth to. Then *Ottomans* ſonne *Orchanes* fought amongſt the *Grecians*, and in divers Chriſtian parts of Europe, & conquered *Pruſia*. Then *Amurathes* their thirde Emperour, made great conqueſt in Europe about *Conſtantinople*, & fought many battels thereabout: and ſo foorth, with their ſucceſſours, who never reſted, making all their battels among the Chriſtians in *Aſia minor, Bulgaria, Walachia, Servia, Boſna, Croatia, Illyria, Hungaria, Epirus, Peloponneſus*, and at length, in *Spaine* and *Italie*, with marveilous ſucceſſe, in ſo much that *Mahomet*, their ninth Emperour ſubverted the two Chriſtian Empires of *Conſtantinople* and *Teapezonda*, and wanne twelue Chriſtian kingdomes, with two hundreth Chriſtian Cities, beſide divers Ilandes, as yet ſhal finde *lib.* 2. *Romanorum Principum Ioan. Baptiſtæ Egnatij*, and in other hiſtories.

[*q*) Of this diſſolution of the heaven, the earth, and all the elements, read 2. Pet. 3.10. & next chapter hereof: So that by the Scriptures, apparantly heauen and earth ſhall be one habitacle of Gods Saints and holy ſervants.

(*r*) The booke of life and predeſtinate ſalvation, is effectuallie regiſtred with God in his mercie, whereof, although by faith the Spirit of God giveth our Spirit a feeling, yet the ſame is not effectuallie regiſtred in our conſcience, I meane, our conſcience beareth no teſtimonie that we haue any merite of life in vs. But the other bookes of Gods iuſt iudgement, are indeed effectually regiſtred in our conſcience, which beareth the feareful record of all our iniquities, whereby the conſciences of the wicked do teſtifie their owne damnation. Wiſd. 17.10.11. while as contrarilie, the regiſter book of Gods mercy exeemeth the godly from their deſerved damnation, and promoteth them freelie to life everlaſting.

(*ſ*) By workes here are we iudged and iuſtified, & not by faith onely, as alſo *Iames* 2.24. teſtifieth, meaning hereby that of liuelie faith, and of the good works that followeth therevpon man is iuſtified, & not of that dead faith, that is by it ſelfe alone, with out any good workes: otherwiſe were the words of Paul, (Rom. 3.28.) expreſſe contrarie to this text, and to *Iames*, for ſaith *Paul, We are iuſtified by faith, without the workes of the Law*, that is to ſay, not without good works whatſoeuer, but meaning, that wee are iuſtified by liuely faith, with ſuch ſmall good works, as our weake nature will ſuffer that faith to produce, although it be without the preciſe workes that the Law requireth: and for confirmation of this interpretation, and vnion of theſe texts, ye ſhall finde both *Iames* and *Paul* agree in divers places, that faith without workes is a dead faith, and ſerueth nothing to juſtification. And againe, they agree both, that all workes (how good ſoeuer they ſeeme) that proceed not from faith, are euill. And ſo it is no difference to ſay with Saint *Paul.* We are juſtified by fruitfull faith, or faith that produceth good workes, although not the workes that the Law requireth, or to ſay with *Iames*, and here with Saint *Iohn*, we are iuſtified by faithfull workes, ſeeing a working faith, and faithfull workes are inſeparable, and none can haue the one without the other. So for concluſion, theſe workes by the which heere wee are iudged, are to be eſteemed good or euill, not in themſelues, or in ſo farre as they ſatisfie the Law [for ſo were all workes euil and imperfect] but in ſo farre foorth, as they haue, or want faith adoiyned with them, they are accounted good or euil onlie.

[*t*] To caſt death or hel, or ſuch vnliuelie things, into this eternall fire, cannot be meaned here: But for as much as death and hell, by Sathans procurement is brought vpon mankind, and ſo Sathan is authour thereof: Therefore *metonymicè*, death and hell, are taken for the deuill and al damned ſpirits. The like is mentioned in divers places of the Scripture, as in Eſa. 28.15.18. whereby their league and bond made with death and hell, is doubtleſlie meant, the giuing of themſelues ouer vnto the Deuill, and all iniquitie, and in Eſay 38.18. where, in ſaying, that hell and death ſhall not praiſe God, is meant,

that the deuill and damned ſpirits ſhall not praiſe God: and ſo in Oſea, 13.14. O *death, I will bee thy death, and O hell, I will be thy deſtruction*, What els is this, than the victorie of Chriſt ouer death and hell, by treading down the head of Sathan, & caſting him into the eternal torment of death and hell, who hath beene the entiſer and onely author of death and hell. So then here [verſe 13.] the Deuil and his inſtruments, vnder the tearmes of death and hell, doth exhibite to judgment al thoſe whom he hath deſtroyed by his tentations, as the world is ſaid before (vnder the tearmes of ſea) to haue exhibited al that haue died by it, including betwixt them all that euer died, for all men either die by corporall griefe of body, or ſpirituall grief of mind. So then here (verſe 14.) this death and hell, that are eternally caſt into the lake of fire, can be no other than the deuil & damned ſpirits, who entiſed man to death and hell, and therefore are caſten into that eternall and vnquencheable fire of *Gehenna*, agreeing with verſe 10. preceeding.

CHAP. XXI. THE ARGVMENT.

The divine Prophet Saint Iohn, hitherto from the firſt to the laſt comming of Chriſt, prophecying the mutabilities of this world, wherein the Chriſtian Church hath neuer had earthly reſt: now in this Chapter, and beginning of the next, concludeth, and endeth all theſe their definit and temporal miſeries, by aſſuring them of infinite and eternall ioy & reſt, in that new world, and heauenly habitation, which here he deſcribeth, not as it is, for that no eye hath ſeene, nor eare hath heard, nor heart can conſider the ioyes thereof. (1. Cor. 2.9.) but after a certain metaphoricall compariſon, moſt proper for our ſenſes and capacitie, is that heauenly habitation deſcribed. So that vnder the literall glory thereof, obiected here to our capacitie, and worldly ſenſes, is not only meaned an infinit & endles glory, exceeding all humane capacitie and iudgement, but alſo by the ſymmetries thereof, diverſe myſteries of the God hood, and grounds of religion are typicallie expreſſed.

Paraphraſtical expoſition.

THen I foreſaw that the heauens and the earth ſhould be (*a*) renewed to their vttermoſt perfection: for this imperfect eſtate of the heauens & the earth that now is, ſhall diſſolue and melt away, and there ſhall be no ſea, or (*b*) raging inundations

The text.

ANd I ſaw a new heauen, & a new earth, for the firſt heauen, and the firſt earth were paſſed away, and there was no more ſea.

of worldly troubles.

2 And I *John*, forefaw and behelde that (*c*) fpirituall *Jerufalem* (euen our eternall felicitie, and glory in Gods Majeftie) vouchfaued, and (*d*) fent downe of God, from the higheft heauens, prepared to decore and attire true Chriftians with, for their (*e*) fpoufe Chrift Iefus.

3 And I heard the mightie voyce of God his eternal decree from heauen faying, Behold, from hence-foorth the Tabernacle and dwelling of God, fhall be with men, and by (*f*) Chrift fhall hee dwell in them, and they in him, & they fhall be his people, and God himfelfe dwelling with them, fhall be their God for euer.

4 And he fhall put an (*g*) end now to all their miferies, and they fhall bee no more fubject to death, nor to forrow, nor to mourning, nor to any manner of paine, for all fuch carnal pafsions of this our former worlde fhall bee abolifhed.

5 And (*h*) God who fitteth in ful authoritie among them, hath promifed, faying, Behold (faith he) I fhall now renewe and make perfect all thefe imperfections, and he (*i*) commanded me to write, and affure all men, that thefe words are faithfull and true.

6 And he faide vnto me, now are all (*k*) things done and perfected: I am the firft and the laft, euen the firft before all beginning, and the laft after all ending: to him that hath thirfted and zealed after trueth & juftice, wil I now freely of my mercie, giue to drinke and receiue of my liuely fountaine, euerlafting life.

7 He that hath ouercome his earthly tentations, fhall in this heauenlie world poffeffe, and inherite all joyfull things: and I will be his God and Father, and he fhall be (*l*) adopted one of my fonnes.

8 But contrarily, the fearefull doubter, and the vnbeleeuer, and the execrable and abominable liuers, cruell murtherers, filthy whoremongers, execrable forcerers, abominable Idolaters, & all teachers, and authorifers of lies, fhall haue their

2 And J John faw the holy citie, new Jerufalem come downe from God out of heauen, prepared as a bride, trimmed for her hufband.

3 And J heard a gret voice out of heauẽ, faying, Behold, the tabernacle of God is *with men, and he wil dwell with them, and they fhall be his people, and God himfelfe fhall bee their God with them.*

4 And God fhal wipe away all teares from their eies: & there fhal be no more death, neither forrow, neither crying, neither fhal there be any more pain, for the firft things are paffed.

5 And he that fate vpon the throne, faid, Behold, J make all thinges new, and hee faid vnto me, Write, for thefe things are faithful and true.

6 And he faid vnto me, It *is done, J am Alpha and Omega, the beginning and the end, J will giue to him that is a thirft, of the well of the water of life freely.*

7 He that ouercommeth fhall inherite all things, and I wil be his God, and he fhal be my Sonne.

8 But the fearful and unbeleeuing, and the abominable, & murtherers, & whoremongers, and forcerers, & idolaters, part & portion, in that endles depth of hell, which burneth eternally with vnquencheable fire, and bitter torment, which is the euerlafting (*m*) death of the foule.

9 And there came vnto mee, the (*n*) laft of the feuen Angels, which had power to execute the feuen laft plagues and he talked with me, faying, Come, and I wil let thee fee the fpiritual eftate of the glorified Church, who is the true Spoufe and Bride of the immaculate Lambe, Chrift Iefus.

10 And hee ravifhed me away in the Spirit, to a great and hie contemplation, wherein, hee let me to vnderftand that eternall felicitie in GOD, and (*o*) holy habitation, which from the higheft heauens, God had appointed to beftow and fend downe vnto his elect, and holy Church.

11 And that happie habitacle, was decored with the glory of God, and the fhining beautie thereof exceeded all precious ftones, euer flourifhing in all felicitie, and waxing greene as Iafper, but pure, and cleare, from all fpots of vncleanneffe, as fhining Cryftall.

12 And it was ftronglie environed about, with the true and invincible (*p*) Apoftolical doctrin, and forenent euerie elect people & nation, there was fet an open and patent gate, and ready acceffe, whereat Gods Angels, (*q*) Apoftles, and Minifters, called, & conducted feuerallie by name, euerie (*r*) tribe, kinred, and nation of Chrift his fpiritual Ifraelites to enter.

13 Toward the holie elect people of the *Orient*, was the (*s*) knowledge of the Trinitie fet patẽt as a threfold entrie to this felicitie. To the elect people of the *North* was the fame knowledg of Gods trinitie a threfold entrie to felicitie the elect people of the *South* alfo entred in by the fame Trinitie. And finallie, the elect people of the *Weft*, and all the airthes of the worlde had no other entrie to felicitie, but by the knowledge and profefsion of this Trinitie.

14 And that ftrong bulwarke and (*t*) wall of

& al liars fhal haue their part in the lake, which burneth with fire & brimftone, which is the fecond death.

9 And there came vnto me, one of the feuen Angels, which had the feuen vials, full of the feuen laft plagues, and talked with me, faying Come: J wil fhew thee the bride, the Lambes wife.

10 And hee caried me away in the fpirit, to a great & hie mountaine, and hee fhewed me the great citie, holy Jerufalem, defcending out of heauen, from God.

11 Hauing the glorie of God, & her fhining was like vnto a ftone, moft precious, as a Jafper ftone, cleare as cryftall.

12 And had a great wall and hie, and had twelue gates, and at the gates twelue Angels, & the names written, which are the twelue tribes of the children of Jfraell.

13 On the Eafte part there wer *three gates* and *on the North fide three gates, on the South fide three gates,* and *on the Weft fide three gates.*

14 And the wal of

Gods trueth and true doctrine, wherewith that holy habitation, was defended and environed was euen that ſelfe ſame doctrine, whereof the ſtrong foundations were laide before in this worlde, twelue folde by Chriſt Ieſus his twelue Apoſtles, euery one by name in their ſeuerall calling.

the citie had twelue foundations, and in them the names of the Lambes twelue Apoſtles.

15 And the Angell who reuealed theſe thinges to me, repreſented by the (*u*) ſymbole of a golden met-wand, that he was to mette and expone by meaſures, the figuratiue ſymmetrie of that holie habitation, of the entries thereof, and of the walles enuironing the ſame.

15 And he that talked with me, had a golden reede to meaſure the citie withal and the gates thereof, and the wall thereof.

16 And this our heauenlie habitation and eternall felicitie in God, was (*x*) ſquared and foure-foldlie deſcribed by the foure Evangeliſtes: and the indwellers are to enioy a like ſelicitie in the Father as in the Sonne, and the Angell by the ſymbole of metting repreſented vnto mee the infinite roome and ſufficient dwelling places that were therein, (*y*) even that there were twelue thouſande times 12000 ſtadges, repeated or multiplied twelue thouſande times, anſwering to the infinite number of Gentiles and to the twelue thouſande of euerie tribe of *Iſraell*, mentioned Apoc. 7. all theſe ſhall participate (*z*) ſo much felicitie in the Father, as they doe in the Sonne, and alike in the Son, as they doe in the holy Spirite,

16 And the citie lay foure ſquare, and the length is as large as the bredth of it, and he meſured the citie with the reede, twelue thouſand furlongs, and the length and the bredth, and the height of it are equall.

17 And the Angel by the figure of metting repreſented to me, that, that ſtrong bulwarke and wall of the Apoſtolicke doctrin, raiſed vp twelue courſes high by theſe twelue Apoſtles, is now as it were raiſed vp (*aa*) twelue times ſo high, euen 144. courſes vp, ſpeaking theſe things after the manner and compariſon of men which are meaned in a ſpirituall and Angelicall meaning.

17 And he meaſured the wal thereof, a hundreth fourtie and foure cubites by the meaſure of man, that is, of the Angel.

18 And the building of this fortreſſe and wall of the Apoſtolik doctrin, was firme, pure & bright, euer flouriſhing & waxing green as the *Jaſper* & the happie habitation of Saintes within the ſame, was more glorious than goulde, and more gliſtering and pure than the cleare glaſſe.

18 And the building of the wall of it, was of Iaſper, and the citie was pure golde, like vnto cleare glaſſe,

19 And the foundations of that ſtrong walle of this holie habitacle, was decored with the Prophets, Martyrs, Profeſſours, Preachers, interpreters, and other precious people of the primitiue Church which as (*bb*) precious ſtones, God hath gathered out of al countries, by his twelue Apoſtles, to foũd this eternall Sanctuarie with: for of theſe 12. Apoſtles ſome one from the *Indies* gathered in al theſe green *Iaſpers*, who abod euer green and vnwithered from the true faith, to found with theſe the firſt foundation of this heauenly habitacle: Another Apoſtle to make the ſecond courſe of this foundation, broght in thoſe cœleſticall *Saphires*, who, indued with heauenly hewes and qualities reſpected no earthly ſhewes, The third Apoſtle for the third courſe of this foundation, gathered in from among the *Ægyptians*, the victorious and fiery *Chalcedonies*, euen the zealous profeſſours and victorious Teachers of trueth: ſome fourth Apoſtle, to builde the fourth foundation, euen from the *Scythians* and *Bactrians*, brought in the greene and glancing *Smaragdes*, euen the flouriſhing, vnwithered, and ſyncere Chriſtians of thoſe countries.

19 And the foundationes of the wall of the citie were garniſhed with all manner of precious ſtones, the firſt foundation was *Jaſper, the ſecond of Saphir, the thirde of a Cha'cedonie, the fourth of a Smaragde.*

20 For the fift courſe of this foundation, ſome Apoſtles gathered in from among the *Arabians*, their meeke lowelie and chaſte people, comparable to the ſimple and chaſte nature of the *Sardonyx*. The ſixte Apoſtle gathered in theſe mollified people in whoſe ſoft heartes are ingrauen the ſeales of ſaluation, euen the continual memorie of Chriſts paſsion, repreſented by the bloodie and fleſhly colour of the *Sardius*. The ſeuenth Apoſtle, for the ſeuenth ground and foundation, the wiſe and conſtant *Chryſolites*: The eight Apoſtle for his courſe and foundation, gathered in al diligent Chriſtians and peacemakers comparable to the Pretious *Beryll.* the ninth Apoſtle to found his courſe with, gathered in the patient people, and reſtrayners of their affections, repreſented by the nature of the *Topaze*. The

20 The fift of a Sardonyx: the ſixt of a Sardius, the ſeuẽth of a Chryſolit: the eight of a Beryl: the ninth of a Topaze, the tenth of a Chryſopraſus, the eleuenth of a Jacinct, the twelfth an Amethyſt.

tenth Apoſtle, broght into the tenth courſe of this foundation, the golden greene *Chryſopraſus*, euen thoſe Chriſtians, that renouncing avarice, glorie in that golden treaſure of heauen that neuer withereth nor decayeth. The eleuenth Apoſtle to found his eleuenth courſe with, gathered in the golden purple and princelie *Hyacinthes*, euen thoſe magnificent and princely Profeſſors, who being richly decored with ſpirituall benefites, ouercame all tentations: Finallie, for the twelfth courſe, and to compleate the foundation of this holie worke, the twelfth Apoſtle brought in, euen from *Armenia*, the pure and temperate *Amethyſtes*, to wit, thoſe Chriſtians, who deteſting gluttonie & drunkennes, are indued with ſobrietie and temperance.

21 And theſe twelue Apoſtles, teachers of the way of life, ſtood as precious pearles, decoring theſe entries, euen euerie Apoſtle, was in euery entrie of that heauenlie habitacle, as a precious pearle, & the (*cc*) patent paſſages or ſtreets of that glorious habitacle, are more glorious than the pure gold, and more bright than the ſhining of glaſſe.

22 And there was no diſtinct Temple, in that heauenlie habitacle, for vniverſally, and ouer it all, the (*dd*) Maieſtie of the Almighty God, & of Chriſt Ieſus, is in ſtead of their Temple.

23 So that therefore, this holy habitation needeth no light of (*ee*) ſpiritual Teachers, nor ſeculare lawes, for the glory of God ſhineth in it, and Chriſt Ieſus is their lanterne of light.

24 And the elect people & nations, ſhal enioy the ful fruition of that light, and all that were kinges of the earth, ſhall aboue all worldlie pompe, (*ff*) glorifie that glory.

25 In all this eternall day time, ſhall neuer the elect be barred out from this felicity, & there ſhal be no obſcurity of (*gg*) night, or darknes of errours there.

26 And the only glory and felicitie of all the elect nations, ſhall be eſtabliſhed in it.

21 And the twelue gates, were *twelue pearles, and euery gate* is *of one pearle, & the ſtreet of the city* is *pure gold, as ſhining glaſſe.*

22 And I ſaw no Temple therin: for the Lord God Almightie, and the Lamb are the Temple of it.

23 And the citie hath no neede of the Sunne, neither of the Moone to ſhine in it, for the glory of God did light it, and the Lamb is the light of it

24 And the people which are ſaued, ſhall walk in the light of it, and the kings of the earth ſhall bring their glory & honor vnto it.

25 And the gates of it ſhall not bee ſhut by day: for there ſhall bee no night there.

26 And the glory & honor of the Gentiles, ſhal be broght vnto it.

27 And there ſhal enter

27 Into this glorious habitation, ſhall no wicked or filthy creature haue entreſſe, neither any abominable ſinners, or deceitfull lyers, but onely they which are predeſtinat to ſaluation, and regiſtred in the booke of life of our Sauiour Chriſt Ieſus.

into it, none vncleane thing, neither whatſoeuer worketh abomination, or lies: but they which are written in the Lambs booke of life.

Notes, Reaſons, and Amplifications.

[*a*] This renewing of the heauen and earth, is without al doubt, literally meant: for ſeeing this is the time of eternall reſt, al heauenly motions, and earthly viciſſitudes muſt haue an end: yea, ſeeing the motions of the Spheares, Planets, and ſtarres, were made for diſtinctiõ of times, Gen. 1.14. as alſo forth from this day ſhall be no time, Apo. 10.6. Therefore ſhall their eſtate without all queſtion, be renewed: yea, and the eſtate of euery creature, for ſaith *Paul*, Rom. 8.21.22. *The Creature alſo ſhal bee deliuered from the bondage of corruption, into the glorious libertie of the ſonnes of God: for we know that euery Creature groneth with vs alſo, and traueleth in paine together, vnto that preſent.* Of this renovation, read 2. Pet. 3.10. and Eſay 65.17. and Eſay 66. 22. and Apoc. 20.11. and note *q*. thereof.

[*b*] Becauſe this deſtruction of the world ſhall be with fire, (as teſtifieth 2. Pet. 3.10.) It may be that the Sea ſhal be vtterly dried vp, & ſo the *Sea* here to be literally taken: but alwayes, in the figuratiue ſenſe, we are ſure that all tumultuous troubles, ſhall ceaſe at that day for euer, and that the *Sea* doth ſignifie tumults, and tumultuous people, is ſaid in our note *a*, Apoc. 13.

[*c*] This *Ieruſalem*, called new *Ieruſalẽ* & the ſpouſe of the Lamb muſt therefore be the renewed Church, euen both the glorified Church, and the felicitie thereof: for that to both the glorified Church who are the inhabitants, and to that felicitie, in which they dwel, doth the whole ſubſequent proprieties of this Ieruſalem agree: euen as the name of a citie is not onely taken for the towne, but alſo *metonymicè*, for the inhabitants thereof.

(*d*) For that God is the ſupream height, all that proceedeth frõ him, is ſaid to deſcend, and ſo this deſcending of ſpirituall Ieruſalem, is not meant locally, but figuratiuely, that both the godlie their ſelues, and their eternall felicitie haue their originall, and flowing from the Maieſtie of God, and ſo deſcende from him.

[*e*) Seing by the Apoc. 12.1. and note *a*, thereof, and by our 22. propoſition, the Church militant, is Chriſts ſpouſe, farre rather is ſhe now his ſpouſe, when ſhe is become triumphant: yea, before was ſhee in a part divorceable for her offences, but now, for her confirmed puritie, and eſtabliſhed chaſtitie, al wholly indivorceable.

(*f*) For this end hath Chriſt taken on him our nature, & is become *Emmanuel* (Eſay 7.14.) that is to ſay, God with vs, that we might now bee with God, and ſo by this text, God ſhall be with man, and man ſhal be with God, in this holy habitation and eternall felicitie.

(*g*) This end of all miſeries, yeeldeth comfortable occaſion to all Gods ſeruants, to endure patiently, temporall and definite troubles, knowing hereby, that now ſhall follow the rewarde of infinit and eternall felicitie.

(*h*) Here ſaith the text, *he who ſate in the throne, ſpake*, to wit, that ſame Deitie, and divine eſſence. But for aſmuch, as both the perſon of the Father, and alſo of the Sonne, are meaned to ſit in this throne, in the next chapter, verſe 1.3. therefore it is to be reaſoned whether this be the perſon of the Father, or of the Sonne, that here ſpeaketh to S. *Iohn.* Concerning this wee ſay, that although it was the perſon of Chriſt, who ſitting in this (throne, lately in the end of the former chapter, iudgeth the world, as both our beleefe, and Apoc. 14.14. do teſtifie, yet here is there a tranſlation made of perſons, and it is the Father that now ſpeaketh. For not onely now hath that iudgement ſeat (attributed in the Scriptures to Chriſt) taken full effect, but alſo Chriſt hitherto hauing perfected his functions of incarnation, teaching redemption, mediation, and of damning, & throwing down the wicked, doeth now deliuer vp the kingdome vnto God the Father, that he may be all in all, as teſtifieth *Paul* 1. Cor. 15.24.25.28. And againe on the other part, now are here introduced the offices, and functions of God the Father, and Creator, in newe creating and renewing the worlde. So that here the functions of the Sonne ending, and the functions of the Father of newe againe beginning, this muſt needes be the Father, that here ſheweth, that he is to renew the world, ſeeing that is his function and office. And to confound the perſons, or yet their offices, is contrary to the godly Symbole of *Athanaſius*, ſaying, *neque confundentes perſonas, neque Subſtantiam ſeparantes*, as who would ſay, the holie ſpirite had redeemed vs, or that Chriſt either firſt created, or now at laſt reneweth the world ſimply & altogether in his own perſon: Indeed Chriſt is meant with the Father, and the Father by Chriſt (but not Chriſt ſimplie) to creat the world, Heb. 1.2.9. conferred with *Pſal.* 102.26. & Col. 1.16. & the firſt article of our beleefe and ſo this muſt not be the perſon of Chriſt, but of the Father that ſpeaketh this & for more proofe that this is the father, he calleth vs afterward, verſ. 7. his ſonnes, whereas Chriſt calleth vs not ſo, yea not *ſeruantes*, but *friendes*, Ioh. 15.15. for both Chriſt and we are called Sonnes to God the Father, hee naturall, and we adopted.

[*i*] It is ſaid by *Paul* (2. Cor. 11.14.) that Sathan may take the ſimilitude, of an Angell of light, but that he may profeſſe himſelfe God, by taking his

ſimilitude, we read not, neither poſsiblie dare he, leaſt ſo the moſt godlie Prophets (meaning nothing but godlines) ſhould be abuſed & worſhip Sathan. Then followeth, that this being verily God, that here apeareth, & teſtifieth, the words of this book to be true and faithfull, there can therefore bee no doubt of the authoritie thereof. But if any will ſay, that though Sathã dare not faine the ſimilitude of God, yet durſt *Cerinthus* or other heretickes, *Chiliaſts*, fain Gods viſions. In contratrie of that conceiued ſuſpition, the heauenly ſtile, the preciſe performance, the harmonie with other Scriptures, the infallible notes, names, numbers, & dates of times confirme the veritie of this book, and divine certaintie thereof. The ſame veritie hereof, that here is approoued by God, was before intimate by the Angel Apoc. 19.9. note *e*, and is hereafter repeated. Apoc. 22.6.

[*k*] This is the ſame thing that is meant, by the latter ſound of the ſeuenth trumpet. Ap. 11.15. and out powring of the ſeuenth Vial. Apoc. 16.17. and theſe are the performances of the Angels oath and vow made in Daniell. 12.7. and Reuel, 10.6. of which our 14. propoſition reckeneth the dates.

(*l*) That this is God the Father that ſaith this, is ſaid at length, note *h*, preceeding.

[*m*] The firſt death being the bodily death, the ſecond death, muſt be the eternall death of ſoule and body, as is ſaid, Apoc. 20, note *i*, and Apoc. 2. note *k*.

[*n*] This one of theſe ſeuen Angels, both here, and Reuel. 17.1. mentioned, appeares to be the laſt of theſe ſeuen Angels, for ſuch reaſones as are contained there, Reuel. 17. note *a*, thereof.

[*o*] What this *Hieruſalem* and deſcending thereof meaneth, read the former notes hereof *c*, and *d*.

(*p*) The Apoſtolical doctrin, being indeed a ſtrong wall againſt all errors, is therefore iuſtlie here meant by this wall, which hereafter verſe 27. debarreth out all vncleane and abominable creatures and lyers: Further, that bold and conſtant Preachers and their doctrine, is called a braſen wall: reade Ieremie 1.18. And for the more euident token, that this wall can meane no other thing nor the Apoſtolike doctrine, the ſame hath twelue foundations, which are intituled by the names of the twelue Apoſtles: ver. 14. and 19. following: and ſo it being a ſolide wal to debarre out the wicked, hath not the leſſe threefold ports in it, by the knowledge of the Trinitie to receiue in the elect at all quarters as followeth verſe. 12. and note *ſ*, hereof pointing out ſo by number, names & circumſtances, that it doth meane the Apoſtolicke doctrine.

(*q*) Seeing miniſters and Paſtors euen before their glorification are called angels, as is ſhewed. Reuel. I. note *x*, therof: how much more then may the twelue, Apoſtles now after their glorificatiõ be meant by theſe Angels, the number of twelue, and other circumſtances following ſo well agreeing.

(*r*) Vnder the number of twelue tribes of Ifrael, (beaufe they fometime were the only peculiar people of God,) are now figuratiuely included the whole elect tribes of al the people and nations of the world, who now alfo are become Gods people. So that as teftifieth *Paul*, Gal. 3. 7. all the faithfull are the fonnes of Abraham, and confequently, all the faithfull are accounted the twelue tribes, in thefe figuratiue and propheticall fpeeches.

[*f*] This defcription of this fpiritual *Hierufalem*, & heauenly habitation, by three portes in each of the foure airthes, making in the whole, twelue, hath in it a perfect harmonie & reprefentation of our abfolut & eternal felicitie, wherof the entries & ports is the knowledge of the Trinitie, written & teftified by the foure Evangeliftes, and patently preached, made open, and difperfed through all the world, by the twelue Apoftles.

(*t*) What this wall is, reade before in the note *p*, preceeding,

(*u*) Nothing is more common amongft the Prophets, & in propheticall vifions, than to reprefent their figuratiue matters by an outward and eternall action, as is to be feene in *Ieremies* yokes. chap. 27 and 28. and *Ezechieles* gefture. cap. 4. and cap. 5. and in the Angels metting, *Exech.* 40.3, and diuers other places. And this is done for our weake capacitie, which imprinteth no verbal document fo deeplie in minde as we doe an actuall gefture: So here the Angell by this externall gefture of metting, willeth vs deepely to imprint in minde, all the fubfequent fymmetrie of this holie habitacle, as containing in it an holie harmonie, with the chiefe grounds of Chriftian trueth.

(*x*) Proceeding here further to the defcriptiõ of that eternal felicitie, in which we are to dwel for euer: for our capacitie, the fpirite of God fetteth downe the fame in fuch forme and fhape, as moft liuelie pointeth out the Doctrine of the Godhood, and teachers thereof in this world, by whome this heauenly felicity hath bene fquared and founded in vs in this life, that afterwarde wee might be citizens in the ioyfull habitacle of euerlafting life. Now therfore faith he here, this citie is fet in fquare or four cornered: meaning no doubt, hereby, that this our heauenly habitacle and eternal felicitie, wherein now we are to dwel and the whole doctrine pertinent there vnto, was fquared and four-foldly defcribed by the foure Evangels of *Matthew, Marke, Luke*, & *Iohn*: for where before, note *p*, the twelue Apoftles by their fhort and prefent doctrine, are made the firft layers of the twelue foundations of this work, doth it not verie harmonically agree, that the 4. Evangelifts by their long lafting and written teftimonies, haue builded out the fquares and four corners of that work, to the finall perfection thereof. If this were not the caufe, why this raifed fquare, & *Cubique* figure wer rathereft here of al other figurs chofen: then had either the Spirit of God chofen the round figure, as of all folide bodies the moft perfect figure: or then had he chofen the *Pyramid Triangulare*, which is the moft fimple, firme, and conftant figure: but thefe partlie not agreeing with the conuenient forme of a citie, and partlie, not bearing the ſymbole of the vnite and æquall Trinitie and of the foure Enangeliſtes, and other ſymmetries needful are therefore here refuſed, and the raiſed ſquare or *Cubique* figure choſen.

(*y*) The whole elect Iſraelites being reckened to a hundreth fourtie foure thouſande to wit twelue thouſande, of euerie Tribe, and the elect Gentiles to be innumerable, [Apoc. 7.4.9. note *f*, thereof, and Apoc. 14.1. notes *a*, and *b*, thereof.) here ſetteth he down a citie almoſt of infinite roome, to containe them into, according to Chriſte his ſaying, Ioh. 14.2. *In domo Patris mei manſiones ſunt multæ. In my Fathers houſe are manie dwelling places*, for the length of this Citie being twelue thouſand ſtadges, the breadth twelue thouſande ſtadges, and the height twelue thouſande ſtadges, as here the texte meaneth, the whole citie muſt containe twelue thouſande times twelue thouſande ſtadges, multiplied by twelue thouſande which is, 1728000000000. Cubique ſtadges: where as one cubique ſtadge, [that is a ſtadge in length and a ſtadge in breadth, and a ſtadge in height] were ſufficient to hold one whole region of people. Hereby then is meaned, that this heauenly habitacle and eternall fœlicitie is infinite, and aboundantlie capable of all the elect number. Of the preciſenes of numbers in ſuch like accomptes, we haue ſaid ynough Apoc. 7 note *f*, and Apoc. 14. notes *a*, and *b*, thereof and other places. And as for this number in particular it hath his originall from twelue for the twelue, Apoſtles, and then from a thouſand anſwering to the ſubdiſciples of euery Apoſtle: and then againe, theſe are multiplied cubiquelie to repreſent the greate increaſe of Gods ſeruants to the worlds ende and conſequently, of the agreable manſions, prouided in heauen for them.

z Seeing the whole body of this citie repreſenteth moſt properly that whole felicitie, that we are eternallie to enioy in the whole vnite Godhead, the three ſeueral dimenſions of longitude latitude and altitud of that citie muſt repreſent theſe ſeuerall meaſures and portions of ioy, which we haue in the Father, in the Sonne, and in the holie Spirite, and conſequentlie, the being of the longitude, of the latitude and of the altitude of that holy habitation all equal by this text doth repreſent that portion of ioy that we haue in the Father, that portiõ of ioy that we haue in the Sonne, and that felicitie that we enioy in the holie Spirite to be equall and alike. So that although euery one of vs inioy not alike felicitie with other, yet, as God in his three perſons is equall, and alike in him ſelfe, ſo of all his three perſons giueth he alike preſence and fruition: The father ſhal not be abſented from vs more then the Sonne, nor the Sonne no more than the holie Ghoſt: we ſhall ſee all three in one, we ſhall worſhippe all three in one, and finallie of all the three perſons of that vnite Godhood ſhall we haue ful ioy, comfort and fruition, and that alike of the one as of the other, even as the perſons are alike in Godhood, and one in ſubſtance.

(*t u x y z*) Of all thefe texts and former notes thereof is gathered the marueilous harmonie and accord in all pointes, betwixt God and this his holie *Hierufalem*. For God who created man according to his own image, here alfo defcribeth his glorified churches fymmetrie to his owne fimilitude. God is one, fo here by one onlie fpirituall *Hierufalem*, he reprefenteth his Church. There be three equal perfons of the Deitie, Father, Sonne, & holie Ghoft, fo be there here of this *Hierufalem* three equal dimenfions of longitude, latitude, and altitude: None of the three perfons of the Deitie is feparable from other, fo none of thefe three dimenfions of a citie, or of any folide bodie, can be feparable one frõ another, for then fhould it become a fuperfice, and no folide bodie. The three perfons of the Deitie and their functions, cannot be confounded: fo are not thefe three dimenfions confounded, for the length is not the breadeth, nor the breadth is not the height: Like as in *Athanafius* Creede, the perfon of the Father, is not the perfon of the Sonne. Neither is the perfon of the Son the perfon of the holy Ghoft, and yet the Father, Sonne, & holy Ghoft, are one onely God: as here the length, bredth, & height, maketh one only Citie. Further, the doctrine of the deitie is witneffed and fourfoldlie fet down by the foure Evangelifts: So here is this *Hierufalem* fet downe in quadrat, or fourfquare. Twelue Apoftles difperfed in all airths, hath founded this doctrine, & there by enter we to the knowledge of God: So on twelue foundations ftandeth this *Ierufalem*, and twelue entries there be thereunto in feverall airthes. Finally, the Deitie is vniverfall, infinite, and infinitely extended over al his workes. So in this *Hierufalem*, is there infinite felicitie, and aboundant roome for ftowing of all her inhabitants.

[*aa*] The Apoftolike doctrine, amongft thefe of the Church militant, being by worldly impediments holdẽ low, is now amõgft thefe of the triumphant Church, raifed to the full and perfect height, fignified here by the full increafe of 12. multiplied by 12. quadratly making 144. courfes of height.

[*bb*] Of all that hereafter followeth, & is fpoken of thefe precious ftones, wherewith are founded the 12. foundations of this ftrong wall of the Apoftical doctrin, this one thing is the fumme & fummar meaning, that as a diligent mafter of worke, will fend his quarriors, mafons, and builders in all parts, to fearch, dreffe, forme, & hew ftones proper and meet for their building, & glorious for the decorement thereof. So hath our God in this world fent his holy Apoftles, through *Inde*, through *Arabie*, through *Egypt*, through *Armenie*, and throughout all the countries of the world, to gather in, teach, & inftruct thefe holy & precious people, fome Prophets, fome Martyrs, fome teachers, fome interpreters, fome hauing one gift and vertue, & fome another, of whom God had decreed to build vp this holy and eternall habitacle, & firft to lay by them the twelue foundations of the Apoftolical & militant Church in this world, which thereafter in the world to come, euen in this triumphant Church, &

holy *Ierufalem*, ſhould ſhine eternally, and vtter al their gliſtring vertues: as faith Dan. 12.3. *Qui autem docti fuerint, fulgebunt ſicut ſplendor firmamenti, & qui ad iuſtitiam erudiunt multos, quaſi ſtellæ in perpetuas æternitates. And they that be wiſe, ſhal ſhine at the brightnes of the firmament, & they that turne many to righteouſnes, ſhal ſhine as the ſtars, for euer & euer.* In the ſeparat hewes, natures & qualities of theſe ſtones, in the Paraphraſe, I haue followed the receiued opinion of writers, leaning alwayes more to this their ſummar effect and meaning.

[*cc*] By theſe ſtreets & paſſages, is ſignified libertie (amongſt other felicities) ſo that we haue no maner of thirlage, as tied or addicted to any one place: yea, go where we will, we goe alwayes in God, and our wayes and ſtreets are beautifull and glorious, & importeth to vs full felicitie.

[*dd*] Of this temple, and how here, & in ſome other places, it meaneth, the Maieſtie of God, we haue ſhewed in the diſcourſe of our 20. Propoſition.

[*ee*] How the Sun is taken for the ſpiritualitie, and the Moone or the temporalitie, read the note *h*, Apoc. 6.

[*ff*] This is no worldly glory, that they ſhall bring thither, for that heavenly habitacle, that hath no need of the glorious light of the Sunne, and Moone, verſe 23. hath farre les need of worldly glory. It is therefore hereby meant, that theſe, who were godly Princes in this world, ſhall lift vp all their felicitie, glorie, and eſtimation from the pleaſures of the world, and fixe the ſame whollie vpon that felicitie and glorie, which there they ſhall haue in God.

[*gg*] This night that ſhall not be found, there agreeth both literallie, and figuratiuely: for literallie there ſhall be no night after the latter day, ſeing diſtinctions of time, by night and day, & by ſommer and winter ſhall ceaſe, as is lately ſaide in the note *a*, hereof, and Apoc. 10.6. Againe, figuratiuelie, there ſhall bee no obſcuritie of errours there, for that no vncleane thing, no lyar, nor abominable perſon abideth there, as hereafter, verſ 27. is teſtified, of no night to be here, read alſo Revel. 22. verſe 5.

CHAP. XXII. THE ARGVMENT.

The Spirit of God in the former Chapter, hauing deſcribed the glory of the glorified Church, vnder the compariſon and forme of a Citie, now proſecuteth out in this chapter, to declare how that citie and heauenly citizens thereof are furniſhed of ſuch ſpiritual and liuelie drinke, meat, and medicaments, as ſhall conſerue eternall life, felicitie and health, vſing here theſe earthly tearmes, for our capacitie, in expreſsing of heauenly matters. Thereafter followeth a concluſion of the proprieties of that glorie, & of this whole prophecie, with a confirmation, yet againe of the veritie hereof, ending finally with a threatning againſt all the wicked, and a terrible proteſtation againſt all the eikers and pairers of this book, and then contrarily commending the faithfull Churches to God.

Paraphraſticall expoſition.

AND the Angell let mee to vnderſtand: how (*a*) God the holy Spirit (flowing as a riuer, to ſatisfie al that thirſted after his ſpiritual gifts) proceedeth from the perſons of God the Father, and of God the Sonne, who ſate there in ſupreme authoritie.

2 And amidſt the wayes & paſſages of all that glorified congregation, and with theſe flowing waters and graces of the holy Ghoſt was alſo (*b*) Chriſt Ieſus conjunct, who is the author of our life and ſaluation: who continuallie, and euer recentlie yeelded, and exhibited the fruite and fruition of all his heauenly vertues, in number and meaſure ſufficient, to ſatisfie euery man, of the (*c*) twelue tribes of ſpirituall Iſraelites: to wit, ſuch as had obſerued and imbraced the doctrine of the (*c*) twelue Apoſtles: and the ſmalleſt benefites, and lighteſt vertues that floweth from Chriſt, ſerued ſome way to the comfort, joy, and continuall health of euery elect people, & nation.

3 And there ſhall bee no accurſed thing in that holie habitacle, for the throne, and eternall ſeate of God the (*d*) Father, and of Chriſt Ieſus his Son, ſhall be there, and all theſe his ſeruants ſhall now moſt perfectly ſerue him.

4 And they ſhall haue the full fruition and enjoying of his face and preſence, and with conſtant (*e*) courage, and bold faces, ſhall they profeſſe his name.

5 And there ſhall bee no darkeneſſe there, neither of (*f*) night, nor of errours, and no neede ſhall they haue of temporall nor eccleſiaſtical teachings, for GOD himſelfe, in Trinitie and Vnitie, ſhall lighten them, and with him ſhall they raigne, for euermore.

6 And the Angel againe aſſured me, that theſe wordes were faithfull and true: for the Almightie

The text.

AND he ſhewed me a pure riuer of water of life, cleare as cryſtal, proceeding out of the throne of God, and of the Lamb.

2 In the midſt of the ſtreet of it, and of either ſide of the riuer, was the tree of life, which bare twelue maner of fruits, and gaue fruit euery moneth, & the leaues of the tree ſerued *to heal the nations with.*

3 And there ſhall be no more curſe, but the throne of God, and of the Lambe, ſhal bee in it, and his ſeruants ſhal ſerue him.

4 And they ſhall ſee his face, and his name ſhall bee in their foreheades.

5 And there ſhall be no night there, & they neede no candle, neither light of the ſunne, for the Lord God giueth them light & they ſhall raigne for euermore.

6 And he ſaid vnto me, Theſe wordes are faithfull and true: and the Lord God of the holy Prophets, ſent his Angel to ſhew vnto his ſeruantes, the thinges which muſt ſhortly be fulfilled.

God, Lord ouer all the holy Prophets, had ſent him who is his holy Angel, to ſhewe and publiſh vnto his ſeruants vpon earth theſe thinges, which muſt beginne ſhortly, and bee performed amongeſt them.

7 Beholde, it ſhall not bee (*g*) long (ſaith Chriſt) till my comming, in reſpect of my eternall abiding, happie ſhall he be, who obſerueth, and doeth according vnto the wordes written in this booke of Prophecie.

8 And I the writer hereof, am *John*, the Apoſtle and Evangeliſt, who ſpiritually ſaw and heard all thee things and when I had heard and ſeene the ſame, (*h*) I fel down proſtrat at the feet of the Angell, who ſhewed mee theſe things, to haue worſhipped him.

9 But he rebuked me, ſaying, Beware thou do it not, for I am but a ſeruant of Gods, as thou art, and as one of thy brethren, which are Prophets, and am one of them which obſerue the wordes of God, contained in this booke, worſhip none but God onelie.

10 And hee commanded me not to (*i*) conceale, or vtterlie to ſeale vp the words of the prophecie, contained in this booke, for that the time of the performance thereof, commeth dayly on, and approcheth.

11 Now, he that doth wrong, (*k*) proceed (if he liſt) to doe wrong, and hee who is a filthie ſinner proceed (if he liſteth) in his filthie ſinnes: and contrarily, let the righteous man continue in his righteouſneſſe, and the holy man in his holineſſe.

12 For behold (ſaith God) I come ſhortly, and that not emptie, but bearing with me euery mans reward, to recompence them therewith, according to their faithfull (*l*) workes.

13 I am the firſt and the laſt, the beginning and the ending, euen the firſt before all beginning, and the laſt after all ending.

14 Bleſſed are they that (*m*) keepe the commandements of this their God, that they may obtaine a joyfull part and portion in Chriſt Ieſus, the true Author of life: and that they by the true knoweledge

7 Behold, J come ſhortly bleſſed is *he that kepeth the wordes of the prophecie of this book.*

8 And I am Iohn, which ſawe and heard theſe things and when I had heard and ſeene I fell downe to worſhip before the feete of the Angell, which ſhewed me theſe things.

9 But he ſaide vnto me, ſee thou doe it not: for J am thy fellow ſeruant, and of thy brethren the Prophets, & of them which keepe the words of this booke worſhip God.

10 And he ſaid to me, Seale not the wordes, of the prophecie of this booke: for the time is at hand.

11 He that is vniuſt, let him be vniuſt ſtil: & he which is filthy, let him be filthy ſtil: & he that is righteous, let him be righteous ſtill: and he that is holy let him be holy ſtill.

12 And behold I come ſhortly and my reward is with me, to giue euery man according as his worke ſhall be.

13 J am Alpha & Omega, the beginning & the ending, the firſt & the laſt.

14 Bleſſed are *they, that doe his commandements, that their right may bee in the tree of life, and may enter in*

of the Trinitie, according to the Apoſtolike doctrine: may enter to eternall felicitie.

15 But forth of the ſame, ſhal be debarred all Apoſtatike (*n*) revolters to their filthineſſe, as dogges to their vomit, & all inchanters, whoremongers, murtherers, and Idolaters, and finally, all that teacheth lyes, or loued and delighted therein.

16 I (ſaith the Lord Ieſus) aſſure all men, that I haue ſent my holy Angell, to teſtifie theſe things vnto you all my holy Churches, and Congregations. I am (in my Deitie) that [*o*] root and foundation, whereof *David*, and all the faithfull hath ſprung, and (in my humanitie) I am of that (*o*) poſteritie, and of-ſpring that hath ſprung of *David*, and am that morning [*p*] ſtarre, & Son of juſtice, whoſe bright Evangell lightened the whole world.

17 And all godlie [*q*] Spirits, and Chriſts holy Church, and chaſt ſpouſe, wiſhed him to come, and ye that heare, pray for his comming, and let euery man that zealeth, and thirſteth after ſpirituall graces, approach, and come vnto him: For what faithfull man ſoeuer lifteth to come, ſhall freelie, & in his mercy receiue of him, and drinke vp as water the aboundant graces of the holy Ghoſt, to life eternall.

18 Now therefore, I proteſt vnto euery man that heareth the wordes of this booke of prophecie, that if anie man ſhall pretend to adde any thing further vnto this book, God ſhall adde his torment, and bring vpon that man all the plagues that are contained in this booke.

19 And likewiſe, if any man ſhall mean to pare or diminiſh of the words of this holie propheticall booke. God ſhall vtterly take away all his portion and part, out of the regiſter booke of eternal life, & out of this holy habitacle of eternall felicitie, and from enioying of theſe joyes and (*r*) benefits herein promiſed for the faithfull.

20 The Lord Ieſus, who teſtifieth theſe things, hath promiſed, that hee ſhall come moſt ſpeedily. Euen ſo muſt it be, O Lord Ieſus, come therfore ſpeedilie.

through the gates into the Citie.

15 For without ſhall be *dogges and inchanters and whormongers and murtherers, and Idolaters, and whoſoeuer loueth or maketh lyes.*

16 I Ieſus haue ſent mine Angel, to teſtifie vnto you theſe thinges in the Churches: J am the roote and the generation of David, and the bright morning ſtarre.

17 And the ſpirite and the bride ſay, Come, And let him that heareth, ſay come, and let him that is a thirſte, come: and let whoſoeuer wil, take of the water of life freelie.

18 For I proteſt vnto euery man, that heareth the wordes of the prophecy of this booke, if any man ſhall adde vnto theſe things, God ſhal adde vnto him the plagues that are written in this booke.

19 And if any man ſhall diminiſh of the words of the booke of this prophecy, God ſhal take away his part out of the book of life, and out of the holy city, & frõ thoſe things which ar writen in this book.

20 He which teſtifieth theſe things, ſaith, Surely J come quickly, Amen, euen ſo, come Lord Jeſus.

21 The grace of our

21 The (*f*) grace of this our Lord Jeſus Chriſt, be with al you his faithful for euer. So be it.	*Lord Jeſus Chriſt,* be *with you all, Amen.*

Notes, Reaſons, and Amplifications.

[*a*] This pure and cryſtalline riuer of liuely waters, is taken in the ſcriptures for the ſpirit of God and graces thereof, which our ſoules ever thirſte and drinke, as is plainely expounded in *Iohn* 7.38.39. ſaying, *Fluminæ de ventre eius fluent aquæ vivæ: hoc autem dicebat de ſpiritu quem accepturi erant credentes in eum, &c. Out of his bellie ſhal flow rivers of water of life: This ſpake he of the ſpirite, which they that beleeved in him ſhould receaue.* And for confirmation that this river is the holy ſpirit, he ſaith here, that it proceeds of the throne of God the Father, and of God the Sonne, vſing the proper tearme of *proceeding*, for that the holy Spirite is ſaid to proceed from the perſon of the Father and of the Sonne, as ſaith the godly *Athanaſius* in his Symbole: *Spiritus ſanctus à Patre & Filio, non factus, nec creatus, nec genitus, ſed procedens: The holy Ghoſt, not made, nor created, nor begotten of the Father and of the ſonne, but proceeding from them.*

(*b*) This tree of life that groweth in the middeſt of this glorified congregation, & among theſe flowing waters of Gods ſpirit, & yeeldeth continuall fruit, wherevpon the godly feedeth to life eternall, muſt needs be Chriſt Iesus the author of our life and ſalvation, of whom the tree of life in earthly Paradiſe, did beare the type: and who hath his reſidence among his elect, and is conjoyned with the holy ſpirit in Deitie, and yeeldeth to theſe his elect, continuall benefites, and ſpirituall graces and vertues, whereon they ſpiritually doe feede and eternally liue.

(*c*) Becauſe the citie by the number of ports, foundatiõs, roomes, and ſymmetries thereof, hath bene before accommodate to the number of twelue, correſpondent to the twelue tribes of ſpiritual Iſraelites, & twelue Apoſtles: Therfore agreeable to that ſame nũber of twelue, he here alſo deſcribeth their ſpiritual food, making thereby the citie, & the inhabitants, & their ſpiritual food, and furniture, all correſpondent & conformablie agreeing together in one, with the twelue Apoſtles, of whom they haue receiued their compariſon, and of whom in truth, both is that heauẽly habitacle & proprieties thereof: to wit, our reſurrection, iuſtification, immortalitie, & eternall felicitie deſcribed: as alſo, the number of all faithfull, are by them and their ſucceſſors, collected and gathered in: As thirdly, theſe ſpirituall vertues flowing here from Chriſt, are diſtributed to every faithfull, according as they are more or leſſe of the number of the twelue tribes of ſpirituall Iſraelites, or hath imbraced more or les the doctrin of the twelue Apoſtles: and therefore, not only firſt the ſymmetrie of the cities and ſecondly, the roome and inhabitants thereof, but alſo thirdly, the furniture, euen theſe fruits, benefites and graces that here floweth from Chriſt, vnto thẽ, are all reckened, after the number of twelue: and ſo their monethly increaſes, meaneth no other thing, but firſt, this alluſion to the number of the twelue Apoſtles And ſecondly, that theſe fruits, and benefites, are euer continuall freſh, & recent: for (as we haue ſaid, note *a*, of the former chapter & in other parts) ther is no diſtinctiõ of times, by years, moneths or dayes, neither in heauẽ, neither after the latter day any wayes.

[*d*] Here he (as alſo verſe 1.) ſpeaketh only of two perſõs of the Godhood: to wit, the Father and the Soone ſitting in this throne, although inſeparablie the whole Deitie in Trinitie & Vnitie here ſitteth: but theſe two perſons, the Father & the Sonne are onely here mentioned, becauſe the iudicial matters of this throne, and the preſent functions belonging thereto, at this time concerned chieflie the offices of God the Father, and of God the Sonne. Of God the Son, in that here he ſitteth, now lately iudging the whol world generally, which chiefly appertaineth to him, as ſaith our Beleefe, *inde venturus eſt indicare vivos & mortuos, frõ thence he ſhal come to iudge the quick and the dead.* Of God the Father, in that hee here ſitteth reforming and renewing the world, as he teſtifieth in the text ſaying, *Ecce, nova facio omnia, beholde, I make al things new*, which belongs to his office, as being Creator of all, as teſtifieth our beleefe, calling him, *Creatorem cœli & terra, maker of heauen & earth*, This approoveth the note *h*, of the former chapter.

(*e*) This name of God was promiſed before (Apoc. 3.12) to be writtẽ vpon the elect, where the firſt mention is made of this new *Hieruſalem*: and now at this time ſince they are come to that *Hieruſalem* that promiſe here taketh effect. This (contrarie all wayfaring and worldly Profeſsions) meaneth ſo conſtant & vnchangeable a Profeſsion of the name of God, that it ſhall never be deleated nor forgotten out of their heartes, mindes, and continual thoughts, but euer ſhal abide ſo rooted in them, that with conſtãt courage and bold faces, ſhall they profeſſe that name for ever, & therefore is that name of God ſaid here to be in their foreheads.

(*f*) Read of this night before, Apoc. 21.25. & note *gg*, thereof.

[*g*] For three cauſes, are times which appeare long to men, called ſhort in the ſcriptures. Firſt, to moue the elect to patience. Secondly, in reſpect of the contrair purpoſes & matter, as here the definite time of Chriſts bodily abſence in this world, in reſpect of the infinite time of his eternall preſence in that everlaſting world, is called but ſhort. Thirdly, in reſpect of God who ſpeaketh this, in whoſe ſight 1000. yeares are but as one daye (2. Pet. 3.8.) this definite time is but ſhort: conferre herewith Apoc. 17.10, and note *t*. thereof.

(*h*) Marveilous is the forcible tentations of Idolatrie, & wrong worſhipping. Sundrie times hath ſaint Iohn ſeene the perſons of the Deitie in theſe viſions:

and although we read, that he hath then fallen and bene proſtrate with that glorious maieſtie of Gods preſence (Apoc. 1.17) yet finde wee not, that he proſtrated himſelfe voluntarilie to worſhip God, but contrarily, this is the ſecond time, that voluntarily he proſtrates & bowes down himſelfe to commit wrong worſhipping: and that even although hee was before forbidden of the Angels ſelfe, according to the proverb, *Nitimur in vetitum ſemper cupimuſque negata. Forbiddẽ things of men are chooſed, & moſt we ſue that is refuſed.* The divine Prophet Apoſtle & Evangeliſt *Iohn*, makes notable here to vs by his owne two-folde example, the great infirmitie in mankind on the one parte, and the terrible force of Idolatrous tentations on the other part. To the effect we wretched ſinners neither truſt overmuch in our owne ſtrength, neither yet be over-ſtack & ſlouthful againſt ſuch tentations: Seeing this holy and beloved ſervant of God ſo farre overcome, read more hereof, Revel. 19.10. and note *f*, thereof.

(*i*) In that here the Prophet is forbidden to ſeale & cloſe vppe theſe myſteries, it is not ſimplie meant, that he ſhould write all ſo plaine, that no part thereof were in any wiſe ſealed or kept ſecret: for contrarilie of the ſeven thunders (Revel. 10.4.) it is commanded to ſeale them, and not to write them plainely. The meaning therefore of this text, is not ſimplie, but in compariſon of other Prophets, to wit, that ſaint Iohn who was neerer to the time of the performance, ſhould not ſeale vp theſe things ſo obſcurely as other Prophets did, who were further from that time of performance, and therefore ſaith this text here, *ne ſignaveris, &c. tempus enimprope eſt, Seal not theſe words, &c. for the time is at hand.* And cõtrarily is ſaid, in *Dan. 8.26. tu ergo viſionem ſigna, quia poſt multos dies erit, Seal you vp the viſion, for it ſhal be after many daies*, & *Daniel* 12.4 9, *Claude ſermones & ſigna librum, vſque ad tempus ſtatutum, Shut vp the wordes and ſeale the booke, til the ende of the time.* And ſo is Iohn here indeed a great deale more plaine, ample, and orderly nor is *Daniel*: *Daniel* having ſealed his book for a long time, & *Iohn* having onely concealed this book from Antichriſtians, during the Antichriſts raigne, which time he calleth but ſhort. Apo. 17.10. ſaying, *Oportet eum breve tempus manere, He muſt continue a ſhort ſpace*, agreeable with theſe words here ſpoken, of the revealing of this book, which are, *tempus enim prope eſt, for the time is at hand*, & ſo in the end of the Antichriſts raign (which ends in our daies) ſhould theſe myſteries be throughly revealed. This texte maketh expreſſely againſt them, that eſteeme this book to be ſo cloſed, that it ſhall never be vnderſtood till Chriſt come, & conſequently, it ſhould be ſuperfluous.

(*k*) This is spokẽ Ironically in mockage of the wicked, & thereafter ſpeaketh he exhortingly to the godly to cõfirme them. This Ironical ſpeech if vſed oft in the ſcripture, for the more aſhaming of vs from ſin, as (beſides many other examples) the like is to be found in Eccleſ. 11.9. where the ſpirit of God by *Salomon* mocks the licentious youth, in bidding them tauntinglie followe their luſtes & hearts deſire, but know, ſaith he, certainly, that of al theſe things God ſhall craue accompt in iudgement. The like is ſaid even here in the next verſe, *For behold* [ſaith he] *I ſhal come ſhortly & my rewarde with mee, to giue to every man according to his workes*. So this threatning concluſion inferreth, that the former bidding of the wicked continue wicked, is altogether Ironicall, and contrarilie, that the bidding of the godly continue godly, is altogether allowablie and exhortingly ſpoken to them.

(*l*) How there works are not ſimply meant of only works, but of workes that proceed of Faith, or liuely Faith that produceth good workes, and ſo Faith to be the ground of Iuſtification, read Apoc. 20. note *ſ*, thereof.

(*m*) The vulgare tranſlation, containeth here theſe words, *Beati qui lavant ſtolas ſuas in ſanguine Agni. Bleſſed are they that waſh their garments in the blood of the Lambe*. The moſt allowable tranſlation bearing, *Beati qui ſeruant eius mandata, Bleſſed are they that keepe his commandements*, as the literal difference hereof is ſmall in the original greeke, ſo is the meaning of both alike in ſubſtance, for theſe that do keepe the commandements of Chriſt Ieſus ar theſe ſelfe ſame, that figuratiuely are ſaid, Apoc. 7.14. to waſh, their garments in the blood of Chriſt, and therefore ſet we the plaineſt in the paraphraſe, as an interpretation to both.

[*n*] To dogs of filthines in luſt, inſatiable gluttonie, vomiting & returning to their vomit, are compared all ſuch filthie perſons as how oft ſoeuer, the Spirit of God ſtriueth to purge thẽ of their ſinnes, yet will they euer revolt and returne to them againe, til at length the ſpirit of God renounce them, and leaue them in their owne filth. Of theſe ſaith *Salomon, Proverb*. 26.11. *Sicut canis quire vertitur ad vomitum ſic imprudens qui iterat ſtultitiam ſuam, As a dog who turneth againe to his vomit, ſo is the fooliſh man who turneth to his fooliſhnes.* And in 2. Perer. 2.22. It is ſaide of theſe revolters, & back-turners to their ſinnes. *Canis reverſus ad vomitum ſuum, & ſus lota in volut abro luti. The dog is returned to his vomit, and the ſow that was waſhed to the wallowing in the mire*. This generall doggiſh propertie of back-turning, and continuall revolting to ſinne, without amendement, is here ſet formoſt, & is to be vnderſtood, common to the inchanters, whoremongers, murtherers, idolaters, & lyers after ſpecified, wherethrough, here they are debarred from God: otherwiſe, if they repented their ſins, & returned not euer thereunto, as dogs to their vomit, they might be receiued among the faithfull number, as Apoc. 14. note *d*, is declared.

[*o*] In ſimple tearms Chriſt here, & Apo. 5.5. ſetteth out his two natures very liuely, *I am* (ſaith he) *the root & the generatiõ of David.* This apparant contrariety making Chriſt the roote whereof *David* is ſprung, & again Chriſt to be a branch, who is ſprung of *David*, is doubtleſlie to be vnderſtood

diverſly. The firſt of Chriſts deitie, in reſpect wherof, he was before *David*, as he teſtifies ſaying [Iohn. 8.58.) *antequàm Abraham fieret, ego ſum, Before Abraham was I am*: yea, ſeeing God by Chriſt created the whole worlde, [Coloſ. 1.16. and Iohn 1.3.] therefore, Chriſt in his Deitie, is the very roote and originall, whereof all things, chieflie godlie men, as *Abraham* and *David* hath their beginning and being. The ſecond of Chriſtes humanitie, in reſpecte whereof, Chriſt contrarilie is ſprung of *Abraham* and of *David*, Matthew. 1.1. and ſo is Chriſt perfitelie both Davids roote, and alſo his branch.

(*p*) For many cauſes is Chriſt called the morning ſtar: for as the morning ſtar endeth the nights darknes, and beginneth the day light. So doth Chriſt at his comming put end to all errors. And doth begin and yeeld to vs that eternall light and day, which never decaieth: of which is ſaid before, verſe 5. *Et nox vltra non erit illic, & non egebunt lumine lucernæ, neque lumine Solis, quoniam Dominus Deus illuminabit illos. And there ſhal be no night there, & they need no candle, neither the light of the Sun, for the Lord God giueth thẽ light.* For that ſame cauſe, is he likewiſe called, *Sol Iuſtitiæ, the ſunne of iuſtice*, that riſeth to the world, Malach. 4.2. Alſo Chriſt is called this morning ſtar, to ſhewe thereby, that the ſtar that ſhould riſe of *Iacob* (Numb. 24.17) was performed in his perſon. Moreouer, becauſe Chriſts nativitie & comming was repreſented by an extraordinary ſtar, ſeene in the Eaſt Orientall. And it is frequent in the ſcriptures, to call the thing ſignified, after the name and the ſtyle of the ſigne it ſelfe: as where Chriſt is called the Lambe, &c. Therefore is Chriſt called here the beautifull, orientall, riſing, or morning ſtarre.

(*q*) This ſpirit that craveth Chriſts latter comming, is neither the damned ſpirits, for that they thereby are to receiue no gaines but paine and eternal torment. Neither yet is it ſimplie and in itſelfe the ſpirit of God, for that holy ſpirit in himſelfe hath no yearning, nor other paſsions, but hath the preſent eſtate of all things agreeable to his will and ordinance, lacking nothing longer nor he craveth it or yearneth after the ſame. This therfore is the ſpirit of the holy, elect and choſen number, or the ſpirite of God in them, that craveth here Chriſts comming, & yearneth for the ſame. For ſeeing all the good creatures of God long for that day, that they may haue libertie, freedome & reſt in God, as ſaith *Paul* Rom. 8.21.22. Therefore, far rathereſt of all doe theſe godly ſpirits that God hath created and breathed in man, yearne and deſire Chriſts comming, that from henceforth they may be conjoyned with God their original, and haue the full fruition of his preſence, as ſaith alſo S. *Paul* (Phil. 1.23.) of himſelfe & his godly ſpirit, *deſiderium habens diſſolvi, & eſſe cum Chriſto*, by this deſire that *Pauls* ſpirit had to be with Chriſt before the latter day, and without the body, importeth he a far greater deſire that our ſpirits haue in yearning for the latter day, that body & ſoul altogether may be with Chriſt our Lord and Saviour.

(*r*) Although the greeke word may here ſeeme to meane alike. *Theſe men that are written*, or, *Theſe things that are written*: & alſo that this booke may either meane indifferently here the booke of life, or this preſent booke of Apocalyps: yet by reaſon of the like wordes contained in the former verſes, ſpeaking directly of the plagues written in this booke of Apocalyps. It muſt needs be, that contrarily theſe ſpoken of here, are the benefites and good things written alſo in this ſame booke of Apocalyps.

(*ſ*) Here endeth Saint *Iohn*, this holy Prophecie of Apocalyps: imploring the grace of Gods ſpirit through Ieſus Chriſt, to abide with all the faithfull. Here end we alſo that interpretation thereof, which God by that ſelfeſame grace hath made the faithful of theſe our latter daies to vnderſtand, in diſcoverie of Gods enemies, and revealing of his trueth, that his Church being purged from Antichriſtianiſme, may from hencefoorth abide pure and holy, and readie decked as a comely bride, waiting the ſudden comming of her Lord and bridegrome Chriſt Ieſus: To this God in Trinitie and Vnitie, who here giueth theſe graces to vs, do we therfore render eternal praiſe, honour and glory, for ever and euer.

Amen.

Rom. cap. 15. verſe. 4. 5. 6.

4 *Whatſoeuer things are written afore time, are written for our learning that we through patience and comfort of the ſcriptures might haue hope.*

5 *Nowe the God of patience and conſolation, giue you that yee be like minded one towards another, according to Chriſt Ieſus.*

6 *That yee with one mind and with one mouth may praiſe God, euen the Father of our Lord Ieſus Chriſt.*

To God onelie wiſe, be praiſe through Ieſus Chriſt for euer. *Amen. Rom.* 16.27.

To the miſlyking Reader Whoſoeuer.

Svch is the miſerie of theſe latter dayes, through ſelfe loue, and coldneſſe of charitie (as Chriſt foretelleth) that euery man envying other, aſpires by reproching his neighbours godly works, and good exerciſes, to get himſelfe a name, ſo that no mans works, how godly ſoeuer, lacketh ſome one or other deteſtable calumniator, and envious reprocher: Neither looke I more, than others more learned to eſcape the ſame. To thee therefore, whoſoeuer miſliker hereof, & opponer againſt this work, I haue thought good to direct this gentle admonition, crauing thee earneſtlie, in the feare of God, Read ouer, ponder well, print all in mind, cõfer places, leaue off partialitie. So truſt I in God, you ſhal find in the ſame, nothing either greatly worthy of blaſphemous reproach, or able juſtlie to be refelled: For firſt, as touching the opponing againſt theſe preſents, by blaſphemous reproches, becauſe that raylings, reproches, & invectiues, exceeds the boũds of brotherly loue, & Chriſtiã charity, I truſt you, nor no Chriſtiã profeſſour, wil borrow theſe manifeſt darts of the deuill, & armour of Sathan to fight againſt any man, euẽ not againſt a reprocher, & farre leſſe againſt me who reproch no man that euer wrote hereof: but iudging them all to haue done faithfullie as they could, doe therefore but ſimple without iniury to thẽ, ſet out this my preſent work, confirming the ſame by cõference of ſcripture expreſlie, (as the Lord knoweth) to glorifie God therby, & to profite thee, O Reader, in repelling of errours, & diſcouering the trueth, without partiall loue to Proteſtant, or hatred to Papiſt, as to this hour, neuer hauing receiued worldly iniurie of Papiſt, nor fleſhly cõmoditie for being Proteſtãt: yea, I ſpare not in this work, to blame the negligence of ſuch profeſſed Proteſtants, as in theſe days ar waxed cold, careleſſe & without charity: as again, I ſpare not to purge and excuſe the Pope and Papiſtes of ſome textes in this booke which others applie, againſt them: namely in that (contrarie the opiniõ of many learned) I purge the Pope from being the Apoſtat of the fift trumpet: And whereſoeuer I accuſe the Pope in this work, by admoniſhing of his errors & thretning the decay of his ſect, ſeat & kingdom: I do nothing more thereinto, thã the very euidence of the text, harmony of other ſcriptures, and the truth of God enforceth me vnto. This doing then being offenſiue to no good man: I am ſure no profeſſed Chriſtiã, will by blaſphemous reprochings, raylinges, publike ſlaunders, priuat backbitings, or ſuch other dartes of the Deuil oppone againſt this booke. But if ſecondly, leauing theſe Ethnick invectiues & detractiõs, thou thinkeſt by pretẽded arguments, to refute this work and forme of interpretatiõ therein contained: be ſure, that canſt thou not iuſtly, nor by no good reaſons perform, ſeeing the grounds & ſubſtance hereof are proued, both by ſure teſtimonies & very probable arguments: by ſure teſtimonies (I ſay) both of ſcripturs interpreting ſcripturs: whereof, our whole paraphraſtical expoſition, & diuers notes are compoſed & made vp, And alſo by the teſtimony of vnſuſpect writers, euẽ of Papiſts where they wrote of themſelues, & of ethnicks, where they wrot of themſelues and of theſe vnſuſpect hiſtories (verily credible againſt thẽſelues) is the hiſtoricall application of this work compoſed, & diuers amplifications amongſt the nots, made vp. Then reſteth our propoſitions & the reaſons that are among the nots, & theſe I ſay we proue by very probable arguments: yea euen by arguments as neceſſary, as true religion (which falles not vnder the capacitie of man, nor necefsitie of mans reaſon) will ſuffer & permit: For theſe argumẽts are founded vpon the moſt euident reaſons, cleareſt tokens, & plaineſt ſcriptures, & thereafter euery argument follow one another, from the plainer text, to the obſcurer text, till the whole be diſcouered. So then of our propoſitions, Paraphraſtical expoſition, Hiſtorical application, Notes, Reaſons & Amplifications: & conſequentlie, of this our whole work, nothing is mine, or of my only aſſertion, to God be the glory, true teſtimonie & reaſon yeldeth al. Think not therfore to refute the verity, for ſtrõg & mighty is the protector therof, And wheras I haue ſaid that our Paraphraſtical expoſition, is not only groũded vpon the preſent text, but cõpoſed of the conference of many texts & ſcriptures interpreting others: that haue I purpoſly, & by al means poſsible preaſed to do, expreſly, becaus many interpreting the preſẽt text only by it ſelf, & according to that ſenſe, that to them ſeemed moſt proper in that place, they haue gretly erred, for to others another ſenſe doth ſeem more proper, & ſo infinit & repugnant interpretations do ariſe: For as in diſciphering of ciphered letters, he who judged ſuch a mark or character to ſignifie ſuch a letter, for that it ſo agrees & ſpels in ſome one place, he may be deceiued, and therfore he muſt tak that character for ſuch a letter as it ſpels & agrees with, not only in that one place, but in al places where it is foũd throghout that whol Epiſtle. Euẽ ſo in prophecies, who by an only text judged ſuch a prophetical terme or phraſe to haue ſuch a ſenſe, for that it ſo appears ther to agree, he may be deceiued: & therfore he muſt take that prophetical terme in ſuch a ſẽſe, as not only doth agre with that one preſent text, but with divers other texts where the ſam terme is to be found, & ſo ſhal al interpreters, agre, & no mã vary frõ other: for ſcriptures interpreting ſcripturs, furniſh one only veritie, & thogh ſomtime divers interpretations yet not diſagreing nor repugnãt, but in a maner al on: & after this maner of interpretatiõ I proceed, ſo far as God hath givẽ me judgmẽt, memory, & ſight of ſcriptures, referring the reſt to whõſoeuer God bleſſeth with mo talẽts, & greter benefits. If thẽ thou O Chriſtiã reader herof, firſt laying aſide blaſphemie, invectiues, & reproches againſt me: ſecõdly, caſting off al captious ſophiſms aganſt the truth, doth now think this form of Paraphraſtical interpretatiõ by ſcripturs interpreting ſcripturs, after the maner of diſciphering of letters, & this form of hiſtoricall application by approved, and vn-ſuſpect writers, and this forme of gathering our groundes by moſt probable indices and argumentes, to be the perfiteſt way to diſcover the myſteries of this holy *Revelation*, (as in trueth it is) & yet doth think that I haue not ſo accuratly followed that order my ſelfe, as I haue heere preſcribed it to other: then & in that caſe, I am not only cõtented, but earneſtly I craue and exhort thee, and all other, ſuſpecters hereof, whom God hath indued with greater benefits, to put hand to work, & follow this moſt perfect order, more accuratelie then I may attain to: for verely, I do acknowledge mine owne imperfection, euen my tender memorie, and hard vtterance: But ſure I am, and experience ſhall declare, that whoſoeuer by this moſt perfect way, and maner of expoſition ſhall proceed, he ſhal ſal vpon the ſame interpretation in ſubſtance, that I haue heere ſet downe, & though now thou diſagreeſt, yet then ſhalt thou accord herewith in one veritie, without any ſubſtantiall difference in effect or matter. I grant indeed, and am ſure, that in the ſtile of words & vtterance of language, we ſhal greatly differ: for therein I do judge my ſelfe iuferior to all men: So that ſcarcely in theſe high matters, could I with long deliberation, find words to expreſſe my mind: but this imperfection, ſeeing God perchance ſendeth it, to make the greater perſection in humilitie, and to beate downe vaine glory, and that *Moſes*, and many great men haue lacked the eloquence of tongue: I truſt charitable Chriſtians ſhal not diſdain me therefore, but rather amend the ſame in their own editions. Finally, by this my brotherlie admonition, I hope you will not giue place vnto thy firſt conceiued wrath: ſo that howſoeuer you at this firſt approching diſdained theſe preſents: yet now at thy more reſolued revewing hereof, I hope, & alſo proteſte, that in the feare of God, you caſting of all blaſphemous reproches againſt me, & al captious ſophiſmes, againſt the trueth, muſt either yeeld to the aſſured veritie of this worke, or then thy ſelfe ſet out the better expoſition heereof, vpon theſe, or more ſurer grounds: And I beſeech the Almightie, the God of trueth and vnitie, to bleſſe that thy godly enterpriſe, and to vnite it with his truth: So ſhall we alſo pleaſe one another, and abide vnited in him, who in Trinitie is vnited, & abideth bleſſed for euer and euer. So, fare you well in the Lord, and grow in him, from rankor & reproch to Chriſtian charitie: foreuer, *Amen*.

Your louing Friend in the Lord, I.N.

A
RESOLVTION,
of certaine doubts, proponed by
well-affected brethren, and needfull to be explained

in this Treatiſe.

AS *we are commanded by the Spirit of God to ſeparate our ſelues from all diſputers contentiouſlie, by ſtrife of words.* 1. Tim. 6.4.5. *ſo are we boũd and commanded with gentlenes and meekenes, to inſtruct all that are doubtfull minded, that they may know the trueth.* 2. Tim. 2.23.24.25.26. *And ſeeing there are certaine wel-affected brethren, who not in the ſpirit of arrogancie and contention, but in all ſobrietie and meekenes, haue craved of me the reſolution of ſome doubts, ariſing vpon my Treatiſe of the Revelation: therefore, for diſcharging my duetie, J haue thought good to write a Reſolution of their doubts, and to inſert the ſame in this Treatiſe vpon the Revelation, for the better ſatisfaction of their reaſonable deſire, and inſtruction of others, meeke and zealous perſons, whom the like doubts might hinder. As to the contentious and arrogant reaſoners, I leaue them to the mercie of the Lord.*

That the ſpace betuixt one yeare of Jubilie, and the next yeare of Iubilie is 49. yeares preciſely, and not 50. yeares as ſome do ſuppoſe.

FOr the clearer vnderſtanding of this queſtiõ, there are herein (as in al numeratiõ) two things to be conſidered: the parts nũbred & maner of numeration. The parts that ar to be numbred or accoũted are three, the two extreme terms, & the midſt. In a who Iubilie ſpace, the firſt yeare, is the firſt tearme, or extremitie, and the laſt yeare, which is called the year of Iubilie, is the laſt tearme and the yeares that are interiected betwixt theſe (how many ſoeuer they be) are called the middle yeares. There be alſo three manners of reckening, or accounting, The firſte, that neglecting and excluding both the extremities, accounteth onlie the middle interiected yeares, or vnities, as amongſt vs, our Lawyers in the account of the ſixe daies, that goe betwixt euerie citation, and ſummonds of the letters of foure formes, they neither account the firſt daye of the ſummondes, neither the nexte day nor any daye, vpon which they doe ſummond, but leauing out the extreames, they recken onlie the ſix middle whole daies, vpon the which no citation, or ſummonds falleth. As for example, If the firſt ſummonds be execute vnpo Tuiſday, it is not Lawfull to execute the next ſummonds, before the next Tuiſday and this they call a ſummonds of ſix daies, while as neuertheles, it accompliſheth, &

filleth out the whole weeke, & this forme of account, is rare, and according thereto, the Iubilie hath but 48. yeares. The ſecond forme of account is *incluſiuè*, where the midſt with both the extreame tearmes is included, and reckened, and this forme is more rude and vulgar, than preciſe, and vnder this forme, are expreſſed all computations, that are made by ordinall numbers, ſuch as firſt, ſecond, third, &c. As for example, a child is new borne, yet euen then is the firſt yeare of his age: or a Prince is new crowned, and euen then goe out the edicts dated with the firſt yeare of his raigne, yet hath he not raigned one yeare, while that time twelue month.

Item, among the Muſitians, the eight voice, or octaue aboue *de-ſol-re*, is called *de-la-ſol-re*, and the octaue aboue *de-la-ſol-re*, is called *de-la-ſol*, yet from *de-ſol-re*, to *de-la-ſol*, there are not twiſe eight, or ſixteene voices or harmonical notes, but fourteene, alanerlie, and yet is that ſpace called two *octaves*.

Item, among the Papiſticall cleargie the ſame weeke day that the Epiphanie of *Vp-hallimeſe* fell vpon, in the next weeke was called *octava Epiphaniæ*, that is, the eight day from *Vp-hallimeſe*. Alſo, that ſame weeke day that was the feaſt of the tabernacles, when it returned againe in the next weeke, was called (Levit. 23 39.) *octava feſti tabernaculorum*, the eight day from the feaſt of the tabernacles, while as nevertheles, it is vnpoſsible that betuixt one tuiſday, and the next tuiſday, or one wedneſday and the next wedneſday moe daies than ſeven can be contained. Chriſt aroſe from death the third day, yet lay hee not the more three daies dead, yea, ſcarſe two, as ſhal be ſaid hereafter: and after this count *incluſiuè*, the Iubilie is the fiftieth yeare.

The third forme of numeration and accounting *excluſiuè*, and which is moſt exact, and moſt agreeable with reaſon, is where the midſt with one of the extreme termes is included and reckened, and the other terme excluded. Vnder this forme of reckening, falleth the true account of al reckening by cardinal numbers that trulie inquire how manie, and not of what order, as for example, To the queſtion, how many yeares a Iubilie intervall or ſpace betuixt the ſounding of one Trumpet, and the ſounding of the next doth comprehend: *Moſes* doeth moſt truly anſwere, after this third forme in theſe words. (Levit. 25.8.9.) *Thou ſhalt number vnto thee* (to wit, from the former ſound of the Iubilie trumpet) *ſeven Sabbothes or weekes of yeares, even ſeven times ſeven yeares, and the ſpace of the ſeven Sabbothes of the yeares will be vnto thee nine and fourtie yeares: Then thou ſhalt cauſe to blowe the Trumpet of the Iubilie the tenth day of the ſeventh moneth*, &c. So that here is directly anſwered, that 49. yeares are interponed, betuixt the ſounde of the trumpet of the one Iubilie, vnto the ſound of the next. Likewiſe, vnto the queſtion: *quotus ſit Iubileus*? Of what order is the year of Iubilie? *Moſes* immediatly after in the tenth verſe maketh anſwer that it is the fiftieth year: As whẽ one would ſay together and in one ſpeech, a Childe of two yeare olde, died the thirde yeare [while perchance he lived not two or three daies after the two yeares expired.] Item, Chriſt lay dead and buried two daies, the third day he roſe againe, while as he onely laye dead thirtie nine houres. For, from the Fryday at the Iewiſh nine houres, (which to vs is three after noone) that Chriſt gaue vppe the Ghoſt, till Sunday at the Sun-riſing that Chriſt aroſe, are no moe houres interjected, which nevertheles, becauſe they participate with a part of the Frydaie, whole Saterday, and beginning of the Sunday, therefore is Chriſt ſaid to ariſe the third day by the former ſecond maner of account: Whereas, by the third maner of moſt exact count he remained not two daies dead. By the like reaſon, the Iubilie by the ſecond maner of account, and by ordinall number falleth in the fiftieth yeare: but by the third maner of true account, it containeth onely fourtie nine yeares betuixt trumpet and trumpet. Neither is this any repugnance to ſay a Iubilie falleth in the fiftieth yeare, and is but 49. from trumpet to trumpet, more than to ſay of Chriſt, he lay only but two daies dead, & aroſe the third: for as the two daies that Chriſt lay dead containe the evening of the firſt day, and morning of the third, and whole ſecond day: So a Iubilie interval or ſpace containeth the laſt halfe of the firſt Eccleſiaſticall yeare, even the harveſt and Winter quarter thereof, and the firſt halfe of the fiftieth and laſt Eccleſiaſticall yeare, even the ſpring and Sommer quarter thereof, together with fourtie eight whole and vndivided Ec-cleſiaſticall yeares.

Who thẽ doth not evidently ſee that fourtie eight whol years with two halfe yeares, doe make iuſtlie the ſpace of fourtie nine civill yeares onely: and yet to participate and take part of fiftie Eccleſiaſticall years? For the civill yeares, which the firſt Patriarches obſerved from the beginning, differ nothing from the Eccleſiaſticall yeares, which God appoynted afterwards to the Iſraelites, Exod. 12. except onely that the Eccleſiaſticall yeare began at the Equinoctiall of the ſpring, and the civill or Iubilie yeares had their beginning at the harveſt Equinoctiall.

Theſe things well vnderſtood, yee ſhall eaſilie conſider a Iubilie ſpace (that is, betweene trumpet and trumpet) not to containe fiftie yeares, but onely fourtie nine yeares preciſely: As *Theodorus Bibliander, Ioſephus Scaliger, Paulus Cruſius, Iunius*, and the learnedſt neotericks do acknowledge. Alſo, the ancient Iewes, as well as the modernes, are found by their tables, cycles, and periodicall numbers of yeares, ever and at all times to haue obſerued their Iubilie to be 49. yeares *excluſiuè*, that is, excluding the one terme, according to the ſaid third manner of account; or to be the fiftieth yeare *incluſiuè*, that is, including and reckening of both the termes by the ſecond manner of account: which commeth all to one effect, and agreeth to our aſſertion.

But who beleeveth that a Iubilie intervall, even the ſpace betuixt trumpet and trumpet, doeth comprehend fiftie yeares *excluſiuè*, by the third manner of exact and preciſe counting, hee alutterlie erreth: for it containeth onely 49. yeares preciſely, as wee ſhall prooue by theſe reaſons.

In all reckenings of Feſtivall and Sabbaticall times, the terme that is celebrated and kept holy, is ever a part of the Sabbaticall number or reckening, and is included thereinto. As in memorie of the weeke of daies, the eight day is not celebrated or kept holie, but the ſeventh day, which is included within the weeke, and is a part thereof.

Likewiſe, in the weeke of yeares, the eight yeare, which is excluded, is not celebrated, but the ſeuenth yeare which is included within the ſaid weeke: But the Iubilie intervall, or ſpace, is a time moſt Sabbaticall: for the Iubilie is a Sabboth of Sabboths, as is evident by the text of Leviticus, 25.8. therefore the Iubilie yeare, which is the obſerued and celebrated tearme of a Iubilie intervall, muſt not fall out of the ſeuen weekes of yeares, or 49. yeares, but within the ſame.

The ſecond reaſon, that if in this analogie, betwixt Sabbaticall dayes, Levit, 23. and Sabbaticall yeares, Levit 25. the feſtiuall day of the firſt fruits, which is called by *Moſes*, the fiftieth day, from the ſhake offering, be preciſely to be holden for the laſt of the 49 dayes, it muſt follow by the like reaſon, that the yeare of Iubilie, which *Moſes* alſo calleth the fiftieth yeare, muſt preciſelie be holden for the laſt of 49. yeares, but that firſte is true, therefore this alſo: for from the morne after that Sabboth day of the ſhake offering (Levit. 23.15,16.) vntill the morne after the Sabboth of the offering of firſt fruits, there are numbred ſeuen Sabboths, or, ſeuen weekes fullie: but theſe ſeuen weekes counting preciſelie and *excluſiuè*, muſt needs be (by the ſaid third maner of counting) onely fourtie nine daies, and yet the laſt and hindmoſt of theſe 49 daies is called by the number of ordour, and by the ſecond manner of counting *incluſiuè*, the fiftieth day in order. Therefore alſo the yeare of Iubilie, which he likewiſe calleth the fiftieth yeare in ordour, is alſo to be holden by iuſt account and *excluſiuè* for the laſt and hindmoſt yeare of fourtie nine.

The third reaſon, that if the ſpace of a Iubilie contained fullie fiftie yeares, the yeare of the Iubilie would oft befall betuixt two Sabbatical yeares, and ſo there ſhould occurre then within eight yeares, three yeares of reſt, in which the ground ſhould be vntilled: to wit, one Iubilie, and two Sabbatical yeares: yea, and more (which is a great abſurditie) oftentimes two of theſe yeares of reſt would fall both together: but that ſuch an inconvenient at anie time befel, it cannot be prooved, neither by the holie write, nor Iudaick hiſtories. Moreover, if at any time two yeares of reſt had bene to haue fallen together, would not God, when (Levit. 25.20.21.22.) he anſwereth to them that were doubtfull what they ſhould eate in that one only yeare, in which they did ſowe nothing, haue rather anſwered them what they ſhould eate in theſe two yeares, in which they were to ſowe nothing.

But ſeeing theſe inconveniences never occurred, therefore neither is the Iubilie interval of fiftie yeares, but of fourtie nine yeares only: For this our Iubilie of fourtie nine yeares ever falleth vpon a Sabbaticall yeare, becauſe the ſeptenarie number doeth divide and meaſure fourtie nine, but not fiftie.

Fourthlie, if the ſpace of a Iubilie containe fullie fiftie yeares, it would neceſſarily follow that in each ſeventh Iubilie, the firſt yeare, the eight yeare, the fifteenth, the 23. the 29. the 36. the 43. and the 50. yeare it ſelfe ſhould al be Sabbatical & yeares of reſt, and ſo that fiftie years Iubilie ſhould ſometimes containe eigh Sabbatical years of reſt. But ſeeing it appeareth evidently by the text, that the true Iubilie neither containeth moe Sabboths or weekes of yeares than ſeven, neither any moe years of reſt to be contained within the ſpace of a true Iubilie than ſeven: therfore the true Iubilie ſpace is not of fiftie yeares, which bringeth the ſaid inconvenient, but is of 49, years allanerlie, of which ever the ſeventh, 14. the 21. the 28. the 35. the 42. and the ſelfe Iubilie yeare 49. are ever and vnchangeably Sabbaticall and years of reſt, abiding ſeven in number only, as the text doth appoint, all inconvenients and abſurdities being remooved.

Fiftly, ſeeing in that ſelf-ſame ſeventh moneth of *Tiſhri* (Deut 31.10. Levit. 25.9.) both the Sabbatical yeares, and the Iubilie yeares begin their account, and vpon that ſame moneth alſo endeth their account: as alſo ſeeing the Iubilie interval is compoſed of ſeven Sabboths or weeks of years, and by the ſound of the trumpet is deſigned, it is certaine, that the beginning of the firſt year of the firſt Sabbatical weeke, with the beginning of the firſt yeare of the Iubilie intervall, doe fall both at once in that ſame moneth, and alſo that the end of the ſeventh year of the ſeventh Sabbatical weeke, together with the end of the hindmoſt yeare of the ſaid Iubilie, doe fall alſo together in the ſaid moneth of *Tiſhri.*

But betuixte the beginning of the firſte yeare of the firſte weeke, and the end of the ſeventh yeare of the ſeventh weeke, it is vnpoſsible that any moe yeares can be interjected than 49. therfore the ſpace of a true Iubilie, muſt neceſſarily be of no mo years than fourtie nine.

Sixtly, ſeeing 400. and 90 yeares, or *Daniels* ſeuentie weekes of yeares may iuſtly be called the great Iubilie (as wee haue ſaid in our fift reaſon of our fift propoſition) for theſe weeks do expreſſe all the myſteries both of the common Sabbatical years, & common Iubilies, for as the common Iubilie brought in the looſing & remiſsion frõ bodily ſervitude: ſo did this great Iubilie bring in remiſsion of ſinnes and freedome from ſpiritual ſervitude: as that Iubilie brought in the reſt from bodily labours, ſo this Iubibilie hath brought in reſt to our ſoules, by finiſhing the wickedneſſe, ſealing vppe our ſinnes, and reconciling our iniquities, and bringing in our everlaſting righteouſneſſe, even the Meſsias (Daniel 9.24.25.) As theſe Iubilies at the appoyntment of GOD, had their beginning in the firſt yeare of the poſſeſsion of the land of Canaan, and were to be obſerved vnto the time of the extreme deſolation thereof: even ſo this Iubilie, at the command of the Angel *Gabriel*, began at the out-going of the command, to reedifie & repaire Ieruſalẽ againe, & endeth at the warfare, deſtruction, & horrible deſolation of the temple, of the towne, and of the whole kingdome of the Iewes.

Further, as the ſound of the trumpet [Levit. 23.24. Numb. 29.1.] vpon the firſt day of the ſeventh moneth, ſignifieth the vulgare Iubilie to containe ſeven Sabboth or weekes of yeares, even ſo the ſound of the trumpet ſounded vpon the tenth daye of the ſeventh moneth, betokeneth this great Iubilie to containe ten times ſevẽ Sabbothes or weeks of years, or ten cõmon Iubilies. But this great Iubilie of 400. and nintie yeares, neither doth containe ten nor nine, nor any whole number of quinquagenaries or fifties: for 50. doth not divide nor meaſure 400. & 90. but contrarily, fourtie nine, doe preciſely meaſure and divide foure hundreth and nintie, tenr. etimes: Therefore, the true leſſer Iubilie, whom the greater doeth ten times containe is a Iubilie of 49. yeares, and not of fiftie.

Seventhlie, from the yeare of the dedication of the Temple, to the end of the captivitie, wee finde not in the Scripture, that the people obſerved their Iubilie and Sabbaticall reſtes, as God had commanded: but contrarily, laboured the landes thoſe yeares with out intermiſsiõ. Now the ſumme of theſe years by the moſt learned is reckened, to be 490. years, which containeth alanerly but 70. yeares of Iubilie and Sabbatical reſts, if Iubilie yeares fall vpon Sabbaticall yeares, as our Iubilie of 49. yeares ever doeth.

But if the Iubilie contained 50. yeares, then would the ſaid 490 yeares, containe 78. or moe Iubilie and Sabbaticall reſts, which is not of veritie, for the ſaid ſpace of four hundreth and 90. yeares, that the people neglected the yeares of reſt, containeth only 70. yeares of reſt: for the which the people & the land were puniſhed with as many yeares deſolation, euen 70. yeares, till the land had fulfilled her Sabboths, [2. Chron. 36.21, Ezra 1.1] and therefore it followeth that the Iubilie reſts are not accounted according to the fiftieth yeare, but according to the 49. yeare onlie: and conſequentlie, the Iubilie intervall, or ſpace contained, 49. yeares preciſely and not 50.

Laſtly, not onely the tables, computations, and reckenings of the Iewes, whereof *Ioſephus Scaliger* maketh mention, and their continual cuſtome, as *Tremellius* (who alſo was a Iew) teſtifieth in his notes vpon the 25. of Levit. but alſo manie and diverſe texts of holy Scripture doeth evidently prooue that not only the reckening of the Iubilie, but of other feſtivall times, was reckened amongſt the Iewes *incluſiuè*, comprehending both the extreame tearmes, and that, that forme of reckening was cuſtomably vſed among them in all times by-gane, as where the Sabboths or feaſtes of the Tabernacle, whereof the firſt was obſerved vpon the fifteenth day of the moneth *Tiſhri*, the next, the 22. of the ſame moneth, that firſt Sabboth is called the firſt day [Levit. 23.39.] and that next Sabboth is called the eight day, and ſo both the extreme tearmes are included in the reckening: Otherwiſe, it could not be poſsible, that one Sabboth from another, or the 22. day from the 15. day could be called or accounted the eight day, in reſpect that by iuſt calculation *excluſiuè*, one Sabboth frõ his former Sabboth muſt be but ſeven daies different. Likewiſe, how were it poſsible, that the latter feaſt of the weekes could be ſaid to be obſerved on the fiftieth day after the firſt feaſt of the weekes, ſeeing both the feaſtes fall vpon the ſame day of the weeke, and haue whole weekes of daies interjected, vnleſſe that both the feaſts or feſtivall termes were *incluſiuè* reckened: therefore, by the like reaſon in all points, it were not poſsible to the latter Iubilie to be called the fiftieth yeare from the former Iubilie, ſeeing both fall vpon a Sabbatical yeare, and haue ſeven Sabbothes or weekes of yeares interjected betweene them, vnles that alſo both the extreme termes or yeares of Iubilie, together with the fourtie eight midle interiected years were al numbered together, making fiftie *incluſiuè*, which is preciſlie fourty nine *excluſiuè*, and agreeth to our Iubilie, which was taken to prooue.

The ſecond reſolution.

That the yeare of God 71. and conſequently each 49. yeare thereafter are Iubilie yeares, and not the years of Chriſts birth, as ſome ſuppoſe, nor of Chriſts paſsion, as others.

HAuing ſufficiently hitherto prooved, that a Iubilie intervall, or ſpace betweene trumpets, containeth preciſely 49. yeares *excluſiuè*: it reſts now to prooue what yeares of Chriſt, the yeare of Iubilie falleth vpon: for manie wreſte the ſame violentlie, ſome to the yeare of Chriſts birth, ſome to the yeare of his baptiſme and teaching, and ſome to the yeare of his paſsion: but here to the true reckening will not agree, which being deduced from the beginning of the world, to the yeare that the people of Iſrael divided & tooke poſſeſsion of the land of Canaan, and from that yeare to the end of the captivitie and beginning of *Daniels* ſeventie weekes, and from that time againe, to the deſtruction of Ieruſalẽ, at the which *Daniels* ſeuentie weeks did end, doth make all theſe periods to fall vpon Iubilie yeares, and doeth no wiſe admit that either Chriſts birth, baptiſme, or paſſion ſhould be a yeare of Iubilie: for the true figure of the Iubilie was not ſo much a ſigne of our libertie in Chriſt Ieſus, while as we are in the miſerable perplexities of this world, or yet a memoriall to the Iewes of their poſſeſsion of the land of Canaan to the finall deſolation thereof, vnder

Titus and *Vefpafian* in *Anno Chrifti* 71. as it is chiefly a figure of our fpiritual and eternall Canaan, wherein Chrift Iefus fhall at his feconnd comming infeft and poffeffe vs reallie. For in trueth & by iuft account, it is found that as the obferved Iubilies began at the poffefsion of the land of Canaan, and ended at the faid deftruction of Ierufalem in *An.* 71. even fo wee finde that by the faid iuft account, thefe fame Iubilies though vnobferved to haue begunne at the creation of the world, and appear to end at Chrifts fecond comming, as a figure thereof, as further I intend; God willing, to fet out fometime in a litle table of chronologies: but in the mean time I remit the reader to *Iofephus Scaliger de emendat. temporum*, whom I cannot but follow in fo far as he prooues and demonftrates, and no otherwife: for both by him, and by the Iewes tables and accounts it is found, that the faid yeare of the deftruction of *Ierufalem*, was the yeare of Iubilie: and it agreeth very wel with reafon that the myfterie of the obferued Iubilies, which tooke the beginning at the firft poffefsion of the land of *Canaan* fhould ende and be referred to the hindmoft loffing and deftruction thereof, and confequently, that 71. yeare to be the yeare of Iubilie, advife with *Daniel* in that matter, who beginneeth the reckening of his tenne Iubilies or feventie weekes, at the going out of the commandement to build vp and repaire *Ierufalem* againe & endeth the fame weekes or Iubilies at the faid deftruction, and defolation, bringing in, in the meane time, the death of the *Mefsias*, and his abrogating of all facrifices, [verfe 26.27.] not in the end of a weeke, and farre leffe of a Iubilie, but almoft in the middeft thereof: for he faith [chap. 9.24.] *feuentie weekes*, (which we haue lately prooued to be ten Iubilies) *are appointed over thy people and ouer thine holie citie.* And thereafter, verfe 26. *And the people of the Prince that fhal come, fhal deftroy the citie, and the Sanctuarie, and the end thereof fhal be deftruction, & after the end of the war, perpetual defolation.* So in this deftruction & war is the end of thefe ten Iubilies and feventie weeks, and after that end, a perpetual defolation appoynted. But it is obiected, that the yeare of Iubilie is a figure of libertie, & therefore fhould rather end in Chrift, who is the author of our libertie, than at the deftruction & fubverfion of *Ierufalem* and captivitie of the Iewes. I anfwere, that the Iubilies and their figure doe not as yet end, but only their outwarde obfervation: neither did they begin vnder *Iofhua*, but from the beginning of the world. For like as many Iubilies preceeded the yeare of the poffefsion of the land of Canaan, which Iubilies were not celebrat nor obferved: even fo alfo, many Iubilies haue fucceeded after the deftruction of Ierufalem, and loofing of the land of Canaan, and fhal ftil fucced till Chrifts fecond comming: in the end of which reft, peace & eternal libertie fhal be brought in to the elect people, which were in vaine to looke for in this world. Herefore in Chrifts fecond comming, and not in his firft comming is this point of the myfterie to be perfited.

The third Refolution.

How, and for what caufes, (Propof 7.) both the laft feale, and firft Viall, or trumpet, doe begin at the deftruction of Hierufalem, in Anno 71. *& not the laft feale to end before the trumpets and Vials do begin.*

SEeing each one of the former feales, follow feuen yeares after other, fome thinke ftrange, why I giue not alfo feuen yeares, to the feuenth and laft feale, making it to end, in Anno 78. before I beginne the trumpets or Vials? I anfwere; I cut not off frõ the laft feale his progreffe be it feven years or longer more than from the former feales: but I make thefe feven yeares to be common to both the laft feale, and to the firft trumpet or Viall, becaufe I finde apparantly that the Spirit of God makes the laft feale and the firft trumpet or Viall to beginne at once, at the downe-cafting of the fire from the Altar. [Apoc. 8.5.) which was the fire of Gods revenge and vengeance vpon the Iewes, Anno 71. Why this is done, and not rather the trumpets or vials to begin Anno 78. thefe feeme to be reafons.

Firft, to the effect that the coherence and connexion, betwixt the hiftoricall part bygane of the feales, and propheticall part to come of the trumpets or Vials, might bee the more fure and firme: therefore, hath the Spirit of God, annexed both their beginnings, to one, and the felfe-fame memorable action, or epocha to wit, to the fnbverfion of *Hierufalem*, for the connexion and coherence betwixt the laft feale, and firft trumpet, would not be fo certain, if the one did begin after the fuppofed end of the other, as when they do convene and agree in fome memorable action common to both.

Secondly, if the feales were ended, and the fubverfion of *Hierufalem* were bypaft, anie time before the trumpets began, then fhould there remaine no memorable action about the yeare of Chrift, 78. from the which the trumpets or vials could be begun to be numbered, as from a fure epocha, and therefore it may bee that the Spirit of God hath thought that account more fure and certain for vs that beginneth at that notable yeare of the fubverfion of *Ierufalem*, thã at that, which would begin at the wafte & idle feventie eight yeere.

But fome will obiect that the text feemeth rather to conioyne the beginning of the feuenth feale, with the Angels receiuing of the trumpets and preparation of themfelues to blow: therefore, thirdly, to that I anfwere, that the preparation and action of blowing the trumpets were both at once, and not fo far fundrie as feven yeares, otherwife why would the text knit vp together the beginning of this preparation, which is an vnprofitable thing, with fo notable an epocha, as the fubverfion of Hierufalem, and not rather knit together therewith the beginning of the trumpets or vials, and of the notable actions to be done and fall out in them, if fo it were not that in that fame houre the feven Angels began to prepare their trumpets and vials, and the firft alfo to blowe his trumpet or powre out his viall: chieflie, feeing that fourthly (Apoc. 8. in the neereft verfes fift and fixt) the preparatiõ of the Angels to blow is immediatly ſubioyned after the thunders and lightenings of the wrath and vengeance of God, which fell on Hieruſalem.

Likewiſe, alſo in the neereſt ſixt and ſeventh verſes, there is ſubioyned to the Angels preparation that immediatly the firſt blewe the trumpet, neither is there in the meane time any action or delay of time, or ſo much as one worde in the text inſert betweene the preparation & the firſt blowing of the trumpet; Wherfore [by my ſimple iudgement] it may very well be inferred, that after the deſtruction of Hieruſalem, the ſeven Anqels prepared themſelues to blow the trumpets, and the firſt did even then blow, and not that the delay of preparation can be gathered by the text to be long, far les to be of ſeven years ſpace. Fiftly, that this delay of preparation, was either litle, or none at al, may be gathered by the laſt two verſes of the 15. chap, & two firſt verſes of the 16. chap. wherein alſo there is no action or delay, interponed betweene the Angels preparation or receiuing of their vials (which are all one with the trumpets, as our ſecond propoſition proueth) and outpowring of the firſt Viall: for inſtantlie after the Angels haue receiued their Vials, in the penult verſe of the 15. chapter, and the effect of the Vials is deſcribed in the laſt verſe thereof, immediatlie thereafter (chap. 16. verſe 1.) commandement is given to the ſeven Angels to powre out their Vials, and inſtantly verſe 2. the firſt powreth out his Viall.

Sixtly, ſeeing the ſeals began to be opened in Anno Chriſti 29, as is prooved by our 8. propoſition, then if (as is ſuppoſed) that after the end of the ſeventh ſeale, to wit, in the yeare of Chriſt 78. the firſt trumpet or Viall did beginne, this intervall would be of 49. yeares, even the ſpace of an whole Iubilie, & ſo the ſame according to the accuſtomed manner, and as we haue ſhewed in the third reaſon of our ninth propoſitiõ, the ſpirit of God would rather haue expreſſed them vnder the names of trumpets and Vials which comprehend whole Iubilies, than vnder the names of ſeales or ſealed weeks: and therefore theſe trumpets or vials could not beginne in the yeare of Chriſt 78. but in the yeare of Chriſt 71. making ſo from the beginning of the ſeales to the beginning of the trumpets or vials onelie 42. interjected yeares, which are ſixe ſeales completed, and ſeven ſeales begun.

Seventhly, if the firſt trumpet did begin in the year of Chriſt, 78. it behooveth to follow that the ſecond trumpet did beginne in the yeare of Chriſt 323. the third trumpet in the yeare of Chriſt 568, &c. and ſo neither would the trumpets or vials end at the yeare of Iubilie, neither begin thereat: neither [which is the greateſt abſurditie of all] could they iuſtlie be called trumpets, ſeeing they would not be diſtinguiſhed by the ſound of trumpets, as the Iubilies are.

Eightlie, if the trumpets did not begin in *Anno* 71. but in *An.* 78. and then confequently that everie one of the reft were remooved forwarde feven yeares fpace, it fhould then come to paffe, that manie of the actions which the text afcribeth to the latter trumpets, fhould contrarily fall within the former trumpets, which were a great abfurditie.

Ninthly, feeing by the Aoc. 15. penult. verfe, & our annotation *f*, there-vpon, S. *Luke* the Evangleift, hath written the whole actions of the Church almoft vnto the yeare of Chrift 70. in his booke of the acts of the Apoftles, & the reft of the actions from thenceforth to the end of the world, he hath cõmitted the declaratiõ therof to the 7. angels, by his delivering to thẽ of their 7. vials in a vifion: therfore it is more apparent that thofe Angels inftantly after the booke of the acts of the Apoft. in the 71. yeare of Chrift, did begin to powre foorth their propheticall vials, and to found their trumpets, than to haue over-lopen an whole 7. years waft & idle. Laftly, though the Spirit of God afsigneth a begining of time to the opening of the feals of our falvation, yet, who can iuftly afsigne an end to the progres thereof before the end of the world, for the white horfe yet rideth dailie fhoting his arrows & conquering in the elect, Apoc. 19.11.19. the red horfe daily perfecuteth thẽ: the black horfe bringeth hunger, dearth & other calamities among thẽ: the pale horfe herefies & errors, &c. And not that the progres of the firft feale ended at the opening of the fecond, in an. 36. but only his firft week, neither at the opning of the third feale ended the fecond, &c. And fo, feing the feals haue not properly an ende, it were very improper to make the 78. yeare [which is the end of the fevẽth week, or the end of the firft week of the 7. feal) to be the end of the very 7. feal it felf, or yet to begin the new account of the trumpets & vials at an endles end of the 7. feal. Now therefore, haue we explained this doubt fufficiently, as well that the 71. yeare of Chrift, as why the 71. yeare of Chrift is a common beginning to both the laft of the feven feales, and firft of the feven trumpets and vials.

The fourth Refolution.

That the fourth kingdom in Daniel, is the monarchy of the Romans *and not the fmal divided kingdoms of the* Seleucians, *and* Syrians, *as fome of late doe fuppofe.*

Let not the followers of *Tremellius* take it in evill part, nor think that I derogate to the authority of fo great a man in fo far, as in this I lift not to aplaud vnto his fentence, but rather to the opinions of *Calvin, Oecolampadius*, and the ancients, for many reafons force me therevnto. Firft, the vniverfal confent and authoritie of all the learned that haue written before *Tremellius*, who all with one vniforme accord, do refer and apply both the fourth iron Kingdome of the image (Dan. 2) & kingdome of the fourth beaft

[Dan. 7.] vnto the *Roman* monarchie, & do ſhew thẽ to agree in all their notes and tokens. Secondly, becauſe that the *Seleucians* & *Syrians*, who ſucceeded to *Alexander*, & *Alexander* alſo himſelfe are ſaid to raign as *Grecians*, & not to brook a diſtinct or divers kingdom frõ the *Greciã* Monarchie, which is prooved moſt manifeſtly by the 8. chap of Dan. where the two-horned ram is expreſly interpreted verſe 20. The kings of the *Medes* & *Perſians*, and the Goat is interpreted the whol kingdom of the *Grecians*, to wit, not only *Alexander* the great, but even alſo his whole ſucceſſors in theſe words, verſ. 21. *And the goat is the king of Grecia, and the great horne that is betweene his eies, is the firſt king* (to wit Alexander) *and where that it is broken, and four ſtood vp for it, are foure kingdomes that ſhal ſtand vp of that nation but not in his ſtrength.* Herewith agreeth Dan. chap. 7. verſ. 6. in theſe words: *After this I beheld, and loe, there was an other like a Leopard, which had vpon his back foure wings of a foule, the beaſt had alſo foure heads, and dominion was giuen him.* And in the beginning of the book of Maccabees, the ſame is made plain in theſe words, chap. 1. verſ. 5. *And when Alexander had raigned over regions, nations and kingdomes, they became tributaries vnto him. After theſe things he fell ſicke, and knew that he ſhould die. Then he called for the chief of his ſervants which had bene brought vp with him of children and parted his kingdom among thẽ, while he was yet aliue. So Alexander had raigned twelue years when he died, and his ſeruants raigned everie one in his roome, and they all cauſed themſelues to be crowned after his death, and ſo did their children after them many years.* Seeing then the *Seleucians* & kings of *Syria* whoſoever ſucceeded to *Alexander*, are holden for *Grecian* Kings, & the *Grecians* were *Daniels* third beaſt or Monarchie, it followeth therefore neceſſarily, that the *Seleucians* & *Syrians* were not Daniels fourth beaſt or Monarchy, but only a part of the third. Thirdly, the years of the raigne of the *Seleucians* & *Syrians*, through all the booke of the Macchabees are dated and ſtiled by the yeares of the Grecian reigne, & one & the ſelfe-ſame dating or *æra* doeth ſerue for both. Therfore vnder the ſelfſame kingdom & monarchy of the *Greciãs* are they cõprehẽded & are no fourth kingdome, &c. Fourthly, no man, as I beleue, wil deny, but that *Daniels* fourth beaſt (cha. 7) that aroſe out of the ſea, & had ten horns, & a mouth ſpeking great things, & words againſt the higheſt, & which ſuppreſſed the Saints of the moſt high for a time, times, & halfe a time, is that ſelfe-ſame beaſt which S. Iohn *Apoc.* 13. & 17. deſcribeth riſing likewiſe out of the ſea, & hauing ten hornes & a mouth ſpeaking great things & blaſphemy againſt God, & making war againſt his Saints, & overcomming them, & having power 42. moneths (which we prooue in our 15. prop. to be all one, with a time, times & halfe a time) but this beaſt of the revelation raigned in the daies of S. Iohn, and this metrapolitan citie reigned evẽ thẽ over the whol world, Apoc. 17. ver, laſt: but the *Seleucian* & *Syrian* kingdoms were extinguiſhed before the comming of Chriſt, as witneſſeth *Tremellius* himſelfe: Therefore the fourth beaſt, neither in *Daniel* nor in the Revelation can ſignifie the litle decaied kingdomes of the *Seleucians* & *Syrians*. To this argument I adde the inconvenient that would follow, if this fourth beaſt in the Revelatiõ ſhould ſignifie the kingdom of the *Seleucians* & *Syrians* then paſt and by-gone, then were the whole Apocalypſe made of none vſe, but alluterlie in vaine, as an idle & ſuperfluous hiſtorie of thinges by-gone, ſet out vnder myſteries, and not to be a prophetical premonition of things to come But to the effect that the trueth of this matter may be the more deeplie riped out, let vs weigh & conſider the chief notes & tokens contained in the text of *Daniel*, whether the ſame agree rather to the *Seleucians* & *Syrians*, or to the *Romãs*, & whether *Daniels* quadriparted image & four beaſts are to be extinguiſhed in Chriſts firſt comming, or in Chriſts ſecond comming. Fiftlie, then it is manifeſt, that this Image and theſe beaſts (at the leaſt the laſt of them) are after a maner to abide vntil the latter daies. For *Daniel* 2 28. ſaith, *But there is a God in heauen that revealeth ſecrets, & ſhewed the king Nebuchadnez ar what ſhal be in the latter daies.* But the latter daies, are the daies of the Antichriſt. 1. Iohn. 2.18. and ſince Chriſt, while as the Romans do reigne and the kingdomes of the *Syrians* & *Seleucians* long time ſince were extinguiſhed. Therefore, the fourth kingdome or Monarchie is not of the *Seleucians* & *Syrians*, but of the *Romans*. I know that ſome will miſconſtrue theſe latter daies, for the daies before Chriſt: but I appeale to the Angels conſtruction of the like viſion, Dan. 7.17.18. Where it is ſaid, that the four beaſts are foure kingdoms which ſhal riſe out of the earth, & they ſhal take the kingdome of the Saints of the moſt High, and poſſeſſe the ſame for ever, even for ever & ever.

Sixtly, Dan. 7.9.10.13.17.22.25.26.27. it doth evidẽtly appear that there ſhal ſome relicts or ſhew of the fourth beaſt & monarchy abide & remaine while the throns be ſet, & the anciẽt of daies ſit in iudgement, and the booke be opened, and till the ſonne of man come in the cloudes of heauen, and till he get all dominion and honour, and kingdoms, that all people, nations and tongues may ſerue him, and till his everlaſting kingdome come, which never ſhall haue an end: which doubtleſſe, is the day of Iudgement and Chriſt his ſecond comming.

And further, by the ſaid texts, the blaſphemous horne is to make warre againſt the Saints, and to prevaile againſt them, vntill the ancient of daies come and giue iudgement to the Saints of the moſt high, & that the time approach that the Saints poſſeſſe the kingdome. And the ſaid horne ſhall ſpeake wordes againſt the moſt high, and ſhall conſume the Saints of the moſt high, and thinke that hee may change times and lawes, and they ſhall be given into his hand vntil a time, times, and halfe a time, but the iudgement ſhall ſit, and they ſhall take away his dominion, to conſume and deſtroy it vnto the end. And the kingdome and dominion, and the greatnes of the kingdom vnder the whol heauen, ſhal be giuen to the holy people of the moſt high, whoſe kingdome is an euerlaſting kingdome, and all power ſhall ſerue him. Which all is doubtleſly meant of the eternall kingdome of Chriſt and his Elect, which is not in this world, nor can bee long before the day of iudgement, and blaſt of the laſt trumpet, Apoc. 11.15.17.18. But long before the day of iudgement, euen before Chriſts firſt comming, the kingdomes of the *Seleucians*, and *Syrians* decayed, therefore, their kingdome is not the fourth Monarchie, or fourth beaſt. Seuenthly, theſe foure beaſts or Monarchies ſhall take vnto them the kingdome of the Saints of the moſt hie, and ſhall poſſeſſe the ſame for euer, euen for euer and euer, [Dan. 7.17.18.21.] But who in theſe ages, ſince Chriſts natiuitie, haue taken vp, and poſſeſſed the kingdome ouer all Chriſtians, and Saints of God, except the Romans? It is therefore the Romans, that haue brooked the fourth beaſts Monarchie, & not the *Seleucians*, or *Syrians*: But ſome wil here object, that this poſſeſsion of the kingdome of the Saints, *for euer*, is but meant while Chriſts firſt comming: for all the ceremonies of the Leviticall Lawe, were commanded by *Moſes* to be obſerued for euer and yet were all abrogated, and annulled at Chriſts firſte comming. I anſwere hereto, that the ſingle terme [*for euer, in ſæculum*] is oftentimes found to end in Chriſts firſt comming, but we read not, but that whẽ the ſame is doubled *in ſeculum ſeculi*, or *in ſecula ſeculorũ*, that is to ſay, *forever & forever*. But then in that caſe it abideth ſo long as this worlde ſtandeth, even to Chriſts ſecond comming, far rather here, where the ſame terme is tripled in thoſe words: *And they ſhall poſſeſſe the kingdome for ever, yea, even for ever and ever*.

Eightly, it appeareth evidently that Chriſt Ieſus being the litle ſtone that was cut off the hil, at his firſt comming ſhould not become the great mountaine that filleth the whole earth, til neere his ſecond comming, at which time he ſhould ſmite the image vpon the feet, and breake the ſame (Dan. 2.31. &c.) and ſhould ſtand for ever and deſtroy all other kingdomes: for while as the Chriſtiã Church is but yet militant, & Chriſt that litle ſtone but as yet increaſing to a mountain, it is not apparant that he ſhal deſtroy the Image till his ſecond comming, at which time he ſhall become a great mountaine, filling the whole earth, and deſtroying all other kingdomes, and from thence he and his Church ſhal raigne and triumph eternallie, for the which it is written, Pſal. 109.1. *The Lord ſaid vnto my Lord, ſit at my right hand till I make thine enemies thy footeſtoole*. And Heb. 10.13. *He ſitteth for ever at the right hand of God, and from hencefoorth tarieth till his enemies be made his footeſtoole*. For if before Chriſts firſt comming this image of the wicked kingdomes and fourth beaſt had bene alreadie extinguiſhed, why would God haue bidden our Lord to tary and await for that deſtruction of his enemies, which alreadie was begun and paſt. Therefore this fourth kingdome of the image and beaſtes is to abide almoſt to Chriſts ſecond comming, and conſequently it is the *Roman* Empire yet extant, and not the decayed *Seleucian* and *Syrian* kingdomes. Ninthlie, the fourth beaſt by his wicked and litle horne, made warre againſt the Saints of the moſt high, and ouercame them, and they were giuen into his hand, for a time, times and halfe a time (Dan. 7.21.25) which wee haue proued prop. 16. to be the ſpace of 1260. years. But neither the *Seleucians*, nor *Syrians*, neither far leſſe *Antiochus Magnus* (if he be called the litle horne) did altogether raigne longer than 245, yeares. Therefore neither is the kingdome of the *Seleucians*, and *Syrians*, to be holden for the fourth beaſt or Monarchie: neither this *Antiochus*, for the litle horne thereof, but rather the Romane Monarchie: for that beaſt, and the Biſhope of *Rome* for this horne

Tenthly, the third kingdom (to wit of the Grecians) impired over the whole earth [Dan. 2.39.] and nevertheleſſe, the fourth beaſt or monarchie was much ſtronger than it, [Dan. 2.40. and 7.7.19.23.] but the *Seleucian* and *Syrian* kingdomes, were not ſo ſtrong as the *Grecian* kingdome, even of *Alexander* himſelfe (incaſe yee would diſtinguiſhe theſe from the Grecians) neither did they raigne over the whole earth, but were a great deale feebler than the kingdome of *Alexander* (Dan 8.22) where it is ſaid *that the great horne was broken and foure did ſtande vp for it, which are foure kingdomes which ſhal ſtande vp of that nation, but not in his ſtrength.* But the *Roman* Empire was a great deale ſtronger than the *Grecian* Empire, it is therefore the Empire of the *Romans*, and not the ſmal kingdome of the *Seleucians* and S*yrians*, that is to be eſtabliſhed for the fourth beaſt or monarchie.

Elevently, in the daies of theſe kingdoms ſhal the God of heaven raiſe vp the eternall kingdome of Chriſt, and of his Saints [Dan. 2.44.]. But long after the kingdome of the *Seleucians* and *Syrians* was extinguiſhed , muſt wee recken Chriſts kingdome to begin, whether yee count by his firſt comming, or by his ſecond comming. But contrarily, the *Roman* impiring, Chriſt firſt came in the world, and thoſe alſo not vtterly aboliſhed, ſhall Chriſt returne again in his ſecond comming. Wherfore it is ſaid of the *Roman* Antichriſt, *that the Lord Ieſus ſhal conſume him with the ſpirit of his mouth, and ſhal aboliſh with the brightneſſe of his comming* [2. Theſ. 2] Even not long before his ſecond cõming as in divers parts of the *Reuelation* is moſt manifeſt. Then ſeeing in the daies of the Empire of the *Romanes*, Chriſtes kingdome ſhould ariſe [the *Seleucians* and *Syrians* long before decayed] therefore the *Romans* and not the *Seleucians* and *Syrians* are to be reckened for this fourth beaſt or monarchy.

Twelfthly, ſeeing this fourth tyrannical Empire deſcribed by *Daniel* in his 2. & 7. chapters is not allutterly to be rooted out, til Chriſts ful kingdome firſt come. And feeing his full kingdome is not to be efteemed of this worlde, as Iohn. 18.36. teftifyeth, faying, *My kingdome is not of this world, &c.* And that artickle of the Lords prayer, *Let thy kingdome come*, And Apoc, 11.15. fpeaking of the latter day, faith, & *the feventh Angel blew the trumpet and there were great voyces in heaven, faying, the kingdoms of this world are our Lords, and his Chrifts, and he fhal raigne for euermore.* For Sathan, through all the Scriptures is called the prince of this world, and therefore vnto the end of the world there muft abide fome dregs & remanent of this fourth beaft or Empire: But the kingdomes of the *Seleucians*, *Syrians*, and of *Antiochus* evanifhed and decayed many ages fince, while as there yet abideth a fhew and remnant of the Romane Empire, both with the Emperours of Germanie, and with the Popes of *Rome*. Therefore it is moft fure that the title of the fourth kingdome or Monarchie in *Daniel*, ought iuftly to be attributed to the Roman Empire, and not to the fmal kings of *Seleucia* and *Syria*, which was taken to prooue.

Fift Refolution.

That the litle horne in Daniel *chap. 7. doeth fignifie the Romane Antichrift, and not* Antiochus *properly, as fome fuppofe.*

LEt the twelue reafons which we haue lately fhewed be confidered, by the which we haue prooved the fourth beaft, and all his properties, hornes and members to be referred to the Romane Monarchie, and not to the kingdomes of the *Seleucians*, and *Syrians*. There are indeed certaine notes and tokens fhewed in that chapter that doe agree both to the Roman Antichrift, and to *Antiochus*, yet truely and properly to the Antichrift: but to *Antiochus* they cannot be attributed except by a fimilitude and comparifon: wherefore iuftlie we call *Antiochus* to be the fymbol and figure of the Antichrift. But many things are herein fpoken, that cannot be iuftly referred to *Antiochus*, though fome do violently wrefte and throwe the fame.

By the Apoc. chap. 11.2.3. Wee are taught, that the prophane and idolatrous Gentiles, fhall poffeffe the externall and vifible Church, and fhall tread downe the true latent Church fourtie two Moneths, and that in that meane time the teftaments and witneffes of God fhall prophecie cladde in fack-cloth, for all the fame daies of 1260. And chapter 12.6. that the true fpoufe of Chrift, even his holy Church fled then in the wilderneffe 1260. daies, and againe (verfe 14.) that fhee flew away to the defert to her place, where fhe was nourifhed for a time, times, and halfe a time, from the prefence of the Dragon.

And by the 13. chapter, that the Antichriftian and blafphemous mouth of the beaft was opened with blafphemies againft God, to blafphem his name & his tabernacle, & thefe that dwelt in heauen: and that it was given him to make warre with the Saints, and to overcome them, and power alſo given him over al people, tribes, tongues and nations 42. moneths, and that all they that dwelt vpon earth, adored the beaſt, &c.

Now ſince all theſe actions are one (or at the leaſt) doe conveene and accorde in one purpoſe, therefore, alſo they muſt neceſſarilie conveene in one date or time: And that date by our 16. propoſition is prooved to be 1260. years, and the ſame is confirmed by the Apoc. 23.3. by the which it is manifeſtly ſaid, that the Antichriſtan ſeate ſhall adiudge and condemne to death, and ſhall behead the witneſſes of Ieſus, and them that haue not adored the Antichriſt, nor received his marke, but haue lived godly with Chriſt the 1000. yeares of the Antichriſtian raigne: by the which it is certaine that a time, times, and halfe a time, or a thouſand, two hundreth, threeſcore daies, are not three yeares and an halfe, as many doe miſconſtrue them to be, but to be 1260, years, as we haue prooved.

But now (to returne to *Daniel* again,) that litle horne is ſaid here to haue a mouth ſpeaking great things, & to vtter words againſt the moſt high, and to be more mighty then the reſt, and to conſume the Saints of the moſt high, & to think that he ſhal change times and Lawes, and that they ſhall be giuen ouer in his hand, for a time, times, and halfe a time. Behold how the Empire of this horne in *Daniel*, and the empire of the Roman Antichriſt, in the Apocal. doth moſt perfitely agree, both in their actiõs, & in their cõſent of time: The date therfore of the time, that that horn ſhal reign, muſt be 1260, years: neither would the old *Chaldeãs*, of *Daniels* time: nor the *Greciãs* of S. *Iohns* time, deny this form of account if they were aliue: for with them the name of the leſſer time, was taken for a yeare, and the name or terme of the greater time was taken for a yeare of yeares: or 360. yeares. And that forme of account, was vulgar, and familiar among them, though now to vs, they ſeeme myſteries. But ſeeing, not only *Antiochus* himſelfe, but all his predeceſſours, from *Alexander* the great, & al his ſucceſſours, vnto *Philip*, the laſt king of *Syria*, and *Aſia*, did no longer all reigne, than 245. yeares. And that the Pope or Roman Antichriſt, hath now reigned 1260. yeares. Therefore to this Antichriſt, and not to that *Antiochus*, muſt that little horne in *Daniel*, be attributed, And where it may be obiected, that by the books of the *Maccabees*, it doth appeare, that *Daniel* did meane of *Antiochus* I put a great diſtinction betwixt thoſe things that Daniel doeth ſpeak in one place of *Antiochus* properly, & in other places as properly of the Antichriſt. For I denie not, but that theſe things that *Daniel* doth ſpeake concerning that little horne, which waxed great toward the South, and towarde the Eaſt, and grew vppe vnto the Hoſt of Heauen, and tread downe of them, and did exalt himſelfe againſt the Prince of princes, and againſt his Saints, and which tooke away the daylie Sacrifice, 2300. euenings and mornings, (that is to ſay, naturall dayes) mentioned *Dan.* chap. 8. verſ. 9.10.11.12.13.14.23.24.25.26. & manie of theſe things which are ſpoken by *Daniel*, in his eleuenth chapter, doe properlie appertaine to *Antiochus*, and are exponed in the bookes of the *Maccabees*: for in theſe places, *Daniel* ſheweth definitly *Antiochus* riſing, his deeds & actions, with their times & dates: but contrarilie, I conſtantlie affirme theſe things that are ſpoken by *Daniel*, in the 2. and 7. chapter of the fourth kingdome, and fourth beaſt, and of thoſe that aroſe thereof, and of theſe time, times, and halfe a time, and of theſe things that are ſpoken by *Daniel*, in the 12. chapter, to be ſpoken of the riſing of the Antichriſt, of his reigne, and of the dates and times of his reigne, and of actions that were to be accompliſhed, after the birth of Chriſt: and neither to bee ſpoken of *Antiochus* properlie, neither yet to be exponed and performed in the books of the *Maccabees*. I feare greatlie, leaſt ſome that defend too wilfullie, that opinion, wold contend, and ſay alſo that all things ſpoken by *Daniel*, were accompliſhed before Chriſts paſſion, vnleſſe that Chriſts owne wordes gaue expreſſe cleare teſtimonie in the contrair, Mat. 24.15.16.17. Marke 13.14. Luke 21.20. ſhewing that ſome Prophecies of *Daniel* remained to be accompliſhed after Chriſts dayes.

The ſixt Reſolution.

That the Popes kingdome, both ſpiritual and temporal began in the dayes of Sylveſter *the firſt, betwixt the yeares of God* 300. *and* 330.

THogh this be proued in our 36. propoſition, yet ſome zealous brethren, vnderſtanding that the Pope acclaimeth the title of both the ſwords, and not finding great mention thereof in *Euſebius, Ruffinus, Socrates, & Theodoretus*, haue willed me to write more particularlie, and amply of the Popes power, vſurped ouer ſecular princes and kings, and likewiſe ouer the remanent Chriſtian Churches. The Papiſticall cleargie beginneth the account of their kingdome at the Calends of Ianuarie, the yeare of Chriſt 313. and from thence euen vnto this day, counteth by fifteenes, which they call *Indictiones pontificiæ*.

Againe, it is founde, that the Emperour *Conſtantine* remooued the imperiall ſeate from *Rome* to *Conſtantinople*, about the yeare of Chriſt, 320. As ſaieth *Nauclerus* or 330. as others write, or thereby: and ſo left the towne of *Rome*, at the leaſt, to Pope *Syluester*, if he gaue not the ſame to him. Againe, there is found into the *Decretals diſtinct.* 96. a gift and donation granted by *Conſtantine* to Pope *Sylveſter*, giving him the towne of Rome, the threefolde crowne, the Pontifical robe of ſcarlet and purple, the preheminence ouer other Churches, and divers other priviledges, which ſeeme to haue bene done about the 316. yeare of Chriſt. And Paul. 2. Theſ. 2. foreſheweth that the Antichriſt ſhall not be revealed till he who holdeth, hold ſtill a while, and then be remooved, or a diviſion and tranſlation made: which apparantly tooke effect, whẽ *Conſtantin* divided the Empire, & removed to *Conſtantinople* and left *Rome*. Theſe & other divers reaſõs make vs to begin the Popes kingdom about theſe times & dates.

But forſomuch as ſome ſuſpect the ſaid donation to be falſe and fained: amongſt whom is *Laurentius Valla*. Let vs therefore weigh his arguments, and we ſhall finde his chiefe reaſons to be, that it was not likely that *Conſtantine* would haue raiſed *Sylveſter* to ſo hie a preheminence of honour and power vpon the ſuddaine, ſeeing he received no poſſeſsion and deliverie thereof in hand. The ſecond reaſon, becauſe that *Euſebius* and remanent Hiſtoriographers aboue writtẽ, makes no mentiõ thereof. Thirdlie, he thinketh it to be vnlikelie, that *Conſtantine* or any other King would at any time haue given the like, or that *Sylveſter* would haue accepted the ſame.

Fourthly, VALLA doeth obiect the ſaid gift to be of ſo rude a ſtile and ſo barbarous Latine, that by reaſon thereof he iudgeth it to be fained and falſlie deviſed by ſome ignorant vnlearned man, and vnskilfull of eloquent Latine.

As to the firſt argument, the not-taking of actuall poſſeſsion and publike deliverie or infeſtment prooveth not the gift the more to be falſe. *Conſtantins* leauing of the towne of *Rome*, and *Sylveſters* dwelling thereinto was enough for the firſt poſſeſsion, till time ſhould offer occaſion, with leaſure and lenitie to creepe further in outward poſſeſsion. Which outwarde poſſeſsion and publike infeſtment, had both bene more reproachful and ſlanderous to *Conſtantine* to haue given, and more dangerous to *Sylveſter* to haue taken vpon the ſuddaine and violently, then ſoftlie at leaſure to creepe thereunto: for all ſuddaine promotions are envied and hated. Anent the ſeconde argumente, all Hiſtorigraphers knowe not all that is true, and ſome write not all that they know, but purpoſely conceale ſecret things, therefore all is not falſe whereof they omit to write. And yet the ſame is not omitted by other Hiſtoriographers, who (I muſt ſay) are worthie of credite. For whenſoever an Hiſtoriographer writeth againſt his affection in favours of the thing which he hateth, then is he the more to be credited: But *Hermannus Gigas, Malleolus de nobilitate*, and *Nauclerus lib. 2. generat. 11.* hating this gift and making it be a miraculous voice and hand-write to be a venome and poyſon to the Church of God: nevertheleſſe (in favours of the trueth of the gift) make it to be truely given, and not forged and falſe, otherwiſe, if it were fained, it had bene better and more eaſie to them to haue improoved it as falſe, than to let it ſtand as true, and to forge as falſe a lie, to make that gift thereby odious.

As to the third argument, whoſoever ſhall weigh and conſider well the actions of all Chriſtian Princes, and ſhall conferre the ſame with the holy Scriptures, they ſhall finde it moſt likelie, yea verie true, that both Princes haue giuen, and the Papiſt cleargie hath taken the ſubſtance of the rents of all kingdomes.

What Chriſtian kingdome is not heavilie rent and broken by their owne kings, and is diſtributed to the Papiſtical Church and cleargie? Shall it not be ſaid to be likelie to be true, or rather abſolutelie to be true? ſeeing the holie Scripture forewarneth the ſame to come. Apoc. 17. in theſe words, *The ten hornes are tenne kings, which as yet haue not receiued their kingdome, but ſhall receiue power as kings at one houre with the beaſt: theſe haue one counſel, and ſhal giue their ſtrength and power vnto the beaſt. &c.* Thereafter it is ſaid, *For God hath put in their to fulfil his wil, and to doe with one conſent, for to giue their kingdome vnto the beaſte, vntil the words of God be fulfilled.* Ye ſee then that Chriſtian kings muſt giue their kingdomes to the beaſt, Shall we then thinke that the Emperour ſhal be free from this decree of God? Nay truelie, Chieflie ſeeing that *Paul* hath expreſly foretold, that he muſt be remoued, and taken out of the way, and his kingdome divided, & the Antichriſt placed there, 2. Theſſ. 2. in theſe wordes, *And now, ye know what doth withhold that he may be revealed in his time: for the myſtery ef iniquity now doth worke, only he which holdeth now, muſt hold ſtil, til he be divided or remoued, and then ſhal that man of ſinne be revealed*, &c. It is ſaid Apoc. 13.12. of this Antichriſtian beaſt, That *he did all that the firſt beaſt* [to wit, the Emperour] *could do before him, &c.* And Apoc. 20.4. it is ſaid of the Papiſticall cleargie, *And I ſaw the ſeats, and they ſat vpon them, and iudgment was giuen vnto them, &c.* Whereby it is manifeſt, that not only a kinglie power was given by kings, but alſo an imperiall power was giuen by Emperours, at the pleaſure and appointment of God. Neither was it apparant that any Emperour would rather doe the ſame, than he who was the firſt Chriſtian, and baptized Emperour, and who was (as is alledged) miraculouſlie cured of his leproſie, and who firſt of all did tranſport himſelfe, and his ſeat out of the way, and dwelling in *Conſtantinople*, left *Rome* to the Pope. Neither doth it appeare, that any of the Popes would rather accept the ſaide gift, than he who after thirtie three Biſhops of *Rome*, that were martyred, was the firſt that purchaſed to himſelfe, & his after-commers freedome, and ſafetie in all time comming. Wherfore both *Conſtantine*, blinded with a ſuperſtitious ſimplicitie, might haue bin the ſame man, that would haue giuẽ the ſaid gift: as alſo *Sylveſter* might haue bin the ſame man, & ſo givẽ to auarice, that he wold haue accepted the ſaid gift, ſeeing that not only the holy Scripture teſtifieth, & forewarneth, that Chriſtian Princes ſhall be ſuch, but alſo the rich rents, & patrimonies, of the Papiſticall Church and cleargie, granted to them by the princes of al countries, doe proue and teſtifie, that the princes were ſuch men in effect.

Concerning the fourth argument, The rudenes of the ſtile and barbaritie of the *Latine*, were litle to the purpoſe, to improoue the gift, though that ſame Latin exemplar, had bene the principall gift: but farre leſſe doth it worke againſt the originall gift, which was written in Greeke language, as teſtifieth *Bartholemæus Picernus, de monte arduo*, who alſo tranſlated the ſame out of Greek, in better Latin. I haue ſeen alſo other tranſlations thereof in Latin. Therfore this Argument taken frõ the vnlearnednes, & barbaritie of the ſpeech, is without any force, ſeeing contrare a gift written in Greeke, the obiection of the barbaritie, of the tranſlation thereof in Latin is no matter. Theſe are *Vallaes* chiefe Arguments, who in the reſt, preaſſeth more through eloquence and ornat ſpeech, to perſwad than to proue or demonſtrat.

Laſtlie, ſeing in the end of his oration, he objecteth many crimes to the Pope, to proue that his tyrannie ſhould bee bridled: by theſe ſharpe & bitter invections, he appeareth to haue oppugned againſt the trueth of this gift, more for particular hatred he had againſt the Pope, than for any certaintie he had, that this gift was falſe or forged: which appeareth more wicked, troubleſome, & noyſome, to the true Church of God, than falſe or fained in it ſelfe. Read *Genebrard lib. 3. chronogr.* about the yeare of Chriſt 315. where he proueth the trueth of this gift, by teſtimonies of Ethnicks, or Iewiſh Rabbins, and by Kings, and Emperours, confirming the ſame. Neither doe I affirme it ſimple to bee true, but leaueth the ſame doubtfull, and vncertaine: for alwayes bee it true or falſe, it may iuſtlie be the period, or epocha, of the popes kingdome, ſeeing ſome of his cleargie do reckon therefra, and others do make their reckoning by indictions not farre therefra, as we haue ſhewed in our 36. propoſition. So all theſe periods of *Conſtantines* tranſporting of his ſeat, [be it Anno Chriſti 320. as *Nauclerus* affirmeth, or 330. or further, as others make it] and the period of the Papiſticall Indictions, Anno 313. and the period of *Conſtantines* gift, Anno 316. fall all in *Sylveſter* the firſt his dayes, ſo neere together, that it is no matter, from which of them wee rathereſt make the reckoning of the Popes ſeculare kingdome. As to his ſpirituall kingdome & power, vſurped ouer Chriſtian Churches, we can not begin that account at the time that they challenged, and claimed the ſupremacie ouer their brethren, while as yet they were but feeble, and lacked power to tyranize. For if we would begin at that time. *Lucius*, Pope of *Rome*, in his firſt Epiſtle, to the Biſhops of *France*, & *Spaine*, in An. Chriſti, 256. writeth, that the Roman church is the Apoſtolik head & mother of all Churches. Itẽ *Bartholomæus, Carranza Mirandenſis*, writeth in his book, *de ſumma conſiliorum*, that *Marcellinus* Pope of *Rome*, about the year of Chriſt 300. being convict by *Caius*, & *Innocencius*, deacons, & by *Vrbanus, Caſtorius, & Iuvenalis* prieſts, & by 28. witneſſes moe, & at laſt in a Synod of 300. Biſhops, and 30. prieſts, that he had ſacrificed, & offered incenſe in the Temple of *Iſis, & Veſta. Marcellinus*, his head ouerſpred with duſt, & aſhes, crieth out with a loud voice, I haue ſinned before you, & outht not to ſtand in the order of prieſts: for auarice hath corrupted me with gold, & they ſubſcribed to his conviction, and condemned him outwith the citie, ſaying, of his owne mouth, he is condemned,

& of his owne mouth, being accurſed, he hath receiued excommunication to death: for no man yet euer could judge ouer his Biſhop, or a prelat ouer his prieſt: for the firſt, and principall ſeat, cannot be adjudged by no man. Thou art the iudge, & of thy ſelfe ſhal be condemned: yet in preſence of vs, thou art the iudge & thou art the offender. Hitherto *Carranza*, ſpeaketh of *Marcellinus*. Again the ſame *Carranza*, ſpeaking of *Marcellus*, who was pope, about the year of Chriſt 309. writeth, that there is an epiſtle of his, written to the Biſhops of the province, of *Antiochia*, & extant in the decretals (2.46 *ad Romanam*) where firſt he diſcourſes of the primacie, of the Roman church, vnto the which al Biſhops ſhould haue their recourſe, as vnto their heade, according to the iniunctiõs of the Apoſtles, & ſhould appeal thereunto, that therfra they may receiue their ſentẽces & decrees, frõ whõ they haue receiued their orders, & cõſecratiõ: vnto the which ſeat, the ſeat of *Antiochia*, that ſomtime was the firſt, hath yelded & givẽ place

By thẽ it is appointed, & ordained that no Synodal convention ſhal be, without the authority of that ſame ſeat. Hitherto ſpeakes *Carranza*. But al this time, euẽ frõ the 256. year, to the 313. year of Chriſt, the popes wer but feble, & lacked tẽporal power, till *Sylveſter* (who began in the 313. year of God, to account their kingdome by indictiõs) did purchaſe ſecular authoritie, & power of *Conſtantin*, & thẽ, as ſaith *Carranza*, he made ordinãces for the liberty of the clergie, that no laick mẽ, ſhuld pretẽd to impute any crime to any of the clergie, & that none of the clergie ſhuld enter in iudgment, before a laik iudge, for any maner of cauſe, but before the church. Moreover, he ſaith, that *Sylveſter* made a ſynodal convention in *Rome*, in whoſe ſecond action, *Can.* 18. are written theſe words. *Nemo iudicabit primã ſedem, quoniam omnes ſedes a prima ſede iuſtitiam deſiderant temperari: neque ab Auguſto, neque ab omni clero, neque à Regibus, neque à populo Iudex iudicabitur, Et ſubſcripſerunt 284. Epiſcopi. 45. Deaconi, & duo ſequentes, Auguſtus Conſtantinus, & Mater eius Helena. Et fixit Canonem hunc Sylveſter Epiſcopus in Vrbe Roma, & omnibus Epiſcopis aſperſit & ſcripſit. Actum in Traianis thermis 3. Calend. Iulij Domino Conſtantino Auguſto tertiò & Priſco Conſule*: That is to ſay, *No man ſhall iudge over the firſt ſeate, for all ſeats doe deſire to haue iuſtice modified by the firſt ſeate, and neither by the Emperour, neither by the whole cleargie, neither by kings, neither by the people ſhal the iudge be iudged. And 284 Biſhops, 45. Deacons, and theſe two following, Auguſtus Conſtantinus and his mother Helen all ſubſcribed this Canon, and Sylveſter Biſhoppe ſet vp the ſame in the Towne of Rome, and wrote and ſpread it among all the Biſhops. Dated & done at Traians hote-bathes, the third day before the Calends of Iulie, the Lord Conſtantine Auguſtus the third time, and Priſcus being Conſuls.*

At this time therefore, we may juſtlie begin the Popes ſpiritual kingdome, which from this foorth, piece & piece, and ſoftlie increaſed craftilie, and grewe at length to that tyrannie, that the true profeſſours were martyred, and the true Church conſtrained to lurke, and lie inviſible. And whereas *Ruffinus* writeth, that it was ſtatuted in the Counſell of *Nice*, *vt Epiſcopus Romanus, ſubvrbicarum eccleſiarum ſollicitudinem gerat, That the Biſhop of Rome take a care and ouerſight of the Churches, in and about Rome*: It followeth not therefore, that onely this care and credit was committed vnto him: for it is another matter, to deſire a King or Biſhop to take care ouer his owne familie, and another thing, to bid him take care alanerlie ouer his owne familie: for there are manie Princes who while as they neglect their owne families, and families that are neereſt them, wil nevertheles be ſaid and accounted to haue preheminence and rule over all the families and maiſters of the families of the whole realme. Neither hee that chargeth a Prince with this particular care, doth the more diſtract from him the generall care: ſo now haue we diſcourſed ſufficiently the time that both the Papiſtical kingdomes, temporal and ſpiritual had their powerful beginning, which after my ivdgement are meeteſt to be accounted from the 313. yeare of God, and to proceed by the Indictions, as the Papiſtical cleargie themſelues do recken.

The ſeventh Reſolution.

That the Pope during his foreſaid reignes, hath poſſeſſed and corrupted the outward and viſible face of the Church, and hath perſecuted God his true Church, and made the ſame to lurke and become latent and inviſible all theſe daies.

For the better reſolution of this queſtion, who they are that haue poſſeſſed the outward Corte of the Temple, and doe tread downe the holy citie 42. moneths, Apoc. 11.2. I remit the reader to the notes, [*a, b, c,*] of the ſaid 11. chapter, where it is ſhewed that the vtter court of the Temple, is the outward and viſible face of the Church, with the externall rites and gliſtering ſuperſtitions thereof. The holy citie is the true latent Church and ſpouſe of Chriſt. The 42. moneths, are 1260. yeares of the Antichriſts reigne. But forſomuch as ſome doe thinke that the outward and external face of the viſible Church was not at that time corrupted with any externall rites or vaine ſuperſtitions: therfore, it is thought good that they be expremed in particular.

Theſe then that were deviſed about that time, were the Biſhops ſtafes, the Archbiſhops cloke or pale, the Cardinals croſſe and remanent ornaments (as they call them) of the Cleargie. Item, their Holie-water, their Chriſmes vſed in baptiſme, the ſhaving of their heades, their golden and ſilver veſſels in the Church, their albs and corporals of linnen for the altars, their conſecration of the altar, the obſervation of daies, the publike faſts foure times in the yeare called *Angariæ*, the faſting in Lent from fleſh, the choiſe of meates, the pretended chaſtitie of the Cleargie, the celebration of Maſſes in memorie of martyrs, and vpon their graues, the adoration of the croſſe, and many other ſuperſtitious dotings, which all were inſtituted and deviſed, and aroſe about the 313 yeare of Chriſt, together with the pride of Popes, & the immunitie of his cleargy frõ ſecular powers, & ſtile of primacie vſurped by the Romanh Biſhop. This proud vſurped primacie, with the glorious heape of other ſuperſtitions externallie gliſtering, are called the outwarde court of the Temple, that is to say, the externall and viſible face of the Church, and conſequently the authors, deviſers, intertainers & followers of theſe outward and glorious ſuperſtitions, to wit, the Pope & his cleargie, are theſe whom the text calleth Gentiles, to whom the outward court of the Temple was giuen, and who tread downe the holie citie, &c. But leaſt that ſome ſhould yet thinke, that the Papiſt Church, was not as yet become ſo Antichriſtian, nor the outward face of their Church, ſo corrupt in doctrine, but that good mẽ might haue adioyned them thereunto. I will declare beſides the former externall rites, three horrible, and deteſtable Antichriſtian tokens, introduced about *Sylveſters* dayes, amongſt the Papiſticall cleargie. The firſt is, the proud ſupremacie, mentioned 2. Theſſ, 2. by the which the Antichriſt ſhall extoll himſelfe aboue all that is called God. This primacie, and proud ſupremacie, we haue taught at length in diverſe places, that the Biſhop of *Rome*, doth claime and challenge to himſelfe, euer ſince the 313. yeare of God. The ſecond token forewarned by *Paull*, 1. Tim. 4.3. is the abſtinence from marriage, and choiſe of meats, which not only the Roman Biſhops *Teleſphorus, Calixtus, Fabianus, Melchiades*, & *Sylveſter* the firſt, haue appointed, and inſtituted: but alſo the Synodall counſels, conveened in this *Sylveſters* dayes, to wit, *Neoceſarienſe conſilium; Elibertinum*, and *Arelatenſe ſecundum*, and manie moe Counſels after, that haue approoved, and euer ſince *Sylveſter* the firſt his dayes, vnto theſe our dayes, are mightily defended. To theſe may be added, the third Antichriſtian token, which aroſe about the ſaid *Sylveſters* dayes, to wit, the abuſe and corrupting of both the Sacraments, for not onely the Popes of theſe dayes, but alſo their whole cleargie [in whoſe handes was the externall and viſible miniſterie of the Church] in ſtead of the Supper of the Lord to be celebrated of pure and ſimple bread & wine, to the behoue of Gods Saints liuing vpõ earth, did appoint the maſſe, of bread and wine, mixt with water, and ordained the ſame to be celebrated vpon the graues, and in memorie of men dead, & deceaſed (as teſtifieth, *Bartholomæus Carranza*, ſpeaking of Pope *Felix)* & in ſtead of pure, & vnmixt, water only to haue bin miniſtred in the ſacrament of Baptiſme, they adioyned in theſe daies conſecrat (or rather execrat) chriſme, like as ſhortly there after they eiked to ſpitle & ſalt. Who then of good conſcience and preciſe chriſtiã, wold haue ſuffred [by other ſuperſtitions] the holy Sacrament of Baptiſme, this wiſe to be polluted with chriſme, & the Supper of the Lord to be degenerate, and turned ouer to the maſſe, & to be celebrated for dead mẽ, & to be polluted with the relicts of dead mẽ [althogh martyrs] & to be mixed with water? Further, what godly mã could haue ſuffred, or allowed the choſe of meats, which Chriſt hath commanded vs to eat indifferentlie with thankſgiuing, euẽ of al that is ſold in the market, 1. Cor. 10.25? Or who wold comport with the forbidding of mariage to the clergie alluterlie, & to laick perſons, to the fift degree of affinitie, or conſanguinitie. Finallie, what godly perſon could comport, or beare with the pride, & primacie of the Biſhops of *Rome*, then be gun? Neither was the Synodall counſels altogether pure & clean, & free from theſe corruptions: yea, who can ſay, that the firſt coũſell of *Nice* (which was the pureſt, & moſt godly counſel, that we read of) was ſincere, & vpright in al the ſeuentie articles thereof, that was ſaid to haue bene ſtatuted at that time, ſeing we finde in thoſe 20. articles, that only are extant, ſome ſuperſtitious rites introduced, euẽ the obſervation of daies, to wit, of peace & ſuperſtitous pẽnances (as they cal thẽ) which afterward moſt abſurdlie were called, ſatisfactions? Likewiſe, in the general counſel, holdẽ at *Conſtantinople*, in the fift Canon therof, it is ſtatute, that the Biſhop of *Conſtantinople*, ſhal haue the honor of the primacie, after the Biſhop of *Rome*, becauſe that it is new *Rome* (*diſt*. 21. *Conſtantinopolitanæ civitatis*) in the counſel at *Chalcedon*, *Can.* 14. it is ſtatut that a Nun who marieth, after ſhe receiued orders ſhal be accurſed [25.*q*.1. *deaconiſſa non*] Itẽ *can.* 15. It is ordained, that neither a Virgin dedicat to God, neither a Monke may be ioyned in marriage, vnder the pain of excõmunication (27 *q*.1. *Si quæ Virgo*) Theſe & divers other abuſes, & ſuperſtitious rits, creeped in, in the outward & viſible face of the Church, in *Sylveſter*, the firſts dayes, & about the firſt beginning of the papiſtical kingdom. And as to the increaſe of their errours of idolatrie, artolatrie, transſubſtantiation, pardõs, indulgẽce, remiſsiõ of ſins, which they haue al broght in ſince, there is no mã ignorãt therof, that knoweth the hiſtories The greateſt queſtion conſiſted in their firſt errors, by the which in the beginning of their kingdome, they did pollute & corrupt the outward and viſible Church, ever more and more vntill our daies, during the ſaide 1260. years of their reigne. This ſarre concerning there poſſeſsing & brooking of the outward and viſible Church, and polluting of the ſame, all the daies of their raigne, which was the firſt part of the former reſolution: reſts to ſhewe the ſecond part, to wit, that the Biſhops of *Rome* haue, during this time, perſecuted the true Church of God, and made the ſame to lurke, and become inviſible.

There is none that haue read hiſtories, who will denie, but that there haue bene manie in the midſt of the Popes reigne, about the 1000. yeare of Chriſt, who haue opponed againſt diverſe of the Popes errours, and abuſes, and were therefore perſecuted, and diverſe martyred by the Pope. But manie men ſuppoſe, that in the beginning of his reigne, none made oppoſition againſt his errours, and none therefore was perſecuted by him, & his cleargie, at that time, as they ſuppoſe: but the contrare may be ſeen by hiſtories, & ſpeciallie in the book called *Catalogus teſtium veritatis*, where in An. Chriſti 300. *Lactãtius* did inveigh againſt images, ſuperſtitious rites & ornaments of the Churches, the Biſhops of *Rome*, contrarily, by edicts, & excomunications, defẽded the ſam.

Antonius Eremite, about the year of Chriſt 350. opponed againſt the worſhipping of Saints, reliques, & againſt the regular ordors. *Hilarius*, about the yeare of Chriſt 360. writeth againſt humane traditions, and againſt free-will, and plainly withſtood Pope *Leo* thereinto. *Vigilantius*, about that time, inveighed againſt the worſhipping of departed ſaints, and their reliques, and againſt compelled chaſtitie, and therefore was hee purſued, and condemned by the Pope, for an Heretik. *Epiphanius*, about the yeare of Chriſt, 370. plainlie inveighed againſt the worſhipping of *Marie*, and of the ſaints, and publikelie teare and rent a vaile, in the which the image of Chriſt was painted. *Auguſtine*, about the yeare of our Lord 380. inveighed againſt rites and ceremonies, againſt forbearing of meates and againſt adoration of ſaints, and authoriſeth the ſcripture aboue all traditions, and writeth ſoundlie of iuſtification. *Ierome*, about the yeare of God, 390. inveigheth againſt the primacie of the Biſhops of *Rome*, and remouing thence, to *Hieruſalem* in his old age, writeth in his Epiſtles, and calleth *Rome*, *Babylon*. About the yeare of our Lord 400. *Chryſoſtomus* inveigheth againſt the interceſſion of ſaints, and humane traditions. Againſt theſe three, *Ierome, Auguſtin, & Chryſoſtom*, thogh the Popes of theſe dayes durſt not much gainſtand, yet more craftilie. the Pope & his clergie ſenſine haue perſecuted their works, & books I meane, haue ſo altered, & perverted them, that they now ſpeak alluterlie in fauours of the Pope, againſt their firſt editions, as teſtifie ancient copies, yet extant, and *Index expurgatorius*. In the days of *Clodowick* king of *France*, the Biſhops of *France* & *Germany* wrote to the Pope *Anastaſius*, againſt his primacie & ſuperiority.

About the yeare of Chriſt 500. *Alachis* king *of Lombards* diſdained and inveighed greatly againſt the filthie luſt of the cleargie, vnder pretext of Chaſtitie, for the which they ſtirred vp warres againſt him. In the meane time, he made his vowe that he ſhould fill an whole Well full of their ſtones and privie members. About the 700. yeare of Chriſt the *Beſarites* inveighed againſt Images, and therefore did the Pope condemne them for heretickes, they having no other errour.

About the yeare of Chriſt, 715. *Clemens Scotus* by his preaching, and *Adelbertus Gallus* by his writes opponed againſt *Bonifacius* the Legate of

Pope *Gregorie* the fecond, and oppugned againft the Popes Primacie and tyrannie, for the which they were both excommunicated, and *Adelbertus* being apprehended was caft in prifon into a monaftrie, where he died miferablie.

Claudius Alticidionenfis, otherwife *Claudius Clement Scotus, anno* 800. inveighed againft their pretented chaftitie, the Popes primacie, and againft adoration of Images and reliques, and therefore was perfecuted of the Pope, and excommunicated for an hereticke.

Methodius Illyricus, anno 860. fet out vnto the *Sclaues & Vandals*, the fcriptures in their vulgare language, and draue out the Roman fuperftitions and Latine language forth of thefe Churches, & thought to haue done the like in *Bavaria*, but being perfecuted and chafed by the Priefts, he fled away.

About the 1000. yeare of Chrift, *Otho* Bifhop of *Conftance*, defended againft Pope *Gregorie* the feventh, the mariage of the cleargie, and therefore was perfecuted and deprived, and excommunicated by the Pope.

About the 1050. yeare of our Lord, *Nicetas* fuftained the argument, that the Church-men fhould marrie, againft whome Pope *Leo* wrote, and purfuing him, condemned him for an hereticke.

From this time foorth the hiftories make mention of diverfe dailie till thefe our daies, who all fuftained and defended the trueth, and being perfecuted therefore, and diverfe of them apprehended and martyred, did therefore lurke and lye darned for fafetie of their liues, fo that they were fcarcely feene, and therefore it is faid in the Revelation of the true Church and fpoufe of Chrift. Apoc. 12.6.14. *That fhee fled into the wilderneffe, where fhee hath a place prepared of God that they fhould feede her there 1260. daies.* And againe. *But to the woman were giuen two wings of a great Eagle, that fhe might flee vnto the wilderneffe into her place where fhe is nourifhed for a time, times, and halfe a time, from the prefence of the ferpent.* And therfore iuftly haue we affirmed in the latter part of the former Refolution, that the Pope and his cleargie during the 1260. yeeres of his raigne hath perfecuted the true and latent Church of God, and made them to lurke and become invifible. The Lord now put his true Church to freedome, and make it to triumph fullie over all the enemies. *Amen.*

FINIS.

HEREAFTER FOLLOWETH CERTAINE NOTABLE

Prophecies agreeable to our purpoſe, extract out of the books of SIBYLLA, whoſe authorities neither being ſo authentik, that hitherto wee coulde cite any of them, in maters of Scriptures, neither ſo prophane that altogether we could omit them: We haue therefore thought very meet, ſeuerally, & apart, to inſert the ſame here, after the end of this worke of holy Scripture, becauſe of the famous antiquitie, approued veritie, & harmonicall conſentment thereof, with the Scriptures of God, and ſpeciallie with the 18. Chapter of this holie *Revelation.*

OVT OF THE SECOND BOOKE OF

the Oracles of Sibylla, *beginning at this part according to Caſtalio his Latine tranſlation.*

—hominum tum denique ſæclum,
Exiſtet decimum, quando qui ſydera fecit, &c.

Faithfullie engliſhed this way.

T*Here ſhal of men the tenth age then ariſe,*
When God who did the heauens high deviſe,
At whoſe preſence, the earth doth quake and mooue,
Of (a) *Jmagerie extinguiſh ſhal the loue:* Apoc. 17.9 & 23. pro.
And ſhal of Rome raiſed on her (b) *hilles ſeuen,*
The people ſhake. Then beis the ſtrength downe dreuen Apo. 18.17
Of her riches, whill (c) *Vulcane in his ire* Apo. 17.16 and 18.8.9
Oft ſhal on her ſend furious flames of fire.

(*a*) Hereby ſhe noteth the inſeparable Idolatrie of *Rome* in ſo much as *Rome*, and her idolatrie, ſhal ſtand together and fall together.

(*b*) Of theſe ſeuen hilles of *Rome*, read Apoc. 17.9. and our 23. propoſition.

[*c*] Here agreeable with the Revelation, doth ſhe forewarne the finall deſtruction of *Rome*, to be by fire, vſing here poetically the name of *Vulcan* expreſlie, for mocking of them, and their idol Gods, for ſhee in all her books, acknowledgeth only the true God.

OVT OF THE FIFT BOOK AT THIS
Part. De cœlo veniet ſydus magnum in mare magnum, &c.

From heauen there fell a great [a] *ſtar in the* [b] *ſea,* Apoc. 8.8.
And burnt it vp although both great and hie:
So beis burnt vp, and ſhall no longer ſtand, Apo. 18. &c
Thou [c] *Babylon of Italie the land:* Propoſ. 23.
Who murthereth Saints, and many faithfull Jow, Apo. 17.6. and
Temples tread downe that did the trueth avow, 18.24.
Yet thou O miſchant ſuffer ſhall great paine, Apo. 11.2.
As thou deſerueſt deſert ſhall thou remaine:
Remain ſhall thou deſert for euermore, Apo. 17.16 and 18.
Thy natiue ſoile henceforth thou ſhalt abhore: &c
For thou in [d] *poyſon hath delighted thee,*
Mother of whoredome and adulterie: Apoc. 17.5.
And bougerie gainſt nature nouriſhing, Apoc. 18.7.
A towne of eaſe in filth moſt flouriſhing:
Wicked, vniuſt, vnworthie to be named,
Ah towne of Latins vtterlie defamed:
As [e] *widow wood, and Viper venemous,*
Vpon thy banks thou ſhalt ſit dolorous: Eſay 47.7.8.9.
And [f] *Tiber flood for thee his ſpouſe ſhall mourne,* Apoc. 17.6. and
Thy mind is mad, thy heart for blood doth burne: 18.24.
Thou doeſt misknow Gods minde and might, for why Eſay 47.7.8.6. &c.
Onely I am (ſayeſt thou) and who but I? Apoc. 18.7.
Eternal God, now wrack ſhall thee and thine,
In all the earth ſhal not reſt but ruine, Eſay 47. &c
A monument or yet a marke of thee, Apo. 18 &c
Which thou hadſt when God gaue proſperitie,
Thou miſchant now ſit ſolitar alone, Apo. 19.20
Hurld ſyne to hel with many grieuous grone:
Where thou ſhalt bide burning both bone and lyre,
Into that lake of furious flamming fire.

(*a*) This great ſtarre, Apoc. 8.8. is called a mountaine of fire, and is there prooved to meane, the multitude of fierce and fiery people that beganne alreadie the firſt deſtruction of Rome.

(*b*) The ſea is both prooved, Apoc. 8.8. note *d* thereof, to meane *Italie*, as alſo here ſubſequently doeth follow.

[*c*] How this *Babylon* is *Rome*, we haue proved Prop. 23. and confequently, muft be of the land of *Italie*, as here both marveilouflie & plainly this Prophetefſe affirmeth.

[*d*] How much the people of *Rome* and all *Italie*, are inclined to poifoning and privie murthers, all men knowe in fo much as vulgarlie we call poifon, an Italian poffet.

(*e*) Becaufe as teftifieth both Efay 47.7. and Revel. 18.7. that fhe is a Queene and Ladie, fornicating with manie vowers, and thinketh her felfe no widowe, therefore at once fay they there, as *Sibylla* doeth here, fhe fhal be made a widowe and defolate of all her vowers.

(*f*) This is the flood that goeth through the towne of Rome, bearing the name of *Tiber* to this day.

OVT OF THE SAME FIFT BOOK
at this Part,
Italia infœlix deferta infleta manebis.

O Curfed and vnhappie Jtalie,
Unmeind or mournd for, barren fhalt thou be:
To ground as greene, as wildernes vnwrought, Apoc. 18.2.
To woods wild and bufhes beis thou brought.
And thereafter faith fhee
Stay and reftraine ye [a] *mother murtherers,*
Your hands and hearts, O filthie furtherers,
(b) *To bougrie, like beafts abufing boyes,*
Who women (c) *chafte with villanie annoyes:* Apoc. 17.5.
And maketh whores by your perfwafion,
Overcomd by fubtil folistation:
In thee the Sonne doth with his mother mell,
The daughter wife doth with her Father dwell,
And Kings their (d) *mouthes polluted haue in thee,*
Jn thee is found with beafts bougerie:
Sit dolorous downe, deplore thy paine with pitie,
Thou leacherous feat fuppreft, and finful citie: Ap.18. &c

(*a*) Becaufe that *Nero*, who was Emperor of *Rome*, murthered his mother, therefore here Sybilla ftyleth the whol Romanes after him, Mother-murtherers.

(*b*] Who could more viuely deſcribe by preſent hiſtory, the filthie vices of *Rome*, then here Sibylla doeth by her marveilous prophecie, for proofe whereof, read their owne hiſtories.

(*c*] To wit, chaſt *Lucretia*, whom *Sextust Tarq. Aruns* ſonne to *Tarq. Superbus.* the king of *Rome*, deforced filthily againſt her will.

[*d*] In mouing of Kings and Emperours to make oathes of maintenance, ſeruice and obedience to her, euen againſt God and his trueth.

OVT OF THE VII. BOOKE AT THIS PART

Roma ferox animi, poſt fracta Macedonis arma, &c

Rome fierce in mind, when [*a*] *Grecians ar down driven,*
Shal ſtellifie thy ſelfe vp to the heauen.
But when thou thinkſt thee in the higheſt hight,
God ſhal tread downe thy ſturdie ſtrength and might: Ap.18.7.8
Then ſhal J ſay when thou art dongen downe,
Sometime on hie reſounded thy renowne:
The (b) *ſecond time, O Rome againe to thee,*
Hereafter yet I wil ſpeake ſecondlie.

(*a*) *Alexander* King of *Macedonia*, Monarch of the world being dead, and the Monarchie of the Grecians being ended, then vp-ſtart the Romans, and become monarch of the whole earth, and thereafter by the Pope they vſurped alſo the power of heauen.

[*b*] For the greater emphaſis ſhe doubles the word [ſecondly] meaning thereby, of a firſt and of a ſecond deſtruction of Rome, according to the which, ſhee here pronounceth a firſt, and hereafter a ſecond prophecie.

OVT OF THE VIII. BOOK AT THIS VERSE

Par tibi de cœlo ſublimis vertice Roma, &c.

O Rome vpraiſed now with thy toppes hie,
The like ruine from heauen ſhall fall on thee:
Plaine beis thou made, downe ſhal thy toppes turne, Ap. 18.21.
And flaming fire all whollie ſhal thee burne:
Farre ſhalt thou flit into an vncouth land, Ap. 17.16. & 18.8.9.
Thy riches ſhal be reft out of thine hand: Ap. 18.17.

Jn thy wall-ſteds ſhall Wolues and Toddes convene, Apo. 18.2.
Waſte ſhalt thou be, as thou had neuer bene: Iere. 50.39.
Where then ſhal be thy Oracles devine,
What golden gods ſhal keepe or ſaue thee ſyne: Iſa. 47.2.13.
What God, I ſay, of copper or of ſtone,
Where then ſhal be the conſultation
Of thy Senate? What helps thy noble race,
Of Saturne, Ioue, or Rhea in this cace?
Whoſe ſenſles ſoules and Idoles thou before,
Religiouſlie did worſhip and adore:
Of whoſe greene graues, vnhappie Crete avants,
Their figures dead vp doeſt thou ſet like Sants.

Then followeth immediatly of the Pope.

When rung in thee hath (a) *fifteen Emperours,*
Of all the world that hath bene conquerours:
Then comes a King (b) *a manifolde crowne to beare,*
Whoſe name ſhal be to (c) PONTI *very neare.*
His wicked (d) *foote the world ſhal viſitate,*
Great gifts and goods then ſhal he impetrate:
Huge heapes of golde he ſhal haue into treaſure,
With ſilver hid, and money without meaſure:
(e) *Diſcovered things he ſhal looſe and remit,*
Of (f) *Magick art, wel ſhal he know and wit*
The myſteries and ſecret ſorcerie,
The mightie God he makes (g) *a babe to be:*
And downe he ſhal tread all true worſhippin, Dan. 7.25c
And at chiefe heads of errours firſt begin Apoc. 11.2
His myſteries to all he ſhal expone, 2. Theſſ. 2.9.10.
Then comes the time of mourning and of mone:
For in a (h) *roape his ending ſhal he make,*
The potent people, and the towne ſhal wrack:
Thus ſhal they ſay, we knew and hoped ay,
Some-time on vs ſhould come this bitter day:
The Fathers olde, and babes ſhal mourne for thee,
Beholding then thy dolorous deſtinee:
On Tiber banke lamenting ſore thy caſe,
Sadde ſhal they ſit, with many loud alace.

[*a*] Hauing before in the beginning of the fift booke,

amongſt other Emperours, deſcribed the fifteenth Emperour *Adrianus*, with many of the ſame tokens wherewith here the Antichriſt is deſcribed, *Sybilla* here ſheweth, that this wicked Antichriſt ſhall come after fifteene Emperors, expreſly thereby to put difference betwixt him and *Adrianus*, and not meaning thereby, that immediatly after the fifteenth Emperour, came the wicked Antichriſt.

[*b*] *Adrianus* Emperour was before deſcribed with a ſiluer crowne, and to be a good man, but where here ſhe deſcribeth this wicked man to weare a manifolde crowne called in Greeke, πολλικρανος, ſhee thereby expreſſeth and points out the Pope moſt liuely, who ever ſince *Conſtantine* the Emperours daies, beareth a threefold crowne, as the gift contained in the *Decret. diſt*. 96. proports.

[*c*] Where ſhee deſcribes *Adrianus* Emperour, ſhe ſaith not that his name is neere vnto *Pontus*, which is to ſay, neere to the Sea, but plainly ſaith ſhee, he ſhall haue the name of the Sea: for where *Adrianus* was borne, the ſea there is called the *Adriaticke* or *Adriane* Sea, but here ſaith ſhee, this wicked man ſhall haue a name neere to *Ponti*, meaning thereby plainly of the Pope, who is ſtyled *Romanus Pontifex*, or *Summus Pontifex*.

(*d*) With his wicked foot doth he viſit the world, both in treading downe Kings and Princes vnder his ſeete, as he did to *Fredericke Barbaroſſa* and *Henrie* the Emperours: as alſo in ſetting vp his foot to be viſite, ſeene and kiſſed as a reliqne and publike ſpectacle to the whole world.

[*e*] Even ſinnes diſcovered by auricular confeſsion ſhal he preſume to looſe and remit.

[*f*] Hereof read Platina the Popes owne ſecretarie, together with other true hiſtories, who make mention of 22. Necromantick Popes, and of eight Atheiſts, beſide others of horrible natures.

(*g*) By repreſenting God by ſtones, and carued ſtockes of bony Baby Ieſus, and ſuch like Idoles to the poore people to play them with, & many other vanities, where by he mocketh God, and would make God to ſeeme as an ignorant childe.

[*h*] This is vnperformed as yet, and ſeemes to meane of the laſt Pope, who ſhall make ſuch a tragicall end, his people, ſeate and towne being vtterly wracked.

And ſhortly after followeth.

Lament ſhal you, and mourne, laying aſide
Thy [a] *purple weed, imperiall robes of pride:*
And into ſackcloth ſitting ſorrowfull,
Repeat ſhalt thou thy plaints pitifull:
O royall Rome, thou bragging Prince but Peere,
Of Latine land the onely daughter deere: Apo. 18.19
Thy pride but pompe ruyned ſhall remaine: Apo. 18.14 21.
Thou (b) *once trode downe ſhall neuer riſe againe:*
For gone ſhall be the glore of that armie,
That beareth the (c) *Eagles in their enſeignie:*
To whence then ſhall thy forces foreward fair,
What land with armes ſhall helpe the any mair:
Yea, none now bound is to thy ſeruitude,
O miſhant, &c.

(*a*) How both Popes and Emperours of Rome, haue their robe royall of Purple and skarlet, reade the Decret, diſt. 96. as is ſaid in the note (*g*) Apoc. 17.

(*b*) Shee ſpeaketh not here of the wrackes that Rome ſuſtained in times paſt, by the *Gothes*, and *Vandales*, for all theſe wrackes were repared: but here ſhee ſpeakes of that greate wracke that is yet to come which neuer ſhall be repaired as it is teſtified Apoc. 18.14.21.22.23.

(*c*) The Roman Emperours beares the double diſplaied Eagle in their badge.

And hereafter followeth.

Then ends the world, then comes the latter light,
Then God ſhall come to iudge his folke aright:
But firſt ſhall fall on Rome but reſiſtance, Apoc 14.8.14.
Of Gods wrath the woſull vengeance. Apoc. 17.6. and
A wofull life a bloudie time ſhall be, 18.24.

Oh, people rude, Oh land of crueltie,
Thou little lookeſt, nor doth regarde aright,
How poore and bare thou firſt came in the light:
That to the like againe you ſhould returne,
And laſt before a dreadfull iudge ſhould murne.

IN ſummar concluſion, if thou O *Rome* aledges thy ſelfe reformed, and to beleeue true Chriſtianiſme, then beleeue Saint *John* the Diſciple, whome Chriſt loued, publikely here in this Reuelation proclaiming thy wracke, but if thou remaine Ethnick in thy priuate thoghts, beleeuing, the old Oracles of the *Sibyls* reuerently keeped ſomtime in thy *Capitol*: then doth here this *Sibyll* proclame alſo thy wracke. Repent therefore alwayes, in this thy latter breath, as thou loueſt thine Eternall ſaluation. *Amen*.

FINIS.

MATHEMATICAL INTRODUCTION

Enrique A. González-Velasco

References to the *Constructio* are given by Article number so that any edition can be consulted. For the same reason, references to the *Descriptio* are by Book and Chapter, but with a numbered item added when possible. References to *De Arte Logistica* and the *Rabdologiæ* are by page numbers, first to the Latin editions and, after a semicolon, to the translations in this volume.

The photographs of Napier's Radical Table from the *Constructio* (p. 412), the first page of his 1614 table of logarithms from the *Descriptio* (p. 417), and his abacus from the *Rabdologiæ* (p. 459) are reproduced by courtesy of the National Library of Scotland.

A portrait of John Napier (1550–1617) owned by the Scottish National Portrait Gallery. It is believed to be a near contemporary copy of the one owned by Edinburgh University. Indeed it may have been painted by the same artist at or about the same time since commissions for multiple copies of portraits were not uncommon then. However, there are some key differences compared to those owned by Edinburgh University and Lord Napier: JN *instead of* IN *either side of the shield,* ÆTATIS SVC 66 *instead of* ÆTATIS 66 *in the top right corner and John Napier of Merchiston spelt out across the top. It was purchased by the Scottish National Portrait Gallery from a Mr. Morris in 1974 but its previous history appears to be unknown. Oil on canvas, 43½” × 39½”.* (*Courtesy of the Scottish National Portrait Gallery.*)

NAPIER'S QUEST TO REMOVE THE LABOUR AND TEDIUM OF CALCULATIONS

When Andro Hart of Edinburgh published a book entitled *Mirifici Logarithmorum Canonis Descriptio* [**27**] (Description of the Wonderful Table of Logarithms) in 1614 what we would now call the mathematics world was taken by surprise. Even its title was a mystery because this was the first appearance in print of the word *logarithms*. In this small book of 147 pages, its author and inventor, identified as Ioanne Nepero (known today as John Napier although he never signed his name this way) Baron of Merchiston, provided the world with an unexpected and very powerful solution to a long-standing problem.

Astronomy is a science that goes back to antiquity and, since very early days, it required the elaboration of what we now call tables of sines and other trigonometric functions. They were not our modern sines and they were not functions then but lengths, measured in some arbitralily chosen units which varied from author to author. For instance, in the Hellenic world, Ptolemy divided the diameter of a circle into 120 parts, and constructed a table of chords of central angles. In this scheme, and referring to the figure below,

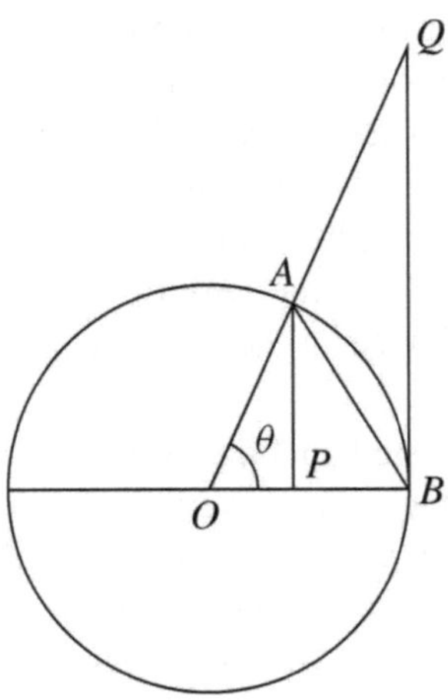

he found that if $\theta = 72°$ the length of the chord AB is 72 parts plus 32 sixtieths of a part plus 3 sixtieths of a sixtieth (he was a proponent of the sexagesimal system, in support of which he gave ample reasons). Later, the Hindus replaced the concept of chord of an angle with that of half-chord of twice the angle, represented in the figure by AP. For instance, the astronomer

Aryabhata divided the radius OA into 3438 parts (for his reasons to chose this apparently strange number see [**14**, p. 27]) and found that the half-chord of $71\frac{1}{4}^\circ$ is 3256. Abu'l Wafa, working in Baghdad in the tenth century, was the first to select the present-day value of 1 for the radius, and therefore this was the half-chord of 90°. However, no one else would make the same choice for a very long time. In Latin-writing scientific Europe, and due to linguistic happenstance, the half-chord of 90° was called the *sinus totus* (whole sine). But the Europeans, at a time when the decimal point was unheard of, chose a huge number for the *sinus totus*, typically 10,000,000. Then the *sinus*, to distinguish it from our modern sine, of 30° was 5,000,000 and those of many angles in the first quadrant were seven-figure numbers. They were entered as such in popular tables of the fifteenth century. Other trigonometric lengths, introduced in the Islamic world and imported into Europe, were the cosine OP and the tangent QB.

The long-standing problem mentioned above was that computations involving such large numbers were extremely time-consuming, tedious beyond belief, and subject to all kinds of errors. This didn't make the task of astronomers easy or appealing. A solution was desperately needed and this is precisely what Napier's book provided. Today we are familiar with logarithms and their use, even if our logarithms are different from the ones provided by the tables in the *Descriptio*, but to the readers of Napier's first mathematical book such a wonderful solution must have been totally unexpected.

It is quite likely that, at this period, a number of people engaged in computation were using some form of mechanical aid that saved them time and increased accuracy. Napier had already developed a method of calculation with some rods, that would soon become known as "Napier's bones," which are essentially portable multiplication tables and will be described below. These were popular at the time the *Descriptio* was published, and thus the invention of mechanical devices to help with computations would have been less surprising.

Napier had been interested in facilitating the computations required in astronomy and other sciences since very early in his life. At the time of his death he left a large collection of manuscripts that were probably his first writings on the subject. However, his attention was then diverted to the development and application of what he would later call logarithms and his earlier writings remained unpublished until the nineteenth century, when they appeared under the title *De Arte Logistica* [**25**]. A description of the entire work is contained in [**40**] and a more recent one in [**15**, pp. 155–178]. Napier stated at the very beginning of Chapter 1 that *Logistica* "is the art of correct calculation."

While the contents of this work does not fall within the limits of this Introduction's title, we should remark on a couple of points that show how Napier's thought was ahead of his time. In Chapter VI of Book I, "*De Quantitatibus Abudantibus et Defectivis*" [**25**, p. 17;], which Hawkins translated as "Positive and Negative Numbers," he dealt with negative quantities in a completely natural manner. First Napier stated:

> Negative numbers are less than zero and carry the minus sign before them. They are always preceded by this sign $-$ which is called the minus sign.

Then he gave the following example from financial mathematics:

> For instance, if a man's wealth is estimated, and his debts exceed his assets by 100 crowns, then his wealth is properly expressed -100 crowns, with the sign in front. This is pronounced 'minus one hundred crowns' and it always signifies loss or deficiency.

Had his contemporaries been able to read these words, many would have been shocked by the naturalness with which Napier worked with quantitites that could not possibly exist, although they had already been used outside Europe, in much earlier times, in several works in Chinese, Sanskrit and Arabic.

In the rest of the chapter, Napier discussed arithmetic with positive and negative numbers, including the statement

> that a positive number with an even index [that is, an even positive number] must have two roots, one positive, the other negative. But negative numbers have no roots.

He elaborated further on this point in Chapter I of Book III, "*De Logistica Geometrica*" (The Logistics of Geometry). Napier found that when taking the root of a defective number [**25**, p. 85; 877]

> the roots are neither abundant nor defective, and they are called trifling or imaginary [*nugacia* $\approx$ nonsensical].

Shortly after this assessment, Napier, who at the time used $\sqcup$ as the symbol for square root, cautioned his readers as follows [**25**, p. 86; 877]:

> In dealing with imaginary cases, you are strongly advised not to put the minus sign ($-$) in front, for the sign must be inserted between the root sign and the number.
>
> For if, instead of putting $\sqcup{-9}$ (which makes nonsense) you put $-\sqcup 9$, then you make a big mistake, for this signifies a minus quantity, the square root of nine. But the square root of nine, in this case abundant ($\sqcup 9$) gives a

pair of answers +3 (abundant) and −3 (defective). And so the product will be negative. And so, he who writes −⊔9 in place of ⊔−9, instead of an imaginary and impossible quantity of no significance, introduces a pair of numbers with opposite signs. Beware of this error, into which many people fall.

So Napier was aware of imaginary numbers, although he considered them nonsense and developed no arithmetic for them, unlike Rafael Bombelli who published the arithmetic of what we now call complex numbers in 1572 [**4**]. At the time of writing *De Arte Logistica* Napier thought that [**25**, p. 85; 877]

the future may show how much credit will accrue to this art, and to other mathematicians.

Returning now to our main quest, Napier began the Preface of the *Descriptio*, by describing his interest in correct calculation as follows:

Seeing there is nothing (right well beloued Students in the Mathematickes) that is so troublesome to Mathematicall practise, nor that doth more molest and hinder Calculators, then the Multiplications, Diuisions, square and cubical Extractions of great numbers, which besides the tedious expence of time, are for the most part subject to many slippery errors. I began therefore to consider in my minde, by what certaine and ready Art I might remoue those hindrances.

PART I. LOGARITHMS

We don't know the particular sequence of events that led Napier to the discovery of logarithms, but his interest in simplifying computations may have been redirected during a visit by Dr. John Craig, later James VI's physician. Craig had been dean for several years at Frankfurt an der Oder, where he met and was tutored by Paul Wittich (c. 1546–1586) in 1576. Craig must have learned from Wittich an ingenious method to easily and indirectly perform the tedious multiplications required by astronomical work, a method on which he left annotations in his own copy of Copernicus' *De revolutionibus* [**12**, pp. 103–106]. It rests on the use of the trigonometric identity

$$\sin A \cdot \sin B = \tfrac{1}{2}[\cos(A - B) - \cos(A + B)],$$

discovered by Johannes Werner and contained in Book IV of his *De triangulis sphaericis*, written before 1515 but not published until 1907 [**2**, Props. 2 & 10 for the case $A + B < 90°$ and Props. 5 & 13 for $A + B > 90°$]. For a history and use of this and similar identities see [**43**].

This method, known by the almost impossible name *prosthaphæresis*, from the Greek words for addition (προσθέσις) and subtraction (ἀφαίρεσις), is applied as follows. Suppose that we want to find the product of two large numbers a and b. First we use tables of sines (*sinuses*) to find numbers A and B such that, at least approximately, $a = \sin A$ and $b = \sin B$ (if it is necessary to multiply or divide a and/or b by a power of 10, this is easily done). Then obtain the values $A - B$ and $A + B$ and find their cosines (either from a table of cosines if there is one or find their sines and use trigonometry), subtract the cosines as shown in the formula, and divide by 2. The result is the product ab except, possibly, for a power of 10. Thus, the problem of multiplication of large numbers was reduced to the simpler tasks of addition and subtraction. The division by 2 is easy, but repeated application of this step when performing many multiplications becomes tiresome and is a drawback of the method.

It is possible that the information on prosthaphaeresis acted as a stimulus for Napier to redouble his efforts in computation, but prosthaphaeresis seemed to be a dead end and he worked on a different concept. The initial idea used by Napier, but only to a point as we shall see, had been well-known for some time and had been very clearly explained in 1544 in the *Arithmetica integra* of Michael Stifel (1487–1567) [**41**, Book III, p. 237]. He gave two parallel lists of numbers, containing powers of 2 in the bottom row and the exponents that generate those powers in the top row:

0.	1.	2.	3.	4.	5.	6.	7.	8.
1.	2.	4.	8.	16.	32.	64.	128.	256.

Below this small table he explained how it can be used to facilitate some computations:

> Just as from the addition (in the top row) of 3 and 5 there comes 8, thus (in the bottom row) from the multiplication of 8 and 32 there comes 256. But 3 is the exponent of eight itself, & 5 is the exponent of the number 32. & 8 is the exponent of the number 256. In the same manner in the top row, from of the subtraction of 3 from 7, remains 4, so in the bottom row from the division of 128 by 8, there comes 16.

In today's notation we would write these operations as $8 \times 32 = 2^3 \times 2^5 = 2^8 = 256$ and $128/8 = 2^7/2^3 = 2^4 = 16$. In this way, multiplication and division are reduced to the much simpler operations of addition and subtraction of exponents.

However, these explanations and examples by Stifel must have been considered by many as an entertaining if somewhat childish game, but totally impractical for real computations. For instance, it is impossible to multiply 24 by 119 using this method because the table does not contain these numbers. Successive powers of 2 are just too far apart. Of course, mathematicians must have quickly realised that replacing the number 2 by a number very close to 1 would solve this problem immediately. However, this solution creates an even worse problem: how to compute successive powers of such a new number.

It was the genius of Napier that would overcome this problem. The *Descriptio* gave us the finished table of logarithms at the end of two books on the mathematics of the subject. Book I is on definitions, including that of logarithm as Definition 6, theorems, a description of the table, and examples of the use of the logarithms. It would be pointless to reproduce the definition of logarithm at this point without a certain amount of preparatory work, but all this will be done below. Book II is on application of the logarithms to trigonometry, including Napier's own contributions to spherical trigonometry (see [**15**, pp. 83–92]). What the *Descriptio* did not provide is an explanation of how Napier arrived at his logarithms. Instead, in a note at the end of Chapter II of Book I, we find the following modest statement:

> Now by what kinde of account or method of calculating they may be had, it should here bee shewed. But because we do here set down the whole Tables, and all his *Logarithmes* with their Sines to euery minute of the quadrant: therefore passing ouer the doctrine of making *Logarithmes*, till a fitter time, we make haste to the vse of them: that the vse and profit of the thing being first conceiued, the rest may please the more, being set forth hereafter, or else displease the lesse, being buried in silence.

However, the tables of logarithms proved to be a huge success immediately after the publication of the *Descriptio*, to the extent that they were already used in China by the 1640s. Therefore, after Napier's death, his son Robert decided to publish some earlier writings of his father to explain their construction. Robert wrote in the preface to the published work that [**26**, p. A2]:

> it has been made plain to me by unmistakable proofs, that the most skilled in the mathematical sciences consider this new invention of very great importance, and that nothing more agreeable to them could happen, than if the construction of this Wonderful Canon, or at least so much as might suffice to explain it, go forth into the light for the public benefit.

Thus, Napier's manuscript on the construction of his logarithms appeared posthumously in print with the title *Mirifici Logarithmorum Canonis Constrvctio* (Construction of the Wonderful Table of Logarithms) [**26**].

THE RADICAL TABLE

The construction of the logarithms was a long and involved process, which Napier tackled in several separate stages. This had the unintended but welcome benefit of keeping error accumulation within each stage.

To describe the creation of the tables of logarithms, as detailed in the *Constructio*, we return first to Stifel's table and note that the exponents in the top row are in arithmetic progression, while the powers in the bottom row are in geometric progression, specifically they are numbers of the form 2^n. The proposal to replace them with numbers of the form r^n where r is very close to 1 was temporarily dismissed above, on the grounds that computation of the values of r^n might be too involved. However, Napier did not dismiss it because he noted that cleverly choosing r would make the computations easy. In fact, he stated [Art. 13]:

> Those geometrical progressions alone are carried on easily which arise by subtraction of an easy part of the number from the whole number.

He gave a fuller explanation of these words in Article 14. Interpreted in today's terminology, what Napier proposed is to choose r of the form $r = 1 - 10^{-k}$, where k is a positive integer. Then, if a is any number and ar^n is known,

$$ar^{n+1} = ar^n(1 - 10^{-k}) = ar^n - ar^n 10^{-k}.$$

In this way the computation of the next term of the progression is accomplished by means of a simple subtraction rather than multiplication. Furthermore, the term $ar^n 10^{-k}$ is easily obtained by a shifting of the decimal point.

Napier knew about the use of the decimal point. It had been introduced before in several parts of the world and its use had been explained by Simon Stevin (1548–1620), of Bruges, in his book *De Thiende* ("The Tenth" in Flemish) of 1585. But his notation for decimal fractions was exceedingly cumbersome, and so almost no use was made of Stevin's version of decimal fractions. Napier's consistent and clear use of it in the *Constructio* may have contributed to its subsequent popularity.

Napier chose $a = 10{,}000{,}000$ because his goal at this point was computing with sines, and the most popular tables of sines at the end of the sixteenth century took the whole sine (*sinus totus*) to be 10,000,000. Then, he took $r < 1$ so that the terms of the progression ar^n were parts of the whole sine. Finally to facilitate the computations he selected

$$r = 1 - 0.0000001 = 0.9999999.$$

Having selected a and r, Napier continued the computation of terms of the progression in Article 16 as follows:

> Thus from radius, with seven ciphers added for greater accuracy [that is, with seven zeros after the decimal point], namely, 10000000.0000000, subtract 1.0000000, you get 9999999.0000000; from this subtract .9999999, you get 9999998.0000001; and proceed in this way as shown on the side [the table shown below was originally printed on the left of this paragraph], until you create a hundred proportionals, the last of which, if you have computed rightly, will be 9999900.0004950.

In this manner Napier constructed what he called the

First table.

$$\begin{array}{r} 10000000.0000000 \\ 1.0000000 \\ \hline 9999999.0000000 \\ .9999999 \\ \hline 9999998.0000001 \\ .9999998 \\ \hline 9999997.0000003 \\ .9999997 \\ \hline 9999996.0000006 \end{array}$$

to be continued up to

9999900.0004950

In today's notation, what Napier did was use the formula

$$10^7 r^{n+1} = 10^7 r^n r = 10^7 r^n (1 - 10^{-7}) = 10^7 r^n - r^n$$

with $n = 0, \ldots, 99$ to obtain all but the first line in the following list:

$$\begin{aligned} 10000000 \times 0.9999999^0 &= 10000000 \\ 10000000 \times 0.9999999^1 &= 9999999 \\ 10000000 \times 0.9999999^2 &= 9999998.0000001 \\ 10000000 \times 0.9999999^3 &= 9999997.0000003 \\ 10000000 \times 0.9999999^4 &= 9999996.0000006 \\ &\;\cdot\;\cdot\;\cdot\;\cdot\;\cdot\;\cdot\;\cdot\;\cdot \\ 10000000 \times 0.9999999^{100} &= 9999900.0004950. \end{aligned}$$

Notice that these numbers are only approximate. For instance, to obtain $10000000 \times 0.9999999^3$ Napier subtracted .9999998 from 9999998.0000001 instead of subtracting .99999980000001, but he had already stated in Article 6 that "fractions following the period [such as .0000001] may then be rejected, without any sensible error."

Napier did not comment on the obvious fact that continuing in this manner might be too time-consuming. The numbers obtained in the first table remain very close to 10000000, the whole sine, so that this method is too slow to cover the sines of angles in the first quadrant coming down to zero. What Napier did was simply to begin another table [Art. 17], but replacing $r = 0.9999999$ with

$$r = 0.99999 = 1 - 0.00001$$

to obtain his

Second table.

$$\begin{array}{r} 10000000.000000 \\ \underline{100.000000} \\ 9999900.000000 \\ \underline{99.999000} \\ 9999800.001000 \\ \underline{99.998000} \\ 9999700.003000 \\ \underline{99.997000} \\ 9999600.006000 \end{array}$$

&c., up to

9995001.222927

In other words, using now the equation $10^7 r^{n+1} = 10^7 r^n - 100 r^n$,

$$\begin{aligned} 10000000 \times 0.99999^0 &= 10000000 \\ 10000000 \times 0.99999^1 &= 9999900 \\ 10000000 \times 0.99999^2 &= 9999800.001 \\ 10000000 \times 0.99999^3 &= 9999700.003 \\ 10000000 \times 0.99999^4 &= 9999700.006 \\ &\cdots\cdots\cdots \\ 10000000 \times 0.99999^{50} &= 9995001.222927 \end{aligned}$$

Unfortunately, the last entry is incorrect; it should be 9995001.224804 (there is a recent reconstruction of all of Napier's tables in [**35**]). This calculation error was discovered by Jean-Baptiste Biot, who reported it in a review article of 1835 of Mark Napier's *Memoirs of John Napier of Merchiston* [**29**]. After endlessly praising John Napier's work, Biot stated that there is indeed [**1**] "a very small error in the last term of the second progression which he forms for preparing the calculation of his table."

In common with the first table, the numbers in the second table also stay close to 10000000, so that Napier decided (but again without giving a reason) to begin again [Art. 18], this time using the ratio

$$r = 0.9995 = 1 - \frac{1}{2000},$$

and computed 21 entries:

$$10000000 \times 0.9995^0 = 10000000$$

$$. \quad . \quad . \quad . \quad . \quad . \quad . \quad .$$

$$10000000 \times 0.9995^{20} = 9900473.57808.$$

But now he introduced a change in his procedure in order to speed things up. With the numbers on the right he made a first column of a third table, but keeping only four decimal digits for each entry. Then, from each of these entries, he computed 68 additional values using the ratio $r = 0.99$, which he placed horizontally from each first column entry. This resulted in a third table that had 69 columns [Art. 20]. The numbers in each column are 99 hundredths of those in the previous column. The complete arrangement is shown below, but for reasons of space we have omitted Napier's "*Third Column.*"

PROPORTIONALS OF THE THIRD TABLE.

First Column.	*Second Column.*	*Thence 4th, 5th, &c., up to*	*69th column.*
10000000.0000	9900000.0000	&c., up to	5048858.8900
9995000.0000	9895050.0000	&c., up to	5046334.4605
9990002.5000	9890102.4750	&c., up to	5043811.2932
9985007.4987	9885157.4237	&c., up to	5041289.3879
9980014.9950	9880214.8451	&c., up to	5038768.7435
&c., continuously to	&c., descending to		finally to
9900473.5780	9801468.8423	finally to	4998609.4034

Note that the last entry of the 69th column is 4998609.4034, which is about half of the whole sine. Thus the sines in the third Table are those of arcs from 90° to 30°, and this is enough for now. Later in the *Constructio* Napier would devise two methods to deal with arcs smaller than 30°.

The most cursory examination of this third table shows that it is useless for the computations that Napier had in mind, for these obvious reasons. The first is that, with 21 rows and 69 columns, there are very few entries in it. The second is that it is a strange mixture, for the ratio coming down each colum is 0.9995 while the ratio moving right in each row is 0.99. Finally, these entries are sines but they could not be found in any tables of Napier's time because those tables only listed sines that are whole numbers, while these have four decimal digits each.

Napier did not despair and did not discard his first three tables. Instead, he used them as stepping stones in the construction of his final table in the *Descriptio*. But in order to do that he needed a new idea. In 1915, Lord Moulton distinguished three stages in Napier's discovery of the logarithms [**24**]. The first was his salvaging of the idea of the two progressions, which led to the construction of the three tables described above. We are about to unveil the second and third stages.

The fact that Napier was repeatedly subtracting a proportional part of each sine that he computed to obtain the next, always the same proportional part, must have led him to switch from an arithmetical to a geometrical point of view. Thus, in Article 24 he started with a segment ST, of length 10000000, to represent the *sinus totus*, and on this segment chose points 1, 2, 3, 4, ...

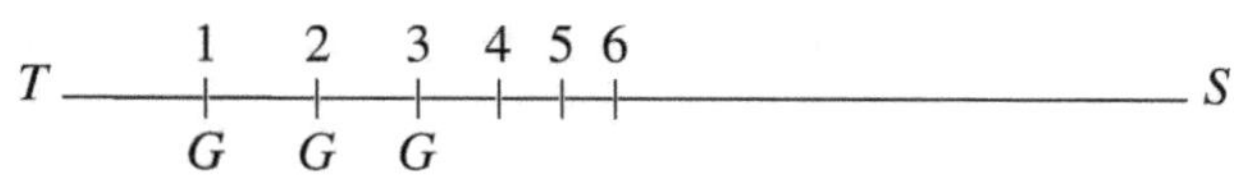

such that the sines TS, $1S$, $2S$, $3S$, $4S$, ... are in continued proportion

$$\frac{1S}{TS} = \frac{2S}{1S} = \frac{3S}{2S} = \frac{4S}{3S} = \cdots,$$

such as those of the first table, in which case this ratio is 0.9999999. Then he imagined a point G moving from T to S with decreasing velocity and in such a way that G traverses each of the segments $T1$, 12, 23, 34, ... in equal times. Napier added that the stated sines

> are said to decrease geometrically, because in equal times they are diminished by unequal spaces similarly proportioned.

Hence his choice of the letter G.

Regardless of the geometric interpretation just described, the fact is that the sines in Napier's third table are still laden with decimals and cannot be found in tables. Napier's next idea, and this was a giant leap in his time, was to use the concept of continuity. To put it simply, if, for example, the point G were to move continuosly on TS, with a decreasing velocity, from table entry 9890102.4750 to table entry 9885157.4237 it is bound to hit some whole numbers that are table sines.

In Article 26 Napier changed the notation for the moving point from upper-case G to lower-case g, and then gave a general definition of what he would later call logarithm. But it is easier to begin with an example before the general definition. Assume that $1S$, $2S$, $3S$, $4S$, … are the sines in the first table, and consider a coordinate infinite half-line bi (shown in the next figure) with $b = 0$. Now imagine a new point a moving on this half-line from b to the right with constant velocity, the same as the initial velocity of g at T. We choose the coordinates $1, 2, 3, 4, \ldots$ (not shown in the figure below) to be the points on bi at which a arrives as g arrives at the points labeled $1, 2, 3, 4, \ldots$ on TS. Therefore, as g passes through the left endpoints of the sines $1S$, $2S$, $3S$, $4S$, … of the first table in equal amounts of time, a passes through the coordinates $1, 2, 3, 4, \ldots$, which are the exponents that generate these sines, in the same amounts of time.

Napier did not give a preliminary example but considered at once the case of an arbitrary sine:

> Let the line *TS* be radius, and dS a given [arbitrary] sine in the same line: let g move geometrically from T to d in certain determinate moments of time. Again, let *bi* be another line, infinite towards i, along which, from b, let a move arithmetically [with constant velocity] with the same velocity as g

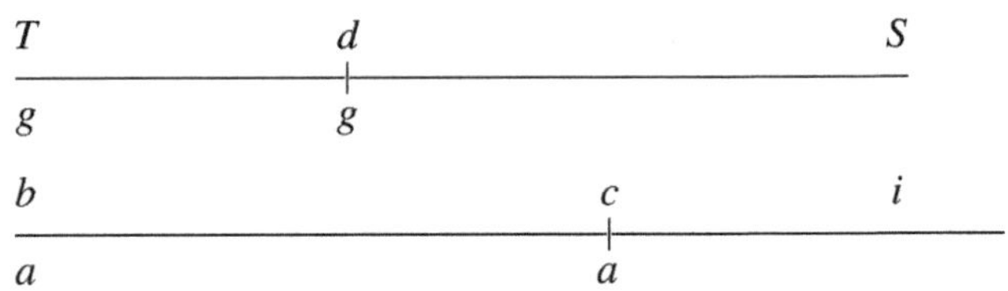

> had at first when at T; and from the fixed point b in the direction of i let a advance in just the same moments of time [as it takes g to move to d] to the point c. The number measuring the line *bc* is the logarithm of the given sine dS.

Thus if the line *bi* is as described above, if dS is an arbitrary sine—possibly a whole number chosen from a published table—if the points g and a start their motions with identical initial velocities, and if a arrives at c in the same amount of time as g arrives at d, the distance *bc*—not necessarily an integer in this general case—represents what we could call the generating exponent of the sine dS.

Actually, Napier did not use the word logarithm in the *Constructio*. He called the distance *bc* the "artificial number" of the sine dS. It must have been later, when he was ready to publish the *Descriptio*, that he introduced the word now in use. He may have realised that, for instance, $4S$ is obtained from *TS* after four multiplications by the ratio 0.9999999, in the case of the first table. Thus, the artificial number of $4S$ is 4, the number of ratios needed to obtain this sine from the whole sine. Using the Greek words *λόγων* (*logon*) = ratio and *ἀριθμός* (*arithmos*) = number, he concocted the word *logarithm* or number of ratios. This is the prevailing theory, but it has also been argued [**42**, p. 192] that Napier took the word logarithm directly from the *The Sand-Reckoner* of Archimedes. Regardless of its origins, "logarithm" was the word used instead of "artificial number" in the first translation of the *Constructio*, and we shall use it from now on.

Napier began the computation of the logarithms of table sines as follows [Art. 27]:

> Whence nothing is the logarithm of radius.

That was the end of the easy part of his task. Thinking today of that continuous motion of the point g that Napier described, we know that differential equations are involved. However, there were no differential equations in Napier's time, and the consequence of this lack is that Napier was unable to compute any other logarithms beyond this first one.

This did not prevent him from going forward. If computing the logarithms was impossible, he settled for the next best thing: finding bounds for the logarithms of table sines. He was able to give them as follows [Art. 28]:

> Whence also it follows that the logarithm of any given sine is greater than the difference between radius and the given sine, and less than the difference between radius and the quantity which exceeds it in the ratio of radius to the given sine. And these differences are therefore called the limits of the logarithm.

As shown in the figure below, Napier denoted the given sine by dS and the quantity that exceeds it in the ratio TS/dS, in current notation, by oS.

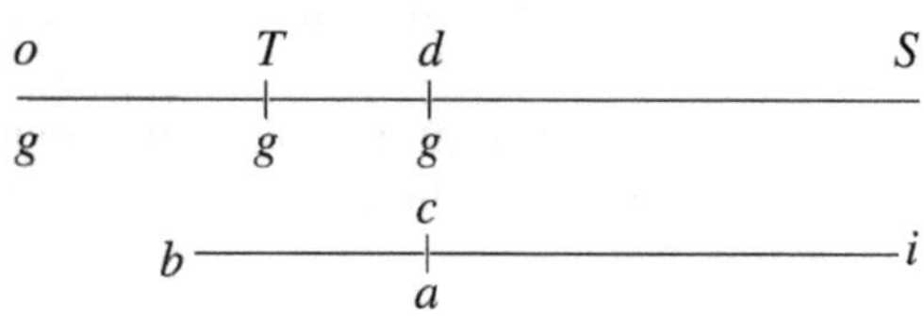

Napier clarified his statement in narrative form by saying that oS is to TS as TS is to dS, and we can express this as

$$\frac{oS}{TS} = \frac{TS}{dS}.$$

Recall that according to the motions of the points g and a, "oT, Td, and bc are distances traveled in equal amounts of time," as Napier stated. Since the velocity of g at T is the same as that of a at b, and since the velocity of g is decreasing but that of a is constant,

$$oS - TS = oT > bc\ [= \log dS] > Td = TS - dS.$$

This is the restatement of Article 28 in current notation. We have taken the liberty of parenthetically using the abbreviation log for "logarithm of," which Napier never used.

As an additional liberty, we can denote the given sine dS by s and if we notice that $oS = TS^2/s$, then we can reverse and rewrite the preceding inequalities as

$$TS - s < \log s < \frac{TS^2}{s} - TS.$$

Then Napier decided to modify the upper limit by stating [Art. 29]:

> that radius being multiplied by into the less limit and the product divided by the given sine, the greater limit is produced, as in the following example.

Thus, we can replace the previous inequalities with

$$TS - s < \log s < \frac{TS(TS - s)}{s}.$$

Notice the difference between Napier's statements in Articles 28 and 29 and the equivalent inequalities in present-day notation. Mathematicians of Napier's time, and for thousands of years before, wrote their results in narrative form because that was the only way available to them and they thought nothing of it. This limitation is now considered as one of the reasons why mathematics developed very slowly up to and including the first quarter of the seventeenth century. Later in that century some necessary notation to abbreviate the writing started to become available, mainly through the pioneering efforts of Descartes, Newton, and Leibniz, and the writing of mathematics improved significantly.

The example that Napier referred to in Article 29 is that of the first sine of the first table after the whole sine. In this case [Art. 30], $s = 9999999$, $TS - s = 1$ and

$$\frac{TS(TS - s)}{s} = \frac{10000000}{9999999} = 1.00000010000001.$$

Disregarding the last decimal digit on the right, we have

$$1 < \log 9999999 < 1.0000001,$$

and since the upper and lower limits are so close to each other, "they or anything between them may be taken as the true logarithm" [Art. 31]. Napier split the difference and took

$$\log 9999999 = 1.00000005.$$

In Article 32 Napier showed how easily the logarithms of all the sines of the first table can then be obtained. As the point g moves through those sines in equal times, the point a moves through their logarithms in equal steps. These steps are equal to the difference of the first two logarithms

$$\log 9999999 - \log 10000000 = 1.00000005 - 0 = 1.00000005.$$

Therefore [Art. 33], recalling the second and third sines in the first table after the whole sine, we obtain $\log 9999998.0000001 = 2 \log 9999999 = 2.00000010$, $\log 9999997.0000003 = 3 \log 9999999 = 3.00000015$, "and so of the rest." In particular, although Napier did not state this, for the last sine of the first table we have

$$\log 9999900.0004950 = 100 \log 9999999 = 100.000005$$

or, from the inequalities $1 < \log 9999999 < 1.0000001$ of Article 30 we would obtain

$$100 < \log 9999900.0004950 < 100.0000100.$$

Unfortunately, the inequalities of Articles 28 and 29 do not work as well for sines further away from 10000000. If, for instance $s = 8000000$, the stated inequalities would give

$$10000000 - 8000000 < \log\ 8000000 < \frac{10000000(10000000 - 8000000)}{8000000},$$

that is, $2000000 < \log\ 8000000 < 2500000$. The difference between these limits is too large to be able to accurately select $\log 8000000$. So Napier had to search for better inequalities that would produce tighter limits for a logarithm. But, before that, he put down the basic result that he used to compute the logarithms of the sines of the first table as follows [Art. 36]:

> The logarithms of similarly proportioned sines are equidifferent.

He would use this direct consequence of the motions of the points g and a over and over again. In our current notation, we can express this as follows: if s, t, u, and v are sines such that

$$\frac{t}{s} = \frac{v}{u},$$

then

$$\log t - \log s = \log v - \log u.$$

Then Napier found better limits for a logarithm in Article 39:

> The difference of the logarithms of two sines lies between two limits; the greater limit being to radius as the difference of the sines to the less sine, and the less limit being to radius as the difference of the sines to the greater sine.

This can be shown with reference to the next figure, in which dS is the greater sine and eS is the lesser sine. Napier selected points V and c such that ST is

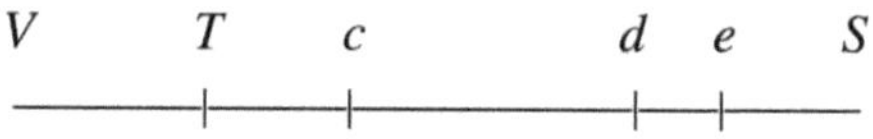

to TV as eS is to the difference of the sines *de* and such that TS is to Tc as dS is to *de*. We can write the reciprocals of these proportions as

$$\frac{TV}{TS} = \frac{de}{eS} \qquad \text{and} \qquad \frac{Tc}{TS} = \frac{de}{dS}.$$

Adding 1 to both sides of the first equation, subtracting from 1 both sides of the second, simplifying in both cases, and comparing results, we obtain

$$\frac{VS}{TS} = \frac{dS}{eS} = \frac{TS}{cS}.$$

From these equations, for which Napier himself did not provide any explanation other than they follow "from the nature of proportionals," we draw two conclusions. Ignoring first the middle term, it is clear that $TS^2 = VS \cdot cS$. Also, by the proposition in Article 36, $\log dS - \log eS = \log TS - \log cS$, and since $\log TS = 0$, we have

$$\log cS = \log eS - \log dS.$$

Using now the limits already obtained after Article 28, namely

$$TS - s < \log s < \frac{TS^2}{s} - TS,$$

with $s = cS$ and recalling that $TS^2 = VS \cdot cS$, we find that

$$TS - cS < \log cS < VS - TS,$$

that is,

$$Tc < \log eS - \log dS < VT.$$

Having obtained these new limits, Napier ended Article 39 with the words *quod erat demonstrandum*, even though these limits do not correspond to the verbal description stated above. Notice, however, that at the begining of the proof, Napier showed that TV is to ST as the difference of the sines is to eS, which is equivalent to the statement "the greater limit being to radius as the difference of the sines to the less sine." Similarly, the fact that Tc is to TS as de is to dS is equivalent to "the less limit being to radius as the difference of the sines to the greater sine."

We would, of course, prefer to have these limits in today's notation rather than in verbal form. If s and t denote the two sines and if $s < t$, the proportions stated at the beginning of the proof of Article 39 can be written as

$$Tc = TS\frac{t-s}{t} \qquad \text{and} \qquad VT = TS\frac{t-s}{s}.$$

Thus we can rewrite the result of Article 39 as

$$\frac{TS(t-s)}{t} < \log s - \log t < \frac{TS(t-s)}{s}.$$

Napier himself gave these limits in Article 40, but again in verbal form.

We do not describe his first example because he immediately found a better use for this theorem, computing the logarithms of whole numbers that are not in the first table but near some table entries [Art. 41]. These inequalities, as opposed to those of Article 28, involve two logarithms, so that one of the sines can be a whole number and the other a table entry.

In his second example, Napier chose $s = 9999900$ and noticed that the first table sine nearest and larger than s is $t = 9999900.0004950$ and, as shown in Article 33 (page 407),

$$100 < \log t < 100.0000100.$$

In this case, $TS(t-s) = 4950$, and then the inequalities in Article 40, in the form stated above, give the approximate inequalities

$$0.00049500495002 < \log s - \log t < 0.00049500495005.$$

These inequalities are not in the *Constructio*, where Napier just stated that the difference of these logarithms would be 0.0004950. Hence, from the limits for $\log t$,

$$100.0004950 < \log s = \log 9999900 < 100.0005050.$$

Then Napier observed that [Art. 42]

> sine 9999900, the first proportional of the Second table, was shown in the preceding example to lie between the limits 100.0005050 and 100.0004950; necessarily (by 32) the logarithm of the second proportionals will lie between the limits 200.0010100 and 200.0009900; and of the third proportional, between the limits 300.0015150 and 300.0014850 &c. And finally, the logarithm of the last sine of the Second table, namely 9995001.222927, is included between the limits 5000.0252500 and 5000.0247500.

For each second table sine a logarithm can then be selected between the stated limits.

However, when Napier set out [Art. 43] to find the logarithms of the sines that are whole numbers, and thus not in the second Table of proportions, "but near or between them," he used a new method. Let s be a given sine that is a natural number and is near or between second table sines, and let t be the second table sine that is nearest to s. To simplify the explanation of the method, assume that $t > s$, but this is not really necessary and Napier did not do it. Now choose a number u such that

$$\frac{u}{TS} = \frac{s}{t}$$

which Napier called a *fourth proportional*. Necessarily, u is within the limits of the first table because $s \approx t$ implies that $u \approx TS$, and then its logarithm can be computed by the method of Article 40. Then, by Article 36 and the fact that $\log TS = 0$, we have

$$\log u = \log s - \log t.$$

Then limits for the logarithm of s can be found from those for the other two logarithms.

Napier chose the example of $s = 9995000$. This is near the last sine of the second table, $t = 9995001.222927$, for which he had already shown [Art. 42] that

$$5000.0247500 < \log t < 5000.0252500.$$

In this case, the fourth proportional is

$$u = TS\,\frac{9995000}{9995001.222927} = 9999998.7764614.$$

Then, replacing s with $u = 9999998.7764614$ and t with 9999999, the limits of Article 40 are

$$\frac{TS(9999999 - u)}{9999999} \approx 0.2235386$$

and

$$\frac{TS(9999999 - u)}{u} \approx 0.2235386.$$

Therefore,

$$\log u - \log 9999999 \approx 0.2235386.$$

But we have shown above [Art. 30] that $1 < \log 9999999 < 1.0000001$, and then

$$1.2235386 < \log u < 1.2235387.$$

Since $\log s = \log u + \log t$, the limits found for the two logarithms on the right show that

$$5001.2482886 < \log s < 5001.2487887,$$

and Napier, who erroneously gave the last digit on the right as an 8, was able to conclude:

> Whence the number 5001.2485387, midway between them, is (by 31) taken most suitably, and with no sensible error, for the actual logarithm of the given sine 9995000.

But 9995000 was chosen above because it is the second sine in the first column of the third table. Since the logarithms of the first two sines in this column are known, and since all these sines are in continued proportion, their logarithms are equidifferent by Article 36. Then all their logarithms can easily be computed [Art. 44]. Next, Napier used the method of the fourth proportional to find the logarithm of 9900000, which is near the last sine in the first column of this third table [Art. 45]. He chose 9900000 because it is the first sine in the second column, and then, looking at the first two sines of the first and second columns, we know that

$$\frac{9895050}{9900000} = \frac{9995000}{10000000}.$$

By Article 36, $\log 9895050 - \log 9900000 = \log 9995000 - \log 10000000$, and then log 9895050 can be found because the other three logarithms in this equation are already known. Since the sines in the second column are in continued proportion, their logarithms are equidifferent and they are easily found from the first two. Similarly, once the logarithms of the sines in the first two columns are known, the fact that the sines in each row of the third table are also in continued proportion implies that their logarithms are equidifferent. In this manner, Napier was able to compute the logarithms of all the sines in the third table [Art. 46].

At this point, Napier constructed a fourth table containing the third table sines and their logarithms in parallel columns, as shown in the next figure,

RADICALIS TABVLÆ

Columna prima.		*Columna secunda.*	
Naturales.	Artificiales	Naturales.	Artificiales
10000000.0000	0	9900000.0000	100503.3
9995000.0000	5001.2	9895050.0000	105504.6
9990002.5000	10002.5	9890102.4750	110505.8
9985007.4987	15003.7	9885157.4237	115507.1
9980014.9950	20005.0	9880214.8451	120508.3
&c. usq. ad	usque ad	usque ad	usque ad
9900473.5780	100025.0	9801468.8423	200528.2

	Columna 69ª.	
	Naturales.	Artificiales.
& cæteri usque ad	5048858.8900	6834225.8
	5046334.4605	6839227.1
	5043811.2932	6844228.3
	5041289.3879	6849229.6
	5038768.7435	6854230.8
	& tandem	usque ad
	4998609.4034	6934250.8

reproduced from page 28 of the *Constructio*. He called this *The Radical Table* [Art. 47]. Here he called the original third table sines *Naturales*, and their logarithms he called *Artificiales*. Notice that these are given with just one decimal digit in the Radical Table, although he had used seven in all previous computations to avoid error accumulation. In spite of this precaution it must be said that these logarithms are not totally correct due to the erroneous last sine in the second table, an error that trickles down to the very end of this Radical Table (the last logarithm of the 69th column should be 6934253.4 rather than 6954250.8), but these small errors did not detract from the usefulness of Napier's tables in the seventeenth century.

The Radical Table, being just the third table with the logarithms of its sines, is still unsuitable for computation for the same reasons already given above: it has too few entries, the sines are not whole numbers, and the vertical sine ratio is not the same as the horizontal sine ratio. But for Napier this was just the beginning of his computations, hence the name *radicalis* for this table. From these roots would grow the final table published in the *Descriptio*.

THE TABLE OF LOGARITHMS

Thus, at this point, Napier still faced the third and main part of his task [Art. 50]: to find the logarithms of all natural numbers that are sines near or between the entries in the Radical Table, these sines to be obtained from published tables. He would refer to the result as the Principal Table of logarithms. Given a common table sine s, the procedure that he used to find its logarithm consists of finding first the closest sine t in the Radical Table (in his examples he chose $t > s$), and then compute $TS(t - s)$. Since speed is essential from this point on, he abandoned the inequalities of Article 40 and just divided the product $TS(t - s)$ by the easiest number between s and t. If we denote this number by n, the inequalities of Article 40 are then replaced by the approximation

$$\log s - \log t \approx \frac{TS(t - s)}{n}.$$

Napier gave two examples, the first of which is to compute the logarithm of $s = 7489557$. The nearest sine in the Radical Table is $t = 7490786.6119$, and then, if we choose the easiest number between s and t to be 7490000, we obtain $TS(t - s)/n \approx 1640.1$ (the decimal point was not printed in the *Constructio*, but this quotient is not quite right anyway). Thus, taking $\log t = 2889111.7$ from the Radical Table gives

$$\log s \approx \log t + 1640.1 = 2889111.7 + 1640.1 = 2890751.8.$$

> But since the principal table admits neither fractions nor anything beyond the point, we put for it 2890752, which is the required logarithm.

As we have seen, the Radical Table was nothing but a stepping stone on which Napier would stand to produce the 90 pages of logarithms published in the *Descriptio*. It must have taken quite some time and effort to pick many sines one by one from common tables of sines and compute their logarithms by the method just described. There is evidence that he began his work much earlier than the publication date of 1614. To begin with, the words which he started the dedication of his second mathematical book, the *Rabdologiæ* of 1617:

> To perform calculations is a difficult and lengthy process, the tedium of which deters many from the study of mathematics. I have always tried, with such strength and talent as I possess, to expedite the process. It was with this end in view that I produced in former years my canon of logarithms, at which I had labored over a long period.

There is corroborating evidence in a letter written by Johannes Kepler to Petrus Crugerus on 9th September 1624, in which he stated that [**19**, p. 47]

> a certain Scotsman in a letter written to Tycho a. 1594 already gave hope of that wonderful Canon.

There can be little doubt that this "certain Scotsman" was Dr. John Craig, James VI's physician. We have already suggested that Craig, a friend of Napier and an accomplished mathematician himself, had learned about some new ways to perform the tedious multiplications required by astronomical calculations during a stay in the continent. On his return to Scotland, he paid a visit to Napier and must have told him about these advances. The English antiquarian Anthony à Wood left anecdotal evidence of Napier's response in his *Athenæ Oxonienses* of 1691 [**48**]:

> Which hint *Neper* taking, he desired him at his return to call upon him again. *Craig*, after some weeks had passed, did so, and *Neper* then shew'd him a rude draught of that he called, *Canon mirabilis Logarithmorum.*

This account is rather too fantastic to be accepted literally. For one thing, Napier referred to logarithms as *numeros artificiales* in the *Constructio* [Art. 22], and didn't use the word "logarithms" until the publication of the *Descriptio* in 1614. It is also very unlikely that he had a complete table to show Craig, as we know how long it takes to complete. However, it is probable that he described to Craig his project of that Wonderful Canon, about which Craig later wrote to Tycho.

Returning to Napier's work, the completion of the Principal Table was not the end of it. Since we know that it suffices to find the logarithms of the sines of arcs in the first quadrant, it remains to find those of sines of small arcs that are not in the Radical Table. To complete this task Napier proposed two methods, and both of them are based on the theorem in Article 36.

According to this theorem, if s is a sine not embraced within the limits of The Radical Table.

$$\frac{s}{2s} = \frac{5000000}{10000000}$$

implies that

$$\log s - \log 2s = \log 5000000 - \log 10000000 = \log 5000000.$$

Then

$$\log s = \log 2s + \log 5000000 = \log 2s + 6931469.22,$$

where $\log 5000000$ has been obtained from the Principal Table but before rounding it as a whole number. If $2s$ is within the limits of the Radical Table, its logarithm is found first by the method of Article 50, and then this equation will provide $\log s$. Napier, who didn't write equations as we do, expressed this procedure, and went beyond, as follows [Art. 51]:

Therefore, also, 6931469.22 will be the difference of all logarithms whose sines are in the proportion of two to one. Consequently the double of it, namely 13862938.44, will be the difference of all logarithms whose sines are in the ratio of four to one; and the triple of it, namely 20794407.66, will be the difference of all logarithms whose sines are in the ratio of eight to one.

Immediately, he also showed that [Art. 52]

All sines in the proportion of ten to one have 23025842.34 for the difference of their logarithms.

Thus, we can find $\log s$ from $\log 10s$ if $10s$ is within the limits of the Radical Table. Then Napier combined these last two results as follows [Art. 53]:

Whence all sines in a ratio compounded of the ratios two to one and ten to one, have the difference of their logarithms formed from the differences 6931469.22 and 23025842.34 in the way shown in the following

Short Table.

Given Proportions of sines.		Corresponding Differences of Logarithm.	Given Proportions of sines.		Corresponding Differences of Logarithm.
Two	to one	6931469.22	8000	to one	89871934.68
Four	”	13862938.44	10000	”	92103369.36
Eight	”	20794407.66	20000	”	99034838.58
Ten	”	23025842.34	40000	”	105966307.80
20	”	29957311.56	80000	”	112897777.02
40	”	36888780.78	100000	”	115129211.70
80	”	43820250.00	200000	”	122060680.92
A hundred	”	46051684.68	400000	”	128992150.14
200	”	52983153.90	800000	”	135923619.36
400	”	59914623.12	1000000	”	138155054.04
800	”	66846092.34	2000000	”	145086523.26
A thousand	”	69077527.02	4000000	”	152017992.48
2000	”	76008996.24	8000000	”	158949461.70
4000	”	82940465.46	10000000	”	161180896.38

This allowed him to state a procedure to find the logarithms of all sines that are outside the limits of the Radical Table [Art. 54]:

> This is easily done by multiplying the given sine by 2, 4, 6, 8, 10, 20, 40, 80, 100, 200, or any other proportional number you please, contained in the short table, until you obtain a number within the limits of the Radical table.

Then find the logarithm of this number by the method of Article 50 and add the corresponding difference from the preceding table. For example, Napier considered 378064, which is outside the limits of the Radical Table. But $20 \times 378064 = 7561280$ is within these limits, and the method of Article 50 shows that its logarithm is 2795444.9. To this value add 29957311.56, which is the difference shown in the preceding table for a sine in the proportion of 20 to one. This gives

$$\log 378064 = 2795444.9 + 29957311.56 \approx 32752756.$$

The second method provided by Napier to find the logarithms of sines outside the limits of the Radical Table is based on a trigonometric identity that he gave in verbal form as follows [Art. 55]:

> As half radius is to the sine of half a given arc, so is the sine of the complement of the half arc to the sine of the whole arc.

We omit the proof, but if we recall that his sine (*sinus*) is the product of the radius R and our present-day sine, this statement can be formulated as

$$\frac{R/2}{R\sin\frac{1}{2}\alpha} = \frac{R\sin\left(90^\circ - \frac{1}{2}\alpha\right)}{R\sin\alpha},$$

which, simplifying and regrouping, becomes the familiar identity $\sin\alpha = 2\sin\frac{1}{2}\alpha\cos\frac{1}{2}\alpha$. Then, using the theorem in Article 36,

$$\log\frac{R}{2} - \log\left(R\sin\tfrac{1}{2}\alpha\right) = \log\left[R\sin\left(90^\circ - \tfrac{1}{2}\alpha\right)\right] - \log\left(R\sin\alpha\right),$$

a result that Napier stated in verbal form [Art. 57] but giving the terms in a slightly different order. This equation can be used to evaluate $\log\left(R\sin\frac{1}{2}\alpha\right)$ if the other three *sinuses* are in the Principal Table. Since the sines of all arcs not smaller than 45° are within the limits of the Radical Table, Napier continued as follows [Art. 58]:

> From the logarithms of all [sines of] arcs not less than 45 degrees, given by hypothesis, you can obtain (by 57) the logarithms of all the remaining [sines] of arcs decreasing down to 22 degrees 30 minutes. From these, again, may be had in like manner the logarithms of [the sines of] arcs down to 11 degrees 15 minutes. And from these the logarithms of [the sines of] arcs down to 5 degrees 38 minutes. And so on, successively, down to 1 minute.

At this point all that remains is the elaboration of the logarithmic table itself, for which Napier gave detailed instructions beginning with [Art. 59]:

> Preapare forty-five pages, somewhat long in shape, so that besides margins at the top and bottom, they may hold sixty lines of figures.

The publisher, however, found this instruction impossible to follow, so he split each proposed page in two. The first page of the published table is reproduced here from the 1614 edition of the *Descriptio*. Hart scrupulously followed the

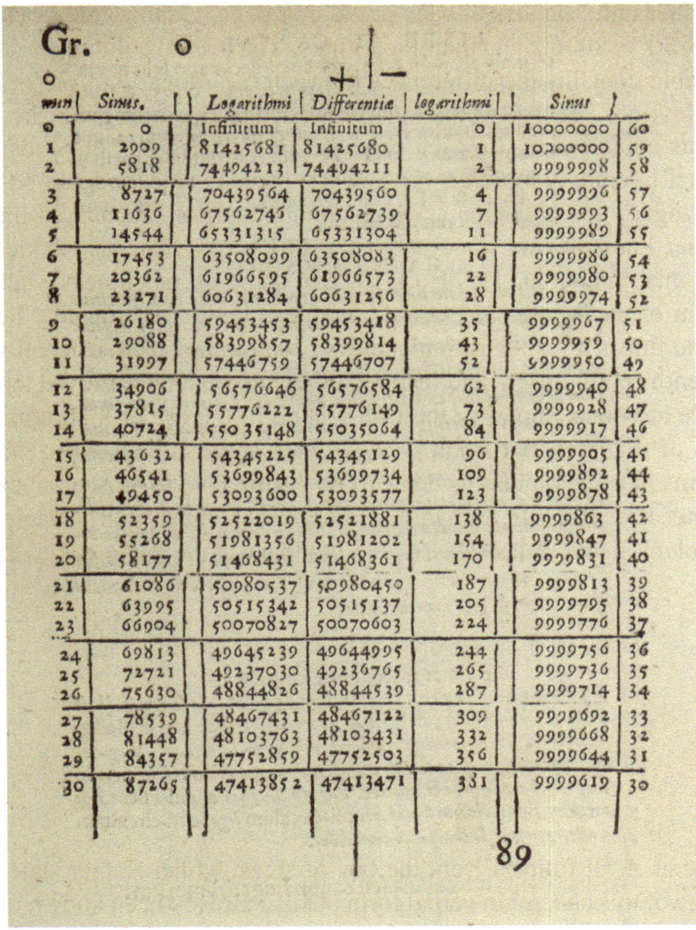

Gr. 0

0 + | −

min	Sinus.	Logarithmi	Differentiæ	logarithmi	Sinus	
0	0	Infinitum	Infinitum	0	10000000	60
1	2909	81425681	81425680	1	10000000	59
2	5818	74494213	74494211	2	9999998	58
3	8727	70439564	70439560	4	9999996	57
4	11636	67562745	67562739	7	9999993	56
5	14544	65331315	65331304	11	9999989	55
6	17453	63508099	63508083	16	9999986	54
7	20362	61966595	61966573	22	9999980	53
8	23271	60631284	60631256	28	9999974	52
9	26180	59453453	59453418	35	9999967	51
10	29088	58399857	58399814	43	9999959	50
11	31997	57446759	57446707	52	9999950	49
12	34906	56576646	56576584	62	9999940	48
13	37815	55776222	55776149	73	9999928	47
14	40724	55035148	55035064	84	9999917	46
15	43632	54345225	54345129	96	9999905	45
16	46541	53699843	53699734	109	9999892	44
17	49450	53093600	53093577	123	9999878	43
18	52359	52522019	52521881	138	9999863	42
19	55268	51981356	51981202	154	9999847	41
20	58177	51468431	51468361	170	9999831	40
21	61086	50980537	50980450	187	9999813	39
22	63995	50515342	50515137	205	9999795	38
23	66904	50070827	50070603	224	9999776	37
24	69813	49645239	49644995	244	9999756	36
25	72721	49237030	49236765	265	9999736	35
26	75630	48844826	48844539	287	9999714	34
27	78539	48467431	48467122	309	9999692	33
28	81448	48103763	48103431	332	9999668	32
29	84357	47752859	47752503	356	9999644	31
30	87265	47413852	47413471	331	9999619	30

89

rest of Napier's instructions. Thus, the first column of this page contains the arcs from 0° 0′ to 0° 30′ and the second column lists their sines. Napier stated: "[Erasmus] Reinhold's common table of sines [**33**, ff. 114–132], with these values." Napier's values coincide with Reinhold's up to and including 88°, but his subsequent values are identical to those of Fincke's table [**10**, pp. 172–173], as originally discovered by Glowatzki and Göttsche [**13**]. Their logarithms are computed by the methods explained above. The logarithms of the sines of the arcs from 0° 31′ to 0° 60′ are on the second page, not reproduced here. The last table page contains on its first three colums the arcs from 44° 30′ to 44° 60′, and their sines and their logarithms. Then the table continues on the right side of this last page with arcs increasing from 45° 0′ to 45° 30′ from bottom to top on the seventh column. Their sines and their logarithms are displayed in the sixth and fifth columns, respectively, and the table is constructed in this manner until, on arrival back at the first page, it shows the sines of arcs from 89° 30′ to 89° 60′ and their logarithms. The central column on each page of the table contains the differences between each logarithm on its left and that on its right. These are the logarithms of the sine of a given arc and the sine of its complement (on each page the arcs located on both ends of each horizontal line are complementary), which is its cosine, and can be used to find the logarithms of tangents, which are quotients of sine and cosine.

Napier knew at this stage about the little error that has trickled down through this final table. He also knew that some of his choices, such as the logarithm of the whole sine, could have been better. Fortunately he persevered and finished his logarithmic tables because they were of great value to seventeenth century users and inspired others to prepare improved tables.

There are no examples in the *Constructio* of the use of logarithms in the solution of triangles, but Book II of the *Descriptio* is devoted to the use of logarithms in trigonometry, and it will be informative to present an example. In Chapter II of Book II, and with reference to the next figure, Napier posed this problem, if $AB = 26302$, $BC = 57955$, and if the angle at C is 26°, find

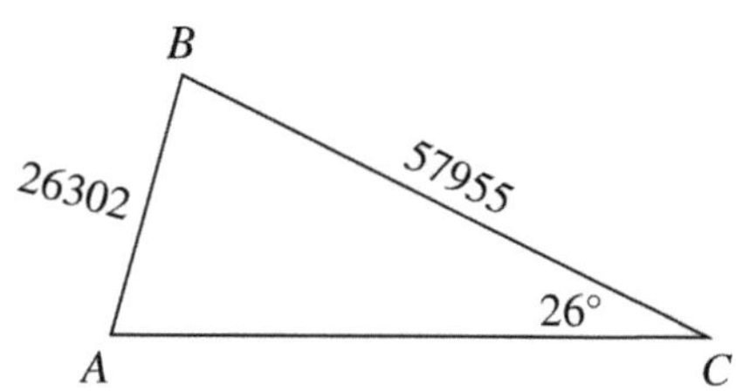

the angle at A. It follows from the law of sines, which Napier stated at the beginning of this chapter in verbal form but had already been known for many centuries, that

$$\frac{\sin A}{57955} = \frac{\sin C}{26302},$$

and then, since the logarithms of similarly proportioned sines are equidifferent

[Art. 36 in the *Constructio*, whose counterpart in the *Descriptio* is Proposition 1 of Chapter II in Book I],

$$\log(\sin A) - \log 57955 = \log(\sin C) - \log 26302.$$

From this equation we could find $\log(\sin A)$, and then $\sin A$ and A itself, if 57955 and 26302 were sines in the logarithmic table. But they are not. While each of them is between two sines on the first page of the table, corresponding to zero degrees, these two sines are too far apart to select a good approximation. For example, 26302 is between the table sines 26180 and 29088, which is too wide an interval.

Napier had already suggested a way out of this difficulty earlier in the *Descriptio*, when he described several ways to accurately find log 137 [Art. 2 of Chapter IV in Book I]. His suggestion, applied to the problem stated above, amounts to replacing the preceding equations with

$$\frac{\sin A}{5795500} = \frac{\sin C}{2630200}$$

and

$$\log(\sin A) = \log(\sin C) + \log 5795500 - \log 2630200.$$

The sine 2630200 is between the table sines 2627505 and 2630312, on the first page for 15 degrees, whose logarithms are listed as 13365493 and 13354817. Napier chose $\log 2630200 \approx 13354921$, and also $\log 5795500 \approx 5454707$ (actually, he did not use this notation, writing instead "Logarithm of AB, which is 13354921–00," where -00 means that the two "last figures" of the sine whose logarithm is 13354921 "are to be cut off"). Since $\log(\sin 26°)$ is given in the table as 8246889, we obtain

$$\log(\sin A) \approx 8246889 + 5454707 - 13354921 = 346675.$$

The logarithm closest to this number in the table is $346683 = \log(\sin 75°)$, and then $A \approx 75°$.

It would be interesting to know how Napier selected the approximated logarithms of 2630200 and 5795500 used in this example, but he did not explain. Strangely enough, he did not use the approximation that he advocated in Article 50 of the *Constructio*. In fact, if we set $s = 2630200$ and $t = 2630312$, then the easiest divisor between s and t is $n = 2630300$ and

$$\log s - \log t \approx \frac{TS(t-s)}{n} \approx 426.$$

From this and the table value $\log t = 13354817$, $\log s \approx 13355243$ rather than Napier's chosen value 13354921. Similarly, letting $s = 5795183$ and

$t = 5705500$, the easiest divisor between these numbers is $n = 5795300$, and then

$$\log s - \log t \approx \frac{TS(t-s)}{n} \approx 547.$$

With the table value $\log s = 5455577$ this gives $\log t \approx 5455030$, and then

$$\log(\sin A) \approx 8246889 + 5455030 - 13355243 = 346676.$$

This gives the same approximation for A as obtained by Napier.

NEW LOGARITHMS BY HENRY BRIGGS

Returning to the *Constructio*, Napier knew that modifications to his table were necessary. He was also aware that his life was coming to an end. Thus, he finished Article 59 in the following way:

> We have computed this Table to each minute of the quadrant, and we leave the more exact elaboration of it, as well as the emendation of the table of sines, to the learned to whom more leisure may be given.

In a final Article 60 he acknowleged that the table of logarithms is faulty in some places, and advised the learned, who may have "plenty of pupils and computors," to publish a more exact table. However, the outline that he described for a new table is very similar to what he had already done, from a First table to a Radical Table. The difference was to begin with a whole sine of 100,000,000, that is, "with eight cyphers after the unit instead of seven only."

However, at the very beginning of an Appendix, Napier made a much more important proposal: to construct a new system of logarithms in which zero is adopted as the logarithm of unity. By the time he published *Constructio* Napier had come to realise that logarithms are useful to simplify computations with numbers, and that these numbers need not be sines of then current tables. The choice that he made abandons the large whole sine and aims at a further simplification. It should be clear that some of the choices he made in building his published table represent a hindrance in the computation of logarithms, Indeed, if $s = tu$, then

$$\frac{s}{t} = \frac{u}{1},$$

and, since the logarithms of similarly proportioned sines are equidifferent [Art. 36], $\log s - \log t = \log u - \log 1$. That is

$$\log tu = \log t + \log u - \log 1.$$

The problem is that 1 is a number very close to S in the top segment TS on page 404, and then its logarithm is a very large number on the bottom line bi. Napier computed it to be 161180896.38, as shown by the last entry in the short table of Article 53, which is the difference $\log 1 - \log 10^7$. This value is somewhat incorrect because of the errors already mentioned, but the true value is very close to this figure. Thus, the logarithm of a product is not the sum of the logarithms of the factors in Napier's original scheme, and one must remember and use the logarithm of 1 in order to find the logarithm of a product. The same is true for a quotient: if $s = t/u$, then $\log s - \log 1 = \log t - \log u$, so that

$$\log \frac{t}{u} = \log t - \log u + \log 1.$$

But if $\log 1 = 0$ these problems disappear and, in particular, we get the usual product rule as known today: $\log tu = \log t + \log u$. Then, if we continue to use current notation and if n is a positive integer, repeated application of the product rule shows that $\log s^n = n \log s$, which we could call the power rule. Then, if $t = s^{m/n}$, where m is also a positive integer, we have $s^m = t^n$. Taking logarithms of both sides and using the power rule gives $m \log s = n \log t$, so that

$$\log s^{\frac{m}{n}} = \frac{m}{n} \log s.$$

We conclude that, if r is a rational number, $\log s^r = r \log s$, which easily deals with the computation of logarithms of powers and roots. Thus, accepting Napier's proposal that $\log 1 = 0$, leads to these very convenient computational features that his original logarithms lack.

Napier was aware of these computational advantages, which is why he proposed the change to $\log 1 = 0$. In the second section of the Appendix to the *Constructio*, he clearly stated these advantages one by one, and then summed them up as follows (we have completed Napier's statement in brackets):

> Finally any common number being formed from other common numbers by multiplication, division, [raising to a power] or extraction [of a root]; its Logarithm is correspondingly formed from their Logarithms by addition, subtraction, multiplication by 2, 3, &c. [or division by 2, 3, &c.]: whence the only difficulty is in finding the Logarithms of the prime numbers; and these may be found by the following general method.

Napier was well aware of the fact that at his age (he was 64 when he published the *Descriptio*) he would never be able to elaborate the new table of logarithms, but he wanted to leave as many hints as possible to whoever would accept this task. One of them was that only the logarithms of prime numbers need be computed, for then the logarithm of any composite number (according to the fundamental theorem of arithmetic, such a number can be factored uniquely as a product of primes) can be found from those of its prime factors by using the product rule.

To obtain the logarithms of primes he proposed two methods: the method of the fifth roots and the method of square roots. To explain the beginning of each of these methods let's accept a second proposal that Napier made in the opening paragraph of the Appendix, that 10,000,000,000 be taken as the logarithm of 10 or that of 1/10. Say that we take $\log 10 = 10{,}000{,}000{,}000$. Dividing this logarithm by 5 ten consecutive times we obtain the logarithms 2000000000, 400000000, 80000000, 16000000, 3200000, 640000, 128000, 25600, 5120, and 1024. According to the rules that follow from the choice of log 10, these are the logarithms of $\sqrt[5]{10}$, the fifth root of this root, and so on. In the same manner, dividing the logarithm 1024 by 2 ten consecutive times we obtain 512, 256, 128, 64, 32, 16, 8, 4, 2, and 1. These are the logarithms of $\sqrt{1024}$, $\sqrt[4]{1024}$, and so on.

Beyond these logarithms, Napier was content with stating, near the end of the first part of the Appendix:

> From these, once computed, there may then be formed both the proportionals of other Logarithms and the Logarithms of other proportionals.

Happily enough, Napier was able to witness the beginning of the construction of a new table of logarithms based on these guidelines. His collaborator was Henry Briggs (1561–1631). A graduate of the University of Cambridge, he became the first Geometry Reader in 1596 at Gresham College in London, holder of a chair endowed by Sir Thomas Gresham in that very same year. He was deeply impressed by Napier's logarithms, as shown by what he wrote to his friend James Ussher on 10th March 1615 [**30**, p. 36 of the collection of three hundred letters at the end of the volume]:

> *Napper*, Lord of *Markinston*, hath set my Head and Hands a Work, with his new and admirable Logarithms. I hope to see him this Summer if it please God, for I never saw [a] Book that pleased me better, or made me more wonder.

Briggs did indeed travel to Merchiston in the summer of 1615, and his first encounter with Napier was described in a rather fantastic manner by the astrologer William Lilly (1602–1681) in his autobiography of 1667–1668 [**22**, p. 236]. Lilly stated that when Briggs was brought into Napier's chamber,

> almost one quarter of an hour was spent, each beholding the other almost with admiration, before one word was spoke.

Briggs, who remained at Merchiston for about a month, had also noticed some shortcomings with Napier's logarithms, and had been independently thinking of ways to solve these problems. Years later, in [the Preface] to the Reader of his book *Arithmetica Logarithmica sive Logarithmorvm Chiliades Triginta* (Logarithmic Arithmetic or Thirty Thousands of Logarithms) of 1624, he stated [**7**, p. A3]:

> When I explained their [the logarithms'] doctrine to my auditors in London, publicly at Gresham College; I noticed that it would be very convenient in the future, if 0 were kept for the Logarithm of the whole Sine (as in the *Canone Mirifico*) while the Logarithm of the tenth part of the same whole sine ... were 10000000000.

This proposal can be written today as $\log 10^7 = 0$ and $\log 10^6 = 10^{10}$, and the reason for these choices must have been that they simplify the computation of $\log 10s$ from $\log s$. Indeed,

$$\frac{10s}{s} = \frac{10^7}{10^6}$$

implies that

$$\log 10s - \log s = \log 10^7 - \log 10^6.$$

Then, with the choices for the last two logarithms,

$$\log 10s = \log s - 10^{10}.$$

This is a simpler computation than the one using Napier's value of $\log 10^6 = 23025842.34$, provided by the difference $\log 1 - \log 10^6 = 138155054.04$ from Article 53 and the previously found value $\log 1 = 161180896.38$.

Briggs had previously sent a letter with his proposal to Napier, and while staying at Merchiston he learned of Napier's better proposal, which he described as follows [**7**, p. A3]:

> But [Napier] recommended to therefore make this change, that 0 should be the Logarithm of unity, and 10000000000 [that of] the whole sine: which I could not but acknowledge was most convenient by far.

The *Constructio* was published after Napier met Briggs at Merchiston, so if this recollection is correct Napier later changed this suggestion that $\log 10^7 = 10^{10}$ to the published version that either $\log 10$ or $\log 1/10 = 10^{10}$.

So Briggs, who had already computed some logarithms based on his own proposal of $\log 10^7 = 0$ and $\log 10^6 = 10^{10}$, put them aside. He began a new table based on $\log 1 = 0$ and, modifying Napier's second proposal, on $\log 10 = 10^{14}$. More precisely, if $10^y = x$ Briggs took $\log x$ to be $10^{14}y$, while today we take y to be the decimal logarithm of x. Thus, many of today's decimal logarithms contain a decimal point. By making his logarithms 10^{14} times as large, Briggs avoided the use of the still unfamiliar decimal point without giving up accuracy.

Briggs journeyed to Edinburgh again in 1616 and showed Napier how he was recomputing the logarithms. The first thousand logarithms appeared first in a privately printed booklet of 16 pages, *Logarithmorvm Chilias Prima* [**8**], whose first page does not contain a date or the author's name. It was published in London in 1617, just after Napier's death, for on this page Briggs told his readers that these logarithms are different from those

> whose Illustrious inventor, of always cherished memory, published in his *Canone Mirifico*.

The last 15 pages of this booklet contain the table of logarithms, but no explanation is provided about how they were obtained (a recent reconstruction of this table can be found in [**36**]).

Briggs remained at Gresham College for 23 years before moving to Oxford, where he became the first Savilian professor of geometry, a chair endowed by Sir Henry Savile (1549–1622), warden of Merton College. Briggs remained at Merton College for the rest of his life, and in 1624 he published the already mentioned *Arithmetica Logarithmica*, containing the new logarithms of the integers from 1 to 20,000 and 90,000 to 100,000 (there is an explanation for the large gap in his table, see p. 429 below), and in this volume he described how the logarithms were computed.

To begin this computation, Briggs used one of the two methods suggested in the Appendix to the *Constructio*: the method of the fifth roots and the method of square roots. According to the first method, which was admitted to be more difficult but considered more exact, if $\log 10 = 10^{10}$ then

$$\log \sqrt[5]{10} = \frac{1}{5} \log 10 = \frac{1}{5} 10^{10} = 2 \times 10^9,$$

and so on for the logarithms of other successive fifth roots of 10. Briggs, however, chose the method of the square roots. We know that he was an incredible calculator, but he must have sensed the enormity of the task ahead of him and chose the simpler method. So he computed successive square roots of 10, beginning with

$$10^{1/2} = \sqrt{10} = 3.16227766016837933199889354,$$

so that $\log 3.16227766016837933199889354 = 0.5$. We have used, and will continue to use, current exponential notation, unknown to Briggs. He denoted the square root of 10 by $l.10$, the *latus* (side) of 10, that is, the side of a square of area 10. Then the fourth root of 10 would be $ll.10$, the eighth root $l.(8)10$, and so on. By successive extraction of square roots, Briggs evaluated $10^{1/4}$, $10^{1/8}$, continuing down to

$$10^{1/2^{53}} = 1.000000000000000255638298640064 70$$

and

$$10^{1/2^{54}} = 1.00000000000000012781914932003235.$$

He gave a complete table showing the results of this labour in Chapter 6 of the *Arithmetica Logarithmica* [**7**, p. 10], of which we reproduce here the first six and the last four rows. Except for the top row—which contains only the

	10		1,000
1	31622,77660,16837,93319,98893,54		0,50
2	17782,79410,03892,28011,97304,13		0,25
3	13335,21432,16332,40256,65389,308		0,125
4	11547,81984,68945,81796,61918,213		0,0625
5	10746,07828,32131,74972,13817,6538		0,03125
.	.		.
.	.		.
.	.		.
51	10000,00000,00000,10225,53194,56025,921	L	0,00000,00000,00000,44408,92098,50062,61616,94526
52	10000,00000,00000,05112,76597,28012,947	M	0,00000,00000,00000,22204,46049,25031,30808,47263
53	10000,00000,00000,02556,38298,64006,470	N	0,00000,00000,00000,11102,23024,62515,65404,23631
54	10000,00000,00000,01278,19149,32003,235	P	0,00000,00000,00000,05551,11512,31257,82702,11815

number 10 and its logarithm—the first column of this table shows the value of the exponent n in $10^{1/2^n}$, the second shows the value of the full power $10^{1/2^n}$, and the third the value of the exponent $1/2^n$. Thus, the third column shows, below the logarithm of 10, the logarithms of the 54 weird numbers on the second column. These logarithms are easy to compute, as each one is half of the preceding one.

Some remarks are in order about this somewhat strange table. To begin with, there are missing zeros in some of the entries in the last column and more than 14 digits in others. For example, Briggs had chosen $\log 10 = 10^{14}$, but here it appears as 1,000. Briggs used commas to help counting spaces, but if we read the first comma in the third column as a decimal point, these are today's decimal logarithms, beginning with $\log 10 = 1.000$. The Briggsian logarithms are just 10^{14} times larger. But the powers (successive square roots) of 10 in the second column should contain a decimal point after the first digit, as shown in the three powers of 10 evaluated above before the table. However, Briggs did not use the decimal point in this table, in spite of Napier's pioneering use of this device in the *Constructio*. In what follows we use the decimal point for these numbers to avoid possible confusion, but retain the Briggsian first comma for their logarithms.

It must be pointed out that Briggs made a mistake in his computation of $10^{1/4}$. As Bruce has pointed out in his third note on Chapter 6 of the *Arithmetica Logarithmica* [**7**], it should be 17782,79410,03892,2801**2,25421,19519**, with the corrected digits in boldface. This mistake propagates through the entire table, but the incorrect digits are far on the right, and they move farther and farther to the right as the error trickles down. The last two entries in the second column are almost in complete agreement with values obtained today using a computer.

Of course, the goal that Briggs had in mind was to compute the logarithms of whole numbers, so did he need to go so far in finding logarithms of numbers laden with decimals? We don't know how far Napier proposed to carry out his proposal of finding logarithms by root extractions. Fortunately Briggs persevered, for this allowed him to make an observation that led him to discover a highly original method to find the logarithms of whole numbers. Then this strange table became just an intermediate stage in his journey, just as the Radical Table was an intermediate stage in Napier's computation of logarithms, as published in the *Descriptio*.

Let us take a closer look at the last four rows of the previous table, whose second-column entries were labeled L to P by Briggs for easy reference. His observation was that, if we look at these numbers from the bottom up—that is, in the order P, N, M, L—they [**7**, p. 11] "all increase almost equally, with the ratio of the distance between each other." This can be explained more clearly with the use of current notation. If we write $P = 1 + p$, where $p = 0.0000000000000001278191493200323$5, we notice that $N = 1 + 2p$, $M \approx 1+4p$, and $L \approx 1+8p$. We have equality for N, but only approximations for M and L, and the approximation for L is less acccurate than that for M, so Briggs decided to go no further. As for the logarithms of these numbers, the fact that $p \ll 1$ means that any power of p is much smaller that p, and this allows us to write

$$\log N = \log(1 + 2p) \approx \log(1 + p)^2 = 2\log(1 + p),$$

$$\log M \approx \log(1 + 4p) \approx \log(1 + p)^4 = 4\log(1 + p),$$

and

$$\log L \approx \log(1 + 8p) \approx \log(1 + p)^8 = 8\log(1 + p).$$

Therefore, if a number of the form $1 + q$ is "near the numbers L M N & P," then

$$\log(1 + q) = \log\left(1 + \frac{q}{p}p\right) \approx \frac{q}{p}\log(1 + p).$$

Briggs wrote mathematics in narrative form, as it had been written for millenia. He did not write any equations or approximations, as we have done, and simply stated that "the Logarithm of this [number; a number near the L to P range] is easily found, by the rule of proportion" [**7**, p. 11], which he called the "golden rule" (*auream regulam*). As an example he considered the number $X = 1.00000{,}00000{,}00000{,}01$, for which

$$\frac{q}{p} = \frac{10000000000000000}{12781914932003235},$$

and found that $\log X = 0{,}000000000000000004342944819032518 04$ (the last three digits are in error, as shown by Bruce in his fifth note on Chapter 6 of the *Arithmetica Logarithmica* [see **7**]).

The basic idea to compute the logarithm of a whole number is to bring it, or some of its prime factors, down to the L to P range by continued square-root extractions. Before we embark on such a folly, we must realise that it would take 53 square root extractions of 2, the first integer whose logarithm is wanted, to bring it to the L to P range as

$$2^{1/2^{53}} = 1.00000000000000007695479593116620.$$

This task (performed today by a computer) is too time-consuming, and Briggs had a better idea.

In Chapter VII, in which Briggs began finding ways to shorten the computations, he approached that of log 2 indirectly as follows. First, since $2^{10} = 1024$ [**5**, p. 13],

$$\log 2 = \frac{\log 1024}{10},$$

and

$$\log 1024 = \log 1000 + \log 1.024 = 3 \log 10 + \log 1.024.$$

Thus, it suffices to compute the logarithm of 1.024 (since Briggs was reluctant to use the decimal point, he wrote this number as $1\underline{024}$), which is closer to 1 than 2. Then it takes 47 (as opposed to 53) square root extractions of 1.024 to obtain

$$1.024^{1/2^{47}} = 1.00000000000000016851605705394977,$$

a number in the L to P range, and "the Logarithm of that number is found by the golden rule to be 0,000000000000000073185593690623933 6." [**7**, p. 13] Therefore (recall that we retain Briggs' first comma for logarithms, as stated on page 426),

$$\begin{aligned}\log 1.024 &= (2^{47})(0{,}0000000000000000731855936906239336)\\ &= 0{,}01029995663981195,\end{aligned}$$

and then

$$\log 2 = \frac{3{,}0000000000000000 + \log 1.024}{10} = 0{,}30102999566398 1195,$$

as given in the *Arithmetica Logarithmica* [**7**, p. 14].

Next Briggs used the proportion 2 is to 1 as 10 is to 5 and the properties of logarithms to obtain $\log 5 = 0{,}69897000433601 8805$, and near the end of Chapter VII listed the numbers whose logarithms are easily computed from those of 2 and 5:

> From the multiplication of Two alone by itself & into its factors, 4. 8. 16. 32. 64. &c. likewise of Five by itself & into its factors, 25. 125. 625. 3125. &c, Two in factors of Five, 250. 1250. 6250. &c. Two into Ten 20. 200. 2000. 40. 400. 80. 800. &c.

The last logarithm that Briggs computed by extraction of square roots was that of the number 3, "whose Logarithm is most conveniently found from the Logarithm of Six." It is clear that he had in mind the fact that $3 = 6/2$, but it is also a fact that 6 is farther away from 1 than 3, and root extraction alone would take longer for 6 than for 3. Thus, Briggs decided to find first the logarithm of

$$\frac{6^9}{10^7} = 1.0077696,$$

which is very close to 1. He did this again by the golden rule, obtaining 0,00336125345279269, and then

$$\log 6 = \frac{7{,}0000000000000000 + \log 1.0077696}{9} = 0{,}77815125038364363.$$

At this point, it must have been evident to Briggs that the elaboration of an entire table of logarithms by square-root extractions would be so time-consuming as to render the task impossible. However, he made an additional discovery that saved the day and he announced at the beginning of Chapter VIII that [**7**, p. 15] "the annoyance of so much labour is greatly diminished through Differences." Thus he found a way to compute square roots, not by laborious, slow extractions, but by the simple computation of some differences of numbers. It was a great breakthrough, but one that cannot be explained briefly (see [**14**, pp 110–114], and for a complete description of the 1624 edition of the *Arithmetica Logarithmica* and a reconstruction of the tables see [**38**]). Suffice it to say that, propelled by his considerable talent and a prodigious capacity for computation, Briggs was able to use this new method to compute the logarithms of the numbers 1 to 1000 in *Logarithmorvm Chilias Prima*.

But Briggs was far from done, and the *Arithmetica Logarithmica* includes a number of chapters unveiling new mathematical techniques to enable the computation of logarithms. Particularly noteworthy is Chapter XII [**7**, pp. 24–32], in which he developed a method of second-order differences that is better known today as *Newton's forward difference method* because it was later rediscovered by Newton. This is the method that Briggs used to compute the logarithms of the numbers from 1 to 20,000 and 90,000 to 100,000. But the method of Chapter XII is not applicable to the range 20,000 and 90,000, and Briggs proposed a new method in a longer Chapter XIII. He intended to compute the logarithms of the missing chiliads, as we know from a letter that he wrote to John Pell on 25th October 1628 [**3**, p. 774]:

> My desire was to have these Chiliades that are wantinge betwixt 20 and 90 calculated and printed, and I had done them all almost by my selfe, and by some frendes whom my rules had sufficiently informed, and by agreement the busines was conveniently parted amongst us; but I am eased of that charge and care by one Adrian Vlacque ...

The large gap in the *Arithmetica Logarithmica* of 1624 had been filled in a second edition of 1628, published without any notice to Briggs by Adriaan Vlacq (c. 1600–1667) of Gouda, Holland. It is unclear exactly who computed these logarithms, although the author may have been Ezechiel de Decker, a Dutch surveyor, perhaps assisted by Vlacq (these matters are discussed in [**37**], which also contains a reconstruction of the table). However, one change that displeased Briggs is that the accuracy of these logarithms was reduced from 14 to 10 digits, probably for the sake of speed in publication.

Frustrated by his inability to properly complete the table in the *Arithmetica Logarithmica*, Briggs devoted all his attention to a new work, the *Trigonometria Britannica.* He worked on this project for the rest of his life but had to leave it unfinished. His friend Henry Gellibrand (1597–1636), a professor of astronomy at Gresham College, was able to complete it and had it published by Adriaan Vlacq in Gouda in 1633.

THE QUESTION OF PRIORITY

At about the same time as Napier was elaborating his table of logarithms, Jost Bürgi (1552–1632), a Swiss clock and instrument maker working at the court of several rulers of the Holy Roman Empire in Prague, had a similar idea of simplifying computations by constructing two sequences, one geometrical and one arithmetical. It appears that he was indirectly inspired by Stifel's table, but chose 1.0001 as his initial ratio close to 1, rather than Napier's 0.9999999.

If we write

$$B = 100000000 \times 1.0001^{R/10},$$

Bürgi called R the "red number" (*rote Zahl*) corresponding to the "black number" (*schwarze Zahl*) B. These are the colours in which these numbers were printed in his tables *Aritmetische vnd Geometrische Progress Tabulen* of 1620 [**9**]. An introduction (*Unterricht*) in which Bürgi explained his methods was not printed with the table, but was later reproduced in the twentieth century [**11**].

What Bürgi gave us was a 58-page table that is very useful for computational purposes. It is definitely more refined and extensive than Stifel's and uses a non-integral ratio, 1.0001. But conceptually his work is at the level of Stifel's. What Bürgi did not give us was an abstract concept of logarithm with a general definition, as Napier did. Using the continuous motion of the points g and a, one on TS and the other on the half line bi, Napier defined the logarithm of an arbitrary sine dS. Furthermore, Napier, unable to exactly compute the logarithm of an arbitrary sine, proved theorems finding error bounds for it. Thus, conceptually his work was much superior to Bürgi's.

Whenever Bürgi's work is discussed the question of priority is likely to arise. It has been frequently stated that Bürgi had his idea of logarithms as early as 1588, which would give him priority in invention, if not in publication. Unfortunately, most of these claims do not offer any supporting evidence, and those that do give no proof. In a lecture delivered in 1872, Rudolf Wolf stated that [**47**, p. 13]:

> Reimarus Ursus had already pointed out in a publication from the year 1588, that Bürgi had a means to make his extraordinary calculations easier,

but gave no specific reference with title and page number. Wolf's statement was picked up, almost verbatim, by Erwin Voellmy in 1948 [**46**, pp. 17–18]. He gave an incomplete reference to Wolf, also without title or page number. Then, in 1992, Lutstorf and Walter gave the following variant of these statements in the opening paragraph of a work on Bürgi's *Progress Tabulen* [**23**, p. 1]:

> We know from contemporary evidence*), that already in 1588 Bürgi had at his disposal a means to simplify complicated calculations,

where the *) is a footnote mark that gives the following reference: Raimarus Ursus Dithmarsus, *Fundamentum astronomicum*. Strassburg, 1588. Once more, no page number is given, but at least there is a title. Since 1992 there has been a deluge of extrapolating statements, mostly online, suggesting or claiming that Bürgi knew about logarithms in 1588.

Nicolaus Reimarus, nicknamed Ursus, (1551–1600) did indeed publish a small book entitled *Fvndamentvm Astronomicvm* in 1588 [**32**]. A line by line examination of this work, has revealed eight references to Bürgi, the most extensive on the fifth page (not numbered) of the *Epistola Dedicatoria* and the others on ff. 5^v, 8^v, 9^r, 19^v, 22^r, 22^v, and 37^r. None of these references are accompanied by a statement about Bürgi having a means to simplify complicated calculations. The *Epistola Dedicatoria* makes it clear that the context is the Doctrine of Triangles, and Reimarus states that this "very profound but formerly intricate doctrine" was given to him by his teacher (*Præceptoris*) Jost Bürgi, exceedingly praising him before announcing that in this volume he is passing the method along to us. This makes sense, for at that time Bürgi was working on his never-published tables of sines, and the method that Reimarus passed along to us in f. 9^r is about the computation of sines. This is made abundantly clear by the subtitle of this book: *Nova Doctrina Sinvvm et Triangvlorvm.*

Benjamin Bramer, who lived in Bürgi's house from 1603 to 1611, stated in 1630 that [**5**, p. 5]

> my dear brother-in-law and preceptor Jobst Bürgi had calculated more than twenty years ago a beautiful table of progressions ...

(the complete paragraph was copied by Tropfke in [**45**, footnote 589, p. 146]). This means that Bürgi had calculated his table or part of it before 1610. But not in 1588, which is more than forty years before 1630, and not in 1594, when Napier had already shown something to John Craig.

In a recent paper, Denis Roegel has described the notion that Bürgi had began work on his logarithms in 1588 as "wishful thinking," and concluded for a variety of reasons that [**34**, p. 8]

> Bürgi's tables were probably conceived around 1605–1610, after he had finished his work on sines.

It seems a fair conclusion that Napier has both the priority in the invention of the logarithms as well as the priority in publication.

PART II. MECHANICAL DEVICES

John Napier published his second mathematical book in 1617, the year of his death. Like the *Descriptio*, it was devoted to simplifying computations with large numbers, but the methods proposed in this new book are radically different from the logarithms of his previous volume. These methods are already suggested by its Greek-inspired main title: *Rabdologiæ* [**28**], constructed from the Greek words *ῥάβδος* (*rabdos*) = rod and *λογία* (*logia*) = study, science, or collection. The purpose and content of the volume is then greatly clarified and expanded by the author in his *Dedicatory Epistle* to "the most illustrious Alexander Seton, Earl of Dumferline, Lord of Fyfe & Urquhart, &c. Supreme Chancellor of the Kingdom of Scotland."

In the opening paragraph Napier referred to his original logarithms, and then stated that he had discovered a greatly superior species of logarithms. However, because of the poor state of his health, he decided to leave the actual computation of the new canon to his learned friend Dr. Henry Briggs.

"In the meantime," he informed the reader in the second paragraph, he had devised three other methods to shorten calculations, which Napier briefly described as follows:

> The first of these, which I call RABDOLOGIÆ, uses *rods with numbers on them*: The second, which for multiplication is the fastest of all, uses strips arranged on a box; for this the name *promptuary for Multiplication* will not be inappropriate. The third and last uses *location Arithmetic* and is performed on a chess-board.

Therefore, this book describes the use of three mechanical aids to computation, and Napier justified his decision to publish, beyond the obvious reason of getting priority in publication, by the fact that the rods had already become so popular as to be "in common use both at home and abroad." If this was true when Napier wrote the *Dedicatory Epistle*, it is a reasonable inference that the rods had been in use for many years before 1617, that Napier had used them himself in the elaboration of his logarithmic table and that the rods had been invented precisely for the purpose of minimising errors in the construction of that table.

The organisation of the *Rabdologiæ* is somewhat strange, consisting of two books, or parts, one appendix, and then another book, as if Napier had just thought of a new important topic after the appendix. The first book is devoted to the description and use of the computing rods, the second to the use of the rods in geometry and mechanics, the appendix is devoted to the multiplication promptuary, and the third book to location arithmetic and the chess-board. Each of the books and the appendix are then divided into chapters.

While the *Constructio* is a true mathematical work, in the sense that it contains proved theorems—albeit without that label—on the properties of logarithms, the *Rabdologiæ* is in the nature of what would now be called a user's manual. It describes the three devices that Napier invented and simply tells us how to use them. There is no theory to explain, and it is as a how-to book that we describe it here.

NAPIER'S BONES

In Chapter I of the First Book Napier described the making and the marking of some rods, beginning as follows (in quotations from the *Rabdologiæ* we use the translation by Richardson included in this volume) [**28**, p. 2; 659]:

> Construct from silver, ivory, boxwood or some similar solid material a number of square rods . . . They should all be the same length, that is, three fingers more or less. The width of each should be one tenth of the length, enough to contain comfortably two arithmetical figures, and the height should be equal to the width.

The last statement means that the rod's cross-section is a square, as shown in the left part of the next figure, from which we learn the general appearance of each rod. The length of each rod proposed by Napier is about three fingers,

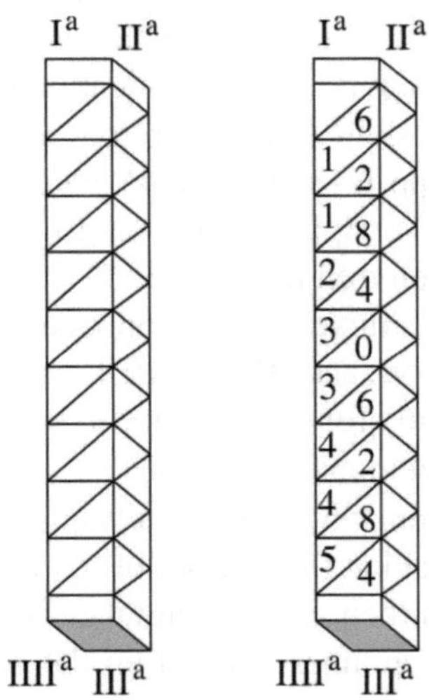

which is about 57mm since the width of a finger is about 19mm (the rods in the set at the *Museo Arquelógico Nacional* in Madrid, which are made of ivory, are 52mm long and their cross section is 5mm × 5mm [**44**, p. 57]). Since the rods were frequently made of bone or ivory, they soon became known as Napier's bones, a popular name that seems to have appeared in print for the first time in 1618. Thomas Bretnor, a mathematics teacher who used to publish a yearly almanac, cited his "loue to the *Mathematickes*" while giving an account of recent publications in the field in the 1618 issue. One of these was "an excellent Treatise of *Rabdologia*, or Arithmeticall Rods, inuented and published by the learned Lord of *Merchiston*, which Instruments or Rods (commonly called *Napiers Bones*) are made by *Nathaniell Gosse* in Hosiers lane." This statement appears in the penultimate page of this small booklet with unnumbered pages [**6**].

Now draw nine squares on each of the four faces, as shown in the figure, with an additional rectangle (half a square) at each end as a margin. Divide each square in two by a diagonal line from the lower left corner to the upper right corner. As the figure shows, Napier labelled the four faces of each rod as I, II, III and IIII (actually, I^a to IIII^a [**28**, p. 3; 660]), but this was only to help his exposition and should not actually be drawn on the rods. The order of these labels is such that, if a rod is lying on the table with face I up, then II is to the right, III down, and IIII is to the left.

In the top square of each face, write down a single digit in the right triangle. Napier called this digit a *simple*. On each of the eight squares below the top one, write down the multiples of the simple in order, from the second to the ninth, with the unit's digit in the right triangle and the tens digit in the left triangle of each square. This is illustrated on the right side of the figure above if the simple on a face is 6. Thus the collection of rods for all simples from 1 to 9 can be considered as a portable multiplication table.

The next three figures show a set of ten rods with the complete markings suggested by Napier [**28**, pp. 6–8; 662–664]. Each rod is shown flattened out, with faces I to IIII from left to right. Abandoning the use of the Greek language in these figures, Napier referred to each rod by its Latin name, *virgula*. The first of these contains the simples 0 and 1 on faces I and II, respectively, and then the simples 9 and 8 on faces III and IIII, respectively. The last two are upside down, but this is only for convenience in selecting rods as needed, and does not affect the computations in any way.

There are some additional markings that Napier proposed, which are suggested in his figures. Imagine that the set of ten rods is stored in a box and that the rods are stood on their ends when taken out of the box. It would be convenient to have some numbers carved on the square ends (two on each end) to indicate what simples are to be found on the four faces [**28**, pp. 9–10; p. 664]. One easy way to do this, which is shown in a photograph of an actual set (not reproduced here), is to draw a diagonal on the top end separating the faces containing the simples 0 and 1 of the first rod, for instance. Then the two triangles on the top can be labeled 0 and 1, appropriately. Turning the rods on their other end, a diagonal can be drawn on the square end to separate the faces containing the simples 9 and 8, and similar markings made.

Now let's notice some relevant features of the markings made on these ten rods. The first is that the simples on opposite faces of each rod add up to nine. Thus, $0 + 9 = 1 + 8 = 9$ for the first rod, $1 + 8 = 4 + 5 = 9$ for the seventh rod, and so for the others. The second is that each digit is a simple in exactly

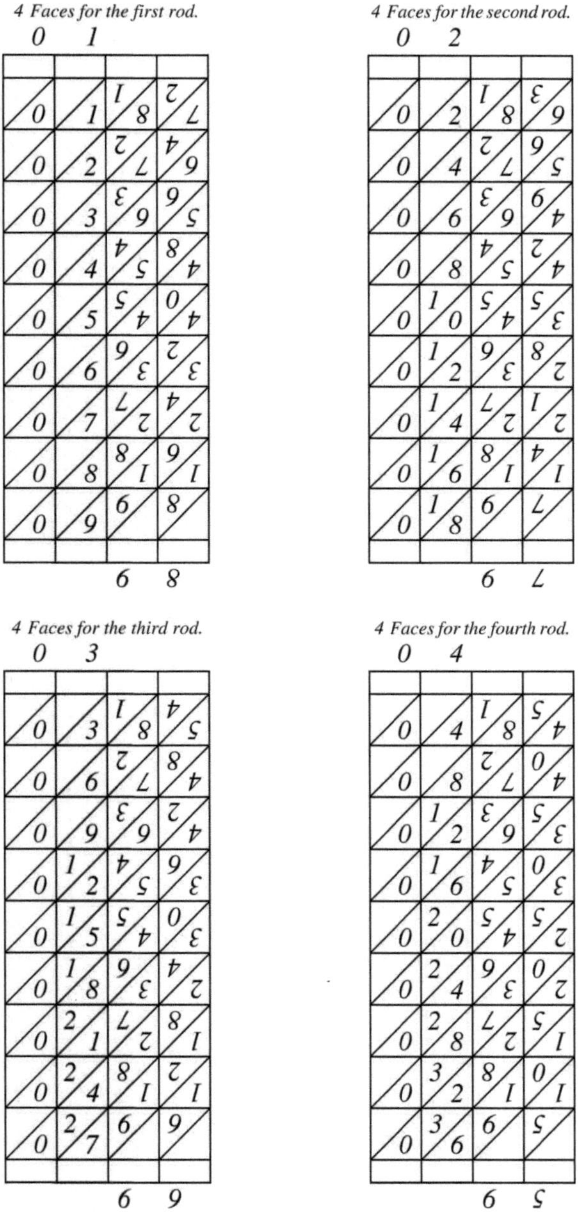

four of the 40 faces. For example, 3 is a simple on face II of the third, the sixth, and the eighth rods, and on face I of the tenth rod; while 5 is a simple on face IV of the fourth, seventh, ninth, and tenth rods. The third is that each simple is paired with every other simple on some rod. For instance, 2 is paired

4 Faces for the fifth rod.

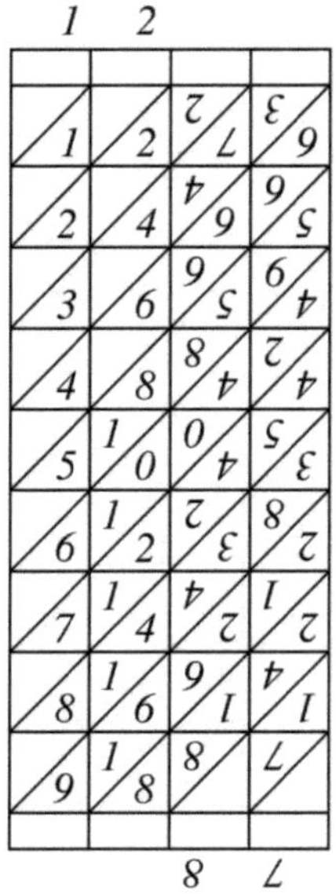

4 Faces for the sixth rod.

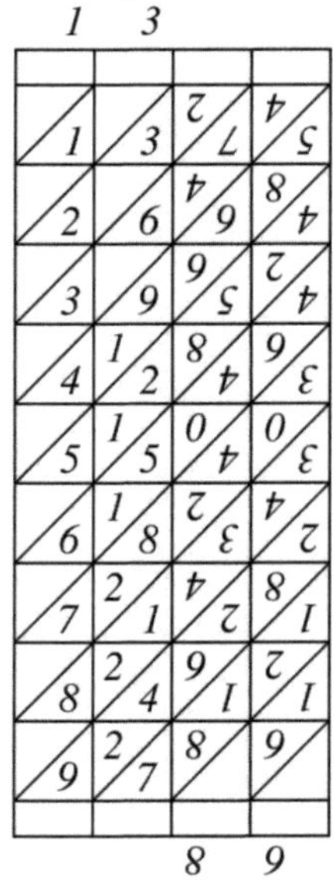

4 Faces for the seventh rod.

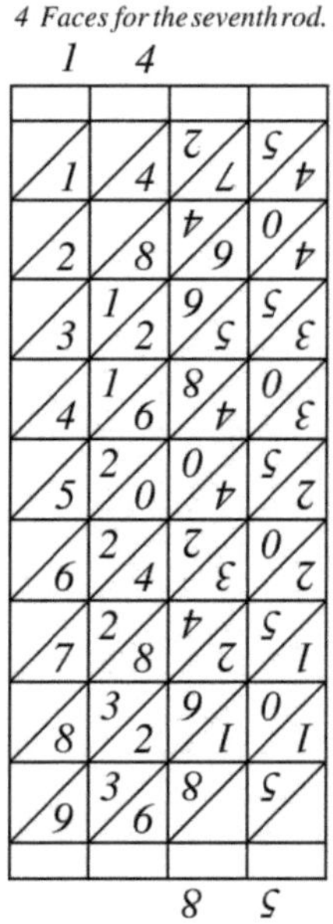

4 Faces for the eighth rod.

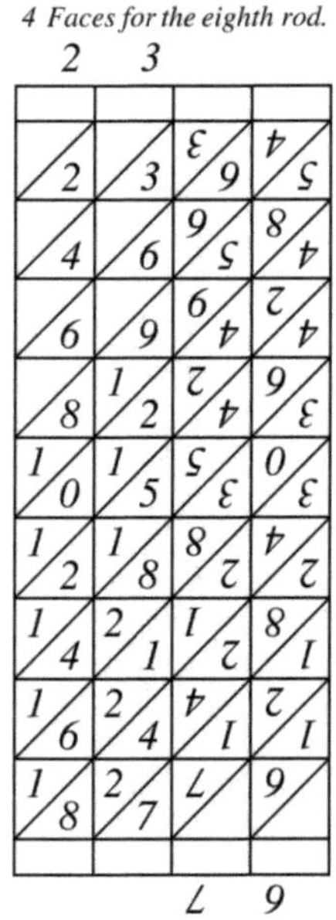

with 0–7–9 in rod two, with 1–7–8 in rod 5, with 3–6–7 in rod 8, and with 4–5–7 in rod 9. This is important because, as we shall see when we learn to display numbers with rods, it sets some limitations on the numbers for which ten rods are sufficient.

4 Faces for the ninth rod.

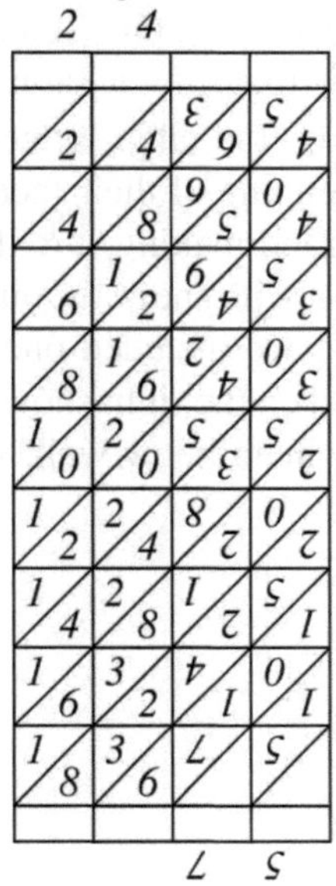

4 Faces for the tenth rod.

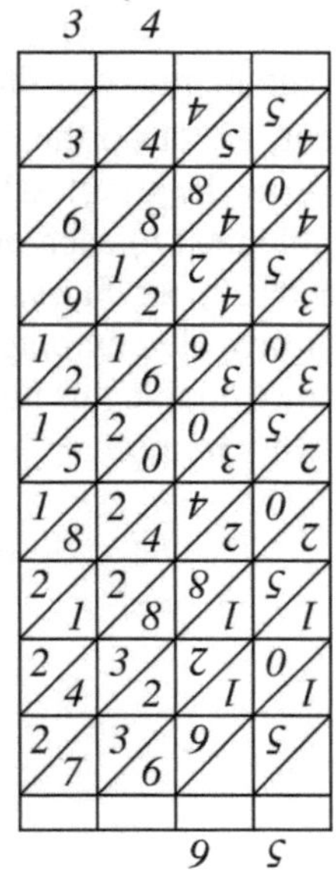

With these end markings, Napier concluded the first part of his explanations, and stated [**28**, p. 10; 664]: "So much for constructing the rods. I now turn to their use."

Napier did this in a number of propositions in Chapter II of the First Book, but the word proposition did not have in his time the meaning that it does now, and it is neither necessary nor desirable to quote and address all of them one after the other. The first proposition is the statement of a problem [**28**, p. 10; 664]: "Setting up on the table a given number with its multiples in order." As his first example, Napier took "the year of the Lord 1615 as the number to be set up on the table with its multiples."

Before we accomplish this task, it must be remarked that this choice of the year 1615 is very probably significant. In 1614 Napier had published the *Descriptio*, and it is only natural that he would start work on the *Rabdologiæ* next, having put the *Constructio* aside while waiting for confirmation of the success of the *Descriptio*. True, some of these these mechanical devices were in use throughout Europe, according to Napier's own testimony, but their users must have been trained by previous users, and the need for written explanations was evident. Napier provided these explanations in the volume under consideration, and his reference to the year 1615 cannot be coincidental. The most obvious reason for his choice of this date is that he was writing Chapter II of the First Book. The book was published in 1617, a relatively short time after its beginning, because there is no long table to compute, as in the *Descriptio*. No doubt, the explanations to use these devices were very fresh in Napier's mind, as he had probably been using these mechanical aids for years to compute many of the entries in his logarithmic table.

In any event, to set up the number 1615 on the table, select four of the ten rods that show the simples 1, 6, 1 and 5 on one of their faces, and place them on the table next to each other in this order, making sure these simples are shown on the upper faces. The top row will show the number 1615. Napier did not provide any diagram, but we have included one below. The second row shows double this number. The rod on the right shows $2 \times 5 = 10$ and, as those who are used to multiplying by hand know well, we can write

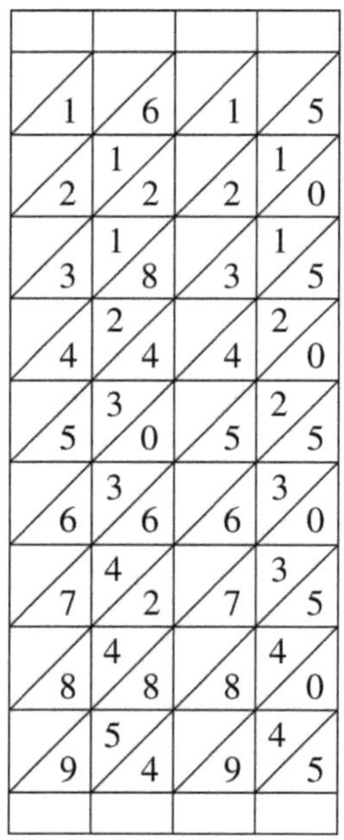

down the 0 but the 1 is the tens digit in the product and must be "carried" to the next column on the left to be added to the 2 that is there. Napier himself talked about "rhomboids" made of triangles on adjacent rods. The first rhomboid on the right contains the numbers 1 and 2, and these must be added to make 3. The next rhomboid to the left shows a blank, that is a 0, and a 2, and their sum is 2, Finally, the last rhomboid, spanning the first two rods on the left, contains the numbers 1 and 2, which are added to make 3. Thus, twice 1615 is 3230.

In Proposition 3 Napier noticed that the sum of the digits in any rhomboid, before or after a "carry" from the right, could exceed 9. In that case [**28**, p. 11; 665], "subtract ten and retain only the remainder." But, of course, the tens digit from the 10 must be "carried" to the rhomboid on the left. Napier gave a new, elaborate example to show these operations, the number 166,702,498, but this is not really necessary here. For example, in row 7 of the diagram under discussion the rhomboid on the right contains the digits 3 and 7, whose sum is 10. We retain the 0, and the 1 is carried to the next rhomboid on the left, where it is added to 2 to obtain 3. Finally, the last rhomboid on the left gives $4+7=11$, so that we retain 1 and carry the other 1 to the last triangle on the left. Thus, $7 \times 1615 = 11305$. It should be clear what to do in cases, as that of Napier's new example, in which several consecutive carries are necessary.

This use of the rods together is reminiscent of the *lattice* method of multiplication, very popular in Europe in the Middle Ages after it was imported from Hindu and Arabic sources. For comparison, here is an example of the method from an anonymous and untitled Languedocian manuscript on commercial mathematics, fully described in [**39**]. It shows the product of 345

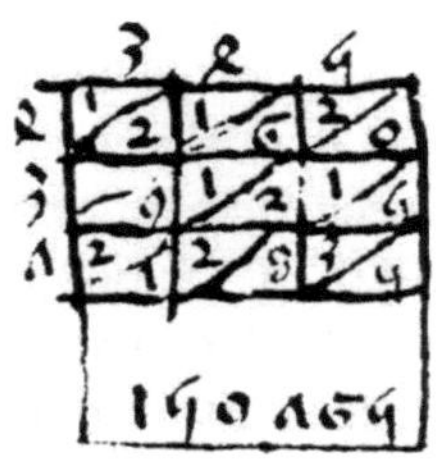

by 437, shown on the top row and left column otside of the lattice, respectively, and the result is 150765 (the Hindu numerals that we use today have evolved from previous forms), obtained by adding diagonally. For instance, $5+3+8=16$, and then we write the 6 as the second digit of the result and carry 1, which added to $0+1+2+2+1$ gives 7, the third digit of the product, and so on.

This may be the best place to point out something that Napier had stated at the outset. For anyone who wonders why he made the choices shown above for the simples on the ten rods, this is Napier's own answer [**28**, p. 8; 661]:

> The arrangement given above offers the greatest flexibility. Although other arrangements are possible, none is superior to this, and none can express more numbers with so few rods.

Ten rods are sufficient to display all numbers smaller than 11,111, but this number itself cannot be displayed on the table because it has five 1's, while only four of the ten rods contain the simple 1. Numbers higher than 11,111 can be displayed with ten rods but not all of them, certainly not those containing five equal digits. Other exceptions are numbers containing two digits, each repeated four times, and numbers containing three digits, repeated a total of 10 times. For example, it is impossible to display a number containing the digits 1 and 3, each four times. Such a representation would need to use rods one, five, six, and seven face up showing the simple 1, but then rod six contains the simple 3 facing to the right, and that leaves only three other rods to show 3 face up, an insufficient number. Similiarly, a number containing the digits, 1, 1, 1, 2, 2, 2, 3, 3, 3, 3 cannot be displayed on the table. That would require rods three, six, eight, and ten for the four 3's, then rods two, five, and nine for the three 2's, leaving only rods one, four, and seven unused, but rod four does not contain the simple 1. However, these limitations are more theoretical than practical. The rods are not expensive to make in most kinds of materials, so that sets of 20 or 30 are perfectly affordable. These sets have similar limitations, but they apply now to such large numbers that they have little, if any, practical significance.

Napier concluded Chapter II with the remark that the rods were invented for more difficult arithmetical operations than addition and subtraction, which are within the reach of any novice, and so he began with multiplication.

First, Napier acknowledged the usual terminology, multiplier, multiplicand, and multiple, but introduced the new term *quotumus* to identify the line of the rods that is being used [**28**, p. 15; 668]. For instance, the line on the rods that contains the multiple 11305 of 1615 in the previous diagram has a quotumus of 7. Then Napier began his example of multiplication as follows [**28**, p. 16; 668]:

> To multiply the year of our Lord 1615 by 365, the former number should be set up on the rods, and the latter written on a piece of paper as in the diagram.

365	365
4845	8075
9690	9690
8075	4845
589475	589475

Taking the figures of 365 as quotumi in order, the left part of the diagram also shows their products by 1615, as obtained from the rods on the table. Note that these products (4845, 9690, and 8075) have been written one below the other on a diagonal so that their leftmost digits are under 3, 6 and 5, respectively. Then, their sum, 589475, is the product of 1615 by 365. Alternatively, as in the righthand half of the diagram above, the digits of 365 can be taken as quotumi in reverse order. Their products by 1615, written on a diagonal with their rightmost digits under under 5, 6, and 3, respectively, naturally yield the same sum.

Napier provided a method to check the result of a multiplication under the heading ANOTHER WAY OF MULTIPLYING [**28**, p. 17; 669]. If we take the four rods displaying 1615 on the table and turn them together so that the sides previously facing up now face down, the top row would now display the number $8384 = 9999 - 1615$, whose third, sixth, and fifth multiples are 25152, 50304, and 41920. Write these on the diagonal as shown below, as well as the multiplier 365, and add them together to obtain 3060525. This is

```
 25152
  50304
   41920
     365
--------
 3060525
```

the value of the product $365(8384 + 1)$. Now subtract it from 3650000, which is 365 followed by as many zeros as the number of rods on the table, to obtain 589475. This is the same as the product of 365 by 1615 obtained above. This works because this product can be written as $365[10000 - (8384 + 1)] = 3650000 - 365(8384 + 1)$.

After ending Chapter III with some mnemonic verses Napier stated the general procedure of division first, at the beginning of Chapter IV, and then gave examples. However, the general procedure is somewhat long, and it may be rather disconcerting to read it without the help of an example, so we try to mix the procedure and example together. Our objective is to divide a number, the dividend (589475), by another, the divisor (365) [**28**, p. 19; 670]. Write down the dividend on paper, as in the diagram below, and display the divisor on the rods as shown above for the number 1615. Now, looking at the rods, find the largest multiple of the divisor that does not exceed the first (leftmost) three digits of the dividend (365 is the largest multiple of itself not exceeding 589). Write this multiple under the dividend and subtract it from those first digits (589), writing the remainder (224) above the dividend, and writing the quotumus of the selected multiple (1) to the right of the dividend:

```
     0
  182
  54
224
589475 (1615
365
2190
  365
  1825
```

Then we repeat this procedure, but now find the largest multiple of the divisor that does not exceed the remaider just obtained with one extra added digit from the dividend (2190 is the largest multiple of 365 that does not exceed 2244, where the last 4 is the fourth digit of the dividend from the left). The new multiple (2190) is written under the previous one (365) but lined up with the extended remainder (2244), and then is subtracted from it. The new remainder (54) is written above the previous one, and the quotumus of the new multiple (6) is placed next to the 1 in the quotient. Next, the largest multiple of the divisor that does not exceed the new enlarged remainder (547) is the divisor itself (365). This new multiple is written under the previous one, but lined up with the last remainder. A new remainder (182) is written above the previous one and the quotumus of the new multiple (1) is put in the quotient. Finally, the multiple of the divisor that does not exceed the new extended remainder (1825) is 1825 itself, whose quotumus is 5. This allows us to complete the division as shown in the diagram.

This quotient is exact and has no final remainder, but Napier gave a second example, 861094 divided by 432, that does [**28**, p. 20; 671]. Since the method is the same, only the diagram is provided and the reader will easily follow the complete process:

```
   118
  141
 402
429            118
861094 ( 1993 -----
               432
432
3888
 3888
  1296
```

Once the last remainder, 118, is obtained, there is no way to divide it by 432, so that the fraction $\frac{118}{432}$ is just written as such to the right of the whole part of the quotient.

This method of division is essentially the same as that of long division that has been in fashion for the last four centuries, but the arrangement of the results above and below the dividend is different. This was the most widely used method of division in Napier's time (he had already used this very same example and arrangement in *De Arte Logistica* [**25**, p. 38; 840]) and for the previous sixteen centuries. At one time, it became known as the *galley method* because the arrangement of numbers on the writing surface reminded some people of a galley being rowed.

After his second example, Napier included some remarks in favour of decimal arithmetic, very uncommon at the time of his writing the *Rabdologiæ* [**28**, pp. 21–22; 671–672]. He stated that in the second example above the quotient is 1993,273 (using now a decimal comma rather than a decimal point), which means 1993 whole units and 273 thousandths $\left(\text{or } \frac{273}{1000}\right)$.

This chapter ends, as the previous one did, with some verses to help the reader's memory with the rules of the operation.

Then Napier turned his attention to the extraction of roots and, although it is possible to obtain them with the rods previously described, he announced that he had designed a plate (actually two, one for square roots and one for cube roots) to facilitate the extractions [**28** p. 23; 673]. These are rectangular plates constructed of the same material as the previous rods, and of the same length

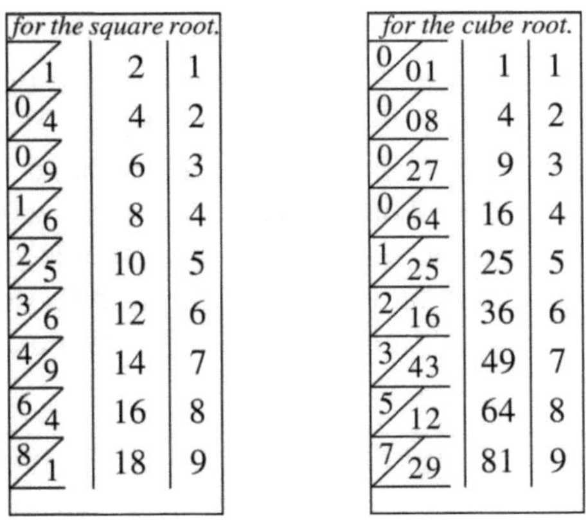

and thickness, but they are one third as broad as they are long. Each top face is divided into three columns, as shown in the accompanying illustration. The left-hand column of the square-root plate contains the squares of the first nine integers. The latter are on the right-hand column, while the central column contains their products by 2. The cube-root plate also contains the first nine integers on its right column, while the central column contains their squares and the left-hand column contains their cubes.

The square root algorithm was then outlined by Napier as follows (as in the case of division we merge in his first example for added clarity) [**28**, p. 27; 675]: write down the number whose square root is to be extracted (117716237694) and put a dot between each pair of digits, starting from the right (11.77.16.23. 76.94.). Under this modified radicand draw two horizontal lines far enough apart to contain the root to be found. This is shown in the next diagram.

The process of root extraction begins by finding a number (in this case 3) whose square (9) does not exceed the number in the radicand before the first

```
                90
          54895
        67
    21
  2
11.77.16.23.76.94.
------------------
  3  4  3  0  9  8
------------------
9
256
  2049
     617481
       5489504
```

dot on the left (11), and writing this number between the two lines for the solution, under the first dot. Square this number (9), subtract it from the number before the first dot (11), and write the remainder (2) over it. Now double the digit already found of the root we are seeking (6), and to continue the extraction select a regular rod containing this simple and place it adjacent to the square root plate and on its left, as shown in the left part of the next figure (Napier did not provide any diagrams here). Write down the remainder

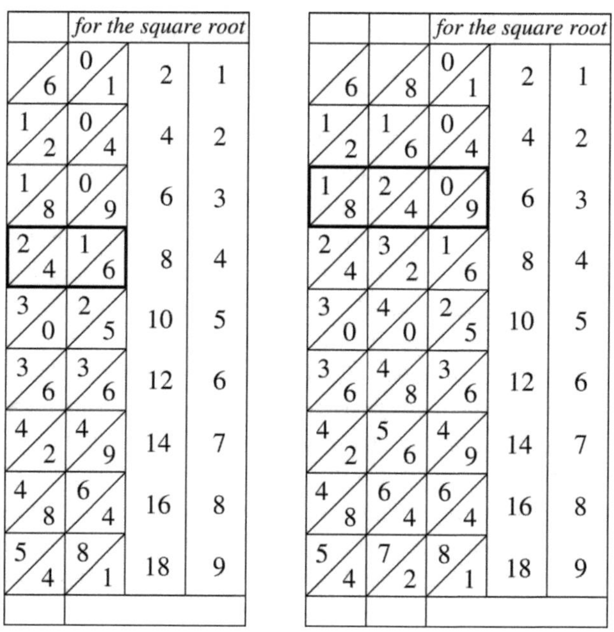

already found (2) extended with the next two digits between dots (277), and use the displayed rod and square root plate to find, on the left, a number not exceeding it (256, with 5 in a rhomboid as $1+4$). Then enter the quotumus of the row containing this number (4) as the next digit of the root we seek under the next dot.

Thinking that the whole procedure is clear from what has been said, Napier quickly summed up the rest as follows [**28**, p. 31; 675]:

> Now repeat this operation as many times as there are dots, remembering the rule that double the last quotumus must be added each time between the previous double and the plate.

The last statement means that a rod with a simple that is double the last quotumus (8) must be inserted between the previous rod (the one with the simple 6) and the square root plate, as shown on the right part of the figure.

We can provide a more detailed summary of the necessary steps for each iteration of the procedure in this way:

(*a*) Square the portion of the root obtained so far.

(*b*) Subtract this square from an equivalent portion of the radicand.

(*c*) Write the remainder on top.

(*d*) Extend this remainder with the next two digits between dots.

(*e*) Enlarge the rod display by inserting a new rod, whose simple is twice the last found digit of the root.

(*f*) In the enlarged display, find the number not exceeding the extended remainder and write it below.

(*g*) Write the quotumus of the line from which we obtained the last number as the next digit of the root.

We had obtained the first two digits of the root, 34. Then, if we separate with semicolons the steps in each iteration, the remaining ones are:

1. $34^2 = 1156$; $1177 - 1156 = 21$; 2116; insert a rod in the display with simple 8; 2049, highlighted and shown in the right part of the previous figure as $1(2+8)(0+4)9$, with quotumus 3. (Here is how this works using modern inequalities. For the next approximation to the root we need to find the largest digit d such that $340 + d \leq \sqrt{117716}$. Squaring both sides and rearranging, $(680 + d)d \leq 117716 - 340^2 = 2116$, and so $d = 3$.)

2. $343^2 = 117649$; $117716 - 117649 = 67$; 6723; insert a new rod with simple 6 (as shown in the left part of the next figure); any line in the enlarged rod display shows a number that exceeds 6723, so we enter 0 as the next digit of the root (as advised by Napier in his general procedure under the heading Caution II [**28**, p. 31; 678]).

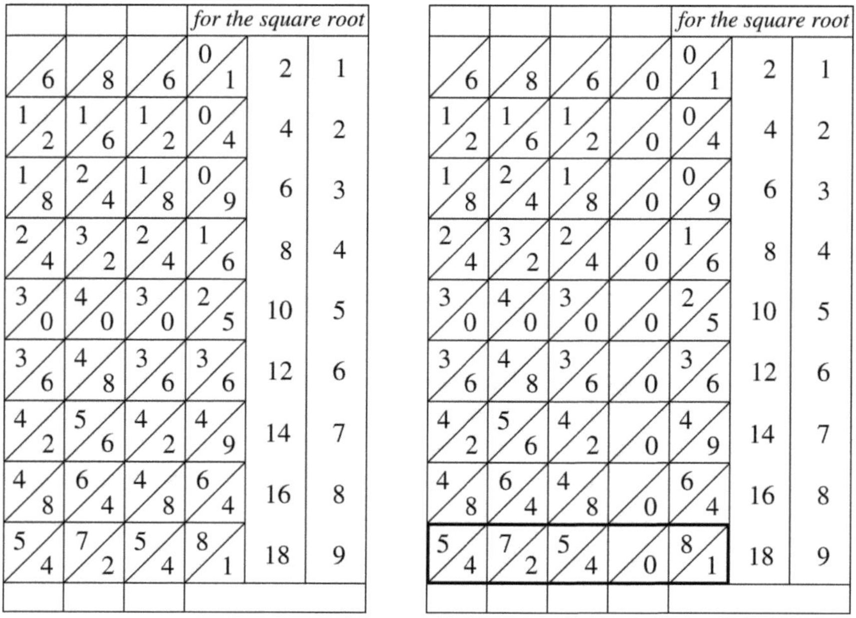

3. $3430^2 = 11764900$; $11771623 - 11764900 = 6723$; 672376; insert a new rod with simple 0, as shown in the right part of the figure above; 617481 with quotumus 9.

4. $34309^2 = 1177107481$; $1177162376 - 1177107481 = 54895$; 5489594; but now we find something new because $2 \times 9 = 18$ is not a simple, and what we do, as Napier stated in general under the heading Caution I [**28**, p. 31; 675], is to insert a rod for the simple 8, while the 1 is added to the rod for 0 on its left. This rod is then removed and one for the total, $0+1=1$, is inserted in its place (as shown in the last of the rods-plus-plate displays); 5489504 with quotumus 8.

This concludes the computation of the integral part of the square root of 117716237694 as 343098, and leaves the remainder $5489594 - 5489504 = 90$. The method just outlined is essentially the well-known manual square root algorithm but rearranged in the galley style favoured in Napier's time.

As before, the chapter ends with some verses as an aid to memory.

Napier considered the extraction of cube roots next [**28**, p. 29; 656], which is similar to that of square roots but considerably more complex. Write down the radicand and use dots to split it into groups of three digits each starting on the right, then draw two lines under it as for the square root. Find a number whose cube does not exceed the first group on the left, and write it under the first dot between the two lines, subtract it from the first group of the radicand

					for the square root		
6	8	6	1	8	0/1	2	1
1/2	1/6	1/2	2	1/6	0/4	4	2
1/8	2/4	1/8	3	2/4	0/9	6	3
2/4	3/2	2/4	4	3/2	1/6	8	4
3/0	4/0	3/0	5	4/0	2/5	10	5
3/6	4/8	3/6	6	4/8	3/6	12	6
4/2	5/6	4/2	7	5/6	4/9	14	7
4/8	6/4	4/8	8	6/4	6/4	16	8
5/4	7/2	5/4	9	7/2	8/1	18	9

and write the remainder on top. But now the process becomes too involved for easy words. Three times the first digit of the root must be set up on the rods, while three times its square root is also set up on different rods. Some are placed on the right and some on the left of the cube root plate. Some number from the rods on the left must be written on a piece of paper and separate operations will take place there. Some of the trial computations are to be kept and some to be rejected, but Napier provided not a single illustration of the rods to help, though there are two diagrams in a note by Hawkins [**17**, pp. 26–27 of his translation of *Rabdologiæ*]. However, the chapter ends with the expected mnemonic verses.

It is fair to say that to understand the procedure of cube root extraction using an example it is absolutely necessary to have a set of rods and use them as suggested by Napier. This is left to those whose interest, energy and available time are equal to the task. Aware of the difficulties of the case, Napier provided in Chapter VIII "A shortcut in extracting cube roots" [**28**, p. 35; 656], but for many this shortcut may prove to be more confusing than the original method.

Chapter IX, the last of the First Book, deals with "The Rule of Three, direct & inverse." This refers to problems in which three numbers in a proportion are given and the fourth is sought. These are well-known to modern readers and there are no additional examples of the use of the rods.

In the Second Book both the rods and the rule of three are implicitly used to solve problems in geometry (Chapters I to VII) and to find the weights and volumes of metals (Chapter VIII). But, again, there is nothing new about this use of the rods. Therefore we do not describe the Second Book, and move on to the very interesting Appendix between this and the next and last book of the *Rabdologiæ*.

THE HIGH-SPEED PROMPTUARY

The title of this Appendix is THE HIGH-SPEED PROMPTUARY FOR MULTIPLICATION, and in the opening paragraph of the Preface Napier stated his reason for this apparently capricious location of an appendix [**28**, p. 91; 713]:

> Although this promptuary for Multiplication was the last of all my discoveries, it does not deserve to come last in this book. With its assistance, all multiplications, however long and difficult, can be carried out easily and quickly.

This is a correct assessment of the usefulness of this device. Napier called this artifact a PROMPTUARY and, according to his assembly instructions, it can be viewed as a repository or dispenser that promptly issues some of a large collection of marked ivory strips that can be used for multiplication.

Before we describe the box that houses the strips, it is well to begin with an explanation of how the strips are made and marked, which Napier did in Chapter I. He chose his promptuary to contain 200 strips to be able to multiply numbers less than 10,000,000,000. Each of them is about one finger ($\approx$ 19mm.) wide and eleven fingers long. Napier continued [**28**, p. 92; 713]:

> At one end there should be a greater margin of two-thirds of a finger, and at the other a lesser margin of one-third, and the space between the two margins should be divided exactly into ten small squares.

The strips will be of two kinds, 100 one finger thick and 100 thin, as thin as is feasible without causing them to break easily. The thick strips are to be engraved with numbers as explained below, and the thin strips will contain perforations rather than numbers. The greater margin of each strip, thick or thin, would show a simple corresponding to that strip.

Each original rod described by Napier in the First Book contains a multiplication table on each face for the simple at the top of that face, filling nine vertical squares. Napier must have thought this arrangement inefficient, and created a more compact one for his promptuary strips. Each strip is designed to accomodate ten identical multiplication tables, rather than one, for the simple

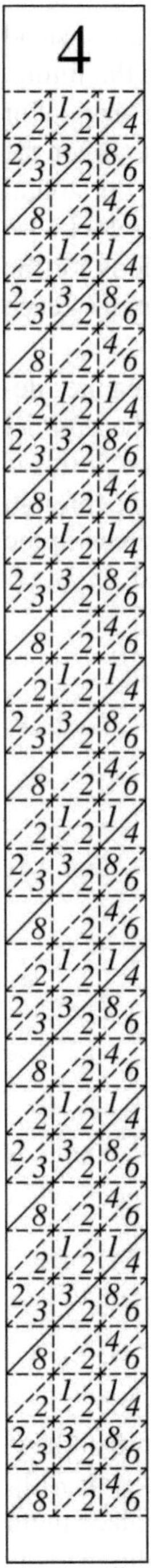

in its greater margin. To make this possible the strips are three times as wide as the original rods, and each multiplication table fills one of the ten small squares mentioned in the last quotation, each of which is subdivided into nine even smaller squares to contain the entries of the multiplication table. This is shown, for the simple 4, in the reproduction of this strip shown on the left, but the arrangement of the digits in each small square needs additional explanation. Napier provided it through the following figure [**28**, p. 94; 714], which shows the multiplication table for the simple *a*, in which its products by 2, 3, ..., 9 are denoted by *bb*, *cc*, ..., *ii*. For instance, if $a = 4$ these products are 08, 12, ..., 36. The units digit of each of these products is located below the solid main diagonal, and these are arranged from top to bottom and left to right. The product of $a = 4$ by 7 is $gg = 28$, since *g* is the seventh letter in $a, b, \ldots, g$, so 8 is in the lower left corner in this arrangement, as shown in any of the squares of the thick strip for the simple 4. The tens digit of each product, if any, is located above the solid main diagonal, also arranged from left to right and top to bottom. For instance, the 2 of $gg = 28$ is in the third triangle from the top in the left column of this square for the simple 4.

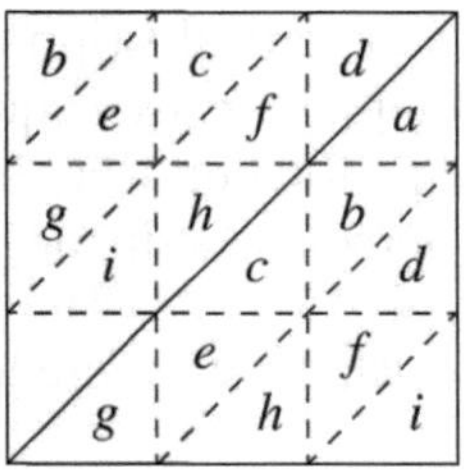

It is convenient to compress the multiplication table for a simple in this manner, but how is a user to keep track of things? What is the seventh letter of the alphabet? Must we remember the position of *gg* in the master square shown above? No, because Napier made this easy for us through the use of the thin strips. If the thick strips are located vertically, as on this page, the thin strips are used horizontally and lying over the vertical strips. Here is a portion of the horizontal strip for the simple 7. It also contains ten squares (of which the four closest to the simple are shown in this figure) of the

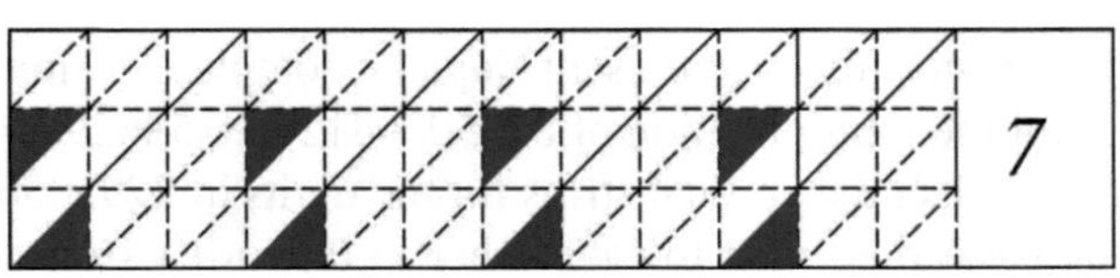

same size as those of the vertical strips, and also subdivided into nine smaller squares. There are no numbers engraved on the horizontal strips, but instead some triangular perforations are made on them, shown in black in the figure. The location of the perforations depends on the simple for each strip. For example, since the seventh letter of the alphabet is *g* the perforations for the simple 7 are made in the *gg* positions. Note that the ten squares on a horizontal strip, as well as the simple on its wide margin, are rotated counterclockwise from their positions on the vertical strips, so that the *gg* perforations for the simple 7 are made on triangles 3 and 6 counting down the left column of each master square.

When a horizontal strip is placed over a vertical strip in the manner described, with a master square exactly on top of a master square, the product of their two simples is shown through the perforations. For instance, if this is done using the vertical strip for 4 and the horizontal strip for 7, we see through the perforations the digits 2 and 8, that is $4 \times 7 = 28$. The grey areas in the next figure merely show the surface on which the strips are located.

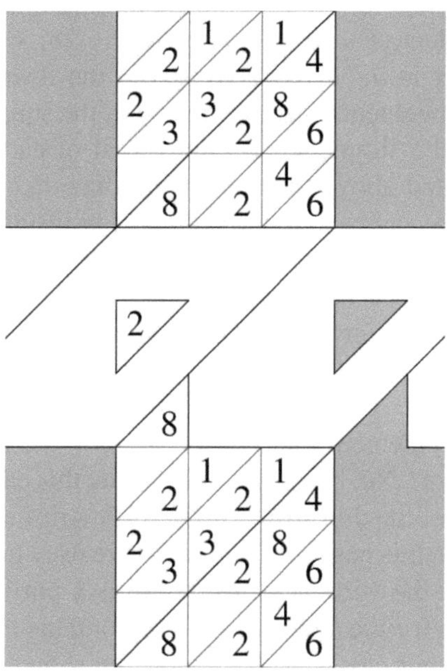

We are now ready to present Napier's complete example of the use of these strips for the multiplication of 8795036412 by 3586290741 [**28**, p. 104; 720]. First we select the vertical strips for the digits in 8795036412 and place them on the table right next to each other in this order from left to right. This is

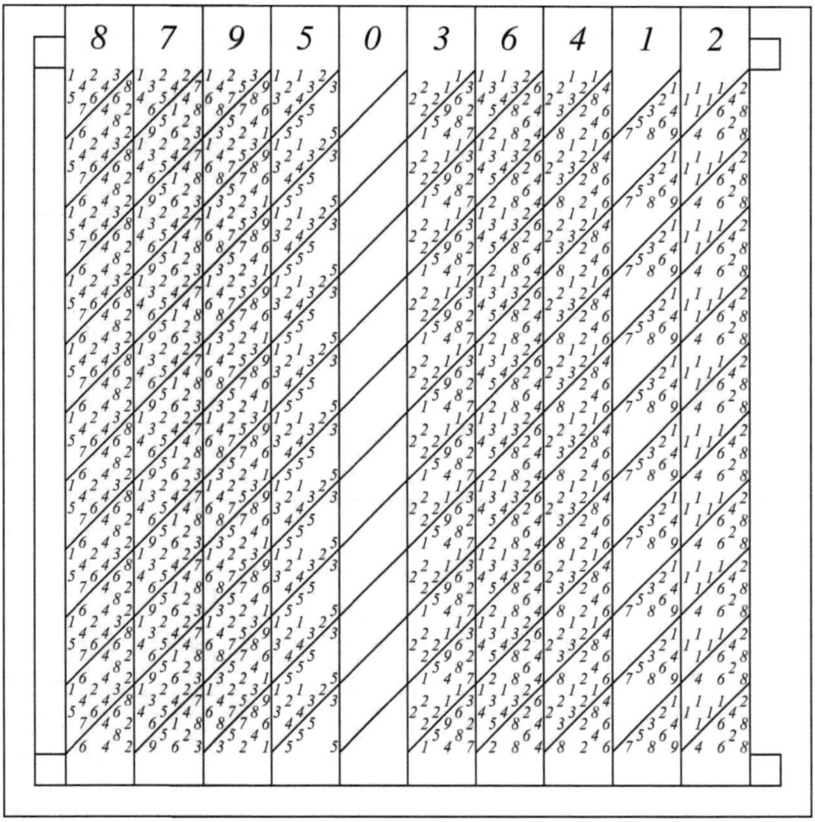

shown in the figure above, which is a portion of Napier's published figure (the complete figure can be seen in the translation of the *Rabdologiæ* included in this volume, p. 721), but we have added some vertical lines to show the boundaries between strips. Now we select the horizontal strips for the digits in 3586290741 and place them over the vertical strips in this order from top to bottom, as shown on the subsequent figure. Again, we have added horizontal lines to a portion of Napier's published figure to show the boundaries between strips. The product of each simple on a horizontal strip by the complete multiplicand is shown through the perforations on the horizontal strip.

Since the original numbers may not be easy to see in either Napier's published figure or in the one included here, we have reproduced below it those digits for the simple 3 without the distraction of the perforations and the diagonal lines (a photographic reproduction of this multiplication performed with the strips of the only extant old promptuary can be seen in [**44**, p. 64]). The product of 3 by the multiplicand is obtained in the same way as if we used the original

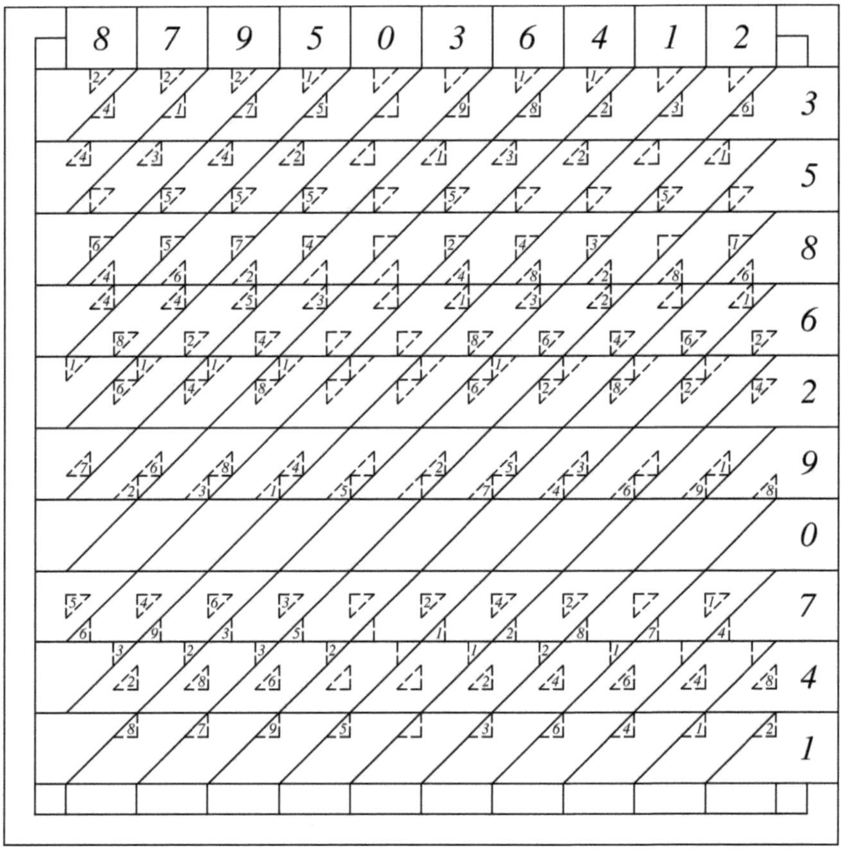

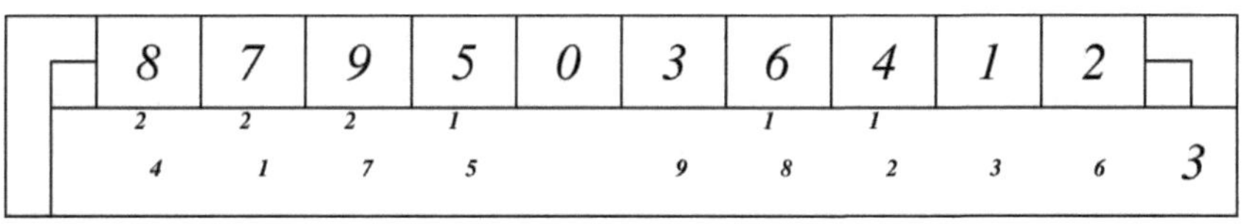

rods, the digits are added diagonally down and from right to left (the diagonal lines are there to facilitate this addition), and it may be necessary to carry to the column on the left. Thus, the result for the simple 3 is 26385109236. When the entire multiplier is involved the procedure is the same but the diagonals longer. Napier carefully outlined the details of the complete procedure in the following manner [**28**, p. 106; 722]. To begin write down 2, the product of the simple 1 and the right-most digit of the multiplicand, which is shown between the lower right angle and the first diagonal on its left in the figure. Then write 9 as the sum of the numbers 1 and 8 between the first and second diagonal lines in the lower right of this figure. Between the second and third diagonals you will find 4, 4 and 4, which add up to 12, so write 2 as the third digit of the product, carrying 1 "in your mind." Adding this 1 to the figures between the third and fourth diagonals, which are 6, 6, 7 and 1, makes 21. Then 1 is the next digit of the product, and 2 is carried to the next diagonal. The entire procedure is now quite clear and it is not necessary to reproduce Napier's remaining instructions. When these are followed to their conclusion, the product so obtained is 31541557651113461292 [**28**, p. 108; 724].

This is quite similar to what we would do today, if we were inclined to multiply two such large numbers by hand using only pencil and paper. We would separately find the products of each simple by the multiplicand, starting with the lower simple of the multiplier and ending with the upper one, and writing the results on a diagonal as shown below.

```
          8795036412
        35180145648
       61565254884
      00000000000
     79155327708
    17590072824
   52770218472
  70360291296
 43975182060
26385109236
--------------------
31541557651113461292
```

Having written these rows on the diagonal, we obtain the result adding by columns, while Napier, constrained to a square arrangement by the display of the strips, had to add by diagonals. Different rite, same result.

Napier admitted [**28**, p. 108; 724] that the promptuary is only for multiplication, and that division by a number can only be performed as multiplication by its reciprocal. Well, not quite, because there were no tables of reciprocals. Today the sine of an arc and the secant of its complement are reciprocals, and there were tables of sines and secants in Napier's time, but one has to deal with the fact that the *sinus totus* was not one. In fact, its value was a choice made independently by each table's author.

What Napier did was to define the *extreme relate* of a number as the number whose ratio to the whole sine (which varied according to which table was used) is the same as that of the whole sine to the given number. For instance, if we denote the given number by a, its extreme relate by ER, and if we take the whole sine to be 10^7, as in the *Constructio*, this means that:

$$\frac{ER}{10^7} = \frac{10^7}{a}$$

If a is any number between zero and 10^7 then a is the *sinus* of some arc α in the first quadrant, or $a = 10^7 \sin\alpha$. Then:

$$ER = \frac{10^7}{\sin\alpha} = \frac{10^7}{\cos(90^\circ - \alpha)} = 10^7\,''(90^\circ - \alpha)$$

The quantity on the right hand side was known as the *secantium* of $90^\circ - \alpha$ in Napier's time, and the values of the *secantium* were included in some published tables. This made it possible to easily find ER from a: α is found in a table from its *sinus*, a, and then $ER =$ *secantium* $(90^\circ - \alpha)$ is also obtained from a corresponding table.

In his first example of division [**28**, p. 111; 726], Napier used the tables by Johan Philip van Lansberge (1561–1632), whose choice of the whole sine was precisely 10^7 [**20**].

> To divide 8795036412 by 27884 using LANSBERGE's Tables and applying the above rule of multiplication, you must multiply 8795036412 by the extreme relate of 27884, which is 3586290741.

The product is already known to us from the previous example on multiplication: 31,541,557,651,113,461,292. This must be divided by 10^{14} to obtain the desired quotient, which Napier gave as:

$$315415\,\frac{57651113461292}{100000000000000}$$

without using the decimal point that he favoured so much in the *Constructio*.

It must be pointed out that Napier's example seems to contain an error. Indeed, a computer can easily show that the extreme relate of 27884 is

$$\frac{10^{14}}{27884} = 3586286042.17$$

when rounded to two decimal digits. But Napier's example is not so much in error as badly expressed, because an examination of his next example shows that 27884 is a rounded version of the intended divisor, 27883.963465, whose extreme relate after rounding is as given by Napier. He must have rounded all his figures for publication after performing the correct computations.

Of course, if the given divisor is larger than 10^7 Lansberge's table is no longer sufficient, and a table of sines and secants based on a larger *sinus totus* must be procured to find the extreme relate.

As a final remark on the example given above, it should be said that the divisor cannot be found exactly in Lansberge's table, but its place is between two table entries, so Napier must have used some interpolation procedure that he did not describe.

Napier proposed to construct the promptuary [**28**, pp. 98–99; 716–717] as a box whose cross-section parallel to the base is a square. This is necessary for the strips to lie on its top, as already shown. This box is reproduced next (a

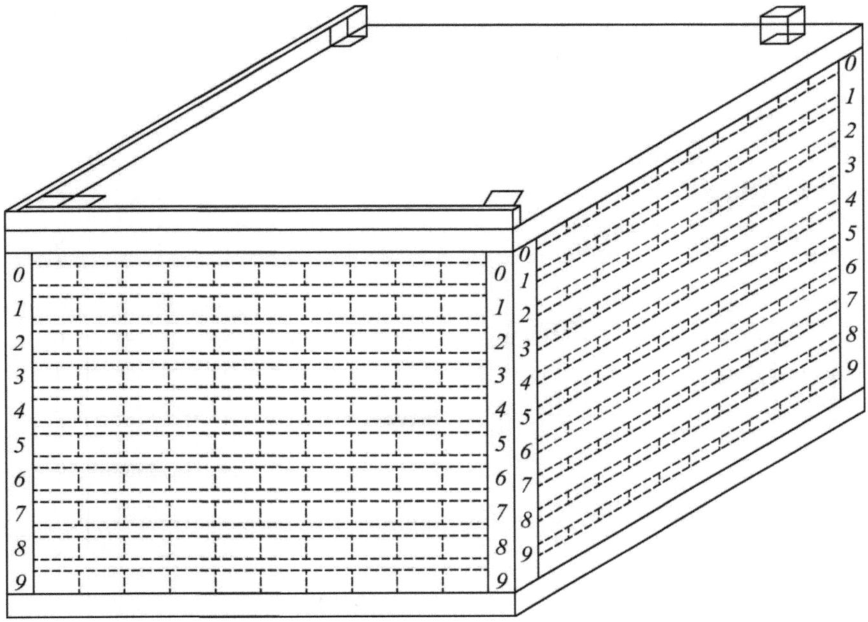

different and more accurate reproduction with a better perspective can be seen in [**21**, p. 44]). Note the raised edges on two sides of the top to keep the strips aligned. The front vertical side of the box, the one closest to the user, shows the tops of the vertical strips stored inside the box, ten for each of the digits 0 to 9. When needed, they are pulled out toward the user and then placed on the top. Similarly, the horizontal strips are stored inside the box with their tops shown on the vertical face to the right of the user. The empty spaces in the box, where the strips have been removed to make up the numbers 8795036412 and 3586290741, are marked in black in the next two views of the front and the right-hand sides of the box. This is a rather intricate construction, and it is not known whether or not it was implemented in Napier's time. The only known extant early promptuary, which possibly dates from the seventeenth century [**18**, pp. 61–62], is housed at the *Museo Arqueológico Nacional* in Madrid, but it was made along very different lines. The ten vertical or horizontal strips for a given simple are just stored in a drawer. This

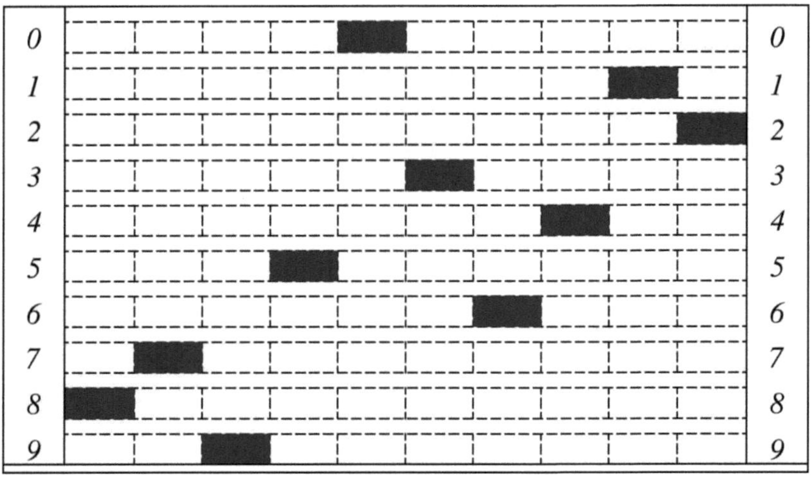

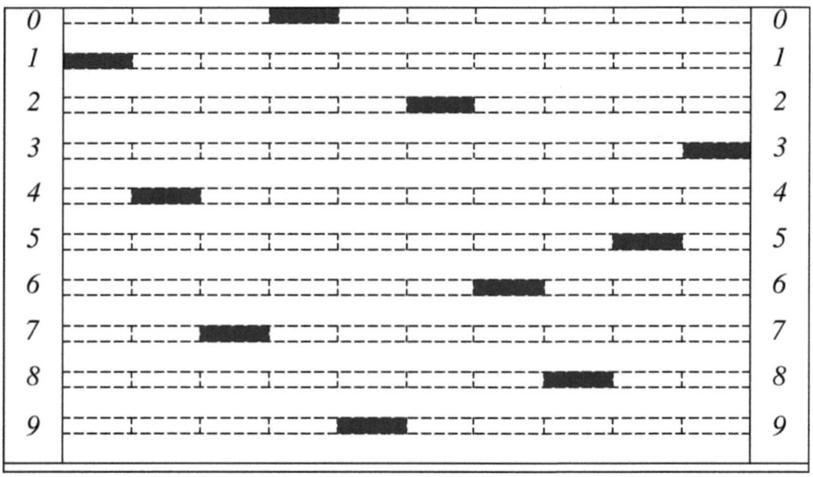

device was first described in [**31**] and then in [**44**]. The first of these works contains a brief explanation of the promptuary's operation [pp. 77–80], but there is a much more extensive one in [**16**].

THE CHESS-BOARD ABACUS

The Promptuary was not the last of Napier's efforts in his quest to simplify computations. In the Preface to the last part of the *Rabdologiæ*, he stated [**28**, p. 113; 727]:

> While working in my spare time on these short methods & seeking ways in which the labour & tedium of calculation might be removed, I developed (not only my LOGARITHMS, my RABDOLOGY, my PROMPTUARY for Multiplication, & other things) but also a method of Arithmetic on a flat surface.

He called this surface a chess-board (*scacchia*), as it is not too dissimilar from it, and the different operations are performed by moving counters on the board as in playing chequers. Napier thought of this process "as more of a lark than a labour" [**28**, p. 114; 727] but warned the reader about a small difficulty. He did not express it in this way, but the fact is that, in today's terms, operations on the board are performed in binary arithmetic,

> so that one must begin by expressing ordinary numbers in the new [binary] form and end by reducing them to common form.

Napier referred to this new form of computation as location arithmetic, and to explain it he constructed a column or measuring rod of powers of 2 [**28**, p. 116; 728] which he labelled, from bottom to top, a to q. He omitted

q.	32768
p.	16384
o.	8192
n.	4096
m.	2048
l.	1024
k.	512
i.	256
h.	128
g.	64
f.	32
e.	16
d.	8
c.	4
b.	2
a.	1

the letter j, which until late in the 18th century was regarded not as a separate letter but as a scribal variant of the letter i. A number in local notation is given by letters. For example, *abe* represents the common number $1+2+16=19$. To represent the ordinary number 1611 (Napier referred to it as "the year of our LORD 1611," so he may well have been writing this at that time), two methods are possible, by subtraction and by division [**28**, p. 117; 729].

Using the first method, subtract the largest number on the rod that does not exceed 1611 from it, which gives $1611 - 1024 = 587$. Do the same for this remainder and for any subsequent remainders until this is no longer possible. Thus, $587 - 512 = 75$, $75 - 64 = 11$, $11 - 8 = 3$, $3 - 2 = 1$, and $1 - 1 = 0$. The numbers that are subtracted in this process are represented by the letters l, k, g, d, b and a, so that 1611 becomes *abdgkl* in local notation.

Using the second method [**28**, p. 118; 729], if the given ordinary number is odd, place a counter in position a, subtract 1 from the number, and divide the remainder by 2 (if the given ordinary number is even just divide it by 2, and do not place a counter in position a). If this half is odd, subtract 1 from it and place a counter in position b; otherwise place nothing. Continue this process until the last quotient is 1. For example, 1611 is odd, subtract 1 and place a counter in position a; $1610/2 = 805$, subtract 1 and place a counter in position b; $804/2 = 402$, place no counter in position c; $402/2 = 201$, subtract 1 and place a counter in position d; $200/2 = 100$ and $100/2 = 50$, so place no counters in positions e and f; $50/2 = 25$, subtract 1 and place a counter in position g; $24/2 = 12$ and $12/2 = 6$, so place no counters in positions h and i; $6/2 = 3$, so place a counter in position k; $2/2 = 1$, so place a counter in position l. This terminates the process and the local representation of 1611 is again given by the counters as *abdgkl*.

Next, Napier explained the conversion of a location number into an ordinary number [**28**, p. 120; 730]. This is rather obvious by the addition of the numbers corresponding to the given letters. Alternatively, the division method just explained can be reversed by doubling, but this is not really necessary.

The representation of a number in local arithmetic is not unique, for it is possible to place more than one counter in a given position, as it happens frequently during computation. For example, *abbdeffg* is the local notation for the ordinary number $1+2+2+8+16+32+32+64 = 157$, but the local notation can be *abbreviated*, as Napier referred to this process [**28**, p. 124; 732]:

> replace the two bs with a single c. The two fs should be replaced by a single g, but as there is a g already there, the two gs must be replaced by a single h. Thus the number [*abbdeffg*] in abbreviated form is *acdeh*, but it is still the same number.

Conversely, a location number can be *extended* so that there are no "blank spaces" in it. For instance, *acdeh*, which is equivalent to *a.cde..h* with the dots marking the positions where there are no counters, can be extended to *abbccddeefg* by replacing some counters with two in the next lower position. The process should be clear enough and there is no need to provide the details here.

Finally, Napier proceeded to computation. Addition and subtraction are straightforward [**28**, p. 125; 733]. To add two numbers write them in abbreviated form, put all their letters together in alphabetic order, and abbreviate the result, For example, the sum of *acdeh* and *bcfgh* is *abccdefghh* or, in abbreviated form, *abhi*. To subtract two numbers write the minuend in extended form and the subtrahend in abbreviated form, then remove the letters of the subtrahend from those of the minuend, and abbreviate the result if necessary. For instance, subtracting *bcfgh* from *abhi*, which is *abccdefghh* in extended form, gives the remainder *acdeh*.

Multiplication and division are more conveniently carried out in two dimensions, on some sort of chess-board shown below with corners marked by

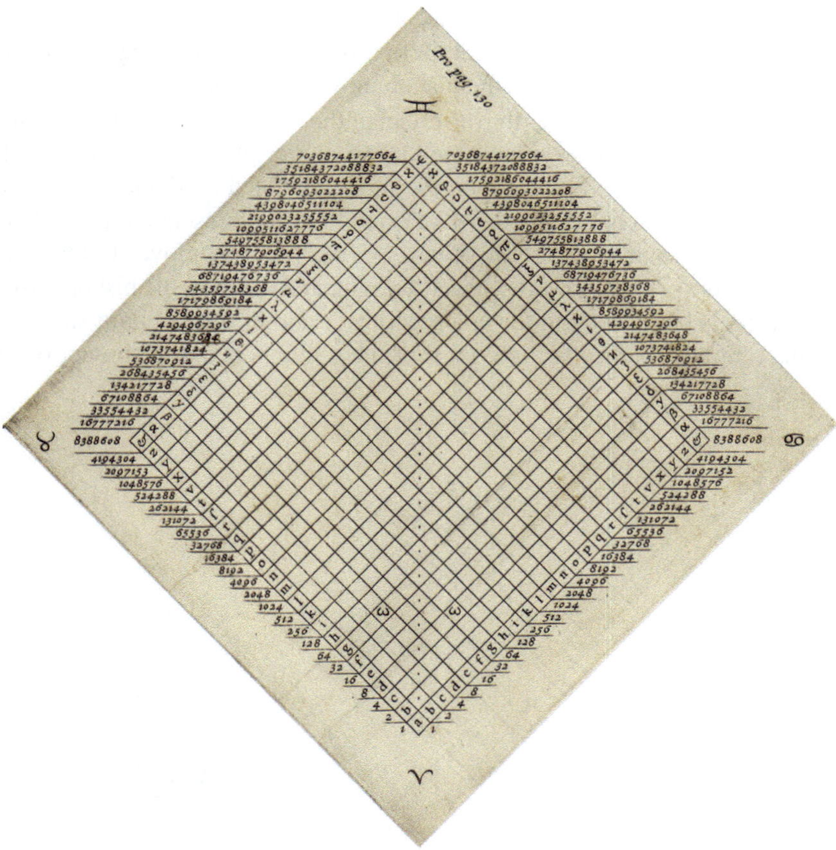

the symbols ♈ ♉ ♊ ♋ (the signs of the Zodiac for Aries, Taurus, Gemini and Cancer, in this order) [**28**, p. 129; 734]. Napier intended the board to be shown diagonally, as it is here (a modern reconstruction of this figure can be seen in the translation of *Rabdologiæ* included in this volume [p. 721]), but in the 1617 edition this and similar diagrams showing the board were printed rotated

45° counterclockwise. On the margins we have the powers of 2, as in the measuring rod, but this time they go beyond 32768 to permit computations with larger numbers. This made it necessary to continue with Greek letters on the margins to denote additional positions after running out of Latin characters.

Napier began his explanations by stating [**28**, p. 131; 736]:

> Numbers are expressed & manipulated by moving counters here and there over the squares on the board.
>
> Motion or progress on the board is of two types: Direct & diagonal.

Direct motion is parallel to the sides of the board, either ♈♉ or ♈♋, as the rook moves in chess. Diagonal motion is parallel to the diagonals of the board, either ♈♊ or ♉♋, as the bishop moves in chess.

However, before we describe the effect of the motion of a counter, we must explain what numbers are represented by a counter placed on the board. If, for example, we place a counter on the square denoted by the left ω (Napier labeled these two ω squares just to give examples), it represents four numbers: $g = 64$ as the rook moves to the left, $d = 8$ as the rook moves to the right, $k = 512$ as the bishop moves to the left or right, and $d = 8$ as the bishop moves down (this is better seen in the enlarged lower corner of the board shown in the next figure). Notice that the third of these numbers is the product of the

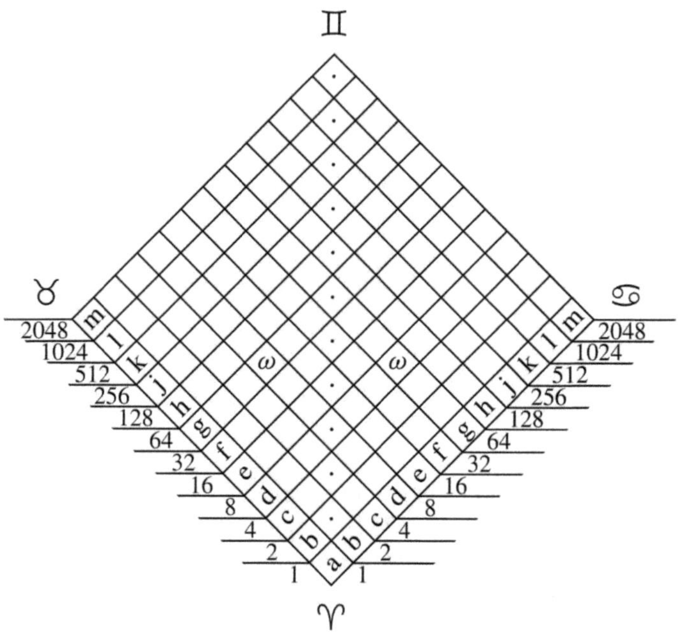

first two ($512 = 64 \times 8$), and that the fourth of these numbers is the quotient of the first two ($8 = 64/8$).

We define the value of a counter placed on the board as that of the corresponding letter or common number to its right or left as the bishop moves. For instance, the value of a counter on any of the two ω squares is $k = 512$. Then moving the counter up one square as the rook moves, to the right or left, doubles its value to $l = 1024$ [**28**, p. 133; 737]. On the other hand, moving the counter up one square as the bishop moves quadruples its value to $m = 2048$. Downward motion halves the value as the rook moves or divides it by four as the bishop moves. Notice, in particular, that the value of counters placed in the positions indicated by the dots on the central vertical line on the board are the squares of the powers of two: 2, 4, 8, etc. To sum up, we have here the seeds of three operations: multiplication, division, and square root extraction.

Multiplication is taken up next [**28**, p. 137; 739], but there is a limit to how large the multiplicand and the multiplier can be because they have to be placed along the margins ♈♉ and ♈♋, respectively. For instance, and this was Napier's first example, to multiply 19 by 13 temporarily place a counter on, or mark with chalk (*cretâ* was Napier's Latin word), the letters *abe* on the left margin, since this is the local arithmetic representation of $19 = 1+2+16$, and over the letters *acd* on the right margin, since this is the local arithmetic representation of $13 = 1 + 4 + 8$. Then place a counter on the board at each intersection of the direct rows and columns of the numbers so marked, as shown in the next figure. As we have seen above, each counter represents

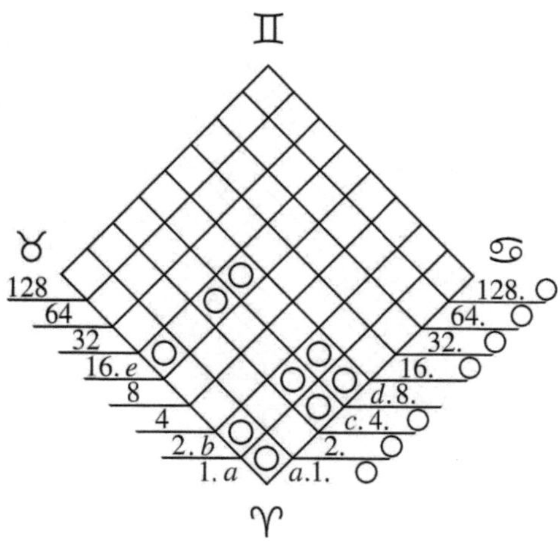

the product of two numbers. For instance, the top counter in this arrangement represents the product of 16.e by d.8, and the result is 128, as shown on the margin as the bishop moves. What remains is to add these results. To do this, notice that the three lowest counters are at levels 1, 2, and 4. Move these counters to the right margin, next to *a*.1, 2, and *c*.4, as shown on the right margin. Next, the two counters at level *d*.8, which make 16, are replaced by a single counter at level 16. The next two counters, at level 16 are replaced by a single counter on the margin at level 32, and the remaining two counters, at levels 64 and 128 are moved to the margin. The complete product is the sum of these marginal numbers: $1 + 2 + 4 + 16 + 32 + 64 + 128 = 247$.

As a second, more involved example, Napier considered the product of 1206 and 604. These numbers are represented by the letters *l*, *h*, *f*, *e*, *c*, *b* for the first and *k*, *g*, *e*, *d*, *c* for the second (contrary to his previous practice, Napier gave these letters in reverse alphabetical order). He marked the first number on the right margin and the second on the left margin, as well as placing counters on the intersections of the direct rows corresponding to the stated letters, as shown in the next figure.

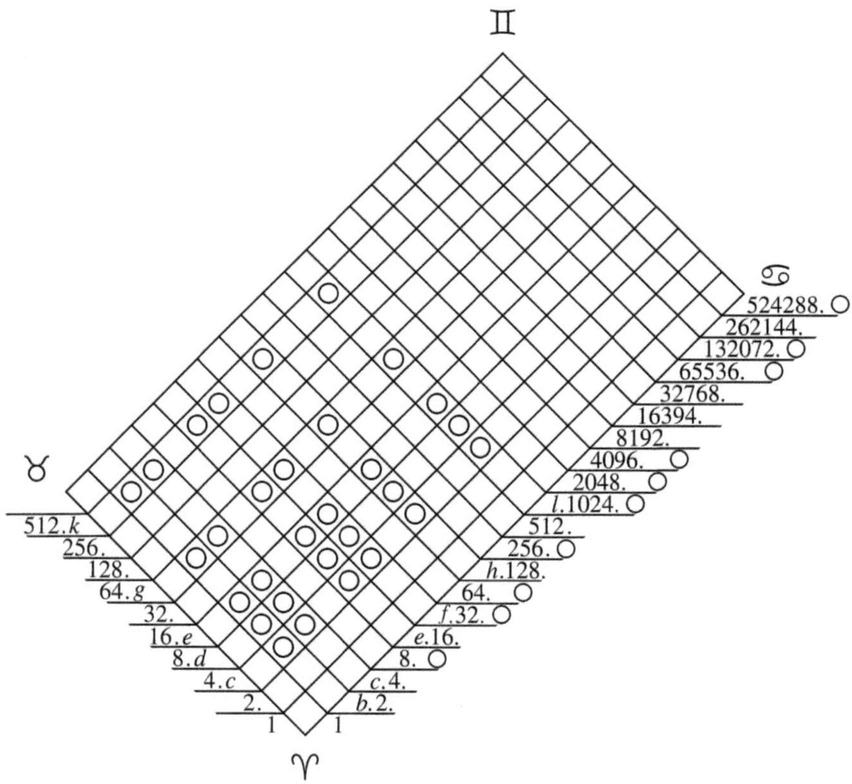

To begin the multiplication, the single counter at level 8 is moved to the right margin, then the two counters at level 16 are replaced with a single counter at level 32. Now there are three counters at level 32, so leave one there and replace the other two with a counter at level 64. Again, we find three counters at this level, so leave one there and replace the other two with one at level 128. That gives us four counters at level 128. We replace them with two counters at level 256, which makes a total of five at this level. Leave one there, at the margin, and replace the other four with two at level 512. The rest of the process is clear so there is no need to provide the details to the end. The resulting product, looking at the marginal counters on the right, is $8 + 32 + 64 + 256 + 1024 + 2048 + 4096 + 65536 + 131072 + 524288 = 728424$.

This procedure is reversed for division [**28**, p. 144; 743]. As a first and simple example, to divide 250 by 13 mark the dividend on the ♈♋ margin by placing counters on the numbers 128, 64, 32, 16, 8, and 2 (which add up to 250), and then mark the divisor with chalk marks on the ♈♉ margin on

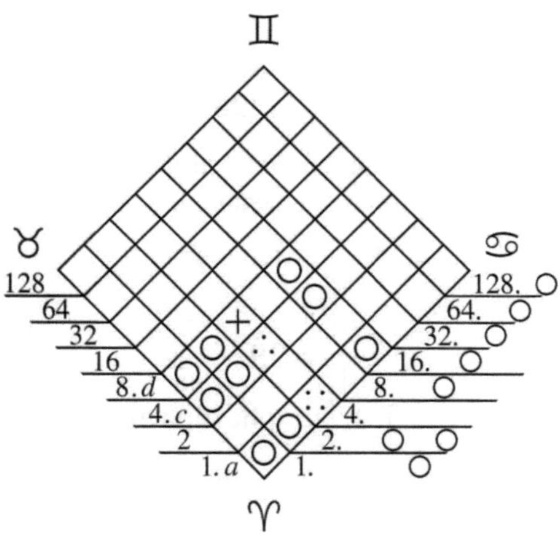

the numbers 8, 4, and 1. Since the two top numbers in the divisor and the dividend are 8 and 128 [**28**, p. 145; 743],

> go up from 8 in the left margin by the rook's move & across 128 in the right margin by the bishop's move, & from the square where they meet, place a row of counters parallel to the divisor.

Make sure to place counters only on the direct rows corresponding to the numbers 8, 4, and 1 of the divisor, as shown in the figure. Napier called this row of counters a *congruous segment*, meaning that its total value ($128+64+16 = 208$) is smaller than the dividend. In that case, this total is subtracted from the dividend to obtain 42, the first remainder, which is marked with marginal counters on the numbers 32, 8, and 2 (that is, imagine the marginal counters on 128, 64, and 16 removed, although this cannot be done physically in the figure).

To continue the division, we might want to place a counter at the point of intersection of the direct row from 8 on the ♈♉ margin and the diagonal row from 32, the largest number in the remainder, on the ♈♋ margin. Then we might want to place a row of counters parallel to the divisor but only on the direct rows corresponding to 8, 4, and 1. These positions are indicated in the figure by the symbols + ∴ and ::, but no counters are placed there because their total would be $32 + 16 + 4 = 52$, which exceeds the first remainder (42). Instead, the counters are placed on the row that is right below this one and parallel to it and the divisor. This is the new congruous segment, and the total value of its counters is $16 + 8 + 2 = 26$. Since $42 - 26 = 16$, this is the second remainder and a single marginal counter is left on the right at this level, and all others are removed.

The new congruous segment is easily seen to contain the three counters in the direct row next to the ♈♉ margin, and its total value is $8 + 4 + 1 = 13$. Thus, the third and last remainder is $16 - 13 = 3$, which is represented by two marginal counters at levels 2 and 1 [**28**, p. 146; 744].

> Then when added together, the numbers in the right-hand margin toward which the individual congruous segments point (that is 16, 2, and 1) form the correct quotient 19.

The complete result is:

$$\frac{250}{13} = 19 + \frac{3}{13}$$

which is corroborated by the first multiplication example given above to show that $19 \times 13 = 247$.

After this Napier gave another example, which shows that the dividend and the divisor can be placed on the same margin. To divide 728424 by 1206, place counters on the ♈♋ margin next to the numbers 524288, 131072, 65536, 4096, 2048, 1024, 256, 64, 32, and 8, whose sum is 728424. Then mark with chalk on the same margin the numbers 1024, 128, 32, 16, 4, and 2, which add up to the divisor. Nothing else is new, so Napier described the rest of the process very briefly [**28**, pp. 147–148; 744–745]:

> Find the first congruous segment of these as I have described. It will stand directly above the number 512 [it is parallel to the divisor, which is now on the ♈♋margin], which appears in the lower left margin. Subtract the value of this segment from the dividend, which leaves a remainder to be noted. Subtract from it its own congruous segment, and you will be left with one remainder after another and eventually with nothing. In the calculation five congruous segments appear that point toward the numbers 512, 64, 16, 8, 4. These added together produce 604, the required quotient.

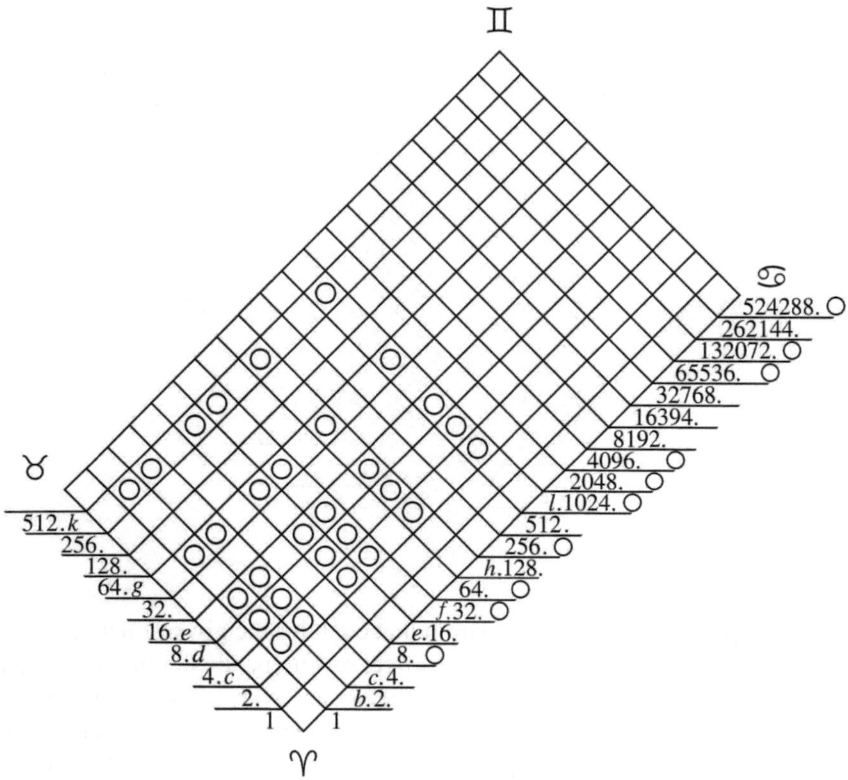

The last chapter on the use of the chess-board [**28**, p. 148; 745] deals with square-root extraction. We have already stated that the value of counters placed on the vertical diagonal line *a* to ψ, marked with dots on the board, are the squares of powers of two. The process of square-root extraction begins by marking the radicand with counters on either margin, and then by placing a counter on the highest dotted square that can be subtracted from the radicand. This is shown in the next figure for the radicand $1238 = 1024 + 128 + 64 + 16 + 4 + 2$, which was Napier's first example, and subtracting the value of this counter (1024) from the radicand leaves marginal counters on 128, 64, 16, 4, and 2 to denote the first remainder (214).

Napier called this first counter on the dotted line the head of the gnomons, and defined a *gnomon* as a collection of counters, arranged in the shape of an elbow, that added below the head of the gnomons to a single counter or to a square formed by counters results in a larger square. This is an incomplete definition and, of course, impossible to understand without an example, but to do that it is best to resort to modern notation. Denote an arbitrary square on the chess-board by (m, n), where m is the number on the ♈♉ margin for the

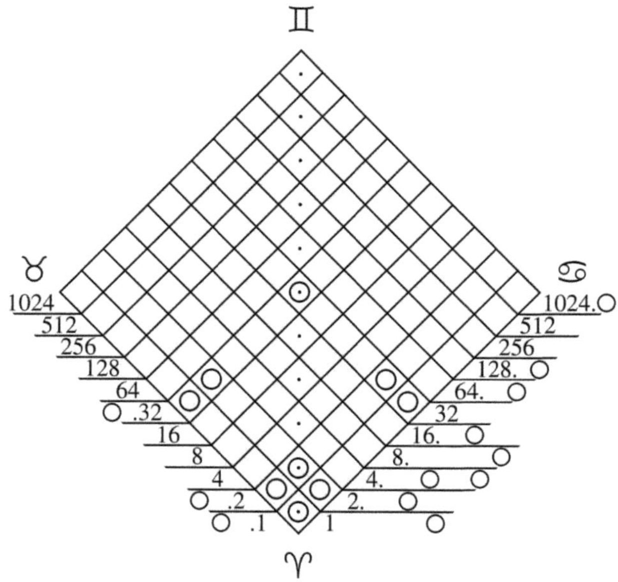

direct row perpendicular to this margin that contains the given square and n is the number on the ♈♋ margin for the direct row perpendicular to this margin that contains the same square. Thus, the square with a dot at level 1024, where we have already placed a counter to begin the root extraction of 1238, is denoted by $(32, 32)$. If we place three additional counters on the squares $(32, 16)$, $(16, 16)$, and $(16, 32)$, these three counters form a gnomon because, together with the original counter, they make a square and they are below the head of the gnomons. Two other possible gnomons at this stage consist of counters on $(32, 8)$, $(8, 8)$, and $(8, 32)$, or on $(32, 2)$, $(2, 2)$, and $(2, 32)$. This last one is shown in the figure.

Napier gave an additional rule to complete the definition of a gnomon stated above: a gnomon added to a single counter must contain 3 counters, producing a square with 4 counters; a gnomon added to such a square must contain 5 counters, producing a square with 9 counters; and so on. For instance, the figure shows a gnomon with counters on $(32, 1)$, $(2, 1)$, $(1, 1)$, $(1, 2)$ and $(1, 32)$ to be added to the square $(32, 2)$, $(2, 2)$, $(2, 32)$, and $(32, 32)$. There is an additional rule that Napier followed but did not write down: the counters of each gnomon can only be placed in direct rows containing counters of the previous square. Thus, the last described gnomon cannot contain a counter on $(1, 4)$ instead of $(1, 2)$.

Now, we have seen three possible gnomons to be added to the counter on $(32, 32)$ above, but only one of them is shown in the figure. Napier gave an additional definition from which the choice can be made [**28**, p. 149; 746]:

> The largest gnomon that can be subtracted from the remaining counters in the margin & for which there is space vacant on the board is called a *congruous gnomon*.

Neither (32, 16), (16, 16), (16, 32) nor (32, 8), (8, 8), (8, 32) are congruous because their values, that is the added values of their counters, are $512+256+512 = 1280$ and $256 + 64 + 256 = 576$, respectively. Each of these totals is larger than the remainder $1238 - 32^2 = 214$, but the total value of (32, 2), (2, 2), (2, 32) is $64+4+64 = 132 < 214$. Only a congruous gnomon can be placed on the board, and the difference between the previous remainder and the total value of the current congruous gnomon, in this case $214-132 = 82$, is the second remainder. For this we leave marginal counters on 64, 16, and 2.

The next gnomon is (32, 1), (2, 1), (1, 1), (1, 2) and (1, 32), already described above, and has total value $32 + 2 + 1 + 2 + 32 = 69 < 82$. Thus, it is a congruous common and the third and last remainder is $82 - 69 = 13$, for which we leave marginal counters on 8, 4, and 1. Finally, the last square built from the head of the gnomons and the three gnomons is made up of three direct rows, and whether these are viewed as parallel to ♈♉ or parallel to ♈♋ makes no difference in interpreting Napier's concluding statement in this example [**28**, p. 152; 748]:

> Take one counter from each of the three rows of this square, move it straight down its row onto the lower margin. The counters will land on the numbers 32, 2, and 1, and their total, 35, is the square root we require.

Well, not quite, because of the little matter of the remainder. $35^2 \neq 1238$ but $35^2 + 13 = 1238$. Napier gave a last example, to find the square root of 2209, but it adds nothing to our knowledge of the procedure.

We do not need to use the board in our time, but out of simple curiosity we could ask a couple of questions: why does this method of root extraction work? and was its accuracy limited in Napier's time because we so easily reach the bottom of the board, as in the stated example, and there is no way to continue?

The answer to the first question is based on the symmetry in the selection of the gnomons. To start with, the head of the gnomons is on a dotted square and is itself the square of $f = 32$, which is 1024 and falls short of the radicand, 1238. Adding the first congruous gnomon, and remembering the process of multiplication, we see that the four counters now on the board represent the product of $bf = 34$ by itself, that is, $34^2 = 1156$, which is closer to but also smaller than 1238. Adding one more congruous gnomon, the counters on the board represent the product of $abf = 35$ by itself, which is $35^2 = 1225$ and even closer to 1238. Thus, the addition of congruous gnomons to the board, symmetric about the dotted line, constitutes a method of successive approximations to the true square root of 1238.

But the method comes to a halt as soon as we run out of room at the bottom of the board, and it is easy to imagine the frustration of early seventeenth-century calculators with the stated result: $\sqrt{1238} \approx 35$. Napier did not elaborate on a way to obtain a more accurate result but we can do so, although it involves extra work. Start by multiplying 1238 by 10000 and then use any of the methods of Chapter II to write the product in local arithmetic as $fgiklopqtvxy\&$. No counters have been placed on the right margin of the next figure to represent this number because there is no room for all of

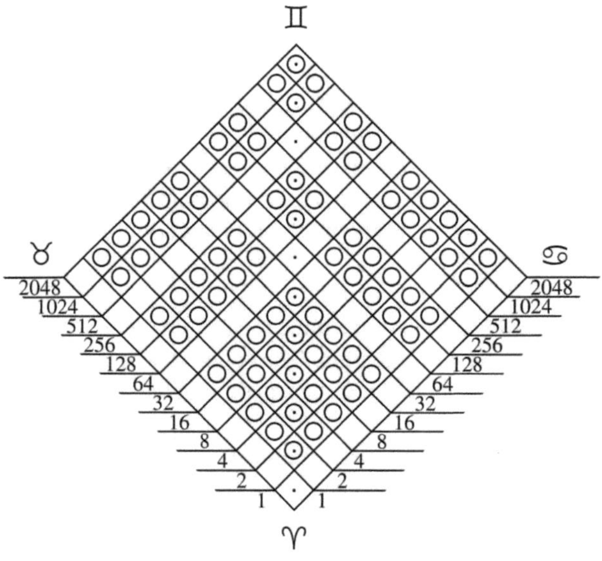

them. In any event, the square-root extraction process begins by placing a counter on (2048, 2048) because 2048 is the largest marginal number whose square can be subtracted from the radicand, 12380000. Then, successive congruous gnomons are constructed as already explained and can be seen in the figure. The square formed by all these counters represents the product of $bcdefhilm = 2 + 4 + 8 + 16 + 32 + 128 + 256 + 1024 + 2048 = 3518$ by itself, which is the approximation provided by the chess-board for $\sqrt{12380000}$. Therefore, dividing the last two numbers by 100, $\sqrt{1238} \approx 35.18$, which is a better approximation than 35.

What we have described shows very clearly that Napier succeeded in his quest to remove the tedium of calculations, giving us an abundance of choices to do so. We know that the bones were widely used during his lifetime, and for a couple of centuries thereafter. There were detractors about the usefulness of these rods, since they are nothing but portable multiplication tables, and all calculators knew their multiplication tables. Some thought that they could perform their calculations more speedily by hand, and that may have been true for some, but it is hardly believable that that could be so in general. However, speed is not the only use of the rods because the multiplication tables that they contain were etched in bone, and that is the second aspect of their usefulness. Their use greatly reduces the possibility of "slippery errors," as Napier termed them, for most calculators.

There can be no argument that the promptuary greatly increases the speed with which the multiplication of large numbers could be accomplished. This can be considered a true calculating machine, as Hawkins called it [**16**], the first one ever built. However, it is unclear how often it was used in Napier's time, or how many were built, and the only known extant early promptuary is the one in Madrid. Nor is the speed with which multiplications can be completed in doubt because Tee and Hawkins had one built at the University of Auckland in 1979, following Napier's design, and then Hawkins publicly performed the multiplication of two ten-digit numbers in less than one minute [**21**, p. 59]. Both the promptuary and the abacus for local arithmetic were largely ignored by the mathematical community in the early seventeenth century. The main reason was Napier's own invention of logarithms at almost the same time as he published a description of his mechanical inventions. Logarithms made these mechanical devices obsolete at the very moment they were born. The use of logarithmic tables spread very rapidly throughout the world because they provided the best possible answer to the problem of performing complicated calculations, becoming an everyday necessity for this purpose. Napier's mechanical devices may not have shared the limelight with logarithms, but his genius was soon appreciated worldwide. Through his logarithms Napier became a member of that select group of thinkers who changed the face of mathematics forever.

BIBLIOGRAPHY

1. **Biot, Jean-Baptiste**, "Mémoire sur J. Napier de Merchiston, contenant sa généalogie, sa vie, le tableau des temps oú il a vécu, et une histoire de l'invention des logarithmes. Par Mark Napier.—Extrait en trois articles," Journal des Savants, March 1835, pp. 151–162 and 257–270. Abridged English translation as "Memoir of John Napier, Baron of Merchiston. By J. B. Biot," pp. 81–94 and 161–178 in Robert D. Thomson, *Records of general science*, Vol. III, Taylor and Walton, London, January 1836. The quoted statement is on p. 173.
2. **Björnbo, Axel Anthon**, "Ioannis Verneri De Triangulis Sphaericis, Libri quatuor". *Abhandlungen zur Geschichte der Mathematischen Wissenschaften*, **24**, 1–184, 1907.
3. **Black, Adam & Charles, eds.**, *The Encyclopædia Britannica. A Dictionary of arts, sciences, and general literature*, Ninth Ed., Vol. XIV (KAO – LON), Edinburgh, 1882.

The main item in this photograph is the Promptuary held in the Museo Arqueológico Nacional *in Madrid. It was exhibited in 1876 at the* South Kensington Museum (*later the* Victoria and Albert Museum) *in London where it was catalogued as a device "for making mathematical calculations". It was largely forgotten until Erwin Tomash re-identified it as an original Promptuary in 1988. It is not laid out as per the instructions and drawing given by Napier but its method of working is exactly as he described it. It first entered the Museum*'s *records in 1867, coming from the* Biblioteca Nacional de España *who in turn received it from the royal collection of Philip V (1683–1746). Before that it probably belonged to the* Academia de Matemáticas *founded by Philip II (1527–1598). It is thought that it was constructed in Milan, probably in the early 17th century. It is certainly the only one known to exist in the world, other than the one built by the University of Auckland in 1979. In front and to the left are a set of Napier's Bones and the box to store them in. This box is housed inside the Promptuary itself, the very top rectangular piece of wood sliding to one side to reveal a recess for holding it. There are 60 rods, 59 of which still exist, an unusually large number since a set normally contains 20 rods (courtesy of the* Museo Arqueológico Nacional, *Madrid*).

4. Bombelli, Rafael, *L'Algebra. Opera di Rafael Bombelli da Bologna, diuisa in tre libri, con la quale ciascuno da se potrà venire in perfetta cognitione della teorica dell'aritmetica. Con vna tauola copiosa delle materie, che in essa si contengono. Posta hora in luce à beneficio delli studiosi di detta professione*, Giouanni Rossi, Bologna, 1572 and 1579. The complete manuscript is contained in Codice B. 1569 of the *Biblioteca comunale dell'Archiginnasio* in Bologna. It was first edited and printed by Ettore Bortolotti as *Rafael Bombelli da Bologna. L'Algebra. Prima edizione integrale*, 1966.

5. Bramer, Benjamin, *Beschreibung eines sehr leichten Perspectiv: vnd grundreissenden Instruments auff einem Stande*, Johan Wessel, Cassel, 1630

6. Bretnor, Thomas, *Bretnor, 1618.* [*A newe almanacke*] *and prognostication, for 1618. Being the second from leape-yeare*, 1617. No publisher or location is mentioned on the title page. There is a modern reprint by ProQuest EEBO Print Editions, undated.

7. Briggs, Henry, *Arithmetica logarithmica sive logarithmorvm chiliades triginta, pro numeris naturali serie crescentibus ab vnitate ad* 20,000: *et a* 90,000 *ad* 100,000, William Jones, London, 1624. Latin transcription with complete (except for the Preface) annotated translation into English by Ian Bruce as *Briggs' Arithmetica logarithmica*, http://www.17centurymaths.com. Second edition: *Arithmetica logarithmica, sive logarithmorvm chiliades centvm, pro numeris naturali serie crescentibus ab vnitate ad* 100000, Petrus Rammasenius for Adriaan Vlacq, Gouda, Holland, 1628; reprinted (except for the table) by Georg Olms Verlag, Hildesheim and New York, 1976.

8. Briggs, Henry, *Logarithmorvm chilias prima*, London, 1617. There is a black and white photographic reproduction by Peter Monta in http://www.pmonta.com/tables/logarithmorum-chilias-prima/index.html.

9. Bürgi, Jost, *Aritmetische vnd geometrische Progress Tabulen, sambt gründlichem Vnterricht, wie solche nützlich in allerley Rechnungen zugebrauchen vnd verstanden werden sol*, Paul Sessen, Löblichen Universitet Buchdruckern, Prague, 1620.

10. Fincke, Thomas, *Thomæ Finkii flenspurgensis geometriæ rotvndi libri XIIII*, Basileæ: Per Sebastianvm Henricpetri, 1583.

11. Folta, Jaroslav and Luboš Nový, "Zu Bürgi's Anleitung zu den Logarithmentafeln." *Acta Historiae Rerum Naturalium Necnon Technicarum*, **4**, 97–126, 1968.

12. Gingerich, Owen Jay, *The book nobody read: Chasing the Revolutions of Nicolaus Copernicus*, Walker Publishing Company, New York, and William Heinemann, London, 2004.

13. Glowatzki, Ernst and Helmut Göttsche, *Die Tafeln des Regiomontanus: ein Jahrhundertwerk*, "Algorismus," Vol. 2, Institut für Geschichte der Naturwissenschaften, Munich, 1990.

14. González-Velasco, Enrique A., *Journey through mathematics. Creative episodes in its history*, Springer-Verlag New York, 2011.

15. Havil, Julian, *John Napier: life, logarithms, and legacy*, Princeton University Press, Princeton, New Jersey, 2014.

16. Hawkins, William Francis, "The first calculating machine (John Napier, 1617)," *New Zealand Mathematical Society Newsletter*, Supplement to Newsletter Number 16, 1–23, December 1979; abridged version reprinted in [**18**, 37–51].

17. **Hawkins, William Francis**, *Napier's mathematical works*, PhD thesis, University of Auckland, 1982. This three-volume thesis does not have global page numbering. It was published through University Microfilms International (now called Proquest Dissertations and Theses).

18. **Hawkins, William Francis, Erwin Tomash and Michael R. Williams**, "The promptuary papers," *Annals of the History of Computing*, **10** (1988) 35–67. This article contains two papers, listed here separately as [**16**] and [**44**], with an introduction by Williams.

19. **Kepler, Johannes**, *Joannis Kepleri astronomi opera omnia*, Vol. VI, ed. by Charles Frisch, Heyder & Zimmer, Frankfurt and Erlangen, 1866.

20. **Lansberge, Johan Philip van**, *Triangvlorvm geometriæ libri qvatvor; in quibus nouâ & perspicuâ methodo, & λ'ποδείχει tota ipsorum triangulorum doctrina explicatur*, Franciscum Raphelengium, Leiden, 1591.

21. **Ligonnière, Robert**, "John Neper, premier pionnier de l'aide au calcul: 7. Le principe du 'promptuarium'," *Ordinateurs*, 44–48, October 19, 1987; and 8. "Deux exemplaires du promptuarium," *Ordinateurs*, 56–59, November 2, 1987.

22. **Lilly, William**, *William Lilly's history of his life and times, from the year 1602 to 1681*, J. Roberts, London, 1715; reprinted for Charles Baldwin, 1822; facsimile reproduction of this edition by Kessinger Publishing Company, Kila, Montana, 2004, 2007. There are additional recent editions.

23. **Lutstorf, Heinz Theo and Max Walter**, *Jost Bürgi's «Progress Tabulen»* (*Logarithmen*), Vol. 28 of *Schriftenreihe der ETH-Bibliothek*, ETH-Bibliothek, Zürich, 1992.

24. **Moulton, John Fletcher (Lord Moulton)**, "The invention of logarithms, its genesis and growth," in *Napier tercentenary memorial volume*, ed. by Cargill Gilston Knott, Longmans, Green and Company, London, 1915, pp. 1–32; reprinted by Nabu Press, Berlin, 2010.

25. **Napier, John**, *De arte logistica. Joannis Naperi Merchistonii baronis libri qui supersunt*, Ballantyne et Hughes, Edinburgh, 1839. This is a volume published by the Bannatyne Club that has a long introduction by Mark Napier. It reproduces a copy by Robert Napier of John Napier's untitled notes on this subject. The present title was selected by Mark Napier. English translation by William Francis Hawkins [**17**]. This translation, somewhat modified to conform to the Latin text published by Mark Napier, is reprinted in this volume, pp. 813–941.

26. **Napier, John**, *Mirifici logarithmorvm canonis constrvctio; Et eorvm ad natvrales ipsorum numeros habitudines; vnà cvm appendice, de aliâ eâque præstantiore logarithmorum specie condenda, quibis accessere propositiones ad triangula sphærica faciliore calculo resolvenda: vnà cum annotationibus aliquot doctissimi D. Henrici Briggii in eas & memoratam appendicem*, Andreas Hart, Edinburgh, 1619; reprinted by Bartholomæum Vincentium, Lyons, 1620. English version as *The Construction of the wonderful canon of logarithms by John Napier Baron of Merchiston. Translated from Latin into English with notes and a catalogue of the various editions of Napier's works, by William Rae Macdonald, F.F.A*, William Blackwood & Sons, Edinburgh and London, 1889 (reprinted in this volume, pp. 754–808); facsimile reproduction by Dawsons of Pall Mall, London, 1966; also in The Classics of Science Library, New York, 1997. There are additional recent editions. Modern English translation with notes by Ian Bruce as *The Construction of the wonderful canon of logarithms. and the relations of these to the natural numbers; concerning another kind of logarithms to be put in place*, http://www.17centurymaths.com.

27. **Napier, John**, *Mirifici logarithmorum canonis descriptio, ejusque usus, in utraque trigonometria; ut etiam in omni logistica mathematica, amplissimi, facillimi, & expeditissimi explicatio*, Andreas Hart, Edinburgh, 1614; reprinted 1619; reprinted by

Bartholomæum Vincentium, Lyons, 1620; reprinted by Francis Maseres in *Scriptores Logarithmici*, Vol. 6, London, 1807, pp. 475–624. Translated into English by Edward Wright as *A Description of the admirable table of logarithmes: with a declaration of the most plentifvl, easy, and speedy use thereof in both kindes of trigonometrie, and also in all mathematicall calculations*, Nicholas Okes, London, 1616 (reprinted in this volume, pp. **??**–647); facsimile reproduction by Theatrum Orbis Terrarum, Amsterdam and Da Capo Press, New York, 1969 (this was made from a defective copy lacking the first two pages of the table of logarithms, and also some pages of commendatory poems). Second edition of Wright's translation printed for Simon Waterson, London, 1618. Modern English translation with notes by Ian Bruce as *The Description of the wonderful canon of logarithms, and the use of which not only in trigonometry, but also in all mathematical calculations, most fully and easily explained in the most expeditious manner*, http://www.17centurymaths.com.

28. **Napier, John**, *Rabdologiæ, sev nvmerationis per virgulas, libri dvo: cum appendice de expeditissimo mvltiplicationis promtvario. Quibus accessit & arithmeticæ localis liber vnvs*, Andreas Hart, Edinburgh, 1617; an incomplete English translation, with commentary by the author can be found in [**17**]; complete English translation by William Frank Richardson as *Rabdology*, Charles Babbage Institute Reprint Series for the History of Computing, Vol. 15, The MIT Press, Cambridge, Massachusetts, 1990 (reprinted in this volume, pp. 652–749).

29. **Napier, Mark**, *Memoirs of John Napier of Merchiston, his lineage, life, and times, with a history of the invention of logarithms*, William Blackwood, Edinburgh; and Thomas Cadell, London, 1834.

30. **Parr, Richard**, *The Life of the most reverend father in God, James Usher, late Lord Arch-Bishop of Armagh, Primate and Metropolitan of all Ireland. With a collection of three hundred* LETTERS, *between the said Lord Primate and most of the eminentest persons for piety and learning in his time, both in England and beyond the seas.* Printed for Nathanael Ranew, at the Kings-Arms in St. *Pauls* Church-Yard, London, 1686.

31. **Picatoste, Felipe**, *Abaco neperiano ó rabdológico del Siglo XVII, que se conserva en el Museo Arqueológico Nacional*, Museo Español de Antigüedades, Madrid, Vol. IX, 1878, pp. 51–80.

32. **Reimarus, Nicolaus**, *Fundamentvm astronomicvm: id est. nova doctrina sinvvm et triangvlorvm*, Bernhardus Iobin, Argentorati [Strasbourg], 1588.

33. **Reinhold, Erasmus**, *Primvs liber tabvlarum directionvm discentibvs prima elementa astronomiæ necessarius & utilissimus. His insertvs est canon fecundus ad singula scrupula quadrantis propagatus. Item nova tabvla climatum & parallelorum, item umbrarum. Appendix canonvm secvndi libri directionum, qui in Regiomontani opere desiderantur*, Apvd Hæredes Vlrici Morhardi, Tübingen, 1554.

34. **Roegel, Denis**, "Bürgi's «Progress Tabulen» (1620): logarithmic tables without logarithms," http://locomat.loria.fr/buergi1620/buergi1620doc.pdf, 10 January 2013.

35. **Roegel, Denis**, "Napier's ideal construction of the logarithms," http://http://locomat.loria.fr/napier/napier1619construction.pdf, 12 September 2012.

36. **Roegel, Denis**, "A Reconstruction of Briggs' *Logarithmorum chilias prima* (1617)," http:// locomat.loria.fr/briggs1617/briggs1617doc.pdf, 11 January 2011.

37. **Roegel, Denis**, "A Reconstruction of De Decker-Vlacq's tables in the *Arithmetica logarithmica* (1628)," http://locomat.loria.fr/vlacq1628/vlacq1628doc.pdf, 16 November 2014.

38. **Roegel, Denis**, "A Reconstruction of the tables of Briggs' *Arithmetica logarithmica* (1624)," http://locomat.loria.fr/briggs1624/briggs1624doc.pdf, 30 November 2014.

39. **Sesiano, Jacques.**, "Une Arithmétique médiévale en langue provençale," *Centaurus*, **27**, 26–75, 1984. The reproduced figure is on f. 29^r of the original manuscript and on p. 34 of Sesiano's paper.

40. **Steggall, John Edward Aloysius**, "A Short account of the treatise 'De arte logistica'," in *Napier tercentenary memorial volume*, ed. by Cargill Gilston Knott, Longmans, Green and Company, London, 1915, pp. 145–161; reprinted by Nabu Press, Berlin, 2010.

41. **Stifel, Michael**, *Arithmetica integra*, Iohannis Petreium, Nuremberg, 1544. German translation by Eberhard Knobloch and Otto Schönberger as *Michael Stifel vollständiger Lehrgang der Arithmetik*, Verlag Königshausen & Neumann, Würzburg, Germany, 2007.

42. **Thomas, W. R.**, "John Napier," *Mathematical Gazette*, **19**, 192–205, 1935.

43. **Thoren, Victor Eugene**, "Prosthaphaeresis revisited," *Historia Mathematica*, **15**, 32–39, 1988.

44. **Tomash, Erwin**, "The Madrid promptuary," in [**18**, pp. 52–67].

45. **Tropfke, Johannes**, *Geschichte der Elementar-Mathematik in systematischer Darstellung*, Part II, Leizig, Veit & Comp., 1903.

46. **Voellmy, Erwin**, *Jost Bürgi und die Logarithmen*, Nr. 5 of *Beihefte zur Zeitschrift «Elemente der Mathematik»*, Birkhäuser Verlag, Basel, 1948.

47. **Wolf, Rudolf**, *Johannes Keppler und Jost Bürgi. Vortrag, gehalten den 4. Januar 1872 auf dem Rathhaus in Zürich*, Druck und Verlag von Friedrich Schulthess, Zürich, 1872.

48. **Wood, Anthony à**, *Athenæ Oxonienses. An Exact history of all the writers and bishops who have had their education in the most ancient and famous University of Oxford, from the fifteenth year of King Henry the Seventh, Dom. 1500, to the end of the year 1690. Representing the birth, fortune, preferment, and death of all those authors and prelates, the great accidents of their lives, and the fate and character of their writings. To which are added, the fasti or annals, of the said university, for the same time*, Vol. I. Printed for Thomas Bennet at the Half-Moon in *S. Pauls* Churchyard, London, 1691. New edition as *Athenæ oxonienses. An Exact history of all the writers and bishops who have had their education in the university of Oxford. To which are added the fasti, or annals of the said university. A New edition, with additions, and a continuatiom by Philip Bliss, Fellow of St. John's College*, Vol. II, London, 1815. There are modern reprints of both editions. The quoted statement is in column 469 of the 1691 edition and in columns 491–492 of the 1815 edition.

DESCRIPTIO

Translated by Edward Wright

All the sines and logarithms in Wright's translation of 1616 are shorter by one figure than those originally published by Napier in 1614. For example, in Napier's table the sine of $0°45'$ is 130896 but appears as 13090 in Wright's table. Wright also used a decimal point on some pages of his table, so that the logarithm of the sine of $89°15'$, listed in Napier's table as 857, is 85.7 in Wright's table. In this volume Wright's translation has been kept unchanged, including typos, but Napier's 1614 table is reproduced here instead of Wright's.

R. A. Sampson pointed out in Cargill Gilston Knott's *Napier tercentenary memorial volume* (p. 186) that a poem by Thomas Bretnor (fl. 1607–1618, an almanac maker and self-described "professor of the mathematicks and student in physicke") was missing from most copies of the 1616 English edition of the *Descriptio*. The present authors have not been able to find a copy of the *Descriptio* including this poem, nor have they been able to trace the copy exhibited at the Napier Tercentenary Exhibition in 1914 that was lent by the now closed Royal Observatory. Therefore it has been inserted here, as published by Knott, but using the long *s*. This poem proclaims that the achievements of the Greeks have been surpassed by some modern mathematicians, with Napier as the greatest of them all. However, many of Napier's contemporaries would have been scandalised by such a claim, which could well explain why this page is missing from most copies of the *Descriptio*.

The photographs of the title page of the *Descriptio* (p. 477) and its first English translation (p. 479) are reproduced by courtesy of the National Library of Scotland.

MIRIFICI

Logarithmorum
Canonis descriptio,

Ejusque usus, in utraque
Trigonometria; ut etiam in
omni Logistica Mathematica,
Amplissimi, Facillimi, &
expeditissimi explicatio.

Authore ac Inventore,
IOANNE NEPERO,
Barone Merchistonii,
&c. Scoto.

EDINBURGI,
Ex officinâ ANDREÆ HART
Bibliopólæ, CIↃ. DC. XIV.

A
DESCRIPTION
OF THE ADMIRABLE TABLE OE LOGARITHMES:

WITH

A DECLARATION OF

THE MOST PLENTIFVL, EASY, and speedy vse thereof in both kindes of Trigonometrie, as also in all *Mathematicall calculations.*

INVENTED AND PVBLISHED IN LATIN BY THAT Honorable L. IOHN NEPAIR, Baron of *Marchiston*, and translated into English by the late learned and famous Mathematician *Edward Wright.*

With an Addition of an Instrumentall Table to finde the part proportionall, inuented by the Translator, and described in the end of the Booke by HENRY BRIGS *Geometry-reader at* Gresham-*house in London.*

All perused and approued by the Author, & published since the death of the Translator.

LONDON,
Printed by NICHOLAS OKES.
1616.

TO THE RIGHT HONOVRABLE AND RIGHT WORSHIPFVLL COMPANY OF MERCHANTS of London trading to the Eaſt-Indies, SAMVEL WRIGHT

wiſheth all proſperitie in this life, and happineſſe in the life to come.

YOur fauours towards my deceaſed Father, and your imployment of him in buſineſſe of this nature, but chiefely your continuall imployment of ſo many Mariners in ſo many goodly and coſtly ſhips, in long and dangerous voyages, for whoſe vſe (though many other wayes profitable) this little booke is chiefly behoouefull: may chalenge an intereſt in theſe his labours. This *Book* is noble by birth, as being deſcended from a Noble Parent, & not ignoble by educatiõ, hauing learned to the learned, and experience of the common ſort, famous for knowledge and practiſe in the Mathematickes: whoſe care thereof was ſo great, to ſend it abroad with the true reſemblance of his worthy father, and ſufficient knowledge of the Engliſh tongue to inſtruct our Countrey-men, that hee procured the Authors peruſall of it: who after great paines taken therein, gaue approbation to it, both in ſubſtance and forme, as now I preſent it vnto you. I am the bolder thus to do, in regard it is not vnknowne to many men, that my ſaid father ſpent a great part of his time in ſtudy of the Art of Nauigation, and had gathered much vnderſtanding by his owne practiſe in ſome voyages to ſea with the right Honourable the Earle of *Cumberland* deceaſed: whereupon he publiſhed a painful worke diſcouering errours committed by Mariners in that Art, with corrections and ready wayes for reformation therof. So that I thinke it is out of doubt, that his iudgement therein was great. And ſeeing hee not onely gaue much commendation of this worke (and often in my hearing) as of very great vſe for Mariners: but alſo to help the want of thoſe that could not vnderſtand it in Latine, tranſlated the ſame into Engliſh, and added thereto an inſtrumentall Table to finde the part proportional, whereof alſo the noble Author approued well. I doubt not but it is apparant enough that he eſteemed of it, and intended to haue recommended it as a booke of more then ordinary worth, eſpecially to Sea-men. But ſhortly after he had it returned out of *Scotland*, it pleaſed God to call him away afore he could publiſh it, or but write a deſcription of the ſaid inſtrumentall Table which he had deuiſed, therefore hee left the publiſhing of it to me, as an inheritance, and the ſaid deſcription to his learned and kind friend Mr. *Henry Brigges*, who hath performed it accordingly. All which I humbly preſent vnto you, hoping you ſhall receaue as much profite by the vſe of it, as there hath been learning, care, and paines beſtowed in the penning and fitting it thus to you-hands.

TO THE MOST NOBLE AND HOPE-*FVLL PRINCE,*

CHARLES:

ONELY SONNE OF the high and mightie IAMES by the grace of God, King of great Brittaine, France, and Ireland: Prince *of Wales: Duke of Yorke and Rothesay: Great Steward of Scotland: and Lord of the Islands.*

MOST NOBLE PRINCE,

SEEING there is neither study, nor any kinde of learning that doth more acuate and stirre vp generous and heroicall wits to excellent and eminent affaires: and contrariwise that doth more deiect and keepe downe sottish and dull mindes, then the Mathematickes. It is no maruell that learned and magnanimous Princes in all former ages haue taken great delight in them, and that vnskilfull and slothfull men haue alwayes pursued them with most cruell hatred, as vtter enemies to their ignorance and sluggishnesse. Why then may not this my new inuention (seeing it abhorreth blunt and base natures) seeke and flye vnto your Highnesse most noble disposition and patronage? and especially seeing this new course of LOGARITHMES *doth cleane take away all the difficultie that heretofore hath beene in mathematicall calculations, (which otherwise might haue beene distastfull to your worthy towardnesse) and is so fitted to helpe the weaknesse of memory, that by meanes thereof it is easie to resolue moe Mathematical questions in one houres space, then otherwise by that wonted and commonly receiued manner of Sines, Tangents, and Secants, can bee done euen in a whole day. And therefore this invention (I hope) will bee so much the more acceptable to your Highneſſe, as it yeeldeth a more eaſie and ſpeedy way of accompt. For what can bee more delightfull and more excellent in any kinde of learning then to diſpatch honourable and profound matters, exactly, readily, and without loſſe of either time or labour. I craue therefore (moſt gracious Prince) that you would (according to your gentleneſſe) accept of this gift (though ſmall, and farre beneath the height of your deſeruings, and worth) as a pledge and token of my humble ſeruice: which if I vnderſtand you doe, you ſhall (euen in this regard onely) encourage me that am now almoſt ſpent with ſickneſſe, ſhortly to attempt other matters, perhaps greater then theſe, and more worthy ſo great a Prince. In the meane while, the ſupreame King of Kings, and Lord of Lords long defend and preſerue to vs the great lights of great Brittaine, your renowned parents, and your ſelfe the noble branch of ſo noble a ſtemme, and the hope of our future tranquilitie: to him be giuen all honour and glory.*

Your Highneſſe moſt
deuoted Seruant,

IOHN NEPAIR.

The Authors Preface to the Admirable Table of Loga-rithmes,

SEEING there is nothing (right well beloued Students in the Mathematickes) that is ſo troubleſome to Mathematicall practiſe, nor that doth more moleſt and hinder Calculators, then the Multiplications, Diuiſions, ſquare and cubical Extractions of great numbers, which beſides the tedious expence of time, are for the moſt part subiect to many ſlippery errors. I began therefore to conſider in my minde, by what certaine and ready Art I might remoue thoſe hindrances. And hauing thought vpon many things to this purpoſe, I found at length ſome excellent briefe rules to be treated of (perhaps) hereafter. But amongſt all, none more profitable then this, which together with the hard and tedious Multiplications, Diuiſions, and Extractions of rootes, doth alſo caſt away from the worke it ſelfe, euen the very numbers themſelues that are to be multiplied, diuided, and reſolued into rootes, and putteth other numbers in their place, which performe as much as they can do, onely by Addition and Subtraction, Diuiſion by two, or Diuiſion by three: which ſecret inuention, being (as all other good things are) ſo much the better as it ſhall be the more common, I thought good heretofore to ſet forth in Latine for the publique vſe of Mathematicians. But now ſome of our Countreymen in this Iſland well affected to theſe ſtudies, and the more publique good, procured a moſt learned Mathematician to tranſlate the ſame into our vulgar Engliſh tongue, who after he had finiſhed it, ſent the Coppy of it to me, to bee ſeene and conſidered on by my ſelfe. I hauing moſt willingly and gladly done the ſame, finde it to bee moſt exact and preciſely conformable to my minde and the originall. Therefore it may pleaſe you who are inclined to theſe ſtudies, to receiue it from me and the Tranſlator, with as much good will as we recommend it vnto you. *Fare yee well.*

THE PREFACE
TO THE READER
By *Henry Brigges*.

GENTLE Reader, ſeeing I haue publickly taught the meaning & vſe of this booke at *Greſham houſe*, and haue had ſome charge about this Impreſſion committed vnto me, both by the Honourable Authour the L. of *Marchiſton*, and by my very good frend Mr. *Edward Wright* the Tranſlator. And ſeeing the one who hath moſt right, and is beſt able to commend it, is ſo farre abſent, and the other hath made a moſt happy change of this place and life for a better: thou maiſt happily expect that I ſhould write somewhat that may giue ſome taſte of the excellent vſe of it to thoſe who by reason of the diſtance of place, or other occasions, cannot come to heare me. In a word therefore I will bee bold to ſet downe mine opinion, writing nothing but that which I hope I ſhall alwayes be able and willing to make good and maintaine. There hath been for many former ages euen vnto this preſent, a very great deale of time and expences beſtowed by moſt induſtrious, learned, and worthy men about the doctrine of Triangles, and the making of the Tables of *Sines*, *Tangents*, and *Secants*, that by the helpe of them we may attaine to the knowledge and vſe of the Mathematickes, and eſpecially of Aſtronomie and Nauigation, as namely by *Hipparchus*, *Ptolomey*, *Theon*, *Regiomontanus*, *Copernicus*, *Reinoldus*, *Finkius*, *Lansbergius*, *Clauius*, *Adr. Romanus*, *Joach. Rheticus*, *Valent Otho*, and *Pitiſcus*. All theſe, and diuers others, to their exceeding praiſe, and the great eaſe & contentment of all ſuch as ſet themselues to the ſerious ſtudie of the Mathematickes, haue laboured much, and ſome of them beſtowed very great coſt, both of their owne eſtate, & alſo from the liberall contribution of ſundry great Princes vpon the maintenance of diuers men, who for many yeares together haue wholly employed themſelues to calculate theſe Tables. Yet notwithſtanding this little Table of Logarithmes being firſt begun, and finished by the charge and paines of the honourable Authour alone, may for exactneſſe and certainty compare with all thoſe Tables, and for eaſe and expedition go very farre beyond them, for all Trigonometricall operations, eſpecially Sphæricall, and for the making of the Tables of *Proſthaphæreſes* for the Planets. Which conſiderations may iuſtly warrant the Title of *The Admirable Table of Logarithmes*. But beſides all this, there is an other very excellent and admirable vſe of this Table, which is not at all furthered by the other Tables formerly mentioned, nor can (for any thing I know) be any other way performed, but with very great paines and loſſe of much time: and that is in numbers continually proportionall, hauing any two numbers giuen with their diſtance, or with the number of meane proportionals betwixt them, at one operation to find any one of thoſe meane proportionals, or any one of the numbers, whithout the giuen numbers at any diſtance aſſigned. And becauſe theſe things may to ſome ſeeme obſcure, giue me leaue to explaine then by an *example*. Let the two giuen numbers be 1. and 3000. and let there be ſuppoſed foure meane proportionals betwixt them. If of theſe foure I deſire that which is neereſt to the leſſe extreame, that meane, (becauſe here the leſſe extreme is an vnity) is called the ſurſolide root of the other extreame, to wit, of 3000. and that, or any other root may farre more eaſily be had by theſe Logarithmes then by any rule or other way. But the finding of any root is but one particular meane proportionall, to wit, the next meane to the vnity: and this way is generall, giuing as eaſily the third or fourth meane as the firſt. And not onely where the one extreame is an vnity, but betwixt any two numbers aſſigned. For example, if the giuen extremes bee 19 and 738. and there ſhall be betwixt theſe two, ten meane proportionals: by this Table we may finde the 7th or 8th, or any other aſſigned, from the leſſe or from the greater number: or if they be continued further either diminiſhing vnder 19. or increaſing aboue 738. we may finde any of them for any diſtance aſſigned, as the fifth and ſixth in the same proportion aboue 738, or vnder 19. And thus hauing two extremes giuen, and the number of meane proportionals betwixt them, we may finde any, for any aſſigned diſtance within or without. In like ſort, hauing a proportion aſſigned in numbers, and a third number giuen, we may from that third number find an extreme: betwixt which and that third number ſhall bee any ſet number of meane proportionals, keeping the proportion aſſigned one from another. For example, if 73£ be yearely ſo to be increaſed, as that $\frac{1}{16}$ be ſtill to bee added vnto the former yeares summe, and I would know what is the whole at the end of ſeuen yeares: here the proportion aſſigned is 16 to 17 the third number, or the beginning of the progreſſion is 73, the meane proportionals are ſixe: I would know the other extreme, to wit, the ſeuenth from the beginning, and by this Table of *Logarithmes* I find it to bee 111£, 11s, 9d$\frac{696}{1900}$, which perhaps by curious ſearch, after the laborious ordinarie way, will bee found too great by $\frac{1}{11}$ of a peny, or thereabouts. And thus we ſee the admirable vſe of theſe *Logarithmes*, not onely in the doctrine of Triangles (which I account to bee farre the moſt excellent part, and which may by other Tables be performed as exactly, but nothing ſo ſpeedily, or with the like eaſe) but alſo in all our common accounts of ordinary proportionall numbers: wherein wee may not expect the same exactneſſe which we may attaine vnto by rule, and by long tedious practiſe, because this table is but ſmall, and the numbers neuer exceed the eighth place; but wee may ſafely truſt to it to performe all things without ſenſible errour: or to performe truly, ſo farre as can be expreſſed in 7 figures. And if it ſhall pleaſe God (who beſides his other mercies hath granted this honour vnto the Authour, to begin and thus farre to accompliſh this admirable worke) further to grant vnto him life and competent ſtrength, I doubt not we ſhall haue the worke ſo enlarged and perfected, that we may vſe it, both with greater eaſe, & with exactneſſe vnto the 10th place. And thus commending theſe things to thy conſideration, and vs and all our honeſt ſtudies to the Lords bleſſing, I euer reſt a louer of all them that loue the Mathematickes,

H. Brigges.

In praiſe of the neuer-too-much praiſed Worke and Authour the L. of *Marchiſton.*

W*Hat, like our bodies,* ſoules *rare excellence.*
Our bodies bound, yet haue thereof no ſight.
(*Enwomb'd with* clouds *of* Myſtery *from ſence*)
Is here (*well* borne, *and* ſhap't) *produc'd to light*;
This skill, *ſince first men knew*, *ſtill lay unknowne.*
As if ſome meere Impoſſibilitie
Had ſtood twixt It *and how it might be ſhowne.*
But now it lookes like ſelfe-Facilitie!
How happy *and* acute *were his* Wits *eyes,*
That for the Mathematicks *found this Key,*
To ope the lockes *of all their* Miſteries,
That from all eyes ſo long concealed lay.
It was at hand, *and yet it was* vnſeene.
Inuiſible, *and yet was* cleere *to* wit
As it could wiſh, or as it could haue beene
In Art *or* Nature; *yet* Art *miſt of* It.
From whence a queſtion *may ariſe* (*perchance*)
Whether, or *no*, This *do extenuate*
The Authors *merit? No, it doth aduance*
His praiſe the more, the leſſe he toil'd for that.
For who with eaſe *hath done what* none *ere could*
Is moſt like God *in* workes *of* rareſt skill:
This argues He *can do what ere he would*
In Art with eaſe, *if he had but a* Will (*then,*
* Wright (*ſhip-wright? no*; *ſhip-right, or righter* M. *Wrights*
when wrong ſhe goes) *to this, with eaſe, will make* Tract of
Thy Rules *to make the* ſhip *run rightly, when* Nauigati-
She thwarts the Maine *for* Praiſe *or* profits *ſake.* on.
If after-times, *that ſtill ſhall bleſſe his* name,
Shall ſeeke more eaſe *than, in his* eaſineſſe,

To worke by Figures, *he muſt make* Art *lame*
(*So leſſe deſir'd*) *with* Eaſes *great exceſſe*!
For his Rules *are ſo* firme *and* facill *too*,
As makes Art *laugh their* quick-diſpatch *to waigh*
With Tangents *and with* Secants *much-a-do*,
And Enuy *with that* eaſe *to pine away*.
O that great Lords *no worſe would vse their leaſure*
In ſeuerall kinds, *then* (*kindly*) *were they* Great:
*But they make ſmall thẽselues w*th *too-great pleſure*:
So, great-Lords *th'are not*, *nor their* Counterſet.
Scotland, *two Miracles of* Men, *this* Age
In thee affoords the world, *to future yeares*:
Bucanan. *The* Tutor *of our* Rulers *Pupillage*,
And this rare Lord, *a* Lead-ſtarre *to his* Peeres.
The ground of whose iuſt praiſes *is ſo ſure*,
That it will beare more Fame *then* Fames *right* wing:
Birth, Grace, *and* Art, *and* all *ſurpaſſing pure*,
Makes him more good *then* great, *although a* king.
Then great-good Lord, *liue euer in my* Lines,
By thy iuſt lauds *that shall then* (*dead*) *reuiue*,
Vnill the Sunne *forſake the heauenly* Signes,
And in the Signes *of thy* worth *euer liue*.
To light the world *through* them, & them *through* thee,
And me through both, *to* Fame, & *that through me*!

By the unfained louer and admirer of his Art and matchleſſe vertue,

Iohn Dauies *of Hereford*.

In the iust praiſe of this
Booke, Authour, and
Tranſlator.

A Rts, in thẽſelues, haue ſuch diuine Perfection,
As Human reaſon *cannot alwaies ſee*;
Yet God *all good, to man giues ſuch direction*
As hidden things *ſometimes* discouered *bee*:
    What many men *and* ages *could not finde*,
    Is, at the laſt, by ſome one *brought to mind.*
This noble Author *first due* honour *gaue*
To him from whom true honours doe proceed, *
Who now to him doth graciouſly vouchſaue:
Beſides his Stile, *much* honour *for his* meed,
    By bringing Him *thus clearly to reueale*
    Such profit *both to* Church & Comon-weale.
This little Booke (*to let the other paſſe*)
As Title *ſhewes, is truly admirable*;
Th'inuention *rare, for* practiſe *nothing leſſe*,
Briefe, eaſie, plaine, *and paſſing* delectable.
    What earſt was hard *and* tedious *to vnfold*,
    Here how to find with eaſe, *is* plainly *told.*
The toyleſome Rules *of due* Proportion
Done here by Addition *and* Subtraction,
By Bipartition *and* Tripartition,
The ſquare *and* cubicke rootes *extraction*:
    And ſo, all questions Geometricall,
    But with most ease Triangles-ſphæricall.
The vſe *is great in all true* Meaſuring
Of Lands, Plots, Buildings, *and* Fortification,
So in Astronomie *and* Dialling,
Geography *and* Nauigation.
    In theſe *and* like, *yong ſtudents ſoone may gaine*,
    The skilfull too, may ſaue cost, time, & paine.
In Latine *to the* world *it first appear'd*

* By his worke on the *Reuelation*, first printed in *Ano*. 1593 And againe in *Ano* 1611.

Mr *Wright* detected & corrected many errors in the vulgar Nauigation.

Strange vnto them to whom that tongue *is ſtrange*:
But he who carſt our Nauigation *clear'd*,
From that ſtrange tongue to Engliſh *it did change*,
That famous, *learned*, Errors *true Corrector*,
England, *great* Pilot, Mariners Director.
Whoſe care thereof was ſuch, that he obtain'd
The Authors Approbation, *and withall*,
He, for the helpe of Practiſers *ordain'd*,
A way to finde the part Proportionall:
The vse whereof too-timely death *deny'd*,
Which famous Briggs *hath learnedly ſupply'd.*
Thus haue you here the quinteſſence *of* Art,
Fitted to hand by men of rareſt skill,
Whoſe euerlaſting praiſes *in each part*
So farre extend that here couclude I will.
And ſay; *For* Matter, Author, *and* Tranſlator,
Nere had theſe Arts so good a Demonſtrator,
Pulchra hæc facilia.

Ri. Leuer.

To the worthily honoured
Authour and Tranſlator

Pull off your Laurel rayes, you learned Greekes,
Let Archimed and Euclid both give way,
For though your pithie ſawes have paſt the pikes
Of all opponents, what they e'er could ſay,
And put all modern writers to a ſtay.
Yet were they intricate and of ſmall uſe,
Till others their ambiguous knots did looſe.

And bonnets vaile (you Germans) Rheticus,
Reignoldus, Oſwald, and Iohn Regiomont,
Lanſbergius, Finckius, and Copernicus,
And thou Pitiſcus, from whose cleerer font
We ſucked have the ſweet from Helleſpont:
For were your labours ne'er compoſ'd ſo well,
Great Napier's worth they could not paralell.

By thee great Lord, we ſalve a tedious toyle,
In reſolution of our trinall lines,
We need not now to carke, to care, or moile,
Sith from thy witty braine ſuch ſplendor ſhines,
As dazels much the eyes of deepe Diuines.
Great thy inuention, greater is the praiſe,
Which thou unto thy Nation hence doſt raiſe.

Nor are we less oblig'd to thee good Wright,
By whoſe induſtrious paines are vulgar made
Not onely thoſe, but things that in the night
Haue ear'ſt lyen hid, and couered in the ſhade,
Till thou into the Ocean ſeas didſt wade,
And there foule errors didſt discouer more,
Than any witty Mariner before.

And though thy fatall threed be cut in twaine,
Thy pilgrimage come to a common end,
Yet as a patterne do thy workes remaine,

To ſuch braue ſpirits as do ſtill intend
Their deereſt lives in ſacred Arts to ſpend,
As Heriots, Gunter, Briggs and Torperley
Who to the world rare ſecrets can diſplay,
But ſuch is now th' ingratitude of time,
That hees the wiſeſt which doth loweſt clime.

THOMAS BRETNOR, Mathem.

A VIEW OF THIS BOOKE.

- In *Logarithmes* we confider
  - The *Logarithmes* themfelues,
    - Nature
      - Vpfpringing, *Booke* 1, *chap*. 1.
      - Affections *chap*. 2.
    - Table
      - Defcription *Chap*. 3.
      - Vfe thereof, *Chap*. 4.
  - Their vfe
    - Generall, *Chap*. 5.
    - Particular in triangles
      - Right-lined.
        - Right-angled *Booke* 2, *Chap*. 5.
        - Oblique-angled *Chap*. 2.
      - Sphæricall, *Chap*. 3.
        - quadrantall, *cha*. 4.
        - not-qadrantall.
          - mixt, *ch*. 5.
          - pure. *cha*. 9

R. L.

Some faults haue efcaped in printing of the Table, which the practifer (if it pleafe him to take the paines) may eafily correct afore he vfe the Table, after this manner, or what elfe he fhall find.

Deg.	*mi*			Deg.	*m*		
0	10	*Diff.*	5839981	62	18	*Sine*	885392
0	11	*Sine*	3200	28	34	*Diff.*	607955
1	21	*Sine*	23560	61	1	*Sine*	874761
1	60	*Log.*	3355282	60	47	*Log.*	136071
2	1	*Log.*	3346986		60	*Sine*	874620
4	46	*Log.*	2487733	29	59	*Diff.*	549978
85	0	*Sine*	996195	60	20	*Log.*	140504
83	44	*Sine*	994025		21	*Log*	140339
	3	*Sine*	992652		7	*Log.*	142668
8	33	*Diff.*	1894833	57	30	*Sine*	843391
81	30	*Sine*	989015		14	*Sine*	840882
11	18	*Sine*	195946	33	59	*Log*	581692
	24	*Log.*	1621220	54	5	*Log.*	210818
78	42	*Log.*	19576	36	28	*Sine*	594355
13	45	*Sine*	237686	53	1	*Diff.*	283560
	46	*Sine*	237968		7	*Diff.*	287193
	47	*Sine*	238251	52	15	*Log.*	234850
	48	*Sine*	238533	38	26	*Log.*	475452
	49	*Log.*	1432062		59	*Log.*	463474
	50	*Log.*	1430880	51	0	*Sine*	777146
15	5	*Sine*	260224		4	*Sine*	777878
16	26	*Diff.*	1220955		19	*Sine*	780612
21	6	*Sine*	359997	38	49	*Log.*	467079
22	4	*Diff.*	902930		44	*Log*	468889
	50	*Log.*	946616	39	13	*Sine*	632255
23	2	*Log.*	938366				
24	58	*Diff.*	764429				
64	2	*Sine*	899049				
27	4	*Sine*	455027				
	11	*Sine*	456839				
	20	*Diff.*	659953				
	31	*Sine*	462007				
62	30	*Sine*	887011				

A DESCRIPTI-ON OF THE ADMIRABLE TABLE OF LOGARITHMES, WITH THE MOST PLEN-TIFVL, EASY, AND READY Vſe thereof in both kindes of *Trigonometrie, as alſo in all Ma-thematicall Accounts.*

THE FIRST BOOKE.

CHAP. I.
Of the Definitions.

A LINE *is ſaid to increaſe equally, when the poynt deſcribing the ſame, goeth forward equall ſpaces, in equall times, or moments.* *1. Definition.*

Let *A* be a poynt, from which a line is to be drawne by the motion of another poynt, which let be *B*.

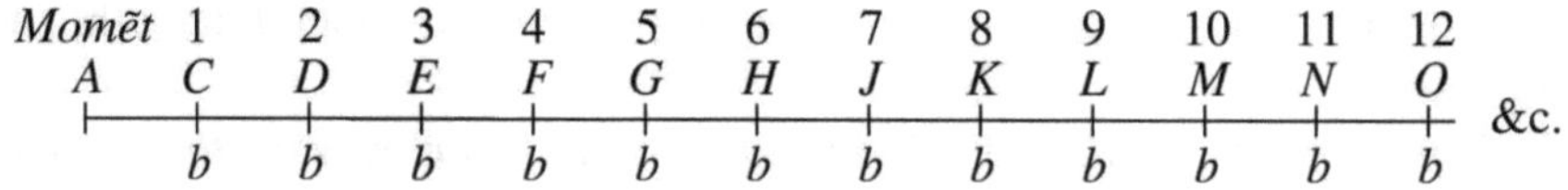

Now in the firſt moment, let *B* moue from *A* to *C*. In the ſecond moment from *C* to *D*. In the third moment from *D* to *E*, & ſo forth infinitely, deſcribing the line *ACDEF*, &c. The ſpaces *AC*, *CD*, *DE*, *EF*, &c. And all the reſt being equall, and deſcribed in equall moments (or times.) This line by the former definition ſhall be ſaid to increaſe equally.

Therefore by this increaſing, quantities equally differing, muſt needes be produced, in times equally differing. *A Corollary or conſequent.*

As in the Figure before, *B* went forward from *A* to *C* in one moment, and from *A* to *E* in three moments. So in ſixe moments from *A* to *H*: and in 8 moments from *A* to *K*. And the differences of thoſe moments, one and three, and of theſe 6 and 8 are equall, that is to ſay two.

So alſo of thoſe quantities AC, and AE, and of theſe, AH, and AK, the differences CE, and HK are equall, and therefore differing equally, as before.

2. *Definition.* *A Line is ſaid to decreaſe proportionally into a ſhorter, when the poynt deſcribing the ſame in æquall times, cutteth off parts continually of the ſame proportion to the lines from which they are cut off.*

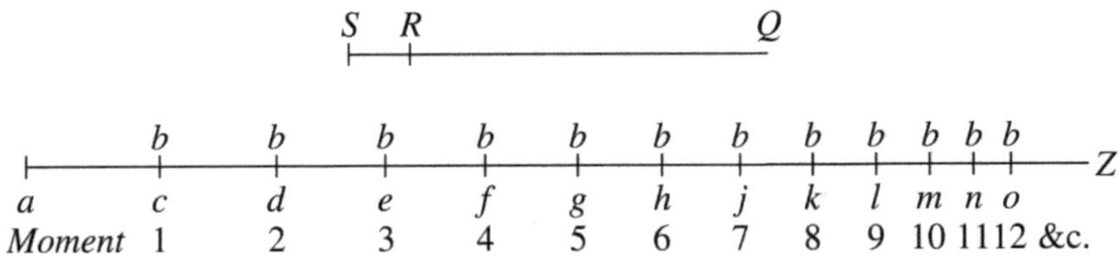

For examples ſake. Let the line of the whole ſine aZ be to bee diminiſhed proportionally: let the poynt diminiſhing the ſame by his motion be *b*: and let the proportion of each part to the line from wch it is cut off, be as QR to QS. Therefore in what proportion QS is cut in *R*, in the ſame proportion (by the 10 of the *6* of *Euclid*) Let aZ be cut in *c*. and ſo let *b*. running from *a* to *c* in the firſt moment, cut off *ac* from aZ, the line or ſine cZ remaining.

And from this cZ let *b* proceeding in the ſecond moment, cut off the like ſegment, or part, as QR to QS: and let that bee *cd*, leauing the ſine. dZ. From which therefore in the third moment, let *b* in like manner, cut off the ſegment *de*, the ſine eZ being left behinde. From which likewiſe in the fourth moment, by the motion of *b*, let the ſegment *cf* be cut off, leauing the ſine fZ. From this fZ in the fifth moment, let *b* in the ſame proportion cut off the ſegment *fg*, leauing the ſine gZ, and ſo forth infinitly. I ſay therfore out of the former definition, that here the line of the whole ſine aZ, doth proportionally decreaſe into the ſigne gZ, or into any other laſt ſine, in which *b* ſtayeth, and ſo in others.

1 *Corolary.* *Hence it followeth that by this decreaſe in equall moments (or times) there muſt needes alſo bee left proportionall lines of the ſame proportion.*

For what continuall proportion there is before of the ſines to be diminiſhed, *az*, *cz*, *dz*, *ez*, *fz*, *gz*, *hz*, *iz*, and *kz*, &c. and of the ſegments cut off from them, *ac*, *cd*, *de*, *ef*, *fg*, *gh*, *hi*, and *ik*, there muſt needes be alſo the ſame proportion of the ſines remaining, that is, *cz*, *dz*, *ez*, *fz*, *gz*, *hz*, *iz*, and *kz*, as may manifeſtly appeare by the 19 *Prop. 5.* and 11. *Prop. 7, Euclid.*

3. *Def.* *Surd quantities, or vnexplicable by number, are ſaid to be defined, or expreſſed by numbers very neere, when they are defined or expreſſed by great numbers which differ not ſo much as one vnite from the true value of the Surd quantities.*

As for example. Let the ſemidiameter, or whole ſine be the rational number 10000000 the ſine of 45 degrees ſhall be the ſquare root of 50,000,000,000,000, which is ſurd, or irrationall and inexplicable by any number, & is included between the limits of 7071067 the leſſe, and 7071068 the greater: therfore, it differeth not an vnite from either of theſe. Therefore that ſurd ſine of 45 degrees, is ſaid to be defined and expreſſed very neere, when it is expreſſed by the whole numbers, 7071067, or 7071068, not regarding the fractions. For in great numbers there ariſeth no ſenſible error, by neglecting the fragments, or parts of an vnite.

Equall-timed motions are thoſe which are made together, and in the ſame time. 4. *Def.*

As in the figures following, admit that *B* be moued from *A* to *C*, in the ſame time, wherin *b* is moued from *a* to *c* the right lines *AC* & *ac*, ſhall be ſayd to be deſcribed with an equall-timed motion.

Seeing that there may bee a ſlower and a ſwifter motion giuen then any motion, it ſhall neceſſarily follow, that there may be a motion giuen of equall ſwiftneſſe to any motion (which wee define to be neither ſwifter nor ſlower.) 5. *Def.*

The Logarithme *therfore of any ſine is a number very neerely expreſsing the line, which increaſed equally in the meane time, whiles the line of the whole ſine decreaſed proportionally into that ſine, both motions being equal-timed, and the beginning equally ſwift.* 6. *Def.*

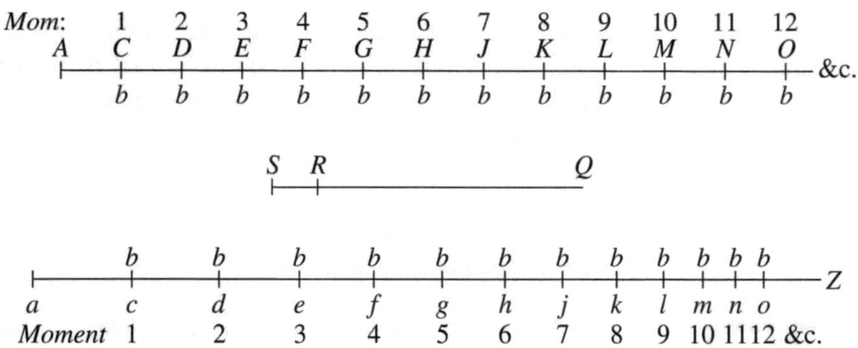

As for example. Let the 2 figures going afore bee here repeated, and let *B* bee moued alwayes, and euery where with equall, or the ſame ſwiftneſſe wherewith *b* beganne to bee moued in the beginning, when it was in *a*. Then in the firſt moment let *B* proceed from *A* to *C*, and in the ſame time let *b* moue proportionally from *a* to *c*, the number defining or expreſſing *AC* ſhal be the *Logarithme* of the line, or ſine *cZ*. Then in the ſecond moment let *B* bee moued forward from *C* to *D*. And in the ſame moment or time let *b* be moued proportionally from *c* to *d*, the number defining *AD*, ſhall bee the *Logarithme* of the ſine *d* Z. So in the third moment let *B* go forward equally from *D* to *E*, and in the ſame moment let *b* be moued forward proportionally from *d* to *e*, the number expreſſing *AE* the *Logarithme* of the ſine *cZ*. Alſo in the fourth moment, let *B* proceed to *F*, and *b* to *f*, the number *AF* ſhall be the *Logarithme* of the ſine *fz*. And keeping the ſame order continually (according to the former definition) the number of *AG* ſhall be the *Logarithme* of the ſine *gz*. *AH* the *Logarithme* of the ſine *hz*. *AI* the *Logarithme* of the ſine *iz*. *AK* the *Logarithme* of the ſine *kz*, and ſo forth infinitely,

cõſequẽt. *Therefore the Logarithme of the whole ſine* 1000000 *is nothing, or* 0: *and conſequently the Logarithmes of numbers greater then the whole ſine, are leſſe then nothing.*

For ſeeing it is manifeſt by the definition, that the ſines decreaſing from the whole ſine, the *Logarithmes* increaſe from nothing: therfore contrariwiſe the numbers which yet we call Sines, increaſing vnto the whole ſine, that is to 1000000, the *Logarithmes* muſt needs decreaſe to 0. or nothing: and by conſequent the *Logarithmes* of numbers increaſing aboue the whole ſine 1000000, which wee call *Secants*, or *Tangents*, and no more ſines, ſhall be leſſe then nothing.

Therefore we call the Logarithmes of the ſines Abounding, becauſe they are alwayes greater then nothing, and ſet this marke + *before them, or elſe none. But the Logarithmes which are leſſe then nothing, we cal Defectiue, or wanting, ſetting this marke* − *before them.*

It was indeed left at libertie in the beginning, to attribute nothing, or 0. to any ſine or quantitie for his Logarithme: but it was beſt to fit it to the whole ſine, that the Addition or Subtraction of the Logarithme which is moſt frequent in all Calculations, might neuer after be any trouble to vs.

Chap. II.

Of the Propoſitions of Logarithms.

Propoſ. 1. T*HE Logarithmes of Proportionall numbers and quantities are equally differing.*

As for example. The Lo[ga]rithmes of the proportionall ſines, namely *cz*, which is to *ez*, as *hz* is to *kz*, are reſpectiuely the numbers defining *AC*, *AE*, *AH*, *AK*, (as is manifeſt by the 6 Definition.) Now *AC*, and *AE* differ by the difference *CE*, and *AH* and *AK* by the difference *HK*. But by the firſt definition and his Corolarie *CE* and *HK*, are equall: therefore the *Logarithmes* of the foreſaid proportional ſines are equally differing. And ſo in all proportionals.

For what affections and ſymtomes the *Logarithmes* haue gotten in their firſt beginning and generation, the ſame muſt they needes retaine and keepe afterwards.

But in their beginning and generation, they are indued with this affection, and this law is prefcribed vnto them, that they bee equally differing, when their fines or quantities are proportionall (as it appeareth by the definition of a *Logarithme*, and of both motions, and fhall hereafter more fully appeare in the making of the *Logarithmes*.) Therefore the *Logarithmes* of proportional quantities are equally differing.

Of the Logarithmes of three proportionals, the double of the fecond or meane, made leffe by the firft, is equall to the third. *Propof.* 2.

Seeing that by the firft propof. the difference of the *Logarithme* of the firft and fecond, is equall to the difference of the *Logarithmes* of the fecond and third, that is, the fecond made leffe by the firft, is equall to the third, leffe by the fecond: therefore the fecond being added to both fides of the equation twice, the fecond, or the double of the fecond made leffe by the firft, fhall come forth equall to the third, which was to bee proued.

Of the Logarithmes of three proportionals, the double of the fecond, or middle one, is equall to the fumme of the extremes. *Propof.* 3.

By the fecond Propofition next going before, the double of the fecond, made leffe by the firft, is equall to the third. To both the equall fides adde the firft, and there fhall arife the double of the fecond equall to the firft and third, that is, to the fumme of the extremes, which was to bee demonftrated.

By the fecond Propofition next going before, the double of the fecond, made leffe by the firft, is equall to the third. To both the equall fides adde the firft, and there fhall arife the double of the fecond equall to the firft and third, that is, to the fumme of the extremes, which was to bee demonftrated.

Of the Logarithmes of foure proportionals, the fumme of the fecond and third, made leffe by the firft, is equall to the fourth. *Propof.* 4.

Seeing by the firft Propofition of the *Logarithmes* of 4 proportionals, the fecond made leffe by the firft, is equall to the fourth leffe by the third: adde the third to both fides of the equality, and the fecond and third made leffe by the firft, fhall bee made equall to the fourth, which was propounded.

Of the Logarithmes of foure proportionals, the fumme of the middle ones, that is, of the fecond and third, is equall to the Logarithme of the extreames, that is to fay, the firft and fourth. *Propof.* 5.

By the 4 Propofition going afore the 2 & third made leffe by the firft, were equall to the fourth: to both fides of the equality adde the firft, and the fecond more by the third fhall bee made equall to the fourth, more by the firft, which was to be demonftrated.

Of the Logarithmes of foure continuall proportionals, the triple of either of the middle ones, is equall to the fumme of the further extreame, and the double of the neerer. *Propof.* 6.

By the ſecond Propoſition, the double of the ſecond made leſſe by the firſt, is equall to the third; and by the third Propoſition the double of this, that is, the fourefold of the ſecond made leſſe by the double of the firſt, ſhall be equall to the ſumme of his extreames, that is, the fourth more by the ſecond. Now, if from both ſides of the equality you ſubtract the ſecond, the triple of the ſecond made leſſe by the double of the firſt, ſhall be made equall to the fourth. Againe, to the ſides of this equality adde the double of the firſt, and there ſhall ariſe the triple of the ſecond, equall to the fourth, more by the double of the firſt, which wee vndertooke to proue.

An Admonition.

Hitherto we haue ſhewed the making and ſymptomes of *Logarithmes*; Now by what kinde of account or method of calculating, they may be had, it ſhould here bee ſhewed. But becauſe we do here ſet down the whole Tables, and all his *Logarithmes* with their Sines to euery minute of the quadrant: therfore paſſing ouer the doctrine of making *Logarithmes*, til a fitter time, we make haſte to the vſe of them: that the vſe and profit of the thing being firſt conceiued, the reſt may pleaſe the more, being ſet forth hereafter, or elſe diſpleaſe the leſſe, being buried in ſilence. For I expect the iudgement and cenſure of learned men hereupon, before the reſt raſhly publiſhed, be expoſed to the detraction of the enuious.

CHAP. III.

Containing the deſcription of the Table of Logarithmes and of the ſeuen Columnes thereof.

1 *Section.* *THE firſt Columne is expreſly of the Arches increaſing from* 0 *to* 45 *degrees, and is alſo vnderſtood to bee of their remainders to a ſemicircle.*

2 *Section.* *The ſeuenth columne is of arches decreaſing from a quadrant to* 45 *degrees, and is alſo vnderſtood to bee of their remainders to a ſemicircle.*

3 *Section.* *So the Arches of the one columne are the complements of the Arches of the other anſwering ouer-againſt them.*

4 *And in the firſt is expreſſed the leſſe ſharpe angle of any right-lined right-angled triangle.*

5 *But in the ſeuenth ouer againſt it, is placed the greater ſharpe angle of the ſame right-angled triangle.*

6 *In the ſecond columne are the ſines of the arches of the firſt columne.*

And thefe are the leffe legges fubtending the leffe angle of a right angled 7
triangle, whofe Bafe, or Hypotenufe is the whole fine.

In the fixth columne are the fines of the arches of the feuenth columne. 8

And thefe are the greater legges fubtending the greater fharpe angle of the 9
fame right-angled triangle, whofe Hypotenufe is the whole fine.

Hence it followeth, that of the whole fine, and the fine of the fecond columne, 10
and the fine of the fixth columne anfwering ouer-againft the fame, there is made
a triangle that is equiangled, and like to any right-angled right-lined triangle.

The third columne containeth the Logarithmes of the arches and fines 11
towards the left hand.

Which are the Logarithmes of the proportion of the leffe legge of a right- 12
angled triangle, to the Hypotenufe of the fame.

And they are alfo the Logarithmes of the complements of the arches and 13
fines towards the right hand, which we call Antilogarithmes.

The fift columne containeth the Logarithmes of the arches and fines towards 14
the right hand.

Which are the Logarithmes of the proportion of the greater legge of a 15
right-angled triangle, to the Hypotenufe of the fame.

They are alfo the Antilogarithmes *of the arches and fines towards the left* 16
hand, or the Logarithmes of the complements.

Laftly, the fourth or middle columne containeth the differences betweene 17
the Logarithmes of the third and fifth columnes. And fo this columne is two-
fold, Abounding and Defectiue.

Thofe differences are Abounding, which arife out of the fubtraction of the 18
Logarithmes of the fifth columne from the Logarithmes of the third columne.

But the differences arifing by fubtraction of the Logarithmes of the third 19
columne out of the Logarithmes of the fifth columne, are Defectiue, which
therefore are leffe then nothing.

The Abounding differences are called the differentiall numbers of the arches 20
towards the left hand.

And are the Logarithmes of the proportion of the leffe legge of a right- 21
angled triangle, to the greater legge of the fame.

And are alfo the Logarithmes of the Tangents of the left hand arches. 22

But the defectiue Differences, are called the differentiall numbers of the 23
right hand arches.

And are the Logarithmes of the proportion of the greater legge of a right- 24
angled triangle, to the leffe legge of the fame.

And are alfo the Logarithmes of the Tangents of the right-hand arches. 25

Alfo euery left hand arch, and the remainder thereof to a femicircle, is 26
called the arch of the complement of the arches, fines, & right hand Loga-
rithmes, and of the Defectiue differentials.

27 *And contrariwiſe euery right hand arch, and the remainder thereof to a ſemicircle, is called the arch of the complement of the arches, ſines and left hand Logarithmes, and of the Abounding differentials.*

Admonitions.

28 *HEre it is to be noted, that if you make the Logarithmes of the third columne Defectiue, ſetting before them this marke, – they ſhall bee made the Logarithmes of the Hypotenuſes or Secants of the right hand arches of the ſeuenth columne.*

29 *And theſe alſo ſhall bee made the Logarithmes of the proportion of the Hypotenuſe of a right angled triangle to the leſſe legge of the ſame.*

30 *And if you make the Logarithmes of the fifth columne Defectiue, they ſhall bee the Logarithmes of the Hypotenuſes, or of the Secants of the left hand arches of the firſt columne.*

31 *The ſame ſhall alſo be the Logarithmes of the proportion of the Hypotenuſe of a right-angled triangle to the greater legge of the ſame. But becauſe the ſines onely, and their arches, and the Logarithmes with their Differentials, are ſufficicient for attaining the knowledge of right-lined triangles, and for the knowledge of ſphæricall triangles, the arches onely with their Logarithmes and Differentials are ſufficient without regard of the ſines. Therefore we haue excluded the Tangents, and the Hypotenuſes, or Secants, out of the Table: and in ſphærical triangles we will haue the ſines alſo not regarded; yet we will ſhew you by the way, that you may, if you liſt, vſe them all readily enough in right-lined triangles, but not in ſphæricall.*

Chap. IV.

Of the vſe of the Table, and of the numbers thereof.

Sect. 1 *THE Sines, Tangents and Secants being preciſely found in their Tables, to finde their Logarithmes as preciſely.*

By the 11 and 14 Section of the third chapter, the Sine giuen being found in the ſecond, or ſixith columne of our Table, the Logarithme thereof ſhall bee found in the third or fifth columne of the ſame line.

So therefore, the Logarithmes of the Sines that are in the table are exactly had. And the numbers of the Tangents and Secants being found in their owne Tables, you haue their arches.

And the arches being knowne, our Table giueth you the *Logarithmes* of the Tangents, or the differentials with their ſignes or marks in the middle columne, by the 22 and 25 Sect And the *Logarithmes* of the Secãts reciprocally in the third & fifth columnes; yet ſetting before them this ſigne – by the 28 and 30 Sect. Therefore the *Logarithmes* of the Sines, Tangents and Secants that are in the Tables, are thus had.

Examples of Sines.

I Seeke the *Logarithme* of the ſine 694658. I finde that ſine preciſely in the ſecond columne, anſwering to the arch 44 degrees, 0 min. & in the ſame line of the third columne, there ſtandeth ouer-againſt it, the *Logarithme* 364335 which I ſought. Alſo let the *Logarithme* of the ſine 721357 bee ſought. This ſine ſhall bee found anſwering to the arch 46 degr. 10 min. and neere adioyning thereto 326620. the Logarithme thereof that was ſought.

Examples of Tangents.

LEt the *Logarithme* of the Tangent 218645 bee ſought. To this Tangent there anſwereth in his Table the arch of 12 degr. 20 min. and to this arch in the middle columne of our Table, anſwereth the Logarithme, or differentiall abounding 1520306 which was ſought. Alſo if you ſhal ſeeke the Logarithme of the Tangent 4573629. you ſhall finde in the Table of Tangents his arch 77 degr. 40 min. and the ſame differentials of this arch in our Table, but yet defectiue, that is, −1520306.

Examples of Secants.

TO the Secant 1811801 there anſwereth in the Table of Secants, the arch 56 degr. 30 min. and to this arch in our Table agreeeth reciprocally −594321 the defectiue Logarithme of the Secant 1811801, aboue written. So you ſhall find −271425, the Logarithme of the Secant 1311834. & of the ſecant 1396059. you ſhall find the Logarithme −333653.

To æſtimate the Logarithmes of the numbers giuen, and not found in the Tables of the Sines, Tangents, ond Secants.

Seeke the number that is moſt like the number giuen in the ſecond or ſixt columne of our Table, whether it be ten fold, an hundred fold, a thouſand fold, 10000 fold, 100000 fold, 1000000 fold: or if you will in the Tables of Tangents and Secants: and note the arche hereof. For the Logarithme thereof taken out of our Table, is that you ſeek for: yet keeping in minde, or for memory ſake, ſetting downe in cyphers, the number of the places or figures of the multiplicitie. As if the Logarithme of the number 137 bee ſought, which is not found in the Tables, you ſhall finde among the Sines 1454. 13671. and 137156. And among the Tangents 1370305. but among the Secants, the number 1370505 which is likeſt of all to the number giuen, if the laſt foure figures toward the right hand be vnderſtood to be blotted out. Therefore let the Logarithme of this Secant 1370305. and of his arch 43 degr. 8. min. be ſought out by the former Section, or by the 28 and 30 Sections of the third chapter, and it ſhall bee found −315033, which is alſo taken for the Logarithme of the number giuen 137 remembring, notwithſtanding, that the 4 laſt figures are to be cut off, or for memory ſake to be noted thus expreſly −315033 −0000 Likewiſe if by the Tangent aboue expreſſed, 1370505. you ſhall ſeeke the Logarithme of the number 137 by the arch of that Tangent 53 degr. 53. min. ſhall be found by the 25 Section in the middle columne −315179, the Logarithme of that Tangent 1370505 which becauſe it exceedeth 137 the number giuen by foure places; or figures. Therefore −315179 −0000 ſhall be the Logarithme of the number giuen 137; yet this Logarithme if ſo much leſſe exact by how much 1370505 is more vnlike to the number 1370000, or the 10000 fold of the number giuen. But this error exceedeth not $\frac{505}{10000}$. Laſtly, if you ſhall seeke the Logagarithme of the number giuen 137 by the Sine aboue written 137156. that ſhall bee found to bee 1986633 – 000 by this & the 11 Section of the third chapter. In like manner you ſhall work by the ſigne + when the number of the figures of the quantitie giuen, exceedeth the number of the figures of the ſine that is likeſt thereto, which ſeldome happeneth. As if the Logarithme of the number (or diſcreet quantitie) 232702 bee ſought for, you ſhall finde in the Table, the ſine 2327 moſt like thereto; but it wanteth two figurs. Therfore to the Logarithme hereof found in the Table (by the 11 Sect. chap. 3) which is, 6063128. let be added two cyphers, the ſign: + being put betweene, and it ſhall be made 6063128 +00 for the Logarithme of the number 232702. which was ſought for.

But the beſt way of eſtimating Logarithmes, is that whereby they were *firſt* made, wherof we ſhall ſpeake in another place.

3 *Therfore as in the firſt Section going afore, ſimple and pure Logarithmes are giuen: ſo in this Section next going before by putting cyphers to them, they become impure.*

4 *To adde Logarithmes of like ſignes, is to giue the ſumme of them both, with their ſigne common to them both.*

As by the Addition of −56312 to −73495. there ſhall come forth −129807. Alſo 4216 being added to +5392, there comes forth 9608. So 3219 −00 added to 4360 −000 make 7579 – 00000.

5 *To adde the Logarithmes of vnlike ſignes, is to giue the difference of them with the ſigne of the greater number.*

As of the Addition of −210 to 332 is prodnced +122.

Alſo of the addition of −210 to 192, comes forth −18.

So −210 + 000 added to 332 – 00 are 122 + 0.

Alſo −210 − 000 added to 192 + 00, are −18 − 0.

Of two Logarithmes this is properly ſaid to bee the Defectiue of that, and that the Abounding of this: when they haue both number and cyphers common, or the ſame, and all the ſignes + and − altogether contrary. 6

As of the Abounding Logarithme 56312, the defectiue is −56312. Alſo of the Abounding Logarithme 56312 − 00 the Defectiue is −56312 + 00. So of the Abounding Logarithme 56312 + 00, the Defectiue is, 56312 − 00.

To ſubtract an abounding Logarithme, is to adde his defectiue. 7

As to ſubtract the abounding Logarithme 56312 out of −73495, is the ſame as to adde his defectiue which (by the ſixith Section.) is −56312 to the ſame −73495, and they ſhall bee made (by the 4 going before) −129807. So to ſubtract 56312+00 out of −73495−000 is the ſame as to adde −56312 + 0 to 73495−000, and they are made (by the 4 and 5 Sect. going before) −129807−00000.

To ſubtract a defectiue is to adde his aboundant. 8

As to ſubtract a defectiue −4216 out of +5392, is the ſame that it is to adde 4216 to 5392. and (by the fourth Section) to bring forth 9608. So it is the ſame to ſubſtract −4216 + 00 out of 5392 + 0, that it is to adde 4216 − 00 to 5392 + 0. and to bring forth 9608 − 0.

To increaſe or diminiſh a Logarithme in number, his former value remaining, is to adde to it, or ſubtract from it, any of the Logarithmes following, as 2302584 + 0, *or* 4605168 + 00, *or* 6907753 + 000, *or* 9210337 + 0000, *or* 11512921 + 00000, *ſinifying nothing at all.* 9

As let the Logarithme bee 3916 − 0 whereto if you adde any of them, as for example ſake, 2302584 + 0, there ſhall bee made thereof 2306500 greater in number, but in value altogether the ſame that 3916 − 0 is: for the quantity or numerall value of this Logarithme 3916 − 0 (by the 12 and 13 Sections following of this Chapter) is 996092, from which take onely the laſt figure, as −0, ſignifieth, and it ſhall be made 99609. And the numeral value of that Logarithme 2306500 (by the 12 and 13 Sections following of this chapter) is alſo 99609 the ſame that was before.

An example of diminiſhing.

Let the Logarithme 2545177 bee to be diminiſhed, from which if you ſubtract 2302584+0, there is left 242593−0 of the ſame value that this former 2545177 was. For the value of the ſimple and pure Logarithme 242593 is ten fold the value of either of them. Their values therefore are equall each to other. For the addition of the Logarithme 2302584 + 0, ſignifieth nothing elſe, but that the value of the number whereto it is added, is to be diuided into ten parts, and that one cypher is to bee added to this tenth part: but the ſubtraction of the ſame ſignifieth that the value of the Logarithme from whence it is ſubtracted, is made tenne fold more, and that one cypher is caſt away from this ten fold. There remaineth therfore the ſame value in both of them.

So 46051684 + 00 added, ſignifieth that two cyphers are added to the hundreth part of the value: and being ſubtracted, it ſignifieth that two cyphers are caſt away from the hundreth fold, and ſo of the reſt aboue expreſſed.

An Admonition.

But becauſe the addition and ſubtraction of theſe former numbers may ſeeme ſomewhat painfull, I intend (if it ſhall pleaſe God) in a ſecond Edition, to ſet out ſuch Logarithmes *as ſhal make thoſe numbers aboue written to fall vpon decimal numbers such as* 100,000,000, 200,000,000, 300,000,000, &c. *which are eaſie to bee added or abated to or from any other number.*

10 *If therefore you ſhall adde to a Logarithme that is diminiſhed by ſome cyphers, or ſhall ſubtract from a* Logarithme *increaſed by cyphers, any of the* Logarithmes *aboue written that containe ſo many cyphers, there ſhall out of an impure* Logarithme *bee produced, a pure one of the ſame value.*

As in the firſt example going before, let the impure Logarithme 3916 − 0 bee to bee purged from his cypher and ſigne −, adde therefore thereto 2302584 + 0 there ſhall thereof be made, as before, 2306500, the pure Logarithme of his former value. So from the impure Logarithme 6358447 + 00 if you ſubtract 4605168 + 00, (which containeth as many cyphers) there ſhall remaine the pure Logarithme 1753278, and of the ſame value, whereof that former impure Logarithme was.

11 *If to a* Logarithme *that is Defectiue in number, you ſhall adde any of the foreſaid* Logarithmes *of the ninth Section, that is greater in number, there ſhall come forth a* Logarithme *of the ſame value Abounding in number.*

As to the Logarithme −2859527 − 0000 adde any of the numbers of the ninth Section, that is greater in number. As for example, 4605168 + 00, and there ſhall bee made thereof 1745641 − 00 of the ſame value, and Abounding in number.

12 *You may giue the Sines, Tangents, and Secants, or any numerall values whatſoeuer, of the* Logarithmes *that are found in our Table by the* 11. 14. 22. 25. 28. 30. *Section of the* 3 *Chapter, whether they be pure or impure.*

As to the Logarithme of 36 degrees, 40 minutes 515572, in the third columne, anſwereth his ſine: 597159 in the ſecond columne, & to the Defectiue therof −515572 there anſwereth in the Table of Secants, 1674597, the Secant of 53 degrees, 20 minutes.

Alſo to the Differentiall Logarithme 295079 in the fourth columne, anſwereth the Tangent 744472 in his Table, and to his Defectiue −295079 anſwereth 1343233 the Tangent of 53 degrees and 20 minutes. So of the Logarithme 220493 in the fifth columne, the numerall value in the ſixith columne is 802123, that is the Sine of 53 deg. and 20 min. and the numerall value of the Defectiue thereof, that is −220493 is the Secant 1246691, agreeing to 36 degrees and 40 minutes.

An example of impure Logarithmes.

LEt the value of the impure Logarithme 9780 − 0 bee to bee ſought out; to this number, there anſwereth in our Table the Sine 990268, from which take the figure next the right hand (as −0 doth ſhew) & they ſhall be made 99027, the value of the Logarithme 9780−0 which was ſought. So the value of the Logarithme 2545177+00 is 7845900, becauſe that to the pure Logarithme 2545177 there anſwereth in our table the Sine 78459. Alſo of the Logarithme 34914 − 00 found in the fourth columne at 46 degrees, the value ſhall be 10355, becauſe the Tangent of 46 degr. is 1035530. So of the Logarithme −635030 − 00 found in the third columne at 32 degrees, the value is 18871, becauſe the Secant of the complement of 32 degrees, that is of 58 degrees, is 1887080, whoſe two laſt figures next the right hand 80, are to be blotted out for −00 adioyned to the Logarithme.

To eſtimate the numerall values of the Logarithmes *giuen, and not found in our Table.* 13

For common meaſuring, it is ſufficient for the moſt part, to take for the Logarithme giuen, the numeral value of the Logarithme in the Table, that comes neereſt that, which is giuen. But if you deſire to come neerer the marke, increaſe or diminiſhin number the Logarithme giuen, by the 9 Section of this chapter, his former value remaining vntill it be either found in the Table, or become like enough to ſome Logarithme in the Table, and the value of this Logarithme found by the former Section, is that which we ſeek for. As for example, let the value of this Logarithme 2314972 + 0 bee ſought, to which there is none found like or neere enough in the Table; but if you ſubtract from it 2302584 +0, there ſhal be left 12388 almoſt, to which vnder 81 degr. there ſhall be found one that is neere, and like enough to it, that is, 12388, the Sine whereof 987688 found by the former Section, is the value of the Logarithme Propoſed 2314972 + 0 which was ſought for.

An Admonition.

FOr this and the 2 Sect, of the chapter, we would haue you admoniſhed, that the Logarithmes of the numbers giuen, & contrariwiſe the numerall values of the Logarithmes giuen, when they are not found in the Table, are moſt exactly giuen by the way, by which the Logarithmes are made or reſolued, which is that you defcend from the fine giuen by meanes. Geometrically proportionall, vntill you come to the next leffe fine in the Table. Likewife from the Logarithme heereof, in the Table, that you defcend alfo by as many agreeable meanes Arithmeticall: and the laft of thefe fhall be the Logarithme of the firft of them, and contrariwife by refolution that you defcend from the Logarithme giuen by Arithmeticaall meanes to the next leffe Logarithme in the Table, and from the value of this in the Table likewife, that you defcend, by as many meanes Geometricall and agreeable: and the laft of thefe fhall bee the numerall value of the firft of thofe Logarithmes. But what Arithmeticall equalitie of difference agreeth and is fitting to euery continued Geometricall proportion, is a matter of no meane skil to finde out. Wherefo re of thefe (if God will) we fhall intreate hereafter more at large, when we fhall handle the making of Logarithmes.

CHAP. V.

Of the moft ample vfe of the Logarithmes, and ready practife by them.

Probleme

OF the Logarithmes of three proportionals, the middle Logarithme being giuen, and one extreame to finde the other extreame, or his proportionall, or arch by one doubling, or fubtraction onely.

Seeing that by the fecond Propofition, Chap. 2. the double of the middle (Logarithme) made leffe by one of the extreames, is made equall to the other; Therefore from the double of the middle Logarithme giuen, fubtract the giuen Logarithme of the extreame, and there fhall remaine the Logarithme of the extreame that was fought for: which being found in the third, fourth, or fifth columne of the Table, you haue the arch anfwering thereto in the firft and feuenth columne, and the Sine in the fecond or fixith, and their Secants or Tangents in their Tables, by the third Chapter, Section 1. 2. 6. 8. 11. 14. 22. 25. 28. 30. for the extreame that was fought for. *Example.*

LEt the firft proportionall giuen, bee 1000000, and the fecond 707107: let the third be fought for, which commonly is found by multiplying the middle number by it felfe, & diuiding this fquare by the firft But we find it eafilier by doubling the Log: of the middle number 346573, and by fubtracting from this double (w[ch] is 693147) the Logarithme of the firft, which is 0, & there remaineth 693147, the Logarithme fought for, whofe arch you fhall finde to be 30 degrees, and the Sine adioyning thereto 500000, which is the proportionall number fought for. Therefore 1000000. 707107. 500000, are three proportionall numbers, the laſt whereof wee found onely by doubling, and ſubtraction, which wee promiſed. Alſo let there bee two proportionall numbers giuen, the firſt 1056256, & 766045 the ſecond, or at leaſt their Logarithmes −54730, and 266515. The third you ſhall thus finde: From the double of this laſt 533030 ſubtract −54730, and by the 8 Section of the 4 chapter, there is bronght forth 587760, the Logorithme of 33 degrees. 45 minutes, the ſine whereof 555570 is the third proportionall number that was ſought for.

Out of the Logarithmes of three proportionals, the extreame Logarithmes being giuen, to finde the middle Logarithme and his proportionall and arch, by one addition onely, and diuiſion by two. *P*

Seeing by the third Propoſition of the ſecond chapter, the double of the middle Logarithme is equall to the ſumme of the extreames, therefore adde the Log. of the extremes, and diuide the product by 2, & there ſhall come forth the Logarithme of the middle proportionall number: and thereby the middle proportionall, and the arch thereof, is knowne in the columnes, and by the Sections, as before. *As for Example.*

LEt the extremes 1000000 and 500000 bee giuen, and let the meane proportionall be ſought: that commonly is found by multiplying the extreames giuen, one by another, and extracting the ſquare root of the product. But we finde it eaſilier thus; We adde the Logarithmes of the extreames 0 and 693147, the ſumme whereof is 693147 which we diuide by 2 & the quotient 346573 ſhall be the Logar. of the middle proportionall deſired. By which the middle proportionall 707107, and his arch 45 degrees are found as before. Alſo let the extremes giuen bee 1056256 and 555570, their Logarithmes are −54730 and 587760. The ſumme of theſe put together, is 533030 by the 5 Sect. Chap. 4. which we diuide by two, and the quotient is 266515, the Logarithme and his arch 50 degr. and the ſine or meane proportionall ſought for is 766044. found by addition onely, and diuiſion by two.

Out of the Logarithmes of foure proportionals, three being giuen, or their arches, to find the fourth Logarithme with the ſine and arch thereof, by one addition onely, and ſubtraction. *Prob.* 3.

In this probleme wee alwayes make the thing demanded the fourth, ſo that as the firſt of the numbers giuen is to the ſecond, ſo is the third to the number demanded. And ſeeing the ſumme of the Logarithmes of the ſecond and third of the numbers ſo placed, diminiſhed by the Logarithmes of the firſt, is equall to the Logarithme of the fourth, by the 4. Prop. Chap. 2. Therefore adde the Logarithmes of the ſecond and third, and from the ſumme of them take the Logarithme of the firſt, and there ſhall remaine the Logarithme of the fourth proportionall number demanded, and thence the fourth number it ſelfe, and the arch thereof.

For examples ſake.

AS 766044 is to 984808: ſo let 500000 be to the fourth proportionall which wee ſeeke for. This they commonly finde by multiplying the ſecond and third, and diuiding the product by the firſt. But you may find it more eaſily thus: Adde the Logarithme of the ſecond 15309, and of the third 693147, the ſumme whereof ſhall be 708456: out of which ſubtract the Logarithme of the firſt, which is, 266515, and there ſhall remaine 441941, the Logarithme of the fourth, whoſe ſine 642788 is the fourth proportionall deſired, and the arch thereof 40 degrees. The ſame would come forth if (the ſines being neglected) their three arches onely were giuen 50 degrees, 80 degrees, and 30 degrees. For out of the Logarithmes of the arches 80 degrees, and 30 degrees, ſubtract the Logarithme of 50 degrees, there ſhall remaine the Logarithme of 40 degrees: and ſo the arch it ſelfe 40 degrees, ſhall be knowne without the ſines, or their multiplication or diuiſion, according as we promiſed in the beginning.

Another Example.

AS the Tangent of 43 degrees is to the Sine of 57 degrees, ſo let the Tangent of 35 degrees bee to a fourth Sine vnknowne, whoſe arch without regard either of Sines or Tangents, we ſhall thus finde: Wee adde the Differential Logarithme of 35 degrees, that is, 356378 found in the middle columne to the Logarithme of 57 degr. that is 175937 placed in the fifth columne from the product, that is, 532316, wee ſubtract the Differentiall of 43 degrees, which is 69870, and there remaineth 462446, the Logarithme of the fourth (Sine) which being found in the third columne, by the 11 Section of the third chapter, you ſhall finde cloſe by it in the firſt columne 39 degrees 2 minutes almoſt, which is the arch of the fourth proportionall, or Sine neglected.

Thus the arches of proportionall numbers are found without their Sines, Tangents, Secants, or any proportionall numbers whatſoeuer.

Which ſo ſhort a way of working, doth helpe very much for meaſuring the angles of plaine triangles, and for the whole *Trigonometrie* of ſphærical triangles, as in his proper place ſhall appeare.

Prob. 4. *Of foure numbers in continuall proportion, the extremes being giuen, or their arches, to finde any of the middle numbers, or any of their arches, onely by diuiding by three, inſtead of the troubleſome extracting of the cubicke root.*

Seeing that in the Logarithmes of theſe, the triple of any middle one, is equall to the ſumme of the extreme remoued, and the double of the next extreame, by the ſixith Propoſition of the ſecond chapter. Therefore adde the double of either extreame Logarithme to the Logarithme of the extreame remaining, and diuide the product by three, and there ſhall come forth the

Logarithme of the middle proportionall next the former extreame, and after the fame manner, the other meane proportionall alfo.

As for examples fake.

LEt the firft extreame be 402925, and the laft, 1056256, the meane proportionals are fought for, which without extraction of the cubicke roote you fhall thus finde. The Logarithme of the numbers giuen are 909005, and −54730: to the double of that 1818010, adde this, and the fumme fhall bee 1763280, which diuided by three, bringeth forth 587760 the Logarithme, whofe Sine 555570 is the firft meane proportionall fought for. Alfo in like manner to the double of this −54730, which is −109460, adde that 909005, and the product will bee 799545, which diuided by three, bringeth forth 266515 the Logarithme, whofe Sine 766044 is the later meane which was fought for. Thefe therefore are foure continuall proportionals 402925, 555570, 766044 and 1056256.

Another example.

Let the extreames given bee 1414213, and 500000: the firft of thefe being found in the Table of Secants, the Logarithme thereof in our Table is −346573, and the Logarithme of 500000 is 693147 to the double whereof, 1386294 adde −346573, the fumme fhall be 1039720, which diuided by 3, will be 346573 the Logarithme of the meane proportional next the leffe extreme, which is 707107. So to the double of −346573, which is −+693147, add 693147, and there fhall be made thereof nothing, which diuided by 3, maketh alfo 0, the fine and the value whereof is 1000000 for the remaining and greater meane proportionall. Thefe foure therefore are continually proportionall, 1414213. 1000000, 707107. 500000.

The Conclufion.

NOw out of this that is already deliuered, let the learned iudge how great benefit the *Logarithmes* bring them; feeing that by the addition and fubftraction of them, and by diuiding by 2 and 3. and by other eafie additions, or fubtractions, multiplication, diuifion: the extraction of the fquare and cubicke rootes, and all the great toyle of calculating is auoided, a generall tafte whereof we haue giuen in this Booke.

But in the booke following we fhall treate of their proper and particular vfe in that noble kinde of *Geometrie* which is called *Trigonometrie*.

The end of the firft Booke.

THE SECOND BOOKE.

Of the excellent vſe of the admirable Table of Logarithmes in Trigonometrie.

Chap. I.

SEEING that *Geometrie*, is the Art of meaſuring well, and meaſuring belongeth to Magnitudes, and Magnitudes are Figures, (at leaſt in power) and a Figure is either a Triangle, or Triangled, and that which is triangled, is compounded, or made of Triangles: which, and whoſe parts, being meaſured, that figure alſo, and all the parts thereof will bee meaſured. It is therefore certaine, that the Arithmetical ſolution of any Geometricall queſtion, dependeth on the doctrine of Triangles. A Triangle is either right-lined or ſphærical.

Of Right-lined Triangles.

Propoſ. 1. *The three Angles of a right-lined Triangle, are equall to two right Angles.*

Therefore if two angles be giuen, take the ſumme of them out of 180 degrees, and there will come forth the third angle. Alſo one angle being taken out of 180 degr. the ſumme of the other two remaines.

A Right-lined Triangle is either right-angled, or obliquangled. In right angled triangles wee call the ſides that are about the right angle, Leggs; and that which ſubtendeth the right angle wee call the *Hypotenuſe*.

Propoſ. 2. *In a right-angled triangle the Logarithme of the legg, is equall to both the Logarithme of the angle oppoſite thereto, and the Logarithme of the Hypotenuſe.*

Seeing it is manifeſt out of the principles of *Trigonometrie*, that either legg hath the ſame proportion to the ſine of the angle oppoſite therto, that the *Hypotenuſe* hath to the whole ſine: and (by the fifth Propoſition of the ſecond chapter of the 1 book) the Logarithmes of the ſecond and third of theſe foure proportionals are equall to the Logarithmes of the firſt and fourth: and the Logarithme of the fourth is 0, or nothing, by the Corolarie of the ſixith definition of the firſt Chapter, and firſt Booke.

Therefore (as before) the Logarithme of the legge is equall to the fumme of the Logarithme of the angle which it fubtendeth, and the Logarithme of the Hypotenufe.

Therefore any two of the Hypotenufe, the legg, and angle which it fubtendeth, being giuen: the third, and thence all the reft of the parts of a right angled triangle will bee knowne. *A confequent.*

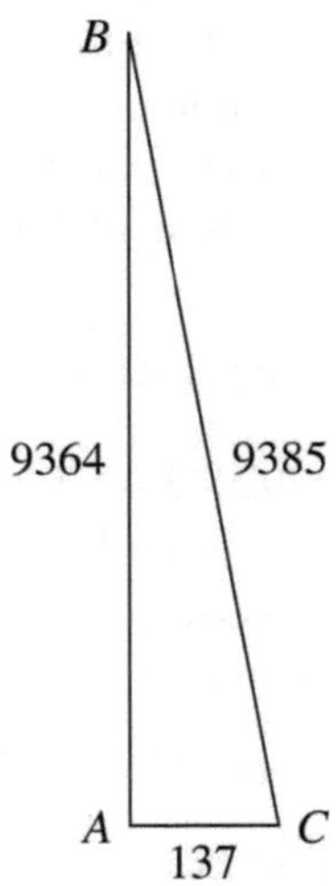

Becaufe thefe three, with the whole fine, make foure prorportionals, it is certaine that any of them may be put in the fourth place, and fo be found out by the 3 *Probl.* of the 5 *Chap.* of the 1 Booke.

As in the triangle *ABC* right angled in *A*.
let the Hypotenufe *BC* be giuen 9385 with the leg *AB* 9384 The oblique angles *C* and *B* are fought for. Therefore out of the Logar. of *AB* 63587 – 00 take the Logar. of *BC* 63480 – 00. There remaines 107. The Logarithme of the Angle *C*, whereto there anfwereth in the Table 89 degrees, $9\frac{3}{4}$ for the angle *C*, and oueragainft it 0 degr. $50\frac{1}{4}$ for the complement therof, namely the Angle *B*.

Contrariwife, if the Angle *C* bee giuen, with the legge of the right angle *AB*, and the Hypotenufe *BC* be fought for.

Out of the Logar. of *AB*.	+63587 – 00
Take the Logar. of the angle *C*	+107
And there will come forth	+63480 – 00

the Logarithme of *BC* 9385 the Hypotenufe that was fought for.

Thirdly, if *BC* and the angle *C* being giuen, and *AB* be fought for,

Adde the Logarithme of *BC*	+63480 – 00
to the Logar. of the angle *C*	+107
And there will be brought forth	63587 – 00

the Logarithme of the number 9384, anfwering to the legge *AB*, which was fought for.

No otherwife is the legge remaining *AC* found by the angle *B*. (which is the complement of the angle *C*) already knowne. And fo all the parts of this right-angled triangle are knowne.

In a right angled triangle the Logarithme of any legge is equall to the fumme of the Differentiall of the oppofite angle, and the Logarithme of the leg remaining. *Propof.* 3.

Seeing it is manifeft out of the common doctrine of Triangles, that either legge hath the fame proportion to the Tangent of the angle oppofite thereto, that the other hath to the whole ſine: and ſeeing that (by the 5 Propoſition of the ſecond chapter of the firſt Booke) of theſe foure proportionals, the Logarithmes of the middle ones, that is to ſay, the Differential of the angle, and the Logarithme of the legge including it, are equall to the Logarithmes of the legge ſubtending the ſame, and of the whole ſine, (which is 0, or nothing) therefore the Logarithme of the legge is equall to the ſumme, &c. as before.

Corolarie. *Therefore of the legges of the right angle, and the angle oppoſite to one of them, any two being giuen, the third is knowne* (*by this Prop.*) *and therfore all the other parts of the right angled triangle by the former Propoſition.*

Becauſe theſe three, with the whole ſine, doe make foure proportionals, it is certaine that euery one of them may be placed in the fourth place, and bee found out by the third *Prob.* of the 5 *Chap.* of the firſt Booke.

As in the triangle going before ABC, right angled at A: the leggs AB being giuen 9384, and AC 137, let the angle B be ſought out.

From the Logarithme AC	$+4292453-00$
Subtract the Logar. of AB,	$+63587-00$
And there will come forth	$+4228866-00$

the Differentiall of the angle B, 0 deg. 50. 11. which was ſought for.

But if the legg AC be giuen 137, and the angle B, 0 degr. $50'$. $11''$, the legg AB, ſhall be thus found.

Out of the *Logarithme* of AC	$+4292453-00$
Subtract the Differential of the Angle B	$+4228866-00$
the number comming thereof	$+63587-00$

is the Logarithme of the number 9384 which is the legg ſought for AB.

Thirdly, the legg AB being giuen 9384, and the angle B, 0 deg. $50'$. $11''$. that the legg AC may be found,

Adde the Logar. of the leg AB	$+63587-00$
to the Differ. of the angle B	$+4228866$
and there will come forth	$+4292453-00$

the Logarithme of 137 the legg AC, which was ſought for.

The Hypotenuſe BC is found by the former Propoſition. Alſo the angle C is knowne, becauſe it is the complement of the angle B, already knowne. And *ſo* by this, and the former Propoſition, by any ſide, and any other part of a right-angled triangle giuen, all the other parts thereof are made knowne.

You haue therfore the knowledge of right-angled right-lined triangles accompliſhed: Now of oblique angled triangles.

Of Right-lined Triangles, efpecially obliquangled.

CHAP. II.

I*N any Triangle: the fumme of the Logarithmes of any angle and fide inclofing the fame, is equall to the fumme of the Logarithmes of the fide, and the angle oppofite to them.* *Propof.* 4

Becaufe, *there is the fame proportion of all the fides to the fines of the oppofite angles*: and fo the product of the right fine of any angle, & any fide including the fame, is equall to the product of the fide fubtending the former angle, and the fine of the angle fubtended by the firft fide. Therefore by the 5. *Prop.* 2. *Chap.* 1. *Booke.* the fumme of the Logarithme, &c. is equall, as before.

Therefore, of two angles whatfoeuer of a kinde giuen, and their fubtendants: if three be giuen, any fourth will be knowne, and thence all the other parts of the triangle. *Corolarie.*

For of thefe foure proportionals, any that is fought for, may be put in the fourth place, & be found out by the third *Prob. Chap.* 5. *Book* 1.

As of the obliquangled Triangle ABC, let AB be giuen 26302, and BC 57955, and the angle C 26 degrees: and let the angle A be fought for, which is thus found.

Adde the Logarith. of BC	$+545471-0$
To the Logar. of C. 26. deg.	$+824689$
And there will bee made	$+1370160-0$
From thence take the Logarithme of AB,	$+1335492-0$
There remaines	$+34668-0$

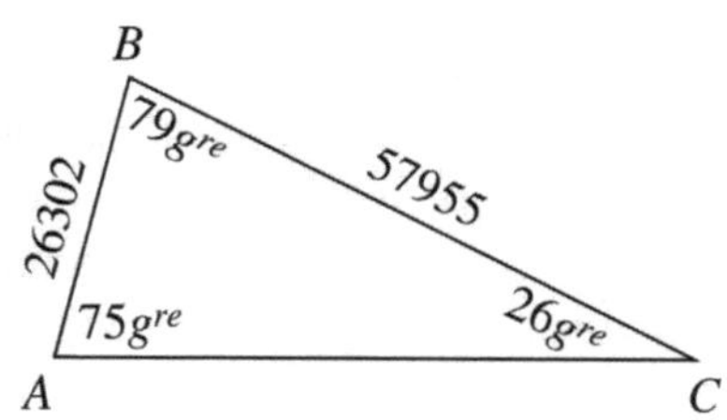

the Logarithme of 75 degrees, and a little more wch is the angle A fought for: if A appeare to be an acute angle, otherwife 105 deg. (by the 1 and 2 Sect. chap. 3. book. 1.) if it appeare to be an obtufe angle.

Contrariwife, if the angle A bee giuen 75 degr. and the angle C, and the fide BC as before, and AB be fought for.

Adde the Logarithme of *BC*	+545471 −0
to the Logar. of the angle *C*	+824689
they will be made as afore	+1370160 −0
From which take the Logar. of the angle *A*	+34668
There will come forth	+1335492 −0

the Logarithme of the ſide *AB*, and the number thereof 26302, which was ſought for.

The angles *A* 75 degr. and *C* 26 deg. being now found, the angle *B* ſhal be 79 deg (by the 1. *Prop.* of this book:) out of which being now found, the ſide oppoſite thereto *AC* 58892 is no otherwiſe found then the ſide oppoſite thereto (*AB*) was lately found by the angle *C*. Therfore now all the parts of this obliqueangled triangle are knowne.

In the obliquangled triangles, we call them legs which are about any angle, & the baſe which ſubtendeth the ſame.

Prop *In obliquangled triangles, the Logarithme of the ſumme of the legges, ſubtracted from the ſumme made of the Logarithme of the difference of the legs, and the Differentiall of halfe the ſumme of his oppoſite angles, leaueth the Differentiall of halfe the difference of the ſame.*

Becauſe *as the ſumme of the legges is to the difference of the legges; ſo is the Tangent of halfe the ſumme of their oppoſite angles to the Tangent of halfe the difference of the ſame*: Therfore they are proportionall, and by the 1 *Prop.* 2. *Chap.* 1 *Book.* the differences, or exceſſes of their Logarithmes are equall. Therfore (by the 4. *Prop.* 2. *Chap.* 1. *book*) we muſt neceſſarily conclude as before.

Therefore by two legs, and the angle contained betweene them, are knowne by this Propoſition, the other oppoſite angles, and thereby the other ſide, by the Propoſition going before.

For the Logarithme of the ſumme of the leggs being ſubducted out of the ſumme made of the Logarithme of the difference of the leggs, and the Differentiall of halfe the ſumme of the oppoſite angles put together, there ſhall come forth the Differentiall of halfe the difference of the ſame angles; which halfe difference being added to the halfe ſumme aforeſaid, there ſhall come forth the greater angle; and being ſubtracted, the leſſe.

As in the foreſaid *Obliquangled* triangle *ABC*

Let there be giuen *AB* one legg		26302
	BC th'other leg	57955
	B the angle contained	

betweene them, 79 degrees, and let the other angles *A* and *C* be ſought for.

The ſumme of the legges *AB*, and *BC* is 84257, the Logar. thereof 2473882, and the difference of the ſame *AB* and *BC* is 31653 the Logar. thereof 3452921. And ſeeing the angle *B* is giuen 79 deg. (by the firſt of this book) the ſumme of the angles A and C will be 101 degr. and halfe the ſumme 50 deg. 30′.

the Differentiall wherof is	−193177
Which being added to the Logarithme of the difference of AB, and BC	+3452921
There wil ariſe	+3259744
Out of which ſubtract the Logarithme of the ſumme of the legs AB and BC	+2473882
There will remaine	+785862

the Differential of 24 deg. 30′, which are the halfe-difference, of the angles A and C that were ſought for.

Therfore adde this half-differ.	24 deg. 30′.
to the halfe ſumme	50 deg. 30′.
And they will make	75 deg. 0′.

for the angle A, the greater of the angles ſought for.

And out of the	50 deg. 30.
Subtract the ſame	24 deg. 30.
And there will remaine	26 deg. 0.

for the angle C, the leſſe of the angles ſought for.

In obliquangled triangles, *the true baſe is alwayes, either the ſum of the caſes* (& *then the difference of the caſes is called the Alterne baſe*:) *or the true baſe is the difference of the caſes*, & *then we call the ſumme of the caſes, the Alterne baſe.* *A Definitiõ*

As in the Triangle ABC.

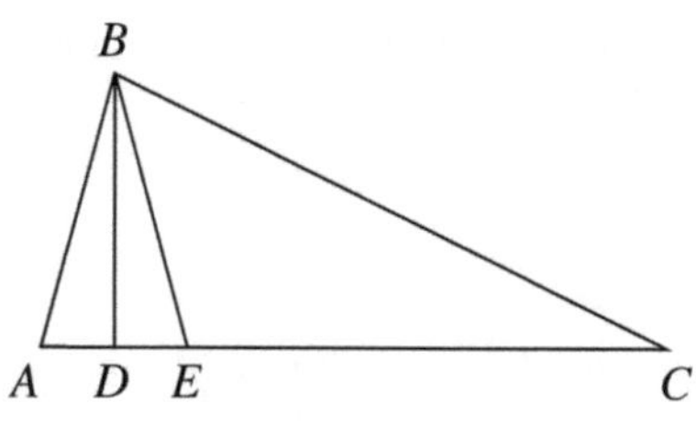

The leſſe caſe is AD
The greater caſe is DC
The ſum of the caſes AC, is the true baſe

And in this triangle take the leſſe caſe AD,
or that which is equall to it DE,
from the greater caſe DC,
and there will remaine EC,
the difference of the caſes which we call the *Alterne baſe*.

Contrariwiſe, in the triangle EBC.

The leſſe caſe is *DE* (Whereto *AD* is equall)
The greater caſe is *DC*
The difference of the caſes *EC* is the true baſe

But the ſum of the caſes, that is, *AC*, wee call the *Alterne baſe.*

Propoſ.6. *In obliquangled triangles, the ſumme of the Logarithmes of the ſumme, and difference of the legs is equall to the ſumme of the Logarithmes of the true and Alterne baſe.*

Becauſe the true baſe is to the ſumme of the leggs, as the difference of the leggs is to the *Alterne baſe*; therfore (by the 5. *Prop.* 2. *Chap.* 1. *Book.*) we neceſſarily conclude, that the Logarithmes of the baſes are equall to the Logarithmes of the ſumme, and difference of the leggs, as before.

A Corolar *Therfore, of an obliquangled triangle, whoſe ſides are giuẽ,* 2 *right angled triangles are made, whoſe Hypotenuſes are knowne with one of the legges of either of them, which (by the ſecond of this book) makes knowne all the other parts of the obliquangled triangle.*

For the Logarithme of the ſumme of the leggs being added to the Logarithme of the difference of the leggs, and the Logarithme of the true baſe being taken from hence, there will come forth the Logarithme of the Alterne baſe (by the 4. *Prop*. of the 2 *Chap*. and 3 *Problem* of the 5 *Chap*-of the 1 *Book.*) Therfore halfe the ſumme of theſe baſes is the greater caſe, and the halfe-difference is the leſſe caſe.

As in the former triangle *ABC*,
Let the ſides be giuen, that is to ſay,
The legge *AB* 26302
The legge *BC* 57955
and the true baſe *AC* 58892 and let the reſt bee ſought for.

The ſumme of the leggs is, 84257	
the Logarithme therof is	+2473882
The diffrence of the legs is, 31653	
and the Logar: thereof is,	+3452921
Adde theſe Logarithmes together, and they will make,	+5926803
From wch take the Log: of the true baſe *AC*	+2831930
There remaines	+3094873
the Logarithme of the number of the Alterne baſe *EC*	45286
which adde to the number of the true baſe *AC*.	58892
And there remaines	104178
The halfe whereof *DC* is the greater caſe.	52089
Subtract one out of another, viz.	
Out of the true baſe *AC*	58892
Take the Alterne baſe *EC*	45286

And there remaines –13606
The halfe whereof AD. is the leſſe caſe. 6803

Therefore of the right-angled triangle ADB, the Hypotenuſe AB, and one of the legs AD being found: and of the right angled triangle BDC the Hypotenuſe BC, and the leg DC being found (by the ſecond of this chapter) the angles of the right angled triangle at A, and B, and C, are known, and by by conſequent, alſo all the parts of the obliquangled triangle propoſed, are manifeſted by the premiſes. Neither ſhould you doe otherwiſe if the ſides of the triangle EBC, were giuen, and the other parts were ſought. For out of the legges, and the true baſe EC, the Alterne baſe AC is knowne, and out of theſe both caſes, and the reſt, as before.

The Concluſion.

NOW therefore, you haue the doctrine of all right-lined triangles perfected & accompliſhed, which if it ſeeme ſomewhat toyleſome in finding out the Logarithmes of variable right-lines; yet in calculating the motions of the planets, (in which the excentricities of the Orbs, the diſtances of the Auges & Apogæs the diameters of the Epicycles and other right lines, remaine the ſame, and invariable) their Logarithmes being once exactly ſet downe, ſhall alwayes ſerue after-wardes without any changing, with maruailous facilitie and certaintie.

Now, there followes the *Sphæricall* triangles, which are moſt hard, as they are commonly deliuered by others; but by our Logarithmes they are the moſt eaſie of all.

Of Sphæricall Triangles
CHAP. III.

IN Sphæricall triangles, the angle that is neereſt in quantitie to a quadrant, and the ſides ſubtending the ſame, are doubtful whether they be of the ſame, or of a diuers kind, except the account, or poſition bewray the ſame. Sentences

But euery one of the two oblique angles, is of the ſame kinde with the ſides ſubtending the ſame. Therefore knowing of what kinde the one is, it appeareth alſo of what kinde the other is. 2

If any angle of a triangle bee neerer to a quadrant then the ſide ſubtending the ſame: two ſides thereof ſhall be of one kinde, and the third leſſe then a quadrant. 3

But if any ſide of a triangle be neerer to a quadrant then the angle ſubtended thereby, two angles thereof ſhall bee of the ſame kinde, and the third greater then a quadrant. 4

5 *A Sphæricall triangle, is either quadrantall or not.*

6 *A quadrantal, is that whoſe ſide or angle is equall to a quadrant: whereby we teach, that the knowledge of a quadrantal that is not right angled may as eaſily be gotten, as if it were right angled.*

7 *A quandrantall triangle, is either manifold, or ſingle.*

8 *A manifold quadrantall, is either three right angled, or two right angled.*

9 *A three right angled triangle, is that, whereof euery part is equall to a quadrant.*

10 *Therefore euery triangle, each of whoſe three parts not being oppoſite, are equall to a quadrant, is three right angled.*

11 *A two right angled triangle, is that wherof two angles onely, and the ſides ſubtending them, are ſeuerally equall to a quadrant.*

12 *In euery two right angled triangle, the oblique angle, is equall to his ſubtending ſide.*

13 *Euery triangle, whereof any part is equall to a quadrant, and any oblique angle, equall to his ſubtendant, is two right angled.*

14 *Euery triangle hauing any two parts ſeuerally equall to a quadrant, and the third unequal, is two right angled.*

15 *All the reſt are called ſingle quadrantals.*

Of Single Quadrantals.
CHAP. IV.

1 *A Single Quadrantall, is that whereof one part onely is equal to a quadrant, and the other fiue parts are not quadrants.*

2 *Of theſe fiue parts which are not quadrants, thoſe three which are furtheſt remoued from the right angle, or the ſide that is a quadrant, we turne into their complements, and retaining the old order, we bring them all fiue into a circular, or quinquangled ſituation, wee call them Circulars.*

Firſt let the triangle BPS be right angled in B, the fiue oblique parts therof which are not quadrãts are theſe, BP one of the ſides about the right angle: P one of the oblique angles: PS the ſide ſubtending the right angle: S the other oblique angle: SB the other ſide about the right angle, for which we (for the eaſier calculation) take the ſide it ſelfe BP: the complement of the angle P: the complement of the ſide PS: the complement of the angle S: and the ſide it

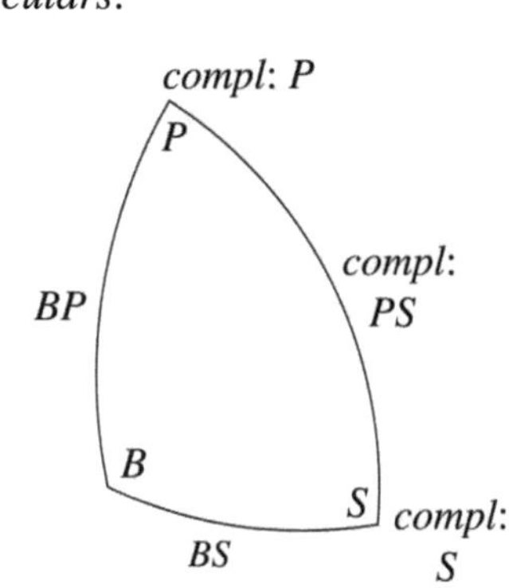

ſelfe SB, and keeping their naturall ſituation, we place theſe fiue parts in order, as in the margine, and we call them Circulars.

Likewiſe ſecondly, let SPZ be a quadrantall ſingle triangle, not right angled (made of the centers of the Sunneriſing, the pole and the zenith) quadrantall in the ſide ZS, his fiue parts not quadrants are theſe, Z one of the angles compaſſed by the quadrant ſide: PZ the diſtance of the pole from the zenith: P the angle ſubtended by the quadrant: the ſide PS the diſtance of the pole from the Sunne: And laſtly S one of the angles, about which the quadrant is: inſtead wherof we for our eaſier account do take the angle Z or PZS, being adioyned to a quadrant, and is the arch of the Sunnes diſtance from the North. The complement of PZ, which is the eleuation of the pole: the complement of the angle P, or of the angle ZPZ which is the difference aſcentional (that is) the difference of the time of the Sunnes riſing or ſetting from ſixe a clocke.

The complement of the ſide PS, which is the declination of the Sunne: and the angle it ſelfe S, or PSZ, which wee call the angle of the Sunnes poſition (that is in reſpect of the Pole and Zenith.) Theſe fiue parts we place in a circular or pentagonall ſite, as in the margine, and we call them Circulars, neither ſhall there be made any other circular parts of the former right angled triangle BPS, if you make P the Pole: S the Sunne: B the Northpoint: for the ſide BP will be the eleuation of the pole: the cõplement of P the diffrence aſcenſionall: the complement of PS, the declination of the Sun: the complement of S the angle of the Sunnes poſition: and laſtly, BS the Azimuth of the Sunne, which are altogether the ſame circular parts that were before, and placed in the ſame ſite towards the left hand that the other was towards the right.

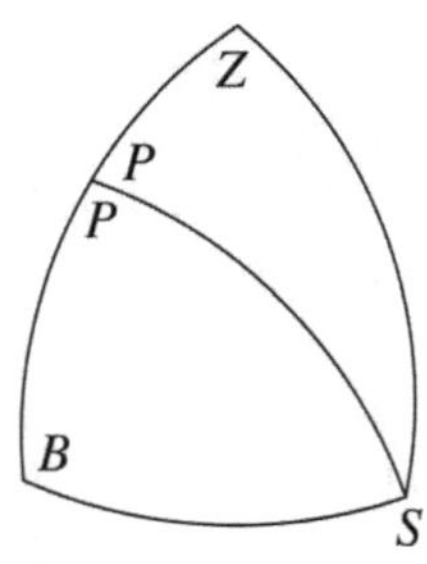

And ſo in all quadrantals, as well right-angled, as not.

Hence it is that there bee many triangles, not conformable in their naturall parts, which in theſe Circular parts, doe altogether agree, and are reſolued by this our methode of Circulars. *Corolarie.*

As it clearely appeareth in the two former triangles BPS, and PZS ioyned together: In which all the naturall parts (beſides PS and BS of the former, and PS, and PZS of the latter, do altogether differ, but all the Circular parts agree, as is aforeſaid.

4 *This vniformitie of the Circular parts, moſt manifeſtly appeareth in right-angled triangles made on the ſuperficies of a globe, of fiue great circles, the firſt whereof cutteth the ſecond, the ſecond the third, the third the fourth, the fourth the fifth: and laſtly, the fifth the firſt, at right angles. But all the other Sections ſhall bee made at oblique angles.*

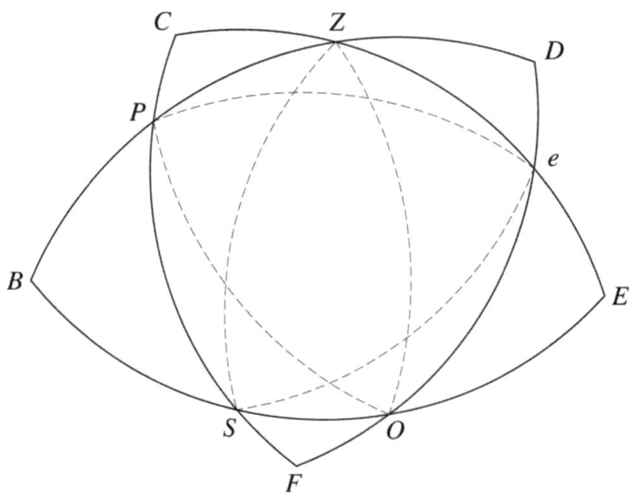

For example ſake: the meridian of any region DB cuts the Horizon BE in the poynt B. The Horizon BE cuts the circle EC, which compaſſeth about the Sunne (that is to ſay, which is drawne about the Sun as his pole) in the poynt E. The circle EC, which compaſſeth the Sunne, cuts the meridian of the Sun CF in the poynt C. The meridian of the Sun CF cuts the Equator FD in the poynt F. And laſtly, the Equator FD cuts the meridian (of that region DB) in the poynt D. And all theſe fiue Sections in the poynts B, E, C, F, D, are made orthogonally and at right angles. The other Sections in the points Z. P. S. O. Q being made at oblique angles. There will alſo bee made of theſe Sections fiue right-angled triangles, PBS, SFO, OEQ, QDZ, and ZCP, the naturall parts whereof, although they differ, and are varied in each ſeueral triangle, yet the fiue Circular parts are the ſame that were before without any difference.

5 *The ſame vniformitie of the circular parts appeareth alſo in quadrantals that bee not right angled, made vpon the superficies of a Globe out of fiue poynts, the firſt whereof is diſtant from the ſecond, the ſecond from the third, the third from the fourth, the fourth from the fifth, and the fifth from the firſt by diſtances and arches equall to a quadrant; but the other diſtances of the poynts bee vnequall to a quadrant.* As in the ſame figure the poynts P from Q, Q from S, S from Z, Z from O, and O from P, are diſtant by ſpaces equall to

a quadrant. But P from Z, Z from Q, Q from O, O from S, and S from P, are diſtant each from other by arches which are not quadrants. There will alſo bee made out of theſe diſtances, fiue quadrantals not right angled, PZQ, ZQO, QOS, OSP, and SPZ, whereof although the naturall parts differ, yet the circular parts remaine the ſame vnchangeable here as before, that is to ſay: BP the eleuation of the pole: the complement of BPS, or SPZ, the difference aſcentionall: the complement of PS, which is SF, the declination of the Sunne: the complement of PSB, which is PSZ, the angle of the Sunnes poſition: BS the Azimuth of the Sunne, which doe indifferently agree to all the former triangles, and not to thoſe alone, but alſo to all triangles which do ariſe of the other interſections of theſe ten arches drawn forth, to whole circles, which becauſe they are many, and confuſed, we here let them paſſe, it is ſufficient to haue warned by this abridgement, that all the confuſion of the naturall parts, and of their rules is auoyded, and taken away by theſe few Circular parts, and their onely rule.

Of the fiue circular parts three alwayes come in queſtion: whereof the two firſt, are giuen, the third is ſought for. 6

And of theſe three, one is in the middle, and two are the extreames which are either ſet about the middle, or oppoſite to it. 7

For example ſake. Let the three parts Propoſed in the queſtion be theſe: the Azimuth of the Sunne BS: the eleuation of the Pole BP: and the aſcenſionall difference the complement of BPS, whereof the eleuation of the pole is called the middle, and the other two extreames are called neighbours vnto it, or ſet about it. But if the three parts coming in queſtion were: the declination of the Sun: the complement of PS: the eleuation of the Pole BP: and the angle of the Sunnes poſition PSZ, the eleuation of the pole ſhal be called the middle one, as before, but the declination of the Sunne, and the angle of the Sunnes poſition, ſhall bee called the extreames, remoued from the middle, or oppoſite to it. The like reaſon is in the other fiue.

The Logarithme of the middle one is equall to the Differentials of the extremes ſet about it, or to the Antilogarithmes of the oppoſite extreames. 8

This Theorem is proued by induction of all the three parts or triplicities which can be made, and come into queſtion of the fiue circular parts of the former right-angled quadrantall BPS. But wee omit the triplicities of the latter triangle not right-angled PZS. becauſe all the circular parts thereof are altogether the ſame in quantitie which were in the former (by the 3. 4. and 5. of this chapter.) Now therefore of the fiue circular parts of the right-angled triangle BPS, (which are BS, or the Azimuth of the Sunne riſing; the complement of BSP, or the angle of the Sunnes poſition: the complement of SP, or the declination of the Sunne: the complement of SPB, or the difference aſcenſinall: and PB, or the eleuation of the pole) The 3 which come in queſtion of extremes, ſet about the middle one, are either firſt BS, the complement of BSP, and the complement of SP: or ſecondly the complement of BSP, the complement of SP, and the complement of SPB: or thirdly, the complement of SP, the complement of SPB, and PB: or fourthly, the complement of SPB, PB, and BS: or fiftly, PB, BS, and the complement of BSP.

But becauſe in all theſe triplicities, *the Tangent of one of the extreames is to the right ſine of the middle one, as the whole ſine is to Tangent of the other extreame* (as it is manifeſt out of the common demonſtrations of *Trigonometrie*) therefore by our demonſtrations of the 5. *Prop.* of the 2. *Chap.* 1. *Book. the Logarithmes of the middle ones* (which are the Logarithme of the middle one onely, by the Corollarie of the ſixt definition of the firſt cha. 1. Book) *are equall to the Tangents of both the extremes.* But *the Logarithmes of the Tangents of theſe extremes are the Differentials of the ſame* (by the 22 and 25 *Sect. Chap.* 3. *Booke* 1.) Therefore *the Logarithme of the middle one only, is equall to the Differentials of the extremes ſet about it*, as we ſaid in the former part of the *Theorem.* The confirmation of the latter part followeth.

Therefore of the ſame fiue circicular parts the three which come into queſtion of the extreames oppoſite to the middle one, are either, firſt PB the complement of BSP, and the complement of SPB: or ſecondly, BS the complement of SP and PB: or thirdly, the complement of BSP, the complement of SPB and BS: or fourthly, the complement of SP, PB, and the complement of BSP: or fiftly, and laſtly the complement of SPB, BS and the complement of SP.

But in all theſe triplicities, or fiue caſes, *the right ſine of the complement of one of the extreames is to the right ſine of the middle one, as the whole ſine is to the right ſine of the complement of the other extreame* (which is more largely demonſtrated by *Regiomontanus*, *Copernicus*, *Lansbergius*, *Pitiſcus*, and others: then that it can be repeated in this abridgement) therfore by our demonſtrations (the 5. *Prop.* 2. *Cha.* 1. *Book the Logarithmes of the complements of theſe extreames, are equall to the Logarithmes of the middle ones*, that is (as is aforeſaid) to the Logarithme of the middle one onely.

But *the Logarithmes of the complements of theſe oppoſite extreames, are the Antilogarithmes of the very ſame parts*, out of the definition, *Sect.* 13. and 16. *Chap.* 3. *Book* 1. Therfore it followeth in theſe caſes, that *the Logaaithme of the middle one only is equal to the Antilogarithmes of his oppoſite extremes*, as the latter part of the Theorẽ affirmeth. Therfore *the whole Theorem is manifeſt.* Beſide this proofe now made by induction of all the caſes which can happen, the ſame Theorem may bee alſo clearely perceiued by the 4 and 5 of this chapter, in the figure whereof, the like conſtitution of the circular parts doth argue the ſimilitude of the analogie of the ſame. So that whatſoever may be truly ſaid of any middle one, and his extreames ſet about, or oppoſite, the ſame cannot iuſtly be denied of the other foure middle ones and their extreames reſpectiuely ſet about, or oppoſed.

A generall Conſequent.

H*Ence it followeth in ſingle Quadrantals, that* out of any two parts giuen, 9
any third ſhall be found. For *alwayes* either the middle one is ſought for, & his Logarithme is found by adding the Differentials of the giuen extreames ſet about, *or* one of the extremes is ſought for & his Differential ariſeth out of the ſubtraction of the Differentiall of the other extreme giuen out of the Logarithme of the middle one already knowne, as in the fiue former triplicities of a right-angled triangle of the Theorem going before, and as many of a not-right-angled triangle: *or elſe* the middle one is ſought for, and his Logarithme commeth forth by adding the Antilogarithmes of the oppoſite extreames giuen. *Or laſtly*, one of the oppoſite extreames is ſought for, and his Antilogarithme is found by ſubtraction of the Antilogarithme of the other oppoſite extreme already giuen out of the Logarithme of the middle one already knowne. *As in the fiue latter caſes of the right angled triangle of the Theorem going before, and as many of a not-right-angled triangle. But to euery one of theſe Logarithmes, Antilogarithmes and Differentials already found, there anſwere two arches of diuers kindes. Therefore out of the kinde of the arch ſought for, knowne* by the 2. 3. or 4. of this chapter, or elſe by poſition, *the true arch it ſelfe ſhall be made knowne.*

As in the former example of the ſeuenth Section of this Chapter, three parts of the queſtion are Circular, *The Azimuth of the Sunne, the Eleuation of the Pole, the Difference Aſcenſionall*: that is, in the right-angled triangle *BPS*, the parts are *BS* and *PB*, and the complement of *SPB*: or elſe in the not-right-angled triangle quadrantall *PZS*, the parts are *PZS*, the complement of *PZ*, and the complement of *SPZ*, of which three let the extreames ſet about be giuen, that is, *The Azimuth of the Sunne riſing BS, or PZS*, 70 *degrees*, and *the difference Aſcenſionall the complement of SPB, or the complement of SPZ*, 16 deg. $24'\ 27''$ and the middle part *PB* be ſought, or the complement of *PZ*, which is *the Eleuation of the Pole*.

Let the differentiall therefore of the complement of *SPZ* 16 degr. $24'\ 27''$

	+1222618
Bee added to the Differentiall of *BS*, or *BZS* 70 degr.	−1010683
And there will come forth	+211935

the Logarithme of *BP* 54 deg. for the eleuation of the pole ſought for.

An Admonition.

BEſides the *Eleuation of the Pole* thus now found, there is alſo found by the ſame maner of working.

2 *The Azimuth of the Sunne* by the eleuation of the pole, and the angle of the Sunnes pofition giuen.

3 *The Angle of the Suns Pofition* out of the Azimuth of the Sun, and his declination giuen.

4 *The declination of the Sunne* out of the angle of the Sunnes pofition, and the difference Afcenfionall giuen.

5 *The Difference Afcenfionall* out of the Declination of the Sunne, and Eleuation of the pole giuen.

The fecond Example.

LEt the *Azimuth of the Sun rifing* bee giuen *BS*, or *PZS* 70 degr. and the *Eleuation of the Pole*, 54. *degr.* which is *PB*, or the compleplement of *PZ*: and let the *Difference Afcenfional* be fought, that is, the cõplement of *SPB*, or the complement of *SPZ*. And becaufe here likewife the extreame parts are fet about the middle part, therefore

Take the Differentiall of the Suns Azimuth, *BS*, or *BZS* 70 deg.

which is	−1010683
Out of the Log. of the eleuation of the Pole *BP*, 54. deg.	+1211935
And there will come forth	+1222618

the Differentiall of *SPB* 16 deg. 24′ 27″ the arch of the *Afcenfionall difference* fought for.

An Admonition.

IN imitation of this example there is found

2 *The Declination of the Sunne* out of the difference Afcenfionall, and the Eleuation of the pole giuen.

3 *The Angle of the Sunnes pofition* out of the declination of the Sunne, and Difference Afcenfionall giuen.

4 *The Azimuth of the Sunne* out of the angle of the Sunnes pofition, and the declination of the fame giuen.

5 *The Eleuation of the Pole* is had out of the Sunnes Azimuth, and the angle of the Suns pofition giuen

Alfo contrariwife there is found

6 *The Difference Afcenfionall* out of the declination of the Sunne, and the angle of the Sunnes pofition giuen.

7 *The declination of the Sunne* out of the angle of the Sunnes pofition, and his Azimuth giuen.

8 *The Angle of the Sunnes Pofition* is had out of the Azimuth of the Sunne, and the Eleuation of the pole giuen.

9 *The Azimuth of the Sunne* out of the Eleuation of the pole, and the Difference Afcenfionall giuen.
10 And laftly, *the eleuation of the Pole* is found out of the difference Afcenfionall, and the Declination of the Sunne giuen.

The third Example.

IN the latter example of the fame 7th Section of this chapter, thefe three circular parts of the queftion are propounded, *The Declination of the Sunne*, *the Eleuation of the Pole*, and the *Angle of the Sunnes pofition*. Thefe in the right angled triangle *BPS*, are the complements of *PS*, *BP*, and the complement of *BSP*. And in a not-right-angled quadrantall *PZS*, they are the complement of *PS*, *ZP*, and *ZSP*. Of which three let the oppofite extreames be giuen, that is, *the Declination of the Sun*, which is the complement of *PS* 11 deg. $35'$. $51''$. and *the angle of the Sunnes pofition*, which is the complement of *BSP*, or *ZSP* 34 degr. $19'$. $21''$. almoft. And let the middle part *BP* be fought, or the complement of *ZP*, *which is the eleuation of the Pole*. Therefore

Let the Antilog. of the complement of *PS* 11 degr. $35'$. $51''$. which is	+20627
be added to the Antilog of *BS*, 34. degr. $19'$. $21''$.	+191308
and there will come forth	+211935

the Logarithme of *BP* 54 degr. for the *Eleuation of the pole* that was fought for.

An Admonition.

BEfides the *Eleuation of the Pole* now firft found after this manner, you may by the fame practife haue
2 *The Azimuth of the Sunne* out his declination and the difference Afcenfionall giuen.
3 *The angle of the Sunnes pofition* out of the difference Afcenfionall, and the Eleuation of the the pole giuen.
4 *The Declination of the Sunne* out of the Eleuation of the pole, and Azimuth of the Sun giuen.
5 And laftly, you fhall finde the *Difference Afcentional* out of the Azimuth of the Sunne, and the angle of the Sunnes pofition giuen.

The fourth Example.

LEt *the Declination of the Sunne* bee giuen the complement of *SP* 11 degr. $35'$. $51''$. And the *Eleuation of the pole BP*, or the complement of *PZ*

54 degrees. And let *the angle of the Sunnes poſition* the complement of BSP, or PSZ be ſought for. And here likewiſe, becauſe the extreame parts are oppoſed to the middle, therefore

Out of the Log. of BP 54 deg.	$+211935$
take the Antilog. of the complement of PS 11. deg. $35'$. $51''$.	$+20627$
and there will remaine	$+191308$

the Antilogarithme of the complement of BSP. 34. deg. $19'$ $21''$. almoſt, the angle of the *Poſition of the Sunne* ſought for.

An Admonition.

BEſides the angle of the Sunnes poſition found out by this firſt practiſe, there is found

2 By the ſame practiſe *the Declination of the Sunne*, out of the difference Aſcenſionall, and the Azimuth of the Sunne giuen.

3 The *Difference Aſcenſionall* is found out of the Eleuation of the pole, and angle of the Sunnes poſition giuen.

4 *The Eleuation of the pole* is found by the Sunnes Azimuth, and his declination giuen.

5 *The Azimuth of the Sunne* is found, out of the angle of the Sunnes poſition, and the difference Aſcenſionall.

6 In a contrary order, *The Angle of the Suns Poſition* is found by the Azimuth of the Sun, and the difference Aſcenſionall giuen.

7 *The declination of the Sunne* is had out of the angle of the Sunnes poſition, and the eleuation of the pole giuen.

8 *The Difference Aſcenſionall* is found out of the Sunnes declination, and Azimuth giuen.

9 *The Eleuation of the Pole* is had out of the Difference Aſcenſionall, and the angle of the Suns poſition being giuen.

10 And laſtly, *The Azimuth of the Sunne* is found by the Eleuation of the pole, and the Suns declination giuen.

And ſo in imitation of theſe foure examples, thirtie ſeuerall queſtions of Circular parts in a right-angled quadrantall, and as many in a not right-angled quadrantall, are reſolued by this general Conſequent, by the benefit of one Addition or Subtraction onely. But for the vnderſtanding of the latter part of this Conſequent, of the kindes of arches, ſee the 3. 4. 5 and 6 Examples of the Chapter following.

Of Not-Quadrantals mixt.
CHAP. V.

HItherto hath beene taught the doctrine of quadrantall Sphæricall triangles: there followeth now the doctrine of Sphæricall triangles not quadrantall.*

1 *A not quadrantall is a Sphæricall triangle, whereof neither ſide nor angle is a quadrant.*

2 *A not-quadrantall is reduced to two quadrantals, if from the top either a perpendicular or a quadrant arch be let downe to the baſe (extended as need ſhal be)*

The perpendicular falles within the triangle, if the angles at the baſe bee both of one kinde; but it falles without if they bee of diuers kindes, and 3
contrariwiſe.

The quadrant arch falles without the Triangle if the legs be of one kinde; but within if they bee of diuers kindes, and contrariſe. 4

Out of the ſixe parts of a not quadrantall three giuen only, are ſufficient to get knowledge of the reſt, except of the three giuen, whereof one is oppoſit 5
to the other, the third be neerer to a quadrant, then the other giuen of the ſame kinde: for in this caſe it is required alſo, that the kind of the part which is oppoſite to the third be alſo giuen, that the other parts may be knowne.

Examples of this caſe are the 4 and 6 examples following.

The three parts giuen are either mixt or pure.

They are mixt whereof one is of a diuers kinde from the other two: As when 6
two ſides and any angle are giuen, or two angles with any ſide.

In mingled parts giuen, if from that tearme of the ſide giuen, in whoſe other 7
terme is the angle giuen, a perpendicular or a quadrant arch ſubtending that
angle, fall to the baſe, the not-quadrantall triangle ſhall be reduced to two 8
quadrantals that may be known by the 9 Sect. of the 4 chap of this booke.

And therefore, the parts of a not-quadrantall, becauſe they are all one with the parts, or remainders of theſe parts to a ſemicircle, are eaſily known, the kindes of the parts being yet firſt knowne by the ſecond, third, and fourth Section of the third chapter of this booke, or elſe by poſition.

An Example of two ſides, and the angle betweene them giuen.

AS for vſe and exerciſe ſake, let there be a *Sphærical* triangle not quadrantal deſcribed on the ſuperficies of the *Primum Mobile* PZS repreſenting the pole, the zenith, and the Sunne; whereof there be ſix parts.

The ſide PZ, which is the diſtance of the pole from the zenith, or the complement of the poles eleuation.

The ſide ZS, the diſtance of the zenith and Sunne, or the complement of the Sunnes height.

The ſide PS, the diſtance of the pole and the Sun, or the complement of the declination of the Sunne from the Equator.

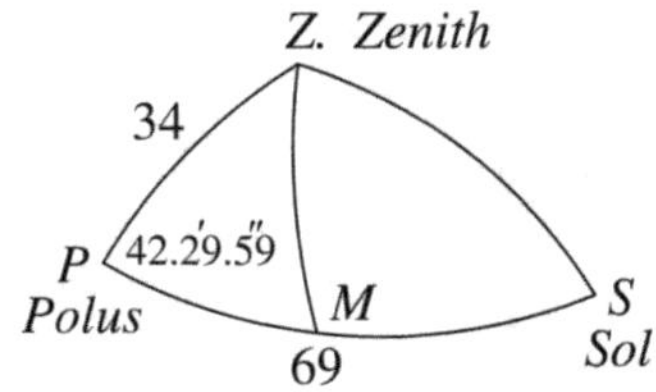

The angle ZPS, the houre of the day, or the degrees of the Equator.

The angle PZS, which is the Azimuth of the Sunne from the North.

The angle PSZ, which is the angle of the ſituation and poſition of the Sunne to the pole and zenith.

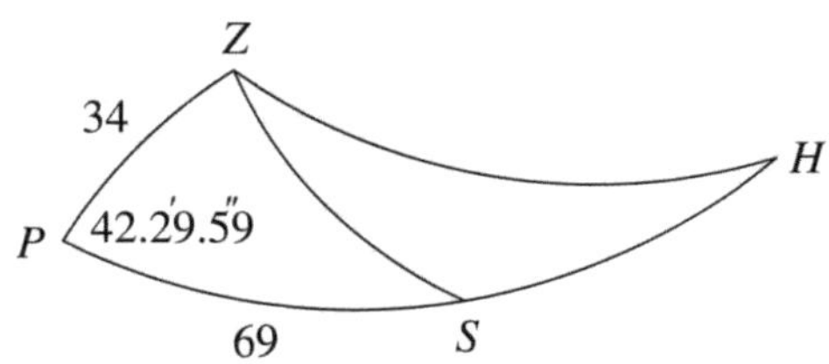

Of theſe ſix parts let any three be giuen, partly angles, partly ſides. *For example ſake.*

The houre angle ZPS 42 degr. $29'$. $59''$. which ſheweth two of the clock afternoone 49. 59. and 56.

And the ſide PZ 34. *the complement of the eleuation of the Pole.*

And the ſide PS 69 *the complement of the declination of the Sunne.*

Out of which that the other three partes may bee gotten: from Z the end of the ſide PZ that was giuen, let the perpendicular ZM, or rather (if you will) the quadrant ZH be drawne downe, ſubtending the angle ZPS, and reducing the not-quadrantall Propoſed PZS into two triangles quadrantall in the angle M, which are PMZ, ZMS, as in the firſt figure: or if you bee delighted with varietie, let them be reduced to two triangles quadrantall in the ſide ZH, which are ZHP, and ZHS, as in the ſecond figure. All the parts of which quadrantals you ſhall get by the ninth Section of the 4th Chapter of this booke.

For by hauing PZ giuen	34. deg.
and ZPM, or ZPS	42 deg. $29'$. $59''$
You may find the perpend. $Z.M$	22 deg. 11. 47
And the angle PZM	52 deg. 46. 38
And the ſide PM	26 deg. 26. 29

Which *PM* being taken out of *PS*	69 deg.
There remaines *MS*	42 deg. 33'. 31''.

Now the ſide *MS*, and the perpendicular *ZM* being knowne, you may (by the ſaid ninth Section of the fourth chapter of this booke,) finde out

The angle oppoſite to the perpendicular *MSZ*, or that which was ſought for *PSZ* 31. deg. 6'. 5''.

And the ſide that was ſought *SZ*.	47 deg.
And the angle *MZS*	67 de. 38'. 11''
Which being added to *PZM*	52 de. 46. 38
Makes *PZS* the angle ſought	120 de. 24. 49

You haue therefore three parts which you ſought for, found by helpe of the perpendicular *ZM* of the former figure.

You may alſo finde the ſame by helpe of the quadrant *ZH* in the latter figure.

For hauing *PZ* giuen	34 deg.
And *ZPS*, or *ZPH*	32 deg. 29'. 59''.

You may finde by the ſame 9. Sect. of the 4. chap. of this booke,

The angle *ZHP*	22 deg. 11'. 47''.
And the angle *PZH*	142 deg. 46'. 38''.
And the ſide *PH*	116 deg. 26'. 29''.
Out of wch *PH* ſubtract *PS*	69 deg.
there remaines the ſide *SH*	47 deg. 26'. 29''.

Which ſide *SH* being now had, together with the angle *ZHP* 22 deg. 11'. 47''. you may alſo (by the ſaid 9 Section of the 4 chap. of this book) finde out

The angle *HSZ*	148 deg. 53'. 55''.
And the remainder therof to a ſemicircle, the angle *PSZ*	31 deg. 6'. 5''.
that was ſought for.	
And the ſide *SZ*	47 deg.
that was ſought for.	
And laſtly the angle *HZS*	22 deg. 21'. 49''.
which being taken out of *HZP*.	142 deg. 46'. 38''.
There remaines *PZS*	120 deg. 24' 49''.

the other angle that was ſought for, in all poynts, as before.

An Admonition.

IN imitation of this example, nine diuerſe queſtions may be reſolued both of this, and any triangle. For by the *Eleuation of the Pole*, and *the houre of the day*, and *the declination of the Sunne that day*, being giuen, there is had, as afore:

1 *The Azimuth of the Sunne.*

2 *The height of the Sunne.*

3 *The angle of pofition of the Sunne*: alfo by hauing *the declination of the Sunne, the angle of the Sunnes pofition, and his height* giuen, you haue

4 *The Sunnes Azimuth.*

5 *The Eleuation of the Pole,*

6 *The houre, or houre arch.*

Alfo if you haue *the height of the Sunne, his Azimuth*, and *the height of the pole* giuen, there is found,

7 *The houre of the day.*

8 *The declination of the Sunne.*

9 *And the angle of the Sunnes Pofition.*

9 *And the angle of the Sunnes Pofition.*

The fecond example of two angles giuen, and the fide betweene them.

THe angles in the figures going afore, being giuen, to wit,

The houre angle ZPS		42 deg. 29'. 59''.
& the azimuth of the fun PZS.		120 deg. 24' 49''.
with the fide between them, being *the compl. of the poles eleuation*	*PZ*	34 deg.

The other 3 parts are fought out. For as afore,

Hauing firft giuen	*ZM.*	22. deg. 11'. 47''.
And	*PM.*	26 deg. 26'. 29''.
And the angle	*PZM.*	52 deg. 46'. 38''.
wch being taken out of	*PZS.*	120 deg. 24' 49''.
there being left remai.	*MZS.*	67 deg. 38'. 11''.

By which *MZS*, and *ZM*, already known, there fhall at length be found,

The fide *ZS* 47 deg. the fide fought for.

And the angle *ZSM*, or *ZSP*. 31 deg 6'. 5''. the angle fought for.

And the fide *MS*. 42. deg. 33'. 31''. which being added to *PM*. 26. deg. 26'. 29''. the fide remaineth *PS* 69 deg. which was fought for.

And thefe you haue by meanes of the perpendicular of the former figure. In like manner you may finde the fame by helpe of the quadrant of the latter figure. For they are found by the ninth Section of the fourth chapter of this booke.

by hauing giuen the angle	$PHZ.$	22, de. $11'$. $47''$.
And the angle	$PZH.$	142. de. $46'$. $38''$
Out of which the angle giuen PZS being giuẽ		120 de. $24'$. $49''$
There remaines	$SZH.$	22 de. 21. 49

which together with the angle $PHZ.$ now knowne, all the reſt of the parts are brought forth. viz.

$PZ.$	34 deg.
$ZPS.$	42 deg. $29'$. $59''$.
$PS.$	69 deg.
$PSZ.$	31 deg. 6. 5.
$SZ.$	47 deg.
$ZSH.$	148 deg. 53. 55.
$SH.$	47 deg. 26. 29.

An Admonition.

IN imitation of this example, nine diuerſe queſtions of this and of any other triangle, are reſolued.

For the houre of the day, the Eleuation of the pole, and the *ſuns azimuth* being giuen, there is had,

1 *The declination of the ſunne,*

2 *The angle of the ſunnes poſition.*

3 *The height of the ſunne.*

Alſo, *the houre of the day, the declination of the ſunne*, and *angle of the ſunnes poſition* being giuen, there is had,

4 *The height of the ſunne,*

5 *The ſunnes azimuth,*

6 *The height of the pole.*

Alſo, *the angle of the ſunnes poſition, the height of the ſunne*, & his *azimuth* being giuen, there is had,

7 *The height of the pole,*

8 *The houre of the day,*

9 *The declination of the ſunne.*

The third example of two ſides giuen, whereof that which is neereſt to a quadrant, ſubtendeth the angle giuen.

IN the figures afore, let there be giuen

The ſide	$PZ.$	34 deg:
And that which is neerer then it to a quadrant,	$ZS.$	47 deg.
With that angle which this ſide ſubtendeth	$ZPS.$	42 de. $29'$. $59''$

By the 9 Sect. of the 4. chap. of this Booke, let there be ſought out

The ſide	ZM	22 deg. 11. 47
And the angle	PZM	52 deg. 46. 38.
And the ſide	PM	26 deg. 26 29.
And in like manner you may haue	ZSM	31 deg. 6. 5.

or ZSP the angle ſought for: which is moſt certainly known (by the 2 Sent. 3. chap. of this book) to be leſſe then a quadrant, namely, that it is 31 deg $6'$. $5''$. ad that it is not 148 deg. $53'$. $55''$

Alſo you may haue the angle	MZS	67 deg. $38'$. $11''$
which being added to	PZM	52 deg. $46'$. $38''$
makes the other angle which was ſought for.	PZS	120 de. 24. 49.

Laſtly, you ſhall haue the caſe	MS	42 deg. $33'$. $31''$
Which being added to	MP	26 deg. 26. 29
makes the ſide which was ſought for.	PS	69 deg.

No otherwiſe (if you will) you may find out the ſame by helpe of the quadrant ZH of the latter figure.

The fourth example, of two giuen ſides, whereof that which is leſſe neere a quadrant ſubtendeth the angle giuen, and that which is neereſt, ſubtendeth an angle of the kinde giuen onely.

Admit there be giuẽ the ſides	ZS	47 deg.
And that which is not ſo neere a quadrant	PZ	34 deg.
with that angle which this ſubtendeth	ZSP	31 deg. $6'$. $5''$

And let it be giuẽ, that the angle wch ZS ſubtendeth, that is the angle SPZ, is by kind leſſe then a quadrant: therfore the perpendicular ZM being let down

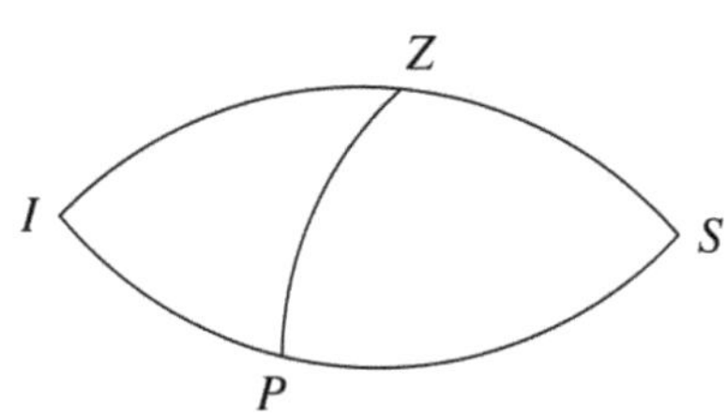

from Z to the baſe PS (as before) or the quadrant ZI (as here) ſubtending the giuen angle ZSP. By the 9 Sect. of the 4 chap of this booke, let the other parts be gotten (as for exercife and varieties fake) by the quadrant of this figure, *ZI*, you may get

The angle	*ZIS*	22 deg. 11′. 47″.
And	*IZS*	157 deg. 38. 11.
And	*SI*	132 deg. 33. 31.
and in like maner you may haue	*IPZ*	137 deg. 30. –1.
& by confequent the angle	*SPZ*	42 deg. 29. 59.

that was fought for.

Becaufe by the giuen pofition, it is expreſly declared to be leffe then a quadrant: otherwife except the kinde therof were giuen, it would be vncertaine (by the 1. Sect. of the 3. Chap. and the 5 fect. of this chap.) for it might otherwife haue been 137 deg. 30′. 1″.

So alfo fhall you haue the angle	*IZP*.	37 deg. 13′. 22
which being taken out of	*IZS*.	157 deg. 38. 11
there remaines the other angle fought.	*PZS*.	120 deg. 24. 49
To conclude, you fhall alfo haue	*IP*.	63 deg. 33 31
Which being taken out of	*IS*.	132 deg. 33. 31
There remaines	*PS*.	69 deg.

the fide that was fought for.

You fhall alfo hit the fame markes (as it were) if you feeke the account or number of the parts, by helpe of the perpendicular *ZM* of the firft figure.

An Admonition.

BY imitation of the third example going before, and this fourth, there are refolued eighteene diuerfe queftions of this and any other triangle. For (as it is in the 3. example) *the height of the pole*, *the height of the Sun*, & *houre of the day*, being giuen, there is found,

1 *The Azimuth of the Sunne.*

2 *The angle of the Sunnes pofition.*

3 *The declination of the funne.*

Alfo, the *Eleuation of the pole*, *the height of the funne*, and *the angle of the funnes pofition* being giuen, (as in the fourth example) there are found,

4 *The Azimuth of the funne*,

5 *The houre of the day*,

6 *The declination of the funne.*

Alfo, *the height of the funne*, *the declination of the funne*, and *houre of the day being giuen*, there is found,

7 *The angle of the funnes pofition*,

8 *The Azimuth of the funne*,

9 *The Eleuation of the pole.*

Alſo, *the height of the ſunne, the declination of the ſunne, and Azimuth of the ſunne* being giuen there is found,

10 *The angle of the ſunnes poſition,*
11 *The houre of the day,*
12 *The Eleuation of the pole.*

Alſo, *the declination of the ſunne, the Eleuation of the pole,* and *the angle of the ſunnes poſition* being giuen, there is found.

13 *The Azimuth of the ſunne,*
14 *The height of the ſunne,*
15 *The houre of the day.*

Alſo, *the declination of the ſun, the height of the pole, & azimuth of the ſun being giuen,* you haue

16 *The houre of the day,*
17 *The angle of the ſunnes poſition,*
18 *The height of the ſunne.*

The fifth example of two angles giuen, the neerer whereof to a quadrant is ſubtended by the ſide giuen.

IN the Triangle *PZS* of the firſt Diagram, Let the angle *PSZ* bee giuen	31. 6′. 5″
And *SPZ* which is neerer then it to a quadrant	42. 29′. 59″.
With the ſide *ZS* ſubtending the ſame	47. 0″.
Out of which *PSZ*, and *SZ*, is found the perpendicular *SM* (By the 9th of the 4th of this booke.)	22. 11′. 47″.
And the other parts of the quadrantal, *SZM*, to wit, *MZS*.	67, 38′, 11″,
And the ſide *MS*.	42. 33′. 31″

As alſo by this perpendicular, with the angle *ZPS* being giuen, or the angle *ZPM*, all the partes of the quadrantall *ZMP* are found.

As firſt the ſide ſought for *PZ*. For this is moſt certainely knowne (by the ſecond Sentence of the firſt Chapter of this book) to be leſſe then a quadrant, namely, that it is 34, and not 146.

Then we haue	*PZM.*	52, 46′, 38″,
W^ch being added to	*SZM.*	67, 38′, 11″,
there is made the angle	*PZS.*	120, 24′, 49″,
Laſtly, there is alſo had	*PM.*	26, 26′, 29″,
W^ch being added to	*MS.*	42, 33′, 31″,
There is made the other ſide,	*PS.*	69, 0′. 0″,

that was ſought for.

You may alſo finde out theſe parts otherwiſe (if you wil) by the two quadrantals of the figure next going before, *ZIS*, and *ZIP*.

The ſixth example of two giuen angles, whereof that which is not neereſt a quadrant is ſubtended by the ſide giuen, and that which is neereſt, is subtended by a ſide, whoſe kinde onely is giuen.

OF the Triangle *PZS* of the firſt figure let there be giuen,

The angle,	*ZPS*,	$42, 29', 59''$,
And that which is not ſo neere to a quadrant as it	*ZSP*.	$31, 6'. 5''$,
With the ſide ſubtending the ſame	*PZ*.	$34. 0'. 0''$,

And let it be knowne that the ſide *ZS*, ſubtending the angle *ZPS*, is leſſe then a quadrant.

By theſe thus giuen, let the perpend: be ſought *ZM*. $22. 11'. 47''$ for.

And the other parts of the quadrantall *PZM*.

To wit the angle	*PZM*.	$52, 46', 38''$,
And the ſide	*PM*.	$26, 26, 29$,
As alſo by this perpendicular, together with the angle giuen	*ZSM* or *ZSP*.	$31, 6'. 5''$

Let all the parts of the quadrantal *ZMS* be ſought,

As firſt the deſired ſide *ZS* $47. 0'. 0''$. becauſe that by poſition giuen, it is expreſſely declared to bee leſſe then a quadrant, otherwiſe it might haue been here 133. For (by the firſt Chap 3 and 5 of this booke) it is vncertaine, except the kinde thereof be expreſſely giuen,

Then the angle	*MZS*	$67. 38'. 11''$
which added to the angle	*MZP*	$52. 46. 38$
Maketh the angle	*PZS*.	$120. 24'. 49''$

which was deſired.

Laſtly, there is alſo obtained	*SM*.	$42, 33', 31''$
Which added to the ſide	*PM*.	$26, 26, 29$
Maketh the baſe deſired	*PS*.	$69, 0', 0''$

You may alſo moſt eaſily get the ſame parts out of the two quadrantals *PHZ*, and *SHZ* of the ſecond figure.

An Admonition.

BY imitation of the 5 example going afore, and this ſixith, 18 ſeuerall queſtions of this and any other Triangle are reſolued.

For (as in the fifth example) out of the *Angle of the poſition of the ſunne, the houre of the day*, and *height of the ſunne* being giuen, is gotten,
1 *The Eleuation of the pole,*
2 *The Azimuth of the ſunne,*
3 *The declination of the ſunne,*

Alſo (as in this ſixith example) by *the houre of the day, the angle of poſition of the ſunne*, and *height of the pole* being giuen, there is gotten
4 *The height of the ſunne,*
5 *The ſunnes Azimuth,*
6 *The declination of the ſunne.*

Alſo by *the houre of the day, the azimuth of the ſunne*, and *the height of the ſunne* giuen, there be gotten,
7 *The declination of the ſunne,*
8 *The angle of the ſunnes poſition,*
9 *The height of the pole.*

Alſo by *the houre of the day, the Azimuth of the ſunne*, and *the ſunnns declination* giuen, you haue,
10 *The height of the ſunne,*
11 *The angle of the ſunnes poſition,*
12 *The height of the pole.*

Alſo by *the Azimuth of the ſunne, the angle of the ſunnes poſition*, and *the declination of the ſun* giuen, there bee gotten,
13 *The height of the pole,*
14 *The houre of the day,*
15 *The height of the ſunne.*

Alſo by *the azimuth of the ſunne, the angle of poſition of the ſunne*, and *height of the pole* giuen, there be gotten,
16 *The declination of the ſunne,*
17 *The houre of the day,*
18 *The height of the ſunne.*

And ſo by the method of this Canon onely, foure and fiftie ſeuerall queſtions of the ſame triangle, not being quadrantall, are reſolued. The reſt ſhall bee reſolued hereafter.

By theſe therefore it is manifeſt, that of two angles, and their ſubtending ſides, three being giuen the Logarithme of the fourth of them at the leaſt, ſhal be made known euen without any deſcription at all of the quadrantals. For out of the ſum of the Logarithmes of the angle and ſide adioyning thereto being giuen, ſubtract the Logarithme of the third thing that is giuen, and thence ſhall come the Logarithme of the fourth that was ſought for; and that fourth it ſelfe ſhall alſo be made knowne if the kinde thereof be not vnknowne.

As may be perceiued by the third, fourth, fifth, and ſixth examples going before. For of the angles of the baſe, ZPS, and ZSP, and of their ſubtending leggs ZS, and ZP. let three bee giuen, (for examples ſake)

The legs ZS. 47 deg. and his Logar.	312858
And —- ZP. 34 deg. and his Logar.	581261
with the agle adioying, ZPS. $42, 29', 59''$ & his Logar.	392172
added to the Logar. of ZP laſt mentioned,	
Their ſumme is,	+973433

(which is the Logarithme of the ſecret and ſuppreſſed perpendicular ZM, or of the angle ZHS, or ZIP.)

Out of which ſubtract the Logar. of ZS	+312858
There remaineth the Log. of the 4th ZSP. that was ſought for.	+660575

Therefore the ſame fourth it ſelfe ZSP. will be $31.6', 0''$. Becauſe (by the ſecond Section of the third chapter) it is proued to bee leſſe then a quadrant.

Now contrariwiſe,

There being giuen ZS 34 deg. and his Logar.	+581261
And ZS 47 deg. and his Logar.	+312858
wth the angle adioyning thereto ZSP, $31, 6', 0''$.	& his Log. 660575
added to the Logarithme of the ſide laſt mentioned, ZS, The ſumme is	+973433
Out of which take the Log. of ZP	+581261
There will emaine the Logarithme of the 4th	+392172

tht was ſought for: that is, ZPS, whoſe arc (by the 1 *Sect.* 3. *Chap.*) is vncertaine whether it be $42, 29', 59''$, or $137, 30', 1''$, except it be knowne by poſition giuen whether it be greater or leſſe then a quadrant.

Of not-quadrantals which be pure.
Chap. VI.

Hitherto wee haue ſpoken of intermingled parts giuen: now follow ſuch parts as are pure.

They are pure when the three parts giuen are of the ſame kinde, and they are either three ſides giuen, and the angles are ſought for: or the three angles giuen, and the ſides are ſought for.

An Admonition.

6 *Although the pure parts are the former in regard of their ſimplicitie, yet for their difficultie they do worthily take the latter place.*

In Sphæricall Triangles.

3 *Halfe the baſe, and halfe the difference of the legs being taken together, and the Logarithme thereof, and the Logarithme of the difference of them being added together; and out of that ſumme, the ſumme of the Logarithmes of the legs being ſubducted, the halfe of that which remaineth is the Logarithme of halfe the verticall angle.*

Becauſe *Regiomontanus* in the ſecond chapter of his fifth booke of Triangles, and others do teach, that *as the rectangle comprehẽded vnder the right ſines of the legs, is to the ſquare of the whole ſine: ſo the difference of the verſed ſine of the baſe, and difference of the legs is to the ve'ed ſine of the verticall angle.* Seeing alſo as tat difference is to this verſed ſine: ſo is the rct-angle made of the right ſines of the ſumne, and difference of the halfe baſe, and halfe difference of the leggs, to the ſquare of the right ſine of halfe the verticall angle (for this laſt rectangle is to that difference of the verſed ſines, and this laſt ſquare to that verſed ſine in a 5000000^{fold} proportion, the whole ſine being 10000000) therefore it ſhall follow, that *as the rectangle contained vnder the right ſines of the leggs, is to the ſquare of the whole ſine, ſo ſhall the rectangle made of the right ſines of the ſumme and difference of the halfe baſe and halfe difference of the leggs, be to the ſquare of the right ſine of halfe the verticall angle.* And by conſequent (out of the Corolarie of the ſixt Definition of the firſt Chapter, & the fourth Propoſition of the ſecond Chapter, and third Problem of the fift Chapter firſt Booke) *The ſumme of the Logarithmes of the leggs ſubtracted out of the Logarithmes of the ſumme and difference of the halfe baſe, and halfe difference of the leggs, leaueth the double of the Logarithme of half the verticall angle*, as is abouesaid.

4 *Secondly, halfe the baſe, and halfe the aggregate of the legs being taken together, and the Logarithme thereof, and the Logarithme of the difference of them being added together; & out of that ſumme, the ſumme of the Logarithmes of the legs being ſubtracted, the halfe of that which remaineth is the Antilogarithme of halfe the verticall angle.*

For the ſumme of the Logarithmes of the ſumme and difference of the halfe baſe, and halfe ſumme of the legs of this Propoſition, hath no other proportion to the ſumme of the Logarithmes of the ſumme and difference of the halfe baſe, and halfe difference of the legs of the former Propoſition, then the double of the Antilogarithme of halfe the verticall angle here, hath to the double of the Logarithme of the ſame halfe verticall angle before: The demonſtration whereof belongeth to another place.

An Admonition.

IN Sphæricall triangles alſo, we take the true & Alterne baſe in the ſame ſence as before in right lined triangles, that is, the one for the ſum, and the other for the difference of the caſes. 5

Thirdly, the Differentiall of the aggregate, and the Differential of half the difference of the legs being added together, and out of the ſumme thereof the differentiall of halfe the true baſe being ſubducted, there will remaine the Differentiall of halfe the Alterne baſe. 6

The fundamentall reaſon hereof is, becauſe that *as the Tangent of the true halfe baſe is to the Tangent of halfe the ſumme of the legs, ſo is the Tangent of halfe the difference of the legs to the Tangent of the Alterne halfe baſe.* For the Logarithmes of Tangents are the Differentials of their arches, by the 22 and 25 Sect. 3. Chap. 1. Booke.

And therfore ſhall that equalitie of the Logarithmes or Differentials follow this analogie of the Tangents (by the 4 Prop. 2. Chap. 1. Booke.) But becauſe the readers hereof wil perhaps require of me the demonſtration of this fundamentall analogie, or proportion of Tangents (hitherto vnknowne) I will here therefore ſhew the ſame, ſo farre forth as the ſhortneſſe of this abridgement will permit.

Let the Sphære therefore $AFPG$ lye vpon the flat ſuperficies $HIKQ$, that they may touch each other in the common poynt A: from which by the center of the Sphære $\odot$, let the right line $A\odot P$ be raiſed, cutting the vpper halfe of the Sphære in the poynt P, and ſo $A\odot P$ ſhall bee perpendicular to the plaine or flat $HIKQ$. Then from the angle A, let be deſcribed vpon the ſuperficies of the Sphære, the triangle $A\lambda\gamma$ ſharpe angled in γ, or $A\lambda\beta$ blunt angled in β, and the ſemicircles $A\lambda P$ and $A\gamma P$, or $A\beta P$ being drawne forth, taking λ for the pole, according to the diſtance $\lambda\gamma$ or $\lambda\beta$ which is equall thereto, draw the circle $\epsilon\delta\beta\gamma$, cutting λP in ϵ, and λA in δ and $A\beta\gamma$ in the poynts β and γ. From the poynt λ to the Arch $A\beta\gamma$ let downe the perpendicular Arch $\lambda\mu$. Here therefore $A\lambda$, ſhall be the greater leg, and $\lambda\gamma$ or $\lambda\beta$ the leſſe leg, $A\gamma$ and $A\beta$ the baſes, the one true, the other alterne, $A\delta$ the difference of the legs, and $A\epsilon$ the ſumme of the legs, becauſe $\lambda\epsilon$ and $\lambda\delta$ by the conſtruction, are equall to the leſſe leg $\lambda\gamma$ or $\lambda\beta$. This being done, and ſuppoſing P to be in ſtead of an eye, or ſome lightſome body, from the ſame P to the flat lying vnder $HIKQ$, let downe the beame $P\gamma$, cutting the flat in c, and the beame $P\beta$ cutting the flat in b: and becauſe $\gamma\beta A$ are in the ſame plaine or circle with the lightſome body P, their ſhadows c b A ſhall be in the ſame right line.

Likewiſe from the ſame poynt P to the ſame plaine, let downe the beame $P\epsilon$, cutting the plaine in e, and the beame $P\delta$ cutting the plaine in d, and becauſe ϵ δ A are in the ſame plaine and circle with the lightſome body P: therefore their ſhadowes e d A ſhall be in the ſame ſtreight line. Moreouer, becauſe

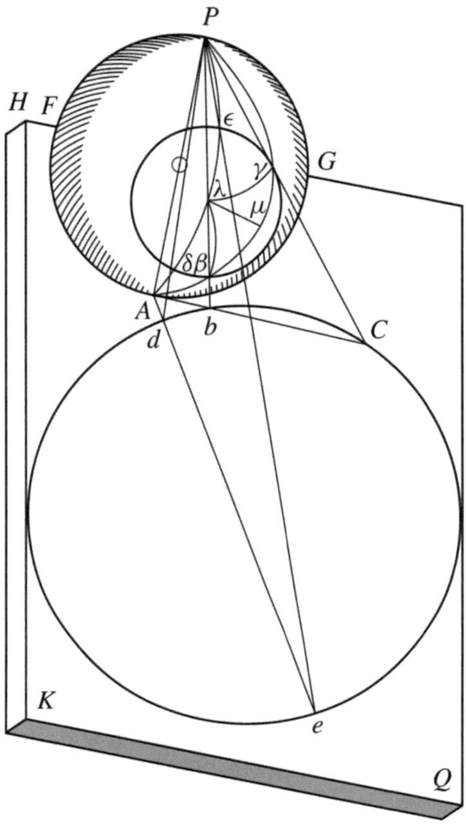

$P\odot A$ is perpendicular to the plaine, therefore the triangles PAd and PAe, and PAb and PAc are right angled in A: and therfore alſo Ad is the Tangent of the angle $AP\delta$ or APd & Ae the tangent of the angle $AP\epsilon$ or APe. So alſo Ab is the tangent of the angle $AP\beta$, or APb and Ac is the Tangent of the angle $AP\gamma$ or APc ſuppoſing PA to be the gnomon of the whole ſine. And because Ad is the Tangent of the angle $AP\delta$, and $AP\delta$ is halfe of the angle $A\odot\delta$, (by the 20 *Prop.* 3. *Eucl.* becauſe this angle is in the center, and that in the circumference) therefore Ad is the tangent of halfe the angle $A\odot\delta$, or which is the ſame of halfe the Arch $A\delta$, which is the halfe difference of the legs. Likewiſe becauſe Ae is the tangent of the angle $AP\epsilon$, and the angle $AP\epsilon$ in the circumference is halfe the angle $A\odot\epsilon$ in the center, therefore Ae is the tangent of halfe $A\odot\epsilon$, or of halfe the arch $A\epsilon$, which is the halfe ſumme of the legs. In like manner in the baſes true and alterne, Ab ſhall bee the Tangent of the angle $AP\beta$, or of halfe the angle $A\odot\beta$, or of halfe the arch $A\beta$ which is the one halfe baſe: and Ac ſhall be the Tangent of the angle $AP\gamma$, or of halfe the angle $A\odot\gamma$, or of halfe the arch $A\gamma$, which is the other halfe bafe. Now feeing it hath been fhewed that *Ab* is the Tangent of one of the halfe bafes, and *Ac* the Tangent of the other halfe bafe, and that *Ad* is the Tangent of half the diffrence of the legs, and *Ae* the Tangent of halfe the fumme of the legs. I fay, that

As Ab the Tangent of the true halfe bafe, is to Ae the Tangent of the halfe fumme of the legs:

So is Ad the tangent of the halfe difference of the legs, to Ac the tangent of the altern halfe bafe.

Or contrariwife, by making the true bafe of the alterne.

As Ac the tangent of the true halfe bafe, is to Ae the Tangent of the halfe fumme of the legs:

So is Ad the tangent of the halfe difference of the legs, to Ab the tangent of the alterne halfe-bafe.

Which I proue thus: If the poynts *b c d e* be in the fame circle, then as *Ab* is to *Ae*, fo is *Ad* to *Ac*. And contrariwife, as wee faid euen now (by 36 *Prop*. 3 and 16 *Prop*. 6. *Eucl.*) But the poynts *b c d e* fall in the fame circle. For the fhadow of any circle defcribed in the fuperficies of a fphære comming from a lightfome body in the fame fuperficies which is not in the circumference of the circle, maketh a circle perfectly round in the plaine perpendicular to the right line which goeth from the lightfome body by the center of the fphære, as it is manifeft out of the Optickes, & making of the *Aftrolabe*: and by *Apollonius* in his 1 book of Conick fections prop. 5. But here the circle $\delta\beta\gamma\epsilon$ is defcribed in the fuperficies of the Sphære, and the lightfome body *P* is without the circumference of the circle, and the right line that goeth from the fame by the center (that is $P\odot A$ is perpendicular to the plaine) therefore the fhadow of that circle which falleth vpon the poynts *d b c e*, is neceffarily circular, and perfectly round. Therefore

As Ab is to Ae, fo is Ad to Ac.

And contrariwife, that is,

As the tangent of the true halfe bafe, is to the tangent of the halfe fumme of the legs:

So is the tangent of the halfe difference of the legs, to the tangent of the alterne halfe bafe. And by confequent, the Differential of the true halfe bafe, fubtracted out of the fumme of the Differentials of the halfe fumme and halfe difference of the legs, is equall to the Differentiall of the Alterne halfe bafe, which things we vndertooke to demonftrate.

Therefore three fides of a Sphæricall triangle being giuen, any one of the angles is had three wayes. 7

The firft way is, *That you make any fide the bafe* (*efpecially that which* 8 *commeth neereft a quadrant*) *then*

Adde halfe the baſe and halfe the difference of the legs together, and to the Logarithme thereof adde the Logarithme of the difference of them; out of which ſumme ſubduct the ſumme of the logarithme of the legs: and the halfe of the remainder is the Logarithme of an arch, which being doubled is the verticall angle. And ſo the reſt.

As of the triangle PZS, let the ſides PZ be giuen 34 deg. and ZS 47 deg. and SP 69 degr. let the angles bee ſought out, and firſt the angle PZS cõming neereſt a quadrant, which SP 69 deg. (that is, the ſide nereſt a quadrant) ſubtendeth. Therefore let this SP 69 be made the baſe: then

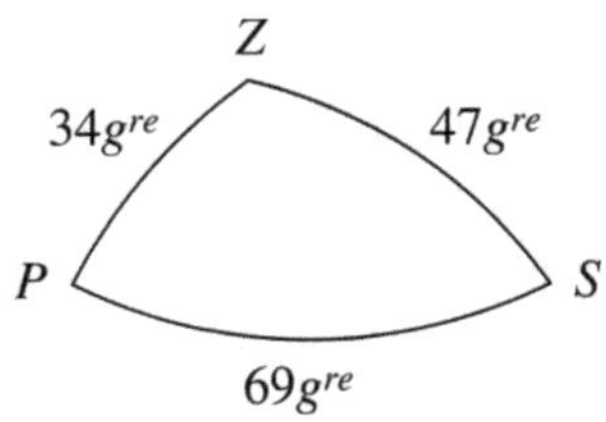

Adde $\frac{1}{2}$ the baſe PS	34 deg. 30′	
To $\frac{1}{2}$ the difference of the legs PZ and ZS	6 deg. 30	
The ſumme is 41 deg. 0′. the Logar.		421504
The differ. is 28 deg. 0′. the Logar.		756147
The ſumme is		+1177651
Adde the Logar. of the leg PZ 34 deg. 581260 / To the Log. of the legg. ZS 47 deg. +312858	ſumme	+894118
Subtract the ſame out of the former ſumme, the reſt is		283533
whereof take the halfe		+141766

W^ch is the Logar. of 60 deg. 12′. 24″ $\frac{1}{2}$. and being doubled is 120 deg. 24′ 49″. the verticall angle PZS that was ſought for.

No otherwiſe you may (if you will) finde out the other angles but they ſhall be found more eaſily by 9 *Chap*. 5. of this booke, becauſe (by 2 *Sent*. 3 *chap*.) they are of a certaine and knowne kinde.

9 The ſecond way is, *that any ſide (eſpecially that which is neereſt a quadrant) being made the baſe, you adde halfe the baſe, and halfe the ſumme of the legs together, and to the Logarithme thereof adde the Logarithme of the difference of them: out of which ſumme ſubduct the ſumme of the Logarithme of the legges, and the halfe of the remainder is the Antilogarithme of an arch, which being doubled, is the verticall angle, and ſo the reſt.*

As of the ſame Triangle PZS

Adde halfe the baſe PS. 34 deg. 36.		
To halfe the ſum of the legs PZ, and ZS.	40 deg. 30′.	
The ſumme is 75 deg. 0′. the Logar.		2258295
The Differ. is 6 deg. 0′ the Logar.		34668
The ſumme is		+2292963

Adde the Logar. of the leg. PZ 34 deg. 581261 To the Log. of the leg ZS. 47 deg. +312858,	fumme is	+894119
Subtract the fame out of the former fumme, The reft is		1398844
whereof take the halfe		+699422

W[ch] is the Antilog. of 60 deg. 12'. 24''. $\frac{1}{2}$ and being doubled is 120 deg. 24'. 49''. the verticall angle PZS fought for.

The other angles, although you may finde after this manner, yet you fhall finde them more eafily by 9 *Chap* 5 of this booke. For by the fecond fentence of the third Chap. they are of a knowne kinde.

The third way is, that any fide being put for the bafe, you adde the Differential of halfe the fumme of the legs, to the Differentiall of halfe the difference of the legs, and fubtract from the product the Differentiall of the true halfe bafe, and there fhall come thereof the Differential of the alterne halfe bafe. The fumme of which halfe bafes is the greater cafe, and the difference the leffe cafe, diftinguifhing two right-angled triangles, which do make knowne both their owne parts, and all the parts of the triangle Propofed (*by* 9 *chap.* 4. *and* 8 *chap.* 5. *of this booke.*)

As the fides of the triangle propounded PZS being giuen, as before, let the angles at the bafe ZPS, and ZSP be fought for.

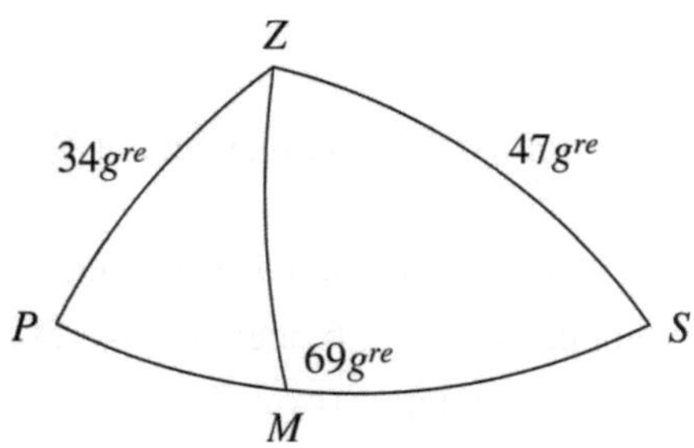

Half the fum of the leggs PZ, & ZS is 40 deg. 30. the Differ.		157730
Half the difference of them is 6 deg. 30. the Differ		2172121
W[ch] added together, bring forth		2329851
From which fubtract the differ: of halfe the true bafe PZ.	viz. 34. deg. 30'.	375012
And there will remaine		+1954839

Which is the Differential of 8 deg. 3'. 31''. halfe the alterne bafe.

Adde therefore together the two halfe bafes, to wit,

Halfe the true bafe	34 deg. 30'. 0''.
and half the altern bafe	8 deg. 3'. 31''.
And they make	42 deg. 33'. 31''.

the greater cafe *MS*.
Take the one out of the other, refts 26 deg. 26'. 29''.
the leffe cafe *PM*.

Therefore by helpe of thefe cafes, you haue now two triangles right angled at *M*, that is *PMZ*, and *SMZ*, which do lay open both the perpendicular *ZM*, and the verticall angles *PZM* and *SZM*: or (if you will) *PZS* (by 9 *Chap*. 4 and 8 *Chap*. 5 of this book) But thefe things being omitted, let vs return to the angles of the bafe *ZPS*, *ZSP* which were fought for.

Adde the Differential of the cafe *PM*. 26, 26' 29'' already found (by 9 *Sect*. 4 *chap*.)	+698552
To the Differentiall of the complement of *PZ*, which is 56 degrees	393771
There will come forth	304781

Which is the Logarithme of the complement of the angle *ZPS*, which complement is 47 deg. 30' 1''.

Likewife,

Adde the Differentiall of the greater cafe *SM* 42 de. 33'. 31'' already found by the 9 *Sect*.	+85324
To the Differentiall of the complement of *SZ*, which is 43 degrees	+69870
There will come forth	+155194

Which is the Logarithme of the complement of the angle *ZSP*, which complement is 58 deg. 53', 55''.

But here remember, that not the parts *PZ* 34, and *ZPS*, or *SZ* 47, and *ZSP*, but their complements, that is, 56 degr. and 47, 30', 1''. and 43 deg. and 58 53', 55''. are here called circular parts, (by 2 *Chap*. 4. of this booke) Therefore the true angle fought for *ZPS*, is 42, 29', 59''.

And *ZSP* is 31, 6', 5''.

As it alfo manifeft by 8 *Sect*. *Chap* 5 of this booke.

Another example of the fame triangle.

THe fame triangle *PZS* being placed otherwife, let *ZS* be the bafe, and the fides being giuen, as before, let the angle *PZS* bee fought for. Therefore.

Halfe the fum of the legs *SP* and *PZ* is	51 deg. 30' the Differ-	228865
Half the difference of them is 17 de. 30'. the Diff.		+1154234
Which added together, bring forth the Differ.		+925369

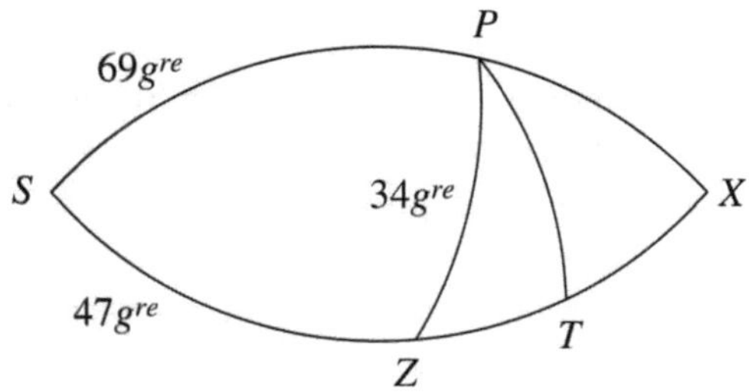

From wch fubtract the Differentiall of halfe the true bafe *SZ*, viz. of } 23 de. 30   +832840

And there will remaine   +92529.

W^ch^ is the Differential of 42 deg. $21'$. $11''$. halfe the alterne bafe.

Adde therefore together the two halfe bafes, to wit,

Halfe the true bafe	23 deg. $30'$, $0''$.
And half the alterne bafe	42 deg. $21'$, $11''$,
And there wil come forth the greater cafe *ST*.	65 deg. $51'$, $11''$,

Take the difference of them 18 deg. $51'$, $11''$, the leffe cafe *TX*, or *TZ*.

Adde therefore the Differentiall of the the fame *TZ*, 18, $51'$, $0''$, viz, }   +1074520

And there will remaine   +92529.

To the differentiall of the complement of *ZP*, which is 56 degrees, }   393771

And from thence will arife   +680749

Which is the Logarithme of the complement of the angle *PZT*. 59 deg $35'$. $11''$

Of which angle *PZT*, feeing the angle fought for *PZS* is the remainder to a femicircle (which alwayes happeneth when the alterne bafe is greater then the true) the angle *PZS* muft needs be 120 degrees, $24'$. 49, otherwife if the true bafe exceed the alterne bafe, the angles *PZT*, and *PZS* fhall bee all one.

An Admonition.

YOu haue now three true wayes to finde out the angles by the fides giuen, by euery one whereof three feuerall queftions of this and any other triangle are refolued.

For by *the eleuation of the pole*, *the height of the Sunne*, *and declination of the* fi*unne* being giuen, they that doubt are fatisfied concerning the queftion: whereby either

1 The Azimuth of the Sunne,
2 The angle of the ſituation and poſition of the Sunne: or
3 The houre of the day is demanded.

Hitherto wee haue found the angles by the ſides.

It remaineth to finde the ſides by the angles.

11 *In any Sphæricall Triangle the ſides may bee changed into angles, and the angles into ſides: yet taking firſt for any one angle and his ſubtending ſide, the remainders of them to a ſemicircle.*

For example ſake.

Let QRT be a triangle, whoſe angles let bee Q 47, R 111, and T 34.

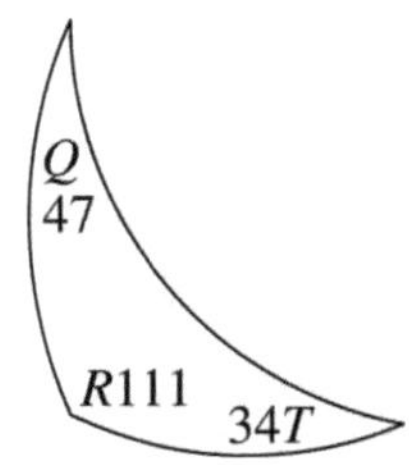

Let vs firſt take for any angle, as for R 111, his remainder to a ſemicircle, which is 69 degrees.

I ſay that theſe angles 47^{d}. 69^{d}. 34^{d}. may bee changed into ſides, & the triangle PZS going afore, and heere now againe expreſſed ſhall be made.

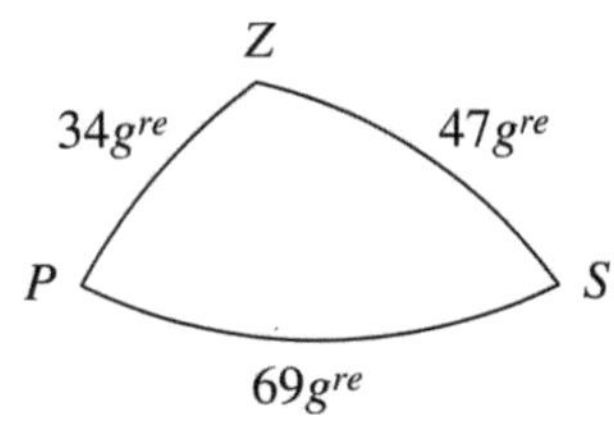

Wherein PZ is 34 degrees,
ZS is 47 degrees,
And PS is 69 degrees,

As alſo of the angles of this repeated triangle ſhall mutually bee made the ſides of the other:

For the angle of this triangle ZSP $31, 6^{\prime}, 5^{\prime\prime}$. is the ſide of the other triangle QR.

And the angle of this ZSP $42, 29^{\prime}, 59^{\prime\prime}$. is the ſide of the other RT.

And of the third angle of this triangle which is SZP $120, 24^{\prime}, 49^{\prime\prime}$, the remainder to a ſemicircle that is $59, 35^{\prime}, 11^{\prime\prime}$, is the ſide QT of the other triangle.

The demonſtration whereof *Bartholomew Pitiſcus*, *Adrian Metius* and others ſet downe, therefore I thinke it no need to repeate the ſame in this ſhort Treatiſe.

12 *From whence it followeth, that the three angles of a Sphæricall triangle being giuen, the ſides are found by an eaſie converſion.*

As in the former triangle QRT, let the angles be giuen Q 47, R 111, and T 34, let the ſides be ſought.

For any one angle, for example ſake, (as before) for R 111, let the remainder thereof to a ſemicircle 69 degrees be taken.

Then 47, 69, 34 being ſet for ſides (as was done in the former triangle PZS, by any of the three wayes aboue written, ſeeke his angles, and you ſhall finde,

Againſt the ſide 47, the angle 42. 29′. 59″.

And againſt the ſide 34 the angle 31, 6′, 5″.

And againſt the ſide 69. (which wee put for 111) you ſhall finde the angle 120, 24′, 49″. Therfore in the triangle propounded, *QRT*.

For the ſide *RT* ſubtending the angle *Q* 47, ſet downe 42 29′ 59″.

And for the ſide *QR* ſubtending the angle *T* 34, ſet downe 31. 6′, 5″.

But for the ſide *QT* ſubtending the angle *R* 111. ſet downe 59, 35′, 11″.

Which are the remainder of 120 deg. 24′, 49″. to a ſemicircle, becauſe before you tooke for 111 his remainder to a ſemicircle, that is 69. And ſo by conuerſion you ſhal finde the ſides by the angles.

An Admonition.

OVt of this finding of the ſides by the angles giuen, three diuers queſtions of this and any other triangle whatſoeuer are reſolued.

As in the triangle *PZS* out of *the houre of the day, the azimuth of the Sunne, and the angle of the ſite or poſition of the Sunne*, this propoſition going before, giueth ſatisfaction to the queſtion, whereby either

1 The height of the pole,

2 The height of the Sunne, or

3 The declination of the Snnne is demanded.

Therefore (out of the 8 *ſ*ect. of the former 5 chap. and the 7 & 13 *ſ*ect. of this booke) you haue the *ſ*olution of 60 *ſ*euerall que*ſ*tions, wᶜʰ fall into any triangle: neither can there any moe varieties then the*ſ*e ari*ſ*e out of the manifold compoſition of any three parts.

You haue therfore a perfect & ab*ſ*olute doctrine of triangles, as well *Sphæricall* as *Plaine*.

THE CONCLVSION.

N*Ow therfore it hath been ſufficiently ſhewed that there are Logarithmes, what they are, and of what vſe they are: for with helpe of them we haue both demonſtratiuely ſhewed and taught by examples of both kindes of Trigonometrie, that the Arithmeticall ſolution of any Geometricall queſtion may moſt readily bee performed without trouble of Multiplication, Diuiſion, or extraction of roots. You haue therfore the admirable Table of Logarithmes that was promiſed, together with the moſt plentifull vſe thereof, which if* (*to you of the learneder ſort*) *I ſhall by your letters vnderſtand to be acceptable to you, I ſhall be incouraged to ſet forth alſo the way to make the Table. In the meane time, make vſe of this ſhort Treatiſe, and giue all praiſe and glory to God the high Inuenter and guider of all good workes.*

The end of the Treatiſe.
Now followeth the Table or Canon of Logarithmes.

Deg. **0**

0 + | −

min	Sine	Logarithm	Difference	logarithm	Sine	
0	0	Infinity	Infinity	0	10000000	60
1	2909	81425681	81425680	1	10000000	59
2	5818	74494213	74494211	2	9999998	58
3	8727	70439564	70439560	4	9999996	57
4	11636	67562746	67562739	7	9999993	56
5	14544	65331315	65331304	11	9999989	55
6	17453	63508099	63508083	16	9999986	54
7	20362	61966595	61966573	22	9999980	53
8	23271	60631284	60631256	28	9999974	52
9	26180	59453453	59453418	35	9999967	51
10	29088	58399857	58399814	43	9999959	50
11	31997	57446759	57446707	52	9999950	49
12	34906	56576646	56576584	62	9999940	48
13	37815	55776222	55776149	73	9999928	47
14	40724	55035148	55035064	84	9999917	46
15	43636	54345225	54345129	96	9999905	45
16	46541	53699843	53699734	109	9999892	44
17	49450	53093600	53093577	123	9999878	43
18	52359	52522019	52521881	138	9999863	42
19	55268	51981356	51981202	154	9999847	41
20	58177	51468431	51468361	170	9999831	40
21	61086	50980537	50980450	187	9999813	39
22	63995	50515354	50515173	205	9999795	38
23	66904	50070827	50070603	224	9999776	37
24	69813	49645239	49644995	244	9999756	36
25	72721	49237030	49236765	265	9999736	35
26	75630	48844826	48844539	287	9999714	34
27	78539	48467431	48467122	309	9999692	33
28	81448	48103763	48103431	332	9999668	32
29	84357	47752859	47752503	356	9999644	31
30	87265	47413852	47413471	381	9999619	30

89

Deg. 0

0 + | −

min	*Sine*	*Logarithm*	*Difference*	*logarithm*	*Sine*	
30	87265	47413852	47413471	381	9999619	30
31	90174	47085961	47085554	407	9999593	29
32	93083	46768483	46768049	434	9999566	28
33	95992	46460773	46460312	461	9999539	27
34	98901	46162254	46161765	489	9999511	26
35	101809	45872392	45871874	518	9999482	25
36	104718	45590688	45590140	548	9999452	24
37	107627	45316714	45316135	579	9999421	23
38	110536	45050041	45049430	611	9999389	22
39	113445	44790296	44789652	644	9999357	21
40	116353	44537132	44536455	677	9999323	20
41	119262	44290216	44289505	711	9999289	19
42	122171	44049255	44048509	746	9999254	18
43	125079	43813959	43813177	782	9999218	17
44	127988	43584078	43583259	819	9999181	16
45	130896	43359360	43358503	857	9999143	15
46	133805	43139582	43138686	896	9999105	14
47	136714	42924534	42923599	935	9999065	13
48	139622	42714014	42713039	975	9999025	12
49	142531	42507833	42506817	1016	9998984	11
50	145439	42305826	42304768	1058	9998942	10
51	148348	42107812	42106711	1101	9998900	9
52	151257	41913644	41912499	1145	9998856	8
53	154165	41723175	41721986	1189	9998811	7
54	157074	41536271	41535037	1234	9998766	6
54	159982	41352795	41351515	1280	9998720	5
56	162891	41172626	41171299	1327	9998673	4
57	165799	41006643	41005268	1375	9998625	3
58	168708	40821746	40820322	1424	9998577	2
59	171616	40650816	40649343	1473	9998527	1
60	174524	40482764	40481241	1523	9998477	0
						min
						Deg.
						89

89

Deg. 1

1

+ | −

min	Sine	Logarithm	Difference	logarithm	Sine	
0	174524	40482764	40481241	1523	9998477	60
1	177433	40317483	40315909	1574	9998426	59
2	180341	40154899	40153273	1626	9998374	58
3	183250	39994918	39993239	1679	9998321	57
4	186158	39837448	39835715	1733	9998267	56
5	189066	39682421	39680633	1788	9998212	55
6	191975	39529765	39527922	1843	9998157	54
7	194883	39379407	39377508	1899	9998101	53
8	197792	39231274	39229318	1956	9998044	52
9	200700	39085307	39083293	2014	9997986	51
10	203608	38941441	38939368	2073	9997927	50
11	206517	38799612	38797479	2133	9997867	49
12	209425	38659767	38657573	2194	9997806	48
13	212333	38521858	38519603	2255	9997745	47
14	215241	38385824	38383507	2317	9997683	46
15	218149	38251613	38249233	2380	9997620	45
16	221057	38119183	38116739	2444	9997556	44
17	223965	37988481	37985972	2509	9997491	43
18	226873	37859471	37856896	2575	9997425	42
19	229781	37732105	37729464	2641	9997359	41
20	232689	37606339	37603631	2708	9997292	40
21	235597	37482135	37479359	2776	9997224	39
22	238505	37359458	37356613	2845	9997155	38
23	241413	37238269	37235354	2915	9997085	37
24	244321	37118532	37115546	2986	9997014	36
25	247229	37000208	36997150	3058	9996943	35
26	250137	36883272	36880142	3130	9996871	34
27	253045	36767690	36764487	3203	9996798	33
28	255953	36653428	36650151	3277	9996724	32
29	258861	36540448	36537096	3352	9996649	31
30	261769	36428748	36425320	3428	9996573	30

88

Deg. **1**

1

\+ | −

min	Sine		Logarithm	Difference	logarithm		Sine	
30	261769		36428748	36425320	3428		9996573	30
31	264677		36318272	36314768	3504		9996496	29
32	267585		36209009	36205427	3582		9996419	28
33	270493		36100924	36097264	3660		9996341	27
34	273401		35994000	35990261	3739		9996262	26
35	276308		35888207	35884388	3819		9996182	25
36	279216		35783520	35779620	3900		9996101	24
37	282124		35679917	35675935	3982		9996019	23
38	285032		35577380	35573316	4064		9995937	22
39	287940		35475892	35471745	4147		9995854	21
40	290847		35375415	35371184	4231		9995770	20
41	293755		35275935	35271619	4316		9995685	19
42	296663		35177444	35173042	4402		9995599	18
43	299570		35079909	35075420	4489		9995512	17
44	302478		34983320	34978743	4577		9995424	16
45	305385		34887652	34882987	4665		9995336	15
46	308293		34792895	34788141	4754		9995247	14
47	311200		34699029	34694185	4844		9995157	13
48	314108		34606036	34601101	4935		9995066	12
49	317015		34513899	34508872	5027		9994974	11
50	319922		34422606	34417486	5120		9994881	10
51	322830		34332140	34326926	5214		9994787	9
52	325737		34242484	34237176	5308		9994693	8
53	328645		34153629	34148226	5403		9994598	7
54	331552		34076549	34071050	5499		9994502	6
54	334459		33978246	33972650	5596		9994405	5
56	337367		33891701	33886007	5694		9994307	4
57	340274		33805893	33800100	5793		9994208	3
58	343181		33720820	33714927	5893		9994109	2
59	346088		33636464	33630471	5993		9994009	1
60	348995		33552817	33546723	6094		9993908	0
								min
								Deg.
								88

88

Deg. 2 **2**

+ | −

min	Sine	Logarithm	Difference	logarithm	Sine	
0	348995	33552817	33546723	6094	9993908	60
1	351902	33469860	33463664	6196	9993806	59
2	354809	33387588	33381289	6299	9993703	58
3	357716	33305993	33299590	6403	9993599	57
4	360623	33225056	33218549	6507	9993495	56
5	363530	33144770	33138158	6612	9993390	55
6	366437	33065128	33058410	6718	9993284	54
7	369344	32986107	32979282	6825	9993177	53
8	372251	32907712	32900779	6933	9993069	52
9	375158	32829923	32822881	7042	9992960	51
10	378064	32752740	32745588	7152	9992850	50
11	380971	32676149	32668887	7262	9992740	49
12	383878	32600139	32592866	7373	9992629	48
13	386785	32524706	32517221	7485	9992517	47
14	389692	32449837	32442239	7598	9992404	46
15	392598	32375526	32367814	7712	9992290	45
16	395505	32301761	32293934	7827	9992175	44
17	398412	32228539	32220596	7943	9992060	43
18	401318	32155852	32147793	8059	9991944	42
19	404225	32083692	32075516	8176	9991827	41
20	407131	32012045	32003751	8294	9991709	40
21	410038	31940909	31932496	8413	9991590	39
22	412944	31870276	31861743	8533	9991470	38
23	415851	31800141	31791487	8654	9991349	37
24	418757	31730492	31721716	8776	9991228	36
25	421663	31661332	31652434	8898	9991106	35
26	424570	31592644	31583623	9021	9990983	34
27	427476	31524424	31515279	9145	9990859	33
28	430382	31456672	31447402	9270	9990734	32
29	433288	31389371	21379975	9396	9990608	31
30	436194	31322524	31313001	9523	9990482	30

87

Deg. **2**

2

\+ | −

min	Sine		Logarithm	Difference	logarithm		Sine	
30	436194		31322524	31313001	9523		9990482	30
31	439100		31256121	31246471	9650		9990355	29
32	442006		31190158	31180380	9778		9990227	28
33	444912		31124626	31114719	9907		9990098	27
34	447818		31059521	31049484	10037		9989968	26
35	450724		30994841	30984673	10168		9989837	25
36	453630		30930577	30920277	10300		9989706	24
37	456536		30866722	30856290	10432		9989574	23
38	459442		30803277	30792712	10565		9989441	22
39	462348		30740230	30729531	10699		9989307	21
40	465253		30677578	30666744	10834		9989172	20
41	468159		30615317	30604347	10970		9989036	19
42	471065		30553442	30542335	11107		9988899	18
43	473970		30491949	30480704	11245		9988761	17
44	476876		30430834	30419451	11383		9988623	16
45	479781		30370090	30358568	11522		9988484	15
46	482687		30309715	30298053	11662		9988344	14
47	485592		30249702	30237899	11803		9988203	13
48	488498		30190049	30178104	11945		9988061	12
49	491403		30130749	30118661	12088		9987918	11
50	494308		30071797	30059565	12232		9987775	10
51	497214		30013193	30000817	12376		9987631	9
52	500119		29954933	29942412	12521		9987486	8
53	503024		29897014	29884347	12667		9987340	7
54	505929		29839424	29826610	12814		9987193	6
54	508834		29782165	29769203	12962		9987045	5
56	511740		29725236	29712125	13111		9986897	4
57	514645		29668628	29655367	13261		9986748	3
58	517550		29612331	29598920	13411		9986598	2
59	520455		29556358	29542796	13562		9986447	1
60	523360		29500706	29486992	13714		9986295	0
								min
								Deg.
								87

87

Deg. **3**
3

min	Sine	Logarithm (+)	Difference (+ | −)	logarithm (−)	Sine	
0	523360	29500706	29486992	13714	9986295	60
1	526265	29445354	29431487	13867	9986143	59
2	529170	29390307	29376286	14021	9985989	58
3	532075	29335565	29321389	14176	9985835	57
4	534980	29281122	29266791	14331	9985680	56
5	537884	29226973	29212486	14487	9985524	55
6	540789	29173115	29158471	14644	9985367	54
7	543694	29119548	29104746	14802	9985209	53
8	546598	29066270	29051309	14961	9985050	52
9	549503	29013273	28998152	15121	9984891	51
10	552407	28960557	28945276	15281	9984731	50
11	555312	28908117	28892675	15442	9984570	49
12	558216	28855951	28840347	15604	9984408	48
13	561120	28804057	28788290	15767	9984245	47
14	564024	28752430	28736499	15931	9984081	46
15	566928	28701071	28684975	16096	9983917	45
16	569832	28649975	28633714	16261	9983752	44
17	572736	28599142	28582715	16427	9983586	43
18	575640	28548570	28531976	16594	9983419	42
19	578544	28498247	28481485	16762	9983251	41
20	581448	28448177	28431246	16931	9983082	40
21	584352	28398354	28381253	17101	9982912	39
22	587256	28348782	28331510	17272	9982742	38
23	590160	28299459	28282015	17444	9982571	37
24	593064	28250377	28232761	17616	9982399	36
25	595967	28201535	28183746	17789	9982226	35
26	598871	28152930	28134967	17963	9982052	34
27	601775	28104561	28086423	18138	9981877	33
28	604678	28056428	28038114	18314	9981701	32
29	607582	28008524	27990033	18491	9981525	31
30	610485	27960848	27942178	18670	9981348	30

86

Deg. **3**
3 + | −

min	*Sine*	*Logarithm*	*Difference*	*logarithm*	*Sine*	
30	610485	27960848	27942178	18670	9981348	30
31	613389	27913400	27894552	18848	9981170	29
32	616292	27866180	27847153	19027	9980991	28
33	619196	27819184	27799977	19207	9980811	27
34	622099	27772408	27753020	19388	9980631	26
35	625002	27725848	27706278	19570	9980450	25
36	627905	27679504	27659752	19752	9980268	24
37	630808	27633374	27613439	19935	9980085	23
38	633711	27587457	27567338	20119	9979901	22
39	636614	27541753	27521449	20304	9979716	21
40	639517	27496257	27475767	20490	9979530	20
41	642420	27450968	27430291	20677	9979343	19
42	645323	27405885	27385020	20865	9979156	18
43	648226	27361003	27339950	21053	9978968	17
44	651129	27316323	27295081	21242	9978779	16
45	654031	27271843	27250411	21432	9978589	15
46	656934	27227563	27205940	21623	9978398	14
47	659837	27183476	27161661	21815	9978207	13
48	662739	27139581	27117573	22008	9978015	12
49	665642	27095878	27073676	22202	9977822	11
50	668544	27052373	27029976	22397	9977628	10
51	671447	27009057	26986465	22592	9977433	9
52	674349	26965926	26943138	22788	9977237	8
53	677251	26922980	26899995	22985	9977040	7
54	680153	26880218	26857035	23183	9976843	6
54	683055	26837639	26814257	23382	9976645	5
56	685957	26795243	26771661	23582	9976446	4
57	688859	26753027	26729244	23783	9976246	3
58	691761	26710988	26687003	23985	9976045	2
59	694663	26669126	26644939	24187	9975843	1
60	697565	26627442	26603052	24390	9975640	0
						min
						Deg.
						86

86

Deg. **4**

4

+ | −

min	*Sine*	*Logarithm*	*Difference*	*logarithm*	*Sine*	
0	697565	26627442	26603052	24390	9975640	60
1	700467	26585929	26561335	24594	9975437	59
2	703369	26544587	26519788	24799	9975233	58
3	706270	26503416	26478411	25005	9975028	57
4	709172	26462418	26437207	25211	9974822	56
5	712073	26421589	26396171	25418	9974615	55
6	714975	26380927	26355301	25620	9974408	54
7	717876	26340428	26314593	25835	9974200	53
8	720777	26300094	26274050	26044	9973991	52
9	723678	26259923	26233669	26254	9973781	51
10	726579	26219913	26193448	26465	9973570	50
11	729480	26180067	26153390	26677	9973358	49
12	732381	26140377	26113487	26890	9973145	48
13	735282	26100842	26073738	27104	9972931	47
14	738183	26061465	26034146	27319	9972717	46
15	741084	26022244	25994709	27535	9972502	45
16	743985	25983176	25955424	27752	9972286	44
17	746886	25944260	25916290	27970	9972069	43
18	749787	25905496	25877308	28188	9971851	42
19	752688	25866884	25838477	28407	9971633	41
20	755588	25828423	25799796	28627	9971414	40
21	758489	25790110	25751262	28848	9971194	39
22	761389	25751942	25722872	29070	9970973	38
23	764290	25713920	25684727	29293	9970751	37
24	767180	25676043	25646527	29516	9970528	36
25	770090	25638310	25608570	29740	9970304	35
26	772991	25600722	25570757	29965	9970079	34
27	775891	25563273	25533082	30191	9969854	33
28	778791	25525966	25495548	30418	9969628	32
29	781691	25488798	25458152	30646	9969401	31
30	784591	25451769	25420894	30875	9969173	30

85

Deg. **4**

4

+ | −

min	Sine		Logarithm	Difference	logarithm		Sine	
30	784591		25451769	25420894	30875		9969173	30
31	787491		25414876	25383772	31104		9968944	29
32	790391		25378119	25346785	31334		9968715	28
33	793291		25341498	25309933	31565		9968485	27
34	796191		25305013	25273216	31797		9968254	26
35	799090		25268662	25236632	32030		9968022	25
36	801990		25232442	25200178	32264		9967789	24
37	804889		25196355	25163857	32498		9967555	23
38	807789		25160399	25127666	32733		9967320	22
39	810688		25124571	25091602	32969		9967085	21
40	813587		25088870	25055664	33206		9966849	20
41	816486		25053298	25019854	33444		9966612	19
42	819385		25017853	24984170	33683		9966374	18
43	822284		24982533	24948610	33923		9966135	17
44	825183		24947340	24913177	34163		9965895	16
45	828082		24912272	24877868	34404		9965655	15
46	830981		24877326	24842680	34646		9965414	14
47	833880		24842502	24807613	34889		9965172	13
48	836778		24807799	24772666	35133		9964929	12
49	839677		24773219	24737841	35378		9964685	11
50	842575		24738761	24703138	35623		9964440	10
51	845474		24704420	24668551	35869		9964194	9
52	848372		24670196	24634080	36116		9963948	8
53	851271		24636090	24599726	36364		9963701	7
54	854169		24602100	24565487	36613		9963453	6
54	857067		24568228	24531365	36863		9963204	5
56	859965		24534473	24497359	37114		9962954	4
57	862863		24500829	24463463	37366		9962703	3
58	865761		24467298	24429679	37619		9962452	2
59	868659		24433880	24396008	37872		9962200	1
60	871557		24400578	24362452	38126		9961947	0
								min
								Deg.
								85

85

Deg. 5

5

+ | −

min	Sine	Logarithm	Difference	logarithm	Sine	
0	871557	24400578	24362452	38126	9961947	60
1	874455	24367384	24329003	38381	9961693	59
2	877353	24334302	24295665	38637	9961438	58
3	880250	24301329	24262435	38894	9961183	57
4	883148	24268467	24229316	39151	9960927	56
5	886045	24235712	24196303	39409	9960670	55
6	888943	24203064	24163396	39668	9960412	54
7	891840	24170523	24130595	39928	9960153	53
8	894737	24138089	24097900	40189	9959893	52
9	897634	24105760	24065309	40451	9959632	51
10	900531	24073540	24032827	40713	9959370	50
11	903428	24041422	24000446	40976	9959107	49
12	906325	24009408	23968168	41240	9958844	48
13	909222	23977495	23935990	41505	9958580	47
14	912119	23945685	23903914	41771	9958315	46
15	915016	23913978	23871940	42038	9958049	45
16	917913	23882373	23840067	42306	9957782	44
17	920809	23850867	23808292	42575	9957515	43
18	923706	23819460	23776615	42845	9957247	42
19	926602	23788153	23745038	43115	9956978	41
20	929498	23756943	23713557	43386	9956708	40
21	932395	23725832	23682174	43658	9956437	39
22	935291	23694818	23650887	33931	9956165	38
23	938187	23663900	23619695	44205	9955893	37
24	941083	23633080	23588601	44479	9955620	36
25	943979	23602355	23557601	44754	9955346	35
26	946875	23571725	23526695	45030	9955071	34
27	949771	23541190	23495883	45307	9954795	33
28	952667	23510748	23465163	45585	9954518	32
29	955563	23480399	23434535	45864	9954240	31
30	958458	23450143	23403999	46144	9953962	30

84

Deg. 5
5 + | −

min	Sine		Logarithm	Difference	logarithm		Sine	
30	958458		23450143	23403999	46144		9953962	30
31	961354		23419980	23373556	46424		9953683	29
32	964249		23389908	23343203	46705		9953403	28
33	967144		23359927	23312940	46987		9953122	27
34	970039		23330036	23282766	47270		9952840	26
35	972934		23300235	23252681	47554		9952557	25
36	975829		23270525	23222686	47839		9952274	24
37	978724		23240903	23192778	48125		9951990	23
38	981619		23211368	23162956	48412		9951705	22
39	984514		23181920	23133220	48700		9951419	21
40	987408		23152560	23103572	48988		9951132	20
41	990303		23123287	23074010	49277		9950844	19
42	993198		23094100	23044533	49567		9950555	18
43	996092		23064999	23015141	49858		9950266	17
44	998987		23035985	22985836	50149		9949976	16
45	1001881		23007056	22956615	50441		9949685	15
46	1004775		22978212	22927478	50734		9949393	14
47	1007669		22949449	22898421	51028		9949100	13
48	1010563		22920769	22869446	51323		9948807	12
49	1013457		22892172	22840553	51619		9948513	11
50	1016351		22863658	22811742	51916		9948218	10
51	1019245		22835227	22783013	52214		9947922	9
52	1022139		22806878	22754366	52512		9947625	8
53	1025032		22778609	22725798	52811		9947327	7
54	1027926		22750420	22697309	53111		9947028	6
54	1030819		22722311	22668899	53412		9946729	5
56	1033713		22694283	22640569	53714		9946429	4
57	1036606		22666333	22612316	54017		9946128	3
58	1039499		22638461	22584140	54321		9945826	2
59	1042392		22610667	22556041	54626		9945523	1
60	1045285		22582951	22528019	54932		9945219	0
								min
								Deg.
								84

84

Deg. 6

6

+ | −

min	Sine	Logarithm	Difference	logarithm	Sine	
0	1045285	22582951	22528019	54932	9945219	60
1	1048178	22555313	22500075	55238	9944914	59
2	1051071	22527752	22472207	55545	9944609	58
3	1053964	22500267	22444414	55853	9944303	57
4	1056857	22472859	22416697	56162	9943996	56
5	1059749	22445527	22389055	56472	9943688	55
6	1062642	22418272	22361490	56782	9943379	54
7	1065534	22391091	22333998	57093	9943069	53
8	1068426	22363984	22306579	57405	9942759	52
9	1071318	22336951	22279233	57718	9942448	51
10	1074210	22309991	22251959	58032	9942136	50
11	1077102	22283104	22224757	58347	9941823	49
12	1079994	22256290	22197627	58663	9941509	48
13	1082886	22229549	22170570	58979	9941194	47
14	1085778	22202881	22143585	59296	9940879	46
15	1088669	22176285	22116671	59614	9940563	45
16	1091561	22149762	22089829	59933	9940246	44
17	1094452	22123308	22063055	60253	9939928	43
18	1097344	22096925	22036351	60574	9939609	42
19	1100235	22070612	22009717	60895	9939290	41
20	1103126	22044368	21983151	61217	9938970	40
21	1106017	22018195	21956655	61540	9938649	39
22	1108908	21992090	21930226	61864	9938327	38
23	1111799	21966054	21903865	62189	9938004	37
24	1114690	21940086	21877571	62515	9937680	36
25	1117580	21914186	21851344	62842	9937355	35
26	1120471	21888355	21825185	63170	9937029	34
27	1123361	21862590	21799091	63499	9936703	33
28	1126252	21836892	21773064	63828	9936376	32
29	1129142	21811261	21747103	64158	9936048	31
30	1132032	21785698	21721209	64489	9935719	30

83

Deg. **6**

6 + | −

min	Sine		Logarithm	Difference	logarithm		Sine	
30	1132032		21785698	21721209	64489		9935719	30
31	1134922		21760199	21695378	64821		9935389	29
32	1137812		21734767	21669613	65154		9935058	28
33	1140702		21709400	21643912	65488		9934727	27
34	1143592		21684100	21618278	65822		9934395	26
35	1146482		21658865	21592708	66157		9934062	25
36	1149372		21633695	21567202	66493		9933728	24
37	1152261		21608586	21541756	66830		9933393	23
38	1155151		21583540	21516372	67168		9933057	22
39	1158040		21558557	21491050	67507		9932721	21
40	1160929		21533639	21465793	67846		9932384	20
41	1163818		21508781	21440595	68186		9932046	19
42	1166707		21483986	21415459	68527		9931707	18
43	1169596		21459254	21390385	68869		9931367	17
44	1172485		21434585	21365373	69212		9931026	16
45	1175374		21409979	21340423	69556		9930685	15
46	1178263		21385434	21315533	69901		9930343	14
47	1181151		21360949	21290702	70247		9930000	13
48	1184040		21336524	21265931	70593		9929656	12
49	1186928		21312160	21241220	70940		9929311	11
50	1189816		21287855	21216567	71288		9928965	10
51	1192704		21263609	21191972	71637		9928618	9
52	1195592		21239423	21167436	71987		9928271	8
53	1198480		21215297	21142959	72338		9927923	7
54	1201368		21191230	21118540	72690		9927574	6
54	1204255		21167222	21094179	73043		9927224	5
56	1207143		21143273	21069877	73396		9926873	4
57	1210031		21119381	21045631	73750		9926521	3
58	1212918		21095546	21021441	74105		9926169	2
59	1215806		21071769	20997308	74461		9925816	1
60	1218693		21048049	20973231	74818		9925461	0
								min
								Deg.
								83

83

Deg. **7**

7

min	Sine	Logarithm +	Difference	− logarithm	Sine	
0	1218693	21048049	20973231	74818	9925461	60
1	1221580	21024385	20949209	75176	9925106	59
2	1224467	21000779	20925245	75534	9924750	58
3	1227354	20977230	20901337	75893	9924393	57
4	1230241	20953738	20877485	76253	9924036	56
5	1233128	20930302	20853688	76614	9923678	55
6	1236015	20906922	20829946	76976	9923319	54
7	1238901	20883595	20806256	77339	9922959	53
8	1241788	20860323	20782620	77703	9922598	52
9	1244674	20837106	20759038	78068	9922236	51
10	1247560	20813945	20735512	78433	9921874	50
11	1250446	20790838	20712039	78799	9921511	49
12	1253332	20767785	20688619	79166	9921147	48
13	1256218	20744785	20665251	79534	9920782	47
14	1259104	20721828	20641935	79903	9920416	46
15	1261990	20698946	20618674	80272	9920049	45
16	1264876	20676106	20595465	80642	9919682	44
17	1267761	20653321	20572308	81013	9919314	43
18	1270647	20630588	20549203	81385	9918945	42
19	1273532	20607906	20526148	81758	9918575	41
20	1276417	20585278	20503146	82132	991820	40
21	1279302	20562701	20480194	82507	9917832	39
22	1282187	20540176	20457293	82883	9917459	38
23	1285072	20517703	20434444	83259	9917086	37
24	1287957	20495281	20411645	83636	9916712	36
25	1290841	20472909	20388895	84014	9916337	35
26	1293726	20450587	20366194	84393	9915961	34
27	1296610	20428316	20343543	84773	9915584	33
28	1299494	20406096	20320942	85154	9915206	32
29	1302378	20383925	20298389	85536	9914828	31
30	1305262	20361806	20275887	85919	9914449	30

82

Deg. **7**

7 + | −

min	Sine	Logarithm	Difference	logarithm	Sine	
30	1305262	20361806	20275887	85919	9914449	30
31	1308146	20339737	20253435	86302	9914069	29
32	1311030	20317717	20231031	86686	9913688	28
33	1313914	20295746	20208675	87071	9913306	27
34	1316798	20273822	20186365	87457	9912923	26
35	1319681	20251947	20164103	87844	9912540	25
36	1322564	20230120	20141888	88232	9912156	24
37	1325447	20208341	20119720	88621	9911771	23
38	1328330	20186611	20097600	89011	9911385	22
39	1331213	20164931	20075530	89401	9910998	21
40	1334096	20143301	20053509	89792	9910610	20
41	1336979	20121717	20031533	90184	9910221	19
42	1339862	20100180	20009603	90577	9909832	18
43	1342744	20078689	19987718	90971	9909442	17
44	1345627	20057245	19965880	91365	9909051	16
45	1348509	20035846	19944086	91760	9908659	15
46	1351392	20014494	19922338	92156	9908266	14
47	1354274	19993189	19900636	92553	9907873	13
48	1357156	19971931	19878980	92951	9907479	12
49	1360038	19950718	19857368	93350	9907084	11
50	1362920	19929552	19835802	93750	9906688	10
51	1365802	19908432	19814281	94151	9906291	9
52	1368683	19887357	19792805	94552	9905893	8
53	1371564	19866327	19771373	94954	9905494	7
54	1374446	19845341	19749984	95357	9905095	6
54	1377327	19824400	19728639	95761	9904695	5
56	1380208	19803504	19707338	96166	9904294	4
57	1383089	19782652	19686080	96572	9903892	3
58	1385970	19761844	19664865	96979	9903489	2
59	1388851	19741081	19643694	97387	9903085	1
60	1391731	19720362	19622566	97796	9902681	0
						min
						Deg.
						82

82

Deg. **8**

8 + | −

min	*Sine*		*Logarithm*	*Difference*	*logarithm*		*Sine*	
0	1391731		19720362	19622566	97796		9902681	60
1	1394612		19699687	19601482	98205		9902276	59
2	1397492		19679054	19580439	98615		9901870	58
3	1400373		19658464	19559438	99026		9901463	57
4	1403253		19637917	19538479	99438		9901055	56
5	1406133		19617413	19517562	99851		9900646	55
6	1409013		19596952	19496687	100265		9900237	54
7	1411893		19576535	19475856	100679		9899827	53
8	1414772		19556160	19455066	101094		9899416	52
9	1417652		19535827	19434317	101510		9899004	51
10	1420531		19515538	19413611	101927		9898591	50
11	1423410		19495290	19392945	102345		9898177	49
12	1426289		19475084	19372320	102764		9897762	48
13	1429168		19454918	19351734	103184		9897347	47
14	1432047		19434794	19331190	103604		9896931	46
15	1434926		19414711	19310686	104025		9896514	45
16	1437805		19394669	19290222	104447		9896096	44
17	1440684		19374668	19269798	104870		9895677	43
18	1443562		19354708	19249414	105294		9895257	42
19	1446441		19334787	19229068	105719		9894837	41
20	1449319		19314908	19208763	106145		9894416	40
21	1452197		19295072	19188501	106571		9893994	39
22	1455075		19275275	19168277	106998		9893571	38
23	1457953		19255517	19148091	107426		9893147	37
24	1460831		19235798	19127943	107855		9892723	36
25	1463708		19216118	19107833	108285		9892298	35
26	1466586		19196477	19087761	108716		9891872	34
27	1469463		19176875	19067727	109148		9891445	33
28	1472340		19157313	19047732	109581		9891017	32
29	1475217		19137792	19027777	110015		9890588	31
30	1478094		19118310	19007861	110449		9890159	30

81

Deg. 8

8 + | −

min	Sine	Logarithm	Difference	logarithm	Sine	
30	1478094	19118310	19007861	110449	9890159	30
31	1480971	19098865	18987981	110884	9889729	29
32	1483848	19079459	18968139	111320	9889298	28
33	1486724	19060091	18908334	111757	9888866	27
34	1489601	19040761	18928566	112195	9888433	26
35	1492477	19021469	18908835	112634	9887999	25
36	1495353	19002215	18889141	113074	9887564	24
37	1498229	18982999	18869485	113514	9887128	23
38	1501105	18963822	18849867	113955	9886692	22
39	1503981	18944682	18830285	114397	9886255	21
40	1506857	18925581	18810741	114840	9885817	20
41	1509733	18906517	18791233	115284	9885378	19
42	1512608	18887489	18771760	115729	9884938	18
43	1515484	18868498	18752323	116175	9884498	17
44	1518359	18849543	18732921	116622	9884057	16
45	1521234	18830625	18713556	117069	9883615	15
46	1524109	18811744	18694227	117517	9883172	14
47	1526984	18792899	18674933	117966	9882728	13
48	1529859	18774090	18655674	118416	9882283	12
49	1532734	18755318	18636451	118867	9881838	11
50	1535608	18736581	18617262	119319	9881392	10
51	1538482	18717882	18598111	119771	9880945	9
52	1541356	18699218	18578994	120224	9880497	8
53	1544230	18680589	18559911	120678	9880048	7
54	1547104	18661995	18540862	121133	9879598	6
54	1549978	18643437	18521848	121589	9879148	5
56	1552852	18624915	18502869	122046	9878697	4
57	1555725	18606428	18483924	122504	9878245	3
58	1558599	18587975	18465013	122962	9877792	2
59	1561472	18569557	18446136	123421	9877338	1
60	1564345	18551174	18427293	123881	9876883	0
						min
						Deg.
						81

81

Deg. **9**

9

+ | −

min	Sine	Logarithm	Difference	logarithm	Sine	
0	1564345	18551174	18427293	123881	9876883	60
1	1567218	18532826	18408484	124342	9876427	59
2	1570091	18514511	18389707	124804	9875971	58
3	1572964	18496231	18370964	125267	9875514	57
4	1575837	18477984	18352253	125731	9875056	56
5	1578709	18459772	18333576	126196	9874597	55
6	1581581	18441594	18314933	126661	9874137	54
7	1584453	18423451	18296324	127127	9873677	53
8	1587325	18405341	18277747	127594	9873216	52
9	1590197	18387265	18259203	128062	9872754	51
10	1593069	18369223	18240692	128531	9872291	50
11	1595941	18351214	18222213	129001	9871827	49
12	1598812	18333237	18203765	129472	9871362	48
13	1601684	18315294	18185351	129943	9870897	47
14	1604555	18297384	18166969	130415	9870431	46
15	1607426	18279507	18148619	130888	9869964	45
16	1610297	18261663	18130301	131362	9869496	44
17	1613168	18243851	18112014	131837	9869027	43
18	1616038	18226071	18093758	132313	9868557	42
19	1618909	18208323	18075533	132790	9868087	41
20	1621779	18190606	18057338	133268	9867616	40
21	1624649	18172924	18039177	133747	9867144	39
22	1627519	18155273	18021047	134226	9866671	38
23	1630389	18137654	18002948	134706	9866197	37
24	1633259	18120067	17984880	135187	9865722	36
25	1636129	18102511	17966842	135669	9865246	35
26	1638999	18084987	17948835	136152	9864770	34
27	1641868	18067495	17930859	136636	9864293	33
28	1644738	18050034	17912913	137121	9863815	32
29	1647607	18032604	17894997	137607	9863336	31
30	1650476	18015207	17877114	138093	9862856	30

80

Deg. 9

9

+ | −

min	Sine		Logarithm	Difference	logarithm		Sine	
30	1650476		18015207	17877114	138093		9862856	30
31	1653345		17997839	17859259	138580		9862376	29
32	1656214		17980503	17841435	139068		9861895	28
33	1659082		17963198	17823641	139557		9861413	27
34	1661951		17945922	17805875	140047		9860930	26
35	1664819		17928677	17788139	140538		9860446	25
36	1667687		17911463	17770433	141030		9859961	24
37	1670555		17894281	17752759	141522		9859475	23
38	1673423		17877128	17735113	142015		9858989	22
39	1676291		17860006	17717497	142509		9858502	21
40	1679159		17842915	17699911	143004		9858014	20
41	1682027		17825852	17682352	143500		9857525	19
42	1684894		17808820	17664823	143997		9857035	18
43	1687761		17791817	17647322	144495		9856544	17
44	1690628		17774843	17629849	144994		9856053	16
45	1693495		17757899	17612406	145493		9855561	15
46	1696362		17740985	17594992	145993		9855068	14
47	1699229		17724100	17577606	146494		9854574	13
48	1702095		17707244	17560248	146996		9854079	12
49	1704962		17690418	17542919	147499		9853583	11
50	1707828		17673622	17525619	148003		9853087	10
51	1710694		17656856	17508349	148507		9852590	9
52	1713560		17640118	17491106	149012		9852092	8
53	1716426		17623408	17473890	149518		9851593	7
54	1719292		17606726	17456701	150025		9851093	6
54	1722157		17590073	17439540	150533		9850593	5
56	1725022		17573448	17422406	151042		9850092	4
57	1727887		17556851	17405299	151552		9849590	3
58	1730752		17540283	17388220	152063		9849087	2
59	1733617		17523744	17371169	152575		9848583	1
60	1736482		17507234	17354146	153088		9848078	0
								min
								Deg.
								80

80

Deg. 10
10

+ | −

min	Sine	Logarithm	Difference	logarithm	Sine	
0	1736482	17507234	17354146	153088	9848078	60
1	1739347	17490751	17337150	153601	9847572	59
2	1742211	17474296	17320181	154115	9847066	58
3	1745075	17457869	17303239	154630	9846559	57
4	1747939	17441470	17286324	155146	9846051	56
5	1750803	17425098	17269435	155663	9845542	55
6	1753667	17408754	17252573	156181	9845032	54
7	1756531	17392438	17235738	156700	9844521	53
8	1759394	17376149	17218929	157220	9844010	52
9	1762258	17359887	17202147	157740	9843498	51
10	1765121	17343652	17185391	158261	9842985	50
11	1767984	17327444	17168661	158783	9842471	49
12	1770847	17311263	17151957	159306	9841956	48
13	1773710	17295109	17135279	159830	9841440	47
14	1776573	17278982	17118627	160355	9840924	46
15	1779435	17262882	17102001	160881	9840407	45
16	1782298	17246809	17085401	161408	9839889	44
17	1785160	17230762	17068827	161935	9839370	43
18	1788022	17214742	17052279	162463	9838850	42
19	1790884	17198749	17035757	162992	9838329	41
20	1793746	17182783	17019261	163522	9837808	40
21	1796608	17166843	17002790	164053	9837286	39
22	1799469	17150929	16986344	164585	9836763	38
23	1802331	17135041	16969924	165117	9836239	37
24	1805192	17119179	16953529	165650	9835714	36
25	1808053	17103342	16937158	166184	9835189	35
26	1810914	17087531	16920812	166719	9834663	34
27	1813774	17071746	16904491	167255	9834136	33
28	1816634	17055987	16888195	167792	9833608	32
29	1819495	17040254	16871924	168330	9833079	31
30	1822355	17024542	16855678	168869	9832549	30

79

Deg. **10**
10

\+ | −

min	Sine		Logarithm	Difference	logarithm		Sine	
30	1822355		17024547	16855678	168869		9832549	30
31	1825215		17008866	16839458	169408		9832019	29
32	1828075		16993210	16823262	169948		9831488	28
33	1830935		16977579	16807090	170489		9830956	27
34	1833795		16961973	16790942	171031		9830423	26
35	1836654		16946392	16774818	171574		9829889	25
36	1839513		16930836	16758718	172118		9829354	24
37	1842372		16915305	16742642	172663		9828818	23
38	1845231		16899799	16726590	173209		9828282	22
39	1848090		16884317	16710561	173756		9827745	21
40	1850949		16868860	16694557	174303		9827207	20
41	1853808		16853428	16678577	174851		9826668	19
42	1856666		16838021	16662621	175400		9826128	18
43	1859524		16822638	16646688	175950		9825587	17
44	1862382		16807280	16630779	176501		9825046	16
45	1865240		16791946	16614893	177053		9824504	15
46	1868098		16776636	16599030	177606		9823961	14
47	1870956		16761351	16583191	178160		9823417	13
48	1873813		16746090	16567375	178715		9822872	12
49	1876670		16730853	16551583	179270		9822227	11
50	1879527		16715640	16535814	179826		9821781	10
51	1882384		16700451	16520068	180383		9821234	9
52	1885241		16685286	16504345	180941		9820686	8
53	1888098		16670145	16488645	181500		9820137	7
54	1890954		16655028	16472968	182060		9819587	6
54	1893810		16639934	16457313	182621		9819037	5
56	1896666		16624864	16441681	183183		9818486	4
57	1899522		16609817	16426072	183745		9817934	3
58	1902378		16594794	16410486	184308		9817381	2
59	1905234		16579794	16394922	184872		9816827	1
60	1908090		16564818	16379381	185437		9816272	0
								min
								Deg.
								79

79

Deg. 11

11

+ | −

min	Sine	Logarithm	Difference	logarithm	Sine	
0	1908090	16564818	16379381	185437	9816272	60
1	1910945	16549865	16363862	186003	9815716	59
2	1913800	16534935	16348365	186570	9815160	58
3	1916655	16520028	16332890	187138	9814603	57
4	1919510	16505144	16317438	187706	9814045	56
5	1922365	16490283	16302008	188275	9813486	55
6	1925220	16475445	16286600	188845	9812926	54
7	1928074	16460630	16271214	189416	9812366	53
8	1930928	16445837	16255849	189988	9811850	52
9	1933782	16431067	16240506	190561	9811243	51
10	1936636	16416320	16225185	191135	9810680	50
11	1939490	16401596	16209886	191710	9810116	49
12	1942344	16386895	16194610	192285	9809551	48
13	1945197	16372216	16179355	192861	9808986	47
14	1948050	16357559	16164121	193438	9808420	46
15	1950903	16342924	16148908	194016	9807853	45
16	1953756	16328311	16133716	194595	9807285	44
17	1956609	16313720	16118545	195175	9806716	43
18	1959462	16299151	16103395	195756	9806147	42
19	1962314	16284604	16088266	196338	9805577	41
20	1965166	16270079	16073159	196920	9805006	40
21	1968018	16255576	16058073	197503	9804434	39
22	1970870	16241095	16043008	198087	9803861	38
23	1973722	16226636	16027964	198672	9803287	37
24	1976574	16212198	16012940	199258	9802712	36
25	1979425	16197782	15997937	199845	9802137	35
26	1982276	16183388	15982955	200433	9801561	34
27	1985127	16169016	15967994	201022	9800984	33
28	1987978	16154665	15953053	201612	9800406	32
29	1990829	16140336	15938133	202203	9799827	31
30	1993679	16126028	15923233	202795	9799247	30

78

Deg. **11**
11

+ | −

min	Sine		Logarithm	Difference	logarithm		Sine	
30	1993679		16126028	15923233	202795		9799247	30
31	1996530		16111742	15908355	203387		9798667	29
32	1999380		16097477	15893497	203980		9798086	28
33	2002230		16083232	15878658	204574		9797504	27
34	2005080		16069008	15863839	205169		9796921	26
35	2007930		16054805	15849040	205765		9796337	25
36	2010780		16040623	15834261	206362		9795753	24
37	2013629		16026462	15819502	206960		9795168	23
38	2016478		16012322	15804764	207558		9794582	22
39	2019327		15998203	15790046	208157		9793995	21
40	2022176		15984105	15775348	208757		9793407	20
41	2025025		15970028	15760670	209358		9792818	19
42	2027874		15955972	15746012	209960		9792228	18
43	2030722		15941936	15731373	210563		9791638	17
44	2033570		15927921	15716754	211167		9791047	16
45	2036418		15913926	15702154	211772		9790455	15
46	2039266		15899951	15687573	212378		9789862	14
47	2042114		15885996	15673012	212984		9789268	13
48	2044962		15872062	15658461	213591		9788674	12
49	2047809		15858148	15643949	214199		9788079	11
50	2050656		15844254	15629446	214808		9787483	10
51	2053503		15830371	15614953	215418		9786886	9
52	2056350		15816518	15600489	216029		9786288	8
53	2059197		15802685	15586044	216641		9785689	7
54	2062043		15788871	15571617	217254		9785090	6
54	2064889		15775077	15557210	217867		9784490	5
56	2067735		15761303	15542822	218481		9783889	4
57	2070581		15747559	15528463	219096		9783287	3
58	2073427		15733824	15514112	219712		9782684	2
59	2076272		15720109	15499780	220329		9782080	1
60	2079117		15706414	15485467	220947		9781476	0
								min
								Deg.
								78

78

Deg. **12**

12 + | −

min	Sine	Logarithm	Difference	logarithm	Sine	
0	2079117	15706414	15485467	220947	9781476	60
1	2081962	15692738	15471172	221566	9780871	59
2	2084807	15679082	15456896	222186	9780265	58
3	2087652	15665445	15442639	222806	9779658	57
4	2090497	15651828	15428401	223427	9779050	56
5	2093342	15638230	15414181	224049	9778442	55
6	2096180	15624651	15399979	224672	9777833	54
7	2099030	15611092	15385796	225296	9777223	53
8	2101874	15597552	15371631	225921	9776612	52
9	2104718	15584031	15357484	226547	9776000	51
10	2107562	15570530	15343356	227174	9775387	50
11	2110405	15557048	15329246	227802	9774773	49
12	2113248	15543585	15315155	228430	9774159	48
13	2116091	15530141	15301082	229059	9773544	47
14	2118934	15516715	15287026	229689	9772928	46
15	2121777	15503308	15272988	230320	9772311	45
16	2124620	15489920	15258968	230952	9771693	44
17	2127462	15476551	15244966	231585	9771075	43
18	2130304	15463200	15230981	232219	9770456	42
19	2133146	15449868	15217014	232854	9769836	41
20	2135988	15436554	15203064	233490	9769215	40
21	2138830	15423259	15189133	234126	9768593	39
22	2141671	15409982	15175219	234763	9767970	38
23	2144512	15396724	15161323	235401	9767347	37
24	2147353	15383484	15147444	236040	9766723	36
25	2150194	15370262	15133582	236680	9766098	35
26	2153035	15357059	15119738	237321	9765472	34
27	2155876	15343874	15105911	237963	9764845	33
28	2158716	15330708	15092102	238606	9764217	32
29	2161556	15317560	15078310	239250	9763589	31
30	2164396	15304430	15064535	239895	9762960	30

77

Deg. **12**
12

+ | −

min	Sine		Logarithm	Difference	logarithm		Sine	
30	2164396		15304430	15064535	239895		9762960	30
31	2167236		15291319	15050779	240540		9762330	29
32	2170076		15278226	15037040	241186		9761699	28
33	2172916		15265150	15023317	241833		9761067	27
34	2175755		15252092	15009611	242481		9760435	26
35	2178594		15239052	14995922	243130		9759802	25
36	2181433		15226030	14982250	243780		9759168	24
37	2184272		15213025	14968594	244431		9758533	23
38	2187111		15200038	14954955	245083		9757897	22
39	2189949		15187068	14941333	245735		9757260	21
40	2192787		15174116	14927728	246388		9756623	20
41	2195625		15161182	14914140	247042		9755985	19
42	2198463		15148266	14900569	247697		9755346	18
43	2201300		15135367	14887014	248353		9754706	17
44	2204137		15122485	14873475	249010		9754065	16
45	2206974		15109621	14859953	249668		9753423	15
46	2209811		15096774	14846447	250327		9752781	14
47	2212648		15083944	14832957	250987		9752138	13
48	2215485		15071132	14819485	251647		9751494	12
49	2218322		15058337	14806029	252308		9750849	11
50	2221158		15045559	14792589	252970		9750203	10
51	2223994		15032799	14779166	253633		9749557	9
52	2226830		15020056	14765759	254297		9748910	8
53	2229666		15007330	14752368	254962		9748262	7
54	2232502		14994620	14738992	255628		9747613	6
54	2235337		14981927	14725632	256295		9746963	5
56	2238172		14969251	14712288	256963		9746312	4
57	2241007		14956592	14698960	257632		9745660	3
58	2243842		14943950	14685649	258301		9745008	2
59	2246677		14931325	14672354	258971		9744355	1
60	2249511		14918717	14659075	259642		9743700	0
								min
								Deg.
								77

77

Deg. 13

13

+ | −

min	Sine		Logarithm	Difference	logarithm		Sine	
0	2249511		14918717	14659075	259642		9743700	60
1	2252345		14906126	14645812	260314		9743045	59
2	2255179		14893551	14632564	260987		9742389	58
3	2258013		14880993	14619332	261661		9741733	57
4	2260847		14868452	14606116	262336		9741076	56
5	2263680		14855927	14592916	263011		9740418	55
6	2266513		14843419	14579732	263687		9739759	54
7	2269346		14830928	14566564	264364		9739099	53
8	2272179		14818453	14553411	265042		9738439	52
9	2275012		14805995	14540274	265721		9737778	51
10	2277844		14793553	14527152	266401		9737116	50
11	2280676		14781128	14514046	267082		9736453	49
12	2283508		14768719	14500955	267764		9735789	48
13	2286340		14756325	14487878	268447		9735124	47
14	2289172		14743947	14474817	269130		9734459	46
15	2292004		14731585	14461771	269814		9733793	45
16	2294835		14719239	14448740	270499		9733126	44
17	2297666		14706909	14435724	271185		9732458	43
18	2300497		14694595	14422723	271872		9731789	42
19	2303328		14682297	14409737	272560		9731120	41
20	2306159		14670015	14396766	273249		9730450	40
21	2308949		14657749	14383810	273939		9729779	39
22	2311819		14645498	14370868	274630		9729107	38
23	2314649		14633263	14357941	275322		9728434	37
24	2317479		14621044	14345029	276015		9727760	36
25	2320309		14608841	14332132	276709		9727085	35
26	2323138		14596654	14319250	277404		9726409	34
27	2325967		14584483	14306384	278099		9725733	33
28	2328799		14572328	14293533	278795		9725056	32
29	2331625		14560189	14280697	279492		9724378	31
30	2334454		14548066	14267876	280190		9723699	30

76

Deg. 13

13

\+ | −

min	Sine	Logarithm	Difference	logarithm	Sine	
30	2334454	14548066	14267876	280190	9723699	30
31	2337282	14535958	14255069	280889	9723020	29
32	2340110	14523866	14242277	281589	9722340	28
33	2342938	14511789	14229500	282289	9721659	27
34	2345766	14499727	14216737	282990	9720977	26
35	2348594	14487680	14203988	283692	9720294	25
36	2351421	14475648	14191253	384395	9719610	24
37	2354248	14463632	14178533	285099	9718926	23
38	2357075	14451631	14165827	285804	9718241	22
39	2359902	14439645	14153135	286510	9717555	21
40	2362729	14427674	14140457	287217	9716868	20
41	2365555	14415718	14127793	287925	9716180	19
42	2368381	14403777	14115143	288634	9715491	18
43	2371207	14391851	14102507	289344	9714802	17
44	2374033	14379941	14089887	290054	9714112	16
45	2376859	14368046	14077281	290765	9713421	15
46	2379684	14356166	14064689	291477	9712729	14
47	2382509	14344301	14052111	292190	9712036	13
48	2385334	14332451	14039547	292904	9711343	12
49	2388159	14320616	14026997	293619	9710649	11
50	2390983	14308796	14014461	294335	9709954	10
51	2393808	14296991	14001939	295052	9709258	9
52	2396632	14285200	13989430	295770	9708561	8
53	2399456	14273424	13976935	296489	9707863	7
54	2402280	14261662	13964453	297209	9707165	6
54	2405104	14249915	13951986	297929	9706466	5
56	2407927	14238182	13939532	298650	9705766	4
57	2410750	14226464	13927092	299372	9705065	3
58	2413573	14214761	13914666	300095	9704363	2
59	2416396	14203072	13902253	300819	9703660	1
60	2419219	14191398	13889854	301544	9702957	0
						min
						Deg. 76

76

Deg. **14**

14 + | −

min	Sine		Logarithm	Difference	logarithm		Sine	
0	2419219		14191398	13889854	301544		9702957	60
1	2422041		14179738	13877468	302270		9702253	59
2	2424863		14168092	13865095	302997		9701548	58
3	2427685		14156461	13852737	303724		9700842	57
4	2430507		14144844	13840392	304452		9700135	56
5	2433329		14133242	13828061	305181		9699428	55
6	2436150		14121654	13815743	305911		9698720	54
7	2438971		14110081	13803439	306642		9698011	53
8	2441792		14098522	13791148	307374		9697301	52
9	2444613		14086977	13778870	308107		9696590	51
10	2447434		14075447	13766606	308841		9695879	50
11	2450254		14063931	13754355	309576		9695167	49
12	2453074		14052429	13742117	310312		9694454	48
13	2455894		14040940	13729891	311049		9693740	47
14	2458714		14029465	13717679	311786		9693025	46
15	2461533		14018004	13705480	312524		9692309	45
16	2464352		14006557	13693294	313263		9691593	44
17	2467171		13995124	13681121	314003		9690876	43
18	2469990		13983705	13668961	314744		9690158	42
19	2472809		13972300	13656814	315486		9689439	41
20	2475628		13960909	13644680	316229		9688719	40
21	2478446		13949532	13632559	316973		9687998	39
22	2481264		13938168	13620450	317718		9687277	38
23	2484082		13926818	13608354	318464		9686555	37
24	2486900		13915482	13596272	319210		9685832	36
25	2489717		13904159	13584202	319957		9685108	35
26	2492534		13892850	13572145	320705		9684383	34
27	2495351		13881554	13560100	321454		9683657	33
28	2498168		13870272	13548068	322204		9682931	32
29	2500984		13859004	13536049	322955		9682204	31
30	2503800		13847749	13524042	323707		9681476	30

75

Deg. 14

14

min	Sine		+ Logarithm	Difference	− logarithm		Sine	
30	2503800		13847749	13524042	323707		9681476	30
31	2506626		13836508	13512048	324460		9680747	29
32	2509431		13825280	13500066	325214		9680017	28
33	2512248		13814066	13488097	325969		9679287	27
34	2515064		13802865	13476141	326724		9678556	26
35	2517879		13791678	13464198	327480		9677824	25
36	2520694		13780504	13452267	328237		9677091	24
37	2523509		13769343	13440348	328995		9676357	23
38	2526324		13758195	13428441	329754		9675623	22
39	2529138		13747061	13416547	330514		9674888	21
40	2531952		13735940	13404665	331275		9674152	20
41	2524766		13724833	13392796	332037		9673415	19
42	2537580		13713739	13380939	332800		9672677	18
43	2540393		13702658	13369094	333564		9671938	17
44	2543206		13691590	13357262	334328		9671199	16
45	2546019		13680535	13345442	335093		9670459	15
46	2548832		13669493	13333634	335859		9669718	14
47	2551645		13658464	13321838	336626		9668976	13
48	2554458		13647448	13310054	337394		9668233	12
49	2557270		13636445	13298282	338163		9667490	11
50	2560082		13625454	13286521	338933		9666746	10
51	2562894		13614476	13274772	339704		9666001	9
52	2565706		13603511	13263035	340476		9665255	8
53	2568517		13592559	13251310	341249		9664508	7
54	2571328		13581620	13239597	342023		9663761	6
54	2574139		13570694	13227896	342798		9663013	5
56	2576950		13559781	13216208	343573		9662264	4
57	2579760		13548880	13204531	344349		9661514	3
58	2582570		13537992	13192866	345126		9660763	2
59	2585380		13527117	13181213	345904		9660011	1
60	2588190		13516255	13165572	346683		9659258	0
								min
								Deg. 75

75

Deg. 15

15

+ | −

min	Sine		Logarithm	Difference	logarithm		Sine	
0	2588190		13516255	13165572	346683		9659268	60
1	2591000		13505406	13157943	347463		9658505	59
2	2593809		13494570	13146326	348244		9657751	58
3	2596618		13483746	13134720	349026		9646996	57
4	2599427		13472934	13123126	349808		9656240	56
5	2602236		13462135	13111544	350591		9655484	55
6	2605045		13451348	13099973	351375		9654727	54
7	2607853		13440573	13088413	352160		9653969	53
8	2610661		13429810	13076864	352946		9653210	52
9	2613469		13419060	13065327	353733		9652450	51
10	2616277		13408322	13053801	354521		9651689	50
11	2619084		13397596	13042286	355310		9650927	49
12	2621891		13386883	13030783	356100		9950165	48
13	2624698		13376182	13019291	356891		9649402	47
14	2627505		13365493	13007810	357683		9648638	46
15	2630312		13354817	12996341	358476		9647873	45
16	2633118		13344153	12984883	359270		9647108	44
17	2635924		13333502	12973438	360064		9646342	43
18	2638730		13322863	12962004	360859		9645575	42
19	2641536		13312237	12950582	361655		9644807	41
20	2644342		13301623	12939171	362452		9644038	40
21	2647147		13291022	12927772	363250		9643268	39
22	2649952		13280432	12916383	364049		9642498	38
23	2652757		13269854	12905005	364849		9641727	37
24	2655562		13259288	12893638	365650		9640955	36
25	2658366		13248734	12882282	366452		9640182	35
26	2661170		13238191	12870936	367255		9639408	34
27	2663974		13227660	12859601	368059		9638633	33
28	2666777		13217141	12848278	368863		9637858	32
29	2669580		13206633	12836965	369668		9637082	31
30	2672383		13196137	12825663	370474		9636305	30

74

Deg. 15
15

+ | −

min	Sine		Logarithm	Difference	logarithm		Sine	
30	2672383		13196137	12825663	370474		9636305	30
31	2675186		13185653	12814372	371281		9635527	29
32	2677989		13175181	12803092	372089		9634748	28
33	2680792		13164721	12791823	372898		9633969	27
34	2683595		13154273	12780565	373708		9633189	26
35	2686397		13143837	12769318	374519		9632408	25
36	2689199		13133413	12758082	375331		9931626	24
37	2692001		13123000	12746856	376144		9630843	23
38	2694802		13112599	12735641	376958		9630059	22
39	2697603		13102210	12724438	377772		9629275	21
40	2700404		13091833	12713246	378587		9628490	20
41	2703205		13081468	12702065	379403		9627704	19
42	2706005		13071114	12690894	380220		9626917	18
43	2708805		13060771	12679733	381038		9626129	17
44	2711605		13050440	12668583	381857		9625341	16
45	2714405		13040120	12657443	382677		9624552	15
46	2717204		13029812	12646314	383498		9623762	14
47	2720003		13019515	12635195	384320		9622971	13
48	2722802		13009229	12624086	385143		9622179	12
49	2725601		12998955	12612988	385967		9621387	11
50	2728400		12988692	12601901	386791		9620594	10
51	2731198		12978441	12590825	387616		9619800	9
52	2733996		12968201	12579759	388442		9619005	8
53	2736794		12957972	12568703	389269		9618209	7
54	2739592		12947755	12557658	390097		9617413	6
54	2742389		12937549	12546623	390926		9616616	5
56	2745186		12927354	12535598	391756		9615818	4
57	2747983		12917171	12524584	392587		9615019	3
58	2750780		12906999	12513580	393419		9614219	2
59	2753577		12896838	12502586	394252		9613418	1
60	2756373		12886689	12491603	395086		9612617	0
								min
								Deg.
								74

74

Deg. **16**

16 + | −

min	Sine		Logarithm	Difference	logarithm		Sine	
0	2756373		12886689	12491603	395086		9612617	60
1	2759169		12876551	12480630	395921		9611815	59
2	2761965		12866423	12469667	396756		9611012	58
3	2764761		12856306	12458714	397592		9610208	57
4	2767556		12846200	12447771	398429		9609403	56
5	2770351		12836105	12436838	399267		9608598	55
6	2773146		12826021	12425915	400106		9607792	54
7	2775941		12815948	12415002	400946		9606985	53
8	2778735		12805886	12404099	401787		9606177	52
9	2781529		12795835	12393206	402629		9605368	51
10	2784323		12785795	12382323	403472		9604559	50
11	2787117		12775766	12371450	404316		9603749	49
12	2789911		12765748	12360587	405161		9602938	48
13	2792704		12755741	12349734	406007		9602126	47
14	2795497		12745745	12338891	406854		9601313	46
15	2798290		12735760	12328059	407701		9600499	45
16	2801082		12725785	12317236	408549		9599685	44
17	2803874		12715821	12306423	409398		9598870	43
18	2806666		12705868	12295620	410248		9598054	42
19	2809458		12695926	12284827	411099		9597237	41
20	2812250		12685995	12274044	411951		9596419	40
21	2815041		12676075	12263271	412804		9595600	39
22	2817832		12666166	12252508	413658		9594781	38
23	2820623		12656267	12241754	414513		9593961	37
24	2823414		12646379	12231010	415369		9593140	36
25	2826244		12636501	12220275	416226		9592318	35
26	2828994		12626633	12209550	417083		9591495	34
27	2831784		12616776	12197835	417941		9590672	33
28	2834574		12606929	12188129	418800		9589848	32
29	2837364		12597093	12177433	419660		9589023	31
30	2840153		12587267	12166746	420521		9588197	30

73

Deg. 16

16 + | −

min	Sine	Logarithm	Difference	logarithm	Sine	
30	2840153	12587267	12166746	420521	9588197	30
31	2842942	12577452	12156069	421383	9587371	29
32	2845731	12567647	12145401	422246	9586544	28
33	2848520	12557853	12134743	423110	9585716	27
34	2851308	12548069	12124094	423975	9584887	26
35	2854096	12538296	12113455	424841	9584057	25
36	2856884	12528533	12102825	425708	9583226	24
37	2859674	12518780	12092204	426576	9582395	23
38	2862459	12509038	12081593	427445	9581563	22
39	2865246	12499306	12070992	428314	9580730	21
40	2868033	12489585	12060401	429184	9579896	20
41	2870819	12479874	12049819	430055	9579061	19
42	2873605	12470174	12039247	430927	9578225	18
43	2876391	12460484	12028684	431800	9577389	17
44	2879177	12450804	12018130	432674	9576552	16
45	2881963	12441134	12007585	433549	9575714	15
46	2884748	12431474	11997049	434425	9574875	14
47	2887533	12421824	11986522	435302	9574036	13
48	2890318	12412184	11976004	436180	9573196	12
49	2893103	12402554	11965495	437059	9572355	11
50	2895888	12392934	11954996	437938	9571513	10
51	2898672	12383324	11944506	438818	9570670	9
52	2901456	12373724	11934025	439699	9569826	8
53	2904240	12364134	11923553	440581	9568982	7
54	2907023	12354554	11913090	441464	9568137	6
54	2909806	12344984	11902636	442348	9567291	5
56	2912589	12335425	11892192	443233	9566444	4
57	2915371	12325876	11881757	444119	9565596	3
58	2918153	12316337	11871330	445007	9564747	2
59	2920935	12306808	11860912	445896	9563898	1
60	2923717	12297289	11850503	446786	9563048	0
						min
						Deg.
						73

73

Deg. **17**

17 + | −

min	Sine		Logarithm	Difference	logarithm		Sine	
0	2923717		12297289	11850503	446786		9563048	60
1	2926499		12287780	11840104	447676		9562197	59
2	2929280		12278280	11829713	448567		9561345	58
3	2932061		12268790	11819331	449459		9560492	57
4	2934842		12259310	11808958	450352		9559639	56
5	2937623		12249840	11798594	451246		9558785	55
6	2940403		12240379	11788239	452140		9557930	54
7	2943183		12230928	11777893	453035		9557074	53
8	2945963		12221487	11767556	453931		9556217	52
9	2948743		12212056	11757228	454828		9555360	51
10	2951523		12202634	11746908	455726		9554502	50
11	2954302		12193222	11736597	456625		9553643	49
12	2957081		12183820	11726295	457525		9552783	48
13	2959860		12174427	11716001	458426		9551922	47
14	2962630		12165044	11705716	459328		9551061	46
15	2965416		12155671	11695440	460231		9550199	45
16	2968194		12146308	11685173	461135		9549336	44
17	2970972		12136954	11674914	462040		9548472	43
18	2973750		12127610	11664665	462945		9547607	42
19	2976527		12118276	11654425	463851		9546742	41
20	2979304		12108952	11644194	464758		9545876	40
21	2982081		12099637	11633971	465666		9545009	39
22	2984857		12090332	11623757	466575		9544141	38
23	2987633		12081036	11613551	467485		9543272	37
24	2990409		12071749	11603353	468396		9542403	36
25	2993185		12062472	11593164	469308		9541533	35
26	2995960		12053204	11582983	470221		9540662	34
27	2998735		12043945	11572810	471135		9539790	33
28	3001510		12034696	11562646	472050		9538917	32
29	3004284		12025456	11552490	472966		9538043	31
30	3007058		12016225	11542341	473884		9537169	30

72

Deg. **17**
17 + | −

min	*Sine*		*Logarithm*	*Difference*	*logarithm*		*Sine*	
30	3007058		12016225	11542341	473884		9537169	30
31	3009832		12007004	11532202	474802		9536294	29
32	3012606		11997792	11522071	475721		9535418	28
33	3015380		11988589	11511948	476641		9534541	27
34	3018153		11979396	11501835	477561		9533664	26
35	3020926		11970212	11491730	478482		9532786	25
36	3023699		11961037	11481633	479404		9531907	24
37	3026472		11951872	11471545	480327		9531027	23
38	3029244		11942716	11461465	481251		9530146	22
39	3032016		11933569	11451393	482176		9529264	21
40	3034788		11924431	11441329	483102		9528382	20
41	3037559		11915303	11431274	484029		9527499	19
42	3040330		11906184	11421227	484957		9526615	18
43	3043101		11897074	11411188	485886		9525730	17
44	3045872		11887973	11401157	486816		9524844	16
45	3048643		11878881	11391134	487747		9523958	15
46	3051413		11869798	11381119	488679		9523071	14
47	3054183		11860724	11371113	489611		9522183	13
48	3056953		11851659	11361115	490544		9521294	12
49	3059723		11842603	11351125	491478		9520404	11
50	3062492		11833557	11341144	492413		9519514	10
51	3065261		11824520	11331171	493349		9518623	9
52	3068030		11815492	11321206	494286		9517731	8
53	3070798		11806473	11311249	495224		9516838	7
54	3073566		11797463	11301300	496163		9515944	6
54	3076334		11788461	11291358	497103		9515050	5
56	3079102		11779468	11281424	498044		9514155	4
57	3081869		11770484	11271498	498986		9513259	3
58	3084636		11761509	11261580	499929		9512362	2
59	3087403		11752543	11251670	500873		9511464	1
60	3090170		11743586	11241768	501818		9510565	0
								min
								Deg.
								72

72

Deg. **18**
18

+ | −

min	Sine	Logarithm	Difference	logarithm	Sine	
0	3090170	11743586	11241768	501818	9510565	60
1	3092936	11734638	11231874	502764	9509666	59
2	3095702	11725699	11221988	503711	9508766	58
3	3098468	11716768	11212109	504659	9507865	57
4	3101234	11707846	11202239	505607	9506963	56
5	3103999	11698933	11192377	506556	9506061	55
6	3106764	11690029	11182523	507506	9505158	54
7	3109529	11681133	11172676	508457	9504254	53
8	3112294	11672246	11162837	509409	9503349	52
9	3115058	11663368	11153006	510362	9502443	51
10	3117822	11654499	11143183	511316	9501536	50
11	3120586	11645638	11133367	512271	9500629	49
12	3123349	11636786	11123559	513227	9499721	48
13	3126112	11627943	11113759	514184	9498812	47
14	3128875	11619109	11103967	515142	9497902	46
15	3131638	11610283	11094182	516101	9496991	45
16	3134400	11601466	11084405	517061	9496080	44
17	3137162	11592658	11074637	518021	9495168	43
18	3139924	11583858	11064876	518982	9494255	42
19	3142686	11575067	11055123	519944	9493341	41
20	3145448	11566285	11045378	520907	9492427	40
21	3148209	11557511	11035640	521871	9491512	39
22	3150970	11548746	11025910	522836	9490596	38
23	3153731	11539989	11016178	523802	9489679	37
24	3156491	11531240	11006471	524769	9488761	36
25	3159251	11522500	10996763	525737	9487842	35
26	3162011	11513768	10987062	526706	9486923	34
27	3164770	11505045	10977369	527676	9486003	33
28	3167529	11496330	10967683	528647	9485082	32
29	3170288	11487624	10958004	529620	9484160	31
30	3173047	11478926	10948332	530594	9483237	30

71

Deg. 18

18

min	Sine		+ Logarithm	Difference	− logarithm		Sine	
30	3173047		11478926	10948332	530594		9483237	30
31	3175805		11470237	10938669	531568		9482314	29
32	3178563		11461556	10929013	532543		9481390	28
33	3181321		11452883	10919364	533519		9480465	27
34	3184079		11444219	10909723	534496		9479539	26
35	3186837		11435563	10900090	535473		9478612	25
36	3189594		11426915	10890464	536451		9477685	24
37	3192351		11418275	10880845	537430		9476757	23
38	3195108		11409644	10871234	538410		9475828	22
39	3197864		11401021	10861630	539391		9474898	21
40	3200620		11392406	10852033	540373		9473967	20
41	3203375		11383800	10842444	541356		9473035	19
42	3206130		11375202	10832862	542340		9472103	18
43	3208885		11366612	10823287	543325		9471170	17
44	3211640		11358030	10813719	544311		9470236	16
45	3214395		11349456	10804158	545298		9469301	15
46	3217150		11340891	10794605	546286		9468366	14
47	3219904		11332334	10785059	547275		9467430	13
48	3222658		11323785	10775520	548265		9466493	12
49	3225412		11315244	10765988	549256		9465555	11
50	3228165		11306711	10756462	550249		9464616	10
51	3230918		11298186	10746944	551242		9463677	9
52	3233671		11289670	10737434	552236		9462737	8
53	3236423		11281162	10727931	553231		9461796	7
54	3239175		11272662	10718436	554226		9460854	6
54	3241927		11264170	10708948	555222		9459911	5
56	3244679		11255686	10699467	556219		9458968	4
57	3247430		11247210	10689993	557217		9458024	3
58	3250181		11238742	10680526	558216		9457079	2
59	3252932		11230282	10671066	559216		9456133	1
60	3255682		11221830	10661613	560217		9455186	0
								min
								Deg.
								71

71

Deg. **19**

19 + | −

min	*Sine*	*Logarithm*	*Difference*	*logarithm*	*Sine*	
0	3255682	11221830	10661613	560217	9455186	60
1	3258432	11213386	10652167	561219	9454239	59
2	3261182	11204950	10642728	562222	9453291	58
3	3263931	11196522	10633296	563226	9452342	57
4	3266681	11188102	10623871	564231	9451392	56
5	3269430	11179690	10614453	565237	9450441	55
6	3272179	11171286	10605042	566244	9449490	54
7	3274927	11162889	10595637	567252	9448538	53
8	3277675	11154500	10586239	568261	9447585	52
9	3280423	11146119	10576849	569270	9446631	51
10	3283171	11137746	10567466	570280	9445676	50
11	3285918	11129381	10558090	571291	9444720	49
12	3288665	11121024	10548721	572303	9443764	48
13	3291412	11112675	10539359	573316	9442807	47
14	3294159	11104334	10530004	574330	9441849	46
15	3296906	11096000	10520655	575345	9440890	45
16	3299652	11087674	10511313	576361	9439931	44
17	3302398	11079356	10501977	577379	9438971	43
18	3305144	11071046	10492648	578398	9438010	42
19	3307889	11062744	10483326	579418	9437048	41
20	3310634	11054449	10474010	580439	9436085	40
21	3313379	11046162	10464702	581460	9435122	39
22	3316123	11037883	10455401	582482	9434158	38
23	3318867	11029612	10446107	583505	9433193	37
24	3321611	11021348	10436819	584529	9432227	36
25	3324355	11013092	10427538	585554	9431260	35
26	3327098	11004843	10418263	586580	9430293	34
27	3329841	10996602	10408995	587607	9429325	33
28	3332585	10988368	10399733	588635	9428356	32
29	3335327	10980142	10390479	589663	9427386	31
30	3338069	10971923	10381231	590692	9426415	30

70

Deg. **19**
19

+ | −

min	Sine		Logarithm	Difference	logarithm		Sine	
30	3338069		10971923	10381231	590692		9426415	30
31	3340811		10963712	10371990	591722		5425444	29
32	3343553		10955509	10362756	592753		9424472	28
33	3346294		10947313	10353528	593785		9423499	27
34	3349035		10939125	10344307	594818		9422525	26
35	3351776		10930944	10335092	595852		9421550	25
36	3354516		10922771	10325884	596887		9420575	24
37	3357256		10914606	10316682	597924		9419599	23
38	3359996		10906448	10307486	598962		9418622	22
39	3362736		10898298	10298297	600001		9417644	21
40	3365475		10890156	10289115	601041		9416665	20
41	3368214		10882021	10279940	602081		9415685	19
42	3370953		10873894	10270772	603122		9414705	18
43	3373691		10865774	10261610	604164		9413724	17
44	3376429		10857661	10252454	605207		9412742	16
45	3379167		10849555	10243304	606251		9411760	15
46	3381905		10841457	10234161	607296		9410777	14
47	3384642		10833366	10225024	608342		9409793	13
48	3387379		10825282	10215893	609389		9408808	12
49	3390116		10817206	10206770	610436		9407822	11
50	3392852		10809137	10197653	611484		9406836	10
51	3395588		10801075	10188542	612533		9405849	9
52	3398324		10793021	10179438	613583		9404861	8
53	3401060		10784974	10170340	614634		9403872	7
54	3403795		10776934	10161248	615686		9402882	6
54	3406530		10768902	10152162	616740		9401891	5
56	3409265		10760877	10143082	617795		9400900	4
57	3411999		10752860	10134009	618851		9399908	3
58	3414733		10744850	10124942	619908		9398915	2
59	3417467		10736847	10115881	620966		9397921	1
60	3420201		10728852	10106827	622025		9396926	0
								min
								Deg.
								70

70

Deg. **20**

20 + | −

min	Sine		Logarithm	Difference	logarithm		Sine	
0	3420201		10728852	10106827	622025		9396926	60
1	3422934		10720865	10097781	623084		9395931	59
2	3425667		10712885	10088741	624144		9394935	58
3	3428400		10704912	10079707	625205		9393938	57
4	3431133		10696945	10070678	626267		9392940	56
5	3433865		10688984	10061654	627330		9391941	55
6	3436597		10681030	10052636	628394		9390942	54
7	3439329		10673085	10043626	629459		9389942	53
8	3442060		10665147	10034622	630525		9388941	52
9	3444791		10657216	10025624	631592		9387939	51
10	3447522		10649292	10016632	632660		9386937	50
11	3450253		10641375	10007646	633729		9385934	49
12	3452983		10633465	9998666	634799		9384930	48
13	3455713		10625562	9989692	635870		9383925	47
14	3458442		10617667	9980725	636942		9382919	46
15	3461171		10609779	9971764	638015		9381913	45
16	3463900		10601898	9962810	639088		9380906	44
17	3466629		10594024	9953862	640162		9379898	43
18	3469357		10586157	9944920	641237		9378889	42
19	3472085		10578297	9935984	642313		9377880	41
20	3474813		10570444	9927054	643390		9376870	40
21	3477540		10562598	9918130	644468		9375859	39
22	3480267		10554760	9909213	645547		9374847	38
23	3482994		10546929	9900302	646627		9373834	37
24	3485721		10539104	9891396	647708		9372820	36
25	3488447		10531286	9882496	648790		9371806	35
26	3491173		10523474	9873601	649873		9370791	34
27	3493899		10515669	9864711	650958		9369775	33
28	3496624		10507871	9855827	652044		9368758	32
29	3499343		10500080	9846949	653131		9367740	31
30	3502075		10492295	9838076	654219		9366722	30

69

Deg. 20
20

min	Sine	Logarithm	+ | − Difference	logarithm	Sine	
30	3502075	10492295	9838076	654219	9366722	30
31	3504799	10484516	9829209	655307	9365703	29
32	3507523	10476745	9820349	656396	9364683	28
33	3510247	10468981	9811495	657486	9363662	27
34	3512971	10461225	9802648	658577	9362640	26
35	3515694	10453476	9793807	659669	9361618	25
36	3518417	10445734	9784972	660762	9360595	24
37	3521140	10437999	9776143	661856	9359571	23
38	3523862	10430271	9767320	662951	9358546	22
39	3526584	10422550	9758503	664047	9357521	21
40	3529306	10414836	9749693	665143	9356495	20
41	3532027	10407129	9740889	666240	9355468	19
42	3534748	10399429	9732091	667338	9354440	18
43	3537469	10391735	9723298	668437	9353411	17
44	3540190	10384047	9714510	669537	9352382	16
45	3542910	10376366	9705728	670638	9351352	15
46	3545630	10368692	9696951	671741	9350321	14
47	3548350	10361024	9688179	672845	9349289	13
48	3551070	10353362	9679412	673950	9348257	12
49	3553789	10345706	9670650	675056	9347224	11
50	3556508	10338057	9661894	676163	9346190	10
51	3559227	10330415	9653144	677271	9345155	9
52	3561945	10322780	9644400	678380	9344119	8
53	3564663	10315152	9635662	679490	9343082	7
54	3567380	10307531	9626930	680601	9342045	6
54	3570097	10299916	9618204	681712	9341007	5
56	3572814	10292308	9609484	682824	9339968	4
57	3575531	10284707	9600770	683937	9338928	3
58	3578247	10277113	9592062	685051	9337887	2
59	3580963	10269526	9583360	686166	9336846	1
60	3583679	10261946	9574664	687282	9335804	0
						min
						Deg. 69

69

Deg. **21**
21

+ | −

min	*Sine*	*Logarithm*	*Difference*	*logarithm*	*Sine*	
0	3583679	10261946	9574664	687282	9335804	60
1	3586395	10254372	9565973	688399	9334761	59
2	3589110	10246804	9557287	689517	9333717	58
3	3591825	10239243	9548607	690636	5332673	57
4	3594540	10231688	9539932	691756	9331628	56
5	3597254	10224140	9531263	692877	9330582	55
6	3599968	10216598	9522599	693999	9329535	54
7	3602682	10209063	9513941	695122	9328488	53
8	3605395	10201534	9505288	696246	9327440	52
9	3608108	10194012	9496642	697370	9326391	51
10	3610821	10186496	9488001	698495	9325341	50
11	3613533	10178987	9479366	699621	9324290	49
12	3616245	10171484	9470736	700748	9323238	48
13	3618957	10163988	9462111	701877	9322186	47
14	3621669	10156498	9453491	703007	9321133	46
15	3624380	10149015	9444877	704138	9320079	45
16	3627091	10141538	9436268	705270	9319024	44
17	3629802	10134067	9427664	706403	9317969	43
18	3632512	10126603	9419066	707537	9316913	42
19	3635222	10119145	9410473	708672	9315856	41
20	3637932	10111694	9401886	709808	9314798	40
21	3640642	10104249	9393305	710944	9313739	39
22	3643351	10096811	9384730	712081	9312680	38
23	3646060	10089379	9376160	713219	9311620	37
24	3648768	10081953	9367595	714358	9310559	36
25	3651476	10074533	9359035	715498	9309497	35
26	3654184	10067120	9350481	716639	9308434	34
27	3656892	10059713	9341931	717782	9307371	33
28	3659599	10052312	9333386	718926	9306307	32
29	3662306	10044918	9324847	720071	9305242	31
30	3665012	10037530	9316313	721217	9304176	30

68

Deg. 21

21

min	Sine	Logarithm +	Difference	− logarithm	Sine	
30	3665012	10037530	9316313	721217	9304176	30
31	3667718	10030148	9307784	722364	9303109	29
32	3670424	10022773	9299261	723512	9302042	28
33	3673130	10015404	9290744	724660	9300974	27
34	3675835	10008041	9282232	725809	9299905	26
35	3678541	10000685	9273726	726959	9298836	25
36	3681246	9993335	9265215	728110	9297766	24
37	3683951	9985991	9256729	729262	9296695	23
38	3686655	9978653	9248238	730415	9295623	22
39	3689359	9971322	6Ɛᄅ5753	731569	9294550	21
40	3692062	9963997	9231273	732724	9293476	20
41	3694765	9956678	9222798	733880	9292401	19
42	3697468	9949366	9214326	735037	9291326	18
43	3700170	9942060	9205865	736195	9290250	17
44	3702872	9934760	9197406	737354	9289173	16
45	3705574	9927466	9188952	738514	9288096	15
46	3708276	9920178	9180503	739675	9287018	14
47	3710977	9912896	9172059	740837	9285939	13
48	3713678	9905620	9163620	742000	9284859	12
49	3716379	9898350	9155186	743164	9283778	11
50	3719080	9891086	9146757	744329	9282697	10
51	3721780	9883828	9138333	745495	9281615	9
52	3724480	9876577	9129915	746662	9280532	8
53	3727179	9869332	9121502	747830	9279448	7
54	3729878	9862093	9113094	748999	9278363	6
54	3732577	9854860	9104691	750169	9277278	5
56	3735275	9847633	9096293	751340	9276192	4
57	3737973	9840412	9087900	752512	9275105	3
58	3740671	9833192	9079512	753685	9274017	2
59	3743369	9825988	9071129	754859	9272928	1
60	3746066	9818785	9062752	756033	9271839	0
						min
						Deg.
						68

68

Deg. 22

22

+ | −

min	Sine		Logarithm	Difference	logarithm		Sine	
0	3746066		9818785	9062752	756033		9271839	60
1	3748763		9811589	9054381	757208		9270749	59
2	3751460		9804399	9046015	758384		9269658	58
3	3754156		9797215	9037654	759561		9268566	57
4	3756852		9790036	9029296	760740		9267474	56
5	3759548		9782863	9020943	761920		9266381	55
6	3762243		9775696	9012595	763101		9265287	54
7	3764938		9768535	9004252	764283		9264192	53
8	3767633		9761380	8995914	765466		9263096	52
9	3770327		9754231	8987581	766650		9262000	51
10	3773021		9747088	8979253	767835		9260903	50
11	3775715		9739950	8970929	769021		9259805	49
12	3778408		9732818	8962610	770208		9258706	48
13	3781101		9725693	8954297	771396		9257606	47
14	3783794		9718574	8945989	772585		9256506	46
15	3786486		9711461	8937686	773775		9255405	45
16	3789178		9704354	8929388	774966		9254303	44
17	3791870		9697353	8921196	776157		9253200	43
18	3794562		9690158	8912809	777349		9252097	42
19	3797253		9683069	8904527	778542		9250993	41
20	3799944		9675986	8896250	779736		9249888	40
21	3802635		9668908	8887977	780931		9248782	39
22	3805325		9661836	8879709	782127		9247676	38
23	3808015		9654770	8871446	783324		9246569	37
24	3810704		9647709	8863187	784522		9245461	36
25	3813393		9640654	8854933	785721		9244352	35
26	3816082		9633605	8846683	786922		9243242	34
27	3818771		9626562	8838438	788124		9242131	33
28	3821459		9619525	8830198	789327		9241020	32
29	3824147		9612494	8821963	790531		9239908	31
30	3826834		9605468	8813732	791736		9238795	30

67

Deg. 22
22

min	Sine	Logarithm +	Difference	− logarithm	Sine	
30	3826834	9605468	8813732	791736	9238795	30
31	3829521	9598448	8805506	792942	9237682	29
32	3832208	9591434	8797285	794149	9236568	28
33	3834895	9584426	8789069	795357	9235453	27
34	3837581	9577424	8780859	796565	9234337	26
35	3840267	9570427	8772653	797774	9233220	25
36	3842953	9563436	8764452	798984	9232103	24
37	3845638	9556451	8756256	800195	9230985	23
38	3848323	9549472	8748065	801407	9229866	22
39	3851008	9542498	8739878	802620	9228746	21
40	3853692	9535530	8731696	803834	9227625	20
41	3856376	9528567	8723518	805049	9226504	19
42	3859060	9521610	8715345	806265	9225382	18
43	3861743	9514659	8707177	807482	9224259	17
44	3864426	9507713	8699013	808700	9223135	16
45	3867109	9500773	8690854	809919	9222010	15
46	3869791	9493839	8682700	811139	9220884	14
47	3872473	9486911	8674551	812360	9219758	13
48	3875155	9479988	8666405	813583	9218631	12
49	3877837	9473071	8658264	814807	9217504	11
50	3880518	9466160	8650128	816032	9216376	10
51	3883199	9459254	8641996	817258	9215247	9
52	3885880	9452354	8633870	818484	9214117	8
53	3888560	9445460	8625749	819711	9212986	7
54	3891240	9438571	8617632	820939	9211855	6
54	3893919	9431688	8609520	822168	9210723	5
56	3896598	9424810	8601412	823398	9209590	4
57	3899277	9417938	8593309	824629	9208456	3
58	3901955	9411071	8585210	825861	920732	2
59	3904633	9404210	8577116	827094	9206185	1
60	3907311	9397354	8569026	828328	9205049	0
						min
						Deg.
						67

67

Deg. 23

23

+ | −

min	Sine	Logarithm	Difference	logarithm	Sine	
0	3907311	9397354	8569026	828328	9205049	60
1	3909989	9390504	8560941	829563	9203912	59
2	3912666	9383660	8552861	830799	9202774	58
3	3915343	9376821	8544785	832036	9201635	57
4	3918020	9369988	8536714	833274	9200496	56
5	3920696	9363160	8528647	834513	9199356	55
6	3923372	9356337	8520584	835753	9198215	54
7	3926048	9349520	8512525	836995	9197073	53
8	3928723	9342708	8504470	838238	9195931	52
9	3931398	9335902	8496420	839482	9194788	51
10	3934072	9329101	8488374	840727	9193644	50
11	3936746	9322306	8480333	841973	9192499	49
12	3939420	9315516	8472296	843220	9191353	48
13	3942093	9308731	8464263	844468	9190207	47
14	3944766	9301952	8456236	845716	9189060	46
15	3947439	9295178	8448213	846965	9187912	45
16	3950112	9288410	8440195	848215	9186763	44
17	3952784	9281647	8432181	849466	9185614	43
18	3955456	9274890	8424171	850718	9184464	42
19	3958128	9268138	8416167	851971	9183313	41
20	3960799	9261392	8408167	853225	9182161	40
21	3963470	9254651	8400171	854480	9181009	39
22	3966140	9247915	8392179	855736	9179856	38
23	3968810	9241185	8384192	856993	9178702	37
24	5971480	9234460	8376209	858251	9177547	36
25	3974149	9227741	8368231	859510	6176391	35
26	3976818	9221027	8360257	860770	9175235	34
27	3979487	9214319	8352288	862031	9174078	33
28	3982155	9207616	8344322	863294	9172920	32
29	3984823	9200918	8336360	864558	9171761	31
30	3987491	9194226	8328403	865823	9170601	30

66

Deg. 23
23 + | −

min	Sine	Logarithm	Difference	logarithm	Sine	
30	3987491	9194226	8328403	865823	9170601	30
31	3990159	9187539	8320450	867089	9169440	29
32	3992826	9180857	8312501	868356	9168279	28
33	3995493	9174181	8304558	869623	9167117	27
34	3998159	9167510	8296619	870891	9165955	26
35	4000825	9160844	8288684	872160	9164792	25
36	4003491	9154183	8280753	873430	9163628	24
37	4006156	9147528	8272827	874701	9162463	23
38	4008821	9140878	8264905	875973	9161297	22
39	4011486	9134233	8256987	877246	9160131	21
40	4014150	9127593	8249073	878520	9158964	20
41	4016814	9120959	8241164	879795	9157796	19
42	4019478	9114330	8233259	881071	9156627	18
43	4022141	9107706	8225358	882348	9155457	17
44	4024804	9101087	8217461	883626	9154286	16
45	4026467	9094473	8209568	884905	9153115	15
46	4030130	9087865	8201679	886186	9151943	14
47	4032792	9081262	8193794	887468	9150770	13
48	4035454	9074664	8185913	888751	9149597	12
49	4038115	9068071	8178036	890035	9148423	11
50	4040776	9061483	8170163	891320	9147248	10
51	4043437	9054901	8162295	892606	9146072	9
52	4046097	9048324	8154431	893893	9144895	8
53	4048757	9041752	8146571	895181	9143718	7
54	4051416	9035185	8138715	896470	9142540	6
54	4054075	9028623	8130863	897760	9141361	5
56	4056734	9022066	8123015	899051	9140181	4
57	4059392	9015514	8115172	900342	9139001	3
58	4062050	9008968	8107334	901634	9137820	2
59	4064708	9002427	8099500	902927	9136638	1
60	4067366	8995891	8091670	904221	9135455	0

min *Deg.* 66

66

Deg. **24**

24 + | −

min	*Sine*	*Logarithm*	*Difference*	*logarithm*	*Sine*	
0	4067366	8995891	8091670	904221	9135455	60
1	4070023	8989360	8083844	905516	9134271	59
2	4072680	8982834	8076022	906812	9133087	58
3	4075337	8976313	8068204	908109	9131902	57
4	4077993	8969797	8060389	909408	9130716	56
5	4080649	8963286	8052578	910708	9129529	55
6	4083305	8956780	8044771	912009	9128342	54
7	4085960	8950280	8036969	913311	9127154	53
8	4088615	8943785	8029171	914614	9125965	52
9	4091269	8937295	8021377	915918	9124775	51
10	4093923	8930810	8013587	917223	9123584	50
11	4096577	8924330	8005801	918529	9122392	49
12	4099231	8917855	7998019	919836	9121200	48
13	4101884	8911385	7990241	921144	9120007	47
14	4104537	8904920	7982467	922453	9118814	46
15	4107189	8898460	7974697	923763	9117620	45
16	4109841	8892005	7966931	925074	9116425	44
17	4112493	8885555	7959169	926386	9115229	43
18	4115144	8879110	7951411	927699	9114032	42
19	4117795	8872670	7943657	929013	9112835	41
20	4120446	8866235	7935908	930327	9111637	40
21	4123096	8859804	7928161	931643	9110438	39
22	4125746	8853379	7920419	932960	9109238	38
23	4128395	8846959	7912681	934278	9108038	37
24	4131044	8840544	7904947	935597	9106837	36
25	4133693	8834134	7897217	936917	9105635	35
26	4136341	8827729	7889491	938238	9104432	34
27	4138989	8821329	7881769	939560	9103228	33
28	4141637	8814934	7874051	940883	9102024	32
29	4144285	8808544	7866337	942207	9100819	31
30	4146932	8802159	7858627	943532	9099613	30

65

Deg. 24

24 + | −

min	Sine	Logarithm	Difference	logarithm	Sine	
30	4146932	8802159	7858627	943532	9099613	30
31	4149579	8795779	7850921	944858	9098406	29
32	4152226	8789404	7843219	946185	9097198	28
33	4154872	8783033	7835520	947513	9095990	27
34	4157518	8776667	7827825	948842	9094781	26
35	4160163	8770306	7820134	950172	9093572	25
36	4162808	8763959	7812456	951503	9092362	24
37	4165453	8757599	7804764	952835	9091151	23
38	4168097	8751253	7797085	954168	9089939	22
39	4170741	8744912	7789409	955503	9088726	21
40	4173385	8738575	7781736	956839	9087512	20
41	4176028	8732243	7774067	958176	9086297	19
42	4178671	8725916	7766402	959514	9085082	18
43	4181313	8719594	7758741	960853	9083866	17
44	4183955	8713277	7751084	962193	9082649	16
45	4186597	8706965	7743431	963534	9081432	15
46	4189239	8700657	7735782	964875	9080214	14
47	4191880	8694354	7728137	966217	9078995	13
48	4194521	8688056	7720496	967560	9077775	12
49	4197162	8681763	7712859	968904	9076555	11
50	4199802	8675475	7705226	970249	9075334	10
51	4202442	8669192	7697597	971595	9074112	9
52	4205081	8662913	7689970	972943	9072889	8
53	4207720	8656639	7682347	974292	9071665	7
54	4210359	8650370	7674728	975642	9070441	6
54	4212997	8644106	7667113	976993	9069216	5
56	4215635	8637846	7659501	978345	9067990	4
57	4218273	8631591	7651893	979698	9066763	3
58	4220910	8625341	7644289	981052	9065535	2
59	4223547	8619096	7636689	982407	9064307	1
60	4226183	8612856	7629093	983763	9063078	0
						min
						Deg.
						65

65

Deg. 25
25

+ | −

min	Sine	Logarithm	Difference	logarithm	Sine	
0	4226183	8612856	7629093	983763	9063078	60
1	4228819	8606620	7621500	985120	9061848	59
2	4231455	8600389	7613911	986478	9060618	58
3	4234090	8594163	7606326	987837	9059387	57
4	4236725	8587942	7598745	989197	9058155	56
5	4239360	8581725	7591167	990558	9056922	55
6	4241994	8575513	7583593	991920	9055688	54
7	4244628	8569306	7576023	993283	9054454	53
8	4247262	8563103	7568456	994647	9053219	52
9	4249895	8556905	7560893	996012	9051983	51
10	4252528	8550712	7553333	997379	9050746	50
11	4255161	8544523	7545776	998747	9049508	49
12	4257793	8538339	7538223	1000116	9048270	48
13	4260425	8532160	7530674	1001486	9047031	47
14	4263065	8525985	7523129	1002856	9045791	46
15	4265687	8519815	7515588	1004227	9044551	45
16	4268318	8513650	7508051	1005599	9043310	44
17	4270949	8507489	7500517	1006972	9042068	43
18	4273579	8501333	7492987	1008346	9040825	42
19	4276209	8495181	7485460	1009721	9039582	41
20	4278838	8489034	7477937	1011097	9038338	40
21	4281467	8482892	7470418	1012474	9037093	39
22	4284096	8476754	7462902	1013852	9035847	38
23	4286724	8470621	7455389	1015232	9034600	37
24	4289352	8464493	7447880	1016613	9033353	36
25	4291979	8458369	7440374	1017995	9032105	35
26	4294606	8452250	7432872	1019378	9030856	34
27	4297233	8446135	7425373	1020762	9029606	33
28	4299859	8440025	7417878	1022147	9028356	32
29	4302485	8433919	7410386	1023533	9027105	31
30	4305111	8427818	7402898	1024920	9025853	30

64

Deg. 25
25

+ | −

min	Sine		Logarithm	Difference	logarithm		Sine	
30	4305111		8427818	7402898	1024920		9025853	30
31	4307736		8421722	7395414	1026308		9024600	29
32	4310361		8415630	7387933	1027697		9023347	28
33	4312986		8409543	7380456	1029087		9022093	27
34	4315610		8403460	7372982	1030478		9020838	26
35	4318234		8397382	7365512	1031870		9019582	25
36	4320858		8391308	7358045	1033263		9018326	24
37	4323481		8385239	7350582	1034657		9017069	23
38	4326104		8379174	7343122	1036052		9015811	22
39	4328726		8373114	7335665	1037449		0014552	21
40	4331348		8367059	7328212	1038847		9013292	20
41	4333970		8361008	7320762	1040246		9012031	19
42	4336591		8354962	7313316	1041646		9010770	18
43	4339212		8348920	7305873	1043047		9009508	17
44	4341833		8342883	7298434	1044449		9008245	16
45	4344453		8336850	7290998	1045852		9006982	15
46	4347073		8330822	7283566	1047256		9005718	14
47	4349693		8324798	7276138	1048660		9004453	13
48	4352312		8318778	7268713	1050065		9003187	12
49	4354931		8312763	7261292	1051471		9001921	11
50	4357549		8306752	7253874	1052878		9000654	10
51	4360167		8300746	7246459	1054287		8999386	9
52	4362785		8294744	7239047	1055697		8998117	8
53	4365402		8288747	7231639	1057108		8996848	7
54	4368019		8282754	7224234	1058520		8995578	6
54	4370635		8276765	7216832	1059933		8994307	5
56	4373251		8270781	7209434	1061347		8993035	4
57	4375867		8264801	7202039	1062762		8991762	3
58	4378482		8258826	7194648	1064178		8999489	2
59	4381097		8252855	7187260	1065595		8989215	1
60	4383712		8246889	7179875	1067014		8987946	0
								min
								Deg.
								64

64

Deg. **26**
26 + | −

min	*Sine*	*Logarithm*	*Difference*	*logarithm*	*Sine*	
0	4383712	8246889	7179875	1067014	8987940	60
1	4386326	8240927	7172493	1068434	8986664	59
2	4388940	8234970	7165115	1069855	8985388	58
3	4391554	8229017	7157740	1071277	8984111	57
4	4394167	8223068	7150368	1072700	8982833	56
5	4396780	8217124	7143001	1074123	8981555	55
6	4399392	8211184	5135637	1075547	8980276	54
7	4402004	8205249	7128277	1076972	8978996	53
8	4404616	8199318	7120920	1078398	8977715	52
9	4407227	8193391	7113566	1079825	8976433	51
10	4409838	8187469	7106216	1081253	8975151	50
11	4412449	8181551	7098868	1082683	8973868	49
12	4415059	8175638	7091524	1084114	8072584	48
13	4417669	8169729	7084183	1085546	8971299	47
14	4420278	8163824	7076845	1086979	8970013	46
15	4422887	8157923	7069510	1088413	8968727	45
16	4425496	8152027	7062179	1089848	8967440	44
17	4428104	8146135	7054851	1091284	8966152	43
18	4430712	8140247	7047526	1092721	8964864	42
19	4433320	8134364	7040205	1094159	8963575	41
20	4435927	8128485	7032887	1095598	8962285	40
21	4438534	8122610	7025572	1097038	8960994	39
22	4441140	8116739	7018260	1098479	8959702	38
23	4443746	8110873	7010952	1099921	8958410	37
24	4446352	8105011	7003647	1101364	8957117	36
25	4448957	8099153	6996345	1102808	8955824	35
26	4451562	8093299	6989045	1104254	8954530	34
27	4454167	8087450	6981749	1105701	8953235	33
28	4456771	8081605	6974456	1107149	8951939	32
29	4459375	8075764	6967166	1108598	8950642	31
30	4461978	8069927	6959879	1110048	8949344	30

63

Deg. **26**

26

min	Sine		Logarithm +	Difference | −	logarithm		Sine	
30	4461978		8069927	6959879	1110048		8949344	30
31	4464581		8064095	6952596	1111499		8948045	29
32	4967184		8058267	6945316	1112951		8946746	28
33	4469786		8052443	6938039	1114404		8945446	27
34	4472388		8046623	6930765	1115858		8944146	26
35	4474990		8040808	6923495	1117313		8942845	25
36	4477591		8034997	6916228	1118769		8941543	24
37	4480192		8029190	6908964	1120226		8940240	23
38	4482792		8023387	6901703	1121684		8938936	22
39	4485392		8017589	6894446	1123143		8937632	21
40	4487992		8011795	6887191	1124604		8936327	20
41	4490591		8006005	6879939	1126066		8935021	19
42	4493190		8000219	6872690	1127529		8933717	18
43	4495788		7994437	6865444	1128993		8932406	17
44	4498386		7988660	6858202	1130458		8931098	16
45	4500984		7982887	6850963	1131924		8929789	15
46	4503582		7977118	6843727	1133391		8928479	14
47	4506179		7971353	6836494	1134859		8927169	13
48	4508776		7965592	6829265	1136327		8925858	12
49	4511372		7959835	6822039	1137796		8924546	11
50	4513968		7954083	6814817	1139266		8923234	10
51	4516563		7948335	6807597	1140738		8921921	9
52	4519158		7942591	6800380	1142211		8920607	8
53	4521753		7936851	6793166	1143685		8919292	7
54	4524347		7931115	6785955	1145160		8917976	6
54	4526941		7925383	6778747	1146636		8916659	5
56	4529535		7919655	6771542	1148113		8915341	4
57	4532128		7913932	6764340	1149592		8914023	3
58	4534721		7908213	6757141	1151072		8912704	2
59	4537313		7902498	6749945	1152553		8911385	1
60	4539905		7896787	6742752	1154035		8910065	0
								min
								Deg.
								63

63

Deg. 27

27

min	Sine	Logarithm	+ | − Difference	logarithm	Sine	
0	4539905	7896787	6742752	1154035	8910065	60
1	4542497	7891080	6735562	1155518	8908744	59
2	4545088	7885377	6728375	1157002	8907422	58
3	4547679	7879678	6721191	1158487	8906099	57
4	4540270	7873983	6714010	1159973	8904776	56
5	4552860	7868292	6706832	1161460	8903452	55
6	4555450	7862605	6699657	1162948	8902127	54
7	4558039	7856923	6692486	1164437	8900802	53
8	4560628	7851245	6685318	1165927	8899476	52
9	4563216	7845571	6678153	1167418	8898149	51
10	4565804	7839901	6670991	1168910	8896821	50
11	4568392	7834235	6663832	1170403	8895492	49
12	4570979	7828573	6656676	1171897	8894163	48
13	4573566	7822915	6649523	1173392	8892833	47
14	4576153	7817261	6642373	1174888	8891502	46
15	4578739	7811611	6635225	1176386	8890171	45
16	4581325	7805965	6628080	1177885	8888839	44
17	4583911	7800323	6620938	1179385	8887506	43
18	4586496	7794685	6613799	1180886	8886172	42
19	4589081	7789051	6606663	1182388	8884838	41
20	4591665	7783422	6599531	1183891	8883503	40
21	4594249	7777797	6592402	1185395	8882167	39
22	4596833	7772176	6585276	1186900	8880830	38
23	4599416	7766558	6578152	1188406	8879492	37
24	4601999	7760944	6571031	1189913	8878154	36
25	4604581	7755334	6563913	1191421	8876815	35
26	4607163	7749728	6556797	1192931	8875475	34
27	4609744	7744126	6549684	1194442	8874134	33
28	4612325	7738528	6542574	1195954	8872793	32
29	4614906	7732934	6535467	1197467	8871451	31
30	4617486	7727344	6528363	1198981	8874108	30

62

Deg. 27
27

min	Sine		Logarithm	+ | − Difference	logarithm		Sine	
30	4617486		7727344	6528363	1198981		8870108	30
31	4620066		7721757	6521261	1200496		8868765	29
32	4622646		7716174	6514162	1202012		8867421	28
33	4625225		7710596	6507067	1203529		8866076	27
34	4627804		7705022	6499975	1205047		8864730	26
35	4630382		7699452	6492886	1206566		8863383	25
36	4632960		7693886	6485800	1208086		8862035	24
37	4635538		7688324	6478717	1209607		8860687	23
38	4638115		7682766	6471637	1211129		8859338	22
39	4640692		7677212	6464560	1212652		8857989	21
40	4643268		7671662	6457485	1214177		8856639	20
41	4645844		7666116	6450413	1215703		8855288	19
42	4648420		7660574	6443344	1217230		8852936	18
43	4650995		7655035	6436277	1218758		8852583	17
44	4653570		7649500	6429213	1220287		8851230	16
45	4656145		7643969	6422151	1221817		8849876	15
46	4658719		7638442	6415094	1223348		8848521	14
47	4661293		7632919	6408039	1224880		8847165	13
48	4663866		7627400	6400987	1226413		8845809	12
49	4666439		7621885	6393938	1227947		8844452	11
50	4669012		7616374	6386893	1229481		8843095	10
51	4671584		7610867	6379850	1231017		8841737	9
52	4674156		7605363	6372809	1232554		8840378	8
53	4676727		7599863	6365771	1234092		8839018	7
54	4679298		7594367	6358735	1235632		8837657	6
54	4681869		7588875	6351702	1237173		8836295	5
56	4684439		7583387	6344672	1238715		8834932	4
57	4687009		7577903	6337645	1240258		8833569	3
58	4689578		7572422	6330620	1241802		8832205	2
59	4692147		7566945	6323598	1243347		8830841	1
60	4694716		7561472	6316578	1244894		8829476	0
								min
								Deg. 62

62

Deg. **28**

28

+ | −

min	Sine	Logarithm	Difference	logarithm	Sine	
0	4694716	7561472	6316578	1244894	8829476	60
1	4697284	7556003	6309561	1246442	8828110	59
2	4699852	7550538	6302547	1247991	8826743	58
3	4702419	7545076	6295535	1249541	8825375	57
4	4704986	7539618	6288526	1251092	8824007	56
5	4707553	7534164	6281520	1252644	8822638	55
6	4710119	7528714	6274517	1254197	8821268	54
7	4712685	7523268	6267517	1255751	8819898	53
8	4715250	7517826	6260521	1257305	8818527	52
9	4717815	7512388	6253528	1258860	8817155	51
10	4720380	7506954	6246538	1260416	8815783	50
11	4722944	7501524	6239550	1261974	8814408	49
12	4725508	7496097	6232564	1263533	8813034	48
13	4728071	7490674	6225581	1265093	8811659	47
14	4730634	7485255	6218601	1266654	8810285	46
15	4733197	7479840	6211624	1268216	8808907	45
16	4735759	7474428	6204649	1269779	8807530	44
17	4738321	7469020	6197676	1271344	8806152	43
18	4740882	7463616	6190706	1272910	8804773	42
19	4743443	7458216	6183739	1274477	8803394	41
20	4746004	7452819	6176774	1276045	8802014	40
21	4748564	7447426	6169812	1277614	8800633	39
22	4751124	7442037	6162853	1279184	8799251	38
23	4753683	7436651	6155896	1280755	8797869	37
24	3756242	7431269	6148942	1282327	8796486	36
25	4758801	7425891	6141991	1283900	8795102	35
26	4761359	7420517	6135063	1285474	8793717	34
27	4763917	7415146	6128096	1287050	8792332	33
28	4766474	7409779	6121152	1288627	8790946	32
29	4769031	7404416	6114211	1290205	8789559	31
30	4771588	7399057	6107273	1291784	8788171	30

61

Deg. **28**
28

\+ | −

min	Sine	Logarithm	Difference	logarithm	Sine	
30	4771588	7399057	6107273	1291784	8788171	30
31	4774144	7393701	6100337	1293364	8786782	29
32	4776700	7388349	6093404	1294945	8785393	28
33	4779255	7383001	6086474	1296527	8784003	27
34	4781810	7377657	6079547	1298110	8782613	26
35	4784365	7372316	6072622	1299694	8781222	25
36	4786919	7366979	6065700	1301279	8779830	24
37	4789473	7361646	6058781	1302865	8778437	23
38	4792026	7356316	6051863	1304453	8777044	22
39	4794579	7350990	6044948	1306042	8775650	21
40	4797132	7345668	6038036	1307632	8774255	20
41	4799684	7340349	6031126	1309223	8772859	19
42	4802236	7335034	6024219	1310815	8771462	18
43	4804787	7329723	6017315	1312408	8770065	17
44	4807338	7324415	6010413	1314002	8768667	16
45	4809888	7319111	6003514	1315597	8767267	15
46	4812438	7313811	5996618	1317193	8765868	14
47	4814988	7308514	5989723	1318791	8764468	13
48	4817537	7303221	5982831	1320390	8763068	12
49	4820086	7297931	5975941	1321990	8761665	11
50	4822635	7292645	5969054	1323591	8760263	10
51	4225183	7287363	5962170	1325193	8158860	9
52	4827731	7282084	5955288	1326796	8757456	8
53	4830278	7276809	5948409	1328400	8756051	7
54	4832825	7271538	5941533	1330005	8754646	6
54	4835371	7266270	5934659	1331611	8753240	5
56	4837917	7261006	5927787	1333219	8751833	4
57	4840462	7255746	5920918	1334828	8750425	3
58	4843007	7250489	5914051	1336438	8749016	2
59	4845552	7245236	5907187	1338049	8741607	1
60	4848096	7239987	5900326	1339661	8746197	0
						min
						Deg.
						61

61

Deg. **29**

29

\+ | −

min	*Sine*		*Logarithm*	*Difference*	*logarithm*		*Sine*	
0	4848096		7239987	5900326	1339661		8746197	60
1	4850640		7234742	5893468	1341274		8744787	59
2	4853184		7229500	5886612	1342888		8743376	58
3	4855727		7224262	5879759	1344503		8741964	57
4	4858270		7219027	5872906	1346119		8740551	56
5	4860812		7213795	5866059	1347736		8739137	55
6	4863354		7208567	5859213	1349354		8737722	54
7	4865895		7203342	5852368	1350974		8736307	53
8	4868436		7198121	5845526	1352595		8734891	52
9	4870977		7192903	5838686	1354217		8733475	51
10	4873517		7187689	5831849	1355840		8732058	50
11	4876057		7182478	5825014	1357464		8730640	49
12	4878596		7177271	5818182	1359089		8729221	48
13	4881135		7172068	5811353	1360715		8727801	47
14	4883674		7166868	5804526	1362342		8726381	46
15	4886212		7161672	5797701	1363971		8724960	45
16	4888750		7156480	5790879	1365601		8723538	44
17	4891287		7151291	5784059	1367232		8722116	43
18	4893824		7146106	5777242	1368864		8720693	42
19	4896361		7140924	5770427	1370497		8719269	41
20	4898897		7135746	5763615	1372131		8717844	40
21	4901433		7130572	5756806	1373766		8716418	39
22	4903968		7125401	5749999	1375402		8714992	38
23	4906503		7120234	5743195	1377039		8713565	37
24	4909037		7115070	5736392	1378678		8712138	36
25	4911571		7109909	5729591	1380318		8710710	35
26	4914105		7104752	5722793	1381959		8709281	34
27	4916638		7099598	5715997	1383601		8707851	33
28	4919171		7094448	5709204	1385244		8706420	32
29	4921703		7089301	5702413	1386888		8704989	31
30	4924235		7084158	5695625	1388533		8703557	30

60

Deg. 29
29

+ | −

min	Sine	Logarithm	Difference	logarithm	Sine	
30	4924235	7084158	5695625	1388533	8703557	30
31	4926767	7079018	5688839	1390179	8702124	29
32	4929298	7073882	5682056	1391826	8700691	28
33	4931829	7068749	5675275	1393474	8699257	27
34	4934359	7063620	5668496	1395124	8697822	26
35	4936889	7058494	5661719	1396775	8696386	25
36	4939418	7053372	5654945	1398427	8694949	24
37	4941947	7048253	5648173	1400080	8693512	23
38	4944476	7043138	5641404	1401734	8692074	22
39	4947004	7038026	5634637	1403389	8690636	21
40	4949532	7032918	5627873	1495045	8689197	20
41	4952059	7027814	5621111	1406703	8687757	19
42	4954586	7022713	5614351	1408362	8686316	18
43	4957113	7017615	5607593	1410022	8684873	17
44	4959639	7012521	5600838	1411683	8683431	16
45	4962165	7007430	5594085	1413345	8681988	15
46	4964690	7002342	5587334	1415008	8680544	14
47	4967215	6997258	5580586	1416672	8679100	13
48	4969740	6992177	5573840	1418337	8677655	12
49	4972264	6987099	5567095	1420004	8676209	11
50	4974788	6982025	5560353	1421672	8674762	10
51	4977311	6576954	5553613	1423341	8673314	9
52	4979834	6971886	5546875	1425011	8671866	8
53	4982356	6966822	5540140	1426682	8670417	7
54	4984878	6961761	5533407	1428354	8668968	6
54	4987399	6956704	5526677	1430027	8667518	5
56	4989920	6951650	5519949	1431701	8666067	4
57	4992441	6946600	5513224	1433376	8664615	3
58	4994961	6941553	5506500	1435053	8663162	2
59	4997481	6936509	5499778	1436731	8661708	1
60	5000000	6931469	5493059	1438410	8660254	0
						min
						Deg.
						60

60

Deg. **30**

30

+ | −

min	Sine		Logarithm	Difference	logarithm		Sine	
0	5000000		6931469	5493059	1438410		8660254	60
1	5002519		6926432	5486342	1440090		8658799	59
2	5005038		6921399	5479628	1441771		8657344	58
3	5007556		6916369	5472916	1443453		8655888	57
4	5010074		6911342	5466206	1445136		8654431	56
5	5012591		6906319	5459498	1446821		8652973	55
6	5015108		6901299	5452792	1448507		8651514	54
7	5017624		6896282	5446088	1450194		8650055	53
8	5020140		6891269	5439387	1451882		8648595	52
9	5022656		6886259	5432688	1453571		8647134	51
10	5025171		6881253	5425992	1455261		8645673	50
11	5027686		6876250	5419298	1456952		8644211	49
12	5030200		6871250	5412605	1458645		8642748	48
13	5032714		6866254	5405915	1460339		8641284	47
14	5035227		6861261	5399227	1462034		8639820	46
15	5037740		6856271	5392541	1463730		8638355	45
16	5040253		6851285	5385858	1465427		8636889	44
17	5042765		6846302	5379177	1467125		8635423	43
18	5045277		6841323	5372499	1468824		8633956	42
19	5047788		6836347	5365822	1470525		8632488	41
20	5050299		6831374	5359147	1472227		8631019	40
21	5052809		6826405	5352475	1473930		8629549	39
22	5055319		6821439	5345805	1475634		8628079	38
23	5057829		6816476	5339137	1477339		8626608	37
24	5060338		6811516	5332471	1479045		8625137	36
25	5062847		6806560	5325808	1480752		8623665	35
26	5065355		6801607	5319147	1482460		8622192	34
27	5067863		6796657	5312488	1484169		8620718	33
28	5070370		6791710	5305831	1485879		8619243	32
29	5072877		6786767	5299177	1487590		8617768	31
30	5075384		6781827	5292525	1489302		8616292	30

59

Deg. **30**
30

+ | −

min	Sine		Logarithm	Difference	logarithm		Sine	
30	5075384		6781827	5292525	1489302		8616292	30
31	5077890		6776890	5285874	1491016		8614815	29
32	5080396		6771956	5279225	1492731		8613338	28
33	5082901		6767026	5272579	1494447		8611860	27
34	5085406		6762099	5265934	1496165		8610381	26
35	5087911		6757175	5259291	1497884		8608901	25
36	5090415		6752255	5252651	1499604		8607420	24
37	5092919		6747338	5246013	1501325		8605939	23
38	5095422		6742424	5239377	1503047		8604457	22
39	5097925		6737513	5232743	1504770		8602975	21
40	5100427		6732606	5226112	1506494		8601492	20
41	5102929		6727702	5219482	1508220		8600008	19
42	5105430		6722802	5212855	1509947		8598523	18
43	5107932		6717905	5206230	1511675		8597037	17
44	5110431		6713011	5199607	1513404		8595551	16
45	5112931		6708120	5192986	1515134		8594064	15
46	5115431		6703232	5186367	1516865		8592577	14
47	5117930		6698348	5179751	1518597		8591089	13
48	5120429		6693467	5173137	1520330		8589600	12
49	5122927		6688589	5166525	1522064		8588110	11
50	5125425		6683714	5159914	1523800		8586619	10
51	5127922		6678842	5153305	1525537		8585127	9
52	5130419		6673974	5146699	1527275		8583635	8
53	5132916		6669109	5140095	1529014		8582142	7
54	5135412		6664247	5133493	1530754		8580649	6
54	5137908		6659388	5126892	1532496		8579155	5
56	5140403		6654532	5120293	1534239		8577660	4
57	5142898		6649680	5113697	1535983		8576164	3
58	5145393		6644831	5107103	1537728		8574668	2
59	5147887		6639985	5100511	1539474		8573171	1
60	5150381		6635142	5093921	1541221		8571673	0
								min
								Deg.
								59

59

Deg. **31**

31 + | −

min	Sine		Logarithm	Difference	logarithm		Sine	
0	5150381		6635142	5093921	1541221		8571673	60
1	5152874		6630302	5087332	1542970		8570175	59
2	5155367		6625465	5080745	1544720		8568676	58
3	5157859		6620631	5074160	1546471		8567176	57
4	5160351		6615801	5067578	1548223		8565675	56
5	5162843		6610974	5060998	1549976		8564173	55
6	5165334		6606150	5054420	1551730		8562671	54
7	5167825		6601329	5047844	1553485		8561168	53
8	5170315		6596512	5041271	1555241		8559664	52
9	5172805		6591698	5034700	1556998		8558160	51
10	5175294		6586887	5028130	1558757		8556655	50
11	5177783		6582079	5021562	1560517		8555149	49
12	5180271		6577275	5014997	1562278		8553643	48
13	5182759		6572474	5008434	1564040		8552136	47
14	5185246		6567676	5001873	1565803		8550628	46
15	5187733		6562881	4995313	1567568		8549119	45
16	5190220		6558089	4988755	1569334		8547609	44
17	5192706		6553300	4982199	1571101		8546096	43
18	5195192		6548514	4975645	1572869		8544588	42
19	5197667		6543731	4969093	1574638		8543077	41
20	5200162		6538951	4962543	1576408		8541565	40
21	5202646		6534174	4955994	1578180		8540052	39
22	5205130		6529400	4949447	1579953		8538538	38
23	5207614		6524629	4942902	1581727		8537024	37
24	5210097		6519862	4936360	1583502		8535509	36
25	5212580		6515098	4929820	1585278		8533993	35
26	5215062		6510337	4923282	1587055		8532476	34
27	5217544		6505580	4916742	1588833		8530958	33
28	5220025		6500826	4910213	1590613		8529440	32
29	5222506		6496075	4903681	1592394		8527921	31
30	5224986		6491327	4897151	1594176		8526402	30

58

Deg. 31
31

min	Sine		Logarithm	+ | − Difference	logarithm		Sine	
30	5224986		6491327	4897151	1594176		8526402	30
31	5227466		6486583	4890624	1595959		8524882	29
32	5229946		6481842	4884098	1597744		8523361	28
33	5232425		6477103	4877573	1599530		8521839	27
34	5234904		6472367	4871050	1601317		8520317	26
35	5237382		6467634	4867529	1603105		8518794	25
36	5239860		6462904	4858010	1604894		8517270	24
37	5242337		6458177	4851493	1606684		8515745	23
38	5244814		6453453	4844978	1608475		8514220	22
39	5247290		6448732	4838465	1610267		8512694	21
40	5249766		6444014	4831954	1612060		8511167	20
41	5252241		6439299	4825444	1613855		8509639	19
42	5254716		6434588	4818937	1615651		8508111	18
43	5257191		6429880	4812432	1617448		8506582	17
44	5259665		6425175	4805929	1619246		8505052	16
45	5262139		6420473	4799427	1621046		8503522	15
46	5264612		6415774	4792927	1622847		8501991	14
47	5267085		6411078	4786429	1624649		8500459	13
48	5269557		6406385	4779933	1626452		8498927	12
49	5272029		6401695	4773439	1628256		8497394	11
50	5274501		6397008	4766947	1630061		8495860	10
51	5276972		6392324	4760456	1631868		8494326	9
52	5279443		6387643	4753967	1633676		8492791	8
53	5281913		6382965	4747480	1635485		8491255	7
54	5284383		6378290	4740995	1637295		8489718	6
54	5286852		6373618	4734512	1639106		8488180	5
56	5299321		6368949	4728031	1640918		8486641	4
57	5291789		6364283	4721552	1642731		8485102	3
58	5294257		6359620	4715074	1644546		8483562	2
59	5296725		6354961	4708599	1646362		8482022	1
60	5299192		6350305	4702126	1648179		8480481	0
								min
								Deg. 58

58

Deg. **32**
32

min	Sine	Logarithm +	Difference +/−	logarithm −	Sine	
0	5299192	6350305	4702126	1648179	8480481	60
1	5301659	6345652	4695655	1649997	8478939	59
2	5304125	6341002	4689186	1651816	8477297	58
3	5306591	6336354	4682717	1653637	8475854	57
4	5309056	6331709	4676240	1655459	8474310	56
5	5311521	6327067	4669785	1657282	8472765	55
6	5313985	6322428	4663322	1659106	8471219	54
7	5316449	6317792	4656861	1660931	8469673	53
8	5318913	6313159	4650402	1662757	8468126	52
9	5321376	6308529	4643944	1664585	8466579	51
10	5323839	6303902	4637488	1666414	8465031	50
11	5326301	6299278	4631034	1668244	8463482	49
12	5328763	6294657	4624582	1670075	8461932	48
13	5331224	6290039	4618131	1671908	8460381	47
14	5333685	6285424	4611682	1673742	8458830	46
15	5336145	6280812	4605235	1675577	8457278	45
16	5338605	6276203	4598790	1677413	8455725	44
17	5341065	6271597	4592347	1679250	8454172	43
18	5343524	6266994	4585906	1681088	8452618	42
19	5345983	6262394	4579467	1682927	8451064	41
20	5348441	6257797	4573030	1684767	8449509	40
21	5350898	6253203	4566594	1686609	8447953	39
22	5353355	6248612	4560160	1688452	8446396	38
23	5355812	6244024	4553728	1690296	8444838	37
24	5358268	6239439	4547298	1692141	8443280	36
25	5360724	6234857	4540859	1693988	8441721	35
26	5363179	6230278	4534442	1695836	8440161	34
27	5365634	6225702	4528017	1697685	8438600	33
28	5368088	6221129	4521594	1699535	8437039	32
29	5370542	6216559	4515172	1701387	8435477	31
30	5372996	6211992	4508752	1703240	8423915	30

57

Deg. **32**
32

+ | −

min	*Sine*		*Logarithm*	*Difference*	*logarithm*		*Sine*	
30	5372996		6211992	4508752	1703240		8433915	30
31	5375449		6207427	4502333	1705094		8432352	29
32	5377902		6202865	4495916	1706949		8430788	28
33	5380354		6198306	4489501	1708805		8429223	27
34	5382806		6193750	4483088	1710662		8427658	26
35	5385258		6189197	4476676	1712521		8426092	25
36	5387709		6184647	4470266	1714381		8424525	24
37	5390159		6180100	4463858	1716242		8422957	23
38	5392609		6175556	4457452	1718104		8421389	22
39	5395058		6171015	4451048	1719967		8419820	21
40	5397507		6166477	4444646	1721831		8418250	20
41	5399855		616 942	4438245	1723697		8416679	19
42	5402403		6157409	4431845	1725564		8415108	18
43	5404851		6152879	4425447	1727432		8413536	17
44	5407298		6148352	4419051	1729301		8411963	16
45	5409745		6143828	4412656	1731172		8410390	15
46	5412191		6139307	4406263	1733044		8418816	14
47	5414637		6134789	4399872	1734917		8407241	13
48	5417082		6130274	4393483	1736791		8405666	12
49	5419527		6125762	4387096	1738666		8404090	11
50	5421972		6121253	4380711	1740542		8402513	10
51	5424416		6116747	4374327	1742420		8400935	9
52	5426859		6112244	4367945	1744299		8399357	8
53	5429302		6107744	4361565	1746179		8397778	7
54	5431745		6103246	4355186	1748060		8396199	6
54	5434187		6098751	4348809	1749942		8394619	5
56	5436629		6094259	4342433	1751826		8393038	4
57	5439070		6089770	4336059	1753711		8391456	3
58	5441510		6085284	4329687	1755597		8389873	2
59	5443950		6080800	4323316	1757484		8388290	1
60	5446390		6076319	4316947	1759372		8386706	0
								min
								Deg.
								57

57

Deg. 33

33 + | −

min	Sine		Logarithm	Difference	logarithm		Sine	
0	5446390		6076319	4316947	1759372		8386706	60
1	5448829		6071841	4310579	1761262		8385121	59
2	5451268		6067366	4304213	1763153		8383536	58
3	5453707		6062894	4297849	1765045		8381950	57
4	5456145		6058425	4291487	1766938		8380363	56
5	5458583		6053958	4285126	1768832		8378776	55
6	5461020		6049494	4278766	1770728		8377188	54
7	5463456		6045033	4272408	1772625		8375599	53
8	5465802		6040575	4266052	1774523		8374009	52
9	5468328		6036120	4259698	1776422		8372419	51
10	5470763		6031668	4253346	1778322		8370828	50
11	5473198		6027218	4246994	1780224		8369236	49
12	5475632		6022771	4240644	1782127		8367644	48
13	5478066		6018327	4234296	1784031		8366051	47
14	5480499		6013886	4227950	1785936		8364457	46
15	5482932		6009448	4221605	1787843		8362862	45
16	5485364		6005013	4215262	1789751		8361266	44
17	5487796		6000580	4208920	1791660		8359670	43
18	5490228		5996150	4202580	1793570		8358073	42
19	5492659		5991723	4196241	1795482		8356476	41
20	5495090		5987299	4189904	1797395		8354878	40
21	5497520		5982878	4183569	1799309		8353279	39
22	5499950		5978460	4177236	1801224		8351680	38
23	5502379		5974044	4170904	1803140		8350080	37
24	5504808		5969631	4164573	1805058		8348479	36
25	5507236		5965221	4156244	1806977		8346877	35
26	5509664		5960814	4151917	1808897		8345274	34
27	5512091		5956409	4145591	1810818		8343671	33
28	5514518		5952007	4139267	1812740		8342067	32
29	5516944		5947608	4132944	1814664		8340463	31
30	5519370		5943212	4126623	1816589		8338858	30

56

Deg. 33
33

+ | −

min	Sine		Logarithm	Difference	logarithm		Sine	
30	5519370		5943212	4126623	1816589		8338858	30
31	5521795		5938829	4120314	1818515		8337252	29
32	5524220		5934438	4113996	1820442		8335646	28
33	5526645		5930050	4107680	1822370		8334039	27
34	5529069		5925665	4101366	1824299		8332431	26
35	5531493		5921273	4095043	1826230		8330822	25
36	5533916		5916893	4088731	1828162		8329212	24
37	5536338		5912516	4082421	1830095		8327602	23
38	5538760		5908142	4076113	1832029		8325991	22
39	5541182		5903771	4069807	1833964		8324380	21
40	5543603		5899402	4063501	1835901		8322768	20
41	5546024		5895036	4057197	1837839		8321155	19
42	5548444		5890673	4050895	1839778		8319541	18
43	5550864		5886313	4044594	1841719		8317927	17
44	5553283		5881955	4038294	1843661		8316312	16
45	5555702		5877600	4031996	1845604		8314696	15
46	5558120		5873248	4025700	1847548		8313079	14
47	5560538		5868899	4019405	1849494		8311462	13
48	5562956		5864552	4013111	1851441		8309844	12
49	5565373		5860208	4006819	1853389		8308226	11
50	5567790		5855867	4000529	1855338		8306607	10
51	5570206		5851529	3994241	1857288		8304987	9
52	5572622		5847193	3987953	1859240		8303367	8
53	5575037		5842860	3981667	1861193		8301746	7
54	5577452		5838530	3975383	1863147		8300124	6
54	5579866		5834203	3969101	1865102		8298501	5
56	5582280		5829878	3962819	1867059		8296877	4
57	5584693		5825556	3956539	1869017		8295253	3
58	5587106		5821237	3950261	1870976		8293628	2
59	5589518		5816920	3943984	1872936		8292002	1
60	5591929		5812606	3937709	1874897		8290376	0
								min
								Deg.
								56

56

Deg. 34

34

min	Sine	Logarithm	+ | − Difference	logarithm	Sine	
0	5591929	5812606	3937709	1874897	8290376	60
1	5594340	5808295	3931435	1876860	8288794	59
2	5596751	5803987	3925163	1878824	8287121	58
3	5599161	5799681	3918892	1880789	8285493	57
4	5601571	5795378	3912623	1882755	8283864	56
5	5603981	5791078	3906355	1884723	8282234	55
6	5606390	5786780	3900088	1886692	8280603	54
7	5608798	5782485	3893823	1888662	8278972	53
8	5611206	5778192	3887559	1890633	8277340	52
9	5613614	5773902	3881297	1892605	8275708	51
10	5616021	5769615	3875036	1894579	8274075	50
11	5618427	5765330	3868776	1896554	8272441	49
12	5620833	5761048	3862518	1898530	8270806	48
13	5623239	5756769	3856261	1900508	8269170	47
14	5625644	5752493	3850006	1902487	8267534	46
15	5628049	5748219	3843752	1904467	8265897	45
16	5630453	5743948	3837500	1906448	8264259	44
17	5632857	5739680	3831249	1908431	8262621	43
18	5635260	5735414	3824999	1910415	8260982	42
19	5637663	5731151	3818751	1912400	8259343	41
20	5640066	5726891	3812505	1914386	8257703	40
21	5642468	5722634	3806261	1916373	8256062	39
22	5644869	5718379	3800017	1918362	8254421	38
23	5647270	5714127	3793775	1920352	8252779	37
24	5649670	5709878	3787538	1922343	8251136	36
25	5652070	5705631	3781296	1924335	8249492	35
26	5654469	5701387	3775059	1926328	8247847	34
27	5656868	5697145	3768822	1928323	8246202	33
28	5659266	5692906	3762587	1930319	8244556	32
29	5661666	5688670	3756354	1932316	8242909	31
30	5664062	5684436	3750122	1934314	8241262	30

55

Deg. **34**
34

			+ | −			
min	*Sine*	*Logarithm*	*Difference*	*logarithm*	*Sine*	
30	5664062	5684436	3750122	1934314	8241262	30
31	5666459	5680205	3743891	1936314	8239614	29
32	5668856	5675976	3737661	1938315	8237965	28
33	5671252	5671750	3731433	1940317	8236316	27
34	5673648	5667527	3725206	1942321	8234666	26
35	5676043	5663306	3718980	1944326	8233015	25
36	5678438	5659088	3712756	1946332	8231363	24
37	5680832	5654872	3706532	1948340	8229711	23
38	5683226	5650659	3700310	1950349	8228058	22
39	5685619	5646449	3694090	1952359	8226405	21
40	5688012	5642241	3687871	1954370	8224751	20
41	5690404	5638036	3681653	1956383	8223096	19
42	5692796	5633834	3675437	1958397	8221440	18
43	5695187	5629635	3669223	1960412	8219784	17
44	5697578	5625438	3663010	1962428	8218127	16
45	5699968	5621244	3656799	1964445	8216469	15
46	5702358	5617052	3650588	1966464	8214810	14
47	5704747	5612863	3644379	1968484	8213151	13
48	5707136	5608676	3638171	1970505	8211491	12
49	5709524	5604492	3631965	1972527	8209831	11
50	5711912	5600311	3625761	1974550	8208170	10
51	5714289	5596132	3619557	1976575	8206508	9
52	5716686	5591956	3613355	1978601	8204846	8
53	5719072	5587782	3607154	1980628	8203183	7
54	5721458	5583611	3600954	1982657	8201519	6
54	5723844	5579443	3594756	1984687	8199854	5
56	5726229	5575277	3588559	1986718	8198188	4
57	5728613	5571114	3582364	1988750	8196522	3
58	5730997	5566953	3576169	1990784	8194855	2
59	5733381	5562795	3569976	1992819	8193188	1
60	5735764	5558639	3563784	1994855	8191520	0
						min *Deg.* 55

55

Deg. 35

35 + | −

min	Sine	Logarithm	Difference	logarithm	Sine	
0	5735764	5558639	3563784	1994855	8191520	60
1	5738147	5554486	3557594	1996892	8189851	59
2	5740529	5550336	3551405	1998931	8188182	58
3	5742911	5546188	3545217	2000971	8186512	57
4	5745292	5542043	3539031	2003012	8184841	56
5	5747672	5537900	3531846	2005054	8183170	55
6	5750052	5533760	3526662	2007098	8181498	54
7	5752432	5529622	3520479	2009143	8179825	53
8	5754811	5525487	3514298	2011189	8178151	52
9	5757190	5521354	3508118	2013236	8176477	51
10	5759568	5517224	3501939	2015285	8174802	50
11	5761946	5513096	3495761	2017335	8173126	49
12	5764323	5508971	3489585	2019386	8171449	48
13	5766700	5504849	3483410	2021439	8169772	47
14	5769076	5500729	3477236	2023493	8168094	46
15	5771452	5496612	3471064	2025548	8166416	45
16	5773827	5492497	3464892	2027605	8164737	44
17	5776202	5488385	3458722	2029663	8163057	43
18	5778576	5484275	3452553	2031722	8161376	42
19	5780950	5480168	3446386	2033782	8159695	41
20	5783324	5476063	3440219	2035844	8158013	40
21	5785697	5471961	3434054	2037907	8156330	39
22	5788069	5467861	3427890	2039971	8154647	38
23	5790441	5463764	3421728	2042036	8152963	37
24	5792812	5459669	3415566	2044103	8151278	36
25	5795183	5455577	3409406	2046171	8149593	35
26	5797553	5451488	3403248	2048240	8147907	34
27	5799923	5447401	3397090	2050311	8146220	33
28	5802292	5443317	3390934	2052383	8144532	32
29	5804661	5439235	3384779	2054456	8142844	31
30	5807030	5435156	3378626	2056530	8141155	30

54

Deg. 35

35 + | −

min	Sine	Logarithm	Difference	logarithm	Sine	
30	5807030	5435156	3378626	2056530	8141155	30
31	5809398	5431079	3372473	2058606	8139465	29
32	5811766	5427005	3366422	2060683	8137775	28
33	5814133	5422933	3360172	2062761	8136084	27
34	5816499	5418864	3354024	2064840	8134393	26
35	5818865	5414797	3347877	2066920	8132701	25
36	5821230	5410733	3341731	2069002	8131008	24
37	5823595	5406671	3335586	2071085	8129314	23
38	5825959	5402612	3329443	2073169	8127620	22
39	5828323	5398555	3323300	2075255	8125925	21
40	5830687	5394501	3317159	2077342	8124229	20
41	5833050	5390449	3311019	2079430	8122532	19
42	5835412	5386400	3304880	2081520	8120835	18
43	5837774	5382353	3298742	2083611	8119137	17
44	5840136	5378308	3292605	2085703	8117439	16
45	5842497	5374266	3286470	2087796	8115740	15
46	5844858	5370226	3280335	2089891	8114040	14
47	5847218	5366189	3274202	2091987	8112339	13
48	5849578	5362154	3268070	2094084	8110638	12
49	5851937	5358122	3261939	2096183	8108936	11
50	5854295	5354093	3255810	2098283	8107234	10
51	5856653	5350067	3249683	2100384	8105531	9
52	5859010	5346043	3243557	2102486	8104827	8
53	5861367	5342021	3237431	2104590	8102122	7
54	5863724	5338002	3231307	2106695	8100417	6
54	5866080	5333985	3225184	2108801	8098711	5
56	5868436	5329970	3219061	2110909	8097004	4
57	5870791	5325958	3212940	2113018	8095296	3
58	5873145	5321948	3206820	2115128	8093588	2
59	5875499	5317940	3200700	2117240	8091879	1
60	5877852	5313935	3194582	2119353	8090170	0
						min
						Deg.
						54

54

Deg. 36

36

min	Sine	Logarithm	+ | − Difference	logarithm	Sine	
0	5877852	5313935	3194582	2119353	8090170	60
1	5880205	5309932	3188465	2121467	8089460	59
2	5882558	5305932	3182350	2123582	8086749	58
3	5884910	5301935	3176236	2125699	8085038	57
4	5887262	5297940	3170123	2127817	8083326	56
5	5889613	5293947	3164011	2129936	8081613	55
6	5891964	5289957	3157900	2132057	8079899	54
7	5894314	5285969	3151790	2134179	8078185	53
8	5896664	5281984	3145682	2136302	8076470	52
9	5899013	5278001	3139575	2138426	8074754	51
10	5901361	5274020	3133468	2140552	8073038	50
11	5903709	5270042	3127363	2142679	8071321	49
12	5906056	5266066	3121259	2144807	8069603	48
13	5908403	5262092	3115155	2146937	8067885	47
14	5910750	5258121	3109053	2149068	8066166	46
15	5913096	5254152	3102952	2151200	8064446	45
16	5915442	5250186	3096853	2153333	8062726	44
17	5917787	5246222	3090754	2155468	8061005	43
18	5920132	5242261	3084657	2157604	8059283	42
19	5922476	5238302	3078561	2159741	8057561	41
20	5924820	5234346	3072466	2161880	8055838	40
21	5927163	5230392	3066372	2164020	8054114	39
22	5929505	5226441	3060280	2166161	8052389	38
23	5931847	5222492	3054188	2168304	8050664	37
24	5934189	5218545	3048097	2170448	8048938	36
25	5936530	5214601	3042008	2172593	8047212	35
26	5938871	5210659	3035919	2174740	8045485	34
27	5941211	5206720	3029832	2176888	8043757	33
28	5945551	5202783	3023746	2179037	8042028	32
29	5945890	5198848	3017660	2181188	8040299	31
30	5948228	5194916	3011576	2183340	8038569	30

53

Deg. **36**
36

+ | −

min	Sine		Logarithm	Difference	logarithm		Sine	
30	5948228		5194916	3011576	2183340		8038569	30
31	5950566		5190986	3005493	2185493		8036838	29
32	5952904		5187059	2999412	2187647		8035107	28
33	5955241		5183134	2993331	2189803		8033375	27
34	5957578		5179211	2987251	2191960		8031642	26
35	5959914		5175291	2981173	2194118		8029909	25
36	5962250		5171373	2975095	2196278		8028175	24
37	5964585		5167457	2969018	2198439		8026440	23
38	5966919		5163544	2962943	2200601		8024705	22
39	5969253		5159633	2956868	2202765		8022969	21
40	5971586		5155724	2950794	2204930		8021232	20
41	5973919		5151818	2944722	2207096		8019494	19
42	5976251		5147914	2938650	2209264		8017756	18
43	5978583		5144012	2932579	2211433		8016017	17
44	5980915		5140113	2926510	2213603		8014278	16
45	5983246		5136216	2920442	2215774		8012538	15
46	5985577		5132322	2914375	2217947		8010797	14
47	5987907		5128430	2908309	2220121		8009056	13
48	5990237		5124540	2902244	2222296		8007314	12
49	5992566		5120653	2896180	2224473		8005571	11
50	5994894		5116768	2890117	2226651		8003828	10
51	5997222		5112886	2884056	2228830		8002084	9
52	5999549		5109006	2877995	2231011		8000339	8
53	6001876		5105128	2871935	2233193		7998593	7
54	6004202		5101253	2865877	2235376		7996847	6
54	6006528		5097380	2859819	2237561		7995100	5
56	6008853		5093509	2853762	2239747		7993352	4
57	6011178		5089641	2847706	2241935		7991604	3
58	6013502		5085775	2841651	2244124		7989855	2
59	6015826		5081911	2835597	2246314		7988105	1
60	6018150		5078050	2829544	2248506		7986355	0
								min
								Deg.
								53

53

Deg. **37**

37 + | −

min	Sine		Logarithm	Difference	logarithm		Sine	
0	6018150		5078050	2829544	2248506		7986355	60
1	6020473		5074191	2823492	2250699		7984604	59
2	6022796		5070334	2817441	2252893		7982852	58
3	6025118		5066479	2811391	2255088		7981100	57
4	6027439		5062627	2805342	2257285		7979347	56
5	6029760		5058777	2799294	2259483		7977593	55
6	6032080		5054929	2793247	2261682		7975838	54
7	6034400		5051084	2787201	2263883		7974084	53
8	6036719		5047241	2781156	2266085		7972328	52
9	6039038		5043401	2775113	2268288		7970572	51
10	6041357		5039563	2769071	2270492		7968815	50
11	6043675		5035727	2763029	2272698		7967057	49
12	6045992		5031894	2756989	2274905		7965299	48
13	6048309		5028063	2750949	2277114		7963540	47
14	6050625		5024234	2744910	2279324		7961780	46
15	6052940		5020408	2738873	2281535		7960020	45
16	6055255		5016584	2732836	2283748		7958259	44
17	6057570		5012762	2726800	2285962		7956497	43
18	6059884		5008942	2720764	2288178		7954735	42
19	6062198		5005125	2714730	2290395		7952972	41
20	6064511		5001310	2708696	2292614		7951208	40
21	6066824		4997497	2702663	2294834		7949443	39
22	6069136		4993687	2696632	2297055		7947678	38
23	6071448		4989879	2690602	2299277		7945912	37
24	6073759		4986073	2684573	2301500		7944146	36
25	6076069		4982270	2678545	2303725		7942379	35
26	6078379		4978469	2672518	2305951		7940611	34
27	6080688		4974670	2666492	2308178		7938842	33
28	6082997		4970873	2660467	2310406		7937073	32
29	6085306		4967079	2654444	2312635		7935303	31
30	6087614		4963287	2648421	2314866		7933533	30

52

Deg. **37**

37 + | −

min	Sine		Logarithm	Difference	logarithm		Sine	
30	6087614		4963287	2648421	2314866		7933533	30
31	6089922		4959497	2642399	2317098		7931762	29
32	6092229		4955710	2636378	2319332		7929990	28
33	6094536		4951925	2630358	2321567		7928218	27
34	6096842		4948142	2624338	2323804		7926445	26
35	6099147		4944361	2618319	2326042		7924671	25
36	6101452		4940582	2612301	2328281		7922896	24
37	6103756		4936806	2606284	2330522		7921121	23
38	6106060		4933032	2600268	2332764		7919345	22
39	6108364		4929260	2594252	2335008		7917569	21
40	6110667		4925490	2588237	2337253		7915792	20
41	6112970		4921723	2582224	2339499		7914014	19
42	6115272		4917958	2576211	2341747		7912235	18
43	6117573		4914195	2570199	2343996		7910456	17
44	6119873		4910435	2564189	2346246		7908676	16
45	6122173		4906677	2558179	2348498		7906896	15
46	6124473		4902921	2552170	2350751		7905114	14
47	6126772		4899168	2546163	2353005		7903332	13
48	6129071		4895417	2540156	2355261		7901550	12
49	6131369		4891668	2534150	2357518		7899767	11
50	6133667		4887921	2528145	2359776		7897983	10
51	6135964		4884177	2522141	2362036		7896198	9
52	6138261		4880435	2516138	2364297		7894413	8
53	6140557		4876695	2510136	2366559		7892627	7
54	6142853		4872957	2504134	2368823		7890841	6
54	6145148		4869222	2498134	2371088		7889054	5
56	6147442		4865489	2492135	2373354		7887266	4
57	6149746		4861758	2486136	2375622		7885477	3
58	6152030		4858029	2480138	2377891		7883688	2
59	6154323		4854302	2474140	2380162		7881898	1
60	6156615		4850578	2468144	2382434		7880108	0
								min
								Deg.
								52

52

Deg. 38

38

min	Sine	Logarithm	+ | − Difference	logarithm	Sine	
0	6156615	4850578	2468144	2382434	7880108	60
1	6158907	4846856	2462149	2384707	7878317	59
2	6161198	4843136	2456154	2386982	7876525	58
3	6163489	4839418	2450160	2389258	7874732	57
4	6165781	4835702	2444167	2391535	7872939	56
5	6168070	4831989	2438175	2393814	7871145	55
6	6170259	4828278	2432184	2396094	7869350	54
7	6172648	4824569	2426193	2398376	7867555	53
8	6174936	4820862	2420203	2400659	7865759	52
9	6177224	4817158	2414215	2402943	7863963	51
10	6179512	4813456	2408227	2405229	7862166	50
11	6181799	4809756	2402240	2407516	7860368	49
12	6184085	4806058	2396254	2409804	7858569	48
13	6186371	4802363	2390269	2412094	7856770	47
14	6188656	4798670	2384285	2414385	7854970	46
15	6190940	4794979	2378301	2416678	7853169	45
16	6193224	4791290	2372318	2418972	7851368	44
17	6195508	4787603	2366336	2421267	7849566	43
18	6197791	4783919	2360355	2423564	7847764	42
19	6200074	4780237	2354375	2425862	7845961	41
20	6202356	4776557	2348396	2428161	7844157	40
21	6204638	4772880	2342418	2430462	7842352	39
22	6206919	4769205	2336441	2432764	7840547	38
23	6209199	4765532	2330465	2435067	7838741	37
24	6211479	4761861	2324489	2437372	7836935	36
25	6213758	4758192	2318514	2439678	7835128	35
26	6216037	4754525	2312539	2441986	7833320	34
27	6218315	4750860	2306565	2444295	7831511	33
28	6220593	4747198	2300593	2446605	7829702	32
29	6222870	4743538	2294621	2448917	7827892	31
30	6225146	4739880	2288650	2451230	7826082	30

51

Deg. 38

38 + | −

min	Sine	Logarithm	Difference	logarithm	Sine	
30	6225146	4739880	2288650	2451230	8703557	30
31	6227422	4736224	2282680	2453544	7824271	29
32	6229698	4732571	2276711	2455860	7822459	28
33	6231973	4728920	2270743	2458177	7820647	27
34	6234248	4725271	2264775	2460496	7818834	26
35	6236522	4721624	2258808	2462816	7817020	25
36	6238796	4717979	2252842	2465137	7815205	24
37	6241069	4714336	2246876	2467460	7813390	23
38	6243342	4710695	2240911	2469784	7811574	22
39	6245614	4707056	2234946	2472110	7809758	21
40	6247885	4703419	2228982	2474437	7807941	20
41	6250156	4699785	2223020	2476765	7806123	19
42	6252426	4696153	2217058	2479095	7804304	18
43	6254696	4692523	2211097	2481426	7802485	17
44	6256966	4688895	2205136	2483759	7800665	16
45	6259235	4685269	2199176	2486093	7798845	15
46	6261503	4681645	2193217	2488428	7597024	14
47	6263771	5678024	2187259	2490765	7795202	13
48	6266038	4674405	2181302	2493103	7793380	12
49	6268305	4670788	2175345	2495443	7791557	11
50	6270572	4667173	2169389	2497784	7789733	10
51	6272838	4663561	2163435	2500126	7787909	9
52	6275103	4659951	2157481	2502470	7786084	8
53	6277368	4656343	2151528	2504815	7784258	7
54	6279632	4652737	2145575	2507162	7782432	6
54	6281895	4649133	2139623	2509510	7780605	5
56	6284158	4645531	2133672	2511859	7778777	4
57	6286420	4641931	2127721	2514210	7776949	3
58	6288682	4638334	2121772	2516562	7775120	2
59	6290943	4634739	2115824	2518915	7773290	1
60	6293204	4631146	2109876	2521270	7771460	0
						min
						Deg. 51

51

Deg. **39**

39

+ | −

min	Sine	Logarithm	Difference	logarithm	Sine	
0	6293204	4631146	2109876	2521270	7771460	60
1	6295464	4627555	2103929	2523626	7769629	59
2	6297724	4623966	2097982	2525984	7767797	58
3	6299983	4620379	2092036	2528343	7765965	57
4	6302242	4616794	2086091	2530703	7764132	56
5	6304501	4613211	2080146	2533065	7762299	55
6	6306759	4609630	2074202	2535428	7760465	54
7	6309016	4606052	2068259	2537793	7758630	53
8	6311273	4602476	2062317	2540159	7756794	52
9	6313529	4598902	2056376	2542526	7754958	51
10	6315784	4595330	2050435	2544895	7753121	50
11	6318039	4591760	2044495	2547265	7751283	49
12	6320293	4588192	2038555	2549637	7749445	48
13	6322547	4584627	2032617	2552010	7747606	47
14	6324800	4581064	2026679	2554385	7745766	46
15	6327053	4577503	2020742	2556761	7743926	45
16	6329305	4573944	2014806	2559138	7742085	44
17	6331557	4570387	2008870	2561517	7740244	43
18	6333808	4566832	2002935	2563897	7738402	42
19	6336059	4563279	1997000	2566279	7736559	41
20	6338310	4559728	1991066	2568662	7734716	40
21	6340560	4556179	1985133	2571046	7732872	39
22	6342809	4552632	1979200	2573432	7731028	38
23	6345058	4549088	1973269	2575819	7729183	37
24	6347309	4545546	1967338	2578208	7727337	36
25	6349553	4542006	1961408	2580598	7725490	35
26	6351800	4538468	1955478	2582990	7723642	34
27	6354046	4534932	1949549	2585383	7721794	33
28	6356292	4531398	1943621	2587777	7719945	32
29	6358537	4527866	1937693	2590173	7718096	31
30	6360782	4524336	1931766	2592570	7716246	30

50

Deg. **39**

39 + | −

min	Sine		Logarithm	Difference	logarithm		Sine	
30	6360782		4524336	1931766	2592570		7716246	30
31	6363026		4520808	1925839	2594969		7714395	29
32	6365270		4517282	1919913	2597369		7712544	28
33	6367513		4513758	1913988	2599770		7710692	27
34	6369756		4510236	1908063	2602173		7708839	26
35	6371999		4506717	1902140	2604577		7706986	25
36	6374241		4503200	1896217	2606983		7705132	24
37	6376482		4499685	1890295	2609390		7703277	23
38	6378722		4496172	1884373	2611799		7701422	22
39	6380962		4492661	1878452	2614209		7699566	21
40	6383201		4489152	1872531	2616621		7697710	20
41	6385440		4485645	1866611	2619034		7695853	19
42	6387678		4482140	1860692	2621448		7693995	18
43	6389916		4478637	1854773	2623864		7692137	17
44	6392153		4475136	1848855	2626281		7690278	16
45	6394390		4471637	1842937	2628700		7688418	15
46	6396626		4468140	1837020	2631120		7686549	14
47	6398862		4464646	1831105	2633541		7684687	13
48	6401097		4461154	1825190	2635964		7682835	12
49	6403332		4457664	1819276	2638388		7680973	11
50	6405566		4454176	1813363	2640813		7679110	10
51	6407799		4450690	1807450	2643240		7677246	9
52	6410032		4447206	1801537	2645669		7675382	8
53	6412264		4443724	1795625	2648099		7673517	7
54	6414496		4440244	1789714	2650530		7671652	6
54	6416728		4436766	1783803	2652963		7669786	5
56	6418959		4433290	1777893	2655397		7667919	4
57	6421189		4429816	1771983	2657833		7666051	3
58	6423419		4426344	1766074	2660270		7664183	2
59	6425648		4422875	1760166	2662709		7662314	1
60	6427876		4419408	1754259	2665149		7660445	0
								min
								Deg.
								50

50

Deg. **40**

40

+ | −

min	*Sine*		*Logarithm*	*Difference*	*logarithm*		*Sine*	
0	6427876		4419408	1754259	2665149		7660445	60
1	6430104		4415943	1748353	2667590		7658575	59
2	6432331		4412480	1742447	2670033		7656704	58
3	6434558		4409019	1736542	2672477		7654833	57
4	6436785		4405560	1730637	2674923		7652961	56
5	6439011		4402103	1724733	2677370		7651088	55
6	6441236		4398648	1718829	2679819		7649215	54
7	6443461		4395195	1712926	2682269		7647341	53
8	6445685		4391743	1707022	2684721		7645466	52
9	6447909		4388293	1701119	2687174		7643591	51
10	6450132		4384845	1695216	2689629		7641715	50
11	6452355		4381399	1689314	2692085		7639838	49
12	6454577		4377955	1683412	2694543		7637960	48
13	6456799		4374514	1677512	2697002		7636082	47
14	6459020		4371075	1671613	2699462		7634204	46
15	6461240		4367638	1665714	2701924		7632325	45
16	6463460		4364203	1659816	2704387		7630445	44
17	6465679		4360770	1653918	2706852		7628564	43
18	6467898		4357339	1648021	2709318		7626683	42
19	6470116		4353910	1642124	2711786		7624802	41
20	6472333		4350483	1636228	2714255		7622920	40
21	6474550		4347058	1630332	2716726		7621037	39
22	6476766		4343635	1624437	2719198		7619153	38
23	6478982		4340214	1618542	2721672		7617269	37
24	6481198		4336795	1612648	2724147		7615384	36
25	6483413		4333378	1606755	2726623		7613498	35
26	6485628		4329963	1600862	2729101		7611612	34
27	6487842		4326550	1594970	2731580		7609725	33
28	6490055		4323139	1589078	2734061		7607837	32
29	6492268		4319730	1583187	2736543		7605949	31
30	6494480		4316323	1577296	2739027		7604060	30

49

Deg. 40

40

+ | −

min	Sine		Logarithm	Difference	logarithm		Sine	
30	6494480		4316323	1577296	2739027		7604060	30
31	6496692		4312919	1571407	2741512		7602170	29
32	6498903		4309517	1565518	2743999		7600280	28
33	6501114		4306116	1559629	2746487		7598389	27
34	6503324		4302717	1553740	2748977		7596498	26
35	6505533		4299320	1547852	2751468		7594606	25
36	6507742		4295925	1541964	2753961		7592713	24
37	6509950		4292532	1536077	2756455		7590819	23
38	6512158		4289141	1530191	2758950		7588925	22
39	6514365		4285752	1524305	2761447		7587031	21
40	6516572		4282365	1518420	2763945		7589136	20
41	6518778		4278980	1512535	2766445		7583240	19
42	6520984		4275597	1506651	2768946		7581343	18
43	6523189		4272216	1500767	2771449		7579446	17
44	6525394		4268837	1494884	2773953		7577548	16
45	6527598		4265460	1489001	2776459		7575650	15
46	6529801		4262085	1483119	2778966		7573751	14
47	6532004		4258712	1477237	2781475		7571851	13
48	6534206		4255341	1471356	2783985		7569951	12
49	6536408		4251972	1465476	2786496		7568050	11
50	6538609		4248605	1459596	2789009		7566148	10
51	6540809		4245240	1453717	2791523		7564246	9
52	6543009		4241877	1447838	2794039		7562343	8
53	6545208		4238516	1441960	2796556		7560439	7
54	6547407		4235157	1436082	2799075		7558535	6
54	6549606		4231800	1430205	2801595		7556630	5
56	6551804		4228445	1424328	2804117		7554724	4
57	6554001		4225092	1418451	2806641		7552818	3
58	6556198		4221741	1412575	2809166		7550911	2
59	6558394		4218392	1406699	2811693		7549004	1
60	6560590		4215044	1400823	2814221		7547096	0
								min
								Deg.
							49	49

Deg. **41**

41 + | −

min	Sine		Logarithm	Difference	logarithm		Sine	
0	6560590		4215044	1400823	2814221		7547096	60
1	6562785		4211698	1394947	2816751		7545187	59
2	6564979		4208354	1389072	2819282		7543277	58
3	6567173		4205012	1383197	2821815		7541367	57
4	6569367		4201672	1377323	2824349		7539457	56
5	6571560		4198334	1371450	2826884		7537546	55
6	6573753		4194999	1365578	2829421		7535634	54
7	6575945		4191666	1359707	2831959		7533721	53
8	6578136		4188335	1353836	2834499		7531808	52
9	6580326		4185006	1347966	2837040		7529894	51
10	6582516		4181679	1342097	2839582		7527980	50
11	6584705		4178354	1336228	2842126		7526065	49
12	6586894		4175030	1330358	2844672		7524149	48
13	6589082		4171708	1324489	2847219		7522233	47
14	6591270		4168388	1318620	2849768		7520316	46
15	6593458		4165070	1312752	2852318		7518398	45
16	6595645		4161754	1306884	2854870		7516840	44
17	6597831		4158440	1301017	2857423		7514561	43
18	6600016		4155128	1295150	2859978		7512642	42
19	6602201		4151818	1289284	2862534		7510722	41
20	6604386		4148510	1283418	2865092		7508801	40
21	6606570		4145204	1277553	2867651		7506879	39
22	6608753		4141900	1271688	2870212		7504957	38
23	6610936		4138598	1265824	2872774		7503034	37
24	6613118		4135298	1259960	2875338		7501111	36
25	6615300		4132000	1254097	2877903		7499187	35
26	6617481		4128703	1248233	2880470		7497262	34
27	6619661		4125408	1242370	2883038		7495336	33
28	6621841		4122115	1236507	2885608		7493410	32
29	6624021		4118824	1230645	2888179		7491484	31
30	6626200		4115535	1224783	2890752		7489557	30

48

Deg. **41**
41

+ | −

min	Sine		Logarithm	Difference	logarithm		Sine	
30	6626200		4115535	1224783	2890752		7489557	30
31	6628379		4112248	1218922	2893326		7487629	29
32	6630557		4108963	1213061	2895902		7485700	28
33	6632734		4105680	1207201	2898479		7483771	27
34	6634911		4102399	1201341	2901058		7481842	26
35	6637087		4099120	1195482	2903638		7479912	25
36	6639263		4095843	1189623	2906220		7477981	24
37	6641438		4092567	1183763	2908804		7476049	23
38	6643612		4089293	1177904	2911389		7474117	22
39	6645786		4086021	1172045	2913976		7472184	21
40	6647959		4082751	1166187	2916564		7470251	20
41	6650132		4079483	1160329	2919154		7468317	19
42	6652304		4076217	1154472	2921745		7466382	18
43	6654476		4072953	1148615	2924338		7464447	17
44	6656647		4069691	1142759	2926932		7462511	16
45	6658817		4066431	1136904	2929527		7460574	15
46	6660987		4063173	1131049	2932124		7458637	14
47	6663156		4059917	1125195	2934722		7456699	13
48	6665325		4056663	1119341	2937322		7454761	12
49	6667493		4053410	1113487	2939923		7452822	11
50	6669661		4050159	1107633	2942526		7450882	10
51	6671828		4046910	1101780	2945130		7448941	9
52	6673994		4043663	1095927	2947736		7447000	8
53	6676160		4040418	1090074	2950344		7445058	7
54	6678326		4037175	1084222	2952953		7443116	6
54	6680491		4033934	1078370	2955564		7441173	5
56	6682655		4030695	1072518	2958177		7439229	4
57	6684818		4027458	1066667	2960791		7437284	3
58	6686981		4024223	1060816	2963407		7435339	2
59	6689144		4020990	1054966	2966024		7433394	1
60	6691306		4017759	1049116	2968643		7431448	0
								min
								Deg.
								48

48

Deg. **42**

42 + | −

min	Sine		Logarithm	Difference	logarithm		Sine	
0	6691306		4017759	1049116	2968643		7431448	60
1	6693468		4014529	1043266	2971263		7429501	59
2	6695629		4011301	1037416	2973885		7427553	58
3	6697789		4008075	1031567	2976508		7425605	57
4	6699949		4004851	1025718	2979133		7423657	56
5	6702108		4001629	1019870	2981759		7421708	55
6	6704267		3998409	1014023	2984386		7419758	54
7	6706425		3995191	1008176	2987015		7417807	53
8	6708582		3991974	1002329	2989645		7415856	52
9	6710739		3988759	996482	2992277		7413905	51
10	6712895		3985546	990636	2994910		7411953	50
11	6715051		3982335	984790	2997545		7410000	49
12	6717206		3979126	978944	3000182		7408046	48
13	6719361		3975919	973099	3002820		7406092	47
14	6721515		3972714	967254	3005460		7404137	46
15	6723668		3969511	961409	3008102		7402181	45
16	6725821		3966310	955565	3010745		7400225	44
17	6727973		3963110	949720	3013390		7398268	43
18	6730125		3959912	943876	3016036		7396311	42
19	6732276		3956716	938032	3018684		7394353	41
20	6734427		3953522	932189	3021333		7392394	40
21	6736577		3950330	926346	3023984		7390435	39
22	6738726		3947140	920504	3026636		7388475	38
23	6740875		3943951	914661	3029290		7386515	37
24	6743024		3940764	908819	3031945		7384554	36
25	6745172		3937579	902977	3034602		7382592	35
26	6747319		3934396	897135	3037261		7380629	34
27	6749465		3931215	891294	3039921		7378666	33
28	6751611		3928036	885453	3042583		7376702	32
29	6753757		3924859	879613	3045246		7374738	31
30	6755902		3921684	873773	3047911		7372773	30

47

Deg. 42
42

+ | −

min	Sine	Logarithm	Difference	logarithm	Sine	
30	6755902	3921684	873773	3047911	7372773	30
31	6758047	3918511	867934	3050577	7370807	29
32	6760191	3915339	862094	3053245	7368841	28
33	6762334	3912169	856255	3055914	7366874	27
34	6764477	3909001	850416	3058585	7364907	26
35	6766619	3905835	844577	3061258	7362939	25
36	6768760	3902671	838739	3063932	7360970	24
37	6770901	3899509	832901	3066608	7359001	23
38	6773041	3896348	827063	3069285	7357031	22
39	6775181	3893189	821225	3071964	7355061	21
40	6777320	3890032	815388	3074644	7353090	20
41	6779459	3886877	809551	3077326	7351118	19
42	6781597	3883723	803714	3080009	7349145	18
43	6783734	3880571	797877	3082694	7347173	17
44	6785971	3877421	792041	3085380	7345199	16
45	6788007	3874273	786205	3088068	7343225	15
46	6790143	3871127	780369	3090758	7341250	14
47	6792278	3867983	774534	3093449	7339274	13
48	6794413	3864841	768699	3096142	7337298	12
49	6796547	3861701	762865	3098836	7335322	11
50	6798681	3858563	757031	3101532	7333345	10
51	6800814	3855426	751197	3104229	7331367	9
52	6802946	3852291	745363	3106928	7329388	8
53	6805078	3849158	739529	3109629	7327409	7
54	6807209	3846027	733696	3112331	7325429	6
54	6809340	3842898	727863	3115035	7323449	5
56	6811470	3839770	722029	3117741	7321468	4
57	6813599	3836644	716196	3120448	7319486	3
58	6815728	3833520	710363	3123157	7317504	2
59	6817856	3830398	704530	3125868	7315521	1
60	6819984	3827278	698698	3128580	7313537	0
						min
						Deg. 47

47

Deg. **43**
43

+ | −

min	Sine		Logarithm	Difference	logarithm		Sine	
0	6819984		3827278	698698	3128580		7313537	60
1	6822111		3824160	692866	3131294		7311553	59
2	6824237		3821044	687035	3134009		7309568	58
3	6826363		3817929	681203	3136726		7307583	57
4	6828489		3814816	675372	3139444		7305597	56
5	6830614		3811705	669541	3142164		7303610	55
6	6832738		3808596	663711	3144885		7301623	54
7	6834861		3805488	657880	3147608		7299635	53
8	6836984		3802382	652050	3150332		7297647	52
9	6839107		3799278	646221	3153057		7295658	51
10	6841229		3796176	640392	3155784		7293668	50
11	6843350		3793075	634562	3158513		7291678	49
12	6845471		3789976	628732	3161244		7289687	48
13	6847591		3786879	622903	3163976		7287695	47
14	6849711		3783784	617074	3166710		7285703	46
15	6851830		3780691	611246	3169445		7283710	45
16	6853949		3777600	605418	3172182		7281716	44
17	6856067		3774510	599589	3174921		7279722	43
18	6858184		3771422	593760	3177662		7277728	42
19	6860301		3768336	587932	3180404		7275733	41
20	6862417		3765252	582104	3183148		7273737	40
21	6864533		3762170	576277	3185893		7271741	39
22	6866648		3759090	570450	3188640		7269744	38
23	6868762		3756011	564622	3191389		7267746	37
24	6870876		3752934	558795	3194139		7265748	36
25	6872989		3749859	552968	3196891		7263749	35
26	6875102		3746786	547142	3199644		7261749	34
27	6877214		3743714	541315	3202399		7259748	33
28	6879325		3740644	535489	3205155		7257747	32
29	6881436		3737576	529663	3207913		7255746	31
30	6883546		3734510	523838	3210672		7253744	30

46

Deg. **43**

43 + | −

min	Sine	Logarithm	Difference	logarithm	Sine	
30	6883546	3734510	523838	3210672	7253744	30
31	6885656	3731446	518013	3213433	7251741	29
32	6887765	3728383	512187	3216196	7249737	28
33	6889874	3725322	506361	3218961	7247733	27
34	6891982	3722263	500536	3221727	7245729	26
35	6894089	3719206	494711	3224495	7243724	25
36	6896196	3716150	488886	3227264	7241718	24
37	6898302	3713096	483061	3230035	7239711	23
38	6900408	3710044	477236	3232808	7237704	22
39	6902513	3706994	471411	3235583	7235697	21
40	6904617	3703946	465587	3238359	7233689	20
41	6906721	3700899	459762	3241137	7231681	19
42	6908824	3697854	453938	3243916	7229672	18
43	6910927	3694811	448114	3246697	7227662	17
44	6913029	3691770	442291	3249479	7225651	16
45	6915131	3688730	436467	3252263	7223639	15
46	6917232	3685692	430644	3255048	7221627	14
47	6919332	3682656	424821	3257835	7219614	13
48	6921432	3679622	418999	3260623	7217601	12
49	6923531	3676590	413177	3263413	7215588	11
50	6925630	3673559	407355	3266204	7213574	10
51	6927728	3670530	401533	3268997	7211559	9
52	6929825	3667503	395711	3271792	7209543	8
53	6931922	3664478	389889	3274589	7207527	7
54	6934018	3661454	384067	3277387	7205511	6
54	6936114	3658432	378245	3280187	7203494	5
56	6938209	3655412	372423	3282989	7201476	4
57	6940303	3652394	366602	3285792	7199457	3
58	6942397	3649377	360780	3288597	7197438	2
59	6944491	3646362	354958	3291404	7195418	1
60	6946584	3643349	349136	3294213	7193398	0
						min
						Deg.
						46

46

Deg. **44**
44

+ | −

min	Sine	Logarithm	Difference	logarithm	Sine	
0	6946584	3643349	349136	3294213	7193398	60
1	6948676	3640338	343315	3297023	7191377	59
2	6950767	3637329	337494	3299835	7189355	58
3	6952858	3634321	331673	3302648	7187333	57
4	6954949	3631315	325852	3305463	7185310	56
5	6957039	3628311	320032	3308279	7183287	55
6	6959128	3625308	314211	3311097	7181263	54
7	6961216	3622307	308390	3313917	7179238	53
8	6963304	3619308	302570	3316738	7177213	52
9	6965392	3616311	296750	3319561	7175187	51
10	6967479	3613315	290930	3322385	7173161	50
11	6969565	3610321	285110	3325211	7171134	49
12	6971651	3607329	279290	3328039	7169106	48
13	6973736	3604338	273469	3330869	7167078	47
14	6975821	3601349	267649	3333700	7165049	46
15	6977905	3598362	261829	3336533	7163019	45
16	6979988	3595377	256009	3339368	7160989	44
17	6982071	3592394	250190	3342204	7158958	43
18	6984153	3589412	244370	3345042	7156927	42
19	6986235	3586432	238550	3347882	7154895	41
20	6988316	3583454	232731	3350723	7152863	40
21	6990396	3580478	226912	3353566	7150830	39
22	6992476	3577504	221093	3356411	7148796	38
23	6994555	3574531	215274	3359257	7146762	37
24	6996634	3571560	209455	3362105	7144727	36
25	6998712	3568590	203635	3364955	7142691	35
26	7000789	3565622	197816	3367806	7140655	34
27	7002866	3562656	191997	3370659	7138618	33
28	7004942	3559691	186178	3373513	7136581	32
29	7007018	3556728	180359	3376369	7134543	31
30	7009093	3553767	174541	3379226	7132504	30

45

Deg. **44**
44 + | −

min	*Sine*	*Logarithm*	*Difference*	*logarithm*	*Sine*	
30	7009093	3553767	174541	3379226	7132504	30
31	7011167	3550808	168723	3382085	7130465	29
32	7013241	3547851	162905	3384946	7128425	28
33	7015314	3544895	157087	3387808	7126385	27
34	7017387	3541941	151269	3390672	7124344	26
35	7019459	3538989	145451	3393538	7122303	25
36	7021530	3536038	139632	3396406	7120261	24
37	7023601	3533089	133814	3399275	7118218	23
38	7025671	3530142	127996	3402146	7116175	22
39	7027741	3527197	122178	3405019	7114131	21
40	7029810	3524253	116359	3407894	7112086	20
41	7031879	3521311	110541	3410770	7110041	19
42	7033947	3518371	104723	3413648	7107995	18
43	7036014	3515432	98904	3416528	7105949	17
44	7038081	3512495	93086	3419409	7103902	16
45	7040147	3509560	87268	3422292	7101854	15
46	7042213	3506626	81450	3425176	7099806	14
47	7044278	3503694	75632	3428062	7097757	13
48	7046342	3500764	69824	3430940	7095708	12
49	7048406	3497835	64006	3433829	7093658	11
50	7050469	3494908	58178	3436730	7091607	10
51	7052532	3491983	52360	3439623	7089556	9
52	7054594	3489060	46543	3442517	7087504	8
53	7056655	3486139	40726	3445413	7085452	7
54	7058716	3483219	34908	3448311	7083399	6
54	7060776	3480301	29090	3451211	7081345	5
56	7062836	3477385	23273	3454112	7079291	4
57	7064895	3474470	17455	3457015	7077236	3
58	7066953	3471557	11637	3459920	7075181	2
59	7069011	3468645	5818	3462827	7073125	1
60	7071068	3465735	0	3465735	7071068	0
						min
						Deg.
						45

45

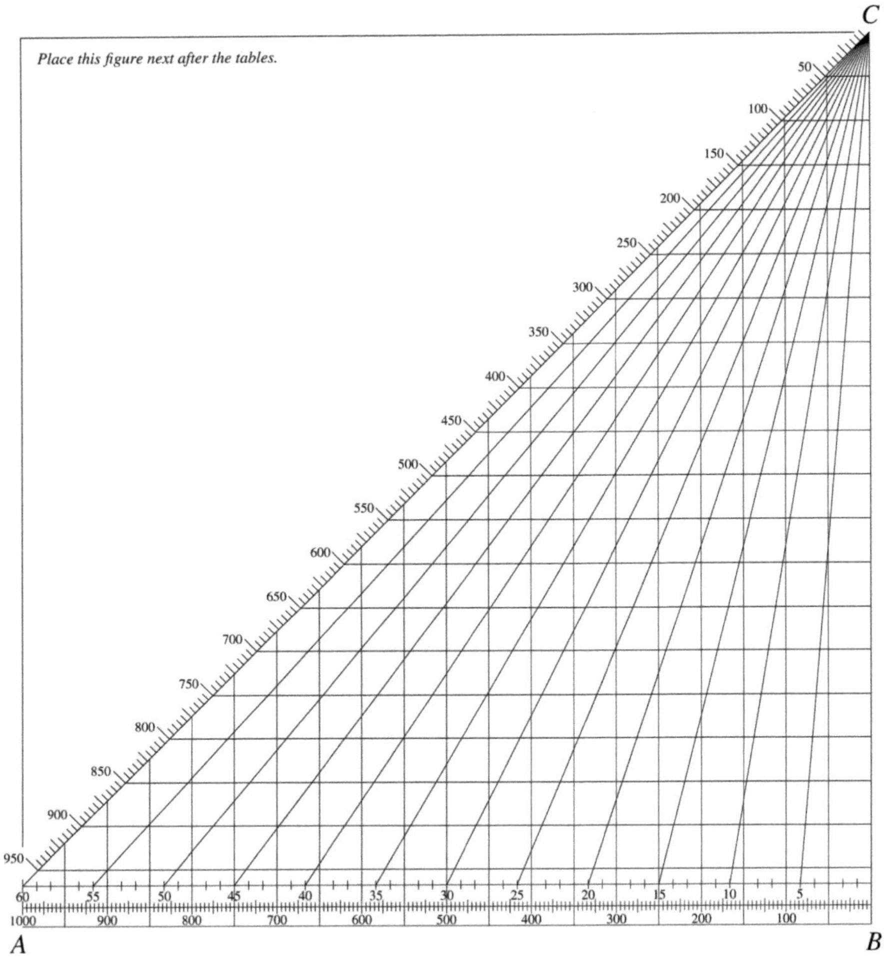
C
Place this figure next after the tables.
50
100
150
200
250
300
350
400
450
500
550
600
650
700
750
800
850
900
950
60
55
50
45
40
35
30
25
20
15
10
5
1000
900
800
700
600
500
400
300
200
100
A
B

THE VSE OF THE TRI-
angular Table for the finding of
the part Proportionall, penned
by Henry Brigges.

THE compendiouſneſſe of theſe Tables of *Logarithmes*, cannot be without ſome defect, which is to be ſupplied (as in all other Tables) by the part proportionall: that whereas ſomtimes the number we deſire, is not to be found in the Table, we may by the difference of that number, the number in the Table neereſt vnto it, and the proportionall part anſwerable to that difference, haue our deſire, ſo neere as may be, or is needfull. This proportionall part is had three wayes.

1. Either by the Rule of Proportion, commonly called the Golden Rule: Or

2. By the helpe of this Table of Logarithmes. Or

3. By this Triangular inſtrumentall Table.

Of which three wayes the firſt is moſt exact, and the laſt moſt eaſie, but not ſo exact as the other two. All theſe wayes hauing three numbers giuen, do helpe vs to finde the fourth proportionall number. The manner of the operatation is beſt ſhewed by example.

Pag. 81. lin. 15. [p. 545, line 22 in this volume] we haue found that 141766 is the Logarithme of halfe the angle *PZS*. which number I ſeeke in the Table, and finde it not; but the two *Logarithmes* next vnto it, are 141834 and 141667, which anſwere to 12 & 13 minuts aboue 60 degrees: ſo that it is apparant, that the arke which we ſeeke is 60: $12'$ and more. Now to finde how much more is to be added to this number, I take the difference of the tabular Logarithmes, 167 being firſt of the three numbers, which before I ſaid muſt be giuen, and the difference of the two anſwerable tabular arkes $60'$ ſeconds, which is the ſecond giuen number, and theſe two may fitly be termed *Tabular differences*. The third giuen number which we call the *Occurring difference*, is the difference of the former of the two Tabular Logarithmes, and of this Logarithme 141766 which we ſeeke for: which difference is 68. Theſe three giuen numbers 167 $60'$. 68 do helpe vs to the part proportionall, to be added to 60: $12'$, by euery one of the three former wayes.

1. By the golden rule I multiply the third 68, by the ſecond 60, and the product 4080 being diuided by the firſt 167, giueth in the quotient 24 $\frac{1}{2}$ almoſt, ſo that the arke anſwering to halfe the angle *PZS*, is found to be 60: $12'$, $24''\,\frac{1}{2}$.

2 By this Table of Logarithmes thus, I take the Logarithmes of thefe three giuen numbers, fo neere as the bare Table will affoord, without any further curious fearch (becaufe in this cafe I need not feeke to be exact; and for the fame caufe I cut off the two laft figures in euery Logarithme) the Logarithmes I find to be 17928. 5109. 3943. But becaufe the numbers in the Table, to which the Logarithmes are adioyned, are *Sines*, and I haue found thefe Logarithmes anfwerably, as if the giuen numbers were 167. 600. 681, It is therefore apparant, that the fourth proportionall, which I feeke for, muft likewife increafe aboue the totall *Sine*; fo that his Logarithme is leffe then nothing, and the number anfwering to it is not to be found in this Table. Therefore by the 4. *chap.* 1. *lib.* the 9 *Sect. pag* 18. I doe adde 23025 (cutting off here alfo the two laft figures) to the Logarithme of 680: therefore the third number being 68, and not 680. his Logarithme by this new increafe is 28134. And becaufe by the 5 *prop.* 2. *chap. lib.* 1. the Logarithmes of the two middle numbers are equall to the Logarithmes of the two extreame proportionals, therefore out of 32077, the fumme of the two middle Logarithmes I take 17928, the Logarithme of the firft, and there remaineth 14149. the Logarithme of the fourth proportionall, which I defire: to the which, in the Table, 243 doth anfwere. Therefore I fay, as before, that the proportionall part to bee added to the arke found, is $24' \frac{3}{10}$. For though the fourth proportionall 243 be of 3 figures, yet wee may prefently difcerne that the caufe of this greatneffe in this laft proportionall is becaufe the one of the two middle numbers is too great, which fhould be onely 60 and 68, and that if here we cut off the laft figure, then will the foure numbers keepe due proportion 167. $60''$. 68. $24' \frac{3}{10}$.

3 The third way of finding this proportionall part, is by this triangular table which was drawne by Mr. *E. Wright*; and becaufe at his death he left no defcription of it, nor manner how to vfe it, at the requeft of fome friends, I make bold to fupply his place fo well and plainly as I can.

You fee then in this triangle, three forts of lines, fome paralell to the bafe *AB*, others perpendicular vnto it, and the third fort drawne all from the verticall angle *C*, vnto equall partes of the bafe. Thefe laft may be called *Diagonall* lines.

You fee likewife the bafe *AB* diuided into 60 equall parts, and a line vnder it, and paralell vnto it into 1000.

In like fort, the perpendicular line *CB* is diuided into 1000, and vpon the interfections of the paralells, and the Diagonall *CA* are fet the fame numbers that are fet vpon the other ends of the paralels, in the perpendicular *CB*. Amongft thefe numbers we muft place our three giuen numbers, and by helpe of the lines we fhall amongft the fame, finde out Geometrically, the fourth proportionall, which we defire.

For Example.

Take the ſame numbers wee had before 167, and 60. the two Tabular Differences, and 68 the Occurring difference. Of theſe there are two which are differences of Logarithmes, to wit, 167, and 68, the firſt and the third: theſe being of one kinde, or homogeneall, are to haue like ſituation in the Triangle. And the ſecond being Homogeneall to the fourth, which is ſought for, is to bee placed on a differing ſide from the other two, vpon which differing ſide the fourth proportionall is to bee found. As here I take 167, and 68, and count them from the poynt C in the Diagonall line CA, and ſuppoſing a perpendicular line to bee drawne from the end of the leſſe number till it cut the paralell line drawne from the end of the greater number, by this poynt of Interſection I drawe an imaginarie Diagonall line from the poynt C, till it cut the Baſe AB, and counting from B to that Diagonall, I finde 24, and about $\frac{1}{2}$, which is the part proportionall I deſire, as in the former operations.

But becauſe 167 and 68 are ſuch ſmall numbers, and fall ſo neere the angle, therefore the concourſe of the paralell and perpendicular is not ſo exactly diſcerned, and the whole operation is more troubleſome and vncertaine. It is therefore conuenient in ſuch caſes, to take the double, or treble of both theſe giuen numbers, or the halfes, or any like parts of them both: and to enter the Table with theſe other numbers, in ſtead of the former, proceeding in all thing as before; then ſhall we, when the numbers reach neerer vnto the Baſe, finde the poynt of concourſe, and the part proportionall more exactly and eaſily then before. As if we take 835 and 340 the quintuples of the firſt and third numbers, wee ſhall more plainely and diſtinctly finde the fourth proportionall to be 24 $\frac{1}{3}$.

In like ſort, page 52, line 36, I would finde the differentiall anſwering to $16^\circ{:}\overset{\prime}{24}{:}\overset{\prime\prime}{27}$. and becauſe the Table extendeth but to minuts, I muſt finde the part proportionall anſwering to $\overset{\prime\prime}{27}$. Here the firſt giuen number is the Tabular differẽce of minuts $\overset{\prime\prime}{60}$: the third number is the Occurring Homogeneall difference $\overset{\prime\prime}{27}$. The ſecond giuen number is the Tabular difference of the Differentials 1074. Theſe three I place thus: The firſt $\overset{\prime\prime}{60}$. and the third $\overset{\prime\prime}{27}$, being *Homogeneall* vnto it, I place vpon the baſe AB: and becauſe the ſecond 1074 is too great for the table, & if one figure be cut away 107 remaining wil fall vpon the Diagonall line CA (on the which it is to be placed) too neere to the angle C, therefore I take 537, the halfe of the ſecond, and place it vpon the Diagonall line CA, drawing a paralell from that poynt, till it meete with the Diagonall comming from $\overset{\prime\prime}{27}$, and from the concourſe of that paralell with this Diagonal, I draw a perpendicular vpwards, till it cut the Diagonall line of $\overset{\prime\prime}{60}$. or the line CA, and I finde that the diſtance of this interſection from C, is about 240, which is the halfe of the fourth proportionall, becauſe 537 is the halfe of the ſecond giuen number *Homogeneall* vnto this fourth. I ſay therefore that the part proportionall anſwering vnto 27″, is 480, which being ſubducted from 1223101, the differentiall of 16°:24′, there remaineth 1222621 for the differentiall of 16°:24′, 27″, which differeth ſomewhat from that differentiall which is ſet downe in the booke: the reaſon whereof is, becauſe in ſo ſmall a Table it is impoſſible to diſcerne all the ſeuerall parts, the omiſſion of which will make no ſenſible difference in any worke. If in ſeeking the ſquare roote, or cubicke roote of 19, I would find the Logarithme of 190000, ſeeking this number amongſt the *Sines*, I cannot finde it, but at 10°:57′ I finde 189952, which is leſſe then the giuen number by 48. This is the *Occurring difference*. The tabular difference *Homogeneall* to this, is 286. The other tabular difference of the Logarithmes is 1502: theſe differences are giuen, which being placed in due order, the firſt of them is 286. the ſecond 1502, the third 48, and that the firſt & third number may fall neerer to the baſe, I double them both, ſo haue I 572, and 96, theſe I place, in ſtead of the other giuen numbers, vpon the perpendicular *CB*, or on the Diagonall *CA*. Likewiſe becauſe the ſecond number 1502 is too great, I take the halfe of it, ſo haue I 751, which I place vnder the baſe *AB*, vpon the line diuided into 1000: and from that point draw a diagonall, till it meete with the paralell of 96, the third number, and from the poynt of concourſe with that paralell, I draw a perpendicular, till it croſſe the paralell of 571 the firſt number. By this laſt concourſe of the perpendicular and paralell, I draw an other diagonall, cutting the line vnder *AB*, in which the ſecond number was counted, and the parts 125 betwixt that poynt and the end of the line towards *B*, being doubled, (becauſe the ſecond being Homogeneall to this, was halfed) are the fourth proportionall required, which may bee 250. and this proportionall being taken away from the tabular Logarithme 1660982, leaueth 1660732 for the Logarithme of 190000.

The ſame may be performed, if the firſt and third be placed on the baſe; and the ſecond vpon the ſide line, thus; Draw two Diagonals from thoſe two points of the baſe, & from the concourſe of the Diagonall of the third with the paralell of the ſecond, draw a perpendicular vpward, till it meete with the Diagonall of the firſt; the paralell paſſing by that point, ſhall in the line *CB* ſhew the fourth proportionall.

Thus may wee inſtrumentally come ſomewhat neere to that which wee deſire, eſpecicially if vpon a faire large Paſte-boord wee make a great Triangle, curiouſly diuided and lined, according to this patterne.

FINIS.

Errata in the Treatife.[94]

Pag. 15.	lin. 28.	make it Tangẽts	1370505
	lin. 29.	Make it number	1370305
Pag. 19.	lin. 25.	reade	4605168–00
Pag. 21.	lin. 33.	reade	34914–00
Pag. 29.	lin. 10.	reade	693147
Pag. 45.	lin. 4.	reade ZSP	
Pag. 72.	lin. 30.	reade ZP.	
Pag. 75	lin, 17.	reade halfe the aggregate.	

[94] This is the original *Errata* list, but it must be modified as follows to conform to the pagination in this volume:

Pag. 503.	lin. 36.	make it Tangẽts	1370505
Pag. 504.	lin. 1.	Make it number	1370305
Pag. 506.	lin. 5.	reade	4605168–00
Pag. 507.	lin. 13.	reade	34914–00
Pag. 511.	lin. 21.	reade	693147
Pag. 521.	lin. 19.	reade ZSP	
Pag. 539.	lin. 17.	reade ZP.	
Pag. 541	lin, 5.	reade halfe the aggregate.	

RABDOLOGIÆ

Translation and notes by William Frank Richardson

Small additions and modifications to the translation by Richardson [in brackets], as well as the addition of notes 9, 27, 28 and 47, have been made by Enrique González-Velasco. Note that the present typesetting has been formatted to be as close as possible to the style of the original Latin publication.

The photograph from the title page of the *Rabdologiæ* (p. 651) is reproduced by courtesy of the National Library of Scotland.

RABDOLOGIÆ,

SEV NVMERATIONIS PER VIRGULAS LIBRI DVO:

Cum APPENDICE de expeditissimo MVLTIPLICATIONIS PROMPTVARIO.

Quibus accessit & ARITHMETICÆ LOCALIS LIBER VNVS.

Authore & Inventore IOANNE NEPERO, *Barone* MERCHISTONII, &c. SCOTO.

EDINBVRGI,

Excudebat *Andreas Hart*, 1617.

RABDOLOGY,
OR CALCULATION WITH RODS

IN TWO BOOKS:

With an APPENDIX on the high-speed PROMPTUARY FOR MULTIPLICATION.

To which is added ONE BOOK on LOCATION ARITHMETIC.

Author & Inventor JOHN NAPIER, *Baron of* MERCHISTON, &c. A SCOTSMAN.

EDINBURGH,
Published by *Andro Hart*, 1617.

To the most Noble [*Sir*]

ALEXANDER SETON

EARL OF DUNFERMLINE, LORD OF FYFE & URQUHART, &c.

SUPREME *Chancellor*
OF THE KINGDOM OF
Scotland,

Greetings.

TO perform calculations (Noble Sir) is a difficult and lengthy process, the tedium of which deters many from the study of Mathematics. I have always tried, with such strength and talent as I possess, to expedite the process. It was with this end in view that I produced in former years my *canon of Logarithms*, at which I had laboured over a long period. In this I abandoned natural numbers and the more difficult operations which are carried out by means of them and substituted others which achieve the same result by simple addition, subtraction, and division by two or three. I have now discovered a greatly superior species of *Logarithms*[1] and have decided (if God will grant me life and health for a little longer) to publish the method of creating them and also their use, but because of the poor state of my bodily health I have left the actual computation of the new canon to men versed in this type of pursuit and principally to my dear and learned friend D. HENRY BRIGGS, public professor of geometry at LONDON.[2] In the meantime, however, for the benefit of those who prefer to operate with the natural numbers as they are, I have worked out three other short methods of calculation. The first of these, which I call RABDOLOGY, uses *rods with numbers* on them. The second, which for multiplication is the fastest of all, uses strips arranged on a box; for this the name *promptuary for Multiplication* will not be inappropriate. The third and last uses *location Arithmetic* and is performed on a chess-board.

I had two reasons for making my book about the manufacture and use of the rods available to the public. The first was that the rods have found favour with so many people that they could almost be said to be already in common

[1] Base 10.

[2] Henry Briggs (1561-1630) became the first professor of geometry at Gresham House, London, in 1596.

use both at home and abroad. The other was that it was brought to my attention that you had kindly advised me to publish them, lest they be published under someone else's name and I be forced to sing with *Virgil*,

'Tis I who wrote these little lines, etc.[3]

Such friendly counsel from your highness must carry the greatest weight with me, and certainly without it this tiny work on rods (to which I have subjoined the other two methods) would never have seen the light of day.

The thanks which students of Mathematics owe for these books, (Noble Sir), must go entirely to you, who are not so much their patron as their second father, especially as I know full well that you hold my rods in such esteem that you have had a set made not from any ordinary material but from silver.

So I ask you, (Noble Sir), to receive this little work, such as it is, with forbearance. Though it is unworthy of such a Maecenas, take it up and nurture it as your own child.[4] I pray earnestly to God that you, as a patron of Justice and equity, may long be preserved in the best of health for the benefit of our people and our Country.

Your highness's
devoted servant

JOHN NAPIER
BARON OF MERCHISTON.

[3] The lines referred to are not in his published works but are from Donatus's *Life of Virgil*.

[4] Gaius Maecenas (c. 70–8 B.C.) was Virgil's patron and became the archetype of all literary patrons.

TO THE MOST WORTHY AUTHOR.

Arithmetic has now become a game.
Gone is the tedium of calculation.
And *Logarithms*, *Chess-board*, *Rod*, *and Strip*,
Confirm once more great the NAPIER's reputation.

TO THE READER OF THE RABDOLOGY.

The number'd rods will help you do your sums
With *quotumus* and *Multiple* display'd.
Extract a root, divide, or multiply,
Using with confidence their novel aid.

PATRICK SANDYS.

TO THE READER.

Would you divide or multiply, that so
A quotient or a product you may know?
Would you extract a Cubic *root* or square
And thus a given figure's Side declare?
Would you a measurement of size secure?
This method, reader, is both swift and sure.

ANDREW JUNIUS.

[CHAPTER CONTENTS.
AND USE OF THE WHOLE
WORK.

First book of Rabdology,] *the use of the Rods in general.*

CHAP. I.

[*Second book of* Rabdology,] *the use of the Rods in Geometry & Mechanics with the help of Tables.*

Appendix.
The Multiplication Promptuary.

Book on local Arithmetic.

THE END.

The tasks that fill'd beginners with dismay
This little book has banish'd clear away.

RABDOLOGY
THE FIRST BOOK
THE USE OF THE RODS
in general.

CHAPTER I.
How to Make, & inscribe the Rods.

RABDOLOGY *is the Art of Calculating by means of Rods with numbers on them.*

These rods are square and movable and are inscribed with the multiples of simple numbers. They are for performing easily and quickly the more difficult operations of ordinary Arithmetic. I shall therefore consider how to Make and use these Rods.

HOW THEY ARE MADE.

CONSTRUCT from silver, ivory, boxwood or some similar solid material a number of square rods: ten for numbers less than 11111 (five places), twenty for numbers less than 111111111 (nine places), thirty for numbers less than 1111111111111 (thirteen places).

They should all be the same length, that is, three fingers more or less.[5] The width of each should be one tenth of the length, enough to contain comfortably two arithmetical figures, and the height should be equal to the width. The four faces or sides should be filed to right angles with such accuracy that, however the rods are joined together, they should look like a single flat tablet. When the rods are ready they should be divided along each edge into ten equal parts but in such a way that there are nine full parts in the middle while the upper half of the tenth part forms the upper margin and the other, lower half the lower margin. Then let straight lines be drawn from edge to edge at the points that mark the divisions so as to divide each face of each rod into nine small square portions (not counting the margins), and let each square be divided into two by diagonals drawn from the lower left to the upper right corner, as in the figure below. The rods are then ready for writing in the numbers.

The first step is to set the rods in front of you and mark the faces of each

[5] 1 finger = 3/4 inch = 19 mm.

I^a, II^a, III^a, and IV^a (or similar marks). The first face is the one confronting you, the second faces right, the third faces down to the ground, and the fourth faces left. These marks are purely to assist your memory and my exposition and should be erasable.

Shape of a Rod.

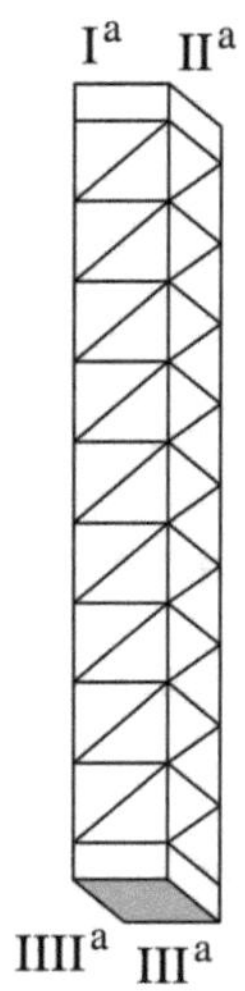

Second, observe that the first figure on each face, which is to be written in the right-hand portion of the first square, is a single figure and is called the simple, the figures which follow in the second square are twice that figure, those in the third square are three times, in the fourth four times, and so through the remaining multiples up to and including nine times. If any of these multiples consists of a single figure, it is to be written on the right-hand side of its square; if it consists of two, the right-hand one is to be written on the right-hand side of the square and the left-hand one on the left-hand side.

Third, note that the third face of each rod is always opposite to the first and the fourth to the second, and that the simples on each face are opposed not merely in the sense that one is on the upper and the other on the lower face or that one is on the right-hand and the other on the left-hand face but also in that they are at opposite ends, one at the top and the other at the bottom of the rod, and that these two opposed simples always add up to nine. So hereafter I will call those numbers *opposed* whose sum contains no figure except 9, for these are the only ones that are opposed on the rods. So much by way of general observation. The inscription of the RODS individually is carried out as follows.

In the lower right of each square on the first face of the first four rods write the cipher 0. Then turn these four rods over in such a way that the head of each is where the foot was before and that the face that before was uppermost is now underneath, and write on each one the number nine and its multiples, to wit 9, 18, 27, 36, 45, 54, 63, 72, 81, in the fashion described above under general method.

Then similarly on the second face of the first rod and the first face of the fifth, sixth, and seventh rods write unity with its multiples, to wit 1, 2, 3, 4, 5, 6, 7, 8, 9, in a row from top to bottom. Then turn the rods over and write on each the number eight with its multiples, that is 8, 16, 24, 32, 40, 48, 56, 64, 72.

Third, on the second face of the second and fifth rods and the first face of the eighth and ninth write the number two with its multiples, that is 2, 4, 6,

8, 10, 12, 14, 16, 18. Then turn the rods over and write on each the number seven with its multiples, to wit 7, 14, 21, 28, 35, 42, 49, 56, 63.

Then on the second faces of the third, sixth, and eighth rods and the first face of the tenth write the number three and its multiples, that is 3, 6, 9, 12, 15, 18, 21, 24, 27. Then turn them over and write on each the number six and its multiples, to wit 6, 12, 18, 24, 30, 36, 42, 48, 54.

Lastly, on the second face of the fourth, seventh, ninth, and tenth rods write the number four and its multiples, to wit 4, 8, 12, 16, 20, 24, 28, 32, 36. Then turn them over and inscribe the number five and its multiples, to wit 5, 10, 15, 20, 25, 30, 35, 40, 45, keeping always to the general principles set down above.

The following are FIGURES *of the above ten Rods with their four faces unrolled and flattened out for easier reading.*

You now have ten rods, each inscribed with its own individual set of numbers. The arrangement given above offers the greatest flexibility. Although other arrangements are possible, none is superior to this, and none can express more numbers with so few rods, for not only can they express every number without exception below 11111 (five places), but they can also express all numbers below 10000000000 (elven places) except those in which, and in the opposites of which, one figure occurs five times or any two figures occur eight times or any three figures occur ten times. If you desire to work with all numbers without exception below 111111111 (nine places), add to these ten rods another ten similar ones, and these twenty rods will express not only every number without exception below the aforesaid 111111111 (nine places) but also every number below 100000000000000000000 (twenty-one places) except those in which (and in the opposites of which) one figure occurs nine times or any two figures occur fifteen times or any three figures occur niineteen times. If it is desired to go even further to numbers of thirty places, you can add to these a third decade of rods constructed in the same way as the previous ones, These thirty rods will express not only every number without exception below 1111111111111 (thirteen places), but also every number below 1000000000000000000000000000000 (thirty-one places) except those in which, and in the opposites of which, one figure occurs thirteen times or any two figures occur twenty-two times or any three figures occur twenty-eight times.

4 Faces for the first rod.

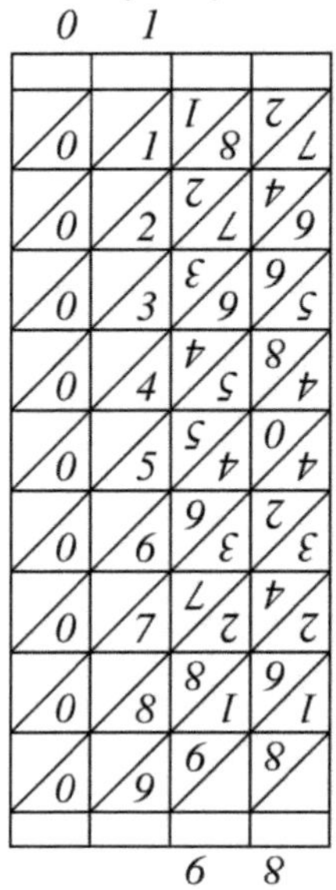

4 Faces for the second rod.

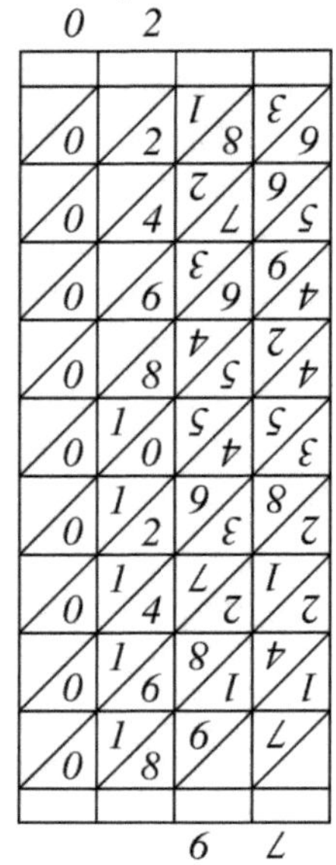

4 Faces for the third rod.

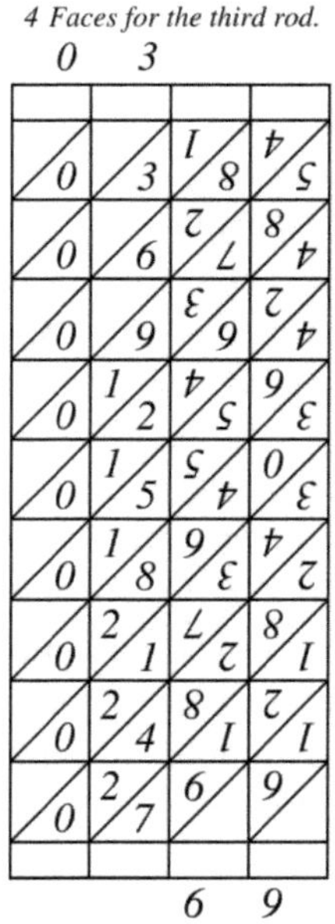

4 Faces for the fourth rod.

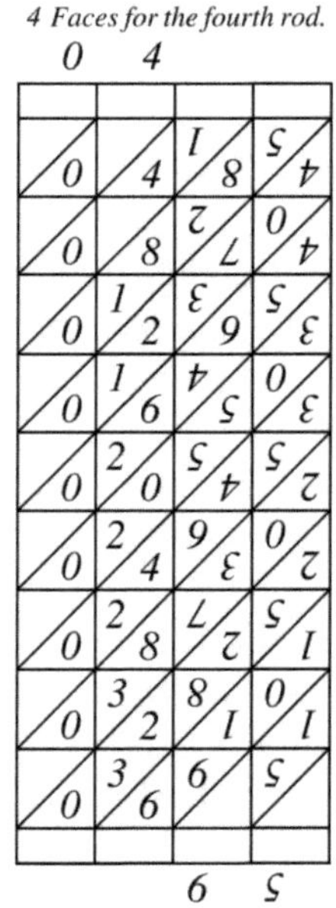

4 Faces for the fifth rod.

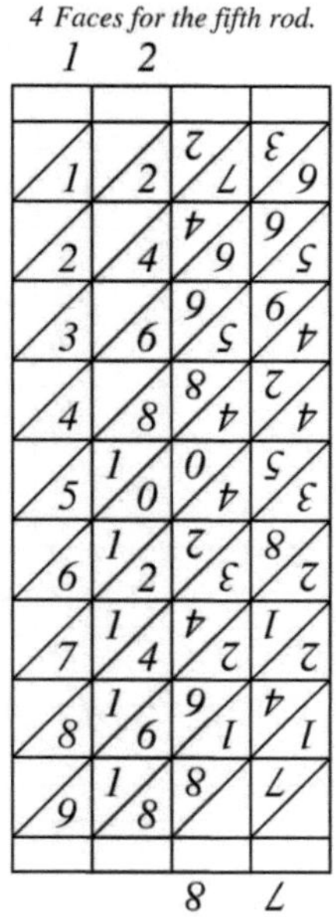

4 Faces for the sixth rod.

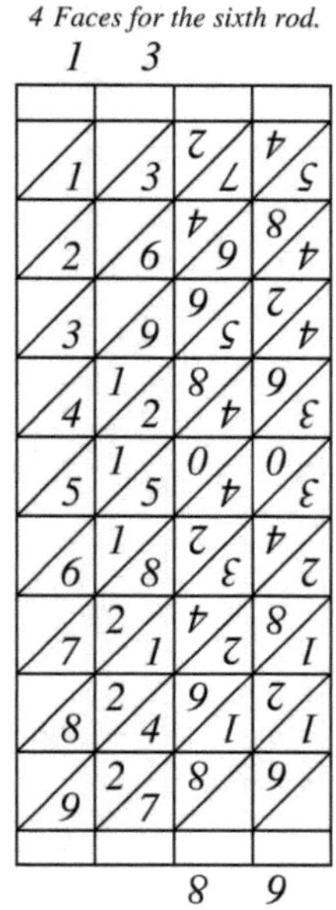

4 Faces for the seventh rod.

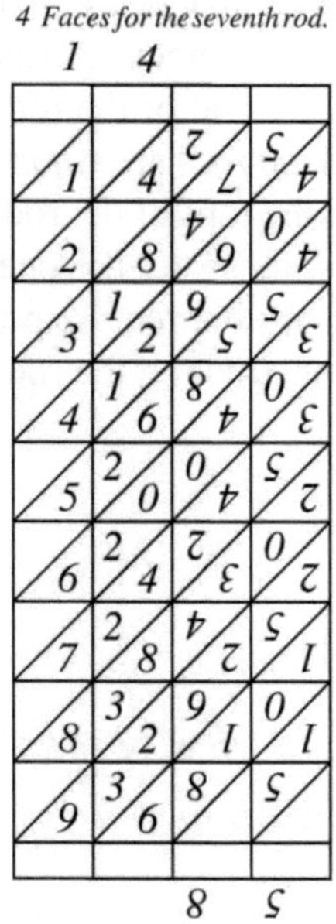

4 Faces for the eighth rod.

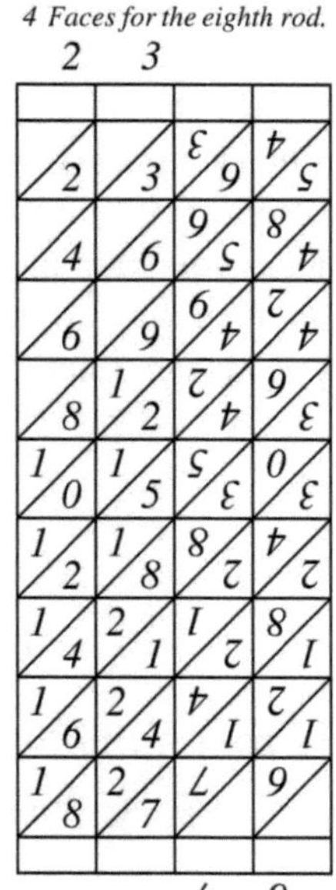

4 Faces for the ninth rod.

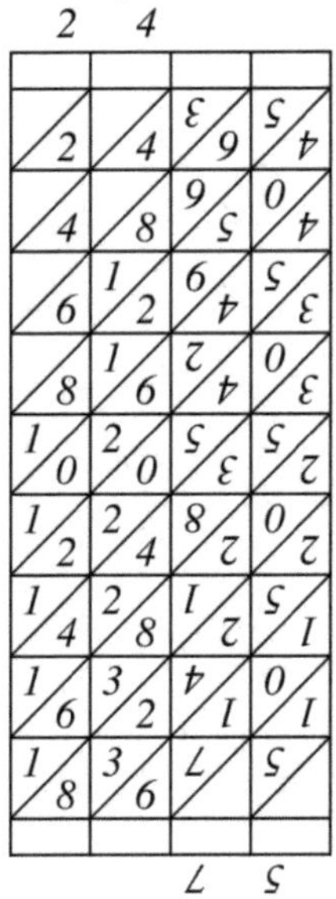

4 Faces for the tenth rod.

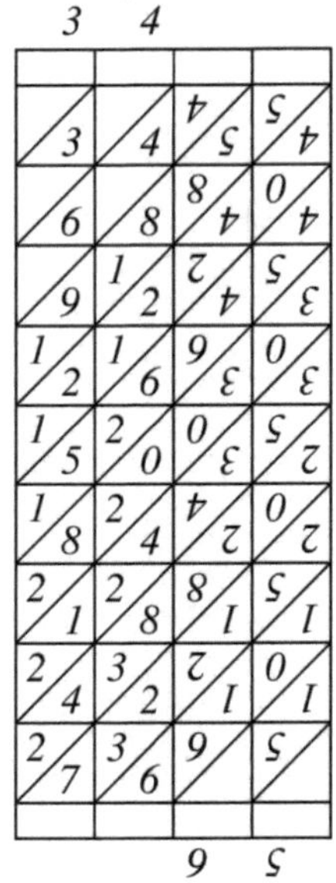

Furthermore, so that the rods which you need may be the more easily and quickly selected from the whole heap, the simple number inscribed in the first square of each face should be carved also on the end of the rod at the top of each face, for thus when the box is opened and the rods are stood on end while still in a bundle, you will see at first glance on the tops of the rods which ones have the numbers you want or their opposites: these are the ones that, either that way or the other way up; will serve your present requirements, and they are the ones you must extract. So much for constructing the rods. I turn now to their use.

Chapter II.

From numbers to Rods, and vice versa.

First Proposition. Problem I.

Setting up *on the table a given number with its multiples in order.*
Take the year of our Lord 1615 as the number to be set up on the table with its multiples. From the whole pile take four rods, of which one should have one written on the top, one the number six, the third one, and the fourth the number five. Place in front of you the faces corresponding to

these figures, and in the top squares of the four rods you will see the simple number facing you, in the second squares twice the given number, in the third squares (that is, the third interval between lines) three times the given number, in the fourth four times, in the fifth five times, and so on up to nine times, which you will find in the ninth interval between lines.

PROPOSITION II.

WHEN *the the rods are set up, single places are marked off by single diagonals. Hence, two figures in the same rhomboid are of the same place and must therefore be added together.*

Thus when the year of our LORD 1615 is set up along the top line (as above), the figures that denote twice that number appear automatically in the second line in four places, In the first rhomboid of.that line appear the numbers 2 and 1 (which are added to make three), in the second rhomboid appears the number 2, in the third 2 and 1 (which again are added to make three), and finally in the end space 0. So we get 3230 as exactly twice the aforesaid year.

PROPOSITION III.

WHEN *the total in any place is greater than nine, then subtract ten and retain only the remainder. When the total is less than nine, retain it intact. For the case where the total is exactly nine, see the next proposition.*

To illustrate this, set up the number 166702498 on the rods [(using the first proposition above)]. In the first rhomboid from the left in the ninth interval you will find 9 and 5, whose sum is 14. So take away ten and keep in mind four. In the seventh rhomboid of the seventh interval you will find 8 and 6, whose sum is 14. So again reject ten and remember four. In the above examples the sum is greater than nine; in those that follow it is less. In the first rhomboid of the third interval are found 3 and 1: their sum, 4, is less than nine and is therefore to be remembered. In the first space (the one on the far left) of the sixth interval stands nothing.[6] So nothing is kept in mind. In the sixth rhomboid of the second interval stands four; four is less than nine and is therefore to be remembered. In the fifth rhomboid of the fourth interval stands a cipher, or nothing. Nothing is therefore to be remembered. And so on.

PROPOSITION. IV.

WHEN *a total has been kept in mind and the next rhomboid to the right has a total greater than nine and the one following that a total less than*

[6] The space is a triangle, not a rhomboid.

nine, the number one must be added to the remembered total before it is written down.[7] *If rhomboids with a total of exactly nine intervene before the one whose total is greater than nine, they should each be represented by a cipher, but if rhomboids with a total of nine are followed by one whose total is less than nine, they should each be represented by a nine.*

In the first example under proposition 3 above the number four was kept in mind from the first RHOMBOID of the ninth interval. The next rhomboid to the right has a total greater than nine [namely 4 and 6, (which make 10)].[8] So instead of the four that was kept in mind, five is to be written down. Again, in the first rhomboid of the third interval the figure 4 was kept in mind (see the thid example above); the next rhomboid to the right has a total of nine in it, and the one to the right of that has more than nine (8 + 2); so instead of four and nine the figures 5 and 0 must be written down. Again, for the vacant space at the extreme left of the sixth interval (see the fourth example above) nothing was kept in mind; but the next rhomboid to the right contains nine and so does the next one, and the one after that contains more than nine (6 + 4 or 10); so the nought will have to be increased by unity, and instead of writing down 099, you will have to write down 100. On the other hand, for the seventh rhomboid of the seventh interval four was kept in mind (see the second example above), and as the next rhomboid to the right is less than nine[, namely 3 + 5, which make 8], the integer four is to be written down without increase. Again, for the sixth rhomboid of the second interval four was kept in mind in the fifth example, and this rhomboid is followed by two nines and then by six, which is less than nine, so the number four is to be written down along with the two nines thus: 499. Finally, in the sixth example nothing was kept in mind for the fifth rhomboid of the fourth interval. This rhomboid is followed by three nines and then by the number two, which, being less than nine, indicates that the zero which was kept in mind is to be written down with the three nines after it without increase thus: 0999. And so on.

PROPOS. V. PROBL. II.

TO FIND *and write down the desired multiple* (*up to nine times*) *of a given simple.*

Since Proposition 3 indicates when the figures on the rods are to be reduced by ten and when not and Proposition 4 indicates when they are to be increased by unity and when not and there are no other differences between the numbers

[7] Note that Napier reads the figures of the multiple from left to right.

[8] Napier has ignored a rhomboid of 9 between them.

on the rods and those to be written down, it is easy to deduce from the figures on the rods what figures should be written down, if only by copying the following examples. To write down the multiples of the first example, the year of our LORD 1615. Let the rods be set up so that 1615, the simple, appears in the first interval (see problem 1 above).[9] In the second interval we find 2 and 1, [2,][10] 2 and 1, 0, which are 3230 or twice the said year. In the third we find 3 and 1, 8, 3 and 1, 5, which are 4845 or three times the said year. In the fourth 4 and 2, 4, 4 and 2, 0; these are 6460 for four times. In the fifth 5 and 3, 0, 5 and 2, 5, which are 8075 for five times. In the sixth 6 and 3, 6, 6 and 3, 0, which are 9690 for six times. In the seventh 7 and 4, 2, 7 and 3, 5, which are 11305 for seven times. In the eighth space 8 and 4, 8, 8 and 4, 0, which are 12920 for eight times the given year. Finally, in the ninth space are 9 and 5, 4, 9 and 4, 5, which are 14535 for nine times the given year. Likewise, the simple of the second number that we set up on the rods, 166702498, will stand in the first or top interval. In the second stands twice this number, which is read off and written down thus: 333404996. From the third interval three times the given number is written down thus: 500107494. From the fourth four times is written down thus: 666809992. From the fifth five times the given number is written down thus: 833512490. From the sixth interval six times the number is read off and written down thus: 1000214988. From the seventh seven times the same number is read off and written down thus: 1166917486. From the eighth eight times the given number is to be transcribed thus: 1333619984. Lastly, nine times the given number is written down thus from the ninth interval: 1500322482. With a little practice you will learn to read off and write down these and similar numbers either forward or backward. Reading off and writing down multiples is the only difficulty in performing Arithmetic by Rods.

WARNING ON ADDITION AND SUBTRACTION.

Since these Rods were invented for the purpose of the more difficult operations of Arithmetic (these being Multiplication, Division, and the Extraction of square and cube roots) and since Addition and Subtraction sums are within the capabilities of every novice, I shall therefore omit these and begin with Multiplication.

[9] Napier did not provide a diagram, but one can be found in th mathematical introduction to this volume.

[10] Napier's printer accidentally omitted this figure.

CHAPTER III.
Multiplication.

THE TERMS MULTIPLIER, multiplicand, and multiple are familiar from ordinary arithmetic. I use the term quotumus (quotulus I might have called it) to refer to the single number that contains unity as many times as the multiple on the rods contains its own simple. This means that if the intervals were numbered in order, the quotumus would be the same as the number indicating that interval.

For the easier multiplication of numbers it is expedient that the simple and all the multiples shown by a given set of rods should consist of an equal number of figures (either intrinsically or by writing an additional cipher in front). This means that the left-hand margin will be justified, as it were, and will form a straight line like the one on the right.

Set up one of the numbers to be multiplied together (preferably the larger) upon the rods (as in chapter 2, *problem* 1). *Write down the other on paper and draw a line under it. Then, choosing any figure on the paper, regard it as a quotumus, and write down on the paper the corresponding multiple from the rods in such a way that the right-hand (or left-hand) figures of all the multiples should form a diagonal line slanting to the right (or left) in the same order as the quotumi that designate them. Then the multiples so set out are added up in the normal way, and the product of multiplication will result.*

To multiply the year of our Lord 1615 by 365, the former number should be set up on the rods, and the latter written on a piece of paper as in the [margin]. Three times, six times, and five times the number on the rods are to be taken in order as quotumi, as indicated by the figures of the number 365 written on the paper. So three times the number 1615, which, written down from the rods, is 4845, and six times (which is 9690) and five times (8075) are to be written in a diagonal line beneath their respective quotumi 3, 6, 5 either beginning from each respectively as in the first display or ending with each respectively as in the second display. It makes no difference which pattern you follow, provided that the lefthand figures, when justified, progress in a straight line diagonally in the same order as the aforesaid indices or quotumi. When these multiples have thus been set out in order, they will be added in the usual way, and the required number 589475, the product of multiplication, will result.

365	365
4845	8075
9690	9690
8075	4845
589475	589475

The same result will be achieved by writing 1615 on the paper and setting up 365 on the Rods. The multiples of 365 are then the simple (365), six times (2190), the simple again (365), and five times (1825), as is indicated by the figures 1615. These are then justified by the addition of a cipher on the left of the two simples, set out diagonally as you see here, and added up, resulting in the same product (589475) as above.

```
1615
-------
0365
 2190
  0365
   1825
-------
 589475
```

Another Way of Multiplying.

ALTERNATIVELY *and as a test of the preceding multiplication, turn over the entire table of Rods* "en bloc," *and you will find at the head of the table the number* 8384, *the opposite of the original number* (1615). *Now write down diagonally three times, six times, and five times this number, that is,* 25152, 50304, *and* 41920, *with the lesser of the multiplicands,* 365, *directly underneath. Add all these up as in the* [*margin*], *and they will give* 3060525. *Subtract this from* 3650000 (*that is, from the same multiplicand increased by as many ciphers as there are Rods on the table or numbers in the other multiplicand*), *and the remainder is* 589475, *the same number as above. To fix this dual form of multiplication more firmly in the memory, I have seen fit to add the following verses.*

```
25152
 50304
  41920
    365
-------
3060525
3060000
-------
 589475
```

Set up the larger number on the rods.
The smaller shows which multiples to take.
Obliquely write them down beneath the line.
And, added up, your answer they will make.
To check your answer, first invert the rods.
The smaller shows the multiples once more.
Write them obliquely with the smaller too.
And sum them by addition as before.
Now to the smaller add as many noughts:
As there are rods before you on the board.
Subtract the second sum you found above.
And thus the self-same answer you'll record.

Chapter IV.
Division.

WRite down on paper the number to be divided or cut, and set up the divisor or sector on the rods (as in chapter 2, problem 1). Select the largest multiple that can be subtracted from the left-hand side of the dividend; it will be either equal to it or nearly equal but less. Subtract this multiple, whatever it is, from the left-hand side of the dividend. Then write the multiple below the dividend, the remainder above, and the quotumus of this multiple on the far side of the semicircle that marks the quotient. Next, select from the table of rods another multiple that is either equal to or as close as possible to but less than the left-hand figures of the remainder. Write this below and subtract it from the remainder written above, writing the remainder from this operation above the previous one and the quotumus of this multiple after the previous quotumus. Repeat this second step over and over again until the right-hand figure of the last multiple stands directly under the right-hand figure of the dividend. Then the quotumi beyond the semicircle are the desired quotient, and the number left over (if any) is the numerator of the remaining fraction and the divisor is its denominator. I shall illustrate all this by examples.

Dividing the number 589475 by 365. [First] write the former number on a piece of paper (as in the [margin]), and set up the latter at the top of the rods.

```
      0
    182
    54
224
589475(1615
365
2190
   365
   1825
```

Among all the multiples of 365 the simple 365 itself is the closest to but less than 589 (the first figures of the dividend). The figure 365 should therefore be written under the 589 and subtracted from it. The remainder 224 is to be written above the 589, and the quotumus (or index) of the simple, namely unity, is to be placed in the quotient. Second, In the sixth interval of the set of rods you will find 2190, which is six times the divisor and is nearest to but less than the number 2244, which appears above the dividend.[11] The number 2190 should therefore be written directly under the number 2244 and subtracted from it. This gives a remainder of 54, which is to be written above, while the quotumus of this multiple, 6, is to be added to the quotient. [Third (repeating the second step)] look for the multiple nearest to but less than the number 547. This is once again the simple itself, 365. This is subtracted from 547 to give a remainder of 182, which is written above, and the index of the simple, which is unity, is to be placed in the quotient. Lastly [fourth], the multiple closest to the number

[11] The 224 appears above the dividend; the other 4 is the next figure of the dividend itself.

1825 is to be sought, and a figure equal to this, 1825, will be found in the fifth interval. This number should be written beneath the other and subtracted from it, which leaves a remainder of nothing. So 0 should be written above and the figure 5 added to the quotient. Thus the desired quotient is 1615.

Another Example.

DIviding the number 861094 by 432. Write the former on a piece of paper as in the [margin], and set up the latter on the rods. The multiple of the latter nearest to but less than the number 861 is the simple number 432 itself, which, when subtracted from 861, gives a remainder of 429. The quotumus, 1, is to be placed in the quotient. The nearest multiple less than 4290 occurring on the rods is 3888, which is nine times. This is subtracted from the surviving number 4290, which gives a remainder of 402, and the quotumus, 9, is placed in the quotient. The nearest multiple below 4029 is again 3888 or nine times. This is subtracted from 4029, which gives a remainder of 141, and the quotumus, 9, is put in the quotient. The multiple of the divisor on the rods nearest to but less than the surviving number 1414 is three times, or 1296. This is subtracted from 1414, which leaves 118, and the index of the triple, namely the quotumus 3, is placed in the quotient. Thus the whole quotient is 1993 with a remainder of 118, which supplies the numerator of the remaining fraction, whose denominator is the divisor itself, 432. The quotient is arranged like this: $1993\frac{118}{432}$.

```
    118
   141
  402
 429
 861094 (1993 118/432
 432
 3888
  3888
   1296
```

Excursus on Decimal Arithmetic.

PErhaps the difficulty of working with fractions that have various denominators is not to your taste, and you prefer the ones whose denominators are always 10, 100, 1000, etc. (denoted thus by the learned mathematician *Simon Stevin* in his Decimal ARITHMETIC:[12] ① firsts, ② seconds, ③ thirds), in which there is the same facility of working as with whole numbers. If so, you

[12] Simon Stevin (1548-1620) was a Flemish mathematician who became commissioner of public works and quartermaster general of the army. He did not invent decimal fractions but helped popularize them by the thorough account that he gave of them in 1585 in his pamphlet *La Thiende* (The Tenth), to which Napier here refers.

```
          64
         136
        316
      118.000
     141
    402
   429
   861094.000 (1993.273
   432
   3888
    3888
     1296
   ------
       864
       3024
        1296
```

can, after finishing the common division, end with a period (as in the [margin]) and add to the dividend or to the remainder one cypher for tenths, two for hundredths and three for thousandths (or more if you wish) and then proceed by the above method.[13] In the last example, which I have repeated here with the addition of three ciphers, the quotient is 1993.273, which denotes 1993 units and 273 thousandths, or $\frac{273}{1000}$ or (following Stevin)

$$1993,\overset{\prime}{2}\,\overset{\prime\prime}{7}\,\overset{\prime\prime\prime}{3}.$$

The final remainder, 64, is ignored in this decimal Arithmetic as of insignificant value. The same procedure is followed in similar examples. To fix these rules of division, both common and decimal, more firmly in your memory, learn the following verses.

FOR EITHER.

Set the divisor up; go on to choose
The highest multiple that suits your need.
Its quotumus within the quotient place;
Remainders write above; and so proceed
Right through the dividend. And if the last
Remainder as a decimal you'd show,
Augment the dividend by several noughts,
And so proceed as far as you would go.

FOR COMMON DIVISION.

Set the divisor up; from dividend
To take the highest multiples proceed
Slantwise; do this as often as you can;
Then from the quotumi the quotient read.

FOR DECIMAL DIVISION.

Set the divisor up; from dividend
Enlarged by noughts (as many as you need)

[13] This is the earliest published reference to the now familiar decimal point whose use was apparently invented by Napier.

To take the highest multiples proceed
Slantwise; do this as often as you can;
Then from the quotumi the quotient read.

NOTE.

It is therefore clear that the operator, or logist, has nothing to do but to place the multiples slantwise and add them for multiplication or subtract them for division. The most difficult part of the operation is the computation of the multiples themselves, and they are automatically displayed with the utmost clarity when the rods are correctly set out.

CHAPTER V.
The Plate for extracting Roots.

IN extracting roots the principal difficulty lies in the multiplications and divisions that occur in the course of the operation. The process can be performed quite easily by means of the rods alone, but this requires that the multiple of the divisor and the square (or cube) of the same multiple must be calculated separately and subtracted separately.[14] To avoid this double subtraction and at the same time to facilitate the finding of the numbers especially required (that is, the simple quotumi or roots and their doubles, squares, or cubes) by placing them all in the same line, I have designed a plate on which these numbers are inscribed. I shall now briefly describe its construction and then consider its use.

CONSTRUCTION OF THE PLATE.

MAke from the same Material as the rods a rectangular plate of the same length and thickness as the rods and one third as broad as it is long with each face (one for square roots, the other for cube roots) smooth and polished. Each face is to be divided into three columns. On the face for square roots

[14] In the long method of extracting square roots it is necessary. when finding each figure of the root after the first, to find by trial and error a value for x such that if r is the approximation to the root so far found. $20rx + x^2$ shall be equal to, or as large as possible but less than, the remainder to that point. The "multiple of the divisor" to which Napier refers is the $20rx$ in this formula. Napier's plate for square roots, when placed to the right of a rod (or rods) whose simple is $2r$, automatically calculates $20rx + x^2$ for x equal to 1 through 9 and so greatly assists the above process.

the left-hand column should be divided into nine small squares bisected diagonally. They should be drawn with clear lines that correspond exactly to the lines on the rods.

The first, or topmost, of these squares is inscribed with the figures 0, 1, the second with the figures 0,4, the third 0, 9, the fourth 1,6, the fifth 2, 5, the sixth 3, 6, the seventh 4, 9, the eighth 6, 4 and lastly the ninth 8, 1, in other words, with the square numbers. In the second column on the same face and in the first square thereof is inscribed 2, in the second 4, in the third 6, in the fourth 8, in the fifth 10, in the sixth 12, in the seventh 14, in the eighth 16, in the ninth 18, in other words, the even numbers. In the third, or right-hand, column of this face the nine figures 1, 2, 3, 4, 5, 6, 7, 8, 9, go down in order. The face for the extraction of square roots is then complete.

for the square root.		
/1	2	1
0/4	4	2
0/9	6	3
1/6	8	4
2/5	10	5
3/6	12	6
4/9	14	7
6/4	16	8
8/1	18	9

for the cube root.		
0/01	1	1
0/08	4	2
0/27	9	3
0/64	16	4
1/25	25	5
2/16	36	6
3/43	49	7
5/12	64	8
7/29	81	9

The [other] face (for cube roots) has three columns like the other, except that the first, or lefthand, column is big enough to take three figures and its first, or top, square is inscribed 0, 01, the second 0, 08, the third 0, 27, the fourth 0, 64, fifth 1, 25, the sixth 2, 16, the seventh 3, 43, the eighth 5, 12, the ninth 7, 29, in other words, the cubic numbers going down the strip in order. The second column on this face contains the square numbers 1, 4, 9, 16, 25, 36, 49, 64, 81, in order going down. The third column on this face, like that on the other face, has the nine figures 1 ,2, 3, 4, 5, 6, 7, 8, 9, in order going down. With the addition of the headings *square roots* on the top of the first face described above and *cube roots* on the top of the second, the construction of the strip is complete.

I now turn to the use of the strip in conjunction with the rods.

Chapter VI.
How to extract square roots.

B*Eginning always from the right-hand side of the right-hand figure of the given number whose root is to be extracted, mark off its figures in pairs by means of dots, and underneath these draw two lines far enough apart to contain the root. The process of extraction starts from the left-hand side and moves toward the right. Extract the square root of the numbers to the left of the first dot (the one farthest to the left) or, if they are not a perfect square, the square root nearest to but less than their exact square root. Place this root between the lines under the first dot, and then subtract its square from the numbers to the left of the first dot and note the remainder directly above them. Next, set up twice this root on the rods, and to their right place the square-root plate. From the rods and the plate select the multiple that is either equal to or closest to but less than the numbers as far as the second dot (in other words, the largest multiple that you can subtract from these). Subtract it from them, and note the remainder directly above; then set down the quo tum us of this multiple (which you will find in the right-hand column of the plate on the same line) between the lines and beneath the second dot. This is the second figure of the root. Now repeat this operation as many times as there are dots, remembering the rule that double the last quotumus must be added each time between the previous double and the plate.*

Caution I.

A*T this point it must be observed that if the number that is twice the quotumus consists of two figures, the rod corresponding to the figure on the right is to be inserted, and the figure on the left is to be added to that on the rod already there, which is then removed and replaced by the rod corresponding to their total.*

Caution II.

I*T must also be observed here that if you can subtract none of the multiples nor even the simple that is supposed to be subtracted from the figures preceding the dot you have reached, then a cipher* 0 *is to be placed beneath that dot as quotumus and everything else left unchanged.*

Example.

EXtracting the square root of the number 117716237694. Mark the number off with dots, and draw lines under it, as in the [margin]. Take the figures

```
                90
        54 89 5
       67
   21
  2
11.77.16.23.76.94.
------------------
  3  4  3  0  9  8
------------------
9
2 56
 20 49
     61 74 81
       5 48 95 04
```

as far as the left-hand dot, namely 11, extract the square root that is nearest but smaller, namely 3, and write this under the first, or left-hand, dot. Subtract its square, 9, from 11, which gives a remainder of 2. Write this remainder above. [Second,] double this root, which gives 6, find the rod labeled 6, and lay the square root plate beside it. Look on the rod and plate for the multiple that is nearest to but less than the remainder up to the second dot, 277. You will find this to be 256, which is four times, and beside it in the same line in the third column stands its quotumus, 4. Subtract 256 from 277, which gives a remainder of 21. Cancel the previous remainder, and write 21 beside it above the number, and to represent four times write the quotumus, 4, under the second dot. [Third,] double this four to give 8, which you will find in the middle column of the plate on the same line. Insert the 8 rod between the plate and the first rod, and then from the rods and plate thus repositioned choose the multiple closest to but less than the remainders as far as the third dot, namely 2116. This multiple is 2049. Subtract 2049 from 2116, which gives 67. Write this remainder above the third dot, and the quotumus for three times, namely 3, below it. [Fourth,] twice the quotumus three is 6. So place the 6 rod between the preceding rods and the plate, and with the plate back in position, look for the multiple that is nearest to but less than the remainder as far as the fourth dot, which is 6723. You will not find one that can be subtracted from 6723 (see caution 2 above), so leaving the remainder intact, place 0 below the fourth point as the fourth figure, and go on to the fifth. [Fifth,] twice 0 is 0, so interpose the 0 rod between the last (or 6) rod and the plate. Then the multiple that is nearest to but less than the remainder as far as the fifth dot, namely 672376, is nine times, which is the number 617481. Subtract this from 672376, which gives a remainder of 54895, and to represent nine times, add 9 as the fifth figure of the root. [Sixth,] to represent 18, which is twice the 9 just found, insert the 8 rod on the right-hand side, and add unity to the preceding 0 rod. This will become 1 instead of 0, so remove the 0 rod, substitute the 1 rod, and then replace the plate. Then look for the multiple that is nearest to but less than the remainder as far as the sixth dot. This remainder is 5489594 and the multiple is eight times or 5489504. Subtract this from 5489594, which gives a remainder of 90. Thus the required root is 343098, and 90 is left over as the final remainder. For other examples the method is the same.

So that the special instructions for extracting square roots may stick more firmly in your mind, I have adjoined the following verses:

THE FIRST OPERATION
IN EXTRACTING SQUARE ROOTS.

Mark off the figures from the right in pairs.
The largest square of those the plate doth show
Subtract from the first group.
Remainder write above the group, and quotumus below.

The second and subsequent operations.

Insert the rods for twice the quotumus beside the plate.
Of multiples they show select the largest that does not surpass
The figures to the point you've reached, and so subtract it as before.
Remainder write above the group and quotumus below.

CHAPTER VII.
How to extract cube roots.

FRom the right-hand side of the right-hand figure of the number whose cube root is to be extracted, mark off with dots groups of three figures, and draw underneath two lines far enough apart to write the cube root in. Starting from the figure or figures to the left of the lefthand dot and working toward the right, find with the help of the cube root plate either the exact cube root or the number that is closest to but less than the true cube root. Write this root (which is a single figure) between the lines and underneath its own dot. Then subtract its cube from the figures that precede the first (or left-hand) dot, and write the remainder above. Next set up three times this root on the rods, and put it aside. Take three times the square of this same root, set this figure up on the rods, and place them to the left of the cube root plate. Take the rods that were set aside, and place them to the right of the plate, which is now in between.[15] *From the rods to the left and the plate,*[16] *is choose the*

[15] The rods to the left of the plate and the figures in the first column of the plate calculate $300r^2x + x^3$ (where r is the approximation to the root already found) for x equal to 1 through 9. The supplementary calculation performed "somewhere else on the paper" works out $30rx^2$ and adds it to the other total.

[16] By "the plate" Napier means the left-hand column of the plate.

multiple that is nearest to but less than the figures as far as the second dot, and write it down somewhere else on the paper. Draw a line above it, and over its right-hand figure, write its quotumus and then to the left of this the square of the quotumus so that the figures on the paper appear in the same order as on their line of the plate. Beneath each figure of the square, write the multiple that appears beside that figure on the rod to the right in such a way that the last figure stands directly below its own figure (which is also its quotumus). This means that these multiples are written down in a diagonal line. Now add them up, subtract their sum from the figures as far as the second dot, and write the remainder above. Finally, below the second dot and between the two lines write down as the second figure of the root the quotumus that earlier was noted down on the right.[17] *This concludes the operation at the second dot. Repeat it unaltered at the following dots right through to the end.*

CAUTION I.

IN the operation at every dot, note that if not even the smallest multiple from the rods to the left of the plate can be subtracted from the remainder to that dot, a cipher 0 should be placed under that dot in the root. Everything else is left unchanged.

CAUTION II.

IF the sum to be subtracted from the figures as far as a certain point cannot be so subtracted, add up the multiples indicated by the line above on the plate, and subtract their sum instead.

Example of extracting a Cube root.

FInding the cube root of the number 22022635627. Mark off the number with dots, and draw lines under it as [shown below]. From the figures preceding the first, or left-hand, dot (that is, 22) extract the cube root that is nearest to but less than the true one (it does not have an exact cube root). This is seen from the plate to be 2, which should be placed between the lines under the first dot as the first quotumus.[18] Next subtract the cube of this (which the plate shows to be 8) from the figures as far as the first dot, that is, from 22. This leaves 14, which is to be written above these figures. That ends the operation at the first

[17] By "on the right" Napier means above the right-hand figure of the supplementary calculation.

[18] By "as the first quotumus" Napier means as the first figure of the required root.

070
14
22.022.635.627.
2 8 0 3
8
13952
70635627

819
11529
6
48
16389

648
10112
24
36
13952

dot. [Second,] select from the rods the one that corresponds to three times the quotumus already found [, which is 6], and place it on the right-hand side of the plate. Find also the rods for three times the square of the first quotumus [, which is 12], and place them on the left-hand side of the plate. From the left-hand rods and the plate, select the multiple that is nearest to but less than the figures as far as the second dot, namely 14022. This is 11529 (nine times). Note this multiple down somewhere else on the paper (as in the [margin]), and above its right-hand figure, 9, write its quotumus 9 and draw a line between the two. To the left of the quotumus, write down its square, 81, so that the order of the figures is the same as on the plate. Beneath the 1, write its multiple from the rod to the right, [which] is [the simple] 6, and beneath the 8 write its multiple from the rod to the right [which is] 48. These three multiples, written slantwise below the line and added up ([as in the margin]), make 16389. As this cannot be subtracted from the figures up to the second dot as given above (14022), the number 9 is to be rejected, and (by [the second] caution) instead of 819 are to be taken the figures next above them on the plate, which are 648. In accordance with these take the eight-times figure from the left-hand rods, which is 10112, and from the right-hand rod the four-times figure [which is] 24 and the six-times figure [which is] 36. In accordance with these take the eight-times figure from the left-hand rods, which is 10112, and from the right-hand rod the four-times figure [which is] 24 and the six-times figure [which is] 36. Write these down in a diagonal line ([as in the margin]) and add them up to give 13952. Subtract this from 14022, which leaves 70 as the remainder to the second dot. For the quotumus at the second dot take the right-hand figure of the chosen figures 648, [which is] 8, and write it between the lines under the second dot. Third, set up on the rods the triple of the quotumi found so far ([namely] 28), [which is] 84, and place them to the right of the plate. Work out three times the square of 28 either in the usual way or by means of the short method that follows (the answer is 2352). Set this up on the rods and position them to the left of the plate. Of all the multiples and the simple generated jointly by the rods to the left of the plate (the least of which is 235201) none will be found that can be subtracted from the figures as far as the third dot, that is, from 70635. By [the first] caution, therefore, the remainder is left untouched and the cipher 0 placed beneath the third dot as the third quotumus of the root. This completes the operation at the third dot. [Fourth,] to the right of the plate, place three times the preceding quotumi [(namely 280) which is 840], and to the left three times the square of 280 [which is] 235200, with the plate between them. From the multiples on the left, choose the one that is nearest to but less than the figures as far as the fourth dot, 70635627; the multiple is 70560027, which is three times. Write this multiple below a line, the quotumus 3 above its right-hand figure, and the square of the quotumus [, which is] 9, above the line to the left of the quotumus. Beneath the 9, write the figure to the right of 9 on the plate, namely 7560. Add these two multiples, as in the [margin], to give a total of 70635627, which is then subtracted from the figures as far as the fourth dot. There is no remainder. Of the figures 93, therefore, the one on the right [, namely 3,] should be placed below the fourth and last dot as being the fourth and last quotumus of the root.

$$\begin{array}{r} 93 \\ \hline 70560027 \\ 7560 \\ \hline 70635627 \end{array}$$

So the complete and exact cube root of 22022635627 is 2803. The method is the same for other numbers.

To fix this circular system and method of extracting cube roots more firmly in the mind, use the following verses.

THE FIRST OPERATION IN EXTRACTING CUBE ROOTS.

Mark off the figures from the right in threes.
The largest cube of those the plate doth show
Subtract from the first group. Remainder write
Above the group and quotumus below.

THE SECOND AND SUBSEQUENT OPERATIONS.

Set up three times the root beside the plate
Upon the right, and twice its square must go
Upon the left. The largest then select
Of all the multiples the plate doth show,
Which from the numbers to the point you've reached
Can be subtracted. Write it down elsewhere
Beneath a line, its quotumus above.
Beside the quotumus inscribe its square
Upon its left. Directly underneath

Write down the multiples this square will show.
Next, everything beneath the line add up,
And note the total in the bottom row.
Upon Subtract it as above. Remainder write
Above the dot and quotumus below.

CHAPTER VIII.
A shortcut in extracting cube roots.

HEre is a method for finding three times the square of a root from the cube root and three times the square of a previously established part of it by an easy shortcut.

For example, in [the third operation] above we had the cube root 28 (albeit incomplete). And before that in [the second operation] we had three times the square of the previous part of it, namely 12,[19] which stood at the head of the rods on the left. We require three times the square of the whole number 28. To find this, first find three times the square of the right-hand figure of the root, which in this case is 192. Also, multiply this right-hand figure by the figures to the left of it, and add a cipher, which in this case gives 160. Third, take half this figure, 80, and add 0 to it, which gives 800. Fourth, take three times the square of the previous part of the root, which appears on the rods to the left of the plate in the previous operation, in this case 12. Add two ciphers to this, which gives a figure of 1200. Add these four numbers, as in the [margin], which gives 2352, and this is three times the square of 28. So by this easy shortcut you have the figure to place to the left of the plate in order to find the fourth quotumus of the root as above. The General Rule is sufficiently apparent from this specific example.

192
160
800
1200
2352

Warning.

SO far in this work I have used the words simple, multiple, and quotumus invariably in their strict senses. I have used simple for that which, when multiplied by a quotumus, produces a multiple. I have used multiple for that which, when divided by a simple, produces a quotumus. I call a quotumus that

[19] That is, $2^2 \times 3$.

which multiplies a simple to produce a multiple or arises from the division of a multiple by its simple. Furthermore, multiples and quotumi (which appear rather frequently in this short work) are always in the same place in every operation, so that two times is always in the second line across the rods, three times in the third, four times in the fourth, and so on to nine times, which you will find in the ninth line.

However, the quotumi 2, 3, 4, 5, etc., up to 9 that denote these multiples not only occur expressly in their own places in the right-hand column of the plate but also contained tacitly in the ordinal numbering of the intervals. Only in the position of the simple is there a distinction, for the position of its right-hand figure or figures always varies according to the different operations being performed. Sometimes all the figures of the simple, whether right-hand or left-hand, are found at the tops of their own rods, as in multiplication and division. Sometimes the right-hand figure is found in the same interval of the third column as its multiple, and the rest on the tops of the rods, as in the extraction of square roots using the appropriate plate.[20]

And finally, sometimes its two right-hand figures are found in the same interval of the middle column as its multiple, and the rest of the figure on the tops of the rods, as in the extraction of cube roots by means of the appropriate plate.[21] I thought I should point this out at this stage, as upon a proper understanding of it depend not only the methods of all the operations in this little work but also the extraction of fourth roots and higher. I have now said enough and more than enough about extracting roots. It remains to discuss the rule of proportion commonly known as the rule of three. The use of this rule is indeed golden not only in Arithmetics but also in Geometry and Mechanics, as I shall show in [the next tract].

Chapter IX.
The Rule of Three, direct and inverse.

IN the direct Rule of Three the second and third numbers must be multiplied together, and the product divided by the first. This is done with the aid of

[20] Thus when the rods and plate are set up for [the second] step in Chapter 6, the multiple 256 is the result of multiplying the simple 64 by the quotumus 4, which means that the 4 in the right-hand column of the plate is both the quotumus and the last figure of the multiple.

[21] Thus when the rods to the left of the plate and the plate itself are set up for [the second] step in Chapter 7, the multiple 11529 is the result of multiplying the simple 1281 by the quotumus 9. The figures 12 of this simple appear on the heads of the rods, and the figures 81 are in the middle column opposite the quotumus 9.

the rods by allowing the figures of the second number to designate multiples of the third number (or vice versa), then writing them diagonally and adding them up, and then subtracting in a diagonal line from the product multiples of the first number that are either equal to or nearest to but less than the remainder. The quotumi of these numbers written in order are the required fourth number.

EXAMPLE.

IF 12 months contain 365 days, 27 months contain proportionally how many days? From the table of rods it appears that the third number 27 multiplied by [three, six, and five] (the figures [3, 6, and 5] of the second number) gives respectively 81, 162, and 135, or alternatively that the second number 365 multiplied by [two and seven] (the figures [2 and 7 in 27]) gives respectively 730 and 2555. Either of these sets of multiples, when written down diagonally and added up, makes 9855. In dividing this number by 12, eight times the divisor 96, twice the divisor 24, and the simple divisor 12, are one by one subtracted in a diagonal line from the number 9855, and these quotumi combine to give the quotient 821, which is the required fourth number. There is a remainder of three-twfelfths or one-fourth of a day. By decimal ARITHMETIC this fourth comes out as $821,\overset{\prime}{2}\overset{\prime\prime}{5}$, which means [821, and] $\frac{25}{100}$ and is the same thing.

```
 365              27
----            ----
81              730
162      or     2555
 135            ----
----            9855
9855

    3
   1
  2
9855 (821 3/12
12
96
 24
  12
```

INVERSE.

In the inverse Rule of Three the first and second numbers must be multiplied together, and the product divided by the third, by using the rods in the above way, that is, by addition and subtraction.

EXAMPLE.

If 27 workmen built a tower in 365 days, 12 workmen will take how many days to build a similar tower? The rods will show the same answer as before.

The numbers are the same, and the method of working the same. The only difference is that the first and third quantities have changed places. They will therefore build this tower in $821\frac{1}{4}$ days. Similar sums are worked out likewise.

A Shortcut in the Rule of Three,

FArsighted Arithmeticians take great care in constructing their tables that, whenever the rule of Three has to be applied by means of numbers taken from them, the divisor should always (or nearly always) be a figure consisting of unity followed by a number of ciphers, and for this reason they give the figure for the whole sine in this form.[22] I have seen to it that this is the case also in my tables below. The reason for this is that when the divisor is 10, 100, 1000, and so on we can remove the tedium of division and reduce that of multiplication by the following shortcut. Division is accomplished by counting the number of ciphers in the divisor and cutting off that number of figures from the right-hand side of the dividend. Now since this dividend had previously to be constructed by multiplication, the process of multiplying should begin on the left and cease before one gets to the figure on the extreme right, for there is no point in establishing through multiplication figures that soon have to be deleted and cut off through division.

An Example of this shortcut.

A circle whose diameter is 100000 units has a circumference of approximately 31416 units. How long is the circumference of a circle whose diameter is 635 units? Finding six times, three times, and five times the number 31416 and cutting off the useless figures on the right, we have 18849., 0942.., and 157..., which are equalized on the left with the addition of ciphers, [as in the *multiplication*] CHAPTER,[23] arranged (as in the [margin]) and added up (we ignore the four figures in the right-hand column) to give 1994 or 1995 as the required fourth term.[24] If, however, you want to calculate this fourth term

635	
1884	9.
094	2..
15	7...
1994	8

[22] For Napier's own explanation of the term *whole sine*, see page 725. The expression was apparently extended from unity in a set of sine tables (sin 90°) to unity in in any sort of table such as those in Book 2 of the present work. Decimal notation would, of course, render the extra noughts unnecessary.

[23] Hence the initial 0 in 0942.

[24] If the four right-hand figures are simply ignored, the answer is 1994. The addition of the fourth figure from the right, as in the [margin], shows that strictly the answer should be rounded up to 1995.

635	
1884	96
094	248
15	7080
1994	$\overset{\prime}{9}\overset{\prime\prime}{1}\ \overset{\prime\prime\prime}{6}\ \overset{\prime\prime\prime\prime}{0}$

exactly rather than easily, the multiplication must be carried right through as in the other examples, and the product will be 1994,$\overset{\prime}{9}\overset{\prime\prime}{1}\ \overset{\prime\prime\prime}{6}\ \overset{\prime\prime\prime\prime}{0}$, which (by the rules of decimal Arithmetic) is equivalent to $1994\frac{9160}{10000}$ or $1994\frac{916}{1000}$ as the required fourth term. This, by common abbreviation, is equivalent to $1994\frac{229}{250}$. Similar sums are worked out likewise.

To fix the rule of three, both direct and inverse, more firmly in your memory, here are some verses.

For the Direct Rule.

Set up the third upon the number'd rods.
The figures of the second will now show
Which multiples to add in line aslant.
Add them and write their total down below.
Set up the first, and from the sum just found
Deduct the highest multiples. Proceed
As often as the dividend allows.
Then from the quotumi your answer read.

For the Inverse Rule.

Set up the first upon the number'd rods.
The figures of the second now will show
Which multiples to add in line aslant.
Add them and write their total down below.
Set up the third, and from the sum just found
Deduct the highest multiples. Proceed
As often as the dividend allows.
Then from the quotumi your answer read.

[First book of Rabdology
The use of the Rods
in general
the end.]

RABDOLOGY
THE SECOND BOOK
The use of the Rods in Geometry and Mechanics.
with the Help of Tables.

CHAPTER I.
Description of the Tables that Follow.

I RETAIN the terms column and row, which are normally used with reference to tables. A series of numbers ascending or descending vertically I call a *column*, and a series of numbers in a horizontal line from right to left or left to right a *row*.

Each column and each row refers to the polygon or other entity whose name appears in it. Thus, for instance, if the first column refers to triangles, the second to squares, the third to pentagons, and so on. Then also the first row refers to triangles, the second to squares, the third to pentagons, and so on.

The tables show the sides, quadratrices, and diameters of polygons and the sides, cubatrices, and diameters of regular solids, together with the weights and volumes of metals.

The quadratrix of a figure is the square root of its area, or the side of the square equal in area to the figure.

The cubatrix of a solid is the cube root of its volume, or the side of the cube equal in volume to the solid.

Each table contains two different quantities: the first contains the sides and quadratrices of polygons, the second quadratrices and diameters, the third diameters and sides, and similarly for the rest, as you will see.[25] From top left to bottom right of each table runs a diagonal line of rectangles containing names (polygons, regular solids, or metals) and the figure [one thousand].

In each table, groups of four numbers found one in each corner of a rectangle are in proportion. Thus in [the first] table the figures 1520, 2450, 525, and 846 are in the four corners of a rectangle and are in proportion: 1520 is to 2450 as 525 is to 846. Likewise, if the first and fourth numbers in a group of four are equidistant from one of the [thousand] squares and the second and

[25] As in Napier's first table.

third numbers are equidistant from either the same or any other [thousand] square, the numbers are in proportion. So 502 is to 525 as 1904 is to 1991.[26]

This means that an almost infinite number of proportions arise on the tables from various similar combinations, but to avoid confusion, I am interested only in those that have one thousand as their first term, for the reason given above under the shortcut in the rule of three. Henceforth it is only of these that I shall speak.

Chapter II.
How to Find the Sides and quadratrices of Polygons with [the first] Table.

THIS TABLE *(like the two that follow) contains the names of the prime polygons in most frequent use. The names descend diagonally accompanied by the figure* 1000. *They are triangle, square, pentagon, hexagon, heptagon, octagon, nonagon, and decagon.*[27]

---

[26] In [the First table], 502 and 1991 are both one square away (horizontally or vertically) from a 1000 square, and 525 and 1904 are both two squares away.

[27] The numerical values of the table entries are obtained from formulas that were known in Napier's time. He didn't mention any, but we can obtain, for example, the one for the first table, in current notation. In the next figure, s is the side of a regular polygon,

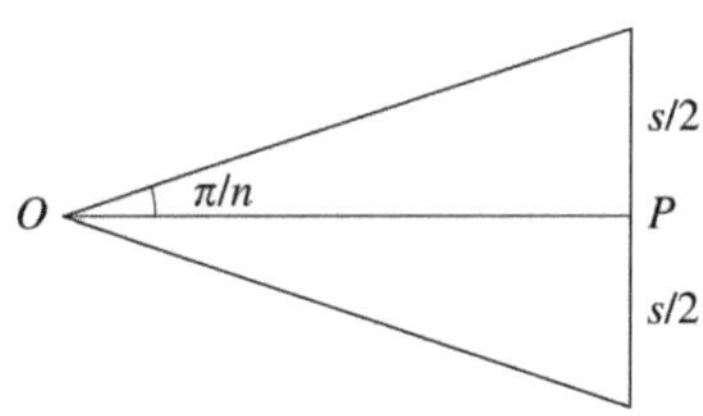

represented by the vertical segment, O is the centre of the polygon, OP is perpendicular to s, and if n is the number of sides of the polygon the angle shown above OP is π/n because twice this angle is $2\pi/n$. It is clear that

$$OP = \frac{s/2}{\tan(\pi/n)},$$

and then the area of the isosceles triangle in the figure is

$$\frac{1}{2}s\frac{s/2}{\tan(\pi/n)} = \frac{s^2}{4\tan(\pi/n)}.$$

The area of the polygon is n times this value, and then the quadratrix q is

$$q = \frac{s}{2}\sqrt{\frac{n}{\tan(\pi/n)}}.$$

First table sides & quadratrices of polygons.

Triangle 1000	1520	1991	2450	2896	3344	3771	4217
658	*Square* 1000	1312	1612	1904	2196	2487	2769
502	762	*Pentagon* 1000	1231	1456	2019	1895	2119
408	620	812	*Hexagon* 1000	1182	1364	1539	1721
345	525	687	846	*Heptagon* 1000	1154	1301	1455
299	455	495	733	867	*Octagon* 1000	1128	1261
265	402	528	650	769	887	*Nonagon* 1000	1118
237	361	472	581	687	793	895	*Decagon* 1000

The purpose of this TABLE *is to find the sides and quadratrices of* POLYGONS. *Any number in the table can be either a side or a quadratrix. If it is a side, it is the side of the polygon on the same row; if it is a quadratrix, it is the quadratrix of the polygon in the same column.*

For example, the number 1456, which appears in the pentagon row and the heptagon column, can be either a side or a quadratrix.[28] If it is a side, it is the side of a pentagon; if it is a quadratrix, it is the quadratrix of a heptagon. Similarly the number [one thousand], which appears in both the pentagon row and the pentagon column, can be either the side of the pentagon or the quadratrix of the same figure.

Numbers in the same column are the sides of polygons having the same quadratrix, and this quadratrix is the second number in the same column.

Thus 867 is the side of an octagon, and 1456 (which is found in the same column) is the side of a pentagon equal in area to the octagon and having the same quadratrix 1904, the second number in the same column.

[28] The values in the table appear here as they were printed in the *Rabdologiæ* of 1617, but Robinson's translation has replaced two very wrong values with corrected ones. The fact is that, redoing the table with a modern computer and rounding correctly, we notice that 33 of Napier's 56 entries are incorrect, although only 20 of them are wrong by more than one unit. For instance, the entries in the pentagon row should be:

502 762 1000 1229 1453 1675 1895 2114.

Numbers in the same row are the quadratrices of polygons having the same side, and this side is the second number in the same row.

Thus 687 is the quadratrix of a pentagon, and 1301 (which is found in the same row) is the quadratrix of a nonagon, and they share the side 525, which is the second number in the same row.

Special Theorems concerning Proportions.

Theor. I.

As one thousand is to the given side of a specified polygon, so the second number in the column of that polygon is to the quadratrix of that polygon.

Example.

As 1000 is to 315, the given side of a pentagon, so 1312 (second number in the pentagon column) is to 413, the required quadratrix of the pentagon. See Problem 1 below.

Theo. II.

As one thousand is to the given quadratrix of a specified polygon, so the second number in the row of that polygon is to the side of the same polygon.

Example.

As 1000 is to 413, the given quadratrix of a pentagon, so 762 (the second number in the pentagon row) is to 315, the required side of the pentagon. See Problem 2 below.

Theor. III.

If two polygons are equal in area (or have the same quadratrix), then as 1000 *is to the given side of the first, so the number at the intersection of the column of the first and the row of the second is to the side of the second.*

Example.

Let there be two polygons of equal area, a pentagon whose side is 315, and a triangle whose side is required. As 1000 is to the given side 315, so 1991 (the number at the intersection of the pentagon column and the triangle row) is to the required side of the triangle [which is] 627. See problem 3 below.

Theor. IV.

If two polygons have the same side, then as 1000 *is to the given quadratrix of the first, so the number at the intersection of the row of the first and the column of the second is to the quadratrix of the second.*

Example.

Let there be two polygons with the same side, a pentagon whose quadratrix is 413, and a triangle whose quadratrix is required. As 1000 is to 413, the given quadratrix, so 502 (the number at the intersection of the pentagon row and the triangle column) is to the required quadratrix of the triangle, 207. See the fourth problem below.

Problems illustrating the above material.

Probl. I.

Given the side of a specified polygon, to give its quadratrix.

Example.

Let there be a pentagon with a side of 315. By [the first] theorem, 1000 is to 315 as 1312 (the second number in the pentagon column) is to the required quadratrix of the pentagon. By the shortened form of the rule of Three, three times, once, and five times the number 1312, or once, three times, once, and twice the number 315, added diagonally and with the three right-hand figures of the product cut off produce 413 as the required quadratrix of the pentagon whose side is 315.[29]

Probl. II.

Given the quadratrix of a specified polygon, to give its side.

Example.

Let there be a pentagon with a quadratrix of 413. By [a] theorem, 1000 will be to the given number 413 as 762 (the second number in the pentagon row) is to the required side. Add diagonally seven times, six times, and twice the number 413 (or four times, once, and three times the number 762) and cut off the last three figures of the total, and you will get 315 as the required side of the given pentagon whose given quadratrix is 413.

Probl. III.

Given two polygons of equal area (which therefore have the same quadratrix) and the side of one of them, to give the side of the other and the quadratrix of each.

---

[29] Napier's phrase "add diagonally" means to write down in a diagonal line and add. Compare Chapter 3 of Book I.

EXAMPLE.

Let there be two polygons of equal area (and therefore having the same quadratrix): a pentagon whose side is 315 and a triangle whose side is to be found. By theorem 3, 1000 is to 315 as 1991 (the number at the intersection of the pentagon column and the triangle row) is to the required side of the triangle. Following the shortened method, add diagonally three times, once, and five times the number 1991 (or once, nine times, nine times, and once the number 315), cut off the three right-hand figures of the total, and you will get 627 as the required side of the triangle. [The first] problem above gives the quadratrix, [namely] 413.

PROBL. IV.

Given two polygons of equal side and the quadratrix of the first, to find the quadratrix of the second and the side of each.

Refer back to the EXAMPLE of [the fourth] THEOREM, in which a pentagon whose quadratrix is 413 and a triangle whose quadratrix is required have equal sides. By that theorem, as 1000 is to 413, so 502 (the number at the intersection of the pentagon row and the triangle column) is to the required quadratrix of the triangle. Following the shortened method, add diagonally four times, once, and three times the number 502 (or five times, nought times, and twice the number 413), cut off the three righthand figures of the total, and you will get 207 as the required quadratrix of the triangle. [The second] problem above gives the side of each, namely 315.

CHAPTER III.
How to Find the Quadratrices and Diameters of Polygons with [the second] Table.

BESIDES the names of the common polygons, [this] Table contains also their quadratrices and diameters, and as circles also possess quadratrices and diameters, I include the circle also among the polygons in this table on the grounds that it is a polygon with an infinite number of sides. The term polygons therefore includes the circle, and the phrase the diameters of the polygons refers to the diameters of circles and the greater diameters of the other polygons, which are the diameters of the circles circumscribed around the polygons. I omit the lesser diameters, which are those of the circles inscribed within the polygons, as being less useful: their function is performed by the quadratrices.

Thus every number on this Table can be taken either as the quadratrix or as the diameter of some polygon. If as quadratrix, it is the quadratrix of the polygon in the same row, and if as diameter, it is the diameter of the polygon in the same column.

Numbers in the same row are the diameters of polygons with the same quadratrix, and this quadratrix is the number at the far right of the same row.

Numbers in the same column are the quadratrices of polygons of the same diameter, and this diameter is the number at the bottom of the same column.

Second table quadratrices of polygons & diameters of their circumscript circles.

Triangle 1000	806	739	707	689	678	670	665	643	570
1241	*Square* 1000	917	877	855	841	832	825	798	707
1353	1090	*Pentagon* 1000	957	932	917	907	900	872	771
1414	1140	1045	*Hexagon* 1000	974	959	947	940	909	806
1451	1169	1073	1026	*Heptagon* 1000	984	972	965	933	827
1476	1188	1090	1043	1016	*Octagon* 1000	989	982	950	841
1492	1203	1103	1056	1029	1011	*Nonagon* 1000	992	959	850
1504	1212	1112	1063	1036	1019	1008	*Decagon* 1000	967	857
1555	1253	1149	1100	1072	1054	1042	1034	*Circle* 1000	886
1755	1414	1297	1240	1209	1189	1176	1167	1128	1000

Proportions in [the second] TABLE.

THEOR. I.

AS one thousand is to the given quadratrix of a specified polygon, the number at the bottom of the column of that polygon is to the diameter of the same polygon.

THEOR. 2.

If two polygons have the same diameter, then as one thousand is to the quadratrix of the first, so the number at the intersection of the column of the first and the row of the second is to the quadratrix of the In second.

THEOR. 3.

As one thousand is to the given diameter of a specified polygon, so the number on the far right of the row of that polygon is to the quadratrix of the same polygon.

THEOR. 4.

If two polygons have the same quadratrix, then as one thousand is to the diameter of the first, so the number at the intersection of the row of the first and the column of the second is to the diameter of the second.

[*Warning*.]

No examples are given of the above theorems, as they are made sufficiently clear in part by comparison with the similar rationale of [the first] table and in part by the examples that illustrate the following problems.

Problems Illustrating the use of [the second] TABLE.

Prob. 1.

G*Iven the quadratrix of a specified polygon, to find the diameter of the same polygon.*

Example.

A triangle has a quadratrix of 800. Find the diameter of the same triangle, that is, the diameter of the circle circumscribed around the triangle. From [the first] theorem it follows that as 1000 is to 800 (the given quadratrix of the triangle), so 1755 (the number at the bottom of the triangle column) is to the required diameter of the same triangle. Following the shortened form of the rule of three, set up the number 1755 on the rods, take the multiple that represents eight times that, add two ciphers (because the given quadratrix is 800), and then remove the three right-hand figures, which gives 1404. This is the required diameter of the triangle whose quadratrix of 800 was given.

Probl. 2.

Given the diameter of a specified polygon, to find the quadratrix of the same polygon.

EXAMPLE.

A triangle has a diameter of 1404. Find the quadratrix of the same triangle. From theorem 2, it follows that as 1000 is to 1404 (the given diameter of the triangle), so 570 (the number at the far right of the triangle row) is to the required quadratrix of the same triangle. So diagonally add five times, seven times, and nought times 1404 (or once, four times, nought times, and four times 570) to give 800280, cut off the three right-hand figures of this number, and the result is 800. This is the required quadratrix of the triangle whose diameter of 1404 was given.

PROBL. III.

Given two polygons of the same diameter and the quadratrix of one of them, to find the quadratrix of the other and the diameter of each.

Example.

Let there be two polygons of the same diameter, a circle whose quadratrix is 1205, and a heptagon whose quadratrix is to be found. From theorem 3 it follows that as 1000 is to 1205 (the given quadratrix of the circle), so 933 (the number at the intersection of the circle column and the heptagon row) is to the required quadratrix of the heptagon. So add diagonally nine times, three times, and three times the number 1205 (or once, twice, nought times, and five times the number 933) to give 1124265, cut off the three right-hand figures, and the result is 1124. This is the required quadratrix of the heptagon. You can, if you wish, hunt out the common diameter of the circle and heptagon by the method of Probl. 1; it is approximately 1359.

Probl. IIII.

Given two polygons with the same quadratrix and the diameter of one of them, to find the diameter of the second and the quadratrix of each.

Example.

Let there be two polygons, a nonagon and a circle, that possess the same quadratrix and so are equal in area. From 1302 as the diameter of the nonagon, find the diameter of the circle. By theorem 4, as 1000 is to 1302, the given diameter of the nonagon, so 959 (the number at the intersection of the nonagon row and the circle column) is to the required diameter of the circle. Add diagonally nine times, five times, and nine times the number 1302; or once, three times, nought times, and twice the number 959, to give 1248618, cut off the three right-hand figures, and the result is 1249. This is the required diameter of the circle. You can, if you wish, get the common quadratrix of the nonagon and the circle by [Probl.] 2; it is approximately 1107.

CHAPTER IV.
How to find the diameters and sides of Polygons with [the third] Table.

THe third Table contains the diameters and sides of polygons and their ratios to each other. Hence, every number in this table can be taken as either the diameter or the side of some polygon. If as diameter, it is the diameter of the polygon in the same row, and if as side, it is the side of the polygon in the same column.

Third table sides of polygons & diameters of their circumscript circles.

1000	866	707	588	500	434	383	342	309
1154	*Triangle* 1000	817	676	577	501	442	394	357
1414	1225	*Square* 1000	832	707	614	541	483	437
1700	1472	1202	*Pentagon* 1000	850	738	650	580	525
2000	1732	1414	1176	*Hexagon* 1000	868	765	684	618
2304	1995	1629	1355	1152	*Heptagon* 1000	881	786	712
2614	2264	1848	1537	1307	1134	*Octagon* 1000	891	807
2929	2537	2071	1722	1462	1271	1122	*Nonagon* 1000	904
3236	2802	2288	1903	1618	1404	1239	1107	*Decagon* 1000

Numbers in the same column are the diameters of polygons with the same side, and this side is the number at the top of on each column.

Numbers in the same row are the sides of polygons with the same diameter, and this diameter is the first number in each row.

Proportions in [the third] TABLE.
Theor. 1.

AS *one thousand is to the given diameter of a specified polygon, so the number at the top of the column of that polygon is to the side of the same polygon.*

Theor. 2.

As one thousand is to the given side of a specified polygon, so the first number in the row of that polygon is to the diameter of the same polygon.

Theor. 3.

If two polygons have the same side, then asone thousand is to the diameter of the first, so the number at the intersection of the column of on row the first and the row of the second is to the diameter of the second.

Theor. 4.

If two polygons have the same diameter, then asone thousand is to the side of the first, so the number at the intersection of the row of the first and the column of the second is to the side of the second.

The examples that illustrate the following problems illustrate the above theorems as well.

Problems illustrating the use of [the third]Table.

Prob. I.

G*Iven the greater diameter of a specified polygon, to find the side of the same polygon.*

Example.

A pentagon has a diameter of 536. Find its side. By [the first] theorem it is apparent that as 1000 is to 536, the given diameter, so 588 (the on top number in the pentagon column) is to the required side of the same pentagon. [Therefore] (following the shortened form of the rule of three), take five times, eight times, and eight times the number 536 (or five times, three times, and six times the number 588), add them diagonally to give 315168, and remove the last three figures, which leaves 315 as the required side of the pentagon whose greater diameter was given as 536.

Probl. II.

Given the side of a specified polygon, to find its greater diameter.

Example.

A pentagon has a side of 315. Find its diameter. By theorem 2 it is apparent that as 1000 is to the given side 315, so 1700 (the first number in the pentagon row) is to the required diameter of the pentagon. Then three times, once and five times the number 1700 (or once, seven times, nought times and nought times the number 315) being added diagonally produce the total 535500 and this being reduced by rejecting the three figures on the right, the result is 536, which is an approximation to the required diameter of the pentagon whose side was given as 315.

Probl. 3.

Given two polygons with the same side and the diameter of the first, to find the diameter of the second and the common side of each.

Example.

Let there be two polygons with the same side, the first a pentagon and the second a triangle. Given that the diameter of the pentagon is 536, find the diameter of the triangle. It is apparent (from [the third] theorem) that as one thousand is to 536 (the given diameter of the pentagon), so 679 (the number at the intersection of the pentagon column and the triangle row) is to the required diameter of the triangle. So five times, three times, and six times the number 679 (or six times, seven times, and nine times the number 536), being added diagonally and reduced by canceling the three right-hand figures, produce 364 as an approximation to the required diameter of the triangle. If in addition you require the common side of each, you will find it by [the first] problem to be 315, as above.

Probl. 4.

Given two polygons of the same diameter and the side of the first, to find the side of the second and the common diameter of each.

Example.

Let there be a pentagon and a triangle of the same diameter, and let the side of the pentagon be given as 315. Find the side of the triangle. By [the fourth] theorem it is apparent that as 1000 is to 315 (the given side of the pentagon), so 1472 (the number at the intersection of the pentagon row and the triangle column) is to the required side of the triangle. So add diagonally three times, once, and five times the number 1472 (or use 1472 instead as the multiplier) to give 463680, cut off the three right-hand figures, and the result is 464 as an approximation to the required side of the triangle. If in addition you require the common diameter of each, you will find it by problem 2 to be 536.

WARNING.

THese and the following Tables have innumerable other uses. Some of these focus specifically on particular numbers, as in cutting a given number as accurately as possible (through the extreme and mean ratio using the num-

bers 618, 1000, *and* 1618 *from* [*the third*] *table*).[30] *Other miscellaneous uses are derived from the above theorems.* (*For instance, if a triangle and a pentagon have the same side, the pentagon and a heptagon the same quadratrix, and the heptagon and a nonagon the same diameter and if a single side, quadratrix, or diameter of any of the four is known, all the others may be found for all the rest.*) *Anyone with a modicum of intelligence can fathom these uses from the foregoing. I have no room in this brief summary to describe them all, and in any case the purpose of these excursions into Arithmetic and Geometry is merely to teach the use of the rods by means of them.*

[*So far we have taught how to find the sides, quadratrices and diameters of polygons. It remains to find the sides, cubatrices, and diameters of the five regular solids and the sphere, as set out by the following three tables.*]

CHAPTER V.
How to Find the Sides and Cubatrices of the Five Regular Solids with [the fourth] TABLE.

THe fourth Table ([*which*] *is the first to concern Cubic measure*), *contains the sides and cubatrices of five regular solids. Every number in the table can be taken either as the side or as the cubatrix of a certain regular solid. If as side, it is the side of the regular solid in the same row*; *if as cubatrix, it is the cubatrix of the regular solid in the same column.*

Numbers in the same column are the sides of regular solids with the same cubatrix, and the cubatrix is the middle number in the same column.

Numbers in the same row are the cubatrices of solids with the same side, and this side is the number in the middle of the same row.

Proportions in Table 4.

THEOR. I.

AS one thousand is to the given side of a specified regular solid, so the same number in the middle of the column of that solid is to the cubatrix of the same regular solid.

[30] *Mean ratio* is Euclid's name for the golden ratio (1:1.618) (*Elements*, B. 6, Prop. 30).

Fourth table sides & cubatrices of the five regular solids.

Tetra. 1000	1587	2040	2689	4088
630	*Octa.* 1000	1285	1694	2575
490	778	*Cube.* 1000	1318	2003
372	590	759	*Icosa.* 1000	1521
245	388	499	658	*Dodeca.* 1000

Theor. II.

As one thousand is to the given cubatrix of a specified regular solid, so the middle number in the row of that solid is to the side of the same solid.

Theor. III.

If two regular solids are equal (that is, if they have the same cubatrix), then as 1000 is to the given side of the first, so the number at the intersection of the column of the first and the row of the second is to the side of the second.

Theor. IV.

If two regular solids have the same side, then as 1000 is to the given cubatrix of the first, so the number at the intersection of the row of the first and the column of the second is to the cubatrix of the second.

Problems illustrating the use of the fourth Table.

Prob. I.

Given the side of a specified regular solid, to give the cubatrix of the same solid.

Example.

An Octahedron has a side of 452. Find its cubatrix. By [the first] theorem, as one thousand is to 452 (the given side of the Octahedron), so 778 (the middle number in the Octahedron column) is to the required cubatrix of the Octahedron. Take four times, five times, and twice the number 778 (or seven times, seven times, and eight times the number 452), add them diagonally, reduce the total by canceling the three right-hand figures, and you are left with 352, which is an approximation to the required cubatrix of the Octahedron whose side was given as 452.

PROBL. II.

Given the cubatrix of a specified regular solid, to find the side of the same solid.

EXAMPLE.

An octahedron has a cubatrix of 352. Find its side. By theorem 2, as [one thousnd] is to 352 (the given cubatrix of the octahedron), so 1285 (the number in the middle of the octahedron row) is to the required side of the same Octahedron. So take three times, five times, and twice the number 1285 (or vice versa using 1, 2, 8, and 5 as multipliers), add them diagonally, reduce the total by canceling the three right-hand figures, and you are left with 452, which is the required side of the Octahedron whose cubatrix was given as 352.

PROBL. III.

Given two regular solids that are equal (that is, have the same cubatrix) and the side of the first, to find the side of the second and the common cubatrix of each.

Example.

Let there be two equal solids, an octahedron and an icosahedron. Given that the side of the octahedron is 452, find the side of the icosahedron. By theorem 3, as [one thousand] is to 452 (the given side of the octahedron), so 590 (the number at the intersection of the octahedron column and the icosahedron row) is to the required side of the icosahedron. So take four times, five times, and twice the number 590 (or five times, nine times, and nought times the number 452), add them diagonally, reduce the total by canceling the three right-hand figures, and you are left with 267 as an approximation to the required side of the icosahedron. In addition, the common cubatrix of each ([which is] 352) is found by [Problem] 1.

Probl. IV.

Given two regular solids with the same side and the cubatrix of the first, to find the cubatrix of the second as well and the common side of each.

Example.

Let there be two regular solids with the same side, an octahedron and an icosahedron. Given that the cubatrix of the octahedron is 352, find that of the icosahedron. By theorem 4, as one thousand is to 352 (the given cubatrix of the octahedron), so 1694 (the number at the intersection of the octahedron row and the icosahedron column) is to the required cubatrix of the icosahedron. So take three times, five times, and twice the number 1694 (or vice versa), add them diagonally, reduce the total by canceling the three right-hand figures, and you are left with 596 as the required cubatrix of the icosahedron. In addition, the common side of each is found by Problem 2 (452), as above.

CHAPTER VI.

How to find the cubatrices and diameters of regular solids and the sphere with [the fifth] TABLE.

THis Table contains the cubatrices and diameters of regular solids. As spheres also have a cubatrix and a diameter, I include spheres among the regular solids in this table. In this chapter, therefore, the term regular solids must be understood to include spheres, and the phrase diameters of regular solids to include the diameter of a sphere as well as those of the other regular solids, the greater diameter (I omit other less useful diameters) is that of the sphere circumscribed around the regular solid.

Every number in this table can be taken either as the cubatrix or as the diameter of a regular solid. If as cubatrix, it is the cubatrix of the regular solid in the same row, and if as diameter, it is the diameter of the regular solid in the same column.

Numbers in the same column are cubatrices of regular solids with the same diameter, and this diameter is the number at the bottom of the same column.

Numbers in the same row are diameters of solids with the same cubatrix, and this cubatrix is the number at the right-hand end of the same row.

Proportions in [the fifth] TABLE.

Theor. 1.

AS one thousand is to the given cubatrix of a specified regular solid, so the number at the bottom of the column of that solid is to the diameter of the same solid.

Fifth table cubatrices of the five regular solids & the diameters of their circumscribed spheres.

Tetra. 1000	727	693	577	560	496	400
1376	*Octa.* 1000	953	794	769	683	550
1443	1049	*Cube* 1000	833	807	716	577
1732	1260	1201	*Icosa.* 1000	970	860	693
1487	1300	1238	1031	*Dodeca.* 1000	887	715
2015	1465	1396	1163	1127	*Sphere* 1000	806
2499	1817	1732	1443	1399	1241	1000

Theor. 2.

As one thousand is to the given diameter of a specified regular solid, so the number at the right-hand end of the row of that solid is to the cubatrix of the same solid.

Theor. 3.

If two regular solids have the same diameter, then as one thousand is to the given cubatrix of the first, so the number at the intersection of the column of the first and the row of the second is to the cubatrix of the second.

Theor. 4.

If two regular solids have the same cubatrix, then as one thoussnd is to the given diameter of the first, so the number at the intersection of the row of the first and the column of the second is to the diameter of the second.

Problems illustrating the use of [the fifth] TABLE.

Probl. I.

GIven the cubatrix of a specified regular solid, to find the diameter of the same solid.

Example.

AN Octahedron has a cubatrix of 352. Find its diameter. By theorem 1, as one thousand is to 352 (the given cubatrix of the Octahedron), so 1817 (the number at the bottom of the Octahedron column) is to the required diameter. So take three times, five times, and twice the number 1817 (or vice versa), add them diagonally, reduce the total by canceling the three right-hand figures, and you have 639, which is the required diameter of the octahedron whose cubatrix was given as 352.

Probl. II.

Given the diameter of a specified regular solid, to find the cubatrix of the same solid.

Example.

An Octahedron has a diameter of 639. Find its cubatrix. By theorem 2, as [one thousand] is to 639 (the given diameter of the Octahedron), so 550 (the number on the right-hand end of the Octahedron row) is to the required cubatrix of the Octahedron. So take five times, five times, and nought times the number 639 (or alternatively six times, three times, and nine times the number 550), add them diagonally, reduce the total by canceling the last three figures, and you have 352 as an approximation to the required cubatrix of the Octahedron whose diameter was given as 639.

Probl. III.

Given two regular solids with the same diameter and the cubatrix of the first, to find the cubatrix of the second and the common diameter of each.

Example.

Let there be two solids with the same diameter: an Octahedron and an Icosahedron. Given that the cubatrix of the Octahedron is 352, find that of the Icosahedron. By theorem 3, as one thousand is to 352 (the given cubatrix of the Octahedron), so 1260 (the number at the intersection of the octahedron column and the icosahedron row) is to the required cubatrix of the Icosahedron. So take three times, five times, and twice the number 1260 (or once, twice, six times, and nought times the number 352), add them diagonally, reduce the total by canceling the three right-hand figures, and you have 444 as an approximation to the required cubatrix of the icosahedron. The common diameter of each, which is 639, is found by problem 1.

Probl. 4.

Given two regular solids with the same cubatrix and the diameter of the first, to find the diameter of the second and the common cubatrix of each.

Example.

Let there be two regular solids with th same cubatrix: an octahedron and an icosahedron. Given that the diameter of the octahedron is 639, find that of the icosahedron. By theorem 4, as one thousand is to 639 (the given diameter of the octahedron), so 794 (the number a the intersection of the octahedron row and the icosahedron column) is to the required diameter of the icosahedron. So take six times, th ee times, and nine times the number 794 (or vice versa), add them diagonally, reduce the total by canceling the three right-hand figures, and you have 507 as the required diameter of the icosahedron. You can find the common cubatrix of each (352) by problem 2 as above.

CHAPTER VII.
How to find the diameters and sides of the five of regular solids with [the sixth] Table.

THe sixth Table contains the greater diameters and the sides of regular solids and their ratios to each other. Every number in this table can be taken as either the diameter or the side of some regular solid; if as diameter, it is the diameter of the solid in the same row, and if as side, it is the side of the regular solid in the same column.

Numbers in the same column are the diameters of regular solids with the same side, and this side is the number at the top of the column.

Numbers in the same row are the sides of regular solids with the same diameter, and this diameter is the first number in the same row.

Proportions in Table 6.

THEOR. I.

AS one thousand is to the given diameter of a specified solid, so the number at the top of the column of that solid is to the side of the same solid.

Sixth table sides of the five regular solids & the diameters of their circumscribed spheres.

1000	817	707	577	521	357
1225	*Tetra.* 1000	966	707	643	437
1414	1035	*Octa.* 1000	817	742	505
1732	1414	1225	*Cube* 1000	909	618
1902	1555	1347	1099	*Icosa.* 1000	679
2802	2287	1981	1618	1437	*Dodeca.* 1000

THEOR. II.

As one thousand is to the given side of a specified regular solid, so the first number in the row of that solid is to the diameter of the same solid.

THEOR. III.

If two regular solids have the same side, then as one thousand is to the diameter of the first, so the number at the intersection of the column of the first and the row of the second is to the diameter of the second.

THEOR. IV.

If two regular solids have the same diameter, then as one thousand is to the side of the first, so the number at the intersection of the row of the first and the column of the second is to the side of the second.

Problems illustrating the use of [the sixth] TABLE.

Probl. 1.

G*Iven the diameter of a specified regular solid, to find the side of the same solid.*

Example.

An Octahedron has a diameter of 639. Find its side. By theorem 1, as one thousand is to 639 (the given diameter of the octahedron), so 707 (the top number in the octahedron column) is to the required side of the octahedron. So take six times, three times, and nine times the number 707 (or seven times, nought times, and seven times the number 639), add them diagonally, reduce the total by canceling the three right-hand figures, and you have 452 as an approximation to the side of the octahedronm whose dismeter was ivem as 639.

Probl. 2.

Given the side of a specified regular solid, to find the diameter of the same solid.

Example.

An octahedron has a side of 452. Find its diameter. By theorem 2, as one thousand is to 452 (the given side of the octahedron), so 1414 (the first number in the octahedron row) is to the required diameter of the same octahedron. So take four times, five times, and twice the number 1414 (or vice versa), add them diagonally, reduce the total by canceling the three right-hand figures, and you have 639 as the required diameter of the octahedron whose side was given as 452.

Prob. III.

Given two regular solids with the same side and the diameter of the first, to find the diameter of the second and the common side of each.

Example.

Let there be two regular solids with the same side: an octahedron and an icosahedron. Given that the diameter of the octahedron is 639, find that of the icosahedron. By theorem 3, as one thousand is to 639 (the given diameter of the octahedron), so 1347 (the number at the intersection of the octahedron column and the icosahedron row) is to the required diameter of the icosahedron. So take six times, three times and nine times the number 1347 (or vice versa), add them diagonally, reduce the total by canceling the three right-hand figures, and you have 861 as an approximation to the required diameter of the icosahedron.

If in addition you require the common side of each, you can find it by problem 1 [and discover that] it is 452.

Probl. IV.

Given two regular solids with the same diameter and the side of one, to find the side of the second and the common diameter of each.

Example.

Let there be two regular solids with the same diameter: an octahedron and an icosahedron. Given that the side of the octahedron is 452, find the side of the icosahedron. By theorem 4, as one thousand is to 452 (the given side of the octahedron), so 742 (the number at the intersection of the octahedron row and the icosahedron column) is to the required side of the Icosahedron. So take four times, five times, and twice the number 742 (or alternatively seven times, four times, and twice the number 452), add them diagonally, reduce the total 335384 by canceling the three last figures, and you will have 335 left as the required side of the icosahedron. You can find the common diameter of each (639) by problem 2.

CHAPTER VIII.
How to find the weights and volumes of Metals.

SO far I have shown how to find quickly and easily the dimensions of plane and solid figures. I now propose to discuss the weights and volumes of Metals and Stones and their mutual ratios (which Mechanics frequently use) by means of [*the seventh*] *Table.*

Description of [the seventh] Table.

THe seventh Table contains the names of Metals and Stones, each with its own figure [*one thousand*], *in a diagonal line from top left to bottom right. It also contains the weights of metals and stones in drams and their volumes in cochlears.* The dram *is, as always, one eighth of an ounce.* The cochlear *is used here by me as a liquid measure denoting the amount displaced by a piece of gold weighing* 10 *drams from a jar full to the brim of liquid.*[31] *The dram varies from province to province, which means that my cochlear varies as well, but the numbers of drams and cochlears given in the table and their mutual ratios remain constant.*[32]

Thus each number in this Table can be taken either as the weight in drams or as the volume or capacity in cochlears of some metal or stone. If as drams, it is the weight of the metal or stone in the same column; if as cochlears, it is

[31] Latin cochlear means spoonful. As defined here it equals 2 cc.

[32] In Apothecaries' weight, 1 dram = 1/8 ounce = 3.89 g.

Seventh table volumes and weights of *Metals & Stones.*

Gold 1000	747	644	599	470	409	387	155	106	100
1340	*Mercury* 1000	862	803	630	548	518	207	142	134
1554	1160	*Lead* 1000	931	730	635	601	241	165	155
1670	1247	1075	*Silver* 1000	785	683	646	258	178	167
2127	1588	1369	1274	*Bronze* 1000	870	823	329	227	213
2446	1826	1574	1465	1150	*Iron* 1000	946	380	261	245
2585	1929	1663	1548	1215	1057	*Tin* 1000	402	276	259
6451	4830	4147	3875	3038	2630	2487	*Marble* 1000	688	645
9433	7042	6060	5616	4405	3830	3622	1453	*Stone* 1000	943
10000	7463	6435	5990	4700	4088	3868	1549	1060	1000

the volume of the metal or stone in the same row.

Numbers in the same column are the volumes in cochlears of metals or stones of the same weight, and the number of drams in this weight is the number at the bottom of the same column.

Numbers in the same row are the weights in drams of metals and stones of the same volume, and the number of cochlears in this volume is the number on the right-hand end of the same row.

Proportions in [the seventh] Table.

Theor. I.

As one thousand is to the volume in cochlears of a specified metal or stone, so the number at the bottom of the column of that metal is to the weight in drams of the same metal.

Theor. 2.

As one thousand is to the weight in drams of a specified metal or stone, so the number at the right-hand end of the row of that metal is to the volume in cochlears of the same metal.

Theor. 3.

If two pieces of metal or stone are of the same weight, then as one thousandf is to the volume in cochlears of the first, so the number at the intersection of the column of the first and the row of the second is to the volume in cochlears of the second.

Theor. 4.

If two pieces of metal or stone are of the same volume, then as one thousand is to the weight in drams of the first, so the number at the intersection of the row of the first and the column of the second is to the weight in drams of the second.

PROBLEMS illustrating the use of Table 7

Prob. 1.

Given the volume in cochlears of a specified metal or stone, to find its weight in drams.

Example.

The mold for a silver statue measures 562 cochlears in volume. Find how many drams the statue will weigh. By theorem 1, as one thousand is to 562 (the given volume in cochlears), so 5990 (the number at the bottom of the silver column) is to the required weight in drams. So take five times, six times, and twice the number 5990 (or vice versa), add them diagonally, reduce the total by canceling the three right-hand figures, and you have 3366 as the required weight in drams of the statue whose volume was given as 562 cochlears.

Probl. 2.

Given the weight in drams of a specified metal or stone, to find its volume in cochlears.

Example.

A silver statue weighs 3366 drams. Find its volume in cochleal's. By theorem 2, as one thousand is to 3366 (the weight of the given statue in drams), so 167 (the number at the right-hand end of the silver row) is to the required volume in cochlears.

So take once, six times and seven times the number 3366 (or vice versa), add them diagonally, reduce the total by canceling the three right-hand figures, and you have 562 as the required volume in cochlears of the statue whose weight was given as 3366 drams.

PROBL. III.

Given the same weight of two metals or stones and the volume in cochlears of one of them, to find the volume in cochlears of the other and the weight in drams of each.

Example.

Two small objects, such as the columns of a model of a siege engine or other familiar article, are of the same weight. One is made of tin and has a volume of 551 cochlears; the other is made of bronze and its volume is to be found. By theorem 3, as one thousan is to 551 (the given volume in cochlears of the tin specimen), so 823 (the number at the intersection of the tin column and the bronze row) is to the required volume in cochlears of the bronze specimen. So take eight times, twice, and three times the number 551 (or five times, five times, and once the number 823), add them diagonally, reduce the total by canceling the three right-hand figures, and you have 453 as the required volume in cochlears of the bronze specimen.

You can find their common weight by the method of problem 1 [and find that] it is 2131 drams.

Probl. 4.

Given the same volume of two metals or stones and the weight in drams of one of them, to find the weight in drams of the second and the volume in cochlears of both.

Example.

Suppose that a lump of tin has been made as above into a small specimen of a machine and weighs 2131 drams. Find the weight of bronze that will exactly fill the same mold. By theorem 4, as one thousand is to 2131 (the given weight in drams of the tin specimen), so 1215 (the number at the intersection of the tin row and the bronze column) is to the required weight in drams of bronze to be poured into the mold. So take twice, once, three times, and once the number 1215 (or once, twice, once, and five times the number 2131), add them diagonally, reduce the total by canceling the three right-hand figures, and you have 2589 as the required weight in drams of the bronze specimen.

You can find the common volume of both by the method of problem 2 [and find that] it is 551 cochlears.

WARNING.

THe above simple examples of Theorems and Problems depend on equalities. In addition, there are many derived from these that depend on inequalities, such as the solution of the following problem.

Suppose a model of a machine that is made of tin and weighs 2131 drams. The full-size machine is to be cast in bronze, and its volume is one thousand times the volume of the model. What will be the weight of the machine?

The answer is that if the bronze machine is to have the same capacity as the tin model, its volume will be 551 cochlears, and its weight a mere 2589 drams (as in problem 4 above). But [by hypothesis] it is one thousand times the size of the model. It will therefore have one thousand times the volume and one thousand times the weight, in other words, a capacity of 551000 cochlears and a weight of 2589000 drams.

However, the length, diameter, and other linear measures of the machine will be not one thousand times but only ten times those of the model, as is apparent from Euclid's [*Elements*], book 5, definition 10, and book 11, proposition 33.

It is no part of my purpose to treat these matters at greater length here. Let me therefore leave these and similar problems, which depend on the preceding simple theorems, and here make an end of Rabdology.

To God alone be all the praise and glory,

Amen.

Appendix. THE HIGH-SPEED PROMPTUARY FOR MULTIPLICATION.

PREFACE.

Although this promptuary for Multiplication was the last of all my discoveries, it does not deserve to come last in this book. With its assistance, all multiplications, however long and difficult, can be carried out easily and quickly. Divisions too can all be performed by means of it, but they must first be converted into multiplications with the help of sines, tangents, and secants or else the Tables in [*this*] *book. That was why I decided that the appropriate place for this treatise was after* [*the second*] *book, where accordingly I have placed it. It begins with the construction of the apparatus.*

CHAPTER I.

How to make the strips for the promptuary.

THe strips should be made of ivory or any other solid white material. For multiplying numbers less than 100000, you will require one hundred of them, and more (or less) in proportion to the size of the numbers you wish to multiply. I have chosen two hundred strips, with which one can multiply numbers less than 10000000000. These two hundred strips should be one finger wide and eleven fingers long.[33] At one end there should be a greater margin of two-thirds of a finger, and at the other a lesser margin of one-third, and the space between the two margins should be divided exactly into ten small squares. One hundred of the strips should be one quarter of a finger thick; the other one hundred should be only half as thick as that (but more if the material requires it). Each of the one hundred thicker strips is to be placed before you in such a way that it points toward your chest with the greater margin farther from you and the lesser margin closer. These strips I call *vertical* strips. Each of the thinner strips, on the other hand, should lie with the wider margin to the right and the narrower to the left, in other words, at right angles to the others. These strips I therefore call *horizontal* strips.

[33] 1 finger = 3/4 inch = 19 mm.

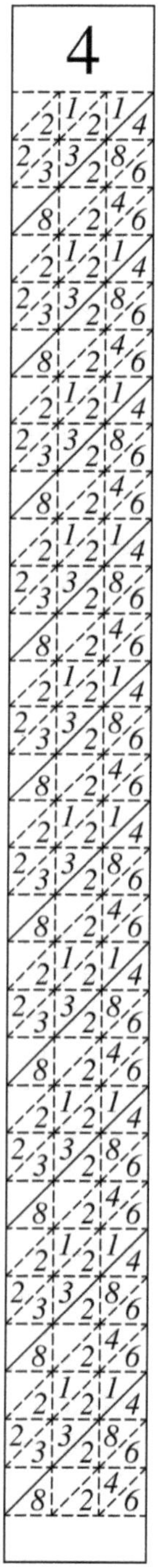

On all the strips when arranged as above should be drawn clear diagonal lines from the lower left to the upper right angle of each square, thus dividing each square in half to form two triangles. Then divide the length and breadth of each square into three, and draw lines through the opposite points of division. These lines, which should be erasable, divide each square into nine small squares, each one of which must now be divided diagonally into two by erasable lines parallel to the clear diagonal line previously marked. This cuts each square into two triangles, which I will henceforth call *places*.

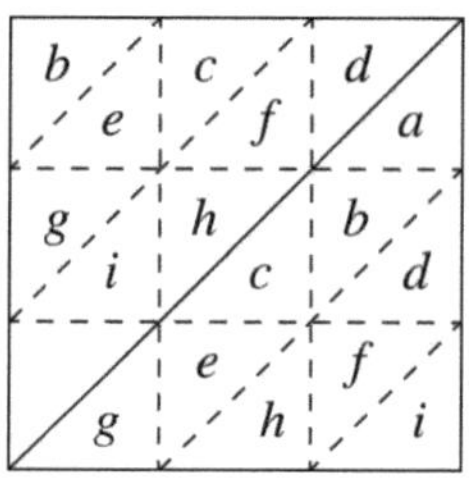

Each large triangle now contains nine places. For the exposition that follows, these should be marked (erasably) with the nine letters [$a, b, c, d, e, f, g, h, i$, as shown in the next example]. After the clear and the erasable lines have been drawn as above, one of the digits should be written or carved in the larger margin of each strip. On ten of the one hundred thicker strips and ten of the thinner ones write (indelibly) the cipher 0. On another ten of the thicker ones and the same number of the thinner ones write (indelibly) the figure 1. On another ten of the thicker ones and the same number of the thinner ones write (indelibly) the figure 2. On another ten of the thicker ones and another ten of the thinner ones write in the larger margin the figure 3. And so on for four, five, and the rest up to nine inclusive until everyone of the two hundred strips has a figure written in its larger margin.

The features described so far are common to the thicker (vertical) and thinner (horizontal) strips. Now I must consider them separately. Multiples are inscribed on the thicker rods as follows. In all the places on the one hundred thicker rods that correspond to the letter a, write with indelible ink the same simple figure as inscribed earlier in the larger margin.

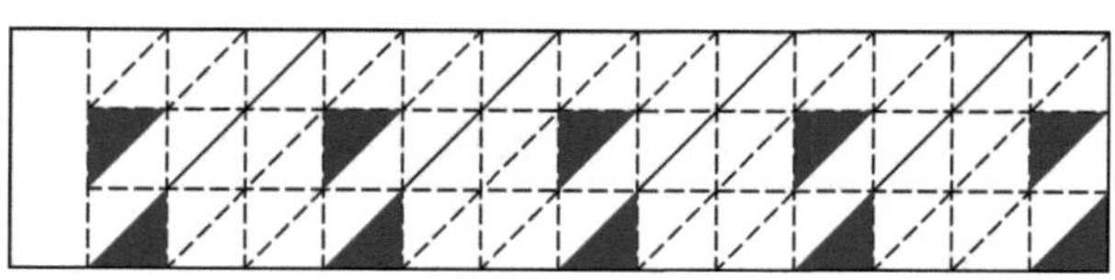

In the places that correspond to the letter *b* in the left triangle, write the left-hand figure of the number that is twice the one that appears in the upper margin, and in the places that correspond to the letter b in the right triangle, write the right-hand figure of the same number. In the places which correspond to the letter *c* on the left, inscribe the left-hand figure of the number that is three times the number in the margin, and in the places on the right inscribe the right-hand figure of the same number. And so proceed, using indelible ink, to write in the remaining multiples up to nine times inclusive. If any multiple (twice, three times, or whatever) has no left-hand figure or if any righthand figure is a cipher, leave that place vacant or fill it up with a cipher.

Example.

To write the multiples on a number 4 strip, first inscribe (in permanent form) the simple number 4 in the places marked *a*. Twice 4 is 8. So inscribe 8 in place *b* on the right, but leave place *b* on the left vacant, as the number 8 has no left-hand figure. Three fours are 12, and the number 12 is inscribed as follows: put unity in place *c* on the left and 2 in place *c* on the right. Four fours are 16, which is inscribed by putting 1 in place *d* on the left and 6 in place *d* on the right. Five fours are 20, which is inscribed by putting 2 in place *e* on the left and nothing in place *e* on the right. Six fours are 24, which is inserted by putting 2 in place *f* on the left and 4 in place *f* on the right. Seven fours are 28, which is inserted by inscribing 2 in place *g* on the left and 8 in place *g* on the right. Eight fours are 32, which is inscribed by putting 3 in place *h* on the left and 2 in place *h* on the right. Finally, nine fours are 36, which is inserted by inscribing 3 in place *i* on the left and 6 in place *i* on the right. All these figures are to be inscribed in some permanent way. The writing of multiples on the number four strip is thus complete, and it should appear as in the figure. Proceed thus for the multiples on the other number four strips and all the figures on the hundred thicker (or vertical) strips. When all are complete, rub out all the faint and erasable lines and letters on the strips, leaving only the simple figures in the margins, the multiples, and the diagonal line down the middle of each of the larger squares, as can be seen on the number four strip or the other strips shown in [the penultimate example of this Appendix].

So much for the inscription of multiples on the one hundred thicker strips; we turn now to the one hundred thinner strips.

THe thinner or horizontal strips serve to block out the multiples on the thicker strips that you do not need and to reveal through windows or holes the ones that you do. These strips are therefore referred to also as *excised* or *perforated* strips. The required perforations are made as follows. No cuts at all are made in the strips that have a cipher in their righthand or larger margin. In the strips that have unity inscribed in the larger margin, holes should be cut in the places corresponding to the letter *a*. In strips inscribed with the number two, holes should be bored in the places corresponding to *b* on the left and on the right. In strips inscribed with the number three, holes should be bored in all the places corresponding on either side to the letter *c*. In strips inscribed with the number four, holes should be bored in all the places corresponding to the letter *d*. In those inscribed with the number five, holes should be bored in all the places marked *e*. In those inscribed with the number six all the places marked *f* should be cut out. In those inscribed with the number seven all the places corresponding to the letter *g* should be cut out. In those inscribed with the number eight, holes should be bored in all the places corresponding on either side to the letter *h*. And finally, in the strips inscribed with the number nine all the places marked *i* on the left and on the right should be cut out. Now you have all one hundred of the thinner [number seven] strips correctly perforated. Compare [the preceding] figure showing a number seven strip correctly cut. This done, rub out all the faint and erasable letters and lines on the horizontal strips, leaving only the diagonals that divide the larger squares in two and the figures inscribed in the right-hand margin, as can be clearly seen in [the newest] diagram [in this Appendix].

All two hundred of the strips are now complete. The construction of the Box follows.

CHAPTER II.

How to Make the Box that holds the strips.

TO build the Box you need four columns, two boards, and two guide pieces. The columns are two thirds of a finger square and about five fingers long. The boards are eleven and one third fingers square. One of them forms the base and the other the top of the box, and each must have four holes bored through it, each of which is one third of a finger square and one third of a finger from the edge of the board. The four columns must then be inserted through the holes in such a way that the boards lie at right angles to the columns. This means that the minimum distance between the holes, and between the columns which go through them, whether in the upper or the lower board, is ten fingers, so that there is room for exactly ten strips side by side at both the bottom and the top. The space between the boards, or the length of the columns between them, is equal to the thickness of ten vertical plus ten horizontal strips, so that there is room for just twenty such strips one on top of the other between the boards. Finally, the length of the two guide pieces should be equal to the edge of the boards, and their thickness should be one third of a finger, the same as the distance between the holes and the nearest edges of the board, thus enabling them to be glued like walls between the ends of the columns and the edge of the board: one of them goes on the left-hand edge, and the other on the edge nearest to you. The width, or height above the board, of each guide piece should be equal to thickness of one thicker plus one thinner strip. If any part of the columns sticks up above these guide pieces, trim it off.

After the component parts of the Box have been glued together in this way, the eight exterior faces of the columns must be divided lengthwise into ten equal parts, and each of these must be divided into two unequal parts. The larger part, equal to the thickness of one of the thicker strips, is underneath, and the smaller, equal to the thickness of one of the thinner strips, is on top. Then in the lowest of the larger divisions on the front and back of the box, write the figure nine. Moving up the four columns, skip over the smaller division, and in the next large division, write the figure eight. In the third large division of each column, write the figure seven, and so proceed upward, writing in the larger divisions on the front and back of the box the figures six, five, four, etc., down to [a cipher] inclusive. This done, turn to the right and left faces of the box and write the figure nine in each of the lowest of the smaller divisions. Moving up the box, skip over the larger division, and in the next smaller division on each face, write the figure eight. In the smaller divisions above these on the same faces write seven, and thereafter, proceeding alway upward, the number six, five, and so on down to [a cipher] inclusive. This completes the construction of the box and the placing of the numbers on the columns. This is how the strips are to be inserted into the box.

Place the box so that one guide piece faces to the left and the other toward your chest. Now lay the ten vertical trips caarrying the figure nine on the bottom board between the two nines facing you. The face with the numbers on it should point upward, and the blank side downward, and the wider marging should be at the back of the box with the narrower margin af the front. Vertical strips inserted this way are said to be *correctly laid in.* Then take the ten thinner or horizontal strips that have the number nine on them and lay them accross the others betweeen the two nines. The wider margin should be at the righthnd side and the narrower to the left, with the inscribed face pointing upward and the blank face downward. Horizontal strips inserted in this way are said to be *correctly laid in*. Next take ten vertical strips bearing the figure eight and lay them in correctly on top of the previous ones and between the two eights facing you. Then on top of these lay correctly (that is crosswise) the ten horizontal strips bearing the figure eight between the eights on the right-hand side of the box. Third, on top of these horizontal strips correctly lay in the ten vertical strips that bear the figure seven, placing them between the two sevens on the front of the box. On top of these again correctly lay in the ten horizontal strips that bear the figure seven, placing them between the two sevens on the columns on the right-hand side. Fourth, on top of these correctly lay in the ten vertical strips that bear the figure six, placing them between the two sixes on the columns facing you. Upon these again correctly lay in the ten horizontal strips that bear the figure six, placing them between the two sixes on the columns on the right-hand side. Similarly, lay in the vertical strips marked five, four, three, etc., from the front and the horizontal strips marked five, four, three, etc., from the right, correctly placing them between their own figures on the columns, proceeding alternately to the ciphers. The box is then full. To the box when thus filled I give the name *promptuary*. Its construction is now complete. At this point I adjoin a picture of it.

[*Near this place is inserted the promptuary diagram denoted by the letter.* A.]

Chapter III.

Easy Multiplication with the promptuary.

I *now examine the use of the promptuary, which is mainly for Multiplication. To perform multiplication, the figures of the multiplicand and multiplier must be set up in a certain way on the top of the Box. The Multiplicand is set up as follows. Start with the first* [*or*] *right-hand figure of the multiplicand. Select from the box the vertical strip bearing this figure that lies directly under the right-hand side of the top of the box, and place it on the top of the box on that side. For the second position*[34] *I select from the box the strip bearing the second figure*[35] *of the multiplicand that lies directly under this position, and*

[34] That is, immediately to the left.
[35] That is, second from the right.

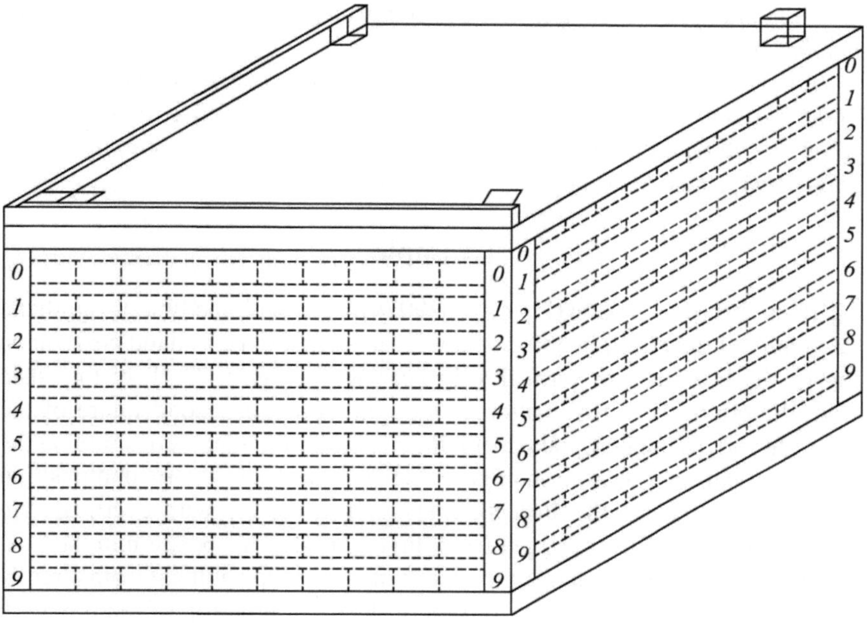

place it on the top in that position. Similarly, fill the third, fourth, fifth, etc., positions on the top of the box with vertical strips bearing the third, fourth, fifth, etc., figures of the multiplicand, taking them from the box directly beneath these respective positions, until the multiplicand is complete. If there are vacant places on the left, fill them up with strips bearing the cipher so that the top of the box may be more tightly filled up. You now have the multiplicand set up on the top.

It remains to set the multiplier up on the top, and this is done as follows. Start with the first [*or*] *right-hand figure of the multiplier. Select from the box the horizontal strip bearing this figure that lies directly under the part nearest to you of the top of the box, and place it over the vertical strips on top of the box in that position. For the second position,*[36] *select from the box the horizontal strip bearing the second figure*[37] *of the multiplier that lies directly under this position, and place it horizontally on top of the vertical strips in this position. Similarly, fill the third, fourth, fifth, etc., positions on top of the box with horizontal strips bearing the third, fourth, fifth, etc., figures of the multiplier, taking them from the box directly beneath these respective positions,*

[36] That is, the figure immediately beyond.

[37] From the right.

until the multiplier is complete. Fill up any empty spaces with the required number of strips bearing the cipher 0.

You now have both multiplier and multiplicand correctly set up on the top of the box, and scattered over the area you also have the figures of the product when they are multiplied together. These must now be added up to produce a single sum, which is exactly the required multiple.

Example.

TO multiply 8795036412 by 3586290741, start with the first or righthand figure of the multiplicand, namely 2. In the first or right-hand position on top of the box place a vertical strip bearing the figure 2. Take this strip from directly under this position in the box. For the second figure of the multiplicand, [namely 1,] take from directly under the second position on the box a vertical strip bearing the figure one and place it in the second position on top of the box. Fill the third position on top of the box with a vertical strip bearing the figure four, taking it from directly under this position in the box. Fill the fourth position on top of the box with a six strip, taking it from directly under this position in the box. Fill the fifth position on top of the box with a three strip, taking it from directly under this position in the box. The sixth position on top of the box should be occupied by a cipher strip taken from directly under this position. The seventh position should be occupied by a vertical five strip taken from directly under this position. The eighth and ninth positions should be occupied respectively by a nine strip and a seven strip, each taken from under the appropriate position. Finally, in the tenth position on top of the box, place a vertical strip bearing the figure eight and taken from beneath this position in the box. Always make sure that these vertical strips (and the horizontal ones that follow) are taken from directly below the position they will occupy on top of the box and that they are correctly placed on that top. You now have the vertical strips denoting the multiplicand correctly disposed on the top of the box and displaying all their multiples, both those that you need for your calculation and those that you do not, as in [the accompanying] diagram. In this diagram, as [well as in the last one in this appendix], I have marked in black the empty spaces in the box where vertical and horizontal strips have been removed.

[*Near this place is inserted the multiplicand diagram denoted by the letter.* B.]

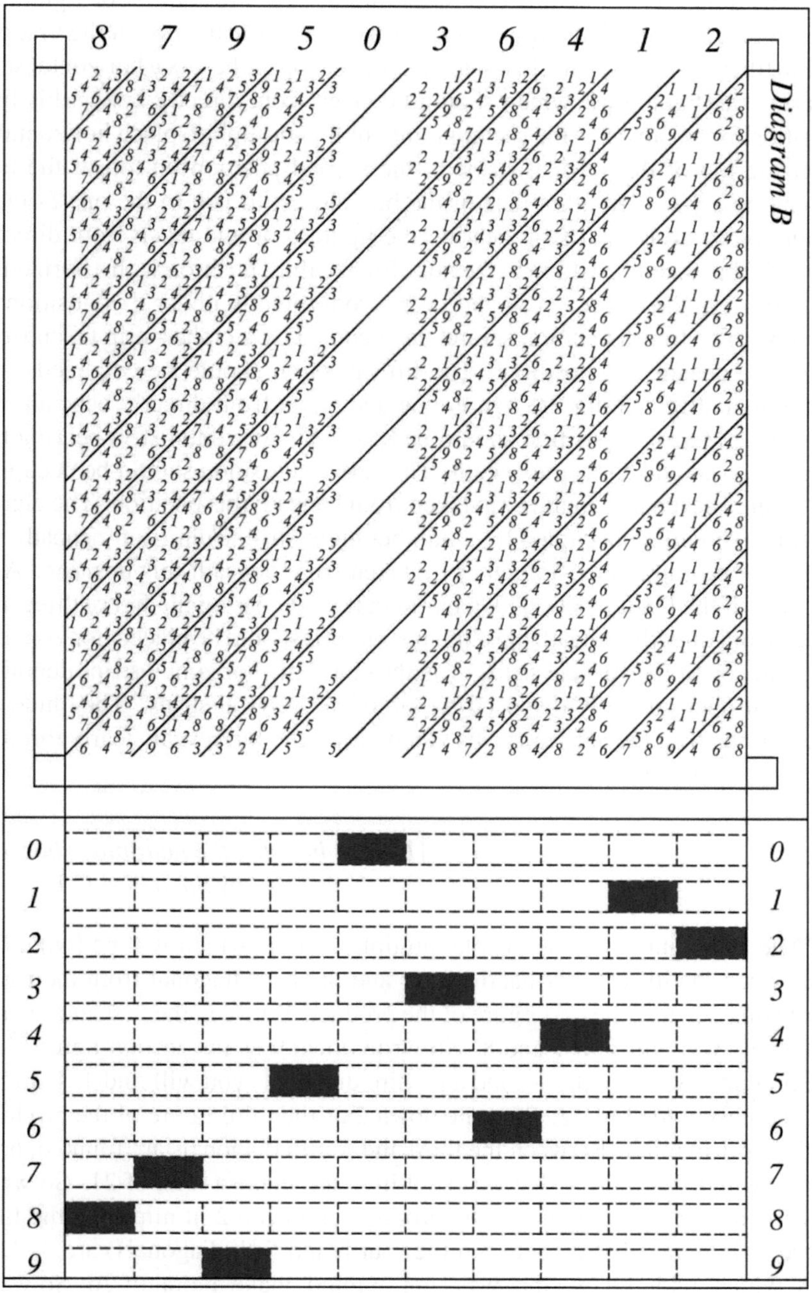
8 7 9 5 0 3 6 4 1 2
Diagram B
0
1
2
3
4
5
6
7
8
9

THis done, one must proceed to the multiplier. For its first or righthand figure, place horizontally over the top of the vertical strips in the first position (that is, the position closest to you) on the top of the box a horizontal strip bearing the figure one, which should be taken from directly below this first position in the box. For the second figure of the multiplier, place horizontally over the top of the vertical strips in the second position on top of the box (counting always from the face of the box closest to you to the one farthest from you) a horizontal strip bearing the figure four and taken from directly beneath this second position in the box. Fill the third position with a horizontal strip bearing the figure seven and taken from directly under this position in the box. The fourth position should be occupied by a cipher-strip taken from directly beneath this position. The fifth position should be occupied by a horizontal strip bearing the figure nine and taken from directly beneath this position. The sixth position should be held by a horizontal strip bearing the figure two and taken from beneath. The seventh position should be occupied by a figure seven horizontal strip taken from beneath this position. The eighth position should be occupied by a number eight strip taken from beneath. In the ninth position place a number fiv taken from beneath this position. And finally, fill the tenth position by a horizontal strip bearing the figure three and taken from directly beneath the tenth position. It should be placed horizontally over the top of the vertical strips. You thus have the horizontal strips denoting the multiplier correctly disposed on top of the vertical strips. They hide all the multiples on the vertical strips you do not need but reveal clearly through their holes the ones you do need.

[*Here is inserted the multiplier diagram denoted by the letter.* C.]

THe first figure of these visible multiples is two, which will be found between the lower right-hand corner and the first diagonal from the right. Write this down as the first figure of the product. Between the first and second diagonals you will find 1 and 8. So write down 9 as the second figure[38] of the product. Between the second and third diagonals you will find 4, 4, and 4, which make a total of 12. So write down 2 as the third figure of the product, and keep 1 in mind. Between the third and fourth diagonals are found 6, 6, 7, and 1, which, with the 1 that was kept in mind, make a total of 21. So write down 1 as the fourth figure of the product, and keep 2 in mind. In the fifth place or interval (that is, between the fourth and fifth diagonals) are 3, 4, 1, 8, 8, which, with the two that was kept in mind, make a total of 26. So write

[38] From the right.

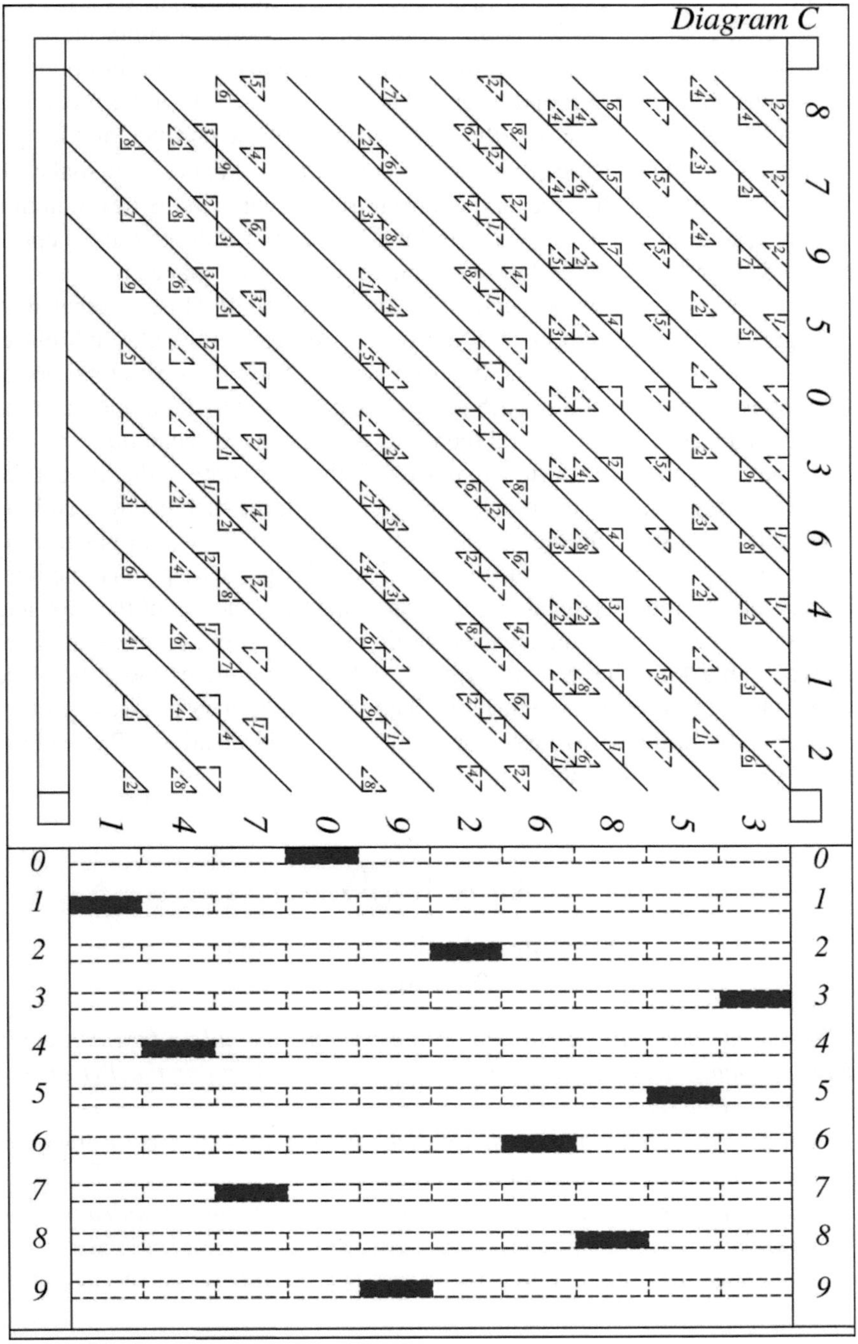
Diagram C

down 6 as the figure for the fifth place, and keep 2 in mind. In the sixth interval are 2, 2, 2, 2, 9, 1, 4, which, with the two that was kept in mind, make 24. So write down 4 as the figure in the sixth place, and keep two in mind. In the seventh interval are found figures that, with the two from the preceding diagonal, make 23. Write 3 as the seventh figure of the product, and keep 2 in mind. The figures in the eighth interval plus the one kept in mind make 41. Write one in the eighth place, and add four to the figures in the next column. This next total is 51, so 1 is written down in the ninth place, and 5 kept in mind. 5, together with the figures in the tenth interval, makes 61. Write 1 as the tenth figure of the product, and keep six in mind. This, together with the figures in the eleventh interval, makes 55. Write 5 in the eleventh place, and add 5 to the numbers in the twelfth interval. These numbers total 36, of which 6 is to be written in twelfth position and 3 added to the numbers in the thirteenth interval. Keeping to this common ARITHMETICAL technique, you will find that the figure in the thirteenth place is 7, the fourteenth 5, the fifteenth 5, the sixteenth 1, the seventeenth 4, the eighteenth 5, the nineteenth 1, and the twentieth and last 3. So the answer that we sought to our multiplication is 31541557651113461292. Write this on a piece of paper to help you remember it, then take all the horizontal and vertical strips from the top of the box, and return them to their previous places within the box. These empty spaces are marked in black in [the shown] diagram, and each is directly under the place where the strip stood on top of the box. Thus the promptuary is always prompt to your hand for the performance of many other multiplications. Proceed the same way in other examples.

CHAPTER IV.

Division with the Promptuary & Tables.

DIvision cannot be performed on the promptuary unless first converted into multiplication. This conversion is done by changing the divisor into its own extreme relate.

The extreme relate is the number whose ratio to the mean relate is the same as that of the mean relate to the given number.

The mean relate is always unity with a number of ciphers added to it on the right.[39]

Thus in some authors' tables the mean relate is 1000 (four places), as in

[39] The mean relate is the mean proportional.

[the second] book of my RABDOLOGY above. In others it is 100000 (six places), as in Pitiscus's manual of sines, tangents, and secants.[40] In other authors again it is 10000000 (eight places), as in LANSBERGE's canon of sines, tangents, and secants.[41] But though it differs from author to author, it is always denoted by unity and some ciphers, and in this form it is commonly known as the whole sine.

Example.

Where the mean relate or whole sine is 1000 and the given number is 125, its extreme relate will be 8000, because the ratio of 8000 to 1000 is the same as that of 1000 to 125.

Thus the mean relate or whole sine is always the mean proportional between some number and its extreme relate.

Furthermore, the product of a number and its extreme relate is equal to the square of the whole sine, or mean relate, (as I call it).

So 8000 multiplied by 125 is the same as the square of 1000, the mean relate; both equal 1000000.

In tables these extreme relates usually appear diametrically opposite the given numbers or are written in conspicuous places so that when one is found the other can be found immediately.

In Pitiscus's canon the given number and its extreme relate can be found as follows. Find the given number in the column of sines (which is the first column), and the other will be found opposite to it in the column of secants (which is the last column). Or find the given number in the second column (that of tangents), and the other will be found opposite to it in the second-to-last column on the page. Or else find the given number in the third column, and the other will be found opposite to it in the third-to-last column.

LANSBERGE, on the other hand, deals with given numbers and their extreme relates as follows: either the given number is among the sines of arcs and the other is among the secants of the complements of these arcs [42] or the given number is among the tangents of arcs and the other among the tangents of their complements.[43] In [the second] Book of my RABDOLOGY I placed the pairs of extreme relates (one given and the other required) in the same di-

[40] Bartolomaeus Pitiscus (1561–1613) was a German clergyman who published a systematic textbook on trigonometry in 1595. He invented the word trigonometry and used it as the title for his textbook.

[41] Philip van Lansberge (1561–1632) was a Protestant clergyman in the Netherlands. He published *The geometry of triangles* in 1591. Book 2 of this work contains tables of sines, tangents, and secants, which were used by Kepler in his calculations.

[42] Now known as cosecants.

[43] Now known as cotangents.

agonal line equidistant from the middle diagonal bearing the number [one thousand].[44] In [the first] Table, for instance, 658 and 1520 are extreme relates, so are 502 and 1991, so are 408 and 2450, so are 702 and 1312. So for all the other extreme relates in this book.

If, therefore, you are given a number to be divided by another number, you must convert the division into multiplication as follows.

Multiply the given dividend by the extreme relate of the given divisor. Form a fraction by writing beneath this product the square of the mean relate (or else cut off from its right-hand side as many figures as there are ciphers in the square). The result is the quotient of the set sum.

Example.

TO divide 8795036412 by 27884 using LANSBERGE's tables and applying the above rule of multiplication, you must multiply 8795036412 by the extreme relate of 27884, which is 3586290741, and the product is 3154155765-1113461292, under which you must write the square of the mean relate (that is, the square of the whole sine) which, from LANSBERGE, is 100000000000000 (fifteen places). One writes the resultant fraction either $\frac{31541557651113461292}{100000000000000}$ or, distinguishing the whole numbers from the fraction, $315415\frac{57651113461292}{100000000000000}$ or, omitting the fraction, 315415. This is the required quotient.

Another Example.

To divide 8795036412 by 27883963465 using a Table whose whole sine (or mean relate) is 10000000000 (eleven places), using my promptuary, you must multiply 8795036412 by the extreme relate of the number 27883963465, which is 3586290741. This will give as a product 31541557651113461292, under which is placed the square of the mean relate, which is 100000000000000000000 (21 places). This gives the fraction $\frac{31541557651113461292}{100000000000000000000102}$. Proceed thus in all cases of division, converting them with the help of Tables into multiplication, and then the promptuary will give the desired quotient with the greatest of ease. Having completed the construction and use of the promptuary, let us now proceed to location Arithmetic.

TO GOD BE THE PRAISE.

[44] The whole sine. See note 22.

LOCATION ARITHMETIC,

as performed on a CHESS-BOARD.

BOOK ONE.

PREFACE.

W*Hile working in my spare time on these short methods and seeking ways in which the [labour and] tedium of calculation might be removed, I developed not only my* LOGARITHMS, *my* RABDOLOGY, *my* PROMPTUARY *for Multiplication and other things but also a method of Arithmetic on a flat surface. As this performs all the more difficult operations of common arithmetic on a chessboard, it might well be described as more of a lark than a labour, for it carries out addition, subtraction, multiplication, division, and [even] the extraction of roots purely by moving counters from place to place. There is one small difficulty in working with it, and that is that the numbers it uses differ from ordinary numbers, so that one must begin by expressing ordinary numbers in the new form and end by reducing them to common form. Either process is simple enough even during the course of the calculation, and overall my method is easier and more accurate than almost any other short method of Arithmetic. I therefore resolved not to bury it in silence nor (as it is so short) to publish it on its own but to subjoin it to my Rabdology after the aforementioned promptuary for the benefit of the studious and the scrutiny of the learned.*

LOCATION ARTHMETIC

FIRST CHAPTER.

Description of the Rod for location in a straight Line.

LOCATION ARITHMETIC *is performed by means of counters set out according to certain rules either along a line or within a square.*

Linear arithmetic uses counters set out along a line or a rod or one edge of a chessboard.

Let there be a Rod a, b, c, ..., etc., divided into equal parts, one for each counter or number that you desire it to hold. If, then, you want it to hold 16 counters or 16 numbers, you will divide it into 16 parts, in which case the sixteenth number will be 32768, and this rod will calculate all numbers less than 65536, which is sufficiently large for ordinary use. On the other hand, if you divide it into 24 parts, it will take 24 counters and numbers, of which the twenty-fourth will be 8388608, and then the rod will calculate all numbers less than 16777216. If, however, you want to work with larger numbers (such as sines, tangents, and secants) make the rod 48 fingers long and divide it into the same number of parts to accommodate 48 counters and 48 numbers, the last being 140737488355328. Such a rod will calculate all numbers less than double this (that is, less than 281474976710656). The rod that I have taken as my example and shown here contains only 16 parts. The divisions of the rod are designated as follows. The first section has the letter a and the number 1, the second is b and its number 2, the third c and its number 4, the fourth d and its number 8, the fifth e and its number 16, the sixth f and its number 32, and so on. In other words all the sections of the rod proceeding upward in the usual numerical order are denoted by the letters of the alphabet in order and are assigned values by a continual process of doubling [as in the accompanying diagram]. The parts a, b, c, d, e, f, etc., are called *positions*.

Rod. &c.

q.	32768
p.	16384
o.	8192
n.	4096
m.	2048
l.	1024
k.	512
i.	256
h.	128
g.	64
f.	32
e.	16
d.	8
c.	4
b.	2
a.	1

CHAPTER II.
How to Change ordinary numerals into location numerals.

THe rod so inscribed is used first to change ordinary into location numerals and last to reduce location numerals to their ordinary form .

Ordinary numerals may be changed into the location (or alphabetical) variety either by subtraction or by division by two.

The method by subtraction involves deducting from the given number whichever number on the rod is nearest to but less than the given number from the remainder whichever number on the rod is nearest to but less than the remainder, and so on until the given number is entirely used up. The numbers subtracted are noted by placing counters on them on the rod or, if you prefer, by writing their letters on a piece of paper so as to keep a record of them. The counters on the rod or the letters on the paper will represent the given number in local notation.

TO express the number of the YEAR OF OUR LORD 1611 in local notation, from 1611 subtract 1024, the number on the rod that is closest to but less than it. The remainder is 587. From this, subtract the number on the rod that is less than but closest to it, namely 512. The remainder is 75. From this, subtract the nearest smaller number 64. The remainder is 11. From this subtract 8; the remainder is 3. From this subtract 2; the remainder is 1. Subtract 1; the remainder is [nothing]. So by placing counters on the numbers 1024, 512, 64, 8, 2, and 1 on the rod, or by noting on a piece of paper the location letters *l*, *k*, *g*, *d*, *b*, *a*, we have changed the number 1611 into local notation.

The other method, using division by two, is as follows: If the given number is odd, place a counter in position a, subtract one, and divide the remainder by two. Otherwise, if the given number is even, no counter is to be placed in position a. Next, if this half is odd, subtract one, and place a counter in position b. If it is even, place nothing. Third, divide this half by two, and if this half of a half is odd, subtract one, and place a counter in position c. Otherwise, place nothing. Fourth, divide by two again, and for an odd result place a counter in position d. Otherwise, place nothing. So proceed through the rest of the spaces, always dividing by two and if the result is odd, subtracting one and placing a counter in that location, but if the result is even, placing nothing, until the given number is reduced to unity, for which a counter is placed in the position reached. Then these counters on the rod or these letters on a piece of paper will represent the given number in local notation.

Represent the number 1611 by counters or letters in local notation. As the number is odd, subtract one and place *a* counter in position *a*. Then divide

1610 by two. This gives 805, which is odd. So subtract one and place a counter in position *b*. Then divide 804 by two. This gives 402, which is even. So no counter is placed in position *c*. Then divide 402 by two. This gives 201, which is odd. So subtract one, and place a counter in position *d*. Divide 200 by two. This gives 100, which is even. Hence, no counter is placed in position *e*. Divide 100 by two. This gives 50, which is even. So no counter is placed in position *f*. Divide 50 by two. This gives 25, which is odd. So mark position *g* with a counter and, subtracting one, divide 24 by two. This gives 12, which is even. So position *h* must remain empty. Divide 12 by two. This gives 6, which is even. So position *i* must remain empty. Divide 6 by two. This gives 3, which is odd. So mark position *k* with a counter and, subtracting one, divide 2 by two, which at last will give unity, for which position *l* is to be marked with a counter. Thus by means of counters placed beside the numbers 1, 2, 8, 64, 512, and 1024 on the rod, or by means of the letters *a*, *b*, *d*, *g*, *k*, *l*, you have by continually dividing by two expressed the number 1611 in local notation.

CHAPTER III.

How to reduce location numerals to ordinary numerals.

THe reduction of numbers in local and alphabetical notation to their ordinary form is accomplished in two ways: by addition and by doubling.

By addition means collecting together the numbers in the positions marked by counters or denoted by letters of the alphabet and adding them up. This total is the required number that the letters or counters indicate.

Given a number expressed in local notation as *a*, *b*, *d*, *g*, *k*, *l*, find the number. The numbers corresponding to these positions are *a* 1, *b* 2, *d* 8, *g* 64, *k* 512, *l* 1024. Add these up, and the total sum is 1611, which is the required number designated in local notation by counters in positions *a*, *b*, *d*, *g*, *k*, *l*.

The method for returning from local to ordinary notation by doubling is as follows. For the position that denotes the last (or highest) number, double unity. To this add one if you find a counter in the penultimate position, but if not, add nothing. Double this (whether or not one has been added to it), and add one if the antepenultimate position is marked by a counter. Otherwise, add nothing. Double this again, and add one if the position before the antepenultimate is marked by a counter. Otherwise, add nothing. Keep on dou-

bling, adding unity whenever you strike a counter, until you arrive at the first position, marked **a**. *The number finally resulting from this continual doubling and addition of unity down to position a is the required number concealed beneath the counters and letters of local notation.*

The Rod *with a*
synopsis of the examples.

Rod. &c.

Rod. &c.		First Example.	Second.	Third.	Fourth.
q.	32768				
p.	16384				
o.	8192				
n.	4096				
m.	2048				
l.	1024	1611 (*l*)	1 (*l*)	(*l*) 1024	(*l*) 1
k.	512	587 (*k*)	3 (*k*)	(*k*) 512	(*k*) 3
i.	256		6		6
h.	128		12		12
g.	64	75 (*g*)	25 (*g*)	(*g*) 64	(*g*) 25
f.	32		50		50
e.	16		100		100
d.	8	11 (*d*)	201 (*d*)	(*d*) 8	(*d*) 201
c.	4		402		402
b.	2	3 (*b*)	805 (*b*)	(*b*) 2	(*b*) 805
a.	1	1 (*a*)	1611 (*a*)	(*a*) 1	(*a*) 1611

GIven a number expressed in local notation as *l*, *k*, *g*, *d*, *b*, *a*, find the number by this method. For *l* double unity, which makes 2. To this, add one for the counter on *k*, which makes 3. Double this again, which makes 6, but add nothing, as position *i* is vacant. Double the 6, which makes 12, but again add nothing, as position *h* is vacant. Double this, which makes 24, and add one for the counter found in position *g*. Double 25, which makes 50, as position *f* is vacant. Double again, which gives 100, as position *e* is vacant. Double again, which makes 200, and add one because of the counter on *d*. Double 201, which gives 402 for the vacant position *c*. Double again, which makes 804, and add one for the counter on *b*. Finally, double 805, which gives 1610. This must be increased by unity because of the counter on *a*. The number 1611, which thus arises for position *a*, is the required number concealed by the [calculation and the letters *a*, *b*, *dg*, *g*, *k*, *l*; as from the Rod and the examples as can be seen by the decription of the synopsis above.]

CHAPTER IV.

Abbreviation & extension.

FRom notation and reduction I pass to computation. This depends entirely on position, abbreviation, and extension.

Position means that the components of numbers in local notation are to be marked with counters as required.

Abbreviation means that two counters standing in a certain position are to be replaced by one in the next higher position.

Extension means that a single counter in a certain position is replaced by two in the next lower position.

This means that neither abbreviation nor extension alters the value of the number.

For example, to abbreviate the number *abbdeffg*, replace the two *b*s with a single *c*. The two *f*s should be replaced by a single *g*, but as there is a *g* already there, the two *g*s must be replaced by a single *h*. Thus the number in abbreviated form is *acdeh*, but it is still the same number.

[Likewise,] extend the number *acdeh*. This number appears on the rod (with dots representing the blank spaces) as *a.cde..h*. It is to be extended so that there are no blank spaces in it, which is done as follows. Take away the counter on *h* and replace it in turn by *gg*, *gff*, *gfee*, *gfedd*, *gfedcc*, *gfedcbb*, or finally *gfedcbaa*, each time changing the last counter into two counters in the next lower position. These are all the same and all equivalent to *h*. So anyone of them added to *acde* is equivalent to *acdeh*. Hence, *abccddeefg* has the same value as *acdeh* but is in extended form.

CHAPTER V.

Addition & subtraction with a short method for changing into and out of local notation.

ADdition is performed simply by writing abbreviated numbers together on a piece of paper or marking them side by side with counters on the rod and then abbreviating the result.

Thus if *acdeh* is to be added to *befgh*, they are written together, which gives *abccdefghh*, and then abbreviated to give *abhi*. And so on.

Subtraction is performed by removing a number in abbreviated form from a number in extended form and (if necessary) abbreviating the result.

Thus *befgh* is subtracted not from *abhi* but from its extended equivalent *abccdefghh*, which gives *acdeh* as the required remainder.

Such additions and subtractions supply us with a short and easy method of reducing ordinary numbers to local notation and numbers in local notation to ordinary numbers by means of [*the following*] *table.*

Reduction Table.

	1	10	100	1000	10000	100000	1000000
1	*a*	*bd*	*cfg*	*dfghik*	*eiklo*	*fhklqr*	*gkprstv*
2	*b*	*ce*	*dgh*	*eghikl*	*fklmp*	*gilmrs*	
3	*ab*	*bcde*	*cdfi*	*defhikm*	*efilnop*	*fghiknqt*	
4	*c*	*df*	*ehi*	*fhiklm*	*glmnq*	*hkmnst*	
5	*ac*	*bef*	*cefghi*	*dhikn*	*egikpq*	*fioqrst*	
6	*bc*	*cdef*	*degk*	*efgikln*	*fgkmopq*	*ghiklorv*	
7	*abc*	*bcg*	*cdefhk*	*degikmn*	*efginr*	*fgklmoqsv*	
8	*d*	*eg*	*fik*	*giklmn*	*hmnor*	*ilnotv*	
9	*ad*	*bdeg*	*chik*	*dfiko*	*ehiklmnpr*	*fhikmnoqrtv*	

To reduce 3783 to local notation, look first in the table for 3000. You will find it in the space at the intersection of row 3 and column 1000; it is *defhikm*. Similarly, look for 700 at the intersection of 7 and 100; it is *cdefhk*. Look for 80 at the intersection of 8 and 10; it is *eg*. Lastly, look for 3, for which you will find *ab* in the space at the intersection of 3 and 1. Add these four expressions as above, and the result will be *abeghklm* for the number 3783.

To reduce *abeghklm* to ordinary numerals, the number must first be extended to the form *abcddefgghhikkm*.

Subtract from this the largest possible location numeral on the table, *defhikm* (which represents 3000). The remainder is *abcdgghk*. Extend this (as in Chap. 4) to *abcdeefghk*. Subtract the largest possible location numeral on the table, *cdefhk* (which represents 700). The remainder is *abeg*. Subtract the number *eg* on the table (which represents 80), and the remainder is *ab* (which in the table corresponds to 3).

The numbers so found—3000, 700, 80, and 3—are conjoined to give 3783, which is the number that corresponds to *abeghklm* in local notation.

Warning.

Multiplication and division also can be performed on this Rod, but as compared with the single dimension of the Rod, these operations are carried out much more clearly and easily by using placements in two dimensions, as on a chess-board, which has horizontal and vertical axes. Let us therefore turn our attention to placement in two dimensions and see how multiplication and division, the extraction of roots are carried out by means of it.

Chapter VI.

Description of the board for location in two dimensions.

The location *of numbers in two dimensions is their designation by means of counters placed on the squares and lines of a chess- or checker-board or any similar flat surface with squares on it.*

Let there be be a square Board ♈♉♊♋, ♈ marking the corner closest to you, ♉ the corner on your left, ♊ the corner farthest from you, and ♋ the corner on your right. Let the side ♈♉ be divided into as many equal parts as you wish, such as 18, 24, or more, according to the number of counters you wish the board to have room for. In [the next] figure I have divided it into 24. Let the side ♈♋ and the other sides be divided into the same number of parts. Draw lines from side ♈♉ to side ♊♋ and from side ♈♋ to side ♉♊ through the points marking the divisions, and you will have a table divided horizontally and vertically into 576 small squares. Mark the right side from ♈ to ♋ and from ♋ to ♊ with letters of the alphabet and with numbers each double the last as on the rod. When the Latin alphabet gives out, proceed with the Greek alphabet. Do the same on the left side from ♈ to ♉ and from ♉ to ♊, leaving on each side a margin sufficient to accommodate the numbers, just as I instructed for making the Rod.

[*Here is inserted the* DIAGRAM *of the* CHESS-BOARD, *or* TWO-DIMENSIONAL ABACHUS, denoted by *the signs* ♈♉♊♋.]

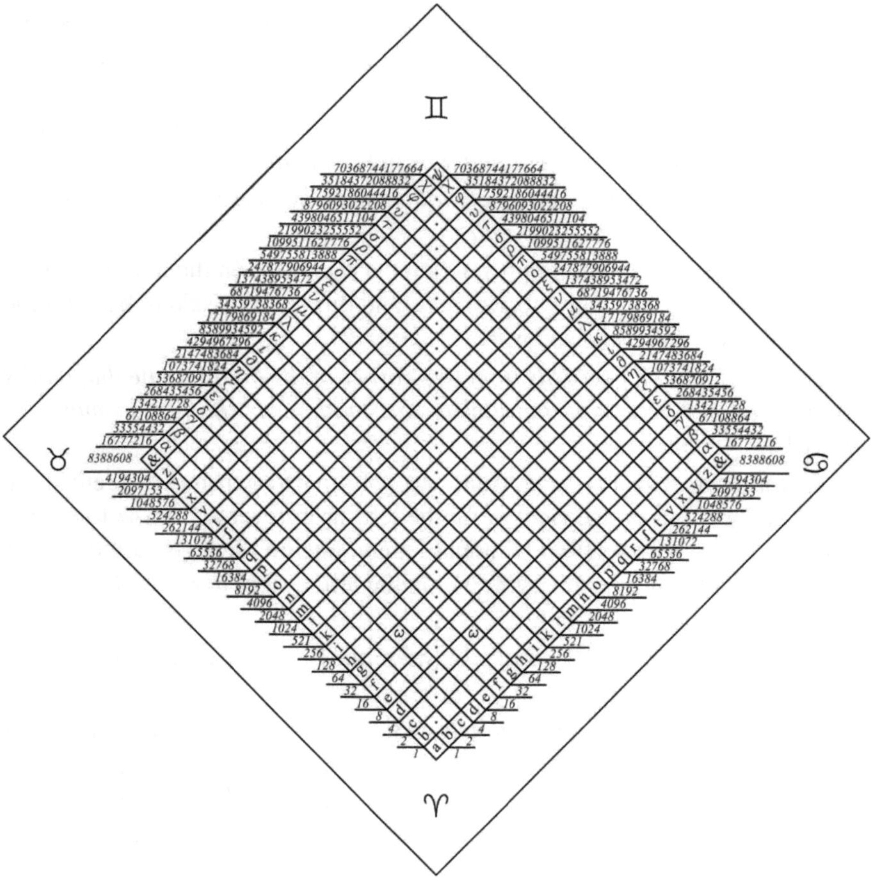

CHAPTER VII.
The two-dimensional movement of counters on the Board.

NU *mbers are expressed and manipulated by moving counters here and there over the squares on the board.*

Motion or progress on the board is of two types: Direct and diagonal.

Direct motion is parallel to the sides, as the rook moves in chess.[45]

As for instance from *a* to &, from *b* on the left to α on the right, from *c* on the left to β on the right, from *d* on the left to γ on the right, [and so on thereafter.]

Or alternatively from *b* on the right to α on the left, from *c* on the right to β on the left, from *d* on the right to γ on the left, or in the opposite direction, from & to a, α to *b*, β to *c*, γ to *d*, etc., to the right or to the left, and up or down.

Hence, direct motion is of two types: either parallel to the lines ♈♋ *and* ♉♊ *or at right angles to these and parallel to the lines* ♈♉ *and* ♊♋.

Furthermore, these two types of motion always intersect at some common point, which must be noted carefully.

For instance, direct motion from *d* on the right to γ on the left and from *g* on the left to ζ on the right intersect at ω, and this space is said to be common to *d* right and *g* left. And so on.

Diagonal motion proceeds from one corner of the board to the diagonally opposite corner or else is parallel to this motion, as the bishop moves in chess.[46]

As for instance from *a* to ψ, from *b* to χ, from *c* to φ from *d* to υ, etc., both letters being either on the right or on the left; or in reverse from ψ to *a*, etc.; or alternatively between two identical letters, one on the right and the other on the left, as from *b* on the right to *b* on the left, *c* right to *c* left, *d* right to *d* left, or left to right; and so on.

So diagonal motion also is of two types, one between identical letters and one between different letters.

Movement between likes occurs when we progress from right to left [*as far as* ♋ *or to the left toward* ♉, *or in reverse we march from* ♉ *to* ♋].

Movement between unlikes occurs when we ascend from [♈ *to* ♊, *or descend from* ♊ *to* ♈], *as is clear from the preceding examples.*

---

[45] Napier calls the rook "the tower-bearing elephant" (*elephantis turriferi*).

[46] Napier calls the bishop the archer (*sagittiferi*).

CHAPTER VIII.

Axioms and corollaries
for each type of motion on the board.

AXIOM I.

IN direct motion (like the rook's move) the value of each space is double that of the one before, whether to the left or to the right.

So from *a* to *b* on either the right or the left the increase between the squares is twofold, for square *a* has a value of 1, and square *b* a value of 2. Similarly, if you ascend from *b* on either the right or the left, the next square *c* will have a value of 4, [which is] twice two. Similarly for the other squares going up and the reverse, coming down.

Axiom 2.

All squares lying diagonally between two identical letters have the same value as the number noted in either margin and are understood to be marked (at least potentially) by the same letters.

Thus all squares lying between *l* and *l* are understood to be marked with an *l* and to have the value 1024.

From the two types of motion, direct like the rook and diagonal like the bishop, and from the axioms just given follow a number of Corollaries which are given below.

Coroll. 1.

[*In the first place,*] *a counter moving diagonally between similar letters changes neither its letter nor its numerical value, and hence this motion is deservedly called equivalent.*

Corol. 2.

[*Second*], *whereas the diagonal motion of a counter to the right or left (like the bishop in chess) does not change its value, diagonal ascent multiplies its value by four. Thus the value of each square is four times that of the square joined diagonally beneath it.*

Corol. 3.

[*Third*], *the squares on the diagonal* ♈♊ *or aψ ascend by alternate letters and by numbers each of which is four times the last and all of which are perfect squares. These squares should be marked with dots for the purpose of extracting square roots.*

Thus *a*, 1; *c*, 4; *e*, 16; *g*, 64; *i*, 256; etc. up to ψ.

Corol. 4.

[*Fourth*], *the squares on the diagonal line* $b\chi$ *ascend by alternate letters and by numbers each of which is four times the last but none of which is a perfect square.*

So *b*, 2; *d*, 8; *t*, 32; II, 128; *k*, 512; and so on up to χ.

Corol. 5.

[*Fifth*], *the squares from c to* φ; *from e to* τ, *from g to* ρ, *etc., proceed like the squares in line* $a\psi$ *except that they begin from the number given in the margin.*

Corol. 6.

[*Sixth*], *the squares from d to* υ, *fromb f to* σ, *from h to* π, *and so on alternately proceed like the squares in the line* $b\chi$ *except that each begins from the number given in the margin.*

Corol. 7.

[*Seventh*], *the multiplication of two numbers one of which is in margin* ♈♋ *and the other in margin* ♈♉ *produces the number in the common square or angle at the intersection by direct motion, and this number is indicated by the two identical letters to right and left by diagonal motion.*

Thus the result of multiplying *d* 8 by *g* 64 is *k* 512, these being the letter and number of the square or common angle between *d* and *g*, which I have marked with the sign ω. And so on.

Corol. 8.

[*Eighth*], *to any counter placed on a square there correspond three numbers and the three letters that belong to them. Two of these lie under that counter by direct motion, one of these being found to the right and one to the left. The third is found to either the right or left by the diagonal motion of the bishop in chess and is marked by identical letters and numbers in the two margins.*

So to a counter on square ω there correspond by the rook's move in chess the numbers and letters *d* 8 and *g* 64, and the third number 512 with its letter *k* is found in either margin to the right or left proceeding by the bishop's move.

Corol. 9.

[*Ninth*], *in the operation of multiplying, the last of these three numbers (the one indicated by the bishop's move to the right or left) is the product of the other two, one of which is the multiplier and the other the multiplicand. In the operation of division the same number is the dividend, and the other two (in the lower margins as indicated by the rook's move) are the divisor and the quotient.*

[So in the previous example, of the three numbers *d* 8, *g* 64 and *k* 512, *k* 512 is the product of 8 and 64, one of which is the multiplier and the other the multiplicand. However, in division the third number, *k* 512, is the dividend and of the remaining two one is the divisor and the other the quotient.

Reminder.]

In accordance with these corollaries counters placed on the board are variously moved, extended, and abbreviated, and while retaining the former value, they produce various shapes such as rectangles (or oblongs), squares, etc. These shapes are used in the processes of multiplication, division, and the extraction of square roots, as will be apparent from what follows.

CHAPTER IX.
Multiplication.

IN multiplication it is necessary that the multiplier and multiplicand should each individually be less than twice the middle number (by which I mean the number at the angles ♉ or ♋). On the board in figure 18 twice the middle number is 16777216. *Use counters or chalk to mark one of the two numbers to be multiplied in the lower margin on the right (♈♋) and the other in the lower margin on the left (♈♉). Mark them not within the area of the board but on the actual numbers beside the letters. Then starting from these two numbers as marked with counters or chalk in the margins, proceed to mark with counters all the common angles on the board, taking great care not to leave any out. These counters will then mark out a rectangular shape on the board, and the rectangle denotes the desired multiple or required product, which can be retrieved by abbreviation, translation, and reduction.*

To multiply 19 (which in translation is *abe*) by 13 (which in translation is *acd*), mark with counters or chalk in the lower left-hand margin ♈♉ the letters *a*, *b*, *e* (or their numbers 1, 2, 16) and in the right-hand margin ♈♋ the letters *a*, *c*, *d* (or their numbers 1, 4, 8), as in the figure [below]. Mark on the board with counters all the common angles between the left-hand marks *a*, *b*, *e* (or 1, 2, 16) and the right-hand ones *a*, *c*, *d* (or 1, 4, 8). They will produce

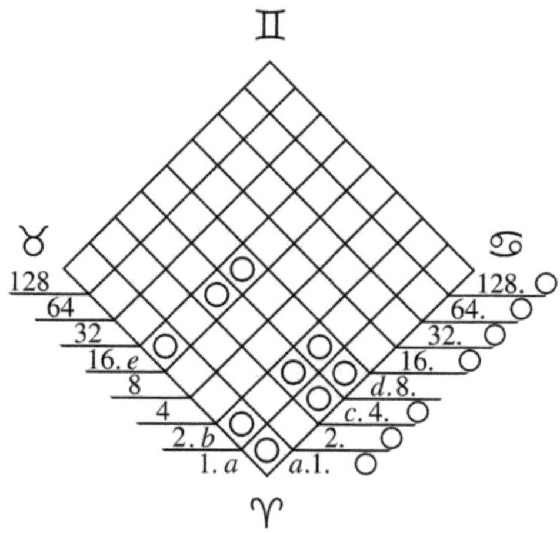

a rectangular figure as in [the near] figure.[47] Now remove the counters in the margin and rub out the marks denoting multiplier and multiplicand. Then abbreviate the total represented by the rectangle on the board by shifting its counters as follows. Transfer the counter on the board in the common angle between 1 and 1 to 1 in the right-hand margin, and the counter between 2 and 2 to 2 in the right-hand margin. Again, for the counter between 4 and 4 let a single counter be placed in the same margin on 4. Instead of the counters between 8 and 8, which are to be removed, let a single counter be placed on the board between 16 and 16. There are now three counters on the board between 16 and 16, for which let a single counter be placed in the aforesaid margin at 16 and another on the board between 32 and 32. As this is the only one on this row, it must be transferred to the margin at 32.

There is another counter on the board, between 64 and 64, which (as it is the only one on this row) I transfer to the margin at 64. Lastly, there is a counter on the board between 128 and 128, and as this is the only one, I transfer it to the margin at 128. So from the counters in the margin at 128, 64, 32, 16, 4, 2, and 1 we derive 247 as the required product of multiplying 19 by 13. It is much easier to follow all this using movable counters on a large board than from the ones immovably printed on the board in the diagram. You should therefore use the former when learning how to do it.

[47] This and the remaining figures in this Book are shown here in the position that Napier intended for the board's use, with "♈ marking the corner closest to you." However, in the original 1617 publication of *Rabdologiæ* they appear rotated 90° counterclockwise, probably for a better fit in the very narrow page width that was used a the time.

Another Example.

To multiply 1206 by 604, mark the former number (which is expressed by the letters *l*, *h*, *f*, *e*, *c*, *b* or the numbers 1024, 128, 32, 16, 4, 2) by counters or chalk in the right margin, and mark the latter (which is expressed by the letters *k* , *g*, *e*, *d*, *c* or the numbers 512, 64, 16, 8, 4) in the left.

Place counters in all the common angles, as in [the] figure [below]. Remove the marginal marks. Now abbreviate by shifting the movable counters on the

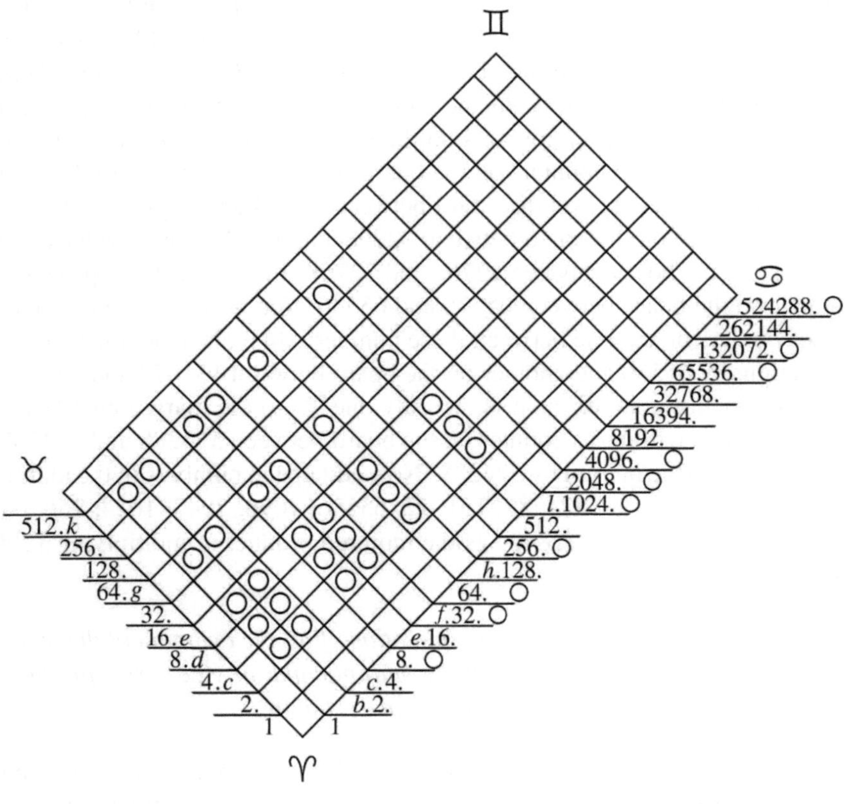

proper board (for the printed ones in [the] figure cannot be moved) as follows. Transfer the single counter on the board between 8 and 8 to the right margin at 8. For the two counters between 16 and 16, place one on the board between 32 and 32.

There are now three counters on the board between 32 and 32. Instead of them, place one counter in the same margin at 32 and another on the board between 64 and 64. There are now three counters on the board between 64 and 64. Instead of them, place one in the margin at 64 and another on the board between 128 and 128. You now have four counters on the board between 128 and 128. For these, place two counters on the board between 256 and 256. You now have five counters on the board between 256 and 256. Instead of them, place one counter in the margin by 256 and two on the board between 512 and 512. You now have four counters on the board between 512 and 512. Remove these, and place in their stead two counters on the board between 1024 and 1024. You now have five counters on the board between 1024 and 1024. Instead of them, place one in the margin by 1024 and two on the board between 2048 and 2048. You now have five counters on the board between 2048 and 2048. Remove these, and place in their stead a single counter in the margin by 2048 and two on the board between 4096 and 4096. You now have three counters on the board between 4096 and 4096. Instead of these, place one in the margin by 4096 and another on the board between 8192 and 8192. You now have four counters on the board between 8192 and 8192. In their stead, place two counters on the board between 16384 and 16384. You now have four counters on the board between 16384 and 16384. Instead of these, place two on the board between 32768 and 32768. You now have two counters on the board between 32768 and 32768. Instead of them, place one counter on the board between 65536 and 65536. You now have on the board three counters on the line between these numbers. In their stead, place one in the margin by 65536 and another on the board between 131072 and 131072, and transfer this counter (as it is the only one[48]) to the margin by 131072. Last of all you will find a counter on the board between 524288 and 524288. Transfer this to the margin by 524288. Now add up the numbers in the margin marked by counters on 524288, 131072, 65536, 4096, 2048, 1024, 256, 64, 32, and 8, and you have 728424 as the required product of multiplying 1206 by 604.

It follows from this that the multiplication of all the counters of the multiplier by each individual counter of the multiplicand, or vice versa, produces rows of counters that I call segments of the rectangle.

Thus in the rectangle in the previous example the line of counters stretching up from the *k* on the lower left in accordance with the rook's move is called a segment of that rectangle.

The line of counters going up from *g* is another segment of the same rectangle.

Similarly, the horizontal line of counters proceeding in accordance with the rook's move toward *l* on the right is one of the segments of the same rectangle.

[48] That is, in its row.

So also the line that stretches towards h, and others similarly.

Chapter X.
Division.

In division the point of intersection of the bishop's move (by equivalent motion) from the highest counter of the dividend and the rook's move from the highest counter of the divisor designates the square from which a row of counters proceeds parallel to the divisor, and this row is called a segment.[49] *It is congruous if its total is less than the remainder of the dividend. Otherwise, the row lying next under it must be taken as the congruous segment.*

This will become clear from the examples of division that follow.

This is how division is carried out. Mark the dividend with counters in either margin and the divisor with chalk marks (to distinguish it from the dividend) in either the same or the other margin. Then set up their congruous segment on the board, subtract it (that is, its numerical value) from the dividend, and leave counters denoting the remainder in the margin. Then subtract from these their congruous segment and again place counters to denote the remainder, and from each remainder, subtract its congruous segment over and over again until at length nothing is left or else a number less than the divisor (which is placed to one side to indicate the final remainder). The numbers on the side in the other margin, added together, will give you the correct quotient.[50]

[So] to divide 250 by 13, mark the dividend 250 on the right margin by placing counters on the numbers 128, 64, 32, 16, 8, and 2, and the divisor 13 on the lower left margin by marks against 8, 4, and 1. Find the first congruous segment of these as follows: go up from 8 in the [left] margin by the rook's move and across from 128 in the right margin by the bishop's move, and from the square where they meet, place a row of counters parallel to the divisor.[51] This row points toward 16 and is the first congruous segment. Subtract it from the dividend. which leaves counters beside 32, 8, and 2 to denote the first remainder.[52] The point of intersection between the largest of these 32 (by the bishop's move) and the largest number of the divisor 8 (by the rook's move)

---

[49] Counters are placed not on every square of the row but only on those that correspond to the counters marking the divisor.

[50] These numbers arc the ones toward which the lines denoting the congruous segments point.

[51] See note 49.

[52] $250 - 208 = 42$.

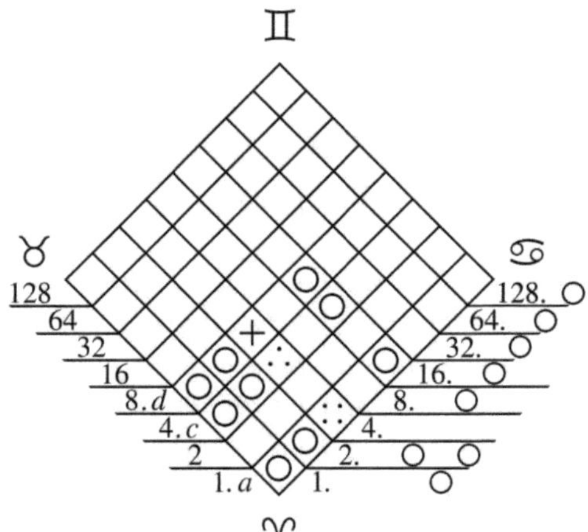

comes at $+$, and so the segment of the divisor is $+\ \therefore$ and :: in [the] figure. But because the value of this[53] exceeds the stated remainder,[54] this incongruous segment must be rejected. Instead we take the next lower row of counters, which points toward 2, and these three counters are the congruous segment, whose subtraction from the stated remainder will leave as the second remainder a counter on 16.[55] Now find the congruous segment between this counter and the largest number of the divisor. It points toward 1. There are three counters, whose value is subtracted from the single counter at 16 on the right denoting the second remainder, and there are left as the third and last remainder a counter on the right at 2 and another at 1. These indicate three as the final remainder, which is set aside.

Then when added together, the numbers in the right-hand margin toward which the individual congruous segments point (i.e., 16,2,1) form the correct quotient, 19.

Another Example.

To divide 728428 by 1206, designate in the right margin the number to be divided by placing counters on 524288, 131072, 65536, 4096, 2048, 1024, 256, 64, 32, and 8, and designate the divisor in the same margin (if you wish) by placing marks or letters beside the numbers 1024, 128, 32, 16, 4, and 2. Find the first congruous segment of these as I have described. It will stand directly above the number 512, which appears in the lower left margin. Subtract the

[53] That is, 52.
[54] That is, 42.
[55] $42 - 26 = 16$.

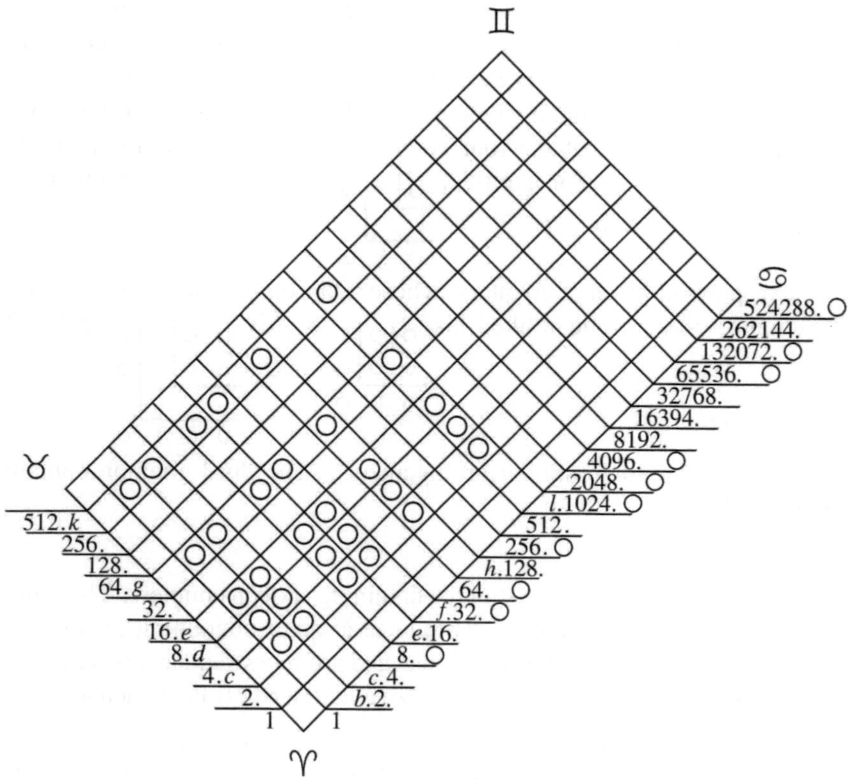

value of this segment from the dividend, which leaves a remainder to be noted: Subtract from that its own congruous segment, and you will be left with on remainder after another and eventually with nothing. In this calculatIon five congruous segments appear that point toward the numbers 512, 64, 16, 8, 4. These added together produce 604 the required quotient. This is illustrated by [the] figure [in the second example for multiplication,which is also inserted here].

CHAPTER XI.
The extraction of square roots.

ON the row marked with dots (from a to ψ) place a counter on the square of highest value that can be subtracted from the number whose square root is to be found. This counter is called the head of the gnomons or the head of the square.[56] *Its numerical value is to be increased by the gnomons.*

The name gnomon is used here to denote a row of counters that, when added to a single counter or to a square, produces a larger square.[57]

The process starts from a single counter.[58] to which is added a gnomon of three counters, so that the four counters stand one at each corner of a square, like [this]  [or like this] 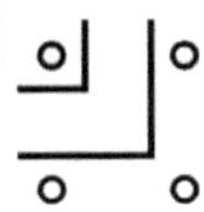 or some similar arrangement.

The next gnomon has five counters, whose addition produces a square with nine counters in it, arranged like [this  or this 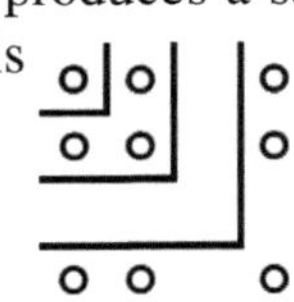 or this][59]

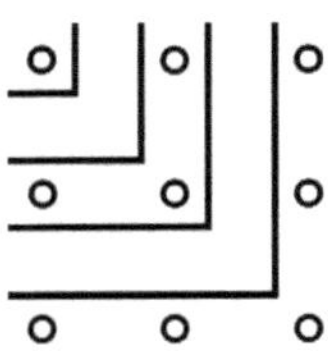 [or some similar arrangement]. The third gnomon contains seven counters and produces a square containing sixteen counters. The fourth gnomon contains nine counters and produces a square containing 25 counters. The fifth gnomon contains eleven counters and produces a square containing 36 counters. Thus the square increases in size with the addition of each gnomon.

The largest gnomon that can be subtracted from the remaining counters in the margin and for which there is space vacant on the board is called a congruous gnomon.

It follows that the congruous gnomon always occupies the next lower or next but one lower space than that occupied by the highest counter in the margin.

So much by way of introduction. Now this is how square roots are extracted. Mark by counters in either margin the number whose square root is required. From it subtract the value of the counter that I have called the head of the gnomons (but leave this counter on the board), and leave this remainder noted by counters in the margin. From this remainder, subtract the value of the three counters that make up the first congruous gnomon (but leave the gnomon on the board), and leave counters in the margin to denote this second remainder.

[56] It forms the top corner of all the square patterns that the calculation produces on the board.

[57] The "row" is *L* shaped.

[58] Napier's head of the gnomons.

[59] Note that the counters of the gnomon are added to the board in such a way as to maintain horiontal and vertical lines.

From the second remainder, subtract the value of the five counters that make up the second congruous gnomon (but leave the gnomon on the board), and leave counters in the margin to denote this third remainder. Subtract the third congruous gnomon to get a fourth remainder. Proceed thus to get a fifth remainder, a sixth, etc., until at length there is no remainder or a remainder less than even the smallest gnomon. The counters remaining on the board mark out a perfect square. Move a counter down each individual row into either margin, and these will denote the exact square root required.

Example.

To find the square root of 1238, denote this number in either margin (the right, say) by means of counters on the numbers 1024, 128, 64, 16, 4, and 2, as in [the] figure [below]. Place a counter in the square (marked with a dot)

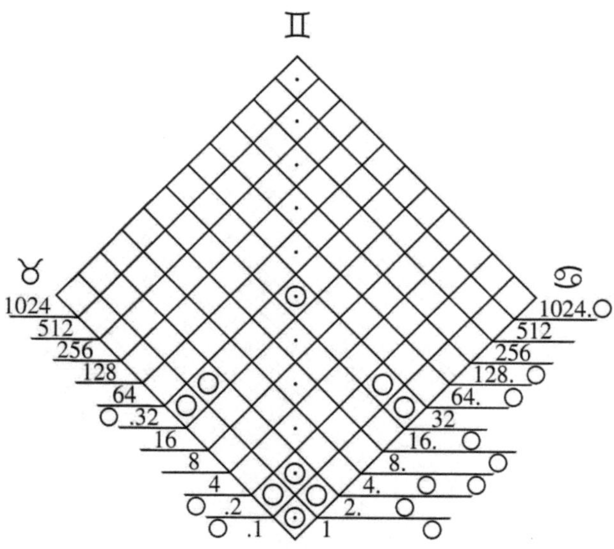

that has the value 1024. This counter is the head of the Gnomons. Leaving this on the board, subtract its value from the aforesaid counters in the margin. This leaves counters on 128, 64, 16, 4, and 2 which denote the first remainder. Place on the board the three counters that make up the first congruous gnomon ([as can be seen]), and subtract their value[60] from the first remainder, which leaves counters in the margin on numbers 64, 16, and 2 to denote the second

[60] $64 + 64 + 4 = 132$.

remainder. Place on the board the five counters that make up the second congruous gnomon, and without removing them from the board, subtract their value[61] from the second remainder. This leaves counters on 8, 4, and 1, which, added together, make 13 as the third and last remainder. Take one counter from each of the three rows of this square, move it straight down its row onto the lower margin. The counters will land on the numbers 32, 2, and 1, and their total, 35, is the square root we require.

Another Example.

To extract the square root of 2209, denote this number in either margin (the right, say) by placing counters on 2048. Place a counter in the square (marked with a dot) that has the value 1024. This is the head of the gnomons. Subtract this from the counters in the margin, which leaves counters on 1024, 128, 32,

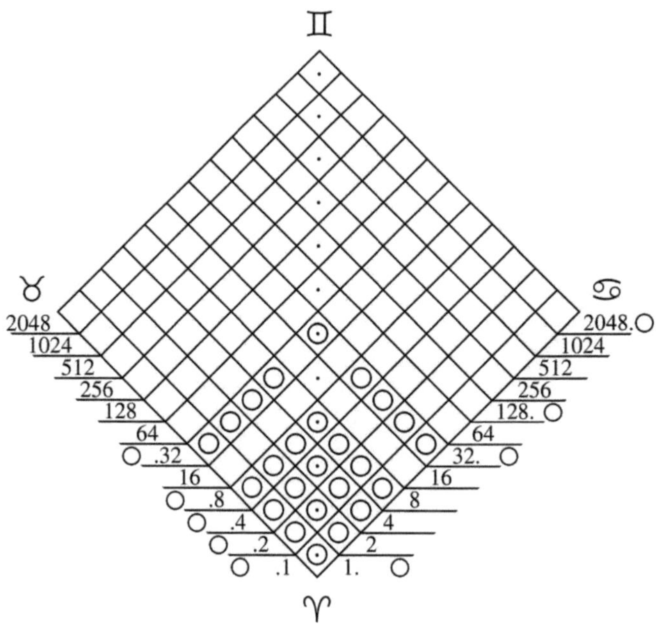

and 1 to denote the first remainder. Place the first congruous gnomon on the board, and subtract its value from the first remainder, which leaves counters on 512, 64, 32, and 1 to denote the second remainder. Place the second congruous gnomon on the board and subtract its value from this second remainder. This leaves the third remainder, denoted by counters in the margin on 256, 16,

[61] $32 + 32 + 2 + 2 + 1 = 69$.

and 1. From this third remainder subtract the value of the third congruous gnomon. This leaves the fourth remainder, denoted by counters in the margin on numbers 64, 16, 8, 4, and 1. From this fourth remainder, subtract the value of the fourth congruous gnomon, and the last remainder is nothing. Place counters in the margin on the numbers toward which the individual rows of the square point. They will go on the numbers 32, 8, 4, 2, and 1, which, when added up, give 47, the required root. This can be seen in [the] figure, so far as a picture allows, but as I advised earlier it is much easier to gain an exact understanding of the movements of the counters if you use a larger board and movable counters. Here I end my work on LOCATION ARITHMETIC.

To GOD alone be all the praise and honour.

THE END.

CONSTRUCTIO

Translation and note by William Rae Macdonald

Macdonald's notes (16 pages in small print in the original) are not included in this edition. For an explanation of the remark in brackets at the end of Article 17 of the *Constructio* see the Mathematical Introduction to this volume, pages 401–402. Note also that Briggs did not use the decimal point in his remarks on Napier's Appendix, except in his third example. Elsewhere, the decimal part of a number was just underlined, as in 25118865.

The photograph of the title page from the *Constructio* (p. 753) is reproduced by courtesy of the National Library of Scotland.

MIRIFICI LOGARITHMORVM CANONIS CONSTRVCTIO;

Et eorum ad naturales ipſorum numeros habitudines;

VNÀ CVM

Appendice, de aliâ eâque præſtantiore Logarithmorum ſpecie condenda.

QVIBVS ACCESSERE

Propoſitiones ad triangula ſphærica faciliore calculo reſolvenda: *Unà cum Annotationibus aliquot doctiſſimi* D. HENRICI BRIGGII, *in eas & memoratam appendicem.*

Authore & Inventore *Ioanne Nepero*, Barone *Merchiſtonii*, &c. Scoto.

EDINBVRGI,

Excudebat ANDREAS HART.

ANNO DOMINI 1619.

THE CONSTRUCTION OF THE WONDERFUL CANON OF LOGARITHMS;

And their relations to their own natural numbers;

WITH

An Appendix as to the making of another and better kind of Logarithms.

TO WHICH ARE ADDED

Propositions for the solution of Spherical Triangles by an *easier method: with Notes on them and on the above-mentioned Appendix by the learned* HENRY BRIGGS.

By the Author and Inventor, *John Napier*, Baron *of Merchiston*, &c., in Scotland.

Printed by ANDREW HART,
OF EDINBURGH;
IN THE YEAR OF OUR LORD, 1619.

Translated from Latin into English by William Rae Macdonald, 1888.

TO THE READER STUDIOUS OF THE MATHEMATICS,

GREETING.

SEVERAL *years ago (Reader, Lover of the Mathematics) my Father, of memory always to be revered, made public the use of the Wonderful Canon of Logarithms; but, as he himself mentioned on the seventh and on the last pages of the Logarithms, he was decidedly against committing to types the theory and method of its creation, until he had ascertained the opinion and criticism on the Canon of those who are versed in this kind of learning.*

But, since his departure from this life, it has been made plain to me by unmistakable proofs, that the most skilled in the mathematical sciences consider this new invention of very great importance, and that nothing more agreeable to them could happen, than if the construction of this Wonderful Canon, or at least so much as might suffice to explain it, go forth into the light for the public benefit.

Therefore, although it is very manifest to me that the Author had not put the finishing touch to this little treatise, yet I have done what in me lay to satisfy their most honourable request, and to afford some assistance to those especially who are weaker in such studies and are apt to stick on the very threshold.

Nor do I doubt, but that this posthumous work would have seen the light in a much more perfect and finished state, if God had granted a longer enjoyment of life to the Author, my most dearly loved father, in whom, by the opinion of the wisest men, among other illustrious gifts this showed itself pre-eminent, that the most difficult matters were unravelled by a sure and easy method, as well as in the fewest words.

You have then (kind Reader) in this little book most amply unfolded the theory of the construction of logarithms, (here called by him artificial numbers, for he had this treatise written out beside him several years before the word Logarithm *was invented,) in which their nature, characteristics, and various relations to their natural numbers, are clearly demonstrated.*

It seemed desirable also to add to the theory an Appendix as to the construction of another and better kind of logarithms (mentioned by the Author in the preface to his Rabdologiae) *in which the logarithm of unity is* 0.

After this follows the last fruit of his labours, pointing to the ultimate perfecting of his Logarithmic Trigonometry, namely certain very remarkable propositions for the resolution of spherical triangles not quadrantal, without dividing them into quadrantal or rectangular triangles. These propositions, which are absolutely general, he had determined to reduce into order and successively to prove, had he not been snatched away from us by a too hasty death.

We have also taken care to have printed some Studies on the above-mentioned Propositions, and on the new kind of Logarithms, by that most excellent Mathematician Henry Briggs, *public Professor at* London, *who for the singular friendship which subsisted between him and my father of illustrious memory, took upon himself, in the most willing spirit, the very heavy labour of computing this new Canon, the method of its creation and the explanation of its use being left to the Inventor. Now, however, as he has been called away from this life, the burden of the whole business would appear to rest on the shoulders of the most learned* Briggs, *on whom, too, would appear by some chance to have fallen the task of adorning this Sparta.*

Meanwhile (Reader) enjoy the fruits of these labours such as they are, and receive them in good part according to your culture.

Farewell,

ROBERT NAPIER, Son.

THE CONSTRUCTION OF *THE WONDERFUL CANON* OF LOGARITHMS; (HEREIN CALLED BY THE AUTHOR THE ARTIFICIAL TABLE)

and their relations to
their natural
numbers.

1. A LOGARITHMIC TABLE *is a small table by the use of which we can obtain a knowledge of all geometrical dimensions and motions in space, by a very easy calculation.*

IT is deservedly called very small, because it does not exceed in size a table of sines; very easy, because by it all multiplications, divisions, and the more difficult extractions of roots are avoided; for by only a very few most easy additions, subtractions, and divisions by two, it measures quite generally all figures and motions.

It is picked out from numbers progressing in continuous proportion.

2. *Of continuous progressions, an arithmetical is one which proceeds by equal intervals; a geometrical, one which advances by unequal and proportionally increasing or decreasing intervals.*

Arithmetical progressions: 1, 2, 3, 4, 5, 6, 7, &c.; or 2, 4, 6, 8, 10, 12, 14, 16, &c. Geometrical progressions: 1, 2, 4, 8, 16, 32, 64, &c.; or 243, 81, 27, 9, 3, 1.

3. *In these progressions we require accuracy and ease in working. Accuracy is obtained by taking large numbers for a basis; but large numbers are most easily made from small by adding cyphers.*

Thus instead of 100000, which the less experienced make the greatest sine, the more learned put 10000000, whereby the difference of all sines is better expressed. Wherefore also we use the same for radius and for the greatest of our geometrical proportionals.

4. *In computing tables, these large numbers may again be made still larger by placing a period after the number and adding cyphers.*

Thus in commencing to compute, instead of 10000000 we put 10000000.0000000, lest the most minute error should become very large by frequent multiplication.

5. *In numbers distinguished thus by a period in their midst, whatever is written after the period is a fraction, the denominator of which is unity with as many cyphers after it as there are figures after the period.*

Thus 10000000.04 is the same as $10000000\frac{4}{100}$; also 25.803 is the same as $25\frac{803}{1000}$; also 9999998.0005021 is the same as $9999998\frac{5021}{10000000}$, and so of others.

6. *When the tables are computed, the fractions following the period may then be rejected without any sensible error. For in our large numbers, an error which does not exceed unity is insensible and as if it were none.*

Thus in the completed table, instead of 9987643.8213051, which is $9987643\frac{8213051}{10000000}$, we may put 9987643 without sensible error.

7. *Besides this, there is another rule for accuracy; that is to say, when an unknown or incommensurable quantity is included between numerical limits not differing by many units.*

Thus if the diameter of a circle contain 497 parts, since it is not possible to ascertain precisely of how many parts the circumference consists, the more experienced, in accordance with the views of Archimedes, have enclosed it within limits, namely 1562 and 1561. Again, if the side of a square contain 1000 parts, the diagonal will be the square root of the number 2000000. Since this is an incommensurable number, we seek for its limits by extraction of the square root, namely 1415 the greater limit and 1414 the less limit, or more accurately $1414\frac{604}{2828}$ the greater, and $1414\frac{604}{2829}$ the less; for as we reduce the difference of the limits we increase the accuracy.

In place of the unknown quantities themselves, their limits are to be added, subtracted, multiplied, or divided, according as there may be need.

8. *The two limits of one quantity are added to the two limits of another, when the less of the one is added to the less of the other, and the greater of the one to the greater of the other.*

Thus let the line *abc* be divided into two parts, *ab* and *bc*. Let *ab*

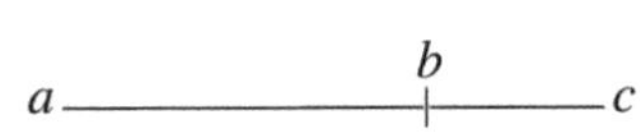

lie between the limits 123.5 the greater and 123.2 the less. Also let *bc* lie between the limits 43.2 the greater and 43.1 the less. Then the greater being added to the greater and the less to the less, the whole line *ac* will lie between the limits 166.7 and 166.3.

9. *The two limits of one quantity are multiplied into the two limits of another, when the less of the one is multiplied into the less of the other, and the greater of the one into the greater of the other.*

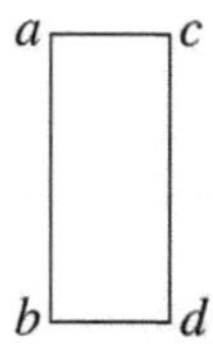

Thus let one of the quantities *ab* lie between the limits 10.502 the greater and 10.500 the less. And let the other *ac* lie between the limits 3.216 the greater and 3.215 the less. Then 10.502 being multiplied into 3.216 and 10.500 into 3.215, the limits will become 33.774432 and 33.757500, between which the area of *abcd* will lie.

10. *Subtraction of limits is performed by taking the greater limit of the less quantity from the less of the greater, and the less limit of the less quantity from the greater of the greater.*

Thus, in the first figure, if from the limits of *ac*, which are 166.7 and 166.3, you subtract the limits of *bc*, which are 43.2 and 43.1, the limits of *ab* become 123.6 and 123.1, and not 123.5 and 123.2. For although the addition of the latter to 43.2 and 43.1 produced 166.7 and 166.3 (as in 8), yet the converse does not follow; for there may be some quantity between 166.7 and 166.3 from which if you subtract some other which is between 43.2 and 43.1, the remainder may not lie between 123.5 and 123.2, but it is impossible for it not to lie between the limits 123.6 and 123.1.

11. *Division of limits is performed by dividing the greater limit of the dividend by the less of the divisor, and the less of the dividend by the greater of the divisor.*

Thus, in the preceding figure, the rectangle *abcd* lying between the limits 33.774432 and 33.757500 may be divided by the limits of *ac*, which are 3.216 and 3.215, when there will come out $10.505\frac{857}{3215}$ and $10.496\frac{2364}{3216}$ for the limits of *ab*, and not 10.502 and 10.500, for the same reason that we stated in the case of subtraction.

12. *The vulgar fractions of the limits may be removed by adding unity to the greater limit.*

Thus, instead of the preceding limits of *ab*, namely, $10.505\frac{857}{3215}$ and $10.496\frac{2364}{3216}$, we may put 10.506 and 10.496.

Thus far concerning accuracy; what follows concerns ease of working.

13. *The construction of every arithmetical progression is easy, not so, however, of every geometrical progression.*

This is evident, as an arithmetical progression is very easily formed by addition or subtraction; but a geometrical progression is continued by very difficult multiplications, divisions, or extractions of roots.

Those geometrical progressions alone are carried on easily which arise by subtraction of an easy part of the number from the whole number.

14. *We call easy parts of a number, any parts the denominators of which are made up of unity and a number of cyphers, such parts being obtained by rejecting as many of the figures at the end of the principal number as there are cyphers in the denominator.*

Thus the tenth, hundredth, thousandth, 10000th, 100000th, 1000000th, 10000000th parts are easily obtained, because the tenth part of any number is got by deleting its last figure, the hundredth its last two, the thousandth its last three figures, and so with the others, by always deleting as many of the figures at the end as there are cyphers in the denominator of the part. Thus the tenth part of 99321 is 9932, its hundredth part is 993, its thousandth 99, &c.

15. *The half, twentieth, two hundredth, and other parts denoted by the number two and cyphers, are also tolerably easily obtained; by rejecting as many of the figures at the end of the principal number as there are cyphers in the denominator, and dividing the remainder by two.*

Thus the 2000th part of the number 9973218045 is 4986609, the 20000th part is 498660.

16. *Hence it follows that if from radius with seven cyphers added you subtract its* 10000000th *part, and from the number thence arising its* 10000000th *part, and so on, a hundred numbers may very easily be continued geometrically in the proportion subsisting between radius and the sine less than it by unity, namely between* 10000000 *and* 9999999; *and this series of proportionals we name the First table.*

First table.

10000000.0000000
1.0000000
9999999.0000000
.9999999
9999998.0000001
.9999998
9999997.0000003

Thus from radius, with seven cyphers added for greater accuracy, namely, 10000000.0000000, subtract 1.0000000, you get 9999999.0000000; from this subtract .9999999, you get 9999998.0000001; and proceed in this way, as shown at the side, until you create a hundred proportion-

.9999997	als, the last of which, if you have computed
9999996.0000006	rightly, will be 9999900.0004950.

to be continued up to

9999900.0004950

17. *The Second table proceeds from radius with six cyphers added, through fifty other numbers decreasing proportionally in the proportion which is easiest, and as near as possible to that subsisting between the first and last numbers of the First table.*

Second table.

10000000.000000
100.000000
9999900.000000
99.999000
9999800.001000
99.998000
9999700.003000
99.997000
9999600.006000
&c., up to
9995001.222927

Thus the first and last numbers of the First table are 10000000.0000000 and 9999900.0004950, in which proportion it is difficult to form fifty proportional numbers. A near and at the same time an easy proportion is 100000 to 99999, which may be continued with sufficient exactness by adding six cyphers to radius and continually subtracting from each number its own 100000th part in the manner shown at the side; and this table contains, besides radius which is the first, fifty other proportional numbers, the last of which, if you have not erred, you will find to be 9995001.222927.

[This should be 9995001.224804 — see note.]

18. *The Third table consists of sixty-nine columns, and in each column are placed twenty-one numbers, proceeding in the proportion which is easiest, and as near as possible to that subsisting between the first and last numbers of the Second table.*

Whence its first column is very easily obtained from radius with five cyphers added, by subtracting its 2000th part, and so from the other numbers as they arise.

First column of Third table.

10000000.00000
5000.00000
9995000.00000

In forming this progression, as the proportion between 10000000.000000, the first of the Second table, and 9995001.222927, the last of the same, is

4997.50000
9990002.50000
4995.00125
9985007.49875
4992.50374
9980014.99501
&c., up to
9900473.57808

troublesome; therefore compute the twenty-one numbers in the easy proportion of 10000 to 9995, which is sufficiently near to it; the last of these, if you have not erred, will be 9900473.57808.

From these numbers, when computed, the last figure of each may be rejected without sensible error, so that others may hereafter be more easily computed from them.

19. *The first numbers of all the columns must proceed from radius with four cyphers added, in the proportion easiest and nearest to that subsisting between the first and the last numbers of the first column.*

As the first and the last numbers of the first column are 10000000.0000 and 9900473.5780, the easiest proportion very near to this is 100 to 99. Accordingly sixty-eight numbers are to be continued from radius in the ratio of 100 to 99 by subtracting from each one of them its hundredth part.

20. *In the same proportion a progression is to be made from the second number of the first column through the second numbers in all the columns, and from the third through the third, and from the fourth through the fourth, and from the others respectively through the others.*

Thus from any number in one column, by subtracting its hundredth part, the number of the same rank in the following column is made, and the numbers should be placed in order as follows:—

Proportionals of the Third Table.

First Column.	*Second Column.*
10000000.0000	9900000.0000
9995000.0000	9895050.0000
9990002.5000	9890102.4750
9985007.4987	9885157.4237
9980014.9950	9880214.8451
&c., continuously to	&c., descending to
9900473.5780	9801468.8423

Third Column.	*Thence 4th, 5th, &c., up to*	*69th column.*
9801000.0000	&c., up to	5048858.8900
9796099.5000	&c., up to	5046334.4605
9791201.4503	&c., up to	5043811.2932
9786305.8495	&c., up to	5041289.3879
9781412.6967	&c., up to	5038768.7435
&c., descending to		finally to
9703454.1539	finally to	4998609.4034

21. *Thus, in the Third table, between radius and half radius, you have sixty-eight numbers interpolated, in the proportion of* 100 *to* 99, *and between each two of these you have twenty numbers interpolated in the proportion of* 10000 *to* 9995; *and again, in the Second table, between the first two of these, namely between* 10000000 *and* 9995000, *you have fifty numbers interpolated in the proportion of* 100000 *to* 99999; *and finally, in the First table, between the latter, you have a hundred numbers interpolated in the proportion of radius, or* 10000000 *to* 9999999; *and since the difference of these is never more than unity, there is no need to divide it more minutely by interpolating means, whence these three tables, after they have been completed, will suffice for computing a Logarithmic table.*

Hitherto we have explained how we may most easily place in tables sines or natural numbers progressing in geometrical proportion.

22. *It remains, in the Third table at least, to place beside the sines or natural numbers decreasing geometrically their logarithms or artificial numbers increasing arithmetically.*

23. *To increase arithmetically is, in equal times, to be augmented by a quantity always the same.*

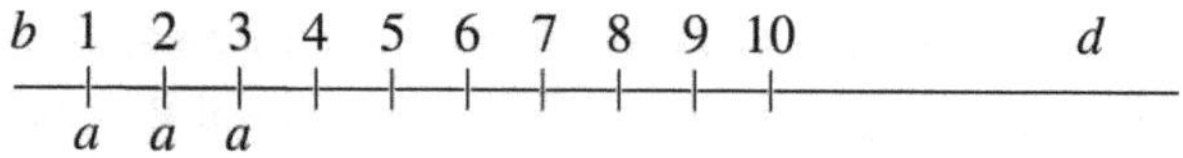

Thus from the fixed point b let a line be produced indefinitely in the direction of d. Along this let the point a travel from b towards d, moving according to this law, that in equal moments of time it is borne over the equal spaces $b1$, 12, 23, 34, 45, &c. Then we call this increase by $b1$, $b2$, $b3$, $b4$, $b5$, &c., arithmetical. Again, let $b1$ be represented in numbers by 10, $b2$ by 20, $b3$ by 30, $b4$ by 40, $b5$

by 50; then 10, 20, 30, 40, 50, &c., increase arithmetically, because we see they are always increased by an equal number in equal times.

24. *To decrease geometrically is this, that in equal times, first the whole quantity then each of its successive remainders is diminished, always by a like proportional part.*

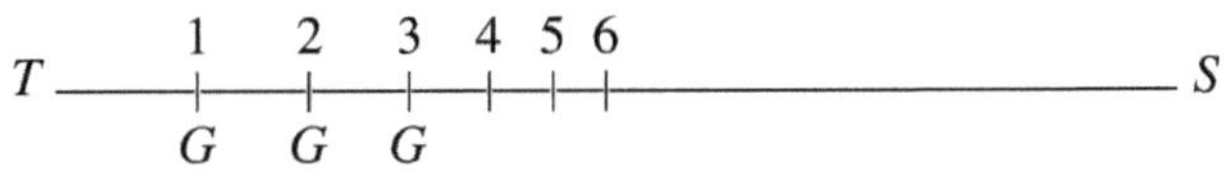

Thus let the line TS be radius. Along this let the point G travel in the direction of S, so that in equal times it is borne from T to 1, which for example may be the tenth part of TS; and from 1 to 2, the tenth part of 1S; and from 2 to 3, the tenth part of $2S$; and from 3 to 4, the tenth part of $3S$, and so on. Then the sines TS, $1S$, $2S$, $3S$, $4S$, &c., are said to decrease geometrically, because in equal times they are diminished by unequal spaces similarly proportioned. Let the sine TS be represented in numbers by 10000000, $1S$ by 9000000, $2S$ by 8100000, $3S$ by 7290000, $4S$ by 6561000; then these numbers are said to decrease geometrically, being diminished in equal times by a like proportion.

25. *Whence a geometrically moving point approaching a fixed one has its velocities proportionate to its distances from the fixed one.*

Thus, referring to the preceding figure, I say that when the geometrically moving point G is at T, its velocity is as the distance TS, and when G is at 1 its velocity is as $1S$, and when at 2 its velocity is as $2S$, and so of the others. Hence, whatever be the proportion of the distances TS, $1S$, $2S$, $3S$, $4S$, &c., to each other, that of the velocities of G at the points T, 1, 2, 3, 4, &c., to one another, will be the same.

For we observe that a moving point is declared more or less swift, according as it is seen to be borne over a greater or less space in equal times. Hence the ratio of the spaces traversed is necessarily the same as that of the velocities. But the ratio of the spaces traversed in equal times, $T1$, 12, 23, 34, 45, &c., is that of the distances TS, $1S$, $2S$, $3S$, $4S$, &c.[*] Hence it follows that the ratio to one another of the distances of G from S, namely TS, $1S$, $2S$, $3S$, $4S$, &c., is the same as that of the velocities of G at the points T, 1, 2, 3, 4, &c., respectively.

[*] It is evident that the ratio of the spaces traversed $T1$, 12, 23, 34, 45, &c., is that of the distances TS, $1S$, 2S, $3S$, $4S$, &c., for

when quantities are continued proportionally, their differences are also continued in the same proportion. Now the distances are by hypothesis continued proportionally, and the spaces traversed are their differences, wherefore it is proved that the spaces traversed are continued in the same ratio as the distances.

26. *The logarithm of a given sine is that number which has increased arithmetically with the same velocity throughout as that with which radius began to decrease geometrically, and in the same time as radius has decreased to the given sine.*

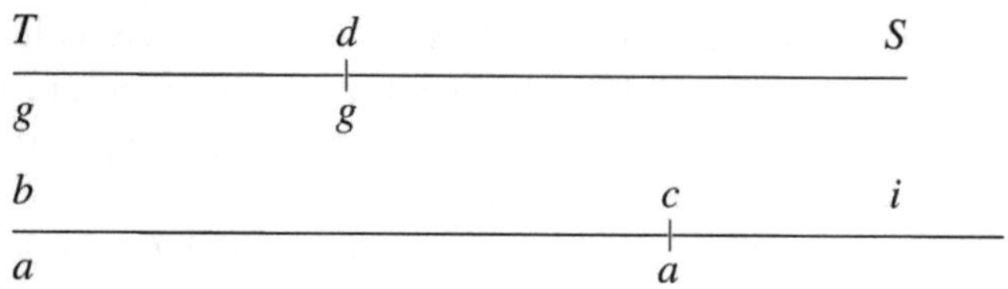

Let the line TS be radius, and dS a given sine in the same line; let g move geometrically from T to d in certain determinate moments of time. Again, let bi be another line, infinite towards i, along which, from b, let a move arithmetically with the same velocity as g had at first when at T; and from the fixed point b in the direction of i let a advance in just the same moments of time up to the point c. The number measuring the line bc is called the logarithm of the given sine dS.

27. *Whence nothing is the logarithm of radius.*

For, referring to the figure, when g is at T making its distance from S radius, the arithmetical point a beginning at b has never proceeded thence. Whence by the definition of distance nothing will be the logarithm of radius.

28. *Whence also it follows that the logarithm of any given sine is greater than the difference between radius and the given sine, and less than the difference between radius and the quantity which exceeds it in the ratio of radius to the given sine. And these differences are therefore called the limits of the logarithm.*

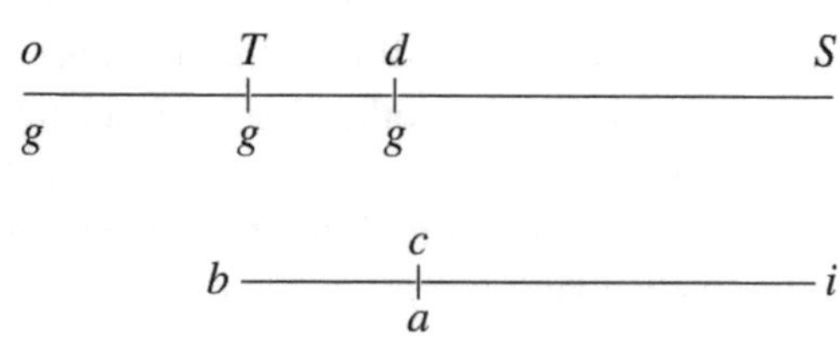

Thus, the preceding figure being repeated, and ST being produced beyond T to o, so that oS is to TS as TS to dS. I say that bc, the logarithm of the sine dS, is greater than Td and less than oT. For in the same time that g is borne from o to T, g is borne from T to d, because (by 24) oT is such a part of o as Td is of TS, and in the same time (by the definition of a logarithm) is a borne from b to c; so that oT, Td, and bc are distances traversed in equal times. But since g when moving between T and o is swifter than at T, and between T and d slower, but at T is equally swift with a (by 26); it follows that oT the distance traversed by g moving swiftly is greater, and Td the distance traversed by g moving slowly is less, than bc the distance traversed by the point a with its medium motion, in just the same moments of time; the latter is, consequently, a certain mean between the two former.

Therefore oT is called the greater limit, and Td the less limit of the logarithm which bc represents.

29. *Therefore to find the limits of the logarithm of a given sine.*

By the preceding it is proved that the given sine being subtracted from radius the less limit remains, and that radius being multiplied into the less limit and the product divided by the given sine, the greater limit is produced, as in the following example.

30. *Whence the first proportional of the First table, which is* 9999999, *has its logarithm between the limits* 1.0000001 *and* 1.0000000.

For (by 29) subtract 9999999 from radius with cyphers added, there will remain unity with its own cyphers for the less limit; this unity with cyphers being multiplied into radius, divide by 9999999 and there will result 1.0000001 for the greater limit, or if you require greater accuracy 1.00000010000001.

31. *The limits themselves differing insensibly, they or anything between them may be taken as the true logarithm.*

Thus in the above example, the logarithm of the sine 9999999 was found to be either 1.0000000 or 1.00000010, or best of all 1.00000005. For since the limits themselves, 1.0000000 and 1.0000001, differ from each other by an insensible fraction like $\frac{1}{10000000}$, therefore they and whatever is between them will differ still less from the true logarithm lying between these limits, and by a much more insensible error.

32. *There being any number of sines decreasing from radius in geometrical proportion, of one of which the logarithm or its limits is given, to find those of the others.*

This necessarily follows from the definitions of arithmetical increase, of geometrical decrease, and of a logarithm. For by these definitions, as the sines decrease continually in geometrical proportion, so at the same time their logarithms increase by equal additions in continuous arithmetical progression. Wherefore to any sine in the decreasing geometrical progression there corresponds a logarithm in the increasing arithmetical progression, namely the first to the first, and the second to the second, and so on.

So that, if the first logarithm corresponding to the first sine after radius be given, the second logarithm will be double of it, the third triple, and so of the others; until the logarithms of all the sines be known, as the following example will show.

33. *Hence the logarithms of all the proportional sines of the First table may be included between near limits, and consequently given with sufficient exactness.*

Thus since (by 27) the logarithm of radius is 0, and (by 30) the logarithm of 9999999, the first sine after radius in the First table, lies between the limits 1.0000001 and 10000000; necessarily the logarithm of 9999998.0000001, the second sine after radius, will be contained between the double of these limits, namely between 2.0000002 and 2.0000000; and the logarithm of 9999997.0000003, the third will be between the triple of the same, namely between 3.0000003 and 3.0000000. And so with the others, always by equally increasing the limits by the limits of the first, until you have completed the limits of the logarithms of all the proportionals of the First table. You may in this way, if you please, continue the logarithms themselves in an exactly similar progression with little and insensible error; in which case the logarithm of radius will be 0, the logarithm of the first sine after radius (by 31) will be 1.00000005, of the second 2.00000010, of the third 3.00000015, and so of the rest.

34. *The difference of the logarithms of radius and a given sine is the logarithm of the given sine itself.*

This is evident, for (by 27) the logarithm of radius is nothing, and when nothing is subtracted from the logarithm of a given sine, the logarithm of the given sine necessarily remains entire.

35. *The difference of the logarithms of two sines must be added to the logarithm of the greater that you may have the logarithm of the less, and subtracted from the logarithm of the less that you may have the logarithm of the greater.*

Necessarily this is so, since the logarithms increase as the sines decrease, and the less logarithm is the logarithm of the greater sine, and the greater logarithm of the less sine. And therefore it is right to add the difference to the less logarithm, that you may have the greater logarithm though corresponding to the less sine, and on the other hand to subtract the difference from the greater logarithm that you may have the less logarithm though corresponding to the greater sine.

36. *The logarithms of similarly proportioned sines are equidifferent.*

This necessarily follows from the definitions of a logarithm and of the two motions. For since by these definitions arithmetical increase always the same corresponds to geometrical decrease similarly proportioned, of necessity we conclude that equidifferent logarithms and their limits correspond to similarly proportioned sines. As in the above example from the First table, since there is a like proportion between 9999999.0000000 the first proportional after radius, and 9999997.0000003 the third, to that which is between 9999996.0000006 the fourth and 9999994.0000015 the sixth; therefore 1.00000005 the logarithm of the first differs from 3.00000015 the logarithm of the third, by the same difference that 4.00000020 the logarithm of the fourth, differs from 6.00000030 the logarithm of the sixth proportional. Also there is the same ratio of equality between the differences of the respective limits of the logarithms, namely as the differences of the less among themselves, so also of the greater among themselves, of which logarithms the sines are similarly proportioned.

37. *Of three sines continued in geometrical proportion, as the square of the mean equals the product of the extremes, so of their logarithms the double of the mean equals the sum of the extremes. Whence any two of these logarithms being given, the third becomes known.*

Of the three sines, since the ratio between the first and the second is that between the second and the third, therefore (by 36), of their logarithms, the difference between the first and the second is that between the second and the third. For example, let the first logarithm be represented by the line *bc*, the second by the line *bd*, the third by the line *be*, all placed in the one line *bcde*, thus:—

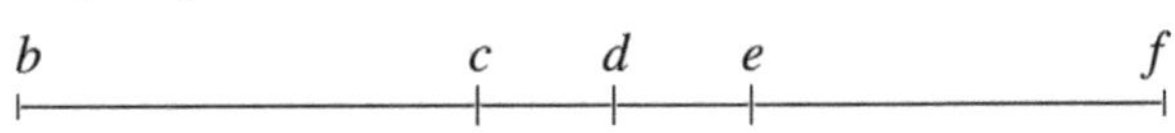

and let the differences *cd* and *de* be equal. Let *bd*, the mean of them, be doubled by producing the line from *b* beyond *e* to *f*, so that bf is double bd. Then bf is equal to both the lines bc of the first logarithm and be of the third, for from the equals bd and df take away the equals cd and de, namely cd from bd and de from df, and there will remain bc and ef necessarily equal. Thus since the whole bf is equal to both be and ef, therefore also it will be equal to both be and bc, which was to be proved. Whence follows the rule, if of three logarithms you double the given mean, and from this subtract a given extreme, the remaining extreme sought for becomes known; and if you add the given extremes and divide the sum by two, the mean becomes known.

38. *Of four geometrical proportionals, as the product of the means is equal to the product of the extremes, so of their logarithms, the sum of the means is equal to the sum of the extremes. Whence any three of these logarithms being given, the fourth becomes known.*

Of the four proportionals, since the ratio between the first and second is that between the third and fourth; therefore of their logarithms (by 36), the difference between the first and second is that between the third and fourth. Hence let such quantities be taken in the line bf as that ba may represent the first logarithm, bc the second, be the third, and bg the fourth, making the differences ac

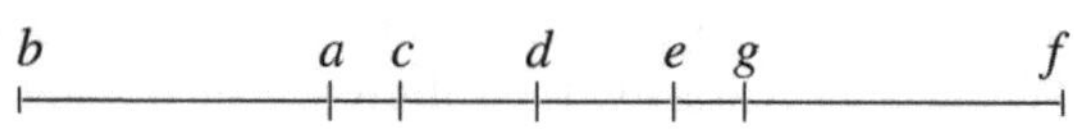

and eg equal, so that d placed in the middle of ce is of necessity also placed in the middle of ag. Then the sum of bc the second and be the third is equal to the sum of ba the first and bg the fourth. For (by 37) the double of bd, which is bf, is equal to bc and be together, because their differences from bd, namely cd and de, are equal; for the same reason the same bf is also equal to ba and bg together, because their differences from bd, namely ad and dg, are also equal. Since, therefore, both the sum of ba and bg and the sum of bc and be are equal to the double of bd, which is bf, therefore also they are equal to each other, which was to be proved. Whence follows the rule, of these four logarithms if you subtract a known mean from the sum of the known extremes, there is left the mean sought for; and if you subtract a known extreme from the sum of the known means, there is left the extreme sought for.

39. *The difference of the logarithms of two sines lies between two limits; the greater limit being to the radius as the difference of the sines to the less sine,*

and the less limit being to radius as the difference of the sines to the greater sine.

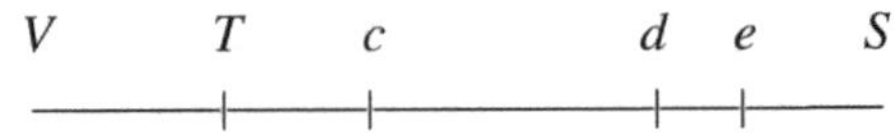

Let TS be radius, dS the greater of two given sines, and eS the less. Beyond ST let the distance TV be marked off by the point V, so that ST is to TV as eS, the less sine, is to de, the difference of the sines. Again, on the other side of T, towards S, let the distance Tc be marked off by the point c, so that TS is to Tc as dS, the greater sine, is to de, the difference of the sines. Then the difference of the logarithms of the sines dS and eS lies between the limits VT the greater and Tc the less. For by hypothesis, eS is to de as TS to TV, and dS is to de as TS to Tc; therefore, from the nature of proportionals, two conclusions follow:—

Firstly, that VS is to TS as TS to cS.

Secondly, that the ratio of TS to cS is the same as that of dS to eS. And therefore (by 36) the difference of the logarithms of the sines dS and eS is equal to the difference of the logarithms of the radius TS and the sine cS. But (by 34) this difference is the logarithm of the sine cS itself; and (by 28) this logarithm is included between the limits VT the greater and Tc the less, because by the first conclusion above stated, VS greater than radius is to TS radius as TS is to cS. Whence, necessarily, the difference of the logarithms of the sines dS and eS lies between the limits VT the greater and Tc the less, which was to be proved.

40. *To find the limits of the difference of the logarithms of two given sines.*

Since (by 39) the less sine is to the difference of the sines as radius to the greater limit of the difference of the logarithms; and the greater sine is to the difference of the sines as radius to the less limit of the difference of the logarithms; it follows, from the nature of proportionals, that radius being multiplied by the difference of the given sines and the product being divided by the less sine, the greater limit will be produced; and the product being divided by the greater sine, the less limit will be produced.

Example.

THUS, let the greater of the given sines be 9999975.5000000, and the less 9999975.0000300, the difference of these .4999700 being multiplied into radius (cyphers to the eighth place after the point being first added to both for the purpose of demonstration, although otherwise seven are sufficient), if you divide the product by the greater sine, namely 9999975.5000000, there will come out for the less limit .49997122, with eight figures after the point; again, if you divide the product by the less sine, namely 9999975.0000300, there will come out for the greater limit .49997124; and, as already proved, the difference of the logarithms of the given sines lies between these. But since the extension of these fractions to the eighth figure beyond the point is greater accuracy than is required, especially as only seven figures are placed after the point in the sines; therefore, that eighth or last figure of both being deleted, then the two limits and also the difference itself of the logarithms will be denoted by the fraction .4999712 without even the smallest particle of sensible error.

41. *To find the logarithms of sines or natural numbers not proportionals in the First table, but near or between them*; *or at least, to find limits to them separated by an insensible difference.*

Write down the sine in the First table nearest to the given sine, whether less or greater. Seek out the limits of the table sine (by 33), and when found note them down. Then seek out the limits of the difference of the logarithms of the given sine and the table sine (by 40), either both limits or one or other of them, since they are almost equal, as is evident from the above example. Now these, or either of them, being found, add to them the limits above noted down, or else subtract (by 8, 10, and 35), according as the given sine is less or greater than the table sine. The numbers thence produced will be near limits between which is included the logarithm of the given sine.

EXAMPLE.

Let the given sine be 9999975.5000000, to which the nearest sine in the table is 9999975.0000300, less than the given sine. By 33 the limits of the logarithm of the latter are 25.0000025 and 25.0000000. Again (by 40), the difference of the logarithms of the given sine and the table sine is .4999712. By 35, subtract this from the above limits, which are the limits of the less sine, and there will come out 24.5000313 and 24.5000288, the required limits of the logarithm of the given sine 9999975.5000000. Accordingly the actual logarithm of the sine may be placed without sensible error in either of the limits, or best of all (by 31) in 24.5000300.

ANOTHER EXAMPLE.

LET the given sine be 9999900.0000000, the table sine nearest it 9999900.0004950. By 33 the limits of the logarithm of the latter are 100.0000100 and 100.0000000. Then (by 40) the difference of the logarithms of the sines will be .0004950. Add this (by 35) to the above limits and they become 100.0005050 for the greater limit, and 100.0004950 for the less limit, between which the required logarithm of the given sine is included.

42. *Hence it follows that the logarithms of all the proportionals in the Second table may be found with sufficient exactness, or may be included between known limits differing by an insensible fraction.*

THUS since the logarithm of the sine 9999900, the first proportional of the Second table, was shown in the preceding example to lie between the limits 100.0005050 and 100.0004950; necessarily (by 32) the logarithm of the second proportional will lie between the limits 200.0010100 and 200.0009900; and the logarithm of the third proportional between the limits 300.0015150 and 300.0014850, &c. And finally, the logarithm of the last sine of the Second table, namely 9995001.222927, is included between the limits 5000.0252500 and 5000.0247500. Now, having all these limits, you will be able (by 3l) to find the actual logarithms.

43. *To find the logarithms of sines or natural numbers not proportionals in the Second table, but near or between them; or to include them between known limits differing by an insensible fraction.*

Write down the sine in the Second table nearest the given sine, whether greater or less. By 42 find the limits of the logarithm of the table sine. Then by the rule of proportion seek for a fourth proportional, which shall be to radius as the less of the given and table sines is to the greater. This may be done in one way by multiplying the less sine into radius and dividing the product by the greater. Or, in an easier way, by multiplying the difference of the sines into radius, dividing this product by the greater sine, and subtracting the quotient from radius.

Now since (by 36) the logarithm of the fourth proportional differs from the logarithm of radius by as much as the logarithms of the given and table sines differ from each other; also, since (by 34) the former difference is the same as the logarithm of the fourth proportional itself; therefore (by 41) seek for the limits of the logarithm of the fourth proportional by aid of the First table; when found add them to the limits of the logarithm of the table sine, or else subtract them (by 8, 10, and 35), according as the table sine is greater or less than the given sine; and there will be brought out the limits of the logarithm of the given sine.

EXAMPLE.

THUS, let the given sine be 9995000.000000. To this the nearest sine in the Second table is 9995001.222927, and (by 42) the limits of its logarithm are 5000.0252500 and 5000.0247500. Now seek for the fourth proportional by either of the methods above described; it will be 9999998.7764614, and the limits of its logarithm found (by 41) from the First table will be 1.2235387 and 1.2235386. Add these limits to the former (by 8 and 35), and there will come out 5001.2487888 and 5001.2482886 as the limits of the logarithm of the given sine. Whence the number 5001.2485387, midway between them, is (by 31) taken most suitably, and with no sensible error, for the actual logarithm of the given sine 9995000.

44. *Hence it follows that the logarithms of all the proportionals in the first column of the Third table may be found with sufficient exactness, or may be included between known limits differing by an insensible fraction.*

For, since (by 43) the logarithm of 9995000, the first proportional after radius in the first column of the Third table, is 5001.2485387 with no sensible error; therefore (by 32) the logarithm of the second proportional, namely 9990002.5000, will be 10002.4970774; and so of the others, proceeding up to the last in the column, namely 9900473.57808, the logarithm of which, for a like reason, will be 100024.9707740, and its limits will be 100024.9657720 and 100024.9757760.

45. *To find the logarithms of natural numbers or sines not proportionals in the first column of the Third table, but near or between them, or to include them between known limits differing by an insensible fraction.*

Write down the sine in the first column of the Third table nearest the given sine, whether greater or less. By 44 seek for the limits of the logarithm of the table sine. Then, by one of the methods described in 43, seek for a fourth proportional, which shall be to radius as the less of the given and table sines is to the greater. Having found the fourth proportional, seek (by 43) for the limits of its logarithm from the Second table. When these are found, add them to the limits of the logarithm of the table sine found above, or else subtract them (by 8, 10, and 35), and the limits of the logarithm of the given sine will be brought out.

EXAMPLE.

THUS, let the given sine be 9900000. The proportional sine nearest it in the first column of the Third table is 9900473.57808. Of this (by 44) the limits of the logarithm are 100024.9657720 and 100024.9757760. Then the fourth proportional will be 9999521. 6611850. Of this the limits of the logarithm, deduced from the Second table (by 43), are 478.3502290 and 478.3502812. These limits (by 8 and 35) being added to the above limits of the logarithm of the table sine, there will come out the limits 100503.3260572 and 100503.3160010, between which necessarily falls the logarithm sought for. Whence the number midway between them, which is 100503.3210291, may be put without sensible error for the true logarithm of the given sine 9900000.

46. *Hence it follows that the logarithms of all the proportionals of the Third table may be given with sufficient exactness.*

For, as (by 45) 100503.3210291 is the logarithm of the first sine in the second column, namely 9900000; and since the other first sines of the remaining columns progress in the same proportion, necessarily (by 32 and 36) the logarithms of these increase always by the same difference 100503.3210291, which is added to the logarithm last found, that the following may be made. Therefore, the first logarithms of all the columns being obtained in this way, and all the logarithms of the first column being obtained by 44, you may choose whether you prefer to build up, at one time, all the logarithms in the same column, by continuously adding 5001.2485387, the difference of the logarithms, to the last found logarithm in the column, that the next lower logarithm in the same column be made; or whether you prefer to compute, at one time, all the logarithms of the same rank, namely all the second logarithms in each of the columns, then all the third, then the fourth, and so the others, by continuously adding 100503.3210291 to the logarithm in one column, that the logarithm of the same rank in the next column be brought out. For by either method may be had the logarithms of all the proportionals in this table; the last of which is 6934250.8007528, corresponding to the sine 4998609.4034.

47. *In the Third table, beside the natural numbers, are to be written their logarithms, so that the Third table, which after this we shall always call the Radical table, may be made complete and perfect.*

This writing up of the table is to be done by arranging the columns in the number and order described (in 20 and 21), and by dividing each into two sections, the first of which should contain the geometrical proportionals we call sines and natural numbers, the second their logarithms progressing arithmetically by equal intervals.

THE RADICAL TABLE.

First column.		*Second column.*	
Natural numbers.	Logarithms.	Natural numbers.	Logarithms.
10000000.0000	.0	9900000.0000	100503.3
9995000.0000	5001.2	9895050.0000	100504.6
9990002.5000	10002.5	9890102.4750	110505.8
9985007.4987	15003.7	9885157.4237	115507.1
9980014.9950	20005.0	9880214.8451	120508.3
&c., up to	up to	up to	up to
9900473.5780	100025.0	9801468.8423	200528.2

	69th column.	
	Natural numbers.	Logarithms.
	5048858.8900	6834225.8
	5046334.4605	6839227.1
	5043811.2932	6844228.3
and the others, up to	5041289.3879	6849229.6
	5038768.7435	6854230.8
	up to	up to
	4998609.4034	6934250.8

For shortness, however, two things should be borne in mind:—First, that in these logarithms it is enough to leave one figure after the point, the remaining six being now rejected, which, however, if you had neglected at the beginning, the error arising thence by frequent multiplications in the previous tables would have grown intolerable in the third. Secondly, if the second figure after the point exceed the number four, the first figure after the point, which alone is retained, is to be increased by unity: thus for 10002.48 it is more correct to put 10002.5 than 10002.4; and for 1000.35001 we more fitly put 1000.4 than 1000.3. Now, therefore, continue the Radical table in the manner which has been set forth.

48. *The Radical table being now completed, we take the numbers for the logarithmic table from it alone.*

For as the first two tables were of service in the formation of the third, so this third Radical table serves for the construction of the principal Logarithmic table, with great ease and no sensible error.

49. *To find most easily the logarithms of sines greater than* 9996700.

This is done simply by the subtraction of the given sine from radius. For (by 29) the logarithm of the sine 9996700 lies between the limits 3300 and 3301; and these limits, since they differ from each other by unity only, cannot differ from their true logarithm by any sensible error, that is to say, by an error greater than unity. Whence 3300, the less limit, which we obtain simply by subtraction, may be taken for the true logarithm. The method is necessarily the same for all sines greater than this.

50. *To find the logarithms of all sines embraced within the limits of the Radical table.*

Multiply the difference of the given sine and table sine nearest it by radius. Divide the product by the easiest divisor, which may be either the given sine or the table sine nearest it, or a sine between both, however placed. By 39 there will be produced either the greater or less limit of the difference of the logarithms, or else something intermediate, no one of which will differ by a sensible error from the true difference of the logarithms on account of the nearness of the numbers in the table. Wherefore (by 35), add the result, whatever it may be, to the logarithm of the table sine, if the given sine be less than the table sine; if not, subtract the result from the logarithm of the table sine, and there will be produced the required logarithm of the given sine.

EXAMPLE.

THUS let the given sine be 7489557, of which the logarithm is required. The table sine nearest it is 7490786.6119. From this subtract the former with cyphers added thus, 7489557.0000, and there remains 1229.6119. This being multiplied by radius, divide by the easiest number, which may be either 7489557.0000 or 7490786.6119, or still better by something between them, such as 7490000, and by a most easy division there will be produced 1640.1. Since the given sine is less than the table sine, add this to the logarithm of the table sine, namely to 2889111.7, and there will result 2890751.8, which equals $2890751\frac{4}{5}$. But since the principal table admits neither fractions nor anything beyond the point, we put for it 2890752, which is the required logarithm.

Another Example.

LET the given sine be 7071068.0000. The table sine nearest it will be 7070084.4434. The difference of these is 983.5566. This being multiplied by radius, you most fitly divide the product by 7071000, which lies between the given and table sines, and there comes out 1390.9. Since the given sine exceeds the table sine, let this be subtracted from the logarithm of the table sine, namely from 3467125.4, which is given in the table, and there will remain 3465734.5. Wherefore 3465735 is assigned for the required logarithm of the given sine 7071068. Thus the liberty of choosing a divisor produces wonderful facility.

51. *All sines in the proportion of two to one have* 6931469.22 *for the difference of their logarithms.*

For since the ratio of every sine to its half is the same as that of radius to 5000000, therefore (by 36) the difference of the logarithms of any sine and of its half is the same as the difference of the logarithms of radius and of its half 5000000. But (by 34) the difference of the logarithms of radius and of the sine 5000000 is the same as the logarithm itself of the sine 5000000, and this logarithm (by 50) will be 6931469.22. Therefore, also, 6931469.22 will be the difference of all logarithms whose sines are in the proportion of two to one. Consequently the double of it, namely 13862938.44, will be the difference of all logarithms whose sines are in the ratio of four to one; and the triple of it, namely 20794407.66, will be the difference of all logarithms whose sines are in the ratio of eight to one.

52. *All sines in the proportion of ten to one have* 23025842.34 *for the difference of their logarithms.*

For (by 50) the sine 8000000 will have for its logarithm 2231434.68; and (by 51) the difference between the logarithms of the sine 8000000 and of its eighth part 1000000, will be 20794407.66; whence by addition will be produced 23025842.34 for the logarithm of the sine 1000000. And since radius is ten times this, all sines in the ratio of ten to one will have the same difference, 23025842.34, between their logarithms, for the reason and cause already stated (in 51) in reference to the proportion of two to one. And consequently the double of this logarithm, namely 46051684.68, will, as regards the difference of the logarithms, correspond to the proportion of a hundred to one; and the triple of the same, namely 69077527.02, will be the difference of all logarithms whose sines are in the ratio of a thousand to one; and so of the ratio ten thousand to one, and of the others as below.

53. *Whence all sines in a ratio compounded of the ratios two to one and ten to one, have the difference of their logarithms formed from the differences* 6931469.22 *and* 23025842.34 *in the way shown in the following*

Short Table.

Given Proportions of sines.	Corresponding Differences of Logarithm.	Given Proportions of sines.	Corresponding Differences of Logarithm.
Two to one	6931469.22	8000 to one	89871934.68
Four ”	13862938.44	10000 ”	92103369.36
Eight ”	20794407.66	20000 ”	99034838.58
Ten ”	23025842.34	40000 ”	105966307.80
20 ”	29957311.56	80000 ”	112897777.02
40 ”	36888780.78	100000 ”	115129211.70
80 ”	43820250.00	200000 ”	122060680.92
A hundred ”	46051684.68	400000 ”	128992150.14
200 ”	52983153.90	800000 ”	135923619.36
400 ”	59914623.12	1000000 ”	138155054.04
800 ”	66846092.34	2000000 ”	145086523.26
A thousand ”	69077527.02	4000000 ”	152017992.48
2000 ”	76008996.24	8000000 ”	158949461.70
4000 ”	82940465.46	10000000 ”	161180896.38

54. *To find the logarithms of all sines which are outside the limits of the Radical table.*

This is easily done by multiplying the given sine by 2, 4, 8, 10, 20, 40, 50, 100, 200, or any other proportional number you please, contained in the short table, until you obtain a number within the limits of the Radical table. By 50 find the logarithm of this sine now contained in the table, and then add to it the logarithmic difference which the short table indicates as required by the preceding multiplication.

EXAMPLE.

IT is required to find the logarithm of the sine 378064. Since this sine is outside the limits of the Radical table, let it be multiplied by some proportional number in the foregoing short table, as by 20, when it will become 7561280. As this now falls within the Radical table, seek for its logarithm (by 50) and you will obtain 2795444.9 to which add 29957311.56, the difference in the short table corresponding to the proportion of twenty to one, and you have 32752756.4. Wherefore 32752756 is the required logarithm of the given sine 378064.

55. *As half radius is to the sine of half a given arc, so is the sine of the complement of the half arc to the sine of the whole arc.*

Let *ab* be radius, and *abc* its double, on which as diameter is described a semicircle. On this lay off the given arc *ae*, bisect it in *d*, and from *e* in the direction of *c* lay off *eh*, the complement of *de*, half the given arc. Then *hc* is necessarily equal to *eh*, since the quadrant *deh* must equal the remaining quadrant made up of the arcs *ad* and *hc*. Draw *ei* perpendicular to *aic*, then *ei* is the sine of the arc *ade*. Draw *ae*; its half, *fe*, is the sine of the arc *de*, the half of the arc *ade*. Draw *ec*; its half, *eg*, is the sine of the arc *eh*, and is therefore the sine of the complement of the arc *de*. Finally, make *ak* half the radius *ab*. Then as *ak* is to *ef*, so is *eg* to *ei*. For the two triangles *cea* and *cie* are equiangular, since *ice* or *ace* is common to both; and *cie* and *cea* are each a right angle, the former by hypothesis, the latter because it is in the circumference and occupies a semicircle. Hence *ac*, the hypotenuse of the triangle *cea*, is to *ae*, its less side, as *ec*, the hypotenuse of the triangle *cie*, is to *ei* its less side. And since *ac*, the whole, is to *ae*

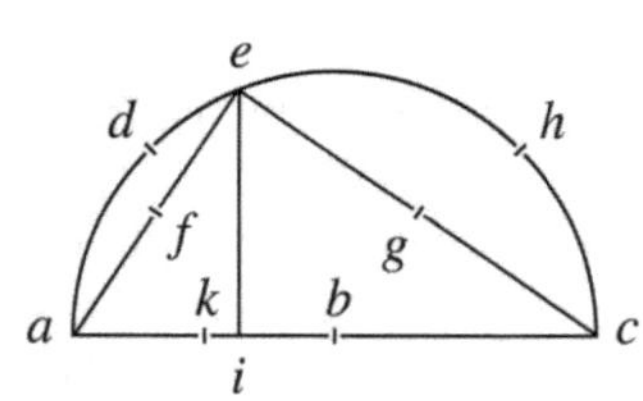

as *ec*, the whole, is to *ei*, it follows that *ab*, half of *ac*, is to *ae* as *eg*, half of *ec*, is to *ei*. And now, finally, since *ab*, the whole, is to *ae*, the whole, as *eg* is to *ei*, we necessarily conclude that *ak*, half of *ab*, is to *fe*, half of *ae*, as *eg* is to *ei*.

56. *Double the logarithm of an arc of* 45 *degrees is the logarithm of half radius.*

Referring to the preceding figure, let the case be such that *ae* and *ec* are equal. In that case *i* will fall on *b*, and *ei* will be radius; also *ef* and *eg* will be equal, each of them being the sine of 45 degrees. Now (by 55) the ratio of *ak*, half radius, to *ef*, a sine of 45 degrees, is likewise the ratio of *eg*, also a sine of 45 degrees, to *ei*, now radius. Consequently (by 37) double the logarithm of the sine of 45 degrees is equal to the logarithms of the extremes, namely radius and its half. But the sum of the logarithms of both these is the logarithm of half radius only, because (by 27) the logarithm of radius is nothing. Necessarily, therefore, the double of the logarithm of an arc of 45 degrees is the logarithm of half radius.

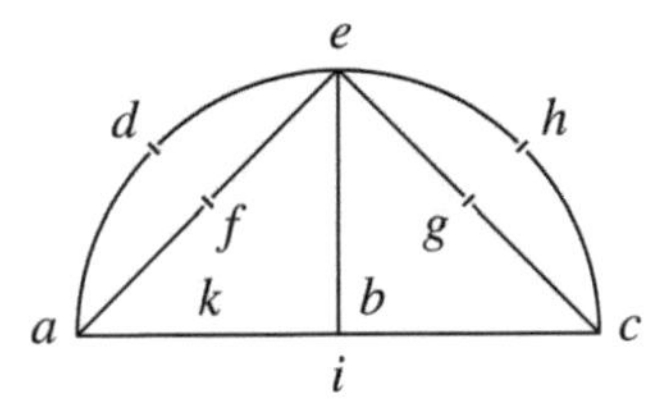

57. *The sum of the logarithms of half radius and any given arc is equal to the sum of the logarithms of half the arc and the complement of the half arc. Whence the logarithm of the half arc may be found if the logarithms of the other three be given.*

Since (by 55) half radius is to the sine of half the given arc as the sine of the complement of that half arc is to the sine of the whole arc, therefore (by 38) the sum of the logarithms of the two extremes, namely half radius and the whole arc, will be equal to the sum of the logarithms of the means, namely the half arc and the complement of the half arc. Whence, also (by 38), if you add the logarithm of half radius, found by 51 or 56, to the given logarithm of the whole arc, and subtract the given logarithm of the complement of the half arc, there will remain the required logarithm of the half arc.

EXAMPLE.

LET there be given the logarithm of half radius (by 51) 6931469; also the arc 69 degrees 20 minutes, and its logarithm 665143. The half arc is 34 degrees 40 minutes, whose logarithm is required. The complement of the half arc is 55 degrees 20 minutes, and its logarithm 1954370 is given. Wherefore add 6931469 to 665143, making 7596612, subtract 1954370, and there remains 5642242, the required logarithm of an arc of 34 degrees 40 minutes.

58. *When the logarithms of all arcs not less than* 45 *degrees are given, the logarithms of all less arcs are very easily obtained.*

From the logarithms of all arcs not less than 45 degrees, given by hypothesis, you can obtain (by 57) the logarithms of all the remaining arcs decreasing down to 22 degrees 30 minutes. From these, again, may be had in like manner the logarithms of arcs down to 11 degrees 15 minutes. And from these the logarithms of arcs down to 5 degrees 38 minutes. And so on, successively, down to 1 minute.

59. *To form a logarithmic table.*

Prepare forty-five pages, somewhat long in shape, so that besides margins at the top and bottom, they may hold sixty lines of figures. Divide each page into twenty equal spaces by horizontal lines, so that each space may hold three lines of figures. Then divide each page into seven columns by vertical lines, double lines being ruled between the second and third columns and between the fifth and sixth, but a single line only between the others.

Next write on the first page, at the top to the left, over the first three columns, "0 *degrees*", and at the bottom to the right, under the last three columns, "89 *degrees*". On the second page, above, to the left, "1 *degree*", and below, to the right, "88 *degrees*". On the third page, above, "2 *degrees*", and below, "87 *degrees*". Proceed thus with the other pages, so that the number written above, added to that written below, may always make up a quadrant, less 1 degree or 89 degrees.

Then, on each page write, at the head of the first column, "*Minutes of the degree written above*"; at the head of the second column, "*Sines of the arcs to the left*"; at the head of the third column, "*Logarithms of the arcs to the left*"; at both the head and the foot of the third column, "*Difference between the logarithms of the complementary arcs*"; at the foot of the fifth column, "*Logarithms of the arcs to the right*"; at the foot of the sixth column, "*Sines of the arcs to the right*"; and at the foot of the seventh column, "*Minutes of the degree written beneath*".

Then enter in the first column the numbers of minutes in ascending order from 0 to 60, and in the seventh column the number of minutes in descending order from 60 to 0; so that any pair of minutes placed opposite, in the first and seventh columns in the same line, may make up a whole degree or 60 minutes; for example, enter 0 opposite to 60, 1 to 59, 2 to 58, and 3 to 57, placing three numbers in each of the twenty intervals between the horizontal lines. In the second column enter the values of the sines corresponding to the degree at the top and the minutes in the same line to the left; also in the sixth column enter the values of the sines corresponding to the degree at the bottom and the minutes in the same line to the right. Reinhold's common table of sines, or any other more exact, will supply you with these values.

Having done this, compute, by 49 and 50, the logarithms of all sines between radius and its half, and by 54, the logarithms of the other sines; however, you may, with both greater accuracy and facility, compute, by the same 49 and 50, the logarithms of all sines between radius and the sine of 45 degrees, and from these, by 58, you very readily obtain the logarithms of all remaining arcs less than 45 degrees. Having computed these by either method, enter in the third column the logarithms corresponding to the degree at the top and the minutes to the left, and to their sines in the same line at left side; similarly enter in the fifth column the logarithm corresponding to the degree at the bottom and the minutes to the right and to their sines in the same line at right side.

Finally, to form the middle column, subtract each logarithm on the right from the logarithm on the left in the same line, and enter the difference in the same line, between both, until the whole is completed.

We have computed this Table to each minute of the quadrant, and we leave the more exact elaboration of it, as well as the emendation of the table of sines, to the learned to whom more leisure may be given.

Outline of the Construction, in another form, of a Logarithmic Table.

60. SINCE *the logarithms found by* 54 *sometimes differ from those found by* 58 (*for example, the logarithm of the sine* 378064 *is* 32752756 *by the former, while by the latter it is* 32752741), *it would seem that the table of sines is in some places faulty. Wherefore I advise the learned, who perchance may have plenty of pupils and computors, to publish a table of sines more reliable and with larger numbers, in which radius is made* 100000000, *that is with eight cyphers after the unit instead of seven only. Then, let the First table, like ours, contain a hundred numbers progressing in the proportion of the new radius to the sine less than it by unity, namely of* 100000000 *to* 999999999.

Let the Second table also contain a hundred numbers in the proportion of this new radius to the number less than it by a hundred, namely of 100000000 *to* 99999900.

Let the Third table, also called the Radical table, contain thirty-five columns with a hundred numbers in each column, and let the hundred numbers in each column progress in the proportion of ten thousand to the number less than it by unity, namely of 100000000 *to* 99990000.

Let the thirty-five proportionals standing first in all the columns, or occupying the second, third, or other rank, progress among themselves in the proportion of 100 *to* 99, *or of the new radius* 100000000 *to* 990000000.

In continuing these proportionals and finding their logarithms, let the other rules we have laid down be observed. From the Radical table completed in this way, you will find with great exactness (*by* 49 *and* 50) *the logarithms of all sines between radius and the sine of* 45 *degrees, from the arc of* 45 *degrees doubled, you will find by* (56) *the logarithm of half radius, having obtained all these, you will find the other logarithms by* 58. *Arrange all these results as described in* 59, *and you will produce a Table, certainly the most excellent of all Mathematical tables, and prepared for the most important uses.*

End of the Construction of the Logarithmic Table.

APPENDIX.

On the Construction of another and *better kind of* LOGARITHMS, *namely* one in which the Logarithm of unity is 0.

AMONG *the various improvements of* Logarithms, *the more important is that which adopts a cypher as the* Logarithm *of unity and,* 10,000,000,000 *as the* Logarithm *of either one tenth of unity or ten times unity. Then, these being once fixed, the* Logarithms *of all other numbers necessarily follow. But the methods of finding them are various, of which the first is as follows:—*

Divide the given Logarithm *of a tenth, or of ten, namely* 10,000,000,000, *by* 5 *ten times successively, and thereby the following numbers will be produced,* 2000000000, 4000000000, 80000000, 16000000, 3200000, 640000, 128000, 25600, 5120, 1024. *Also divide the last of these by* 2, *ten times successively, and there will be produced* 512, 256, 128, 64, 32, 16, 8, 4, 2, 1. *Moreover all these numbers are logarithms.*

Thereupon let us seek for the common numbers which correspond to each of them in order. Accordingly, between a tenth and unity, or between ten and unity (adding for the purpose of calculation as many cyphers as you wish, say twelve), find four mean proportionals, or rather the least of them, by extracting the fifth root, which for ease in demonstration call A. *Similarly, between* A *and unity, find the least of four mean proportionals, which call* B. *Between* B *and unity find four means, or the least of them, which call* C. *And thus proceed, by the extraction of the fifth root, dividing the interval between that last found and unity unto five proportional intervals, or into four means, of all which let the fourth or least be always noted down, until you come to the tenth least mean, and let them be denoted by the letters* D, E, F, G, H, I, K.

When these proportionals have been accurately computed, proceed also to find the mean proportional between K *and unity, which call* L. *Then find the mean proportional between* L *and unity, which call* M. *Then in like manner a mean between* M *and unity, which call* N. *In the same way, by extraction of the square root, may be formed between each last found number and unity, the rest of the intermediate proportionals, to be denoted by the letters* O, P, Q, R, S, T, V.

To each of these proportionals in order corresponds its Logarithm *in the first series. Whence* 1 *will be the* Logarithm *of the number* V, *whatever it may turn out to be, and* 2 *will be the* Logarithm *of the number* T, *and 4 of the number* S, *and* 8 *of the number* R, 6 *of the number* Q, 32 *of the number* P, 64 *of the number* O, 128 *of the number* N, 256 *of the number* M, 512 *of the number* L, 1024 *of the number* K; *all of which is manifest from the above construction.*

From these, once computed, there may then be formed both the proportionals of other Logarithms *and the* Logarithms *of other proportionals.*

For as in statics, from weights of 1 *of* 2, *of* 4, *of* 8, *and of other like numbers of pounds in the same proportion, every number of pounds weight, which to us now are* Logarithms, *may be formed by addition, so, from the proportionals* V, T, S, R, *&c., which correspond to them, and from others also to be formed in duplicate ratio, the proportionals corresponding to every proposed* Logarithm *may be formed by corresponding multiplication of them among themselves, as experience will show.*

The special difficulty of this method, however, is in finding the ten proportionals to twelve places by extraction of the fifth root from sixty places, but though this method is considerably more difficult, it is correspondingly more exact for finding both the Logarithms *of proportionals and the proportionals of* Logarithms.

Another method for the easy construction of the LOGARITHMS of composite numbers, when the LOGARITHMS of their primes are known.

IF *two numbers with known* Logarithms *be multiplied together, forming a third; the sum of their* Logarithms *will be the* Logarithm *of the third.*

Also if one number be divided by another number, producing a third; the Logarithm *of the second subtracted from the* Logarithm *of the first, leaves the* Logarithm *of the third.*

If from a number raised to the second power, to the third power, to the fifth power, &c., certain other numbers be produced; from the Logarithm *of the first multiplied by two, three, five, &c., the* Logarithms *of the others are produced.*

Also if from a given number there be extracted the second, third, fifth, &c., roots; and the Logarithm *of the given number be divided by two, three, five, &c., there will be produced the* Logarithms *of these roots.*

Finally any common number being formed from other common numbers by multiplication, division, [*raising to a power*] *or extraction* [*of a root*]; *its* Logarithm *is correspondingly formed from their* Logarithms *by addition, subtraction, multiplication, by* 2, 3, *&c.* [*or division by* 2, 3, *&c.*]: *whence the only difficulty is in finding the* Logarithms *of the prime numbers*; *and these may be found by the following general method.*

For finding all Logarithms, *it is necessary as the basis of the work that the* Logarithms *of some two common numbers be given or at least assumed*; *thus in the foregoing first method of construction,* 0 *or a cypher was assumed as the* Logarithm *of the common number one, and* 10,000,000,000 *as the* Logarithm *of one-tenth or of ten. These therefore being given, the* Logarithm *of the number* 5 (*which is a prime number*) *may be sought by the following method. Find the mean proportional between* 10 *and* 1, *namely* $\frac{316227766017}{100000000000}$, *also the arithmetical mean between* 10,000,000,000 *and* 0, *namely* 5,000,000,000; *then find the geometrical mean between* 10 *and* $\frac{316227766017}{100000000000}$, *namely* $\frac{562341325191}{100000000000}$, *also the arithmetical mean between* 10,000,000,000 *and* 5,000,000,000, *namely* 7,500,000,000; . . . . .

In all continuous proportionals.

AS *the sum of the means and one or other of the extremes to the same extreme*; *so is the difference of the extremes to the difference of the same extreme and the nearest mean.*

A saving of half the Table of LOGA-RITHMS.

OF *two arcs making up a quadrant, as the sine of the greater is to the sine of double its arc, so is the sine of* 30 *degrees to the sine of the less. Whence the* Logarithm *of the double arc being added to the* Logarithm *of* 30 *degrees, and the* Logarithm *of the greater being subtracted from the sum, there remains the* Logarithm *of the less.*

The relations of LOGARITHMS & their natural numbers to each other.

[A] 1. LET *two sines and their Logarithms be given. If as many numbers equal to the less sine be multiplied together as there are units in the* Logarithm *of the greater*; *and on the other hand, as many numbers equal to the greater sine be multiplied together as there are units in the* Logarithm *of the less*; *two equal numbers will be produced, and the* Logarithm *of the sine so produced will be the product of the two* Logarithms.

2. *As the greater sine is to the less, so is the velocity of increase or decrease of the* Logarithms *at the less, to the velocity of increase or decrease of the* Logarithms *at the greater.*

3. *Two sines in duplicate, triplicate, quadruplicate, or other ratio, have their* Logarithms *in double, triple, quadruple, or other ratio.*

4. *And two sines in the ratio of one order to another order, as for instance the triplicate to the quintuplicate, or the cube to the fifth, have their* Logarithms *in the ratio of the indices of their orders, that is of* 3 *to* 5.

5. *If a first sine be multiplied into a second producing a third, the* Logarithm *of the first added to the* Logarithm *of the second produces the* Logarithm *of the third. So in division, the* Logarithm *of the divisor subtracted from the* Logarithm *of the dividend leaves the* Logarithm *of the quotient.*

6. *And if any number of equals to a first sine be multiplied together producing a second, just so many equals to the* Logarithm *of the first added together produce the* Logarithm *of the second.*

7. *Any desired geometrical mean between two sines has for its* Logarithm *the corresponding arithmetical mean between the* Logarithms *of the sines.*

8. *If a first sine divide a third as many times successively as there are units* [B] *in* A; *and if a second sine divides the same third as many times successively as there are units in* B; *also if the same first divide a fourth as many times successively as there are units in* C; *and if the same second divide the same fourth as many times successively as there are units in* D: *I say that the ratio of* A *to* B *is the same as that of* C *to* D, *and as that of the* Logarithm *of the second to the* Logarithm *of the first.*

9. *Hence it follows that the* Logarithm *of any given number is the number* [C] *of places or figures which are contained in the result obtained by raising the given number to the* 10,000,000,000th *power.*

10. *Also if the index of the power be the* Logarithm *of* 10, *the number of places, less one, in the power or multiple, will be the* Logarithm *of the root.*

Suppose it is asked what number is the LOGARITHM of 2. I reply, the number of places in the result obtained by multiplying together 10,000,000,000 of the number 2.

But, you will say, the number obtained by multiplying together 10,000,000,000 of the number 2 is innumerable. I reply, still the number of places in it, which I seek, is numerable.

Therefore, with 2 as the given root, and 10,000,000,000 as the index, seek for the number of places in the multiple, and not for the multiple itself; and by our rule you will find 301029995 &c. to be the number of places sought, and the LOGARITHM of the number 2.

F I N I S.

SOME REMARKS
BY THE LEARNED
HENRY BRIGGS
On the foregoing APPENDIX.

The relations of LOGARITHMS and their natural *numbers to each other, when the* LOGARITHM *of unity is made* 0.

TWO *numbers with their* Logarithms *being given, if both* Loga- [A]
rithms *be divided by some common divisor, and if each of the given numbers be multiplied by itself continuously, until the number of multiplications is exceeded, by unity only, by the quotient of the* Logarithm *of the other number, two equal numbers will be produced. And the* Logarithm *of the number produced will be the continued product of the quotients of the* Logarithms *and their common divisor.*

		Logarithms.
Let the given numbers be	$2\underline{5118865}$	4
	$3\underline{9810718}$	6
Let the common divisor be		1

The first multiplied by itself 5 times, The second " " 3 " } makes $251\underline{188649}$ / 1000000

Continued Proportionals.			Logarithms.
1	(0)		0
$2\underline{5118865}$	(1)	First power	4
$6\underline{3095737}$	(2)	Second power	8
$15\underline{8489331}$	(3)	Third power	12
$39\underline{810718}$	(4)	Fourth power	16
$100\underline{000000}$	(5)	Fifth power	20
$251\underline{188649}$	(6)	Sixth power	24

Continued Proportionals.		Logarithms.
1	(0)	0
39810718	(1)	6
158489331	(2)	12
630957379	(3)	18
251188649	(64	24

ANOTHER EXAMPLE.

		Logarithms.
Let the given numbers be	316227766	5
	50118724	7
Let the common divisor be		1

The first multiplied by itself 6 times
The second „ „ 4 „ } makes 316227766

		Logarith.			Logarith.
1	(0)	0	1	(0)	0
316227766	(1)	5	50118724	(1)	7
100000000	(2)	10	251188649	(2)	14
100	(4)	20	630957376	(4)	28
1000	(6)	30	316227766	(5)	35
316227766	(7)	35			

It should be observed that if the common divisor be unity, as in both the preceding examples, the product of the given Logarithms *is the* Logarithm *of the number produced, because multiplication by unity does not increase the thing multiplied.*

THIRD EXAMPLE.

		Logarithms.	Quotients.
Let the given numbers be	343	2.53529412	3
	823543	5.91568628	7
Let the common divisor be		84509804	

Number of Places.			
	1	(0)	0
3	343	(1)	2.53529412
6	117649	(2)	5.07058824
8	40353607	(3)	7.60588236
11	3841287210	(4)	10.14117648
18	558545864083284007	(7)	17.74705884
6	823543	(1)	5.91568628
12	678223072849	(2)	11.83137256
18	558545864083284007	(3)	17.74705884

As the quotients of the given LOGARITHMS are 3 and 7, their product is 21, which, multiplied by 84509804 the common divisor, makes 17.74705884 the LOGARITHM of the number produced.

It should be observed that the cube of the second number, and its equal the seventh power of the first (*which some call* secundus solidus), *contain eighteen figures, wherefore its* Logarithm *has* 17. *in front, besides the figures following. The latter represent the* Logarithm *of the number denoted by the same digits, but of which* 5, *the first digit to the left, is alone integral, the remaining digits expressing a fraction added to the integer, thus* $5\frac{5854586408}{10000000000}$ *&c. has for its* Logarithm 74705884. *Again, if four places remain integral,* 3 *must be placed in front of the* Logarithm, *thus* $5585\frac{4586408}{10000000}$ *&c. has for its* Logarithm 3.74705884.

Hence from two given Logarithms *and the sine of the first we shall be able to find the sine of the second.*

Take some common divisor of the Logarithms, (*the larger the better*); *divide each by it. Then let the first sine multiply itself and its products continuously until the number of these products is exceeded, by unity only, by the quotient of the second* Logarithm; *or until the power is produced of like name with the quotient of the second* Logarithm. *The same number would be produced if the second sine, which is sought, were to multiply itself until it became the power of like name with the quotient of the first* Logarithm, *as is evident from the preceding proposition. Therefore take the above power and seek for the root of it which corresponds to the quotient of the first* Logarithm; *thereby you will find the required second sine. Also the* Logarithm *of the power itself will be the continued product of the quotients and the common divisor.*

Thus let the given LOGARITHMS be 8 and 14, and the sine corresponding to the first LOGARITHM be 3. A common divisor of the LOGARITHMS is 2; this gives the quotients 4 and 7. If 3 multiply itself six times, you will have 2187 for the power which, in a series of continued proportionals from unity, will occupy the seventh place, and hence it may, without inconvenience, be called the seventh power. The same number, 2187, is the fourth power from unity in another series of continued proportionals, in which the first power, $6\frac{838521}{1000000}$, is the required second sine. The product of the quotients 4 and 7 is 28, which, multiplied by the common divisor 2, makes 56, the LOGARITHM of the power 2187.

Continued Proportionals.		Logarithms.	Continued Proportionals.		Logarithms.
1	(0)	0	1	(0)	0
3	(1)	8	6838521	(1)	14
9	(2)	16	46765372	(2)	28
27	(3)	24	31980598	(3)	42
81	(4)	32	2187	(4)	56
243	(5)	40			
729	(6)	48			
2187	(7)	56			

It will be observed that these Logarithms *differ from those employed in illustration of the previous Proposition; but they agree in this, that in both, the Logarithm of unity is* 0; *and consequently the* Logarithms *of the same numbers are either equal or at least proportional to each other.*

[B] *If a first sine divide a third,*)

The first must divide the third, and the quotient of the third, and each quotient of a quotient successively as many times as possible, until the last quotient becomes less than the divisor. Then let the number of these divisions be noted, but not the value of any quotient, unless perhaps the least, to which we shall refer presently. In the same manner let the second divide the same third. And so also let the fourth be divided by each.

Thus let the	first	sine	be	2
	second	,,	,,	4
	third	,,	,,	16
	fourth	,,	,,	64

The first, 2. *divides the third,* 16. *four times, and the quotients are* 8, 4, 2, 1. *The second,* 4. *divides the same third,* 16. *two times, and the quotients are* 4, 1. *Therefore* A *will be* 4, *and* B *will be* 2.

In the same manner the first, 2. *divides the fourth,* 64. *six times, and the quotients are* 32, 16, 8, 4, 2, 1. *The second,* 4. *divides the fourth,* 64. *three times, and the quotients are* 16, 4, 1. *Therefore* C *will be* 6, *and* D *will be* 3.

Hence I say that, as A, 4. *is to* B, 2. *so is* C, 6. *to* D, 3. *and so is the* Logarithm *of the second to the* Logarithm *of the first.*

If in these divisions the last and smallest quotient be everywhere unity, as in these four cases, the numbers of the quotients and the Logarithms *of the divisors will be reciprocally proportional.*

Otherwise the ratio will not be exactly the same on both sides; nevertheless, if the divisors be very small, and the dividends sufficiently large, so that the quotients are very many, the defect from proportionality will scarcely, or not even scarcely, be perceived.

Hence it follows that the logarithm) [C]

Let two numbers be taken, 10 *and* 2, *or any others you please. Let the* Logarithm *of the first, namely* 100, *be given; it is required to find the* Logarithm *of the second. In the first place, let the second,* 2, *multiply itself continuously until the number of the products is exceeded, by unity only, by the given* Logarithm *of the first. Then let the last product be divided as often as possible by the first number,* 10, *and again in like manner by the second number,* 2. *The number of quotients in the latter case will be* 100, (*for the product is its hundredth power; and if a number be multiplied by itself a given number of times forming a certain product, then it will divide the product as many times and once more; for example, if* 3 *be multiplied by itself four times it makes* 243, *and the same* 3 *divides* 243 *five times, the quotients being* 81, 27, 9, 3, 1.) *In the former case, where the product is continually divided by* 10, *it is manifest that the number of quotients falls short of the number of places in the dividend by one only. Therefore* (*by the preceding proposition*) *since the same product is divided by two given numbers as often as possible, the numbers of the quotients and the* Logarithms *of the divisors will be reciprocally proportional. But, the number of quotients by the second being equal to the Logarithm of the first, the number of quotients by the first, that is the number of places in the product less one, will be equal to the* Logarithm *of the second.*

Number of Places.		
	1	0
1	2	1
1	4	2
2	16	4
3	256	8
4	1024	10
7	1048576	20
13	1099511627776	40
25	1208925819614	80
31	1267650600228	100
61	16069379676	200
121	25822496318	400
241	66680131608	800
302	107150835165	1000
603	114813014767	2000
1205	131820283599	4000
2409	17316587168	8000
3011	19950583591	10000

Here we see that if we assume the LOGARITHM of 10 to be 10, the number of places in the tenth power is 4, wherefore the logarithm of 2 will be 3 and something over. The number of places in the hundredth power is 31; in the thousandth, 302; in the ten thousandth, 3011; and generally the more products we take the more nearly do we approach the true LOGARITHM sought for. For when the products are few, the fraction adhering to the last quotient disturbs the ratio a little; but if we assume the LOGARITHM of 10 to be 10,000,000,000, and if 2 be multiplied by itself continuously until the number of products is exceeded, by one only, by the given LOGARITHM; then the number of places, less one, in the last product, will give the LOGARITHM of 2 with sufficient accuracy, because in large numbers the small fraction adhering to the last quotient will have no effect in disturbing the proportion.

THE END.

SOME VERY REMARKABLE

PROPOSITIONS FOR THE

solution of spherical triangles

with wonderful ease.

To solve a spherical triangle without dividing it into two quadrantal or rectangular triangles.

Prop. 1. GIVEN *three sides, to find any angle.*

And conversely,

2. *Given three angles, to find any side.*

This is best done by the three methods explained in my work on LOGARITHMS, *Book* II. *chap.* vi. *sects.* 8, 9, 10.

3.

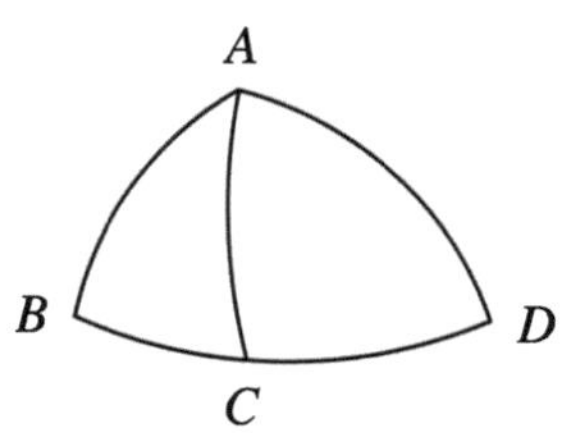

Given the side AD, & the angles D & B, to find the side AB.

Multiply the sine of *AD* by the sine of *D*; divide the product by the sine of *B*, and you will have the sine of *AB*.

4. *Given the side AD, & the angles D & B, to find the side BD.*

Multiply radius by the sine of the complement of *D*; divide by the tangent of the complement of *AD*, and you will obtain the tangent of the arc *C*: then multiply the sine of *CD* by the tangent of *D*; divide the product by the tangent of *B*, and the sine of *BC* will result: add or subtract *BC* and *CD*, and you have *BD*.

5. *Given the side AD, & the angles D & B, to find the angle A.*

Multiply radius by the sine of the complement of *AD*; divide by the tangent of the complement of *D*, and the tangent of the complement of *CAD* will be produced; whence we have *CAD* itself. Similarly multiply the sine of the complement of *B* by the sine of *CAD*; divide by the sine of the complement of *D*, and the sine of *BAC* will be produced;

which being added to or subtracted from CAD, you will obtain the required angle BAD.

6. *Given AD, & the angle D with the side BD, to find the angle B.*

Multiply radius by the sine of the complement of D; divide by the tangent of the complement of AD, and the tangent of CD will be produced; its arc CD subtract from, or add to, the side BD, and you have BC: then multiply the sine of CD by the tangent of D; divide the product by the sine of BC, and you have the tangent of the angle B.

7. *Given AD, & the angle D with the side BD, to find the side AB.*

Multiply radius by the sine of the complement of D; divide the product by the tangent of the complement of AD, and the tangent of CD will be produced; its arc CD subtract from, or add to, the given side BD, and you have BC. Then multiply the sine of the complement of AD by the sine of the complement of BC; divide the product by the sine of the complement of CD, and the sine of the complement of AB will be produced; hence you have AB itself.

Given AD, & the angle D with the side BD, to find the angle A.

This follows from the above, but the problem would require the "Rule of Three" to be applied thrice. Therefore substitute A for B and B for A, and the problem will be as follows:—

Given BD & D, with the side AD, to find the angle B.

This is exactly the same as the sixth problem, and is solved by the "Rule of Three" being applied twice only.

8. *Given AD, & the angle D with the side AB, to find the angle B.*

Multiply the sine of AD by the sine of D; divide the product by the sine of AB, and the sine of the angle B will be produced.

9. *Given AD, & the angle D with the side AB, to find the side BD.*

Multiply radius by the sine of the complement of D, divide the product by the tangent of the complement of AD, and the tangent of the arc CD will be produced. Then multiply the sine of the complement of CD by the sine of the complement of AB, divide the product by the sine of the complement of AD, and you have the sine of the complement of BC. Whence the sum or the difference of the arcs BC and CD will be the required side BD.

10. *Given AD, & the angle D with the side AB, to find the angle A.*

Multiply radius by the sine of the complement of AD, divide the product by the tangent of the complement of D, and the tangent of the complement of CAD will be produced, giving us CAD. Again, multiply the tangent of AD by the sine of the complement of CAD, divide the product by the tangent of AB, and the sine of the complement of BAC will be produced, giving BAC. Then the sum or difference of the arcs BAC and CAD will be the required angle BAD.

11. *Given AD; & the angle D with the angle A; to find the side AB.*

Multiply radius by the sine of the complement of AD, divide the product by the tangent of the complement of D, and you have the tangent of the complement of CAD; CAD being thus known, the difference or sum of the same and the whole angle A is the angle BAC. Multiply the tangent of AD by the sine of the complement of CAD; divide the product by the sine of the complement of BAC, and you will have the tangent of AB.

12. *Given AD, & the angle D with the angle A, to find the third angle B.*

Multiply radius by the sine of the complement of AD, divide the product by the tangent of the complement of D, and the sine of the complement of B will be produced, from which we have the angle required.

Given AD, & the angle D with the angle A, to find the side BD.

This follows from the above, but in this form the problem would require the "Rule of Three" to be three times applied. Therefore substitute A for D and D for A, and the problem will be as follows:—

Given AD & the angle A with the angle D, to find the side BA.

This is the same throughout as problem 11, and is solved by applying the "Rule of Three" twice only.

The use and importance of half-versed sines.

1. GIVEN *two sides & the contained angle, to find the third side.*

From the half-versed sine of the sum of the sides subtract the half-versed sine of their difference; multiply the remainder by the half-versed sine of the contained angle; divide the product by radius; to this add the half-versed sine of the difference of the sides, and you have the half-versed sine of the required base.

Given the base and the adjacent angles, the vertical angle will be found by similar reasoning.

2. *Conversely, given the three sides, to find any angle.*

From the half-versed sine of the base subtract the half-versed sine of the difference of the sides multiplied by radius; divide the remainder by the half-versed sine of the sum of the sides diminished by the half-versed sine of their difference, and the half-versed sine of the vertical angle will be found.

Given the three angles, the side will be found by similar reasoning.

3. *Given two arcs, to find a third, whose sine shall be equal to the difference of the sines of the given arcs.*

Let the arcs be 38° 1′ and 77°. Their complements are 51° 59′ and 13°. The half sum of the complements is 32° 29′, the half difference 19° 29′, and the logarithms are 621656 and 1098014 respectively. Adding these, you have 1719670, from which, subtracting 693147, the logarithm of half radius, there will remain 1026523, the logarithm of 21°, or thereabout. Whence the sine of 21°, namely 358368, is equal to the difference of the sines of the arcs 77° and 38° 1′, which sines are 974370 and 615891, more or less.

4. *Given an arc, to find the Logarithm of its versed sine.* [a]

Let the arc be 13°; its half is 6° 30′, of which the logarithm is 2178570. From double this, namely 4357140, subtract 693147, and there will remain 3663993. The arc corresponding to this is 1° 28′, and the number put for the sine is 25595; but this is also the versed sine of 13°. ⁂

5. *Given two arcs, to find a third whose sine shall be equal to the sum of the sines of the given arcs.*

Let the arcs be 38° 1′ and 1° 28′; their sum is 39° 29′ and their difference 36° 33′, also the half sum is 19° 44′ and the half difference 18° 16′. Wherefore add the logarithm of the half sum, viz. 1085655, to the logarithm of the difference, viz. 518313, and you have 1603968; from this subtract the logarithm of the half difference, namely 1160177, and there will remain the logarithm 443791, to which correspond the arc 39° 56′ and sine 641896. But this sine is equal, or nearly so, to the sum of the sines of 38° 1′ and 1° 28′, namely 615661 and 25595 respectively.

6. *Given an arc & the Logarithm of its sine, to find the arc whose versed sine shall be equal to the sine of the given arc.*

Let the arc be 39° 56′, to which corresponds the logarithm 443791, the sine being unknown. To the logarithm 443791 add 693147, the logarithm of half radius, and you have 1136938. Halve this logarithm and you have 568469. To this corresponds the arc 34° 30′, which being doubled gives 69° for the arc which was sought. This is the case since the sine of 39° 56′ and the versed sine of 69° are each equal, or nearly so, to 641800.

[b] *Of the spherical triangle ABD, given the sides & the contained angle, to find the base.*

LET the sides be 34° and 47°, and the contained angle 120° 24′ 49″. Half the contained angle is 60° 12′ 24½″, and its logarithm 141766. To the double of the latter, namely 283533, add the logarithms of the sides, namely 581260 and 312858, and the sum is 1177651. This sum is the logarithm of half the difference between the versed sine of the base and the versed sine of the difference of the sides; it is also the logarithm of the sine of the arc 17° 56′, which arc we call the "second found," for that which follows is first found.

Halve the difference of the sides, namely 13°, and you have 6° 30′, the logarithm of which is 2178570. Double the latter and you have 4357140 for the logarithm of the half-versed sine of 13°; it is also the logarithm of the sine of the arc 0° 44′, which arc we call the "first found."

The sum of the two arcs is 18° 40′, the half sum 9° 20′, and their logarithms 1139241 and 1819061 respectively. Also the difference of the two arcs is 17° 12′, the half difference 8° 36′, and their logarithms 1218382 and 1900221 respectively.

Now add the logarithm of the half sum, namely 1819061,

either	or
to the logarithm 1218382, and the sum will be 3037443; from this subtract the logarithm 1900221 and there will remain 1137222.	to the logarithm of the complement of the half difference, namely 11307, and the sum will be 1830368; from this subtract 693147 and there will remain 1137221.

Halve the latter and you have the logarithm 568611, to which corresponds the arc 34°30′, and double this arc is the base required, namely 69°.

Conversely, given the three sides, to find any angle. The solution of this problem is given in my work on Logarithms, Book II. chap. vi: sect. 8, *but partly by logarithms and partly by prosthaphæresis of arcs.*

It is to be observed that in the preceding and following problems there is no need to discriminate between the different cases, since the form and magnitude of the several parts appear in the course of the calculation.

Another direct converse of the preceding problem follows:—

[*Given the sides and the base, to find the vertical angle.*]

HALVE the given base, namely 69°, and you have 34° 30′, the logarithm of which is 568611. Double the latter and you have

1137222; corresponding to this is the arc 18° 42′, which note as the second found.

As before, take for the first found the arc 0° 44′, corresponding to the logarithm 4357140.

The complements of the two arcs are 89° 16′ and 71° 18′; their half sum is 80° 17′, and its logarithm 14449; their half difference is 8° 59′, and its logarithm 1856956. Add these logarithms and you have 1871405; subtract 693147 and there remains 1178258. The arc corresponding to this logarithm is 17° 56′, which arc we call the third found.

From the logarithm of the third found, subtract the logarithms of the given sides, namely 581260 and 312858, and there remains 283533; halve this and you have 141766 for the logarithm of the half vertical angle 60° 12′ 24½″. The whole vertical angle sought is therefore 120° 24′ 49″.

Another rule for finding the base by prosthaphæresis.

[*Given the sides and vertical angle, to find the base.*]

NOte the half difference between the versed sines of the sum and difference of the sides, and also the half-versed sine of the vertical angle. Look among the common sines for the values noted, and find the arcs corresponding to them in the table. Then write for the second found the half difference of the versed sines of the sum and difference of these arcs.

Also, as before, take for the first found the half-versed sine of the difference of the sides. Add the first and second found, and you will obtain the half-versed sine of the base sought for.

Conversely — [*given the sides and the base, to find the vertical angle.*]

The first found will be, as before, the half-versed sine of the difference of the sides.

From the half-versed sine of the base subtract the first found and you will have the second found.

Multiply the latter by the square of radius; divide by the half difference between the versed sines of the sum and difference of the sides, and you have as quotient the half-versed sine of the vertical angle sought for.

Of five parts of a spherical triangle, given the three intermediate, to find the two extremes by a single operation. Or otherwise, given the base and adjacent angles, to find the two sides. [c]

(*) OF the angles at the base, write down the sum, half sum, difference and half difference, along with their logarithms.

Add together the logarithm of the half sum, the logarithm of the difference, and the logarithm of the tangent of half the base; subtract the logarithm of the sum and the logarithm of the half difference, and you will have the first found.

Then to the logarithm of the half difference add the logarithm of the tangent of half the base; subtract the logarithm of the half sum, and you will have the second found.

Look for the first and second found among the logarithms of tangents, since they are such, then add their arcs and you will have the greater side; again subtract the less arc from the greater and you will have the less side.

Another way of finding the sides.

ADd together the logarithm of the half sum of the angles at the base, the logarithm of the complement of the half difference, and the logarithm of the tangent of half the base; subtract the logarithm of the sum and the logarithm of half radius, and you will have the first found.

Again, add together the logarithm of the half difference, the logarithm of the complement of the half sum, and the logarithm of the tangent of half the base; subtract the logarithm of the sum and the logarithm of half radius, and you will have the second found.

Proceed as above with the first and second found, and you will obtain the sides.

Another way of the same.

MUltiply the secant of the complement of the sum of the angles at the base by the tangent of half the base.

Multiply the product by the sine of the greater angle at the base, and you will have the first found.

Multiply the same product by the sine of the less angle, and you will have the second found.

[d] Then divide the sum of the first and second found by the square of radius, and you will have the tangent of half the sum of the sides.

Also subtract the less from the greater and you will have the tangent of half the difference of the sides.

Whence add the arcs corresponding to these two tangents, and the greater side will be obtained; subtract the less arc from the greater and you have the less side.

Of the five consecutive parts of a spherical triangle, given the three intermediate, to find both extremes by one operation and without the need of discriminating between the several cases.

(*) OF the angles at the base, the sine of the half difference is to the sine of the half sum, as the sine of the difference is to a fourth which is the sum of the sines

And the sine of the sum is to the sum of the sines as the tangent of half the base is to the tangent of half the sum of the sides.

Whence the sine of the half sum is to the sine of the half difference of the angles as the tangent of half the base is to the tangent of half the difference of the sides.

Add the arcs of these known tangents, taking them from the table of tangents, and you will have the greater side; in like manner subtract the less from the greater and the less side will be obtained.

F I N I S.

SOME NOTES
BY THE LEARNED
HENRY BRIGGS
ON THE FOREGOING PROPOSITIONS.

[a] GIVEN *an arc, to find the logarithm of its versed sine.*

To the end of this proposition ⁂ *I should like to add the following:—*

Conversely, given the logarithm of a versed sine, to find its arc.

Add the known logarithm of the required versed sine to the logarithm of 30°, *viz.*, 693147, *and half the sum will be the logarithm of half the arc sought for.*

Thus let 35791 be the given logarithm of an unknown versed sine, whose arc is also unknown.

To this logarithm add 693147, and the sum will be 728938, half of which, 364469, is the logarithm of $43°\,59'\,33''$. The arc of the given logarithm is therefore $87°\,59'\,6''$, and its versed sine is 9648389.

Again, let a negative logarithm, say −54321, be the known logarithm of the required versed sine. To this logarithm add, as before, 693147, and the sum, that is the number remaining since the sines are contrary, will be 638826, half of which, 319413, is the logarithm of $46°\,36'\,0''$. The arc of the given logarithm is therefore $93°\,12'\,0''$, the versed sine of which is 10558216, and since this is greater than radius it has a negative logarithm, namely −54321.

DEMONSTRATION

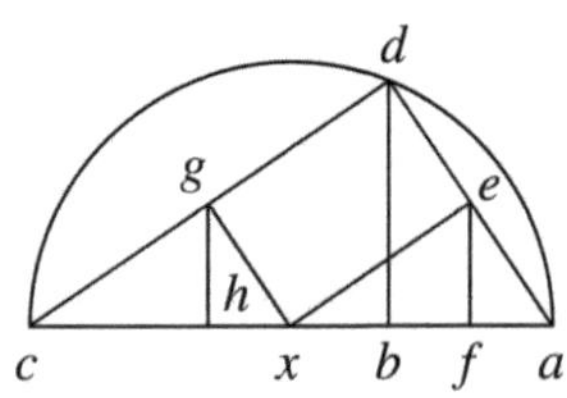

$\left.\begin{matrix} cb \\ ab \end{matrix}\right\}$ versed sine of arc $\left\{\begin{matrix} cd \\ ad \end{matrix}\right.$

$\left.\begin{matrix} xc \\ cg \\ ch \end{matrix}\right\}$ cont. pro- port. $\quad$ $\left.\begin{matrix} xa \\ ae \\ af \end{matrix}\right\}$ cont. pro- port. $\quad$ $\left.\begin{matrix} \frac{1}{2}xc, \text{ sine of } 30°\,0' \\ cg, \text{ sine of } \frac{1}{2} \text{ arc } cd \\ cb, \text{ double of line } ch \end{matrix}\right\}$ cont. pro- port.

Later on I observed that the sixth proposition might be proved in an exactly similar way.

Of the spherical triangle ABD]

In finding the base we may pursue another method, namely:—

Add the logarithm of the versed sine of the given angle to the logarithms of the given sides, and the sum will be the logarithm of the difference between the versed sine of the difference of the sides and the versed sine of the base required. This difference being consequently known, add to it the versed sine of the difference of the sides, and the sum will be the versed sine of the base required.

For example, let the sides be 34°and 47°, their logarithms 581261 and 312858, and the logarithm of the versed sine of the given angle −409615. The sum of these three logarithms is 484504, which is the logarithm of the difference between the versed sine of the base and the versed sine of the difference of the sides.

Now the line corresponding to this logarithm, whether a versed sine or a common sine, is 6160057, and consequently this is the difference between the versed sine of the base and the versed sine of the difference of the sides. If to this you add the versed sine of the difference of the sides, that is 256300, the sum will be the versed sine of the base required, namely 6416357, and this subtracted from radius leaves the sine of the complement of the base, namely 3583643, which is the sine of 21°. Consequently the base required is 69°.

Conversely, given three sides, to find any angle.

If from the logarithm of the difference between the versed sine of the base and the versed sine of the difference of the sides you subtract the logarithms of the sides, the remainder will be the logarithm of the versed sine of the angle sought for.

As in the previous example, let the logarithms of the sides be 581261 and 312858. Subtract their sum, 894119, from the logarithm 484504, and the remainder will be the negative logarithm −409615, which gives the versed sine of the required angle $120°\,24'\,49''$.

Of five parts of a spherical triangle] [c]

This proposition appears to be identical with the one which is inserted at the end, and distinguished like the former by (*). *The latter proposition I consider much the superior. There are, however, three operations in it, the first two of which I throw into one, as they are better combined. Thus:—*

Let there be given
the base 69°,

the angles at the base $\begin{cases} 42^\circ 29' 59'' \\ 31^\circ \ 6' \ 5'' \end{cases}$

73° 36′ 4″	sum.
36° 48′ 2″	half sum.
53° 11′ 58″	complement of $\frac{1}{2}$ sum.
11° 23′ 54″	difference.
5° 41′ 57″	half difference.
84° 18′ 3″	compl. of $\frac{1}{2}$ diff.

			Logarithms.
Proportion 1.	Sine half difference	5° 41′ 57″	23095560
	Sine half sum	36° 48′ 2″	5124410
	Sine difference	11° 23′ 54″	16213641
	Sum of sines		−1757509
Proportion 2.	Sine of sum	73° 36′ 4″	415312
	Sum of sines		−1757509
	Tangent half base	34° 30′ 0″	3750122
	Tangent $\frac{1}{2}$ sum of sides	40° 30′ 0″	1577301
Proportion 3.	Sine $\frac{1}{2}$ sum of angles	36° 48′ 2″	5124410
	Sine $\frac{1}{2}$ diff. of angles	5° 41′ 57″	23095560
	Tangent $\frac{1}{2}$ base	34° 30′ 0″	3750122
	Tangent $\frac{1}{2}$ diff. of sides	6° 30′ 0″	21721272

40° 30′
6° 30′

47° 0′
34° 0′ } sides.

These are the operations described by the Author. But I replace the first two by another, retaining the third.

			Logarithms.
Proportion	Sine compl. $\frac{1}{2}$ sum of angles	53° 11′ 58″	2222368
	Sine compl. $\frac{1}{2}$ diff. of angles	84° 18′ 3″	49553
	Tangent $\frac{1}{2}$ base	34° 30′ 0″	3750122
	Tangent $\frac{1}{2}$ sum of sides	40° 30′ 0″	1577307

ANOTHER EXAMPLE.

Let there be given

the base 47°,

the sides containing it $\begin{cases} 59°35'11'' \\ 31°\ 6'\ 5'' \end{cases}$

90°41′16″	sum.
45°20′38″	half sum.
44°39′22″	compl. of half sum.
28°29′ 6″	difference.
14°14′33″	half difference.
75°45′27″	compl. of half. diff.

			Logarithms.
Proportion 1.	Sine compl. $\frac{1}{2}$ sum of sides	44°39′22″	3526118
	Sine compl. $\frac{1}{2}$ diff. of sides	75°45′27″	312192
	Tan. compl. $\frac{1}{2}$ vert. angle	66°30′ 0″	−8328403
	Tan. $\frac{1}{2}$ sum of angs. at base	72°30′ 0″	−11452329
Proportion 2.	Sine of $\frac{1}{2}$ sum of sides	45°20′38″	3406418
	Sine $\frac{1}{2}$ diff. of sides	14°14′33″	14023154
	Tan. compl. $\frac{1}{2}$ vert. angle	66°30′ 0″	−8328403
	Tan. $\frac{1}{2}$ diff. of angs. at base	38°30′ 0″	2288333

72°30′
38°30′

111° 0′
34° 0′ } angles at the base.

And these relations are all uniformly maintained, whether there be given two angles with the interjacent side or two sides with the contained angle. In each operation the important point is what occupies the third place in the proportion. In the former it is the tangent of half the base, in the latter the tangent of the complement of half the vertical angle. In these examples, if the tangent or the sum of the sines be greater than radius, the logarithm is negative and has a dash preceding, for example −8328403.

Another way of the same]

[d] Then divide the sum of the first and second found by the square of radius, and you will have)

To make the sense clearer, I should prefer to write this as follows:—

Then divide both the first and second found by the square of radius, add the quotients, and you will have the tangent, &c.

This proposition is absolutely true, as well as the one preceding; but while the former may most conveniently be solved by logarithms, the latter will not admit of the use of logarithms throughout, as the quotients must be added and subtracted to find the tangents; for the utility of Logarithms is seen in proportionals, and therefore in multiplication and division, and not in addition or subtraction.

THE END.

DE ARTE LOGISTICA

Translation and notes by William Francis Hawkins

Small additions and modifications to the translation by Hawkins [in brackets], as well as the addition of notes 9, 13, 37, 38, 57, 73 and 77, have been made by Enrique A. González-Velasco. Notes 5, 6, 20 and 24 are by Garry J. Tee, Hawkins' thesis supervisor. Note that the present typesetting has been formatted to be as close as possible to the style of the original Latin publication, which Hawkins could not achieve in 1982 using a typewriter.

DE ARTE LOGISTICA

JOANNIS NAPERI

MERCHISTONII BARONIS

LIBRI QUI SUPERSUNT.

IMPRESSUM EDINBURGI

M.DCCC.XXXIX.

DE ARTE LOGISTICA

JOHN NAPIER

BARON OF MERCHISTON

SURVIVING BOOK.

PRINTED IN EDINBURGH
M.DCCC.XXXIX.

[BOOK ONE.

ON THE CALCULATIONS OF NUMBERS BY ALL KINDS OF COMMON LOGISTICS.]

CHAPTER I.

BASIC CALCULATIONS.

LOGISTICS is the technique of correct calculation.

CALCULATION is a process or operation which derives a required number from several numbers whose properties are given.

The numbers are either named or derived geometrically.

But definitions and notation come first in Logistics, and calculation proceeds from them.

Calculation is either simple or compound.

Simple calculation is that which generates a third number from two given nuubers by means of a single unique operation.

Simple calculation is either basic or derived from a basic operation.

A basic calculation is one in which one number operates on another number once only.

> And so, when any two numbers are given—the whole, or any two of its components—this process leads to the appropriate third number: This will soon be evident in the examples which follow.

This operation is either addition or subtraction.

ADDITION is a basic operation by which several numbers are added to produce their sum.

> For example, 3 and 4 are added and produce 7 as their sum. Likewise, 2, 3 and 4 are added and produce a sum totalling 9.

SUBTRACTION is a basic operation by which one number is taken from another, leaving a remainder.

> As for instance, by taking 4 away from 9 you leave a reminder 5. [Or you can say that 4 is the subtrahend, 9 the minuend, and 5 the remainder. So] if you take, or subtract 3 from 5, you leave remainder 2.

Subtraction is either exact, with no remainder, or inexact.

Inexact subtraction is either a small number taken from a larger number with a remainder greater than zero; or a greater number from a smaller with a remainder less than zero.

> 5 subtract 5 leaves remainder zero. 5 subtract 3 leaves remainder 2, greater than zero. 5 subtract 7 leaves remainder 2, less than zero, or zero minus two.

From this therefore it is clear that negative numbers trace their origin here, surely, in the subtraction of a larger number from a smaller one. This matter is dealt with in its proper place.

From what has been said it is clear that addition and subtraction are related operations and each is a verification of the other.

For we define as verifications only those processes which lead to correct calculations whether the numbers are taken separately or all together.

> As above, a verification of subtraction: 5 subtract 3, remainder 2; Add 2 and 3 giving 5. On the other hand, the verification of addition: 2 and 3 make 5; 5 subtract 3 producing 2 as before. Alternatively, 5 subtract 2 producing 3 [as before].

Besides this there is another internal verification of subtraction. Subtract the remainder from the largest nmber, and this leaves the number previously subtracted.

> For verification: 3 subtracted from 5 leaves 2. Subtract 2 from 5 and restore 3.

And so, when any two nunbers are given (the whole, the part and the corresponding remainder) you get the third number by addition and also by subtraction.

CHAPTER II.

CALCULATIONS DERIVED FROM BASIC OPERATIONS.

UP TO THIS POINT we have dealt with basic calculations; Now follow calculations derived from them.

Derived calculations occur when one number operates on another more than once. And such calculations arise naturally from successive and continuous operations.

And so derived numbers are produced either by several basic operations, or else by the continued application of one basic operation.

Those calculations are derived which, given any two numbers—the whole and one of its parts—lead to a third number which supplements the given part. This will shortly be illustrated by examples.

Multiplication by continued addition, and division by continued subtraction are derived from basic operations.

MULTIPLICATION is therefore the continued addition of one of two given numbers as often as it will go into the other once; and the result is called the multiple.

> To multiply 3 by 5 is equivalent to one of two things—5 continued additions of three, which makes 15; or the addition of five three times, which comes to the same thing. And of the numbers 3 and 5, one is the multiplier and the other the multiplicand. The product 15 is called the multiple.

In this, unity stands to either the multiplier or the multiplicand as the other number stands to the multiple.

> As in the previous example: 1 is to 3 as 5 is to 15; or, 1 is to 5 as 3 is to 15.

The kinds of multiplication are infinite: as duplation, which is the multiplication of a given number by 2; triplation, by 3; quadruplation, by 4; and so on.

The multiplicands are called twofold, threefold, fourfold etc., [the multipliers are 2, 3, 4, etc.,] and the multiples are called double, triple, quadruple, etc.

DIVISION is subtraction of the divisor from the dividend continuously until nothing is left. And the number of subtractions is the required answer.

> Let 15 be divided by 5, 5 is taken from 15 by continued subtraction until nothing remains, and the number of subtractions is three. Therefore the required quotient is 3; 15 is the dividend and 5 is the divisor.

In this operation unity stands to one of the quotient and the divisor as the second stands to the dividend.

> As in the example above; 1 stands to 3 as 5 to 15; or 1 stands to 5 as 3 to 15.

Division is either exact and produces unity, or else it is inexact.

The division of a smaller by a larger number is inexact; and the quotient is a fraction, or broken number less than 1. A greater number divided by a smaller has a quotient greater than 1.

> 10 divided by 10 produces 1 exactly; but 10 divided by 13 produces ten thirteenths parts, and the result is less than unity. Thirdly, 10 divided by 5 produces 2 which is greater than unity.

Further, the division of a greater by a lesser number is either exact or inexact.

It is exact when there is no remainder, and in this case the quotient is an integer.

> As in the previous example 10 divided by 5 yields 2 exactly without remainder.

But division is inexact if a remainder is left which cannot be divided.

> For if 16 is to be divided by 5, you obtain 3 and are left with one unit which remains indivisible. Such a division is said to be inexact.

From this it is seen that fractions owe their origin as much to the division of smaller numbers by larger, as to the inexact division of larger numbers by smaller. This is dealt with in its proper place.

In division the number which is to be divided is called the dividend, the other of the given numbers is the divisor, and the result is called the quotient. Whatever is left over indivisible is called the remainder.

> As in the preceding example 16 is called the dividend, 5 the divisor, and 3 the quotient. The unit left over is the remainder.

The kinds of division are infinite. Bipartition is the division of a given number into two equal parts; tripartition into three; quadripartition into four, and so on. The divisors are 2, 3, 4 etc.; and the quotients are called seconds, thirds and fourths etc.

From the above it is clear that multiplication and exact division are related, and each provides a verification of the other.

> For if you doubt whether 3 multiplied by 5 equals 15, for verification divide 15 by 5, and when the answer comes to 3 you know that your multiplication is correct. Or else divide 15 by 3, and 5 comes out as before. Likewise if you doubt whether 16 divided by 5 produces 3 with

one unit left over undivided; then multiply 3 by 5. This gives I5 and not 16. And so one unit must be added in verification and this shows that the division was inexact by one unit.

Besides this there is another internal verification of division. Surely, on dividing the dividend by the quotient the divisor comes out as before.

For verification that 15 divided by 5 gives 3, divide 15 by 3 and 5 will come out as formerly.

Given any two numbers, one being a factor of the other, then you can find a third number by suitable multiplication and division, which is also a factor.

CHAPTER III.

[CALCULATIONS DERIVED FROM FIRST CALCULATIONS:] OPERATIONS ON POWERS AND ROOTS.

So far I have dealt with calculations derived from basic operations. Calculations derived from a single basic operation follow.

These calculations are such that, from any two given numbers the third can be found—power, index, root.

A power is that number which is divided by some other number as often as necessary until unity is obtained. The number of such divisions is called the index, and the divisor is the root itself.

Thus from the three terms 32, 5 and 2, the number 32 is called the power, because after being divided five times by 2 it becomes unity. The first division gives 16, the second 8, the third 4, the fourth 2 and the fifth finally gives 1. And so of these, 5 is the index and 2 is the root.

The calculations derived from a single basic operation are either radical multiplication by continued multiplication or radical division and the extraction of roots by continued division.

Radical multiplication is the continued multiplication of the given root as many times as the index contains one unit. This produces the required power.

> So 2 is said to be raised to the third power when 8 is produced, because the index three contains three units. The first multiplication from unity gives 2, the second gives 4 and the third 8, which, indeed, is the required power.

The kinds of radical multiplication are limitless. Duplication, which is the multiplication of two equal numbers in turn or a given number with 2 as index; triplication which is a given number with three as index or the multiplication of three equal nmbers.

In these cases the power is said to be duplicated, triplicated, quadruplicated: the indices are two, three, four: the roots are square root, cube root, fourth root.

> Given two for root and two for index then two times two make 4 a square. And index three denotes as many twos and naturally two times two times two make 8 the cube. In the same way index four denotes as many twos and thus two times two times two times two make 16, the fourth power. And so without limit with two as root as in the following table in which the upper line denotes the index and the lower line denotes the power.[1]

	I.	II.	III.	IIII.	V.	VI.	VII.	etc.
1.	2.	4.	8.	16.	32.	64.	128.	etc.

Division of powers is division of a power by its root continuously until unity results, and the number of divisions is the required index.

> Thus the power 8 divided by the root 2 until unity is obtained makes the first division 4, the second division 2 and the third division 1. And from this, 3, the number of divisions, is the index required.

These are the kinds of indices and roots used above in performing radical multiplication.

---

[1] The use of Roman numerals to indicate the index is very significant, since Napier first called his logarithms 'artificial numbers.' Using Napier's artificial numbers:

$$2^{\mathrm{II}} = 4; \quad 2^{\mathrm{III}} = 8; \quad 2^{\mathrm{IV}} = 16; \quad 2^{\mathrm{V}} = 32; \quad 2^{\mathrm{VI}} = 64.$$

Using the current system of notation Napier would write

2℥ = 4; 2ℭ = 8; 2℥℥ = 16; 2ß = 32; 2℥ℭ = 64.

Chapters 3 and 4 show that Napier was in complete command of index notation, and this is strengthened by pages 52–54 of Napier's *Constructio*.

CHAPTER IV.

THE EXTRACTION OF ROOTS.

THE EXTRACTION OF A ROOT when the index is given means finding that number which restores the given power by radical multiplication, and which also divides the power by radical division.

If you want the cube root of 8, then, by using the rules given in [their place (Chapter 3)] you find that it is 2. For two, by radical triplication first produces 2, secondly 4, and thirdly 8, the required power. On the other hand, the same power divided by the root reduces first to 4, then to 2, and by a third division to unity.

The extraction of a root is either exact or inexact.

It is exact when there is no remainder.

As in the previous example.

The root is inexact when there is an irreducible portion left over.

For instance, if the cube root is to be extracted from the power 9, the nearest integer to it is two, which, when raised to the third power gives 8, and not 9. Therefore one unit being left over and the root not being extracted, the extraction is therefore inexact.

Here we have the same kinds of power, root and index as in radical multiplication above. And whatever remainder is left after extraction is called unresolvable.

The result of an inexact extraction is the lower limit; to which, if you add one unit, you get the upper limit. Between these two limits the true, perfect root lies.

In the previous example two came out as the inexact root of nine. If you add one unit it becomes three. And between these two values lies the true and exact cube root of nine.

But geometers, anxious for greater accuracy, prefer to put an index sign before the power itself, rather than to include the power within limits.

In the example above they indicate the cube root of nine thus: $\surd\mathfrak{C}\,9$ and they call it the cube root of nine. But I write it in this way $\llcorner 9$ and call it the cube root of nine. I shall say more about these symbols in the right place.

In this context geometrical or real numbers, which are called irrational and meaningless, have their origin.

In these calculations with radicals some indices are even numbers, others are odd. Some again are prime numbers, that is, only divisible by unity. Other numbers are composite, that is, exactly divisible by some other number.

> Thus the indices 2, 4 and 6 are even; 3, 5, 7 are odd; again indices 2, 3, 5, 7, 11 are primes, and cannot be divided by any number. But 4, 6, 8, 9, 10 etc. are composite and made up of other numbers; 4 from two twos, 6 from two threes or three twos; 8 from two fours, or four twos; 9 from three threes and 10 from two fives.

From these facts a table of radical multiplication and root extraction is made wherein the indices are compounded. For it is easier to extract roots and effect radical multiplication by means of small components than by composite indices.

> For example, a fourth root is extracted from a given power with more difficulty in one operation than if you first extract its square root, and thence extract another square root. To raise a number to the sixth power or to extract the sixth root is not as easy as if you first take the cube, or extract the cube root, and then square the result, or take the square root of the result, for the same reasons already given. A numerical example: The sixth root of 64 is more easily extracted by first extracting the cube root which is 4, and thence extracting the square root, which is 2. Or, first extract the square root of 64 which comes to 8, and then extract the cube root of 8, which is 2 and the required sixth root.

From the foregoing it is clear that the significant features of the multiplications, divisions and extractions of the roots of radicals are subject to two basic criteria. Multiplication can be tested either by division or by root extraction: Division can be tested either by multiplication or by root extraction; and the extraction of roots by multiplication or by division.

> As in the example above: power 32, index 5 and root 2. If you doubt whether 32 is the fifth power of two, divide 32 repeatedly by 2 and you will hit upon 5 as the index. Or extract the fifth root of 32 and you will get 2, the previous root. From this you will agree that 32 is the correct power. Further, if you doubt whether 2 is the root, divide 32 by the root repeatedly, and you will find that the index is 5; or raise 2 to the fifth power and you will obtain 32. Finally, if you wish to show that 5 is the correct index, extract the fifth root of 32, and you will obtain 2. Or raise 2 to the fifth power by repeated multiplication and you will obtain 32.

Thus, from power, index and root, any two of which are given, we can obtain the third number by radical multiplication, division and root extraction.

CHAPTER V.

COMPOUND CALCULATIONS.

So far I have dealt with simple calculations; now follow the rules of compound calculations.

A compound calculation is one which requires several different numbers and several different operations to produce the required result.

Compound calculations and their rules of formation are either the same as for proportion or else otherwise defined.

The rules of proportion. These produce a required quantity from given data by the calculations of simple proportion alone, involving multiplications and divisions.

> For instance, if a man walks four miles in three hours, how many miles will he walk in six hours? Also, if six cows eat three measures of hay in four days, how many cows can be fed on five measures of hay in two days.[2] Also 20 Scottish shillings make one pound, 2 pounds are equal to three marks, and 5 marks are worth one crown. How many shillings are 9 crowns worth? Questions on proportion do not involve the use of addition or subtraction. Multiplication and division are proportionals by reason of their definitions.

In these problems we must pay attention to the position of any quantity, and to the operations needed.

The four rules of position are given below.

First draw a line as shown. A place is now ready below the line for the required quantity and its associated data.

> This follows from the examples given above.

[2] This sentence should end with a '?'.

In the second and third case it is a quicker and easier way to place below the line the smaller of the given numbers together with the required quantity. But whether or not this method is safer is not yet clear.[3]

First example.

6 hours, 4 miles
———
3 hours, how many miles

Second example.

6 cows 5 measures 4 days
———
how many cows 3 measures 2 days

Third example.

20 shillings 2 pounds 5 marks 9 crowns
———
how many shillings 1 pound 3 marks 1 crown

Secondly, if there are two quantities one of which increases as the other decreases, they must be placed side by side on the same side of the line.

> [As in the first example, when there were so many more hours at first, namely three, so much fewer will be the number of miles looked for. Thus, by increasing the number of cows (as in the second example), the number of days will decrease in which they can be fed with the fodder. From which there shall be 3 hours and the number of miles sought, as well as 6 hours and 4 miles; likewise 6 cows and 4 days, and the number of cows sought and 2 days, each put together on the same side of the line.]

Thirdly, two quantities which both increase, or both decrease simultaneously, are inserted on opposite sides of the line.

> [So with a 3 hour increase, also a 4 mile increase will be necessary, and vice versa. Thus with the increase of 6 hours, the increase in the miles is sought out, and vice versa. Also with the increase in the number of cows, it is necessary to increase their fodder, and vice versa, to decrease it with fewer cows. Thus the amount of hay increases with the number of days, and decreases with fewer days. In the third example, with the increase or decrease in the number of shillings, a simultaneous increase or decrease is necessary in the number of pounds equivalent to those shillings. Consequently, the 2 pounds and its equivalent 3 marks also increase or decrease]; this is also true of the 5 marks and the crown to which it is equal. Finally, the required number of shillings must

[3] This is a note inserted by Napier. But the suggestion is not correct. For instance: 6 men make 78 articles in 4 days; how many men make 195 articles in 6 days? (Answer 10).

also change since they must be equal to 9 crowns. And so from each pair, one quantity must be placed below the line, and the other quantity above the line as you may see.

Fourthly, of two quantities of the same kind, one must be above the line, and the other below the line.

[Exactly as in the above examples, separated by the interposed line, 3 hours to 6 hours, and 4 miles to the sought number of miles in the first example: And 6 cows to the number of cows sought, 5 measures of hay to 3 measures, and 4 days to 2 days in the second example: And in the third example 20 shillings to the number sought, and 2 pounds to 1, and 5 marks to 3 marks, and 9 crowns to 1 crown.]

After having studied these rules, this general method of procedure will provide you with a unique solution to all questions of this kind.

Multiply the quantities above the line and then, in turn, multiply the quantities below the line. Then divide the upper multiple by the lower multiple. The quotient will be the answer required.

In this way we can understand all the different rules of proportion by means of a single general method or working.

The scope of this analysis is boundless—the rule of three or the golden rule simple, duplicate, direct and inverse proportion with five quantities and with six—writers hand down the different kinds and their patterns. But the rule of three and multiples of it have not yet attained the synthesis elsewhere all of which you have here in concise form.[4]

This completes the study of proportionals; the rules of non-proportionals are to follow. But these, over and above the calculations in proportion, include additions and subtractions and other mixed calculations which do not conform to the rules of proportion. All of these have therefore been omitted, so that a single book of Algebra may serve us for these topics.

This includes the principal part of all the rules of Arithmetic—alligation, association, the rule of false position, single, double, and many others: likewise the propositions, problems and theorems of Geometry, which play havoc with a memory confused both by the variety and number of topics. These I leave until I deal with Algebra.

[4] This claim is also supported and referred to in Chapter 10 of Book II. "In the three worked exercises of Book I, Chapter 5, the first, second and third rules of proportion were established; and from these a unique and general method of solution was discovered."

CHAPTER VI.

POSITIVE AND NEGATIVE NUMBERS.

So far we have dealt with the calculations of numbers at their source; we now consider their properties.

First, numbers are either positive or negative.

Positive numbers are greater than zero, and carry the sign of addition before them.

These numbers are preceded by nothing, or else by this sign +, which is the sign of addition.

> For instance, if you are debt-free and your assets are assessed at 100 crowns, this sum is written either 100 crowns or thus, +100 crowns with the sign in front; it is pronounced 'an increase of 100 crowns' and it always signifies a favourable balance.

Calculations of this kind can be found in preceding pages, rather than in those which follow.

Negative numbers are less than zero and carry the minus sign before them.

They are always preceded by this sign − which is called the minus sign.

> For instance, if a man's wealth is estimated, and his debts exceed his assets by 100 crowns, then his wealth is properly expressed −100 crowns, with the sign in front. This is pronounced 'minus one hundred crowns' and it always signifies loss or deficiency.

I have shown above that the origin and source of negative numbers arises in the subtraction of a greater from a smaller number.

Positive and negative numbers are added if their prefixed signs are the same, and if their sum is prefaced by the same sign.

> For instance, on adding +3 and +2 they make +5: likewise, on adding −4 and −6 they make −10.

Positive and negative numbers are also added if their preceding signs are different, by prefacing their numerical difference by the sign of the greatest number.

> For instance, on adding +6 and −4 they make +2. Similarly, after adding −6 and +4 they make −2. The sign of the larger number 6 always preceding the difference between 6 and 4.

Subtraction is done by altering the sign of the number to be subtracted, and by adding it to the other number by means of the rules given above.

> To subtract +5 from +8: Change +5 to −5 and add −5 to +8. This makes 3, the required answer. To subtract +8 from −5: Change +8 to −8 and add −8 to −5. This makes −13, the required answer. In the same way −5 from +8 make +13; +5 from −8 make −13; −5 from −8 make −3; +8 from +5 make −3; −8 from +5 make +13; −8 from −5 make +3.

Positive and negative numbers may be multiplied and divided if their signs are the same, by placing the plus sign before the answer. If their signs are different place the minus sign before the answer.

> Multiply +3 by +2, or −3 by −2: This gives a product +6. Divide +6 by +3, or −6 by −3: This gives a quotient +2. If you multiply +3 by −2, or −3 by +2 you get −6 as product. And if you divide +6 by −3, or −6 by +3 you get a quotient −2.

Roots, both positive and negative, with an even index produce a positive power.

> If the root is +2 and the index is 4, the first power gives +2, the second gives +4, the third +8, and the fourth power gives +16, all positive. Similarly with the root −2, the first power gives −2, the second power gives +4, the third −8, and the fourth +16 positive as before.

From this it follows that a positive number with an even index must have two roots, one positive, the other negative. But negative numbers have no roots.

> In the above example, the positive power +16 has its fourth root equally +2 positive and −2 negative. This is clear from the foregoing remarks, and on examination of both. There can be no number, positive or negative, which can be the fourth root of the negative number −16.

Positive roots with an odd index give, on radical multiplication, a positive power, and negative roots produce negative powers.

> The positive root +2, with odd index 5, on radical multiplication gives +32. The first operation gives +2, the second +4, the third +8, the fourth +16 and the fifth +32 a positive power. Similarly, the negative root −2 with index 5 gives −32 on radical multiplication. The first operation gives −2, the second +4, the third −8, the fourth +16 and finally the fifth power is −32, produced by a negative root.

In the same way it follows that a power with an odd index has one root only, a positive root from a positive power, and a negative root from a negative power.

And so in the previous example the positive power $+32$ with index 5 will have a positive root $+2$. Similarly the negative power -32, with the same index, will have a negative root -2. This is clearly seen in the previous example and after studying both roots. The rules of proportionals are not repeated here, because they are made up of multiplications and divisions, and they can be learned in previous chapters.

CHAPTER VII.

FRACTIONS.

So far the first category of numbers has been treated. Now follows the second kind.

In the second category the numbers are either integers or fractions.

We call those numbers integers which have either unity as a denominator, or nothing at all. The calculations produced by integers amongst themselves are dealt with in earlier chapters, but fractions are treated in the following pages.

We call a number a fraction, which has a denominator, differing from unity, placed beneath the numerator.

The denominator is the number placed below the line which indicates the number of parts into which the whole must be divided.

The numerator is a number placed over the line which tells the number of parts to be taken.

For example, the number $3ab$ is an integral quantity. So too, is $\frac{3ab}{1}$ because it is identical (with $3ab$) even though disguised as a fraction. Likewise, $\frac{3ab}{2bc}$, $\frac{5a}{2}$, and $\frac{3a}{2a}$, which is the same as $\frac{3}{2}$, are fractions or fractional numbers, whose upper terms are numerators, and lower terms denominators.

I have shown above that fractions trace their origin to division. Fractional numbers greater than unity come from the inexact division of a smaller number by a greater one.

For 9 divided by 2 produces $4\,\frac{1}{2}$, or, if you wish, $\frac{9}{2}$, and this is greater than unity. Likewise, on dividing 3 by 5 you get $\frac{3}{5}$ as I have shown.

From this, every numerator acts as the number to be divided, but the denominator that of the number dividing it.

> As in the example above, $\frac{3ab}{2ac}$ signifies the same as $3ab$ divided by $2bc$; so also $\frac{3a}{2a}$ has the same meaning as $3a$ divided by $2a$, or more briefly 3 divided by 2. Finally, it means the same as three parts of unity divided into two parts. In the same way, $\frac{3}{4}$ are equal to three-fourths of unity, or three divided by four, which comes to the same thing.

Every number which has a numerator and a denominator is regarded as a fraction, and is computed as such.

Hence it is wise to regard the denominator of an integer as unity. In this way integers and fractions can be dealt with by the rules which apply to fractions.

Fractions can be computed more easily if they are first simplified and reduced, before being increased by subsequent operations.

They are reduced and simplified by dividing composite terms by their greatest common divisor. A greatest common divisor exactly divides both terms, and a larger one cannot be used.

This is found by first dividing the greater term by the lesser, and then by repeatedly dividing the preceding divisor by its remainder, until nothing is left. And the last divisor (ignoring the remainder) is the maximum common divisor sought.

> Thus the greatest common divisor of 55 and 15 is found as follows: divide 55 by 15, remainder 10. Divide 15 by 10, remainder 5. Divide 10 by 5, remainder nil. The greatest common divisor is therefore 5, for it divides 15 three times, and 55 eleven times.

When you reach division by unity, there is no common factor, and both terms are primes, or prime to each other.

> Let the terms be $5a$ and $3a$. On dividing $5a$ by $3a$, the remainder is $2a$. Then divide $3a$ by $2a$ and you have remainder $1a$. This, divided into $2a$ leaves no remainder. And so, $5a$ and $3a$ have no divisor greater than $1a$ or unity. If you divide $5a$ and $3a$ in turn, you get the primes 5 and 3 as explained more fully in its proper place.

Here I must warn you strongly against dividing incommensurable numbers which never terminate. This will emerge clearly in its proper place.

> For the square root of the number ten, which they call its quadratic root, no common measure can ever be found, much less its greatest common measure. But more of this later.

[With the terms that cannot be abbreviated, their abbreviation is by a diminution operation.

By obtaining a maximum common divisor, and with each term divided by it, new abbreviated terms originate; and this operation is called abbreviation.]

CHAPTER VIII.

CALCULATIONS WITH FRACTIONS.

FRACTIONS can be added or subtracted if they have the same denominator.

If the denominators are different they must be converted to the same denominator.

This is effected by dividing both denominators by their highest common factor and by noting the quotients.

Then, by multiplying both terms of the first fraction by the quotient of the denominator of the second fraction we obtain a new first fraction. Then multiply both terms of the second fraction by the quotient of the denominator of the first fraction, and a new second fraction is produced with the same denominator as the first fraction.

> Reduce the fractions $\frac{2}{3}$ and $\frac{7}{9}$ to the same denominator. The highest common factor of denominators 3 and 9 is 3. On dividing the denominators, 1 comes out for the first fraction and 3 for the second fraction. You then multiply both terms of $\frac{2}{3}$ by the second quotient 3, and $\frac{6}{9}$ comes out for the new first fraction. Similarly, on multiplying $\frac{7}{9}$ in the same way by unity, the quotient of the first fraction, $\frac{7}{9}$ comes out with the same denominator as $\frac{6}{9}$.

The new numerators, which now have the same denominator are added and subtracted by keeping their new common denominator. You have the sum by addition, and the residue by subtraction.

> In the example above the new numerators 6 and 7 are added and make 13; and with the common denominator 9 they make $\frac{13}{9}$ as the total after addition. In the same way, if you take $\frac{6}{9}$ from $\frac{7}{9}$ there remains $\frac{1}{9}$ the residue after subtraction.

Fractions are multiplied by dividing each separate pair, one a numerator and the other a denominator, by their highest common divisor. The final quotients of all are recorded. Then by multiplying the quotients of the numerators in turn, the product gives the new numerator. The same is done to the quotients of the denominators, and the product gives the new denominator of the answer required.

> Multiply $\frac{18}{20}$ by $\frac{35}{231}$. First divide 18 and 20 by their highest common divisor which is 2. This gives $\frac{9}{10}$ and $\frac{35}{231}$. Then divide 10 and 35

by 5 giving 2 and 7. But write it down as $\frac{9}{2}$ and $\frac{7}{231}$.[5] Next divide 9 and 231 by 3 which gives 3 and 77; but write it down as $\frac{3}{2}$ and $\frac{7}{77}$. Finally divide 7 and 77 by their highest common divisor 7, and this gives 1 and 11, written down as $\frac{3}{2}$ and $\frac{1}{11}$. This done, multiply the final quotients of the numerators 3 and 1, and do the same with those of the denominators 2 and 11. The first gives numerator 3 and denominator 22. Write down $\frac{3}{22}$ which is the required number. The multiplication of fractions which cannot be simplified is carried out in the same way. Multiply $\frac{2a}{3}$ and $\frac{4}{5}$. First multiply the numerators 2a and 4 giving 8a, the new numerator. Then multiply the denominators 3 and 5, making 15 the new denominator. The answer required is therefore $\frac{8a}{15}$.

By this method of multiplication, fractions of fractions, nay more, fractions of fractions time and again, can be reduced to simple fractions.

Simplify two-fifths of three quarters. This is written $\frac{2}{5}$ of $\frac{3}{4}$ and then, by cancellation $\frac{1}{5}$ of $\frac{3}{2}$. Then by multiplying the numerators and denominators in turn it becomes $\frac{3}{10}$ a unique simplified fraction which is equal to the fraction of a fraction given above. Simplify three-quarters of two-thirds of one-half. This is written $\frac{3}{4}$ of $\frac{2}{3}$ of $\frac{1}{2}$. On cancellation it becomes first $\frac{1}{4}$ of $\frac{2}{1}$ of $\frac{1}{2}$, and then $\frac{1}{4}\ \frac{1}{1}\ \frac{1}{1}$. Finally by multiplying the numerators and the denominators in turn this makes $\frac{1}{4}$ which is equal to $\frac{3}{4}$ of $\frac{2}{3}$ of $\frac{1}{2}$.

Fractions are divided by inversion of the terms of the divisor. The inverted terms are multiplied out as in previous examples of multiplication.

[As in] $\frac{3}{10}$ to be divided by $\frac{3}{4}$. Invert the terms of the divisor. This gives $\frac{4}{3}$. This must be multiplied by $\frac{3}{10}$, giving on a first cancellation $\frac{1}{10}$ and $\frac{4}{1}$, then $\frac{1}{5}$ and $\frac{2}{1}$. Multiply the terms above and below the lines, and this gives $\frac{2}{5}$ the required quotient. This is also a verification of multiplication. [Extraction] is performed by extracting the desired root both from numerator and denominator or from whatever terms there are of the same degree. And they give the terms of the desired root.

The square root of the fraction $\frac{16}{25}$ is required. The square root of 16 is 4, and the square root of 25 is 5. Thus the numerator is 4, and the denominator is 5, and the required square root of $\frac{16}{25}$ is written $\frac{4}{5}$. [Also] the square root of $\frac{3}{4}$ is required. This is more conveniently expressed

[5] According to Napier's verbal rule (on page 18) for multiplying fractions, the reduction of 18/20 to 9/10 should have been followed immediately by the reduction of 35/231 to 5/33.

as $\frac{48}{64}$. When roots cannot be found exactly the greater the number of figures which express the same ratio, the more exactly can roots be calculated. The square root of 48 cannot be evaluated, and so try 49, whose root is 7. Then extract the root of 64 which is 8. The square root of $\frac{3}{4}$ is therefore $\frac{7}{8}$ to a good approximation.

Multiplication and division of radicals, and the rules of proportion, since they consist of repeated multiplications and divisions, can easily be found in previous chapters to which I refer you.

When calculations are finished, mixed fractions, those in which the numerator exceeds the denominator, must be broken down into integers and proper fractions. This is done by dividing the numerator by the denominator, when an integer comes out as the quotient, and the remainder becomes the numerator. The divisor becomes the denominator of the mixed fraction which results.

If, after completing your calculation, $\frac{11}{4}$ comes out proceed as follows. Divide 11 by 4 and you get 2 as quotient, and 3 will remain. And so two whole numbers, and three-quarters of a unit, written $2\,\frac{3}{4}$, are the same as the previous $\frac{11}{4}$, but in a form which is much clearer.

END OF BOOK ONE.

[BOOK TWO.

OF ARITHMETICAL CALCULATIONS.]

CHAPTER I.

THE DEFINITIONS AND NOTATION FOR INTEGERS.

SO FAR I have treated the calculations of quantities common to all kinds of Logistic. Then came their special properties.

In the third place calculations are made with arithmetic numbers or else with algebraic or fictional numbers. And so Logistic is concerned with arithmetic numbers, treated in Books II and III; or with algebraic or fictional numbers which are treated in Book IV.

Arithmetic numbers are defined by real names, which explain how many there are, numerically, or how great they are in size.

Arithmetic numbers are either distinct with a definite number, or compound and defined in relation to other quantities.

And so the Logistic of arithmetic numbers is concerned either with discrete numbers, and is called Arithmetic, treated in Book II; or else it deals with the material quantities involved in Geometry. This is treated in Book III.

ARITHMETIC is therefore the Logistic of discrete quantities by the use of individual numbers.

A number is discrete which measures out its own unique and individual ranking.

A discrete number is either an integer or a fraction. And so Arithmetic deals with numbers of both kinds.

Integers are numbers for which unity is an exact measure.

The spoken names of the integers are those which the local language supplies, just as in English: one, two, three, four, etc.

The written names or symbols are nine in number 1 [one], 2 [two], 3 [three],

4 [four], 5 [five], 6 [six], 7 [seven], 8 [eight], 9 [nine].

These numbers indicate different quantities when put in different places.

Besides these nine symbols or figures there is the circle 0, which signifies nothing at all when in its place. But this 0 is intended to fill up an empty space.

The place-values are counted from right to left. The place-value of the first figure is its nominal value; the place-value of the second digit is tenfold; that of the third digit is one hundredfold; the fourth counts a thousandfold; the fifth counts ten thousand times the nominal value[; the sixth a hundred times a thousandfold; the seventh, a thousand times a thousandfold; the eighth, ten thousand times a thousandfold.] And so on to infinity, progressing all the time by multiples of ten.

> For instance, 7 represents 7, but 70 represents seventy, and 700 represents seven hundred. In the same way 8000 stands for eight thousand, and 60000 stands for sixty thousand. And in the same way 68777 means sixty-eight thousand seven hundred and seventy-seven. Likewise 90630 signifies ninety thousand six hundred and thirty. And so for all the numbers.

From these examples we derive an easy way of describing large numbers. After every third digit a point is inserted. The first point indicates thousands, the second a thousand thousands, the third, thousands of thousands of thousands, the fourth thousands of thousands of thousands of thousands. And so proceed with additional points. The figure in the first place after the point carries its nominal value; the second figure from the point carries ten times its value; and finally, the third figure after the point carries one hundred times its value.

> And so the number 4734986205048205 is punctuated like this: 4. 734. 986. 205. 048. 205 and it is read as follows: Four thousand thousand thousand thousand thousand; seven hundred and thirty four thousand thousand thousand thousand; nine hundred and eighty-six thousand thousand thousand; two hundred and five [thousand thousand; forty-eight thousand, two hundred and five: And so for all the numbers.]

CHAPTER II.

ADDITION AND SUBTRACTION OF INTEGERS.

We have now dealt with the names and symbols for integers. Computation follows; first addition and then subtraction.

In addition, attention must be paid to position, as well as to the operation or practice.

As for position, numbers are entered directly below other numbers. And so, starting from the right, the first figure is entered directly below, in the first place. The second entry is put below in the second place, and the rest directly below in the remaining places. A line is then drawn below the bottom number.

There are three rules for addition.

First, figures which occupy the first place are added together, and the first figure only of this sum must be written in below the line. The remainder, if there is one, must be memorised.

Secondly, this remainder which you have memorised, together with all the figures in the second place, must be added together. The first figure of this sum must be entered below the line in the second place. The rest of this sum, if any, must be memorised. This operation must be repeated through to the last figure of all.

Thirdly, total up the figures in the last column together with the final number in your memory and insert this amount in the place at the extreme left.

The fifth chapter of Genesis gives the years from the creation of [Adam and] the earth as follows: 130 to Sheth; hence 105 to Enosch; hence 90 to Kenan, hence 70 to Mehalalelem, 65 years to Jered; hence 162 to Henoch, likewise 65 to Methusalach, 187 to Lamech, hence 182 to the birth of Noah, 600 from the birth of Noah to beginning of the flood. The question is, when these are added what is the total number of years from the creation [of the earth] to the flood? Set up all the years in columns as shown above. First add up 5, 5, 2, 5, 7, 2 on the extreme right. This gives 26. Write 6 directly below this column and keep 2 in your memory. Secondly, add this 2 to the second column of figures 3, 9, 7, 6, 6, 6, 8, 8 making 55. Write down the first 5 and keep the second in your memory. Thirdly, add this memorised 5 to 1, 1, 1, 1, 1, 6 in the last column of figures, making 16. Enter this number in the left hand places, making a total of 1,656.

130
105
90
70
65
162
65
187
182
600
1656

Nonetheless, addition, strictly speaking, applies to two numbers which produce a third number.

For example, 9754862 and 863556 are to be added. These from the rules given above produce the total 10618418 shown in the margin. You have a verification of this in subtraction which follows this section.

9754862
863556
10618418

In subtraction you pay attention both to the position of a number and to the operation.

Place-value, as in addition, starts to the right. The number to be subtracted is placed on the bottom line, and the number from which the subtraction is to be effected is placed on the middle line above it. A line is drawn separating the top and middle lines, and the remainder is inserted on the top line. Starting from the right, the first figure is written in the first place. The second figure is inserted in the second place, and the remaining figures in order are written into the remaining places.

The working, however, is best begun from the left. But if you want to begin from the right, a single rule applies in both cases.

Whatever figure is on the bottom line is subtracted from the figure immediately above. If there is nothing below, then nothing is subtracted. But if the figure above is smaller than that below, then add 10 to it before subtraction. The integer left over is entered in the top line, provided that the total of the lowest line towards the right does not exceed the total written directly above. But, if it does exceed that total, deduct 1 from the figure, if it is a numeral, or add nine if it is zero. And enter it on the top line.

$$\begin{array}{r} 2690997393 \\ \hline 2738154098 \\ 47156705 \end{array}$$

47156705 is to be subtracted from 2738154098. Write the second number above the first number as shown above, making ten digits corresponding to ten place-values. Accordingly, starting from the left with 27, take away zero (for nothing is written below). There remain 27 reduced by unity since 47156105 below exceed 38154098 on the line above. And so 26 is the remainder which must be written above in place of the previous 27. Now take 4 from 3; but first add 10 to it so that 4 can be subtracted. This leaves 9 which, without change must be written above three, because 7 is less than 8. Next take 7 from 8 leaving unity. This must be reduced by unity because 156705 towards the right exceed 154098 written above it. And so 0 must be written on top. Now take 1 from 1 and 0 remains. But because the 56705 following, exceed 54098 on the line above, instead of 0 the digit 9 must be entered on top. We take 5 from 5 in exactly the same way leaving 0; but 9 must be written on top because the following 6 is greater than the 4 written above it. Next take 6 from 14 leaving 8. But write 7 above because 705 following exceeds 098 above. Then take 7 from 10 leaving 3 which must be entered on top because 0 below is smaller than 9 above it. Now take 0 from 9 leaving 9. This must be entered in the top line because 5 on the bottom line is less than the 8 placed above it. Finally take 5 from 8 leaving 3 and this digit must be entered on top because no more digits are left for consideration. In this way you have the required answer 2690997393.

A verification of this work can be effected by adding 47156705 to 2690997393 and by getting 2738154098 for your answer; or by subtracting 2690997393 from 2738154098 and getting 47156705 as answer. This test can be applied in every case considered.

Here is another example and verification of an addition given above.

863556 is to be subtracted from 10618418. The number left is 9754862 as seen above under addition. It likewise shows if you performed the operation from the right towards the left.

If it is proposed to subtract the greater number from the smaller nevertheless always take the smaller number from the greater by means of the rules given earlier. But place the minus symbol before the remainder, and in this way a defective or negative number emerges.

For instance, we want to take 10618418 from 863556. By the previous rule take the latter from the former and this produces 9754862 as in the example above. This number is converted into the negative or defective number −9754862, and this is the required remainder. I have treated the properties of defective numbers previously in Book I, Chapter 6.

CHAPTER III.

MULTIPLICATION OF INTEGERS.

MULTIPLICATION is effected by a single digit or by a number containing several digits.

Multiplication by a single digit consists of the operation of a single digit on one or more digits.

The multiplication of one digit by another is memorised and learned most readily from the adjacent table.

What is the result of multiplying 7 and 8 together? Look for the greater number on the top row, and look for the smaller number in the left hand column. The junction between the vertical through 8 and the horizontal through 7 indicates 56, which is the required product.

	9	8	7	6	5	4	3	2	1
1	9	8	7	6	5	4	3	2	1
2	18	16	14	12	10	8	6	4	
3	27	24	21	18	15	12	9		
4	36	32	28	24	20	16			
5	45	40	35	30	25				
6	54	48	42	36					
7	63	56	49						
8	72	64							
9	81								

If you have forgotten the product of any two large digits, then subtract each digit from ten, and multiply the differences together. This gives the right hand digit of the product. Next take the smaller difference from the smaller digit; or the greater difference from the larger digit. The numbers are the same in each case and will give the left hand digit of the required product.

8 × 2
7   3
———
5   6

If you have forgotten the product of 7 times 8 or of 8 times 7, then calculate their differences from ten. Multiply the resultant 3 and 2 together to obtain 6. This is the required digit on the bottom right. Next take 2 from 7, or 3 from 8, and 5 remains. This digit is placed at the bottom left. And so 56 is the required multiple of 7 and 8.

The multiplication of several figures by one digit is carried out with due regard to place-value and the operation, proceeding in order from right to left. The multiplicand is placed either above or before the multiplier, and a line is drawn underneath in both cases. The multiplication of several figures by one digit has three rules for its completion.

[The first is,] the right hand figure is multiplied by the given digit. If the multiple is a single digit it is put below the line; but if there are two figures only the digit on the right is entered. The left hand figure is memorised.

[The second is,] the next figure is multiplied by the given digit and the memorised figure is added. If this produces one figure, then enter it below the line. But if there are two figures, only enter the right hand one and commit the left hand one to memory. This must be added as before. The operation is repeated moving towards the left until the final figure is reached.

[The third is,] the multiple of the final figure together with the memorised remainder (if any) are added into one total and entered below the line on the extreme left.

865091372 is to be multiplied by 5[, as in the margin]. First, the right hand figure 2 is multiplied by the digit 5, and this produces 10. The right hand figure 0 is entered in the quotient and the 1 is memorised.

865091372
5
4325456860

865091372 (5
4325456860

Next multiply 7 by 5 which gives 35. Add 1 from your memory, making 36. Enter the right hand 6, and carry the 3 in your memory. To continue: 5 times 3 make 15 and the 3 in your memory makes the total 18. Enter 8, and remember 1. Then follows five times 1 making 5 which makes 6 with 1 from your memory. Enter 6 in the space below, and there is nothing to memorise. Next, five times 9 make 45. Enter 5 and keep 4 in your memory. 5 times 0 follows, and this produces zero. But you must add the 4 in your memory, and the total 4 must be entered. To continue: Five times 5 make 25; enter 5 and remember 2. Then multiply 6 by 5 giving 30, and add the memorised 2 making 32. Write in 2 and remember 3. Next multiply 8 by 5 making 40. To this add 3 from your memory making a final total of 43. Enter both of these figures in the extreme left position of the quotient, And so the required product totals 4325456860.

Multiplication of several figures by several other figures is now left.

This operation has three rules in addition to those given above.

[First, t]he whole multiplicand (as before) is multiplied by every figure in the multiplier, beginning from the right or from the left, as you please.

[Second, t]he first figure of the multiple has its right hand figure placed directly under the figure of the multiplier, and the rest follow in order, towards the left.

[The third rule is,] beneath the separate multiples draw another line, and add up all the multiples into one total. This will be the total multiple and the required product of the operation.

865091372 is to be multiplied by 92105. The figures are set out as shown [in the margin]. Commence the multiplication from whichever side you please, left or right. For this example let us work from the right. Perform the multiplication of the top line by 5. This gives, by the foregoing, 4325456860 which is entered so that the right hand figure 0 is directly below the multiplier 5. The other figures proceed in order to the left. Next the multiplicand is multiplied by 0, which is nothing and which produces nothing. Consequently nothing need be entered from this multiplication by 0. Next perform the multiplication by 1; and this produces its own value 865091372. The right hand digit is entered beneath its multiplier 1, and the other digits in order towards the left. Next double the multiplicand and 1730182744

865091372
92105
4325456860
865091372
1730182744
7785822348
79679240818060

comes out. Its position commences beneath its multiple 2, and from that place the figures are entered in order to the left. Finally perform the multiplication by 9 and this produces 7785822348. The first digit 8 is placed under 9 and from this place in order the digits proceed to the left. At last with all the multiplications completed, draw a horizontal line and add up the columns giving 79679240818060 as the required product. You get exactly the same result from the second method, when you commence the multiplication with the left hand figure of the multiplier 9, and the second lay-out makes this quite clear. When operations are alike, results are also alike.

```
865091372 (92105
----------------
7785822348
 1730182744
   865091372
    4325456860
----------------
79679240818060
```

You will have a verification of this in the following chapter dealing with division: and this exercise is also a verification of the principle of division.

The multiplication of three or more numbers is effected by multiplying the first number by the second; this product is multiplied by the third number; and this is again multiplied by the fourth number and so on.

To multiply 5, 4, 2 and 3. 5 times 4 make 20. Now multiply 20 by 2 giving 40, Now multiply 40 by 3 and this gives 120 the product of all four numbers.

CHAPTER IV.

DIVISION OF INTEGERS.

THE DIVISION of a small number by a larger number cannot be done otherwise than by drawing a line between the dividend above and the divisor below. This produces a fraction less than 1.

Let 3 be divided by 5: this makes $\frac{3}{5}$, which is read three-fifths of one unit or three divided by five; and it is a fraction.

And so fractional numbers arise in this context, as I have shown in Chapter 7, [of fractional quantities in general. The second part of the Arithmetic will deal with this.]

I have indicated earlier that the division of one quantity by an equal quantity produces unity.

Finally, in the division of a greater number by a smaller one the quotient is always greater than unity. And naturally the operation of division stresses this fact.

In the division of a greater number by a smaller one, attention must be paid both to the operation and to place-value. Each of them begins from the left and moves towards the right. It is most convenient to place the left hand figure of the divisor below the left hand figure of the dividend if the latter is smaller. But if not it is placed before the left hand figure of the divisor, followed by other figures (if any), in order, towards the right. Immediately following the divisor insert a parenthesis to indicate the location of the quotient. A line is then drawn underneath.

There are four rules for the operation:

[First,] carefully examine how often the divisor can be taken from the top group of figures on the left. The space below, starting at the parenthesis, is intended for the quotient. Having made your decision, enter the figure after the divisor and parentheses in the space allocated to it.

[Second,] as the divisor is multiplied by the latest figure added to the quotient, the multiple is entered in the debit row. (See Chapter 3).

[Third,] the fresh multiple is subtracted from the row of figures directly above, and these figures are cancelled together with those from which they were subtracted. The remainder is then entered above. (See Chapter 2 of this book).

[Fourth,] these three operations are repeated until the whole dividend is exhausted. The remainder is either zero or a number less than the divisor. No vacant places are left in the quotient up to the extreme right hand figure of the divisor. Finally, on the right hand of the quotient a line is drawn. The remainder is written above, and the divisor is written below the line; and this gives the numerator and denominator of a fraction.

Example: The 366 days of a leap year are divided by the 7 days of the week, to find the number of weeks in a year. The problem is set out in the margin. The two left hand upper figures are 36, and the last figure is 6, as you see. These figures are above empty places for the quotient, next to the parenthesis. You must now ask how many times 7 can be taken from 36. In reply, there are five sevens in 36; so insert 5 in the first vacant space for the quotient. Now multiply 7 by 5, making 35. This must be entered in the debit places, putting 5 under 5, and 3 to the left. Next take 3 and 5 from the 3 and 6 above, which leaves 1 unit to be entered as shown. Then cross out 35 and 36.

  366
7(5
———

   0
 ~~1~~
~~366~~
7(52$\frac{2}{7}$
———
~~35~~
 ~~14~~

Next we deal with the remainder 16, at and above the next vacant place for the quotient. Repeat the previous procedure, and decide how often 7 may be taken from 16. This can be done twice, leaving 2 as you can see. Enter 2 in the next vacant place for the quotient and multiply 2 by 7 giving 14. Enter this below 16, and subtract 14 from it leaving 2. Delete 14 and 16. This leaves 2 days, or better $\frac{2}{7}$ of a week which must be added to the quotient. The full quotient then becomes $52\frac{2}{7}$ weeks in a leap year.

[Another] example.

861094 is to be divided by 432. This is set out in the margin. As you see, 4 is placed before 8 because 4 is less than 8. Next consider how often 432 can be taken from 861. Try 4 into 8, or 43 into 86. Both of these can be taken twice; but 2 cannot be taken from 1 twice. Since this guess fails we take 432 from 861, and this goes once only. We therefore write 1 for the first entry in the quotient, multiply it by 432 and then subtract 432 from 861, leaving 429. Now consider how many times 432 can be taken from 4290. Now 4 goes into 42 nine times and 6 over; further, 69 contains 3 nine times with a sufficiently large remainder to include 2 nine times. And so 9 is placed in the next vacant place as the second figure of the quotient. Next multiply 432 by 9 giving 3888. This must be subtracted from 4290 leaving 402. Next examine how many times 432 can be taken from 4029. Like the previous estimate you will see that this can be done 9 times. Place 9 in the quotient and multiply it by 432, giving 3888 as before. This must be taken from 4029 with remainder 141. Now consider how often 432 can be taken from 1414. This can be done 3 times, and the 3 is placed in the last position for the quotient.

118
~~141~~
~~402~~
~~429~~
~~861094~~
432 (1993$\frac{118}{432}$
~~432~~
~~3888~~
~~3888~~
~~1296~~

Multiply 3 by 432 which gives 1296 to be subtracted from 1414 placed above it. Last of all comes the remainder 118, or $\frac{118}{432}$ which is added to the quotient, and gives a final total of $1993\frac{118}{432}$.

[Third] example [and] verification of the multiplication above.

79679240818060 divided by 865091372. The problem is set out as in the margin. Consider how often 8650 can be taken from 79679: or how often 8 can be taken from 79. You see that this can be done nine times, leaving a sufficient remainder from 6, 5, and succeeding digits.

```
              0
     43254568
    90834594
  182101733
 79679240818060
865091372 (92105
7785822348
 1730182744
   865091372
    4325456860
```

Next calculate 9 times the divisor (865091372 which was placed in the quotient) and which gives 7785822348. This must be inserted in the debit space, and then subtracted from the number above leaving 1821017338. This, in turn, must be divided by 865091372. You will see from these figures ([evidently] 8 into 18 and so on [from all the rest]) that it can only go twice. Put 2 in the quotient and multiply the divisor by 2. This gives 1730182744 which must be taken from 182101733 above, leaving 908345940. From this the divisor can be taken once only, so put 1 in the quotient. Multiply the divisor by 1, which gives 865091372. This is taken from 908345940 and leaves 432545686. But the divisor cannot go even once into this number, so place 0 in the quotient. Then proceed by finding how often the divisor can be taken from 4325456860. By guessing 8 into 43 goes 5 times, we put 5 in the quotient, and multiply the divisor by 5. This gives 4325456860, and it must be taken away from an equal number above, with nothing left as remainder. This division is therefore exact, and the quotient 92105 is an integer. This result verifies the answer obtained in [the penultimate] example for multiplication; and [from that approved. And so for the rest.]

CHAPTER V.

[MISCELLANEOUS] SHORT METHODS OF MULTIPLICATION AND DIVISION.

Multiplication by 10, 100, 1000 or by other numbers made up of unity with as many zeros as you wish is easily effected. This is done by adding as many zeros as are contained on the right hand side of the multiplier.

865091372 multiplied by ten gives 8650913720; multiplied by one hundred gives 86509137200; multiplied by 10000000 gives 8650913720000000.

On the other hand, division by 10, 100, 1000, or by other numbers made up of unity with as many zeros as you wish is also easily done. Cut off as many figures from the right of the dividend as there are zeros in the divisor. The figures cut off are inserted above the line, and those of the divisor below the line. This must be added to the quotient and written as a fraction.

865091372 is to be divided by 100. This gives $8650913\frac{72}{100}$ as the quotient. Similarly, 8650913720000089 is to be divided by 10000000. This produces a quotient of $865091372\frac{89}{10000000}$.

From this it follows that multiplication and division by whatsoever numbers you please, can readily be effected. First, for multiplication: keep the zeros separate until the product of the remaining figures of the multiplier and the multiplicand have been evaluated. Then restore the zeros to the product. For division: cut off the zeros from the divisor, and as many figures from the dividend. Next divide that part of the dividend which is left by the corresponding part of the divisor, draw a horizontal line and insert the fractional remainder.

65294 multiplied by 2300. We take the product of 65294 and 23 giving 1501762. On adding 00 the total multiple becomes 150176200.

~~1~~~~0~~8
~~6~~~~5~~~~2~~|94
23(28
~~4~~~~6~~
~~1~~8~~4~~

[Likewise] 65294 is to be divided by 2300. The lay-out is shown in the margin. To compensate for 00 (taken from 2300) we use the indicator |94 as shown. Next divide 652 by 23, and the quotient, with its fractional remainder is $28\frac{894}{2300}$ or, in its lowest terms $28\frac{447}{1150}$.

Division by two: This lends itself readily to intuition.

For example, 65294 is to be divided by two. Who does not recognise instantly that 3 is the half of 6, that 2 is half of five with 1 over; that this remainder makes 12 with the two following, and that 6 is the half of 12. Then follows 9, whose half is 4 with 1 over (or better, ten). This last makes 14 whose half is 7. Consequently 3, 2, 6, 4, 7 provide the digits for 32647 which is one half of the given number, [as written in the margin.]

65294
32647

Multiplication by five: This follows readily from the previous paragraph, for five is ten divided by two.

For instance, 865091372 is to be multiplied by five. According to the first section above 865091372 multiplied by ten is 8650913720 and one half 8650913720 is 4325456860 and this is five times the given number.

Alternatively, division by five is easily found as twice division by ten.

4325456860 is to be divided by 5. First double the number giving 8650913720. By the rule given [in Section 2] above divide by 10. This gives 86509137 the required fifth part.

[Nor is it difficult] to multiply by nine [by taking] the number away from its tenfold. To multiply by four, take the number away from its fivefold. To multiply by six, add the number to its fivefold. To multiply by two, add the number to its equal. All of these multiplications can be performed most easily by addition or subtraction.

[Also from these, similarly facilitated, we have the triplicate, the sevenfold and the eightfold,] and thus multiplication by each of the nine digits.

[Instead of examples, there shall be the preceding and the following.]

When the multiplier includes all nine digits or the greater part of them; or if it seems that the quotient is likely to be long, it is better to have to hand single multiples. This is best done by the [established] table or by continued addition. Best of all perhaps is the use of my bones referred to in Rabdologiæ.[6] If these are to hand the whole multiplication or division can be accomplished.

For instance, let us see what 92105 multiplied by 865091372 produces. (See the final example in the chapter on multiplication.) Multiply 92105 successively by 2, 3, 4, 5, 6, 7, 8, 9, and tabulate them as shown. So that single multiples shall contain the same number of figures place a 0 before the unit entry. (The tabulated multiples are in the left hand column.) And now for a short method for the multiplication. Single multiples are set in position in the order indicated by the multiplier 865091372. Beneath the digit 8 the eightfold multiple taken from the table commences. Below 6 the sixfold multiple commences, and so on for each of the multiples. Finally, when all are added 79679240818060 emerges. This is the required multiple and it agrees with the results found earlier.

092105	Simple
184210	Double
276315	Triple
368420	4^{ple}
460525	5^{ple}
552630	6^{ple}
644735	7^{ple}
736840	8^{ple}
828945	9^{ple}
921050	Tenfold

```
865091372
736840|276315
  552630|644735
    460525|184210
        828945
          092105
79679240818060
```

Similarly, as a verification of the above example equally for multiplication as for division, 79679240818060 is to be divided by 92105. The exercise is set out as shown [in the margin] (see Chapter 4).

[6] Napier's reference to *Rabdologiæ* must have been written in 1617, after that book was published.

```
          66315
         34263
        12636
       84158
      46894      0
     59952  18421
    7967924081806O
    92105)865091372
    ---------------
    736840  644735
     552630  184210
      460525
       828945
        092105
         276315
```

From 796792 take the next smallest number in the table, namely 736840. Next put 8 the figure corresponding to it in the quotient. This leaves 599524. From this take the sixfold entry of the divisor, which is the next smallest number in the Table 552630. Add 6 to the quotient. This leaves 468940 into which the fivefold divisor 460525 fits. The quotient is now 865 and the dividend 84158. It is not possible to take the simple divisor 092105 from this. Put 0 in the quotient making 8650 and the remainder is 841581. From this take 828945 the ninefold divisor. The quotient is now 86509 with 126368 left. Take away 092105, the simple unit divisor. This gives 865091 and remainder 342630. From this take the threefold divisor 276315 with 663156 left. The quotient is now 8650913. Take the sevenfold multiple 644735 from 663156 leaving 184210. The quotient is now 86509137 and the remainder is equal to the twofold divisor. The quotient is therefore 865091372 which agrees with previous results.

So far I have dealt with simple multiplication and division, and with short methods. Next follow the multiplication and division of radicals.

CHAPTER VI.

RADICAL MULTIPLICATION AND DIVISION OF INTEGERS.

THE OPERATION called multiplication is clearly defined. If you multiply one unit by a radical you get the radical itself; and if you multiply a second time by the radical you get the square (index 2). If you multiply the square by the radical a third time you get the cube (or third power). And again, if you multiply a fourth time by the radical, the quadruplicate or fourth power is produced. And so continue for the fifth and sixth powers according to the nature of the index.

235 is to be raised to the fourth power or quadruplicated. First multiply unity by 235 which produces 235. Then multiply a second time by 235 which gives 55225, the square of the given radical. Multiply a third time by the radical, and this gives 12977875, the cube. Continue

and multiply a fourth time by the radical. This gives 3049800625, the required fourth power of the given radical.

Therefore, starting with unity and multiplying by a root several times produces the same result as multiplying the same number of equal roots in succession.

[As in the example above,] starting with unity multiply four times by the root 235; this gives the fourth power 3049800625. If you multiply 235, 235, 235, 235, in succession the quadruplicate or fourth power emerges[, and so of the others].

Radicals whose indices are compound numbers can more easily be evaluated, as explained above, by means of composite indices.

For instance, it is easier to square 235 making 55225, and to square it again (multiplying by 55225) giving 3049800625, the fourth power of the given number. And so to raise a radical to the fourth power is the same as squaring the square of the given number. In the same way, 10 can be raised to the sixth power. Starting with unity, and using continued multiplication we get—first 10, second 100 (squared), third 1000 (cubed), fourth 10000 (fourth power), 100000 (fifth power) and finally 1000000 (sixth power). But using a shorter method, you can cube the square in order to get the sixth power 1000000. This is a little quicker than previously indicated. It depends on the fact that two multiplied by three equals six.

There is another way of squaring a root, other than by continued multiplication. To do this the root to be multiplied must be divided into two parts or terms. The required square is obtained by taking twice the product of the first term and the second term, and adding the sum of the squares of the first term and second term.[7]

The root 35 is to be squared. Divide it into two parts, first part 30 and second part 5. The product of 30 times 5 makes 150. Double this and you have 300. To this add the square of 30 and the square of 5, that is, 900 and 25. This gives a total of 1225. If you square 35 directly you get the same result. In the same way if the root is increased to 352 make the two parts or terms 350 and 2. The product of 350 and 2 is 700. Double this gives 1400. The square of 2 is 4, and the square of 350 is 122500 by the last section. The sum of these three components is $1400 + 4 + 122500 = 123904$. This is the required root which can be

[7] $(a+b)^2 = 2ab + a^2 + b^2$.

checked by direct multiplication and gives the same result. The problem must be handled in the same way if there are four figures, say 3521. The first part is 3520, which when squared gives 12390400. The square of the second term is 1. The product of the first and second terms, when doubled is 7040. The result is $7040 + 12390400 + 1 = 12397411$. And this is the square of 3521.

There is also another method for cubing a radical. This is by dividing the root into two parts or terms. Then multiply three times each term by the square of the other term and add them. Then add the cube of the first term and the cube of the second term. The sum of these four gives the required cube.[8]

The given root is 35, which is to be cubed. First divide it into two terms 30 and 5. Multiply 3 times 5 by the square of 30, which is 900; then three times 30 by the square of 5, which is 25. The first product is 13500, and the second 2250. Now raise 30 to the third power, giving 27000; and do the same for 5, giving 125. The four sums are 13500, 2250, 27000 and 125. Add them together and we get 42875, which is the required cube of the given root. Test this by the continued multiplication of 35, 35 and 35 in succession, and you will get the same result. In the same way, if the root is extended to 351, divide it into first term 350, and second term 1. From this, by the method shown above, you get four components: 367500, 1050, 42875000, and 1. When added they give 43243551 which is the required cube of the number 351.

There are special rules for finding the power of any index—fourth, fifth, sixth and higher. But because the method of continued multiplication (described in the first section) is quite simple and general enough; and since, on the other hand all these special rules for systematic extraction can be used, I propose, at this point, to omit them.

The rule for division is clear from its very definition. For, if you divide a power by a root until unity appears, the number of divisions is the order of the power, or the required index.

For example, 55225 is to be divided by the root 235. The quotient of the first division is 235, and when this is divided by 235, we find that the second and final quotient is unity. Therefore the index of this divisor is two, and the root is the square root. In the same way the power 12977875 is to be divided by the root 235. After the first division 55225 comes out. The second division produces 235, and the third division gives 1. Hence, the division takes place three times; the index is 3, and the root is called the cube root.

[8] $(a + b)^3 = 3ab^2 + 3a^2b + a^3 + b^3$.

The number of the index, or the kind of root produced comes equally in reducing from a power to unity by division, as in ascending from unity to a power by multiplication. For in both cases the number of operations is the index of the type of root.

In the previous example, just as the power 12977875 divided by the root 235 produces unity on the third division; so, by the continued multiplication of the root 235, after the third operation the same power emerges. And so both methods point to three as the index.

For this reason, the use of radical division is rarely used in calculation, for multiplication performs that function more readily. But radical multiplication is of little value unless the root reproduces the power exactly and vice versa. In these circumstances the given root is an imperfect root of the power.

CHAPTER VII.

RULES FOR EXTRACTION OF ROOTS.

THERE IS A SPECIAL rule for the extraction of every root.

Every rule for extraction consists of the resolution of the power into supplementary parts.

A supplement consists of the difference of two powers of the same order.

For example, 100 and 144 are square powers of the same order. The first is [the square of] 10 and the latter [the square of] 12 and the difference of the squares is 44. This is the true supplement of the given powers. The supplements vary in accordance with variations in the powers and roots.

There is one rule for finding the supplements of squares and for the extraction of square roots; another for cubes and cube roots; and yet another for fourth powers and the extraction of fourth roots; and so on for all cases.

The rules for finding all powers, and the supplement of roots, are contained in our triangular Table, filled with small hexagonal spaces. On the right hand side of the triangle single unit entries are made, but on the left side the numbers increase from unity by increments of one unit as you descend. Each of the hexagonal spaces inside the triangle contains a number which is equal to the sum of the two numbers placed above it.

The triangle ABC has angles A on the left hand side, B at the vertex, and C on the right hand side. When you decide on the maximum order of the root to

TABLE OF SUPPLEMENTS.

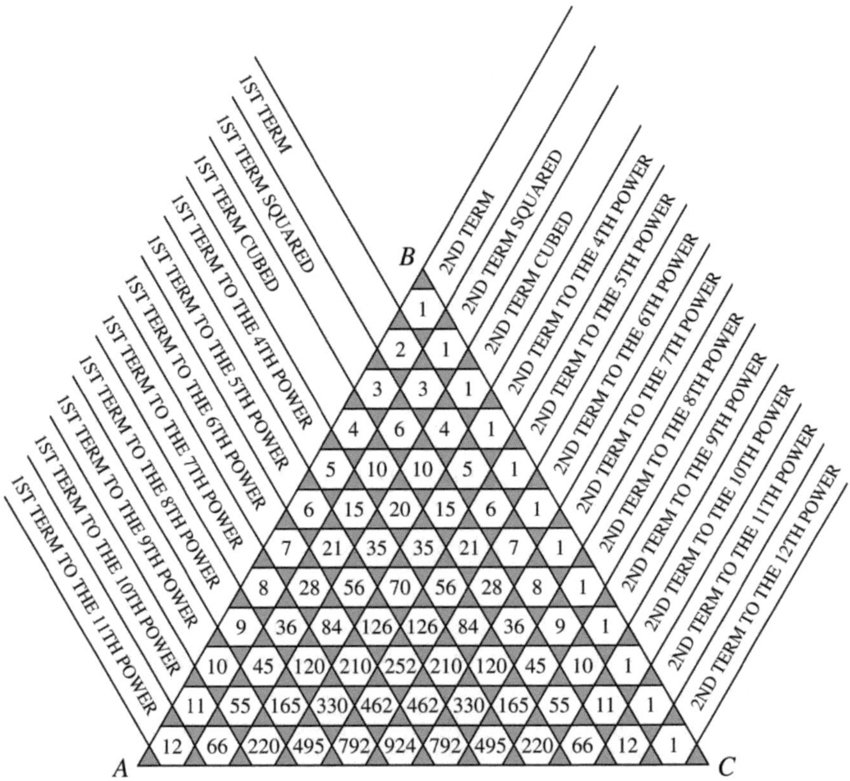

be included in the table, each of the sides of the triangle must contain twice that number plus one, units of length. For example, in order to contain root extraction of order 12, divide each side into 25 equal parts. Starting at A, mark points 1, 3, 5, ..., 21, 23 units of length along AC. Start at A again, and mark points 1, 3, 5, ..., 21, 23 units along AB. Now draw twelve lines parallel to CB joining the points 1, 3, 5, ..., 21, 23 along AC as shown in the diagram. These lines must be extended about two inches to the left outside the triangle ABC as shown. Next, through every second point along AB and CB draw twelve lines parallel to the base AC. You now have a triangle covered with small hexagonal areas of which 12 are on the right hand side, adjacent to the line BC. All 12 of these carry the digit 1. On the left hand side, however, from the vertex B downwards to the angle A, the hexagons carry the digits 1, 2, 3, [etc., as far as] 12. Next, every vacant hexagon within the triangle carries a number which is the sum of the two numbers lying immediately above. For instance, 3 is written in the space immediately below 2 and l; beneath 3 and 3, 6 is entered, and beneath 3 and 1 you can see 4, and so on down to the foot of the table. Finally, the titles are written on the left hand side; 'first term,'[9] above the second hexagon inscribed 2. Above the third hexagon marked 3 write 'first term squared' Above the fourth hexagon write 'first term cubed' And so on for the remaining powers of the first term up to hexagon marked 12 which must be marked 'first term to the eleventh power.' Along the right hand leg BC write above the first hexagon 'second term.' Above the next hexagon 'second term squared;' above the next hexagon 'second term cubed.' And so continue for the remaining powers of the second term up to the twelfth which must be marked 'second term to the twelfth power.'

Every supplement is made up of two parts of the root. One part, called the first term, consists of one or more digits found in the left hand leg of the triangle; the other, called the second term, can be found on the right hand leg.

And these parts of the root, in combination, make up the supplement, as will be seen below.

> The square root of 144 is 12. The two parts of the latter are 1 (which represents 10) and 2 and the supplement 44 is derived from them. The first 1 (or 10) is called the 'first term' and the second 2 is called the 'second term.' Consider the square number 15129, whose square root is 123. The first part is 12 (or better 120) and this is called the 'first term;' whilst the final 3 is called the 'second term.'

From this triangle the rules for finding any supplement can be read out and understood. First you must find the index of the proposed root on the left hand leg of the triangle. From this the required supplement can be read out from the numbers contained in the spaces on the same line as the index or power. But with each separate number the named values obliquely above, both to left and right, must be multiplied together and read out.

> Example:—You are asked to find, from the triangular Table, the rules for finding the supplement of a square, and for the extraction of a square root. In the second line there are two numbers, 2 and 1 and the named values of the terms are to be multiplied. This is expressed as follows: twice the first term times the second term to which the square of the second term is added. The supplement of a square is equal to twice the first term multiplied by the second term added to once the square of the second term[, as will become clear from the following examples].[10]

---

[9] Hawkins denoted the 'first' and 'second terms' by x and y, using the modern notation x^n.

[10] Refer to the definition at lines 4–5 of [Chapter 7] and to the [last] example [above]. $(10+2)^2 = 2\cdot 10\cdot 2 + 10^2 + 2^2 \therefore (10+2)^2 - 10^2 = 2\cdot 10\cdot 2 + 2^2 =$ supplement.

> To find the supplement of a cube look along line three. Here you will find three numbers 3, 3, 1. These numbers must be associated with first and second terms as follows: The supplement of a cube contains three numbers. The first number is three times the first term squared multiplied by the second term. The second number is three times the first term multiplied by second term squared. The third number is the second term cubed.[11] The supplement of a fifth power or the extraction of a fifth root is obtained in the same way. On the fifth line of the triangle you will find five numbers: 5, 10, 10, 5, 1 with their named powers indicated clearly. The supplement of the fifth power therefore contains five different parts. The first part is five times the first term raised to the fourth power multiplied by the seond term. The second part is ten times the first term cubed multiplied by the second term squared. The third part is ten times the first term squared multiplied by the second term cubed. The fourth part is five times the first term multiplied by the second term raised to the fourth power. The fifth part is the second term raised to the fifth power. In this fashion you can find the fourth, sixth and other supplements of the remaining powers in this table.[12]

The root is best extracted by choosing the greatest possible value of the first term, and then collecting together all the parts which contain the second term as parts of the total supplement.

You now have the supplements of all roots so far described. We now discuss the extraction of roots.

CHAPTER VIII.

EXTRACTION OF ROOTS.

EVERY root of a power contains one or more digits.

The table which follows contains powers below the thirteenth for all roots made up of a single digit.

> Find the fifth root of 16809 from the table. The required number can be found on the line containing fifth powers; the nearest number below

[11] $(x+y)^3 = 3x^2y+3xy^2+y^3+x^3 \therefore (x+y)^3-x^3 = 3x^2y+3xy^2+y^3$. Difference of cubes = supplement. This is consistent with the definition of supplement (page 847).

[12] $(x+y)^5 - x^5 = 5x^4y + 10x^3y^2 + 10x^2y^3 + 5xy^4 + y^5$. Difference of fifth powers = supplement of fifth power. And so: difference of nth powers = supplement of nth power. This relationship is used in the next chapter for the extraction of roots.

16809 is 16807. Now look at the head of the column which contains 16807, and you find 7 as the fifth root with remainder 2. The extraction is unresolvable and cannot be found exactly.

TABLE OF POWERS AND ROOTS.[13]

	2	3	4	5	6
Square	4	9	16	25	36
Cube	8	27	64	125	216
Fourth power	16	81	256	625	1296
Fifth power	32	243	1024	3125	7776
Sixth power	64	729	4096	15625	46656
Seventh power	128	2187	16384	78125	279936
Eighth power	256	6561	65536	390625	1679616
Ninth power	512	19683	262114	1953125	10077696
Tenth power	1024	59049	1048576	9765625	60466176
Eleventh power	2048	177147	4194304	48828125	362797056
Twelfth power	4096	531441	16777216	244140625	2176782366

	7	8	9
Square	49	64	81
Cube	343	512	729
Fourth power	2401	4096	6561
Fifth power	16807	32768	58049
Sixth power	117649	262144	531441
Seventh power	823543	2097152	4782969
Eighth power	5764801	16777216	43046721
Ninth power	40353607	134217728	387420489
Tenth power	282475249	1073741824	3486784401
Eleventh power	1977326743	8589934592	31381059609
Twelfth power	13841287201	68719476736	282429536481

The table is built up by means of the continued multiplication of each of the nine digits in successive columns downwards.

At the head of the table you have the digits 2, 3, 4, 5, etc., up to 9 heading the eight columns. Beneath each digit you have its square, its cube, its fourth power and so on up to the twelfth power. Beneath the

[13] For reasons of space, this table is reproduced in two parts.

> digit 2 you have successive entries 4, 8, 16, 32, 64, 128, 256, 512, 1024, 2048 and 4096 with the name of the power given in the first column[; evidently, the duplicate, triplicate, quadruplicate, and so on, and finally the duodecuplicate;] beneath the digit 3 you have the powers 9, 27, 81, 243 and so on in constant ratio downwards. So also, under digit 4 you have 16, 64, 256 and so on. Finally, under 9 you have 81, 729, 6561 and so on. All of these can be found readily in the table.

In the extraction of roots which consist of several digits attention must be paid to place-value as well as to the operation.

Two parallel lines are drawn beneath the given power spaced so that there is plenty of room between them to take the required root. Beneath the extreme right hand digit of the power a point is inserted. From this point to the left other points are inserted; at every second digit for square root; at every third digit for cube root and at every fifth digit for fifth root.

On these points between the lines each single digit of the required root is inserted.

There are two rules for the operation.

First, from the figures of the power which end at the left hand point take away the largest possible power of the required order found in the table. Enter the remainder above and the power (taken from the table) below. Then replace the left hand point with the value of the root found in the table. This then becomes the first figure of the root.

Second, make an informed guess at the supplement required, and insert the new figure at the next point or period. The supplement of the proposed order must be the largest possible, but it cannot exceed the figures above it, since it must be subtracted from the power and the remainder entered above. This second operation must be repeated as many times as there are empty periods on the right. All the figures which fall into place at the points, make up the required root.

> Example: Find the square root of 55225. The lay-out is shown in the diagrams [in the margin]. From 5, located at the left hand point, take away the largest possible square in the table. This is 4, with remainder 1. Cancel the 5 above, and the 4 below. Enter 2 at the first point since it is the square root of 4. The first digit of the root is 2. The next digit must be found and entered in place of the second point. The supplement cannot exceed 152 which is entered above the line. The first term is 2, or allowing place-value 20. If we choose four as the second term the supplement, that is, the first part of the supplement,

55225
· · ·
_____

1
~~5~~5225
· · ·
2
_____
~~4~~

would be 2 times 20 times 4, or 160, which exceeds I52. So we reject four and settle for three as the second term. When 129 is taken from 152 the remainder is 23. Enter 3 in place of the second digit, 129 below and 23 above. Delete 129 and 152. The first term of the root is now 23, or, allowing for place-value, 230. The second operation is now repeated to find a new second term to be entered at the right hand point. The supplement of this new second term, whatever it may be, is, by the first table, twice 230 or 460 multiplied by the second term added to the square of the second term. But if you multiply 6 by 460 it exceeds 2325. We must therefore reject 6 and take five as the second term. Now deduct 2325 from the total left above the line, and there is no remainder. The square root 235 emerges complete and perfect as the square root of 55225.

20
2
40
3
120. First part of the first supplement.
129. Total first supplement.

3
3
9. Second part of the first supplement.

123
55225
2 3
4
129

230
2
460
5
2300. First part of the second supplement.
2325. Total second supplement.

5
5
9. Second part of the second supplement.

Another example of square root extraction.

Find the square root of 164860. The lay-out of the problem is shown in the margin. The square root of 16 placed above the left hand point is 4. Replace the point with 4 and insert its square 16 below. This must be taken from 16 in the dividend. The dividend contains 48 up to the second point. The first term consists of 40 and if we choose one as the second term the supplement will be 81.[14] This cannot be taken from 48,

164860
4 0
16

[14] $2 \cdot 40 \cdot 1 + 1 = 81$.

so we must enter 0 at the middle point. The supplement is also 0 and the remainder is 4860 as shown. The new first term is now 40, and the new second term is 6[, whose supplement is 4836 (clearly six times the double of 400, and six times 6)]. When 4836 is taken from 4860 the remainder is 24, which is unresolvable. The required square root of 164860 is therefore 406, and it is imperfect.

```
 400
 2
 800            6
 6              6
4800. First    36. Second
part of the    part of the
supplement.    supplement.
               4836. Total
               supplement.

       0  24
     164860
     4 0 6
     16
       4836
```

Example of cube root extraction.

Extract the cube root from 12977875. The lay-out for the problem is shown [in the margin]. The greatest cube in the table less than 12 is 8. Take 8 from 12 and leave 4. Delete 12 and write 4 over it. The cube root of eight, namely 2, is inserted in place of the left hand point. The first term is now 20 and the second term and the total supplement are required. The supplement of a cube root (from the triangular table) contains three parts. The first part consists of three times the square of the first term multiplied by the second term.

```
      4
     12977875
      2
      8
 20
 20
 400               20    3
 3                 3     3
1200               60    9
 3                 9
3600. First        540. Second
part of the first  part of the first
supplement.        supplement.
```

If therefore we divide 4977 (the dividend above the line) by three times 400 (equals 1200) this would be an excellent guide to the value of the second term which is generally equal to, or one less, than this value. When 4977 is divided by 1200 we get 4 as the second term. But

```
   3
   3
  ---
   9
   3
  ---
  27.  Third part of the first supplement.
```

```
3600
 540
  27
----
4167.  Total part of the first supplement.
```

```
 4810
12977875
 ·  ·  ·
 2  3
 8
 4167
```

```
  230
  230
 -----
   69
  46
 -----
 52900
     3
 ------
158700
     5
 ------
793500.  First part of the second supplement.
```

```
 230
   3
 ---
 690
  25
 ---
 345
138
-----
17250.  Second part of the second supplement.
```

```
  5
  5
 --
 25
  5
 ---
125.  Third part of the second supplement.
```

```
793500
 17250
   125
------
810875.  Total second supplement.
```

```
 4810
12977875
 ·  ·  ·
 2  3
 8
 4167
  810875
```

the value four gives 5824 as the supplement. Since this exceeds 4977 we must reject 4 and choose 3 as the second term, [whose supplement is 4167 (for 1200 times 3, and three times 3 multiplied by three times 20, plus 3 cubed make the sum 4167)]. Take 4167 from 4977 leaving 810. This makes 810875 up to the third point on the right in the quotient, and gives 230 as the new first term. We must now locate the second term in the space next to the right hand point. This is done in exactly the same way as before and 5 is obtained. This produces a supplement of 810875 [(for three times the square of 230 multiplied by 5, namely 793500, and the triple of 230 multiplied by 25, namely 17250, and the cube of 5, which is 125, make 810875)]. When this supplement is subtracted from the equal number above there is no remainder. The required cube root 235 is an integer and perfect, and when cubed it produces 12977875. The same procedure is used for all root-extractions when the indices are simple integers such as fifth and sixth roots and so on.

Example of root extraction with compound roots.

5522
276   0
524
3049800625
5 5 22
25 552225
525
2204
22084

Extract the fourth root of 3049800625. So that this can be done conveniently the square root must be extracted twice instead of finding the fourth root directly, see Chapter 4, Book I. The square root is extracted by the method and operation shown in the margin. This gives 55225 as the required square root. You must now extract the square root a second time from 55225 as you can see in the margin. You will then find that the square root of 55225 is 235, and this is the required fourth root of the given radical 3049800625. Similar methods can be employed to extract sixth, eighth, ninth and whatever other compound roots you wish to evaluate.

0
23
1
55225
2 3 5
4
129
2325

50	
2	
100	5
5	5
500	25. First supplement.
550	
2	
1100	2
2	2
2200	4. Second supplement.
5520	
2	
1100	2
3	2
2200	4. Third supplement.
55220	
2	
110440	5
5	5
552200	25. Fourth supplement.
20	
2	
40	3
3	3
120	9. First supplement.
230	
2	
460	5
5	5
2300	25. Second supplement.

CHAPTER IX.

ROOT APPROXIMATIONS AND LIMITS.

THE REMAINDER which is left after the extraction of a root, can be converted into a closer approximation, even though the exact root is not resolvable.

The fractions which remain, unless modified in one way or another, for the most part exhibit sensible error.

These sensible errors of a root and its remainder are corrected chiefly in two ways.

[The first method is,] draw a straight line following the imperfect root and enter the remainder above the line. Underneath the line insert the unit supplement.[15] This gives the smaller value of the square root. The greater value is obtained by deducting one from the unit supplement and inserting this figure as the denominator of the fraction. The true value of the root cannot be expressed exactly, but it lies between these two limits.

Extract the square root of 164860. In the second example of the last chapter the square root of the above number was found to be 406 (imperfect) and remainder 24 left over [as in the margin]. Draw a line after 406 and above the line put 24 as shown. From the first term already found (406) find the supplement with unity as its second term. From this 813 is the denominator of the smaller fraction and 812 is that of the greater fraction. And so, without sensible error (especially in applied mathematics) the square root of 164860 can be said to be greater than $406\frac{24}{813}$ and less than $406\frac{24}{812}$.

$406\frac{24}{813}$

$406\frac{24}{812}$

406		
2		
812	1	
1	1	
812	1. Supplement of unity.	

[An example of placing a cube root] between suitable limits.

[Find the cube root of 998.] From the table of cubes, the nearest root is 9 leaving a remainder of 269 for treatment. Now put 9 before the line and 269 above it. The total unit

$9\frac{269}{271}$

$9\frac{269}{270}$

9	9	1
9	3	1
81	27	1. Third part of the supplement.
3	1	
243	27	27. Second part.
1		
243		

[15] That is, the supplement which contains unity as its second term.

supplement is 271, and when unity is deducted this leaves 270 as you can see [in the margin]. Therefore the cube root of the power 998 lies between the limits $9\frac{269}{271}$ (the lower) and $9\frac{269}{270}$ (the larger).

243. First part.
271. Total supplement of unity.

The second method is by multiplying the whole power of the given order by any chosen number of the same order. Then the root of the product of the same order, ignoring the remainder, is divided by the chosen number. This produces a quotient and a fraction which provide the lower limit of the root. If you then add one unit to the numerator of the fraction, you will have the upper limit. The true root lies between these two values.

Find the square root of 50 by using a suitable multiple, say 1000. Since the chosen factor is 1000 its square is 1000000. When this multiplies 50 we get, as the product, 50000000. The lay-out and working are given in the margin. The square root is 7071 and the remainder 959 is ignored. Now divide by the chosen factor 1000. This gives the quotient together with the fraction $7\frac{71}{1000}$ as the lower limit. The upper limit is found by adding one unit to the numerator, giving $7\frac{72}{1000}$. Thus, one or other of these square roots of 50 can be taken without any sensible error.

```
     959
   151
  1
50000000
 7 0 7 1
49
  9849
    14141
```

```
  700
    2
 1400     7
    7     7
 9800    49. First supplement.

 7070
    2
14140     1
    1     1
14140     1. Second supplement.
```

[The preceding cube root example amended in this way.]

Find the cube root of 998 by the multiplication method. Let the chosen multiple be 100 and its cube 1000000. When this is multiplied by 998 we have 998000000 whose cube root is to be extracted [as in the margin]. The root extraction gives approximately 999 with a

```
   997001
  27701
 269
998000000
  9  9  9
729
241299
  26703999
```

```
    90          90
    90           3
  8100         270     81
     3          81      9
 24300         270    729
     9         216
218700.      21870
First part of the first suppl.
```

remainder 997001 which is ignored. But we must divide by the factor 100, and this produces $9\frac{99}{100}$ as the lower limit. When we increase the numerator by one unit we get 10 as the upper limit. Therefore without any sensible error, the cube root of 998 lies between 10 and $9\frac{99}{100}$.

21870.	Second part.	
729.	Third part.	
241299.	Total first suppl.	
990		
990	990	
89100	3	
891	2970	81
980100	81	9
3	2970	729
2940300	2376	
9	240570	
26462700.	First part of the second supp.	
240570.	Second part.	
729.	Third part.	
26703999.	Total second suppl.	

These methods are suited to Applied Mathematics rather than to Pure Mathematics because imperfect roots cannot be evaluated exactly and give approximate results. I pointed this out in Chapter 4 of Book I.

Accordingly geometers place the appropriate root sign before numbers which do not have exact roots. From such signs arise the first kind of geometrical numbers which are called uninomia.

> Do not extract the square roots of numbers like 164860 and 50 because they have no exact location on the number line.[16] Nor can this be done by approximations. Use instead the sign of root extraction placed before the number, like this: $\sqrt{}$ꝗ164860 and $\sqrt{}$ꝗ50; or else thus $\sqrt{}$q164860 and $\sqrt{}$q50; pronounced the square root of the number 164860 and the square root of the number 50. But I use the sign $\sqcup$164860 and $\sqcup$50. I pronounce it the second root of 164860 and the second root of 50. So do not extract the cube or third root of 998 because it has no place in the number line, and don't try to approximate it, but write it like this $\sqrt{}\mathfrak{c}$998; and and pronounce it the cube root of 998. But I use this sign: $\llcorner$998 and I suggest that it is called the third root of 998, and I shall have more to say in its proper place. However, these roots are called uninomia or medialia and are based on the logistics of geometry. The next book will deal with this topic, and it is enough to comment on their origin at this point.[17]

[16] Napier has "quis nullas præcise habent in numeris."

[17] See Book III, Chapter 1: The Logistics of Geometry, p. 875. This also explains Napier's system of symbols for root extraction. For this reason it is included in Book I Algebra.

So far I have dealt with the simple calculations of integers; next follow the rules for composite computation.

CHAPTER X.

THE RULES OF PROPORTION FOR INTEGERS.

THE RULES OF PROPORTION for integers can be satisfactorily and fully explained under the normal operations and basic rules of arithmetic.

For example, in the three worked exercises in Book I, Chapter 5, the first, second and third rules of proportion were established; and from these a unique and general method of solution was discovered.

Nevertheless, some numbers have special methods of treatment or short cuts which speed up calculations. So, for instance, with somewhat larger numbers, a divisor may have zeros inserted on the right hand side. Thanks to this contraction, you can omit nearly all figures on the right or fill up the space with dots. This can be done when you begin a multiplication with the left hand figure of the multiplier.

[As if] (from the table of sines) 10000000 is to an unknown sine as 9925461 is to 7986354, what is the value of the unknown sine? [18]

09925461	1
19850922	2
29776383	3
39701844	4
49627305	5
59552766	6
69478227	7
79403688	8
89329149	9
99254610	10

Now because these numbers have about the same number of digits so 9925461 is multiplied by each of the nine digits and a table of values is constructed [in the margin] (See Book II, Chapter 5). Next begin the multiplication under each single figure of the multiplier 7986354 by choosing the corresponding multiple from the table. Leave out all of the last six figures to the right, since the seven zeros of the divisor 10000000 would have the same effect if taken into account and not omitted. These contracted parts of the multiple, when added, make 79268241. . . . . ., ignoring the six vacant places

[18] The whole sine is 10000000, and this is equivalent to the proportion:

$$\frac{10000000}{9925461} = \frac{7986354}{x}.$$

9925461 is the sine of 83° and 7986354 is the sine of 53°.

on the right. But the seven unused zeros of the divisor require 7926824 as the required integral answer. Therefore, [if 10000000 gives 9925461, it follows that 7986354 will give 7926824.]

```
7986354
69478227
 8932914.
  794036..
   59552...
    2977....
     498.....
      39......
7926824|1......
      10000000
```

In this contracted multiplication, the figures omitted on the right, even if they were all nines, would not increase the answer by more than one unit. With numbers of this magnitude, all of them could properly be neglected, since the sensible error is not even as much as one unit.

Let the points which fill the vacant spaces be replaced by nines (which [is possible by the above]). Nonetheless, the total increase of 5888889 when added to the answer 7926824...... makes the total 7926824688889. When divided by 10000000 this gives [the sum] $7926824\frac{6888889}{10000000}$, which is not even as great as 7926825. So the difference does not exceed the product by one unit. With the largest numbers it is most praiseworthy to use this contracted application of the rule of three.

```
.
..
...
....
.....
......
5888889
```

There is also another short method for this rule which does not require the omission of figures. This is done by drawing a line and putting all the given data of the problem in their proper places above or below the line. I have shown this in the general method proposed in Chapter 5 of Book I. For each pair of numbers, one is placed above the line and serves as a numerator, and the other is placed below the line and serves as a denominator. They are then divided by their highest common factors, until reduced to their lowest terms, and the final figures are noted. Finally, the product of the numerators is divided by the product of the denominators, and this quotient gives the required answer to the problem.

Four builders constructed a wall 6 feet high, 48 yards long, in 42 days. How many days would 5 builders take to make a wall 9 feet high and 50 yards long? By Book I, Chapter 5 all the numbers are placed in their proper places as shown [in the margin]. Numerator 4 and denominator 6 are divided by 2, their greatest factor, giving $\frac{2}{3}$,

Builders	Feet	Yards	Days
4	9	50	42
5	6	48	How many days?

[resulting in $\frac{2 \cdot 9 \cdot 50 \cdot 42}{5 \cdot 3 \cdot 48 \cdot}$]. Next cancel 2 above with 48 below, giving 1 and 24 [resulting in $\frac{1 \cdot 9 \cdot 50 \cdot 42}{5 \cdot 3 \cdot 24 \cdot}$]. Next divide 9 above by 3 below giving 3 above and 1 below [resulting in $\frac{1 \cdot 3 \cdot 50 \cdot 42}{5 \cdot 1 \cdot 24 \cdot}$]. Next divide 50 and 5 by 5, giving 10 above and 1 below [resulting in $\frac{1 \cdot 3 \cdot 10 \cdot 42}{1 \cdot 1 \cdot 24 \cdot}$]. Then divide 10 above and 24 below by their largest common divisor 2. This gives 5 above and 12 below [resulting in $\frac{1 \cdot 3 \cdot 5 \cdot 42}{1 \cdot 1 \cdot 12 \cdot}$]. Finally, divide 42 above and 12 below by their highest factor 6, giving 7 above and 2 below [resulting in $\frac{1 \cdot 3 \cdot 5 \cdot 7}{1 \cdot 1 \cdot 2 \cdot}$]. You now have the easily handled numbers 1, 3, 5, 7 and 1, 1, 2 to multiply instead of the greater numbers at the start of the operation. 1 by 3 by 5 by 7 gives 105, and on the bottom 1 by 1 by 2 equals 2. 105 divided by 2 gives $52\frac{1}{2}$ as the quotient. This therefore represents the number of days required by the question, without laborious divisions and multiplications.

Very many unexpected and unobtrusive processes are covered by the rules of proportion, and they cannot be ignored if advantage is to be taken of their efficiency and speed.

For example, the rule for the reduction of fractions of fractions, repeatedly broken down, comes under the rules of proportion by reason of the definition. This has been treated earlier in a general way, and will be dealt with later on. The same short method used here for cancellation was seen in Book I Chapter 8, and will be seen again later.

This is sufficient for the Arithmetic of integral numbers; fractions or broken numbers remain to be treated.[19]

CHAPTER XI.

FRACTIONS OR FRACTIONAL NUMBERS.

A FRACTION or fractional number is the smallest separate part which defines, measures and enumerates one unit.

For instance, astronomers use the degree as a unit, and 7 minutes make a fraction of a degree. For in this context, one minute is the

[19] Charles Naux, in his "Histoire des logarithmes," Vol. 1, p. 23, referred to one of Kepler's letters dated 1623 ("Opera Omnia," edited by Dr Ch. Frisch, 1868, Vol. VII, Prooemium Editoris, p. 306), in which the above method of contracted multiplication was used 'for the first time.' Napier used this elegant method many years earlier.

smallest separate part which measures one unit and the number 7 enumerates 7 such units. Similarly, 5 shillings make a fraction of a pound, because 1 shilling is the smallest separate part of one pound, which measures the fraction, and the number 5 enumerates 5 such units. In exactly the same way 17 thirtieth parts of unity make a fraction, because one-thirtieth is, in this context, the smallest separate part of one unit. And this fraction measures 17 thirtieths, and the 17 enumerates seventeen of such units.

Some fractions are called vulgar fractions and others are natural or material fractions.

Vulgar fractions have free and unrestricted denominations.

Examples are one half, two thirds, four elevenths, which are produced by division. (Book II, Chapter 4).

Attention must be paid both to denomination and operation in the Arithmetic of vulgar fractions.

Denomination, when spoken, is called pronunciation; but when written it is described as notation.

The denomination of vulgar fractions is expressed by two terms, the numerator, and the denominator.

The denominator is that number which tells into how many equal parts one unit is divided.

The numerator is that number which enumerates how many of these equal parts are included in the fraction.

The numerator is a cardinal number and appears above the line,—and the denominator is called an ordinal number, and is entered below the line.

For instance, if you were to divide one unit, or any article you suggest, into as many parts as you wish, say 11 parts; and if you decide that 5 of these parts are to be chosen—then 11 is the denominator of the fraction, and 5 is its numerator. Of these the first 5 is known by its cardinal number (five); and the 11 which follows is called the ordinal number (eleventh). The whole fraction is pronounced this way: 'five eleventh parts.' But the notation is 5 above the line; and 11 below the line, in this fashion $\frac{5}{11}$, and likewise for other fractions.

We therefore define a fraction or fractional number as one or more equal parts divided by the number of equal parts in one unit.

Moreover, it is clear that the value of a fraction is equal to the number of the numerator divided by the number of the denominator.

The fraction given above $\frac{5}{11}$ is pronounced 'five elevenths of one unit'; and this is exactly the same as 'five divided by eleven'. This was dealt with in Chapter 7 of Book I.

There are certain improper fractions which are neither a part nor parts of unity, but which are clearly parts of fractions. These are called fractions of fractions. These I mark by placing the preposition 'from' between them; but there are some who mark them by omitting the last line or lines.

For instance, two fifth parts of three quarter parts of unity is not, strictly speaking, a fraction of unity, but a fraction of a fraction of unity. It means that I have divided a unit into four parts, and of these three parts are again divided up into five parts, and of these two parts have been taken. This I signify as $\frac{2}{5}$ from $\frac{3}{4}$ but others write it as $\frac{2}{5}\ \frac{3}{4}$. This is pronounced 'two fifths from three fourths' or 'two fifths of three quarters.'

The operations and methods of calculation depend on the denomination of vulgar fractions.

As for method, we reduce an integer to the form of a fraction by putting unity as the denominator, and by cancelling the terms of the fraction as necessary. This is done by dividing them by their highest common factor, which is found by the repeated division of the divisor by its remainder, until zero is produced. In this way we reduce different denominators to a common denominator. We add and subtract the reduced terms in various ways according to the rules laid down in Chapter 7 of Book I.

For example, two a whole number, is reduced to the form of a fraction when it is written $\frac{2}{1}$; or 3 as $\frac{3}{1}$; or 4 as $\frac{4}{1}$; and 5 as $\frac{5}{1}$; for they are the same in value. We also shorten the terms of the fraction $\frac{6}{10}$ by dividing the numerator and denominator by its highest common factor 2, and it becomes $\frac{3}{5}$. In the same way $\frac{35}{49}$, divided by its highest common factor 7 becomes $\frac{5}{7}$. The highest common factor is found in this way: divide 49 by 35 and leave remainder 14; divide 35 by 14 and leave remainder 7; divide 14 by 7 and leave remainder zero; therefore, the highest ommon divisor of the terms 35 and 49 is 7. You can also reduce two fractions with different denominators, say $\frac{11}{132}$ and $\frac{7}{128}$, to the same denominator by the same rules, in this way: divide 132 and 128 by their highest common divisor 4. This gives 33 and 32; multiply 33 by 7, giving 231; multiply 33 by 128, giving 4224. For the given fraction $\frac{7}{128}$ write $\frac{231}{4224}$. [Similarly,] multiply 32 by 11, giving 352; multiply 32 by 132, giving 4224 For the given fraction $\frac{11}{132}$ write $\frac{352}{4224}$. The two given fractions have been reduced to fractions with the same denominator by the prescribed general rules. By the same rules we can subtract the new numerator 231 from the new numerator 352 leaving 121 as remainder. We now have a fraction with its new denominator $\frac{121}{4224}$, and with the residue after subtraction as its new numerator. [Likewise,] reduce $\frac{4}{12}$ and $\frac{7}{15}$ to the same common denominator. Find their sum and their difference. First cancel terms, giving $\frac{1}{3}$ and $\frac{7}{15}$. Next divide the denominators by their highest common divisor 3 and this gives 1 and 5. Next multiply both numerator and denominator of $\frac{1}{3}$ by 5. This gives $\frac{5}{15}$ as the first fraction. Now multiply both numerator and denominator of $\frac{7}{15}$ by 1. This gives $\frac{7}{15}$ as the second fraction [which now added to $\frac{5}{15}$] makes $\frac{12}{15}$ or, on cancelling 3, gives $\frac{4}{5}$. Similarly for the residue after subtraction which gives 2, expressed as $\frac{2}{15}$. [Also,] add 66 and $\frac{2}{3}$ to produce one fraction (or rational number). First write $\frac{66}{1}$ instead of 66. Next change to a common denominator 3, and this gives $\frac{198}{3}$ and $\frac{2}{3}$. Finally we add the numerators and obtain 200. This gives $\frac{200}{3}$ as the required fraction. [Also,] add the fractions $\frac{1}{2}$, $\frac{2}{3}$ and $\frac{3}{4}$. These fractions must first be reduced to common denominator 12. The three fractions then become $\frac{6}{12}$, $\frac{8}{12}$ and $\frac{9}{12}$. Now write the sum of the numerators 23; and the fraction as $\frac{23}{12}$. This fraction is equal to 1 and $\frac{11}{12}$. [Also,] subtract $\frac{8}{12}$ from $\frac{11}{12}$. Taking away $\frac{8}{12}$ from $\frac{11}{12}$ leaves $\frac{3}{12}$. Now cancel numerator 3 with denominator 12 leaving $\frac{1}{4}$.

CHAPTER XII.

MULTIPLICATION AND DIVISION OF SIMPLE AND RADICAL FRACTIONS.

FRACTIONS are multiplied, and even fractions of fractions are repeatedly reduced to simple fractions of fractions by cancellation of numerator and denominator in pairs as shown in Chapter 7 of Book I, until common factors are eliminated. The product of the new terms above provides a new numerator; and that of the terms below a new denominator. These give the multiple, or fraction required.

Multiply the fractions $\frac{1078}{1768}$, $\frac{1955}{2090}$, $\frac{3705}{1449}$. Cancel 2 between 1078 on top, and 1768 below. This gives $\frac{539}{884}$. Likewise cancel 3 from 3705 and 1449. This gives $\frac{1235}{483}$. Next cancel 5 from 1955 and 2090. This gives $\frac{391}{481}$. The position is now: $\frac{539}{884}\ \frac{1235}{483}\ \frac{391}{418}$. Now cancel 539 above and 483 below by the common divisor 7. This gives $\frac{77}{69}$. Next cancel the common divisor 19 in 1235 and 418. We then have $\frac{65}{22}$. Now cancel the highest common divisor 17 between 391 and 884. This produces $\frac{23}{52}$. The position is now: $\frac{77}{52}\ \frac{65}{69}\ \frac{23}{22}$. When this is done cancel [the highest common divisor 11 between] 77 above and 22 below. This leaves 7 above, 2 below. Next cancel 65 above and 52 below by [their highest common divisor] 13. This gives 5 above and 4 below. Finally cancel 23 above and 69 below [by their highest common divisor 23]. This gives 1 above and 3 below. The position now is $\frac{7.5.1}{4.3.2}$. When the numerators and denominators are multiplied this comes to $\frac{35}{24}$. The product of the three fractions is $\frac{35}{24}$. [Also,] multiply the fractions $\frac{3}{4}$ and $\frac{5}{7}$. These cannot be cancelled. Multiply 3 and 5 giving 15 the numerator. Next multiply 4 and 7 giving the new denominator 28. Therefore, the required fraction and multiple is $\frac{15}{28}$. [Likewise,] reduce $\frac{2}{3}$ of $\frac{4}{5}$ of $\frac{6}{7}$ of $\frac{8}{9}$ to its lowest terms. Cancel 6 above and 9 below by the highest common divisor 3. The position now is $\frac{2}{3}\ \frac{4}{5}\ \frac{2}{7}\ \frac{8}{3}$. Further cancellation is impossible between the two lines. The terms are therefore multiplied. [The] product of [the] numerators [is] 128. [The] product of [the] denominators [is] 315. The required fraction is $\frac{128}{315}$ and this is the product of the four fractions given.

Instead of division the terms of the divisor can be inverted and the terms are then multiplied together, as shown in Chapter 7 of Book I.

$\frac{4}{5}$ is to be divided by $\frac{2}{3}$. Keep the first term and invert the second and multiply $\frac{4}{5}$ by $\frac{3}{2}$. Cancel 4 and 2 giving $\frac{2}{5}$ multiplied by $\frac{3}{1}$. This gives $\frac{6}{5}$ or $1\frac{1}{5}$ as the required result.

[Another mixed example.]

Divide $66\frac{2}{3}$ by $2\frac{1}{5}$. This is the same as $\frac{200}{3}$ divided by $\frac{11}{5}$. Invert the second term and multiply $\frac{200}{3}$ by $\frac{5}{11}$. The product is $\frac{1000}{33}$ or $30\frac{10}{33}$ the required result.

[Another example to test multiplication.]

Divide $\frac{15}{28}$ by $\frac{3}{4}$. Invert the divisor and multiply $\frac{15}{28}$ by $\frac{4}{3}$. Now cancel 15 and 3 by 3. This now becomes $\frac{5}{28}$ multiplied by $\frac{4}{1}$. Now cancel 28 and 4 by 4. This then becomes $\frac{5}{7}$ multiplied by $\frac{1}{1}$ or $\frac{5}{7}$.

[The] radical multiplication of fractional numbers is effected by continued multiplication, just as for integers. If you multiply a given fraction by itself, or by an equal quantity, it becomes a square number. If you multiply this square number by the same number again it becomes a cube or cubed number. If you multiply the cube yet again by the same number it becomes a fourth power of the fraction first given. And this process is continued when fifth, sixth and higher powers are produced.

The rooted number $\frac{2}{3}$ is to be multiplied radically to index 5, or raised to the fifth power. First multiply $\frac{2}{3}$ by $\frac{2}{3}$ and this gives $\frac{4}{9}$ the square. Next multiply the square $\frac{4}{9}$ by $\frac{2}{3}$ which gives $\frac{8}{27}$ the cube of the given fraction. Next multiply the cube $\frac{8}{27}$ by $\frac{2}{3}$ which gives $\frac{16}{81}$ the fourth power of the fraction. Finally, multiply $\frac{16}{81}$ the fourth power by $\frac{2}{3}$ giving $\frac{32}{243}$ the fifth power of the fraction.

The radical division of fractions by continued simple division is performed just as for integers using the rules given above. Divide the proposed radical fraction by its given root continuously until unity is left. The number of divisions gives the index or power as previously indicated.

The fraction $\frac{32}{243}$ is to be divided radically by $\frac{2}{3}\cdot$ This serves as a verification of [the preceding] example. The first division gives $\frac{16}{81}\cdot$ The second division gives $\frac{8}{27}\cdot$ The third division gives $\frac{4}{9}\cdot$ The fourth division produces $\frac{2}{3}\cdot$ The fifth division produces unity. The above five divisions show that index five is the power required. This also supports the result obtained in [the preceding] example.

CHAPTER XIII.

ROOTS OF FRACTIONAL NUMBERS

If each term of a fraction reduced to its lowest terms has roots of the required order, they must be extracted just like integers; and they form the numerator and denominator of the required root.

But if each term of the fraction does not have such roots, then the root must be extracted geometrically, or by the methods of practical mathematics.

The cube root of $\frac{54}{16}$ is to be extracted. In its lowest terms this is $\frac{27}{8}\cdot$ Each term clearly contains a cube root – numerator 3, and denominator 2. From this, $\frac{3}{2}$ is the required cube root of $\frac{54}{16}$ or $\frac{27}{8}\cdot$

[Another example.]

If the given fraction were $\frac{290}{40}$ it is reduced to $\frac{29}{4}$. If the square root is required I say that both terms, 29 and 4, do not have square roots (although one of them has a root). It is not therefore possible to extract the perfect arithmetical square root of $\frac{290}{40}$ or $\frac{29}{4}$. To do this you must use geometry or practical (applied) mathematics.

The geometric method is to place the symbol for root extraction before the complete fraction; or else to insert the root sign before each term.

In the above example the fraction $\frac{290}{40}$ or $\frac{29}{4}$ has no true square root. Therefore the radical sign must be placed before the whole fraction, like this $\sqcup\frac{29}{4}$, or before each term like this $\frac{\sqcup 29}{\sqcup 4}$.

Finally, the same root is completely included between limits by the methods of practical (applied) mathematics as shown below.

The method of applied arithmetic is to include the required root, which cannot be evaluated numerically, so that it lies as closely as possible between an upper and a lower limit.

The greater the chosen denomination, and the smaller the difference of the limits, the more precisely and closely is the root defined.

Let the terms be of the desired denomination. Then multiply the number of the desired denomination according to the index and type of root required. Next multiply by the numerator of the given fraction, and divide the product by its denominator. The appropriate root of this quotient is extracted in the same way as for integers, giving an upper and lower limit. The chosen denominator is inserted. This produces two values enclosing the required root, one the upper limit and the other the lower limit.

The square root is to be extracted from the fraction $\frac{290}{40}$ given above. Let the desired denomination be 200 parts. The power of the proposed kind is 40000, a square. Multiply 40000 by 29 the given numerator, and then divide the product by 4, its denominator. This gives 290000 as quotient. The square root, or rather its lower limit is 538. And its upper limit is 539 (by the rules given above for integers). Under these figures insert 200 as denominator. This gives $\frac{538}{200}$, the lower limit. And $\frac{539}{200}$, the upper limit. These differ only by one unit in 200 and very closely enclose the true square root $\sqcup\frac{29}{4}$ of the fraction $\frac{29}{4}$.

[An example of cube root.]

[Likewise,] the cube root of $\frac{2}{3}$ is to be extracted, or rather, since this root cannot be defined in the number line, its upper and lower limits are required. Choose your own denominator, having, say, a thousand parts, between which the cube root of two-thirds or $\frac{2}{3}$ is contained. Multiply the cube of 1000 by 2. This makes 2000000000. Divide this by 3, giving $666666666\frac{2}{3}$. The lower limit of the cube root is 873. The upper limit of the cube root is 874. Insert the chosen denominator 1000. This gives $\frac{873}{1000}$ (lower limit) $\frac{874}{1000}$ (upper limit) of $\llcorner \frac{2}{3}$.[20]

Last of all, fractions whose numerator exceeds the denominator must be restored to whole numbers by dividing the denominator into the numerator as sufficiently well explained at the end of Book I.

For instance, if the fraction $\frac{562}{18}$ emerges from several preceding calculations, the integers in which the final answer must be given are more significant than the fractions. So 562 is divided by 18 which produces 31 as quotient and 4 as remainder; that is $\frac{4}{18}$ or $\frac{2}{9}$. When completely reduced this gives $31\frac{2}{9}$ or 31 numbers and two ninths of one unit.

CHAPTER XIV.

THE RULES OF PROPORTION FOR FRACTIONS.

In the rules of proportion for fractions attention must be paid both to position and to the operation itself. The position of the numerator is generally the same as that of the quantities dealt with in Chapter 5 of Book I. But that of the denominator is reversed. For if the numerator falls below the line by the rule given there, the denominator is reversed and falls above the line. And contrariwise, if the former is above the line, then the latter falls below the line.

If $\frac{3}{7}$ of a gallon of water flows from a cistern in $\frac{5}{12}$ hours, how long does it take for $2\frac{1}{2}$, or $\frac{5}{2}$, gallons to flow from it? First draw a line as shown [in the margin] and enter 5 (the numerator of the first time) above the line. This is according to the general rule of Chapter 5, Book I, since $\frac{5}{12}$ hours and the required hours are quantities of the same kind, which increase and decrease together. In the same way $\frac{5}{2}$ gallons increases and decreases at the same time as the required number of hours. For this reason the numerator 5 is placed

$$\frac{.5.5}{3.}$$

[20] This result with denominator 1000 must have been written by Napier before he had perfected decimal fractions in his *Constructio*. Hawkins, in his translation, transferred the final part of Chapter XIII (starting with "Last of all") to the end of Chapter XI, which is an appropriate place for it.

above the line. But as the first number of gallons $\frac{3}{7}$ increases the fraction of the required time decreases, and vice versa. The denominators of all the entries 7, 12 and 2 are placed on the other side of the line, each one opposite its kind [as in the margin].

$$\frac{7.\ 5.5.}{3.12.2.}$$

The operation is completed as follows. Multiply the top numbers together and obtain the numerator. Do likewise with the bottom numbers and obtain the denominator. Divide the former by the latter and the required answer emerges.

[As in the previous example the upper numbers 7, 5, 5, multiplied together, make the numerator 175; and the lower ones 3, 12, 2, multiplied together, produce the denominator 72;] therefore $\frac{175}{72}$ or by division $2\frac{31}{72}$ hours is the time required for the outflow of $2\frac{1}{2}$ gallons.

$$\frac{7.\ 5.5.}{3.12.2.} \quad \frac{175}{72}$$

The terms in the above example cannot be reduced by cancellation. So here you may recall the second method of Book II Chapter 10, as often as there are entries above and below the line a compound ratio emerges which can perhaps be cancelled.

In the previous example replace the final terms [$\frac{7}{3}$ $\frac{5}{12}$ and $\frac{5}{2}$, which cannot be simplified, by taking these terms $\frac{7}{6}$ $\frac{10}{21}$ and $\frac{15}{8}$, which have composite ratios, and can be simplified]. Rearrange the terms as shown [in the margin.] Divide 7 above and 21 below by their [highest commom measure] 7; this gives $\frac{1}{3}$, [resulting in $\frac{1{\cdot}10{\cdot}15{\cdot}}{6{\cdot}3{\cdot}8{\cdot}}$.] Next, in the same way cancel 10 above and 6 below; this gives [5 above and 3 below, resulting in $\frac{1{\cdot}5{\cdot}15{\cdot}}{3{\cdot}3{\cdot}8{\cdot}}$.] Now cancel 15 above and one of the 3's below; this gives [5 above and 1 below, resulting in $\frac{1{\cdot}5{\cdot}5{\cdot}}{3{\cdot}1{\cdot}8{\cdot}}$, or $\frac{1{\cdot}5{\cdot}5{\cdot}}{1{\cdot}3{\cdot}8{\cdot}}$, or otherwise $\frac{1{\cdot}5{\cdot}5{\cdot}}{6{\cdot}1{\cdot}4{\cdot}}$.] Now multiply 1, 5, 5 together to give 25. In the same way multiply 3, 1, 8 [or 6, 1, 4] to give 24. And so the complete fraction which satisfies the equation is $\frac{25}{24}$ or $1\frac{1}{24}$ in shortened form. And in all similar problems the operations can be effected with speed and precision.

$$\frac{7.10.15.}{6.21.\ 8.}$$

CHAPTER XV.

PHYSICAL OR NATURAL FRACTIONS.

I HAVE NOW DEALT with all the calculations of vulgar fractions, and physical calculations follow in due order.

Physical quantities are those which form a part or parts of a whole natural entity. These physical entities are sub-divided by custom and usage, and the commonly accepted divisor imposed in this way becomes the denominator.

> For it has pleased our financiers to divide one pound in money into 20 parts, neither more nor less, and to give them the name shillings. They then divide the shilling into 12 parts called pence. Similarly physicians have divided the pound weight into 12 parts called ounces; and the ounce into 8 drams; and the dram into 3 scruples[, etc]. Chronologists have divided the year into 12 months; the month into 30 days (or near so); and the day into 24 hours. Astronomers have divided the degree into 60 minutes, the minute into 60 seconds, the second into 60 thirds, and so on.

Physical fractions have one thing in common with vulgar fractions; as often as the numerator exceeds the denominator, the top row is increased by one complete unit. And as often as a larger fraction is taken from a smaller fraction, a complete unit is needed from the top line to make good the deficiency.

> If 14 hours is added to 19 hours, this adds up to 33 hours. But it is not written 33 hours, for under the heading days and hours we write 1 day and 9 hours, because beyond 9 hours there follow 24 hours in a day. Also, when 7 and 9 pence are added we do not write 16 pence. We write 4 pence and in place of the remaining 12 pence 1 is written under the heading shillings. Likewise, 48 minutes added to 15 minutes do not make 63 minutes, but only 3 minutes and 1 degree for the remaining 60 [minutes]. The same thing is done with integers as with vulgar fractions. For when five hundreds are added to eight hundreds, it is neither written nor pronounced thirteen hundred, but one thousand three hundred, like this 1300. And in adding vulgar fractions, whether $\frac{6}{7}$ or six-sevenths and $\frac{5}{7}$ or five-sevenths are added and cancelled in the usual way, you do not say eleven-sevenths but four sevenths and the remaining seven-sevenths are written as one entity like this $1\frac{4}{7}$. Also in subtraction, if 14 hours are to be taken from 1 day and 9 hours (see example above), because 14 hours exceed 9 hours therefore the one day must be resolved into 24 hours, which, with 9 hours makes 33 hours. From this take 14 hours, and there is left, as above, 19 hours as required. 9 pence are to be taken from 1 shilling and 4 pence. Since 9 pence exceeds the value of 4 pence we must change the shilling into 12 pence and add 4 pence to it, giving 16 pence. Now take away the 9 pence and 7 pence are left, the required answer. If 15 minutes are to be taken from 1 degree and 3 minutes the latter must be changed to 63 minutes. From this 15 minutes can be taken, and 48 minutes will be

left. It is exactly the same with integers and vulgar fractions; one unit from the preceding place must answer for a deficiency in the following place and supplement it. And if five hundreds are to be taken from one thousand three hundred, this makes thirteen hundred from which you can now take five hundred, leaving eight hundred over as before. From one and four-sevenths six-sevenths are to be taken. To do this one unit first be converted to seven-sevenths, leaving four-sevenths outstanding, and this gives eleven-sevenths. Six-sevenths from this total leaves five-sevenths as above.

You can therefore see the common harmony in all of these operations. In all cases, fractions and integers, there is equal consideration and regard for their magnitudes whether increasing or decreasing.

Unit integers with tens, hundreds and thousands on the top line; and on the bottom line with tenths, hundredths and thousandths. Minutes, sixty of which make a degree, and degrees (marked with two digits) in ascending order; and seconds which contain sixty parts called thirds in descending order. Likewise the shilling, twenty of which make up a larger coin called a pound; and which in turn contains twelve smaller coins called pence. Equal regard is paid to all such details, and the methods of computation used are similar.

Hence it would be burdensome and unnecessary to develop special techniques for these quantities since their calculations are more easily carried out by integers and vulgar fractions, rather than by special methods and formulae. This is particularly true of astronomy, for which huge tables, including minutes and seconds are required. Further, there is the bother of double figures for each individual subject for which common arithmetic requires only one figure in every place.

This fact is evident to experienced people, for it is easier and quicker for a problem in astronomy to be calculated and resolved by the rules of arithmetic, rather than by the techniques of astronomy. For even beyond the tables mentioned, and the double figures, no matter what their denomination, there is the fatigue involved. For common arithmetic involves a first place-value of one unit; and second place ten units; in the third place one hundred units, and each place further involves the factor ten. But astronomical calculation involves the signs ′ ″ ‴ ⁗, etc. for each single entry, and mostly two figures are involved, and this leads to confusion.

I bring this Arithmetic to a finish having dealt with all the vulgar fractions, except those deliberately left out.

[Moreover] to the BEST and GREATEST GOD; and [to] all his Numbers, Infinite, Immense and Perfect, [may] all praise, honour and glory [be attributed] for ever. Amen.

[END OF BOOK TWO.]

BOOK THREE.

THE LOGISTICS OF GEOMETRY.

CHAPTER I.

NOTATION AND DESCRIPTION OF CONCRETE NUMBERS.

IN THE PRECEDING Book I taught Arithmetic. Here in due order, the logistics of Geometry follow.

GEOMETRY is the Logistics of material quantities by the use of concrete numbers.

A number is called a concrete number insofar as it relates to a material and continuous entity.

For example, if three lines of finger length are drawn thus ________ ________ ________ it is represented by the discrete number $3a$. But when a continuous line of three finger's length is drawn in this fashion ________+________+________ it is called a concrete number. But this is not an exact definition and is subject to further consideration.

Properly, and in their own right we call the roots of numbers which cannot be measured exactly, whether by integers or by fractions, by the name of concrete numbers.

The square or quadratic root of seven lies between two and three. No fraction in the entire range of fractional numbers is found which is equal to, or commensurable with it. It is therefore properly said to be a concrete number. The third, or cube root of ten is not a discrete number, nor does it have a number which is commensurable with it, so it is a concrete number. So also for the infinity of rooted numbers which are commonly called surds or irrational numbers.

The origin of these concrete numbers lies in the extraction of roots of numbers which cannot be evaluated exactly.

I have referred to such matters in Chapter 4 of Book I and Chapter 9 of Book II.

And so the various notations and descriptions of concrete numbers originate in the diversity of their roots.

> For instance, I signify the square root of seven (which is commonly called the quadratic root of 7) thus ⊔7, [pronounced the bipartite root of seven or the square root of seven]. Likewise, instead of the cube root of 10, I write ∟10 and call it the third root of 10. I signify the fourth root of 11 as ⊐11; the fifth root thus □, and the sixth root like this ⊏.

This one figure, divided into distinct parts enables us to remember every kind and variety of root together with its index.

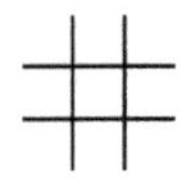

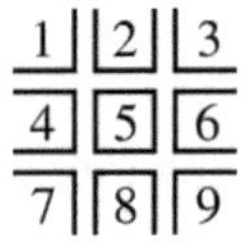

> In the preceding examples the symbols ⊔ ∟ ⊐ □ ⊏ placed before a number denote square, or second root; cube, or third root; fourth root; fifth root; and sixth root. [Just as ┐ is the seventh root; ⊓ the eighth; ┌ the ninth; ┘^{0} the tenth; ┘┘ the eleventh; ┘⊔ the twelfth; ┘∟ the thirteenth; ┘⊐, or ⊣ the fourteenth; ┘□ the fifteenth; ┘⊏ the sixteenth; ┘┐ the seventeenth; ┘⊓ the eighteenth, ┘┌ the ninteenth; ⊔^{0} the twentieth; ⊔┘ the 21st; ⊔⊔ the 22nd; ⊔∟ the 23rd; ⊔⊐, or ⊎ the 24th; etcetera. Likewise ∟$_{0}$ the 30th; ⊐^{0} the 40th; □^{0} the 50th; ⊏^{0} the 60th; ┐^{0}, or ┐$_{0}$ the 70th; ⊓$_{0}$ the 80th; ┌^{0} the 90th; ┘^{00} the 100th. And so on to infinity in the manner of arithmetical figures.]

Because geometric numbers tend to describe a quantity rather than to enumerate it, they are known by their commonly accepted names.[21]

Some of these names are unique; in such cases they are called uninomia. But others have several names.

A uninomium is the same as a unique concrete number, whether it is described accurately or not.

Whence it follows that a uninomium is either a unique simple number, or some root of a unique simple number.

> For instance, 10 is a simple number and it is used by geometers instead of a uninomium. Likewise ⊔10, ∟12, ⊐26 and similar are the roots of numbers, and rightly take on the root implications of uninomia.

Whenever a radical uninomium is the root of an abundant or defective number, whose index is either odd or even, four separate cases ensue. In one case the root is abundant, in another defective, and in yet a third case, one root is abundant and the other defective, giving twin values. And last of all,

[21] Feet, degrees, shillings, etc.

the roots are neither abundant nor defective, and they are called trifling or imaginary.[22]

> I have presented the essence of this great algebraic mystery above, in Chapter 6 of Book I. But, although I know of no one who has produced an answer to this problem, the future may show how much credit will accrue to this art, and to other mathematicians.

In dealing with abundant and defective uninomia it does not matter much whether you place the necessary signs before the roots or between them. It is probably better to place them in front. In uninomia having pairs and in those which are imaginary, the necessary signs must be placed between them.

> [An example of the first case is] └10. By Chapter 6 of Book I this is the same as └+10, an abundant uninomium. [An example of the second case], └−10, is a defective uninomium [(by the same chapter)]. [An example of the third case is] ⊔10. This is the same as ⊔+10 and it signifies an abundant number which when multiplied by itself gives +10. But since +10 signifies ⊔−10 multiplied by itself, so this signifies a defective number as well. To clarify this example ⊔9 or ⊔+9 is equally +3 or −3, as shown previously [in Chapter 6 of Book I]. An example [of the last case is] ⊔−9. This is a pure trifle and signifies nothing, whether abundant or defective; for nine defective has no square root, as shown in Chapter 6 of Book I.

In dealing with imaginary[23] cases, you are strongly advised not to put the minus sign (−) in front, for the sign must be inserted between the root sign and the number.

> For if, instead of putting ⊔−9 (which makes nonsense) you put −⊔9, then you make a big mistake, for this signifies a minus quantity, the square root of nine. But the square root of nine, in this case abundant (⊔9) gives a pair of answers +3 (abundant) and −3 (defective). And so the product will be negative. And so, he who writes −⊔9 in place of ⊔−9, instead of an imaginary and impossible quantity of no significance, introduces a pair of numbers with opposite signs. Beware of this error, into which many people fall.

For the remaining uninomia it makes no difference if you place the symbols + or − between the root signs and the number or if you place them before both. Nor does it change their value by the insertion of a + sign in the empty spaces before or between them (by Chapter 6 of Book I).

[22] The Latin word is *nugacia* (silly, frivolous).
[23] Latin *nugacibus*

So $\sqcup 9$, $\sqcup +9$, $+\sqcup 9$, $+\sqcup +9$ are exactly the same $+3$ or -3. Similarly, $\llcorner 27$, or $\llcorner +27$, [or $+\llcorner 27$,] or $+\llcorner +27$ have the same value $+3$ only. Likewise, $\llcorner -27$, or $+\llcorner -27$, or $-\llcorner 27$ or $-\llcorner +27$ have the same value, -3 only. As for imaginary numbers $\sqcup -9$ and $+\sqcup -9$ imply an equal measure of impossibility. But be sure, in this case, that you do not replace them with $-\sqcup 9$ or $-\sqcup +9$, as I have warned you in in the previous section.

We now deal with the natural interactions of uninomia. The mutual interactions of uninomia follow this section.

Uninomia are of two kinds, they are either mutually commensurable or else incommensurable.

They are commensurable if they contain mutually, numbers which are discrete or absolute.

Every absolute number is commensurable with every other absolute number. Likewise, two uninomia which carry the same root sign; if the simple number of one is divided by the simple number of the other, it produces a number endowed with the same root sign as indicated by the two numbers. They are then said to be mutually commensurable in the same manner indicated by the radical sign.

Therefore 5 and 7, because they are absolute or rational are commensurable with each other. [Likewise,] two uninomia with a root of the same order $\sqcup 8$ and $\sqcup 2$ are considered. The simple number 8 divided by the simple number 2 yields 4. But this carries the sign $\sqcup$ and so gives 2. Consequently $\sqcup 8$ and $\sqcup 2$ are mutually commensurable in the ratio of the root two.

All other uninomia not so reducible are incommensurable.

So $\llcorner 12$ and $\sqcup 2$ have different root signs and so are incommensurable. Likewise, $\sqcup 6$ and $\sqcup 2$ although carrying the same root sign, are incommensurable. This is because 6 and 2 when divided produce 3 which carries the square root sign $\sqcup$. But 12 and $\sqcup 4$ are commensurable because when reduced they are equal to 12 and 2.

And so on.

I could find no more of this geometricall pairt amongst all his fragments.

signed Robert Napier

THE ALGEBRA OF JOHN NAPIER

BARON OF MERCHISTON.

[BOOK ONE.

THE PART OF ALGEBRA CONCERNING NAMES.]

CHAPTER I.

DEFINITIONS, DESCRIPTION AND NOTATIONS.

1. ALGEBRA is the science which leads to solutions of the questions how many? and of what kind?

2. It has two branches,—one classification and definition, the other the solution of equations.

3. Those numbers are defined which are described as either rational numbers, or irrationals.

4. Rational numbers are either absolute numbers, or the parts of such numbers. Arithmetic deals with numbers of this kind.

5. Irrational numbers are the roots of rational numbers which have no roots among the absolute numbers.

6. And these (because they are physical quantities) turn up in Geometry.

7. The positive part of Algebra is that part which produces quantities, and concealed numbers by cans of imaginative suppositions, Book II will deal with them.

8. I shall teach the first part of Algebra, concerning numbers and physical quantities, in this Book I.

9. There are three kinds of defined numbers: Uninomia, plurinomia and universalia. I shall deal with them in order.

10. Uninomia are simple unique numbers, or any root of such a simple number.

11. The roots of numbers are very different, and thanks to the theory and skill of algebra different symbols are used to express them. These symbols are called radical signs.

For instance, the cube root of 6 is written thus √𝔠6. Likewise the square root of 5 is written thus √ʒ5. The rest are expressed as follows:

√ʒ Square root.
√𝔠 Cube root.

√ꝗꝗ	[Square root of the square root.]
√ß	[Supersolid root.]
√ꝗꞇ	[Square root of the cube root.]
√ʃß	[Second supersolid root.]
√ꝗꝗꝗ	[Square root of the square root of the square root.]
√ꞇꞇ	[Cube root of the cube root.]

And so the list continues to infinity.[24]

12. Certain roots and radicals are simple:
Such as √ꝗ, √ꞇ, √ß, √ʃß.

Others are multiple:
Such as √ꝗꝗ, √ꝗꞇ, √ꝗꝗꝗ, √ꞇꞇ, etc.

13. Again, certain of the roots and radicals are called quadratinomiæ named after quadratic whether on its own or in a multiple situation.

For example, square root, [square root of the square root, square root of the cube root, square root of the square root of the square root, and the supersolid root of the square root], etc.

Others do without the quadratic root sign.

Such as cube root, supersolid root, the second supersolid root, the cube [root] of the cube root, the supersolid root of the cube [root], etc.

14. A pair of uninomia operated on by the same root or radical sign are said to be identically rooted; if not they are differently rooted.

15. Every rational number is commensurable with every other rational number. So also a pair of uninomia identically rooted, in which the simple number of one divided by the simple number of the other produces a number of the same order as the radical sign indicates. Such numbers are described as mutually commensurable.

For example, 12 and 2, since they are rational they are commensurable. Likewise, √ꝗ8 and √ꝗ2 are a pair of uninomia identically rooted; from which 8 divided by 2 gives four prefaced by the quadratic root sign ꝗ. This makes 2, proving that √ꝗ8 and √ꝗ2 are commensurable with each other.

Corollary.

16. From this it is clear that the remaining uninomia which cannot be reduced in this way must all be incommensurable.

[24] Supersolid root means fifth root, second supersolid root means seventh root, and so on.

For example, √ʒ6 and √ʒ2 are incommensurable, because 6 divided by 2 gives 3 from which the square root sign is missing. Similarly, √𝔠12 and √ʒ3 are incommensurable because they carry different signs; but 12 and √ʒ4 are commensurable, because on reduction they have the same value as 12 and 2.

CHAPTER II.

ADDITION OF UNINOMIA.

1. Assuming that a given pair of uninomia are commensurable, divide the greater absolute number by the less. Extract the root of this result in accordance with the order of the root symbol, and add one. Multiply this by itself, as often as the power of the radical sign indicates. Next multiply by the absolute number of the smaller term and, before the product insert the original radical sign. This uninomial will then be equal to the sum of the given pair.[25]

The commensurable uninomia √ʒ12 and √ʒ3 are to be added. Divide 12 by 3 and obtain 4. From 4 extract √ʒ, which gives 2. To this add 1, making 3. Now multiply 3 by itself as often as the sign ʒ indicates. This produces 9. Now multiply 9 by 3 (the absolute value of the smaller term) getting 27. Next insert the proper radical sign giving √ʒ27; and this is the aggregate of √ʒ12 and √ʒ3. Likewise √𝔠24 and √𝔠3 (by Chapter 1, Proposition 14) are commensurable. Following the rule these can he added to give √𝔠81. Similarly, the addition of √ʒ$\frac{2}{3}$ to √ʒ2$\frac{2}{3}$ produce √ʒ6.

2. If the given uninomia were incommensurable it is not possible to add them otherwise than by inserting the symbol + which is called the sign of addition.

√𝔠5 and √ʒ3 are to be added, and they make a sum of √𝔠5 + √ʒ3. This is pronounced as follows,—the cube root of five plus the

---

25 In modern notation this is equivalent to the equation:

$$a^{1/n} + b^{1/n} = \sqrt[n]{b\left[\left(\frac{a}{b}\right)^{1/n} + 1\right]^n} \qquad \begin{aligned} n &= \text{power} \\ 1/n &= \text{root} \\ a &> b \end{aligned}$$

a and *b* are commensurable uninomia.

square root of three. Likewise, √ꝗ6 and √ꝗ2 when added make √ꝗ6 + √ꝗ2.

Corollary.

3. From this it is clear that from the addition of incommensurable uninomia, an abundance of binomia and plurinomia arise, so called because they consist of two or more uninomia joined by the addition sign plus. But more of this in due course.

CHAPTER III.

SUBTRACTION OF UNINOMIA.

1. If the given uninomia are commensurable, divide the number of the uninomia, from which the subtraction is to be effected, by the number of the uninomia which is to be subtracted. From the quotient, extract the proper root indicated, and then take away 1. Multiply the remainder by itself as often as the power of the radical sign indicates. Multiply this product by the number of the uninomia to be subtracted, and then insert the original radical sign in front. This then becomes the remainder left by the subtraction of the original pair of uninomia.[26]

Examples.

√ꝗ12 is to be taken from √ꝗ27. Divide 27 by 12 arithmetically giving $\frac{9}{4}$; [whose √ꝗ] is $\frac{3}{2}$. From this subtract 1 giving $\frac{1}{2}$. Multiply this by itself, making $\frac{1}{4}$. Next multiply by 12, giving 3. Now interpose the proper radical sign giving √ꝗ3, which is the remainder after subtracting √ꝗ12 from √ꝗ27. In the same way √𝔠24 taken from √𝔠81 leaves √𝔠3. So too, √ꝗ$\frac{2}{3}$ taken from √ꝗ6 leaves √ꝗ2$\frac{2}{3}$.

[26] In modern notation the formula is as follows:

$$a^{1/n} - b^{1/n} = \sqrt[n]{b\left[\left(\frac{a}{b}\right)^{1/n} - 1\right]^n} \qquad \begin{aligned} n &= \text{power} \\ 1/n &= \text{root} \\ a &> b \end{aligned}$$

a and *b* are commensurable uninomia.

2. If the given uninomia were incommensurable, write them both down with the subtrahend in the final place, and interpose this sign –, which is called the sign of subtraction.

Examples.

$\surd q\,3$ is to have $\surd c\,5$ taken from it. This leaves $\surd q\,3 - \surd c\,5$ which is pronounced, the square root of three lessened by the cube root of five. Likewise $\surd q\,2$ taken from $\surd q\,3$ leaves $\surd q\,3 - \surd q\,2$.

Corollary.

3. From this it is clear that defective remainders, or apotomes arise from binomia and plurinomia in the subtraction of incommensurable uninomia. Apotomes is defined by the mingling of many uninomia under the sign of subtraction.

CHAPTER IV.

EXTRACTION OF ROOTS FROM UNINOMIA.

1. If the root to be extracted is simple, and if it lies within the absolute number of the uninomia, extract the root arithmetically, but keep the former radical sign.

Examples.

The square root is to be extracted from the uninomia $\surd c\,4$. Because the square root is contained within 4, extract the root but keep the former radical sign. This will give $\surd c\,2$. Likewise, $\surd c$ extracted from $\surd q\,\frac{54}{16}$ or $\surd q\,\frac{27}{8}$ will be $\surd q\,\frac{3}{2}$.

2. But if the simple root to be extracted is not found in the absolute number of the uninomia, then insert before the absolute number both the sign of the required root and the former root sign.

Examples.

The root $\surd q$ is to be extracted from the $\surd q\,3$, and this will give $\surd qq\,3$. Likewise $\surd c$ of $\surd q\,5$ makes $\surd qc\,5$.

3. But if a multiple root extraction is to be effected on a uninomia, first carry out the multiplication on one simple root; and then carry out the second; and so extract them all successively (as shown in sections 1 and 2 above).

The root √℥𝔠 is to be extracted from √ß27. By Section 2 extract √℥ and this will give √℥ß27. Then by section 1 extract √𝔠 from it, and this will give √℥ß3. Likewise √℥℥ of √𝔠16 will be √𝔠2. And so √℥℥ of √𝔠10 will be √℥℥𝔠10.

4. In extracting roots from fractions, place the radical sign before a horizontal line which separates the numerator and denominator.

From $\frac{2}{3}$ you are to extract the √℥𝔠, and this will give $\frac{\sqrt{℥\mathfrak{c}}\,2}{\sqrt{℥\mathfrak{c}}\,3}$, or √℥ $\frac{\sqrt{\mathfrak{c}}\,2}{\sqrt{\mathfrak{c}}\,3}$, or √𝔠 $\frac{\sqrt{℥}\,2}{\sqrt{℥}\,3}$, or best of all √℥𝔠 $\frac{2}{3}$; for these are all exactly the same in value.

CHAPTER V.

REDUCTION OF UNINOMIA TO THE SAME ORDER.

Two uninomia have unlike roots. Multiply the absolute number of each one by itself, as often as its unlike partner indicates. Keep both products separate and place before them the root signs of both. You have now converted the uninomia to the same radical, and retained their previous values.

Example.

Let there be two unlike radicals which are to be reduced to the same radical sign. These are √𝔠3 and √℥2. Multiply 3 into itself twice, and 2 into itself thrice. From the former we get 9, and from the latter 8. Now put both the radical signs in front and get √℥𝔠9 and √℥𝔠8. These have like radical signs, and retain their previous value. For √℥𝔠9 has the same value as √𝔠3; and √℥𝔠8 has the same value as √℥2 as Chapter 4 of Book I explains. Likewise, √℥℥2 and √℥𝔠5 are reduced in this way. Multiply 2 into itself thrice and 5 into itself twice, making √℥℥𝔠8 and √℥℥𝔠25. Likewise √℥6 and 2 have the same value as √℥6 and √℥4.

CHAPTER VI.

MULTIPLICATION AND DIVISION OF UNINOMIA.

1. EVERY uninomium without the symbols $+$ or $-$ must be understood to carry the $+$ sign before it.

[$\surd\mathfrak{z}\,10$ is taken for $+\surd\mathfrak{z}\,10$.]

2. Multiplication or division of 2 symbols carrying the same sign, produce the sign of increase; conversely, multiplication or division of two opposite signs produces the sign of decrease.

Examples are given below.

3. So, in the first place, the given uninomia must carry similar radical signs; if not, they must be reduced to the same order. Then multiply or divide number by number, and extract (by Chapter 4) the appropriate root of the product. Finally (by Section 2 of this chapter) multiply or divide the signs and place the appropriate symbol before the product; and this becomes the result of the multiplication or division.

Example.

The uninomia $\surd\mathfrak{z}\,12$ and $\surd\mathfrak{z}\,3$ are to be multiplied. Multiply 12 by 3 and obtain 36, whose square root is 6; and this is the product. Similarly, $\surd\mathfrak{C}\,3$ multiplied by $-\surd\mathfrak{C}\,2$ gives the product $-\surd\mathfrak{C}\,6$ by Section 2 of this chapter. Likewise, $-\surd\mathfrak{C}\,3$ multiplied by $\surd\mathfrak{z}\,2$ which must first be reduced to $\surd\mathfrak{z}\mathfrak{C}\,9$ and $\surd\mathfrak{z}\mathfrak{C}\,8$ by Chapter 5. Then 9 times 8 make 72, whose root $\surd\mathfrak{z}\mathfrak{C}$ (by Chapter 4, Section 2) will be $\surd\mathfrak{z}\mathfrak{C}\,72$. This must be preceded by the $-$ sign, by 2 above, and the product becomes $-\surd\mathfrak{z}\mathfrak{C}\,72$.

Example of Division.

$\surd\mathfrak{z}\,12$ is to be divided by $-\surd\mathfrak{z}\,3$. By a previous section the quotient produced is -2. Likewise, $-\surd\mathfrak{C}\,16$ divided by $\surd\mathfrak{C}\,2$ becomes -2. Finally, $\surd\mathfrak{z}\mathfrak{C}\,72$ divided by $\surd\mathfrak{z}\mathfrak{C}\,9$ makes $\surd\mathfrak{z}\mathfrak{C}\,8$ or otherwise $\surd\mathfrak{z}\mathfrak{C}\,2$, as section 1 of Chapter 4 explains.

Corollary.

4. Hence it is evident that a uninomium squared, cubed, or raised to any order; when its absolute number is squared, cubed, or raised to that order, retains its original root sign.

So $\surd\mathfrak{z}2$ raised to the third power becomes $\surd\mathfrak{z}8$. In the same way $\surd\mathfrak{C}3$ squared yields $\surd\mathfrak{C}9$, and so on.

Corollary.

5. From this it follows that any radical multiplied into itself as often as a quadratinomium indicates, loses both its former root sign, and also its negative sign. But if there is no quadratinomium then only its radical sign is removed.

For example, $-\surd\mathfrak{C}5$ when multiplied into itself thrice makes -5. Similarly, $\surd\mathfrak{z}3$ multiplied into itself twice makes 3. In the same way, $\surd\mathfrak{z}\mathfrak{C}6$ multiplied twice into itself makes $\surd\mathfrak{C}6$, or three times into itself, $\surd\mathfrak{z}6$. The same is true for the whole universe of radicals, and also for the positive parts of algebra. But more of this later.

CHAPTER VII.

PLURINOMIA.

1. Plurinomia consist of two or more uninomia joined by the symbols $+$ or $-$.

2. Some plurinomia are called abundant (described in Chapter 2, Section 3) and others defective, commonly called residual or apotomes, which are dealt with in Chapter 3, Section 3.

3. Lower plurinomia are those in which all the uninomia are the quadratic roots of numbers, either with, or without an absolute number.

As for example, $\surd\mathfrak{z}3 + \surd\mathfrak{z}5 - \surd\mathfrak{z}2 + 5$; or $\surd\mathfrak{z}3 + \surd\mathfrak{z}5 - \surd\mathfrak{z}2$, are called lower plurinomia.

4. All other plurinomia are labelled upper plurinomia.

5. Some plurinomia are described as greater or lesser in accordance with the number of uninomia included.

As for example, the quadrinomium $\surd\mathfrak{z}3 + \surd\mathfrak{z}35 - \surd\mathfrak{z}32 + 5$ is a greater plurinomium than the trinomium $\surd\mathfrak{z}3 + \surd\mathfrak{z}5 - \surd\mathfrak{z}2$, which is itself a greater plurinomium than the binomium $\surd\mathfrak{z}6 - \surd\mathfrak{z}5$.

6. If you convert certain signs in a given plurinomium, but not all, you may change it into a defective or apotomes plurinomium. On the other hand you can change it from a defective to an abundant plurinomium.

Let $\surd\mathfrak{q}5 + \surd\mathfrak{q}3 - \surd\mathfrak{q}2$ be an abundant trinomium. Then its defective will be $\surd\mathfrak{q}5 - \surd\mathfrak{q}3 - \surd\mathfrak{q}2$ or $\surd\mathfrak{q}5 - \surd\mathfrak{q}3 + \surd\mathfrak{q}2$ or $\surd\mathfrak{q}3 - \surd\mathfrak{q}5 + \surd\mathfrak{q}2$ or $\surd\mathfrak{q}3 - \surd\mathfrak{q}5 - \surd\mathfrak{q}2$. Similarly, if the trinomium $\surd\mathfrak{q}5 + \surd\mathfrak{q}3 - \surd\mathfrak{q}2$ is defective, then $\surd\mathfrak{q}5 + \surd\mathfrak{q}3 + \surd\mathfrak{q}2$ will be abundant. From which examples it is clear that the same plurinomium can be both abundant and defective, depending upon the conditions imposed.

7. If a given plurinomium contains two commensurable uninomia of the same sign, add them (by Chapter 2, Section 1), place the sign before the product, and you will have converted a greater plurinomium into a smaller one.

Let the trinomium $\surd\mathfrak{q}12 + \surd\mathfrak{q}3 - 2$, in which $\surd\mathfrak{q}12$ and $\surd\mathfrak{q}3$ are commensurable and have the same sign, be added to make $\surd\mathfrak{q}27$, which, with -2 reduces it to a smaller plurinomium, namely, to $\surd\mathfrak{q}27 - 2$, a binomium.

8. If a given plurinomium contains two commensurable uninomia with different signs, subtract the smaller from the greater (by Chapter 3, Section 1), and place the sign of the greater uninomium before the product, and you will have converted a greater plurinomium into a smaller one.

Let the trinomium $\surd\mathfrak{q}10 + \surd\mathfrak{q}2 - \surd\mathfrak{q}8$, in which $+\surd\mathfrak{q}2$ and $-\surd\mathfrak{q}8$ are commensurable, have its different signs noted. Upon subtraction the difference comes to $\surd\mathfrak{q}2$, before which the sign of the greater uninomium must be placed. This gives $-\surd\mathfrak{q}2$, which, with $\surd\mathfrak{q}10$ completes the reduction to the binomium $\surd\mathfrak{q}10 - \surd\mathfrak{q}2$.

CHAPTER VIII.

ADDITION OF PLURINOMIA.

1. FOR THE ADDITION of plurinomia join all uninomia together with their signs into one single plurinomium. Then, if any of the terms are commensurable, reduce them by Sections 7 and 8 of the previous chapter. From this the sum total is produced.

$\surd q\,3+\surd q\,8$ are to be added to $4-\surd q\,2$. First make $\surd q\,3+\surd q\,8+4-\surd q\,2$; then, because $+\surd q\,8$ and $-\surd q\,2$ are commensurable perform the addition by Section 8 of Chapter 7, which comes to $\surd q\,3+\surd q\,2+4$. Likewise, $\surd q\,5+\surd q\,3$ are to be added to $\surd q\,20-\surd q\,12$. This comes to $\surd q\,5+\surd q\,3+\surd q\,20-\surd q\,12$, and then, by reduction, there will arise $\surd q\,45-\surd q\,3$. [Likewise,] add $\surd c\,16+\surd q\,18$ to $\surd c\,2-\surd q\,2$. This gives $\surd c\,54+\surd q\,8$. [Likewise,] add $\surd c\,54+\surd q\,18-1$ to $\surd q\,2+\surd q\,3$. This makes $\surd c\,54+\surd q\,32+\surd q\,3-1$.

Corollary.

2. From this it is clear that in the addition of an abundant to a defective counterpart, abundant and defective numbers cancel each other and the remaining terms are duplicated.

For example, $12+\surd q\,3$ (abundant) added to its apotomes $12-\surd q\,3$ makes $12+\surd q\,3+12-\surd q\,3$ which is the same as 24 numerically.

CHAPTER IX.

SUBTRACTION OF PLURINOMIA.

1. First change all the signs of [the] plurinomium to be subtracted. This done, the converted plurinomium is added (by the preceding chapters) to the plurinomium from which it was to be taken. And this produces the remainder after subtraction is performed.

For example, $\surd q\,5+\surd q\,3$ is to be subtracted from the binomium $\surd q\,45-\surd q\,3$. Change the signs thus: $-\surd q\,5-\surd q\,3$. Now add them to $\surd q\,45-\surd q\,3$ (by the preceding chapter) and this gives $\surd q\,20-\surd q\,12$. Likewise, $\surd c\,2-\surd q\,2$ is to be subtracted from $\surd c\,54+\surd q\,3$. There will remain $\surd c\,16+\surd q\,3+\surd q\,2$. Similarly, from $\surd c\,54+\surd q\,32+\surd q\,3-1$ the following is to be subtracted; $\surd q\,2+\surd q\,3$. There will remain $\surd c\,54+\surd q\,18-1$.

Corollary.

2. From this it is evident that in the subtraction of a defective from an abundant counterpart, certain abundant or defective numbers are duplicated, but the rest mutually annihilate each other.

For instance, from the abundant binomium $\surd\mathfrak{q}\,13 + 7$ its defective conjugate $\surd\mathfrak{q}\,13 - 7$ is to be subtracted. This gives $\surd\mathfrak{q}\,13 + 7$ $-\surd\mathfrak{q}\,13 + 7$ at first, and then it becomes 14 when the operation is complete.

CHAPTER X.

MLTIPLICATION OF PLURINOMIA.

1. The multiplication of uninomia upon uninomia was dealt with in Chapter 6; but if they included numbers which were mutually commensurable, their magnitudes were reduced as shown in Chapter 7, Sections 7 and 8.

For instance, $\surd\mathfrak{q}\,3 - \surd\mathfrak{c}\,2 + 6$ is to be multiplied by $\surd\mathfrak{q}\,5 - 7$. The product then becomes $\surd\mathfrak{q}\,15 - \surd\mathfrak{qc}\,500 + \surd\mathfrak{q}\,180 - \surd\mathfrak{q}\,147 +$ $\surd\mathfrak{c}\,686 - 42$, which is quite incommensurable, and so it cannot be reduced. Similarly, $\surd\mathfrak{q}\,8 + \surd\mathfrak{q}\,3 - 5$ is to be multiplied by $\surd\mathfrak{q}\,12 -$ $\surd\mathfrak{q}\,2$. This gives $\surd\mathfrak{q}\,96 + \surd\mathfrak{q}\,36$ [(otherwise 6)] $- \surd\mathfrak{q}\,300 - \surd\mathfrak{q}\,16$ [(otherwise -4)] $- \surd\mathfrak{q}\,6 + \surd\mathfrak{q}\,50$. This product (by Chapter 7, Sections 7 and 8) reduces to $\surd\mathfrak{q}\,54 + 2 - \surd\mathfrak{q}\,300 + \surd\mathfrak{q}\,50$.

2. In multiplying an abundant quantity by its conjugate defective it is enough to multiply the abundant part by the defective part, and to multiply the parts common to both into each other.

For example, $\surd\mathfrak{q}\,7 + \surd\mathfrak{q}\,5$ abundant is to be multiplied by $\surd\mathfrak{q}\,7 -$ $\surd\mathfrak{q}\,5$ its conjugate defective. This gives $7-5$ (otherwise 2) for the total product. The transverse multiplications $\surd\mathfrak{q}\,7$ by $-\surd\mathfrak{q}\,5$, and $\surd\mathfrak{q}\,7$ by $+\surd\mathfrak{q}\,5$, come to $-\surd\mathfrak{q}\,35$ and $+\surd\mathfrak{q}\,35$, which cancel each other and so have no effect.

Corollary.

3. If an abundant plurinomium of the lower order is multiplied by its conjugate defective it produces a plurinomium of the lower order which is smaller numerically.

An abundant trinomium of the lower order is multiplied by one of its conjugate defectives, say $\surd\mathfrak{q}\,11 - \surd\mathfrak{q}\,3 + \surd\mathfrak{q}\,2$ multiplied by $\surd\mathfrak{q}\,11 - \surd\mathfrak{q}\,3 - \surd\mathfrak{q}\,2$. This produces the binomium $12 - \surd\mathfrak{q}\,132$,[27]

[27] This depends on the relation $(A - B + C)(A - B - C) = A^2 + B^2 - C^2 - 2AB$. If A, B and C are square roots then A^2, B^2, and C^2 are integers, and only $2AB$ contains the root sign.

which when multiplied by its abundant conjugate $12 + \surd$ʒ$\,132$ creates a uninomium namely, the number 12.

Corollary.

4. If an abundant binomium of the lower order is multiplied by its conjugate defective, then an absolute number is produced.

As already said, if the abundant binomium $12+\surd$ʒ$\,132$ is multiplied into its conjugate defective $12 - \surd$ʒ$\,132$ (by Corollary 2 above), then it produces the number 12.

It must be noted that an irrational binomium can be multiplied by a certain kind of plurinomium, and from this a rational number can emerge. Each of two named cube roots can be squared, and their product can also be added to make an abundant trinomium from an abundant binomium, or a defective trinomium from a like binomium. This trinomium, if it is abundant, is multiplied into the defective binomium; but if it is defective it is multiplied by the abundant binomium, and it will produce a simple number. Otherwise proceed by Proposition 2 of Euclid, Book VIII and look for three numbers in the same ratio as that which exists between the named cube roots; or four numbers in the same ratio as that between the fourth roots; and so on. Then perform the multiplications given above.

Example.

From the abundant binomium $\surd$𝔠$\,6+\surd$𝔠$\,4$ produce the trinomium $\surd$𝔠$\,36+\surd$𝔠$\,24+\surd$𝔠$\,16$. Multiply the latter by the defective $\surd$𝔠$\,6-\surd$𝔠$\,4$ which gives 2. Likewise from the defective $\surd$𝔠$\,6 - \surd$𝔠$\,4$ produce the defective trinomium $\surd$𝔠$\,36-\surd$𝔠$\,24+\surd$𝔠$\,16$. Multiply this by the abundant binomium $\surd$𝔠$\,6 + \surd$𝔠$\,4$ and you get 10.

Another example.

From $\surd$ʒʒ$\,3-\surd$ʒʒ$\,2$ produce the quadrinomium $\surd$ʒʒ$\,27-\surd$ʒʒ$\,18$ $+\surd$ʒʒ$\,12-\surd$ʒʒ$\,8$. Multiply the latter by $\surd$ʒʒ$\,3+\surd$ʒʒ$\,2$, giving 1.[28]

[28] Section 4 depends on $A^3 + B^3 = (A + B)(A^2 - AB + B^2)$ and the companion relation for $A^3 - B^3$. The quadrinomium in Example 2 depends on the relation $A^4 \mp B^4 = (A \mp B)(A^3 \pm A^2B + AB^2 \pm B^3)$. The letters A and B indicate fourth roots in the last case.

CHAPTER XI.

DIVISION OF PLURINOMIA.

1. If the [divisor] is a uninomium, divide each separate uninomium by it (according to Chapter 6). Then join the uninomia which make up the quotient by means of the signs produced.

$\surd\mathcal{Q}\,12 + \surd\mathcal{Q}\,8$ is to be divided by $\surd\mathcal{Q}\,2$. This gives $\surd\mathcal{Q}\,6 + 2$. Similarly, $\surd\mathcal{Q}\,36300 + \surd\mathcal{Q}\,7200 - \surd\mathcal{Q}\,10800 + \surd\mathcal{Q}\,6600 + \surd\mathcal{Q}\,9900$ is to be divided by 12. The quotient then comes to $\surd\mathcal{Q}\,\frac{3025}{12} + \surd\mathcal{Q}\,50 - \surd\mathcal{Q}\,75 + \surd\mathcal{Q}\,\frac{275}{6} + \surd\mathcal{Q}\,\frac{275}{4}$.

2. A plurinomium of the lower order is the divisor. From this plurinomium (using the technique or Chapter 10, Sections 3 and 4), produce a simple number. Using the same plurinomium with which you multiplied, you will multiply both divisor and dividend. Now divide the product by the given simple number, and it will reproduce as quotient the original divisor and dividend.

5 is to be divided by the trinomium of the lower order $\surd\mathcal{Q}\,11 - \surd\mathcal{Q}\,3 - \surd\mathcal{Q}\,2$. If first you multiplied it by its conjugate $\surd\mathcal{Q}\,11 - \surd\mathcal{Q}\,3 + \surd\mathcal{Q}\,2$ it would make $12 - \surd\mathcal{Q}\,132$. When this is multiplied by its conjugate $12 + \surd\mathcal{Q}\,132$ it would make 12 (by 3 and 4 of Chapter 10). Then, using the conjugate trinomium $\surd\mathcal{Q}\,11 - \surd\mathcal{Q}\,3 + \surd\mathcal{Q}\,2$, multiply the dividend which is 5, and this makes $\surd\mathcal{Q}\,275 - \surd\mathcal{Q}\,75 + \surd\mathcal{Q}\,50$. Multiply this again by the previous binomium $12 + \surd\mathcal{Q}\,132$, and it becomes: $\surd\mathcal{Q}\,36300 + \surd\mathcal{Q}\,7200 - \surd\mathcal{Q}\,10800 + \surd\mathcal{Q}\,6600 + \surd\mathcal{Q}\,9900$. This is the new dividend, which, on division by the given number 12 gives, as quotient $\surd\mathcal{Q}\,\frac{3025}{12} + \surd\mathcal{Q}\,50 - \surd\mathcal{Q}\,75 + \surd\mathcal{Q}\,\frac{275}{6} + \surd\mathcal{Q}\,\frac{275}{4}$. This result is consistent with the previous division of 5 by the plurinomium $\surd\mathcal{Q}\,11 - \surd\mathcal{Q}\,3 - \surd\mathcal{Q}\,2$.

> This must be altered, for division can be effected by $6 + \surd\mathfrak{C}\,2$, and by every binomium, as indicated at the end of the previous chapter.

3. If the divisor were chosen from the upper uninomia, it would scarcely ever divide and leave an integral dividend without remainder. And so, between the dividend on the top line, and the divisor below it, a line would have to be drawn after the fashion of arithmetical fractions.

As for example $10 - \surd\mathcal{Q}\,3$ is to be divided by $6 + \surd\mathfrak{C}\,2$. This cannot be done in any other way than by interposing a line to show division, like this $\frac{10-\sqrt{\mathfrak{z}}3}{6+\sqrt{\mathfrak{c}}2}$, which is pronounced $10-\sqrt{\mathfrak{z}}3$ divided by $6+\sqrt{\mathfrak{c}}2$.[29]

Corollary.

4. From this it is evident that irrational fractions have their origin in the division by plurinomia of the upper order.

CHAPTER XII.

ROOTS OF PLURINOMIA.

1. Some roots of plurinomia are easily found, but others are unresolvable. We call those plurinomia resolvable which are not greater than those whose roots they are.

2. We describe those roots as unresolvable which swarm in confusion amongst a multitude of uninomia and roots of plurinomia.

3. If the square root is to be extracted from a given binomium of the lower order, then from the difference of the squares of both uninomia, extract the square root. Add this to the greater uninomium, and subtract it from the same, if they are not commensurable. If the latter is the case, then the root sought is unresolvable. From the halves of these two extract the square root (by Chapter 4). Then connect the two roots by the sign of the original binomium. This will then produce the simple quadratic root of the original binomium.[30]

Example.

The square root is to be extracted from the defective binomium $3-\sqrt{\mathfrak{z}}5$. The squares of the uninomia are 9 and 5, and their difference is 4. The square root of 4 is 2, and this is commensurable with 3,

[29] The denominator $6+\sqrt{\mathfrak{c}}2$ can easily be converted. Multiply by $36-6\sqrt{\mathfrak{c}}2+(\sqrt{\mathfrak{c}}2)^2$ and this gives $6^3+(\sqrt{\mathfrak{c}}2)^3=216+2=218$. The numerator must be dealt with in the same way but here it is impossible to eliminate irrationals. The numerator becomes

$$360+10\sqrt{\mathfrak{c}}4-60\sqrt{\mathfrak{c}}2-36\sqrt{\mathfrak{z}}3+6\sqrt{\mathfrak{zc}}108-4\sqrt{\mathfrak{zc}}32$$

and this is to be divided by 218.

[30] $\sqrt{(a\pm b)}=\sqrt{\frac{1}{2}\left[a+\sqrt{(a^2-b^2)}\right]}\pm\sqrt{\frac{1}{2}\left[a-\sqrt{(a^2-b^2)}\right]}$.

therefore add 3 to 2 making 5. Now subtract 2 from 3 leaving 1. From the halves of 5 and 1 take the square roots, making $\surd\text{Q}\,\frac{5}{2}$ and $\surd\text{Q}\,\frac{1}{2}$. Join these two with the original (minus) sign making $\surd\text{Q}\,\frac{5}{2} - \surd\text{Q}\,\frac{1}{2}$; and this is the square root of the binomium $3 - \surd\text{Q}\,5$. The square root of $\surd\text{Q}\,48 - 6$ is to be extracted. The square root of the difference of the squares is $\surd\text{Q}\,12$, which, added to, and subtracted from $\surd\text{Q}\,48$ makes $\surd\text{Q}\,108$ and $\surd\text{Q}\,12$. The square roots of the halves, joined by the proper sign make $\surd\text{QQ}\,27 - \surd\text{QQ}\,3$, the required root. In similar fashion $\surd\text{Q}\,24 + \surd\text{Q}\,18$ has for its resolvable square root $\surd\text{QQ}\,\frac{27}{2} + \surd\text{QQ}\,\frac{3}{2}$.

4. The roots of all the remaining plurinomia whatsoever are considered to be unresolvable.

Example.

$\surd\text{Q}\,48 + \surd\text{Q}\,28$ lacks an evident root, because the $\surd\text{Q}$ of the difference of the squares, which is $\surd\text{Q}\,20$, is not commensurable with $\surd\text{Q}\,48$ when (by Section 3 above) it ought to be commensurable. Likewise the square or cube root of $\surd\text{C}\,3 + 1$ is unresolvable; and so for the others, excepting binomia of the lower order already mentioned.

5. Unresolvable roots, however, cannot be extracted otherwise than by placing before the root the sign of root extraction, with a period before the plurinomium in question.[31] And this radical with the following period is called the universal sign of root extraction, for it indicates the root of the following universal plurinomium.

As example, the square root of $\surd\text{Q}\,48 + \surd\text{Q}\,28$ is to be extracted. Before this binomium, place the radical sign $\surd\text{Q}$ with this period. This then makes $\surd\text{Q}.\surd\text{Q}\,48 + \surd\text{Q}\,28$, which is pronounced in this way: The square root of the universal radical of the square root of 48 added to the square root of 28. For this signifies that $\surd\text{Q}\,48$ is to be joined with the square root of 28 in a single summation. Then the square root of the total sum must be taken. Likewise, the cube root of $\surd\text{C}\,3 + \surd\text{Q}\,2 - 1$ is to be extracted. This will be expressed: $\surd\text{C}.\surd\text{C}\,3 + \surd\text{Q}\,2 - 1$.

Corollary.

6. From this it is evident that universal radicals have their origin in the extraction of unresolvable roots.

[31] This dot is called the "sign of aggregation".

CHAPTER XIII.

IRRATIONAL FRACTIONS.

1. THAT which can be done for rational fractions in Arithmetic can be accomplished for irrational fractions by Algebra.

In the irrational fractions of plurinomia we operate by Arithmetic insofar as they are fractions; and by Algebra insofar as they are plurinomia and irrational.

For example $\frac{\surd\mathfrak{q}3+2}{\surd\mathfrak{c}3}$ is to be divided by $\frac{\surd\mathfrak{q}5}{\surd\mathfrak{c}2-1}$, which, by Arithmetic, using rational numbers, can be done by cross multiplication of each numerator into its opposite denominator. This can also be done [by Algebra] and yields $\frac{\surd\mathfrak{q}\mathfrak{c}108+\surd\mathfrak{c}16-\surd\mathfrak{q}3-2}{\surd\mathfrak{q}\mathfrak{c}1125}$ and this is the result of the required division.

Likewise, the two following are to be added $\frac{\surd\mathfrak{q}3+2}{\surd\mathfrak{c}3}$ and $\frac{\surd\mathfrak{q}5}{\surd\mathfrak{c}2-1}$. This is done as Arithmetic teaches by multiplying transversely and directly so that a common denominator is produced. But Algebra multiplies in this way. First obtain the common denominator for the first fraction thus: $\frac{\surd\mathfrak{q}\mathfrak{c}108+\surd\mathfrak{c}16-\surd\mathfrak{q}3-2}{\surd\mathfrak{c}6-\surd\mathfrak{c}3}$. And for the second we obtain $\frac{\surd\mathfrak{q}\mathfrak{c}1125}{\surd\mathfrak{c}6-\surd\mathfrak{c}3}$. Then by Algebra, as the rules of Arithmetic demand, add the numbers both of which contain a common denominator, and the result of the addition is then: $\frac{\surd\mathfrak{q}\mathfrak{c}1125+\surd\mathfrak{q}\mathfrak{c}108+\surd\mathfrak{c}16-\surd\mathfrak{q}3-2}{\surd\mathfrak{c}6-\surd\mathfrak{c}3}$.

CHAPTER XIV.

ADDITION AND SUBTRACTION OF UNIVERSAL RADICALS.

1. UNIVERSALS are added by the sign of increase, and are subtracted by the sign of diminution.

In order to add $\surd\mathfrak{q}.10+\surd\mathfrak{q}2$ to $\surd\mathfrak{q}.8-\surd\mathfrak{q}3$, interpose the sign for addition + and it will make $\surd\mathfrak{q}.10+\surd\mathfrak{q}2+\surd\mathfrak{q}.8-\surd\mathfrak{q}3$. Likewise, $\surd\mathfrak{q}.8-\surd\mathfrak{q}3$ is subtracted from $\surd\mathfrak{q}.10+\surd\mathfrak{q}2$; interpose the minus sign − and which makes $\surd\mathfrak{q}.10+\surd\mathfrak{q}2-\surd\mathfrak{q}.8-\surd\mathfrak{q}3$.

2. If the universal square root of a defective binomium of the lower order is added to the universal square root of its abundant conjugate; or else if the

universal square root of an abundant binomium of the lower order is subtracted from the universal square root of its defective conjugate; the result obtained by use of the signs $+$ and $-$ in the proper way will show an abbreviation (as indicated in Chapter 16, Section 5), which follows.

For instance, by adding $\surd Q.10 - \surd Q 2$ to $\surd Q.10 + \surd Q 2$ them will arise $\surd Q.10 + \surd Q 2 + \surd Q.10 - \surd Q 2$. And this can be abbreviated by Chapter 16, Section 5.

CHAPTER XV.

REDUCTION OF DIFFERENT UNIVERSALS TO THE SAME RADICAL SIGN.

1. MULTIPLY each universal plurinomium into itself as often as the order of its companion universal indicates, and then insert the universal of both before each product.

The following are to be reduced to the same universal sign: $\surd QQ.2 - \surd C 3$ and $\surd QC.7 + \surd Q 2$. Multiply the plurinomium $2 - \surd C 3$ into itself thrice; and $7 + \surd Q 2$ into itself twice. This makes $5 + \surd C 1944 - \surd C 5184$, and $51 + \surd Q 392$. Now place the sign $\surd Q$ before both, together with the different signs Q and C which makes $\surd QQC.5 + \surd C 1944 - \surd C 5184$ and $\surd QQC.51 + \surd Q 392$ both reduced to the same universal sign $\surd QQC$.

2. By the same reasoning universals can be reduced to the same radical sign together with their components.

3 and $\surd Q.18 + \surd Q 243$ become $\surd Q.9$ and $\surd Q.18 + \surd Q 243$. [Likewise,] $\surd Q.13 + \surd Q 20$ and $2 + \surd Q 3$.[32] become $\surd Q.13 + \surd Q 20$ and $\surd Q.7 + \surd Q 48$.

Corollary.

3. From this it is evident that a uninomium carrying the universal sign is the same as a rooted particle.

For instance, $\surd Q.9$ and $\surd Q 9$ are exactly the same. Likewise $\surd C 5$ and $\surd C.5$; this is described as the universal sign when it carries the point.

[32] $(2 + \surd Q 3)^2 = 4 + 3 + 4\surd Q 3 = 7 + \surd Q 48 \ \therefore\ 2 + \surd Q 3 = \surd Q.7 + \surd Q 48.$

CHAPTER XVI.

MULTIPLICATION AND DIVISION OF UNIVERSALS.

1. If a universal is to be multiplied by a universal, first equip them with the universal symbol (as in Chapter 15).

2. Next, temporarily forgetting the universal symbol, perform the multiplications and divisions by the rules for uninomia and plurinomia.

3. Finally, place before the product, or quotient, the original universal symbol together with the appropriate sign in front (by Chapter 6, Section 2).

For example, $\surd\mathfrak{C}.5 + \surd\mathfrak{Q}2$ and $\surd\mathfrak{C}.4 - \surd\mathfrak{Q}3$ are to be multiplied together. So multiply $5 + \surd\mathfrak{Q}2$ by $4 - \surd\mathfrak{Q}3$, and this produces $20 + \surd\mathfrak{Q}32 - \surd\mathfrak{Q}75 - \surd\mathfrak{Q}6$, before which place the symbol $\surd\mathfrak{C}$. or $+\surd\mathfrak{C}$, which makes $\surd\mathfrak{C}.20 + \surd\mathfrak{Q}32 - \surd\mathfrak{Q}75 - \surd\mathfrak{Q}6$. Similarly, $\surd\mathfrak{Q}.4 + \surd\mathfrak{C}2$ is to be multiplied by 3 or $\surd\mathfrak{Q}9$, or better still (by Chapter 15, Section 3) use $\surd\mathfrak{Q}.9$. Therefore multiply $4 + \surd\mathfrak{C}2$ into 9, which makes $36 + \surd\mathfrak{C}1458$, by Chapter 6 and Chapter 10. Now place $\surd\mathfrak{Q}$. or $+\surd\mathfrak{Q}$. and this makes $\surd\mathfrak{Q}.36 + \surd\mathfrak{C}1458$. Likewise, $\surd\mathfrak{Q}.10 + \surd\mathfrak{Q}2$ multiplied into $-\surd\mathfrak{Q}.10 - \surd\mathfrak{Q}2$ makes $-\surd\mathfrak{Q}.98$, or $-\surd\mathfrak{Q}98$.

4. If the universal square root symbol is prefixed by the sign + it must be multiplied into the whole universal. But with the minus sign prefixed, delete it and the universal symbol and change the remaining signs into their opposites. The product of the multiplication will then emerge.

$\surd\mathfrak{Q}.2 - \surd\mathfrak{Q}3$ when multiplied by $-\surd\mathfrak{Q}.2 - \surd\mathfrak{Q}3$ makes $+\surd\mathfrak{Q}3 - 2$.

5. But if a universal square root is multiplied into itself, delete the universal root symbol with the sign prefixed, and the product of the multiplication will be left.

$\surd\mathfrak{Q}.10 + \surd\mathfrak{Q}2$ multiplied into itself produces $10 + \surd\mathfrak{Q}2$.

6. But if several universals are to be multiplied or divided by several others, then the total sum produced by the single products will carry the universal symbol indicated by the products.

The universal $\surd\mathfrak{C}.10 + \surd\mathfrak{Q}5 + \surd\mathfrak{Q}.8 - \surd\mathfrak{Q}3$ [is] to be multiplied by $\surd\mathfrak{Q}.3 + \surd\mathfrak{Q}6 - \surd\mathfrak{Q}.4 - \surd\mathfrak{Q}7$ in this way. First reduce $\surd\mathfrak{C}.10 + \surd\mathfrak{Q}5$ with $\surd\mathfrak{Q}.3 + \surd\mathfrak{Q}6$ to the same universal. The former gives $\surd\mathfrak{Q}\mathfrak{C}.105 + \surd\mathfrak{Q}2000$; and the latter becomes $\surd\mathfrak{Q}\mathfrak{C}.81 + \surd\mathfrak{Q}6534$. Now multiply these together, and (by this chapter) it makes:

$\surd QC.8505 + \surd Q13068000 + \surd Q13122000 + \surd Q72037350.$[33] And this is the product of one universal acting on another. The total product has a common universal symbol $\surd QC$ placed before it. In the same way multiply $3 + \surd Q6$ by $8 - \surd Q3$ and place its universal symbol[, which is $+\surd Q.$,] in front. This produces: $+\surd Q.24 - \surd Q27 + \surd Q384 - \surd Q18$. And this is the second part of the product. Thirdly, reduce $-\surd Q.4 - \surd Q7$ with $\surd C.10 + \surd Q5$ to the same universal. This makes $-\surd QC.148 - \surd Q21175$ and also $\surd QC.105 + \surd Q2000$. Multiply these together and they make up the third part of the product: $\surd QC.15540 - \surd Q233454375 + \surd Q43808000 - \surd Q42350000$. Fourthly, multiply $8 - \surd Q3$ by $4 - \surd Q7$. Place the symbol $-\surd Q.$ before the product and it makes: $\surd Q.32 - \surd Q48 - \surd Q448 + \surd Q21$. This is the fourth part of the product. [Whose four parts arise from the multiplication of one unversal into another in some manner; henceforth they are embraced by a single universal sign in some manner, and the whole product is $\surd QC.8505 + \surd Q13068000 + \surd Q13122000 + \surd Q72037350 + \surd Q.24 - \surd Q27 + \surd Q348 - \surd Q18 - \surd QC.15540 - \surd Q233454375 + \surd Q43808000 - \surd Q42350000 - \surd Q.32 - \surd Q448 + \surd Q21$.]

7. An abundant binomium of the lower order carries the universal square root symbol $\surd Q.$, and it is connected by either of the signs + or − to its conjugate defective, which also carries the [same] symbol.

The given universal is as shown: $\surd Q.10 + \surd Q2 + \surd Q.10 - \surd Q2$. This is to be multiplied into itself [and you prefix $\surd Q.$], making $\surd Q.20 + \surd Q392$ which is equal to $\surd Q.10 + \surd Q2 + \surd Q.10 - \surd Q2$ but much shorter.[34] By an equal reasoning $\surd Q.20 - \surd Q392$ is made from $\surd Q.10 + \surd Q2 - \surd Q.10 - \surd Q2$.

---

[33] This perhaps requires some explanation. $A = \surd C.10 + \surd Q5$; $B = \surd Q.8 - \surd Q3$; $C = \surd Q.3 + \surd Q6$; $D = -\surd Q.4 - \surd Q7$. Then, $(A+B)(C+D) = A \cdot C + A \cdot D + B \cdot C + B \cdot D$.
The product of $A \cdot C$ must have the universal symbol $\surd QC$. Multiply $10 + \surd Q5$ into itself twice: $100 + 20\surd Q5 + 5 = 105 + \surd Q2000$.
Multiply $3 + \surd Q6$ into itself three times: $27 + 3 \cdot 3^2 \cdot \surd Q6 + 3 \cdot 3 \cdot 6 + 6 \cdot \surd Q6 = 81 + \surd Q6534$. Therefore [the] product

$$\begin{aligned} A \cdot C &= (105 + \surd Q2000)(81 + \surd Q6534) \\ &= \surd QC.105 \times 81 + 105 \cdot \surd Q6534 + 81 \cdot \surd Q2000 + \surd Q6534 \times 2000 \\ &= \surd QC.8505 + \surd Q72837350 + \surd Q13122000 + \surd Q13068000. \end{aligned}$$

[34] The product is as follows (see Section 5 above) $10 + \surd Q2 + 10 - \surd Q2 + 2\surd Q.10 + \surd Q2 \times \surd Q.10 + \surd Q2 = 20 + 2\surd Q.100 - 2 = 20 + \surd Q392$. Now insert the universal sign $\surd Q$. Therefore $\surd Q.20 + \surd Q392 = \surd Q.10 + \surd Q2 + \surd Q.10 - \surd Q2$.

8. Examples of the division of universals are the same as for plurinomia which were described in Chapter 11 earlier. But you must prefix the quotients of the divisors and dividends with the symbol of the universal root.

From this chapter and from Chapter 11, Section 1, $\sqrt{\mathfrak{C}}.\sqrt{\mathfrak{z}}12 + \sqrt{\mathfrak{z}}8$ is to be divided by $\sqrt{\mathfrak{z}\mathfrak{C}}2$, or what is the same thing, by $\sqrt{\mathfrak{C}}.\sqrt{\mathfrak{z}}2$, giving the quotient $\sqrt{\mathfrak{C}}.\sqrt{\mathfrak{z}}6 + 2$.[35] Likewise, [from this chapter and from Chapter 11, Section 2,] $\sqrt{\mathfrak{z}}.5$ divided by $\sqrt{\mathfrak{z}}.\sqrt{\mathfrak{z}}11 - \sqrt{\mathfrak{z}}3 - \sqrt{\mathfrak{z}}2$ gives a quotient: $\sqrt{\mathfrak{z}}.\sqrt{\mathfrak{z}}\frac{3025}{12} + \sqrt{\mathfrak{z}}50 - \sqrt{\mathfrak{z}}75 + \sqrt{\mathfrak{z}}\sqrt{\mathfrak{z}}\frac{275}{6} + \sqrt{\mathfrak{z}}\frac{275}{4}$. [And so on].

CHAPTER XVII.

EXTRACTION OF THE ROOTS OF UNIVERSALS.

1. Extract the root of a given plurinomium, disregarding the universal symbol, in accordance with Chapter 12. But replace the former universal symbol after the root is extracted.

The square root is to be extracted from $\sqrt{\mathfrak{C}}.3 - \sqrt{\mathfrak{z}}5$, and this (by Section 3 of Chapter 12) will be $\sqrt{\mathfrak{z}}\frac{5}{2} - \sqrt{\mathfrak{z}}\frac{1}{2}$. Now prefix it with the universal symbol $\sqrt{\mathfrak{C}}$. and that will make the required root $\sqrt{\mathfrak{C}}.\sqrt{\mathfrak{z}}\frac{5}{2} - \sqrt{\mathfrak{z}}\frac{5}{2}$. In similar fashion, the cube root of $\sqrt{\mathfrak{z}}.\sqrt{\mathfrak{C}}3 + \sqrt{\mathfrak{z}}2$ will be (by this section and Chapter 12, Section 5) $\sqrt{\mathfrak{z}\mathfrak{C}}.\sqrt{\mathfrak{C}}3 + \sqrt{\mathfrak{z}}2$. Likewise, the sixth root of $\sqrt{\beta}.7 - \sqrt{\mathfrak{z}}48$ is required. First, its quadratic root will be $\sqrt{\beta}.2 - \sqrt{\mathfrak{z}}3$; and then the cube root of this will be $\sqrt{\mathfrak{C}\beta}.2 - \sqrt{\mathfrak{z}}3$, the required root.

2. If you extract the root from several linked universals, or from universals joined with uninomia, then the result is called the universal of universals. But you must place the universal symbol before the extracted root, and draw a line below the whole expression.

The square root of $5 + \sqrt{\mathfrak{C}}2 - \sqrt{\mathfrak{z}}.3 - \sqrt{\mathfrak{z}}2$ is to be extracted. When this is done prefix the symbol of the universal root together with a horizontal line drawn as shown: $\sqrt{\mathfrak{z}}.\underline{5 + \sqrt{\mathfrak{C}}2 - \sqrt{\mathfrak{z}}.3 - \sqrt{\mathfrak{z}}2}$.

[35] $\sqrt{\mathfrak{C}}.\sqrt{\mathfrak{z}}12$ divided by $\sqrt{\mathfrak{C}}.\sqrt{\mathfrak{z}}2$ is equal to $\sqrt{\mathfrak{C}}.\sqrt{\mathfrak{z}}6$; and $\sqrt{\mathfrak{C}}.\sqrt{\mathfrak{z}}8$ divided by $\sqrt{\mathfrak{C}}.\sqrt{\mathfrak{z}}2$ is equal to $\sqrt{\mathfrak{C}}.\sqrt{\mathfrak{z}}4 = \sqrt{\mathfrak{C}}.2$. The answer is therefore $\sqrt{\mathfrak{C}}.\sqrt{\mathfrak{z}}6 + 2$

Corollary.

From this it follows that the effect of the universal symbol on universals extends as far as the line stretches. And if no line is drawn the effect of the universal symbol on a following universal symbol is cut off and ceases.

For instance, $\sqrt{♀}.\underline{60 + \sqrt{♀}16\sqrt{♀}.6 - \sqrt{♀}4}$ indicates the whole of $60 + \sqrt{♀}16 - \sqrt{♀}.6 - \sqrt{♀}4$ a root of four components whose value is $\sqrt{♀}62$. But if the lines were not there, this is how it would go. $\sqrt{♀}.60 + \sqrt{♀}16 - \sqrt{♀}.6 - \sqrt{♀}4$. The first symbol $\sqrt{♀}$. only has power and effect, over $60 + \sqrt{♀}16$. Afterwards $\sqrt{♀}$. has power over the rest, namely, it covers $6 - \sqrt{♀}4$. The situation is as follows: $\sqrt{♀}.\underline{60 + \sqrt{♀}16 - \sqrt{♀}.6 - \sqrt{♀}4}$ has the value $\sqrt{♀}62$[; however] $\sqrt{♀}.60 + \sqrt{♀}16 - \sqrt{♀}.6\sqrt{♀}4$ is the same as 6. [And similarly with similar ones.]

This is enough concerning irrational numbers, and other kinds of irrationals can be treated alike. Uninomia arise through the extraction of the roots of numbers not having a definite place in the number line.[36] (I dealt with these in the first part of this book.) Plurinomia arise from the addition and subtraction of incommensurable uninomia. (I have dealt with these in the second part of this book.) Universalia arise through the extraction of the unresolvable roots of plurinomia. (I have dealt with these in the third and last part of this book.) So also universals of universals spring from universals, and from these again universals ad infinitum. But if you need to make use of these techniques, which happens only rarely, then you can easily find the information in the foregoing pages.

END OF BOOK ONE.

[36] The Latin text reads: *Per extractionem radicum ex numeris non habentibus radices oriuntur uninomia.* (Uninomia arise through the extraction of the roots from numbers which do not have roots.)

BOOK TWO.

[THE POSITIVE OR COSSICK PART OF ALGEBRA].

CHAPTER I.

DEFINITIONS, CLASSIFICATION AND DESCRIPTION OF THE ART.

1. THE SOLUTION of equations by means of assumed values can reveal the required answer correctly, as I have shown in Book I, Chapter 1, Section 7.

2. The values assumed are certain imaginative notions which are marked by unity, which we use in place of quantities and unknown numbers [when we add, subtract, multiply or divide].

3. The assumed numbers and their symbols are as many in number and as diverse in kind as those involved in the actual question of unknown numbers and quantities.

> For example, the symbols and denomination are as follows: 1℞ is called the first unknown; $1a$ called one *a*, is the second unknown; $1b$ called one *b*, is the third unknown; $1c$ called one *c*, is the fourth unknown. And so you proceed through the alphabet.

4. These assumed numbers are commonly called RES and they are of the first order. They represent the unknown number and its measure in every respect.

5. When any of these unknown numbers is multiplied by itself the product which arises is the quadratic power, and it is of the second order.

> For example, 1℞ multiplied by itself makes one quadratic prime, which is written thus, 1♀. Likewise, $1b$ multiplied by itself makes $1b$♀, which is called one *b* to the second power. Similarly, $1a$ multiplied by $1a$ makes $1a$♀, which is called one *a* raised to the second power. And so on.

6. The cube arises when any unknown is multiplied into its quadratic or square. This is a power of the third order.

For example, 1℞ multiplied by 1ʒ makes one cubed, which is written thus, 1ℭ. Similarly, 1*a* multiplied by 1*a*ʒ makes 1*a*ℭ, which is pronounced thus: one *a* cubed. In the same way 1*b* multiplied by 1*b*ʒ makes 1*b*ℭ, [and so on].

7. The fourth power is produced when any unknown is multiplied into its third power. This produces a power of the fourth order.

For example, 1℞ multiplied by 1ℭ makes one fourth power which is written thus, 1ʒʒ. Similarly, 1*a* multiplied by 1*a*ℭ makes 1*a*ʒʒ, which is pronounced thus, one *a* to the fourth power. In the same way 1*b* multiplied by 1*b*ℭ makes 1*b*ʒʒ. Next 1*c*ʒʒ, and so on.

8. The supersolid or fifth power is produced when any unknown is multiplied into its fourth power. It produces a power of the fifth order.

For example, 1℞ multiplied by 1ʒʒ makes 1ß otherwise known as a supersolid. Similarly, 1*a* by 1*a*ʒʒ makes 1*a*ß which is pronounced one *a* supersolid or 1*a* to the fifth power. In the same way we have 1*b*ß, 1*c*ß, and so on.

Corollary.

9. From this it is clear that one particular power arises from powers of other orders arranged in infinite progression.

For example, 1℞ into 1ß makes 1ʒℭ, which is a power of the sixth order. You can examine the rest of the powers in the attached table. In this I have assumed that 1℞ equals 3, that 1*a* equals 2, and that 1*b* is equal to 4. From these values the values of the remaining orders follow in sequence as shown below.

10. Those numbers, whether rational or irrational which bear the cossick symbol, are called cossick numbers.

For example, 6℞, [or] 5*a*, [or] 7*b*ℭ, [or] √ʒ6*b* [or] √ℭ7*a*ʒ are called cossick numbers. But occasionally, the name cossick is taken for any number at all.

11. Any cossick number which is a complete and separate entity, or which assumes a comparable status is termed a simplex.

For example 6*a* is a simplex. [Likewise,] √ʒ3ℭ. [Also,] √ʒ1*ab*.

12. A composite expression is one which contains several simplex numbers connected by the signs [plus or minus].

Power or order 0	[Characters and Examples of orders of the first position]		[Characters and Examples of orders of the first position]		[Characters and Examples of orders of the first position]		&c.
1	1℞	3	1*a*	2	1*b*	4	
2	1ʒ	9	1*a*ʒ	4	1*b*ʒ	16	
3	1𝔠	27	1*a*𝔠	8	1*b*𝔠	64	
4	1ʒʒ	81	1*a*ʒʒ	16	1*b*ʒʒ	256	
5	1ß	243	1*a*ß	32	1*b*ß	1024	
6	1ʒ𝔠	729	1*a*ʒ𝔠	64	1*b*ʒ𝔠	4096	
7	1ſß	2187	1*a*ſß	128	1*b*ſß	16384	&c.
8	1ʒʒʒ	6561	1*a*ʒʒʒ	256	1*b*ʒʒʒ	65536	
9	1𝔠𝔠	19683	1*a*𝔠𝔠	512	1*b*𝔠𝔠	262144	
10	1ʒß	59049	1*a*ʒß	1024	1*b*ʒß	1048576	
11	1ſſß	1177147	1*a*ſſß	2048	1*b*ſſß	4194304	
12	1ʒʒ𝔠	531441	1*a*ʒʒ𝔠	4906	1*b*ʒʒ𝔠	16777216	
13	1ſſſß	1594323	1*a*ſſſß	8192	1*b*ſſſß	67108864	
&c.	&c.		&c.				

[Such as] 6*a* + √ʒ3𝔠. [Likewise,] 5℞ − 2ʒ. [Also,] √ʒ30𝔠 + 3*a* − 4℞*b*.

13. A simplex number which contains a single numeral followed by the symbol for an unknown quantity [is called pure].

[Such as] 5*a*ʒ. [Likewise,] 3𝔠. [Also,] √ʒ2*c*𝔠, etc., are called pure.

14. A simplex numeral which is followed by more than one symbol for an unknown quantity [is called mixed].

[Such] as 5ʒ*ac*, 2℞*ac*, √ʒ1*ab*, √𝔠1*a*ʒ*b*ß*c*, and so on in infinite measure, are called mixed numbers, and their origin will be dealt with in Chapter 5, Sections 2 and 3.

15. Simple rational numbers are those rational numbers which carry the cossick sign together with the proper order of the power or root. Those numbers are irrational which are inconsistent with the cossick root sign.

For instance 6*a*, 5ʒ*a*ʒ and 2℞ are rational numbers. But √ʒ6*a*, √𝔠5ʒ*b*𝔠 and √ʒ𝔠7*b*ß are irrational.

16. Simple numbers whose radical signs are either nothing or similar are said to be similarly rooted. Those which have dissimilar signs are said to have dissimilar roots.

For instance, 2ℨ and 3*a*, or √ℨ3℞ and √ℨ5ℭ, or √ℭ6 and √ℭ2*ab* are similarly rooted. But, √ℨ2ℭ and √ℭ3ℭ, and also √ß1*a* and 5℞*b* are not similarly rooted.

17. Two simple numbers have the same unknown when they carry the very same notation and characteristics. They need not be of the same order.

For instance, 2℞ and √ℨ5ß, 3℞*a*ℨ and √ℭ2ℨ*a*ℭ have the same unknown quantity. But 2℞ and √ℨ1*a*, and 3℞*a*ℨ with 2℞ have different unknowns.

18. Simple numbers with similar radical signs of the same order are said to be of the same order in operations. But they need not have the same unknown quantity.

3*a* and 2*b*, √ℨ2ℭ and √ℨ5ℭ, 2℞ *a* and 5℞ *b*, √ℨ2℞ *a* and √ℨ3*bc* are of the same order in the operations. But 2℞ *a*ℨ and 3*b*, √ℨ2℞ *a*ℨ and √ℨ3*b*, √ℨ2℞ *a* and √ℨ3℞ *a*ℨ, these, and other of its kind have different unknowns. But those which are powers of the same order and the way in which such powers can be renewed by their operations, [are] described below in Chapter 4, Section 5.

19. Two simple numbers of the same unknown and of the same order in operation are commensurable if their uninomia are commensurable, with separate symbols for the unknown, and for the order or power.

For instance, 3ℭ and 2ℭ are commensurable, because 3 and 2 are commensurable (by Chapter 1, Section 15, Book I). Likewise, √ℨ12℞ and √ℨ3℞ are commensurable because √ℨ12 and √ℨ3 are commensurable (by Chapter 1, Section 14, Book I). So also for √ℨ12℞*a*ℨ and for √ℨ3℞*a*ℨ and the same is true for others of the same kind.

CHAPTER II.

ALGEBRAIC ADDITION AND SUBTRACTION.

1. If algebraic numbers, both simple and commensurable, are to be added or subtracted, then add the uninomia of both together; or else take the smaller from the greater; and then place the original cossick sign before the product or residue.

3℞ is to be added to 2℞ making 5℞. Add 4𝔔 to 3𝔔 making 7𝔔. Add 6*a*ℭ to 9*a*ℭ making 15*a*ℭ. Add √𝔔2ℭ to √𝔔8ℭ making √𝔔18ℭ, because √𝔔2 added to √𝔔8 makes √𝔔18 by Chapter 2, Section 1, Book I. √𝔔2℞*a*𝔔 added to √𝔔8℞*a*𝔔 makes √𝔔18℞*a*𝔔. Add (√𝔔2ℭ)/3 to (√𝔔8ℭ)/5 giving (√𝔔242ℭ)/15, because when the uninomia are reduced to the same denomination by Chapter 2, Section 1, Book I, and Chapter 13, Book I we have (√𝔔50)/15 added to (√𝔔72)/15 giving (√𝔔242)/15. Take 3℞ from 5℞ leaving 2℞. Subtract 3*b*𝔔𝔔 from 8*b*𝔔𝔔 giving 5*b*𝔔𝔔. Take √ℭ3ß from √ℭ192ß leaving √ℭ81ß by this paragraph and Chapter 3, Section 1, Book I. Take √ℭ3𝔔*a*𝔔 from √ℭ192𝔔*a*𝔔 leaving √ℭ81𝔔*a*𝔔.[37]

2. If the simple numbers are incommensurable interpose the plus sign + for addition and the minus sign − for subtraction.

Add 3℞ to 2𝔔 making 2𝔔 + 3℞. 4𝔔 added to 2*a*𝔔 makes 4𝔔 + 2*a*𝔔. Add √𝔔5ℭ to √𝔔10ℭ making √𝔔10ℭ + √𝔔5ℭ. 5*a*𝔔*b* added to 7*a*𝔔*b*𝔔 gives 7*a*𝔔*b*𝔔 + 5*a*𝔔*b*. Subtract 3𝔔 from 2*a*𝔔 leaves 2*a*𝔔 − 3𝔔. Take 2*a* from 3*a*𝔔 leaving 3*a*𝔔 − 2*a*. √𝔔3𝔔 from √ℭ12𝔔 leaves √ℭ12𝔔 − √𝔔3𝔔. √𝔔2*a* from √𝔔2*ab* makes √𝔔2*ab* − √𝔔2*a*.[38]

Corollary.

3. Hence it is clear that composite numbers arise from the addition and subtraction of simple incommensurables. The above examples illustrate this fact.

4. Composite numbers are added, subtracted and reduced by the rules applicable to plurinomia. See Chapters 8 and 9 and Chapter 7, Sections 7 and 8 of Book I. What is said there regarding uninomia and plurinomia must be understood to apply to composite and simple numbers.

For example, √𝔔2ℭ*b*𝔔 + 3𝔔 − 2℞ + 1 is to be added to 5ℭ + √𝔔8ℭ*b*𝔔 − 4𝔔 + 3*a* − 6. First add the + sign (by Chapter 8, Section 1, Book I) and this makes √𝔔2ℭ*b*𝔔 + 3𝔔 − 2℞ + 1 + 5ℭ + √𝔔8ℭ*b*𝔔 −

[37] Hawkins presented the translation of these examples on the left of a vertical line. On its right he gave the stated equations in modern notation as follows: $3x + 2x = 5x$, $4x^2 + 3x^2 = 7x^2$, $6a^3 + 9a^3 = 15a^3$, $\sqrt{2x^3} + \sqrt{8x^3} = \sqrt{18x^3}$, $\sqrt{2xa^2} + \sqrt{8xa^2} = \sqrt{18xa^2}$, $\frac{\sqrt{2x^3}}{3} + \frac{\sqrt{8x^3}}{5} = \frac{\sqrt{242x^3}}{15}$, $5x - 3x = 2x$, $8b^4 - 3b^4 = 5b^4$, and $\sqrt[3]{192x^2a^2} - \sqrt[3]{3x^2a^2} = \sqrt[3]{81x^2a^2}$.

[38] Similarly, the resulting expressions in modern notation are $3x + 2x^2$, $4x^2 + 2a^2$, $\sqrt{10x^3} + \sqrt{5x^3}$, $7a^2b^2 + 5a^2b$, $2a^2 - 3x^2$, $3a^2 - 2$, $\sqrt[3]{12x^2} - \sqrt{3x^2}$, and $\sqrt{2ab} - \sqrt{2a}$.

$4\mathfrak{q}+3a-6$.[39] This can be abbreviated (by Chapter 7, Sections 7 and 8 of Book I) and then gives $\sqrt{}\mathfrak{q}18\mathfrak{C}b\mathfrak{q} - 1\mathfrak{q} - 2℞ - 5 + 5\mathfrak{C} + 3a$.[40]

An example of subtraction.

From the last product $\sqrt{}\mathfrak{q}18\mathfrak{C}b\mathfrak{q}-1\mathfrak{q}-2℞-5+5\mathfrak{C}+3a$ the expression $\sqrt{}\mathfrak{q}2\mathfrak{C}b\mathfrak{q}+3\mathfrak{q}-2℞+1$ is to be taken. First, (by Chapter 9, Section 1) change the signs, and it makes $\sqrt{}\mathfrak{q}18\mathfrak{C}b\mathfrak{q} - 1\mathfrak{q} - 2℞ - 5 + 5\mathfrak{C} + 3a - \sqrt{}\mathfrak{q}2\mathfrak{C}b\mathfrak{q} - 3\mathfrak{q} + 2℞ - 1$. This must now be abbreviated and it gives $\sqrt{}\mathfrak{q}8\mathfrak{C}b\mathfrak{q} - 4\mathfrak{q} - 6 + 5\mathfrak{C} + 3a$, as given above.

CHAPTER III.

ROOT EXTRACTION OF SIMPLE NUMBERS.

1. EVERY simple number contains within itself inherent roots, of a nature and character indicated by their signs; beyond these roots there are no others.

For $0\mathfrak{q}\mathfrak{C}$ carries within itself the quadratic root, the cubic root, and lastly the sixth root. But none other.

2. To extract the roots inherent in a simple number, divide the order of the simple number by the order of a natural root, and make a note of the order of the quotient.

For instance, $0\mathfrak{q}\mathfrak{C}\beta a\mathfrak{q}\mathfrak{q}\mathfrak{C}$ has inherent roots, first of which is the quadratic, because the sign $\mathfrak{q}$ is found in both the first and second unknowns.[41] A cube root is also found for the same reason. Finally, there is a sixth root and beyond that nothing. Not even the fifth or supersolid root, because β is not found amongst the signs of the second unknown in this example. Nor is the fourth root contained because $\mathfrak{q}\mathfrak{q}$ is not found in the first unknown in this example.

3. To extract the roots inherent in a simple number, divide the order of the simple number by the order of a natural root, and make a note of the order of the quotient.

---

[39] In modern notation, $\sqrt{2x^3b^2} + 3x^2 + 1 + 5^3 + \sqrt{8x^3b^2} - 4x^2 + 3a - 6$.

[40] In modern notation, $\sqrt{18x^3b^2} - x^2 - 2x - 5 + 5^3 + 3a$. Note that $\sqrt{2x^3b^2} + \sqrt{8x^3b^2} = 3\sqrt{2x^3b^2} = \sqrt{18x^3b^2}$.

[41] $0\mathfrak{q}\mathfrak{C}\beta$ and $a\mathfrak{q}\mathfrak{q}\mathfrak{C}$.

For example, the cube root is to be extracted from 0ℨℭ. The number of the order ℨℭ is 6 which, divided by the number of the cubic order, namely by 3, gives 2 as quotient. This 2 is the sign of the order, and its sign is ℨ. So 0ℨ is the cube root of 0ℨℭ. Similarly, the square root of 0ℨℭ is 0ℭ; and the sixth root of 0ℨℭ is 0℞.

4. A root inherent in a mixed number is extracted when that root is extracted (by the previous paragraph) from each unknown separately.

The cube root is to be extracted from 0ℨℭß*a*ℨℨℭ. First, by the preceding section, the cube root of 0ℨℭß is to be extracted. This is 0ℨß. Then, by the same rule the cube root of *a*ℨℨℭ is to be taken. This is *a*ℨℨ, whence the full cube root of 0ℨℭß*a*ℨℨℭ is 0ℨß*a*ℨℨ. By the same example, the square root is 0ℭß*a*ℨℭ, and its sixth root is 0ß*a*ℨ.

5. If a certain root of a complete simple number is to be extracted then it is not only the absolute number which is involved in the process. The unknown quantity (by Sections 1 and 2 above) is also concerned with the root. But then extract that root from the absolute number, and also the root of any other order indicated by the radical sign.

The cube root is to be extracted from 64ℨℭ. First the cube root of the absolute number is 4. Next, the cube root of the sign ℨℭ is ℨ (by Section 1 above). Therefore the full cube root of 64ℨℭ is 4ℨ. In the same way the square root of 64ℨℭ is 8ℭ; and the sixth root is 2℞. In the same way the square root of √ℭ9ℨ*a*ℨß comes to √ℭ3℞*a*ß.

Corollary.

6. A simple number having both a numeral and an algebraic (or cossick) sign has an inherent root indicated by the total or special radical sign. The simple number is reduced by removing the radical sign, and then extracting the root of the remainder.

The simple number √ℨℭ4ℨß is reduced as follows. Delete the radical sign ℨ and there remains √ℭ4ℨß. From this, by the above method, extract the equivalent root, that is, the square root, and this gives √ℭ2ß as the reduced expression having the same value as √ℨℭ4ℨß. Similarly, √ℨℭ64ℨℭ is reduced as follows. Delete the complete radical sign √ℨℭ, leaving 64ℨℭ. From this, as before, extract the equivalent root, namely the sixth root, giving 2℞ which is the same as √ℨℭ64ℨℭ.

7. If a simple number, a certain root of which is to be extracted, does not carry the natural root in its absolute number and cossick sign, then place the required radical sign before the whole simple number.

For instance, the square root of 4𝕮 is to be extracted. This is √ℨ4𝕮. Likewise, the cube root of 4𝕮 is √𝕮4𝕮; and the cube root of √ℨ3℞ is √ℨ𝕮3℞, Similarly, the square root of 4ℨ*a* is √ℨ4ℨ*a*.[42]

CHAPTER IV.

POWERS AND ROOTS OF SIMPLE NUMBERS.

1. To MULTIPLY a pure number into itself quadratically or cubically, or to any other power, multiply the numerals of each power successively and mark the product with the symbol of its order.

For example, consider the number 0ℨ (whose power is 2) which is to be multiplied into itself cubically whose number is of order 3. Multiply 2 into 3 producing 6. The sign of this order is ℨ𝕮 and so 0ℨ𝕮 is the cube of 0ℨ. Similarly, 0ℨ supersolid (fifth power) when multiplied into itself makes 0ℨ*ß* (the tenth power). Likewise, 0℞ multiplied to the sixth power produces 0ℨ𝕮.

2. To multiply a mixed number into itself to any given power you must multiply each unknown into itself to the required power as indicated in Section 1.

0*ß a*ℨ is to be raised to the sixth power. First, by Section 1, multiply 1*ß* into itself six times. This makes 0ℨ𝕮*ß*. Next multiply *a*ℨ into itself to the sixth power, and this makes *a*ℨℨ𝕮. Consequently the product total 0ℨ𝕮*ß* *a*ℨℨ𝕮 will be the sixth power of 0*ß a*ℨ.

---

[42] Note on the interpretation of 0. Napier frequently uses 0 as an exponent, consistent with the use of other exponents ℨ, 𝕮, ℨℨ, ℨ𝕮 etc. I have shown in Book I, Chapter 3 Arithmetic that Napier was well aware of index notation, and his use of 0 in this and later chapters shows that he was aware of the relation $x^0 = 1$. For instance in Section 2 above we have the expression 0ℨ𝕮*ßa*ℨℨ𝕮 in which zero is used as a placeholder for x^0 thus preserving the true value of the expression $(x^0x^{30}a^{12})$. Professor Steggal of St Andrews University made comment on Napier's use of 0 in these terms (*Napier Memorial Volume* 1915, edited by C. G. Knott):

> The notation is rather puzzling, and it would appear that Napier uses 0, *a*, ℞ indiscriminately, as we use *x*, *y* and *z* . . . and he speaks of extracting a root where indices only are expressed . . . . This seems to involve a suggestion of symbolism, and is, of course, correct, if correctly interpreted.

In a later chapter on DIVISION the 0 is clearly used as $x^0 = 1$.

3. If therefore you wish to multiply a full simple number into itself quadratically, or cubically, or to any other order, first multiply its absolute number arithmetically, and raise its power (by Sections 1 and 2 above) to that order. Then extract the root of the product (by Chapter 3, Sections 5 and 7) as its radical sign (if there is one) indicates.

3ℭ is to be multiplied into itself quadratically. Find the arithmetic product of 3 into itself twice; and the symbol ℭ into itself twice (by Section 1 above) makes 9ℨℭ for the true square of three cubed. In the same way 2ℨℨ raised to the third power makes 8ℨℨℭ. And √ℭ3℞ raised to the second power makes √ℭ9ℨ. Next √ℨ2℞ *a*ℨ*c* is to be raised to the second power. First, 2 and ℞*a*ℨℭ must be squared. This makes 4ℨ*a*ℨℨ*c*ℨ (by Section 2 above) and its root must be extracted according as its root sign. The quadratic root will be (by Chapter 3, Section 5) 2℞*a*ℨ*c* for the true square of this √ℨ2℞*a*ℨ*c*. But indeed, √ℨ2℞*a*ℨ*c* is squared even more easily just by deleting the radical sign (by Chapter 6, Section 6, Book I).

4. If simple numbers differently rooted are to be reduced to a common root, then multiply the rational part of each one into itself as often as the other unlike roots indicate. And before each single product place the common radical sign which embraces all dissimilar powers as well as those which are common to more than one term.

[For example,] √ℭ3℞ and √ℨ2℞ are reduced as follows. Multiply 3℞ into itself twice, and 2℞ into itself three times. They then become 9ℨ and 8ℭ. Now interpose the common radical sign, giving √ℨ*c*9ℨ and √ℨℭ8ℭ, which have the same radical sign but their value is the same as √ℭ3℞ and √ℨ2℞. [Likewise,] √ℨ6ℨ and 2℞ are reduced as follows. First square 2℞ because the other number has the radical √ℨ. We do not multiply 6ℨ because 2℞ does not carry a radical sign. We now have 6ℨ and 4ℨ and we place the radical sign before both √ℨ6ℨ and √ℨ4ℨ. They now have the same radical sign and moreover have the same power in operation. The following √ℨℨ2ℨ*b* and √ℨℭ3*ac* are reduced as follows. Multiply 2ℨ*b* into itself cubically, and 3*ac* into itself quadratically (because their roots differ, one a cube root and the other a square root.) This gives 8ℨℭ*b*ℭ and 9*a*ℨ*c*ℨ. Now place before them the common root sign √ℨ together with the unlike signs ℨ and ℭ giving √ℨℨℭ8ℨℭ*b*ℭ and √ℨℨℭ9*a*ℨ*c*ℨ of the same order and value as before.

[Example] of reduction of more than two terms.

The three following terms are to be reduced to a common radical sign: √ℨ2ℨ, √ℭ3℞, √ℨℨ1*a*. First multiply 2ℨ into itself to the sixth power and not to the twelfth power. The first and third terms have a quadratic root, but the middle term has not. This makes the first term with its assigned radicals √♀*q*ℭ64♀♀ℭ. The second term for a similar reason must be raised to the fourth power, and this makes its product with its own radical sign √♀*q*ℭ81♀♀. Finally the third term raised to the third power with its radical signs included makes √♀*q*ℭ1*a*ℭ. These three now have the same radical sign √♀♀ℭ.

Corollary.

5. Hence it follows that some simple numbers with unlike roots are said to be powers of the same order; and when reduced they are of the same order in operation. But they are not of the same power, for one is of a higher order than the other.

In the above example √♀6♀ and 2℞ are powers of the same order, because when reduced they make √♀6♀ and √♀4♀ which (by Chapter 1, Section 18, Book II) are of the same order in operation. Further √ℭ3℞ and √♀2℞ when reduced become √♀ℭ9♀ and √♀ℭ8ℭ. One of these, [namely,] √♀ℭ8ℭ is of higher degree or order, whilst the other √♀ℭ9♀ is of a lower order.

CHAPTER V.

ALGEBRAIC MULTIPLICATION.

1. Multiplying the symbols of the same unknown is done by adding the numbers of the signs of their order, and noting the product of their absolute numbers.

0*a*♀ is to be multiplied by 0*a*ℭ. The numbers of order *a*♀ and *a*ℭ are 2 and 3, which, when added make 5, and the symbol for order 5 is *a*ß. 0*a*ß is therefore the product found by multiplying 0*a*♀ into 0*a*ℭ. Similarly, 0*b* multiplied into 0*b*ℭ makes 0*b*♀♀; and 0♀ multiplied into 0ß makes 0ʃß.

2. To multiply the pure signs of different unknowns successively write the signs next to each other, always placing the sign of the first unknown (if it is there) before the others.

0*a*℥ is to be multiplied into 0*b*𝔠 which produces 0*a*℥*b*𝔠. When a numeral is included as 6*a*℥ multiplied by 1*b*𝔠 the product is written 6*a*℥*b*𝔠. Similarly 0*a*℥ by 0𝔠 does not produce 0*a*℥𝔠 but 0𝔠*a*℥ with the sign of the first unknown in front. This 0𝔠*a*℥ is pronounced, thus much or nil of the first unknown cubed, multiplied into one squared of the second unknown. On the other hand 0*a*℥𝔠 is an unknown of the second kind (as the Table in Chapter 1 shows) and it is pronounced, one, a to the sixth power.

3. Hence it follows from the pure signs of different unknowns multiplied together mixed numbers result.

As in the example above, the pure number 0*a*℥ when multiplied by the pure number 0*b*𝔠 produces 0*a*℥*b*𝔠 as before; and this, by Chapter 1, Section 14, is a mixed number.

4. For mixed numbers insofar as they represent unknowns are multiplied by Section 1 of this chapter; but insofar as they are different unknowns they are multiplied by the second Section of this chapter.

0℞*a*℥ multiplied by 0𝔠*a* makes 0℥℥*a*𝔠; for multiplying (by Section 1) the same unknown successively namely 0℞ by 0𝔠, and *a*℥ by *a*, giving 0℥℥ and *a*𝔠. And again, ℥℥ multiplied into *a*𝔠 produces (by Section 2 above) 0℥℥*a*𝔠. Similarly, 0℞*a*℥ by 0𝔠*b* makes 0℥℥*a*℥*b*; because 0℞ by 0𝔠 makes 0℥℥ and 0℥℥ by *a*℥ makes 0℥℥*a*℥ (by Section 2); and 0℥℥*a*℥ by 0*b* makes 0℥℥*a*℥*b*.

5. For the successive multiplication of simple numbers: [first] they must have the same root signs, if necessary, by reduction (Chapter 4, Section 4); then multiply the absolute numbers of each in turn; [third] raise the unknowns to the correct power as already described. Finally, extract such a root from the number and sign of the product as its radical sign indicates (Chapter 3) and place before it the appropriate symbol (by Chapter 6, Book I).

2*a*℥ multiplied by 5*a*𝔠 makes 10*a*ß, because 2 by 5 gives 10, and *a*℥ by *a*𝔠 gives *a*ß. Likewise, 2*a*℥ multiplied by 5𝔠 gives 10𝔠*a*℥ Similarly, √℥2℞*a*℥ multiplied by √℥8℞ make 16℥*a*℥, and when the proper square root is inserted, this reduces to 4℞*a*. 2℞*a* is to be multiplied by −√℥3*ab*. First, by reduction these come to √℥4℥*a*℥ and −√℥3*ab*. Then the numbers are multiplied together and the unknowns separately, giving 12℥*a*𝔠*b*, whose square root and proper sign makes −√℥12℥*a*𝔠*b* the product of the operation.

6. If composite expressions are to be multiplied together each simple term of the multiplicand must be multiplied by each term of the multiplier. The

product is reduced by the addition and subtraction of commensurables in the same way as plurinomia are multiplied and reduced by Chapter 7, Sections 7 and 8; and Chapter 10, Section 1 of Book I.

For example, $2a\text{℥} + \surd\text{℥}3\text{℞} - 2\text{℞}\,a - 4$ is to be multiplied by $\surd\text{℥}12\text{ℭ} + 2\text{℥}$. Multiply term by term and you obtain $\surd\text{℥}48\text{ℭ}a\text{℥℥} + 6\text{℥} - \surd\text{℥}48\text{ß}a\text{℥} - \surd\text{℥}192\text{ℭ} + 4\text{℥}a\text{℥} + \surd\text{℥}12\text{ß} - 4\text{ℭ}a - 8\text{℥}$. This can be reduced a little [43] and then becomes: $\surd\text{℥}48\text{ℭ}a\text{℥} - 2\text{℥} - \surd\text{℥}48\text{ß}a\text{℥} - \surd\text{℥}192\text{ℭ} + 4\text{℥}a\text{℥} + \surd\text{℥}12\text{ß} - 4\text{ℭ}a$.[44]

CHAPTER VI.

THE ARRANGEMENT OF ALGEBRAIC EXPRESSIONS.

1. THE INTERVAL between powers is the difference between the powers of simple numbers which carry the same radical sign; or the number by which the highest power exceeds that of the next power in the sequence.

For instance, the interval in the composite expression $\surd\text{ℭ}3\text{℞} - \surd\text{ℭ}2\text{℥}$ is 1; because the number of the power ℞ is 1, and the number of the power ℥ is 2, giving 1 as the difference. Moreover, in this expression $\surd\text{℥}3\text{ℭ} + \surd\text{℥}2\text{℞} - \surd\text{℥}5$ which has the same root sign, the

---

[43] By the addition of $+6\text{℥}$ and -8℥ making -2℥.

[44] Readers may require some help with the reduction of the given product in Section 6. I hope that the presentation of the transition stage may be of value.

$$\underline{2a\text{℥} + \surd\text{℥}3\text{℞} - 2\text{℞}\,a - 4} \times \underline{\surd\text{℥}12\text{ℭ} + 2\text{℥}}.$$

I have used the underline to indicate that the whole of the expression on the left is to be multiplied by the whole expression on the right. The partial products are as shown. Evaluate the terms in reverse order.

$\surd\text{℥}12\text{ℭ} \times 2a\text{℥} = \surd\text{℥}12\text{ℭ} \times \surd\text{℥}4a\text{℥℥} = \surd\text{℥}48\text{ℭ℥℥}$	$(8\sqrt{6})$
$\surd\text{℥}12\text{ℭ} \times \surd\text{℥}3\text{℞} = \surd\text{℥}12\text{ℭ} \times \surd\text{℥}3\text{℞} = \surd\text{℥}36\text{℥℥} = 6\text{℥}$	(24)
$\surd\text{℥}12\text{ℭ} \times -2\text{℞}\,a = \surd\text{℥}12\text{ℭ} \times -\surd\text{℥}4\text{℥}a\text{℥} = -\surd\text{℥}48\text{ß}a\text{℥}$	$(-16\sqrt{16})$
$\surd\text{℥}12\text{ℭ} \times -4 = \surd\text{℥}12\text{ℭ} \times -\surd\text{℥}16 = -\surd\text{℥}192\text{ℭ}$	$(-16\sqrt{16})$
$2\text{℥} \times 2a\text{℥} = 4\text{℥}a\text{℥}$	(16)
$2\text{℥} \times \surd\text{℥}3\text{℞} \times \surd\text{℥}4\text{℥℥} \times \surd\text{℥}3\text{℞} = \surd\text{℥}12\text{ß}$	$(8\sqrt{6})$
$2\text{℥} \times -2\text{℞}\,a = -4\text{ℭ}a$	(-32)
$2\text{℥} \times -4 = -8\text{℥}$	(-32)

These results may be checked by putting $x = 2$ and $a = 1$. Remember that $2\text{℥} = 2x^2$, $12\text{ℭ} = 12x^3$; and $48\text{ß}a\text{℥} = 48x^5a^2$.

interval between √℘3ℭ and √℘2℞ is 2 because the difference between the powers of their orders is 2. Likewise the interval between √℘2℞ and √℘5 is 1, because the number of the power ℞ is 1, and the power of a simple numeral is 0.[45]

2. Nothing or 0 is a false or imaginary simplex of any power which carries radical signs and unknowns of the appropriate order.

0ℭ is an imaginary simplex of order 3; 0℘ an imaginary simplex of order 2, and likewise √℘0ℭ is an imaginary simplex for the square root of some cube. Accordingly, the imaginary simplex for the cube root of some supersolid is fixed as √ℭ0ſß.

3. The same intervals turn up (by 1, 2, 3 or Euclid Book VII) when the greatest common measure dividing it is found. True simplex numbers proceed from the highest power to the lowest by the interval of this measure; but when these are missing imaginary simplex numbers can be used.

For instance, in the given example √℘3ℭ+√℘2℞ − √℘5 there are two different intervals, 2 and 1. And the greatest common measure, by Euclid VII, Proposition 1, of these is unity. It is therefore necessary to proceed from the greatest power √℘3ℭ to the minimum √℘5 by unit intervals. It is done in this way. Subtract a unit from the number of the power of √℘3ℭ, and this yields √℘0℘, an imaginary simplex because the true one is missing. Next, deduct 1 from the power √℘0℘, and this gives the power √℘2℞. Finally take 1 from the last number and you get the minimum power √℘5. You can now arrange the terms of the given example as shown; with equal intervals. √℘3ℭ + √℘0℘ + √℘2℞ − √℘5 [46] Likewise, in the expression 1ℭ − 3℞ − 6 the greatest common measure is 1, but by unit intervals we have 1ℭ + 0℘ − 3℞ − 6 and all the intervals will be the same. But in this sequence √ℭ1℘ℭ − √ℭ3℞ + √ℭ8 the intervals are different, namely 5 and 1 of which unity will be the common measure. So make a progression by subtracting unity each time, and then arrange the orders like this. √ℭ1℘ℭ + √ℭ0ſß + √ℭ0℘℘ + √ℭ0ℭ + √ℭ0℘ − √ℭ3℞ + √ℭ8. But there are intervals of 6 and 4 in the progression √ℭ2ſſß + √ℭ3ſß − √ℭ10℞ and the common measure of 6 and 4 is two. Therefore arrange the

---

[45] In this chapter Napier uses an initial '0' as a 'false or imaginary simplex of any power.' This is difficult to comprehend and lacks coherence. It is advisable to use this symbol as a place-holder for unity ($x^0 = 1$). This is clearly given in *the power of a simple numeral is* 0. The interval, or difference of their order will be 1.

[46] The example is now in decreasing order. $\sqrt{3x^3} + \sqrt{0x^2} + \sqrt{2x} + \sqrt{5}$.

progression in which you have the same twofold intervals as follows: √ℭ2∫∫ß + √ℭ0ℭℭ + √ℭ0∫ß + √ℭ3ß + √ℭ0ℭ − √ℭ10℞.

4. In order to arrange simple numbers of a compound expression correctly proceed as follows: first reduce all of the elements to the same root order by Chapter 4, Section 4. Place those terms of the same order together, with greater powers in front of the smaller powers. Make all the intervals the same by the preceding section. Finally, if possible abbreviate the simple numbers. This can be done by Chapter 3, Section 6 of this book.

2 − √ℭ3℞ + 1Ꝗ + √ℭ2Ꝗℭ. Give each term the same radical sign. This gives √ℭ8 − √ℭ3℞ + √ℭ1Ꝗℭ + √ℭ2Ꝗℭ. Next place the term of highest order in front and the terms of lower order in sequence. This gives √ℭ2Ꝗℭ + √ℭ1Ꝗℭ − √ℭ3℞ + √ℭ8. Next insert terms to provide equal intervals between them. This gives: √ℭ2Ꝗℭ + √ℭ1Ꝗℭ + √ℭ0ß + √ℭ0ꝖꝖ + √ℭ0ℭ + √ℭ0Ꝗ − √ℭ3℞ + √ℭ8. It may now be abbreviated if possible. For √ℭ1Ꝗℭ may be written 1Ꝗ as originally given; and √ℭ0ℭ may be written down as 0℞. This then produces √ℭ2Ꝗℭ + 1Ꝗ + √ℭ0ß + √ℭ0ꝖꝖ + 0℞ + √ℭ0Ꝗ − √ℭ3℞ + 2. In the same way, and for the same reasons the expression 1Ꝗ + √Ꝗ2℞ − 3 is better arranged as follows: 1Ꝗ + √Ꝗ0ℭ + 0℞ + √Ꝗ2℞ − 3 and similar action is appropriate in other cases.

5. Hence it is that the correct arrangement of simple numbers in composite expressions as indicated above, maintains proportional powers. For each intermediate term is the square of the same order or power as the product of the next preceding and succeeding terms.

For instance, in the above example, arranged in order 1Ꝗ [i] + √Ꝗ0ℭ [ii] + 0℞ [iii] + √Ꝗ2℞ [iiii] − 3 [v], the square of the second term is 0ℭ. Now multiply 1Ꝗ into 0℞ which gives 0ℭ. Now multiply √Ꝗ0ℭ into √Ꝗ2℞ which makes 0Ꝗ [, or (what is the same) √Ꝗ0ꝖꝖ.] Likewise the square of the third term is 0Ꝗ and that of the fourth is 2℞. Now multiply 0℞ into 3 which gives 0℞ of the same order as 2℞. Proportional numbers are always associated with similar progressions.

6. Hence it follows that when the smaller powers are arranged in order decreasing, a simple and absolute number must be established in the last place of all, because it does not contain a power of an unknown quantity.

7.[47] In composite numbers, either mixed or with several unknowns, as many places must be allowed as there are different unknowns, or unknowns of

[47] This Section seems largely unintelligible both in its wording and use of symbols.

a different order. And as many symbols of the unknowns must be multiplied and aggregated to produce the second or quadratic order. And each single symbol must be multiplied into those of the second order, and aggregated to complete the third or cubic order. And so continue without end.

[As in] 1Q − 1℞ + 1*a* + 1*a* − 18, there are two unknowns, namely ℞ and *a*. This means that 1℞ + 1*a* are written as unknowns of the first order. Next multiply −1℞ + *a* into itself, producing +Q − 0℞*a* + Q (by Chapter 5);[48] and therefore 1Q − 0℞*a* + 1*a*Q are three simple numbers which make up one particular order, the second or quadratic. Here is the correct arrangement of the composite expression: 1Q − 0℞ *a* + 1*a*Q − 1℞ + 1*a* − 18. By similar reasoning an expression of the third order might be produced.

CHAPTER VII.

DIVISION.

1. To DIVIDE a pure number, the greater by the less, which contain the signs of the same unknown; place the sign of the power of the interval after its simple number or numerator.

0*a*ℭ is to be divided by 0*a*Q. The sign of the interval is *a*℞ which must be inserted in place of a simple number, or after its numerator, as you please. This gives 0*a*℞ or $\frac{0a℞Q}{0}$ as the result of the division. In the same way 0ß divided by 0ℭ gives as quotient 0Q or $\frac{0Q}{0}$.

2. To divide a pure number, the less by the greater, with the same unknown. The quotient will always be a fraction. Place the sign of the power of the interval after the denominator.

0*a*Q is to be divided by 0*a*ℭ. The sign of the interval is *a*, which must be put after the simple number of the denominator. This makes $\frac{0}{0a℞Q}$ for the fractional quotient. In the same way, 0ℭ divided by 0ß gives as quotient $\frac{0}{0Q}$.

3. To divide a pure number by another with a different unknown, draw a line indicating division because the quotient will always be a fraction in this case. Enter the dividend as the numerator and the divisor as the denominator.

[48] −℞ + *a* when squared = 1Q − 2℞*a* + 1*a*Q and not Q − 0℞*a* + *a*Q as in the original. This section is not of great value for the text appears to be in some disorder.

$0a\mathfrak{C}$ is to be divided by $0\mathfrak{z}$. This gives $\frac{0a\mathfrak{C}}{0\mathfrak{z}}$ as quotient. Likewise, if $0\mathfrak{z}$ is to divided by $0a\mathfrak{C}$ the quotient is then $\frac{0\mathfrak{z}}{0a\mathfrak{C}}$.

Corollary.

4. Hence it follows that the division of mixed numbers is to be performed by Sections 1 and 2, since they have the same unknowns; but those with different unknowns must make use of Section 3.

$0\mathfrak{C}a\mathfrak{z}$ is to be divided by $0\mathfrak{z}a$. By Section 1 divide $0\mathfrak{C}$ by $0\mathfrak{z}$ giving $0\mathrm{R}$. Likewise, $a\mathfrak{z}$ divided by a gives a. The complete quotient is therefore $0\mathrm{R}a$. Likewise, $0\mathfrak{C}a\mathfrak{z}$ divided by $0\beta b$ is done in this way. $0\mathfrak{z}$ is divided by 0β and gives $\frac{0}{0\mathfrak{z}}$ (by Section 2). Then $a\mathfrak{z}$ is divided by b and gives the fraction $\frac{a\mathfrak{z}}{b}$ [(by Section 3)]. The total quotient is then $\frac{0a\mathfrak{z}}{0\mathfrak{z}b}$. In the same way $0b\mathfrak{z}c$ when divided by $0bc\mathfrak{z}$ gives as answer $\frac{0b}{0c}$.

5. A fraction can be made from every integer by interposing a line beneath it and inserting 1 as denominator.

For example, 5 is an integer, and from it the fraction $\frac{5}{1}$ can be made. Likewise, $\surd\mathfrak{C}7$ is a complete entity from which the fraction $\frac{\surd\mathfrak{C}7}{1}$ can be formed.

6. Therefore, if a simple number is to be divided by another simple number, first give them similar roots (by Chapter 4, Section 4). Next, divide the simple number of the dividend by the simple number of the divisor. Thirdly, divide the unknown quantity of the dividend by that of the divisor. Fourthly, extract such a root of the product as indicated by the radical sign (by Chapter 3) and this will be the quotient. Finally, place the correct sign in front by Chapter 6 of Book I.

For example, $12b\mathfrak{C}$ is to be divided by $3b\mathfrak{z}$. Divide 12 by 3 making 4. Now divide $b\mathfrak{C}$ by $b\mathfrak{z}$ giving b. So the total quotient is $4b$. Likewise, $\surd\mathfrak{z}20\beta$ is to be divided by $\surd\mathfrak{z}8\mathfrak{C}$. Divide number by number giving $\frac{5}{2}$; and sign by sign, giving $\mathfrak{z}$. From these we get $\frac{5}{2}\mathfrak{z}$ or $\frac{5\mathfrak{z}}{2}$ or $2\frac{1}{2}\mathfrak{z}$(which are the same by 1 above). Now extract the square root and this will give as quotient $\frac{\surd\mathfrak{z}5\mathfrak{z}}{\surd\mathfrak{z}2}$; or $\surd\mathfrak{z}\frac{5}{2}\mathfrak{z}$; or $\surd\mathfrak{z}2\frac{1}{2}\mathfrak{z}$ (and these are the same by Chapter 4, Section 4, Book I). Conversely, $\surd\mathfrak{z}8\mathfrak{C}$ is to be divided by $\surd\mathfrak{z}20\beta$. The quotient will be $\frac{\surd\mathfrak{z}2}{\surd\mathfrak{z}5\mathfrak{z}}$ or, better still $\surd\mathfrak{z}\frac{2}{5}\mathfrak{z}$. Similarly $-\surd\mathfrak{z}12\mathfrak{z}a\mathfrak{C}b$ is to be divided by $\surd\mathfrak{z}3ab$. Divide numbers and signs as already explained and 4ʓ*a*ʓ emerges, whose indicated root is the quadratic or square root −2℞ *a*. When prefaced with the appropriate sign −2℞ *a* emerges as quotient. 4℞ *a* is to be divided by −√ʓ2℞*a*ʓ. First change to the same radical sign and this makes √ʓ16ʓ*a*ʓ and −√ʓ2℞*a*ʓ. Now divide both numbers and signs and this gives 8℞. With the addition of root sign √ʓ and minus sign we get as a quotient −√ʓ8℞.

7. It must be noted that if the quotient of the unknown is a fraction (by Sections 2 and 3 of this chapter) and if the quotient of the absolute number is an integer, the latter can be turned into a fraction by Section 5 of this chapter.

12ℭ is to be divided by 3ß. Divide the numbers and we get 4, an integer. Now divide sign by sign and we get a fraction $\frac{0}{0ʓ}$ or $\frac{1}{ʓ}$. From the integer 4 we get the fraction $\frac{4}{1}$ and so the total quotient is $\frac{4}{1ʓ}$. Likewise, 15*b*ʓ*c* is to be divided by 5*bc*ʓ. The division of numerals gives 3 as quotient; the division of signs produces $\frac{0b}{0c}$ or $\frac{b}{c}$ which is a fraction. From 3 we have the fraction $\frac{3}{1}$ and this gives a total quotient of $\frac{3b}{1c}$, not $3\frac{b}{c}$ [nor $\frac{3b}{0ʓ}$].

8. If composite numbers are to be divided by simple ones, divide, as already explained, each particular simplex of the composite number by this simple divisor; and join the simple numbers of the quotient with the appropriate signs.

For example, 12ʓ−√ʓ2ʓ+6 is to be divided by 2℞. Divide 12ʓ by 2℞ making 6℞. Next divide −√ʓ2ʓ by 2℞ making −√ʓ$\frac{1}{2}$. Finally, divide +6 by 2℞ which gives $\frac{+3}{1℞}$. When joined the total quotient is 6℞ − √ʓ$\frac{1}{2}$ + $\frac{3}{1℞}$.

(If a certain number is to be divided by a composite number of the first order, for instance to be divided by 2℞ − √ʓ3ʓ, proceed as in Chapter 11, Section 2, Book I.)

9. A composite number is to be divided by a composite number consisting of several powers. In this case, the simple numbers of each composite must be arranged in the correct order (by Chapter 6, Section 4). Next divide the simple number of the highest power by the simple number of the highest power of the divisor (by Section 6 of this chapter). The first simple number of the quotient will then be found. Multiply the whole divisor by this number and subtract the product from the whole dividend. Note the remainder, and delete the rest of the dividend. From the remainder choose another dividend, from which, in exactly the same way another simple number is selected as quotient, and

perhaps another remainder is left, until finally there is nothing left to divide or else it is of a lower order than the divisor. When this point is reached arrange each part of the quotient with its sign and make up the full quotient with its final remainder, if there is one.

For example, divide 1𝔔𝔔 + 71𝔔 + 120 − 154℞ − 14ℭ by 6 + 1𝔔 − 5℞. First, by Chapter 6, arrange the problem in this way:

1𝔔𝔔 − 14ℭ + 71𝔔 − 154℞ + 120 (1𝔔
1𝔔 − 5℞ + 6

Next divide 1𝔔𝔔 by 1𝔔 and this gives 1𝔔 as quotient. Enter this in the answer-bracket as shown above. Multiply the whole divisor by 1𝔔 and this makes 1𝔔𝔔 − 5ℭ + 6𝔔, which must be taken from the whole dividend, leaving −9ℭ + 65𝔔 − 154℞ + 120. Make a note of this and delete the rest of the figures by underlining in this way:

− 9ℭ + 65𝔔
1𝔔𝔔 − 14ℭ + 71𝔔 − 154℞ + 120 (1𝔔
1𝔔 − 5℞ + 6
1𝔔 − 5℞ + 6

Divide the remainder by the same divisor as before, and arrange the terms as shown:

+ 20𝔔
− 9ℭ + 65𝔔 − 100℞
1𝔔𝔔 − 14ℭ + 71𝔔 − 154℞ + 120 (1𝔔 − 9℞
1𝔔 − 5℞ + 6
1𝔔 − 5℞ + 6

Finally, divide the remainder in the same way as before, and this will be the arrangement:

+ 20𝔔
− 9ℭ + 65𝔔 − 100℞
1𝔔𝔔 − 14ℭ + 71𝔔 − 154℞ + 120 (1𝔔 − 9℞ + 20
1𝔔 − 5℞ + 6
1𝔔 − 5℞ + 6
− 5℞ + 6

Consequently, the complete quotient is 1𝔔 − 9℞ + 20 and no remainder is left over.[49]

[49] The copy for the first long-division was badly scattered and had to be rearranged. I have arranged the type so that it reads correctly when checked against [the next] example.

[Another] example: arrange 1C − 11Q + 40℞ − 36 divided by 1℞ − 4 in the new style.

```
                    + 12
              + 12℞
         − 7Q
1C − 11Q + 40℞ − 36 ( 1Q − 7℞ + 12  total quotient.
1℞ − 4
         1℞ − 4
              1℞ − 4
```

Finally there are 12 left over[50] [which] is of a lower order than the divisor 1℞ − 4.

10. If the dividend is not integral, or leaves a remainder, it can be divided by the divisor, and the divisor can be placed beneath the remainder, making a fraction. Put this fraction in the quotient, or, if you wish, place the divisor beneath the whole dividend.

For instance, in the last example the true quotient is 1Q − 7℞ + $12\frac{+12}{1\mathrm{R}-4}$, or, if you prefer it $\frac{1\mathrm{C}-11\mathrm{Q}+40\mathrm{R}-36}{1\mathrm{R}-4}$. This is done because the divisor does not go exactly into the dividend, and a remainder is left.[51]

CHAPTER VIII.

EXTRACTION OF ROOTS FROM COMPOSITE NUMBERS.

1. When the square root of a composite number is to be extracted, first rearrange the terms in descending order (see Chapter. 6, Section 4). Next extract the square root from the simple number or numbers of the highest power (see Chapter 3, Sections 5 and 7); and insert a bracket to hold the quotient. Then delete the number with the highest power, without leaving a remainder if possible. Secondly, divide the first term of the remaining

[50] Notice that this has not been underlined or cancelled because the division is inexact.

[51] In the exercises Napier gives substantially the modern method of long division. This method was first used in Florence by Filippi Calandri in the first Italian textbook of Arithmetic printed in 1491 and reprinted in 1518. The book was beautifully printed and the arithmetical operations were very practical, but the problems used were traditional. Calandri's method of Long Division was a full century ahead of its time.

composite expression by twice the quotient (see Chapter 7, Section 9) without leaving a remainder if you can do so. Enter this new quotient after the previous quotient. Take the square of the new quotient from the remaining expression and make a note of the remainder. Perform this second operation a third and a fourth time, or as often as necessary until no remainder is left. Then the whole quotient, with its appropriate signs will be the true and natural root required. But if any but the smallest of remainders is left, then the calculated quotient gives an approximate root and not a true one.

The square root of the following composite number is required: √ʓ4ℭ + 1ʓ − √ʓ576℞ + 144 − 23℞. First, rearrange the expression like this:

1ʓ + √ʓ4ℭ − 1ʓ + √ʓℭ − 23℞ + √ʓ576℞ + 144;

Next, from the first term 1ʓ extract the true root. This gives 1℞ for the quotient, and the rest as remainder like this:

<u>1ʓ</u> + √ʓ4ℭ − 23℞ − √ʓ576℞ + 144 (1℞.

Secondly, double the quotient, giving 2℞ or √ʓ4ʓ. Divide the first term of the remainder, namely √ʓ4ℭ by double the quotient √ʓ4ʓ. This gives a new quotient +√ʓ1℞, with −23℞ − √ʓ576℞ + 144 left over. From the remainder take away the square of √ʓ1℞ (which gives 1℞). The remainder and the quotient will then be as shown below.

− 24℞
<u>1ʓ</u> + <u>√ʓ4ℭ</u> − <u>23℞</u> − √ʓ576℞ + 144 (1℞ + √ʓ1℞
+ √ʓ4ʓ

Repeat the second operation by doubling the quotient giving 2℞ + √ʓ4℞. Then divide the first two terms of those remaining (that is, −24℞ − √ʓ576℞) and this gives a new quotient −12, leaving +144 as remainder. From this take away the square of the new quotient (−144), and no remainder is left over as the computation below shows:

− <u>24℞</u>
<u>1ʓ</u> + <u>√ʓ4ℭ</u> − <u>23℞</u> − <u>√ʓ576℞</u> + <u>144</u> (1℞ + √ʓ1℞ − 12
+ <u>√ʓ4ʓ</u> + <u>2℞</u> + <u>√ʓ4℞</u>

Whence it is clear that 1℞ + √ʓ1℞ − 12 is the true and natural square root of the given composite expression written above, because nothing is left after root extraction.

Another example.

Find the square root of $\sqrt{\mathfrak{C}}\,4\mathcal{Q} - 8 - \sqrt{\mathfrak{C}}\,16\text{℞}$. First set out the problem as shown:

$$\underline{\sqrt{\mathfrak{C}}}\,4\mathcal{Q} - \sqrt{\mathfrak{C}}\,16\text{℞} - 8 \;(\; \sqrt{\mathfrak{C}}\,2\text{℞}$$

Next operate as shown below:

$$\begin{array}{l} \qquad\qquad\qquad\qquad\qquad -\,9 \\ \underline{\sqrt{\mathfrak{C}}\,4\mathcal{Q}} - \underline{\sqrt{\mathfrak{C}}\,16\text{℞}} - \underline{8} \;(\; \sqrt{\mathfrak{C}}\,2\text{℞} - 1 \\ \qquad\qquad + \underline{\sqrt{\mathfrak{C}}\,16\text{℞}} \end{array}$$

From this it is clear that $\sqrt{\mathfrak{C}}\,2\text{℞} - 1$ is an approximate root, and not a true quadratic root of the expression $\sqrt{\mathfrak{C}}\,4\mathcal{Q} - \sqrt{\mathfrak{C}}\,16\text{℞} - 8$ because a remainder -9 is left.[52]

An example of [several] unknowns.

Find the square root of $1\mathcal{Q} + 2\text{℞}a + 1a\mathcal{Q} + 1\text{℞} + 1a - 110$. First, arrange the work as follows (see Chapter 6, Proposition 7):

$$\underline{1\mathcal{Q}} + 2\text{℞}a + 1a\mathcal{Q} + 1\text{℞} + 1a - 110 \;(\; 1\text{℞}$$

Secondly, operate as follows:

$$\begin{array}{l} \underline{1\mathcal{Q}} + \underline{2\text{℞}a} + \underline{1a\mathcal{Q}} + 1\text{℞} + 1a - 110 \;(\; 1\text{℞} + 1a \\ \quad + \underline{2\text{℞}} \end{array}$$

Thirdly, operate as follows:

$$\begin{array}{l} \qquad\qquad\qquad\qquad\qquad\qquad\qquad -\,110\tfrac{1}{4} \\ \underline{1\mathcal{Q}} + \underline{2\text{℞}a} + \underline{1a\mathcal{Q}} + \underline{1\text{℞}} + \underline{1a} - \underline{110} \;(\; 1\text{℞} + 1a + \tfrac{1}{2} \text{ for the} \\ \qquad\;\; \underline{2\text{℞}} \qquad\qquad\quad \underline{2\text{℞}} + \underline{2a} \qquad\qquad\qquad \text{approximate root.} \end{array}$$ [53]

[52] $(4x^2)^{1/3} - (16x)^{1/3} - 8 = (2x)^{2/3} - 2(2x)^{1/3} - 8 = X^2 - 2X - 8 = (X-1)^2 - 9$ where $X = (2x)^{1/3}$.

[53] $x^2 + 2ax + a^2 + x + a - 110 = (x+a)^2 + 2(x+a)\left(\frac{1}{2}\right) + \left(\frac{1}{2}\right)^2 - 110\frac{1}{4} = \left(x + a + \frac{1}{2}\right)^2 - 110\frac{1}{4}$.

Another example of [several] unknowns.

Find the square root of 1Ꝗ + 6℞ a − 7. First, arrange the work as follows (see Chapter 6, Proposition 7):

$$\begin{array}{l} \qquad\qquad\qquad\qquad -\,9a\text{Ꝗ} - 7 \\ \underline{1\text{Ꝗ}} + 6\text{℞}a + 0a\text{Ꝗ} - 7\ (\ 1\text{℞} \end{array}$$

Secondly, operate as follows:

$$\begin{array}{l} \qquad\qquad -\,9a\text{Ꝗ} \\ \underline{1\text{Ꝗ}} + \underline{6\text{℞}a} + \underline{0a\text{Ꝗ}} - 7\ (\ 1\text{℞} + 3a \text{ for the approximate root.[54]} \\ \qquad\ \ \underline{2\text{℞}} \end{array}$$

A more difficult example of one unknown.

Extract the square root of 1Ꝗ − √Ꝗ8Ꝗ − 6℞ + 8 + √Ꝗ32. First, set out the problem as shown below:

$$\underline{1\text{Ꝗ}} - \sqrt{\text{Ꝗ}}8\text{Ꝗ} - 6\text{℞} + 8 + \sqrt{\text{Ꝗ}}32\ (\ 1\text{℞}$$

[Secondly], set it out as follows:

$$\begin{array}{l} \qquad\qquad\qquad\qquad\qquad -\,3 - \sqrt{\text{Ꝗ}}8 \\ \underline{1\text{Ꝗ}} - \underline{\sqrt{\text{Ꝗ}}8\text{Ꝗ}} - \underline{6\text{℞}} + \underline{8} + \underline{\sqrt{\text{Ꝗ}}32}\ (\ 1\text{℞} - \sqrt{\text{Ꝗ}}2 - 3 \text{ for the} \\ \qquad + \underline{\sqrt{\text{Ꝗ}}4\text{Ꝗ}} + \underline{2\text{℞}} \qquad\qquad\qquad\qquad\qquad \text{approximate root.[55]} \end{array}$$

Sixth example.

Find the square root of 1Ꝗ − 0℞ a + 1aꝖ − 1℞ + 1a − 18. First, set out the problem in this way (see Chapter 6, Proposition 7):

$$1\text{Ꝗ} - 0\text{℞}a + 1a\text{Ꝗ} - 1\text{℞} + 1a - 1\ (\ 1\text{℞}$$

Next proceed as shown:

$$\begin{array}{l} \qquad\quad +\,2\text{℞}a \\ -\underline{1\text{Ꝗ}} - \underline{0\text{℞}a} + \underline{1a\text{Ꝗ}} - 1\text{℞} + 1a - 18\ (\ 1\text{℞} - 1a \\ \qquad\quad +\,\underline{2\text{℞}} \end{array}$$

[54] $x^2 + 6ax - 7 = x^2 + 6ax + 9a^2 - 9a^2 - 7 = (x + 3a)^2 - 9a^2 - 7.$

[55] $x^2 - 2\sqrt{2}x - 6x + 8 + \sqrt{32} = x^2 - 2x(\sqrt{2}+3) + (\sqrt{2}+3)^2 - (\sqrt{2}+3)^2 + 8 + 4\sqrt{2} = (x - \sqrt{2} - 3)^2 - (11 + 6\sqrt{2}) + 8 + 4\sqrt{2} = (x - \sqrt{2} - 3)^2 - 3 - \sqrt{8}.$

Thirdly, the layout is shown below:

          + 2℞a   − 18¼
−1ℨ − 0℞a + 1aℨ − 1℞ + 1a − 18 (1℞ − 1a − ½
          + 2℞   + 2℞ − 2a[56]

2. When the cube root is to be extracted the expression must be rearranged correctly (see Chapter 6, Section 4). Then take the cube root from the simple number of the highest order (see Chapter 3, Sections 5 and 7), and place it within the bracket indicating the quotient. Then delete the simple number of the highest power. Secondly, divide the first part of the expression which has not been deleted, by three times the square of the whole quotient. Write the new quotient which arises from this division next to, and after the first quotient. And then multiply three times the square of the new quotient into the first, or leading quotient, and take this away from the previous remainder. This second operation must be repeated until no remainder is left. At this point, the whole quotient with its signs will be the true and innate root which is required. But if even the smallest remainder is left, then the said quotient is called an approximate root and not the true one.

The cube root of the following expression is required. Place it in descending order.

1ℨℭ + 12ß + 60ℨℨ + 160ℭ + 240ℨ + 192℞ + 64 (1ℨ

From ℨℭ extract its cube root 1ℨ and place it in the quotient. Secondly, divide 12ß by three times the square of the quotient, namely by 3ℨℨ. This yields a fresh quotient +4℞. 12ß is then deleted and the position is then as shown:

1ℨℭ + 12ß + 60ℨℨ + 160ℭ + 240ℨ + 192℞ + 64 (1ℨ + 4℞
      + 3ℨℨ

Next multiply the first quotient, that is 1ℨ, by three times the square of the new quotient +4℞. This makes 48ℨℨ which must be taken from 60ℨℨ [etc.], leaving 12ℨℨ + 160ℭ + 240ℨ + 192℞ + 64. From the remainder take the cube of 4℞, which comes to 64ℭ. This leaves 12ℨℨ + 96ℭ + 240ℨ + 192℞ + 64 in this array:

[56] $x^2+a^2-x+a-18 = (x-a)^2-(x-a)+\frac{1}{4}+2ax-16\frac{1}{4} = \left(x-a-\frac{1}{2}\right)^2+2ax-18\frac{1}{4}$ (0 in 0℞a is a true coefficient).

$$
\begin{array}{l}
\qquad\qquad\quad + 12\mathfrak{Z}\mathfrak{Z} + 96\mathfrak{C} \\
\underline{1\mathfrak{Z}\mathfrak{C}} + \underline{12\beta} + \underline{60\mathfrak{Z}\mathfrak{Z}} + \underline{160\mathfrak{C}} + 240\mathfrak{Z} + 192\mathcal{R} + 64 \ (\ 1\mathfrak{Z} + 4\mathcal{R} \\
\qquad + \underline{3\mathfrak{Z}\mathfrak{Z}} + \underline{48\mathfrak{Z}\mathfrak{Z}} + \underline{64\mathfrak{C}}
\end{array}
$$

Thirdly, repeat the second operation, that is, divide the first part of the remainder by three times the square of the quotient, which is $3\mathfrak{Z}\mathfrak{Z} + 24\mathfrak{C} + 48\mathfrak{Z}$. The quotient and remainder are then as shown below:

$$
\begin{array}{l}
\qquad\qquad\quad + \underline{12\mathfrak{Z}\mathfrak{Z}} + \underline{96\mathfrak{C}} + 48\mathfrak{Z} \\
\underline{1\mathfrak{Z}\mathfrak{C}} + 12\beta + \underline{60\mathfrak{Z}\mathfrak{Z}} + \underline{160\mathfrak{C}} + \underline{240\mathfrak{Z}} + 192\mathcal{R} + 64 \ (\ 1\mathfrak{Z} + 4\mathcal{R} + 4 \\
\qquad + \underline{3\mathfrak{Z}\mathfrak{Z}} + \underline{48\mathfrak{Z}\mathfrak{Z}} + \underline{64\mathfrak{C}} \\
\qquad\qquad\quad + \underline{3\mathfrak{Z}\mathfrak{Z}} + \underline{24\mathfrak{C}} + \underline{48\mathfrak{Z}}
\end{array}
$$

Finally, multiply three times the square of the latest quotient, [namely] 48, into the complete antecedent of the quotient namely, $1\mathfrak{Z} + 4\mathcal{R}$, making $48\mathfrak{Z} + 192\mathcal{R}$. Take this from the remainder, that is from $48\mathfrak{Z} + 192\mathcal{R} + 64$, leaving $+64$. From this take the cube of the final quotient, namely $+64$, and there is no remainder. The array then looks as shown below:

$$
\begin{array}{l}
\qquad\qquad\quad + \underline{12\mathfrak{Z}\mathfrak{Z}} + \underline{96\mathfrak{C}} + \underline{48\mathfrak{Z}} \\
\underline{1\mathfrak{Z}\mathfrak{C}} + 12\beta + \underline{60\mathfrak{Z}\mathfrak{Z}} + \underline{160\mathfrak{C}} + \underline{240\mathfrak{Z}} + \underline{192\mathcal{R}} + \underline{64} \ (\ 1\mathfrak{Z} + 4\mathcal{R} + 4 \\
\qquad + \underline{3\mathfrak{Z}\mathfrak{Z}} + \underline{48\mathfrak{Z}\mathfrak{Z}} + \underline{64\mathfrak{C}} \qquad\qquad\qquad\qquad \text{the true} \\
\qquad\qquad\quad + \underline{3\mathfrak{Z}\mathfrak{Z}} + \underline{24\mathfrak{C}} + \underline{48\mathfrak{Z}} \qquad\qquad\qquad \text{cube root.} \\
\qquad\qquad\qquad\qquad\quad + \underline{48\mathfrak{Z}} + \underline{192\mathcal{R}} + \underline{64}
\end{array}
$$

Another example.

Extract the cube root of the following composite number, already arranged in descending order:

$$1\mathfrak{C} - 10\mathfrak{Z} + 31\mathcal{R} - 30 \ (\ 1\mathcal{R}$$

[Second, rearrange it as follows:] [57]

[57] Hawkins added the following explanation about the rearrangement in the main text: "The first calculation is the cube root of $1\mathfrak{C}$, which is $1\mathcal{R}$, to go in the quotient. The second calculation is: divide $-10\mathfrak{Z}$ by three times the square of the quotient, that is, three times $1\mathfrak{Z} = 3\mathfrak{Z}$. This gives $-3\frac{1}{3}$ as the second part of the quotient. This gives $1\mathcal{R} - 3\frac{1}{3}$ as the approximate root, and the arrangement is as shown below:"

$$\begin{array}{l} \quad\quad\quad\; - \tfrac{7}{3}℞ + \tfrac{190}{27} \\ \underline{1ℭ} - \underline{10ʓ} + \underline{31℞} - \underline{30}\;(\,1℞ - 3\tfrac{1}{3} \\ \quad\quad\; \underline{+\,3ʓ} + \underline{\tfrac{100}{3}}℞ - \underline{\tfrac{1000}{27}} \end{array}$$

for the approximate cube root, not the true root, because of the remainder.[58]

3. If you wish to extract the true square root of a composite number which has no natural root of this order, but an approximate one only, then place the universal root sign before the number; and this will indicate the true root even though it cannot be evaluated exactly.

For instance, in the second example of a quadratic root, [namely] $\surd$ℭ4ʓ − $\surd$ℭ16℞ − 8, the true and exact square root is expressed like this: $\surd$ʓ.$\surd$ℭ4ʓ − $\surd$ℭ16℞ − 8.

Likewise when the cube root of 1ℭ − 10ʓ + 31℞ − 30 is to be extracted exactly, the answer is expressed as follows: $\surd$ℭ.1ℭ − 10ʓ + 31℞ − 30.

4. I have not mentioned fourth roots, supersolid roots, and roots of higher orders; partly because they are rarely used, and partly because they can be discussed from what has already been said.

If a fourth root is to be extracted this can be done by a modification of cube root extraction. Firstly, instead of 'cube [root] extraction' read 'fourth [root] extraction.' Secondly, instead of 'three times the square' read 'four times the cube.' Thirdly, instead of 'three times the square multiplied into the first antecedent quotient,' read 'six times the square multiplied into the square of the first antecedent quotient and four times the cube of the new quotient multiplied into the first antecedent quotient' etc. Fourthly, instead of 'cube of the new quotient,' read 'fourth power of the new quotient.' In this way the rule must be adapted to deal with the extraction of a fourth root.

And, indeed, if you wish to adapt the rule to supersolid root extraction, instead of 'three times the square' read 'five times the fourth power'. And instead of 'three times the square multiplied into the first antecedent quotient,' read 'ten times the square multiplied into the cube of the first antecedent quotient, and ten times the cube of the new quotient multiplied into the square of the antecedent; and five times the fourth power of the new quotient into the first antecedent,' etc.' And instead of 'cube of the new quotient' write 'supersolid of the new quo-

58 In an expression of this kind, that is a cubic, only two terms are possible, the term in x and the independent term. Use the expression $(x - A)^3 = x^3 - 3x^2A + 3xA^2 - A^3$. Note that $-3x^2A + 3xA^2 - A^3$ is the supplement. See Arithmetic Book II, Chapter 8 (cube root Example 3).

tient,' etc. And so for the extraction of all higher roots rules can be drawn up.

Example of the rule for fourth roots

$$\underline{1a\text{ƺ}\text{ƺ}} + \underline{4a\mathfrak{C}b} + \underline{6a\text{ƺ}b\text{ƺ}} + \underline{4b\mathfrak{C}a} + \underline{1b\text{ƺ}\text{ƺ}} \;(\; 1a + 1b$$
$$+ \underline{4a\mathfrak{C}b} + \underline{6a\text{ƺ}b\text{ƺ}} + \underline{4b\mathfrak{C}a} + \underline{1b\text{ƺ}\text{ƺ}}$$

for the true cube root.

Example of the rule for fifth roots

$$\underline{1a\text{ß}} + \underline{5a\text{ƺ}\text{ƺ}b} + \underline{10a\mathfrak{C}b\text{ƺ}} + \underline{10\text{ƺ}b\mathfrak{C}} + \underline{5b\text{ƺ}\text{ƺ}a} + \underline{1b\text{ß}} \;(\; 1a + 1b$$
$$\underline{5a\text{ƺ}\text{ƺ}} + \underline{10b\text{ƺ}a\mathfrak{C}} + \underline{10b\mathfrak{C}a\text{ƺ}} + \underline{5b\text{ƺ}\text{ƺ}a} + \underline{1b\text{ß}}$$

the true supersolid root.

5. It is clear from the foregoing that some remainders, after extraction, have no positive sign. All these remainders are called formal. Other remainders which carry the positive sign are called informal.

In Examples 2, 3 and 4 [on the] extraction of square roots all the remainders after the first operation are informal. But in the same examples the final remainders are numbers, and so they are called formal. But in Examples 5 and 6 [on the] square root; and in Section 2 cube root, all the remainders, both first and last, contain positive signs, and so are termed informal.

Sometimes the patterns of the remainders are formable and sometimes reformable; at other times they are quite shapeless and unreformable.

The examples which follow will soon make that clear.

6. Those remainders are formable with which the second part of the rule for extraction can be operated. Then there are no remainders, or else they yield informal remainders less than those before. The operation is then considered compatible with the second part of the rule for extraction of roots.

For instance, in all of the above examples, both quadratic or cubic, all the remainders except the last are called formable because they conform only to the second part of the rules for extraction, and the final remainders then turn out to be informal.

7. Those remainders are reformable if you divide them by a composite number equal to zero (or by an equation equal to 0). Hence, if need be, divide the last outstanding remainders by one or another equation equal to 0. Finally, a remainder, or nothing will be left; which is either formal or formable. The equations so used are called Reformatrices and the work of division itself is termed Reformation.

Example.

An approximate quadratic root was extracted from $1Q-0℞a+1aQ-1℞+1a-18$ in Example 6 of this chapter, Section 1. The root obtained was $1℞-1a-\frac{1}{2}$, and the remainder was informal $+2℞a-18\frac{1}{4}$. Let the composite number $1℞a+1℞-1a-10$ be equal to zero. Divide the above remainder $(2℞a-18\frac{1}{4})$ by this equation and the resultant remainder is $-2℞+2a+1\frac{3}{4}$. By Proposition 6 above it is formable. Therefore the remainder $+2℞a-18\frac{1}{4}$ is said to be reformable. The composite equation $1℞a+1℞-1a-10=0$ is called the reformatrix, and the operation itself is termed reformation.

Another example.

The square root is to be extracted from $1Q+4℞a+1aQ-4℞b-4ab+4bQ+4℞+4a-8b-61$. The approximate root is $1℞+1a-2b+2$ and the informal remainder is $2℞a-65$. Let the equation $1℞a-1ab-1b-5$ be given, and let it be equated to 0. Divide the previous remainder $(2℞a-65)$ by $1℞a-1ab-1b-5$ and the remainder $+2ab+2b-55$ emerges. This is neither formal nor formable and must be divided by another equation equated to 0. For example, use $2ab-3℞-3a+8b-21=0$; and the remainder evolved is $3℞+3a-6b-34$. This is formable, that is, with respect to the nearest previous radical, namely $1℞+1a-2b+2$. And so the remainders $2℞a-65$ and $2ab+2b-55$ are called reformable. The composites $1℞a-1ab-1b-5$ and $2ab-3℞-3a+8b-21$ are called reformatrix equations.

8. Since therefore informal residues may convert to formal, you can change them into conformable by Section 6 of this chapter; or you can change them to reformable by Section 7. You must make a note of all the latest residues, for if there is no remainder, or a formal one, it is good that all the quotients in the expression should be joined together and condensed to give the reformed approximate root. The quotients produced by reformation are always of little worth, and should be avoided.

The approximate square root of the penultimate example was $1℞-1a-\frac{1}{2}$; and the remainder was $+2℞a-18\frac{1}{4}$. Because it was reformable by Section 7, by means of the reformatrix $1℞a+1℞-1a-10$, and, ignoring the quotient, the remainder rose to $-2℞+2a+1\frac{3}{4}$. By Section 7 this was changed into formable. Remainders arise which must be noted as formal, namely $+\frac{3}{4}$ and a quotient with a figure -1 joined to the root mentioned before. When this is simplified it gives 1℞ $-$ 1a $-$ $1\frac{1}{2}$ for the approximate reformed root.

Similarly, the remainder of the last example was first $+$ 2℞a $-$ 65 which was reformed by means of the reforming equation 1℞ $-$ 1ab $-$ 1$b$$-$5. Disregarding the quotient, the remainder rose to 2ab+2$b$$-$55 as I said in Section 7. This again, reformed by the use of a new reformatrix namely, by 2$ab$$-$3℞$-3a$+8$b$$-$21, the remainder, ignoring the quotient, rose to 3℞ $+$ 3a $-$ 6b $-$ 34. But because it is formable by Section 6 it will conform to its approximate root, namely, 1℞ $+$ 1a $-$ 2b $+$ 2 and then the total approximate root, when reformed will be 1℞ $+$ 1a $-$ 2b $+$ $3\frac{1}{2}$. The final remainder must be noted as $-\frac{169}{4}$ or $42\frac{1}{4}$ formal.

9. But if an informal remainder is left after the final rearrangement, due to the failure of the reforming equation, then such remainders are termed deformed or non-reformable.

For instance, the final remainders in Examples 5 and 6 (quadratic root); and Example 2 (cubic root), if no reforming equations were used, would be described as deformed or non-reformable.

10. There are two kinds of roots and their deformed remainders, single and plural. Those deformed expressions are singular which have one simple and pure unknown; or one term of a mixed unknown in the root or quotient, which does not contain an unknown of the same kind either in the quotient or root, nor in the remainder.

For instance, the square root of 1Ꝗ $+$ 6℞a $-$ 7 is 1℞ $+$ 3a, and the remainder is $-$ 9aꝖ $-$ 7. This is therefore a singular root, because it contains only one unknown 1℞. Likewise, the square root of 1ꝖaꝖ $-$ 6℞a $-$ 1a $+$ 8 $=$ 0, is 1℞a $-$ 3, and its remainder is $-$ 1a $-$ 1, which contains the unknown only once.

11. Roots and their remainders are termed plural when several simple numbers with any one unknown are found both in the root and in the remainder.

For instance, the approximate cube root of 1𝔈$-$9Ꝗ$+$36℞$-$80 $=$ 0 is 1℞ $-$ 3; and the remainder is $+$9℞ $-$ 53. In this case, the sign ℞ of the unknown can be found twice.

Likewise, the approximate square root of 1Ꝗ $+$ 1aꝖ $-$ 1℞ $+$ 1a $-$ 18 (which lacks a reforming equation) is 1℞ $-$ 1a $-$ $\frac{1}{2}$; and the remainder is 2℞ a $-$ 18 as explained in Section 6 above. There are two simple numbers attached to the first unknown, 1℞ and 2℞a. And also the second unknown has as many namely 1a and 2℞a.

12. A root therefore has four types: the true root, the formal root, the singular root, the plural root. I shall deal with their uses and extractions later on.

CHAPTER IX.

EQUATIONS AND THEIR SIMPLIFICATION.

1. An equation is an assembly of unknown quantities together with others of equal value; from this equation the value of the unknown is sought.

For instance, if someone imposes 1℞ for a required number or quantity, not knowing its value, and then finds by hypothetical questioning that 3℞ are equal to 21. That expression of equality is called an equation, and from it one thing, or one unknown is worth 7.

2. Between the equal parts of an equation a double line is placed, which is called the sign of equation.

For instance, 3℞ $= 21$ is pronounced three things are equal to 21. From this comes 1℞ $= 7$ which is pronounced one thing is equal to seven.

3. Some equations have one unknown, but other equations have more than one unknown.

$1a\mathcal{Q} + 3a = 10$ is an equation with one unknown. $2\mathcal{Q} - 1a = 6$ is an equation with two unknowns.

4. Some equations are simple because they can be reduced to lower terms of greater clarity and precision. Those equations are perfect which show the greatest clarity and precision.

For instance, 3℞ $= 21$ is a simple equation because it can be reduced to perfect form 1℞ $= 7$. Likewise, $5a\mathcal{Q} = 20$ is a simple equation because it can be reduced to the perfect form $1a\mathcal{Q} = 4$. But $1a\mathcal{Q} = 4$ is a simple equation, although in perfect form, because it can be reduced to its most perfect form $1a = 2$, by a technique which I shall explain later. Likewise, $12\mathcal{Q} + 3a = 6$ is a simple equation because it can be reduced to a more perfect form $4\mathcal{Q} + 1a = 2$.

5. Further, some equations are simple, some quadratic, some cubic, and some of higher degree.

For instance, 3℞ = 27, or 1℞ = 9; likewise 5*b*ʓ = 20, are called simple equations.

6. Some of the simple equations are real, that is, those equating an unknown to a number. Others are radicals, being equal to a number of square, cube, or higher power.

These are real: 3℞ = 21 or 1℞ = 7; 1*a* = 3; 2℞ – √ʓ3 – 1. These are radicals: 2ʓ = 8; 3ℭ = 24; 1*a*ß = √ℭ9 etc.

7. A quadratic equation is one which contains three proportional numbers.

[Such as] 2ʓ + 3℞ = 4, [or] 3℞ = 2ʓ – 4. [Also,][59] 1*a*ʓℭ – 10 = 3*a*ʓ. [Also,] 12 – √ʓ1℞ = 1℞.

8. A cubic equation is one which contains four proportional numbers.

[Such as] 1ℭ – 9ʓ = 24 – 26℞. [Also,] 1ℭ + 0ʓ – 2℞ = 4. [Also,] 1*a*ʓℭ – 2*a*ʓ = 4 is a cubic equation because the terms can be rearranged like this: 1*a*ʓℭ + 0*a*ʓʓ – 2*a*ʓ = 4 (see Chapter 6, Section 4) which gives the required four terms.[60]

9. An equation of the fourth degree is one which contains five proportionals; a supersolid has six; the sixth degree has seven proportionals, and so on to infinity.

Fourth degree: 2ʓʓ – 28ℭ + 142ʓ = 308℞ – 240. Fifth degree: 1*b*ʓß – 4*b*ʓʓʓ + 1*b*ʓℭ – 3*b*ʓʓ – 1*b*ʓ = 12.[61] Sixth degree: 1*a*ʓℭ – 3*a*ß + 2*a*ʓʓ – 6*a*ℭ + 1*a*ʓ = 1*a* + 6.

10. An illusive equation is one which cannot be evaluated and whoever attempts the impossible in an equation of this kind kills his own goose.

For example 1℞ = 3℞ is an illusive equation since it is impossible for one thing to be equal to its triple. Likewise, 1ʓ = 4℞ – 5 is an illusive equation, since no second power can be equal to four of its roots after five is deducted. This is shown below.[62]

11. The exposition is the reduction of a simple equation to a perfect and real equation. That part of a real equation which is equated to one thing or unknown is called the Exponent which solves the problem.

---

[59] The last term of the equation is in error. It should read 1*a*ʓℭ – 10 = 3*a*ℭ [$a^6 - 10 = 3a^3$].

[60] In this case 0 is placeholder only.

[61] The original equation is equal to 12. In such a case the equation is of the 10th degree.

[62] The equation referred to is $x^2 - 4x + 5 = 0$. The roots of this equation are $x = 2 \pm i$. The roots are therefore "illusive".

When the simple equation 3℞ $=21$ is reduced to its most perfect form 1℞ $=7$, then the exponent of either equation is 7, because it is equal to one thing (namely, to 1℞). Likewise, the simple equation 5℥ $=20$ is reduced to its more perfect form 1℥ $=4$, and then to its most perfect form 1℞ $=2$. This operation of reduction is called the exposition and there are 2 exponents to which the unknown is equal. I shall deal with this question of solving exponents later.

12. Every equation, other than an illusive one, has at least one exponent, which may be valid or not.

I shall deal with this matter later, and this notice should be adequate.

13. Valid exponents are those which are placed to one side and marked with the positive sign $+$, and they are always greater than zero. Invalid exponents are those which are set on one side and marked with the negative sign $-$, and such exponents are less than zero.

For instance, in the equation 1℞ $=7$, seven is a valid exponent because ([by] Proposition 1, Chapter 6, Book I) the sign $+$ must be understood. But in the real equation 1℞ $=-7$, the exponent is said to be invalid because it goes against reason, [because the $-$ sign is noted thus, $-$ 7], and is less than [nothing].

14. Some exponents can be described both in number and in kind; others only in number, and yet others in kind. Others are part here, and part there, whilst still more conform to neither.

This may be seen in Chapters 11, 12, and 13 which include effective examples.

15. Every part of an equation controlled by a power symbol and its sign is called Minimal, however many symbols and terms it may have. Moreover its preceding sign is called its leader or Ductrix; but the remaining signs are called Intermediate or neutral.

[As in this] equation: 1ℭ $-3+\sqrt{}$℥$2+\frac{3℞-4}{1℥+1}-\sqrt{}$℥$.6+\sqrt{}$℥1℞ $=0$, [in which] 1ℭ is called minimal, and the sign $+$ is called its leader or ductrix. Likewise 3 is called minimal, and the sign $-$ is its leader or ductrix. Similarly, $\sqrt{}$℥2 is called minimal, and the sign $+$ is its leader or ductrix. Further, the expression $\frac{3℞-4}{1℥+1}$ is called minimal, and the sign $+$ is its leader because its power extends over the whole fraction. But the remaining signs of this fraction are called intermediate or neutral. Similarly, $\sqrt{}$℥$.6+\sqrt{}$℥1℞ is called minimal, and the sign $-$ is its leader or ductrix, because the effect of the total universal sign covers the aggregate value, and the sign $+$ which remains is called intermediate or neutral.

CHAPTER X.

PREPARATION [OF GENERAL] EQUATIONS.

1. Preparation is the reduction of simple equations to an improved form, which can then be further clarified and reduced to real numbers.

> $5a\mathcal{Q} = 20$ is first prepared and becomes $1a\mathcal{Q} = 4$. Next the square root is extracted giving $1a = 2$. The method of preparation is now clear. The extraction of roots will be shown later.

2. Simple equations are prepared and clarified in five ways: by transposition, by abbreviation, by division, by multiplication and by extraction of roots.

> The rules for the various methods follow next with examples.

3. If you transfer an expression from one side of the equation to the other, and then change sign, both sides will remain equal as before. This is called Transposition.

> From the right-hand side of the equation $4\text{℞} - 6 = 5\text{℞} - 20$ transfer -20 to the left-hand side of the equation and then change its sign like this: $4\text{℞} - 6 + 20 = 5\text{℞}$. Now transpose 4℞ and it becomes -4℞ on the other side, $-6 + 20 = 5\text{℞} - 4\text{℞}$.[63] Consider the equation $1\mathcal{Q} - \surd\mathcal{Q}.3\mathcal{Q} - 2 = 3a$. Now transpose $-\surd\mathcal{Q}.3\mathcal{Q} - 2$ into $+\surd\mathcal{Q}.3\mathcal{Q} - 2$. Therefore, it becomes $1\mathcal{Q} = 3a + \surd\mathcal{Q}.3\mathcal{Q} - 2$. Now transpose 3a to the left-hand side which becomes $-3a$, [This gives] $1\mathcal{Q} - 3a = \surd\mathcal{Q}.3\mathcal{Q} - 2$. And the opposite sides remain equal, just as they were before the operation.

4. If all the terms on one side of an equation (by the previous paragraph) are transposed to the other side, the total so compounded will be equal to nothing. The equation is then said to be equated to zero. This equation must be simplified (by Proposition 4, Chapter 2 of this book).

> [In the example above] $4\text{℞} - 6 = 5\text{℞} - 20$,[64] transpose $5\text{℞} - 20$ into $-5\text{℞} + 20$. The equation then becomes $4\text{℞} - 6 - 5\text{℞} + 20 = 0$. And this can be simplified to: $-1\text{℞} + 14 = 0$ which is now equated to zero. [Likewise] $1\mathcal{Q} - \surd\mathcal{Q}.3\mathcal{Q} - 2 = 3a$. If you transpose the left-hand side

[63] Whence $\text{℞} = 14$.

[64] The equation of Section 3.

to the right-hand side we have: $0 = -1\mathfrak{z} + \sqrt{\mathfrak{z}}.3\mathfrak{z} - 2 + 3a$, which is now equated to zero.

5. If the unknown of the highest order has the minus $-$ sign in front, change all the signs of all the terms, and this will produce a clearer equation.

Refer to examples above: [If] $-1\text{℞} + 14$ [is made to be 0], then $-1\text{℞} + 14 = 0$, [and as a consequence] $+1\text{℞} - 14$ [will also be equal to 0], that is, $1\text{℞} - 14 = 0$. [Also, in the same manner, from] $-1\mathfrak{z} + 3a - \sqrt{\mathfrak{z}}.3\mathfrak{z} - 2 = 0$ [we have] $+1\mathfrak{z} - 3a + \sqrt{\mathfrak{z}}.3\mathfrak{z} + 2 = 0$. Similarly, if $-1\text{℞} - 1 + \frac{32}{1a+1} = 0$ then, by changing signs: $1\text{℞} + 1 - \frac{32}{1a+1} = 0$.

6. Divide the term, or terms, of the highest order in an equation by unity with the sign of the unknown. This gives a number by which you must divide the whole expression. This produces a clear equation with unity as the coefficient of the highest order.

Example [of the equation] $2\mathfrak{C} - 8\mathfrak{z} + 6\text{℞} = 0$. Divide $2\mathfrak{C}$, the term of highest order by $1\mathfrak{C}$. This gives 2 as the divisor and we have $1\mathfrak{C} - 4\mathfrak{z} + 3\text{℞} = 0$.

In the equation above, $3\text{℞} - \sqrt{\mathfrak{z}}2\mathfrak{z} - 6 = 0$, the terms of the highest order which include the unknown are $3\text{℞} - \sqrt{\mathfrak{z}}2\mathfrak{z}$, which (by Proposition 5, Chapter 4) are powers of the same order, the order being res or first. [Divide therefore $3\text{℞} - \sqrt{\mathfrak{z}}2\mathfrak{z}$ by 1℞ or (what is the same)] by $\sqrt{\mathfrak{z}}1\mathfrak{z}$. The division gives as quotient $3 - \sqrt{\mathfrak{z}}2$. Divide the whole equation by this quotient (by Proposition 2, Chapter 11, Book I) and this gives the equation: $1\text{℞} - \frac{18}{3} - \frac{\sqrt{\mathfrak{z}}72}{7} = 0$[, which is still a fraction, but is clearer than before, because the sign $\mathfrak{z}$ has been taken away].

[Third example,] $1\text{℞}a + 1a + 1\text{℞} - 31 = 0$. In this case you must aim at removing the mixed sign $1\text{℞}a$. Divide $1\text{℞}a + 1a$ by $1a$, or else divide $1\text{℞}a + 1\text{℞}$ by 1℞ (you can choose either option). In this case let us choose 1℞. When the division is completed the quotient is $1a + 1$, [by which divide the whole equation $1\text{℞}a + 1\text{℞} + 1a - 31 = 0$,] and the equation becomes $1\text{℞} + 1 - \frac{32}{1a+1} = 0$. Although this is a fraction, it is nonetheless clearer than before, since the mixed number, which was formerly obscure, has now been removed.

7. If all the terms in an equation contain the unknown, then divide the whole equation by unity with the lowest power of the unknown. This will produce a clear equation having an absolute number in place of the term of lowest order.

Example, divide $1\mathfrak{C} - 4\mathfrak{z} + 3\text{℞} = 0$ by unity and the power of lowest order, in this case, 1℞. This produces the final equation

$1\text{ʓ} - 4\text{℞} + 3 = 0$. [Likewise] divide $3\text{ʓ} - \sqrt{\text{ʓ}}\,2\,\text{℞} = 0$ by $\sqrt{\text{ʓ}}\,1\text{℞}$, which produces $\sqrt{\text{ʓ}}\,9\text{ℭ} - \sqrt{\text{ʓ}}\,2 = 0$. The final result of this sequence is a number.

8. If some of the terms of an equation are true fractions then multiply the whole equation by the denominators. This will produce an equation which is generally of great clarity.

[In] $\frac{6\text{℞} - 8\text{ʓ}}{1\text{ℭ}+3\text{℞}} + 2 = 0$ $\left[\frac{6\text{℞} - 8\text{ʓ}}{1\text{ℭ}+3\text{℞}}\right.$ is a true fraction that may be abbreviated$\left.\right]$. Multiply the whole equation by the denominator $[1\text{ℭ} + 3\text{℞}]$. This produces $2\text{ℭ} - 8\text{ʓ} + 12\text{℞} = 0$.[65]

[Likewise] $1\text{ʓ} + \frac{2\text{℞}}{3} - \frac{88}{75} = 0$. First multiply by 3 giving: $3\text{ʓ} + 2\text{℞} - \frac{264}{74} = 0$. Next multiply by 75 and this gives $225\text{ʓ} + 150\text{℞} - 264 = 0$. This equation is now integral and free from fractions.

9. If the equation contains one universal radical sign, separate it from the rest of the equation by Section 3. Then multiply each side of the equation by itself as often as the universal sign indicates. This produces an equation of great clarity, without the universal sign.

[Example,] $2\text{ʓ} + 3\text{℞} - \sqrt{\text{ʓ}}.12\text{ℭ} + 4\text{ʓʓ} + 18 = 0$. First, by transposition $2\text{ʓ} + 3\text{℞} = \sqrt{\text{ʓ}}.12\text{ℭ} + 4\text{ʓʓ} + 18$. Now square both sides because the universal sign is $\sqrt{\text{ʓ}}$. This now produces: $4\text{ʓʓ} + 12\text{ℭ} + 9\text{ʓ} = 12\text{ℭ} + 4\text{ʓʓ} + 18$. And by transposition and reduction this gives $1\text{ʓ} = 2$.[66]

[Another example:] $\sqrt{\text{ℭ}}.2\text{℞} - 6 = 3\text{℞}$. Cube both sides, giving $2\text{℞} - 6 = 27\text{ℭ}$[, or $2\text{℞} - 27\text{ℭ} - 6 = 0$].[67]

10. If an equation consists of two universal roots of the same order, separate one from the other by transposition, and multiply each into itself as often as the universal sign indicates. This produces a clear equation, free from universal root signs.

$\sqrt{\text{ʓ}}.2\text{℞} + 5 - \sqrt{\text{ʓ}}.3\text{℞} - 4 = 0$. Separate them by transposition and we have: $\sqrt{\text{ʓ}}.2\text{℞} + 5 = \sqrt{\text{ʓ}}.3\text{℞} - 4$. Now square both sides to remove the root sign: $2\text{℞} + 5 = 3\text{℞} - 4$. By transposition and reduction this gives $1\text{℞} - 9 = 0$.[68]

11. If an equation consists of two dissimilar universal roots, first separate the universals by transposition and then multiply each side into itself as often

[65] $6\text{℞} - 8\text{ʓ} + 2\text{ℭ} + 6\text{℞} = 0$.

[66] In modern terms $x^2 = 2$.

[67] $27x^3 - 2x + 6 = 0$.

[68] In modern terms $x = 9$.

in succession as the power of the universal roots indicates. This produces a clear equation without universal signs.

[Such as] $\sqrt{\beta}.3\text{Q}+6-\sqrt{\text{Q}}.2\text{R}-3=0$. First separate the universals by transposition: $\sqrt{\beta}.3\text{Q}+6=\sqrt{\text{Q}}.2\text{R}-3$. Now raise each side to the second power[69] [and to] the fifth power, [giving] $32\beta-240\text{QQ}+720\text{C}-1080\text{Q}+810\text{R}-243=9\text{QQ}+36\text{Q}+36$. When transposed and simplified this becomes: $32\beta-249\text{QQ}+720\text{C}-1116\text{Q}+810\text{R}-279=0$.[70]

12. If an equation contains two universal quadratic radicals, together with other terms, either simple or uninomiae, separate both universals with their signs, from the other terms. Next square both sides, and the equation then consists of one universal radical which can be eliminated by [Proposition] 9, together with other terms.

This equation $\frac{1}{2}+\sqrt{\text{Q}}.\underline{48\frac{1}{4}+1\text{R}-1\text{Q}}+\frac{1}{2}\text{R}-\sqrt{\text{Q}}.79-\frac{3}{4}\text{Q}=0$ must be transposed as indicated: $\sqrt{\text{Q}}.79-\frac{3}{4}\text{Q}-\sqrt{\text{Q}}.\underline{48\frac{1}{4}+1\text{R}-1\text{Q}}=\frac{1}{2}\text{R}+\frac{1}{2}$. Next square both sides, [giving the equation] $127\frac{1}{4}+1\text{R}-1\frac{3}{4}\text{Q}-\sqrt{\text{Q}}.15247+316\text{R}-460\frac{3}{4}\text{Q}-3\text{C}+3\text{QQ}=\frac{1}{4}\text{Q}+\frac{1}{2}\text{R}+\frac{1}{4}$. After being transposed and simplified we have: $\sqrt{\text{Q}}.15247+316\text{R}-460\frac{3}{4}\text{Q}-3\text{C}+3\text{QQ}=127+\frac{1}{2}\text{R}-2\text{Q}$. Both sides are now squared and finally [(by Proposition 9)] we have: $1\text{QQ}+1\text{C}-47\text{Q}-189\text{R}+882=0$.[71]

13. If an equation consists of three universal quadratic radicals, with no other terms, two of the radicals are separated from the third by transposition. Both sides are then squared, thus producing an equation which contains only one universal quadratic radical. This can be eliminated by the use of [Proposition] 9.

[Consider the equation] $\sqrt{\text{Q}}.3\text{R}+2+\sqrt{\text{Q}}.2\text{R}-1-\sqrt{\text{Q}}.4\text{R}-2=0$. The terms are separated like this: $\sqrt{\text{Q}}.3\text{R}+2+\sqrt{\text{Q}}.2\text{R}-1=\sqrt{\text{Q}}.4\text{R}-2$. Both sides are now squared: $5\text{R}+1+\sqrt{\text{Q}}.24\text{Q}+4\text{R}-8=4\text{R}-2$. This is then simplified to: $\sqrt{\text{Q}}.24\text{Q}+4\text{R}-8=-3-1\text{R}$. Both sides are now squared giving: $24\text{Q}+4\text{R}-8=9+6\text{R}+1\text{Q}$. This finally reduces to $23\text{Q}-2\text{R}-17=0$.[72]

---

[69] $\sqrt{\beta}.9\text{QQ}+36\text{Q}+36=2\text{R}-3$.

[70] $32x^5-249x^4+720x^3-1116x^2+810x-279=0$.

[71] $x^4+x^3-47x^2-189x+882=0$. This is equivalent to $\sqrt{A}+\sqrt{B}+\sqrt{C}=0$ therefore $\sqrt{A}+\sqrt{B}=-C$. Next, $A+B+2\sqrt{AB}=C^2$ therefore $2\sqrt{AB}=C^2-A-B$. Finally, $4AB=(C^2-A-B)^2$ which contains no root signs.

[72] $23x^2-2x-17=0$. The operation is as follows: $\sqrt{A}+\sqrt{B}+\sqrt{C}=0$ therefore $A+B+2\sqrt{AB}=C$. Whence $2\sqrt{AB}=C-A-B$ and $4AB=(C-A-B)^2$. Several misprints [in Mark Napier's transcription] were rectified in the working.

14. If the equation consists of three universal quadratic radicals with a single simplex or uninomia, proceed as follows: transpose two of the universal radicals to one side, with the other universal and the term on the side opposite. Square both sides of the equation. This produces a single quadratic root on both sides which can be eliminated by [Proposition] 12.

[Let the equation $\sqrt{\mathcal{Q}}.2\text{℞} + 3 + \sqrt{\mathcal{Q}}.3\text{℞} - 2 - 2\text{℞} - \sqrt{\mathcal{Q}}.2\mathcal{Q} - 1 = 0$ be transposed as][73] $\underline{\sqrt{\mathcal{Q}}.2\text{℞} + 3 + \sqrt{\mathcal{Q}}.3\text{℞} - 2} = 2\text{℞} + \underline{\sqrt{\mathcal{Q}}.2\mathcal{Q} + 1}$. Square both sides of the equation: $5\text{℞} + 1 + \sqrt{\mathcal{Q}}.24\mathcal{Q} + 20\text{℞} - 24 = 6\mathcal{Q} + 1 + \sqrt{\mathcal{Q}}.32\mathcal{Q}\mathcal{Q} + 16\mathcal{Q}$. Now transpose and simplify where necessary: $\sqrt{\mathcal{Q}}.24\mathcal{Q} + 20\text{℞} - 24 - \sqrt{\mathcal{Q}}.32\mathcal{Q}\mathcal{Q} + 16\mathcal{Q} = 6\mathcal{Q} - 5\text{℞}$.[74] Radical signs can now be eliminated by [Proposition] 12.[75]

15. If an equation is made up of four universal quadratic expressions without any other terms proceed as follows: separate the terms by transposition, two on each side. Now square both sides, and this will produce an equation with two universal signs. These can then be eliminated by [Proposition] 12.

[Let the transposed equation be] $\sqrt{\mathcal{Q}}.\underline{5 - 2\text{℞}} - \sqrt{\mathcal{Q}}.\underline{10 - 1\text{℞}} = \sqrt{\mathcal{Q}}.2\text{℞} + 6 + \sqrt{\mathcal{Q}}.1\mathcal{Q} + 4$. Square both sides which gives: $5\mathcal{Q} - 3\text{℞} + 10 - \sqrt{\mathcal{Q}}.208\mathcal{Q} - 20\mathfrak{C} - 80\text{℞} = 1\mathcal{Q} + 2\text{℞} + 10 + \sqrt{\mathcal{Q}}.8\mathfrak{C} + 24\mathcal{Q} + 32\text{℞} + 96$.[76] This contains two universals which can be eliminated by [Proposition] 12.

16. If a unique universal on one side of an equation is equated to a solitary universal on the other side; whether the universal is alone, or with a uninomium, or with a simple number, or both; then multiply the sides into themselves, according to the nature of the universals. The universal sign is then eliminated. The remaining universals can be eliminated by preceding sections.

[As in this equation] $\underline{\sqrt{\mathcal{Q}}.10 + \sqrt{\mathcal{Q}}.\underline{5\text{℞} - 2}} = \underline{\sqrt{\mathcal{Q}}.3 + \sqrt{\mathcal{Q}}.\underline{3\text{℞} + 1}}$. In this case a major universal is equated to a major universal. When

[73] The published example contains some misprints. Hawkins modified the example to be correct, beginning with the next equation (disregard the underlininig for now).

[74] $\sqrt{24x^2 + 20x - 24} - \sqrt{32x^4 + 16x^2} = 6x^2 - 5x$.

[75] In Section 14 Napier introduced a horizontal line underneath an expression to indicate aggregation (shown above as modified by Hawkins). See Algebra, Book I, Chapter 17, Section 3. This line links the first two quadratic expressions on the left hand side of the equation. If the lines of aggregation are included [as shown in the Latin original] we have: $\underline{\sqrt{\mathcal{Q}}.2\text{℞} + 3 + \sqrt{\mathcal{Q}}.3\text{℞} - 2} = 2\text{℞} + \sqrt{\mathcal{Q}}.2\mathcal{Q} + 1$. Squaring, $2\text{℞} + 3 + \sqrt{\mathcal{Q}}.3\text{℞} - 2 = 6\mathcal{Q} + 1 + \sqrt{\mathcal{Q}}.32\mathcal{Q}\mathcal{Q} + 16\mathcal{Q}$. Simplifying, $\sqrt{\mathcal{Q}}.3\text{℞} - 2 - \sqrt{\mathcal{Q}}.32\mathcal{Q}\mathcal{Q} + 16\mathcal{Q} = 6\mathcal{Q} - 2\text{℞} - 2$ ($\sqrt{3x - 2} - \sqrt{32x^4 + 16x^2} = 6x^2 - 2x - 2$).

[76] Simplify where necessary. $\sqrt{\mathcal{Q}}.208\mathcal{Q} - 20\mathfrak{C} - 80\text{℞} + \sqrt{\mathcal{Q}}.8\mathfrak{C} + 24\mathcal{Q} + 32\text{℞} + 96 = 4\mathcal{Q} - 5\text{℞}$.

we square both sides we obtain 10 + √Q.5℞ − 2 = 3 + √Q.3℞ + 1, [or] 7 + √Q.5℞ − 2 = √Q.3℞ + 1. The universal signs can now be eliminated by [Proposition] 12.

Another example.

[Both sides of √Qß.$\overline{3 + \text{√Q.}\underline{2℞ − 1}}$ = √𝔆ß.$\overline{5 + \text{√Q.}\underline{3℞ − 4}}$ are to be raised to the thirtieth power, giving 18℞ + $\overline{18 + \text{√Q.8𝔆} −}$ 12Q + 6℞ − 1 + √Q.1458℞ − 729 = 21 + 3℞ + √Q.300℞ − 400, or 15℞ − 3 + √Q.8𝔆 − 12Q + 6℞ − 1 + √Q.1458℞ − 729 = √Q.300℞ − 400, whose universal signs cannot be deleted.] [77]

[77] This example is translated directly from the Latin publication. In modern notation, the given equation is

$$\sqrt[10]{3+\sqrt{2x-1}} = \sqrt[15]{5+\sqrt{3x-4}},$$

and raised to the thirtieth power becomes

$$\left(3+\sqrt{2x-1}\right)^3 = \left(5+\sqrt{3x-4}\right)^2.$$

Expanding,

$$27 + 27\sqrt{2x-1} + 9(2x-1) + \left(\sqrt{2x-1}\right)^3 = 25 + 10\sqrt{3x-4} + 3x - 4.$$

Finally, since the cube of a square root is the square root of the cube, simplifying and reordering terms of each side gives the equation

$$18x + 18 + \sqrt{8x^3 - 12x^2 + 6x - 1} + \sqrt{1458x - 729} = 21 + 3x + \sqrt{300x - 400},$$

which is the equation stated by Napier as

18℞ + 18 + √Q.8𝔆 − 12Q + 6℞ − 1 + √Q.1458℞ − 729 = 21 + 3℞ + √Q.300℞ − 400.

The remaining equation is obvious as a global simplification of this one.

However, in spite of the simplicity of the example as is, Hawkins translated it as follows:

√Qß.3 + √Q.2℞ − 1 = √𝔆ß.5 + √Q.3℞ − 4

Both sides are raised to the 30th power. That is, the root sign √Q𝔆ß is to be attached to both sides. To do this, the R.H.S. must be squared:

√Q𝔆ß.25 + 3℞ − 4 + √Q.300 − 400 = √Q𝔆ß.$\underline{21 + 3℞ + \text{√Q.300℞} − 400}$

The L.H.S. must now be cubed and the prefix √Q𝔆ß added; therefore the L.H.S.
= √Q𝔆ß.$\overline{27 + \text{√Q.1458℞} − 729 + 18℞ − 9 + \text{√Q.8Q} − 12𝔆 + 6℞ − 1}$
= √Q𝔆ß.$\overline{18 + 18℞ + \text{√Q.1458℞} − 729 + \text{√Q.8𝔆} − 12Q + 6℞ − 1}$. Since the L.H.S. and R.H.S. are equal their 30th roots must be equal and the root sign √Q𝔆ß can be removed. By transposition and simplification we have: 15℞ − 3 + √Q.$\underline{1458℞ − 729}$ + √Q.$\underline{8𝔆 − 12Q + 6℞ − 1}$ = √Q.$\underline{300℞ − 400}$. These universal signs cannot be eliminated.

Third example.

[Likewise consider] √ℭ.3 + √ʓ.2℞ − 1 = √ℭ.20 − 4℞. Cube both sides and this gives: 3 + √ʓ.2℞ − 1 = 20 − 4℞ [or] √ʓ.2℞ − 1 = 17 − 4℞. You can now eliminate the universal (by [Proposition] 9).

17. The very same propositions by which the elimination of universals was established, can be used to eliminate, transpose and multiply simple irrationals amongst rational expressions.

The equation 12 − ʓ1℞ = 1℞ is to be simplified by [Proposition] 9. By transposition, 12 − 1℞ = ʓ1℞. Now multiply both sides to eliminate the quadratic root. Therefore 1ʓ − 24℞ + 144 = 1℞ and so 1ʓ − 25℞ + 144 = 0. This is now a rational expression. Therefore, what I have indicated in [Propositions] 9, 10, 11, 12, 13, 14, and 15 concerning universals the same must be understood to apply to numbers or expressions with radical signs.

18. What can otherwise be done to simplify equations, over and above the propositions given in this chapter. For instance, the multiplication of simple irrationals for the most part produces more solutions than there should be.

In the previous example 12 − ʓ1℞ = 1℞, when multiplied as shown, produced the equation 1ʓ − 25℞ + 144 = 0, which has two valid solutions, namely 16 and 9. But, actually, the principal equation 12 − ʓ1℞ = 1℞ only has one solution, [namely] 9, as will appear later. The principal equation, unless prepared by [Proposition] 17, can be better and more readily prepared by [Proposition] 20, which follows, as described therein.

19. If the true root of an equation equal to zero is extracted without leaving a remainder, that root will be more concisely expressed when the equation of root and unknown is made equal to 0.

From the equation 1ℭ − 6ʓ + 12℞ − 8 = 0, extract the true cube root. 1℞ − 2 = 0 will be the shortest and most concise equation.

Furthermore, at the bottom of his typewritten page containing the last two examples, he inserted the following:

Note: This is equivalent to $\sqrt{A} + \sqrt{B} + \sqrt{C} + \sqrt{D} = 0$. Therefore $\sqrt{A} + \sqrt{B} = -\sqrt{C} - \sqrt{D}$. By squaring both sides $A + B + 2\sqrt{AB} = C + D + 2\sqrt{CD}$. Therefore $A + B - C - D = 2\sqrt{CD} - 2\sqrt{AB}$.

This is true, but there is no footnote mark anywhere on the page to indicate where this note should be inserted.

Extract the square root of the equation 1℞ − √♀36℞ + 9 = 0. The true root will be (by Chapter 8) √♀1℞ − 3 = 0 which is by far the most exact equation.

20. If the root of an equation equated to 0 can be extracted, it will either be formal or reformed (see Chapter 8, [Proposition] 8 of this book). Change the sign of the remaining terms, and then, from them extract the square, cube, or other suitable root from the remainder. The equations produce these roots (with their changes of sign) together with an approximate and formal root; but a single root, not a quadratic nor a double quadratic [but more briefly equal to 0], and includes the roots (exponents) of the first equation.

Et cætera.

There is no more of his algebra orderlie sett down.[78]

[78] The final sentence was written by Robert Napier when he collected his father's manuscript and wrote out a fair copy for Henry Briggs.

APPENDICES

Brian Rice

An engraving of John Napier (1550–1617) taken from a portrait that Mark Napier recorded in 1839 as being "from the original belonging to Lord Napier, and which has always been in possession of the Napier family". However, the present Lord Napier has confirmed that it is no longer owned by the family and that he has never seen or heard of the original, so it is presumably now lost. (From the frontispiece of Mark Napier (ed.87), De Arte Logistica, *The Maitland Club, 1839; see p. xciii for Mark Napier's quotation).*

APPENDIX 1

Dedicatory Epistle to the 1593 edition of the *Plaine Discovery*

TO THE RIGHT EXCELLENT, HIGH AND MIGHTIE PRINCE, IAMES THE SIXT, KING OF SCOTTES, GRACE AND PEACE, &c.

F*orſomuch (right highe and mightie Prince) as both this our diuine Prophet S.* Iohn, *intreating here moſt ſpeciallie of the deſtruction of the Antichriſtian ſeate, citie and kingdome, doth direct the execution of that great worke of Gods Iuſtice and iuſt iudgement to the Kings of the earth: as alſo the whole Prophets of al ages, haue for the moſt part, directed al their admonitions generally to Kings, princes and gouernors, to the effect that they (as Heads-men) being by holy admonitions forewarned, might (according thereto) holde all the whole body of their commoun wealth in good order: for certaine it is, that the heade being well affected, will of neceßitie miniſtrat health and wholſome humors to the whole body. Therefore, it is likewiſe the dutie of Gods ſeruants in this age, interpreters of Prophecies, as well (according to the example of the Prophets) to incourage and inanimate Princes, to be ready againſt that greate day of the Lords reuenge, as alſo to exhort them generally, to remoue all ſuch impediments in their cuntries and common wealths, as may hinder that work, & procure Gods plagues. For the which cauſes, wee alſo all your M. Subiects, that any waies (how litle ſoeuer) haue addicted our ſtudies vnto theſe propheticall myſteries, doth not onely craue your highnes to abide conſtant and couragious againſt that day of the deſtruction of that Apoſtatik ſeate and citie, in caſe (God willing) it fall in your time: but alſo in the meane time, vntill the reformation of that Idolatrous ſeate, to be preparing and purging your M. own ſeat and kingdome, from all the enemies of that cauſe: yea, and from all others any waies enemies or abuſers of Iuſtice: for verely and in trueth, ſuch is the iniury of this our preſẽt time, againſt both the Church of God, and your M. true Lieges, that Religion is deſpiſed, and Iuſtice vtterly neglected: for what by Atheiſts, Papiſts, and Cold profeſſors, the religion of God is mocked in al eſtates: Againe, for partialitie, prolixitie, dearth, and deceitfulnes of lawes, the poore periſhe, the proud triumphe, and*

Iuſtice is no where to be found. Praying your M. to attend your ſelf vnto theſe enormities, and (without caſting ouer the credite thereof to wrong wreſters of Iuſtice) your M. ſelf, to wit certainly, that iuſtice be done to theſe your true & godly Lieges, againſt the enemies of Gods Church, and their moſt cruell oppreſſors: Aſſuring your M. be concordance of al ſcriptures, that if your M. miniſtrate Iuſtice to them, God the ſupreme Iudge ſhal miniſtrate Iuſtice to you againſt al your enemies: and contrarily, if otherwiſe. Therefore, Sir, let it be your M. continuall ſtudy (as called and charged thereunto by God) to reforme the vniuerſall enormities of your country, and firſt (taking example of the princely Prophet Dauid) to begin at your M. owne houſe, familie and court, and purge the ſame of all ſuſpicion of Papiſts, and Atheiſts or Newtrals, whereof this Reuelation foretelleth, that the number ſhall greatly increaſe in theſe latter daies. For ſhall any Prince be able to be one of the deſtroiers of that great ſeate, and a purger of the world from Antichriſtianiſme, who purgeth not his owne countrie? ſhal he purge his whole country, who purgeth not his owne houſe? or ſhal hee purge his houſe, who is not purged himſelfe by priuate meditations with his God? I ſay therefore, as God hath mercifully begunne the firſt degree of that great worke in your inward minde, by purging the ſame from all apparant ſpot of Antichriſtianiſme, as that fruitfull meditation vpon the 7. 8. 9. and 10. verſes of the 20. Chapter of the Reuelation, which your highnes hath both godly & learnedly ſet forth, doth beare plaine teſtimony, to your M. high praiſe and honour: So alſo wee beſeeche your M. (hauing conſideration of the treaſonable practiſes in theſe preſent daies, attempted both againſt Gods trueth, your authoritie, and the common wealth of this countrie,) to proceede to the other degrees of that reformation, euen orderly from your M. owne perſone till your highnes familie, and from your family to your court. Til at laſt, your M. whol country ſtand reformed in the feare of God, ready waiting for that great day, in the which it ſhall pleaſe God to call your M. or yours after you, among other reformed Princes, to that greate and vniuerſall reformation, and deſtruction of that Antichriſtian ſeat and citie Rome, according to the wordes prophecied, Apoc. 17. ſaying: The ten horns are ten Kings &c. Theſe are they that ſhall hate that harlot, and ſhall make her deſolate and naked, and ſhall eate vp her fleſh and burne her ſelfe with fire: Beſide alſo a warrant and commaund generally giuen to all men. Apoc. 18. ſaying, Rewarde her euen as ſhee hath rewarded you, and giue her double according to her workes, and in the cup that ſhe hath filled to you, fill her the double. And now, becauſe the ſpirit of God, both by all his Prophets generally, and by S. Iohn particularly, commends and directs the execution of Iuſtice to kings and rulers, I truſt no man ſhall thinke, that this our diſcouery (wherein is contained Gods Iuſtice and ſeuere iudgement againſt the Antichriſtian ſeate) can more iuſtly be dedicate vnto any man, than vnto theſe ten Chriſtian Kings, ſometimes maintainers of that ſeat, whome, or whoſe ſucceſſors now both the Prophet promiſes to be executers of that iudgment, as alſo in whoſe kingdomes reformation is already begunne: But becauſe of theſe kingdomes, ſometimes maintainers of that ſeate, and nowe deſiſting therefro, this your M. Realme is vndoubtedly one, as alſo this preſent treatiſe both being written by your highnes ſubiect, and in your M. natiue language, were vnproper to be directed to any of the other Princes. Therefore, of neceßitie I am led (as by the eare) to direct & dedicate theſe primices & firſt fruites of my ſtudy vnto your highnes, wherein, if perchance I ſhould ſeme any waies more preſumpteous then acceptable, I doubt not, but your M. clemencie will pardone that preſumption that comes of neceßitie: But contrarily, if I herein ſhall be found acceptable, (as verely I look for, of your M. humanitie) then certainly, not onely conioyne I vnto the former neceßitie, a voluntary heart, and ſo do offer theſe preſents, both gladly & neceſſarily vnto your highnes, but alſo it ſhall incourage both me and others your M. Lieges, to proceede every man in his own calling, to all kinde of godly workes, and good exerciſes, to the honour of God, ædification of his Church, your highnes renowne, and welfare to your M. realme, when they ſhall finde your clemencie to become the Patrone and protector of all zealous ſtudents, and an allower and accepter of their godly exerciſes: For let not your M. doubt, but that there are within your Realme (als wel as in other countries) godly and good ingynes, verſed & exerciſed in al maner of honeſt ſcience, and godly diſcipline, who by your M. inſtigation might yeelde foorth workes and fruites woorthie of memory, which otherwiſe (lacking ſome mightie Mæcenas to incourage them) may perchance be buried with æternall ſilence. Hoping therefore, that your highnes will be a protector of vs and our godly exerciſes: wee pray and humblie beſeech the Almightie, to be alſo vnto your highnes ſelfe and moſt honourable bedfellowe, the Queenes M. a perpetuall protector of your honourable eſtates and welfare of your perſones, both in body and ſoule, to the quieting of your M. lieges, increaſe of the true church, and honour of God, to whome, in Trinitie and Vnitie bee praiſe for euer. At Marchiſtoun the 29. daye of Ianuar. 1593.

Your highnes moſt humble
and obedient Subject,

IOHN NAPEIR, *Fear of Marchiſtoun*

APPENDIX 2

The origin of the Napier name [79]

Foraſmuch as my entirely-beloved kinsman [...] being deſirous to be informed of their pedigree and deſcent from my houſe; I have declared [...] the truth thereof, and the original of our name, as, by tradition from father to ſon, we have generally, and without any doubt, received the ſame. That one of the ancient earls of Lenox, in Scotland had iſſue three ſons: the eldeſt (Alexander) that ſucceeded him to the earldom of Lenox. The second, whoſe name was Donald; and the third, named Gilchriſt. [...] Alexander, the then king of Scotts,[80] having warrs, did convocate his lieges to the battle; amongſt whom he that commanded was the earl of Lenox, who (keeping his eldeſt ſon at home) sent his ſecond ſon to ſerve for him, with the forces that were under his command. This battle went hard with the Scots; for the enemy preſſing furiouſly upon them, forced them to give ground, untill at last they fell to flat running away; which being perceived by Donald, he pulled his father's ſtandard from the bearer thereof, and valiantly encountering the foe, being well followed by the earl of Lenox's men, he repulſed the enemy, and changed the fortune of the day, whereby a great victory was got. After the battle, as the manner is, every one advancing and ſetting forth his own acts, the king said unto them, "Ye have all done valiantly, but there is one amongst you who hath *Na-pier* [No peer]," and calling Donald into his preſence, commanded him, in regard of his worthy ſervice, and in augmentation of his honour, to change his name from Lenox to Naper; and gave him the lands of Gosford, and lands in Fife, and made him his own ſervant: which discourse is confirmed by ſundry of my old evidences and teſtimonies, wherein we are called Lenox, alias Naper.

[79] John Hutchins, *The history and antiquities of the county of Dorset*, W. Boyer & J. Nichols, 1774, Vol. 2, p. 46.

[80] It is not clear which of the three Scottish kings named Alexander this account refers to. Alexander I reigned 1107–1124, Alexander II reigned 1214–1249 and Alexander III reigned 1249–1286.

APPENDIX 3

The Contract to search for treasure at Fast Castle [81]

Contract Merchiston and Restalrik.

At Edinbruch the [blank] day of Julij, yeir of God im vc foirscoir fourtein yeiris [1594] – It is apointit, contractit, and agreit, betwix the personis ondirwrettin; that is to say, Robert Logane of Restalrige on the ane pairt, and Jhone Neper, fear of Merchistoun, on the uther pairt, in maner, forme, and effect as folowis:- To wit, forsamekle as ther is dywerss ald reportis motiffis and appirancis, that thair suld be within the said Robertis dwellinge place of Fascastell a soum of monie and poiss [treasure], heid and hurdit up secritlie, quilk [which] as yit is on fund be ony man. The said Jhone sall do his utter and exact diligens to serche and sik out, and be al craft and ingyne that he dow, to tempt, trye, and find out the sam, and be the grace of God, ather sall find the sam, or than mak it suir that na sik thing hes been thair; sa far as his utter trawell diligens and ingyne may reach. For the quilk the said Robert sall giff, as be the tenor heirof, giffiis and grantis unto the said Johne the just third pairt of quhatsoewir poiss or heid treasour the said Jhone sall find, or beis fund be his moyan and ingyn, within or abut the said place of Falscastell, and that to be pairtit be just wecht and balance, betwix thaim but ony fraud, stryff, debait, and contention, on sik maner as the said Robert sall heff the just twa partis, and the said Jhone the just third pairt thereof upone thair fayth, truth, and consciens. And for the said Jhonis suir return and saiff bakcumming tharwith to Edinbruch, on beand [without being] spulzeit [plundered] of his said thrid pairt, or utherways hairmit in body, or geir, the said Robert sall mak the said Jhone saiff convoy, and accumpanie him saifflie in maner forsaid bak to Edinburgh, quher the said Jhone, beand saiflie returnit, sall, in presens of the said Robert, cancell and destroy this present contract, as a full discherg of ather of thair pairtis honestlie satisfiet and performit to utheris; and ordanis that na uther discherge heirof but the destroying of this present contract sal be of ony awaill, forse, or effect. And incaiss the said Jhone sal find na poiss to be thair eftir all tryall and utter diligens tane; he referris the satisfactione of his trawell

[81] The reproduction (opposite) and this transcript are taken from the facsimile of the original document reproduced in Mark Napier's *Memoirs of John Napier*, p. 220. The present location of the original document is unknown but it is held neither by the National Library of Scotland nor by the National Records of Scotland.

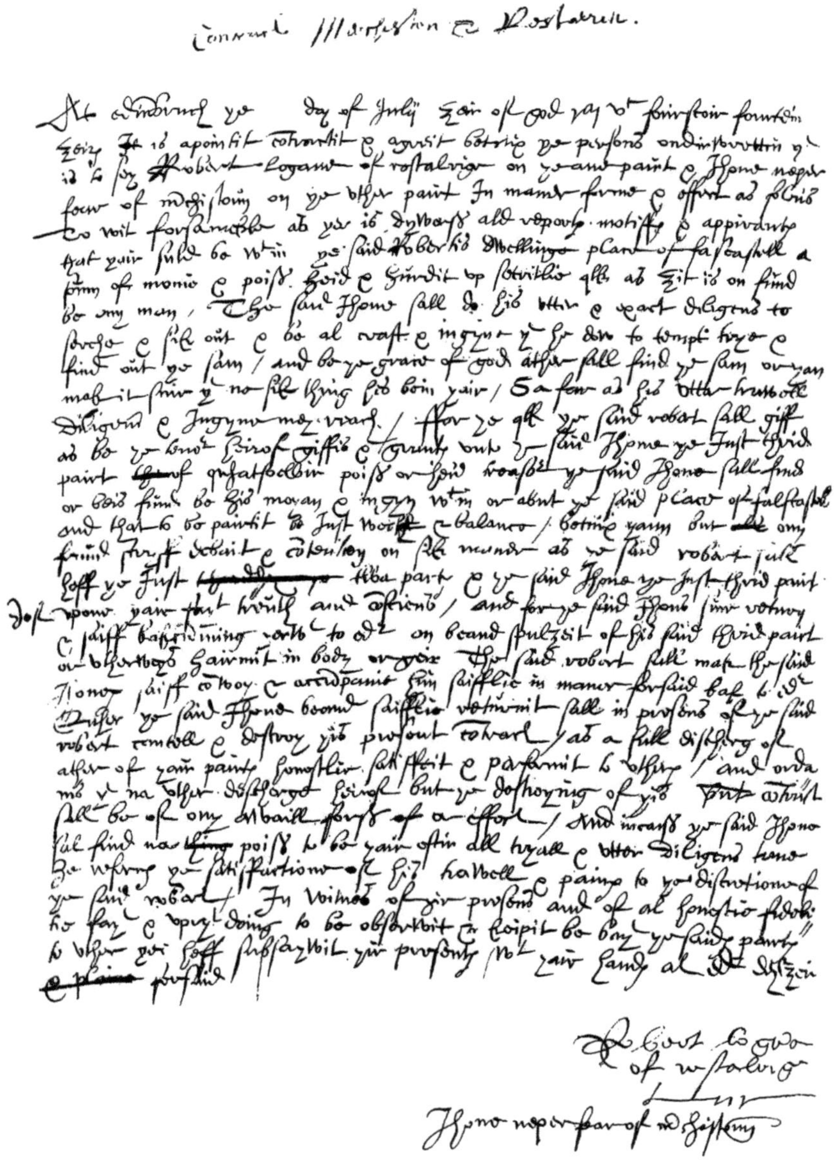
Contract Merchiston & Restalrig.

At edinburgh ye day of July ye zeir of god Jaj vc fourscoir fourtein
zeiris It is apointit contractit & agreit betwix ye persons undirwrittin yt
is to say Robert logane of restalrig on ye ane pairt & Jhone neper
fear of merchistoun on ye uther pairt In maner forme & effect as folowis
to wit forsamekle as yr is dyvers ald reportis motiffis & appirancis
that yair suld be wtin ye said Robertis dwelling place of fastcastell a
soum of monie & poiss hid & hurdit up secretlie qlk as zit is on fund
be ony man, The said Jhone sall do his utter & exact diligens to
serche & sik out & be al craft & ingyne yt he dow to tempt trye &
find out ye sam / and be ye grace of god ather sall find ye sam or yan
mak it sur yt no sik thing hes bein yair / Sa far as his utter travell
diligens & Ingyne may reach / For ye qlk ye said robert sall giff
as be ye tenor heirof giffis & grantis unto ye said Jhone ye Just third
pairt ~~thairof~~ quhatsoevir poiss or hid treasur ye said Jhone sall find
or beis fund be his moyan & ingyn wtin or about ye said place of fastcastell
and that to be pairtit be Just weycht & balance / betuix yaim but ~~all~~ ony
fraud stryf debait & contraversy on sik maner as ye said robert sall
haff ye Just ~~[illegible]~~ twa part & ye said Jhone ye Just third pairt
upon yair faith and conscience / and for ye said Jhonis sur returnyng
& saiff bakcumming wt ye ad on beand spulzeit of his said third pairt
or utherwayis hairmit in body or geir The said robert sall mak the said
Jhone saiff convoy & accumpanie him saiflie in maner forsaid bak to Edr
Quhair ye said Jhone beand saiflie returnit sall in presens of ye said
robert cancell & destroy yis present contract as a full discharg of
ather of yair pairtis honestlie satisfeit & perfurnit to utheris / and orda
nis yt na uther discharge heirof but ye destroying of yis ~~put~~ contract
sall be of ony awaill force or effect / And incaiss ye said Jhone
sal find na ~~thing~~ poiss to be yair efter all tryall & utter diligens tane
he referris ye satisfactioun of his travell & painis to ye discretioun of
ye said robert / In witnes of yir presentis and of al honestie fidelitie
to kep & upricht doing to be observit be baith ye saidis pairtis
to uther yai haff subscrivit yir presentis wt yair handis at Edr
~~[illegible]~~ presentis

Robert Logane of restalrig

Jhone neper fear of merchistoun

and painis to the discretione of the said Robert. – In witnes of thir presens, and of al honestie, fideletie, fayth, and uprycht doing to be observit and keipit be bayth the saidis pairtis to uther, thei heff subscrywit thir presentis with thair handis at Edinbruch, day and yeir forsaid.

Robert Logan
of Restalrige
Johnne Neper, Fear of Merchistoun

APPENDIX 4

Secrett Inventionis [82]

Anno Domini 1596, the 7 of June

Secrett Inventionis, proffitabill and necessary in theis dayes for defence of this Iland, and withstanding of strangers, enemies of Godes truth and Religion.

First the invention, proofe and perfect demonstration, geometricall and alegebricall, of burning mirrour, which, receving the dispersed beames of the sonne, doth reflex the same beames alltogether United and concurring preciselie in one mathematicall point, In the which point most necessarelie it ingendreth fier with an evident demonstration of their error who affirmeth this to be made a parabolik section.

The use of this invention serveth for burning of the enemies shippes at whatsoever appointed distance.

Secondlie the invention and sure demonstration of another mirrour which receving the dispersed beames of any materiall fier or flame yealdeth allsoe the former effect, and serveth for the like use.

Thirdlie the invention and visible demonstration of a peice of Artillery, which, shott, passeth not linallie through the armie destroying onlie those that stand on the randon thereof, and fra them forth flying idly, as otheres doe; but passeth superficially, ranging abrode within the whole appointed place, and not departing furth of the place till it hath executed his whole strength, by distroying those that be within the boundes of the said place.

The use hereof not onlie serveth greatlie against the Armie of the Enemy on land, but alsoe by sea it serveth to destroy, and cut downe, and one shott the whole mastes and tackling of so many shippes as be within the appointed boundes, as well abreid as in large, so long as any strength at all remayneth.

Fourthlie the invention of a round chariot of mettle made of the proofe of dooble muskett, which motion shall be by those that be within the same, more easie, more light, and more spedie by much then so manie armed men would be otherwayes.

The use hereof as well, in moving, serveth to breake the array of the enemies battle and to make passage, as also in staying and abiding within the

[82] Lambeth Palace Library, MS 658, f. 58.

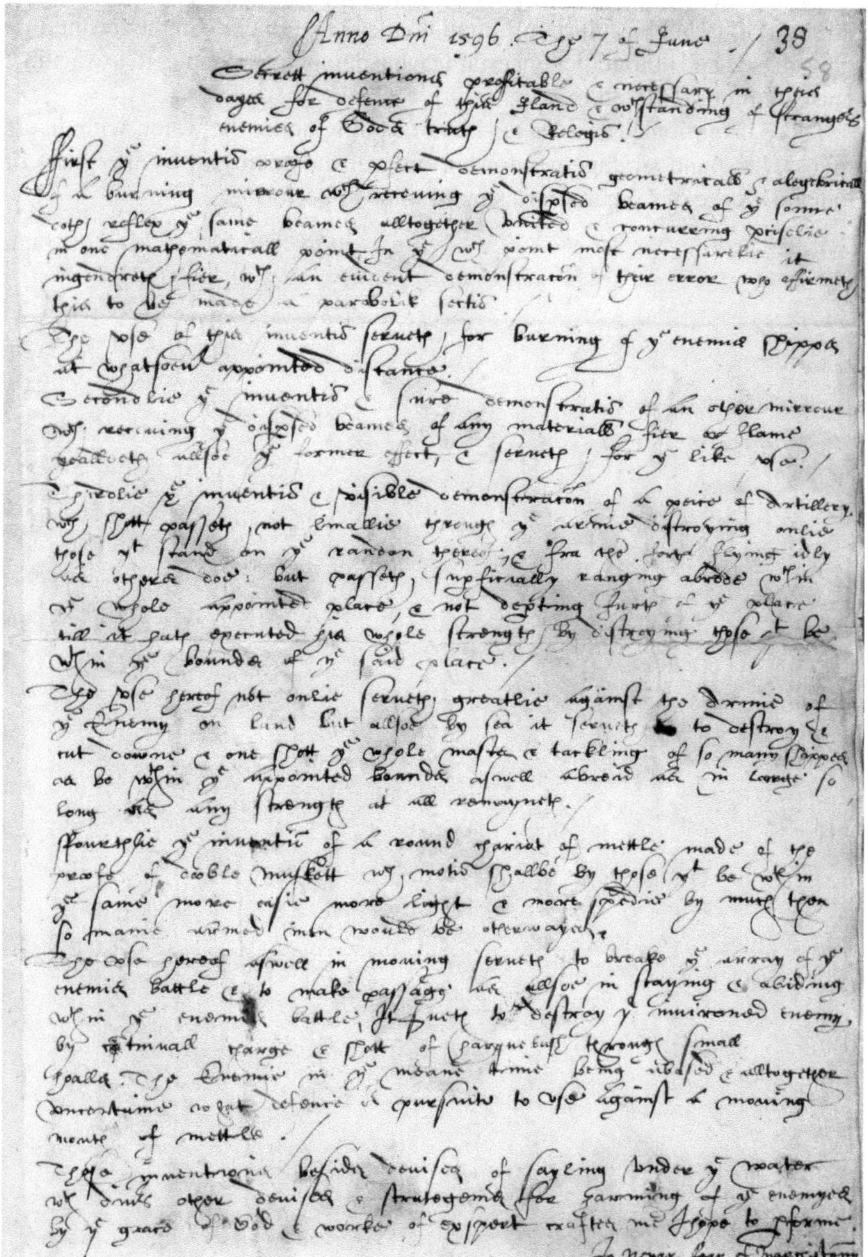

The original manuscript of Napier's Secrett Inventionis *signed by Napier. 11.5" × 8".* (*Courtesy of Lambeth Palace Library.*)

enemies battle, It serveth to destroy the invironed enemy by continuall charge and shott of harquebush through small hoalles. The Enemie in the meanetime being abased and alltogether uncertaine what defence or pursuite to use against a moving mouth of mettle.

These inventiones, besides devises of sayling under the water, with divers other devises and stratagemes for harming of the enemyes, by the grace of God and worke of exspert craftesmen I hope to performe.

Jo. Nepar, fear of Marchistoun.

APPENDIX 5

Gooding and manuring of field land [83]

THE NEW ORDER OF GOODING AND MANURING OF FIELD LAND WITH COMMON SALTS, WHEREBY THE SAME MAY BRING FORTH IN MORE ABUNDANCE, BOTH OF GRASS AND CORN OF ALL SORTS, AND FAR CHEAPER THAN BY THE COMMON WAY OF DUNGING USED HERETOFORE IN SCOTLAND.

Set forth by Archibald Napier, the apparent of Merchistown, conform to the Gift of Office given by the King's Majesty, under the PrivySeal, with advice of the Lord's of Council thereof, and made to him thereanant, of the date, at Holyrood-house, the

[Archibald's pamphlet concludes:]

That no man take upon him this kind of husbandry without licence from the said Archibald, or his deputies, under the pain of ten shillings to be paid him for every acre of land they labour therewith, as well grass as corn, conform to his gift thereupon by his majesty.

Considdering the greit proffite and commoditie that may be redound universallie to this realme be the diligent cair and paines to be taine in laboring, mukking, and manuring of the ground, in sik sort and manner that wes never usit nor frequentit within any pairt of the boundis thereof be anie persoun or persouns of before, and of the greit incres, aslweill of coirnes as grass, as may accress thair throw, and neidful it is that that invention and pratique be useit and exercesid be ane skillfull person, ... and understanding that his heines lovit Archibald Naper, appeirand of Merchinstoun, is ane qualifiet and expert peroun, maist apt and meit for exercising of sik ane commodious industrie and laboure.

[83] The original pamphlet was printed by Robert Waldegrave, a prolific printer in London and then to James VI in Edinburgh, but no copies appear to have survived. It was reprinted in *Archaeologia Scotica or Transactions of the Society of the Antiquaries of Scotland*, William Blackwood, 1818/22, Vol. II, Part 1, p. 154.

APPENDIX 6

Briggs' first meeting with Napier[84]

I will acquaint you with one memorable story related unto me by John Marr, an excellent mathematician and geometrician, whom I conceive you remember: he was servant to King James I and Charles I.

At first, when Lord Napier, or Marchiston, made publick his Logarithms, Mr. Briggs, then reader of the astronomy lecture at Gresham-College in London, was so surprized with admiration of them, that he could have no quietness in himself until he had seen that noble person the Lord Marchiston, whose only invention they were: he acquaints John Marr herewith, who went to Scotland before Mr. Briggs, purposely to be there when these two so learned persons should meet. Mr. Briggs appoints a certain day when to meet at Edinburgh: but failing thereof, the Lord Napier was doubtful he would not come. It happened one day as John Marr and the Lord Napier were speaking of Mr. Briggs, 'Ah! John,' saith Marchiston, 'Mr. Briggs will not now come:' at the very instant one knocks at the gate; John Marr hasted down, and it proved Mr. Briggs to his great contentment. He brings Mr. Briggs into my Lord's chamber, where almost one quarter of an hour was spent, each beholding other with admiration, before one word was spoke: at last Mr. Briggs began.

'My Lord, I have undertaken this long journey purposely to see your person, and to know by what engine of wit or ingenuity you came first to think of this most excellent help unto astronomy, viz. the Logarithms; but, my Lord, being by you found out, I wonder nobody else found it out before, when, now known, it appears so easy.' He was nobly entertained by the Lord Napier; and every summer after that, during the Lord's being alive, this venerable man, Mr. Briggs, went purposely to Scotland to visit him. *Tempora nunc mutantur* [*Times change*].

These two persons were worthy men in their time; and yet the one, viz. Lord Marchiston, was a great lover of astrology, but Briggs the most satirical man against it that hath been known: but the reason hereof I conceive was,

[84] *William Lilly's history of his life and times*, J. Roberts, 1715; Charles Baldwin, 1822, pp. 235–8. William Lilly (1602–1681) was born of humble stock but by marrying his former master's widow secured his fortune. He then devoted himself to astrology, publishing annual almanacs throughout the English Civil War that were hugely popular and becoming close to many influential persons. He wrote a posthumously published autobiography which includes this almost passing mention of Napier's first meeting with Briggs.

Briggs was a severe Presbyterian, and wholly conversant with persons of that judgement; whereas the Lord Marchiston was a general scholar, and deeply read in all divine and human histories: it is the same Marchiston who made the most serious and learned exposition upon the *Revelation of St John*; which is the best that ever appeared in the world.

APPENDIX 7

Kepler's letter to Napier [85]

To the noble and distinguished, Dr. John Napier, Baron of Merchiston, in Scotland. Greetings.

Some years ago, at the beginning of my *Ephemerides*, I started to give my readers further information about the state of the Rudolphine Tables, and to explain the reasons for the delays, which they were complaining about in numerous letters and public documents. Now, noble Baron, I address you, separately from the others, because both the matter itself and your book, entitled *Mirificus logarithmorum Canon*, make it necessary, but also publicly, because those things I am to discuss with you, are known to all readers.

The reason why I have delayed for yet another year, apart from wider causes which have so far got in my way, is because of particular issues which have come together this year. Some of these issues have been talked about publicly, such as wars and comets; others I have spoken about before or touched upon at the beginning of the *Ephemerides* in 1617 and 1619, which appeared in 1618, namely the publication of five books of the *Harmonices Mundi*. This publication alone (not to mention the previous laborious studies) kept me fully occupied for a whole year. Even so, it is finished (may the great harmoniser of the world be praised) despite the roaring, grinding and quite wild rumbling of Bellona with her cannons, pipes and trumpets. So that if on the one hand this frightful goddess had not besieged me at home and abroad, which she is still doing, or on the other hand the delays of scholars had not betrayed me (as happened in the second part of the *Epitomes seu Doctrina Theorica*, which as yet has not even been typeset beyond the first page), those who like to contemplate inwardly the works of God's hands, illumined by the light of the mind, could have had copies of both the *Harmonices* and the *Description of Comets*, at the Autumn markets in Frankfurt, but they have been stuck in Auxburg for three months now.

However, the main cause which this year has got in the way of my writing the Tables, is an entirely new but fortunate calamity concerning that part of the Tables which I had finished long ago. I am referring, Noble Baron, to

[85] Ioanne Kepplero, *Ephemerides novæ motuum coelestium ab anno vulgaris æræ MDCXVII / Ephemeris motuum coelestium ab annum incarnationis verbi MDCXX*, Iohannes Plancvs, 1617–1620, sig. N recto.

your book published in Edinburgh in Scotland five years ago, which I first saw in Prague two years ago. However, I was not able to read it through until last year when I stumbled upon a book by Benjamin Ursin, for a long time an assistant in my household, now astronomer to the Margrave (in which he gave a brief summary of the subject copied from your book) which is why I know about it. However, scarcely had I tried one example, when I discovered, to my great joy, that you had worked on that general function of numbers, a very small particle of which I had been using already for many years, and which I had proposed to make part of my Tables, especially in the matter of parallaxes and of units of duration and delay in eclipses. This very Ephemeris displays an example of this method. I was aware indeed that my method only worked when an arc differed very slightly from a straight line. But this I did not know, that logarithms could be made out of the excesses of secants, which makes this method universal through any length of arc. Therefore, my mind strove before anything else to see whether the logarithms in Ursin's book had been fully worked out. Then, with the help of Janus Gringalletus Sabaudus, an assistant of mine, I ordered him to subtract the thousandth part of the whole sine, to subtract again a thousandth part from the remainder, and to do that more than two thousand times, until there should remain about the tenth part of the whole sine. But with great care I established the logarithm of the whole sine, from which a thousandth part had been subtracted. I began from the unit of that division, which Pitiscus uses most often, that is to say the duodecimal. Having thus established this logarithm, I added it equally to the remainders of all the subtractions. In this way I discovered that there was in essence nothing wrong in these logarithms, but that some slight errors had occurred either in printing, or in that minute distribution of the greater logarithms around the beginning of the quadrant. Incidentally, I wanted you to know this, so that you should understand that it would be to me at least (and to others I think) most welcome, that the methods by which you proceeded (and doubtless you have many and very ingenious ones) should be in the public domain.

Now let us get closer to your Tables. In this month of July, having just been brought to Linz, I at last managed to read as far as folio 28 in a copy of your book. [. . .]

And so that nobody should be in any doubt, my *Ephemeris* has been constructed using this mode; and by right it is inscribed to you, illustrious Baron. Thus your logarithms inevitably become part of the Rudolphine Tables, having been previously reprinted in my printing office; and this is why astronomers will be delighted at my delays. If you have anything of greater use, please let me know as soon as possible. I wish to ask the same of professors of astronomy, some of whom I asked long ago privately, all of whom I now address publicly. Farewell, illustrious Baron, for the sake of our common studies, please value this greeting from a man who stands beneath you. From Linz on the Danube. On the 5th day before the Sextilis Calends in the year 1619.[86]

With the greatest respect for your esteemed rank.

John Keppler.

[86] Sextilis was the sixth month of the Roman year (beginning in March), the equivalent of August. Calends means the first day of the month while V Calends means counting back five days from the beginning of August (including the 1st) which gives 28th July 1619.

APPENDIX 8

Edinburgh in 1598[87]

I rode the reſt of the way [. . .] to *Edenborow*, seated in *Lodouey* [Lodoney – Lothian], (of old called *Pictland*) the moſt ciuill region of Scotland, being hilly and fruitful of corne, but having little or no wood. This city is the ſeat of the King of *Scotland* and the Courts of Iuſtice are held in the ſame. [. . .] This City is high ſeated, in a fruitful ſoyle and wholeſome aire, and is adorned with many Noblemen's Towers lying about it, and aboundeth with many ſprings of ſweet waters. At the end towards the Eaſt is the Kings Pallace ioyning to the Monaſtery of the Holy Croſſe [Holyrood Abbey] which King *David* the firſt built, over which, in a Parke of Hares, Conies [rabbits], and Deare, an high mountaine hangs, called the chaire of Arthur [. . .]. From the Kings Pallace at the Eaſt, the City ſtill riſeth higher and higher towards the Weſt and conſiſts eſpecially of one broad and very faire ſtreet, (which is the greateſt part and ſole ornament thereof), the reſt of the ſide ſtreetes and allies being of poore building and inhabited with very poore people, and this length from the Eaſt to the Weſt is about a mile, whereas the bredth of the City from the North to the South is narrow, and cannot be halfe a mile. At the furtheſt end towards the Weſt, is a very ſtrong Caſtle which the Scots hold unexpugnable. [. . .]. And from this Caſtle towards the Weſt, is a most ſteepe Rocke pointed on the higheſt top, out of which this Caſtle is cut: But on the North & South ſides without the wals be plaine and fruitfull fields of Corne. In the midſt of the aforeſaid faire ſtreete, the Cathedrall Church is built, which is large and lightſome, but little ſtately for the building, and nothing at all for the beauty and ornament.

[87] Fynes Moryson, *An itinerary written by Fynes Moryson Gent. first in the Latin tongue, and then translated by him into English: containing his ten yeeres travell throvgh the twelve domjnions of Germany, Bohmerland, Sweitzerland, Netherland, Denmarke, Poland, Jtaly, Turky, France, England, Scotland, and Ireland*, John Beale, 1617, Part I, Booke 3, Chap V, p. 273.

Moryson (1565/6–1630), the son of a Lincolnshire M.P., was educated at Peterhouse, Cambridge, graduating B.A. in 1584 and M.A. in 1587. He left England in 1591 and spent four years travelling around the Continent to observe local customs, institutions and economies, making extensive notes each day. In 1596 he journeyed with his brother to Jerusalem, Tripoli, Antioch (where his brother died of dysentery), Aleppo, Constantinople and Crete for the same purpose, returning in 1597. Finally, he visited Scotland (where he saw James VI) in 1598 and Ireland. He was an acute observer and his work is full of interesting observations on the local manners, diet and customs of each country and their sources of wealth etc.

[. . .]. The houſes are built of vnpoliſhed ſtone, and in the faire ſtreet good part of them is of free ſtone, which in that broade ſtreete would make a faire ſhew, but that the outſides of them are faced with wooden galleries, built upon the ſecond ſtory of the houſes; yet theſe galleries giue the owners a faire and a pleaſant proſpect, into the said faire and broad ſtreet, when they ſit or ſtand in the ſame. The wals of the City are built of little and vnpoliſhed ſtones, and ſeeme ancient, but are very narrow, and in ſome places exceeding low, in other, ruiued. (Part I, Booke 3, Chap V, p. 273).

APPENDIX 9

Merchiston Tower

The invention of logarithms is widely associated with Merchiston but it is not clear when or even if Napier took full possession of the property before his father, Sir Archibald, died in 1608. It is known that Napier was invested with the Fear of Merchiston on his first marriage in 1572 but it seems unlikely that he actually lived there during his father's lifetime. Napier had a large family (two children by his first wife and ten by his second) so Merchiston would hardly seem large enough for them and the family of his father (six children by his second wife, all of roughly the same age as Napier's own children).

Napier's various publications do not provide any conclusive evidence on this point. The Dedication of his commentary to the *Plaine Discovery* is dated "at Marchistoun, the 29 day of January, 1593", which could indicate that he was living at Merchiston at the time. However, the title page is rather ambiguous since it refers to "Iohn Napeir, L. [Laird] of Marchistoun younger", possibly implying that his father was still residing at Merchiston. As would be expected from their dates of publication, after the death of Napier's father, both the *Descriptio* (1616) and the *Rabdologiæ* (1617) refer to "Iohn Nepair, Baron of Marchiston" in English and "Ioanne Napero, Barone Merchistonii" in Latin on their respective title pages.

The acquisition of Merchiston by the Alexander Napier is poorly documented. James I's original charter for Merchiston has not survived and the Great Seal Register for 1432–8 is lost so its precise terms can only be guessed at. Besides this loss, the printed *Registrum Magni Sigilli* also disguises the fact that what survives is mostly made up of fragments and copies put together after the destruction of the original records sometime before 1580. Therefore all that is known of the original acquisition of Merchiston comes from the Exchequer Rolls. As noted below, there are also gaps in this record which might have provided further information about James I's grant to Alexander Napier. A further major problem in using the Exchequer Rolls is that they were written as a record of business transacted for the king by his officials and were not intended to be a historical source. They answer specific questions regarding money due to the king through his officials; its receipt, its whereabouts and whether any was still owed to the king. The auditing process required scrutiny of documentation, such as warrants and receipts for payments, which the enrolled accounts present in a much summarised form. Therefore what appears in the rolls is merely what the clerks thought necessary to record for financial purposes and is far from comprehensive.[89]

Alexander, the first Laird of Merchiston acquired Merchiston by wadset (mortgage) from James I sometime before 1438 but the king was assassinated in 1437 and the loan was never fully repaid.[90] The account of the king's Receiver south of Forth for 1433–4 shows Merchiston among lands in the environs of Edinburgh belonging to the king.[91] The Receiver's accounts for 1435–6 are missing but in his next surviving account, for 1437–8, the rents of Merchiston are not charged because it is 'pledged by way of sale or gift' by charter to Alexander Napier, Burgess of Edinburgh, and the letter of reversion remains in Edinburgh Castle with Sir William Crichton.[92] As the Receiver had accounted for the rents on 5 June 1434 it is likely that Napier made the loan and the wadset of Merchiston was granted after that date, with a letter of reversion for redemption.

The loan was partly repaid by William Foulis, Keeper of the Privy Seal, from the balance of money advanced to him for purchasing goods in Flanders,

[89] The author is particularly indebted to Dr Athol Murray, former Keeper of the Records of Scotland, for the greater part of this and the succeeding three paragraphs, excepting the footnote below regarding the French embassy of 1435. (See also Athol Murray, The procedure of the Scottish Exchequer in the early sixteenth century in *The Scottish Historical Review*, Oct. 1961, Vol. 40, Part 2, pp. 89–117).

[90] In 1434 Regnault Girard was sent to Scotland by Charles VII of France with a large retinue to negotiate the marriage of James I's daughter, Princess Margaret, to the Dauphin (the future Louis XI) and then escort her to Paris. He arrived in Scotland in January 1435. Louis Barbé states that "Quarters had been secured for the whole party, and those assigned to Girard were in the house of Alexander Napier" (*Margaret of Scotland & the Dauphin Louis*, Blackie & Son, 1917, p. 55), no doubt based on an unreferenced contemporary source held in the Bibliothèque Nationale de France which forms the core of Barbé's work. Helen Armet references Barbé and states that "Alexander Napier was a merchant burgess of Edinburgh, rich enough to lend money to the King and to entertain the French ambassador, Reginald Giraud, when he came to Holyrood. [...] In return for the loan to the King he received a wadset of the lands of Merchiston" (The Tower of Merchiston in *The book of the Old Edinburgh Club*, 1962, p. 1). This almost but doesn't quite link the loan to the French embassy. However, Francis Shennan does conflate the loan and the visit, stating (without any reference but presumably from Armet) that "James I wanted money to entertain him [Giraud] at Holyrood. He borrowed it from Alexander Napier [...] and in return the King pledged the lands of Merchiston" (*Flesh and bones*, Napier Polytechnic of Edinburgh, 1989, p. 2). However, although it is possible that Napier's loan originated with the visiting French embassy, there appears to be no documentary evidence in contemporary sources directly connecting the two.

[91] George Burnett (Ed), *The Exchequer Rolls of Scotland*, HM Register House, 1880, Vol. 4, p. 599.

[92] George Burnett (Ed), *The Exchequer Rolls of Scotland*, HM Register House, 1882, Vol. 5, pp. 63.

on or about 29 June 1435,[93] and a further £20 was paid to Napier by the Custumar of Edinburgh sometime between July 1435 and July 1437,[94] almost certainly before the James I's death on 21 February 1437. Neither the value nor the date of the original loan are known but as the combined repayment of £64 6s 6d was insufficient to clear the whole loan before the period for redemption expired during James II's minority, there seems to have been some compromise whereby Over Merchiston reverted to the crown while Alexander Napier retained Nether Merchiston.[95] This was to be inherited by his son, Sir Alexander Napier, James II's comptroller, in 1454.

Sir Alexander Napier acquired Over Merchiston from the crown, rather than from the city of Edinburgh. On 26 June 1456 a feu charter from the Provost, Bailies and community granted him the lands of Over Merchiston pertaining to the altar of St Catherine in St Giles Kirk, Edinburgh, for a feu-duty of 20 merks annually.[96]

As indicated above it is not known exactly when Merchiston Tower was built, though it is clear that it was sometime during the latter part of the fifteenth century.[97] Merchiston remained the home of the Napier family until it was sold in 1647. The increasingly settled state of Scotland probably encouraged the new owners to turn the tower into a mansion and add extra windows around this time. Merchiston was repurchased by the family in 1752 but they sold it for the second time in 1772 to a family relative. The Napiers repurchased Merchiston for the final time in 1818 and it remained in largely rural isolation until the tower was surrounded by domestic housing built in the 1860s and 70s. In 1911 it was inherited by Col. John (Jack) Scott Napier (1848–1938), the 2nd son of Francis 10th Lord Napier, on the death of his mother.

What subsequently became Merchiston Castle School opened in 1828 and in 1833 they took a lease on Merchiston Tower from the Napier family. They remained at Merchiston as lessees, making numerous additions and alterations to the building to suit their needs, until they purchased the property outright in 1914 from Col. Jack Napier, the last Napier to own the property. The school continued to occupy the tower until 1930, when they moved to new purpose-

[93] George Burnett (Ed), *The Exchequer Rolls of Scotland*, HM Register House, 1880, Vol. 4, pp. 667.

[94] George Burnett (Ed), *The Exchequer Rolls of Scotland*, HM Register House, 1882, Vol. 5, p. 26.

[95] Over Merchiston lies near the top of a ridge marking the edge of the Borough Muir, roughly on the line of the present day Colinton Road, from which the ground slopes down and northwards towards Edinburgh's West End. This is the location of Merchiston Tower. Nether (North) Merchiston is about 50-60 feet lower, roughly along the line of the present day Union Canal.

[96] NRAS (National Register of Archives for Scotland) 156/1/1.

[97] See p. 54 for a more detailed discussion on when Merchiston Tower was built.

built premises. Ownership passed to Edinburgh Corporation in 1935 but the tower remained largely unoccupied and derelict until 1958 when plans were approved to incorporate it into a new technical college which opened in 1964.

Such a variety of owners and uses inevitably resulted in numerous alterations over the generations, so it is difficult to determine the exact appearance and layout of the tower when Napier finalised his work on logarithms there, early in the seventeenth century. However, what is clear is that the so called "Napier Room" on the sixth floor did not exist in Napier's time since the building then only had five floors.[98]

Although Merchiston Tower has been preserved, it is most unfortunate that the modern Napier University, built as Napier Technical College, surrounds this historic tower so unsympathetically with low-grade, concrete buildings dating from the 1960s. Part of the north side of the tower is even bisected by a concrete and glass corridor connecting various parts of the campus. Such insensitive handling of the building would not be allowed today.

The following article provides a description of Merchiston Tower based on the meticulous investigation carried out during the major restoration which commenced in 1958.[99] Much of the supporting detail in the original article has been silently omitted but sufficient has been reproduced here to provide an appreciation of the tower's history, its original form and its extensive restoration.

When the Tower of Merchiston came into the hands of Edinburgh Corporation in 1935 it was the centre of a cluster of attached buildings which to a greater or lesser extent obscured the main tower on all sides except the east. A castellated Gothick building of two storeys and a basement, dating from the early nineteenth century, abutted on the south front. The other buildings had been erected in the later nineteenth century, and comprised a dining hall on the west, a two-storeyed building on the north, and a four-storeyed building in the re-entrant angle of the tower. All these accretions were demolished by the Corporation just before and just after the Second World War, and some emergency repairs, including the strengthening and reslating of the castle roof, were carried out in 1949. The tower was left free-standing, intact but for the

[98] See p. 971.

[99] Stuart Harris, The Tower of Merchiston: Architecture, in *The book of the Old Edinburgh Club*, Published for the Members of the Club, 1962, pp. 13–32. A supplement to this article was subsequently published and, although the main findings remained unaltered, a number of detailed points were amended or added. The plans of the tower were also updated and it is these revised ones that are reproduced here on pp. 56–57. (Stuart Harris, The Tower of Merchiston: a supplementary report, *The book of the Old Edinburgh Club*, Published for the Members of the Club, 1969, pp. 2–9).

numerous openings, fireplaces, and flues which had been cut out of its walls to serve the various attached buildings.

The work of restoration was put in hand in August 1958, under the direction of the Department of the City Architect and with the guidance of the Ministry of Works and the Royal Commission on Ancient Monuments.

The first six months were almost wholly occupied in stripping and examining the structure, and in shoring up weak and dangerous parts. With the exception of the seventeenth-century plaster ceiling on the second floor, which was retained in situ, none of the internal finishings was found to be worthy of preservation; they were all of nineteenth century date, and were mostly copies of earlier work. Their removal exposed numerous structural defects, almost all of them caused by ill-conceived and wretchedly executed alterations carried out principally in the seventeenth and nineteenth centuries. Windows, doorway, fireplaces and flues had been inserted with apparently reckless disregard of the proper abutment of arches or the stability of walls, and in some cases such elementary necessities as lintels had not been provided. Collapse was imminent in many places—notably in the east front, where the arch above the ten-foot wide embrasure of the windows inserted here in the seventeenth century required immediate shoring at a cost of £1500. There was evidence that this arch had been slipping for a very long time, threatening the stability of the entire upper half of the wall on this front. Many other examples, rather less dramatic but equally dangerous, could be quoted, and it is fair to say that, with the exception of the seventeenth-century upperworks above parapet level, all work later than the original was distinguished by its shoddiness and scamped workmanship.

The original masonry is, by contrast, well and substantially built. The tower is founded upon an outcrop of Upper Old Red Sandstone, and the stone used in the building is identical to it. It is similar to the Doddington (Northumberland) stone which is being used in the restoration work, but it is harder, and it has weathered so well over five centuries that there is little difference between the appearance of internal and external work.

The Castle has long been recognised as fifteenth-century tower house built on the L-plan with a wing or jamb on its north side but its original organisation had been so greatly obscured by alterations and additions that accounts such as those by MacGibbon & Ross,[100] and in the Inventory of the Royal Commission

[100] MacGibbon, David and Ross, Thomas, *The castellated and domestic architecture of Scotland from the twelfth to the eighteenth century*, David Douglas, 1889, Vol. 3, and in the Inventory of the p. 263. [MacGibbon and Ross, of the eponymous Edinburgh based architectural practice, documented many of the castles of Scotland in their monumental five-volume work published between 1887 and 1892. They provide the earliest detailed description of Merchiston but to prepare their finely executed drawings of the tower they had to "remove" many of the outbuildings that then existed to reveal the original L plan.]

Merchiston Tower photographed in about 1957, immediately prior to its restoration. The signs of the many additions and alterations that the tower underwent during its long life are clearly visible. (Crown Copyright.)

Merchiston Tower photographed after completion of the restoration in 1964. (*Courtesy of Historic Environment Scotland.*)

on Ancient Monuments[101] were necessarily tentative and incomplete. The present study confirms the fifteenth-century date of the great bulk of the tower, and nothing whatsoever has so far been found to suggest that any earlier work was incorporated in it, but the original form of the castle is now seen to be rather different from what had been supposed. It will be convenient to summarise the main points here before enlarging upon the details.

1. The tower originally contained five floors, with stone vaults at the levels of the second floor and the roof. The principal rooms were on the second floor, with a solar and sleeping apartments above, and storerooms beneath.
2. The turnpike stair [a spiral stair] in the southwest corner led from the second floor to the upper floors and the battlements, and the second and third floors were also connected by a mural stair [an internal stair] in the northeast corner of the main block. Internal access downwards from the principal flat was solely by mural stairs: one in the west wall, connecting the second and first floors, and one in the north wall of the main block, connecting the first floor with the ground floor.
3. A doorway low down in the centre of the south front gave entrance to the store-room flats, but the main entrance to the castle was at second floor level in the south front, immediately to the east of the turnpike stair. This entrance was defended by a drawbridge, and this in turn implies that the castle had some kind of outwork—probably a barmkin [an outer fortification wall] wall enclosing a courtyard—and was a larger fortification than now appears.

The principal entrance is perhaps the most interesting original feature discovered in the course of the work. It has suffered greatly from later alterations, but it has been possible to recover its main outlines. The doorway was about three feet nine inches wide and was surrounded by a moulding of much greater elaboration than the simple broad chamfer characteristic of other openings in the tower, though only an outer cavetto [a hollowed moulding whose profile is in the form of a quadrant of a circle] and a tiny fragment of the next part of the moulding remain. The doorway was set back seven-and-a half inches in a recess some ten feet high and five feet four inches wide, surmounted by a weathered, projecting lintel. This recess was evidently designed to house a drawbridge in the raised position. Traces of a mural chamber or chambers have been found in the wall above, and fixing holes for some kind of pulley

[101] RCAM *An Inventory of the Ancient and Historical Monuments of the City of Edinburgh with the Thirteenth Report of the Commission*, HMSO, 1951. pp. 232–3.

sheaves are visible on the upper ingoes of the recess, but too little remains to enable details of the mechanism for raising the drawbridge to be worked out, though it seems probable that a winch was used. The outer end of the drawbridge must have rested on a platform about ten feet away from the tower and some fourteen feet above the presumed level of the castle courtyard. No direct evidence of outworks has yet come to light, but it may be supposed that this platform was part of the outer barmkin defence, and that it was approached from the courtyard by a flight of steps leading up the inner face of the barmkin wall.

The present roof and all the buildings above the parapet walk (including the so called "Napier Room") are of the late seventeenth century.

Any reconstruction of the external appearance of the castle in the fifteenth century must be speculative in some degree, but the main bulk of the tower, up to and including the battlements, was substantially the same as it is today The roof was probably of a lower pitch than at present, and slabbed in stone; the walls were probably harled [102] and there were perhaps only eight sizable windows, all in the upper half of the castle, and some eleven slit windows lighting the various stairs and the lower rooms. But the most striking difference would lie in the lofty barmkin wall and in the principal entrance with its drawbridge. The line of the barmkin and the shape and extent of the courtyard are as yet unknown, and since the ground round about the castle has been extensively disturbed by later building, are likely to remain so. If the arrangements depicted in late eighteenth-century views of the castle may be taken as a guide, the entrance to the outer defences was probably on the north side, and the barmkin may have enclosed a well which existed about forty feet north of the tower.

It appears that the tower remained in this form without notable alteration for some two hundred years.

A cannon ball found in the east front of the tower may be a relic of these engagements [of 1572]. The shot had apparently smashed one of the heavy corbels of the battlement, bringing down a large part of the parapet, and was found lodged in a mass of loose and shattered rubble in the heart of the wall about two feet below the level of the parapet walk. To all appearances it had come to rest in this position after firing. The damage had been made good in a slipshod fashion: the corbel was not replaced, and the shattered part of the main tower wall was merely faced up internally without consolidation of the hearting. Since the gable of the seventeenth-century upperworks is built off

[102] A rough sort of harling—a smearing of mortar from the joints, covered by limewash—was the usual finish for this kind of rubble work. It greatly enhances the contrast of the dressed stonework at quoins, openings and other features.

the tails of the slabbed paving of the walk, the ball must be of an earlier date; and since the assaults of 1572 are the only military operations recorded in the castle's history, the probability that the ball belongs to this time is high. It measures six inches in diameter, but superficial rust has probably enlarged it slightly, and a calculation based on its weight of 26 pounds suggests that the original diameter was about a quarter of an inch less, which corresponds very well with the probable calibre [103] of the culverin mentioned in Sir William Drury's dispatch of 11th June 1572.

Since there is nothing to suggest that any more extensive changes were made in the sixteenth century, it may be assumed that when John Napier the mathematician lived in it, the tower was still more or less in its medieval form. The identification of the small room in the upperworks above the jamb of the tower as Napier's study is therefore erroneous.

The first important alterations of the tower took place about half-a-century after Napier's death, most probably in the reign of Charles II. Their object was to extend the living quarters and to convert the castle into a mansion house. This involved a major reconstruction of the medieval tower, as well as some extensions above the old roof level. The building was gutted from top to bottom, the stone vaults being removed and the timber floors renewed. Windows were inserted on the first, second, third and fourth floors (the main windows to the latter being dormers in the new roof). The medieval entrance at second-floor level was converted to a window, and a new entrance was formed in the centre of the south front at first floor level.

At various times (not precisely ascertained) in the latter half of the century, the other buildings listed at the beginning of this article were erected around the tower, with further incidental mutilation of its walls

The restoration of the tower, and its adaptation to new uses as a part of the Napier Technical College has been complicated by this sequence of alterations over three centuries. Not only have original features been severely damaged or altogether lost but some of the work which replaced them has itself disappeared or become so unsound as to make rebuilding imperative. The alterations carried out after 1800 being for the most part incidental to the attachment of other buildings, naturally did nothing to enhance the appearance of the tower as a free-standing building, and indeed did much to mar it. In these circumstances a purely conservative preservation has not been feasible, and a bold policy of reconstruction has been adopted.

The remains of each phase of the castle's development have been consid-

[103] B. H. St. J. O'Neill, in *Castles and Cannon*, [Oxford University Press, 1960], gives the following calibres for guns of the time; demiculverin 4 inches, culverin 5½ inches, demicannon 6½ inches, cannon 8 inches.

ered on their merits in relation to the new plan, but preference has been given to the recovery of original fifteenth-century features, the more so since these, as has been noted above, were still in existence in the time of John Napier. Rather than lose all trace of features too fragmentary to retain as they stand, some of the older work has been reconstructed after close study of the fragments and, in some cases, reference to comparable parts of contemporary towers in the Lothians. The resulting amalgam of fifteenth and seventeenth-century features will in fact be of a kind which never existed at any one time in the history of the tower but which will nevertheless be a fair reflection of its past, and—it is hoped—a seemly piece of architecture.

This portrait of Mark Napier (1798–1879) was painted by Colvin Smith (1795–1875), a distinguished Scottish portrait painter who lived in Edinburgh, as did Mark Napier himself. Colvin Smith's Sitters Book shows that it was commissioned by Francis, 10th Lord Napier and 1st Baron Ettrick (1819–1898), who paid £52 10s for it in 1867. It was sold in 1939 by a subsequent Lord Napier to the Scottish National Portrait Gallery for £30. Oil on canvas, 41" × 30". (Courtesy of the Scottish National Portrait Gallery.)

FURTHER READING

ALMOST all the documented facts regarding Napier were recorded by Mark Napier (1798–1879) in his *Memoirs of John Napier of Merchiston: his lineage, life and times with a history of the invention of logarithms* (William Blackwood, 1834). Mark Napier was a descendant of John Napier, being a great-grandson of Francis, 6th Lord Napier. Fortunately for future generations William, 9th Lord Napier (1786–1834), gave his distant cousin full access to the Napier charter chest where the family records were kept.

A lawyer by training and profession, Mark Napier passed advocate at the Scottish bar in 1820 and was appointed Sheriff-Depute for Dumfriesshire in 1844, Galloway being subsequently added to the post. He died in office.[104]

However, he devoted much of his life to historical research and wrote numerous 'Memoirs' of notable Scottish figures from the past. Unfortunately, Mark Napier's works do not make for easy reading. His *Memoirs of John Napier* is prolix and there are innumerable digressions which, though of wider historical interest, have little or no connection with Napier himself or his family.[105] He did not reference his sources as clearly as modern convention dictates and, as Mark Napier himself admitted, he was hardly qualified to discuss the mathematical aspects of his ancestor's life:[106]

> In some respects a philosopher [scientist] would have been the most proper biographer of Napier, particularly in the analysis of his unpublished treatises, to which I can scarcely hope to have done justice beyond the fact of making their contents known.

Mark Napier frequently promoted Napier's actions further than the available evidence allowed.[107] He was also good at wording his speculations as facts, a trait which has had an insidious effect on the work of subsequent historians. Because nearly all of Napier's original manuscripts were lost in a fire, the *Memoirs of John Napier* has been used as the primary source by almost every historian of Napier, and it has therefore defined how both academics and

---

[104] The principal obituaries appeared in *The Scotsman* (24th November 1879), *The Scottish Guardian* (28th November 1879) and *The Journal of Jurisprudence* (1879, Vol. 23, pp. 652–655).

[105] The *Memoirs of John Napier* was reviewed at length and not entirely favourably in *The Quarterly Review*, Vol. 52, No. 104, John Murray, August & November 1834, pp. 443–466.

[106] Mark Napier, *Memoirs of John Napier*, p. ix.

[107] For an example of Mark Napier's excessive promotion of John Napier see fn. 36 on p. 19.

the general public view him. Many of Mark Napier's assertions can be seen in subsequent works, right through to those published in the 21st century. For example, the view of Napier as an academic recluse with overtones of sorcery originates with Mark Napier, not helped by the portraits of John Napier dressed in a black cloak with a long beard, yet this view is contradicted by Mark Napier's own record of John Napier's deep and practical involvement in the worldly affairs of the Kirk, amongst a number of other episodes he recorded.[108]

However, despite its faults, Mark Napier's work is still the definitive source for information relating to Napier and is likely to remain so for the foreseeable future. All subsequent historians and biographers are deeply indebted to him for his research and the records he identified, some of which have now been dispersed or possibly even lost.

During his research Mark Napier discovered in the Napier charter chest mathematical notes by Napier which were subsequently published in Latin as *De arte logistica, Joannis Naperi Merchistoni Baronis libri qui supersunt* (Maitland Club, 1839). This work includes a long and worthwhile introduction by Mark Napier, though it repeats much that appears in his *Memoirs of John Napier*, albeit more succinctly.

Mark Napier also wrote *"The Lanox of auld", an epistolary review of the "Lennox by Willam Fraser" to which is added a postscriptive memorie of the House of Merchiston* (David Douglas, 1877) to demonstrate the descent of the Napiers from the ancient Lennox family.

The first biography of John Napier was written by David Stewart (1742–1829), 11th Earl of Buchan, with the mathematical assistance of Dr. Walter Minto in *An account of the life, writings and inventions of John Napier of Merchiston* (R. Morison & Son/J. Murray/W. Creech, 1778; 2nd edn., R. Morison & Son/C. G. J & J. Robinson/W. Creech, 1787).[109]

[108] This assessment of Mark Napier as a historian is based on Alexander Corrigan, *John Napier of Merchiston's* Plaine Discovery: *A Challenge to the Sixteenth Century Apocalyptic Tradition*, Unpublished Doctoral Thesis, (University of Edinburgh, 2014).

[109] Walter Minto was born in Merse, Scotland, in 1753 but emigrated to America in 1786 where he became Professor of Mathematics and Philosophy at Princeton University. He died there in 1796. A year after his arrival he received a package from the Earl of Buchan containing the first copy he had seen of their joint work on Napier. He replied in a letter dated October 1787 that "I have not had leisure to give Napier an attentive perusal; but in the glance I gave it I discovered a great many typographical errors. These I shall mark down at my leisure which might be printed & distributed among the possessors of the book" (Luther P. Eisenhart, *Walter Minto and the Earl of Buchan*, Proceedings of the American Philosophical Society, Vol. 94, No. 3, 1950, p. 282). Unfortunately no such list of corrigenda appears to exist. Of the four examples of this work which have been examined only one copy, which is dated 1787 (British Library, Shelfmark 10827.h.8), has an errata list but this only has 10 corrections and only one of them relates to the mathematics, the rest being spelling errors.

The Earl of Buchan had to admit that he could discover very little about Napier himself, barely twelve pages out of the whole book, despite enquiring 'without success, amongst all the descendants of this eminent person, for papers or letters, which might elucidate this dark part of his history'.[110] Assuming the Earl of Buchan did approach the contemporary Lord Napier, it is clear that he did not gain access to the family's charter chest.

However, at much the same time Francis, 8th Lord Napier (1758–1823), compiled a summary of the family's charters and private papers to create an early genealogical account of the family.[111] This formed the core of Wood's edition of Douglas's *The Peerage of Scotland* (Archibald Constable, 1813), which provides a good 23-page source for the Napier family from original sources up to that date.

A fuller history of the Napier family can be found in the anonymous (though actually written by Archibald Scott Napier) and scarce *A history of the Napiers of Merchiston ... compiled from old records*, (J & E Bumpus, 1921). This is a useful compilation from many sources including Mark Napier's *Memoirs of John Napier*, works by Mark Napier on the Lennox family, Wood's edition of Douglas's Peerage, manuscripts in the Napier charter chest and others.

Priscilla Napier wrote a number of books on the Napier family into which she married, mostly covering the distinguished generals, admirals and diplomats of the 18th and 19th centuries, but the mediæval and Stuart family are covered in *A difficult country, the Napiers in Scotland* (Michael Joseph, 1972).

A large number of mathematicians from all over the world gathered to

---

A close comparison of the two editions of this work confirms that they are absolutely identical, apart from the date, the publisher's names on the title page and the fact that the Bibliography and the five plates were bound into the two works in different locations. Even the page numbers for pp. 55 & 56 are repeated on the next two pages in both works and it is clear that the second edition was either published using sheets left over from the first printing or printed at a later date from the same formes as the first edition, though the latter option is exceedingly unlikely. The work was given a rather lukewarm review in December 1788 (*The Analytical Review, or history of literature, domestic and foreign*, J. Johnson, Vol II, September to December 1788, p. 413), ten years after the first edition was supposedly published, yet there is no mention of a previous edition being published in 1778.

All these issues, combined with the fact that Minto first received a copy in 1787 and his comment regarding its numerous typographical errors, clearly indicate that the edition of 1778 included a typographic error in the date itself, which should have been 1787. Such a mistake is easy enough to make when typesetting Arabic numerals but the date on the title pages of these two works is set in Roman numerals, so the error must have been within the typesetter's mind as he worked. Despite being catalogued as two separate editions by libraries around the world, they appear to be effectively one and the same edition, published at the same time.

[110] David Stewart, Earl of Buchan, & Walter Minto, *An account of the life, writings and inventions of John Napier of Merchiston*, (R. Morison & Son, 1787, p. 13).

[111] Mark Napier, *Memoirs of John Napier*, p. viii.

mark the tercentenary of the publication of the *Descriptio* at an International Congress held in Edinburgh in July 1914. This led to the publication of Cargill Gilston Knott's *Napier tercentenary memorial volume* (Longmans, Green & Co, 1915) which contains the numerous addresses and essays presented to the Congress. They cover Napier himself, his invention of logarithms, his various works and Merchiston tower, as well as the subsequent development of logarithms. The chapters by Lord Moulton (*The invention of logarithms*), Prof. Hume Brown (*John Napier of Merchiston*), Prof. Eugene Smith (*The law of exponents*) and Prof. G. A. Gibson (*Napier's logarithms and the change to Briggs's logarithms*) can be particularly recommended. There is also a long chapter by Prof. R. A. Sampson on the bibliography of Napier's works.

E. M. Horsburgh's *Handbook of the exhibition of Napier relics and of books, instruments and devices for facilitating calculation* (Royal Society of Edinburgh, 1914; *Handbook of the Napier tercentenary celebration*, Tomash, 1982) catalogues the very comprehensive exhibition which accompanied the Tercentenary Congress. This was a celebration of Napier's pivotal role in the history of calculation and it featured many of his works. However, it also traced the subsequent development of Napier's achievement by displaying many different types of calculating machines, as well as other aids to calculation such as mathematical tables, the abacus and slide rules, planimeters and integraphs, ruled papers and nomograms etc.

E. W. Hobson's, *John Napier and the invention of Logarithms* (Cambridge, 1914) is a useful reconstruction of Napier's work.

William Rae Macdonald published a translation of the *Constructio* under the title *The construction of the wonderful canon of logarithms* (William Blackwood & Sons, 1889). This work also includes a very worthwhile catalogue of all five of Napier's works, detailing the various editions that were published, including those printed elsewhere in Europe, and listing the major bibliographical variations between them. Macdonald also wrote the excellent entry for Napier in Vol. 40 of the *Dictionary of National Biography* (Macmillan & Co, 1894).

Although now outdated for many topics, the erudite and elegantly written eleventh edition of the *Encyclopedia Brittanica*, published in 1911 and the last to be largely compiled in Britain, contains a long article on Napier and a separate one by J. W. L. Glaisher on logarithms.

Gold at Wolf's Crag? An inquiry into treasure at Fast Castle (Oliver & Boyd, 1971) by Fred Douglas takes Napier's manuscript agreement with Sir Robert Logan to search for treasure at Fast Castle and puts it into the context of a much wider political intrigue with possible connections, amongst others, to the search for the Holy Grail. However, though seemingly well researched, no references are provided to the many sources that Douglas obviously consulted.

The most comprehensive description of the method of operating a Promptuary was prepared by William Francis Hawkins and published in *The first calculating machine (John Napier, 1617)* (Annals of the History of Computing, 1988, Vol. 10, No. 1, pp. 37–51) while the re-discovery of the only known original Promptuary and a description of it was provided by Erwin Tomash in the accompanying article, *The Madrid Promptuary* (Annals of the History of Computing, 1988, Vol. 10, No. 1, pp. 52–67).

William Frank Richardson's translation of *Rabdology* (MIT Press, 1990) contains a useful introduction by Robin E. Rider.

D. J. Bryden's *Napier's Bones, a history and instruction manual* (Harriet Wynter Publications, 1992) provides a short but well illustrated description of Napier's Bones and their use.

For Napier's sole religious work Dr. Alexander Corrigan's PhD thesis clearly covers the *Plaine Discovery* in far more detail than is possible in this work.[112] The thesis also includes a comprehensive list of works related to this topic (pp. 13–28) with a commentary on each of them.

There are two relatively modern works on Napier, though neither includes any references. Francis Shennan's *Flesh and Bones, the life, passions and legacies of John Napier* (Napier Polytechnic, 1989) is a modern summary of Napier's life that excludes Mark Napier's digressions. It covers the history of the early Napier's in more detail than was considered necessary in this work.

John Napier: Logarithm John (National Museums of Scotland Enterprises, 2003) by Lynne Gladstone-Millar is another modern but even shorter biography, though it is particularly well illustrated. This work also includes a full explanation of how to use logarithm tables for those that have forgotten, or for those who never learnt how to use them.

However, the most comprehensive of the modern works is Julian Havil's *John Napier: life, logarithms and legacy* (Princeton University Press, 2014), nicely timed to catch the quadcentenary of the publication of the *Descriptio*. It covers Napier exactly as the title indicates; his life, the invention of logarithms (as well as Napier's other mathematical works and the *Plaine Discovery*) and his largely mathematical legacy.

[112] Alexander Corrigan, *John Napier of Merchiston's* Plaine Discovery: *A challenge to the sixteenth century apocalyptic tradition*, Unpublished Doctoral Thesis (The University of Edinburgh, 2014.)

IOANNIS NEPERI
DE MERCHISTON
1617
1842

BIBLIOGRAPHY OF NAPIER'S WORKS

WILLIAM Rae Macdonald's *The construction of the wonderful canon of logarithms* (William Blackwood & Sons, 1889) provides details for all five of Napier's works, including their translations into Dutch, French, German and Italian, and the major bibliographical variations between them. Cargill Gilston Knott's *Napier tercentenary memorial volume* (Longmans, Green & Co, 1915) also contains a long chapter by Prof. R. A. Sampson on the bibliography of Napier's works.

All known Scottish books printed before 1700 were recorded by H. G. Aldis in *A List of books printed in Scotland before 1700*, (Edinburgh Bibliographical Society, 1904; National Library of Scotland, 1970). Over 6,000 items are currently listed in the National Library of Scotland's online version of Aldis. Of these the Library holds about 4,200 titles, some 780 of them being the only known copies.

Napier's works held by the National Library of Scotland have been extracted from the current Aldis online record and are listed below. Nearly all Aldis items are also recorded in the English Short Title Catalogue (ESTC) to which Aldis always gives references when they are available. Each entry ends with a list of libraries known or believed to hold copies (the abbreviations can be identified online) but ESTC provides the most reliable and up-to-date source of library holdings.

1. ***242*** **Napier (J.).** A plaine discouery ... Reuelation of Saint Iohn ... 4to.
Edinburgh: R. Waldegrave, 1593. *STC 18354*; *Dickson & Edmond 206*
NLS holdings: H.36.b.17; H.36.b.17a; Hax.29; Ry.II.f.48
Other locations: D O B T C Yk Folg HN HD NY Y

Facing page. *The large memorial erected to John Napier in the vestibule of St. Cuthbert's, Edinburgh, in 1842. The Latin inscription reads* Somewhere near this place / is buried the body of / John Napier / of Merchiston / who by his marvellous / invention of logarithms / won for himself the everlasting / remembrance of posterity / Died on the day before the Nones of April [4th April] / AD 1617 aged 67 / In honour of this great man / This tablet is erected / AD 1842. *The Romans recorded the dates within each month by counting backwards from certain points called the Calends, Nones and Ides ('Beware the Ides of March' says Shakespeare's soothsayer to Julius Caesar). In April the Nones are the 5th day of the month while 'pridie' or 'prid.' means the day before. Hence the 4th of April. (Photo by Alastair Carew-Cox, courtesy of St. Cuthbert's, Edinburgh).*

2. ***242.1*** **Napier (J.).** A plaine discouery ... Reuelation of Saint Iohn ... 4to. Edinburgh: R. Waldegrave, 1593. *Another issue, with 32pp. of additional matter inserted after S8*
NLS holdings: L.C.1472
Other locations:
3. ***255.3*** **Napier (John).** A plaine discoverie of the whole revelation of Saint John. Newly imprinted. 4to. [Edinburgh: R. Waldegraue, for] J. Norton, [London] [1594]. *STC 18355*
NLS holdings: F.7.g.15
Other locations: B L38 O Cashel Folg HN MIN N NU
4. ***272.5*** **Napier (Archibald).** The new order of gooding and manuring ... field land ... Edinburgh: R. Waldegrave, 1595.
NLS holdings: MS.963, foll.25- (a transcript)
Other locations:
5. ***435*** **Napier (John).** A plaine discovery, of the whole Revelation of S. Iohn ... 4to. Edinburgh: A. Hart, 1611. *STC 18356*
NLS holdings: L.C.515; H.37.d.24 (imperfect)
Other locations: S C C3 E4 Folg HN NY Mich
6. ***436*** **Napier (John).** A plaine discovery, of the whole Revelation of S. Iohn ... 4to. London, for John Norton [A. Hart], 1611. *STC 18356a*
NLS holdings: H.36.b.20
Other locations: C F B O TCD Folg HN HD NY CAL Leighton
7. ***469*** **Napier (John).** Mirifici logarithmorum canonis descriptio ... 4to. Edinburgh: A. Hart, 1614. *STC 18349; sig. m1v blank; in most, though not all, copies C3v and C4r are incorrectly numbered "22" and "23"; on A2r a variant is found in which the comma after "Illustrissimo" is found directly above the "I" of the line below rather than clearly to the left of it*
NLS holdings: H.34.b.22 (imperfect: lacks C2-3, cut out leaving stubs; C4r is correctly numbered "15"); H.34.b.23 (without title page, issued as part of Aldis 539; comma directly above the "I"); Cd.2.1b (missing); UMI 1147:3 (microfilm of Bodleian copy Ash.130(2); C3v and C4r incorrectly numbered)
Other locations: B HN C O E13 G L38 Cashel BO IND MICH Y
8. ***470*** **Napier (John).** Mirifici logarithmorum canonis descriptio ... 4to. Edinburgh: A. Hart, 1614. *STC 18349a; with 'Admonitio' on sig. m1v; C3v-C4r are incorrectly numbered "22" and "23"*
NLS holdings: RB.s.578 (A1 = title can be seen to be conjugate with A4); Ry.II.f.49 (with title page, issued as part of Aldis 539); UMI 1636:07 (microfilm of Edinburgh University copy)
Other locations: B G S E C O TCD HN CH CHI LC Tex
9. ***518*** **Napier (John).** Rabdologiæ, sev nvmerationis per virgulas ... 12mo. Edinburgh: A. Hart, 1617. *STC 18357*

NLS holdings: Ry.III.h.26
Other locations: B E G F C S C16 C19 O TCD Folg HN HD CHI NY Y

10. ***538*** **Napier (John).** Mirifici logarithmorvm canonis constrvctio ... 4to. Edinburgh: A. Hart, 1619. *Part of Aldis 539: see note ad loc.*

11. ***539*** **Napier (John).** Mirifici logarithmorum canonis descriptio ... Accesserunt opera posthuma ... 4to. Edinburgh: A. Hart, 1619. *STC 18350; 1614 sheets of "Descriptio", either issue (Aldis 469, 470), reissued with a new title page (the 1614 title page either being retained or not) and with the addition of Part 2, "Mirifici logarithmorum canonis constructio" first printed in 1619; photograph seen of 1619 general title page without rule above author statement*
NLS holdings: H.34.b.23 (1614 title page removed; m1v blank); Ry.II.f.49 (1614 title page retained; m1v "Admonitio"); H.34.b.23a (1619 parts only, i.e. new general 1619 title page, with conjugate blank, and new Part 2); UMI 1901:9 (microfilm of HN copy; m1v "Admonitio"; C3v and C4r incorrectly numbered; 1614 title page removed)
Other locations (all need to be verified; some may consist of 1619 part only): S C L38 O TCD Folg (impf.) HN BO NY

INDEX

Headings in italics indicate the title of a publication.
Page numbers in italics refer to illustrations
Page numbers followed by n refer to notes

O

P

T

U

V

W

Z